Microbiology

SECOND EDITION

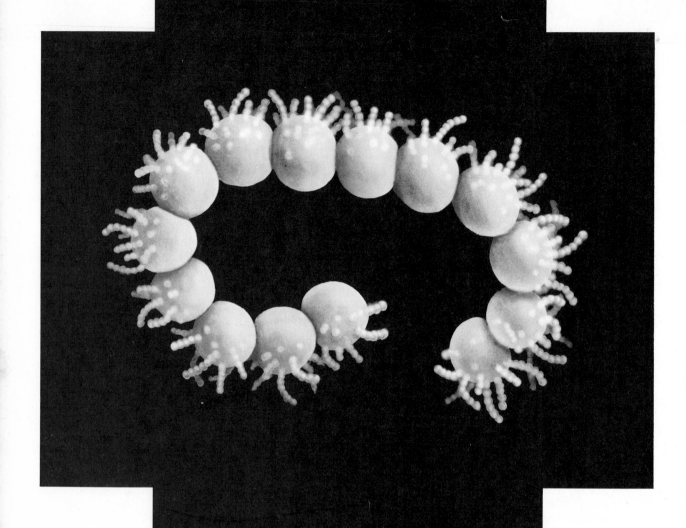

ADDISON-WESLEY PUBLISHING COMPANY

**Reading, Massachusetts / Menlo Park, California
Don Mills, Ontario / Wokingham, England / Amsterdam
Sydney / Singapore / Tokyo / Mexico City
Bogotá / Santiago / San Juan**

Microbiology

Second Edition

Cynthia Friend Norton
University of Maine at Augusta

Sponsoring Editor: Bruce Spatz
Development Editor: Kathe Rhoades
Art Development Editor: Meredith Nightingale
Production Supervisor: Margaret Pinette
Production Coordinator: Susan E. Vicenti
Copy Editor: Deborah Stone
Illustrators: Christine D. Young
Horvath & Cuthbertson
Rossi & Associates
Art Consultant: Joseph K. Vetere
Photo Researcher: Darlene Bordwell
Manufacturing Supervisor: Ann DeLacey
Cover Designer: Marshall Henrichs
The cover shows a model of an Adenovirus.

This book is affectionately dedicated to three fine teachers, each of whom has provided a major inspiration to me in choosing the direction of my professional and personal growth.

John B. Friend, M. A. English
  My first teacher

Edwin R. Frude, B. A.
  Head of the Science Department
  Arms Academy

Galen E. Jones, Ph. D.
  Professor of Microbiology
  University of New Hampshire

Library of Congress Cataloging in Publication Data

Norton, Cynthia F., 1940–
  Microbiology.

  Includes bibliographies and index.
  1. Microbiology.   I. Title.
QR41.2N67 1985   616'.01   85-3909
ISBN 0-201-10997-2

ABCDEFGHIJ-RA-898765

PREFACE

In this new edition of *Microbiology* my goal has been to bring together that selection of core concepts most representative of the increasingly broad field of microbiology. I hope that the ways I have presented them will engage and hold the interest of introductory students from diverse backgrounds, while laying the groundwork for an understanding of the wide-ranging applications of the field.

At the same time, I have chosen to be selective rather than encyclopedic, in order to write a teachable, readable book. I hope that instructors may cover the material within a semester and that students will enjoy and absorb the topics rather than being overwhelmed.

SCOPE AND ORGANIZATION

The text is intended for the type of one-semester or equivalent introductory course in which health issues play a major, but not an exclusive, role. Although microbiological applications are stressed, the student is provided with the solid conceptual framework required for a preprofessional curriculum.

Pedagogically, this text has been designed to equalize the disparities (now found in most classes, it seems) among those students with a weak background in biology or chemistry and those who are better prepared. All common biological and chemical terms are fully defined on first use. The terms are introduced in a careful sequence, so that the student has the opportunity systematically to build a working vocabulary.

The book is organized in five units, moving from general introductory concepts to practical applications. From the beginning, however, concrete examples are freely introduced to help students recognize the everyday encounters we all have with the microbial world.

Part One introduces the science of microbiology by presenting a brief history and statement of the scope and potential of the discipline, a review of basic chemical principles, and a structural analysis and comparison of procaryotic and eucaryotic cells. The next chapters provide an overview of the classification and diversity of the bacteria and the eucaryotic microorganisms, including the helminths. Additional chapters discuss growth conditions and kinetics, cultivation and identification methods, and microbial metabolism. The remainder of the part covers aspects of molecular biology, including a chapter on genetic mechanisms and chapters on the nature and replication of viruses and on the animal viruses, the methods used in studying them, and their contributions to our understanding of the role of oncogenes in cancer.

Part Two focuses on the host-parasite relationship — how our bodies interact with microorganisms. The student becomes familiar with the features that make some microorganisms harmless and those that make others pathogenic. We analyze the body's nonspecific and specific defenses against microbial pathogenicity and encounter some of the effects of malfunctioning defenses. Finally the student is introduced to the challenging field of epidemiology and public health. The concept of a chain of transmission of infectious disease is discussed, and the student is shown some of the types of intervention that can be called into play to reduce the incidence of infectious disease.

Part Three surveys the most significant microbial diseases. The presentation is based on the organ system approach. I chose this approach because it gives learners the opportunity to incorporate the new information with already acquired knowledge of anatomy and physiology, making it easier to learn and retain. The presentation is also structured in such a way that both student and instructor have ready access to information on the microbiology of key pathogenic species.

Part Four brings together the four key approaches to the control of harmful unwanted microorganisms: disinfection, sterilization, and aseptic techniques; antimicrobial chemotherapy; immunization; and hospital infection control. These chapters make a natural unit, since the techniques are often used together in a combined control strategy. I have presented these techniques in this location to enable the instructor to draw effectively on the examples of infectious disease in the previous section. However, each chapter may be taught individually and at other points in the course sequence, if desired.

Part Five, on applied microbiology, concludes the book with chapters on soil and water microbiology and food, dairy, and industrial microbiology.

COURSE SEQUENCES

Different course sequences have been taken into consideration in designing the presentation. Chapter 23, Disinfection and Sterilization, may be placed after Chapter 6, Microbial Growth. Chapter 24, Antimicrobial Drugs, may be placed after Chapter 6, 7, 10, or 13. The disease presentations are self-contained, so that an instructor who wishes to spend only one or two lectures on infectious diseases can assign a sampler, lifting, to give one example, the discussions of influenza from Chapter 18, salmonellosis from Chapter 19, a sexually transmitted disease from Chapter 20, and malaria and AIDS from Chapter 22.

CHANGES IN THIS EDITION

Writing a textbook is a little like making sausage: if you push more stuffing in one end, some is going to fall out the other. Many new topics have been added in this edition. Coverage of the

helminths has been included to round out the discussion of infectious diseases. The chapter on microbial ecology has become a new, more detailed chapter on water and soil microbiology. A new chapter on food microbiology and industrial microbiology concludes the book. I have chosen not to isolate the new techniques of molecular biology, such as the genetic manipulation and monoclonal antibody methods, in a separate chapter, but have allowed them to invade and recombine promiscuously with the rest of the subject matter, as they have in microbiology labs.

This edition includes updates in numerous areas. I have added information on the newly discovered diseases and pathogens such as toxic shock syndrome and AIDS. I have introduced new therapeutic approaches such as third-generation cephalosporins, immunotoxins, and the newer antiviral drugs. I stress modern laboratory techniques that are widely used, especially the automated clinical methods students are most likely to encounter.

The compromises are ones I can live with and hope others can too. The material on biochemical separation techniques was removed from Chapter 2, the discussion of dental microbiology was trimmed, and a number of large reference tables were deleted as their information was felt to be overly detailed. The chapter on wound infections was deleted, with certain essential information being subsumed into other chapters.

However, the greatest economies were achieved through two steps. Essentially every sentence in the book was rewritten for conciseness and clarity. As I rewrote, I constantly questioned "Is this fact absolutely necessary?" Quite often, I discovered less than essential details that could be cut. We believe the result proves once again the validity of Mies van de Rohe's aphorism that "less is more."

UNIQUE FEATURES

A remarkable feature of this edition is its art program, perhaps the most extensive and innovative in the discipline, with some 485 new figures and 350 photographs. The diagrams are designed to emphasize the dynamics of processes, and behind each figure lies a careful analysis of the pedagogic message to be conveyed and a well developed correlation with the text. An organic and realistic style has been employed, as opposed to traditional schematics; people, for example, are drawn in naturalistic form rather than in anatomic outlines.

Certain other features give this book particular distinction: the section on maternal and infant diseases; the chapters on immunization and hospital infection control; the integrated emphasis on epidemiology and disease control; and the special interest boxes.

THEMES

All microbiology books have facts — chemical formulas, disease symptoms, names of microorganisms — but not all have cohesive themes. I believe that it is the themes of this text that make it unique in the field. These themes derive from my own attitudes toward students and teaching microbiology as a discipline and the sciences in general. All my students are adults, preprofessionals who have come to my class to share understandings and gain competence. With this view, I have selected teaching materials that can expand horizons, stimulate discussion, and build the student's professional skills.

With my background, which is in environmental microbiology, I learned to view microorganisms as essential partners in the living world and to view their activities primarily in ecological terms. I carry this approach throughout the book, even the units that deal with infectious diseases. I attempt to lead the students to an understanding that we can live with and perhaps direct and control microbial processes, but that we get nowhere if we try to ignore or blindly eradicate them. I also take care to make students aware of the social, economic, and ethical implications of microbiology and biotechnology, for

although they will be our students for but a few weeks, they will be citizens for life of a complex, rapidly changing world. In addition, I provide them with a core of data with which to make the responsible professional and social decisions that will inevitably be theirs to make.

LEARNING AIDS

Each chapter begins with a brief overview of the major topics. New terms are printed in boldface type and defined when first used. A summary of key points, questions for discussion, and a list of supplementary readings complete each chapter. The special topics boxes in each chapter focus on research breakthroughs, present case studies, or address economic and social implications of a subject area. A comprehensive glossary with pronunciation guide is found at the end of the text.

I have prepared two appendices that I feel will be useful to students. Appendix I presents thumbnail descriptions of the bacterial genera mentioned in the text, and Appendix II contains an outline of bacterial classification based on *Bergey's Manual of Determinative Bacteriology*, 8th edition, and *Bergey's Manual of Systematic Bacteriology* (volume 1).

SUPPLEMENTAL AIDS AVAILABLE

A number of supplements have been designed to increase the usefulness of this text for both instructor and student.

- A student study guide prepared by David Fox of the University of Tennessee is intended to help students learn the vocabulary of each chapter, synthesize the important concepts, and test their understanding of the material.
- An outstanding instructor's resource package prepared by Barbara Pogosian of Golden West College contains a number of features to assist you in planning and teaching. Each chapter contains a list of chapter

objectives, summaries of main points, and lecture outlines. An innovative cross-referencing system shows each of the places in the text where you can find all information pertinent to essential topics. The test questions in each chapter are presented in a variety of formats — multiple choice, fill-in, matching, and essay.

Many adopters of the second edition of *Microbiology* will qualify for free acetate transparencies, color slides, and computerized testing. Please contact your Addision-Wesley representative for details about the complete supplemental package.

ACKNOWLEDGMENTS

An unusual number of people have been involved in the development of this edition. I am grateful to Eric Gordon and Kathe Rhoades for their developmental work on the text; to Meredith Nightingale, Cathy Dorin, and Art Ciccone for their developmental work on the art program; to Darlene Bordwell and Judy Ullman for their assistance in photo research; to my editor Bruce Spatz for his commitment to and overall direction of the project. I have had the help of a number of reviewers, whose names are listed below. I thank them for their criticisms and corrections as well as their very welcome commendations.

Barbara Pogosian
Golden West College

Florence E. Farber
University of New Hampshire

Richard Tilton
University of Connecticut Health Center
John Dempsey Hospital

Robert T. Vinopal
University of Connecticut

Brian Wilkinson
Illinois State University

Bruno Kolodziej
Ohio State University

Carl Warnes
Ball State University

Walter Konetzka
Indiana University

John Holt
Iowa State University

I would also like to thank the following individuals, who participated in our market research surveys.

Alfred Borgaetti
Salem State College

Margaret Heimbrook
University of Northern Colorado

Cynthia Somner
University of Wisconsin

Marjorie Spradling
Western Michigan University

Robert Hairston
Harrisburg Area Community College

Nancy Rubino
SUNY at Binghamton

Nancy Bigley
Wright State University

Isaiah Benathen
Kingsborough Community College

Thomas Kerr
University of Georgia

Walter Appelgren
Northern Arizona University

William Rafaill
Berea College

Violet Schirone
Suffolk County Community College

Dolores Kasarda
Cedar Crest College

Shirley Bishel
Rio Hondo College

Larry Thiery
Willmar Community College

Sylvia Kerr
Hamline University

Barry Teitelbaum
Lansing Community College

Charles Beetz
Parkland College

Josephine Smith
Montgomery County Community College

Jan Van Niel
Everett Community College

Robert Shurer
Villanova University

Remo Morelli
San Francisco State University

Rex Moyar
Trinity University

William Sherwood
Lynchburg College

Jane Voos
William Paterson College

Carolyn Mather
York College of Pennsylvania

Marion Gaffey
Montgomery College

Frederick Schuster
CUNY Brooklyn College

My special appreciation goes to my son Jack, who between the first and second editions learned to put his own worms on his fishhooks. He shares generously with me his observant eye and keen understanding and enjoyment of nature and helps me learn more about how people learn.

East Vassalboro, Maine   C.F.N.

ABRIDGED CONTENTS

DETAILED CONTENTS

PART TWO
HOST AND PARASITE   325

PART THREE
HUMAN INFECTIOUS DISEASES 451

PART FIVE

MICROORGANISMS: THEIR ENVIRONMENTAL AND ECONOMIC IMPACT 793

PART
ONE

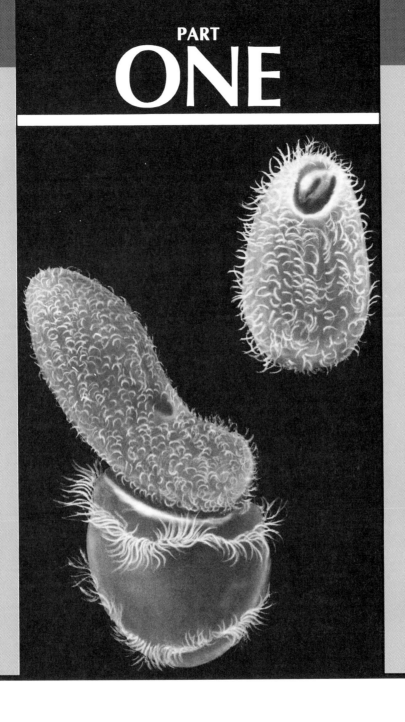

FUNDAMENTALS OF
MICROBIOLOGY

INTRODUCTION TO MICROBIOLOGY

CHAPTER OUTLINE

Welcome to the world of microorganisms. You are beginning an exploration of a remarkable part of the biological world, where you will encounter a variety of tiny life forms, some plantlike, some animal-like, and some unlike anything else you've seen before.

Microorganisms are living things so tiny they can be seen only through a microscope. Although they are so small, they move, nourish themselves, grow, respond to their environments, and reproduce just as do organisms of visible size. When we speak of microorganisms, we are referring to forms of life named bacteria, fungi, algae, and protozoa (Figure 1.1). **Bacteria** are simple one-celled organisms. They inhabit the widest range of environments of any group of living things. Most live off of dead plants and animals and off of their wastes. One group of bacteria called **cyanobacteria,** however, grow from the energy of captured sunlight. These microorganisms, once called blue-green algae, are found in great numbers in water and other moist environments. The **fungi** include yeasts, molds, and mushrooms. Like bacteria, they get their nourishment from dead organic matter. **Algae,** including the seaweeds as well as microscopic forms, use light energy and are very important as food for small aquatic animals. The **protozoa** are one-celled animals. They are classified by their means of locomotion. Most move rapidly and survive by capturing and eating other, smaller microbes such as bacteria. In addition to these microorganisms, we will also be learning about **viruses,** which are quite different from other forms of life. Furthermore, in order to provide a complete picture of the agents that cause human diseases, we will include a few larger animal species, the parasitic **helminths.** These animals, which are composed of many cells, are not considered microorganisms.

As we explore the microbial world, you will discover that microorganisms (in their simplest forms) were in all probability the first living things on earth. By combining and building on their basic biological features, other living things have evolved, that is, developed and changed slowly over time.

From the beginning of life on earth until the present time, microorganisms have been major contributors to the workings of the biological

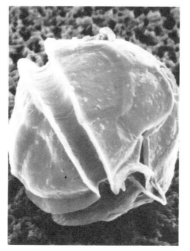

(a) *Bacillus cereus*   (b) *Gonyaulax tamarensis*   (c) *Aspergillus nidulans*

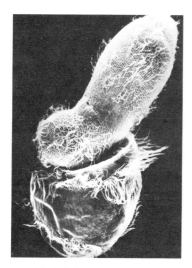

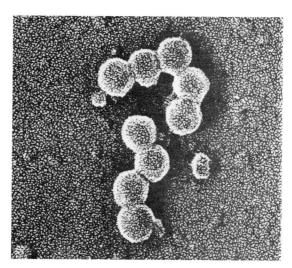

(d) *Didinium nasutum*   (e) *Simian virus 40*

FIGURE 1.1 The Major Microbial Groups
The major microbial groups include: (a) bacteria (represented by *Bacillus cereus*), (b) algae (represented by *Gonyaulax tamarensis,* one of the species responsible for the toxic "red tide"), (c) fungi (represented by *Aspergillus nidulans*), and (d) protozoa (represented by *Didinium nasutum,* at bottom, about to eat *Paramecium multimicronucleatum,* above). Viruses (e) reproduce only within a host cell by unique means. Photo credits: a, (×1443) Z. Skobe, Forsyth Dental Center; b, (×1956) J. M. Sieburth, *Microbial Seascapes.* Baltimore: University Park Press (1979), plate 2.9; c, (×2263) R. G. Kessel and C. Y. Shih, *Scanning Electron Microscopy in Biology.* Berlin: Springer-Verlag (1974); d, (×435) H. S. Wessenberg and G. A. Antipa, Capture and ingestion of *Paramecium* by *Didinium nasutum, J. Protozool.* (1970), 17:250–270; e, courtesy of Jack Griffith.

world. After more complex organisms, such as plants and animals, arose, microorganisms continued to evolve right along with them. All living things, microscopic and macroscopic, developed a complex web of interrelationships.

Although we are rarely made conscious of it, the bacteria, fungi, algae, and protozoa are essential to our lives. They are irreplaceable in fertilizing our soils; aiding the productivity of our fisheries; manufacturing many common foods and beverages, such as cheese, chocolate, some sausages, pickles, breads, wine, and beer; synthesizing life-saving antibiotics; and controlling naturally such destructive insect pests as the gypsy moth.

As comparative newcomers to the planet, human beings have only recently begun to understand and learn how to use the unique talents of these microbes. In the million-year-plus history of our species, scientific inquiry is a very new habit. Microorganisms were first seen by human eyes barely 300 years ago, and serious study began little over 100 years ago. Our studies of the microbial world give us insights into the world of larger plants and animals because microorganisms carry out the same functions as do other life forms, but more simply. Because they are simpler, they are easier to study in depth. Research done with bacteria and viruses helps scientists to understand the working rules of all life on earth.

MICROORGANISMS IN THE BIOSPHERE

The **biosphere** is the part of the global surface that is composed of living things of all types. Microorganisms are an integral part of the biosphere.

LEVELS OF ORGANIZATION

Living things are structured in several levels of organization (Figure 1.2). The primary unit of life is the **cell,** a self-reproducing microscopic struc-

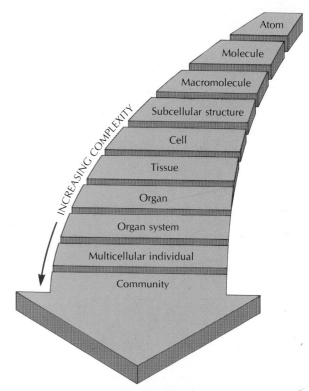

FIGURE 1.2   Levels of Organization
The cell, the fundamental unit of life, can exist only through the systematic organization of simpler, nonliving biochemical substances. Single cells can, in turn, be organized into more complex structures.

ture enclosed by an outer layer called a membrane. There are levels of organization both simpler and more complex than the cell. To understand a cell's function completely, we must first be aware of the simpler levels. The basic chemical materials of which cells are made are atoms, such as carbon and oxygen, and molecules, such as sugars and fats. Small molecules are linked together into macromolecules, such as proteins. All cells have subcellular structures made up of many macromolecules put together. The chemical nature of the macromolecules of a subcellular structure enables it to carry out a certain function, contributing its part to the complex network of activities that constitute a living cell. In contradiction to a law you may have learned in geometry, in biology the whole (the cell) is more than the sum of its parts. Alone, the parts are not alive; when unified in a healthy cell, they live.

Above the cellular level, groups of cells with the same structure and function form **tissues,** such as nerve and muscle; tissues are organized into **organs,** such as the heart and brain. Many animals have **organ systems** composed of interacting organs. Together all these systems make up the individual being. Organisms sharing similar characteristics constitute a species. Organisms within a species interact and cooperate in a variety of ways and thereby constitute a community. Finally, the interactions that take place among groups of organisms — among plants, animals, and microbes — are crucial in enabling each individual organism to survive in the biosphere.

SYMBIOSIS

A **symbiosis** is an interaction between two species. Wherever we look in the biosphere, we will find that microorganisms and macroorganisms are in symbiotic relationships. Their interactions may have positive or negative effects, or none, on one or both partners; the patterns are varied and often shifting. Symbiosis is an important factor in regulating the population size of the partners. Here we will look at just a few examples to illustrate the concept of symbiosis. Our own symbiosis with microorganisms will be developed as one of the main themes of this book.

COOPERATIVE INTERACTION.   Perhaps you have raised a vegetable garden and planted green beans or peas. These two crops are members of a group of plants called legumes. If you pull up a mature legume plant and examine its roots, you will see small whitish nodules on them. Inside each nodule, there is a mass of plant root tissue within which bacteria are growing. The legume and the bacteria have a mutually beneficial sym-

biosis (Figure 1.3). The plant provides sugars to the bacteria as their food source. The bacteria, in turn, draw nitrogen from the air and convert it to the fertilizing compounds that stimulate the plant's growth.

Another example of a cooperative symbiosis takes place in a cow's stomach, which has four chambers. The first chamber, the rumen, is a giant fermentation tank of sorts. The cow coarsely chews its grass or hay, then swallows it into the rumen. Chemically, hay is composed mainly of cellulose, a fiber for which cows and other mammals lack the necessary digestive enzymes. However, the billions of bacteria that live in the cow's rumen do have this enzyme. They dismantle the cellulose into small molecules that the cow can absorb and use. The bacteria also benefit; they use a portion of the cellulose for their own growth. Later, as the stomach contents, including bacteria, pass on into the other parts of the cow's digestive system, the cow actually digests the bacteria, which are rich in proteins. It is from the bacterial cells that the cow gets the protein and other nutrients it needs to thrive on what is otherwise a very non-nutritious diet.

We humans have a cooperative symbiosis with a complex group of microorganisms that live on and in us — our **normal flora.** Normal flora are found on our skin, in our noses, throats, and other body openings, and in our digestive tracts. These microorganisms rarely make us sick; in fact they benefit us, particularly by creating an environment that is inhospitable to harmful microorganisms. The normal flora for human beings will be discussed in detail in Chapter 11, and in Chapters 15 through 20.

NEGATIVE INTERACTIONS.   From almost anyone's perspective, the most important negative interactions we humans have with microorganisms are the **infectious diseases,** that is, diseases caused when microorganisms interfere with our normal body processes. Other animals and plants also contract infectious diseases. We will be discussing infectious diseases at length in

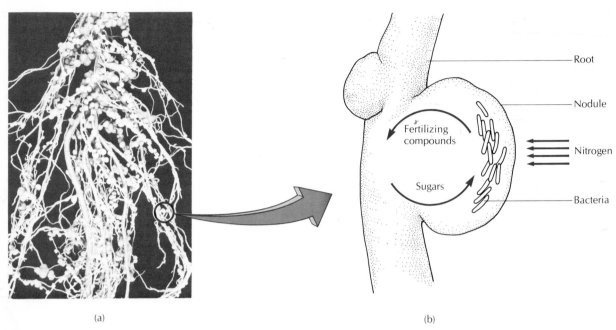

(a)   (b)

FIGURE 1.3   Beneficial Symbiosis
(a) The root system of a plant, showing hundreds of nodules. (Courtesy of the
Nitragin Company.) (b) In this close-up of a root nodule, plant and bacterial cells
exchange essential nutrients. The plant cells make sugars used by the bacteria; the
bacteria fix nitrogen into fertilizing compounds used by the plant cells.

later sections of this book. However, bear in
mind that only a tiny minority of microorganisms — a few dozens of the many thousands of
different types — are harmful.

Microbes can also attack other microorganisms or inhibit their ability to thrive. This
microbial **antagonism** points out the effects microbes have on each other — something you
may never have thought about. In Figure 1.4 you
will see that the fungus *Penicillium* can produce
a chemical that inhibits bacteria from growing
too near it. The *Penicillium* fungus thus gains
exclusive rights to the nutrients in its own small
habitat. Scientists have found hundreds of different chemicals produced by one microbial species
to inhibit others. Some of them, including penicillin, streptomycin, and tetracycline, have been

adapted for human medicine as antibiotics to
control microbial infections, and have saved
millions of lives.

PRODUCTION AND CYCLING
OF NUTRIENTS

Like the rest of us, microorganisms require certain chemical elements to survive. These elements include carbon, nitrogen, oxygen, sulfur,
and phosphorus. Since they are available only in
limited amounts life forms must actively compete for the use of compounds that have these
elements. Any atom that is being used by one
organism is currently unavailable to another. For
this reason, these elements must be efficiently
cycled, or passed from one organism to the next.

Microorganisms play key roles in the chemical steps that keep these important elements in circulation (Figure 1.5). For example, we have already mentioned that bacteria take nitrogen out of the air and provide it to plants. This process is an important step in the **nitrogen cycle** (see Figure 1.3). Microorganisms are also key factors in each of the other steps in the nitrogen cycle.

All plants and animals need oxygen. Free oxygen, such as that we breathe, is a byproduct of **photosynthesis.** This is the process by which plants, algae, and some bacteria capture the energy of the sun and use it to make sugars and other food materials. Water-living microorganisms such as the cyanobacteria and algae produce essentially all the oxygen that is released over the 70% of the earth that is covered by

water. The microbial contribution of oxygen, then, is equal to or greater than that of the more familiar land plants. The nutrients these microbes make feed all aquatic life.

In the **carbon cycle,** green plants, algae, and some bacteria remove carbon dioxide from the air and capture the energy of the sun to convert carbon dioxide into sugars and other nutrients by the process of photosynthesis. Microorganisms, especially bacteria, return carbon dioxide to the air when they decompose dead organisms.

You can see that without the action of microorganisms, the crucial cycles discussed above would be interrupted, and life on earth would gradually cease. The nutrient cycles are discussed in detail in Chapter 27.

In carrying out their role in the nutrient cy-

FIGURE 1.4 A Negative Symbiosis
A colony of the mold *Penicillium* was inoculated on the center of a Petri dish containing a nutrient medium that had first been streaked with bacteria. Both types of microorganisms attempted to grow in the same area, but the mold's production of penicillin prevented the bacteria from growing in the central zone around the mold.

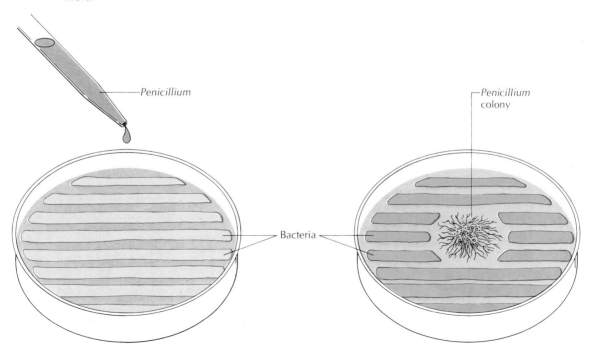

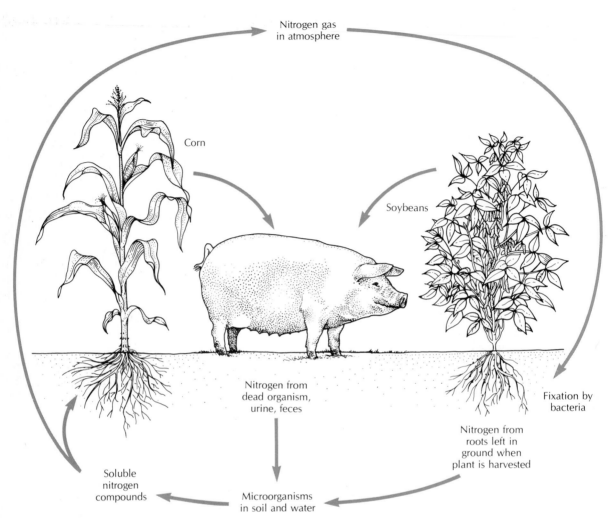

Nitrogen gas
in atmosphere

Corn

Soybeans

Fixation by
bacteria

Nitrogen from
dead organism,
urine, feces

Nitrogen from
roots left in
ground when
plant is harvested

Soluble
nitrogen
compounds

Microorganisms
in soil and water

FIGURE 1.5   Nutrient Cycling
Bacteria in the root nodules of plants fix nitrogen from the air. Animals use the
nitrogen in the plants they eat. Eventually this nitrogen returns to the soil as wastes.
Other microorganisms recycle soluble nitrogen compounds back as plant nutrients.

cles, the bacteria, along with the fungi, perform a humble but absolutely vital function. They **decompose,** or break down, the indigestible parts, waste products, and dead tissues of all the other life forms. In doing so they free up the life-sustaining elements so that they may be used by succeeding generations of organisms. This function keeps us from being overwhelmed by organic waste materials. As we learn to exploit the microbial decomposers better, they are becoming useful tools to help us manage the wastes of our civilization more effectively.

MICROORGANISMS IN EVOLUTION

We have just been sketching a brief impression of the importance of microorganisms in today's world. Let us now step back in time and see what scientists believe about how microorganisms contributed to the evolution of life.

WHAT IS LIFE?

Before we can even consider how life arose, we need some sort of working definition of "life." Philosophers, religious thinkers, and scientists have all tried to define life. The definition scientists agree on is based entirely on what is observable, so we will define life as possessing five observable traits.

Characteristic 1. Each life form is a distinct individual. It is surrounded by boundary layers that protect it, shape, it, and maintain its internal environment. Each life form has a means of communicating with its environment. Even the bacteria, simplest of all life forms, interact extensively with their environments.

Characteristic 2. Each living organism must be able to capture, use, and store some form of energy to carry out its life processes. Briefly, then, life depends on the orderly use of energy. The biosphere provides many sources of energy — for example, light or organic (carbon-containing) compounds. Each organism's place in the biosphere depends on which energy source it uses. Humans can use only organic compounds for energy, whereas microorganisms have a wide variety of energy sources. The microorganisms, in fact, are able to make use of the widest range of energy sources of any major group of living things.

Characteristic 3. Living organisms assemble large and intricate macromolecules necessary for growth, repair, and reproduction. Some of these, such as the proteins, may contain thousands of individual atoms, which are linked together in a unique sequence. We have gotten most of our knowledge of how this is done by studying microorganisms.

Characteristic 4. Each living organism has a set of information, a "blueprint," it uses to direct its growth and reproduction. The information is held in the structure of a macromolecule called deoxyribonucleic acid, or DNA. A DNA blueprint is a "design manual" that tells the organism how to manufacture molecules. It is also an "operations manual" that specifies which jobs are to be carried out and in what order. When live organisms reproduce, they pass on copies of their blueprint to their offspring. Microorganisms reproduce rapidly and efficiently.

Characteristic 5. All living organisms **adapt** (adjust) to changing environmental conditions. They have flexibility in their DNA blueprints, so they can adjust their behavior to changes in the natural environment. Microorganisms as a group are particularly adaptable and many can flourish in rapidly changing or hostile environments.

THE ORIGIN OF LIFE ON EARTH

Although we can identify the characteristics of living things, no one actually knows, of course, how life began on earth. Scientists have attempted to hypothesize what conditions might have been like on earth 3 to 4 billion years ago, and how those conditions could have produced a primitive life form that had a boundary, that used energy, that made macromolecules, that had some form of blueprint, and that was adaptable enough to stay together and reproduce.

Astronomers hypothesize that about 10 billion years ago, a portion of the gas cloud forming our galaxy, the Milky Way, started to condense and heat up. Most of this cloud coalesced to form the sun, which started to give off heat and light. Smaller aggregations became the planets, which at that time were very hot and molten. By roughly 4 billion years ago, the earth's surface had cooled sufficiently that a crust of solid rock was able to form. Geologists have found no signs of living things in the oldest rocks. But in rocks formed only half a billion years later there are structures believed to be the fossils of one-celled life forms. It was during this half-billion-year interval that living organisms probably appeared on the earth (Figure 1.6). At that time the surface of the earth was very hot and was continuously bombarded by lightning discharges and exposed to intense solar radiation. The atmosphere was very thin and was composed largely of hydrogen gas (the most plentiful element) and other elements in combination with hydrogen. As yet there was very little oxygen gas in the atmosphere.

Under these conditions, it is believed that energy discharges bonded together some of the available atoms, more or less at random. When these conditions are recreated today in the laboratory, some of the molecules that appear include amino acids (the building-block molecules for proteins) and other nutrients essential to

FIGURE 1.6 The Clockface of Biological Time
Using a 24-hour time scale, procaryotic life would have appeared relatively early in earth's history, some time before 6 A.M. (color). The first eucaryotes appear at about 4:30 P.M. (color). Multicellular organisms do not appear until evening of that same day, and the human species only a little before midnight.

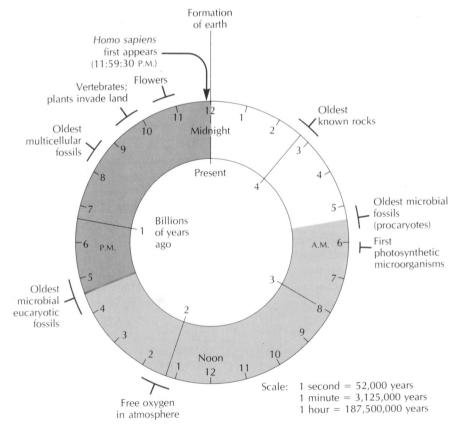

Scale: 1 second = 52,000 years
1 minute = 3,125,000 years
1 hour = 187,500,000 years

present-day life forms. "Hot soup" is the term used to describe the mixture that may have developed as a result of these random chemical reactions.

Some of the fat-like molecules tend to roll up into microscopic hollow spheres, like soap bubbles, spontaneously. This is a possible origin for a cell boundary layer or membrane. Some short simple chains of amino acids have a very inefficient type of energy-using activity. Other molecules tend to attach to and form long strands with others of their kind — a possible origin for a blueprint macromolecule. All these formations are still very far from making up a cell, and we still know essentially nothing of how the first cell arose.

THE FIRST ORGANISMS. The first "cell" was probably a very inefficient, simple (we might even say sloppy) little blob. It probably expanded, rather than grew, by accumulating more of the useful materials from the hot soup, and fragmented, rather than reproduced, when it got too big to be mechanically stable. At first, these "cells" had no competition for nutrients. However, as there came to be more of them, the contents of the hot soup began to be used up faster than they were being formed. The most efficient accumulators and fragmenters spread more rapidly than the others, perhaps reflecting the evolutionary force called **natural selection.** As we see natural selection operating today, the most efficient organisms tend to survive better and are more likely to reproduce themselves, while the less adaptable are more likely to die out. Even at the beginning of evolution, natural selection may have been operating, enabling more functional cells to succeed the earliest blobs.

As we noted, the "hot soup" contained little oxygen, so the first organisms probably were **anaerobes,** organisms that grow in the absence of oxygen. Their cell structure was simple; they contained no organized nucleus. Their descendants today are the bacteria. Organisms that have this type of cell structure are called **procaryotes.**

As time passed, scientists believe, some cells developed the ability to capture solar energy and convert carbon dioxide to nutrients, and photosynthesis began. At first, photosynthesis was carried out anaerobically (without producing oxygen) by organisms whose descendants are the photosynthetic bacteria. In a second major advance, however, some photosynthetic microbes developed the additional ability to use solar energy to split water molecules (H_2O). This led to the release, for the first time, of free oxygen.

The presence of free oxygen caused immense changes. Oxygen-producing photosynthesizers were extremely successful, as are their modern descendants. They converted the global atmosphere from oxygen-free to oxygen-rich in a twinkling — a few tens of millions of years. For human beings, oxygen is a blessing, because our metabolism is designed not only to use it but also to protect us against oxygen's toxic effects. However, in a cell not so protected, oxygen can be lethal, because it reacts with molecules it shouldn't react with and damages them beyond use. The original anaerobic life forms, unable to cope, either died off or came to occupy restricted anaerobic habitats, where they may still be found. When we study anaerobes, we go to deep sediments, marsh mud, the contents of the intestinal tract, the grooves under our gums, and the relatively few other places on earth where oxygen is not found.

As oxygen became plentiful, new **aerobic** or oxygen-using microorganisms became prevalent, sharing the world with the photosynthesizers, and living by using the nutrients the photosynthesizers produced. Both groups became increasingly efficient at using their own energy sources. Their cellular structures underwent changes, with the development of special folded membranes that hold groups of key molecules in an assembly-line precision. Scientists believe that microbes also developed, relatively early, whiplike appendages called **flagella,** which they used for swimming to more favorable spots. Their ability to move through the environment represents a typical adaptation response.

BOX 1.1
Are These Fossils Microorganisms?

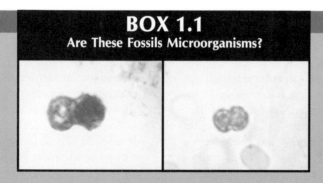

The objects shown in the photomicrograph on the left may just look like blobs to you. However, researchers hypothesize that they are fossil cyanobacteria that are about 3.4 billion years old. How do scientists know this? These rocks, taken from the Swartkoppje chert in Swaziland, Africa, contain a type of carbon compound that strongly suggests vigorous biological activity. The fossils have a regular appearance and size that resembles not only present-day microbes (photo on right) but also relatively recent microbial fossils. Fully 25% of all the fossils observed were lying in pairs, an arrangement that suggests that these cells were capable of reproducing themselves. Their arrangement could further be interpreted as meaning that their environmental conditions were favorable and they were thriving.

Despite the evidence, and these deductions, there remains an element of doubt concerning whether these markings in the rocks actually are microbial fossils. It is remotely possible that they could simply be random, inorganic, crystalline forms. When scientists discuss occurrences that they cannot observe directly, they know that they cannot be certain. This is why they name their ideas hypotheses or theories, not absolute truths.

Photo courtesy of A.H. Knoll and E.S. Barghoon, *Science* 1977; 198:396-398, Fig. 1. Copyright 1977 by the American Association for the Advancement of Science.

Through a period of almost 2 billion years, procaryotes were the sole inhabitants of the planet. During this time they evolved into many different types of microbes.

THE DEVELOPMENT OF EUCARYOTES.   The next advance, which occurred about 1.6 billion years ago, was the appearance of the **eucaryotes,** organisms with a nuclear membrane. Eucaryotic cells are larger and more specialized than procaryotic cells. In addition to having membrane-bounded nuclei, they have numerous other small internal structures called **organelles.** The most favored hypothesis for the origin of intracellular organelles is called **endosymbiosis,** which means "living inside." Briefly, it is proposed that one free-living procaryotic cell ingested others. The relationship was mutually beneficial so it persisted. The inner cells over time lost their independence, and evolved into specialized cell structures found in higher organisms. This hypothesis, and the relevant cell structures, will be discussed in more detail in Chapter 3.

The greater specialization of eucaryotic cells and the fact that they also have the capacity for sexual reproduction are important benefits, because these features promote rapid evolutionary change. Because of these advantages, the pace of evolution accelerated when eucaryotic organisms appeared. All plants and animals, including human beings, are made up of eucaryotic cells. The modern-day microbial eucaryotes we will be studying include fungi, algae, and protozoa.

The next step in evolution was the appearance of **multicellular organisms.** Groups of cells working together enjoy several advantages. One of these is mutual protection. The first multicellular organisms to take advantage of this approach were basically aggregates of like eucaryotic cells. The sponge, the hydra, and the seaweeds are modern-day examples of this plan. Multicellular organisms are recent entrants to the scene; they did not appear in the fossil record until about 0.6 billion years ago.

The second major advantage of multicellularity is the potential for the cells within the organism to specialize in different functions, instead of each having to be able to do everything needed for survival as individuals. As cells aggregated, they also began to acquire differences that enabled groupings of different types of cell to perform specialized functions as tissues and organs. The stomach, for example, contains an epithelial lining, muscle tissues, gland tissue to secrete gastric juice, and nerve tissue to direct the stomach's function. Tissues and organs are found in plants and animals, but not in microorganisms. We will be concerned with human tissues and organs, as we study how they are affected by microorganisms.

A BRIEF HISTORY
OF MICROBIOLOGY

Microbiology, the study of microorganisms, has had a sweeping influence on other sciences, such as agriculture, medicine, biology, and genetics. We will review some of the historical milestones in microbiology, so that you may get a sense of how profoundly microbiological research has changed the quality of human life.

Scientific study of the natural world — study through careful observation — began several thousand years ago in the civilizations that were most advanced at the time, such as in China and Greece. The physical sciences — physics and astronomy — were the first to develop a body of useful, accurate knowledge, followed by the field of chemistry. Biology lagged behind for a long time because there remained a general belief that living things operated under rules different from those that governed nonliving things. Living organisms were believed to be animated by "spirits" that put them outside the **empirical** (based on observation) means of science. Natural processes such as fermentation, decay, and disease were considered mysteries that human reason could not, and should not, attempt to penetrate. A startling discovery was required to change this intellectual climate and start serious investigation.

THE DISCOVERY
OF MICROORGANISMS

Oddly enough, although there were already thousands of different plants and animals to study, it was the wholly unexpected discovery of an entirely new type of living thing, the microorganism, that most spurred scientific curiosity. The discovery was made by a persistent and inquisitive Dutch merchant and amateur scientist named Anton van Leeuwenhoek. He laboriously produced an expertly ground optical magnifying lens, which could magnify up to 200 times (Figure 1.7). Using this simple microscope, he examined all sorts of materials, among them hay infusions, tooth scrapings, blood, and semen. Between 1674 and 1723 Leeuwenhoek sent his observations in detailed letters filled with small sketches to the Royal Society of London. His major discovery was that every natural environment he sampled seemed to be filled with various microscopic creatures in unbelievable numbers. Leeuwenhoek's excitement and his in-

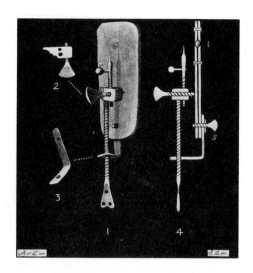
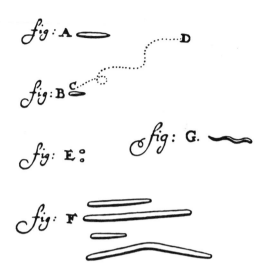

FIGURE 1.7   Leeuwenhoek's Microscope and Drawings
With this device, Leeuwenhoek observed most of the major groups of
microorganisms. The tiny hand-held instrument contained a single, very carefully
ground lens. The fluid specimen was placed in a drop on top of the screw point,
which could be raised or lowered. Then the viewer gazed through the lens at the
droplet in a strong light. Leeuwenhoek's careful drawings (right) created a furor
among his colleagues when he sent them to the Royal Society of London in 1684.
(Courtesy of Dover Publications, Inc.)

sistent emphasis on the vast numbers and wide distribution of a previously unknown form of life sparked a change in attitudes in the scientific community. Scientists became intrigued at the idea that living things, including those invisible to the eye, could lend themselves to scientific investigation. It was at this time that biology began to take its place among the other natural sciences.

THE BATTLE OVER SPONTANEOUS GENERATION

One widely held traditional belief now came under critical scrutiny — that of **spontaneous generation.** Until the second half of the 19th century it was generally assumed by laypersons that the mice and rats that infested people's homes, the maggots that appeared on decaying meat, and the flies found on manure arose spontaneously, without any particular cause — without par-

ents, in other words. This belief assumed that life can arise where there is no pre-existing life. A long sequence of scientific and not-so-scientific experiments were done over the course of two centuries in an effort to prove or disprove the theory of spontaneous generation.

In the 1660s, Francesco Redi, an Italian physician, attacked the maggot issue. In one experiment he placed some fresh meat in tightly covered jars and some in opened jars. He theorized that flies could not reach the meat in the sealed jars to lay eggs and that therefore maggots, the larvae that hatch from fly eggs, would not appear (Figure 1.8). As Redi had anticipated, maggots appeared on the meat in the open jars but not in the sealed jars. This experiment did not convince some critics, who argued that fresh air was required for spontaneous generation. To counter this argument, Redi, instead of sealing jars, covered them with gauze that allowed air in. Once again, no maggots appeared. By excluding the parent flies, Redi showed, there could be no off-spring. Spontaneous generation, at least insofar as it applied to animal pests of a visible size, was discredited.

Some years later the controversy over spontaneous generation erupted again, but the battle-ground had shifted to the mysterious microbes discovered by van Leeuwenhoek. At issue was whether microorganisms arose spontaneously in broths or infusions, or whether they entered from the outside. John Needham had found that when he heated fluids containing nutrients (such as chicken soup) before pouring them into sealed flasks, the solutions developed multi-tudes of microorganisms after they were cooled. Needham claimed that the microbes had gener-ated spontaneously.

The case for spontaneous generation of mi-crobes turned out to be difficult to disprove. First it was necessary to effectively sterilize the broth or other medium under study to remove any mi-crobes present at the beginning of the experi-ment. This was the only way to tell if any new microbes appeared. This is not as simple as it sounds, because microbes may be very difficult to destroy. Needham's heating methods did not completely sterilize his soups. Second, once ste-rility was achieved, it was necessary to keep mi-croorganisms from entering. It was the French chemist Louis Pasteur, who, in 1861 designed the ingenious experiment that showed that when microorganisms grow in a properly ster-ilized broth, they originate from microorganisms introduced from other sources.

Pasteur designed a special piece of glassware to contain his nutrient medium (Figure 1.9). This was a long-necked flask that could be effectively sterilized, yet its design would allow air in. After the broth was placed in the flask, its neck would be heated to melt the glass so it could be bent into a swan-neck shape. Then the whole flask and its contents would be sterilized and cooled. The curve in the neck of the flask would trap any airborne microbes before they could con-taminate the broth, yet the gases in the air would pass in and out without obstruction or alter-

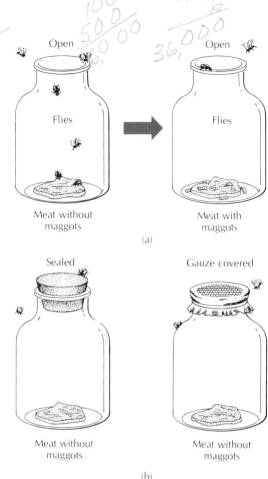

FIGURE 1.8 Redi's Experiment
Francesco Redi observed that when meat was placed in an open jar and left to sit uncovered, maggots eventually appeared on the meat (a). He esigned an experiment that disproved the theory of spontaneous generation of maggots. Redi covered the jars, first with a tight seal and then with a gauze cloth (b). The meat remained free of maggots. Thus, by preventing the parent flies from laying eggs on the meat, Redi showed that there could be no offspring.

ation. Pasteur demonstrated that a medium placed in such swan-necked flasks, then ster-ilized, remained clear and unchanged in any sur-roundings and for any length of time as long as the spout remained intact. If the fragile spout was snapped off, so that airborne bacteria could

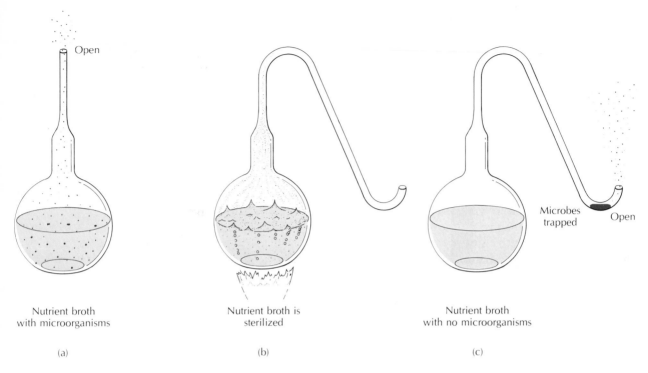

Open

Nutrient broth
with microorganisms

(a)

Nutrient broth is
sterilized

(b)

Microbes
trapped   Open

Nutrient broth
with no microorganisms

(c)

(d)

FIGURE 1.9 Pasteur's Swan-Necked Flask
When an ordinary straight-necked flask was filled with nutrient broth and left open
to the air (a), the broth soon became cloudy with microorganisms. Pasteur, however,
designed a special piece of glassware with a long swan-neck shape. He heated the
broth in the bent-necked flask to sterilize it (b) and then left it to cool. Although the
flask was open to the air, the curve in the neck of the flask trapped incoming
airborne microbes (c), preventing them from contaminating the broth. This photo (d)
shows a group of swan-necked flasks prepared by Pasteur himself. (Courtesy of
Pasteur Institute/Rockefeller University Press.)

drift in, the broth would become cloudy, and would be teeming with microorganisms within one or two days. Some of Pasteur's original swan-necked flasks still exist and remain sterile, as they were when he prepared them over 100 years ago.

It is now generally accepted that in present circumstances life does not generate spontaneously, but develops only from preexisting life. Yet you will recall that only a few pages ago we discussed the fact that many present-day scientists think that spontaneous generation did occur once, if not many times, under the vastly different conditions of the primordial world, about 4 billion years ago. Although we cannot say absolutely that spontaneous generation is not happening somewhere on earth now, there is no evidence that it happens in any setting that we've studied to date.

Once the idea of spontaneous generation was disproved, people acquired a rationale to try to control microorganisms. There is no point in trying to control something that can pop up again spontaneously at any time. But suppose we figure out that once we have killed microorganisms, they will not reappear unless we reintroduce them. Then we can move ahead and develop control measures such as disinfecting, sterilizing, preserving food by canning, and developing aseptic (microbe-free) surgical techniques. And this is precisely what has happened since Pasteur's time.

FERMENTATION

The resolution of the dispute over spontaneous generation opened the way to research on the natural processes of fermentation, decay, and disease. Microbiologists began to try to figure out exactly what microbes (they called them germs) did, how to make them do it in ways that suited human wishes, or, on the other hand, how to make them stop doing it. Although people had used and enjoyed yogurt, cheeses, leavened breads, and alcoholic beverages for countless generations, no one had ever investigated the biological processes involved in their production.

There were two schools of thought. One group of scientists felt that changes such as the fermentation process that turned malted barley into alcohol were strictly chemical; they thought that the microorganisms invariably observed in the fermenting material played no role. The other school were convinced that the microorganisms themselves caused the fermentation. Theodor Schwann, in 1837, was the first to design an experiment that clearly demonstrated that the microorganisms called yeasts play the key role in fermentation. Later, it was shown that different types of organisms could cause different types of chemical change. It was Pasteur, again, who during his investigations into the souring or "sickening" of wine, showed that the cause of the undesirable flavors was the activity of an unwanted microbial species. He developed a technique of heating the grape juice just enough to kill the offending organisms. This process, called **pasteurization,** or partial sterilization by heat, helped to save the French wine industry. Pasteurization was later widely adopted to kill unwanted bacteria in milk and other beverages.

By his work with wines, along with other studies, Pasteur established that all microorganisms were not the same. Rather, he discovered that there are many different specific microbes, each causing specific chemical changes. He also showed that the microbes involved in fermentation are able to live and reproduce entirely anaerobically. Building on his work, chemists studying fermentation developed methods for using microorganisms to produce industrially important chemicals, such as acetone and alcohols, on a large-scale basis. This was the first technological use of microorganisms.

THE GERM THEORY OF DISEASE

The discovery that fermentation is caused by microbial activity suggested to scientists that microbes might also cause changes in plants and animals. In particular, microbes might be the cause of disease. This idea was difficult for many to accept because, since medieval times, dis-

TABLE 1.1 The development of the germ theory of disease	
1530	Gerolomo Frascatoro, in a poem about syphilis, suggested that it is a transmissible disease.
1762	Marc Antony von Plenciz proposed an early statement of the germ theory, that disease was caused by living organisms and each disease had its own causative agent.
1861	Ignaz Philipp Semmelweis concluded that childbed fever was spread by doctors' contaminated hands and instruments, and published his recommendations.
1865	Joseph Lord Lister became convinced of the microbial origin of wound infection. He introduced the use of antiseptics in wound treatment and surgery.
1873	William Budd noted that typhoid outbreaks were caused by contaminated milk or water. He proposed the existence of a transmissible agent, several years before the bacterium was isolated.
1876	Robert Koch isolated the anthrax bacillus in pure culture and started experiments to establish its infectious nature.
1884	Robert Koch published his postulates for proving that a specific type of microorganism caused a specific disease.
1886	Adolf Mayer was the first worker to recognize viruses. He pioneered the study of viral diseases.

eases had been regarded as supernatural in origin, primarily punishments for sin. In 1546, an observant Italian, Gerolomo Frascatoro, argued that diseases could be transmitted by direct physical contact with an infected person, contaminated materials, and infected air. However, his ideas, published in the form of a poem, had no immediate effect on the untrained and superstitious medical practitioners of his time. But more demonstrations of the contagious nature of disease followed, and there was a gradual process of enlightenment. The burning question was "If 'something' is transmitted, what is the nature of that something?"

The **germ theory of disease,** first proposed in 1762, was fully developed in the 19th century, through the combined work of Robert Koch, a German microbiologist, Louis Pasteur, and their students and co-workers. It states that infectious diseases are caused by living microorganisms — and that one type of organism causes one specific type of disease. This theory became the driving force in the development of modern medicine and public health. The term **germ** was originally applied to microorganisms because it means "seed." Now that microorganisms have been more adequately classified, we use more specific terms such as bacterium or virus.

Being able to say that a specific microbe caused a specific disease was a major conceptual leap. Proving this cause–effect relationship, however, requires a series of logical experimental steps. These steps were first developed by Robert Koch, and are called **Koch's postulates.** The criteria established by Koch are still used today in investigating unsolved infectious diseases, such as virus-caused cancers and acquired immunodeficiency syndrome (AIDS) (see Chapter 22).

It was in Koch's laboratories that technical advances such as microbiological staining, the Petri dish, and solid gelled media made **pure culture** techniques possible. Koch and his co-workers learned to isolate one pure strain of microorganism from among a mixture of microorganisms, and maintain or culture it so that its unique properties could be assessed. Koch was studying anthrax, a lethal, highly contagious bacterial disease of domestic animals and humans. At that time anthrax was widespread, causing a financial disaster to European farmers.

In 1876, Koch established four experimental steps to discover the organism responsible for anthrax (Figure 1.10). Koch first isolated a suspect organism from animals sick with anthrax. He showed that the same type of organism was

FIGURE 1.10 Koch's Postulates
Koch, a brilliant young German physician, was the first to establish a series of experimental steps to prove that a specific microbe would cause a specific disease. According to Koch's postulates, bacteria in the blood of an animal that dies from anthrax can be isolated in a pure culture. When samples of the culture are injected into healthy animals, they too sicken and die. Bacteria isolated from the blood of this second group of animals are identical to the bacteria originally isolated. These bacteria will cause disease in any other healthy animals they infect. (Courtesy of Curtis B. Thorne, University of Massachusetts at Amherst.)

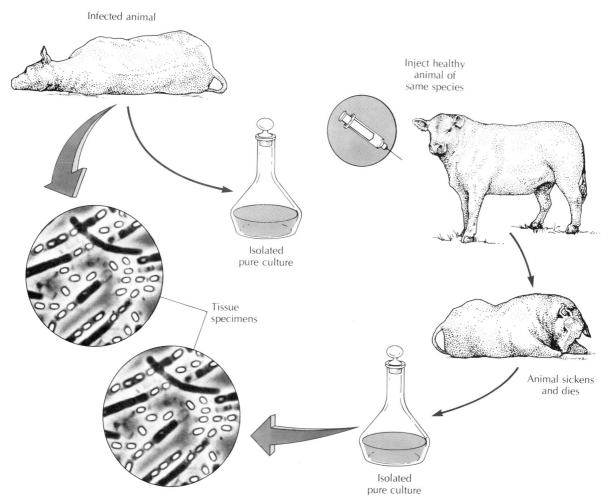

Infected animal

Inject healthy animal of same species

Isolated pure culture

Tissue specimens

Animal sickens and dies

Isolated pure culture

present in the blood of all of a large number of animals with anthrax. As the second step, he got the microorganism to grow in pure culture and established that there was only one type of organism present. In the third step, he inoculated healthy animals with some of this pure culture from his lab and observed that the animals developed anthrax. Fourth, he reisolated organisms from this second generation of anthrax cases and showed they were identical to those in the pure culture he had inoculated into the test animals. By completing these four steps, Koch could logically conclude that the specific bacterial strain he was studying was, indeed, the cause of anthrax.

Other European scientists quickly adopted Koch's methods and his experimental logic. In a brief, exciting period the causative agents of tuberculosis, diphtheria, tetanus, and other major microbial killers were found (Table 1.2).

We should note that Koch's postulates are designed to apply to infectious diseases, ones in which there is a microbial causative agent. We will see in a later chapter that it is much more difficult to establish a specific cause for non-infectious diseases such as heart disease or diabetes. Also, it would be unethical to use Koch's approach in working with any human diseases, such as cancer, where we might be "causing" a disease that is often lethal and for which we have no cure.

THE DEVELOPMENT OF PREVENTIVE MEDICINE

When you don't know where a disease comes from, or what causes it, you are essentially helpless to control it. At the time of the great plague epidemics, people had few choices — they could flee the cities, or they could stay and die. Doctors chose treatments from a large selection of mostly useless and frequently very harmful powders, potions, and salves.

Fortunately, when the cause of a disease is known, ideas for curing and preventing that disease come immediately to mind. The pioneer microbiologists who proved the germ theory also made the initial advances in disease control.

TABLE 1.2 Discovering the causes of bacterial infections		
Date	**Disease**	**Workers**
1876	Anthrax	Koch
1879	Gonorrhea	Neisser
1880	Malaria	Laverans
1880–1884	Typhoid fever	Eberth, Gaffky
1882	Tuberculosis	Koch
1883	Cholera	Koch
1883, 1884	Diphtheria	Klebs, Loeffler
1889	Tetanus	Kitasato
1894	Plague	Kitasato, Yersin
1905	Syphilis	Schaudinn, Hoffman
1906	Whooping cough	Bordet, Gengou

Methods for preventing infectious disease developed along two major lines. One is **immunization,** the process that protects individuals by increasing their resistance to an infectious disease. The other is **sanitation,** a group of methods used to protect communities by treating foods, water, and sewage to prevent the transmission of the microorganisms that cause disease. These two approaches were almost unknown a century ago (Figure 1.11). Now they are the backbone of today's preventive medicine.

The Chinese first observed that people contracted smallpox only once in their lives, and reasoned that smallpox survivors had acquired a permanent resistance to the disease. They developed variolation, a controlled exposure to the smallpox virus in a mild form to cause a mild case of the disease, which would protect the person throughout his or her life. This process frequently backfired, and it appears that about 3% of those who were variolated died. Variolation was introduced into England in 1718 by Lady Mary Wortley Montagu, who had observed the process in her travels in Turkey. Then in 1798, an English country physician, Edward Jenner, observed that those of his patients who had recovered from cowpox, a disease similar to but much milder than smallpox, also seemed to be permanently protected against smallpox. Using the fluid from human cowpox lesions, which contained the cowpox agent, Jenner scratched it into the skin of healthy volunteers. They usually got a localized skin infection, from which they promptly recovered. When the next smallpox outbreak occurred, Jenner observed that his volunteers did not get smallpox. He proved that at the cost of having a small scar at the injection site, people could be protected for life from disfiguring, often lethal, smallpox. The Latin name for cow, *Vacca,* became the basis for the term **vaccination,** denoting a procedure in which an altered disease agent or its products are used to confer immunity. We now routinely administer vaccinations for measles, poliomyelitis, and various other diseases.

Many years later, Pasteur devised methods

FIGURE 1.11 The Transmission of Disease
We tend to take for granted the protection offered by modern sanitation practices that prevent disease transmission through the water and sewage systems of communities. At the turn of the century, however, inadequate sanitation facilities were a major health problem. Here young children in New York, circa 1900, play in filthy gutters near the body of a dead horse. (Photo courtesy of the Library of Congress.)

to weaken infectious agents, so they lost their disease-producing ability and could be safely used in vaccines. This was a necessary advance because in most cases (smallpox being an exception) only the disease agent itself, or its chemical products, will induce the body to develop immunity. Pasteur used his methods to develop immunizations for anthrax and rabies. Because of his work and that of others, vaccinations for diphtheria, tetanus, and other agents were soon developed (Table 1.3).

In the United States we take for granted a clean, safe source of drinking water, but this sanitary measure is relatively new. Many infectious diseases are spread by the use of feces-contaminated water, a fact that was first shown in England for the disease cholera. Major cities

TABLE 1.3 The introduction of vaccination			
Disease	First Immunizing Material Used	Date of First Human Use	Workers
Smallpox	Serum from cowpox sore	1796	Jenner
Anthrax	Weakened bacteria	1881	Pasteur
Rabies	Infected brain tissue	1885	Pasteur
Tetanus	Inactivated toxin	1920	Glenny, Ramon
Diphtheria	Inactivated toxin	1923	Ramon
Polio	Killed virus	1954	Salk
Measles	Live virus	1963	Enders

first began building water and sewage treatment plants a little over a century ago. Other humble but important aspects of sanitation include pest control, garbage disposal, street cleaning, and a variety of simple but immensely valuable activities that prevent the spread of disease. These measures are discussed further in Chapters 14, 27, and 28.

MICROBIOLOGY IN TODAY'S WORLD

Three areas of research have dominated microbiology in this century. The first is microbial physiology — how microbial cells metabolize, grow, and control their vital processes. Investigation of microbial cells has given us valuable information about higher-level cell structures and has contributed to our knowledge of genetics, enzymes, and cellular communication. An invaluable offshoot of research in microbial physiology was the discovery and use of **antibiotics** — antimicrobial substances produced naturally by a microorganism.

A second research area is **virology,** the study of viruses. Because they are so small, viruses are difficult to see; because of their unique character, they are difficult to grow. These factors initially made progress in understanding viruses slow. However, the development of cell-culture methods and the very powerful electron microscope contributed to major advances in virology, especially in diagnosis and prevention of communicable viral diseases.

The third significant research area has been **environmental microbiology,** the study of the essential roles played by microorganisms in the biosphere, recycling crucial nutrients, producing oxygen, and removing wastes. These functions have been clarified through research done in this century, expanding our understanding of our soil, water, and mineral resources.

MICROORGANISMS AND RESEARCH

Microorganisms are ideal research tools. It is relatively simple to maintain tens of millions of them in a very small space. The methods are technically easy and, compared with the cost of maintaining equal numbers of mice, dogs, or geraniums, quite cheap. Microbial populations are fairly stable and one member varies little from another. This certainly is not true of more complex organisms. In addition, being able to use large numbers of experimental organisms simplifies the task of showing that an experimental result is statistically significant. Not to be forgotten is the fact that there are few restrictions on manipulating, altering, or killing microorganisms. Only a few types of research that are potentially risky to human beings require strict precautions.

Many biological researchers, although they might consider themselves biochemists, geneticists, or pharmacologists first, gain important information from their research with microorganisms. For this reason, what you learn from studying microbiology will help you understand exciting new scientific developments in a wide range of fields.

Research with microbes has a wide range of applications. Microbiologists may be involved in such diverse projects as spacecraft sterilization and the improvement of the disposable diaper. You may find a microbiologist out on a clam flat digging up specimens to find out if they are safe to eat, supervising the top-secret yeast cultures used by a major brewery, or field-testing a new vaccine in Central America. Like the microbes themselves, microbiologists engage in a great variety of beneficial activities.

MICROBIOLOGY IN INDUSTRY

Microorganisms play a significant role in industry, either because industries need to control unwanted microbial growth or because they depend on the activities of microorganisms to manufacture their products. Microbial activities help produce billions of dollars worth of industrial goods. We have already noted the use of microorganisms in producing raised bread and other bakery products; fermented dairy products, such as cheese and yogurt; alcoholic beverages; aged meats and sausages; and sauerkraut and other pickled foods. Microbes are also used in industrial fermentation to make commercial acetic acid (used in making vinegar) and pro-

pionic acid (used in making Swiss cheese). They are employed to produce solvents such as acetone, synthetic hormones, enzymes (chemical catalysts), insecticides, antibiotics, and vitamins. Microorganisms are also a potential food source. Scientists have recently cultivated **single-cell protein (SCP)**, a food supplement used in animal feed, which is produced by microorganisms grown on industrial waste.

In industry, unplanned microbial growth can also be a negative factor, causing spoilage problems, or contributing to degradation (breakdown) of certain products. Much research is directed at minimizing losses suffered in this way. The cosmetics industry develops products such as deodorants and mouthwashes that promise to reduce microbial growth on the body; but cosmetics such as hand lotion are often degraded by microbial growth. Microbial degradation causes losses in the lumber and paper industries. Bacteria and fungi rot fabric, corrode glass, leather, and metal, and gum up delicate electronic instruments.

CHANGING PUBLIC-HEALTH EXPECTATIONS

Americans today have a longer life expectancy than at any other time in history. Because of medical advances we have also come to expect that our later years will be relatively free of the crippling diseases and disabilities that routinely afflicted people in earlier eras with blindness and paralysis. We can get ready relief from the pain of bacterially caused sore throats, abscessed teeth, bladder infections — the kind of pain our predecessors simply had to bear.

A comparison of the leading causes of death in 1900 and in 1982 (Table 1.4) shows how far we have come in eradicating infectious diseases. In 1900, the average life expectancy in the United States was estimated at 47.3 years. Most of the leading causes of death were infectious diseases. There were over 400,000 cases of diphtheria a year, and more than 10% of children died in their

| TABLE 1.4 A comparison of the ten leading causes of death in 1900 and 1982 |||||||
| **1900** |||| **1982** |||
Rank	Cause of Death	Deaths per 100,000 Persons		Rank	Cause of Death	Deaths per 100,000 Persons
1	Pneumonia/influenza	202		1	Heart disease	321
2	Tuberculosis	194		2	Cancer	188
3	Diarrhea/enteritis	143		3	Cerebrovascular diseases	69
4	Heart disease	137		4	Accidents	43
5	Cerebral hemorrhage	107		5	Chronic lung disease	25
6	Nephritis	89		6	Suicides, homicides	21
7	Accident	72		7	Pneumonia/influenza	19
8	Cancer	64		8	Infant mortality	17
9	Diphtheria	40		9	Diabetes mellitus	15
10	Meningitis	34		10	Chronic liver disease, cirrhosis	15

first year. Now, by contrast, the combined U.S. life expectancy for both sexes and all races is about 74.4 years and is still rising. Infant mortality is down to less than 1.2%, and there were only two cases of diphtheria in 1984. The discovery and application of sanitation, vaccination, and antibiotics produced this revolution in public health.

The leading causes of death today are the degenerative diseases of aging, accidents, and life-style illnesses, such as lung cancer and diabetes. These are diseases that are not produced by infectious organisms and for which there are no vaccines or preventive drugs. In these more difficult and intractible health problems, infection has become a significant problem in a whole new way — as a complication that may arise as a degenerative or life-style disease worsens. For example, alcoholics with cirrhosis of the liver are frequently susceptible to infection because they have a very poor diet. Largely for this reason, in 1977 infectious disease in the United

States still accounted for about 29 million hospital-occupancy days per year, at a cost of over $4.8 billion. In less developed countries with poor sanitation and inadequate medical care infectious disease remains almost entirely uncontrolled, and is as serious a problem as it was in the United States at the turn of the century.

THE PROMISE OF MICROBIOLOGY

Since its beginning microbiology has been an exciting science in which to be involved. Microbiologists have always been quiet revolutionaries — Pasteur, Lister, and Koch totally shook up the prejudices and practices of the medical profession of their time, and they were certainly not universally thanked for it. Today, the mid-1980s, is another vitally exciting period, when a powerful group of new genetic-research tools allows workers to decode the cell's blueprint and quite literally direct the evolution of microbes, plants, and animals. New foods, medicines, plastics, energy fuels, agriculturally superior plants, and streamlined handling of toxic wastes are a few of the areas in which microbiology is currently making a contribution.

As in Pasteur's time, not everyone is comfortable with these developments. However, we can hope that great benefits will follow, as we learn to integrate our new powers into our society and our value systems.

What advances can we expect in micro-

biology in the years to come? The area that you have probably heard the most about is **genetic engineering,** the manipulation of genetic material in the laboratory setting. Modern biochemical techniques have made it possible to take small amounts of the DNA that provides the blueprint for important substances from an organism and attach these fragments to the DNA in a bacterium. The resulting DNA is called **recombinant DNA.** When bacteria take up and decode this recombinant DNA, they are able to make the desired substance — a nucleic acid or

BOX 1.2
Silks, Brocades, Parasites— A Glimpse into the Past

How did infectious disease touch people's lives in ancient times? We know there were plagues, and we know many children died in infancy. The privileged lives of kings, queens, and the nobly born are about all we have any knowledge of through historical records and archaeology.

The excavation of a 2100-year-old Chinese tomb revealed the beautifully clothed, carefully embalmed body of a noblewoman, probably in her 50s at her death, perhaps the emperor's consort. She lived in a peaceful era in total luxury. Hers was surely a privileged life.

Examination of her body revealed a poorly set fracture of the right arm. Lung x-rays showed tuberculosis scars. She had gallstones that may have caused excruciating pain. She probably walked with a cane.

Her intestines contained three types of microbial parasites: pinworms, whipworms, and schistosomes. She was overweight, had arteriosclerosis (hardening of the arteries), and probably died of a heart attack. If this was the medical history of a privileged person, what must it have been like to be a peasant?

If this woman lived today under even modest circumstances, think how many of her problems could be prevented or relieved. Yet we still do not have the answers for the diseases that perhaps stemmed from her privileged status (and ours) — obesity and heart disease.

Figure of a court attendant, Han Dynasty. Royal Ontario Museum, Toronto, Canada, the George Crofts Collection.

protein. Recombinant-DNA techniques have already made it possible to produce in large quantities a variety of natural substances that would otherwise be very expensive and difficult to produce. For example, scientists have produced insulin, a hormone that is essential for regulating blood sugar levels. They have also isolated interferon, a substance that combats viral infection and that is being investigated in the treatment of cancer. Although not everyone is comfortable with the idea of genetic manipulation, these techniques hold great promise in the alleviation and control of disease. We'll talk more about recombinant DNA techniques in Chapter 8.

We can also expect major advances in our understanding of viruses. As we'll see later in the text, the standard antibiotics do not help in viral diseases. Better antiviral drugs, therefore, are urgently needed, and many promising ones are under development. Shortly, new techniques for identifying viral proteins and products will yield a new generation of much improved viral vaccines. Our study of viruses has also opened up the study of oncogenes — growth-regulating and cancer-inducing genes. Using these insights, researchers are zeroing in on the basic cellular causes of cancer.

Our knowledge of the immune system is constantly expanding, yet we still lack answers as to how immunity is regulated. Many diseases, such as allergies and graft rejection, are caused by misdirected or inadequate immune function. At present we can only treat their symptoms, but progress in immunologic research may soon provide us with the tools to prevent or manage many of these conditions.

The world's population, and its food needs, are continuing to grow. Microbiology, through genetic engineering, may soon contribute hardier, more productive, and more nutritious edible plants, and thriftier, faster-growing domestic animals. We will also be increasing our knowledge of how microorganisms improve soil fertility, work in symbiosis with economically valuable species, and can be raised as foods. New foods, drugs, plastics, and fuels will appear on the market, and improved control methods will improve the quality and shelf life of many products, from yogurt to house paints.

Currently, we are coming to grips with an extremely difficult problem — the safe disposal of toxic chemicals and other dangerous wastes. If you recall that microorganisms are the decomposers of our world, you will understand their potential use in breaking down our wastes.

Microbiology, then, has come a long way. It led the way as the biological sciences emerged in the 19th century. It provided the means for many of the major biological discoveries of the 20th century. We may hope that it will continue to solve health problems and further improve the human condition in the 21st century.

SUMMARY

1. Microorganisms are very small life forms that we study with a microscope. They include the bacteria, algae, fungi, and protozoa.

2. All living things are composed of cellular units. Several levels of organization underlie the complete organism. To thoroughly understand cells, we must analyze atomic and molecular structure and interactions.

3. Microorganisms play an essential role in the biosphere. They live in complex relationships, called

symbioses, with other species. These can be mutually helpful, harmful, or a combination of both.

4. Microorganisms assist in cycling key elements such as carbon and nitrogen, and function as decomposers. Without them, multicellular life would not be possible.

5. All life forms share common characteristics: (1) remaining separate from the exterior environment; (2) using energy; (3) manufacturing macromolecules; (4) maintaining and replicating a blueprint of information; and (5) adapting to changes in the environment.

6. Hypotheses explaining the origin of life have been developed. It is thought that inputs of energy first caused the formation of small, and then more complex, organic molecules. Interaction among these molecules formed aggregations that could grow and divide. At first, such growth was un-regulated or random. Systems eventually developed to regulate the processes. At some point, the aggregates developed all the traits necessary to be defined as living cells. This took place sometime between 4 billion years ago (the age of the oldest rocks) and 3.5 billion years ago (the age of the oldest fossils).

7. The earliest life forms used nutrients from the environment. Later life forms developed the ability for photosynthesis, which led to the accumulation of atmospheric oxygen. At that point the dominant life forms developed aerobic metabolisms. Simple cell forms, the procaryotes, gave rise to more complex cells, the eucaryotes. These, in turn, developed into multicellular organisms.

8. Microbiology as a science is interdisciplinary. Its methods and discoveries depend on and contribute to studies in many other areas, including biochemistry, cell biology, genetics, medicine, pharmacology, public health, and ecology.

9. The discoveries of van Leeuwenhoek, Pasteur, Koch, and many others, and the discovery of pure culture techniques laid the groundwork for studies of the role of microorganisms in fermentation, disease, and natural environmental changes. Bacteria as research tools made possible many of the discoveries of the 20th century.

10. Cancer, immunological malfunction, degenerative and life-style diseases are now among the leading causes of death. Pollution, waste disposal, and resource scarcity are our leading environmental problems. Food shortages and population growth are worsening worldwide. Microbiology is in a pivotal position to contribute toward solving all of these problems.

DISCUSSION QUESTIONS

1. Compare how the five characteristics of life are manifested in a higher plant (such as a tree), in a bird, and in yourself.

2. Compare environmental conditions as they have been hypothesized to have existed on the earth 4 billion years ago and today.

3. Where, would you think, could an anaerobic microorganism survive in today's world?

4. Give some examples of aerobic organisms encountered in daily life.

5. Why will we be studying eucaryotic human cells in a microbiology course?

6. Find 10 examples of microbiological applications in your daily life.

7. What benefits are we still receiving from Louis Pasteur's research of a century ago?

ADDITIONAL READING

Brock T, ed. *Milestones in Microbiology*. Englewood Cliffs, N.J.: Prentice-Hall, 1961.

Curtis H. *Biology*, 4th ed. New York: Worth, 1983.

Fox SW, Dose K. *Molecular Evolution and the Origin of Life*. San Francisco: W.H. Freeman, 1972.

LeChevalier HA, Solotorvsky M. *Three Centuries of Microbiology*. New York: Dover, 1974.

Oparin AI. *The Origin of Life on the Earth*, 3rd ed. New York: Academic Press, 1957.

Woese CR. Archaebacteria. *Scientific American* 1981; 244:98-122.

THE BASIC CHEMISTRY
OF LIFE

CHAPTER OUTLINE

Why begin our study of microbiology with a chapter on chemistry? If you recall our discussion of microorganisms in Chapter 1, you will realize that most of the natural activities of microbes involve a series of chemical reactions. Microbes, like all other living organisms, including human beings, depend on a constant flow of energy to carry out life processes — to repair or build new substances, to eliminate wastes, to reproduce, and simply to stay alive. In this chapter, we will examine how chemical reactions can provide this vital flow of energy. First, however, let us look at the minute particles, the atoms and molecules, that are involved in these reactions.

THE ATOM

All matter is made up of combinations of **elements,** which are substances that cannot be bro-

ken down into simpler substances by ordinary chemical means. There are 92 naturally occurring elements, but of these only six (carbon, oxygen, nitrogen, hydrogen, sulfur, and phosphorus) make up 99% of living matter. The smallest unit of matter within an element is called an **atom;** particles consisting of two or more atoms joined together are called **molecules.** The breaking apart and putting together of these tiny building blocks — and the energy used or released in the process — are essential to all life functions.

SUBATOMIC PARTICLES

Let's look at the atom in greater depth. The atom is composed largely of space containing three subatomic particles: protons, neutrons, and electrons. **Protons** are particles with a positive charge. **Neutrons** are electrically neutral particles with approximately the same mass (quantity of matter in a body) as a proton. **Electrons** are negatively charged particles; their mass is negligible in relation to that of protons and neutrons.

Positive and negative charges attract each other. If positively and negatively charged particles come together, it will take work, or energy, to pull them apart. Alternatively, two similarly charged particles wil repel each other. An atom contains an equal number of protons and electrons. Since the charges on these are of equal magnitude and opposite sign, the atom as a whole has a net charge of zero — it is electrically neutral.

All of the atoms of a particular element have a characteristic number of protons in their nuclei. The number of protons in the nucleus of a single atom is called the **atomic number.** The **atomic weight** of an element is equal to the total number of protons plus neutrons contained in the nuclei of its atoms (Table 2.1).

ISOTOPES

Although each atom of a particular element must contain the same number of protons, the number of neutrons may vary. Therefore atoms of the same element may differ from one another in their atomic weights, but not in their atomic numbers; these different atoms are called **isotopes.** For example, while all carbon atoms have

six protons, one isotope of carbon has six neutrons per atom, while another has eight.

Several of the less common isotopes are **radioactive,** that is, their atomic nuclei are unstable and spontaneously emit energy or radiation. Since even small amounts of radioactivity can be detected by sensitive instruments, radioactive isotopes have a number of valuable uses in medicine and biological research. They may be used to determine the age of fossils; they may be used to destroy harmful cells such as cancer cells; or they may be used as "tracers" to mark the course of many important processes in living organisms.

ATOMIC STRUCTURE

We can visualize an atom as being made up of two basic parts. In its center is a dense, electrically positive nucleus that contains all the atom's protons and neutrons. The electrons are found in the much larger volume outside the nucleus. Diagrams of the structure of atoms and molecules may take different forms (Figure 2.1). Such diagrams are drawn without attention to accurate proportions, as atoms are mostly space and electrons are so small that it would be impossible to make a realistic rendering on a book-sized page. The Bohr model of atoms, named after physicist Niels Bohr, shows a central nucleus

TABLE 2.1 Atomic weight and number for the elements most common to living organisms				
Element	Protons	Neutrons	Atomic Number	Approximate Atomic Weight
Carbon	6	6	6	12
Oxygen	8	8	8	16
Hydrogen	1	0	1	1
Nitrogen	7	7	7	14
Sulfur	16	16	16	32
Phosphorus	15	16	15	31

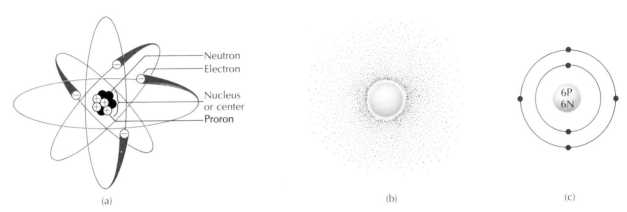

(a)   (b)   (c)

FIGURE 2.1   Atomic Structure
The planetary model (a) shows the electrons orbiting the nucleus as planets circle
the sun. A more modern concept of the atom (b) recognizes that we cannot be sure
exactly where an electron is at any point in time. In this diagram, the nucleus is
surrounded by an electron density cloud, indicating where the electrons are most
likely to be found. The Bohr concept (c) uses concentric circles to emphasize the
relative distances of electrons from the nucleus.

surrounded by electrons that orbit in concentric
three-dimensional shells. Because of its usefulness in visualizing what happens during interactions between atoms, we will use the Bohr
model in depicting most of the processes described in this chapter.

Each concentric shell can contain only a certain maximum number of electrons. For the innermost shell, that number is two. For the next
closest, it is eight. In moving away from the nucleus, greater and greater numbers of electrons
can occupy a shell. Each shell is generally
"filled" before the next shell begins to fill. For
example, oxygen has eight electrons. Two occupy and fill the innermost shell, and the remaining six are in the outer shell, which is not
completely filled. Sodium, with eleven electrons, has two in the first shell, eight in the next,
and one in its outermost shell (Figure 2.2). It is
important to note that the number of electrons
in the outer shell largely determines the chemical properties of an atom. For example, when
the outer shell is filled, the atom tends to be

stable; when the outer shell is incomplete, the
atom is more likely to react with other atoms.

MOLECULES

Although there are only 92 naturally occurring
elements, the enormous diversity of life stems
from the ability of atoms to combine with each
other and form thousands of different molecules.
When a molecule contains at least two different
kinds of atoms, it is called a **compound.** The
atoms in such molecules are held together by
chemical bonds, the forces of attraction between
atoms. Just as there are various ways of depicting
atoms, we can have different representations of
molecules. Let us look at a molecule of glucose,
one of the simple sugars. Each glucose molecule
is composed of 6 atoms of carbon, 12 atoms of
hydrogen, and 6 atoms of oxygen. As a sort of
chemical shorthand, we can represent the 6 carbons, 12 hydrogens, and 6 oxygens as $C_6H_{12}O_6$.
Atoms are represented by their element's sym-

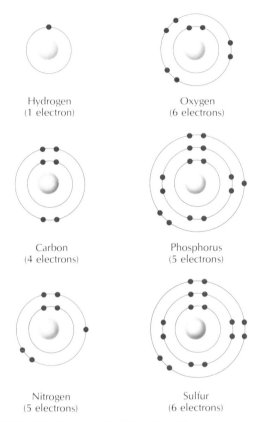

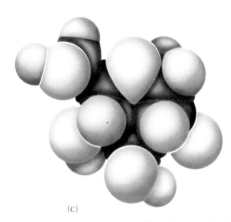

Hydrogen
(1 electron)

Oxygen
(6 electrons)

Carbon
(4 electrons)

Phosphorus
(5 electrons)

Nitrogen
(5 electrons)

Sulfur
(6 electrons)

FIGURE 2.2 Electron Shell Configurations
These are the electron configurations for some of
the elements most common to living organisms. In
each case, the number of electrons in the
outermost shell is indicated.

(a)

(b)

(c)

FIGURE 2.3 Representation of the Glucose Molecule
There are different ways of representing the glucose
molecule; three are illustrated here. In the flat
representation (a), covalent bonds are indicated by
straight lines. In fact, the atoms in most glucose
molecules in water are arranged in a ring structure
(b). The space-filling model (c) gives a rough idea of
the molecule's three-dimensional structure and the
relative sizes of the atoms.

bol, consisting of one or two letters usually derived from the element's English or Latin name.
For example, hydrogen is designated "H," while
sodium is "Na" from the Latin term for sodium,
Natrium.

This chemical shorthand, however, does not
tell us how these 24 atoms are joined together.
We can show the arrangement of atoms in molecules by using straight lines to depict the bonds
between atoms. Other, more complex representations, like space-filling models, show the three-
dimensional structure (Figure 2.3c).

We can use simple arithmetic to determine the molecular weight of a single molecule once we know what atoms it contains. We simply sum the weights of the various atoms that comprise it. The atomic weights of carbon, hydrogen, and oxygen are 12, 1, and 16, respectively (see Table 2.1). Therefore, the molecular weight of glucose is $6 \times 12 + 12 \times 1 + 6 \times 16$, which equals 180.

BONDING

As we have seen, molecules are highly specific arrangements of atoms. The atoms that make up a molecule are held together by chemical bonds. To understand how atoms join together to form bonds, let us return to the concept of electron shells.

The driving force behind the formation of chemical bonds lies in the tendency of each atom to fill its outer shell completely. Remember that each shell can contain a limited number of electrons and that atoms with the maximum number in their outer shell gain a special stability. To achieve this, an atom may give up one of its electrons to another atom or atoms may share one or more electrons with each other. The electrons in the outermost shell, that is, those directly involved with bond formation, are called the **valence electrons.** The **valence** of an atom is the number of electrons the atom must gain or lose in order to achieve a filled outer shell. Note in Table 2.2 that when an element has multiple valences, the most common valence in biological molecules is underlined.

In general, atoms with less than half their maximum number of valence electrons will tend to lose electrons, while atoms with more than half their valence electrons will gain electrons. Let us look at some examples. Oxygen has two electrons in its innermost shell and six in the

TABLE 2.2 Chemical elements in biological systems			
Element	Symbol	Atomic Number	Valence
Hydrogen	H	1	1
Carbon	C	6	4
Nitrogen	N	7	2, 3, 5
Oxygen	O	8	2
Sodium	Na	11	1
Magnesium	Mg	12	2
Phosphorus	P	15	3, 5
Sulfur	S	16	2, 4, 6
Chlorine	Cl	17	1
Potassium	K	19	1
Calcium	Ca	20	2
Iron	Fe	26	2, 3
Iodine	I	53	1

next shell (see Table 2.2). Since the second shell can contain a maximum of eight electrons, oxygen will combine with other atoms in such a way as to acquire two more electrons, filling its outermost shell. Similarly, sodium has eleven electrons in its three shells (two, eight, and one, respectively) and therefore must lose one electron in order to achieve a filled outer shell of eight electrons. Thus, oxygen has a valence of two, sodium a valence of one.

Valence is also a good guide to an atom's combining capacity. Oxygen, with a valence of two, can combine with other atoms to form two chemical bonds, while carbon, with a valence of four, can form four chemical bonds. Of the different bonding relationships that atoms can participate in, we will study three: ionic, covalent, and hydrogen bonding.

IONIC BONDS

Recall that since atoms contain equal numbers of protons and electrons, they are electrically neutral. Therefore if an atom gains or loses electrons, it will become negatively or positively charged, respectively. Such electrically charged particles derived from atoms are called **ions.** Let us see now how ions arise.

Take, for example, the formation of a sodium and a chloride ion from a sodium and a chloride atom. Sodium tends to lose its sole valence electron; with eleven protons and only ten remaining electrons it becomes a positively charged ion. Similarly, chlorine, with seven of eight maximum electrons in its outer shell, tends to gain an electron; with seventeen protons and eighteen electrons, it becomes a negatively charged ion. Since opposite charges attract, sodium and chlorine ions will be attracted to one another, forming the compound sodium chloride. This attraction between positive and negative ions is called an **ionic bond** (Figure 2.4). Ionic bonds are the "glue" that bind together many common nonliving substances such as table salt, baking soda, and concrete.

The positively charged particles that result when atoms lose electrons are called **cations;** the

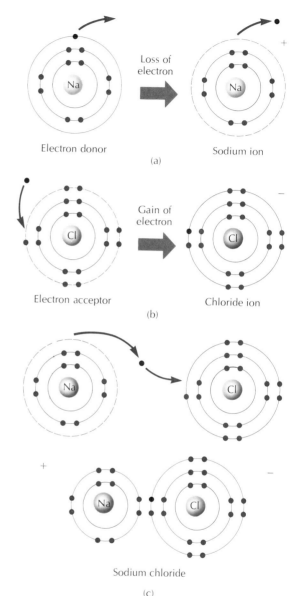

FIGURE 2.4 Ion Formation and Ionic Bonding
(a) Electron donors like sodium may lose the single electron in their outermost orbital, becoming positively charged sodium ions. (b) Electron acceptors like chlorine may gain one electron to complete their outermost orbital, becoming negatively charged chloride ions. (c) An electron donor (sodium) gives its electron to an electron acceptor (chlorine). Ions charged positively (sodium) and negatively (chloride) can form ionic bonds, yielding stable ionic compounds such as sodium chloride.

BOX 2.1
A Bacterium that Carries a Compass

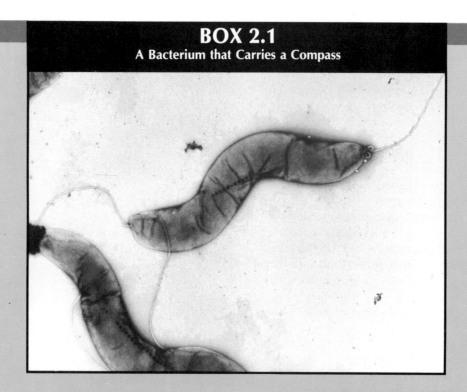

The spiral bacterium in this photomicrograph lives in the waters of a cedar swamp. It swims around by means of the flagella at each end. It is anaerobic, so it needs to move downward, toward the heavy, oxygen-free sediments, not upward, toward the air. Being a primitive organism, it lacks conventional sense organs to find its way. But it has a unique way to distinguish up from down.

Note the series of regular, cubical inclusions aligned with its long axis. These are crystals of magnetite (Fe_3O_4), an iron ore. In a magnetic field, the bacteria swing around to align with the field, like compass needles.

Magnetobacteria live in both the Northern and Southern Hemispheres. The vertical component of the Earth's magnetic field is opposite in the two hemispheres. Not surprisingly, once you come to appreciate the sheer ingenuity that bacteria use in adapting to their environments, we find that New England magnetobacteria and their New Zealand cousins have their internal magnetic poles reversed. Each knows which way is down in its own half of the world.

Magnetic crystals have been found in the abdomens of honeybees, the skulls of homing pigeons, and the brains of dolphins. These mineral crystals seem to be a rather widespread orientation mechanism in the biological world. Humans have been using compasses for only a little more than a thousand years. We seem to have been a bit slow in catching on!

Photo courtesy of D. L. Balkwill, D. Maratea and R. P. Blakemore, *Int. J. Syst. Bacteriol.* (1981) 31:452–455.

negatively charged particles that result when atoms gain electrons are **anions.** It should be noted that an atom can gain or lose more than one electron. For example, calcium can lose two electrons, forming a calcium ion with a positive charge of +2. Oxygen can gain two electrons to form an oxide ion with a negative charge of −2.

COVALENT BONDS

Another way for atoms to achieve stability is by sharing electrons, that is, by forming **covalent bonds.** In covalent bonding, neither atom loses its electrons completely, rather the electrons are shared between the two atoms. Such bonds are stronger and more stable than ionic bonds. Covalent bonds are the type used in the construction of all organic molecules. For example, human muscle fiber and the wood of trees are composed of molecules held together by covalent bonds; they are not only strong but extremely resilient.

An example of covalent bonding can be found in the formation of water molecules (H_2O). Each hydrogen atom needs to gain a *single* electron in order to fill its outer shell, while each oxygen atom needs *two* more electrons to fill its outer shell. In water, each oxygen atom shares one pair of electrons with each of the two hydrogen atoms. Thus, a water molecule contains two single covalent bonds (Figure 2.5).

When two atoms form a chemical bond, they may share only one pair of electrons in a single bond or more than one pair of electrons in double and even triple bonds (Figure 2.6). For example, two hydrogen atoms may complete their outer shells by sharing their single electrons to form a hydrogen molecule, H_2. On the other hand, in a carbon dioxide molecule, CO_2, each oxygen atom shares two pairs of electrons with the central carbon atom. A nitrogen molecule, N_2, contains a triple bond; each nitrogen atom shares three pairs of electrons.

POLAR AND NONPOLAR BONDS

Although atoms may share electrons, they do not necessarily share them equally. In the case of the hydrogen molecule, H_2, both hydrogen atoms have an "equal claim" to the shared electrons. Therefore, all parts of the bond are electrically neutral. This type of covalent bond is called **nonpolar.** On the other hand, in the case of the water molecule, H_2O, the oxygen atom exerts a greater pull on the shared electrons than do the hydrogen atoms. Thus, the electrons in the covalent bond are usually found closer to the oxygen atom than to the hydrogen atom. The region around the oxygen atom will be negatively charged, and that around the hydrogen positively charged. This unequal sharing of electrons between atoms is called a **polar covalent bond** (Table 2.3).

HYDROGEN BONDS

Some chemical bonds possess more energy than others. Bonds that have less energy are more stable than those with high energy. While ionic and covalent bonds are relatively strong, there are other chemical bonds that require little energy to break. One that is of special importance to

FIGURE 2.5 Covalent Bonding in the Water Molecule
One atom of oxygen forms a covalent bond with each of two hydrogen atoms. Each bond is formed by the sharing of a pair of electrons. Of the pair of electrons, one is donated by the oxygen atom and the other by the hydrogen atom. The shared electrons orbit all the atoms so that the oxygen atom has eight electrons to complete its outer shell, while each hydrogen shares the use of two electrons.

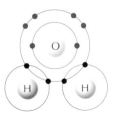

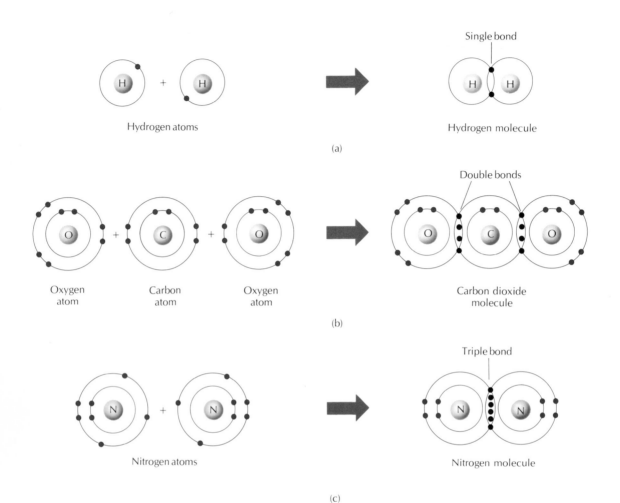

FIGURE 2.6 Single, Double, and Triple Bonds
(a) Two hydrogen atoms share their electrons to form a pair, creating a single covalent bond in a molecule of hydrogen gas. (b) Carbon and oxygen atoms bond to form carbon dioxide. The carbon atom shares two pairs of electrons with each oxygen atom, creating two double covalent bonds. (c) Two nitrogen atoms can form a molecule of nitrogen gas by sharing three pairs of electrons, forming a triple covalent bond.

living organisms is the **hydrogen bond,** a weak attraction between neighboring molecules that have electrons that are not equally shared and the resulting weakly charged regions. A hydrogen bond forms when the positive end of one molecule is attracted to the negative end of the other molecule. Water provides a simple example of hydrogen bonding, and hydrogen bonding in water is intermolecular, that is, between molecules (Figure 2.7).

TABLE 2.3 Some covalent bonds		
Bond	**Relative Polarity**	**Relative Bond Energy**
H—H	Nonpolar	+ +
O=O		+
C—C		+ +
=C=C=		+ + +
=N—H	Polar	+ +
C—O—		+ +
—C=O		+ + + +
—O—H		+ + +

In larger molecules, hydrogen bonding can occur between various portions of the same molecule. Such bonds are significantly weaker than ionic or covalent bonds. In fact, they have only about 1/20 the strength of covalent bonds. Nevertheless, although the effect of any one hydrogen bond is slight, their cumulative effect can be considerable. Some large molecules contain millions of hydrogen bonds, which add considerable strength and stability.

While polar molecules are joined by weak bonds similar to hydrogen bonds, nonpolar molecules also have weak attractions for one another. Polar molecules therefore tend to associate with other polar molecules and nonpolar with nonpolar. In addition, both polar and nonpolar molecules tend to arrange themselves in

FIGURE 2.7 Hydrogen Bonding Between Water Molecules
The water molecule is held together by polar covalent bonds, and thus each water molecule has a negatively charged region around the oxygen atom and positively charged regions around the two hydrogen atoms. Hydrogen bonds form between the positively charged hydrogen atoms of one water molecule and the negatively charged oxygen atoms of another neighboring water molecule.

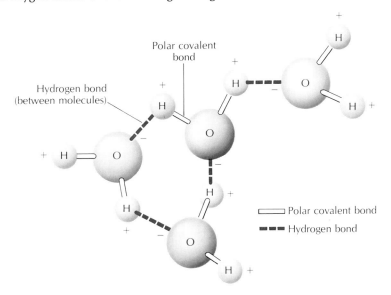

precise spatial patterns with regard to other molecules of similar or differing polarities — an important factor in the structure of living organisms. Polarity proves to be of great importance in the mechanisms by which the various components of living cells are assembled.

CHEMICAL REACTIONS

All biological processes can be viewed at their simplest level as rearrangements of atoms. Such rearrangements, involving the formation and breakage of bonds between atoms, are called **chemical reactions.** Energy changes accompany every chemical reaction. These energy gains and losses are the major influences directing biochemical processes as they take place in living organisms. The energy changes can be divided into two main categories: **exergonic** reactions, which release energy, and **endergonic** reactions, which absorb energy. The two types of reactions arise from the differing amounts of energy stored in chemical bonds. When bonds are broken, energy is released, whereas when bonds are formed, energy is absorbed.

Let us look at a very simple reaction, the formation of water:

$$2H_2 + O_2 \longrightarrow 2H_2O + \text{energy}.$$

The above equation tells us that two hydrogen molecules and one oxygen molecule (the reactants) combine to form two water molecules (the products). The water molecules have a lower energy level than the gas molecules from which they are formed, so that energy is released during the reaction. The energy released represents the difference between the two levels (Figure 2.8a).

A reaction may also occur in the opposite direction, in which the water molecules are split into hydrogen and oxygen molecules:

$$2H_2O + \text{energy} \longrightarrow 2H_2 + O_2.$$

In this case, the products, hydrogen and oxygen, have more energy than the reactants, the water molecules. Energy must be absorbed from the surroundings to make the reaction go (Figure 2.8b). This reaction is endergonic.

THE COLLISION THEORY

We will now look at how chemical reactions occur and the factors that affect the **reaction rate,** that is, how quickly a reaction occurs. The **collision theory** explains reactions in terms of collisions between atoms and molecules. According to this theory, atoms, ions, and molecules are in constant motion and must collide with each other in order to react. Yet, even if two particles collide, there is no guarantee that a reaction will take place.

For example, consider that a can of gasoline may sit unused in the back of a car for months. However, if someone should open the container and light a cigarette nearby, then an explosive reaction would occur — the gasoline would react with oxygen and burst into flames. Since collisions between gasoline and oxygen molecules had been occurring in the container for months, clearly an additional source of energy was needed to cause the reaction. The reaction required additional energy to proceed — in our example the spark from the match. The minimum energy required to start a chemical reaction is known as the **activation energy.** We can think of the activation energy as an energy requirement that must be overcome before a reaction can begin (Figure 2.9). Only those molecules colliding with sufficient energy to cross this barrier will form products. By increasing the temperature, we increase the kinetic energy (the energy associated with motion) of the reacting molecules. Since the molecules move at higher speeds, more collisions will have the required activation energy, and the reaction rate will increase. Any substance that can lower the activation energy needed for a reaction and thus speed up the reaction is called a **catalyst** (Figure 2.10).

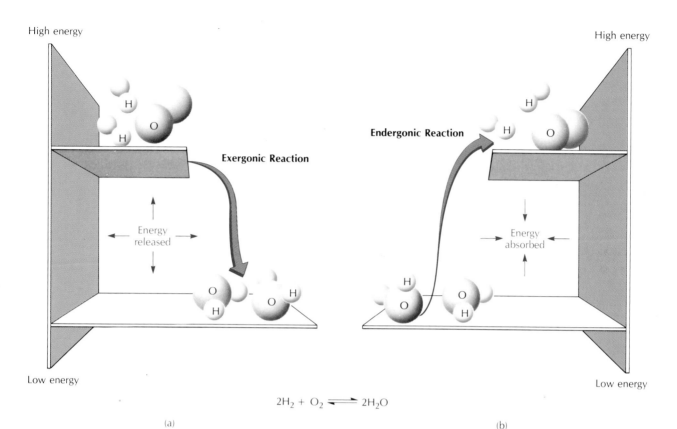

High energy

High energy

Exergonic Reaction

Endergonic Reaction

Energy released

Energy absorbed

Low energy

Low energy

$$2H_2 + O_2 \rightleftarrows 2H_2O$$

(a)

(b)

FIGURE 2.8 Chemical Reactions and Energy
In all chemical reactions, atoms are rearranged into different molecules from the starting materials. For example, (a) the hydrogen atoms in molecules of H_2 gas and the oxygen atoms in molecules of O_2 gas may be rearranged into water molecules. Energy is given off in this exergonic reaction. (b) Conversely, water molecules may be split and H_2 and O_2 gas formed. A great deal of energy is absorbed in this endergonic reaction.

Another factor influencing the reaction rate is the concentration of the reactants. Increasing the concentration means that more molecules are packed into a given space, thus increasing the frequency of collisions between reactants. The greater the number of collisions taking place, the greater the chance that a successful collision will occur. Thus, an increase in the concentration of reacting molecules increases the fre-quency of collisions and consequently the reaction rate.

EQUILIBRIUM

All reactions are theoretically reversible. Once products are formed, they can collide and be turned back into the original reactants. In time, all reactions establish an **equilibrium point** in which the number of conversions from reactants to products is exactly equal to that of products to reactants. Although both forward and backward reactions continue to occur, the net amounts of reactants and products remain constant. We show reversible reactions in the following way:

$$2H_2 + O_2 \rightleftharpoons 2H_2O.$$

In every reversible reaction, one direction is endergonic, or energetically "uphill," and the other is exergonic, or energetically "downhill." Gener-ally the equilibrium will favor the downhill reaction. This means that if the products are more stable than the reactants, there will be more of the products in the equilibrium mixture.

FIGURE 2.10 Catalysts Lower the Activation Energy
A catalyst can promote or accelerate a reaction by making it easier for the atoms to react, thus lowering the activation energy. For example, palladium metal is used as a catalyst to promote the controlled reaction of H_2 and O_2 gases.

FIGURE 2.9 Activation Energy
Before a reaction may occur, the oxygen atoms and hydrogen atoms must be separated from each other so that they are free to combine into water molecules. The activation energy to separate them comes from molecular collisions.

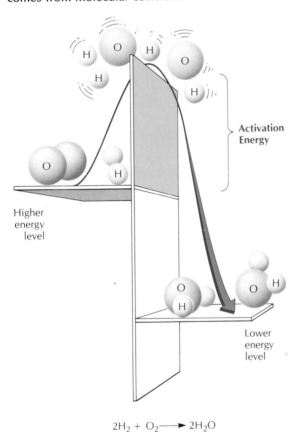

$$2H_2 + O_2 \longrightarrow 2H_2O$$

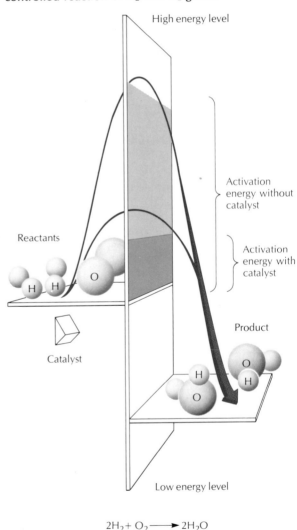

$$2H_2 + O_2 \longrightarrow 2H_2O$$

REACTIONS IN THE LIVING CELL

Although the principles of chemical reactions in living and nonliving things are fundamentally the same, the chemical reactions occurring in living organisms are vastly more complex. The rate of reaction in the living cell depends on both the concentration of reactants and minimum energy requirements. However, cells contain additional factors influencing the rates of reactions: the enzymes. **Enzymes** are large, globular protein molecules that act as specific catalysts; they speed up chemical reactions without being changed themselves. They participate in essentially all the chemical reactions performed by living organisms. Because of the influence of enzymes, cells can carry out specific reactions at high speeds and relatively low temperatures.

Enzymes reduce the time necessary for systems to reach equilibrium; they do not change the equilibrium point of a reaction. We will discuss enzymes and how they work in greater detail in Chapter 7.

Now, let us look at three types of reactions that are particularly important to the living cell: condensation, hydrolysis, and oxidation–reduction.

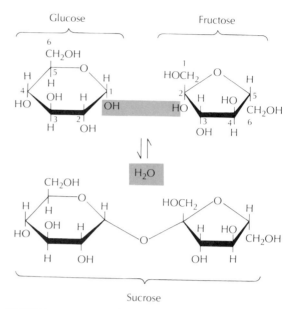

FIGURE 2.11   A Condensation Reaction
One molecule of glucose and one of fructose may react to form sucrose, a larger sugar. A molecule of water is eliminated in the reaction, which is called a condensation.

CONDENSATION REACTIONS

Condensation reactions are reactions in which water molecules are split off during bond formation. When molecules of glucose and fructose join to form sucrose (table sugar, a sweet substance obtained from sugar beets and sugar cane), a molecule of water is removed (Figure 2.11). Condensation is the cell's usual method for building large molecules from small subunits. All condensation reactions are endergonic.

HYDROLYTIC REACTIONS

A **hydrolytic reaction** is essentially the reverse process of condensation. In hydrolysis a molecule of water is added to the reacting molecule as it splits apart. Thus, sucrose can be split into its subunits, glucose and fructose, by the addition of water. Hydrolytic reactions are all exergonic.

Most of the food that we eat is made up of extremely large molecules or **macromolecules.** Inside our bodies, these macromolecules are hydrolyzed into small subunits during digestion. The small subunits can then be reconstructed into the body's own macromolecules through condensation reactions. The equation in Figure 2.12 shows the relationship between hydrolysis and condensation.

OXIDATION–REDUCTION

Earlier, in our discussion of ionic bonding, we discussed the transfer of electrons between atoms. The type of reactions in which electron transfers occur is called an **oxidation–reduction,** or **redox,** reaction. **Oxidation** is the removal of

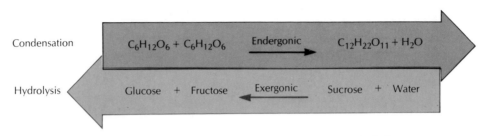

| Condensation | $C_6H_{12}O_6 + C_6H_{12}O_6$ | Endergonic → | $C_{12}H_{22}O_{11} + H_2O$ |
| Hydrolysis | Glucose + Fructose | ← Exergonic | Sucrose + Water |

FIGURE 2.12   The Relationship Between Condensation and Hydrolysis
These two opposite processes are used to build up or to break down large molecules, depending on the current energy and nutrient status of the cell or organisms.

electrons from a substance; oxidation is always exergonic. **Reduction** is the addition of electrons to a substance; it is always endergonic. Therefore in Figure 2.13 sodium is oxidized, and chlorine is reduced. Oxidation and reduction are always coupled in the cell — the electrons removed by oxidation are transferred to reduce another substance. It is through the oxidation of compounds such as glucose that the cell derives its usable energy.

The reactions that occur in the cell — including condensation, hydrolysis, and redox reactions — are collectively known as the cell's **metabolism,** which is the subject of Chapter 7.

FIGURE 2.13   Oxidation and Reduction
When sodium atoms lose electrons, not only do they become ions but they are also oxidized. When chlorine atoms gain electrons they become reduced chloride ions.

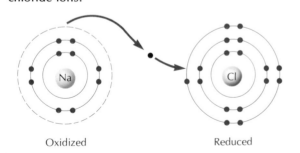

Oxidized   Reduced

WATER

In the rest of this chapter we will look at the compounds that are of fundamental importance to life. Although there are an enormous number of such compounds we can classify them into two broad categories. **Organic** compounds are those that contain carbon; they are found in all living organisms and form the substance of cells and tissues. Compounds that lack carbon are called **inorganic.**

Water, an inorganic compound, makes up the bulk (from 50 to 90%) of living organisms and is by far the most important compound on our planet. Without water, life simply could not exist. Let us look at three properties of water that account for its importance to living systems.

WATER AND THE HYDROGEN BOND

In water, each polar molecule is hydrogen bonded to four others. These hydrogen bonds have a cumulative effect and give water molecules an extraordinary tendency to stick together. The cumulative strength of the hydrogen bonds accounts for the high **surface tension** of water — its capacity to form beads or drops on surfaces such as a windowpane. Water moves by **capillary action;** because water molecules are unwilling to let go of one another, water may be drawn through narrow tubes or vessels. For example, if a thin glass tube is lowered into water, the water adheres to the inner sides of the tube.

Gradually, surface tension pulls a column of water up the tube until it is above the level of water in the container. Similarly, water moves through the tubules in the stem of a plant, the capillaries in the tissues of an animal, and even among the grains of sand and silt in the soil.

The strength of attraction between molecules is directly related to the amount of heat needed to change a liquid to a gas. Thus, a considerable amount of heat is required to break the hydrogen bonds in water in order for it to convert from solid to liquid to gas. Most similarly sized molecules, such as ammonia and methane, are gases at normal temperatures. Because of its relatively high boiling point, water can exist as a liquid over much of the earth's surface. This ability to remain in a liquid state makes water a good environment for many cells.

WATER AS A TEMPERATURE BUFFER

Hydrogen bonding is also responsible for water's ability to resist changes in temperature. While the hydrogen bonds that hold water molecules together are weaker than ionic or covalent bonds, they are much stronger than the forces between many other molecules. Compared with other substances, water requires a greater gain of heat to increase its temperature, and a greater loss of heat to decrease its temperature. Absorbed heat must first be used to break the hydrogen bonds between water molecules, rather than to increase the kinetic energy of the water molecules and thus increase the temperature of the water. Certain cell reactions that take place within a narrow temperature range can occur because of the tendency of water to maintain a constant temperature.

WATER AS A SOLVENT

Water can dissolve most of the compounds important to life. To understand what happens when a substance dissolves, let us look at sugar dissolving in water. Because it is a polar molecule, water is able to form weak bonds with other polar molecules. If sugar, a polar compound, is added to water, the sugar molecules are separated from each other and surrounded by water molecules (Figure 2.14a). When two substances mingle completely so that their molecules are evenly dispersed, we say a **solution** has been formed. The liquid that does the dissolving is called a **solvent,** while the dissolved substance is called the **solute.**

When water dissolves ionic compounds, the compounds tend to separate into cations and anions. These charged particles are then surrounded by water molecules. This process of splitting a compound into ions surrounded by water is called **dissociation.** When we dissolve salt in water, the salt crystals separate into sodium and chloride ions (Figure 2.14b). Since all the sodium chloride is transformed into ions, we say it completely dissociates. Other compounds may only partially dissociate in water. For example, when acetic acid (vinegar) dissolves in water, not all the acetic acid molecules dissociate; the result is a mixture of molecular and ionic forms in solution.

We have seen that water can dissolve both polar and ionic compounds. However, some important compounds, such as fats and oils, do not dissolve in water. These compounds are made up almost entirely of nonpolar bonds. As water is a polar molecule, it cannot break the forces between nonpolar molecules. To dissolve nonpolar substances, we need to use a nonpolar solvent, such as acetone.

ACIDS AND BASES

Water is partially dissociated. At any given moment, about one in every 10 million water molecules will spontaneously dissociate to yield a hydrogen ion (H^+) and a hydroxide ion (OH^-):

$$H_2O \longrightarrow H^+ + OH^-.$$

Many of the chemical reactions of the cell occur

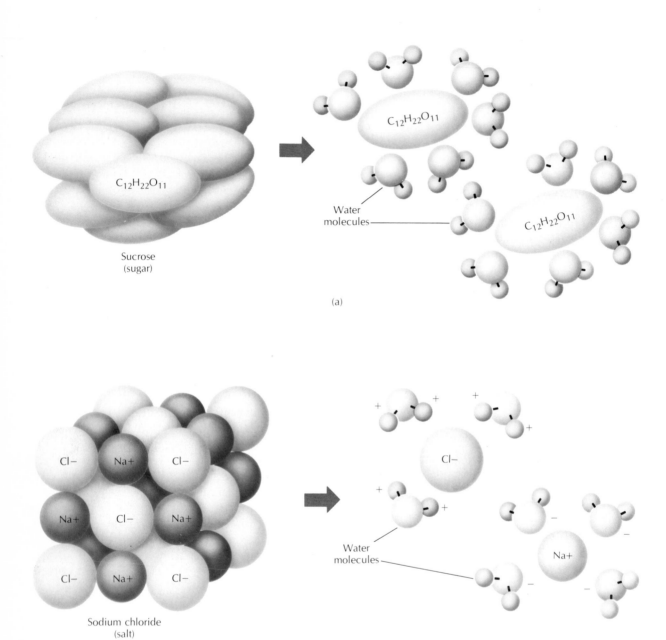

(a)

(b)

FIGURE 2.14 How Substances Dissolve in Water
(a) Sucrose is an example of a polar compound. It dissolves molecule by molecule. When we drop crystals of sucrose (table sugar) in water, each sucrose molecule becomes hydrogen-bonded to many water molecules. The sugar molecules disperse throughout the water. (b) Crystals of sodium chloride, an ionic compound, dissolve ion by ion. Both the positively charged sodium ions and negatively charged chloride ions become surrounded by the polar water molecules.

in water. These reactions are highly sensitive to the concentration of hydrogen ions in the solution. This important factor is often expressed as the **pH** of the solution. In mathematical terms, the pH is the negative logarithm of the hydrogen-ion concentration. In other words, as the hydrogen-ion concentration increases, pH goes down; as the hydrogen-ion concentration decreases, pH goes up.

Because pure water contains an equal mixture of hydrogen ions and hydroxyl ions, it is **neutral.** The pH of such a neutral solution is 7. Any solution that contains more hydrogen ions than pure water is **acidic** and has a pH less than 7; a solution with fewer hydrogen ions than water is **basic** and has a pH greater than 7 (Figure

2.15). In nature acidic environments are more common than basic environments, and many microbial species grow well under relatively acidic conditions.

An **acid** is a compound that dissolves in water, dissociates, and increases the free H^+-ion concentration. Most acids accomplish this by releasing H^+ ions and are therefore called hydrogen-ion donors. For example, hydrochloric acid (HCl) dissociates completely in water into hydrogen and chloride ions. It is called a **strong acid** because all of its hydrogen ions are free and contributing to the acidity of the solution. Compounds that only partially dissociate, such as acetic acid (CH_3COOH), are called **weak acids.**

A **base** is a substance that dissociates in wa-

FIGURE 2.15 The pH Scale
Naturally occurring solutions, household liquids, and laboratory chemicals can all be measured to find their pH, or concentration of hydrogen ions. Solutions with pH values of less than 7 are acidic. Solutions with pH values of more than 7 are basic.

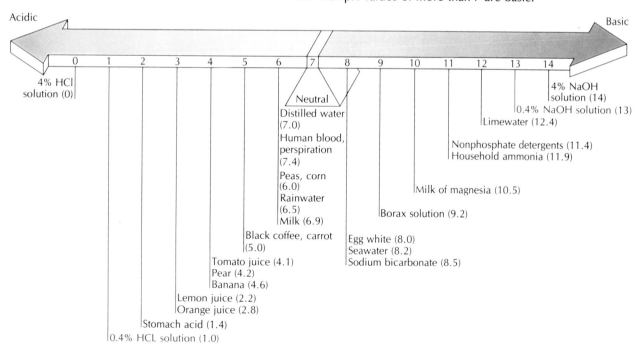

ter and causes a decrease in the hydrogen-ion concentration. Bases function as hydrogen-ion acceptors; these ions bind H⁺ ions. For example, sodium hydroxide (NaOH) dissociates in water into sodium and hydroxide ions (OH⁻). The hydroxyl ions accept hydrogen ions, combining with them to form water molecules, thus decreasing the hydrogen-ion concentration.

Each enzyme in a cell has a unique pH at which it is most effective. In unicellular organisms, the cell's internal pH must be kept within a narrow range or reactions will slow down and the cell will die. Different parts of the human body require different pH values, but fluctuations in pH are always carefully controlled. Living organisms accomplish this through the use of **buffers,** which are combinations of compounds that resist changes in pH, even if a strong acid or base is added.

ORGANIC COMPOUNDS

Carbon atoms normally form four covalent bonds. If these are all single bonds, the resulting molecule has a characteristic tetrahedral shape (Figure 2.16a). Carbon is noted for its ability to bond to other carbon atoms, forming long chains of atoms; however, it can also form branched-chain and ring structures (Figure 2.16b). These long chains that form in many biological molecules are often called "carbon backbones." In addition, carbon can bind to other atoms such as hydrogen, nitrogen, and oxygen, the elements found most frequently in organic compounds.

FIGURE 2.16   Carbon Atoms and Organic Molecules (a) The carbon atom may form four covalent bonds at equal distances from each other as shown in the compound methane (CH₄). (b) Carbon atoms readily bond to each other. They may form straight or branched chains, with single or double covalent bonds. Carbon atoms may also form rings with single or double bonds. Organic molecules are built upon these types of carbon backbones.

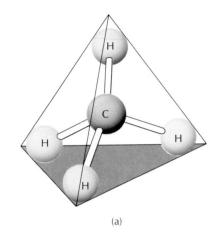

(a)

Double bond — — Single bond

Straight chains

Branched chains

Rings

(b)

The simplest type of organic compound, the **hydrocarbon,** is a compound that contains only carbon and hydrogen. The characteristic structure is a chain or ring of carbon atoms, each with attached hydrogen atoms. Hydrocarbons are found in living organisms in fats and oils, and in fossilized plant and animal remains, such as peat, petroleum, natural gas, and coal.

FUNCTIONAL GROUPS

More complex organic molecules are formed as atoms or groups of atoms of other elements are added to a carbon "backbone." As we look at organic compounds, we can see that they are composed of small groups of atoms, called **functional groups,** which confer specific chemical properties on the parent compound. To illustrate, one functional group consists of an oxygen that is attached to a hydrogen. This –OH combination may be attached to various carbon backbones. The –OH group is called an **hydroxyl group** (Figure 2.17) and the compounds formed are called alcohols. If we designate the remainder of the molecule by the letter R, we can say that alcohols have the general structure R–OH. The hydroxyl group is polar; molecules containing many hydroxyl groups, such as glucose, dissolve readily in water.

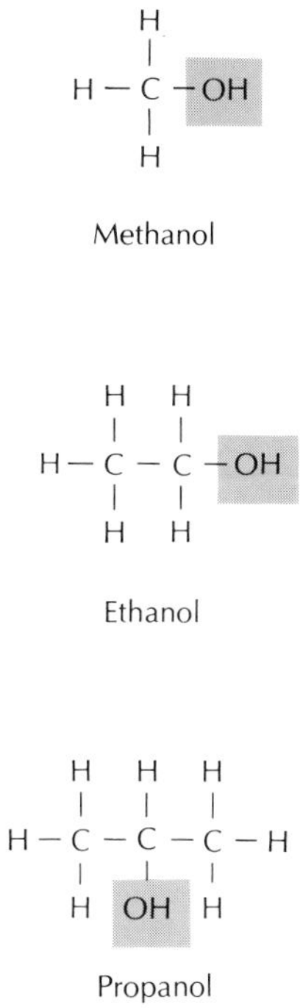

Table 2.4 lists the other common functional groups and the types of compounds that contain them. It should be noted that a molecule may contain more than one functional group. An **amino acid,** for example, is a molecule that contains both an amino (–NH_2) group and a carboxyl (–COOH) group and can be represented as

$$H_2N-\overset{\overset{\displaystyle R}{|}}{\underset{\underset{\displaystyle H}{|}}{C}}-COOH.$$

Large organic molecules are manufactured in the cell by linking together small molecules called subunits or **monomers.** The resulting macromolecule, composed of a sequence of small molecules hitched together by covalent

FIGURE 2.17 Alcohols
The alcohols show the relationship between a carbon backbone and the functional groups that are attached to it. Methanol is composed of a one-carbon backbone and a hydroxyl group. Ethanol has a two-carbon backbone and a hydroxyl group. Propanol has a three-carbon backbone and a hydroxyl group.

bonds, is a **polymer.** For example, proteins are macromolecules made up of hundreds of small amino acids joined together by condensation reactions.

Organic macromolecules are divided into four groups: carbohydrates, lipids, proteins, and nucleic acids. As we shall learn, each of these groups performs different functions in the cell.

THE CARBOHYDRATES

Carbohydrates are a large group of organic compounds predominantly composed of carbon, oxygen, and hydrogen; they include the sugars, the

TABLE 2.4 Functional groups that appear in organic molecules and confer chemical properties

Group	Name	Characteristics	Biochemical Sources
R—OH	Hydroxyl group	Polar, forms hydrogen bonds	Sugars, alcohols
$R-C{\lessgtr}^{O}_{OH}$	Carboxyl group	A hydrogen ion donor or acid	Fatty acids Amino acids
$R-\overset{R}{\underset{}{C}}=O$	Carbonyl group or keto group	Strongly polar, highly reactive group	Sugars, polypeptides Triglycerides
$R-\overset{H}{\underset{}{C}}=O$	Aldehyde group	Even more reactive than keto group	Sugars Aldehydes
$R-\overset{H}{\underset{H}{C}}-H$	Methyl group	Nonpolar grouping	Hydrocarbons Fatty acids Amino acids
$R-N{\lessgtr}^{H}_{H}$	Amino group	Polar, a potential hydrogen ion acceptor or base	Amino acids Nucleic acids Proteins sugars
$R-\overset{O}{\underset{OH}{P}}-OH$	Phosphate group	May ionize one to three times (a strong acid)	Phospholipids Nucleotides Nucleic acids
R—SH	Sulfhydryl group	May form high-energy linkage to carbonyl group; forms disulfide bridges in proteins	Cysteine

starches, and other related compounds. The primary function of the carbohydrates is to act as the principal fuels that drive cell reactions. Secondarily, they serve as energy reserves. The various other functions of carbohydrates include acting as structural elements in cell walls and assisting in the synthesis of fats and/or fatlike substances. Two simple forms of carbohydrates, monosaccharides and disaccharides, are known as sugars, and many have the characteristic sweet taste associated with cane sugar.

The ratio of hydrogen to oxygen atoms in carbohydrates is 2 to 1. Therefore the general formula for a carbohydrate can be expressed as $(CH_2O)_n$, where n represents the number of carbon atoms. For example, the carbohydrate glucose, with six carbons (see Figure 2.3), can be written as either $(CH_2O)_6$ or $C_6H_{12}O_6$.

Monosaccharides. The simple sugars that contain three to seven carbon atoms are called **monosaccharides.** All monosaccharides have specific numbers of carbon atoms. For example, pentoses contain five carbons and hexoses six carbons. Pentoses such as deoxyribose and ribose are found in DNA and RNA, the molecules that carry hereditary information. Perhaps the most familiar hexose is glucose, the main "fuel" of the cell; other nutritionally important hexoses are fructose and galactose.

Monosaccharides contain many alcohol side groups, a factor that makes them highly polar. Therefore they are readily soluble in water. Certain monosaccharides contain other functional groups, such as amino groups, as well.

Disaccharides. The carbohydrates formed from the condensation reaction of two mono-

saccharides are called **disaccharides,** or double sugars. (In Figure 2.11, we saw how the removal of water from the monosaccharides glucose and fructose formed the disaccharide sucrose.) Similarly, the monosaccharides glucose and galactose can combine through condensation to form lactose, or milk sugar. Disaccharides can also be split through a hydrolytic reaction into smaller molecules. (Recall how the addition of water to sucrose caused the molecule to split into fructose and glucose.) The condensation and hydrolytic reactions for a particular disaccharide are usually catalyzed by two different enzymes (Table 2.5).

Polysaccharides. Any sugars not used immediately by the cell are linked together by condensation reactions to form macromolecules. A carbohydrate composed of eight to hundreds or even

TABLE 2.5 Common and not-so-common sugars		
Group	**Name**	**Source**
Hexose	Glucose	Found in almost all organisms
	Fructose	Keto sugar, found in fruits, honey
	Galactose	Found in milk
Pentose	Ribose	In ribonucleic acid (RNA)
	Deoxyribose	In deoxyribonucleic acid (DNA)
	Xylose	From various wood products
	Arabinose	From pine gums
Alcohol	Inositol	Human B-vitamin
sugars	Mannitol	From seaweed polysaccharides
	Xylitol	From xylose; "sugarless" chewing gum
	Sorbitol	From various berries
	Dulcitol	
Amino	Galactosamine	Cartilage
sugars	N-acetyl-glucosamine	Bacterial cell-wall polymer
	N-acetyl-muramic acid	Bacterial cell-wall polymer
	Neuraminic acid	Mucoproteins
Keto sugar	Ascorbic Acid	Vitamin C
Disaccharides	Sucrose (Composed of glucose plus fructose)	Cane syrup
	Maltose (Composed of glucose plus glucose)	Hydrolyzed starch
	Lactose (Composed of glucose plus galactose)	Milk

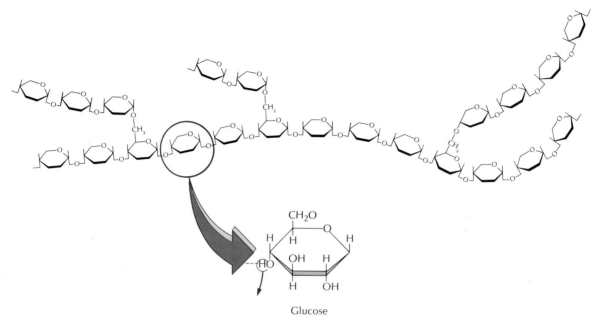

Glucose

FIGURE 2.18 Polysaccharide Chains
The polysaccharide glycogen consists of repeating units of a monosaccharide, glucose, with frequent branches.

thousands of monosaccharides is called a **polysaccharide.** These are more stable than sugars and although as polar, they are less soluble because of their huge size. Polysaccharide chains, which may be either straight or branched, commonly consist of repeating units of a single monosaccharide, such as glucose (Figure 2.18). Like the disaccharides, polysaccharides can be split into simple sugars by hydrolytic reactions.

Polysaccharides have many important functions. Glycogen, found in animal tissues, is stored in certain cells and hydrolyzed for fuel. Some polysaccharides, such as the starches, are easily broken down by human digestive enzymes. Others, such as cellulose and agar, have linkages that our enzymes cannot handle. Carbohydrates are also the major structural material used in the cell walls of microorganisms and plants. Cellulose alone makes up over 50% of the dry weight of the world's biomass.

THE LIPIDS

Probably the lipids most familiar to you are the fats, oils, and waxes. **Lipids** are a diverse group of organic compounds that will not dissolve in water. They are insoluble because they are nonpolar and therefore **hydrophobic** (literally, "water fearing"). The lipids are generally made up of carbon and hydrogen, with only small amounts of oxygen and occasional traces of nitrogen and phosphorus. Important as an energy source and as a component of cell membranes, lipids are essential to living cells.

Triglycerides. The simplest lipids, fats and oils, are formed by combining three fatty acid molecules with a single glycerol molecule (an alcohol with three hydroxyl groups). They are called **triglycerides. Fatty acids** are long hydrocarbon chains, usually with an even number of carbons, ending in a carboxyl group. A fatty acid is **saturated** when each of its carbon atoms is attached to the maximum number of hydrogen atoms. However, when carbon atoms are linked by double bonds, each carbon holds only one hydrogen atom and the fatty acid is **unsaturated.** If two or more double bonds occur, the fatty acid is **poly-unsaturated.** Through a condensation reaction, one fatty acid will combine with each hydroxyl group of glycerol, forming monoglycerides, diglycerides, and triglycerides (Figure 2.19). In the process, an **ester linkage,** or

$$-O-\overset{\displaystyle O}{\overset{\displaystyle \|}{C}}-$$

group, is formed.

FIGURE 2.19 The Structure of a Triglyceride
A triglyceride is formed by condensing a glycerol backbone molecule with three fatty-acid molecules. The hydroxyl groups of the glycerol condense with the carboxyl groups of the fatty acids, forming ester linkages. Three molecules of water are given off. Stearic acid is a saturated fatty acid; oleic acid is an unsaturated fatty acid; linoleic acid is a polyunsaturated fatty acid.

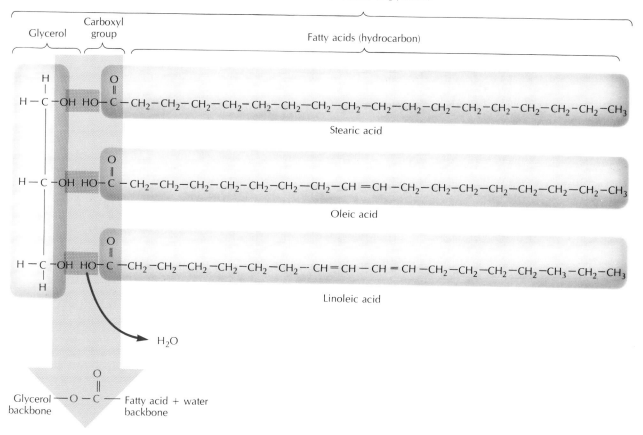

Triglycerides, the familiar dietary fats and oils, are used by plants and animals for storing energy. Gram for gram, lipids contain more calories than any other biological chemical. Because they are insoluble in water, they can be deposited in special areas of the cell or organism.

Phospholipids. The complex lipids composed of glycerol, two fatty acids, and one organic molecule containing a phosphate group are called **phospholipids** (Figure 2.20a). They have an unusual structure composed of both polar and nonpolar regions: the long fatty acids make up the

FIGURE 2.20   Phospholipids
(a) A phospholipid is formed by joining a glycerol molecule bearing a polar organophosphate group to two fatty acids. (b) Phospholipid molecules in water tend to group themselves to form bilayers arranged to create microscopic spherical bodies.

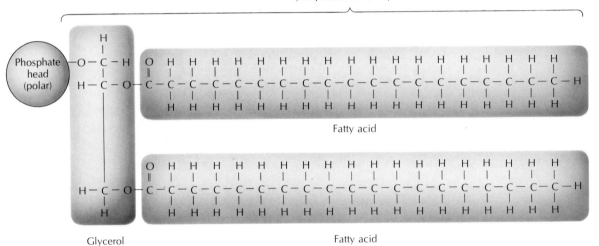

(a)

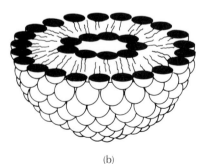

(b)

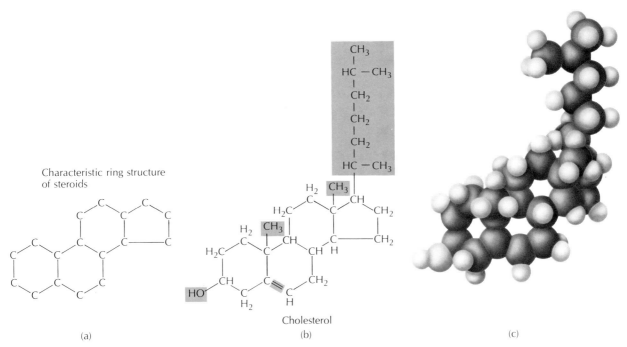

Characteristic ring structure of steroids

Cholesterol

(a)   (b)   (c)

FIGURE 2.21   The Steroids
(a) Steroids are lipids with a characteristic four-ring carbon backbone. (b) A detailed structure for cholesterol, a lipid important in the structure of animal-cell membranes and in digestion. (c) A space-filling molecular model of cholesterol.

nonpolar region, and the phosphate group produces a polar region.

When placed in water, phospholipid molecules will twist and orient themselves in such a way as to place all nonpolar portions in contact only with nonpolar neighbors. All polar portions, on the other hand, will be oriented toward the polar water molecules with which they then form hydrogen bonds (Figure 2.20b). Because of their characteristic behavior, phospholipids are a key component in biological membranes, forming a barrier that separates the contents of the cell from the water-based environment.

Steroids. A third type of lipid with similar solubility characteristics as triglycerides and phospholipids are the **steroids.** They have a characteristic multi-ring structure (Figure 2.21). Those that contain hydroxyl groups in certain positions are called **sterols.** The most well-known sterol is cholesterol, a substance important in human digestion. With other sterols, it is an important component of the membranes of fungi, protozoa, and animal cells. However, sterols are found in only one group of bacteria, the mycoplasmas.

THE PROTEINS

Our third class of organic compounds, the proteins, are crucial to the cell's moment-to-moment functioning. In fact, the synthesis of proteins is one of the most important tasks to which a cell directs its energy. A single cell may contain hundreds of different proteins, each playing a unique role. Some proteins are structural, forming parts of the cell such as membranes, ribosomes, or flagella. In contrast to the rigid structure of many polysaccharides, proteins confer elastic strength. Certain hormones (for example, insulin) and molecules connected with immunity to disease (antibodies) are proteins, as are toxins produced by some microorganisms that cause disease. Proteins are also involved in cell transport and in the transmission of hereditary traits. Perhaps the most crucial role of a cellular protein is as an enzyme; enzymes are the molecules that control each step in the cell's metabolism.

Proteins are macromolecules made up of a sequence of amino acids and containing the elements carbon, hydrogen, oxygen, nitrogen, and sometimes sulfur. They are assembled from 20 naturally occurring amino acids. Each amino acid contains a carbon atom bonded with at least one carboxyl group, one amino group, and a side group called an R group. This R group may take different forms—a single hydrogen atom, a branched chain, or a ring structure (Figure 2.22). It is the R group that lends each of the 20 amino acids its distinctive physical and chemical properties.

Proteins are formed step by step through condensation reactions between the carboxyl group (–COOH) of one amino acid and the amino group (–NH$_2$) of another. The resulting covalent bond is given the special name of **peptide bond** (Figure 2.23). A molecule formed from the condensation of two amino acids is a dipeptide, from three amino acids a tripeptide, and so on. Long chains of amino acids form macromolecules called **polypeptides.**

Many proteins exist as long chains folded on themselves in complex patterns; the end result appears globular. To understand this, we must realize that proteins have several levels of structure, one imposed on the next.

FIGURE 2.22   The Amino Acids
There are 20 common amino acids used in the assembly of proteins. Each bears an amino group and a carboxylic acid group. The individual amino acids have different R groups or side chains (color), which influence their chemical properties. Each amino acid has a name and a three-letter standard abbreviation.

Primary structure.   The simple sequence of amino acids, held together by peptide bonds (Figure 2.24a) is called the **primary structure.** This primary structure determines the fundamental chemical and physical properties of the polypeptide. Thus, any change in the order of linked amino acids may affect the structure and function of the entire molecule.

Secondary structure.   The repeated coiling or zigzag pattern in proteins that results from the formation of hydrogen bonds between amino acids is the **secondary structure.** This two-dimensional arrangement may take the form of either an alpha-helix or a beta-pleated sheet (Figure 2.24b).

Tertiary structure.   The long amino acid chain can next be folded into a three-dimensional conformation. The three-dimensional shape created by the protein's folds and bends is the molecule's **tertiary structure** (Figure 2.24c). The complex forces involved include different interactions between amino acids. For example, sulfhydryl groups (–SH) that are present on certain amino acids may combine with one another to form covalent disulfide (–S–S–) bonds with the removal of the hydrogen atoms. In conjunction with the effects of weak bonds, these interactions may result in a roughly spherical or globular molecule.

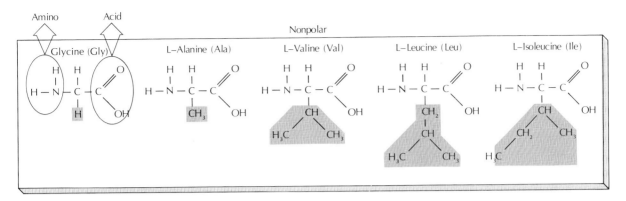

Nonpolar

Amino   Acid

Glycine (Gly)   L–Alanine (Ala)   L–Valine (Val)   L–Leucine (Leu)   L–Isoleucine (Ile)

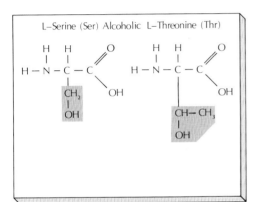

L–Serine (Ser) Alcoholic L–Threonine (Thr)

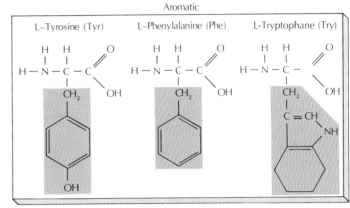

Aromatic

L–Tyrosine (Tyr)   L–Phenylalanine (Phe)   L-Tryptophane (Try)

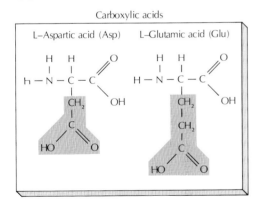

Carboxylic acids

L–Aspartic acid (Asp)   L–Glutamic acid (Glu)

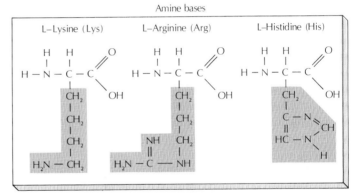

Amine bases

L–Lysine (Lys)   L–Arginine (Arg)   L–Histidine (His)

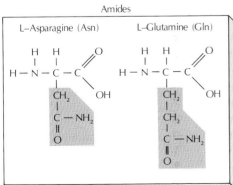

Amides

L–Asparagine (Asn)   L–Glutamine (Gln)

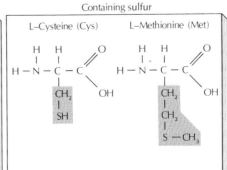

Containing sulfur

L–Cysteine (Cys)   L–Methionine (Met)

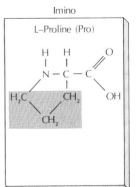

Imino

L–Proline (Pro)

Depending on the folds, the surfaces of some globular proteins have indentations and projections of various sizes and shapes. Many such folded proteins have polar amino acids on their surface and therefore have hydrophilic (water-loving) surfaces — even though their inner surfaces may be strongly hydrophobic (water-fearing). Others may fold so that the hydrophobic regions are on the outer surface. Thus, certain protein molecules interact with the aqueous environment of the cell, while others associate with nonpolar membrane lipids.

Tertiary structure is flexible enough to undergo reversible changes. That is, a protein can change its surface conformation temporarily while performing its function, then return to its original form. These changes are crucial to the operation of enzymes.

Quaternary structure. Some globular proteins take a final **quaternary structure** that is the result of several polypeptides combining into one large molecular complex (Figure 2.24d). A common example is hemoglobin, which includes four interacting protein units. Hemoglobin is made up of two alpha chains and two beta chains. The quaternary structure represents the bonding relationships between these polypeptides.

THE NUCLEIC ACIDS

Nucleic acids are probably the largest molecules in living cells, and they carry the cell's hereditary information. These nucleic acids fall into two basic classes: **DNA** (deoxyribonucleic acid) and **RNA** (ribonucleic acid). Like proteins, they are linear polymers. However, whereas proteins are constructed from amino acids, the basic units of nucleic acids are **nucleotides.** A nucleotide is made up of three connected parts: a nitrogenous base, a five-carbon sugar, and from one to three phosphates (Figure 2.25). In DNA the

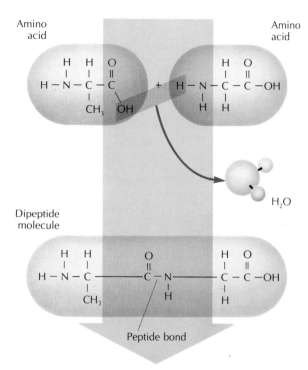

FIGURE 2.23   Peptide Bonds
Two amino acids may condense by forming a single covalent bond called a peptide bond, which joins them into a dipeptide. This is the first step in the formation of the long amino-acid chains called proteins.

FIGURE 2.24   Levels of Structure in Proteins
(a) Chains of amino acids join to form the primary structure of a protein. The backbone is the repeating –N–C–C– of the sequence of amino acids. The R groups of the amino acids are not part of the chain, but stick out to the side. (b) Hydrogen bonding between amino acids can form secondary structures such as the beta sheet or the alpha helix. (c) Secondary structures may fold and coil into compact tertiary structures such as the globular proteins. Most enzymes have this type of structure. (d) Some complex enzymes have a quaternary structure formed by the union of two or more tertiary proteins.

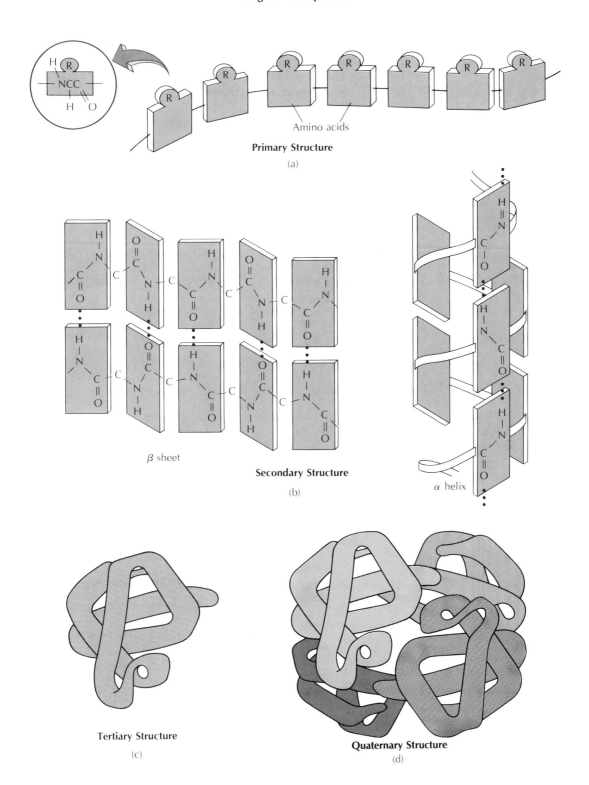

Amino acids

Primary Structure

(a)

β sheet

Secondary Structure

(b)

α helix

Tertiary Structure

(c)

Quaternary Structure

(d)

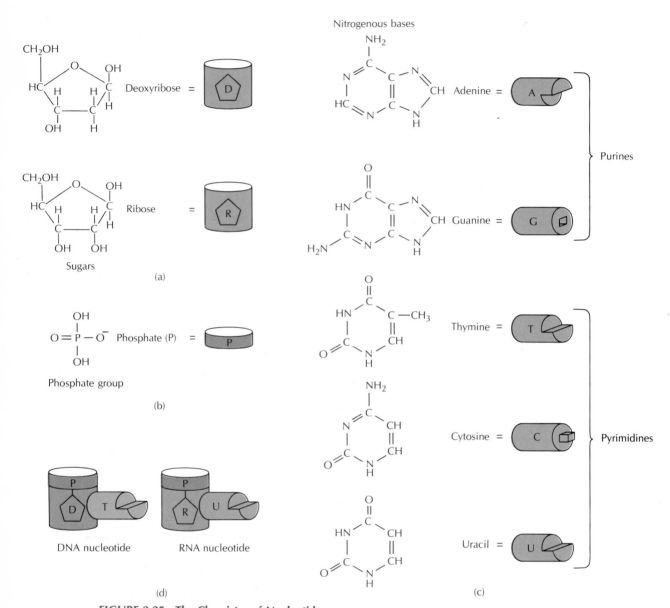

FIGURE 2.25 The Chemistry of Nucleotides
Nucleotides are the building blocks of nucleic acids. DNA and RNA have slightly
different building blocks. (a) All nucleotides contain pentose (five-carbon) sugar
molecules. DNA contains deoxyribose; RNA contains ribose. (b) All nucleotides
contain one to three phosphate groups. (c) There are five nitrogenous bases found
in nucleic acids. Adenine and guanine are purines. Cytosine, thymine, and uracil are
pyrimidines. DNA contains adenine, guanine, thymine, and cytosine; RNA contains
adenine, guanine, uracil, and cytosine. (d) Each nucleotide consists of one
phosphate, one sugar, and one nitrogenous base. A DNA nucleotide will contain
deoxyribose and may contain thymine; an RNA nucleotide will contain ribose and
may contain uracil.

sugar is deoxyribose, and in RNA it is ribose. The nitrogen-containing bases may have either one carbon–nitrogen ring (the pyrimidines) or two such rings (the purines). Of five different bases, each of the nucleic-acid classes contains only four: DNA contains the bases adenine, guanine, cytosine, and thymine; RNA contains adenine, guanine, cytosine, and uracil. A nucleic acid is a long chain of nucleotides.

Nucleotides are assembled into nucleic acids by an enzyme-catalyzed reaction in which the sugar of one nucleotide becomes linked to the phosphate of the next. As a result, a backbone of covalently linked, alternating sugars and phosphates is formed (Figure 2.26). The bases project out at an angle from the chain where they can participate in interactions with other compounds.

The nitrogen-containing bases are strongly polar and form hydrogen bonds with other polar groups. Most importantly, they can bond with each other. Possible pairs of nucleic acids are adenine bonded to thymine, adenine bonded to uracil, and guanine bonded to cytosine (Figure 2.27). These base pairs are said to be complementary. Bases rarely bond to bases other than their normal partners.

DNA. DNA is a double-stranded molecule formed from two chains of nucleotides. The bases pair up to establish the secondary structure of DNA, then the two strands of covalently linked nucleotides twist around each other in a **double helix** (Figure 2.28). The helix forms a "ladder" in which the uprights consist of alternating phosphate groups and deoxyribose rings (sugars). The rungs of the ladder are formed by the paired nitrogenous bases, which are joined by hydrogen bonds.

RNA. RNA, which is synthesized under the direction of DNA, plays its own unique role in protein synthesis. It differs from DNA in several ways: it is predominantly single-stranded; its

sugar (ribose) contains one more oxygen molecule than deoxyribose; one of its bases is uracil instead of thymine; and the molecules are smaller. Both DNA and RNA will be discussed in more detail in Chapter 8.

FIGURE 2.26 Nucleic Acid Structure
The strand of a nucleic acid is formed when the phosphate of one nucleotide condenses to the sugar of the next. The backbone is a repeating structure of alternating sugars and phosphate. The nitrogenous bases are not part of the backbone, but stick out at the side.

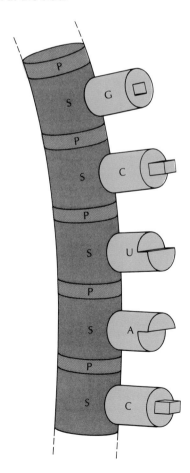

BOX 2.2
Molecular Fingerprints

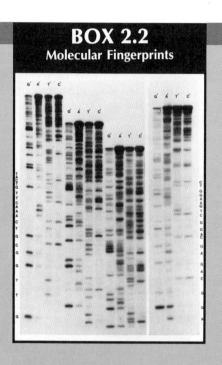

The strange pattern in this photograph is the fingerprint of a DNA molecule. Proteins and nucleic acids are very large, as we've said before. They may contain thousands to millions of sequential units. "Reading" the sequence of a nucleic acid was an impossible chemical task until quite recently. Now we have methods that enable us to rapidly sequence a nucleic acid of tens of thousands of nucleotides, and it is being done routinely. The optimists say that by the time you read this we will have the complete DNA sequence of the human chromosomes.

How is it done? Well, briefly, a section of DNA is isolated. Then it is treated with a hydrolytic enzyme that can cleave the chain only at a specific type of sequence, such as after two adenines and two thymines (–AATT–). The enzyme does its work, and then a mixture of short fragments are left. Those are separated by spotting the mixture on a gel slab, then passing a current of electricity through it. The fragments move one way or the other depending on how many positive or negative charges each has. Afterward the gel is stained and you have a "fingerprint." Any DNA that is identical to this one will have the same fingerprint. Each spot can be cut out and painstakingly analyzed to determine its sequence, then the whole thing is pieced together like a puzzle. Tough work, and slow, but it does the job.

Courtesy of Bio-Rad Laboratories.

Adenosine triphosphate (ATP). The primary energy-carrying molecule of the cell is a nucleotide named **adenosine triphosphate (ATP)**. This molecule consists of adenine and ribose joined to three phosphate groups (Figure 2.29). The bonds that join the two outer phosphate groups to the molecule are considered high-energy bonds because a large amount of energy is released when these bonds are broken by hydrolysis.

When ATP is hydrolyzed to **adenosine diphosphate (ADP)** and inorganic phosphate (P_i), the resulting energy can be used to drive the

FIGURE 2.27   Base Pairing
The nitrogenous bases in nucleic acids can pair with each other by forming hydrogen bonds. Adenine can form two hydrogen bonds with either thymine or uracil. Guanine can form three hydrogen bonds with cytosine. Each base pair contains one purine and one pyrimidine, so all base pairs occupy a comparable space.

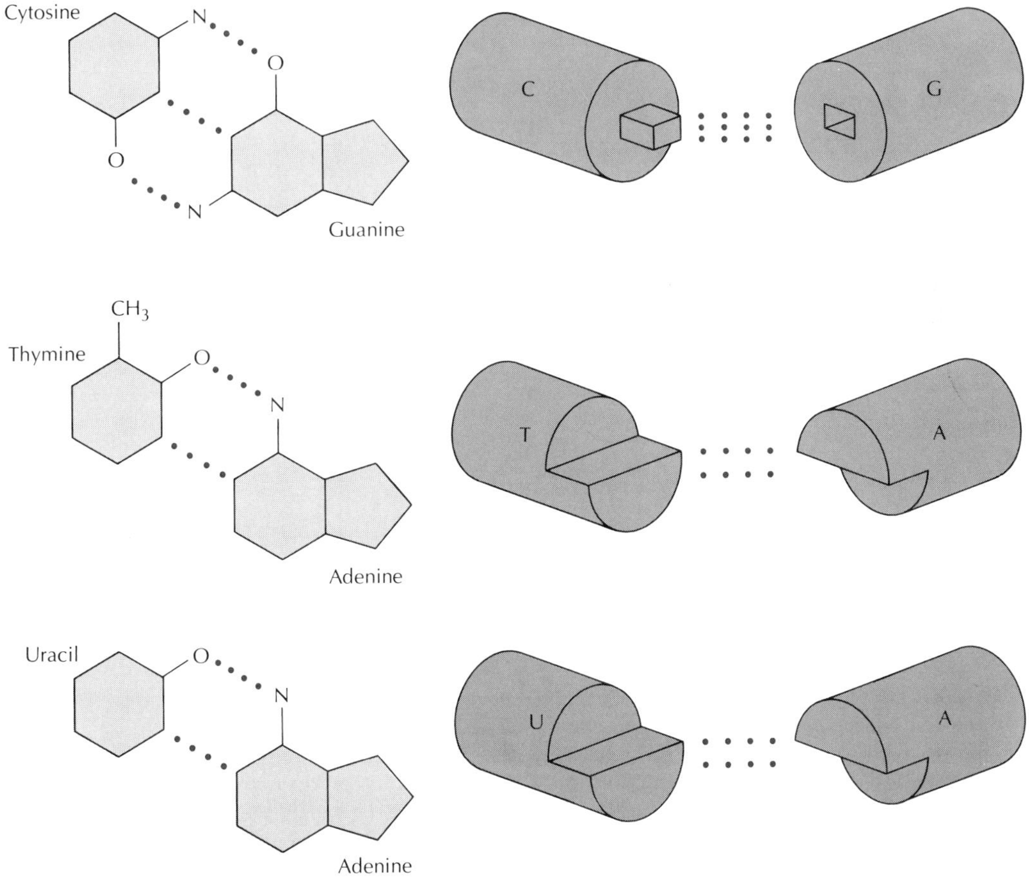

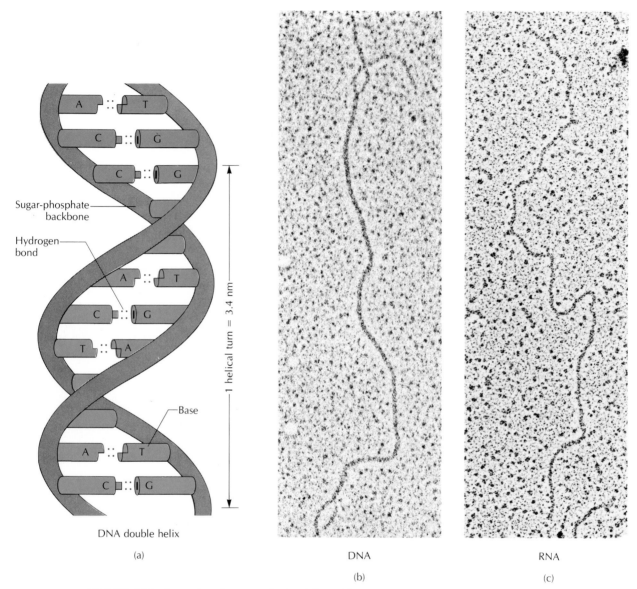

(a)

DNA

(b)

RNA

(c)

FIGURE 2.28   Secondary Structure in Nucleic Acids
(a) DNA forms a secondary structure called a double helix, in which two strands of DNA are joined together by matching and pairing their nitrogenous bases. (b) Electron micrograph of a DNA molecule, 20 Angstrom units in diameter, partly unwound at the top. DNA molecules are often millions of nucleotides in length. (c) Electron micrograph of an RNA molecule. RNA molecules are primarily single-stranded, but may have looped regions of base pairing as seen here. RNA molecules are comparatively smaller than DNA molecules, not usually over a few hundred nucleotides long. (×79,200. Photo credits: Visuals Unlimited/K. G. Murti © 1984.)

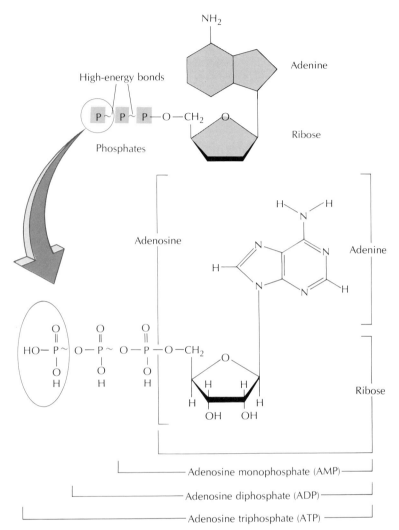

FIGURE 2.29 ATP
ATP is a nucleotide that is used as the primary energy source for cellular work. The molecule has a string of three phosphates. The bonds holding the second and third phosphates are high-energy bonds.

cell's endergonic reactions. Conversely, energy released by exergonic reactions can be used to form ATP from ADP:

$$ATP + H_2O \rightleftharpoons ADP + P_i + energy.$$

Thus ATP acts as an energy carrier, transporting energy from exergonic decomposition reactions to endergonic synthesis reactions.

SUMMARY

1. All matter is composed of atoms of the various elements. Each element has characteristic numbers of protons, neutrons, and electrons arranged in shells. If the numbers of neutrons vary, then the different atomic forms are isotopes and may be radioactive.

2. The number of electrons in an atom's outer shell determines its valence, or combining potential. Because formation of an outer shell with either two electrons in the first shell or eight electrons in the others is stabilizing, atoms form bonds in order to share electrons.

3. Oxidation-reduction reactions form ions (charged particles) that then form ionic bonds. Sharing of electrons between atoms results in covalent bonds. Equally shared electrons yield nonpolar bonds, but unequal sharing gives rise to polarity. These strong bonds hold molecules together. Weak bonds such as hydrogen bonds are responsible for interaction among molecules and for the higher levels of structure of the macromolecules.

4. Most reactions are reversible and subject to considerable variation both in direction and rate. In order for a reaction to occur, its energy needs must be met. Its rate will be enhanced by increases in concentration of the reactants, by increase in temperature, and by the catalytic effect of the specific enzyme.

5. Water is an inorganic molecule, but its peculiar properties are essential to life. Because the polar water molecules are strongly hydrogen-bonded, water has high surface tension and cohesiveness. Water's heat-absorbing capacity renders it a vital climatic regulator that protects cells and organisms from the effects of uncontrolled temperature variation. Water is a polar solvent and as such transports all the critical small chemical substances within and without cells. It causes ionic and polar covalent molecules to dissociate forming free ions. Of these, the hydrogen ion is of special importance since its concentration determines the pH of the aqueous environment.

6. Carbohydrates are organic substances composed of simple and complex sugars, singly, in pairs, or in long polymeric strands. Assembled by endergonic reactions, polysaccharides serve energy storage and structural functions. Disaccharides are the sugars used for transport; the monosaccharides are used for immediate energy utilization. The carbohydrates are polar, and the smaller ones are readily water soluble. Larger carbohydrates can be immobilized for storage.

7. Lipids are a diverse group of compounds. All share high proportions of hydrogen and carbon and a nonpolar, water-insoluble nature. Triglycerides function as energy-storage substances, waxes as surface protection, phospholipids as membrane structural elements, and steroids as hormones.

8. Proteins are informational polymers in which the sequence of the subunit amino acids is critical. Amino acids may be acidic or basic and polar or nonpolar in nature. They are linked by peptide bonds into a primary structure. Hydrogen bonding then directs the primary structure into a helical or flat secondary structure. The helical structure will fold to form a globular protein with unique surface characteristics. Several globular proteins may, in turn, associate to form a functional unit. Some proteins have structural roles but most are enzymatic in nature. They carry out catalysis by means of specific surface interactions with their substrates.

9. Nucleic acids are polymers of nucleotides. These monomers contain a nitrogenous base, a pentose sugar, and phosphate. DNA and RNA are both very large macromolecules, and their bases can interact with other nucleic acids by the formation of base pairs. This mechanism is responsible for their ability to store and transmit genetic information.

10. Single nucleotides such as ATP can serve as energy contributors in driving reactions. Cyclic nucleotides have key roles in modulating cell growth and regulation.

DISCUSSION QUESTIONS

1. Draw or build atomic models of C, H, O, N, P, and S. Which of these tend to gain, lose, or share electrons, and how many? Draw electron dot diagrams of a compound of C and S; H_3PO_4; and nitrogen gas (N_2).

2. Summarize the distinction between ionic, polar covalent, and nonpolar covalent bonds.

3. Describe why the warming of a reaction mixture increases the rate of reactions.

4. What is bond energy?

5. If water were replaced by liquid ammonia in a newly discovered life form, what might be some of the ways in which the new life form would differ from water-based life forms?

6. Distinguish between strong acids, weak acids, salts, weak bases, strong bases, and buffers.

7. Describe how starch, collagen, and DNA all fit the definition of a polymer.

8. Provide four examples of condensation reactions.

9. When radioactive isotopes are introduced into a cell, the cell incorporates them along with the normal isotope into its biochemical products. A substance that contains a radioactive atom can be detected by various types of electronic equipment and is said to be labeled. If C^{14} is introduced, what types of compounds might become labeled? If radioactive P^{32} is introduced? If radioactive sulfur is introduced?

ADDITIONAL READING

Baker J, Allen G. *Matter, energy, and life*, 3rd ed. Reading, Mass.: Addison-Wesley, 1975.

Curtis H. *Biology*, 4th ed. New York: Worth, 1984.

Lehninger A. *Biochemistry*, 2nd ed. New York: Worth, 1975.

Stryer L. *Biochemistry*, 2nd ed. San Francisco: W.H. Freeman, 1981.

Zubay G. *Biochemistry*. Reading, Mass.: Addison-Wesley, 1983.

3

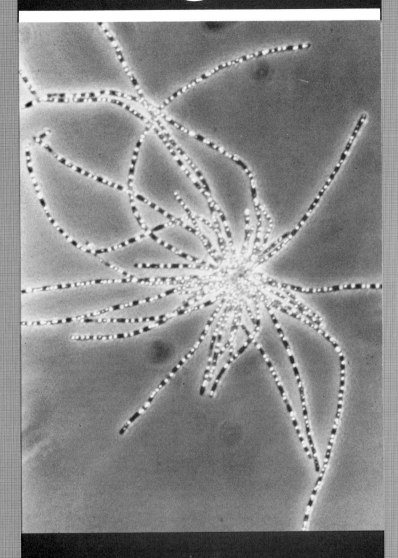

STUDYING PROCARYOTIC AND EUCARYOTIC CELLS THROUGH THE MICROSCOPE

CHAPTER OUTLINE

We saw in Chapter 1 that there are two very different kinds of cells in our biological world: procaryotic and eucaryotic. Now we will systematically examine the structure of these cells. Much of our knowledge of their structure was made possible because of the microscope. Therefore we'll first examine the various types of microscopes, learn what each allows us to see, and familiarize ourselves with the techniques available to study microorganisms. Then we'll turn to the cells themselves to see how the basic cell functions are carried out under these two different structural plans.

MICROSCOPY

Microscopy is the study of small objects or details by the formation of magnified images. Many different methods may be employed to produce the image. It is possible to use either visible light or invisible forms of radiation converted to visible form. Of course, our human eyes and brain are the final receivers of any image, and therefore determine its usefulness. One objective of this chapter is to help you gain skill in interpreting **micrographs,** photographs taken through the microscope.

UNITS OF MEASUREMENT

Microorganisms are measured using the metric system. The standard metric unit is the meter, which is equal to 39.37 inches in the English system. The units used to measure microscopic structures are tiny fractions of the meter. They include:

- the micrometer (μm) = 0.000001 m (useful in measuring microorganisms);
- the nanometer (nm) = 0.000000001 m; 1/1000 μm (useful in measuring cellular structures);
- the angstrom unit (Å) = 0.0000000001 m; 1/10 nm (useful in measuring molecules).

ELECTROMAGNETIC RADIATION (AND HOW IT BEHAVES)

In order to understand how microscopes function, it is useful to know something about light. Light is a small part of a continuous spectrum of electromagnetic radiation, radiant energy that travels in waves. Different colors of light have different **wavelengths** (the distance from one peak to the next). Radiation of a particular wavelength has a characteristic amount of energy, which is inversely proportional to its wavelength. The shorter the wavelength, the higher the energy (Figure 3.1).

There is a continuous **spectrum** of electromagnetic radiation. Different radiation sources emit different parts of the spectrum. An electric stove unit set low, around 300 degrees Celsius, emits **infrared** (IR) radiation — heat. If you turn the stove up to high, it emits not only more IR but also shorter, visible wavelengths, and it glows bright red. An electric light bulb filament at 3000 degrees Celsius gives off the complete visible spectrum. We perceive this as a white blend. The sun has a surface temperature of one million degrees Celsius. It emits the complete electromagnetic spectrum, visible and invisible. We use a variety of devices (light bulbs, cathode-ray tubes, etc.) to generate selected portions of the spectrum.

Certain physical laws affect radiation's behavior. Let us see how these laws apply to visible light. When light strikes an object, four things may happen to the radiation.

- If the object is transparent and light strikes at a right angle, the light will pass through without affecting the object. This is **transmission.**

- If the object is opaque or colored, it will **absorb** some or all portions of the spectrum. The absorbed light energy may heat the object up, may drive chemical reactions such as photosynthesis, or may be given off as **fluorescence** (light of a different wavelength).

- Some light striking an object is **reflected** (bounces off). It does not change the energy content of what it strikes.

- Light rays may be bent as they move from one transparent medium, such as air, to another, such as a glass microscope lens. This bending is **refraction,** and it causes blurring of images.

FIGURE 3.1 The Electromagnetic Spectrum
Electromagnetic radiation exists in a continuous spectrum, of which visible light is a very small part. The energy of radiation increases as the wavelength decreases.

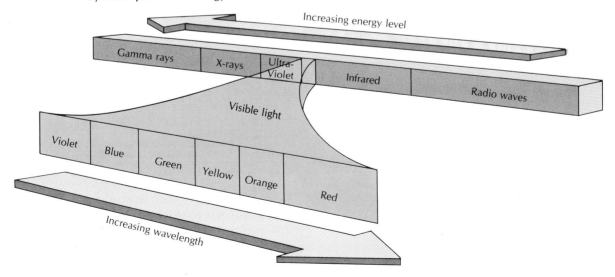

THE COMPOUND LIGHT MICROSCOPE

Light microscopes are the workhorses of the microbiology laboratory. Using them, we study the sizes, shapes, and colors of microorganisms. We can also readily observe the cells of human and other tissues. Light microscopes are (comparatively) inexpensive, easy to operate, and quick to give results.

A **compound microscope** uses the combined magnification of two or more lenses. Depending on the lenses used, its image will vary in magnification, definition, and resolution. Compound light microscopes may be equipped with several different types of **condenser.** The condenser is a device located between the light source and the slide, which aims the light beam at the specimen on the slide. Different types of condensers are designed to give good viewing conditions for certain types of specimens. The parts and operation of compound light microscopes are shown in Figure 3.2.

Magnification, the increase in the diameter of the image, depends on the curvature of the lenses. Total magnification is calculated by multiplying the magnifications of the lenses. **Definition** or clarity is at least as important as size in an image — the image produced by a lens made of poor quality glass may be highly magnified but unclear. The optical limit of a microscope is its **resolving power** — technically, a measure of how close two objects can be and still be perceived clearly as separate objects. Resolution determines the ability to distinguish fine detail. A general rule is the shorter the wavelength of the radiation , the better the resolution. Thus, a light microscope has a maximum resolution of 0.2 μm, while an ultraviolet microscope, using a shorter wavelength, produces a resolution of 0.1 μm. Immersion oil, widely used when viewing bacteria with a light microscope, reduces the light scattering due to refraction (Figure 3.2b) and produces close to the optimal resolution.

THE BRIGHT-FIELD MICROSCOPE

Bright-field microscopes work well for fixed, stained specimens. They are equipped with the Abbé condenser. This focuses and concentrates the light beam for maximum brightness.

THE DARKFIELD MICROSCOPE

A **darkfield microscope** is used for viewing live, moving organisms that cannot be seen by ordinary light microscopes, cannot be stained easily, or are distorted by staining. Microbiologists use a darkfield microscope to spot the rapidly wriggling bacterium that causes syphilis. A darkfield microscope does not focus light directly on the organism. It has a condenser that focuses the light across the slide at a sharp angle, while blocking direct light. Light can enter the objective lens and so reach the eye only if it reflects from the specimen on the slide. The specimen appears brilliantly outlined against a dark background.

THE PHASE-CONTRAST MICROSCOPE

A **phase-contrast microscope** gives superior images of living, naturally moving microorganisms and certain delicate intracellular structures without requiring intrusive fixing and staining procedures that would harm the organism. It can also, however, deliver excellent images of stained specimens. This type of microscope makes use of slight variations in the **refractive index** — the relative velocity at which light passes through different parts of a specimen. The phase-contrast microscope amplifies the differences in velocity into significant differences in light intensity. As the light passes through denser parts of the organism, certain structures stand out and can be examined. Images produced by the three microscopes described above appear in Figure 3.3.

ULTRAVIOLET MICROSCOPES

Some microscopes use ultraviolet radiation; that is, the portion of the electromagnetic spectrum containing wavelengths shorter than those of visible light. Because of their use of shorter-wavelength radiation they have twice the resolving power of standard light microscopes.

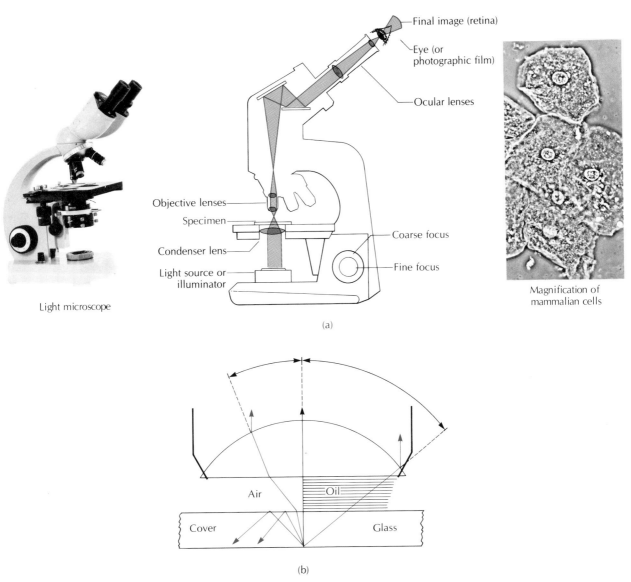

Light microscope

Objective lenses

Specimen

Condenser lens

Light source or illuminator

Final image (retina)

Eye (or photographic film)

Ocular lenses

Coarse focus

Fine focus

Magnification of mammalian cells

(a)

Air

Oil

Cover

Glass

(b)

FIGURE 3.2 The Light Microscope
(a) The light microscope focuses light from a light source through a condenser lens, through the specimen, and through objective and ocular lenses to the eye. It may magnify from 50 to 1000 times. (b) Immersion oil placed between the slide and the objective lens reduces loss and distortion of light. (Photo credits: left, courtesy of Carl Zeiss, Inc., Thornwood, N.Y./BPS; right, courtesy of Department of School Services, Wesleyan University, Middletown, Connecticut.)

However, the human eye cannot see ultraviolet radiation, so we must view the image indirectly, via fluorescence.

When ultraviolet radiation is absorbed by certain substances, these substances immediately re-emit light of a visible color (fluo-

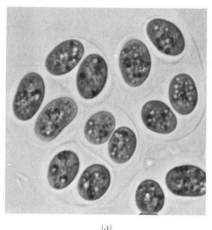
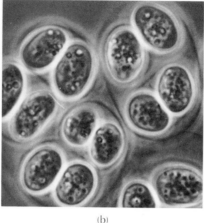
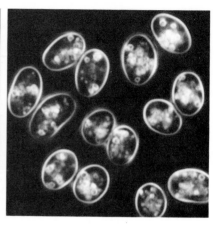

(a)   (b)   (c)

FIGURE 3.3   Typical Images Produced with a Compound Light Microscope
Different techniques with the light microscope: (a) cyanobacterial cells (*Gloecapsa*)
viewed with bright-field illumination (×1568); (b) phase-contrast microscopy (×1736);
(c) darkfield illumination. (×1568. Courtesy of J. Robert Waaland/BPS.)

rescence). Ultraviolet microscopy makes use of fluorescent dyes, such as those in black-light posters and certain kinds of paints. Dyes such as fluorescein and acridine orange absorb ultraviolet light and emit orange, yellow, or greenish light. When such dyes bind to cell surfaces, and the cells are then viewed through an ultraviolet microscope, outlines become visible. Often, fluorescent dyes can be attached to unique chemical structures in cells. This specific attachment has the effect of pinpointing the structures for the microscopist to study — they glow with the color emitted by the dye. Fluorescence microscopy is particularly useful in **immunology,** the study of the immune system and its reactions. We will learn more about this approach in Chapter 12.

THE ELECTRON MICROSCOPE

Some structures are so small that examining them requires the use of microscopes with even shorter wavelengths and higher resolutions. In this situation, **electron microscopes,** which use beams of electrons rather than light, are employed. Electron microscopes are necessary to see individual viruses, for example, or membranes, DNA molecules or other small features of subcellular structures. The resolution of electron microscopes may be more than 100 times greater than that achieved with light microscopes (Figure 3.4).

The electron microscope's radiation source is a stream of electrons emitted by a very hot wire in a vacuum tube. This electron stream is focused by electromagnets as it passes through a vacuum tube onto a specimen. The specimen must be carefully fixed in order to withstand exposure to the vacuum without becoming distorted.

Electron microscopes are of two types. A **transmission electron microscope** (TEM) passes the electrons through specimens and allows the viewer to see details of their internal structures. A specimen is sliced into ultra-thin sections of 0.05 μm or less (Figure 3.5). An average bacterium, which is about 1.5 μm long, could be sliced into 30 sections. The sections are then mounted on metal support grids, which serve as "slides." The image formed by the electrons passing through the specimen is made visible by directing the electrons onto either an electron-sensitive screen for direct viewing or a plate of photographic film for a permanent record.

Several techniques are used to enhance the image produced by transmission electron microscopy. **Shadowing,** which intensifies the surface features of the preparation, involves coating with electron-resistant metals. Staining to increase the contrast between details of the specimen and the background is done by exposing the specimen to metal salts such as osmium tetrox-

ide or ruthenium red. **Freeze-etching** is a technique that peels off outer layers of frozen cells and gives views of inner layers of cell walls and membranes to study their architectural features.

The **scanning electron microscope** (SEM) scans a beam of electrons across the surface of a specimen. It produces highly detailed images of the surface structures of specimens. Because cells relate to their environments via their surface features, the scanning electron microscope can show biologically significant details of microbial structure that may be seen in no other way. The specimen to be viewed is given an electron-dense heavy metal coating. The elec-

FIGURE 3.4 The Electron Microscope
The electron microscope uses beams of electrons emitted by an electron source and focused by magnetic lenses. (Courtesy of Carl Zeiss, Inc., Thornwood, NY/BPS.) The specimen is placed in a vacuum compartment in the microscope tube. The image is shown on a fluorescent screen and recorded on photographic film. Magnifications of up to 50,000× can be achieved. (×4160. Courtesy of Richard Rodewald, University of Virginia/BPS.)

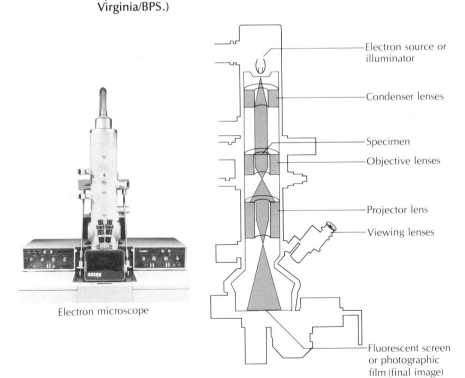

Electron microscope

Electron source or illuminator

Condenser lenses

Specimen

Objective lenses

Projector lens

Viewing lenses

Fluorescent screen or photographic film (final image)

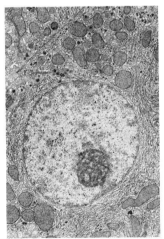

Magnification of rat liver cell

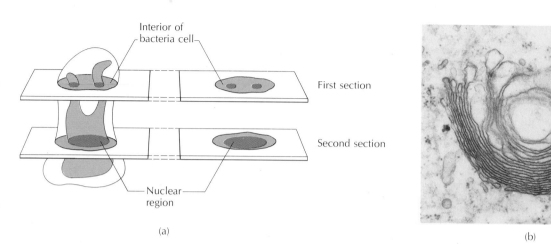

Interior of bacteria cell

First section

Second section

Nuclear region

(a)

(b)

FIGURE 3.5. How Sectioning Works
(a) When a cell is sectioned, the sections may differ depending on how the specimen was positioned. Different sections may show different parts of the same structures. (b) A thin-section electron micrograph through a Golgi body structure in a eucaryotic cell. (Micrograph courtesy of C.J. Flickinger, Department of Anatomy, University of Virginia Medical Center. From *J. Cell Biol.* 49:221-226, 1971, by permission of the Rockefeller University Press.)

FIGURE 3.6   Scanning Electron Micrograph of Mammalian Blood Cells
A scanning electron micrograph of the interior of a small blood vessel, showing the slightly flattened, disk-shaped red blood cells and the rough-surfaced, spherical white blood cells. (×2800. From R.G. Kessel and R.H. Kardon, *Tissues and organs: a text-atlas of scanning electron microscopy.* New York: W.H. Freeman and Co. 1979.)

10μm

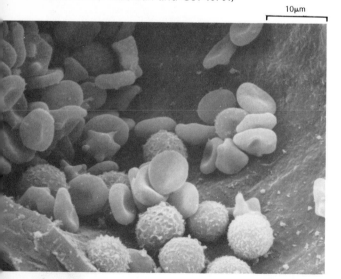

trons are reflected off the surface of the specimen onto a viewing screen. As shown in Figure 3.6, the SEM provides a striking three-dimensional effect.

PREPARING SPECIMENS FOR OBSERVATION

Microorganisms must be prepared for many types of study by **fixing** and **staining.** The general procedure used for bacterial cultures is as follows. A suspension, or **smear,** of material containing the microorganisms is spread on a slide and then air dried. Next, the slide is passed through the flame of a Bunsen burner several times, which causes the microbial material to become bound firmly to the slide and fixed in approximately its natural form.

Staining colors the normally transparent cells. Most biological stains are strongly colored synthetic dyes. They may be acidic, basic, or neutral. The familiar microbiological dyes, such as crystal violet, safranin, and methylene blue, are basic. They combine with acidic cell materials, especially nucleic acids and acidic polysaccharides. Acidic dyes such as eosin are commonly used in staining human tissues. They bind to basic proteins.

Fortunately for microbiologists, preparing and staining single-celled organisms is much simpler than preparing animal tissues. Tissue preparation requires fixation, dehydration, slicing or sectioning, mounting, then differential staining by special techniques. We won't cover tissue techniques here.

There are two basic microbiological staining techniques. In a **simple stain,** a fixed slide of microbial cells is bathed with a single dye, the excess is washed off, and the dried slide observed. All the organisms come out one color. A simple stain reveals shape, size, and grouping of cells.

Differential stains distinguish among different types of bacteria on the basis of their reaction to the staining material. The **Gram stain,** named for a Danish bacteriologist, is the most important differential stain technique in microbiology (Figure 3.7). It separates bacteria into two different color categories, purple and red. The way an organism stains with the Gram reaction is perhaps the single most important piece of information needed in beginning to identify that organism. The reaction has four steps. First, a **primary stain** is applied — the purple stain, crystal or gentian violet. This stain is taken up by all bacterial cells. Second, the smear is covered with a **mordant,** a substance that is added to the dye to make it stain more permanently. In this case iodine is used. It enters the cells and forms large complexes with the crystal violet. Third, a **decolorizing agent** containing alcohol is applied. Decolorizing is the differential step — it removes the purple dye complex from some types of organisms but not from others. Fourth, a **counterstain** of a constrasting red color, safranin, is applied to stain any cells that have been decolorized. Bacteria that retain the first purple stain after decolorizing is attempted are **Gram-positive.** Those that lose the purple color but show the red counterstain are **Gram-negative.** The basis for the different reactions lies in differences in the chemistry of the outer layers of the

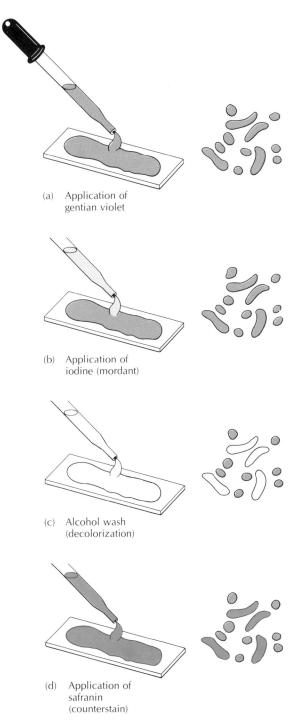

(a) Application of
     gentian violet

(b) Application of
     iodine (mordant)

(c) Alcohol wash
     (decolorization)

(d) Application of
     safranin
     (counterstain)

FIGURE 3.7 Gram Stain Procedure
The Gram staining procedure separates bacteria into the purple-staining Gram-positives and the red-staining Gram-negatives.

cells of different bacteria. We will explain this later in this chapter. The **acid-fast stain** is another common differential staining procedure, used in diagnosing tuberculosis infections.

The **negative stain** is a procedure that colors the background of the slide and leaves the organisms uncolored so they stand out in contrast. Negative stains are often used to make it possible to observe bacterial structures that do not stain well. With a negative stain, overall sizes and shapes stand out as clear areas against a dark background.

PROCARYOTIC CELL STRUCTURE

Having concluded our survey of microscopes, we are ready to turn to the objects we use them to study — cells. We'll begin with the procaryotic cells, which contain fewer and simpler structures than the eucaryotic cells. The bacteria and the cyanobacteria comprise the entire procaryotic group. Procaryotes carry out the same basic functions as eucaryotes: metabolizing food, synthesizing proteins, and storing energy. They are different primarily in structure. All portions of a procaryotic cell are in direct contact with each other; there are no internal compartments within the cell. The cell membrane, also known as the cytoplasmic or plasma membrane, is involved in all the cell's activities. Because of their simplicity, procaryotic cells provide a good starting point for understanding basic cell biology (Figure 3.8).

BOUNDARY LAYERS

Each cell structure exists because it fulfills a function. The functions of a life form arise from the functions of its parts. As we saw in Chapter 1, life forms possess five characteristic traits. Here we'll use these characteristics as a framework to understand the parts of cells. We start with the characteristic that makes a cell a cell — its boundary layers. These protect, shape, and retain the cell contents, and regulate which

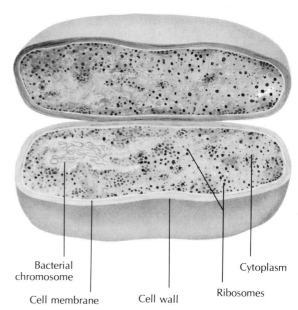

Bacterial chromosome

Cell membrane

Cell wall

Cytoplasm

Ribosomes

FIGURE 3.8   A Typical Procaryotic Cell
The internal structure of a procaryote contains a central mass of DNA (the chromosome), ribosomes, and cytoplasm. The exterior layers are the cell membrane and cell wall.

chemicals enter and leave the cell. Procaryotes have several boundary layers, one inside another. We will proceed from the outside in.

CAPSULES, OR SLIME LAYERS.   Most procaryotes have some type of amorphous external coating. These coatings are usually made up of polysaccharides, but there are a few species with protein coats. A dense, structured layer is referred to as a **capsule,** a loosely associated layer as a **slime layer** (Figure 3.9). Both are often termed the **glycocalyx.** The material is made by the cell that it coats, and may accumulate very thickly. A negative stain is used to reveal these layers for light microscopy, and ruthenium red is used for electron microscopy. The growth environment influences the development of the glycocalyx, which is promoted by the availability of abundant sugar or other carbohydrates. Organisms that live in the human body produce more gly-cocalyx when freshly isolated than they do after long periods of laboratory growth. Glycocalyx producers make shiny **mucoid** or **smooth colonies.** Nonproducers have dry-looking **rough colonies.** Glycocalyx production is also genetically controlled, and the trait can be lost by mutation or gained by genetic exchange. Different strains of a bacterial species may each produce slight chemical variants of the glycocalyx substance. These differences are often useful means of identification in the laboratory.

The glycocalyx is very important to humans. Many bacteria are disease-causing or **pathogenic.** Some dangerous pneumonia bacteria have capsules that protect them against our lungs' normal defense network. In your mouth, slime-producing bacteria turn sucrose (from sweet foods) into **plaque,** the sticky substance that coats tooth surfaces and contributes to tooth de-cay. Bacterial slime is a major industrial nuisance in dairy operations and in paper mills.

On the positive side, capsules frequently stimulate your body to respond defensively, to protect against future infections. We can often deduce the identity of a pathogen by measuring the patient's defenses. Capsular materials are also used to make vaccines to protect against pneumonia and meningitis.

PEPTIDOGLYCAN. Most bacteria are enclosed by rigid outer layers that form the **cell wall,** an important protection against their being killed by osmotic stress. Procaryotes contain a unique substance called **peptidoglycan,** which provides most of the wall's structural strength. Peptidoglycan is a complex polymer that is not found in eucaryotic cells. It is a three-dimensional macromolecule consisting of sugars and chains of amino acids (Figure 3.10). The carbo-

FIGURE 3.9 Capsule Formation
The gram-positive coccus *Leuconostoc mesenteroides* (a) enzymatically converts sucrose into dextran. After two hours' growth in a sucrose medium, the bacterium forms a surface coat (b). After 18 hours, a completed capsule of insoluble dextran many times the diameter of the cell has been formed (c). (Courtesy of B.E. Brooker, *J. Bacteriol.* (1977) 131:288-292, Figures 1, 2 and 3.)

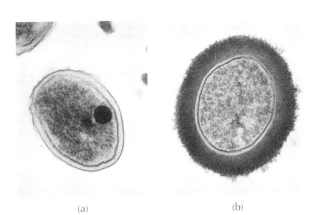

(a)   (b)

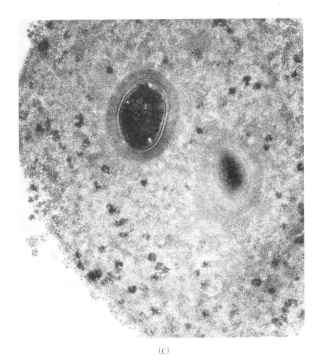

(c)

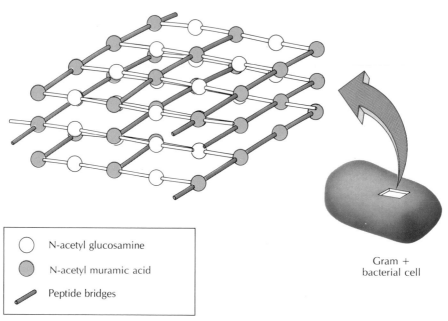

- ○ N-acetyl glucosamine
- ● N-acetyl muramic acid
- ╱ Peptide bridges

Gram +
bacterial cell

FIGURE 3.10   Peptidoglycan
The peptidoglycan molecule is a polymer composed of alternating molecules of the two sugars N-acetyl glucosamine and N-acetyl muramic acid, connected by peptide bridges. A cell wall may be built of several layers of peptidoglycan.

hydrate portion consists of alternating molecules of the amino sugars N-acetyl muramic acid (NAM) and N-acetyl glucosamine (NAG), covalently joined. These chains are then cross-linked through peptide chains. The overlapping of many layers of peptidoglycan creates the three-dimensional wall. Antibiotics such as penicillin stop some bacterial infections because they block peptidoglycan synthesis, arresting the invaders' growth. Not all bacteria are susceptible, however.

GRAM-POSITIVE CELL WALLS.   There are two major cell-wall types (see Figure 3.8), referred to as Gram-positive and Gram-negative because they correlate with (and cause) the Gram-staining reaction. Let us look at the two forms in detail. The structure of Gram-positive and Gram-negative cell walls are shown in Figure 3.11. The Gram-positive cell wall consists of a single layer containing up to 90% peptidoglycan. There are also variable amounts of polysaccharides called **teichoic acids,** polymers of sugar alcohols and phosphate. Some Gram-positives such as the streptococci have cell-wall–associated proteins (M-proteins) that are important in producing disease.

Gram-negative bacteria typically possess a two-layered boundary. This boundary surrounds the **periplasmic space,** which separates the cell wall from the cell proper. The inner wall layer is composed of peptidoglycan, but the layer is much thinner than in the Gram-positive bacteria and less cross-linked. The outer wall layer (called the **outer membrane**) has a double-tracked appearance under the electron microscope. Phospholipids and lipoproteins are found throughout the outer layer, and lipopolysaccharides are found on its exterior-facing surface.

The polysaccharide portions of the outer membrane lipopolysaccharide serve in cell identification in the laboratory. The lipid portion (Lipid A), also called **endotoxin,** causes multiple effects on our circulation when we are infected by gram-negative bacteria. The symptoms include inflammation, fever, muscle pain and headache, and in some cases even fatal shock.

The outer membrane contains pores formed from clusters of protein molecules (**porins**) that allow certain large molecules to pass through.

Proteins in the outer membrane also serve as receptor sites for the attachment of bacterial viruses. Some outer-membrane proteins are toxic to humans.

To summarize, the walls of Gram-negative bacteria differ significantly from the walls of Gram-positive bacteria. These differences are reflected in the Gram-staining reaction, the response to various antibiotics and to environmental challenge, and other characteristics (Table 3.1).

FIGURE 3.11 The Two Types of Bacterial Cell Wall
(a) Gram-positive bacteria have a wall composed of a thick peptidoglycan layer while gram-negative bacteria have a peptidoglycan layer and an outer membrane.
(b) Detail of the gram-positive *Clostridium* and gram-negative *Bacteroides*. (×202,800. Courtesy of V.A. Koch, K. E. Moore, and H.W. McFadden, University of Nebraska Medical Center.)

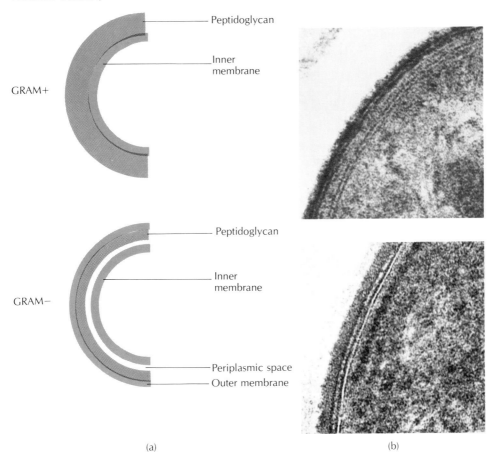

(a)   (b)

TABLE 3.1 Biological characteristics of Gram-positive and Gram-negative bacteria	
Gram-Positive Bacteria	**Gram-Negative Bacteria**
Wall is single layer of peptidoglycan	Wall is typically three layers, with the innermost peptidoglycan; wall-deficient bacteria stain gram-negative
Shape always rigid	Shape may be rigid or flexible
Round, rod-shaped	Many different shapes
Spores formed by some species	Spores rarely formed
Tolerate dry conditions well	Have low tolerance for drying
Growth inhibited by penicillin	Penicillin inhibition not reliable
Wall removed by lysozyme	Wall weakened by lysozyme
Growth often inhibited by dyes	Dyes not generally inhibitory
Enzymes frequently excreted directly	Enzymes retained in periplasmic space, released when wall is disrupted
May secrete protein toxins	Produce endotoxins; secreted toxins uncommon

OSMOSIS.   The cell wall's major function is to provide a mechanical limitation to swelling and shrinking due to the flow of water into and out of the cell. Without this protection, water inflow could burst the cell. This water movement is caused by **osmosis.** As shown in Figure 3.12, osmosis is the movement of water between two solutions separated by a **semipermeable membrane** (a membrane that permits certain molecules and ions to pass through but excludes others). Water tends to flow into a cell when the cell's contents contain a higher concentration of dissolved materials (solutes) than are found in the water outside the cell. This is most often the case for microorganisms living in fresh water. On the other hand, water tends to flow out when the water outside the cells has a higher concentration of solutes. For example, when microorganisms are placed in salty pickling brine, the water inside the cell flows out, because the concentration outside the cell is higher. We will be studying the effects of osmosis on microbial growth and survival in Chapter 6.

CELL-WALL VARIANTS.   Some few bacteria live without cell walls, in environments where osmotic changes are not a problem. A few bacteria have evolved cell walls with other, very different, chemistry and functions. Still others, having had their cell walls chemically removed in the laboratory for study purposes, live in an artificially protected situation. Let us briefly look at some of these cell-wall variants.

L-forms are mutant variants of normal bacteria, which have a defective cell wall. They are found in the tissues of higher organisms, and may cause disease. They grow in erratic, flexible shapes to form small colonies. Because their cell walls are imperfect, they are vulnerable to osmotic destruction. In the laboratory they require highly specialized growing conditions. They must be grown on media that have the same concentration of solute as found within the cell. L-forms lack part or all of their peptidoglycan. Despite this, they survive well in human tissues, because our body provides a controlled osmotic environment.

The **mycoplasmas** are naturally occurring procaryotes that entirely lack a cell wall. Their outer boundary is the cytoplasmic membrane (see next section), stabilized with lipids called sterols, which are like those found in animal cells. The sterols are thought to protect the membrane from damage caused by osmosis. Mycoplasmas are rarely free-living; most live in animal tissues or other protected places. Some cause respiratory and reproductive-system infections; others cause plant diseases.

The **archaebacteria** are a newly classified group of microbes that seem to be closer to the earliest forms of life than to most present-day bacteria. **Halobacteria,** members of the archaebacteria group, inhabit highly saline environments such as the Dead Sea (which contains 29 to 30% salts) and pickling brines. They have a highly specialized protein cell wall that lacks peptidoglycan entirely. The primary function of this unusual cell wall is to capture solar energy. The halobacterial wall does not protect against osmotic lysis (rupture).

Wall-deficient laboratory bacteria may be produced using penicillin, which inhibits peptidoglycan synthesis, or **lysozyme,** an enzyme that catalyzes the breakdown of peptidoglycan. Lysozyme is a normal component of tears, blood, egg white, and other fluids, and helps afford protection against bacteria. Lysozyme removes peptidoglycan; penicillin prevents its replacement. When a Gram-positive organism is exposed to lysozyme, its cell wall is destroyed. The remaining cell cytoplasm surrounded by the cell membrane is called a **protoplast.** Gram-negative bacteria exposed to lysozyme are damaged, but the outer-membrane portion of the wall remains. Such a modified cell — called a **spheroplast** — can still metabolize, but it cannot multiply. Spheroplasts are useful for research on membranes and their transport activities.

In summary, the procaryotic cell wall normally has peptidoglycan, and the wall's mechanical strength is directly related to the amount present. In addition, each strain of bacteria also appears to possess unique individual surface chemicals, such as the teichoic acids and proteins of the Gram-positive bacteria and the lipopolysaccharides of the Gram-negative bacteria. In later chapters we will focus on these surface chemicals and their importance in immunology and diagnostic microbiology. A few bacteria

FIGURE 3.12 Osmosis
To demonstrate osmosis we can place solutions inside membrane bags and suspend them in other solutions. When a starch solution simulating cytoplasm is suspended in a beaker of pure water, water flows into the bag, causing it to swell.

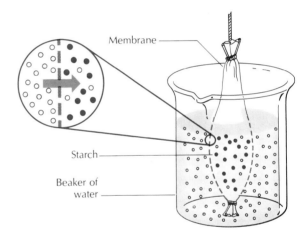

Membrane

Starch

Beaker of water

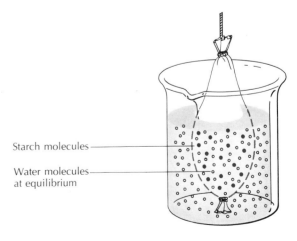

Starch molecules

Water molecules at equilibrium

have cell walls that depart significantly from the two more common cell-wall types.

THE CYTOPLASMIC MEMBRANE. The cytoplasmic membrane in procaryotes is part of the boundary layer. It encloses the **cytoplasm** of the cell. In a procaryotic cell this includes everything inside the cell membrane. The role of the cell membrane is to regulate the inside chemical environment of the cell and receive chemical

BOX 3.1
How to Aim a Magic Bullet

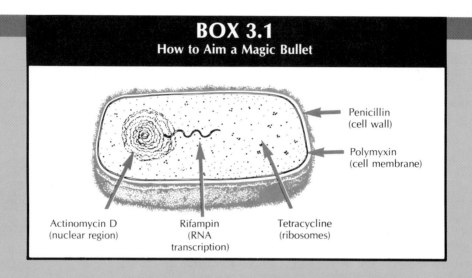

Penicillin (cell wall)

Polymyxin (cell membrane)

Actinomycin D (nuclear region)

Rifampin (RNA transcription)

Tetracycline (ribosomes)

Knowing about cell structure and function can help us begin to understand how many medicines work — what they will and will not do. The lay person often tends to see antibiotics as miracle drugs, good for whatever ails you. This they certainly are not. Antibiotics are toxic chemicals designed to poison cell functions — of bacteria, fungi, or other organisms.

Ideally, antibiotics fulfill the dream of a "magic bullet," a drug that can be introduced safely into the body, killing only the pathogens at which it is aimed, and leaving our normal cells unharmed. The best drugs can be "aimed" at vulnerable spots in the pathogen's cell — points where it differs from human eucaryotic cells. Some examples follow.

Penicillin blocks peptidoglycan synthesis, thus arresting growth of susceptible bacteria. Because there is no peptidoglycan in eucaryotic cells, the drug has no specific effect on human tissues and leaves them unharmed (however, allergy may be a dangerous side effect in some people). Similarly, tetracycline blocks protein synthesis in procaryotes. Rifampin, a new tuberculosis drug, blocks procaryotic RNA synthesis. Polymyxin destroys procaryotic cell membranes.

In the above cases, our cells and the pathogen's are sufficiently different. However, procaryotes make their DNA almost the same way we do. Drugs that block DNA activities are as toxic to humans as they are to bacteria. Thus these drugs cannot be magic bullets, and are not used as antibiotics. Instead, some are used for cancer chemotherapy, where they kill human cells that have gone out of control.

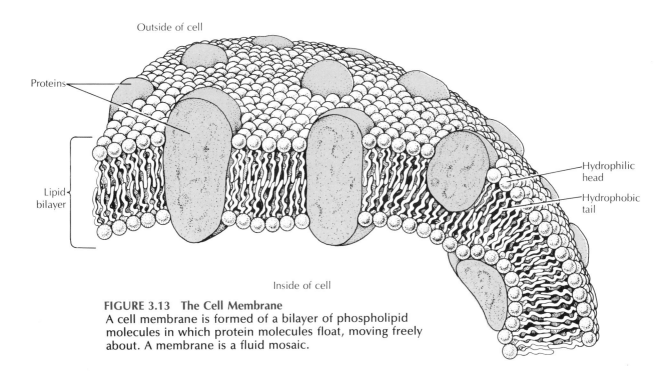

Outside of cell

Proteins

Lipid bilayer

Hydrophilic head

Hydrophobic tail

Inside of cell

FIGURE 3.13 The Cell Membrane
A cell membrane is formed of a bilayer of phospholipid molecules in which protein molecules float, moving freely about. A membrane is a fluid mosaic.

information from the outside environment. In procaryotes, the cell membrane also shares in most other cell functions, including energy capture, energy utilization, protein synthesis, and reproduction.

Biological membranes are approximately 60% protein and 40% phospholipid. A membrane is not a homogeneous structure. Although it has a uniform thickness of 5 to 8 nm, its components have a dynamic arrangement enabling them to move around and rearrange. The phospholipid molecules are arranged in two parallel rows, called a phospholipid bilayer. The protein molecules have a variety of arrangements, often forming clusters to perform individual functions.

The arrangement of the protein and lipid molecules is described in terms of a **fluid mosaic model** (Figure 3.13) Use your imagination for a moment to visualize a soap bubble floating in the sunlight. The simple membrane that encloses the bubble is held together by weak bonds among the soap molecules. Recall the moving

patterns of iridescent color you would see on the bubble's surface. They occur because the soap molecules continuously move around and realign, causing the color patterns to change. The soap bubble demonstrates in a readily visible form what is meant by the term *fluid* when applied to a membrane.

The term *mosaic* reflects the fact that the membrane contains many different proteins embedded in its lipid bilayer structure. Some of these proteins physically stabilize the membrane. Others are enzymes catalyzing the reactions for which the membrane is the site. Proteins are embedded in both of the bilayer surfaces, in the interior of the membrane, and penetrating through the membrane. Cell membranes expand, as additional molecules insert themselves into the mosaic. Many Gram-positive bacterial cells have **mesosomes,** which are infoldings of the membrane. It is unclear what mesosomes do.

A key role of the cytoplasmic membrane is to regulate the entrance and exit of small molecules, such as sugars and amino acids. The membrane is semipermeable; it excludes certain substances from the cell, retains others, and permits yet others to enter and exit. A compound can cross the membrane passively, as water does in osmosis, or be actively assisted in one of the following ways.

Facilitated transport is movement of small molecules across the membrane assisted or facilitated by specific membrane proteins that serve as their carriers. The process may operate across the membrane in either direction. When a substance is more concentrated on one side of a membrane than on the other, there is a difference in concentration called the **concentration gradient.** A substance will move spontaneously by diffusion from an area where it is more concentrated to an area where it is less concentrated. This is called moving *with the concentration gradient.* It does not require energy. Facilitated transport provides for the entrance of certain small polar compounds and exit of some waste products, moving with the gradient only. The membrane protein carriers enable these polar compounds to traverse the nonpolar protein environment of the membrane.

Active transport requires some of the cell's energy. The energy is used to move sugars, amino acids, and ions into the cell *against the concentration gradient.* Active transport depends on membrane-bound transport enzymes, each of which links an energy release from ATP to the transport of a molecule of a specific substance.

CAPTURING, UTILIZING, AND STORING ENERGY

As you'll recall, the second characteristic of life forms is that they use and manipulate energy. Some capture energy from the nonliving environment, such as the energy from sunlight or the energy of reduced inorganic chemicals. Others transform and utilize energy already captured in organic molecules. All store energy in the form of reserve supplies of nutrients. Specific cellular

structures have evolved for each of these processes. They are designed to cope with the special problems of transferring energy and of making and utilizing high-energy molecules.

Any time cells manipulate energy, they are concerned with either making or using molecules of ATP (adenosine triphosphate; review Figure 2.28). ATP is a high-energy molecule that may be made whenever a cell temporarily stores some energy, and used whenever the cell needs energy.

The sun shines on all organisms, but only some can capture the energy from sunlight. Energy capture refers not only to absorbing energy, but also to using the energy to drive reactions that synthesize carbohydrates or other high-energy storage molecules. We all absorb the energy of sunlight to some extent, but only photosynthetic organisms then use the energy effectively. Similarly, some procaryotes can extract useful energy from inorganic chemicals in ways no other forms of life can.

The structures involved in energy capture are contained in the cytoplasmic membrane and other specialized membrane layers. In procaryotic cyanobacteria, for example, a light-capturing apparatus is contained in **lamellae,** layers of parallel folded membrane material. The anaerobic photosynthetic bacteria possess **chromatophores,** which include both folded membranes and sacs of membrane material filled with liquid (Figure 3.14). The "purple membrane" of the halobacteria contains **bacteriorhodopsin,** a light-reactive substance chemically related to the pigment in the human retina. Lamellae are also found in most of the bacteria that use inorganic chemicals as an energy source.

Energy utilization refers to releasing the energy potential of organic molecules and using it to produce ATP. Energy-utilization reactions require complex enzyme pathways. The enzymes that catalyze the major method of ATP synthesis are incorporated in the cytoplasmic membrane. These enzymes are arranged in highly ordered sequences or arrays.

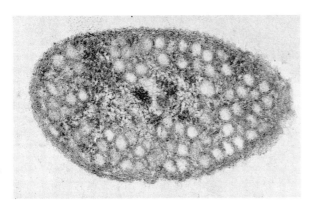

FIGURE 3.14 Bacterial Chromatophores
A photosynthetic bacterium, *Rhodopseudomonas spheroides*. This electron micrograph shows the many oval, lighter-appearing chromatophores containing the organism's photosynthetic apparatus. (×78,000. Courtesy of Dr. Germaine Cohen-Bazire, Director of Research, National Research Service, Paris.)

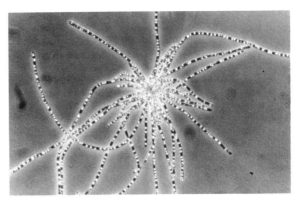

FIGURE 3.15 Sulfur Granules
The brightly reflecting spots within these bacterial filaments are granules of elemental sulfur. This sulfur bacterium, *Thiothrix*, may use the sulfur as an energy reserve. (×160. Courtesy of T.M. Williams. The Pennsylvania State University.)

Energy storage occurs when energy-rich substances are manufactured faster than they are needed. They may then be converted into insoluble reserve materials and deposited in microscopic **inclusion bodies** or **granules** (Figure 3.15). Inclusion bodies serve as a metabolic bank account in that the stored substances can be withdrawn and used for energy when times are lean. Inclusions may contain glycogen, starch, oils, other lipids such as poly-β-hydroxybutyrate (PHB), sulfur, or volutin, a polymer of inorganic phosphate. An organism's storage material is genetically determined, and knowing what it is may help in identification.

PROTEIN ASSEMBLY STRUCTURES

The third characteristic of all living cells is that they synthesize macromolecules. Here we are concerned with their role in synthesizing proteins, which serve as indispensable parts of cell structures and as the enzyme catalysts that direct metabolism. Protein synthesis provides the new molecules the cell needs for growth, replacement, and adaptive shifts in its metabo-

lism. The assembly of a protein requires that a precise sequence of amino acids be covalently linked up. The **ribosome** is the major cell structure in protein synthesis. It provides a surface on which a genetic message and the amino-acid carrier molecules attach and interact.

Procaryotic ribosomes are assembled from two subunits. The smaller (less dense) ribosomal subunit is called the **30S subunit** and the larger one, the **50S subunit.** The numbers 30S and 50S refer to differences in the subunits' physical behavior in a centrifuge. Each subunit contains one large RNA molecule and a mixture of proteins, including the peptide-bond–forming enzymes. The large subunit also contains a small RNA molecule.

A procaryote's cytoplasm always contains some free ribosomal subunits, but in an active cell most subunits are bound into functional complexes called **polysomes.** A polysome is a group of ribosomes actively reading and working on one genetic message. Some polysomes are bound to the cytoplasmic membrane; they are synthesizing proteins that will be exported through the membrane to the outside of the cell.

GENETIC INFORMATION STRUCTURES

As we saw in Chapter 1, the fourth characteristic of life forms is that they have a genetic "blueprint" that directs cellular activities. A simple procaryotic cell needs much less information than we humans do, of course. Still, a conservative estimate is that a bacterium needs at least 2000 individual units of heredity, or **genes.** Genes, more specifically, are segments of DNA specifying a particular protein or polypeptide. The genes of any organism are arranged in **chromosomes,** long strands of DNA. In bacteria the procaryotic DNA molecule bears all the essential genes in one long double helix of DNA, the **bacterial chromosome,** in the form of a closed circle. As the cell grows, new copies of the DNA are made. Sometimes, just prior to division, several chromosome copies may be seen as diffuse, shapeless areas called **nucleoids,** within one cell. Typically, however, a procaryote has just a single chromosome and therefore only one copy of its genetic material. Cells with only one copy of their genes are called **haploid.** Procaryotes characteristically divide by a simple process called binary fission, discussed in Chapter 6.

Most procaryotes also contain accessory DNA structures called **plasmids.** These are small, closed circles in the cytoplasm that contain information that is useful but not essential for growth. Bacteria can lose their plasmids without harm. Plasmids frequently contain information that allows the host bacterium to resist being killed by antibiotics. Even worse (for humans), these **resistance plasmids** or **R-factors** are readily passed from one strain of bacteria to another. Antibiotic resistance in pathogenic bacteria is an ever-worsening problem in modern medicine, and bacterial plasmids are one reason it is spreading.

In Chapter 1 we briefly introduced genetic engineering, the set of laboratory techniques that allows us to use enzymes to isolate specific genes and splice them with genes from other sources. Plasmid DNA is the form most frequently recombined in the lab, then reintroduced into the host bacteria, bearing "foreign" genes from other bacteria, plants, or animals. Once reintroduced, the recombinant plasmid DNA causes the host bacteria to make new proteins. These techniques have been employed to produce a strain of bacteria able to make human insulin, for example.

ADAPTABILITY AND MOVEMENT STRUCTURES

The fifth characteristic of life is that organisms adapt to the rigors of existing in constantly shifting environments. They must be able to adapt to the environment in order to survive. Adaptation may involve variations in enzyme patterns or genetic behavior, some of which are discussed in Chapter 8. It may also involve movement, which requires that the organism have the appropriate appendages. Procaryotes may also adapt by adhering to surfaces to stay in a preferred environment or by forming resting bodies to enable them to survive adverse conditions.

FLAGELLA.   A cell that is capable of independent motion is said to be **motile.** A motile species has the ability to move toward or away from stimuli. Bacteria can be remarkably speedy, in fact. When asked to think of a fast-moving species, you perhaps envision a greyhound or a cheetah. The cheetah, the world's fastest animal, can achieve speeds of about 25 times its own length per second. This is astoundingly fast. Yet one bacterial species can swim 37 times its own length per second. Top honors go to the procaryotes!

Motile procaryotes may have several different means of getting around. The most common is by **flagella** — long, slender protein fibrils attached to one end of the cell. Chemically, a flagellum consists primarily of many units of the protein **flagellin.** Flagellin molecules, which vary slightly from strain to strain of organism, aggregate by a self-assembly process into a tubular form. As they are added, the flagellum grows from the tip. Flagellar growth is very rapid; bacteria may replace their flagella within 10 to 20 minutes.

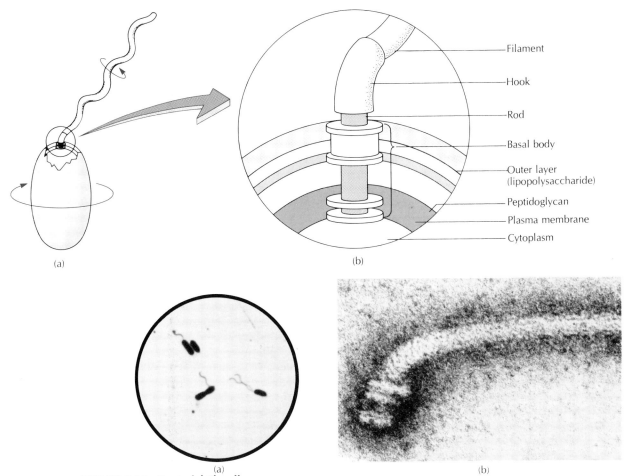

Filament
Hook
Rod
Basal body
Outer layer
(lipopolysaccharide)
Peptidoglycan
Plasma membrane
Cytoplasm

(a)

(b)

(a)

(b)

FIGURE 3.16   Bacterial Flagella
(a) A bacterium with only one flagellum inserted through the cell wall at the pole of
the cell. As the flagellum rotates in one direction, the bacterium moves forward. If
the flagellum rotates in the opposite direction, the bacterium will go backward or
tumble aimlessly. (Courtesy of Centers for Disease Control, Atlanta, GA/BPS.) (b) The
hook and basal body structure that anchor the flagellum into each cell wall and
membrane layer. (Courtesy of M.L. DePamphilis and J. Adler, *J. Bacteriol.* (1971)
105:384-395, p. 388.)

As shown in Figure 3.16, the flagellum is
composed of three parts. The outermost area, the
filament, is attached to a **hook** composed of subunits of a single protein quite different from flagellin. The hook is connected to a **basal body,**
composed of rings bound to a central rod. The
rings on the basal body anchor the flagellum to
the cell wall and cell membrane. Gram-positive

bacteria contain two rings and gram-negatives
contain four. In gram-negatives the outer rings
are attached to the cell wall and the inner rings
to the cell membrane.

The arrangement of the flagella is characteristic of the group to which the bacterium belongs. There are four basic types of arrangements: **monotrichous** (single flagellum at one

pole); **lophotrichous** (two or more flagella at one pole); **amphitrichous** (tufts at both ends); or **peritrichous** (flagella over the entire surface). The flagellar arrangements are shown in Figure 4.3.

Flagellar motion is a coordinated behavior. In action, flagella rotate, probably due to rotation of their basal bodies. When the cell is moving forward (called a **run**) the flagella are rotating mostly in coordinated tufts. When the cell is at rest, or jiggling aimlessly (called a **twiddle**), the flagellar tufts separate into individual flagella.

Flagellar movement often occurs in response to the presence of specific chemicals or in response to wavelengths of light in the environment. Movement in response to chemicals is called **chemotaxis.** Chemotaxis may be positive or negative depending on whether the organism moves toward or away from the chemical stimulus. The bacterial cell receives its sensory input in the form of chemical signals. The cell membrane contains protein receptors that bind the signal substances. Binding appears to alter flagellar rotation. This in turn affects the coordination of the flagella so that the cell moves (rather randomly) toward increasing concentrations of a favorable chemical such as a food source, or away from a harmful chemical.

AXIAL FILAMENTS.   A group of flexible, wriggling bacteria called spirochetes (one of which is the bacterium responsible for syphilis) have a unique motion that is related to the location of their flagella. The flagella are internal. They lie between the cell-wall layers, grouped in **axial filaments** that wrap around the cell in a spiral fashion. When they contract, an undulating motion is produced. Some spirochetes appear to have microtubules (see p. 101) as eucaryotes do.

OTHER FORMS OF LOCOMOTION.   Cyanobacteria and some others move by **gliding.** They have no external structures for motility. In some cyanobacteria there appears to be a layer of parallel fibrils between the outer layers. These may contract to produce the characteristic forward rotating movement of gliders.

Organisms that live in the water often have flotation structures. **Gas vesicles,** which are membranous gas-filled sacs, give organisms a means to move vertically. They can change their buoyancy by increasing or decreasing the amount of stored gas. They are widespread in cyanobacteria and photosynthetic bacteria.

SPORE FORMATION.   Motility enables an organism to move toward favorable environmental stimuli and away from harmful stimuli. Sometimes, however, environmental conditions are so adverse that only those microorganisms that can form specialized survival structures called **endospores** will be able to adapt. An endospore is a resting cell structure formed by some Gram-positive bacteria. It is usually metabolically inactive; is somewhat more resistant than a normal cell to heat, drying, or other adverse conditions; and bears a thickened outer layer. The purpose of endospores is to enhance the chances of survival under difficult conditions. Endospores often survive all but the most determined efforts to kill them. They create quite a problem when we are attempting to disinfect surfaces and materials. Endospores are the reason why autoclaving (sterilizing by steam under pressure) rather than plain boiling is needed for sterilization (see Chapter 23).

Sporulation (spore formation) commences in a **vegetative cell** — that is, a normal, actively growing cell. When the food source the bacterium has been using is exhausted, growth stops. This signals the beginning of a complex series of events, which are shown in Figure 3.17. First, the

FIGURE 3.17   Sporulation
A spore begins to form when one copy of the bacterial chromosome acquires a series of spore coats originating from the cell membrane. As the cortex thickens and matures, the rest of the cell contents break down. Eventually a free spore, exceptionally resistant to heat, drying, and chemicals, is released. (Courtesy of Philipp Gerhardt, Michigan State University.)

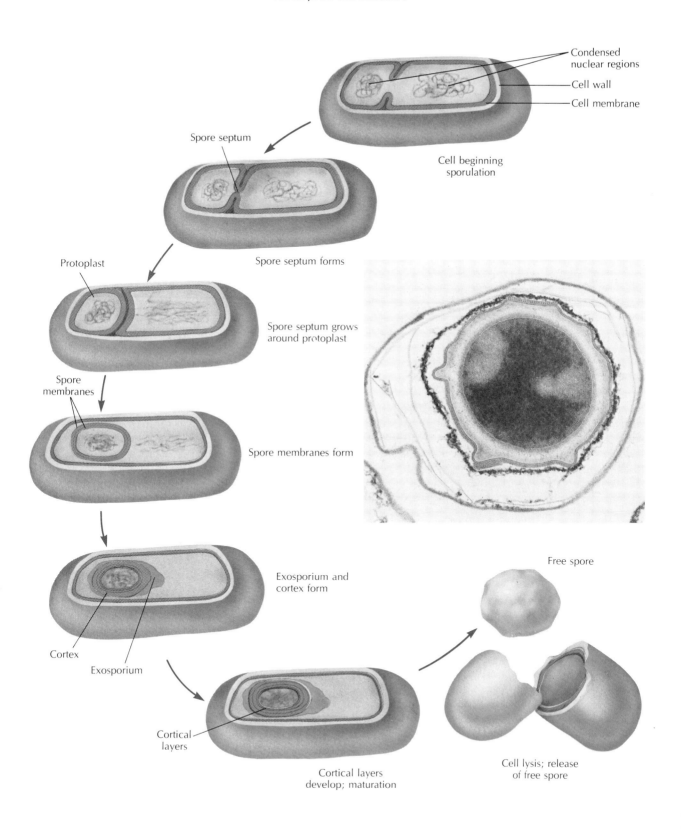

Condensed
nuclear regions

Cell wall

Cell membrane

Cell beginning
sporulation

Spore septum

Spore septum forms

Protoplast

Spore septum grows
around protoplast

Spore
membranes

Spore membranes form

Exosporium and
cortex form

Cortex

Exosporium

Cortical
layers

Cortical layers
develop; maturation

Free spore

Cell lysis; release
of free spore

bacterial chromosome is duplicated. Along with enzymes, nucleic acids, and ribosomes from the cytoplasm, one copy of the chromosome becomes isolated from the rest of the cell's cytoplasm by the ingrowth of the cytoplasmic membrane, forming a **spore septum.** The spore DNA then directs the rest of the sporulation steps from within the new structure, called a **forespore.** Thick coats of peptidoglycan are deposited; first the **cortex,** which is rich in calcium ions and dipicolinic acid, and then one or two outer endospore coats. Dipicolinic acid apparently confers heat resistance. Because of the endospore coats, the endospore (once freed from the vegetative cell) refracts light, resists heat, and excludes many chemicals, including stains.

The endospore contains little water, so its metabolic activity is arrested. Once the endospore matures, the remainder of the vegetative cell usually falls off. Six to eight hours elapse from the time when normal cell growth ceases until the free spore is mature.

An endospore may return to the vegetative state by the process of **germination.** Germination occurs when the endospore contacts certain amino acids, ions, or other chemical signals. The endospore takes on water and swells, loses its heat resistance and refractility, and begins enzymatic activity. Outgrowth occurs as swelling splits the endospore coat and a vegetative cell emerges. Note that one vegetative cell gives rise to only one endospore; the endospore in turn germinates into just one new cell.

Some soil bacteria form resistant cells called **cysts.** Cysts differ from endospores in that they lack dipicolinic acid and are less heat resistant. **Conidia** are structures formed in vast numbers by filamentous bacteria such as the actinomycetes. Their function, quite different from that of endospores, is primarily reproductive.

ATTACHMENT STRUCTURES — THE PILI.   Microorganisms are frequently found attached to a surface in their environment. This might be a leaf, a rock, a soil particle, or a particular type of human tissue. The ability to attach firmly and sometimes highly specifically is very important to the organism's survival because it helps the organism to find nutrients.

Many bacteria have small, tubular surface projections called **pili** or **fimbriae** (Figure 3.18). They are formed by aggregation of protein molecules into a hollow helix. One important function of pili is attachment to surfaces. For example, pili attach bacteria to red blood cells causing **hemagglutination** or clumping of the blood cells. Pili are also the means by which bacteria causing bladder infections and sore throats attach to their respective target tissues. Strains of bacteria without pili cannot cause these infections. Viruses that parasitize bacteria do so by first attaching to the fimbriae of their host cells.

FIGURE 3.18   Pili
The bacterium *Proteus vulgaris* is covered with stiff hairlike pili. The wrinkled surface of its outer membrane is also clearly visible, as is one long, curved flagellum at the right. (×46,769. Courtesy of Stanley C. Holt, University of Texas Health Science Center, San Antonio/BPS.)

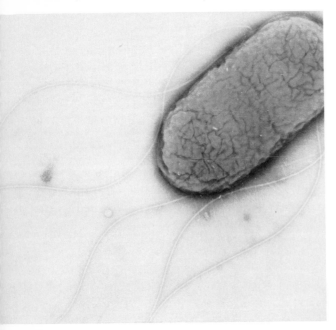

A second function of pili is to facilitate genetic exchange. *Sex pili* attach two bacteria to each other prior to **conjugation,** a process in which DNA is transferred from one cell to another.

Other types of attachment structures are found in aquatic bacteria. These include stalks, holdfasts, and modified adhesive surfaces.

THE EUCARYOTIC CELL

Eucaryotic cells, as we have mentioned before, contain a true nucleus, chromosomes enclosed in a nuclear membrane. They also have numerous smaller membrane-enclosed compartments or bodies called **organelles** within their cytoplasm. The eucaryotic cell is significantly more complex in structure than the procaryotic cell, but still has the same functional requirements. The algae, fungi, and protozoa are the eucaryotic microorganisms. The structure of the eucaryotic cell is shown in Figure 3.19. You'll find it helpful to refer to it as you read the following discussion.

We will discuss eucaryotic cell structure briefly. Our goal is to help you understand the structure of the eucaryotic microorganisms, to compare the eucaryotic and procaryotic cell types, and to be able to understand eucaryotic human cells as they function in defending us against infection and as they are damaged by infectious disease.

BOUNDARY LAYERS

The boundary layers in eucaryotes may include a cell wall and a cell membrane. Some eucaryotes have a cell wall, but others do not. All have a cytoplasmic membrane, which has fewer functions than the analogous structure in procaryotes.

CELL WALLS. Algae, fungi, and higher plants all require a rigid cell wall for osmotic protection. Eucaryotic cell walls contain no peptidoglycan. They use other polysaccharides as building blocks. Fungi, for example, use cellulose and chitin. Alginic acid, carrageenan, and agar add rigidity and tensile strength in the algae. In plants, cell walls composed of cellulose, pectin, and lignin keep out many pathogens. Cell walls are not found in animal cells or in protozoans.

CYTOPLASMIC MEMBRANES. The cytoplasmic membrane of eucaryotes appears to be chemically and structurally similar to the procaryotic membrane. One difference, however, is that the eurcaryotic cell membrane in animals contains complex lipids called sterols that contribute to the strength of the membrane. The eucaryotic cell membrane carries out facilitated and active transport, chemical exchange, and chemical signaling. Membranes may bear receptors for hormones. The membranes of certain protozoan and animal cells can be extensively reshaped to participate in **phagocytosis,** the surrounding and engulfing of particles. The eucaryotic membrane is able to carry out specific adhesion. This provides for active cell-to-cell cooperation.

CAPTURING, UTILIZING, AND STORING ENERGY

ENERGY CAPTURE — CHLOROPLASTS. In photosynthetic eucaryotes, which include the algae and plants, light energy is captured within specialized organelles, the **chloroplasts.** Chloroplasts contain the pigment chlorophyll, as well as enzymes required in photosynthesis (Figure 3.20). Chloroplasts are oval in shape and are enclosed by two complete membrane bilayers. The chlorophyll is contained in flattened membrane disks called **thylakoids.** Thylakoids, in turn, are arranged in interconnected stacks called **grana** and float in the **stroma,** or fluid matrix, of the chloroplast. Chloroplasts have DNA and divide by binary fission.

ENERGY UTILIZATION — MITOCHONDRIA. In eucaryotic cells, energy production involves specialized organelles called **mitochondria.** Mito-

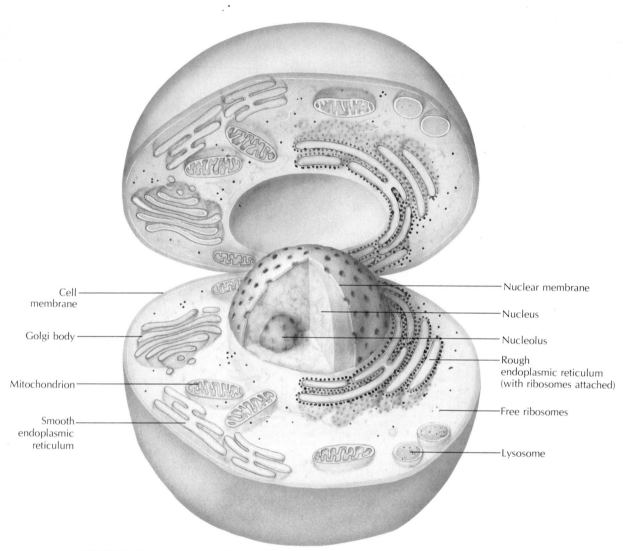

Cell membrane

Golgi body

Mitochondrion

Smooth endoplasmic reticulum

Nuclear membrane

Nucleus

Nucleolus

Rough endoplasmic reticulum (with ribosomes attached)

Free ribosomes

Lysosome

FIGURE 3.19   Eucaryotic Cells
A generalized drawing of a eucaryotic animal cell showing its membrane-enclosed nucleus and diverse organelles.

chondria trap the energy released from the breakdown of food molecules by forming ATP. This is a complex process called **respiration,** which we will describe in Chapter 7. As we saw, procaryotes also respire, but they use the cytoplasmic membrane structure for this purpose.

Mitochondria, like chloroplasts, have two membrane layers. The outer membrane is smooth; the inner membrane is extensively infolded into **cristae** (Figure 3.21). Some of the enzymes for energy utilization are found in the fluid matrix that fills the mitochondrion; others are held in the cristae. Each eucaryotic cell has many mitochondria, but the most actively metabolizing cells have the largest numbers. Mitochondria also have DNA and reproduce themselves by binary fission.

ENERGY STORAGE. Eucaryotic cells frequently contain storage materials. Examples are starch or oil in algae and higher plants and glycogen or fat in protozoans and higher animals. In human

FIGURE 3.20 The Chloroplast
Chloroplasts have external membranes. Within, the photosynthetic pigments are arranged in thylakoid disks stacked in grana. These float in the fluid stroma. Chloroplasts are seen in each cell of eucaryotic photosynthetic microorganisms as well as plants. (×15,120. Courtesy of Dr. Lewis K. Shumway, College of Eastern Utah, San Juan campus.)

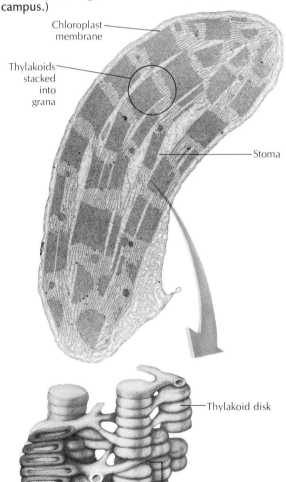

Chloroplast membrane

Thylakoids stacked into grana

Stoma

Thylakoid disk

Granum

FIGURE 3.21 The Mitochondrion
Mitochondria are enclosed by an outer mitochondrial membrane. Their inner membrane is folded into layers called cristae. The cristae bear the components of the ATP-producing system. Many mitochondria are formed in each eucaryotic cell. (Courtesy of Keith R. Porter, University of Colorado.)

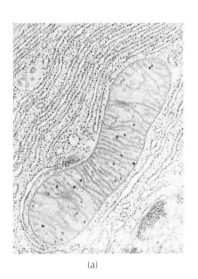

(a)

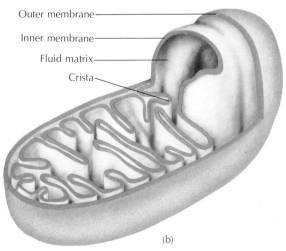

Outer membrane

Inner membrane

Fluid matrix

Crista

(b)

BOX 3.2
Assembling Cells with Parts from Many Sources

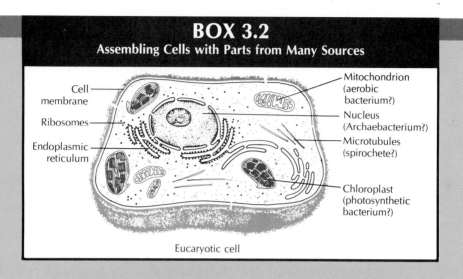

Cell membrane · Ribosomes · Endoplasmic reticulum · Mitochondrion (aerobic bacterium?) · Nucleus (Archaebacterium?) · Microtubules (spirochete?) · Chloroplast (photosynthetic bacterium?)

Eucaryotic cell

The biological nature of a eucaryotic cell doesn't entirely match that of its organelles. The organelles, especially mitochondria and chloroplasts, share many biological traits with procaryotes. A popular hypothesis supported by many investigators proposes that many of the eucaryotic cell's "parts" are derived from procaryotes that came to live within host cells in a symbiotic relationship called endosymbiosis. The eucaryotic chloroplasts are proposed to have descended from free-living photosynthetic bacteria. They carry reminders of their procaryotic past, including an incomplete genetic system with DNA, ribosomes of the procaryotic type, and associated components of the protein-synthesizing system. Like the chloroplast, the mitochondrion contains DNA and procaryote-like protein-synthesizing components, and replicates itself. The endosymbiotic hypothesis proposes that, on one or many occasions, a large procaryotic cell either ingested ("swallowed") or was parasitized by other procaryotes. The internal procaryotes survived intact and multiplied in harmony with the host. Both partners evolved toward a relationship that offered mutual advantage. The host cell provided shelter, nutrients, and an opportunity to reproduce. The internal partner specialized in efficient ATP production and energy capture. It is believed that this positive, stable relationship has persisted to this day.

Some workers also suggest that the microtubular structures of eucaryotes — cilia and flagella, for example — bear a striking resemblance to the microtubule-like structures found in some spirochete bacteria. They propose that microtubules in higher organisms may descend from endosymbiosis of some similar bacteria. However, these ideas are not universally accepted.

So where did the eucaryotic nucleus come from? If the nucleus descends from the "eater," which is not proven, it would be interesting to know which of the early procaryotes was its source. Nucleic acid studies are now suggesting that the *Archaebacteria* seem to be the closest surviving relatives of our cells' nuclei. Inside each eucaryote, there may be several procaryotes — but they are not crying to get out.

beings certain tissues are specialized for this storage function. Adipose (fat) tissue cells enlarge tremendously as they become filled with stored fats. The liver and muscle cells contain deposits of glycogen.

PROTEIN ASSEMBLY AND TRANSPORT STRUCTURES

There are relatively minor differences between the ways procaryotes and eucaryotes actually assemble their protein molecules. There is a major difference, however, in the way the protein molecules are transported inside the cell and to the outside, and in how they are chemically modified after synthesis.

ENDOPLASMIC RETICULUM AND GOLGI COMPLEX. As is true in procaryotes, the basic protein-synthesizing unit in eucaryotes is the ribosome, composed of one large and one small subunit. However, eucaryotic ribosomes are larger than procaryotic ribosomes. The eucaryotic subunits are 40S and 60S; they contain longer RNA molecules and a larger number of proteins per subunit. Polysomes are the basic assembly site, with genetic messages being used to assemble proteins on groups of ribosomes. Some eucaryotic polysomes float freely in the cytoplasm and seem to make internal proteins, whereas other polysomes are attached to the surface of a deeply folding network of internal membranes called the **endoplasmic reticulum** (ER). **Rough ER,** which is ER with associated ribosomes, produces proteins with defined "addresses" or destinations inside or outside the cell. These proteins require some additional processing after assembly. After being synthesized on rough ER, they are first packaged in extensions of the ER lacking ribosomes (**smooth ER**). These extensions then separate from the ER, and migrate to and fuse with **Golgi bodies** — a stack or complex of flattened oval membrane disks (Figure 3.22). Within the Golgi complex, proteins and, some believe, other large molecules, are chemically "sorted." Some are held in reserve for later use. Others are chemically

modified and prepared for secretion outside the cell. Eucaryotic cells frequently secrete large amounts of protein. For example, pancreatic cells secrete the enzymes needed to digest food. Proteins to be secreted are often modified by addition of carbohydrates to form glycoproteins. Vesicles (droplets bounded by membranes) of the Golgi complex pinch off, carrying secretory proteins to the cell membrane, where they discharge their contents to the outside.

The Golgi complex also forms organelles called **lysosomes** and **peroxisomes.** Lysosomes act as internal reservoirs for digestive enzymes, while peroxisomes contain enzymes that produce toxic chemicals to kill captured prey. Both lysosomes and peroxisomes are crucial in the ability of a phagocytic cell to successfully digest captured prey. When the phagocyte is one of our white blood cells and the "prey" is a pathogenic bacterium, the digestive abilities of these small bodies may determine whether or not we recover from the infection.

GENETIC INFORMATION STRUCTURES

There are many differences between procaryotes and eucaryotes in how their genetic information is structured. Recall that procaryotes have a single ring-shaped DNA chromosome. In eucaryotic cells the chromosomes themselves are different, as is their location in the cell. Eucaryotic chromosomes are linear structures, consisting of a DNA helix coiled up with clusters of basic proteins, the **histones.** Most eucaryotic cells have two copies of each type of chromosome. One is obtained from each parent in the course of sexual reproduction, and the two members of the pair usually differ somewhat in the exact details of their genetic instructions. Cells with paired chromosomes are referred to as **diploid.** Other eucaryotic cells, including many microorganisms, have only one copy of their chromosomes and are haploid. All procaryotes, you will recall, are haploid.

NUCLEUS. The eucaryotic cell **nucleus** is a spherical structure containing most of the cell's

DNA. It is separated from the cytoplasm by the double-layered nuclear membrane. The nucleus communicates with the cell cytoplasm via numerous tiny pores in the nuclear membrane. In most eucaryotes, a dense, roughly spherical area, called the **nucleolus,** lies within the nucleus. The nucleolus also contains DNA. Its genes are concerned with the manufacture of ribosomes.

FIGURE 3.22 Endoplasmic Reticulum and the Golgi Complex
Proteins are synthesized when an RNA message, ribosomes, and other factors assemble to form a polysome. When the polysomes are free in the cytoplasm, the protein is released into the cytoplasm when completed. When polysomes are bound to the endoplasmic reticulum, the protein product passes through rough, then smooth endoplasmic reticulum. It is transported to the Golgi apparatus, where it is packaged into vesicles. Some of these fuse with the cytoplasmic membrane to dump their contents exterior to the cell. The inset TEM shows multiple layers of rough endoplasmic reticulum within a cell. (Courtesy of Keith R. Porter, University of Colorado.)

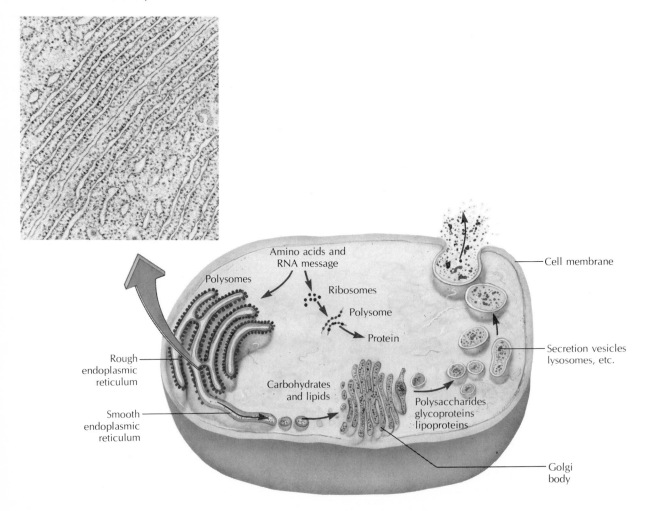

There are other sources of DNA in a cell. One of these may be the genes of infecting viruses, which may become permanently incorporated into the the cell's chromosomal DNA. Both the chloroplasts and the mitochondria of the cell also contain DNA, as previously noted.

The complex structure of the eucaryotic nucleus, plus the need to sort out multiple chromosomes accurately, makes cell division a more complex process for eucaryotes than for procaryotes. Eucaryotic cells divide by mitosis. We'll discuss mitosis in Chapter 5.

ADAPTABILITY AND MOVEMENT. Eucaryotic organisms have developed an amazing diversity of adaptive mechanisms — far too varied to be covered adequately here. We will limit our discussion to the equivalents of procaryotic motility and sporulation found in the eucaryotic microorganisms.

Microfilaments and Microtubules. A eucaryotic cell has a complex internal network of tubular and filamentous structures called a **cytoskeleton,** used both for mechanical support and movement. These structures can move the cell's contents around, rearrange organelles, and change the cell's shape. Two of the four components of the cytoskeleton should be noted. The **microfilaments** are slender contractile protein fibrils that are active in stretching and contracting movements and in cell division. **Microtubules** are hollow protein cylinders. They comprise the spindle during eucaryotic cell division. Equally important, microtubules compose the structures that permit eucaryotic motility. When the structures are few and long they are called **flagella.** When they are numerous, short, and hairlike, they are called **cilia.** Both flagella and cilia are composed of a circle of nine double microtubules surrounding a pair of single microtubules (Figure 3.23). This **"9 + 2" arrangement** is a eucaryotic characteristic found both in flagella and in cilia, which are structurally just short flagella. Many single-celled algae and certain protozoa have one or two flagella.

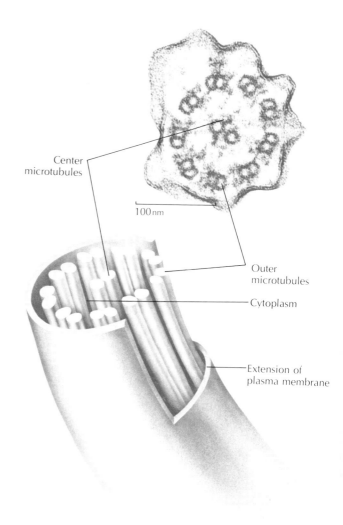

FIGURE 3.23 Eucaryotic Flagella
The flagella and cilia of eucaryotes are composed of nine pairs of microtubules arranged around one central pair of microtubules, and enclosed in an extension of the plasma membrane. (×127,050. Courtesy of W.L. Dentler, The University of Kansas/BPS.)

A uniform covering of short cilia, which exhibit coordinated movement, is seen in protozoa such as *Paramecium* as well as in the ciliated epithelium that lines the human respiratory tract.

TABLE 3.2 Structure and function in procaryotic and eucaryotic cells		
Function	**Procaryotic Structure**	**Eucaryotic Structure**
Boundary layers	Capsule or slime layers	Not commonly found
	Cell wall contains peptidoglycan	Cell wall in algae, fungi contains cellulose and chitin
	Cell membrane is fluid, mosaic type	Basically same as procaryotic
Energy	Lamellae chromatophores	Chloroplasts
Capture		
Utilization	Cell membrane	Mitochondria
Storage	Inclusions of glycogen, starch, volutin, PHB	Inclusions of glycogen, starch, fats, and oils
Protein synthesis, packaging, and transport	30S and 50S ribosomal subunits	40S and 60S ribosomal subunits, endoplasmic reticulum, Golgi complex
Genetic structure	Nucleoid	Membrane-bound nucleus
Organization	Single circular DNA helix	Linear chromosomes, often multiple and in pairs
Accessory DNA	Plasmids	Mitochondrial DNA, chloroplast DNA
Movement	Flagella, hollow protein tubules	Flagella and cilia of "9 + 2" microtubular type; microfilaments
	Gas vesicles	Gas vesicles
Resistant forms	Endospores, cysts	Cysts
Attachment	Pili, stalks	Holdfasts, other structures
Internal vesicles	No structural equivalent	Lysosomes

Eucaryotic flagella and cilia may have a more complex coordination system than the corresponding procaryotic structures.

Spores and Cysts. Many eucaryotic microorganisms form specialized, hard-coated cells. Most fungi, for example, form vast numbers of **spores.** Their spores have a reproductive function, usually becoming airborne for wide distribution. Many protozoans form resting **cysts.** These resistant structures enable the organisms to survive harsh environments, such as drying or a cold winter. Some human pathogens, taken in as cysts with our food, are thus able to survive our stomach's acidity on their way to infect our intestines.

Now that we've completed our description of procaryotic and eucaryotic cell structure, you may wish to review the differences between these types of cells. This information is presented in Table 3.2.

SUMMARY

1. Knowledge of cell structure developed as the microscope improved. Bright-field microscopes allow study of cell shapes and the larger organelles of eucaryotes. We use diverse techniques for the differential staining of cell structures. The most important bacterial stain is the Gram reaction. Darkfield and phase-contrast optics are widely used for the study of unfixed living cells. Ultraviolet microscopy increases resolution and makes use of fluorescent dyes that pinpoint particular structures for study.

2. The electron microscope takes advantage of the increased resolving power of shorter-wavelength radiation. Transmission electron microscopes create images of the internal structures of cells. The specimen is sliced into very thin sections for viewing. Scanning electron microscopes use deflected electrons, creating images of surface details that appear to be three-dimensional.

3. Electron microscopes provided convincing evidence that there are two different types of cell — procaryotic and eucaryotic.

4. The procaryotic cell has a number of surface layers. The glycocalyx is a loosely organized external layer that serves protection and adhesion functions. The cell wall contains a layer of peptidoglycan as a covalently linked molecular enclosure. It provides protection against osmotic stress and confers shape. When the cell wall is weakened or absent, bacteria lose shape and become osmotically fragile. In Gram-positive bacteria the peptidoglycan sac comprises almost the entire cell-wall structure. In Gram-negative bacteria the wall consists of an outer membrane and a thinner peptidoglycan layer. The outer membrane contains a mixture of substances, notably the lipopolysaccharides.

5. Cytoplasmic membranes are fluid mosaics of structural and enzymatic proteins interacting with phospholipid and other lipid components. Membranes regulate what goes into and out of the cell, contain enzymatic and carrier proteins and carry out facilitated and active transport.

6. All of the surface layers contain unique chemical substances that may be used to help us in identifying the specific strain or species of organism.

7. Light-energy capture in procaryotes takes place in lamellae and chromatophores. Energy utilization and ATP production take place in specialized regions of the cytoplasmic membrane. Energy storage involves irregular aggregates of high-energy compounds within the cytoplasm.

8. Protein synthesis occurs on ribosomes. In procaryotes genetic information is contained in the single, circular DNA chromosome. Accessory information is found on extrachromosomal elements called plasmids.

9. Motility in procaryotes is conferred by flagella, hollow tubes formed of the protein flagellin. Flagella rotate in response to chemical stimuli in the environment, allowing coordinated movement (chemotaxis) toward or away from a stimulus.

10. The endospore is a highly refractile, resistant structure formed by some bacteria. Under adverse conditions, spores survive when vegetative cells in the population die. Spores can later germinate and commence active growth.

11. In eucaryotic cells many functions are localized in specialized organelles. These are enclosed in membranes. The cytoplasmic membrane's key function is to regulate permeability and transport. The mitochondrion serves as the site of ATP synthesis; the chloroplast carries out photosynthesis. Protein synthesis takes place on ribosomes. The endoplasmic reticulum and the Golgi complex synthesize proteins and secrete them outside the cell.

12. In eucaryotes, the nucleus is enclosed by a nuclear membrane penetrated by pores. Nuclear genetic information is organized in linear chromosomes. The DNA is enclosed in a regular coating of histones. In most eucaryotic cells, the chromosomes are paired.

13. Eucaryotic structures for movement are formed from either contractile microfilaments or rigid, hollow microtubules. Eucaryotic flagella or cilia are composed of microtubular bundles.

DISCUSSION QUESTIONS

1. Identify the various types of microscopes and explain what each is most useful for viewing.
2. A human red blood cell is 7.5 μm in diameter. Express its size in nanometers and angstrom units.
3. Mitochondria and bacteria are about the same size and have the same type of ribosomal subunits. What might this suggest about a common origin?
4. Compare the structure of the gram-positive and gram-negative cell wall.
5. What are the functions of the procaryotic cell membranes?
6. Why is being able to move important to a microorganism?
7. Compare the structure and characteristics of a bacterial spore with those of the vegetative cell.
8. Describe the process whereby eucaryotic cells produce macromolecules and secrete them to the exterior of the cell.

ADDITIONAL READING

BOOKS

Alberts B, et al. *Molecular biology of the cell.* New York: Garland, 1983.

Lowry AG, Siekevitz P. *Cell structure and function.* 3rd ed. New York: Holt, Rinehart and Winston, 1979.

Morrison P, et al. *Powers of ten.* New York: Scientific American Library, 1982.

REVIEW ARTICLES

Adler J. Chemotaxis in bacteria. *Annu Rev Biochem* 1975; 44:341-356.

Gray MW, Doolittle WF. Has the endosymbiotic hypothesis been proven? *Microbiol Rev* 1982; 46:1-42.

Margulis L, To L, Chase D. Microtubules in Procaryotes. *Science* 1978; 200:1118-1124.

Rothman JE. The Golgi apparatus: two organelles in tandem. *Science* 1981; 213:1212-1219.

Salton, MRJ, Owen P. Bacterial membrane structure. *Annu Rev Microbiol* 1976; 30:451-482.

Silverman M, Simon MI. Bacterial flagella. *Annu Rev Microbiol* 1977; 31:397-420.

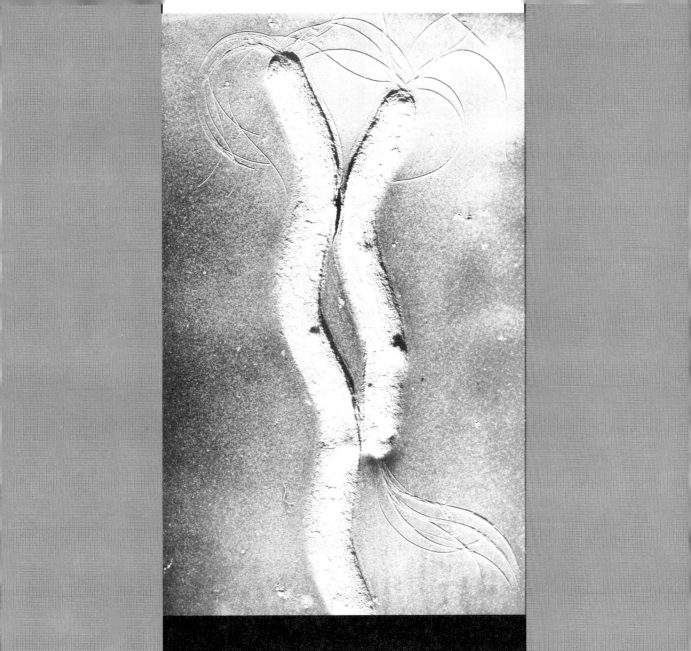

MEET THE
PROCARYOTES

CHAPTER OUTLINE

Microorganisms are amazingly diverse, and endlessly fascinating for the patient observer. In this chapter we will explain what we do and do not know about classifying microorganisms, then we will describe each of the major groups of procaryotes. We will apply the concepts of biochemistry and cell structure we have just acquired by looking at some specific microorganisms and their place in the environment.

BASIC RULES OF CLASSIFICATION

The science of biological classification is called **taxonomy.** By classification we mean grouping things together according to their similarities, and distinguishing them from other groups. Classification satisfies a basic human intellectual drive, whereby we recognize the advantages of generalizing information. Our present

taxonomies are based on those developed by the Swedish scientist Carolus Linnaeus in 1753 and 1758. We have modified his work as we have found new information. Whereas Linnaeus used physical appearance — **morphology** — of the organism as his primary basis for assigning a creature to a particular group, taxonomists today also use techniques from biochemistry and molecular genetics, and take advantage of computerized analysis. Any classification scheme must be regarded as provisional, and certainly the taxonomic schemes for microorganisms are still far from perfect. They are useful but partly arbitrary outlines, constantly being changed as new information comes to light and old criteria are reexamined.

Let us start with the classification groups **genus** and **species.** Each known organism (not only the living ones but also many fossils) has been given a two-part name, a **binomial** designation, an example being the binomial *Escherichia* *coli.* The first portion, capitalized, is the genus name (roughly equivalent to our family surnames) and the second, lower-case part is the species name (equivalent to our first or given names). On a second reference, it is usual to abbreviate the genus name to its first letter, as in *E. coli.* The genus and species are the two classifications you will be using most in microbiology. The names are usually derived from Latin or Greek descriptive terms. A bacterial name must be approved by the custodians of the International Code of Nomenclature for Bacteria. This is to make sure that it is valid and not a duplication of an already named organism. You will find that often microbiologists use common names, such as **pseudomonads** for members of the genus *Pseudomonas* or **staphylococci** for members of the genus *Staphylococcus.* These names, neither capitalized nor italicized, are intended to refer to organisms of the general type of the genus from which the idiomatic name is taken.

Next, we recognize that classifications are **hierarchical** (Table 4.1). That is, groups of closely related species are arranged in genera, groups of similar genera are arranged in families, and groups of families constitute orders. In the case of higher organisms, additional subcategories may be used. As you go up the scale, the groups become progressively larger and the criteria for membership more general. To illustrate, there are several thousand valid bacterial species, but they are all members of four divisions (the comparable group in animals is called a phylum), and the four divisions are all in one kingdom, **Procaryotae.**

We classify hoping to group organisms that are closely related and similar in important characteristics. **Natural taxonomy** aims to bring together, within any group, those organisms that descended from a common ancestor. This type of taxonomy indicates evolutionary relationships. Unfortunately, the goal of a natural taxonomy has so far not been obtained with microorganisms. Their great biological age and the freedom with which they exchange genetic material guarantee that they have extremely complex, often ambiguous, evolutionary interrelationships. Organisms that look very similar may be totally different functionally. How are they actually related, then? Conversely, two very different looking organisms may be very similar genetically, so that it is difficult to escape the conclusion that they are closely related.

TABLE 4.1 Comparing the classification of two kingdoms

A Bacterium		A Human Being	
Kingdom	Procaryotae	Kingdom	Animalia
Division	II	Phylum	Chordata
		Subphylum	Vertebrata
		Class	Mammalia
		Subclass	Eutheria
		Order	Primates
Family	Enterobacteriaceae	Family	Hominidae
Genus	*Escherichia*	Genus	*Homo*
Species	*coli*	Species	*sapiens*

Euglena, a eucaryotic alga, possesses features of both plant and animal cells (it photosynthesizes, as plants do, and captures smaller organisms for food, as animals do). It is vigorously claimed by both botanists and zoologists. The world of microorganisms is full of similar enigmas. In any case, some things can be learned about *Euglena* by considering it as a plant, and other things can be learned by considering it as an animal.

Classification, for all its difficulties, allows us to organize our knowledge, and to formulate sensible questions to ask to get more knowledge. Furthermore, once we have a workable classification scheme, we are in a position to make an identification, that is, to take an unknown organism, study it, and find out what it is. A clinical laboratory's primary job is identification of unknown organisms from patients' specimens. We discuss some of the many microbiological identification techniques in Chapter 6, then again throughout later chapters as we encounter specific pathogens.

THE PROCARYOTIC KINGDOM

The popular Whittaker classification scheme separates all life forms into five kingdoms. Four of the kingdoms are eucaryotic — Plants, Animals, Fungi, and Protists — which leaves one kingdom for the Procaryotes (Figure 4.1).

In this chapter we will examine all the major groups of procaryotes. You will discover that they are surprisingly diverse. Their diversity reflects the fact that each different type of organism is adapted to its own particular place in the environment. This is as true for human disease organisms as it is for soil bacteria. The most important things about each organism are where and how it lives.

The primary reference on bacterial classification is *Bergey's Manual of Determinative Bacteriology*. This reference was first published in 1923. It is now being issued as *Bergey's Manual of Systematic Bacteriology*, starting in 1984, and it has grown from one medium-sized volume to four volumes. This might give you an idea of

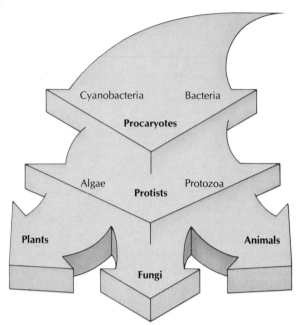

FIGURE 4.1 The Five Kingdoms
Microorganisms are found in three of the five kingdoms — bacteria and cyanobacteria in the kingdom Procaryotae, algae and protozoa in the kingdom Protista, and fungi in the kingdom Fungi.

the rapid expansion in the amount of information available. The order in which the groups of bacteria are presented later in this chapter is based loosely on the newer *Bergey's Manual* system.

THE BACTERIA

The **bacteria** are small procaryotic organisms, predominantly single-celled, but some living in groupings. They are believed to be direct descendants of some of the earliest forms of life, that have survived and prospered. Their diversity is so great that we cannot limit our definition any more than this. Bacteria are found everywhere on the globe, including such unlikely spots as Antarctic ice masses, deep ocean trenches, the Dead Sea, and steaming hot geysers. They play extremely important, though sometimes unexpected, roles in human life. We

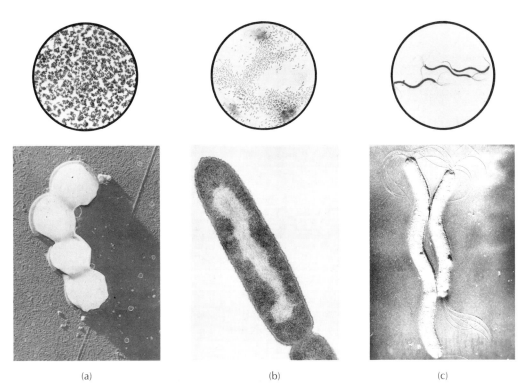

(a)   (b)   (c)

FIGURE 4.2   The Three Basic Shapes of Bacteria
Bacteria occur in three morphological shapes; each basic shape may occur with
variations. These pairs of photomicrographs show (above) many cells of an organism
as it appears through the light microscope and (below) one or more cells viewed
through the electron microscope. (a) The spherical-shaped coccus form (top:
Micrococcus ×480; bottom: *Staphylococcus aureus* ×18,700), (b) the rod-shaped
bacillus form (top: *Azotobacter*; bottom: *Bacillus subtilis* ×21,600), and (c) the
corkscrew-shaped spirillum form (top: *Spirillum volutans* ×296; bottom:
Aquaspirillum bengal ×3465). (Photo credits: 4.2a, (top) General Biological Supply
House, (bottom) Eli Lilly & Co.; c, (top) General Biological Supply House, (bottom)
From Noel R. Krieg, *Int. J. Syst. Bacteriol.* (1974) 24:453–458, Fig. 1, p. 455.)

will see that bacteria are crucial in human
health, nutrition, commerce, industry, agriculture, and the environment. Once you start being
aware of bacteria, you will see their important
contributions all around you in daily life.

CRITERIA USED TO CLASSIFY AND IDENTIFY BACTERIA

SHAPE AND ARRANGEMENT.   If you were asked to
identify an unknown organism in the microbiology laboratory, your first step would be to
prepare a gram-stained slide from which to describe its shape, size, and the groupings in which
it appears. A **bacillus** (plural, bacilli) is a rod-
shaped or sausage-shaped cell; a **coccus** (plural,
cocci) is a spherical cell; and a **spirillum** (plural,
spirilla) is a helical, rigid rod (Figure 4.2). Each of
these three basic forms may occur in numerous
variations — short rods, long rods, slender rods,
and so forth. There are also less common
forms — star shapes, squares, stalked bacteria,
filaments, and who knows what else. The groupings of the cells (clusters, chains, pairs, tetrads)

are also significant. Cell length and/or diameter can be accurately measured to aid in classification. There will be plenty of variation among individual cells in a population, so decriptions are based on a composite picture of the average cell. It is also sometimes important to determine the biochemistry of the cell-wall components. Special structures such as endospores, capsules, and granules are found by special stains, and may help us in identification.

MOTILITY.   The presence, number, and arrangement of flagella or other means of motility are important identification traits. As mentioned in the last chapter, there are two basic patterns, **polar,** with single or clustered flagella at one or both ends, and **peritrichous,** with flagella all over. These and some variations are shown in Figure 4.3.

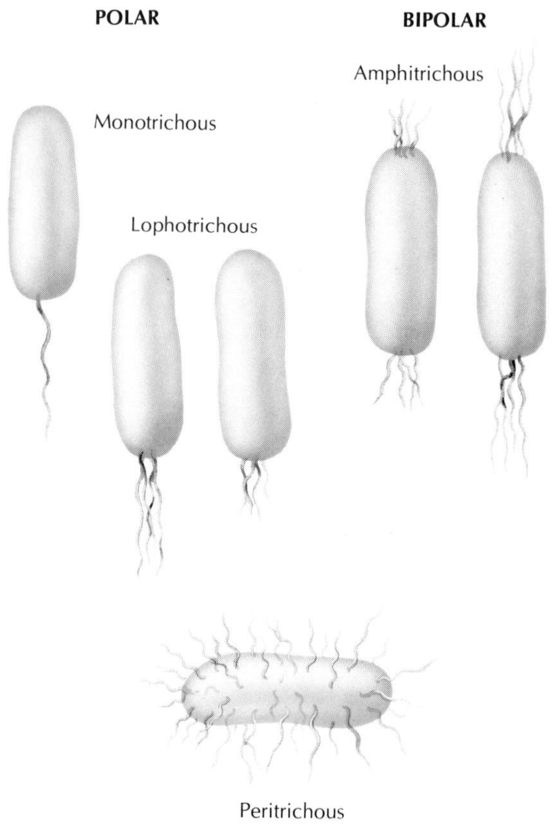

METHOD OF REPRODUCTION.   Most bacteria reproduce by **binary fission,** a simple but orderly process in which one cell enlarges and divides into two cells. Budding, formation of reproductive spores, and fragmentation are other methods. The method of division is important in establishing cell groupings, such as chains or clusters.

NUTRITION.   Our nutritional classification of an organism depends on what carbon sources and energy sources it uses. **Lithotrophs** are organisms that use inorganic carbon sources such as carbon dioxide or carbonate ions for growth. **Heterotrophs** are organisms that require organic carbon, that is, carbon already incorporated into organic compounds such as sugars or amino acids. To subdivide further, we can define where the organism gets its energy. **Photolithotrophs** get their energy from light, while **chemo-lithotrophs** get energy by chemical reactions in which they oxidize certain inorganic compounds.

The heterotrophic microorganisms, our primary concern in this book, fall into two nutritional subgroups. **Saprophytes** use organic compounds they obtain as wastes or residues from

POLAR   **BIPOLAR**

Monotrichous

Lophotrichous

Amphitrichous

Peritrichous

FIGURE 4.3   Arrangements of Bacterial Flagella
Polar flagella are located at one end of the cell. Cells with single flagella are named monotrichous; cells with clusters at one end are lophotrichous. Bipolar cells have two clumps; if the flagella spread over much of both ends they are amphitrichous. Peritrichous cells have flagella evenly distributed over the entire cell surface.

dead organisms. **Parasites** live by invading other cells, or the tissues of a living organism. Many parasites live and reproduce within host cells, and are called **intracellular parasites.** All microbial pathogens are heterotrophic. Some are saprophytic (living off of our wastes) but most are parasitic (living within our cells).

METABOLISM.   Metabolic patterns in bacteria are varied and complex. The energy-yielding portion of metabolism, as we shall see in Chapter 7, works by removing electrons from food materials, then transferring them to other molecules, which serve as **electron acceptors.** One way to classify organisms is on the basis of what substance is used as the final acceptor of electrons in their metabolism. Bacteria that use oxygen or another inorganic electron acceptor are **respiratory.** Bacteria that use organic compounds, such as organic acids, as electron acceptors are **fermentative.** We will discuss respiration and fermentation in more detail in Chapter 7. Note that the terms respiration or respiratory, when applied to microorganisms, apply to metabolism, not to breathing.

Some bacteria grow in the presence of oxygen. They are **aerobes;** if they must respire with oxygen they are **strict aerobes.** These organisms live in such places as water, soil, and on skin surfaces. Other organisms can grow either with or without oxygen. They are called **facultative anaerobes;** they can obtain energy by either respiration or fermentation. Many of these organisms live in places such as the human intestinal contents, where there is little oxygen. **Anaerobes** grow in the absence of oxygen.

Organisms that do not use oxygen and tolerate exposure to it poorly are **strict anaerobes.** These species ferment and sometimes respire, by means that do not use oxygen. Their habitats include the cow's stomach, the mud of swamps, and the area under the flap of the human gums.

DNA AND RNA.   The genetic material of each species is unique. However, the more closely related two species are, the more similar their genomes will be. One commonly performed genetic analysis is the determination of the relative amounts of adenine, guanine, cytosine, thymine, or uracil. This tells us the base composition of a DNA or RNA sample. A determination of the actual base sequence of a nucleic acid is a more powerful analytic tool. Simply put, the more similar the base sequences, the more closely related two organisms are. If DNAs from two organisms are close to identical we have indisputable proof that they are closely related. Table 4.2 sums up our discussion of the various criteria used to classify and identify bacteria.

Let us proceed to discuss the major groupings of the procaryotes. These will be arranged in

TABLE 4.2 Criteria used to classify and identify bacteria					
Microscopic Morphology	**Biochemical Determinations**	**Environmental Relationships**	**Appearance of Growth**	**Metabolic Pathways**	**Molecular Genetics**
Cell shape	Cell-wall materials	Atmosphere required	Location in liquid culture	Carbon sources used	DNA base composition
Cell size	Cellular lipids	pH tolerance	Colonial morphology	Nitrogen sources used	DNA sequences
Groupings	Capsular materials	Temperature and osmotic range	Pigmentation	Sulfur sources used	RNA sequences
Internal structures	Pigments	Symbiosis and pathogenicity	Rate of growth	Fermentation potential	Genetic mapping
Appendages	Cellular storage inclusions	Habitat		Respiratory potential	Amino acid sequences of proteins
Staining reactions		Antibiotic sensitivity		Characteristic end products	

four rather heterogeneous subdivisions, following the outline of *Bergey's Manual*. Dominant criteria for the grouping are the organism's Gram reaction or cell-wall structure, its role in the environment, and the medical or commercial applications of the organism.

GRAM-NEGATIVE BACTERIA OF MEDICAL AND COMMERCIAL IMPORTANCE

Gram-negative bacteria are found in almost every type of natural environment — soil, water, human tissues, and sewage, to name a few. In this section we discuss some common varieties with which you may expect to become more familiar.

THE SPIROCHETES

Spirochetes are long, slender, wavy microorganisms with an unusually flexible cell wall (Figure 4.4a). Because they are so slender, they are often difficult to see with a microscope using a bright-field condenser, and darkfield microscopy may be used. They exhibit an undulating motility using flagella-like axial filaments.

Some highly pathogenic spirochetes include *Treponema*, which causes syphilis and yaws, *Borrelia*, which causes relapsing fever, and *Leptospira* (Figure 4.4b), which causes leptospirosis. Some spirochetes are fermentative, others respiratory. Several, including *Treponema pallidum*, the agent of syphilis, are **obligate parasites,** meaning that they can live only in living organisms.

HELICAL, MOTILE BACTERIA

The true spirilla, in contrast to the spirochetes above, have rigid cell walls, which give them the shape of helices (Figure 4.5). Their flagella, if present, are located on the outside surface. *Spirillum*, *Aquaspirillum*, and *Oceanospirillum* are predominantly aquatic genera, as their names imply. They are usually saprophytic and decompose organic residues. The genus *Campylobacter* contains the major pathogen *C. fetus*, causing reproductive and intestinal infections in human beings and in domestic animals. They are a major cause of human digestive upsets in all age groups. They inhabit the reproductive organs, intestinal tracts, and oral cavities of many animal species. *Bdellovibrio* is a bacterium predatory on other bacteria. Its motile, curved cells attack the host bacterium, penetrate its wall, and multiply in the periplasmic space. A coiled filamentous cell is formed, which subsequently divides into numerous progeny. The host is destroyed when they are released. The bdellovibrios are found in soil, sewage, fresh water, and the ocean.

AEROBIC GRAM-NEGATIVE RODS AND COCCI

There is a large, widely distributed group of heterotrophic bacteria that are strictly respiratory and do not ferment. Most of these bacteria possess the respiratory enzyme **cytochrome oxidase.** This enzyme's presence can be determined by several rapid and simple tests. Organisms with the enzyme activity are said to be **oxidase positive.** This is a very important identifying trait in clinical microbiology.

The genus *Pseudomonas* contains rod-shaped species that bear polar flagella and are active swimmers. They are extraordinarily versatile in their ability to metabolize a wide range of compounds for energy, and to thrive in a variety of environments. The pseudomonads include many water and soil organisms, as well as several serious pathogens of humans (*P. aeruginosa, P. pseudomallei, P. cepacia*) and plants (*P. solanacearum*).

Rhizobium, another genus of aerobic gram-negative rods, is noteworthy for the major contributions it makes to soil fertility. *Rhizobium*, as we saw in Chapter 1, forms symbiotic associations with legume plants such as beans, peas, and alfalfa. The root nodules contain rhizobia,

(a) Spirochete

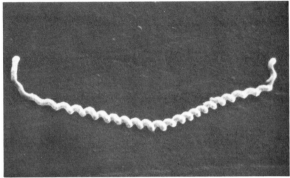

(b) *Leptospira interrogens*

FIGURE 4.4 Spirochetes
(a) A generalized drawing of a spirochete. (b) *Leptospira interrogans*, showing the slender, flexible shape. This species also has hooked ends. It causes the urinary-tract disease called leptospirosis, which affects humans, dogs, and rodents. (Courtesy of Dr. Dorothy L. Patton, Department of Medicine, University of Washington, Seattle.)

(a) Spirillum

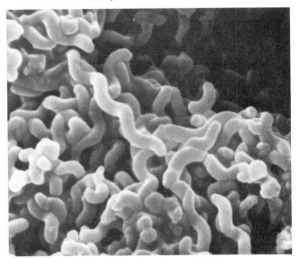

(b) Unidentified spirillum, called the Type 1 *Campylobacter*-like organism.

FIGURE 4.5 Spirilla
(a) A generalized drawing of a spirillum. (b) A scanning electron micrograph of *Campylobacter*-like organisms cultured from the human intestine. Note the regular, helical shape. (Courtesy of Dr. Dorothy L. Patton, Department of Medicine, University of Washington, Seattle.)

bacteria which incorporate gaseous nitrogen from the air into organic "fixed" nitrogen compounds, which may be absorbed by the plant (review Figure 1.5). Legume crops significantly increase the available nitrogen in soils. The discovery of the soil-enriching character of leguminous crops led to the introduction of crop rotation, a major agricultural advance in which a nitrogen-fixing crop such as alfalfa is planted in alternate seasons with nitrogen-demanding crops such as corn. It is useful to recall that it is the bacteria that do the actual enriching.

The genus *Azotobacter* also contains nitrogen-fixers. They are free-living, not symbiotic, and inhabit soil, water, and leaf surfaces. Their contribution to total soil fertility is less than that of the rhizobia.

Agrobacterium has attracted a great deal of interest because it forms abnormal growths or tumors, called crown galls, on certain types of plants (Figure 4.6). The tumor-inducing factor in the bacterium is genetic material within a plasmid. As we saw in Chapter 3, plasmids can be induced experimentally to transfer genes to host cells. The *Agrobacterium* plasmid has been successfully used to introduce valuable new genetic traits into the tobacco plant. It may become quite useful in the genetic engineering of improved plant strains.

Three other Gram-negative, aerobic, rod-shaped pathogens may be considered with the above organisms. Members of the genus *Brucella* cause brucellosis or undulant fever. This disease, rare since the introduction of milk pasteurization, is transferred to humans from infected domestic animals. *Bordetella* is the organism that causes whooping cough, a serious infection of small children. *Francisella* causes tularemia, a severe circulatory disease, in human beings and animals.

AEROBIC GRAM-NEGATIVE COCCI.   The aerobic Gram-negative cocci, which tend to arrange themselves in paired cells called diplococci (Figure 4.7), are respiratory. The oxidase test is generally positive, which distinguishes these organisms from other cocci. The genus *Neisseria* contains two important human pathogens: *Neisseria gonorrhoeae* causes the common sexually transmitted disease gonorrhea, and *N. meningitidis* causes a severe form of bacterial meningitis. *Moraxella* is a normal inhabitant of the human body, but may cause disease in de-

bilitated individuals. *Acinetobacter*, which is oval to rod-shaped, is atypically oxidase negative and moves in a twitching or jumping pattern. It is a serious hospital-acquired pathogen.

GRAM-NEGATIVE, FACULTATIVELY ANAEROBIC RODS

The Gram-negative, facultatively anaerobic rods are capable of both fermentative and respiratory metabolism. They can convert back and forth

FIGURE 4.7   Aerobic Gram-Negative Diplococci (a) Diplococci arise when two cells remain attached after binary fission. (b) The species of the genus *Neisseria* live in close association with humans, sometimes causing infection. Note the doubled arrangement evident in this scanning electron micrograph. (From Yoshii Z, et al. *Atlas of scanning electron microscopy in microbiology.* Tokyo: Igaku-Shoin, Ltd (1976).)

(a) *Diplococci*

FIGURE 4.6 *Agrobacterium tumefaciens* This scanning electron micrograph shows the crown-gall bacillus infecting a carrot cell by first attaching itself via fibrils to the cell surface. (Courtesy of Ann G. Matthysse and the American Society for Microbiology.)

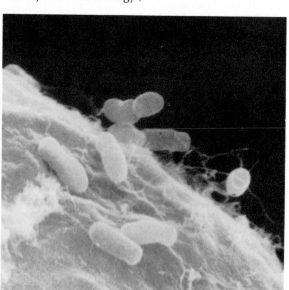

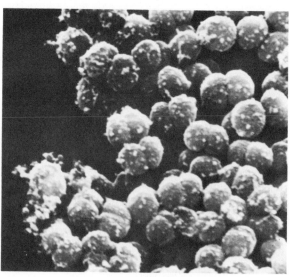

(b) *Neisseria*

BOX 4.1
Versatility — The Key to Success

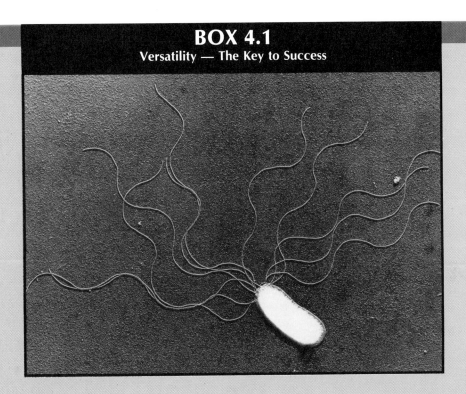

Pseudomonas is a very undemanding microorganism. Most species require no specific vitamins or amino acids and can develop well with only a single organic compound as their sole source of carbon and energy. They are not very fussy about what that compound should be, either. One species studied was quite contented with any of 127 different substances offered. The upper limit was set not by the organism's capacities, but by the investigators' running out of ideas of things to offer it.

As a result of their adaptability, *Pseudomonas* species are found almost everywhere. They thrive in soil, water, and marine environments. Some live on and in both plants and animals. They can be found in spoiling foods, fresh butter, clay suspended in kerosene, donkey feces, chlorinated swimming pools, carnations, and rotted onions. I could go on, but you see what I mean. Some species move freely from one of these environments to another. *Pseudomonas aeruginosa* may enter a hospital unit in the tap water, but promptly take up residence on a patient's burned skin. Versatility and adaptability have combined in a bacterial group that is a force to be reckoned with.

(×11,250. SEM by Gary Gaard, in Diane A. Kuppels and Arthur Kelman, Isolation of pectolytic fluorescent pseudomonads from soil and potatoes, *Phytopathology* (1980) 70: 1110–1115.)

from one metabolic pattern to the other as environmental conditions change. Almost all Gram-negative, facultatively anaerobic rods give a negative oxidase test. If flagella are present, they are usually peritrichous.

The **enterobacteria** are a closely related cluster of species that inhabit the large intestine of mammals. This group contains the most familiar and most thoroughly studied of all bacteria, *Escherichia coli* (Figure 4.8), and its close relatives. Because it has been so extensively used in basic research, we probably know more about *E. coli* than about any other life form. There are several pathogens in the enterobacterial group,

so identification and classification techniques for them have been brought to a very sophisticated and effective level. Some enterobacteria (*Escherichia, Enterobacter, Citrobacter, Proteus*) are components of the normal flora or resident microbial population in the mammalian intestine. Some of this group may also act as **opportunists,** organisms capable of causing disease, but only in an already weakened host. In fact, *E. coli* is the most frequently reported agent detected in the infections of hospitalized patients in the United States. A few strains of *E. coli* are true pathogens and cause very severe diarrhea in healthy individuals. *Salmonella* and *Shigella* cause epidemic diseases, including typhoid fever and bacterial dysentery. *Klebsiella* may cause pneumonia. *Yersinia pestis* causes bubonic plague. *Erwinia* is found in association with plants; some *Erwinia* species are harmless and some are pathogenic. The enterobacteria are discussed in greater detail in Chapter 19.

The **vibrios** are mostly curved rods. Some are pathogens, like *Vibrio cholerae* (Figure 4.9), which causes cholera, and *V. parahaemolyticus,* *V. alginolyticus,* and *V. vulnificans,* which cause wound infections.

Photobacterium is **luminescent** — it produces yellowish light when actively respiring. Most luminescent bacteria are marine; they are usually associated with fish and provide the light source for fish luminescent organs and markings (Figure 4.10). The light-generating enzyme, called **luciferase,** is similar to that of fireflies.

Hemophilus influenzae is the most common cause of childhood meningitis. It also causes pneumonia and ear infections.

GRAM-NEGATIVE, TRULY ANAEROBIC BACTERIA

By definition, an anaerobe lives without oxygen. Most tolerate exposure to oxygen poorly because it is toxic to them. Anaerobes have a restricted choice of environments in our oxygen-permeated world. In humans, strict anaerobes

FIGURE 4.8 The Enterobacteria
(a) A generalized drawing of a bacillus of the enterobacterial type. (b) An electron micrograph of a thin section through *Escherichia coli.* (Electron micrograph, ×40,480. Courtesy of Stanley C. Holt, University of Texas Health Science Center, San Antonio/BPS.)

(a) A bacillus

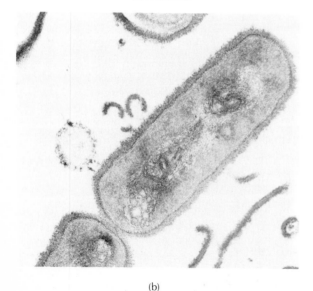

(b)

(a) Vibrio

(b) *Vibrio cholerae*

FIGURE 4.9 Vibrios
(a) A generalized drawing of a vibrio. (b) Many cells of *Vibrio cholerae* adhering to mucus strands. (Scanning electron micrograph, ×3680. Courtesy of M.N. Guentzel, D. Guerrero, and T. V. Gay, Division of Life Sciences, University of Texas at San Antonio.)

FIGURE 4.10 Fish Luminescent Organs
The luminescent spots are specialized structures containing photoluminescent bacteria. Luminescent markings help fish identify members of their own and other species, and lure prey. (Courtesy of D. Powell, Monterey Bay Aquarium/BPS.)

inhabit the intestine, invade the tissues, and colonize microenvironments such as the groove between the teeth and the gums. The rod-shaped genera *Fusobacterium* and *Bacteroides* are numerous in human oral flora and feces. When they escape from their normal sites into the abdominal cavity or tissues, *B. fragilis* and *B. melanogenicus* cause very dangerous abscesses and postsurgical infections. They also frequently infect bite wounds.

THE RICKETTSIAS AND CHLAMYDIAS

There are two types of organisms that in a way are "deficient" bacteria. The **rickettsias** and the **chlamydias** both are clearly cellular, procaryotic organisms. However, each has lost some essen-

tial capacity for an independent free-living existence. As a result, these organisms live as **obligate intracellular parasites.** This means that they can live and multiply only within and at the expense of an intact, eucaryotic animal cell. Most can be cultivated and studied only in live animals and tissue cultures.

The rickettsias, including the genera *Rickettsia, Bartonella*, and others, have a typical gram-negative cell wall, cell membrane, and cytoplasmic components. *Coxiella*, an exception, may stain gram-positive. Rickettsias multiply by binary fission. Their cellular deficiency seems to be an imperfect permeability layer — simply, they are leaky, and lose key cell components. Therefore, they need to live inside another cell (Figure 4.11a), from which equivalent

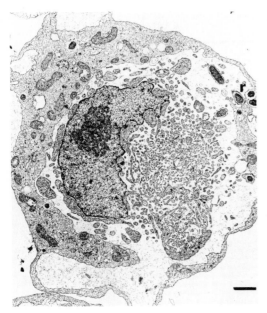

(a)   *Rickettsia rickettsii*

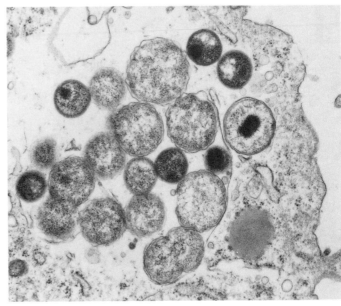

(b)   *Chlamydia psittaci*

FIGURE 4.11   Rickettsias and Chlamydias
(a) Cells of *Rickettsia rickettsii* inside a disintegrating fibroblast cell. The rickettsias are the dark-cored, membrane-covered bodies. (Courtesy of David J. Silverman and Charles L. Wissman, Jr., *R. rickettsia* cytopathology, *Infection and Immunity* (1979) 26:714.) (b) Multiple elementary bodies of *Chlamydia psittaci* maturing within the cytoplasm of an infected host cell. (Transmission electron micrograph, ×15,795. Courtesy of R. C. Cutlip, National Animal Disease Center, Ames, Iowa.)

compounds can leak back in. The parasitic rickettsias are typically transmitted to humans by the bites or feces of arthropods such as ticks or lice. *Rickettsia prowazekii* causes typhus and *R. rickettsii* causes Rocky Mountain spotted fever.

The genus *Chlamydia* contains organisms with cell walls of the Gram-negative type. They appear to have a deficient energy metabolism, and thus must depend on the host for their ATP supply. They have a reproductive cycle in which two cell types alternate. One type, the **elementary body,** is a small, rigid, infective form. It is taken into the host cell by phagocytosis, where it then develops into the second type, a larger, thin-walled, **initial body** (Figure 4.11b). This continues to enlarge, then divides to organize new daughter elementary bodies. Eventually the

cell ruptures and releases these daughter bodies, which may then proceed to infect surrounding host cells.

The genus contains only two species, *C. trachomatis* and *C. psittaci. Chlamydia trachomatis* causes common sexually transmitted diseases such as nongonococcal urethritis, the eye infections inclusion conjunctivitis and trachoma, and pneumonias. *Chlamydia psittaci* causes psittacosis or parrot fever, a form of pneumonia transmitted from infected birds of the parrot family to humans.

THE MYCOPLASMAS

Mycoplasmas are the tiniest bacteria, ranging from 125 to 250 nm in diameter. They are at about the limits of resolution for the light microscope. Their shape is highly variable, or **pleomorphic,** and filamentous shapes are often seen. Although mycoplasmas reproduce by binary fission, they also produce elementary bodies and spherical reproductive bodies, which arise by fragmentation of filaments.

Their most notable structural feature is the lack of a peptidoglycan-containing cell wall. The outer cell boundary of a mycoplasma is a three-layered membrane, and the cells stain gram-negative. Cholesterol is an essential component of the membrane; it is usually required for growth. Colonies on solid laboratory media are very small, sometimes looking like tiny fried eggs, or like a film with spots of precipitated matter. *Mycoplasma* lives in the body cavities and fluids of animals. In human beings, *M. pneumoniae* (Figure 4.12) causes an atypical pneumonia. *Ureaplasma* causes sexually transmitted diseases that may lead to infertility, spontaneous abortion, or infections of newborn infants.

Bacteria of the genus *Acholeplasma* lack cholesterol in their membranes and are more susceptible to damage caused by osmosis. Some of these organisms are free-living while others are parasitic. Other recently discovered genera include *Spiroplasma* and *Thermoplasma,* mycoplasmalike plant pathogens. Some but not all investigators consider these latter genera to belong with the Archaebacteria (page 128).

GRAM-POSITIVE BACTERIA

There are comparatively fewer Gram-positive bacteria than Gram-negative ones, but some are significant disease agents. The Gram-positive bacteria as a group are too diverse to generalize about, so we must look at them individually.

THE GRAM-POSITIVE COCCI

Some Gram-positive cocci are aerobic, and others are facultatively anaerobic. Bacteria of the aerobic genera *Micrococcus* and *Staphylococcus* primarily occur in grapelike clusters, tetrads (fours) or cubical packets (Figure 4.13a). Both respire, but only *Staphylococcus* ferments. They contain the enzyme **catalase,** which breaks down hydrogen peroxide to oxygen and water. Most are animal normal flora, and some are soil and water forms; most are quite salt tolerant. *Micrococcus* species are often brightly colored with yellow to red carotenoid pigments. *Staphylococcus aureus* causes skin, respiratory, and severe wound infections in human beings. *Staphylococcus epidermidis* is a normal skin organism and occasional opportunist (Figure 4.13b).

The genus *Streptococcus* contains gram-positive cocci that characteristically grow in

FIGURE 4.12 Mycoplasmas
Scanning electron micrograph showing filamentous and pleomorphic cells of *Mycoplasma pneumoniae.* (×11,000. Courtesy of Michael G. Gabridge, Bionique Laboratories, Inc.)

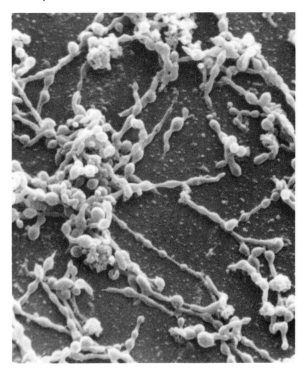

(a) Staphylococcus

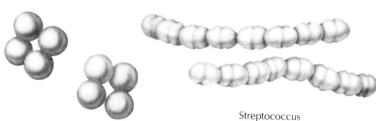

Packets of 4

Streptococcus

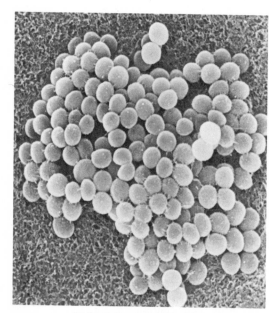

(b) *Staphylococcus epidermis*

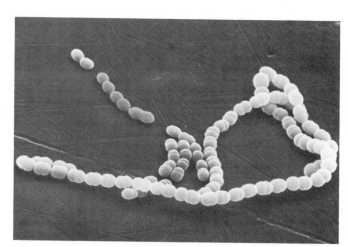

(c) *Streptococcus*

FIGURE 4.13   Gram-Positive Cocci
(a) After binary fission, some cocci remain associated in clusters (staphylococci). Others form pockets of four (tetrads) and yet others form chains (streptococci). (b) A scanning electron micrograph of *Staphylococcus epidermis*. (From Yoshii Z, et al., *Atlas of scanning electron microscopy in microbiology*. Tokyo: Igaku-Shoin, Ltd. (1976)). (c) A scanning electron micrograph of a chain of *Streptococcus*. If you look closely you will see that each coccus is about to divide, elongating the chain. (Courtesy of D. L. Shungu, J. B. Cornett, and G. D. Shockman, Autolytic mutants of *S. faecium*, J. Bacteriol. (1979) 138:601, Fig. 1a.)

chains (Figure 4.13c). Their metabolism is very different from that of the staphylococci above. Streptococci are fermentative, do not respire, and have no catalase. Streptococci also produce lactic acid as an end product of carbohydrate metabolism. They are universally found in the normal oral flora and bowel contents of warm-blooded animals. They are numerous in milk, and cause it to go sour if unchecked. *Strep-*

tococcus pyogenes is one of our most serious bacterial pathogens, causing "strep throat," rheumatic fever, kidney disease, and many other problems. *Streptococcus agalactiae* causes septicemia ("blood poisoning") and meningitis in infants.

Hemolysis, the ability to destroy red blood cells, is an important identification trait for the streptococci and some other bacteria. In addition, chemical analysis of specific cell-wall proteins, teichoic acids, and polysaccharides may be used to identify the streptococci. *Streptococcus* is discussed further in Chapters 17 and 22.

THE ENDOSPORE-FORMERS

True heat-resistant, refractile endospores are effective survival mechanisms for the relatively few species of bacteria that produce them. Two genera are important, *Bacillus* and *Clostridium*. The aerobic genus *Bacillus*, widely distributed in soil and on vegetation, thrives in dry or dusty environments, where their spores give them a strong survival advantage. Many *Bacillus* species can grow, less effectively, without oxygen. *Bacillus* species mostly are strongly Gram-positive and actively motile with peritrichous flagella. A few are pathogenic, including *B. anthracis*, which causes anthrax in animals and human beings; *Bacillus cereus* (Figure 4.14), which sometimes causes food toxicity; and *B. thuringensis*, which is pathogenic to many insect larvae. The genus *Bacillus* is also noteworthy for the many industrially valuable species it contains. Some strains produce commercially valuable extracellular enzymes, such as the protein-digesting enzymes used in laundry products. Others yield antibiotics, including bacitracin and polymyxin. *Bacillus thuringensis* or BT is an effective bioinsecticide when used against insect pests such as the gypsy moth and spruce budworm.

Clostridium species are mostly strict anaerobes that inhabit animal intestinal tracts, animal wastes, and sewage. These Gram-positive rods are also found in waterlogged (airless) soils, black muds, and decomposing organic material. Clostridia are strongly fermentative.

It takes an unusual event to create the specialized conditions under which clostridia may cause human disease. A puncture wound may lead to tetanus, and consumption of improperly canned foods may produce botulism. In fact, *C. tetani* and *C. botulinum* become pathogens when human beings come in contact with the protein toxins that the bacteria excrete while growing anaerobically. Other clostridia are commercially valuable as producers of organic solvents such as acetone and isopropanol. *Clostridium pasteurianum* promotes fertility of anaerobic soils because it is a nitrogen fixer.

GRAM-POSITIVE, NON–SPORE-FORMING RODS

The *Lactobacillus* species have very exacting nutritional requirements for amino acids and vitamins. This restricts them to nutritionally

FIGURE 4.14 *Bacillus*
(a) A single cell of *Bacillus* drawn to show its rather boxy shape. (b) *B. cereus* customarily forms chains. The cells are several times larger than other bacilli, such as *E. coli*. (Scanning electron micrograph. Courtesy of E. G. Afrikian, G. St. Julian, and L. A. Bulla, Jr., *Appl. Microbiol.* (1973) 26:934.)

(a) Bacillus

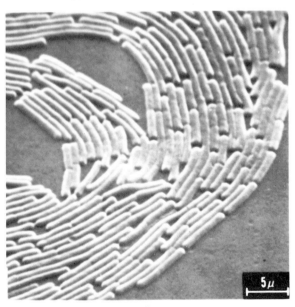

5μ

(b) *Bacillus cereus*

complete environments, such as foods and intestinal contents. Their anaerobic, fermentative metabolic activities are the basis of time-honored methods of food preparation. Lactobacilli are indispensible in making cheeses (*L. casei*), yogurt, sauerkraut, kefir (*L. bulgaricus*), pickled vegetables, and sourdough bread. The lactobacilli are also universally found in digestive tracts, and in the oral and vaginal flora of human beings. As part of the normal flora in these areas, they have poorly understood, but apparently positive, roles in maintaining health.

The genus *Listeria* contains catalase-positive rods. *L. monocytogenes* is an uncommon but serious pathogen, especially in newborn infants, in whom it causes a form of meningitis.

The **coryneform group** of bacteria are notable for their club-shaped, irregular, small rods, often showing pronounced intracellular granules. They are aerobic and nonmotile. They include animal and plant pathogens, and nonpathogenic forms. Human beings carry nonpathogenic diphtheroids (that is, organisms that resemble the diphtheria organism in appearance) such as *Corynebacterium xerosis* and *C. pseudodiphtheriticum* among their normal flora. *Corynebacterium diphtheriae* is the species that is responsible for diphtheria, a life-threatening disease that is now, due to vaccination, very rare.

The genus *Arthrobacter* contains many species of widely distributed soil and water organisms. They characteristically undergo a shape change from rod to coccus as they age. *Gardnerella*, a genus that is believed to be a close relative to *Arthrobacter*, causes vaginal infections.

The genus *Mycobacterium* includes species that are normal human flora, as well as the agents of tuberculosis (*M. tuberculosis*) and leprosy (*M. leprae*). Mycobacteria are identified in part with the acid-fast differential stain. They contain unusual cell-wall lipids. Mycobacteria are somewhat irregular in their cell shape, and grow extremely slowly on laboratory media, forming rough and, in some cases, pigmented colonies. Specialized techniques have been developed to culture and identify many mycobacterial species, but some, such as the leprosy agent, cannot yet be cultured. The antibiotics used to treat mycobacterial infections are generally quite different from those used in other bacterial infections.

There are many other coryneforms. We might note *Bifidobacterium*, which often has branched cells, and is an important colonist of the infant intestine. *Nocardia*, which forms filaments, is found in soil and in animal flora. *Nocardia asteroides* can cause a serious lung infection, and *N. brasiliensis* causes mycetoma, a disease of the feet and hands.

GRAM-NEGATIVE BACTERIA OF ENVIRONMENTAL IMPORTANCE

The procaryotes, as we have seen, do not occupy a separate world from their larger, more complex eucaryotic descendants. Instead, their activities and ours are inextricably intertwined. It bears restating that the procaryotic microorganisms are an intrinsic, though often unrecognized, part of the biosphere. Their contributions to the global environment are of immense significance, even though they are largely invisible to the naked eye. In the following section, you should be pleased to note that practically every microorganism is a beneficial contributor in some way.

THE CYANOBACTERIA

The cyanobacteria were once called blue-green algae and were classified as plants, on the basis of incomplete information. In the early 1960s electron microscopy showed clearly that they are procaryotes. Then the cyanobacteria became accepted as a specialized group of bacteria. It is believed by many biologists that a primitive cyanobacterial cell was the ancestor of the eucaryotic chloroplast.

Often we use the convention of dividing the procaryotes into the cyanobacteria and the "true bacteria" — everything else. This is convenient, although perhaps somewhat arbitrary. The cyanobacteria are generally larger than other bacteria. They are photosynthetic, using light as an energy source. All live in well-lighted environments. Their metabolic patterns are diverse. Overall, in nature, carbon dioxide is the most commonly used carbon source, and most cyanobacteria use no other. Some cyanobacteria also can cope with periods of darkness, growing by using heterotrophic metabolism of organic food sources. Yet others use organic nutrients, but only as sources of carbon and not for energy.

Cyanobacteria use nitrate or ammonia as nitrogen sources. Many cyanobacteria also fix molecular nitrogen (N_2) into amino acids useful as nutrients for higher organisms. They move over solid surfaces by gliding.

STRUCTURES OF THE CYANOBACTERIAL CELL. The cyanobacteria have typical Gram-negative cell walls. Some also produce exterior sheaths that bind groups of cells into long strands, and break apart for reproduction.

Their photosynthetic apparatus consists of light-absorbing pigments built into membranous structures. **Chlorophyll** is the principal energy-capturing pigment. Accessory pigments, the yellow to red **carotenoids** and the red or blue **phycobilins,** further expand the light-absorbing potential. The photosynthetic apparatus (Figure 4.15), is constructed of **thylakoids,** layers of internal membrane connected with the cell membrane. These carry out the "light" reactions of photosynthesis. They can both produce ATP and split water molecules, releasing oxygen into the atmosphere (see Chapter 7). The cytoplasm contains the **carboxysomes,** the sites where the carbohydrates are manufactured. Most cyanobac-

FIGURE 4.15   The Cyanobacterial Cell
(a) A schematic drawing of the structures of the cyanobacterial cell. Photosynthetic membranes are shown in color. (b) A transmission electron micrograph of one cell within a filament of *Anabaena*. This organism is a major contributor to summer algal blooms in fresh water. The filaments also contain nitrogen-fixing heterocysts, which are not shown here. (×4800. Courtesy of William T. Hall, Electro-Nucleonics Labs.)

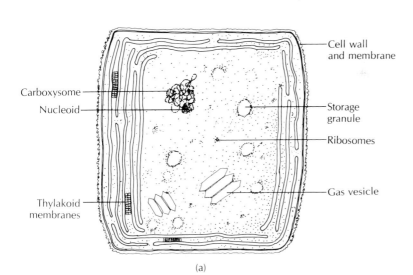

(a)

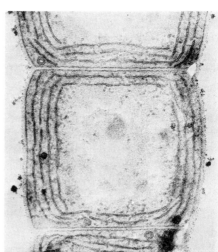

(b)

teria also have storage granules of several types, and gas vesicles for buoyancy.

Nitrogen fixation takes place in specialized cells called **heterocysts.** These are thick-walled structures, interspersed in a filament of regular cells, and connected to them by a peculiar "neck." The interior environment of a heterocyst is oxygen-free, which protects the nitrogen-fixing enzymes from being inactivated. These enzymes incorporate atmospheric nitrogen gas into amino acids that are then used by adjacent cells in the chain, or are excreted into the water to be used by neighboring microorganisms.

CYANOBACTERIA IN THE ENVIRONMENT. Nitrogen-fixing cyanobacteria such as *Anabaena* and *Nostoc* contribute significantly to filling the fixed nitrogen requirements of plant, animal, and microbial life (Figure 4.16). Aquatic floating forms, such as the oscillatorians, are ubiquitous parts of the **phytoplankton** — floating microscopic plant life — and constitute an important base for aquatic food chains. We can find them in fresh water and saltwater, hot springs, and saline lakes. They contribute to productivity by fixing CO_2 into carbohydrates via photosynthesis. Their oxygen production boosts total atmospheric oxygen, continuing the trend that started 3 billion years ago when, it is believed, they converted the earth from an anaerobic to an aerobic habitat. Some cyanobacteria, however, produce odorous substances that may make drinking water unpalatable; others release highly potent neurotoxins poisonous to animals.

Cyanobacteria may also be found in moist, warm soils, such as in greenhouses. Others are desert inhabitants and form crusts on hot, dry soils and over rocks, growing only in the rainy season. In shallow waters they form multicolored mats, often in association with the eucaryotic algae.

Fungi form symbiotic partnerships called **lichens** with photosynthetic microorganisms. In many lichens, cyanobacteria are the photosynthetic partner. Lichens grow on some very

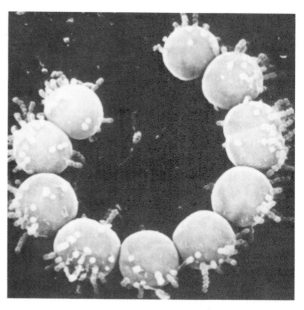

FIGURE 4.16   Cyanobacteria in Nature
These fresh-water cyanobacteria (larger coccoid forms) are producing organic nutrients. The numerous bacteria (chains of small cocci) are adhering to them to utilize these food materials. In nature, cyanobacteria may serve as a primary food source for protozoa, animals, fungi, and bacteria. (Scanning electron micrograph, ×6528. Courtesy of Dr. Hans Paerl.)

inhospitable surfaces, such as rocks, buildings, roof shingles, and Antarctic gravel. Their tissues are intermingled fungal and cyanobacterial cells (Figure 4.17). The photosynthetic cyanobacteria manufacture the food for both partners, while the fungi, with their rigid, resistant cell walls, give the mechanical protection that both need.

THE PHOTOSYNTHETIC TRUE BACTERIA

The photosynthetic true bacteria are found in anaerobic aquatic habitats. It is not easy to find a habitat that is both anaerobic and lighted, so these organisms have a restricted range — below the surface of lakes, on the bottom of shallow salt ponds, and in sulfur springs, to name a few spots.

Their shapes may be rods, cocci, vibrios, spirals, or budding forms. Most reproduce by binary fission and are motile. They are commonly separated into the purple and green groups.

These bacteria differ significantly from the cyanobacteria. The photosynthetic true bacteria are considered by most authorities to be more primitive, and they are likely to have evolved earlier. The distinction is that the photosynthetic apparatus of the true bacteria is able to use light energy to synthesize ATP and make carbohydrates, but cannot use light energy to split water molecules as cyanobacteria do. Thus they produce no oxygen. Most photosynthetic true bacteria are also partly heterotrophic, relying on some organic compounds for growth.

THE CHEMOLITHOTROPHIC BACTERIA

Only among the bacteria do we find the chemolithotrophic mode of nutrition. This term means that the organisms build their cellular materials from inorganic carbon, and at the same time satisfy their energy needs by oxidizing inorganic chemicals. They do not need light or organic carbon. This could be viewed as perhaps the most "independent" way of survival known.

Strictly aerobic chemolithotrophs use reduced nitrogen or sulfur compounds, which they oxidize. The oxidations release energy, and the chemolithotrophs are able to use the energy to make ATP. Many have membranous lamellae containing the necessary enzymes. Nitrifying bacteria may oxidize ammonia (*Nitrosomonas, Nitrosococcus*) or nitrite (*Nitrobacter, Nitrococcus*) (Figure 4.18a) in aerobic, energy-yielding

(a)

Fungal mycelium

Cyano bacterial cells

FIGURE 4.17 Cross-section of a Lichen

(a) A lichen growing on a rock surface. (×0.33. Courtesy of J.M. Cohrader/Photo Researchers, Inc.) (b) An interpretitive drawing of a cross-section showing the structure of the lichen.

(b)

processes to produce nitrate, a valuable plant nutrient. Nitrifiers are common, but rarely numerous, in soil and water. Sulfur-oxidizers such as *Thiobacillus* (Figure 4.18b) utilize reduced inorganic sulfur compounds, releasing sulfate as an end product. Some tolerate extremely acidic or hot conditions. *Sulfolobus* lives in sulfur-rich hot springs, preferring temperatures of 70 to 80 degrees Celsius.

Chemolithotrophy can be viewed as an evolutionary experiment, one that was successful enough that its products have survived. On the deep ocean floor, around 2000 m below the surface, heated chemical-rich water rises up out of rifts in the earth's crust. Dense, vigorous communities of organisms thrive far from the usual primary source of energy for life, the sun. These communities, including clams, mussels, crabs, tube worms, and countless bacteria are supported by chemolithotrophy. In fact, sulfur bacteria appear to be the primary and only producers. They use hydrogen sulfide for energy, fix carbon dioxide to grow, and in turn are being used as food by all the other organisms. Some researchers propose that these rift communities may provide valuable insights into alternative views of how life might have begun on earth.

THE GLIDING BACTERIA

Gliders are common inhabitants of soil, and of areas exposed to animal wastes. They are unicellular, gram-negative rods or filaments. *Cytophaga* is a soil and water organism that lives on cellulose, chitin, and agar. *Beggiatoa* is a filamentous glider found in environments rich in hydrogen sulfide; it accumulates intracellular sulfur granules. *Leucothrix*, also filamentous, lives on seaweeds; only single cells glide, but filaments do not.

FIGURE 4.18   Chemolithotropic Bacteria
(a) *Nitrobacter agilis* uses the oxidation of nitrite to nitrate as an energy-yielding reaction to support its growth. The multilayered membranes within the cell carry out the chemical processes required. (Transmission electron micrograph, ×18,468. Courtesy of Stanley W. Watson, Woods Hole Oceanographic Institute.) (b) *Thiobacillus neopolitanus* oxidizes sulfur compounds to obtain energy. The dark globules are accumulations of elemental sulfur. (Transmission electron micrograph, ×23,360. Courtesy of Jessup M. Shively.)

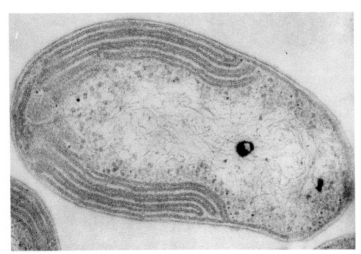

(a)   *Nitrobacter agilis*

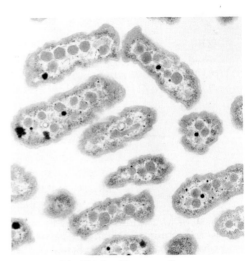

(b)   *Thiobacillus neopolitanus*

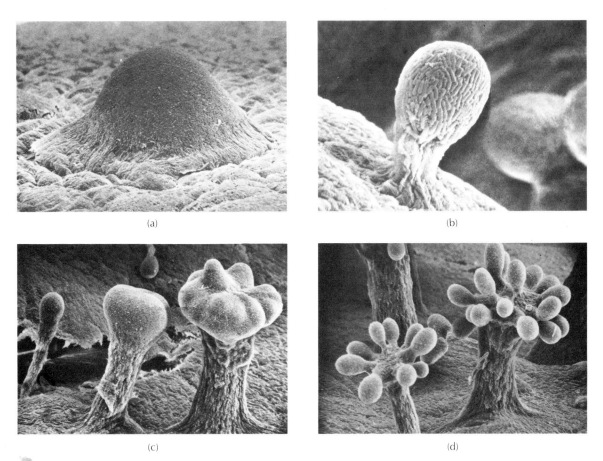

(a)

(b)

(c)

(d)

FIGURE 4.19 *Chondromyces crocatus* Forming a Fruiting Body
(a) Chemical signals emitted by the cells cause them to aggregate by gliding toward a central area. (b) Upward migration of the cells forms a body with a stalk and a head. (c) The developing fruiting body, somewhat obscured by a slime coat. (d) Mature fruiting body containing myxospores. (Courtesy of P. L. Grilione and J. Pangborn, *J. Bacteriol.* 124:1558–1565.)

The **myxobacteria** are gliders, remarkable for their complex developmental cycle. Among the procaryotes they have the most well-developed "life cycle," with different stages, so familiar in eucaryotic plants and animals. Myxobacteria live primarily by breaking down other bacteria and metabolizing their contents. When prey bacteria or other nutrients are scarce, the individual cells, by gliding, converge to form a vegetative colony. This gradually develops into an elevated, complex structure, often brightly pigmented.

The cells signal to each other by emitting chemicals, which coordinate their migratory behavior. Some of the cells within this **fruiting body** become resistant resting cells called **myxospores** (Figure 4.19). The fruiting bodies break open to release the resistant myxospores, which are well equipped to survive until conditions improve. The behavior patterns of *Myxococcus* and *Chondromyces* have been extensively studied, contributing basic knowledge about cell-to-cell communication methods.

THE SHEATHED BACTERIA

There is a diverse assortment of rod-shaped, gram-negative cells that form filaments surrounded by tubular sheaths, often loaded with oxides of iron or manganese. They have a simple sort of life cycle. Within the sheath, binary fission gives rise to motile swarmer cells, which exit from the sheath's ends. Each swarmer may originate a new filament. *Sphaerotilus* and *Leptothrix* (Figure 4.20) are widely distributed in fresh water and saltwater, usually those with significant sewage pollution. An environment a human might regard as unappealing may be just great for a microorganism.

THE BUDDING AND APPENDAGED BACTERIA

These are bacteria in which the cells often become somewhat different at one end than the other, either by developing a bud at one end, or a stalk or holdfast. This group contains aquatic bacteria. **Budding** is a variant of binary fission; it is a form of reproduction in which the bacteria undergo unequal cell divisions. The smaller of the two progeny, the daughter cell, originates from growth of one end of the cell. *Hyphomicrobium* forms long, slender extensions called hyphae; the daughter cell, with a flagellum, develops at the tip, then separates. Appendaged or stalked microorganisms form stalks or adhesive appendages that serve as holdfasts to attach the cell to a solid surface. *Caulobacter* (Figure 4.21) has a stalk that is an extension of the cell. It may attach itself by its stalk to a solid surface, or the stalk may bind to those of other cells, forming a rosette. *Caulobacter* is common in tap water.

THE ARCHAEBACTERIA

The archaebacteria are regarded as having originated in a very early period in biological history. They differ in many significant ways from other bacteria. Some of these key differences are:

- Their cell walls lack peptidoglycan.
- Their membrane lipids are formed from branched fatty acids, not the more common straight-chain type, with an unusual linkage to the alcohol backbone. This type of lipid is very heat stable.
- Their stable RNA sequences are very different from those of all other bacteria, as well as eucaryotes.
- Many live in unusual and extreme habitats such as extremely salty lakes or volcanic springs.
- They have unique metabolic factors assisting their enzymes, found neither in the other bacteria nor in eucaryotes.
- Those that fix carbon into carbohydrate use a method quite different from that used by all other lithotrophs.
- Their enzymes for making DNA are unique.

The **methanogens,** so-called because they produce the gas methane, are receiving intensive study. Methanogens, the first of three major types of Archaebacteria, are strictly anaerobic and frequently are killed by traces of oxygen. Most are chemolithotrophic, and combine CO_2 and H_2 gas to form methane (CH_4). This process yields energy and provides the cell with organic carbon for growth. Some methanogens, however, combine H_2 with simple organic compounds. The natural distribution of the methanogens is restricted by their anaerobic requirement and the fact that almost all need hydrogen gas. Free H_2 gas in nature is usually encountered only if oxygen is unavailable to react with it. Methanogens, then, may be found when special anaerobic isolation techniques are used, in the rumen and intestinal tract of animals, in composting manure, in anaerobic sewage treatment tanks, and in oxygen-free sediments. Commercial usage of methanogens to produce methane gas, a valuable clean fuel, is being explored.

Many researchers consider the **halobacteria** to be the second major group of archaebacteria, as they share the distinctive features of the group as listed above. Halobacteria inhabit shallow, sunlit, highly saline environments. They live in places like the Great Salt Lake, the Dead Sea, and

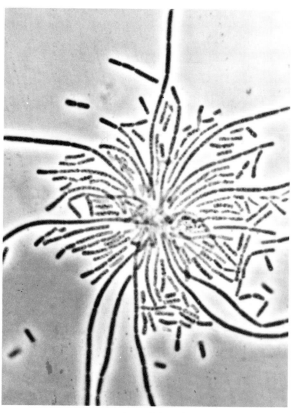

Leptothrix

FIGURE 4.20 A Sheathed Bacterium
Leptothrix, clustered filaments containing many
sheathed cells. The filaments adhere forming a
rosette structure. (Phase-contrast micrograph.
Courtesy of American Society for Microbiology and
Dr. Mulden.)

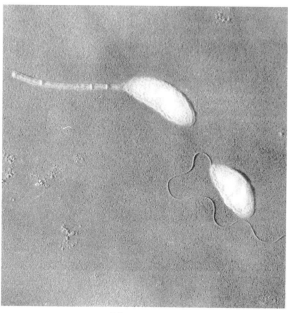

Caulobacter crescentus

FIGURE 4.21 An Appendaged Bacterium
Caulobacter crescentus, showing a stalked cell, and
a swarmer cell with a flagellum. (Shadowed
specimen, ×13,216. Courtesy of Dr. J. S.
Poindexter.)

other bodies of water much saltier than the
oceans. Their enzymes require high salt concentrations for catalytic activity. Highly concentrated yellow and red carotenoid pigments in the
membrane protect against the ultraviolet radiation in sunlight. Halobacteria can capture light
energy and use it to make ATP, but do not carry
out photosynthesis. They produce **bacteriorhodopsin,** a light-sensitive pigment, and incorporate it into their outer layer, the **purple membrane.** This pigment drives an unusual energy-
generating mechanism which establishes a flow
or current of charged ions across the membrane.

This ion current is used to make ATP, thus capturing solar energy in chemical form. A third
group of microorganisms includes the high-
temperature acid-tolerant organisms such as
Sulfolobus and *Thermoplasma*. These also fulfill
the criteria to belong with the Archaebacteria.

GRAM-POSITIVE
FILAMENTOUS BACTERIA

There is a large, diverse group of bacteria characterized by the tendency to form long, branching
filaments that grow over a solid surface to form
a tangled mat, or **mycelium.** They all possess a
Gram-positive type of cell wall, although the actual staining reactions may be unreliable. The
genus *Streptomyces* exemplifies the mycelial
growth pattern. They also produce reproductive

BOX 4.2
The Archaebacteria — Another Kingdom?

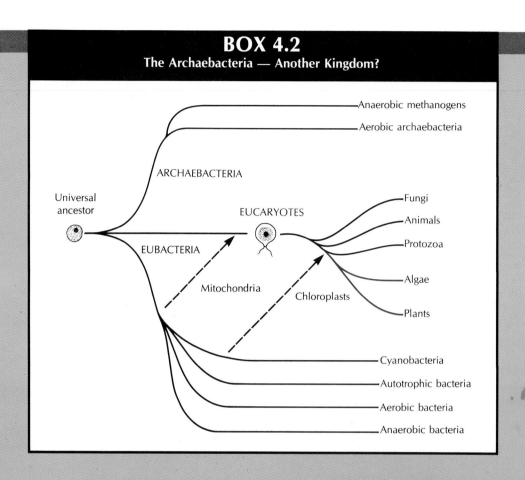

Just so that you will get an idea of how changeable classification schemes can be, let us look over the evidence for the existence of a sixth, previously unsuspected kingdom. Over the past 10 years, experimental evidence has convincingly linked together a scattered group of "oddball" microorganisms — methane producers, halobacteria, and organisms from extremely hot,

acidic environments. They have many common characteristics, presented later in this chapter. The fascinating thing about these organisms is that they are significantly different from the procaryotes in general. All this information is new since the five-kingdom system was proposed. Although most taxonomists are taking a wait-and-see attitude, the scientific evidence is very

convincing that we have found a sixth kingdom. One way we could incorporate these changes into the microbial family tree is shown here. As one of the prime researchers, Carl Woese, said in describing his feelings about these discoveries, "It's comparable to going into the back yard and seeing a creature that's neither a plant nor an animal." That's **really** new!

spores called conidia at certain growth phases. The conidia are produced in large numbers by forming cross-walls in aerial portions of the filaments. The streptomycetes are strict aerobes, and are relatively slow-growing. Streptomycetes are major components of the aerobic bacterial flora of soil, companions and competitors of the eucaryotic fungi. Most are saprophytic and some are actively helpful to humans. The streptomycetes are economically important. They produce hundreds of different antibiotic chemicals, all of which inhibit the growth of other bacteria to some degree. More than 50 of these are actually used. In human medicine, we use streptomycin, chloramphenicol, tetracycline, erythromycin, neomycin, and spectinomycin, among others, to treat bacterial infections. We will discuss antibiotics in depth in Chapter 24.

RECONSIDERING BACTERIA

After examining so many new and bewilderingly diverse groups of bacteria, and reading so many long names, you perhaps feel the need to form some conclusions about what the real significance of bacteria might be. We explored some ideas about microorganisms in the environment in Chapter 1. Now we can examine those concepts with some examples fresh in our minds.

When reading through a survey of the bacteria such as that above, several impressions strike us. The bacteria are certainly diverse in habitat and in the adaptations that suit them to their environment. In fact, bacteria are the only organisms able to live in many extreme environments. It is also clear that most bacteria are actively useful, despite the "bad reputations" they have in the layperson's mind. Only a small minority of bacterial genera produce disease. However, we tend to focus strongly on the occasional species that is pathogenic.

We have seen bacteria contributing as producers that fix carbon and feed heterotrophs or as decomposers that remove and break down waste materials. We have encountered the rhizobia that fix nitrogen to nourish crops and the cyanobacteria that release oxygen to replenish the atmosphere. I hope that you are beginning to be able to see things somewhat from a "bug's-eye view" — to know that the bacterium is doing what it is doing for its own purposes. It is also valuable to become sensitized to the fact that there are many things going on in the biological world that are too small for the naked eye to see.

You may certainly be forgiven for feeling that Linnaeus was on the right track when he lumped all the microorganisms together in a group he named **Chaos!** We have many more names now, but not necessarily more rational order. The taxonomy of the bacteria is far from settled.

SUMMARY

1. Taxonomy is the process of classifying organisms. It involves discerning fundamental similarities and differences, and separating these from superficial resemblances. In the case of the procaryotes, these judgments are often not easy.

2. The discovery of a new organism is followed by its careful examination and description. The results will show either that it is identical to a previously described and named species, or that it is different and should be given a new species name. Each organism bears a binomial designation. All organisms are grouped under a hierarchical system of taxonomy, leading up to the kingdom level. We use the five-kingdom classification here.

3. The procaryotes are divided into four groups. The gram-negative bacteria of medical and commercial importance include many of the organisms we will be studying later, such as the spirochetes, pseudomonads, neisserias, enterobacteria, rickettsias, chlamydias, and mycoplasmas.

4. The Gram-positive bacteria include both rods and cocci, and all the spore formers. Groups of interest include the staphylococci, streptococci, bacilli, clostridia, and mycobacteria.

5. The Gram-negative bacteria of environmental importance include the various lithotrophs — the cyanobacteria, photosynthetic true bacteria, and chemolithotrophic bacteria. The Archaebacteria are also included here.

6. The filamentous streptomycetes and other bacteria that form reproductive spores form a fourth group. They are primarily soil organisms, and strongly resemble the eucaryotic fungi. Many produce useful antibiotics.

DISCUSSION QUESTIONS

1. What are some of the criteria used to classify bacteria? Compare them with criteria used to classify plants or mammals.

2. What are the differences between the cyanobacteria and the photosynthetic true bacteria?

3. Find and list a few microorganisms found in (a) the ocean, (b) soil, (c) the human mouth, and (d) intestinal contents or sewage.

4. Outline the primary differences between pseudomonads and enterobacteria. Refer to *Bergey's Manual*, if it is available, to augment your answer.

5. Describe the major differences between the Archaebacteria and the other procaryotes.

ADDITIONAL READING

Brock TD, Smith DW, Madigan MT. *Biology of microorganisms.* 3rd ed. Englewood Cliffs, N.J.: Prentice-Hall, 1984.

Holt JG, ed. *Bergey's manual of systematic bacteriology.* Vol. 1–4. Baltimore: Williams and Wilkins; Vol. 1, 1984, Vols. 2–4 in preparation. in preparation.

Mayr E. Biological classification: toward a synthesis of opposing methodologies. *Science* 1981; 214:510–516.

Skerman VBD, McGowan V, Sneath PHA, eds. *Approved lists of bacterial names.* Washington, D.C.: American Society for Microbiology, 1980.

Stanier RY, Adelberg EA, Ingraham JL. *The microbial world.* 4th ed. Englewood Cliffs, N.J.: Prentice-Hall, 1976.

Stanier RY, Cohen-Bazire G. Phototrophic procaryotes: the cyanobacteria. *Annu Rev Microbial* 1977; 31:225–274.

Woese CR. Archaebacteria. *Sci Am* 1981; 244:98–122.

MEET THE
EUCARYOTES

In this chapter, we will examine some eucaryotic organisms. Most **algae, protozoa,** and **fungi** are truly microscopic, but others, such as seaweeds and mushrooms, are not. We will also stretch our definition of "microorganism" to look at some medically important animal parasites, the **helminths.** These may have both microscopic and readily visible forms.

As we saw in Chapter 3, eucaryotic cell structure is complex. As a result, the biology of eucaryotes is more elaborate in many ways than that of procaryotes. Their cells grow more slowly and reproduce less rapidly. They divide their highly organized, membrane-enclosed nuclei by the complex process called mitosis. Many reproduce sexually, as well as asexually, and pass through complex life cycles.

ASEXUAL AND SEXUAL REPRODUCTION

There are two means of reproduction in nature. In **asexual reproduction,** new cells or organisms arise from just one parent. This may occur in several different ways. We have already seen how bacteria use binary fission or budding processes in which one cell divides into two. In mitosis, which is used by eucaryotic cells, the parent cell's nuclear membrane pulls apart and its chromosomes, each of which has two duplicate copies, split apart (Figure 5.1). By using a network of microtubules called the spindle, the cell separates the two identical sets of chromosomes into two groups located at each end of the cell. A nuclear membrane then forms around each daughter nucleus. This nuclear division is then frequently followed by a division of the rest of the cell contents. The outcome of asexual reproduction is two offspring that are essentially identical to the parent cell.

Sexual reproduction occurs when a new individual is formed by fusion of sex cells called **gametes** from two parents (Figure 5.2). Typically the parental cells are diploid, containing paired chromosomes. The parental cell undergoes a type of cell division, called **meiosis,** to form gametes. In meiosis only one of each pair of parental chromosomes is placed into each gamete; since gametes contain unpaired chromosomes they are haploid. When the haploid gametes from the two parents fuse, the cell that results, the **zygote,** is again diploid.

Sexually reproducing organisms exhibit **alternation of generations.** This is another way of saying that in the life cycle diploid forms alternate with haploid forms. These chromosome movements are necessary to keep the chromosome number constant from generation to generation; they also continuously create new mixtures of genetic information. The new genetic combinations produced by sexual reproduction pave the way for rapid evolutionary change.

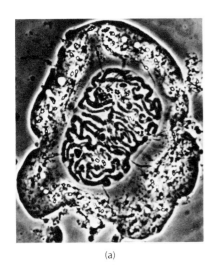

(a)

FIGURE 5.1   Mitosis in a Living Plant Cell
(a) In the center of the cell nucleus are the twisted threads of genetic material. (b) The nuclear membrane disappears and the genetic material (chromosomes) move toward the center. (c) Each chromosome has split and the two groups move apart. (d) In the last stage of mitosis, the two sets of chromosomes begin to spread out and new nuclear membranes form around each nucleus. (Courtesy of William T. Jackson, Dartmouth College.)

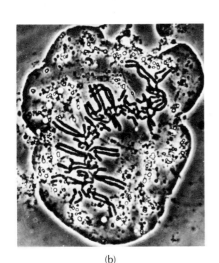

(b)

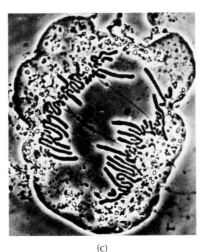

(c)

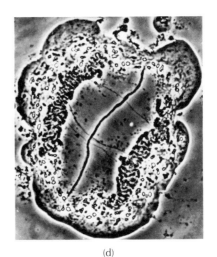

(d)

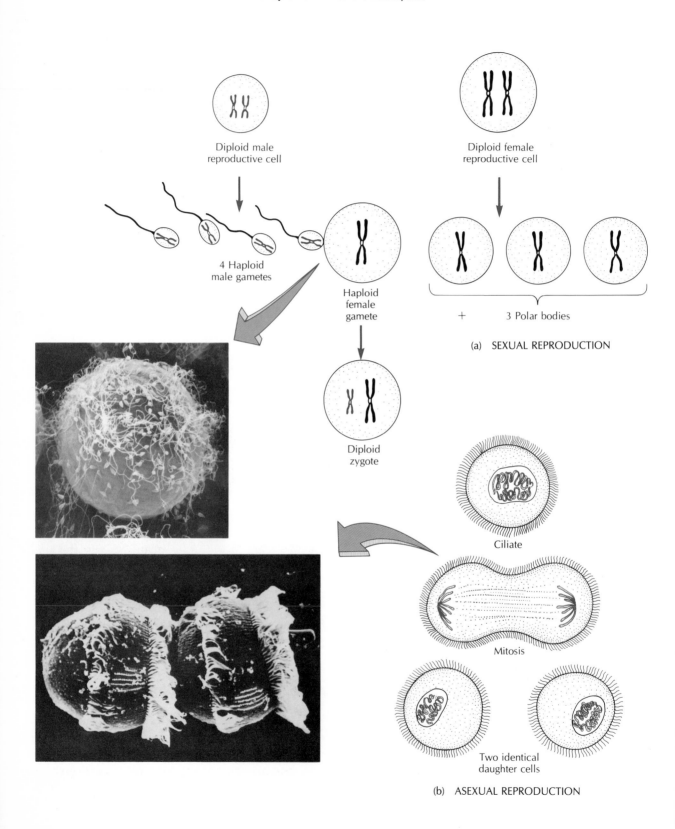

Diploid male
reproductive cell

4 Haploid
male gametes

Haploid
female
gamete

Diploid female
reproductive cell

+   3 Polar bodies

(a)   SEXUAL REPRODUCTION

Diploid
zygote

Ciliate

Mitosis

Two identical
daughter cells

(b)   ASEXUAL REPRODUCTION

THE ALGAE

Algae are members of the kingdom Protista. They are photosynthetic, and thus are valuable because they manufacture much of the annual global production of organic material. Algal cells contain the photosynthetic organelles called chloroplasts and most have cell walls composed of cellulose, pectin, or silica. There are six groups of algae, distinguished primarily on the basis of cell pigments and morphology. They are green algae, euglenids, diatoms, dinoflagellates, brown algae, and red algae. There are both unicellular and multicellular types of algae.

The habitats and contributions of unicellular algae overlap those of procaryotic cyanobacteria. Together, they make up the phytoplankton, the microscopic, floating, photosynthetic life of a body of water. All aquatic animal life depends on these producers for food. They also help replenish global oxygen supplies.

THE GREEN ALGAE

The green algae, or **Chlorophyta,** may have been the ancestors of the land plants. Green algae range from unicellular to colonial and multicellular modes of life and from water to land existence (Figure 5.3). *Chlamydomonas* is a typical unicellular form. *Volvox* grows in spherical colonies containing vegetative and reproductive cells. In *Chlorococcus*, vegetative cells are multinucleate, with each large cell containing many nuclei within one membrane. This type of structure, called a **coenocyte,** forms when a cell undergoes nuclear division several times in succession without undergoing cell division.

Some green algae are fixed in place, and others move. *Ulva*, the sea lettuce, is multicellular; its sheetlike body, called a **thallus,** often grows to several inches in length. The sea lettuce has a specialized attachment structure, a **holdfast,** by which it adheres to submerged rocks and pilings. The unicellular green algae and reproductive cells are motile by means of flagella.

Green algae are extremely widely distributed. They grow on snow and ice and on the shady sides of tree trunks. They grow on the fur and feathers of certain animals and birds giving them a greenish color. They are, however, most common in soil and water habitats.

THE EUGLENIDS

Euglena and its kin are claimed by both botanists and zoologists. We will consider them as algae. Euglenids are unicellular, photosynthetic, and motile (usually by means of two flagella) (Figure 5.4). In these respects, they resemble certain of the green algae. However, they do not have a cell wall, and do have a gullet; they take in and digest particles of organic material. Furthermore, they need organic compounds to live and thus are heterotrophic, which is why the zoologists claim them. Most euglenids are freshwater forms and reproduce asexually.

THE DIATOMS AND DINOFLAGELLATES

Diatoms are **Chrysophyta.** They have single cells encased in a hard, silica-impregnated casing or **frustule** that looks like a tube, petri dish, or

FIGURE 5.2 Sexual and Asexual Reproduction
(a) In sexual reproduction, two separate parent cells, each containing the normal number of chromosomes, produce haploid gametes. When the two gametes fuse, the resulting zygote is again diploid. The scanning electron micrograph shows hundreds of sperm surrounding a single sea urchin egg. One male gamete will penetrate the egg and fertilize it, creating a diploid zygote. (×850. Courtesy of Mia Tegner, Scripps Institute of Oceanography.) (b) In asexual reproduction, a eucaryotic cell divides by the process of mitosis. Two identical sets of chromosomes are pulled to opposite ends of the cell, followed by a division of the rest of the cell contents. The scanning electron micrograph shows two cells of the ciliated protozoan *Didinium nasutum* about to separate after cell division. (×595. Courtesy of E. B. Small, G. A. Antipa, and D. S. Marszalek, The cytological sequence of events in the binary fission of *Didinium nasutum* (O.F.M.), *Acta Protozool.* (1972), 9:275–282.)

box with a lid (Figure 5.5). Diatoms are very beautiful. Their frustules, being almost indestructible, are preserved in vast fossil deposits called diatomaceous earth. This material is useful as an abrasive powder.

Dinoflagellates or **Pyrrophyta** are unicellular and biflagellate. Each cell is armored with interlocking plates (Figure 5.5b). More than 1000 species are known. Most dinoflagellates are photosynthetic; some are also heterotrophic.

Both diatoms and dinoflagellates actively store the liquid lipids called oils. When their cells accumulate in sediments and are fossilized, the oils are transformed into petroleum. Our global petroleum deposits all originate from microbial oil residues. Another contribution of the diatoms and dinoflagellates is that they are ma-

jor marine and fresh-water food producers. They are consumed in vast numbers by schools of small fish such as sardines and herring.

Several species of dinoflagellates synthesize potent toxins. When shellfish such as clams, oysters, and mussels feed, toxic dinoflagellates may accumulate within their digestive organs. When humans eat the shellfish they get a paralytic shellfish poisoning, which is occasionally fatal.

THE BROWN AND RED ALGAE

Brown algae or **Phaeophyta** are true macroscopic seaweeds. They are widely distributed around the coasts of the world's oceans. They have two parts, a thallus and a holdfast. The thallus may

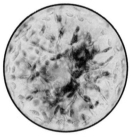

(a) *Eremosphaera*

FIGURE 5.3   Green Algae
Green algae take a wide variety of forms and have adapted to both water and land habitats. (a) *Eremosphaera*, one of the largest unicellular green algae (×222); (b) *Volvox*, a colonial algae made up of thousands of biflagellate cells; (c) *Acetabularia*, a marine algae consisting of a single huge cell that is differentiated into a cap, a stalk, and a "foot." (Photo credits: a, courtesy of Ray F. Evert, University of Wisconsin; b, courtesy of Carolina Biological Supply House; c, courtesy of H. Genthe, Science Photo Library.)

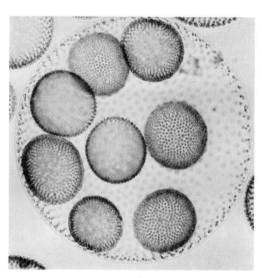

(b) *Volvox*

(c) *Acetabularia*

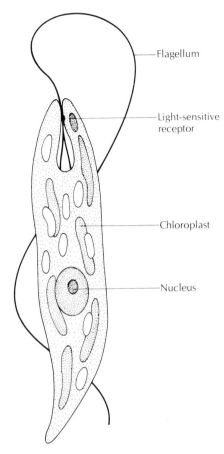

- Flagellum

- Light-sensitive receptor

- Chloroplast

- Nucleus

FIGURE 5.4 Euglenids
Euglena is a free-living unicellular organism with characteristics of both plants and animals. Containing numerous chloroplasts, it is photosynthetic, yet it can also ingest organic materials from the environment. A light-sensitive "eye spot" influences the action of the flagellum and thus the orientation of the cell. (×4640. From R. G. Kessel and C. Y. Shih, *Scanning Electron Microscopy in Biology*. New York: Springer-Verlag (1974).)

be branching or flat. The thallus of a giant kelp such as *Laminaria* or *Macrocystis* may be many meters long and many hundreds of kilograms in weight. Most brown algae attach themselves firmly to rocks or pilings by means of a fibrous, strongly adhesive holdfast. Their habitats include surfaces within the intertidal zone — the area between the high-tide and low-tide lines — and subtidal areas only a few meters below the surface. Their depth is limited by light penetration and the amount of time in each tide cycle that they are able to tolerate exposure to the air (Figure 5.6). We use the brown algae for fertilizer and stock feed. Alginates, polysaccharides obtained from brown algae, are used as food thickeners and stabilizers.

The red algae or **Rhodophyta** are a possible evolutionary link with the ancestral procary-

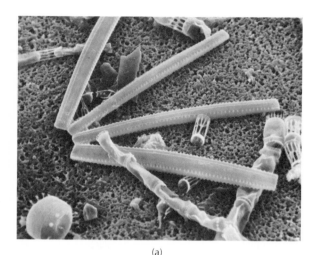

(a)

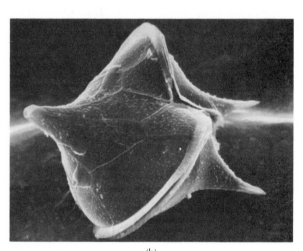

(b)

otic cyanobacteria, since they have the same phycobilin pigments. The phycobilins, accessory photosynthetic pigments, absorb short-wavelength blue light, the part of the visible spectrum that penetrates deepest into water. Red algae can capture sufficient light energy to grow at greater depths than brown algae. They only marginally compete with the brown seaweeds which attach to rock surfaces. Red algae may be unicellular, filamentous, or thallus-producing. The coralline forms deposit calcium carbonate and contribute to building coral reefs in warm seas. We harvest *Chondrus crispus*, the Irish moss, for its polysaccharide carrageenan, a thickener and stabilizer used in gelled puddings, chocolate milk, ice cream, and

FIGURE 5.5   Diatoms and Dinoflagellates
(a) Marine plankton may be collected by towing a fine-meshed plankton net through the water. This sample, obtained in Narragansett Bay, Rhode Island, contains four diatoms: *Skeletonema costatum*, *Thalassionema nitzschioides*, *Detonula confervacea*, and *Thalassiosira nordenskjoldii*. (×740. Courtesy of J. M. Sieburth, *Microbial Seascapes*. Baltimore: University Park Press (1979), plate 2.6.) (b) *Peridinium*, from Florida waters. This dinoflagellate has two flagella, one of which provides forward motility, while the other, housed in the beltlike groove surrounding the cell, creates rotary motion. The cell is enclosed in armorlike plates. (×455. Courtesy of J. M. Sieburth, *Microbial Seascapes*. Baltimore: University Park Press (1979), plate 2.8.)

FIGURE 5.6   Seaweeds Exposed at Low Tide
Two species of brown algae are attached to the upper and middle levels of rock. At the water line mossy red algae are attached. These need a habitat in which they are submerged almost all of the time. The brown algae here can tolerate some exposure to sunlight and air during a portion of the tidal cycle. (Courtesy of J. M. Sieburth, *Microbial Seascapes*. Baltimore: University Park Press (1979).)

dozens of convenience foods. *Porphyra* is an essential ingredient in Japanese dishes such as sushi. *Gelidium* provides the bacteriologist's indispensable agar, used for gelling laboratory media.

Brown and red algae are "grazed" by many marine animals as a food. Their polysaccharides may be used as nutrients by heterotrophic marine bacteria. However, the real significance of the brown and red algae as an ocean food source is less than you might think, because they are largely limited to growing close to shore.

THE FUNGI

The study of fungi is called **mycology.** The members of the kingdom Fungi have rigid cell walls composed of cellulose or chitin. They mainly reproduce by forming asexual or sexual spores, and are classified predominantly by their means of reproduction. Fungi have heterotrophic nutrition. Most fungi are saprophytes, but some are parasitic. Quite a few infect plants and animals. Most fungi are respiratory but some, especially the yeasts, may ferment. The fungi are all aerobic. One of their characteristics is their ability to secrete active digestive enzymes to hydrolyze carbohydrates, proteins, and lipids. These fungal enzymes can cause rapid deterioration of organic materials such as wood, fabrics, and leather in a damp environment or in contact with soil. The fungi, with the bacteria, are decomposers. Ecologically, these organisms are indispensable recyclers. Decomposers release scarce elements from animal and plant remains and return them to the soil. Many fungi tolerate high osmotic pressures and survive in concentrated solutions of salt and sugar. They can cause spoilage of pickles and preserves.

FUNGAL STRUCTURE

Fungi have rigid cell walls and may take several distinctive microscopic shapes. The three basic structural types are **molds, yeasts,** and **fleshy fungi.** Molds form **hyphae,** which are long microscopic filaments. These may or may not have crosswalls, called **septa,** perforated by pores (Figure 5.7). Nonseptate mold hyphae are multinucleate or coenocytic, and all their cytoplasm is continuous. Within the hyphae, cytoplasmic movement carries nutrients from one area to another. Hyphae grow by elongating and branching, and may form a visible mat, called a **mycelium.** Molds growing on a solid surface exhibit a characteristic fuzzy appearance to the naked eye (Figure 5.7b). Their mycelium may be submerged within the soil or a tissue, or it may extend over the surface of the material the fungus is utilizing for growth. Some portions of the mycelium are **vegetative mycelium** and gather nutrients. Other portions are **reproductive mycelium,** producing asexual or sexual spores. In some molds, the reproductive mycelial structures stick up from the surface of the vegetative mat as aerial spore-bearing structures. You can readily see these by examining a moldy piece of bread or orange. The hyphal threads readily penetrate dead plant and animal tissue and, in some cases, enter living cells. This ability can make them highly invasive pathogens.

The **fleshy fungi** include mushrooms, puffballs, and bracket fungi. Their solid structures are formed from highly organized hyphal threads. In a mushroom, part of the reproductive mycelium forms the cap, which bears reproductive spores, as well as the supporting stalk. The vegetative part of the mushroom's mycelium is not readily seen, as it is submerged in the organic matter of soil or wood, which is digested to nourish the fruiting body.

Yeasts are unicellular fungi, usually oval in outline. Yeasts growing on solid media form colonies very similar to bacterial colonies. They reproduce primarily by budding, a highly organized asymmetric form of cell division. When yeasts bud, nuclear division or mitosis is coordinated with the outgrowth of a swelling or bud from one end of the oval parent cell (Figure 5.8). One daughter nucleus migrates into the bud,

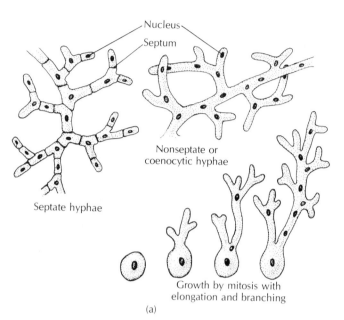

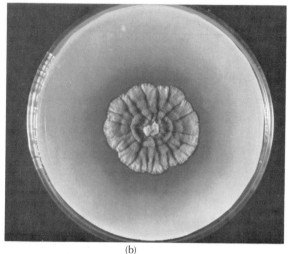

(b)

(a)

FIGURE 5.7 The Structure of Molds
(a) Molds form long filaments called hyphae, which may be separated into cell-like units by septa or exist as one continuous cell with many nuclei. These hyphae grow by elongating and branching at the tips. (b) The mold *Penicillium marneffei* growing on agar shows the characteristic fuzzy mat called a mycelium. (Courtesy of Centers for Disease Control, Atlanta.)

which is then separated from the parent cell by the formation of a new cell wall. The bud then detaches, leaving a bud scar on the parent cell and carrying a birth scar on its own cell wall. Many yeasts, such as the familiar baker's and brewer's yeast *Saccharomyces*, are always unicellular. Others, however, such as the pathogen *Histoplasma*, are **dimorphic.** This means that the organism has two forms. Although it grows as a yeast at body temperature, *Histoplasma* will convert to mycelial growth in a cooler environment such as soil, and may then reproduce by forming spores.

FUNGAL REPRODUCTION

Fungi have evolved a vast repertoire of reproductive structures (Table 5.1). Asexual re-

production is almost continuous in many fungi, while sexual reproduction occurs periodically. We use the means of sexual reproduction, or the lack of it, as the primary criteria for classifying fungi. Fungi are extremely prolific; you may have seen the clouds of thousands of spores that become airborne when a mycelium is disturbed.

All fungal spores are small modified cells. **Asexual spores** (Figure 5.9) arise from portions of the coenocytic cell containing one nucleus. The spores take a variety of microscopic forms, such as arthrospores, sporangiospores, and conidiospores.

Sexual spores are formed by fusion of haploid nuclei from two different parental mating types, referred to simply as plus (+) and minus (−) strains. Many fungi produce chemical sex attractants called **pheromones** to assist in "meeting

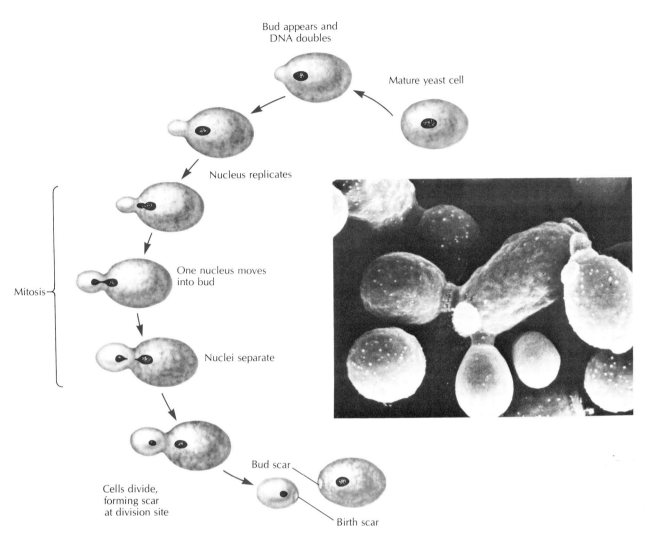

Bud appears and
DNA doubles

Mature yeast cell

Nucleus replicates

Mitosis

One nucleus moves
into bud

Nuclei separate

Cells divide,
forming scar
at division site

Bud scar

Birth scar

FIGURE 5.8 Budding Yeast Cells
A mature, rougher-surfaced yeast cell forms smooth buds. Each bud will contain a
nucleus formed by mitosis. After the bud separates, a collarlike bud scar will be left
on the parent cell; a matching birth scar can be seen on the bud cell. In the
accompanying scanning electron micrograph, buds appear on cells of baker's yeast,
Saccharomyces cereviseae. (From Kenneth Watson and Helen Arthur, *J. Bacteriol.*
(1977) 130: 312–317, Fig. 5.)

their mate." The pheromones guide hyphae of
opposite mating types to extend toward each
other and make contact. After fusion, the diploid
nucleus undergoes meiosis, producing haploid
spores appropriate to the species. Examples of
sexual spores are ascospores, oospores, zygo-
spores, and basidiospores. Some species of fungi
have never been observed to reproduce sexually.

	TABLE 5.1 The major groups of fungi	
Class	Microscopic Morphology	Method of Asexual Reproduction
Lower fungi	Nonseptate mycelial fungi Some unicellular species	Sporangiospores Zoospores Conidiospores
Ascomycetes	Septate mycelial fungi Yeasts	Conidiospores Budding
Basidiomycetes	Septate mycelium May form complex fruiting bodies	Conidiospores (rare)
Deuteromycetes	Yeasts Dimorphic fungi Septate mycelial fungi	Budding Arthrospores Blastospores Chlamydospores

THE CLASSES OF FUNGI

We use a simplified classification of the fungi, placing them in four main categories — the lower fungi, the ascomycetes, the basidiomycetes, and the deuteromycetes. As previously mentioned, note that they are separated primarily on the basis of their mode of reproduction.

THE LOWER FUNGI.   These fungi are found in fresh water, soil, and other moist environments. They have nonseptate mycelium, some unicellular motile forms, and asexual spores produced in a sac called a **sporangium,** from which they burst when mature. Some fungi such as *Blastocladiella* are unicellular through much of their life cycle. They may form large numbers of motile **zoospores** asexually within a thick-walled sporangium. The free zoospores propel themselves by means of their single flagellum (Figure 5.10a).

The **oomycetes** are mildew and blight organisms. Many are highly destructive. *Pythium*, for example, is a pathogen of plants and *Saprolegnia* is a pathogen of fish. Sexual reproduction in the oomycetes involves the fusion of nonmotile gametes; male nuclei contained in separate hyphal branches fuse with eggs contained in a spherical oogonium. The **oospores** that result from this type of fertilization mature into heavy-walled survival spores (Figure 5.10b).

The **zygomycetes** are generally terrestrial fungi that may reproduce through both sexual and asexual processes. In the sexual cycle, two gametangia, one from a parent of each mating type, join and a **zygospore** forms at the point of fusion (Figure 5.10c). *Rhizopus nigricans*, the common black bread mold, also forms asexual

Method of Sexual Reproduction	Habitat	Roles in Environment
Zygospores	Soil	Saprophytes — decay, spoilage
Oospores	Water	Parasites — diseases of plants
Flagellate gametes	Plants	
Ascospores	Foods	Saprophytes — cause fermentation
	Soil	Parasites — disease of plants
	Plants	
Basidiospores	Soil	Saprophytes — decompose wastes
	Plant tissue	Parasites — plant disease
None observed	Animal tissue	Important fungus diseases
	Soil	of human beings
	Plants	

spores in a sporangium. These spherical black balls are just visible when you look at a piece of moldy bread.

THE ASCOMYCETES Ascomycetes are the fungi that form sexual ascospores. Some are mycelial fungi, others are yeasts, and a few such as the morels are fleshy fungi. Human beings derive sustenance and cheer from the products of the most common yeast, *Saccharomyces cerevisiae.* Special yeast strains are employed in bread making and fermenting alcoholic beverages. Other ascomycetes include the fleshy fungi commonly named morels, cup fungi, and truffles. Ascomycetes are widely distributed saprophytes and parasites. They are found in soil, dung, on the roots of plants, on grains, and in the oceans.

The mycelial ascomycetes reproduce asexually when the mycelium forms brushlike co-nidia; these pinch off strings of **conidiospores** (review Figure 5.9). The ascomycete yeasts bud (review Figure 5.8). In sexual reproduction, first the fusion of gametes produces a diploid nucleus. Then meiosis yields four or eight haploid **ascospores;** these are held like peas in a pod in a sac or **ascus** (Figure 5.11). The ascus eventually ruptures to release the mature ascospores.

THE BASIDIOMYCETES Basidiomycetes reproduce sexually by forming microscopic structures called the **basidium** and **basidiospores.** Most of the best-known types of fleshy fungi — mushrooms, puffballs, and bracket fungi — are basidiomycetes. The visible structures are actually portions of the reproductive mycelium — **fruiting bodies.** The rusts and smuts, though less widely known, are among the most destructive agricultural plant pathogens.

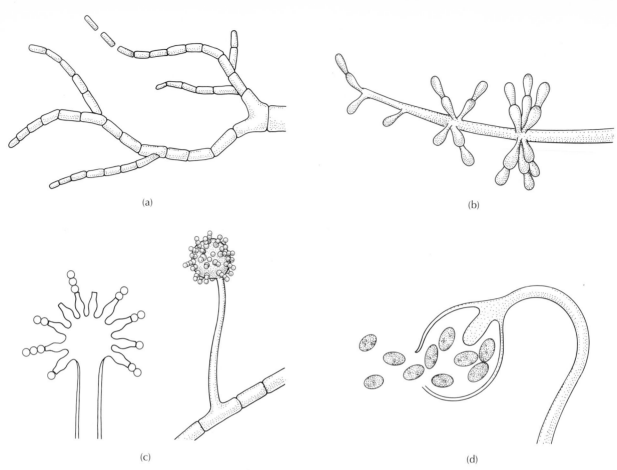

(a)

(b)

(c)

(d)

FIGURE 5.9 Asexual Reproduction in Fungi
(a) The hyphae fragment into single, slightly thickened cells called arthrospores. (b) Buds that form from the parent cell form blastospores. (c) External reproductive cells form chains of conidiospores. (d) A sac (sporangium) at the end of an aerial hypha may enclose hundreds of sporangiospores.

FIGURE 5.10 Reproduction in the Lower Fungi
(a) Motile zoospores of *Blastocladiella emersonii* move by means of a single flagellum arising from a basal body that passes through a large mitochondrion. (From James S. Lovett and W. E. Barstow, Growth and development of *Blastocladiella emersonii, Bacteriol. Revs.* (1975) 39: 345–404, Fig. 3. Reprinted with permission of the publisher and authors.) (b) Mating in the fungus *Achyla ambisexualis* occurs when sperm nuclei, contained in the antheridium, pass through fertilization tubes to the egg nuclei contained within the oospheres. The large spherical structure in the photo is the oogonium, and the dark bodies within contain eggs. (×450. Courtesy of Alma W. Barksdale.) (c) The life cycle of the black bread mold involves both asexual and sexual stages. Initially a spore develops into a mycelium from which rises an aerial hyphae topped by a sporangium. This sporangium is a sac containing hundreds of haploid sporangiospores. On dissemination, mature spores give rise to new mycelial nets. If both + and − mating strains are present, hormones cause the mating tips to come together—the nuclei of the two types fuse, and a zygospore is formed through sexual reproduction.

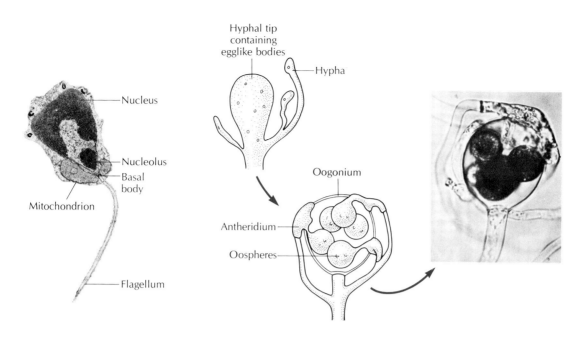

Nucleus

Nucleolus
Basal
body

Mitochondrion

Flagellum

(a) Zoospore

Hyphal tip
containing
egglike bodies
Hypha

Oogonium

Antheridium

Oospheres

(b) Oomycetes

ASEXUAL

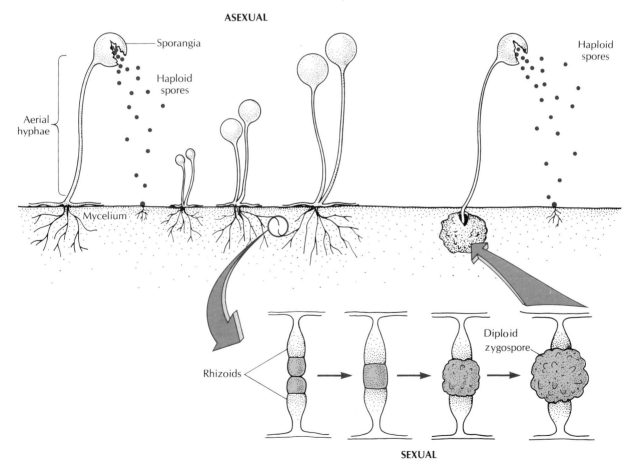

Sporangia

Haploid
spores

Aerial
hyphae

Mycelium

Haploid
spores

Rhizoids

Diploid
zygospore

SEXUAL

(c) Sporangia

147

People are fascinated by mushrooms, by their sudden appearance and rapid growth, by their often fantastic shapes and colors, by the folklore surrounding the few poisonous species, and by the delicate flavors of the edible forms. Mushroom collectors know well that each basidiomycete species has a specific preferred habitat, and develops best in certain types of soil; leaf litter from pine trees, aspens, or oaks; or the tissues of specific trees, shrubs, or grasses. Fruiting is a predictable seasonal phenomenon, usually occurring after a rainy spell.

The fleshy basidiomycetes develop an extensive mycelium, which may infiltrate a radius of many feet of soil or permeate the roots and vascular system of a large tree. The mycelium may be extremely long-lived. Its cytoplasm produces and excretes digestive enzymes to break down organic matter; the enzyme activity may eventually kill a tree. Asexual conidiospores are produced along the mycelium when nutrients are in good supply.

Each mushroom is a sexual reproductive body. It is able to develop rapidly because nutrients are supplied by the large vegetative mycelium underground or within the tree trunk. Sexual reproduction occurs after two haploid mycelia fuse (Figure 5.12). **Dicaryons** (cells with two sep-

FIGURE 5.11 Sexual Reproduction in Ascomycetes
During sexual reproduction in a heterothallic ascomycete, sexual cells pass through a series of cell divisions that leads to the formation of ascospores. Each ascus initially contains two haploid nuclei, one from each mating type. These two nuclei fuse and then divide — first through meiosis and then through mitosis — to yield a total of eight haploid nuclei. These develop into eight ascospores.

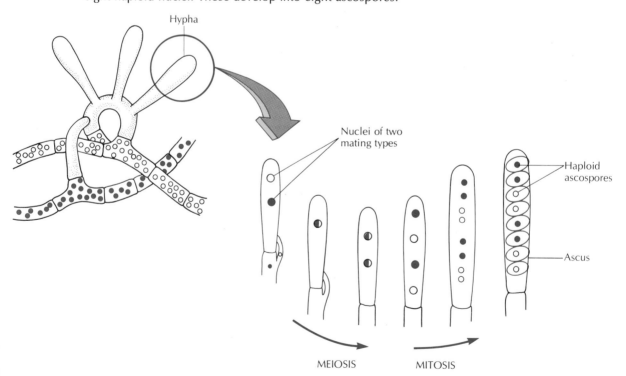

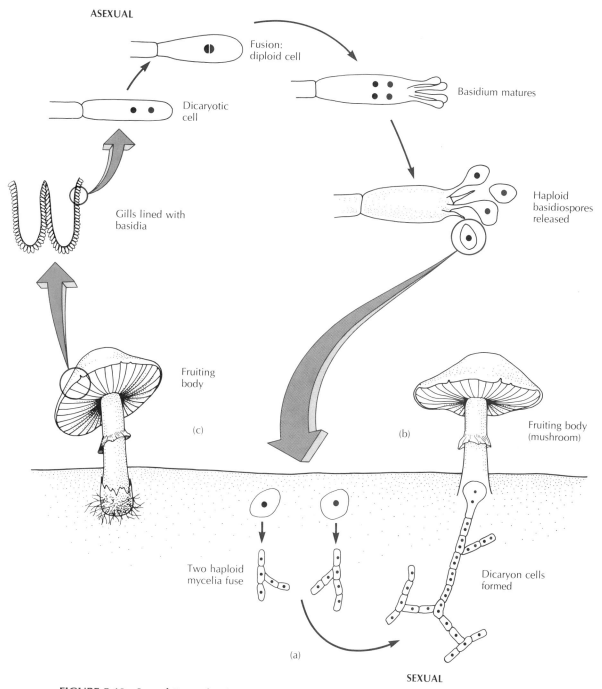

ASEXUAL

Fusion:
diploid cell

Basidium matures

Dicaryotic
cell

Haploid
basidiospores
released

Gills lined with
basidia

Fruiting
body

(c)

Fruiting body
(mushroom)

(b)

Two haploid
mycelia fuse

Dicaryon cells
formed

(a)

SEXUAL

FIGURE 5.12 Sexual Reproduction in the Basidiomycetes
The reproductive cycle of *Agaricus*, a common, edible mushroom involves (a) fusion
of the + and − strain mycelium to yield a dicaryotic mycelium, which develops into
a fruiting body (b). Gills beneath the cap (c) produce spores following nuclear fusion
and meiosis, which recombines the parental genetic material.

arate haploid nuclei) are formed and these proliferate to form the fruiting body. The cap of the mushroom has either gills or pores bearing basidia. As the mushroom ripens, clouds of variously colored haploid basidiospores are formed and released. Each basidiospore, on germination, may initiate the growth of a new mycelium.

The rusts and smuts multiply in soil and on the tissues of woody and soft plants. Black stem rust of wheat causes extensive crop losses. The causative agent *Puccinia graminis* reproduces asexually while attacking wheat as a parasite. However, a sexual cycle involving growth on barberry plants is required for the rust to survive the winter and continue its infestation from year to year. The problem has been partly controlled by eradicating barberries in areas where wheat is cultivated. Another approach is the breeding of wheat varieties genetically resistant to rusts.

THE DEUTEROMYCETES Deuteromycetes are fungal species for which sexual stages have not yet been described. Because of this deficiency, they have also been called the Fungi Imperfecti. Almost all of the human fungal pathogens are found in this group. The deuteromycetes, however, also have their redeeming features. Many valuable drugs including penicillin are produced by these fungi.

The deuteromycetes' asexual reproduction is diverse but highly efficient. *Aspergillus* and *Penicillium* are mycelial fungi that form vast numbers of conidiospores (Figure 5.13). The deuteromycete yeasts such as *Candida* and *Tor-*

BOX 5.1
Requiem for an Elm

Over the past 30 years, the eastern and central United States have suffered the loss of many millions of large and graceful shade trees. Many an Elm Street or Elm Plaza no longer deserves its name. The loss of the elms has greatly increased the harshness of much urban life.

The decline and fall of an elm results from the attacks of succeeding waves of fungi. First comes the lethal Dutch elm disease, caused by the ascomycete *Ceratocystis ulmi*. Many elms are also weakened by the saprophytic ascomycetes *Gnomonia ulmea* and *Systremma ulmi*.

Once the tree is dead, fungi with their cellulose-digesting enzymes will attack, gradually removing it from the landscape. A basidiomycete, the bracket fungus *Pleurotus ostreatus*, will penetrate the bark and wood, fruiting prolifically, using nutrients extracted from digested wood. As limbs fall to the ground, they will be finally destroyed by the dryrot fungus *Merulius lachrymans* and a variety of other soil fungi.

(Photo courtesy of Brown Brothers.)

ulopsis bud. *Candida* may also form **pseudo-hyphae.** These are tubular extensions, consisting of sequences of buds that do not fully separate. Dimorphic deuteromycetes such as *Coccidioides* asexually form chlamydospores, arthrospores, and conidia.

Human fungal diseases, called **mycoses,** are widespread in subtropical and tropical areas. Many fungal infections are chronic and hard to treat effectively. Parasitic fungi growing in human tissue cause a variety of destructive effects. The major mycoses are summarized in Table 5:2.

FIGURE 5.13 Reproduction in the Deuteromycetes
(a) Scanning electron micrograph of young *Penicillium* with a few conidia on each chain. (×436 Courtesy of Centers for Disease Control, Atlanta.) (b) When spores reach a suitable environment, they begin to form the mycelium. Within a short time conidiospores begin to form and eventually bear conidia. The conidia propagate and disseminate the fungus.

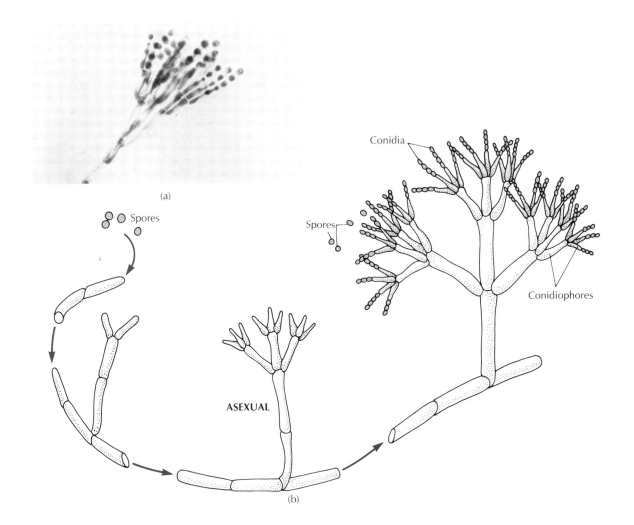

TABLE 5.2 Some fungus diseases of human beings			
Genus Names	**Morphology**	**Disease**	**Clinical Features and Transmission**
Epidermophyton *Microsporum* *Trichophyton*	Mycelial fungi, form conidia	Ringworm of scalp, beard, skin, and groin	Limited, usually to moist areas Some contracted from animals Recurrent in nature Cause inflammation, itching, and peeling of skin layers
Coccidioides	Dimorphic fungi	Coccidiodomycosis	Lung infection, usually mild, oc- casionally severe, caused by fungi in soil
Histoplasma	Dimorphic fungi	Histoplasmosis	Lung infection, mild to severe, transmitted by bat or bird dung
Blastomyces	Dimorphic fungi	North American blastomycosis	Lung infection, mild to severe, life-threatening if spread
Sporotrichum	Dimorphic fungi	Sporotrichosis	Subcutaneous nodules, ulcers, lung involvement; acquired by puncture wound
Cryptococcus	Yeast	Cryptococcosis	Lesions in lung, skin, and cen- tral nervous system; transmitted by pigeon droppings
Candida	Dimorphic fungi	Candidiasis Moniliasis Thrush	Opportunist; causes disease in throat, genitals, and dissem- inated infection

THE PROTOZOA

Protozoa are unicellular heterotrophs classed in the kingdom Protista. Their cell structure carries the unicellular design to its most complex capabilities. They are very diverse in their shape and adaptations. With no cell wall, they have the potential for rapid and flexible movements. In fact, their means of locomotion is the criterion used to classify groups of protozoa (Figure 5.14).

Some protozoans ingest other living organisms and are true predators and consumers. Others take in dissolved nutrients and are saprophytic or parasitic. Most protozoa are harmless free-living components of the aquatic **zooplankton,** the community of tiny floating animals. Certain other protozoa are human patho-gens, causing malaria, sleeping sickness, and other illnesses — some of the world's most severe and poorly controlled infectious diseases. Scientists have described approximately 25,000 species of protozoa. The protozoa are separated into four subgroups — Sarcodina, Mastigophora, Ciliata, and Sporozoa — on the basis of the organelles they use for movement.

THE SARCODINA

The first subgroup, the Sarcodina, move by formation of **pseudopodia,** flexible extensions of the cytoplasm. Pseudopodia advance and retract by active cytoplasmic streaming and stretching of the cell membrane. Pseudopodia are used pri-

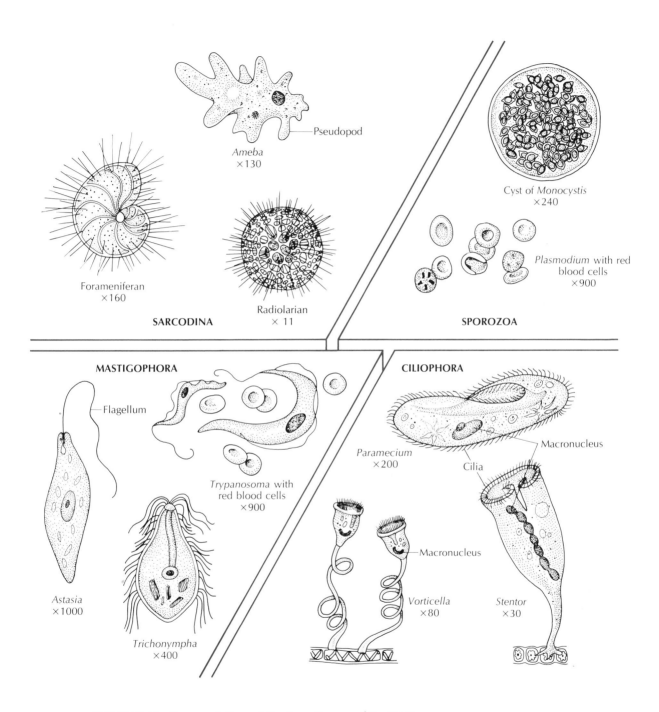

SARCODINA

Pseudopod

Ameba
×130

Forameniferan
×160

Radiolarian
× 11

SPOROZOA

Cyst of *Monocystis*
×240

Plasmodium with red
blood cells
×900

MASTIGOPHORA

Flagellum

Trypanosoma with
red blood cells
×900

Astasia
×1000

Trichonympha
×400

CILIOPHORA

Paramecium
×200

Macronucleus

Cilia

Macronucleus

Vorticella
×80

Stentor
×30

FIGURE 5.14 Representatives of the Four Groups of Protozoa
These eucaryotes are classified on the basis of their means of locomotion, as shown
in these diagrams.

marily as food-capturing devices, but may be used for a type of creeping motion. The **amoebas** are the best-known members of this group of protozoans. They use two or more coordinated pseudopodia to surround and engulf prey organisms that are then digested and assimilated.

All Sarcodina have tough but flexible exterior membranes. True amoebas have no distinct shape but rearrange their cytoplasm continuously. Other sarcodinids such as *Arcella* or *Difflugia* produce protective external shells or skeletons, constructed primarily of inorganic materials excreted or glued together. Although these shells give them a fixed shape, they are able to move by extruding their pseudopods through openings in the casing.

The sarcodinids, like all fresh-water protozoans, must cope with the problem of life in a medium more dilute than their cytoplasm. Water continually enters their cells by osmosis. The unwanted water is collected and evacuated by simple excretory organelles, the **contractile vacuoles.** These are collecting vacuoles that expand as water accumulates within them, then contract, forcing the collected water out of the cell.

The Sarcodina reproduce asexually. Some have a life cycle in which the vegetative cell type or **trophozoite** alternates with a resistant hard-coated cell type — a **cyst** — formed under dry or cold conditions.

True amoebas are mainly free-living aquatic and soil organisms. Some are harmless inhabitants of the gut, but several species cause human infections. The parasitic species *Entamoeba histolytica* causes amebic dysentery, an often chronic, debilitating disease. It is most commonly found in subtropical and tropical countries. *Naegleria fowleri* causes a rare but usually fatal brain disease.

Three types of sarcodinids have their amoeboid cells enclosed in hard shells. The foramenifera produce shells made of calcium salts, resembling miniature snail shells. We find the forameniferan *Globigerina* in enormous numbers in nutrient-rich areas of sea water. When they die, their shells contribute to the forma-

tion of ocean sediments, trapping substantial amounts of calcium and phosphorus. The sediments may eventually be geologically transformed into limestone and chalk. Forameniferan shells compose the chalk of the White Cliffs of Dover. The radiolarians are beautiful microorganisms that form delicate exoskeletons of silica with radially arranged spikes (Figure 5.15). They are also marine and, like the forameniferans, serve as a mainstay food source for larger marine animals and at death contribute their skeletons to bottom sediments. The heliozoa are fresh-water sarcodinids that construct compact casings by making gummy substances used to glue together sand grains.

THE MASTIGOPHORA

The Mastigophora possess flagella, powered by the chemical reactions taking place in an organelle embedded in the cell membrane called a **kinetoplast.** In *Giardia* the flagella are attached only at one end, but in *Trypanosoma*, the flagellum is partly attached to the cell, along the cell's length, while the rest is freely moving beyond the end of the cell. The Mastigophora or flagellates characteristically reproduce asexually. Many form cysts; several undergo cyclic morphologic changes.

Four genera of Mastigophora cause serious human infection. The parasite *Giardia lamblia* causes a diarrheal disease (Figure 5.16a). Giardiasis is the most common protozoan infection in the United States. *Leishmania* species, transmitted by fly bites, cause leishmaniasis, a disease with skin and organ ulcers. *Trypanosoma* species, also transmitted by insects, cause sleeping sickness and Chagas' disease. *Trichomonas vaginalis* (Figure 5.16b) causes a genital infection usually spread by sexual intercourse.

THE CILIOPHORA

The Ciliophora, commonly called ciliates, move by means of bands of cilia. The cilia beat with a coordinated direction and rhythm. A ciliate can respond to chemotactic stimuli and is able to

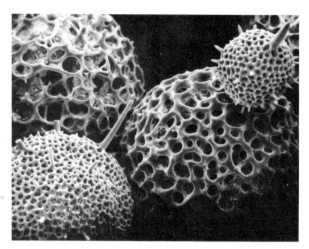

FIGURE 5.15 The Sarcodina
Scanning electron micrograph of the exoskeletons of radiolarians. (×185. Courtesy of J. M. Sieburth, *Microbial seascapes*. Baltimore: University Park Press, 1979, plate 8.27.)

start, stop, or move in any direction. Most ciliates are successful free-living forms; *Paramecium* is a well-known example. They capture other unicellular organisms as prey. They are able to ingest and digest their prey using specialized organelles, the oral groove and gullet. Undigested material is ejected by way of another specialized structure, the anal pore. Some ciliates use another capturing technique employing **trichocysts** — sticky threadlike harpoons that entangle the prey. The suctorians bear clusters of sticky tentacles for trapping prey.

Ciliophora reproduce both asexually and sexually. Each cell has a **macronucleus** and a **micronucleus,** both diploid. The macronucleus directs asexual reproduction, in which the cell divides in half lengthwise. The micronucleus is active in sexual reproduction. Ciliates carry out a sexual reproductive process called conjugation, in which two cells exchange micronuclei.

THE SPOROZOA

Adult sporozoans are nonmotile, although immature forms and gametes are occasionally mo-

tile by means of flagella, flexion, or gliding. The spores that give the group its name are formed following sexual reproduction. The sporozoans are generally parasitic, needing the dissolved nutrients and environmental protection they receive from a host.

Sporozoans often have complex life cycles. Sexual and asexual reproduction may alternate, each taking place in a different host. In a parasite life cycle the **definitive host** is that species in which the adult parasites mature and reproduce sexually. The **intermediate host** is that species in which immature forms develop. Humans may be either type of host, depending on which parasite is causing their infections.

Many sporozoans cause diseases. Malaria, which infects millions worldwide, is caused by several species of *Plasmodium* and transmitted by mosquitoes. In the plasmodian life cycle both humans and mosquitoes are hosts (Figure 5.17).

In malaria, humans are the intermediate hosts; they support asexual reproduction of the parasite. When a human is bitten by an infected mosquito, the parasites are injected. They reproduce first in the liver and then in the red blood cells forming merozoites. When mature, they rupture the red blood cells, releasing more parasites and giving rise to fever and chills.

A mosquito that bites a human who has malaria feeds on infected human blood. It becomes the definitive host for a sexual cycle that takes place in the insect's gut. A new crop of infective oocysts are then transferred to the mosquito's salivary gland. When the mosquito bites a new human host, the sporozoites are injected via the saliva. This step completes the complex life cycle of the plasmodian parasite.

Malaria can be partially controlled by drugs, used both to protect those who are exposed and to treat those who develop the disease. However, getting rid of the mosquito host is a more effective control measure, because then the plasmodia cannot complete their reproductive cycle and the transmission of the disease is arrested. Drainage and spray programs help control the insect vector. Still, every year as many as 300

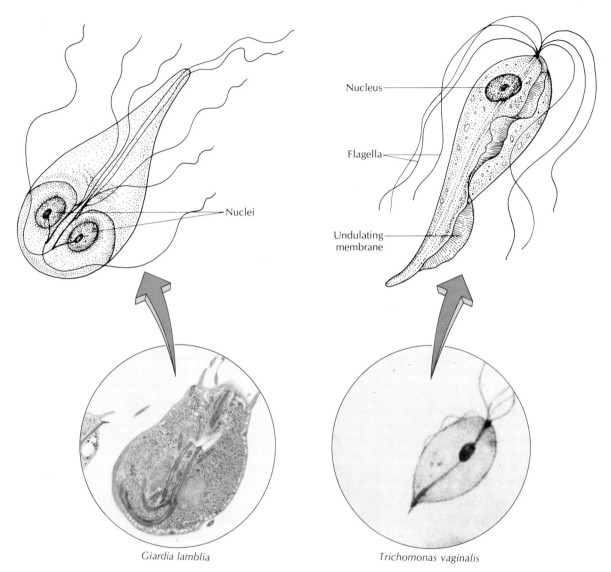

Nuclei

Nucleus

Flagella

Undulating
membrane

Giardia lamblia

Trichomonas vaginalis

FIGURE 5.16   Two Pathogenic Mastigophora
(a) A transmission electron micrograph of *Giardia lamblia*, an intestinal parasite which
can cause gastroenteritis in humans. (×16,055. Courtesy of C. Callaway, Pathology
Division, Centers for Disease Control, Atlanta.) (b) A light micrograph of *Trichomonas
vaginalis*, a urogenital parasite which can cause urogenital infections in humans.
(Courtesy of Centers for Disease Control, Atlanta.)

million people, principally children, contract the
disease, and about 1.5 million die as a result of it.

Toxoplasma gondii, another parasitic sporozoan, causes toxoplasmosis. This infection, although mild or asymptomatic in adults, can
cause severe congenital defects in infants born of
an infected pregnant woman. The cat is the definitive host, and cows, sheep, and humans are

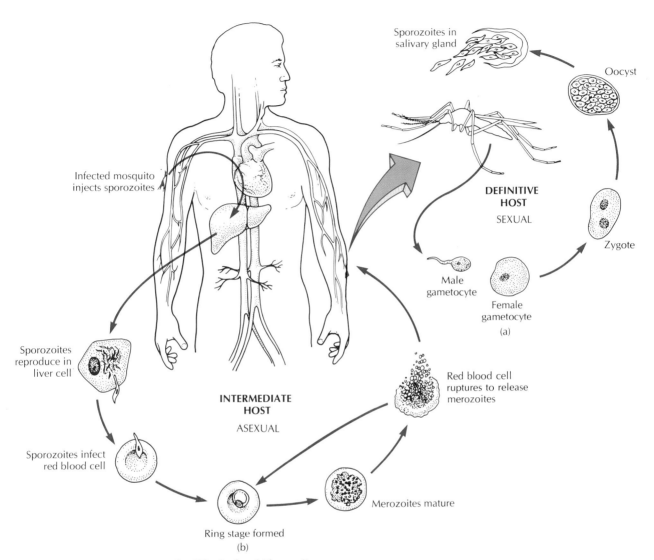

FIGURE 5.17 The Life Cycle of Plasmodium
Many parasites need several hosts to complete their life cycle. (a) The malaria parasite *Plasmodium* reaches sexual maturity in the mosquito, its definitive host. (b) The mosquito injects larval forms into the human intermediate host. The larvae mature through several stages, killing human red blood cells in the process. Then, when another mosquito bites, the insect sucks out larvae that complete the cycle by maturing to adulthood in the insect.

all intermediate hosts. Humans contract the disease by coming into contact with cat feces, or by eating overly rare infected beef or lamb.

Pneumocystis carinii is a sporozoan that causes a severe pulmonary infection in persons with depressed immunity. It is one of the infectious conditions that is often seen in patients with acquired immunodeficiency syndrome (AIDS). The agent is unfortunately not very responsive to drug therapy.

THE HELMINTHS

The **helminths** are not microorganisms, but simple invertebrate animals. Not only are they multicellular but they also have differentiated tissues. Why, then, do we include them in a microbiology book? We do so in order to be able to round out our coverage of the agents of infectious disease. Two groups of helminths, the roundworms and the flatworms, include some infectious parasitic agents. Many of the stages in the helminth life cycles are microscopic. We use many of the usual microbiological techniques and strategies to identify and eradicate the parasitic helminths.

We will study the two groups of helminths in which the parasitic forms are found. The **flatworms** include two subgroups, trematodes and cestodes. In the adult form, the **flukes** or **trematodes** are microscopic or small visible animals. The **tapeworms** or **cestodes** are all macroscopic in the adult form, sometimes reaching a length of many feet if left undisturbed in the host. The **roundworms,** the second major group, include pinworms, hookworms, and similar forms. Some are microscopic and some macroscopic.

Each parasite has its own unique life cycle. We will be studying several of these in later chapters. In humans, infections with adult parasites most commonly involve the digestive tract. However, the liver, blood, muscle, and nervous system are also infected by some forms.

Helminth infections are endemic in subtropical and tropical climates. It is both possible and common for one individual to have several different parasites simultaneously (see Box 1.2). Children with parasites grow more slowly and may not reach their full potentials in strength and intelligence. Adults with parasites are weakened and unproductive. Since parasitic infections are most readily transmitted under conditions of crowding, poor sanitation, and poor nutrition, it is easy to see how these infections contribute to the social cycle of poverty in the less-developed areas of the world.

Helminthic infections are difficult to treat. The infecting agent is an animal, you recall. As such, its cellular biology is very similar to ours. Anything that kills it is likely to be highly toxic to us. There are drugs available, but they must be used very cautiously. Some diseases, such as the fluke infestations, do not respond well to therapy.

STRUCTURAL FEATURES OF HELMINTHS

Unlike the procaryotes, protists, and fungi we have studied so far, helminths have **differentiated tissues.** Their simple digestive tubes, male and female reproductive systems, and primitive nervous systems are formed from differentiated tissues. Because of their parasitic life cycles, the digestive and nervous systems can be very simple — much simpler than they would have to be in an organism adapted for independent life. On the other hand, helminthic reproductive systems are very highly developed. The segments of an adult tapeworm, for example, appear under the microscope to be solidly packed with reproductive organs. Each segment produces thousands of eggs.

REPRODUCTION. The parasitic worms reproduce sexually. Most have complex life cycles. The smallest, least complex form, the **egg,** hatches into an immature, rapidly growing **larva** (plural, **larvae**). Larvae may undergo a sequence of developmental steps or stages before reaching sexual maturity, at which point they are **adults,** and able to produce eggs.

Parasitic helminths need hosts for part or all of their life cycle. Just like the protozoan parasites, individuals of different species may be used as definitive and/or intermediate hosts. Definitive hosts shed eggs or young larvae; intermediate hosts acquire young larvae and shed infectious larvae at a later stage of development. In some cases, humans are the definitive host for a parasitic disease and there is no other host (pinworms). In other cases, we may be only an intermediate host (canine tapeworm). In nature, the canine tapeworm is usually passed back and forth between the definitive host — a canine

BOX 5.2
The Fiery Serpent

One fascinating aspect of medical history is looking at how people's experiences with disease have been expressed in literature, religion, and action.

Dracunculiasis is a disease very few Americans have ever heard of. A parasitic roundworm, *Dracunculus medinensis,* burrows and grows under the skin. It causes fierce inflammation and a burning itch with troublesome subcutaneous ulcers. The parasite may grow to be a yard long. Altogether, it's quite an agent of sheer misery.

The guinea worm, as it is usually called, is found all over the Middle and Far East. Its life cycle involves humans shedding larvae into shallow water, where they develop within *Cyclops,* a small crustacean. Humans then ingest the little crustacean and its larvae with contaminated drinking water. Let's look at some different approaches to dealing with it.

Biblical texts mention both the disease and one form of cure. Many scholars believe that the guinea worm was the fiery serpent sent as a plague on the Israelites in the Old Testament (Numbers 21: 8–11). Its genus name means serpent in Greek and Latin. The biblical writer reported this occurrence as evidence of the wrath of God. Moses, under divine direction, set up a brass image of a serpent on a pole. Sufferers who looked upon the brass serpent were cured.

Traditional contemporary folk medicine has a difficult form of treatment for guinea worm. First, the skin is moistened to induce the worm to stick its head out. The worm's head is caught in a split stick, then the worm is cautiously extracted by winding it up around the stick over the course of many days. Cautiously, because if the worm snaps in two, the patient will end up with a terrible inflammation and ulceration.

In India, guinea worm disease has been a plague for centuries. It is endemic in about 12,000 rural villages, with at least 12.6 million people at risk. India has adopted a modern (by our standards) eradication campaign based on public health principles. Eradication is closely tied to the country's program of providing safe drinking-water supplies to all its rural population. Disinfection, preventive drugs for individuals, and community education are essential components. Progress is coming slowly but steadily. It is expected that at least 10 years will be required for success.

(Courtesy of Dr. E. R. Degginger, FPSA.)

such as a wolf or dog — and an intermediate host, typically the deer. Humans can become intermediate hosts when they eat foods contaminated with dog feces or kiss their dogs.

THE FLATWORMS

The flatworms, as their name implies, have flattened bodies. The trematodes have simple tubular digestive systems with only one opening, the mouth, through which foods enter and wastes leave. The cestodes have no digestive system at all, but absorb small nutrient molecules directly through their outer covering, the cuticle. Flatworms are **hermaphroditic;** that is, each individual has the reproductive organs of both

sexes. Some are self-fertilizing, while others copulate and carry out reciprocal fertilization. Fertilized eggs are released from a genital pore.

THE TREMATODES OR FLUKES These parasites have two suckers, which they use to adhere to the host or to suck its fluids (Figure 5.18a). This is their major means of feeding. Adult flukes release eggs. The eggs develop through up to four different larval stages, often in several intermediate hosts. The larval stages are called **miracidia, redia, cercaria,** and **metacercaria.** Humans may be the definitive host for the adult stages (Figure 5.19a). One of the most common intermediate hosts for trematodes is the snail (Figure 5.19b), although fish and shellfish may also serve as hosts for the larval stages. Water environments such as river bottom mud and rice paddies support their life cycles and facilitate transmitting the parasite to humans. Liver and lung fluke diseases are severe, debilitating infections common in the Far East. Schistosomiasis, caused by a blood fluke, is endemic in Africa and Latin America. In the United States we see swimmer's itch, a mild schistosome disease caused by trematodes whose definitive hosts are waterbirds. When their larvae in pond and lake water mistakenly infect human swimmers, they attempt unsuccessfully to mature in our skin, but quickly die. The outcome is only a mild, brief rash.

FIGURE 5.18   The Fluke
A fluke has suckers, a simple digestive system with a mouth and no anus, and the sex organs of both male and female.

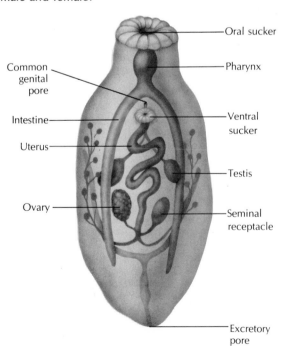

Oral sucker

Common genital pore

Pharynx

Intestine

Ventral sucker

Uterus

Testis

Ovary

Seminal receptacle

Excretory pore

FIGURE 5.19   The Life Cycle of the Fluke
Schistosoma, a trematode, has a life cycle that demonstrates the roles of hosts. (a) The human is the definitive host. The flukes grow to maturity, reproduce, and produce eggs, which are shed in the feces. (b) If feces get into water, the eggs hatch and develop as miracidia, eventually entering snails, where they develop into a tail-bearing cercaria. The snail releases these back into the water. (c) These larvae enter human skin, usually through the feet of a person wading or working in shallow water. Once in the human, they mature to sexually reproducing adulthood, causing damage to liver, lungs, or other organs, depending on the species.

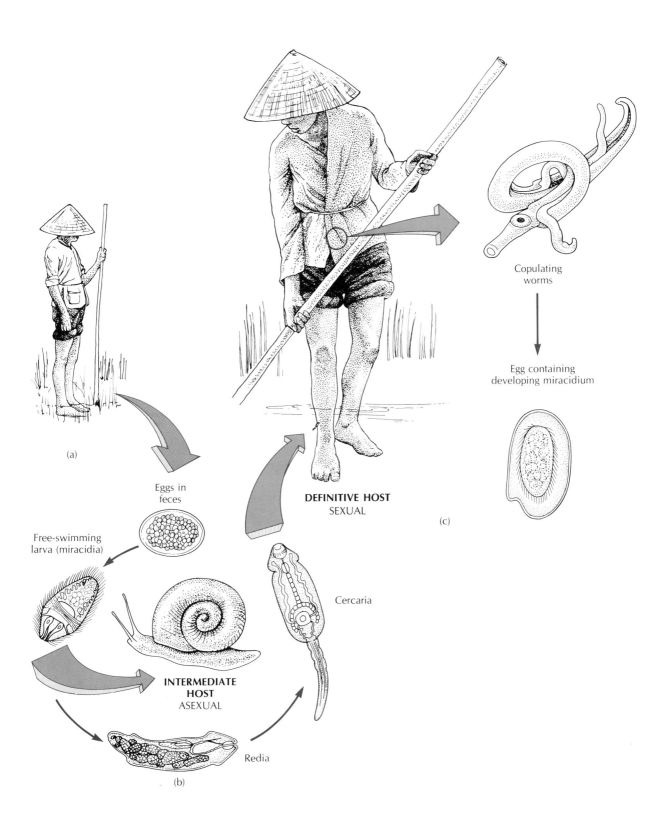

(a)

Eggs in feces

Free-swimming larva (miracidia)

INTERMEDIATE HOST
ASEXUAL

Redia

(b)

Cercaria

DEFINITIVE HOST
SEXUAL

Copulating worms

Egg containing developing miracidium

(c)

THE TAPEWORMS OR CESTODES Tapeworms have immobile adult stages adapted to life as intestinal parasites. Ingested larvae hatch in the mammalian host's intestine, then the head or **scolex** attaches its suckers firmly to the intestinal wall so that it will stay there. Segments called **proglottids** grow from the neck of the scolex. More proglottids are added as time passes, forming the body of the mature worm. As the worm elongates, the largest and oldest proglottids mature into male and female reproductive structures containing ovaries and testes. The eggs are fertilized within the segments of the adult worm and are then discharged (Figure 5.20). Segments of the worm will periodically break off. They are large enough to be visible in the stool. The scolex of a worm may very well remain in place for years, for the entire lifetime of the host, attaining lengths of many feet. See Chapter 19 for further discussion of tapeworm life-cycles. Contrary to folklore, tapeworms are not as debilitating as some other parasites, and tapeworm sufferers do not fade away to a shadow. Tapeworms can be gotten rid of by administering drugs that stun or poison the scolex so that it relaxes its hold. Humans can acquire tapeworms of different genera from many other animals, including dogs, cows, pigs, and deer. Both feces containing eggs, and meat containing larvae, may transmit disease.

THE ROUNDWORMS

The roundworms have cylindrical bodies. Some laypeople tend to confuse them with earthworms, but they are not closely related. Roundworms are somewhat more complex structurally than flatworms (Figure 5.21), having a complete digestive system with both mouth and anus. Most roundworm species are **dioecious,** which means that there are separate male and female individuals. In most cases, the two sexes look different in size and structure. The female, which contains ovaries that produce thousands of eggs, is typically much larger than the male, which contains testes. Roundworms usually have less diverse larval stages and fewer inter-

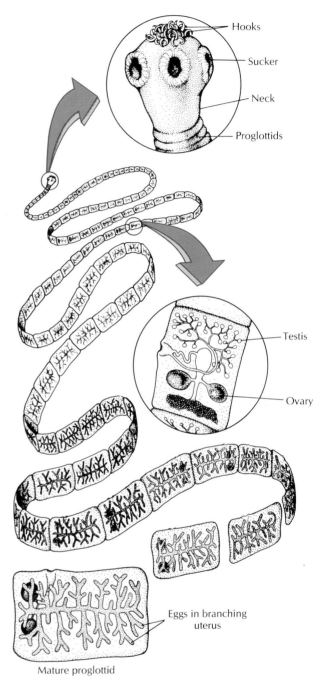

FIGURE 5.20   The Tapeworms
The adult has a head or scolex with hooks and suckers for attachment. Segments or proglottids behind the scolex enlarge and mature progressively. Each is independent and hermaphroditic. At maturity, each will release thousands of eggs into the host's feces.

mediate hosts than flatworms. Some, such as *Enterobius*, the pinworm, have only one host. Among the more serious roundworm diseases are trichinosis, transmitted by the meat of improperly raised, infected pigs, and by game such as bear and walrus meat. Hookworm disease currently affects almost one billion people; it was once extremely common in agricultural areas of the warmer parts of the United States. Although they do not kill humans, hookworms cause continuous slight blood loss, culminating in severe anemia in some cases. Hookworm is transmitted most commonly by feces-contaminated soil, especially through the skin of bare feet.

River blindness, a disease caused by *Onchocerca volvulus*, affects tens of millions of people. In parts of Africa as many as one in three adults has been blinded by this parasite. It is the largest cause of vision loss worldwide.

Parasitic infestations caused by protozoans or helminths, or multiple infections caused by combinations of these, are unquestionably the major infectious disease problems of less-developed countries around the world, claiming millions of lives every year. Tens of millions of children are developmentally disabled, and adults are blinded or too debilitated to work productively. Parasitic diseases remain a largely unconquered area in infectious-disease medicine. Research efforts are increasingly being brought to bear on diseases such as malaria, schistosomiasis, and Chagas' disease. We in the affluent nations are just beginning to look beyond the medical needs of our own comfortable society to those of the impoverished rest of the world.

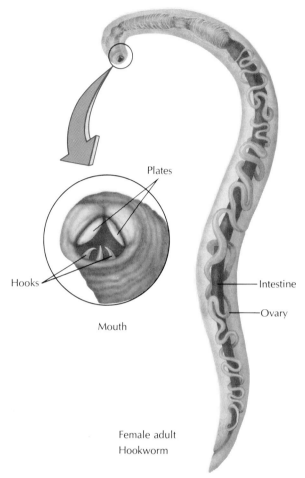

FIGURE 5.21 The Roundworms
The hookworm, a species of roundworm, has a mouth area specialized for adherence and for sucking fluids. The organism has a complete intestinal tract. The female is much larger than the male, with an extensively developed ovary.

SUMMARY

1. The algae are photosynthetic eucaryotes with a true nucleus and subcellular organelles, including chloroplasts, in which photosynthesis takes place. Algae may be capable of motility by flagella. Many are unicellular but some are multicellular, forming an undifferentiated cell mass called a thallus. Others are coenocytic, with many cells merged within one outer boundary. Algae reproduce sexually, and some cells are diploid.

2. Algae are classified on the basis of their photosynthetic pigments, anatomy, and habitat. The unicellular forms (along with the cyanobacteria) make up the phytoplankton. Others live in soil and on damp surfaces. Macroscopic forms, the seaweeds, are harvested as sources of useful polysaccharides, food, or fertilizer.

3. Fungi are eucaryotic and heterotrophic. They possess cellulose or chitin-containing cell walls and are frequently coenocytic. They are heterotrophs; their role in nature is similar to some bacteria in that they function as decomposers and occasionally as pathogens.

4. The classes of fungi are divided on the basis of the means of sexual reproduction. Cell structure may be mycelial, yeastlike, or dimorphic. Both asexual and sexual reproduction occur, and the resulting spore types are both diverse and numerous. In one group, the deuteromycetes, sexual reproduction has not been observed. This group includes all the species that are pathogenic for human beings.

5. Protozoan cells, very complex in design, do not have cell walls. Their shape may be completely changeable, as are the amoebas; maintained by a stiff gel-like exterior layer of cytoplasm, as are the mastigophorans; or enclosed in a mainly inorganic shell or skeleton, as are the foraminiferans. Sexual reproduction is commonly seen. Many form resting bodies called cysts.

6. Most protozoa are free-living components of the aquatic zooplankton, feeding on each other and the phytoplankton and in turn nourishing the larger animals. Others are parasitic and occupy the intestinal tracts and orifices of human beings and animals. Although most of these do no appreciable harm, a few cause diseases characterized equally by their seriousness, wide distribution in certain areas of the globe, and the lack of effective measures to control them.

7. The helminths are multicellular, sexually reproducing animals with complex life cycles. The parasitic groups are roundworms and flatworms. They cause such diseases as pinworm, hookworm, and schistosomiasis.

DISCUSSION QUESTIONS

1. Outline the contributions of the eucaryotic algae to the global environment.

2. Why is the classification of *Euglena* debatable?

3. Fungi, like plants, have cell walls. Why are they given the status of a separate kingdom?

4. What are the types and functions of fungal spores?

5. What effect did the development of sexual reproduction have on the evolution of early eucaryotes?

6. What is meant by alternation of generations? What is a life cycle? Find examples in this chapter.

7. How are the helminths specialized for a parasitic life and unsuited for life on the outside?

ADDITIONAL READING

Alexopoulos CJ, Mims CW. *Introductory mycology.* 3rd ed. New York: Wiley, 1979.

Bold HC, Wynne MJ. *Introduction to the algae: structure and reproduction.* Englewood Cliffs, N.J.: Prentice-Hall, 1978.

Emmons CW, Binford CH, Utz JP, Kwon-Chung KJ. *Medical mycology.* 3rd ed. Philadelphia: Lea and Febiger, 1977.

Grell KG. *Protozoology.* 2nd ed. Heidelberg: Springer-Verlag, 1973.

Noble ER, Noble GA. *Parasitology: the biology of animal parasites.* 5th ed. Philadelphia: Lea and Febiger, 1982.

Yamaguchi T. *Color atlas of clinical parasitology.* Philadelphia: Lea and Febiger, 1981.

6

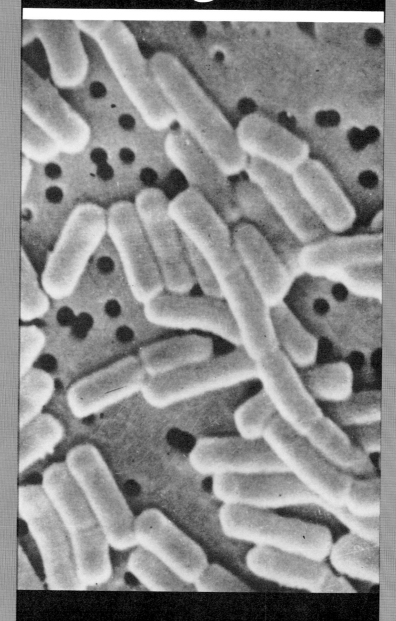

HOW TO GROW
AND STUDY
MICROORGANISMS

CHAPTER OUTLINE

TECHNIQUES FOR ISOLATING
MICROORGANISMS
One Species from a Mixture
Design and Choice of Media
Stock Cultures

PROVIDING THE PHYSICAL GROWTH
ENVIRONMENT
Temperature
Osmotic Pressure
pH of the Environment
Gaseous Atmosphere
Light Exposure

PROVIDING NUTRITIONAL REQUIREMENTS
Carbon Sources
Nitrogen Sources

Other Inorganic Nutrients
Organic Growth Factors

THE GROWTH PROCESS
Binary Fission
Growth Kinetics
The Phases of Growth
Other Growth Patterns

COUNTING MICROORGANISMS
Counting Techniques

IDENTIFYING MICROORGANISMS

AUTOMATION IN CLINICAL MICROBIOLOGY

Microbiologists have developed a group of basic methods they use to isolate, cultivate, identify, and study microorganisms. In this chapter our objective is to become familiar with these methods.

Any organism we might wish to study has its own natural habitat. These habitats are diverse — the human mouth, desert soil, or a marine fish's skin. Its habitat meets the organism's physical requirements (temperature, osmotic pressure, light) and chemical requirements such as carbon and nitrogen sources. If we do not approximately recreate these conditions in the laboratory, the organism will not grow satisfactorily. We adjust the artificial growth medium and the incubation conditions to mirror the natural habitat of the organism as closely as possible.

We use a multistep procedure to grow and study microorganisms. First we isolate a microbial species from the others that share its natural habitat. Then we grow it in culture in the laboratory, using physical and chemical conditions suited to the microorganism's needs. In this chapter we will become familiar with some features of microbial growth — the steps of binary fission and the stages observed as a culture develops through its growth curve.

Once an organism is isolated in a pure culture we may use a variety of identification tests designed to find out in which taxonomic group it fits best. With these tests we determine its genus, species, and often even its subspecies or strain designations. For the clinically important bacteria and fungi, many identification steps are now done by computerized equipment.

TECHNIQUES FOR ISOLATING MICROORGANISMS

Suppose you go out on a lake for a day's fishing, taking a full tackle box. Depending on which type of bait or lure you select, you may catch anything from a sunfish to a salmon, if you're skillful and lucky. Microbiologists may take their sampling bottles to the same lake and recover anything from cyanobacteria to anaerobic, spore-forming bacteria, depending on where they sample and what they do with the samples back in the laboratory. We have an equally wide range of choices when we process a patient sample, such as a throat swab or urine specimen (Figure 6.1) As a general rule, microbiological samples from natural sources yield **mixed cultures** containing many species. We must then isolate from these mixtures the specific types of microbe we want.

Consider this: Soils can contain 2.5 million bacteria per gram, and human intestinal contents, 100 billion. The microbiologist is usually faced with the problem of the needle in the haystack — how to pick out individual species from a mixture that contains astronomical numbers.

ONE SPECIES FROM A MIXTURE

The first step in isolation is to physically separate the organisms so that each can multiply un-

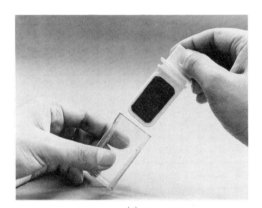

(a)

FIGURE 6.1 A Simple Microbiological Sampling Method
There are many different ways to obtain microbiological samples. (a) A medium-coated paddle is supplied sterile and can be dipped into a urine sample either on location or in the laboratory. (b) The medium picks up a fixed quantity of the urine sample. (c) As colonies develop on the surface of the medium, they can be counted, giving a semiquantitative measurement of the number of bacteria per milliliter of the patient's urine. (Courtesy of Millipore Corporation, Bedford, Massachusetts.)

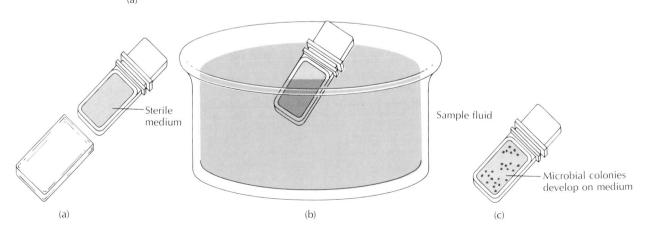

Sterile medium

Sample fluid

Microbial colonies develop on medium

(a)   (b)   (c)

hindered and be studied. Microbiologists often use a solid growth medium containing agar, a polysaccharide gelling agent derived from various red algae. Agar is added at a 1.5 to 2.0% concentration to a nutrient medium. The medium is then boiled to dissolve the agar and autoclaved to heat-sterilize the mixture. When the medium cools to about 45 degrees Celsius the liquid agar solidifies to form a firm gel. Bacteria, fungi, and some other microorganisms grow well on this agar surface when suitable nutrients are included. Aseptic conditions are maintained by enclosing the agar in a dish or test tube. Many laboratories use commercially prepared liquid broth media (without agar) and solid media (with agar).

THE STREAK-DILUTION TECHNIQUE. Most isolation techniques include a method for severely limiting the number of organisms that grow and spacing them out over the available surface area. The most commonly used isolation technique is the **streak-dilution method** or streak plate (Figure 6.2). An inoculating loop or probe made from nichrome wire is sterilized in a flame, then dipped into the material to be sampled. The sample **inoculum** is then applied to the surface of a solid medium in a Petri plate in such a way that only a very small amount is transferred. The inoculum is then streaked out in an orderly pattern. Following the pattern guarantees a progressive thinning of the inoculum so that in some areas on the plate single cells are deposited at a slight distance from each other. The streaked plate is then placed in an incubator at the right temperature and the inoculum is allowed to grow.

After a microbial cell is deposited on an agar surface, it multiplies, eventually forming a visible heap of cells called a **colony,** all derived from a single parent cell. Following incubation, isolated colonies appear in the most thinly inoculated areas. In more densely inoculated areas, the colonies run together and cannot readily be studied.

If we carefully subculture from an isolated colony to fresh, sterile medium, we will obtain a **pure culture.** We define a pure culture as a culture of one type of organism growing in the absence of all other organisms.

We can rarely accurately estimate in advance how many organisms a sample might contain. The streak-plate method allows us to prepare a plate with a range of inoculum densities. If streaking is carefully done, isolated colonies will be obtained every time. The streak-plate method is appropriate to isolate most bacteria and yeasts. A streak plate will not be useful for organisms that produce a spreading growth, such as most mycelial fungi or the extremely motile **swarmer** bacteria, which migrate rapidly over all the surface. Technologists in clinical laboratories routinely use streak plates to isolate suspected pathogens from throat swabs and other body specimens.

THE POUR PLATE METHOD. Another isolation technique is the **pour plate method** (Figure 6.3). Melted agar can be cooled to about 50 degrees Celsius without solidifying. An inoculum can be added to the warm liquid medium, thoroughly mixed in, and the mixture quickly poured into a sterile Petri dish to harden. This process "traps" single cells in the gelled agar, where they develop into isolated colonies. Alternatively, the inoculum can be evenly spread on the agar surface with a glass rod. A standard Petri dish of agar medium can support the growth of no more than about 300 isolated colonies.

DESIGN AND CHOICE OF MEDIA

When we want to isolate one type of organism from a mixture, we must choose or design an appropriate medium. Let us look at the two types of medium most frequently used in microbiology.

SELECTIVE MEDIA. These contain chemicals that inhibit most microorganisms, allowing only certain types to grow. For example, sodium tetrathionite is added to some media used to examine fecal samples. The chemical inhibits growth of

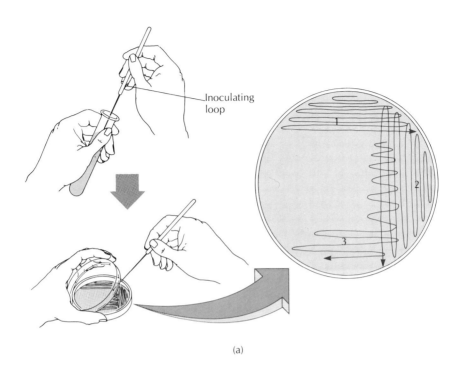

(a)

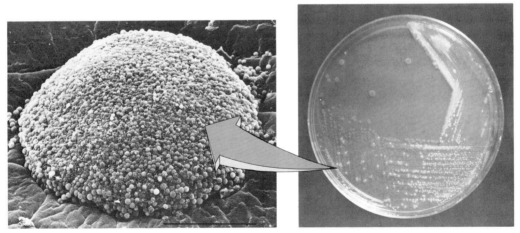

Nonmotile, nonlife *Proteus mirabilis*

(b) Bacterial colonies

FIGURE 6.2   Streaking a Plate
(a) This is a three-zone streaking pattern. First, the heavily loaded inoculating loop is streaked closely on one area of the plate. Then a new area is streaked by moving the loop into the previously inoculated area, then out over clean medium in parallel lines. Last, the loop is moved, without further pickup of inoculum, over the third area. (b) Isolated colonies will develop in the less densely inoculated areas. (Centers for Disease Control, Atlanta.) (c) A scanning electron micrograph of a colony shows that it is a heap of large numbers of bacterial cells. (×260. Courtesy of Drs. T. Eda, Y. Kanda, C. Mori and S. Kimura, from *J. Electr. Microsc.* (1978) 27(2):119–126.)

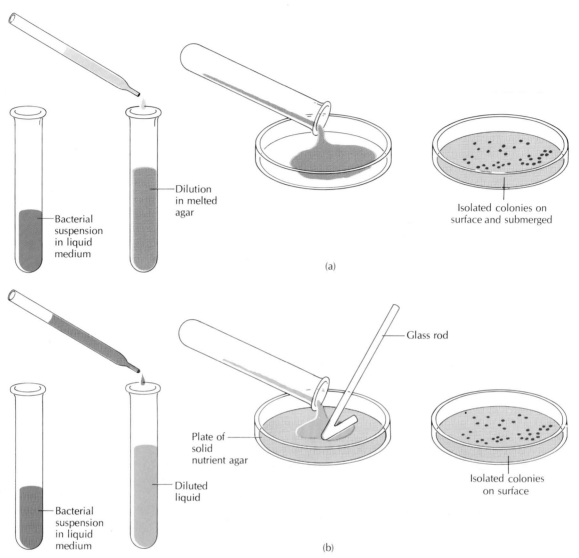

FIGURE 6.3 Pour Plates and Spread Plates.
(a) A diluted bacterial suspension may be placed in melted agar at 45 to 55 degrees Celsius, then poured into an empty, sterile Petri dish and allowed to harden. After incubation, both surface and submerged colonies will be seen. (b) Alternatively, a glass-rod spreader may be used to distribute the inoculum over a firm agar surface. Evenly spaced surface colonies will be seen after incubation.

the normal gut flora but permits the growth of two resistant Gram-negative pathogenic genera, *Salmonella* and *Shigella.* Other selective chemicals include salts, dyes, antibiotics, and specific enzyme inhibitors. A selective medium is chosen when the organism to be isolated is an important but numerically minor member of its microbial community.

DIFFERENTIAL MEDIA. These contain pH indicators, color reagents, or other indicators. These substances act to make the growth of one type of organism look clearly different from the growth of other neighboring colonies. This is because any colony carrying out a particular reaction will change the chemistry of the adjacent medium. After incubation, the reactive colony may itself be a distinctive color or it may be surrounded by a zone of media with an altered color. Nonreactive colonies will grow, but cause no color changes.

Blood agar — a rich formula containing whole blood, with intact red blood cells — is a differential medium. Blood agar is widely used in the clinical laboratory to isolate suspected human pathogens. Freshly prepared, it is bright red and opaque. Some microbial species produce **hemolysins,** enzymes that destroy red-blood-cell membranes. Such a hemolytic colony growing on blood agar will create a visible clear or greenish zone in the agar. This zone marks the area of hemolyzed red blood cells around the colony. A clear zone is designated as **beta hemolysis,** and a greenish zone as **alpha hemolysis.** Nonhemolytic microorganisms cause no changes in the color of the medium; this is sometimes called **gamma hemolysis.** Blood agar is therefore differential, because it visibly distinguishes hemolytic colonies from nonhemolytic neighbors.

STOCK CULTURES

We have seen how a microbiologist obtains pure primary cultures of desired organisms through chosen isolation techniques. Pure cultures may be maintained for future use as **stock cultures.** They can be used as a source of **subcultures** by aseptic transfer of a portion of the growth. Many laboratories maintain their own stock cultures; others purchase them from one of the large service collections. The American Type Culture Collection (ATCC) maintains and provides stock cultures of bacteria, other microorganisms, viruses, and cell-culture lines.

Active microorganisms cannot survive indefinitely in a closed area, such as a test tube. They must be transferred periodically to a fresh medium or they will die. However, subculturing always poses the risk of contamination; some cultures may also develop altered characteristics after months or years of subculturing. For these reasons, long-term preservation in an inert state is preferable for stocks. We may place cultures in a state of suspended animation by freeze-drying or **lyophilizing** them. Lyophilized cultures may be stored indefinitely at room temperature. Simple refrigeration is not a reliable way of preserving most organisms for longer than a few days.

PROVIDING THE PHYSICAL GROWTH ENVIRONMENT

Important aspects of an organism's physical growth environment include the temperature, osmotic pressure, pH, gaseous atmosphere, and light. When we grow an organism in the lab, the growth situation must be adjusted to resemble the microorganism's native habitat. We provide polar bears with air conditioning so they can live in zoos in hot climates. So how can we remove a bacterium from its habitat in a hot sulfur spring at 85 degrees Celsius and expect it to do well in the 37-degree-Celsius incubator designed for our body flora?

TEMPERATURE

In nature, we will find microorganisms surviving and multiplying at temperatures from −10 degrees Celsius to 85 degrees Celsius; a few appear to survive at much higher temperatures in deep-ocean geothermal areas. Each individual species, however, multiplies over a range that rarely exceeds 30 to 40 degrees Celsius. We can separate the microorganisms into three loose groups based on their temperature requirements.

BOX 6.1
First, Catch Your Hare. . .

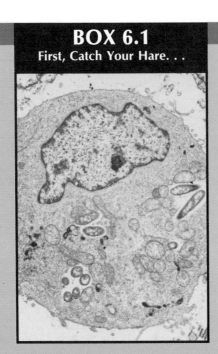

There is an old recipe for jugged hare (a stew with wine) that begins, "First, catch your hare." Many of us can see why it is important to study individual microorganisms, but it is a little hard sometimes to see why there should be so much fuss made about which medium and what formulation is used. However, primary isolation is the equivalent of catching the hare. Our success depends entirely on having the right medium.

In 1975, an outbreak of severe lung infections brought a tragic aftermath to an American Legion convention in Philadelphia. Despite intensive research efforts, months elapsed before the causative bacterium was identified, for lack of a suitable medium. The first media that worked at all were difficult to use and not selective,

so scientists found they could only isolate the *Legionella* organism from normally sterile specimens, such as infected pleural fluid in which it was the only organism. However, isolation from other samples such as sputum was still impossible because the fussy, slow-growing *Legionella* was invariably overgrown by other organisms from the sample — microbial "weeds." Three years went by before a workable semiselective medium was developed so that *Legionella* could be isolated from fluids and tissues containing other organisms. That medium is still being refined and improved. Having it confers an important

benefit — we can make an accurate laboratory diagnosis of legionellosis, and thus provide rational therapy.

We also have a need to be able to isolate *Legionella* from soil and water, its natural habitat. This poses additional, difficult problems of selection, because we must cope with the huge, diverse, and hardy group of soil and water bacteria. Several solid media are available for this purpose. We are now able to identify natural sources of this pathogen, such as air conditioners, standing water, and inadequately chlorinated climate-control systems. These sources can then be cleaned up before an outbreak occurs.

(×4800.) Photo of *Legionella* courtesy of John J. Cardamone, Jr., and P. W. Dowling, University of Pittsburgh School of Medicine/PBS.

The **psychrophiles,** or cold-loving organisms, multiply within the range of −10 degrees to 35 degrees Celsius (Figure 6.4). They include bacteria, fungi, algae, and protozoans. Their natural habitats are the colder waters and soils; most marine organisms are psychrophilic. Since they don't multiply at human body temperatures, psychrophiles aren't pathogenic for human beings. However, some are pathogenic for fish and other aquatic animals. Furthermore, psychrophiles multiply at refrigerator temperatures (4 to 10 degrees Celsius) and cause substantial economic loss from spoilage of fresh foods. We incubate them in refrigerated incubators.

Mesophiles grow best at moderate temperatures, between 20 and 50 degrees Celsius. Mesophiles inhabit warmer soils and waters, animal wastes, and animal tissues. Both our normal flora and our pathogens are mesophiles, growing best over a narrow range from 35 to 40 degrees Celsius, but less well as the temperature rises to 40 degrees Celsius, or 104 degrees Fahrenheit. When a human with an infection runs a fever, his or her elevated temperature works to inhibit the growth of the microbial pathogen, and to promote recovery. We incubate mesophile cultures in a warm incubator, usually at about 35 degrees Celsius.

Thermophiles are heat-loving, and grow between 40 and 85 degrees Celsius, with the most rapid growth usually occurring between 50 and 60 degrees Celsius. At present there are no known eucaryotes that grow at temperatures higher than about 60 degrees Celsius. Hot environments, such as volcanic areas, heaps of decomposing material, and heat-producing industrial processes, are the habitats of thermophiles. Facultative thermophiles, which bridge the mesophile and thermophile temperature ranges, can rapidly spoil cooked foods that are held at warm temperatures for several hours.

Each living organism is restricted to a specific temperature range because of the effects of temperature on its genetic material, its enzymes, and the lipids in its cell membranes.

Cold tends to slow down or stop the action of enzymes, while heat disrupts the weak bonds that maintain an enzyme's secondary and tertiary structures. An overheated enzyme becomes **denatured** and nonfunctional.

Although psychrophiles' enzymes are not at all heat-stable — they are rapidly inactivated at 30 degrees Celsius — they compensate by having very high activity rates at low temperatures. Mesophilic enzymes, like our cellular enzymes, are maximally active at around 35 degrees Celsius and are denatured as the temperature approaches 50 degrees Celsius. A thermophilic organism is adapted to a hot environment because it has the genetic capacity to produce unusually heat-stable enzyme proteins. Thermophilic enzymes are maximally active at 50 to 60 degrees Celsius; some resist denaturation at the temperature of boiling water, as shown by the survival of thermophilic bacteria in superheated

FIGURE 6.4 Temperature Ranges
Each microbial species grows well over its own temperature range, usually about 30 degrees Celsius. Bacteria are grouped in four categories: the cold-loving psychrophiles, medium-range mesophiles, warm-range facultative thermophiles, and heat-loving thermophiles.

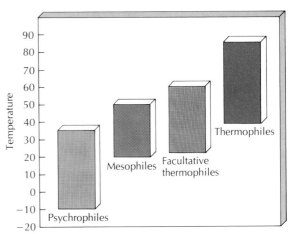

springs at 105 degrees Celsius. The lipids in thermophilic cell membranes are also specially produced to resist "melting" at these high temperatures.

OSMOTIC PRESSURE

As we saw in Chapter 3, cells may gain or lose water by osmosis depending on the relative concentrations of dissolved substances inside and outside the cell (Figure 6.5). When the exterior environment is **hypertonic,** or contains more dissolved materials than the cell's cytoplasm, most types of cell will shrink and become dehydrated. The loss of available intracellular water slows, then stops, cellular activities.

When the exterior environment is **hypotonic,** more dilute than the cytoplasm, water enters the cell. The cell wall's function, as we saw, is to provide a mechanical restraint to this expansionary pressure. If the influx of water exerts greater pressure than the cell wall's mechanical strength can sustain, the cell lyses.

Inorganic salts such as sodium and potassium chlorides, sulfates, and fluorides are the most important solutes affecting osmotic pressure in natural aquatic habitats. **Salinity** is the term used to indicate the concentration of dissolved salts in a fluid. Some organisms tolerate a wide range of salinity, while others, usually marine, tolerate only slight fluctuations. True **halophiles** (salt-lovers) live in environments containing 15 to 30% dissolved salts; they actively require sodium ions to stabilize their unique membrane structure.

The normal flora microorganisms of the human gut are not exposed to elevated salt levels in their natural environment. When feces pass into sewage, and the sewage is discharged into sea water at 3.5% salinity, the gut flora dies off rapidly. However, our normal skin flora is adapted to the salts provided by sweat, and is very salt-tolerant (Figure 6.6). The staphylococci tolerate 10% sodium chloride.

Salts such as sodium chloride and alum have been used for thousands of years to pickle meats

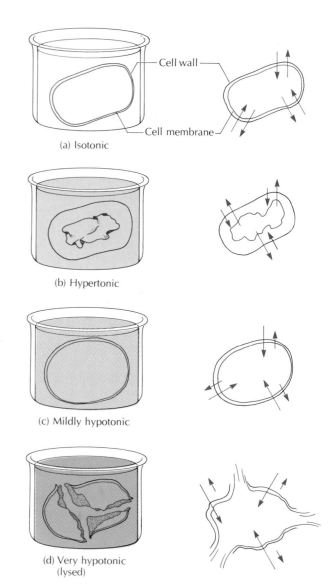

FIGURE 6.5 Effect of Osmotic Pressure on Microbial Cells
(a) When a cell is in an isotonic environment, water flows equally into and out of the cell, and its size does not change. (b) In a hypertonic environment, more water flows out than into the cell, and the cell membrane and cytoplasm shrink, pulling away from the cell wall. (c) In a mildly hypotonic environment, water flows in, exerting pressure on the cell wall. (d) In a very hypotonic environment, the influx of water is so great that the cell wall ruptures, and the cell lyses and dies.

and vegetables, because the salts reduce the available water, inhibit microbial growth, and protect these foods from decomposition. Similarly, high concentrations of sugar preserve fruits. However, these preservative efforts may fail, because osmotically preserved foods often are eventually degraded by microbial growth. **Osmotolerant** organisms, such as the halobacteria and certain yeasts and molds, are the culprits in this type of spoilage. These species are able to grow in concentrated solutions. They are osmotolerant because they can adjust their own internal solute concentration until it is isotonic with that of their environment. Salting and preserving by sugar are now usually supplemented by a heat treatment that absolutely eliminates microorganisms from the food.

pH OF THE ENVIRONMENT

The pH range of natural environments is surprisingly wide. Drainage waters from volcanic soils, mineral deposits, and abandoned mines may be highly acidic, with a pH between 1 and 3. Clean drinking-water sources average about pH 6, the oceans, about pH 8. Alkali lakes may have pH values as high as 10.

Microorganisms are adapted to grow in various portions of the entire pH range. To live successfully in an acidic or basic environment a microorganism must maintain its intracellular pH at about 7.5, independent of the external pH. Because the outside pH is different than the inside pH, the cell experiences a **pH gradient** across its membrane. As we will see in Chapter 7, some organisms are able to use pH gradients as a mechanism for transporting nutrients and a source of energy for motility and ATP synthesis.

Actively metabolizing microorganisms may modify the pH of their environment, as they release either acids or basic ammonia into their environment. In a laboratory medium, these waste products accumulate and, above a certain concentration, eventually arrest the microorganism's growth. This problem must be taken into consideration when preparing media. **Buffers**

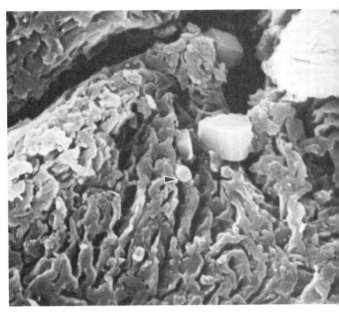

FIGURE 6.6 Salt-tolerant Bacteria
The surface of human skin, showing crystals of salts formed from evaporated sweat and staphylococci, which form part of the normal flora of the skin. These microbes readily tolerate exposure to high levels of salts. (Courtesy of Dr. Robert P. Apkarian, Yerkes Primate Center, Emory University.)

may be included to stabilize the pH of the medium and prevent the organism from inhibiting its own growth. Amino acids and proteins have a buffering effect, and inorganic salts or synthetic organic buffers are also useful.

GASEOUS ATMOSPHERE

We have seen in Chapter 4 that microorganisms in their natural settings need varying amounts of oxygen, carbon dioxide, and other gases. Some must even be protected from oxygen exposure. These gas mixtures must be recreated in the lab to successfully grow the organism.

Strict aerobes require molecular oxygen for growth, but may survive without growth in the absence of oxygen or grow at reduced rates when oxygen is limited. On solid media, oxygen availability is not a problem. In liquid culture, however, aerobes tend to exhaust their dissolved oxygen supplies, using up the dissolved oxygen faster than more can enter the liquid at its surface. Liquid cultures may be aerated by shaking or bubbling; this increases the amount of oxygen entering the fluid and helps to support optimal growth.

Facultative anaerobes have a versatile metabolism that allows them to multiply both with and without oxygen. However, their metabolism is much more efficient when oxygen is present, so aerobic growth is more rapid. We usually incubate facultative anaerobes under conditions that promote fully aerobic growth, unless we are specifically testing them for anaerobic functions such as sugar fermentation. A microbial isolate's ability to ferment individual types of sugar such as lactose, mannitol, or xylose is often determined as part of an identification procedure.

Microaerophiles prefer lowered oxygen concentrations and increased carbon dioxide. This combination is found in their natural habitat, the tissues and body fluids of animals, where there is both a small amount of respired oxygen and a lot of waste carbon dioxide. In the lab, a "candle jar" incubator is a simple device for providing this type of atmosphere. The cultures are placed in a wide-mouthed gallon pickle jar, a candle is inserted and lighted, and the lid put on. As the candle burns out, most of the oxygen will be consumed and replaced by carbon dioxide and water vapor. For large-scale work, or where a more accurately controlled gas mixture is required, labs use incubators supplied with a bottled gas mixture. We use microaerophilic conditions routinely when incubating primary throat cultures, since we are interested in promoting the growth of streptococci. Microaerophilic incubation conditions are also routinely used for gonorrhea cultures.

Strict anaerobes do not grow in the presence of molecular oxygen. Some can tolerate brief exposure to small amounts of oxygen, but others are almost instantaneously killed by minute

traces. Generally speaking, oxygen is toxic to strict anaerobes, and it must be removed and/or excluded if the organism is to grow. This is not an easy technical feat. Anaerobic conditions must be present while we collect the sample, transport it to the lab, and transfer it to culture media. The sample must *at no time* come in contact with the air. Anaerobic cultures are transferred under nitrogen gas, and incubated in airtight jars or incubators. In one method, an incubator–transfer box combination is filled with an oxygen-free atmosphere (Figure 6.7). For small labs, the inoculated cultures may be placed in a sealed jar. Just before sealing the jar, we activate a catalytic reaction that will chemically remove oxygen from the air by combining it with hydrogen gas (review Figure 2.10).

It is critically important that we use correct technique in isolating anaerobic pathogens from human wound infections. Otherwise, the patient's sample may reach the laboratory with all the anaerobes already dead. Mishandled cultures yield no useful information; worse, they may be actually misleading and contribute to an inaccurate diagnosis.

LIGHT EXPOSURE

Photosynthetic bacteria, cyanobacteria, and algae require a light source to provide the energy they use for growth. The red and blue portions of the visible spectrum are the most important energy sources for photosynthesis. However, it is quite all right to use white-light sources in the lab, since white light includes these wavelengths.

Visible light exposure has little effect on most nonphotosynthetic microorganisms. Light exposure may, however, induce pigment formation, a trait used to identify some mycobacteria. The ultraviolet component of sunlight, on the other hand, is lethal to many bacteria and fungi; ultraviolet lights have some usefulness in disinfecting air, small quantities of fluids, and certain types of instruments.

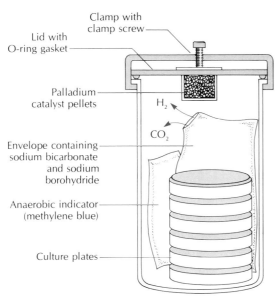

FIGURE 6.7 Anaerobic Incubation
An anaerobic jar is useful for small volumes of anaerobic cultures. Plates or tubes to be incubated anaerobically are placed in this vessel. Foil packets containing hydrogen-generating chemicals are activated, and the jar is then sealed. A catalyst in the jar's lid promotes the formation of water vapor from the oxygen in the vessel and the generated hydrogen. This reaction removes all oxygen from the air sealed in the jar. (Courtesy of BBL Microbiology Systems, Division of Becton Dickinson and Company.)

PROVIDING NUTRITIONAL REQUIREMENTS

In addition to the physical requirements for growth, each organism needs as nutrients chemical elements such as carbon, nitrogen, sulfur, phosphorous, and inorganic ions. These elements are found in nature in a wide variety of compounds, some inorganic and some organic. Each microorganism is adapted to use certain types of compounds as its sources of these nutrients — those available in its natural habitat. This nutrient environment must be recreated or improved on when we lift the organism from its habitat into the artificial environment of the laboratory. An artificial medium is a culture's supply of these elements, in its preferred available forms.

CARBON SOURCES

Cells need carbon compounds for two purposes. First, they use them as building blocks for new organic compounds. Second, heterotrophs also metabolize organic carbon compounds as energy sources. *Pseudomonas*, as we discovered in Chapter 4, has the versatility to use any one of more than a hundred simple organic compounds as its sole carbon and energy source. Although carbohydrates and lipids are preferentially used by many organisms as their carbon and energy sources, the amino acids from proteins can also be metabolized to supply energy for basic growth needs. Laboratory microbiological media may be formulated to provide sugars, starches, and partially hydrolyzed proteins such as peptone and tryptone.

NITROGEN SOURCES

Microorganisms need nitrogen atoms to build the amino groups of the amino acids that make up their cellular proteins, and to form the nitrogenous bases from which they assemble their nucleic acids. A few can get nitrogen from inorganic sources, by reducing nitrate ions to amino groups, or by directly incorporating ammonium ions into amino groups. Other organisms are less self-sufficient and require one, several, or all of the 20 amino acids. A **fastidious** organism is one with complex nutritional requirements, needing many cellular building-block molecules in order to survive in nature or in the laboratory. For example, the fastidious *Lactobacillus* inhabits the human gastrointestinal tract, where its diet is the same as ours — a nutritionally rich mixture. In the lab, hydrolyzed proteins and meat or yeast extracts may supply amino acid requirements.

OTHER INORGANIC NUTRIENTS

Phosphorus is needed in the formation of the nucleotides that make up the nucleic acids. It is commonly encountered in nature and supplied in the lab as inorganic phosphate ion. Preformed nucleotides from meat or yeast extracts provide an additional source. Sulfur is needed to make two of the amino acids, cysteine and methionine. Some microorganisms are able to reduce sulfate ion to sulfhydryl groups to build into the amino acids, while others need cysteine or methionine supplied.

Sodium, potassium, and chloride ions must be supplied to meet an organism's osmotic needs. Trace amounts of metallic ions, such as magnesium, manganese, copper, calcium, and zinc are needed for enzyme function. They may be added to lab media in specific amounts, but usually sufficient metals creep in as impurities in other media constituents. Iron needs are tiny, but deficiencies severely limit microbial growth.

ORGANIC GROWTH FACTORS

Vitamins are organic growth factors. Many microorganisms synthesize vitamins themselves, enough so that we employ them in commercial vitamin production for human diet supplements; these microorganisms don't need an outside supply. Other species, however, need one or more of these growth factors. We commonly use a crude mixed vitamin source called yeast extract, prepared by extracting the water-soluble components of yeast cells, to enrich culture media.

If we wish to cultivate the normal and pathogenic human flora, we usually obtain best results if we add tissue or tissue fractions to the medium. We routinely add whole blood, blood serum, or specific blood factors to the media we use for cultivating human bacteria. For example, media for *Hemophilus influenzae* contain not only hemolyzed blood, but also two specific growth factors, hemin and the coenzyme nicotinamide adenine dinucleotide (NAD). Chopped meat, beef brain/heart extract, and other tissue supplements are occasionally used.

THE GROWTH PROCESS

Let us now look at how microorganisms grow, first individually, by binary fission, then as populations, in characteristic phases.

BINARY FISSION

Unicellular procaryotes, as we saw earlier, usually reproduce asexually by **binary fission,** a process in which one cell divides into two (Figure 6.8). In describing this process, we will look at what happens to three cell components — the cell contents, the genetic material, and the cell wall.

Prior to cell division, the cell contents will roughly double in quantity. Water, ions, small organic molecules, and macromolecules such as enzymes, RNA molecules, and ribosomal subunits are accumulated or synthesized.

The bacterial chromosome must be copied or replicated before cell division occurs; this happens roughly once in each division cycle. Each copy of the chromosome is attached to the cell membrane at an attachment site. As the cell

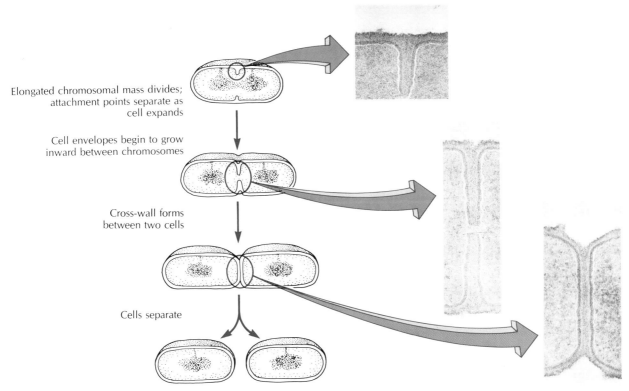

Elongated chromosomal mass divides; attachment points separate as cell expands

Cell envelopes begin to grow inward between chromosomes

Cross-wall forms between two cells

Cells separate

FIGURE 6.8 Binary Fission
Cell division occurs in the zone between two chromosomes or nuclear bodies. New cell-wall material grows inward to form cross-walls between the daughter cells. We can observe the steps involved in the formation of cross-walls through a sequence of transmission electron micrographs. The wall grows inward from both sides and finally closes. As the cell begins to tear apart, small tabs of wall material remain to mark the point where ingrowth originated. (Photo courtesy of I.D.J. Burdett, *J. Bacteriol.* (1980) 137:1395.)

expands, the membrane extends and the two chromosomal copies become separated. Cell division occurs only in the region between two chromosomal attachment sites.

Cell walls must expand if the cell is to grow. The bonds holding the peptidoglycan molecules together are repeatedly broken by enzyme action, making spots where new subunits are added, extending the wall structure. Cocci add new peptidoglycan material only in a central zone (Figure 6.9a). Rod-shaped bacteria add new peptidoglycan as a layer on their interior cell-wall surface, thus elongating the cylinder (Figure 6.9b). Both rods and cocci divide into two daughter cells by forming a cross-wall. The new wall material separates into two layers, which then pull apart. Daughter cells may completely detach, but in some species they remain aggregated in pairs, tetrads, clusters, or chains. These groupings are characteristic identification traits.

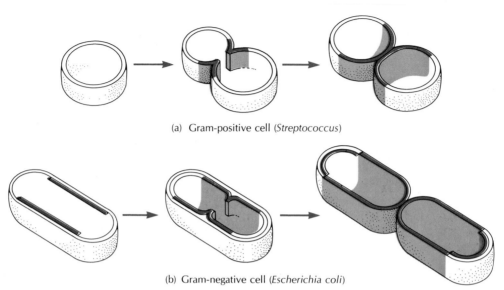

(a) Gram-positive cell (*Streptococcus*)

(b) Gram-negative cell (*Escherichia coli*)

FIGURE 6.9   Areas of Cell-Wall Growth in Bacteria
(a) In a spherical coccus, new cell wall is added at the central zone, where the cross-wall forms during cell division. Each daughter cell keeps the coccus shape.
(b) In a rod-shaped bacillus, new cell wall is added along the cylinder and also forms the cross-wall. The ends of the parent cell are retained almost unchanged in the two daughter cells.

GROWTH KINETICS

All the events in one cell's division cycle are coordinated, but each individual cell in a culture is likely to be dividing on its own timetable. When you look at a slide of an actively growing culture, rod-shaped cells will often be different lengths, since some have just divided, while others are about to divide. Despite this apparent variation, each cell takes just about the same length of time to complete its division cycle. We call this period the **generation time** or **doubling time.** An average generation time of a heterotrophic mesophile growing under optimal conditions is 20 minutes. This means that not only will one cell become two in 20 minutes, but also that the total number of cells in such a population will double in 20 minutes. Although a species' fastest doubling time is probably genetically fixed, actual doubling time frequently changes. This is because growth rates are affected by environmental changes such as temperature or nutrient shifts. Herein lies the explanation why foods spoil more slowly under refrigeration — microbial reproduction is slowed.

In nature, organisms rarely achieve their maximum growth rate, because conditions are rarely perfect. Many microorganisms, especially those from soil and water habitats, have great adaptability. Under unfavorable conditions, their growth slows, and they have greatly extended generation times of days or weeks. The extreme example of this is the complete dormancy of the spore or cyst. Higher organisms such as ourselves have lost this extremely useful ability. When our food runs out, we starve.

EXPONENTIAL GROWTH.   Binary fission has an inevitable numerical result. Every round of division doubles the number of cells in the population. Mathematically, this creates a **geometric progression.** This is defined as a number series in

BOX 6.2
Rapid Growth and a Firm Grip Help a Pathogen Succeed

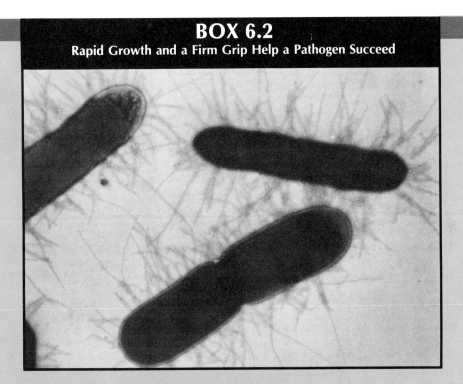

Cystitis (an infection of the bladder) is very common in adult females. *Escherichia coli* is the most common infecting agent. This bacterium, like other cystitis agents, colonizes the normally sterile urinary tract from outside. Why is *E. coli* so well adapted as a bladder pathogen? To find out, investigators built an artificial model of the human bladder. It was supplied with sterile urine, flowing through at the physiologically correct rate. Pure cultures of various bacteria previously isolated from urinary-tract infections were introduced, and their generation time in urine was measured.

The superior adaptation of *E. coli* became clear. In urine it showed generation times of 21 to 22 minutes, compared with other less-common urinary-tract pathogens (*Serratia marcescens*, 34 to 35 minutes; and *Pseudomonas aeruginosa*, 50 minutes). When paired cultures containing *E. coli* and another organism were introduced into the artificial bladder model, *E. coli* outgrew 16 different partners by 50- to 100-fold in 24 hours.

In addition to its speed, *E. coli* can attach firmly to the bladder lining with its **fimbriae.** It is thus perfectly adapted to maintain itself in the bladder despite the tendency to be washed out by the flow of urine.

(×16,425.) Photo courtesy of S. Abraham and E. H. Beachey, V.A. Medical Center, Memphis, Tenessee/BPS.

which each number is derived from the previous one by multiplying it by a certain factor, an exponent. In a **binary** series the exponent is 2. Thus bacterial population growth follows the binary series: $1 \times 2 = 2$; $2 \times 2 = 4$; $4 \times 2 = 8$; $8 \times 2 = 16$; $16 \times 2 = 32$ .... This is called **exponential growth.** Note that each round of division expands the total by the same amount as did all previous increases. Although the number always doubles, recall that we just saw that the time it takes the organism to double might vary. As you can see, if the doubling time is short, bacterial populations can expand at an explosive rate. When exponential microbial growth is occurring in an infection, the person rapidly develops severe symptoms, and treatment must be begun without delay.

Very large numbers are expressed in scientific notation as powers of ten. It is more convenient to say that a certain culture contains 2.5×10^8 cells per milliliter than to write out the number 250,000,000. When analyzing microbial growth graphically, we convert the actual cell numbers to their base-10 logarithms, then plot these values on semilog graph paper. An example of such a plot is shown in Figure 6.10. When the data is handled logarithmically, exponential growth will appear on the graph as a straight line. The actual doubling time of the cells in the culture can be calculated from the slope of the line.

THE PHASES OF GROWTH

In the laboratory, we usually grow a microbial culture in a closed container — a test tube or flask filled with fresh, liquid medium. At least at the start, the inoculum experiences no limitations, and, after a period of adjustment, population growth becomes rapid. As time passes, however, crowding, nutrient exhaustion, and the buildup of waste products take their toll. The **growth curve** of a culture shows its observed growth over time. In a liquid culture in a test tube or flask, the growth curve has four distinct phases.

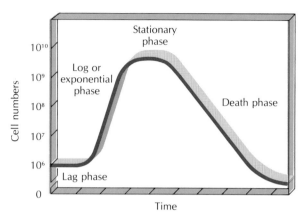

FIGURE 6.10   Typical Growth Curve
When an inoculum is introduced to a container of liquid medium at zero time, the phases of growth succeed each other as shown here. The duration of the phases and the slopes of the exponential growth and death phase lines will vary with the organisms and the conditions. No "average" times can be suggested. However, log phase is usually measured in hours, while stationary and death phases may last several days. Cell numbers are plotted as logarithms to the base 10.

When first introduced into the fresh medium, the inoculated cells adjust to the new environment for a time. Only minimal growth and essentially no cell division occurs during this **lag phase.** The lag phase will be long if the inoculum is added to refrigerated medium, if the inoculum is in poor physiological condition, or if it is being transferred from a different medium. Lag is shortened by using healthy cells and prewarmed medium.

Once the cells have adapted, they start to grow and divide. If base-10 log cell numbers are plotted on a semilog graph, a rising straight line that indicates exponential growth is occurring. This is the **log phase** or **exponential-growth phase.** The generation time is calculated by analyzing the slope of the straight portion of the line. The steeper the angle of the line, the shorter the generation time.

As cell numbers increase, nutrients are consumed and waste material accumulates. This inevitably ends the exponential growth phase. The rising straight line tails off into a horizontal line on the graph. **Stationary phase** is the term for this period in the culture's history. In stationary phase, total cell numbers do not change. Some cells die, releasing nutrients. Others continue to divide by reusing these nutrients. Waste products keep increasing. Most cells simply survive; microscopic observations show that structurally the survivors become progressively more abnormal.

Death phase succeeds stationary phase when cells can no longer find energy sources to maintain life. Aerobes respond to declining energy supplies with **autolysis,** enzymatic self-destruction. Anaerobes accumulate toxic end products that eventually kill them. Death is also exponential. A very few living cells may persist into the tail of the curve, long after the majority have failed.

OTHER GROWTH PATTERNS

A developing bacterial colony contains zones representing each phase of growth. As the colony develops, the initial cell divides and newer cells heap up and push outward; the colony expands radially. Cells left behind in the center quickly suffer nutrient deprivation and waste buildup. The center of a 24-hour-old colony contains cells in a stationary or death phase while its periphery contains young, log-phase cells. Colony centers often collapse due to autolysis of cells in death phase.

While bacterial and yeast colonies grow to a certain size and then stop, mycelial fungi seem to expand their growth indefinitely. The hyphae lengthen only at the tips, but since the cytoplasm in the mycelium is continuous, nutrients that are absorbed at the periphery of the colony will diffuse into the central region. The center of the colony remains well nourished by the nutrients gathered through the extensive mycelium. The colony center is the location where asexual or sexual spore formation takes place.

Sometimes colonies show concentric rings of growth (Figure 6.11). For example, some types

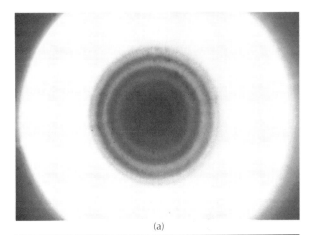

(a)

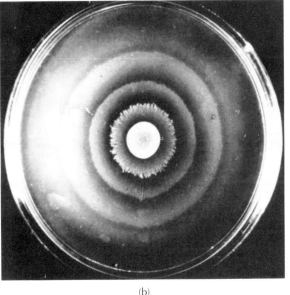

(b)

FIGURE 6.11 Colony Zonation
Many organisms in colonial growth exhibit periodic oscillations in metabolic function. As the colony grows outward, these oscillations may become visible as rings. (a) *Pseudomonas nigrifaciens*, a marine bacterium, produces a blue-black pigment. As colonies develop, zones of pigment production alternate with colorless growth. The outermost ring — the newest growth — is colorless. (From Cynthia Norton.) (b) This swarming colony of a *Proteus* species exhibits the thick, "fuzzy" bands characteristic of periods of active migration. (From H.E. Jones and R.W.A. Park, The Influence of medium composition on the growth and swarming of *Proteus, J. Gen. Microbiol.* (1967) 47:369.)

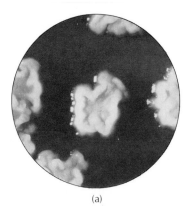
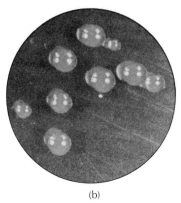
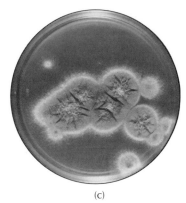

(a)   (b)   (c)

FIGURE 6.12 Colonial Morphology
The morphologies of different microbial species may show highly individual
characteristics that aid in their identification. For example, the rough edges and
flattened folds of one bacterial species (a) contrast with the smooth, raised, circular
appearance of another (b). The fuzzy appearance (c), created by aerial mycelia,
distinguish colonies of the fungi *Aspergillus fumigatus,* grown on agar. (Photo
credits: (a) Drs. R.J. Hawley and T. Imaeda; (b) Curtis B. Thorne, University of
Massachusetts at Amherst; (c) Centers for Disease Control, Atlanta.)

of bacteria have swarming motility, allowing
them to migrate into fresh areas. Such microorganisms form a very diffuse, barely visible film
of growth that spreads over the entire surface of
the medium, far distant from the point of inoculation. The swarming periods may alternate
with periods of less active migration, giving the
colony a banded appearance. **Colonial morphology** is the visible appearance of the colonies
of a microbial species. It may be highly characteristic and a valuable aid in making an identification (Figure 6.12).

COUNTING MICROORGANISMS

We often need to determine the number of cells
and the rate of growth of a culture under differing
environmental conditions. We will attempt not
to duplicate your lab book in giving detailed pro-

cedures. Rather, we will look at how the lab
techniques are based on key concepts about the
biology of microorganisms.

In quantitation, there are two major requirements — accuracy and convenience. Some techniques determine the number of cells; such
counting techniques usually cannot distinguish
living (**viable**) from dead (**nonviable**) microorganisms. Other techniques assess microbial
activity — some aspect of metabolism, or the
ability to reproduce. These techniques miss the
organisms that may be alive, but are inactive
under test conditions that are unfavorable to
them. Some methods lend themselves extremely
well to automated analysis.

COUNTING TECHNIQUES

Some techniques involve counting individual
cells or colonies. The **direct count** involves plac-

ing a microorganism-containing fluid such as milk into the shallow sample well of a special microscope slide. A calibrated cover slide engraved with guidelines is used. You can then count all cells seen in a small area and, by a simple calculation, figure out how many organisms are present per milliliter of the sample. A direct count enumerates both living and dead organisms.

The **plate count** makes use of the pour-plate techniques discussed earlier. A pour plate gives fairly evenly spaced colonies, which may conveniently be counted. In a plate count, we can find out how many microorganisms are able to reproduce. Plate counts have been widely used in water analysis and urinalysis. The count tells us the number of **colony-forming units** in a given volume of sample. This type of data may indicate the degree of contamination of a water sample, or the acuteness of a urinary-tract infection in a patient.

The samples used in plate counts are customarily prepared by **serial dilution,** a basic laboratory technique with many applications (Figure 6.13). To carry out a serial dilution, first 1 ml of the sample is added to 9 ml of sterile diluting fluid — a "blank" — and mixed to form a 1:10 dilution. This type of dilution is called a 10-fold dilution.

A second dilution in the series can next be prepared by taking 1 ml of the 1:10 dilution and adding it to another 9-ml blank, giving a 1:100 dilution. If desired, a third dilution to 1:1000, a fourth dilution to 1:10,000 or any number of additional dilution steps can be added to the series. Measured quantities of each dilution in the series can then be plated on sterile media either by pour plating or by evenly spreading the inoculum over the surface. After incubation, the resulting colonies are counted and reported as colony-forming units (CFUs). One source of error is that a CFU may originate from either a single cell or a clump of microorganisms.

A quicker, alternative approach to colony counting is the **membrane-filter technique.** In this method the sample to be counted is passed through a membrane filter under vacuum (Figure 6.14). The pores of the filter are so small that microorganisms cannot pass through, and they are retained on the filter surface. The filter is next placed on an appropriate medium, which provides nutrients so that each trapped microorganism develops into a colony.

PHOTOMETRIC TECHNIQUES.   Photometry is the process of measuring the intensity of a beam of light. You may have observed already that a liquid culture or a suspension of microorganisms appears cloudy. When a beam of light passes through such a liquid, some of the light is absorbed or scattered. The more organisms present, the cloudier or more **turbid** is the suspension; the more turbid the suspension, the less light gets through. To estimate bacterial numbers by measuring turbidity (called turbidimetry) we use a light source that emits a constant, stable beam of light. We place a suspension of cells to be measured in an optically clear test tube called a **cuvette,** then place the cuvette in the light path. Any light rays that hit the cells are deflected or absorbed, and so do not reach a photoelectric detector on the other end of the light path. The more cells there are, the less light gets through and is detected. (Figure 6.15). Various electronic instruments called colorimeters, spectrophotometers, or nephelometers are employed to take these readings. Both living and dead bacteria deflect light. Growing cultures show steady increases in light absorption that may give a direct measurement of the culture's activity as related to the increase in numbers of cells.

Turbidimetry is now the most common method for detecting and following growth of unicellular microorganisms such as bacteria and yeasts. There are also an increasing number of clinical techniques, based on **fluorometry,** the measurement of the intensity of fluorescence emitted by a sample. Some of these will be discussed in Chapter 12.

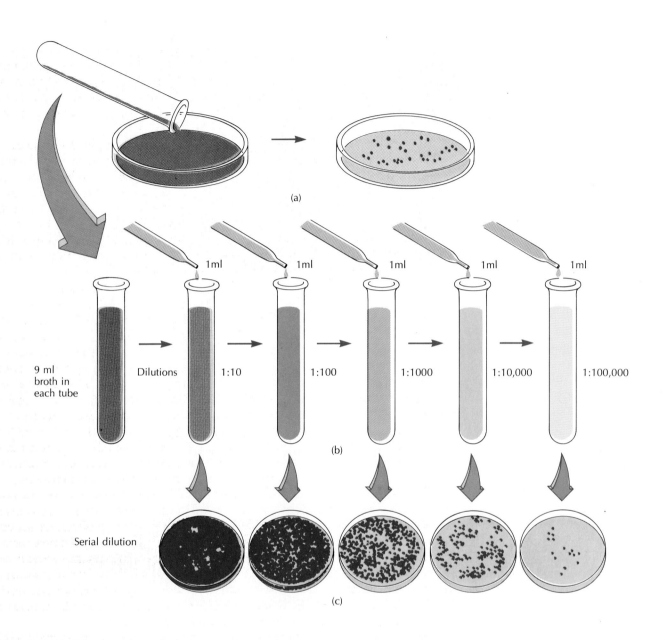

FIGURE 6.13   Plate Count with Serial Dilution
(a) The plate-count technique utilizes a liquid inoculum dispersed in melted agar, which is then poured into a sterile Petri dish and incubated. The number of colonies that develop will depend on the density of microorganisms in the inoculum. (b) To prepare a serial dilution, 1 ml of sample is added to 9 ml of sterile broth or saline to form a 1:10 dilution. This step is repeated as many times as necessary to form further dilutions. (c) One milliliter of each dilution is plated. In the most dilute plates, we can see and count isolated colonies.

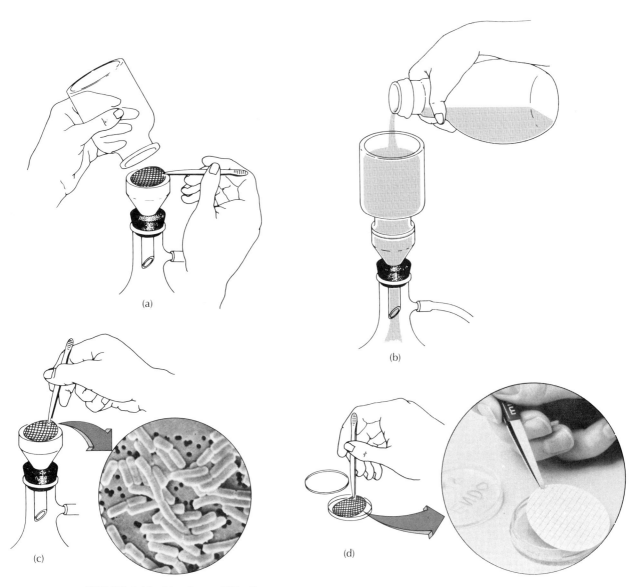

FIGURE 6.14 Membrane Filtration
This technique may be used either for sterilizing fluids or for collecting and counting microorganisms from liquid samples. (a) The presterilized filter is placed onto a porous glass support. The solution holder goes above the membrane filter, which is held in place by a metal clamp. (b) Fluid is poured into the holder. The vacuum causes the fluid to flow through the filter into the lower vessel, where it is aseptically collected. (c) The filter is placed on top of a filter-paper pad moistened with an appropriate medium; the scanning electron micrograph shows bacteria trapped on a filter. (d) The filter pad is placed onto a plate of nutritious medium, and the plate is incubated. (Photos courtesy of Millipore Corporation, Bedford, Massachusetts.)

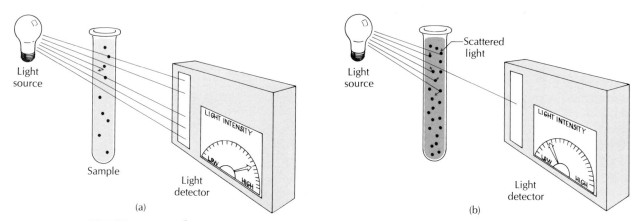

FIGURE 6.15   A Photometer
In a photometer, a light source emits light that passes through a slit and strikes a photoelectric detector. The detector records the intensity of the light that reaches it and displays the value on a dial. When a cuvette containing a bacterial culture is placed in the light path, some of the light is absorbed or scattered and does not reach the detector. As shown in (a) and (b), the light absorbed is generally proportional to the number of cells per milliliter.

OTHER TECHNIQUES.   In some cases it is difficult to count microorganisms by any of the previous methods. For example, the numbers and growth of filamentous organisms cannot accurately be assessed by plate counts or by turbidimetry. The **dry-weight** technique or **total protein analysis** of such a microbial sample can be used to estimate the cell number.

Other techniques directly measure some aspect of cellular metabolic activity. The **cellular ATP** (adenosine triphosphate) **assay** provides an extremely precise measurement of metabolic rate. This very sensitive method makes use of extracts from the luminescent organ of the firefly. These extracts emit light of an intensity exactly proportional to the quantity of ATP provided. A photometer records the intensity of the light emitted. **Gas-exchange** techniques quantitate cellular respiration; they may measure either the amount of oxygen taken up or the amount of radioactive carbon dioxide ($C^{14}O_2$) given off by a culture. The radioactive CO_2 method is used in one popular automated instru-

ment to detect the presence of active microbial growth in clinical blood cultures.

IDENTIFYING MICROORGANISMS

In Chapter 4 we discussed some of the major traits by which bacteria are classified. These are also the characteristics by which we identify them. Identifying unknown cultures is a major activity in many microbiology laboratories, especially those that deal with clinical samples such as throat swabs, urine specimens, and blood cultures. We will be encountering many examples of identification techniques as we discuss individual pathogenic species of microorganisms in later chapters.

Most useful identification schemes work on a process-of-elimination approach (Figure 6.16). With each step in the process, the number of possibilities shrinks until only one possibility is left. Both microscopic and biochemical tests are needed. If the organism has a coccus shape, that rules out bacilli and spirilla. A negative Gram stain rules out Gram-positive organisms. Biochemical tests are usually required to separate members of closely related species. When a group of biochemical tests is usually done simultaneously on an isolate, they are called a **test battery.** Many of these are commercially packaged to facilitate simultaneous inoculation of multiple tests. The biochemical changes that occur are made visible by incorporating color indicators. This permits quick test reading by inspection.

Once test results from an unknown organism are available, we can compare them with standard results for large numbers of other species to determine the closest fit. We can find comparison data in reference books, published charts, and keys. Alternatively, we may feed our data into a computer, which will compare it with species profiles stored in its data bank, and print out an identification based on the best match.

We may report our identifications to the genus, species, or strain, depending on the extent of our testing and our degree of confidence in the data. Since the bacteria "don't read the charts" and undergo constant genetic change, it is not at all uncommon to find yourself working with a strain that falls between the cracks in the neat identification schemes.

In the clinical laboratory, we are usually expected to provide physicians not only with a microbial identification, but also with practical information about the isolate's responses to antibiotics. This data aids the physician in selecting appropriate therapy for the patient's infection. **Antibiotic susceptibility testing** is discussed in Chapter 24.

AUTOMATION IN CLINICAL MICROBIOLOGY

Most clinical laboratories now use automated and computer-assisted systems to carry out much of their routine bacterial identification and to determine the antibiotic susceptibilities

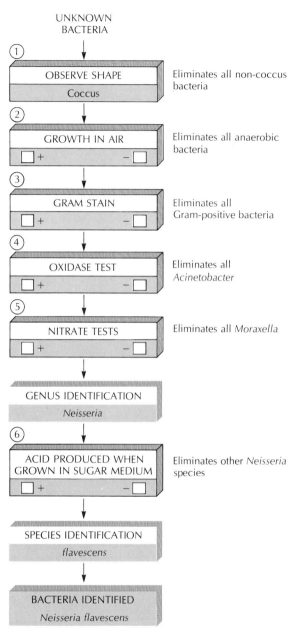

FIGURE 6.16 An Identification Flowchart
A sequence of tests, each allowing the progressive elimination of groups of bacteria, can identify an isolate from a throat culture. In this example, additional tests would be useful to confirm the identity of *Neisseria flavescens*, a harmless member of the normal flora, and to rule out its two pathogenic relatives, *N. gonorrhoeae* and *N. meningitidis*.

of the isolates. One example of a freestanding microcomputer-based system is shown in Figure 6-17.

It is difficult to generalize about this rapidly evolving technology, but we can identify some of its features. A keyboard is used for recording patient identification and sample information, for entering instructions, and for requesting reports. A screen or cathode-ray tube (CRT) is used for checking input and reading results. Samples may be placed in a mechanical sample carrier that advances them past a photometer at timed intervals. This allows the turbidity, color, or fluorescence values of each sample to be read. The data are analyzed and stored in computer memory, and reports may be printed out as needed. Additional features that may be automated include inoculum preparation, the actual inoculation step, and addition of indicator reagents. In large hospitals the computer in the microbiology laboratory is often linked up with units in other areas of the clinical laboratory,

FIGURE 6.17   An Automated System
A representative Sceptor system: At bottom right are tubes of broth with Sceptor panels and report forms. To the left is the automated preparation station. Inoculated broth medium is placed in the preparation station, which accurately rehydrates and inoculates each well of the Sceptor panel with 0.1 ml of inoculated Sceptor broth. Following a given incubation period, the panel is viewed in the Sceptor reader/recorder (bottom center) to detect bacterial growth for identification. Identification and antibiotic susceptibility (MIC/ID) data are analyzed and stored in the data management center (top center and top left). The data may be either displayed on the CRT or turned into hard copy on the printer (top right). (Courtesy of BBL Microbiology Systems, Division of Becton Dickinson and Company.)

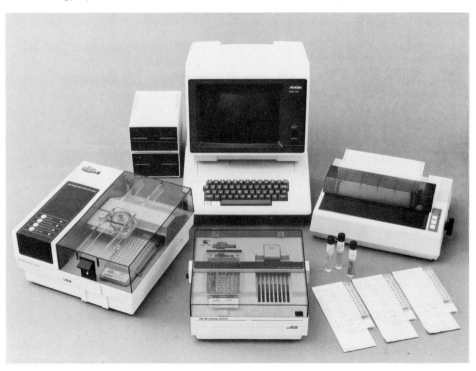

with terminals on other hospital services, and sometimes networked possibly with computers in other hospitals and health agencies locally or nationally.

Automated systems are used largely for routine clinical laboratory analysis. One application is processing large numbers of urine- and blood-culture samples. In this role, the system is programmed to initially detect and report **growth or no growth**; that is, whether the sample is positive or not. Other protocols may be used next to carry out an identification of the microorganism(s) present in the positive samples.

Another common use is the **identification of isolates** transferred as pure cultures from colonies on primary-isolation streak plates, such as from stool, wound, or throat cultures. The pure culture is inoculated into a battery of biochemical tests, these are automatically read, and then the computer compares the test data with the standard microbial-test profiles in its data bank. These profiles are based on previous test records of thousands of known microbial strains. The biochemical behavior of the unknown isolate is compared with that of all agents likely to be recovered from that site (wound, feces, blood).

Automated systems can carry out highly accurate and sophisticated antibiotic susceptibility testing. The minimum inhibitory concentration (MIC) test, discussed in Chapter 24, is based on a serial-dilution technique. It is difficult, tedious, and time-consuming to do manually, but becomes rapid and accurate done mechanically.

Computer data storage and analysis capabilities make possible up-to-the-minute analysis of patterns of infections and infective agents seen within a hospital. This analysis can be very helpful in the continuous campaign, carried out in all hospitals, against hospital-acquired infections.

SUMMARY

1. To acquire accurate, detailed knowledge about microorganisms, it is necessary to isolate them from their natural habitat. Once isolated, they are called pure cultures. They can then be maintained, subcultured, and used for systematic investigation.

2. Successful isolation often first requires a physical process of dilution because most microbial habitats contain vast numbers of individual microorganisms. Pour plates and streak plates with solidified agar are used for this purpose.

3. A selective agent added to a medium may inhibit the growth of a dominant group of flora; a minor component will prosper and become sufficiently numerous for isolation. Many primary isolation media contain dyes or pH indicators to show colonies with certain desired biochemical traits. These are called differential media.

4. Once pure cultures are obtained, they may be subcultured to other media. To maintain a stock culture without loss of its natural characteristics it may be stored under mineral oil, or in lyophilized form. The American Type Culture Collection (ATCC) maintains and supplies stock cultures of microorganisms.

5. The physical growth conditions supplied for a pure culture should approximate the natural habitat. Temperature is an important variable. Each species has a defined growth range, below which metabolism is too slow to produce growth, and above which enzyme losses due to heat denaturation cause death.

6. Cellular organisms maintain a constant internal osmotic pressure. When placed in hypertonic or hypotonic media, they vary in their ability to survive. Loss of water to the exterior reduces rates of cellular metabolism; gain of water results in swelling or bursting.

7. The hydrogen-ion concentration of the environment — its pH — may be as low as 2 and as high as 10. Yet most bacteria maintain an interior pH of about 7.5. Many organisms terminate their own growth by adversely altering the pH of the medium with acid or alkaline end products. Buffering agents may be added to compensate for this tendency.

8. Microorganisms vary widely in the concentration of oxygen and carbon dioxide needed and tolerated. Incubator atmospheres can be adjusted to provide the optimal gas mixture to favor a given group.

9. The nutritional environment must supply all key elements in a useful form. Most microorganisms can use many small organic compounds interchangeably as carbon sources. However, requirements for one or many amino acids are common. These are usually supplied in lab media by hydrolyzed protein, which may also serve as the carbon/energy source. Phosphorus is usually supplied in the form of phosphate ion and sulfur as sulfate ion; however, sulfur-containing amino acids may be required for some strains. Sodium, potassium, and chloride ions are added to fulfill osmotic and transport requirements. Traces of metal ions such as iron, copper, or zinc are also necessary. Vitamins or growth factors are often added in the form of yeast or meat extracts. Nonspecific enrichments such as blood often promote faster or more ample growth.

10. Cell division results from coordinated increases in cell contents, duplication of the chromosome, and formation of membrane and wall partitions. The maximum growth rate is genetically fixed, but it is slowed by suboptimal growth conditions. The observed time needed for a complete round of cell division is the generation time.

11. In the growth curve seen in liquid culture, four sequential phases are observed. The lag phase, just after inoculation, is a period of internal readjustment. Once prepared, cells begin rapid exponential growth that continues through the log phase. As the environment becomes poor in nutrients and rich in waste products, net growth stops and the population moves into stationary phase. Further deterioration of the medium triggers the onset of an exponential death phase.

12. Microorganisms can be quantitated by several means. Actual counting (microscopic or by colonies formed), chemical analysis, and photometric techniques are used.

13. Microorganisms are identified by observing microscopic and macroscopic growth and by carrying out batteries of biochemical tests, then comparing the results with published references.

14. Computer-assisted automated systems are widely used in clinical microbiology to identify organisms, determine antibiotic susceptibilities, and keep records.

DISCUSSION QUESTIONS

1. Design an isolation method for the recovery of thermophilic, anaerobic, nitrogen-fixing bacteria.

2. A microorganism is causing serious mortality in schools of North Atlantic salmon. You wish to isolate and study it. How would you proceed?

3. Explain the biological basis of the antimicrobial preservative effects of salt (brining), acid (pickling), sugar (preserving), refrigeration, and drying as food-storage methods.

4. Strict anaerobes require complete absence of molecular oxygen. Look up methods for

creating laboratory environments that are oxygen-free.

5. Using a standard reference manual, look up the formulations of blood agar, eosin methylene blue agar, Sabouraud's dextrose agar, and Simmons citrate agar. What organisms are supposed to grow on each? Determine the contribution each component makes to the isolation, differentation, and maintenance of organisms.

6. A mesophilic organism is inoculated into a tube of nutrient broth. Describe what is probably happening to an individual cell during each growth phase.

7. The kinetics of the growth curve make periodic subculture of stock cultures necessary. Why? What storage methods avoid this requirement?

8. Find out if there is an automated system available that you may observe. Have someone explain to you how it works and what tasks the system carries out.

ADDITIONAL READING

Gerhardt P, ed. *Manual of methods for general bacteriology.* Washington, D.C.: American Society for Microbiology, 1981.

Holt J, ed. *Bergey's manual of systematic bacteriology,* Vol. 1. Baltimore: Williams and Wilkins, 1984.

Lennette EH, Balows A, Hausler WJ, Jr, Truant JP. *Manual of clinical microbiology,* 4th ed. Washington, D.C.: American Society for Microbiology, 1985.

Mendelson NH. Bacterial growth and division: genes, structures, forces and clocks. *Microbiol Rev* 1982; 46: 341–375.

Tilton R, ed. *Rapid methods and automation in microbiology: Proceedings of the third international symposium.* Washington, D.C.: American Society for Microbiology, 1982.

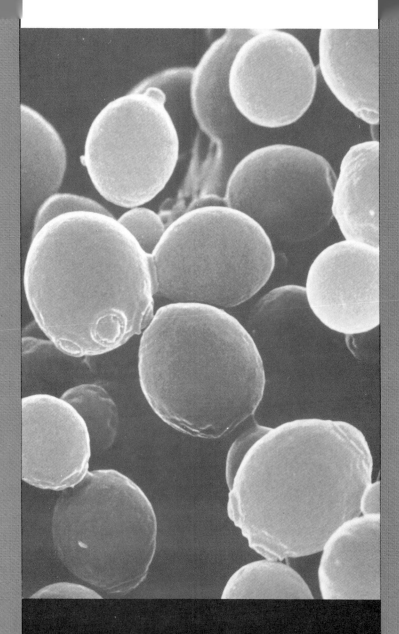

A GUIDE TO
MICROBIAL
METABOLISM

CHAPTER OUTLINE

Every living thing depends totally on energy to do the things that identify it as being alive — growing, moving, and reproducing. Energy will be the focal point of this chapter. Cells use energy for **metabolism,** which is the network of all the chemical reactions occurring in the cell. Microorganisms are fundamentally no different from higher organisms, they just have a wider range of ways to obtain and use energy.

The metabolic network in heterotrophs links two processes, catabolism and anabolism (Figure 7.1). **Catabolism** extracts energy from organic nutrients by breaking them down. Compounds we looked at in the second chapter, such as carbohydrates, fats, and proteins, are useful sources of energy when they are catabolized into smaller pieces. As a general rule, when a cell uses a compound as a source of energy, it will do

so by **oxidizing** it. You will recall that oxidation is the removal of electrons from an atom or compound. While heterotrophs derive their energy from catabolizing organic compounds, lithotrophs either harness the energy of sunlight or oxidize inorganic chemicals.

The energy a cell obtains from catabolism, light, or inorganic chemicals must be stored temporarily to be used later to drive processes that need energy. The compound ATP, which we first learned of in Chapter 2, is the most common and important form of cellular energy currency. Chemical energy is stored in ATP's high-energy bonds.

One metabolic process that demands energy is **biosynthesis,** the formation of organic compounds by living organisms from elements or simple compounds. Biosynthesis requires large

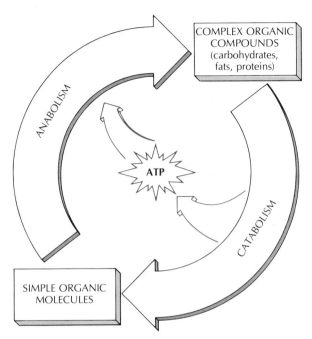

FIGURE 7.1 Cellular Metabolism
The metabolism of a cell is composed of two general processes. Catabolism breaks down complex organic compounds such as carbohydrates into simpler organic molecules, giving off energy, which is captured as ATP. Anabolism puts together complex molecules from smaller units, utilizing the energy stored in the ATP.

amounts of stored energy to make useful products. **Anabolism** is the portion of metabolism that is concerned with biosynthesis.

The reactions of metabolism are all catalyzed by enzymes, so enzymes are crucial to keep the processes of metabolism running smoothly. We will begin this chapter by studying enzymes in greater detail. Then we will examine how various energy sources are utilized to drive ATP synthesis, including the ways microorganisms break down carbohydrates, fats, and amino acids. Finally, we will see how the energy is used to power biosynthesis.

ENZYMES — WHAT THEY DO AND HOW THEY DO IT

An **enzyme** is a protein that catalyzes a chemical reaction. Since we will be investigating enzymes in this section, you may wish to refresh your memory about proteins by referring to Chapter 2. Our objective is to become familiar with the types of enzymes, the kinds of jobs they do, and their locations inside or outside the microbial cell.

THE JOBS DONE BY ENZYMES

Enzymes are classified by the type of chemical reaction they catalyze. There are six functional classes. Enzymes are generally named after the substrate on which they act or the reaction they catalyze. Many enzyme names end in -ase. We will mention the common names for some enzymes. You will be encountering many examples in Chapters 7 and 8 and will find out more about how they fit into the cell's life processes then.

- **Oxidoreductases** catalyze redox reactions (review Chapter 2). In metabolism, useful energy is always obtained by oxidizing an energy-rich (reduced) nutrient. A dehydrogenase, an enzyme that removes hydrogens, is an example (see Step 5, Figure 7.11).

- **Hydrolases** hydrolyze covalent bonds that link monomers into polymers. These enzymes are digestive or degradative. Amylase is a simple hydrolase; it breaks down starch into maltose (Figure 28.13). Other hydrolases break down proteins, lipids, and nucleic acids.

- **Ligases** condense monomers together to form polymers. They are also called **synthetases** and **polymerases.** Their actions are just opposite to those of the hydrolases. The DNA polymerases link nucleotides together to make and repair DNA strands (Figure 8.4).

- **Lyases** remove groups from compounds. When the bonds are broken to remove the group, double bonds form. For example, a lyase removes water from phosphoenol-pyruvic acid in the last step of glycolysis (see Figure 7.11), leaving a double bond in pyruvic acid.
- **Transferases** move functional groups such as amino or phosphate groups from one molecule to another. The transaminases (see Figure 7.17) are examples.
- **Isomerases** reshuffle atoms within molecules without adding or subtracting any. Their activities change polarities, strengths, and energy contents of bonds. An isomerase converts glucose-6-phosphate to fructose-6-phosphate, for example, in the second step of glycolysis (see Figure 7.11).

EXOENZYMES AND CYTOPLASMIC ENZYMES. Does the enzyme customarily do its work inside or outside the cell? Microbial metabolism goes on inside the cell, yet much of an organism's useful nutrient source may be derived from macromolecules found outside the cell in the environment. Many microorganisms excrete **exoenzymes** to function outside the cell's membrane to hydrolyze such external nutrient sources. In Gram-positive bacteria, exoenzymes pass out through the wall. In Gram-negative bacteria, most exoenzymes stay in the periplasmic space between the membrane and the wall. Exoenzymes of decomposer species break down starch or cellulose. Some exoenzymes of pathogens hydrolyze tissue components such as collagen fibers, causing destructive disease symptoms. **Cytoplasmic enzymes** are retained within the cell to catalyze the metabolic processes. Some diffuse freely, while others are embedded in cell structures such as the membrane and ribosomes (Figure 7.2).

As the cell's environment changes, so do demands on cellular metabolism. Some enzymes, such as those that make ATP, are essential under

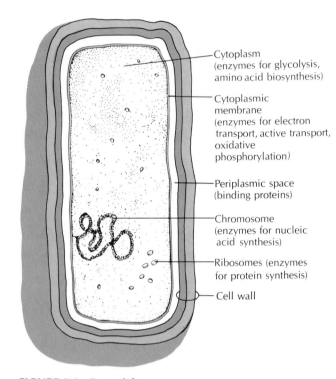

FIGURE 7.2 Bacterial Enzymes
Bacterial enzymes are produced to function in specific locations. In a Gram-negative organism (as pictured), macromolecules are digested in the periplasmic space by hydrolases, which release small molecules that are actively transported across the membrane to the inside. Once in the cytoplasm, substances may be broken down by cytoplasmic enzymes, or routed to nucleic acid or protein synthesis at the designated spots.

Labels in figure:
Cytoplasm (enzymes for glycolysis, amino acid biosynthesis)
Cytoplasmic membrane (enzymes for electron transport, active transport, oxidative phosphorylation)
Periplasmic space (binding proteins)
Chromosome (enzymes for nucleic acid synthesis)
Ribosomes (enzymes for protein synthesis)
Cell wall

all conditions and are always present. They are called **constitutive** enzymes. A few molecules per cell of each constitutive enzyme, replaced on a regular basis, are usually enough.

However, since making enzymes is costly to the cell in terms of cellular energy and materials, microorganisms don't keep supplies of enzymes for which they have no immediate need. **Inducible enzymes** are those that are produced only when the cell has a specific need or opportunity. Inducible enzymes for lactose metabolism are produced by *E. coli* only when the lactose sugar is available in the environment. Once the lactose has been used up, no more enzyme will be produced until the next opportunity to use lactose arises. **Repressible enzymes** are ones that the cell can stop making if it has no immediate need for the enzyme's product. Thus when a cell has an adequate supply of the amino acid tryptophane, it will stop making the enzymes that assemble tryptophane molecules. These shifts in enzyme availability are under genetic control, and will be discussed in Chapter 8.

HOW DO ENZYMES CATALYZE REACTIONS?

Enzymes work by combining specifically with their **substrates,** which are the substances they work on. They may be helped in their action by coenzymes and cofactors, or blocked by inhibitors. In the end, an enzyme is able to catalyze a reaction by lowering energy barriers. Each of these factors in enzyme function will now be discussed. Before you begin this section, you might find it helpful to review the general statements on chemical reactions in Chapter 2.

ENZYME SPECIFICITY. In the healthy, normally functioning cell, enzymes do not randomly alter materials, but are limited to appropriate natural substrates. An enzyme is a three-dimensional globular protein. The substrate binds at the enzyme's **active site.** The active site's intricate surface configuration can only bind substrate molecules with a compatible shape and charge. Some enzymes are highly specific, bind only one unique substrate, and catalyze only a single reaction. Other enzymes, less specific, will bind any of a group of generally similar molecules and catalyze reactions using them.

Basically, an enzyme's role is to combine with its substrate, act on it, and release the product. Each step is (theoretically at least) reversible. This stepwise process is summarized in the **Michaelis–Menton** equation:

$$E + S \rightleftharpoons ES$$
(enzyme) (substrate) (enzyme–substrate complex)

$$\rightleftharpoons E + P$$
(enzyme) (product)

An enzymatic reaction starts when the specific, unique **enzyme–substrate complex** is formed (Figure 7.3). Later, when the substrate(s) have been altered into products, the enzyme releases the products and emerges unchanged, ready to catalyze the next reaction. When we say that enzymatic reactions are reversible, we mean that under certain circumstances it may happen that enzyme could convert products back to sub-

FIGURE 7.3 Enzyme Reaction Model
The enzyme reacts with its substrate by a lock-and-key fit at the enzyme's active site. After the reaction occurs, the substrate has been separated into two products and the enzyme is freed and ready to catalyze the next reaction.

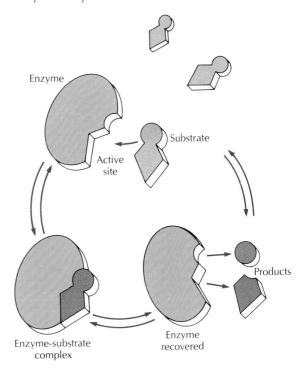

Enzyme

Substrate

Active site

Products

Enzyme-substrate complex

Enzyme recovered

strates again (see Figure. 7.17, for example). However, within an active cell, the product is usually a useful substance which is being consumed about as fast as it is made, so the product does not have a chance to participate in the reverse reaction.

COENZYMES AND COFACTORS.   Many enzymes become functional only when bound to and working with a nonprotein component. There are two different types of molecules that assist enzymes. **Coenzymes** are small, complex organic molecules, frequently consisting of nucleotides bound to vitamins. Table 7.1 shows the

important coenzymes of metabolism. You'll want to refer to it as we explain various reactions. A coenzyme may be loosely or tightly attached to the protein part of the enzyme to make the functional catalytic unit (Figure 7.4). Coenzymes are needed to reversibly accept, hold, or transport any single atoms, electrons, or small groups of atoms that may be substrates for an enzyme reaction. **Cofactors** are inorganic ions such as iron, magnesium, calcium, or zinc. These ions may form the reactive center of an enzyme and assist it to combine correctly with its substrates.

ENZYME INHIBITORS.   Sometimes enzyme specificity doesn't work to the organism's benefit. If an enzyme encounters a natural or human-made substance very similar to its real substrate, the enzyme may accept it as a substrate. The enzyme may bind the incorrect substance equally or more firmly than the real thing (Figure 7.5a). However, since the enzyme would not be able to catalyze any reaction of this substance, the enzyme would be diverted from its normal activities. The inhibitor may dissociate from the enzyme later, allowing it to resume functioning. This type of blocking substance is called a **competitive inhibitor.** Some bacterial infections are treated successfully with sulfa drugs. These are chemicals that act as competitive inhibitors for the enzymes that synthesize folic acid, an essential vitamin. This is the basis for sulfa's therapeutic usefulness.

Noncompetitive inhibitors do not compete with the substrate for the active site. Instead they bind the enzyme irreversibly at another location, inactivating it (Figure 7.5b). Cyanide ions are noncompetitive inhibitors of cytochrome oxidases, which we will meet shortly as key respiratory enzymes. We will also encounter yet a third type of inhibition, beneficial this time, called feedback inhibition.

ENZYMES AND ENERGY REQUIREMENTS.   One outstanding feature of enzymes is their speed, measured in some cases at up to one million individ-

TABLE 7.1 Important coenzymes of metabolism		
Type of Function	Name of Coenzyme	Group Transferred
Electron carriers	Nicotine adenine dinucleotide (NAD)	Two electrons, one proton
	Nicotine adenine dinucleotide phosphate (NADP)	Two electrons, one proton
	Flavin adenine dinucleotide (FAD)	Two electrons, two protons
	Quinones	Two electrons, two protons
	Cytochromes	One electron
Group transfer carriers	Thiamine pyrophosphate (TPP)	Aldehyde
	Pyridoxal phosphate	Amino
	Coenzyme A	Acetyl
	Tetrahydrofolate	I-carbon, such as methyl
Energy donors	Adenosine triphosphate	Phosphate
	Guanosine triphosphate	Phosphate

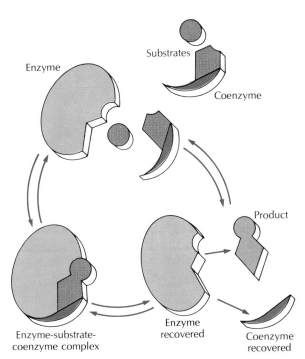

FIGURE 7.4 The Role of a Coenzyme
A coenzyme binds to an active site on an enzyme, bringing a reactant, such as a phosphate group. Once the reactant has been used and incorporated into a product, the coenzyme dissociates. It will then pick up another group and be able to participate in another reaction.

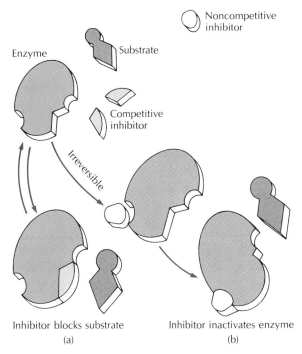

FIGURE 7.5 Enzyme Inhibitors
(a) A competitive inhibitor binds to the same site as the natural substrate. Once bound it blocks the enzyme's binding of substrate. A competitive inhibitor can, however, dissociate so that the enzyme may become functional again. (b) A noncompetitive inhibitor binds irreversibly at a different site on the enzyme, permanently destroying its catalytic ability.

ual catalyses per enzyme molecule per second. This is why a cell needs only a few molecules of each enzyme at any time. How are the energy barriers to such speeds overcome?

Remember that exergonic reactions give off energy, while reactions that consume energy (endergonic) occur only when energy is provided by some other, exergonic, process. Reactions occur between compounds; within compounds each atom is bonded with others. In order for reactions to occur, bonds must first be broken, exposing new combining sites. Breaking these bonds requires energy. Even for an exergonic reaction, such as burning paper (cellulose) with oxygen, a slight input of energy is needed to start

the reaction. This is why paper doesn't burn up spontaneously at room temperature. In fact, paper does oxidize, but imperceptibly slowly. If a sample of paper had been preserved away from acids and microbial enzymes, you would notice little change after the passage of a thousand years. However, consider the difference an enzyme can make. Bacterial cellulase enzymes at work in a cow's rumen efficiently hydrolyze dozens of pounds of hay cellulose per day.

Let's follow up on the example of the breakdown of cellulose by cellulase enzymes in the cow's rumen. **Activation energy** is the name used for the energy that must be supplied to increase collisions between molecules to the point at which bonds break and reaction occurs. Without activation energy, nothing much happens — there is an energy barrier. This barrier is what keeps complex molecules like cellulose stable enough so that they don't spontaneously break down. In the cow's stomach, the bacterial enzymes bind to the cellulose. Enzyme–substrate binding appears to alter the substrate bonds to render them highly unstable and lower their activation energy so that many bonds break. Once some bonds in the cellulose are broken, the enzyme, which is holding the portions, orients them to promote the formation of new bonding combinations in the products, releasing glucose molecules. Because bonds break and reform so much more efficiently in the presence of the catalyst, the reaction goes much faster with than without an enzyme.

CELLULAR CONTROL OF ENZYMES

In cells, enzyme reactions do not stand alone, but are linked in sequences called **metabolic pathways.** A metabolic pathway is a sort of molecular assembly line. A compound, the product of one enzymatic reaction, may be furthur modified when it becomes the substrate for a different enzyme in the next reaction in the pathway. Some examples of metabolic pathways are glycolysis (see Figure 7.11), amino acid synthesis (see Figure 7.18), and the Krebs cycle (see Figure 7.15). A metabolic pathway may provide starting materials for other pathways, regenerate some of its starting materials, and/or yield waste or end products.

Cells may regulate both the production and the activity of their enzymes within these pathways, using several levels of control. Let's adopt the analogy of a home heating system to illustrate this. First, to have heat, you must have a furnace or heater of some sort. Second, the furnace must be hooked up and ready to operate. Third, you must have a thermostat, which senses the need for heat in the house, and turns the furnace on or off accordingly. Fourth, you can improve the efficiency of your heating system by fine-tuning it. Each of these four has its analog in the metabolic situation.

First, the cell must have the genetic information in its DNA in order to be able to make the enzyme. In the heating system analogy this compares to having a furnace. Genes can be gained by gene transfer, or lost by alterations in the DNA. Changes of this sort would be very unusual but significant events in the lifetime of a cell.

Second, the cell's gene may or may not be making the enzyme at any given time. Induction and repression, just mentioned, are genetic functions that turn enzyme production on or off as needed. This is analogous to hooking up or disconnecting the furnace. The response time for induction or repression is measured in minutes.

Third, the enzyme itself, while present, may or may not be active in catalyzing reactions. The activity of existing enzyme molecules can be temporarily inhibited or shut down. In our analogy, we can compare this adjustment to how the furnace is turned on and off in response to the house temperature. As heat (the product) accumulates, the heat-producing unit (the furnace) is shut off by the thermostat. After the heat dissipates, and the house cools down, the thermostat reactivates the furnace. For an example of how this works in a cell, we could observe that a buildup of ATP inhibits certain enzymes, thus slowing down the making of ATP. Once the ATP supply has been depleted somewhat by the cell, ATP synthesis is once again reactivated. A response of this sort takes only a fraction of a second.

This effect (end-product accumulation inhibiting enzyme activity) is called **feedback inhibition** (Figure 7.6). The mechanism for feedback inhibition is an **allosteric** (meaning "other shape") interaction. This type of inhibition occurs in enzymes that have two types of binding sites, the substrate-binding sites and an allosteric site, which binds the product. When the end product of a metabolic enzyme pathway combines with the enzyme at the allosteric site, its binding so changes the whole three-dimensional shape of the enzyme that it can no longer combine with its initial substrate. The blocking molecule stays put until it is removed for cellular use. Once it is gone, the enzyme becomes active again. For example, in the cellular synthesis of the amino acid proline, if excess tryptophane builds up, it inhibits an enzyme in the tryptophane pathway, thus "turning off" production until the excess is used up.

Fourth, the cellular environment fine-tunes enzymatic reactions. Activity rates fluctuate in response to changes in substrate concentration,

temperature, pH, and osmotic conditions. These changes are frequent and rapid.

To summarize, the rates of each enzyme's activities are established by a hierarchy of controls. Because of these controls, the cell's interior functions as a coordinated metabolic machine rather than a random collection of unlinked activities. Rate regulation also enables the cell to adapt to changing needs.

ENERGY AND BIOLOGICAL REACTIONS

Energy is the ability to do work or to drive a process. A cell's energy source allows it to create internal order and maintain its integrity, under the direction of its enzymes. In the next few pages, we will first become familiar with the laws of thermodynamics. These physical laws regulate energy behavior throughout the universe; they also establish the "ground rules" for cellular metabolism. Next, we will examine biological oxidation, the removal of electrons from organic compounds. These electrons are the energy source for heterotrophs. Then we will turn to the synthesis of adenosine triphosphate, or ATP, the primary energy-storing component of metabolism. Then we will see how energy release by oxidation is linked to energy storage by ATP synthesis within cellular membranes.

THE LAWS OF THERMODYNAMICS IN METABOLISM

THE FIRST LAW OF THERMODYNAMICS. This is also called the Law of Conservation of Energy. It states that whenever energy is converted from one form to another, all the energy can be accounted for, and none is either created or destroyed. We saw in Chapter 2 that we can insert energy values into equations for chemical reactions. The First Law says that equations must always balance in terms of the units of energy on both sides of the arrow. If there are 10,000 units of energy on one side, there must be 10,000 units on the other, no matter what form they take. The

FIGURE 7.6 Feedback Inhibition of Enzymes
Feedback regulation depends on the ability of the regulated enzyme to bind two substances — a substrate and an allosteric effector — at different sites. The metabolic pathway's end product is the allosteric effector. If production gets ahead of demand, the allosteric-effector concentration builds up and the effector binds to the enzyme. This binding causes an alteration in the active site of the initial enzyme so that substrate cannot be bound. This process is reversible.

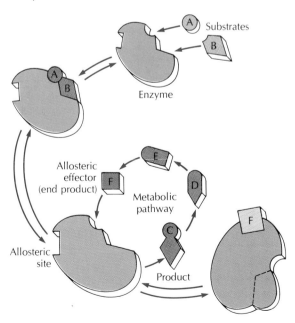

exact energy values of all common compounds can be found in reference tables. So if we want to, we can find the theoretical energy yield from the conversion of any useful nutrient into another.

For example, when glucose is oxidized to carbon dioxide, it releases 686,000 calories per mole in the process. These calories (cal) are not the same calories you count if you're on a diet. They are 1/1000th of the kilocalories (Kcal) used in dietary analysis. A mole (mol) is a chemist's unit, the number of grams equal numerically to a compound's molecular weight. A mole of any substance always contains the same number of molecules (6.16×10^{23}, or Avogadro's number) so we can usefully compare the energy content of a mole of one substance to that of a mole of another. Synthesis of ATP from ADP and inorganic phosphate ion requires about 8000 cal per mole. Simple division tells us that oxidation of one mole of glucose will supply enough energy, theoretically, to put together 85 mol of ATP. It doesn't work that way in real life, however. When we actually measure ATP output from glucose oxidation in healthy live cells, we get values nearer 35 mol. What's the problem?

THE SECOND LAW OF THERMODYNAMICS. The "problem" is the Second Law of Thermodynamics. It states that, in any energy-converting process, some portion of the energy will be inevitably degraded from an ordered, useful form, called **free energy,** into a disordered form unavailable to do work, called **entropy.** In other words, energy may not be destroyed, but it can be (and is) lost to useful purposes. So all biological processes are less than fully efficient. Although cells have had several billion years of evolution in which to work out ways of maximizing their efficiency, and minimizing entropic losses, energy inefficiency is a fact of life.

COUPLING EXERGONIC AND ENDERGONIC REACTIONS

Exergonic reactions such as oxidations or ATP hydrolysis give off energy, whereas endergonic

reactions such as reductions or ATP synthesis absorb energy. Coupling them tightly is life's way of reducing the loss of useful energy. In fact, the emergence of living organisms was dependent on the development of **coupled reactions;** these are reactions in which an exergonic and an endergonic process are linked by sharing an enzyme and/or its products and reactants. To illustrate how this works, let's construct some symbolic reactions, starting with an endergonic process:

$$A + B + \overset{100 \text{ units}}{\text{energy}} \longrightarrow C + D.$$

We will assign an arbitrary value of 100 units for the energy stored in this reaction.

Now let's propose a simultaneous exergonic reaction:

$$X \longrightarrow Y + Z + \overset{200 \text{ units}}{\text{energy}}$$

with an arbitrary value of 200 units of energy released. Now let's also say that both reaction sequences occur at the same site and that they are catalyzed by the same enzyme–coenzyme complex. Combining the two processes

$$A + B + X + \underset{\text{energy stored}}{100 \text{ units}} \longrightarrow$$
$$C + D + Y + Z + \underset{\text{energy released}}{200 \text{ units}}$$

for the coupled reaction, we see that the *net* energy change is

200 units released (exergonic reaction)

−100 units stored (endergonic reaction)
_____
100 units lost as entropy (coupled reaction)

You see, when all is said and done, the coupled process itself is exergonic; it has a net energy loss. More energy must be supplied than is actually conserved; waste is inevitable under the Second Law. Life's only option is tight and efficient energy coupling, which can minimize the waste. Here again we see the drive to evolve closely regulated enzymes in highly structured pathways.

Our next step is to look (separately) at how cells synthesize ATP and how they release and control the energy of electrons.

ATP SYNTHESIS — A KEY ENERGY-STORING PROCESS

ATP is the energy currency of the cell. Like our paychecks, it is earned only to be spent. Its formation stores energy, and its hydrolysis releases energy.

PHOSPHORYLATION. This is a name for reactions in which phosphate groups are added to molecules. Phosphorylation always increases the compound's energy. ATP synthesis (Figure 7.7) is the most important single phosphorylation reaction. ATP is synthesized by adding a third phosphate group to a molecule of ADP (adenosine diphosphate). These are linked together with a covalent bond that stores so much energy that it is often called a high-energy bond. ATP synthesis can take place in two ways. **Substrate-level phosphorylation** takes place in the cytoplasm. The energy released by oxidizing

a substrate is used directly, at the reaction site, to phosphorylate ADP to ATP. This simple process, associated with glycolysis (see steps 6 and 9, Fig. 7.11) and other pathways, is universal. It is considered to be the "primitive" way of transferring energy from food molecules to the useful ATP form.

Oxidative phosphorylation, the second and more efficient way of making ATP, occurs in respiring cells and requires a membrane-bound coenzyme structure. The released energy of electrons drives the formation of the high-energy bonds by enzymes within the membrane structure.

BIOLOGICAL OXIDATION — THE KEY ENERGY-RELEASING PROCESS

The key energy-releasing process in metabolism is oxidation. You'll recall that oxidation is the loss or removal of electrons from a substrate. The substance being oxidized is frequently referred to as the **electron donor.** We will study such electron donors as carbohydrates, fats, reduced coenzymes, or reduced inorganic compounds. The electrons removed from the donor

FIGURE 7.7 ATP
The bond that attaches the third phosphate to the ATP molecule is usually referred to as a high-energy bond. It is unstable, and readily hydrolyzed to release its energy. Energy is stored when ATP is made, and released when ATP is hydrolyzed to ADP and phosphate.

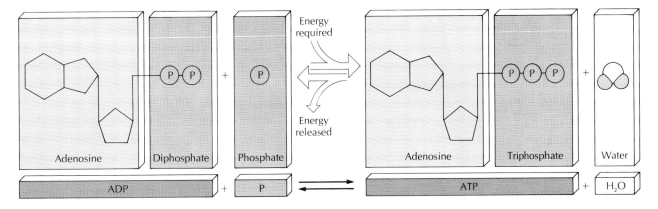

are usually passed to another substrate, called an **electron acceptor.** The latter acquires electrons and as a result is reduced. Some common electron acceptors we will study include oxygen, nitrate or sulfate ion, organic molecules such as pyruvic acid, oxidized coenzymes, and metal ions.

In biological systems, we most frequently deal with electrons as part of hydrogen atoms. A hydrogen atom, you may recall, consists of a nucleus containing one proton, with one orbiting electron (Figure 7.8). When the electron is removed, we have a positively charged hydrogen ion, which is exactly the same as a proton (both terms are used). In many of the oxidations we will study electrons are not the only things lost; whole hydrogen atoms — that is, electrons plus protons — can also be lost. Electrons are never found free in the cell; they are always bound to some molecule or carrier.

Certain coenzymes have a unique and important role in the electron transfers occurring during oxidation and reduction. The key aspect of the behavior of the **electron-transport coenzymes** is that they exist alternately in two states, the oxidized and the reduced states. That is, when oxidized, they can gain electrons and become reduced. On the other hand, if they are reduced, they can lose electrons to a suitable electron acceptor and become reoxidized. Another way to view these coenzymes is as temporary electron acceptors. Often coenzymes pass electrons among themselves, so that reciprocal oxidation and reduction occur. Let us look, for example, at the simple reaction

$$NADH + H^+ + FAD \longrightarrow NAD^+ + FADH_2$$

This shows the interaction of two of the most important electron transport coenzymes. One is called nicotine adenine dinucleotide (NAD). The reduced form of this substance (NADH) appears on the left, and the oxidized form (NAD$^+$) appears on the right. The reduced coenzyme is shown transferring the pair of electrons and the one proton it carries, plus a proton. The recipient is another coenzyme, flavin adenine dinu-

cleotide (FAD), which also has both oxidized and reduced states. In most of the biological oxidations we will be studying, NAD$^+$ accepts a pair of electrons, then passes them on to another acceptor, which may or may not be FAD.

This transfer of electrons is associated with loss of energy. The electrons give up some of their energy in passing from NAD to FAD. Let us see how this happens. Oxidation is exergonic, and a substance has less energy content after it is oxidized. Reduction, on the other hand, is endergonic, and a substance has a higher energy content after being reduced. But oxidation and reduction always occur together, making up a coupled reaction. And coupled reactions, as we just saw, are always exergonic overall. The cell's objective is to get some work out of that released energy before it dissipates to the environment.

ELECTRON-TRANSPORT PATHWAYS.   We have just looked at an isolated example of electron transfer. We saw electrons passing from the oxidized compound to the reduced compound in a redox reaction. If the second compound were now to be oxidized, it too would pass the electrons on to yet another compound. A series of linked redox reactions can pass electrons down a chain of compounds, just as an electrical circuit of interconnected wires can conduct an electrical current, which is also a stream of electrons. The **electron-transport system** or **chain** (Figure 7.9) is a sequence of electron-transport coenzymes that pass electrons from one to the next in a defined order. Electron-transport systems are localized in the cell membrane of the procaryotic species that respire, and in the mitochondria of eucaryotes. The coenzymes of electron-transport systems receive electrons from the oxidation of nutrients and control their flow so their released energy may partially be trapped and used to form ATP. Transferring one pair of electrons from NADH all the way down the chain yields sufficient energy to drive the endergonic synthesis of three molecules of ATP. A typical electron-transport system might be composed of the following coenzymes:

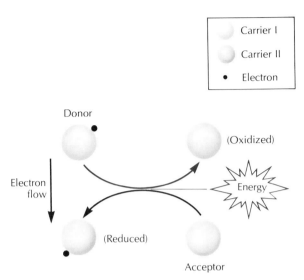

FIGURE 7.8 Electron Transfer
Electrons may be transferred from one coenzyme to
another. Electrons are given up by the reduced
form of the first coenzyme (carrier I), which then
becomes reoxidized. They are picked up by the
oxidized form of the second coenzyme (carrier II),
which then becomes reduced.

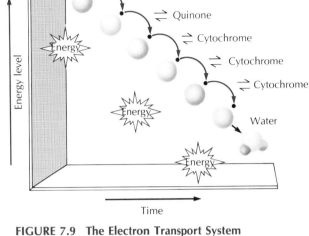

FIGURE 7.9 The Electron Transport System
This linear model of electron transport shows the
sequence of steps in the passage of electrons from
NAD to oxygen. With each transfer, the energy level
of the electrons declines. Each electron finally
unites with a pair of protons ($2H^+$) and reduces an
oxygen atom to H_2O.

- The first coenzyme is NAD^+. On reduction
  it picks up two electrons and one proton
  (H^+) to become the electrically neutral reduced NADH.

- The two electrons are passed from NADH
  to oxidized FAD. In becoming reduced,
  FAD accepts two electrons and a proton
  from NADH and takes on a second proton
  from the environment to become $FADH_2$.
  The transfer is catalyzed by an enzyme.

- Next the paired electrons reduce a **quinone,** a vitamin K derivative. From this
  point on, electrons are passed, but protons
  are not.

- A **cytochrome** next accepts the electrons
  from the quinone. The cytochromes are
  proteins containing heme groups. They are
  chemically similar to chlorophyll and use
  oxidized iron atoms as electron acceptors.

There are many different cytochromes; in
procaryotes there may be no cytochrome,
one, or a sequence of several depending on
the species. In eucaryotes the sequence is
usually cytochrome b, cytochrome c, then
a complex of cytochromes a and a_3.

- At the end of electron transport the electrons have lost some energy. They combine with an inorganic compound, the **terminal electron acceptor,** to form a low-
  energy waste product. If oxygen is the acceptor, the waste product is H_2O. The final
  reaction in this case is catalyzed by a **cytochrome oxidase** enzyme.

Hydrogen peroxide (H_2O_2), which is highly toxic,
may be formed before water. Aerobic organisms
produce the enzyme **catalase** to break down the
H_2O_2. Strict anaerobes usually lack catalase.

This is why exposure to oxygen poisons them. A rapid-slide test for catalase is useful in laboratory identification of some bacterial species.

RESPIRATION

We defined **respiration** originally in Chapter 4 as a pattern of metabolism in which the cell gets electrons from organic compounds and transfers them to an inorganic electron acceptor. Now we are in a position to see where the inorganic acceptor comes into the picture. A respiring cell is one that is transferring electrons via some type of electron-transport system. Respirers usually are also using a metabolic pathway called the Krebs cycle (discussed below) as the source of most of the electrons. However, bacteria vary greatly in the electron-transport mechanisms they possess, and the amount of energy they get from electron transport. Many common bacteria have different or less-complex electron-transport systems than illustrated, and some may stop after the FAD step.

Aerobic respiration occurs when the terminal inorganic electron acceptor is oxygen. It occurs in strict aerobes and facultative anaerobes. Water is the chemical end product. **Anaerobic respiration** occurs when cells use inorganic ions other than oxygen, such as nitrate $(NO_3{}^-)$, to accept electrons. A cell that is able to carry out anaerobic respiration obtains a high energy yield from oxidation of nutrients. Bacterial nitrate and sulfate reduction, occurring as steps in their anaerobic respiration, are essential links in the elemental cycles of nitrogen and sulfur in nature, as we will see in Chapter 27.

Later we will discuss fermentation in detail. For now, however, please note that fermenters, which use organic electron acceptors, are not employing an electron-transport system at all.

CHEMIOSMOSIS — HOW COUPLING REALLY WORKS

How is the energy given off by the electrons of the electron-transport system used to drive the making of ATP? Oddly enough, this extremely important question resisted at least 40 years of investigation. Currently, we are using the **chemiosmosis** model to explain it, and this theory seems to answer most of the questions.

The chemiosmosis model proposes that as electrons move down the electron-transport chain their energy is used to drive a transport mechanism to extrude protons (H^+) through a membrane to the outside (Figure 7.10). In bacteria, protons are extruded through the cell membrane; in eucaryotes they are pushed out through the inner mitochondrial membrane. Thus the medium outside the membrane contains a higher concentration of H^+ than the interior. In other words, a **proton gradient** is created. Like all osmotic gradients, this one has an energy potential, the **proton motive force (PMF).**

At other points in the membrane, specific sites contain membrane-bound ATP-synthesizing enzymes, or ATPases. These ATPase sites are the only channels through which H^+ can reenter the cell, moving by diffusion down the concentration gradient. The PMF potential energy is "harvested" at these sites to phosphorylate ADP to ATP. Phosphorylation occurs only in a structurally intact membrane. If the membrane is damaged so a proton gradient cannot develop, phosphorylation stops even though electron transport continues. Membrane damage, then, uncouples the exergonic and endergonic parts of the process.

Chemiosmosis is not used only in respiration. It is also the critical feature in the trapping of light energy by photosynthetic organisms and the energy-gathering purple membrane of the halophilic bacteria. These organisms use solar energy to pump out the protons.

In summary, the electrons obtained from oxidized nutrients are passed across a sequence of coenzyme carriers and finally to an inorganic acceptor, with a progressive release of energy. This is the aspect of chemiosmosis that we call electron transport. The released energy is used to expel protons to the outside across the membrane. At certain locations on the membrane, these protons reenter, providing energy to make ATP by oxidative phosphorylation.

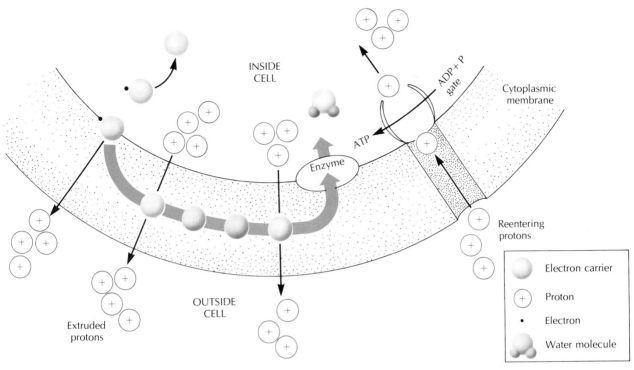

FIGURE 7.10 A Model of Chemiosmotic ATP Production
The electron-transport coenzymes are oriented in the membrane mosaic so they are able to extrude protons. During the passage of two electrons from NAD to oxygen, three pairs of protons may be extruded. The external concentration of protons and positive charges is the proton motive force. This force is captured for ATP synthesis when protons reenter the cell through an ATPase gate.

RELEASING ENERGY FROM NUTRIENTS

As we have just seen, electrons are the primary source of cellular energy. We also know that heterotrophs get their electrons by breaking down organic compounds. It is now time to look at how heterotrophs attack nutrients to extract those all-important electrons. This will involve examination of several enzymatic pathways. Carbohydrates and fats are the cell's preferred fuels and reserve materials. Proteins too may be used as energy sources when they are plentiful or when carbohydrates or fats are not available.

PATHWAYS FOR CATABOLIZING CARBOHYDRATES

We can assume that the starting material of carbohydrate metabolism is glucose, since other carbohydrates are converted into glucose by the cell. Recall that glucose is a six-carbon sugar, with a formula of $C_6H_{12}O_6$.

GLYCOLYSIS. The first stage in the cell's oxidation of glucose may be the nearly universal metabolic pathway called either **glycolysis** or the Embden–Meyerhof–Parnas (EMP) pathway after its discoverers. Glycolysis is a character-

BOX 7.1
The Membrane as an Energy Broker

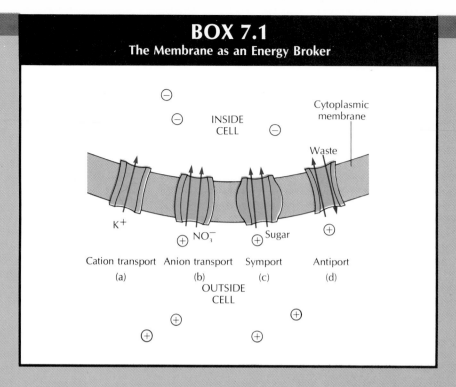

Cation transport (a) — K^+
Anion transport (b) — NO_3^-
Symport (c) — Sugar
Antiport (d) — Waste

INSIDE CELL — Cytoplasmic membrane

OUTSIDE CELL

Membranes do a surprising range of things, many of which seem to involve the flow of energy. The membrane itself, with its clusters of enzymes, appears to act as an energy broker, making ATP here, routing it to drive a process there. We still have a great deal to learn about the actual arrangement of membranes. When you take a cell apart in the lab, to separate out the membrane fragments, what is left of this exceedingly delicate structure unfortunately is rather like what is left of a watch after you hit it with a hammer.

At the moment, we know that proton gradients across a semipermeable membrane have the potential to do work, the proton-motive force. There is also an electrical charge gradient — outside positive, inside negative. This is be-cause lots of positive charges (H^+) accumulate outside, while negative charges, mainly hydroxyl ions (OH^-), remain inside. These two gradients provide the driving force for active transport, which, you recall, requires energy to activate a carrier that transports molecules across a membrane.

Positive ions can enter the cell directly because of the charge gradient, in the presence of a suitable carrier (a). Negative ions can enter with a proton, as a neutral salt (b). Uncharged molecules can also be carried along with a proton (c), in a process called **symport.** By a type of revolving door arrangement (d), the entrance of a proton can be linked to the eviction of an unwanted molecule (**antiport**).

Transport's energy demands can be a problem. En-ergy can only be used once (you can't have your cake and eat it too). Proton-motive force that has been used up in performing transport functions cannot drive oxidative phosphorylation. Fermenting cells have to use some of the scarce ATP they make by their relatively inefficient metabolism to drive the transport of other needed molecules. It's little better than a break-even arrangement.

Flagellar motility is another energy-requiring, membrane-associated function. All the chemotactic receptors are membrane-bound, and directly involved with active transport. Motility also depends on an intact membrane, which must supply the energy for rotation of the flagellum.

istic of *Escherichia coli*, the streptococci, and most other microbial heterotrophs, as well as of eucaryotes. Alternate microbial methods for breaking down glucose include the Entner–Doudoroff pathway, found in pseudomonads, and the pentose phosphate pathway, found in most aerobes.

Glycolysis converts glucose to pyruvic acid, a three-carbon compound, in a nine-step process. It can occur in the absence of oxygen as the first step in fermentation, or in the presence of oxygen as the first stage in respiration.

The nine steps of glycolysis are detailed in Figure 7.11. Follow them as you read the description below.

- In Step 1, a glucose molecule is phosphorylated by transfer of a phosphate from an ATP molecule, activating the glucose.
- In Step 2, the internal arrangement of atoms is shifted.
- In Step 3 a phosphate group from another ATP is added to further activate the molecule. These three steps have stressed the middle bond (between carbons 3 and 4).
- In Step 4 the glucose is split into two three-carbon molecules. Both, as **glyceraldehyde-3-phosphate,** may continue on in the glycolytic pathway.
- Step 5 is a coupling reaction. Glyceraldehyde-3-phosphate is oxidized and the electrons reduce a molecule of NAD$^+$ to NADH. Energy is released, but immediately used to add a high-energy phosphate group to the carbon backbone.
- Step 6 is a transfer reaction. The high-energy phosphate group is moved to ADP, forming ATP. This is the first obvious energy "payoff" of glycolysis.
- Steps 7 and 8 are rearrangements; they convert the remaining low-energy phosphate group to a high-energy group.
- Step 9 is a transfer reaction like Step 6, in which ATP is made. The reaction yields pyruvic acid, and is the second energy payoff.

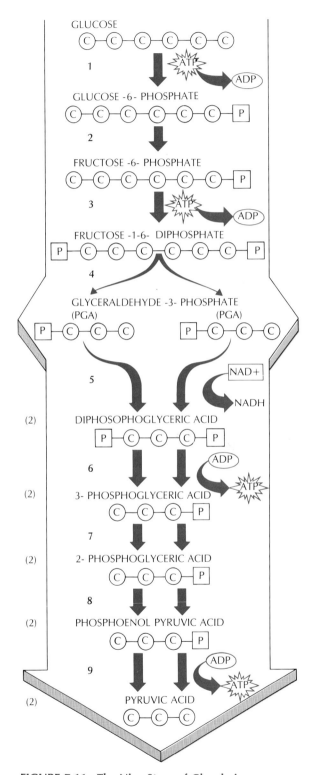

FIGURE 7.11 The Nine Steps of Glycolysis

In summary, three things happen in glycolysis (Figure 7.12). The glucose is partly oxidized, giving up some of its electrons. The glucose also provides some energy directly for phosphorylation. When both halves of the glucose molecule are used, we can see that a net yield of 2ATP is achieved. In addition, the glucose is subdivided into smaller molecules useful for building blocks.

This is the end of glycolysis, but not the end of glucose utilization. There is a lot of unfinished business here! The pyruvic acid still contains, unused, over 90% of the original energy of the glucose molecule. When pyruvic acid undergoes further oxidation, that energy can also be recovered. In addition, the NADH molecules still carry electrons acquired at Step 5. If the cell is able to respire — that is, to use electron transport — these electrons may be used for driving ATP synthesis, and the NAD$^+$ will be regenerated. If it cannot respire, the cell immediately needs to find an alternate means to get rid of the electrons. It must get the coenzyme back into its oxidized state (NAD$^+$), so that glycolysis may keep going. Cells have only so many molecules of NAD; if they are all reduced, then there is a "log jam" that arrests glycolysis at Step 5.

FERMENTATION. Fermentation is the metabolic method cells use so that they can produce ATP (by glycolysis) when they can't respire. Fermentation recycles NAD by using a readily available cellular organic compound as the terminal electron acceptor. This is a comparatively simple metabolic mechanism, which is believed to have evolved early, before the earth's atmosphere had free oxygen. Some bacteria such as the lactobacilli still subsist entirely by fermentation, while higher organisms such as ourselves use it only briefly during intense exertion in oxygen-deprived muscle tissues.

In a fermenting cell, glycolysis is followed by a final step or steps that reoxidize NADH to NAD$^+$. The electrons from the NADH are transferred back to the pyruvic acid. Thus, pyruvic acid, readily available as an end product of glycolysis, becomes the **organic electron acceptor.** In fermentation, the pyruvic acid molecule's oxidation is never completed; just the opposite occurs. The electrons carried by the NADH have no energy value. Fermentation yields only the two units of ATP derived in glycolysis.

There are many different specific pathways of fermentation, as the pyruvic acid may be further modified into various other acids, alcohols, or gases (Figure 7.13). Lactic-acid fermentation is found in bacteria such as some streptococci. They convert pyruvic acid entirely to lactic acid,

FIGURE 7.12 Glycolysis
In this simplified schematic of glycolysis, a molecule of glucose is first activated with phosphate groups, then split to yield two three-carbon molecules. These are oxidized and ATP is formed.

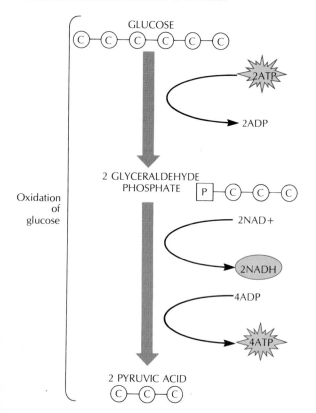

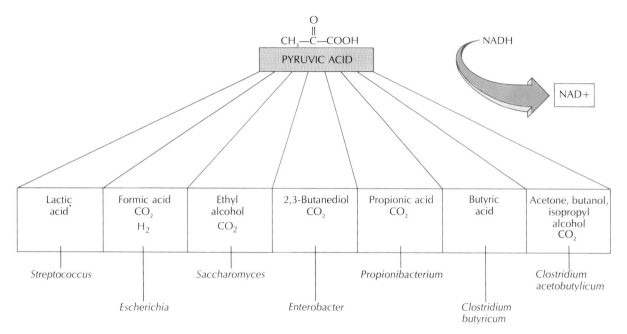

The end products of fermentation in several microorganisms

FIGURE 7.13 Some Pathways of Fermentation
Each species carries out its own genetically programmed fermentation pathway. Most of these are multistep pathways, and produce a mixture of byproducts, only some of which are shown here.

which accumulates, lowers the pH and eventually arrests microbial growth. This fermentation is used in dairy microbiology to produce yogurt and sour cream.

Yeasts, notably *Saccharomyces,* carry out alcoholic fermentation. They convert glucose to a mixture of ethanol and carbon dioxide. Yeast fermentations have been exploited by human beings for thousands of years. We use either or both of the fermentation products. In making bread, when the dough is leavened, bubbles of carbon dioxide expand the dough; the alcohol is driven off in baking. In winemaking, most wines are fermented for alcohol content primarily: the carbon dioxide escapes before bottling. Beers and champagnes make use of both alcohol and carbon dioxide for carbonation.

The pathways of fermentation are genetically determined. We often identify fermentative end products in the laboratory, to assist in identifying an unknown. Fermentation is used to provide many useful industrial products.

COMPLETING THE OXIDATION OF PYRUVIC ACID. Respiring cells have plenty of NAD^+, so they are able to complete the oxidation of pyruvic acid, breaking it down to carbon dioxide via the Krebs cycle. The **Krebs cycle,** also called the **tricarboxylic acid cycle** or the **citric acid cycle,** is a series of metabolic reactions that free up a lot of electrons for electron transport, so their energy may be harvested to make ATP. It is cyclical because the last compound formed is the first to react in the next set of changes. Thus glucose

FIGURE 7.14   A Production Fermenter
The technologist can use this system to grow large batches of microorganisms for the production of any desired fermentation product. The growth conditions are monitored and continuously adjusted to optimum by the microcomputer unit. (Designed and manufactured by New Brunswick Scientific Co., Inc., Edison, N.J.)

metabolism in respiring cells follows glycolysis with the Krebs cycle, electron transport, and oxidative phosphorylation.

Between glycolysis and the Krebs cycle comes a single important, irreversible reaction. The enzyme complex that catalyzes this step uses thiamine pyrophosphate, a coenzyme that carries aldehyde groups, and Coenzyme A (CoA), a coenzyme that carries acetyl groups. Pyruvic acid is again oxidized, then **decarboxylated,** that is, a carbon dioxide group is split off. The remaining two-carbon fragment is an **acetyl** group. The acetyl group is accepted by Coenzyme A, forming acetyl CoA, which then enters the Krebs cycle. The electrons reduce NAD^+ to NADH.

The Krebs cycle has two starting materials, the two-carbon acetyl group and the four-carbon compound oxalacetic acid. Cyclic pathways such as the Krebs cycle regenerate some of their starting materials. Thus although one complete turn of the Krebs cycle gives off two carbons (equivalent to an acetyl group) as carbon dioxide, a waste product it also regenerates one molecule of oxalacetic acid (Figure 7.15).

- Stage 1 bonds the entering acetyl group with the oxalacetic acid molecule, forming citric acid and releasing Coenzyme A. The six-carbon chain is rearranged to prepare it for cleavage.
- Stage 2 oxidizes the six-carbon intermediate; carbon dioxide is released and NAD^+ is reduced. The next product is a five-carbon compound, alpha-ketoglutaric acid.
- Stage 3 oxidizes the five-carbon molecule to a four-carbon compound, succinic acid. The released energy is used to phosphorylate GDP to GTP, a high-energy compound similar to ATP (it has a guanine instead of an adenine). Electrons are removed, and enter the electron-transport system carried by FAD, with a lower energy level than if they had been carried by NAD.
- Stage 4 further oxidizes and rearranges the remaining four-carbon fragment into an oxalacetic acid molecule which can combine with a new acetyl group so that the cycle will continue.

The Krebs cycle gives off two carbon dioxide molecules and provides four pairs of electrons to the electron-transport system per turn. One high-energy molecule of GTP is formed directly. Each acetyl group passing through the Krebs cycle provides, directly or indirectly, an energy equivalent of 12 ATPs.

Let us summarize the energy yield available from respiration when glucose is the electron source. Glycolysis provides two ATPs, two pairs of electrons, and two pyruvic acid molecules. Pyruvic acid oxidation through the Krebs cycle

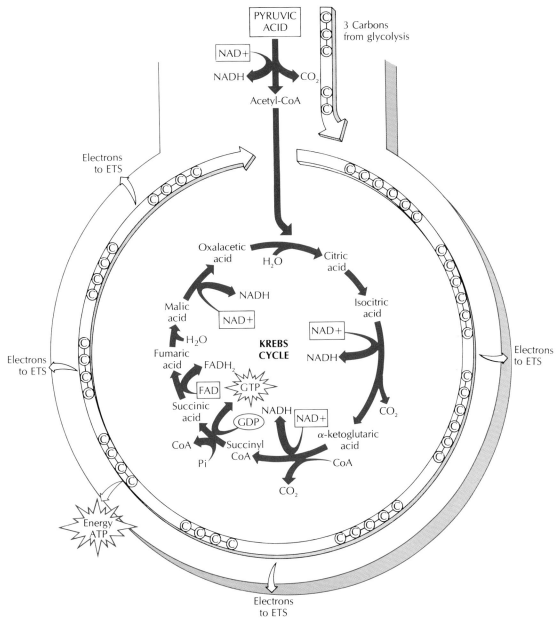

FIGURE 7.15 The Krebs Cycle
In this schematic of the Krebs cycle, two-carbon fragments introduced as acetyl groups join with four-carbon oxalacetic acid. As the cycle progresses, first one, then another carbon is split off as carbon dioxide. Many electrons are passed to electron transport by NAD and FAD. One molecule of ATP is made.

yields four additional pairs of electrons carried by NAD, one pair of electrons carried by FAD, one ATP, and three carbon dioxide molecules per pyruvic acid. You will recall that, depending on the efficiency with which each pair of electrons is carried through the electron-transport system,

the energy release may be sufficient for up to three phosphorylations. Theoretically, one glucose molecule may (at maximum metabolic efficiency) be broken down to provide energy to form 38 high-energy bonds of ATP. When we calculate this efficiency, it comes out to around

BOX 7.2
The Art of Making Home Brew

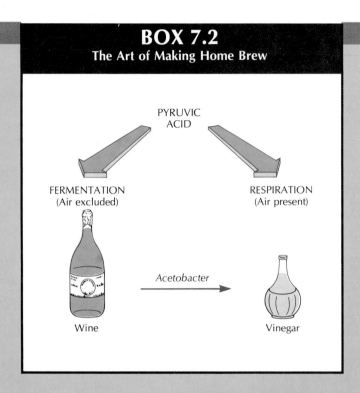

PYRUVIC ACID

FERMENTATION
(Air excluded)

RESPIRATION
(Air present)

Acetobacter

Wine

Vinegar

Many people try to make their own beer or wine at some point or other. They dream that their own home brew will be cheap, and much better tasting than the commercial product. But for every new amateur home-brewer, it is a good bet that there will soon be a jug, barrel, or plastic garbage can filled with non-alcoholic, vilely sour, undrinkable fluid. The Pasteur effect has struck again.

The yeast *Saccharomyces*, a facultative anaerobe, can both ferment and respire, but it prefers to respire. The Pasteur ef-fect is the yeast's prompt shift from anaerobic to aerobic growth when air is introduced. When you are making home brew you want to encourage fermentation, which produces alcohol and carbon dioxide, and to shut off respiration, which uses up the sugars without producing any alcohol. Few novice home-brewers realize how important it is to exclude all air, so their products contain less than the maximum alcohol.

They also tend to use baker's yeast (not a pure culture) and nonsterile condi-tions, so they introduce unwanted organisms. One common contaminant is *Acetobacter*, an aerobic bacterium that promptly oxidizes any alcohol to acetic acid, the sour component of vinegar. Other contaminants produce other unpleasant byproducts, often reminiscent of skunks.

Most fledgling brewers give up in disgust at this point. If they go on, they succeed by mastering three basic micro-biological techniques — controlling the growth conditions, using a good stock culture, and maintaining cleanliness.

40%. In reality, many bacteria lack some of the electron-transport steps, so are much less efficient. Also, in an active cell, many of the intermediate compounds in these pathways may be taken for other uses and not be completely broken down.

We have now seen, from several different perspectives, that organisms can dispose of electrons by more than one route — fermentation, aerobic respiration, or anaerobic respiration. They adjust their metabolic pathways according to environmental conditions. Bacteria, in particular, retain an adaptability in their metabolism that higher organisms have lost. For example, the respiratory bacterial genus *Pseudomonas* uses both aerobic and anaerobic respiration. Similarly, the strictly anaerobic *Clostridium* ferments, but also respires anaerobically when nitrate is available as an electron acceptor. If a species has two genetically possible options, the metabolic path used will be determined by environmental conditions. Respiration is always preferred over fermentation, because the energy yields are as much as 20 times higher.

PATHWAYS FOR CATABOLIZING LIPIDS

Lipids, including fats and oils, are highly reduced compounds, containing two hydrogens for each carbon in the chain, and little oxygen. When a lipid is catabolized it has the potential to yield more pairs of electrons per gram, and thus more energy, than either carbohydrates or proteins. A heterotroph may obtain lipids as nutrients from the tissues of plants and animals, and some microorganisms such as diatoms and dinoflagellates store lipids as reserve-energy materials.

Let us follow the breakdown of a fat in a respiring cell. You may recall that, generally speaking, fats and oils are triglycerides, made up of three fatty acids bonded to a glycerol backbone molecule. First, enzymes called **lipases** separate the fatty acids from the glycerol. Then a stepwise oxidation process begins to break down the fatty acid chain. The oxidation pathway removes two pairs of electrons, one pair by NAD and one by FAD, from the two carbons at the carboxyl end of the long fatty acid chains (Figure 7.16). After this, the two carbons are removed as an acetyl group, available to enter the Krebs cycle. The oxidation process starts again on the next two carbons in the chain. The glycerol backbone is converted into dihydroxyacetone phosphate, which enters glycolysis as glyceraldehyde-3-phosphate at Step 4 (Figure 7.11).

PATHWAYS FOR UTILIZING AMINO ACIDS

Most heterotrophic microorganisms make extensive use of amino acids obtained from plant or animal-tissue proteins as an energy source. They deal with an amino acid as if it is composed of two parts, an amino group and a carbon chain. When the amino group has been disposed of, the carbon chain can be utilized via glycolysis or the Krebs cycle.

Deamination is removal of the amino group. The product is an organic acid. Deaminating glutamic acid gives alpha-ketoglutaric acid, the five-carbon intermediate of the Krebs cycle (Figure 7.17a). Phenylalanine is deaminated to phenylpyruvic acid, a step that is used as the basis of a laboratory test to identify the bacterial species *Providencia*.

The free amino groups form ammonia gas or dissolve as ammonium ion in solution. Ammonium can be toxic because it raises the pH of the medium. Terrestrial animals convert amino nitrogen to urea, a less toxic waste that is important in microbiology because many enteric bacteria can use urea as a nutrient. Since deamination reactions are often reversible, they may also be used to synthesize amino acids.

Transamination is the enzymatic transfer of an amino group from one compound to another. It is a process cells can use either to break down excess amino acids or, in reverse, to synthesize needed amino acids. The coenzyme pyridoxal phosphate carries the amino group. For example, the amino acid glutamic acid donates its amino group to oxalacetic acid. The recipient compound becomes an amino acid, aspartic acid. The

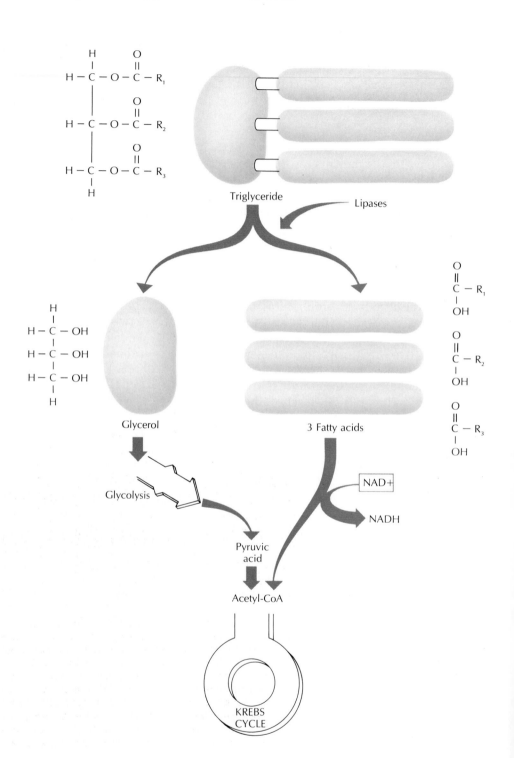

◀ FIGURE 7.16 **Fat Breakdown**
First, the fatty acids are enzymatically separated by a lipase from the glycerol backbone. Then the fatty acids are oxidized. One round of oxidation of an 18-carbon fatty acid removes an acetyl group — two carbons — and retains 16. The acetyl groups go to the Krebs cycle and the glycerol backbone to glycolysis; both are eventually completely oxidized.

FIGURE 7.17 **Amino Acid Conversions**
(a) Deamination removes amino groups. The amino group of glutamic acid may be removed, leaving the organic acid alpha-ketoglutaric acid. (b) Transamination moves amino groups and forms new amino acids at the expense of old ones. In this example, glutamic acid is the amino group donor. Its amino group is transferred to pyruvic acid, yielding alpha-ketoglutaric acid and aspartic acid. Both these reactions are reversible.

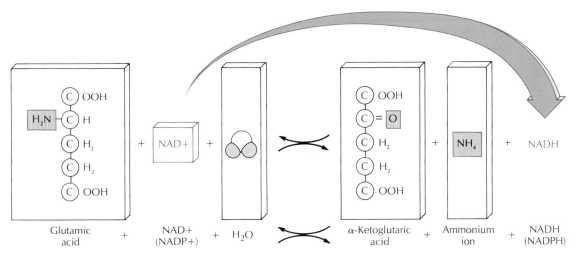

| Glutamic | + | NAD+ | + | H_2O | | α-Ketoglutaric | + | Ammonium | + | NADH |
| acid | | (NADP+) | | | | acid | | ion | | (NADPH) |

(a) Deamination

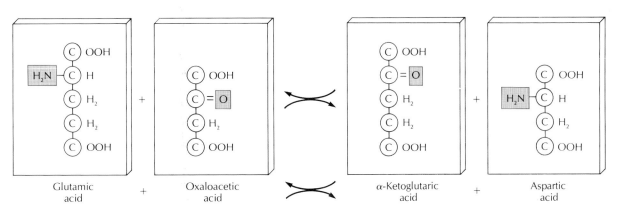

| Glutamic | + | Oxaloacetic | | α-Ketoglutaric | + | Aspartic |
| acid | | acid | | acid | | acid |

(b) Transamination

donor compound, on losing the amino group, becomes an organic acid, oxalacetic acid, which you will recognize as an intermediate of the Krebs cycle (Figure 7.17b).

Note that we are seeing here yet another role for the Krebs cycle — it receives and oxidizes the carbon backbones of amino acids, allowing their energy to be harvested through respiration. In fact, we are now able to see clearly that in respiring cells the Krebs cycle plays the key role in catabolizing all three major organic nutrient groups, carbohydrates, fats, and proteins.

BIOSYNTHESIS

Biosynthesis or anabolism is the cell's way of making new materials. Catabolism, as we saw,

FIGURE 7.18 Biosynthesis — An Overview
The glycolytic pathway and Krebs cycle are the sources of most of the important subunit molecules for the manufacture of the cell's carbohydrates, lipids, proteins, and peptidoglycan molecules.

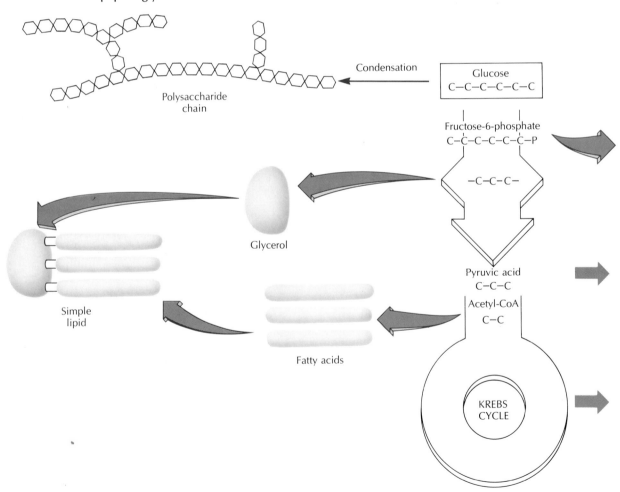

produces ATP and small building-block materials. These are needed for biosynthesis, which consumes large amounts of energy and building block molecules. Biosynthesis often uses different enzymes and pathways than catabolism. It is equally dependent on coupled reactions.

Cells biosynthesize carbohydrates, lipids, proteins, and nucleic acids. The enzymes and coenzymes that make metabolism possible are synthesized by the cell that uses them, as are the protein, lipid, and nucleic acid components of its cellular structures such as membranes or ribosomes. Biosynthesis also provides and replenishes the **cell pool** — the cell's supply of the at least 150 different small building-block compounds needed for cellular growth and reproduction. When we, in an offhand way, have spoken of a substance as being present in a cell, we have been taking biosynthesis for granted. In the section that follows, we will stop taking it for granted, and focus on how cells make new molecules (Figure 7.18).

FIGURE 7.18 **(continued)**

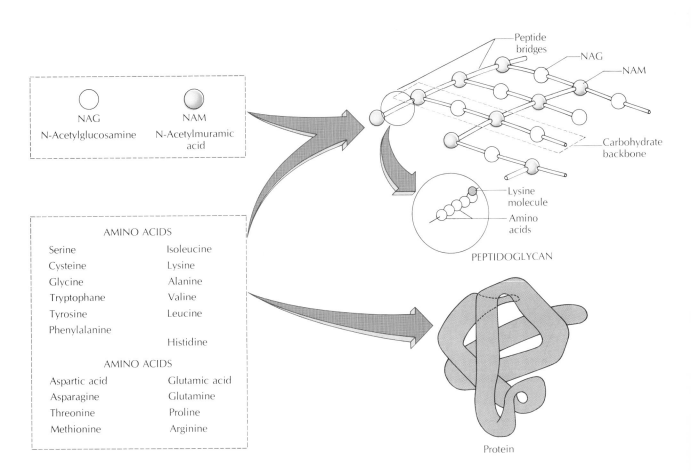

NAG
N-Acetylglucosamine

NAM
N-Acetylmuramic
acid

Peptide
bridges

NAG

NAM

Carbohydrate
backbone

Lysine
molecule

Amino
acids

PEPTIDOGLYCAN

AMINO ACIDS

Serine	Isoleucine
Cysteine	Lysine
Glycine	Alanine
Tryptophane	Valine
Tyrosine	Leucine
Phenylalanine	
	Histidine

AMINO ACIDS

Aspartic acid	Glutamic acid
Asparagine	Glutamine
Threonine	Proline
Methionine	Arginine

Protein

CARBOHYDRATE BIOSYNTHESIS

Microorganisms manufacture both sugars and polysaccharides. The carbon backbones to synthesize glucose and other sugars may be drawn from either glycolysis or the Krebs cycle, and may originate from lipids or amino acids. Once glucose or other simple sugar units have been made, the cell may polymerize them into a variety of polysaccharides. For example, glycogen is a common bacterial energy-storage material. Glycogen is composed of only one type of monomer (glucose) and is thus a **homopolymer.**

The strategy used for glycogen synthesis is rather typical of cellular polymerizations. Any reaction that covalently links two units is strongly endergonic. Both units must first be activated or phosphorylated so they have a high-energy bond. In glycogen synthesis, both the glucose unit to be added and the growing end of the polymer chain are phosphorylated. When the glucose unit is added to the growing polymer a new bond is formed with the release of the activating units. Thus to add one monomer unit (glucose) to a growing polymer (glycogen) the cell must expend the energy equivalent of two ATPs. Clearly, biosynthesis is very energy intensive.

LIPID BIOSYNTHESIS

Lipids are needed primarily to construct cell membranes, but they may also be stored as energy reserves. Triglycerides are synthesized by linking glycerol molecules, derived from glycolysis, with three fatty acids. The fatty acids are synthesized in a stepwise fashion. Remember that fatty acids have even-numbered carbon chains. To synthesize them, a sequence of enzymes add two carbon atoms (an activated acetyl group) at a time to a growing chain. After the acetyl group is added, there are two reduction steps, which add hydrogens. Reduced NADP (a coenzyme closely related to NAD) is the electron donor in fatty acid biosynthesis. NADP is, in fact, the carrier used in most biosynthetic reductions. Fat synthesis stores energy because it absorbs large numbers of electrons that would otherwise be used for ATP synthesis. Carbon dioxide and the B vitamin biotin are additional requirements for fatty acid biosynthesis.

Phospholipids are synthesized from the three-carbon compound 3-glycerophosphate, derived from Step 6 in glycolysis, to which two fatty acids are added. Polar groups such as choline or ethanol are then added to the phosphate.

PEPTIDOGLYCAN BIOSYNTHESIS

Peptidoglycan, a polymer containing both amino sugars and amino acids, is used in building the bacterial cell wall (review Figure 3.10). The two amino sugars N-acetylglucosamine (NAG) and N-acetylmuramic acid (NAM) are both derived from fructose-6-phosphate, from Step 2 in glycolysis. The amino groups are donated by transamination from the amino acid glutamine and the acetyl groups are donated by acetyl CoA. Five amino acids are added one at a time to the NAG and NAM units. Then these complexes are transferred across the cell membrane in an activated form. Once outside the membrane, they are covalently linked to form the repeating network of the polymer.

AMINO ACID BIOSYNTHESIS

A cell needs all 20 amino acids in order to assemble proteins for enzymatic and structural use. The nutritionally self-sufficient bacteria, such as *E. coli*, can synthesize all 20 amino acids from intermediates produced in glucose metabolism. The pathways that result in the synthesis of the 20 compounds are highly complex. There are several major pathways, each of which branches, giving rise to two or more different amino acids. Each branch is catalyzed by a separate group of enzymes, and is controlled by its own feedback-inhibition loop. Note that Krebs cycle intermediates are essential for the manufacture of eight amino acids. This shows again the central importance of the Krebs cycle. Only histidine is synthesized independently of carbohydrate metabolism.

METABOLISM IN LITHOTROPHS

Thus far we've been concentrating on the metabolic processes of heterotrophs, organisms that use organic compounds and thus are dependent on other organisms for their sources of nutrition. Now we will turn to the lithotrophs. Photolithotrophs such as the photosynthetic bacteria and the cyanobacteria, and chemolithotrophs such as the sulfur-oxidizing bacteria, are self-feeders. Lithotrophs are independent because they are able to get the energy to synthesize ATP (and thus other materials) from environmental sources such as light or the electrons of an inorganic chemical. Lithotrophs differ from heterotrophs in that they can satisfy their need for carbon by taking in carbon dioxide and reducing it to carbohydrate and other cell materials. This is a special form of biosynthesis. However, lithotrophs resemble respiratory heterotrophs, in one key feature; their method of converting environmental energy to ATP is based on the chemiosmotic coupling of electron transport to ATP synthesis.

PHOTOSYNTHESIS

Photosynthesis, in the most general terms, is the conversion of light energy to chemical energy, using chlorophyll and other pigments to trap the light energy. Cells may use light energy just to synthesize ATP or to both synthesize ATP and reduce NADP. Photosynthesis also transforms carbon dioxide into carbohydrates, using the energy from the ATP and the reducing power from reduced NADP. The chemical equation for the oxygen-yielding type of photosynthesis is

$$6CO_2 + 6H_2O \longrightarrow C_6H_{12}O_6 + 6O_2.$$

As you will see, this equation presents as a simple one-step operation something that is actually much more complex.

Some true bacteria, the halobacteria, the cyanobacteria, the algae, and the plants are photosynthetic. The equation shown above represents photosynthesis as it occurs in cyanobacteria and eucaryotes.

Those true bacteria that photosynthesize are mostly anaerobic and carry out a simpler form of photosynthesis that does not produce oxygen gas. Some true bacteria switch to heterotrophic nutrition under suitable conditions.

Cyanobacteria, algae, and plants carry out a more complex photosynthetic process, which captures much more light energy, and splits water molecules, giving off molecular oxygen. These organisms are aerobic, and almost all are obligate lithotrophs.

The key feature of all photosynthetic organisms is multiple energy-capturing **photosystems,** clusters of chemicals embedded in membranes. Photosystems are made up of three components. The **chlorophyll** pigments absorb red and blue wavelengths of light. Chemically, the chlorophylls are closely related to heme and the cytochromes. When chlorophyll absorbs light energy, it ejects an electron. The released electron reduces an electron-transport coenzyme.

Electron-transport systems are closely associated with the chlorophyll in the membrane structure. They contain the coenzymes ferredoxin, quinones, and cytochromes and pick up and transfer the electrons evicted from the chlorophyll. The released electrons are highly energetic and may be used to make ATP or reduce NADP to NADPH.

The **antenna** is a group of light-harvesting pigments such as yellow to red carotenoids or red and blue phycobilins. These pigments absorb a broad spectrum of wavelengths of light to augment the light absorption of the chlorophylls.

In looking further into the complex reactions of photosynthesis, we see that they consist of two parts, the light reactions and the dark reactions. The **light reactions** occur only while light is provided. They generate ATP and reducing power. **Dark reactions** fix carbon dioxide into carbohydrate; they depend on the products of the light reactions but do not directly require light energy.

Light reactions can take place only in membrane structures, which ensure the correct spatial orientation of the components. The light reactions coordinate several processes. First, the light energy is used in one photosystem to remove high-energy electrons from the chlorophyll (Figure 7.19a). The electrons reduce $NADP^+$ to NADPH and thus enter the electron transport system. They travel through the system, giving up their energy. The energy is used to pump protons out through the membrane via a chemiosmotic mechanism. As the protons reenter, the energy of the proton gradient is harvested in ATP synthesis. The photosynthetic bacteria, having only one photosystem, can use light only for ATP production.

In the oxygen-producing photolithotrophs, on the other hand, there are two photosystems, one of which carries out the steps just described. Simultaneously, the second photosystem captures additional light energy, used to split the water molecule into hydrogen (which reduces NADP to NADPH) and oxygen. The use of two photosystems thus provides both ATP and reducing power in the form of NADPH.

If this analysis of the light reactions sounds rather like our discussion of respiration earlier in this chapter, it's not accidental. Biological membranes have evolved as efficient energy-coupling structures. Thus the same basic physical and chemical strategies have been adopted by cells as they use membranes to trap energy in both respiration and photosynthesis. The differences should not distract us from the overall similarity.

The **dark reactions** do not use light energy directly. Instead, they use the products of the light reactions. The purpose of the dark reactions is to "fix" carbon — that is, to reduce CO_2 into carbohydrate (Figure 7.19b). Carbon fixation is highly endergonic. Energy is required both to reduce the highly oxidized carbon dioxide and to forge covalent bonds between the carbons to form carbon chains. The dark reactions consume ATP and require a source of electrons. Hydrogen gas and H_2S (hydrogen sulfide) are the electron donors used by the photosynthetic true bacteria, which are all anaerobes. H_2O is the electron donor used by cyanobacteria and eucaryotes.

The **Calvin cycle** is the major metabolic pathway used in fixing carbon. Carbon dioxide becomes fixed when it becomes covalently bonded to a five-carbon sugar, ribulose diphosphate (RUDP), to form a six-carbon intermediate. Further conversions reduce the intermediates, using electrons carried by NADPH and energy input by ATP. The Calvin cycle yields fructose-6-phosphate as its product, and regenerates RUDP, its five-carbon starting material. Fructose-6-phosphate can be converted to glucose; either fructose-6-phosphate or RUDP can be used for polysaccharide synthesis. The glucose can be catabolized for energy (even photosynthesizers need midnight snacks). Some green algae and certain green plants such as the grasses use a different metabolic pathway called the C_4 pathway to synthesize carbohydrates, as an alternative to the Calvin cycle.

To summarize, each molecule of carbon dioxide is fixed via a cyclical pathway in the dark reactions. Each turn of the cycle requires that three molecules of ATP and two molecules of NADPH be supplied by the light reactions. Thus the light and dark reactions are closely linked, and neither runs for long in the absence of the other.

CHEMOLITHOTROPHY

Now let us turn to the unique metabolic process called chemolithotrophy. As previously mentioned, chemolithotrophy is a metabolic pattern in which energy is obtained from the environment through the oxidation of inorganic chemicals, and used to fix carbon dioxide. This nutritional pattern is found only among procaryotes; in certain soil and water bacteria, and some archaebacteria.

Chemolithotrophy can be divided into two processes, just like photolithotrophy. First, the cell must generate ATP and a source of electrons for reducing power. These organisms reduce large amounts of NADPH by oxidizing an inorganic substrate. Reduced forms of hydrogen, sul-

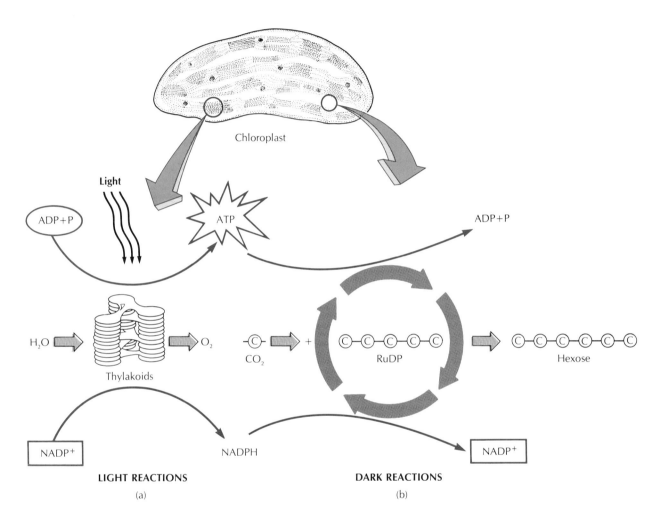

LIGHT REACTIONS

(a)

DARK REACTIONS

(b)

FIGURE 7.19 Photosynthesis
(a) The light reactions. When light strikes a photosystem, shown here located within the thylakoid of a chloroplast, its energy evicts electrons from chlorophyll molecules. These energetic electrons break down water, activate electron transport, and drive ATP synthesis. Light reactions are always membrane-bound, occurring in chloroplasts as shown, or in layers of thylakoid membranes in procaryotes. (b) The dark reactions. In the Calvin cycle, the ATP and the reduced NADP from the light reactions provide the energy needed to attach the carbon or carbon dioxide to a carbon chain and the reducing power to partially reduce it to a carbohydrate. The dark reactions take place in the stroma of the eucaryotic chloroplast or in the cytoplasm of the procaryotic cell. The dark reactions draw on the products of the light reactions.

fur, iron, and nitrogen are among the inorganic electron donors exploited. Next, some of the electrons can pass through the respiratory electron-transport chain to yield ATP. Other electrons will be used for reduction in carbon dioxide fixation. Chemolithotrophs usually have highly developed, specialized membrane structures in which their electron-transport systems are embedded.

Second, the Calvin cycle carries out carbon fixation. The ATP and electrons from the first steps are used to incorporate carbon dioxide into sugars. All the chemolithotrophs use the Calvin cycle to fix carbon, except the methanogens, which are archaebacteria. Their method of carbon fixation is unknown at present.

How should we wrap up our overview of metabolism? Certainly, cells use a lot of different metabolic pathways (believe it or not, we have looked at only a few). Yet all the pathways have a common pattern. All metabolic pathways are "about" the transfer of energy, as carried by electrons (Figure 7.18). We can summarize catabolic patterns as involving the removal of high-energy electrons from a chemical donor, their transfer to an acceptor at a lower energy level, and the capture of part of the energy differential by synthesizing ATP. Every synthetic pathway, on the other hand, involves using some of the ATP (and often some of the electrons) to assemble small molecules into bigger ones to serve the cell's purposes. Although the details may vary, the basic strategy is always the same; it must be. The ability to control energy in this way is one of the characteristics that define life.

SUMMARY

1. Cells are energy-converting systems that capture or extract energy from many sources, then redirect it to their own growth and maintenance needs.

2. Each chemical reaction is catalyzed by a particular, unique enzyme. Enzymes are classified by the type of reaction they perform, their location, and the cellular regulation of their activity rate and synthesis.

3. A catalyst facilitates a reaction, causing it to occur more quickly. An enzyme's catalytic ability depends on the highly specific fit between enzyme and substrate. This fit brings substrate molecules into close proximity, markedly increasing the probability of a reaction. Also, for new compounds to be formed, the original compounds must be altered to expose combining sites; this means that covalent bonds must be broken, a process that requires energy. However, enzymes can alter the chemical environment around the substrate to reduce the activation energy needed to start the reaction.

4. There are several levels of regulation for enzyme activity. Existing enzyme molecules can be con-

trolled or stopped by feedback inhibition. The cell's genetic information may be regulated to start or stop the manufacture of new enzyme molecules.

5. The energy-utilizing mechanisms of cellular life channel energy from catabolism to biosynthesis, using coupled reactions in which the exergonic portion drives the endergonic portion. These processes are always less than completely efficient.

When heterotrophs oxidize organic compounds, electrons are removed and passed to coenzymes such as NAD^+. The electrons are a source of energy that drives phosphorylation. The processes are most efficiently coupled by chemiosmosis with membrane-bound electron-transport systems.

6. Respiration requires a functioning electron-transport mechanism. The electrons are transferred to oxygen (aerobic respiration) or to some other inorganic acceptor (anaerobic respiration). As long as the electron acceptor supply is adequate, cells will respire.

7. The cell derives electrons from carbohydrates, fats, proteins and other organic substances. Glycolysis partially oxidizes and cleaves the glucose molecule to yield pyruvic acid.

8. In anaerobic conditions, facultative and strictly anaerobic bacteria use pyruvic acid as an organic electron acceptor in fermentation. Gases, organic acids, and alcohols are produced as end products.

9. If the cell is respiring, and a supply of NAD^+ is present, pyruvic acid will be decarboxylated to an acetyl group, then oxidized via the Krebs cycle to carbon dioxide.

10. Fats are cleaved to yield glycerol and fatty acids. The glycerol enters the glycolytic pathway; fatty acids are oxidized to remove acetyl groups that enter the Krebs cycle.

11. Proteins are hydrolyzed into amino acids. The amino group is removed by deamination or transamination, leaving an organic acid that directly or indirectly enters the Krebs cycle.

12. Biosynthesis uses up the ATP and reduced coenzymes in producing needed new molecules, such as carbohydrates, fats, and amino acids.

13. Photosynthetic organisms capture light energy. Chlorophyll molecules absorb light energy and emit electrons. As the emitted electrons move through electron-transport systems, chemiosmosis powers ATP synthesis. Simultaneously, light energy may also drive reduction of coenzymes such as NADP. The ATP and the reducing power generated are then used, via the Calvin cycle, to fix carbon dioxide into glucose.

14. Chemolithotrophic organisms derive energy from reduced inorganic compounds, oxidizing them so that the released energy drives phosphorylation and coenzyme reduction. The Calvin cycle carries out carbon fixation in most of the chemolithotrophs.

15. All metabolic pathways have common patterns. High-energy electrons are removed from a chemical donor, transferred to an acceptor at a lower energy level, and the energy differential captured by the synthesis of ATP. Although the sources of electrons, the nature of the final acceptor, and the means of phosphorylation may vary, the basic pattern remains the same.

DISCUSSION QUESTIONS

1. Find examples of the six classes of enzymes in the pathways illustrated in this chapter.

2. Review the different levels of control of enzyme activity.

3. Find some examples of the operation of the laws of thermodynamics in everyday life. (*Hint:* start with your car.)

4. Find some examples of phosphorylation in the pathways we have studied.

5. Describe why a pH gradient across a membrane can serve as a source of energy.

6. Follow one of the carbon atoms in a glucose molecule through glycolysis and one turn of the Krebs cycle. Where does it end up?

7. The Krebs cycle is a pathway for the oxidative breakdown of carboyhydrates.

Krebs-cycle enzymes may be present and extremely important in organisms growing anaerobically without carbohydrates. Why?

8. If an aerobically growing organism is deprived of carbohydrates, its Krebs cycle can be kept operative. Where might it get the intermediates?

9. Facultative anaerobes function both fermentatively and respiratively. What would be happening differently under these two conditions?

10. Describe how light energy is captured and converted to chemical-bond energy.

11. In what ways does photosynthesis in the true bacteria differ from photosynthesis in cyanobacteria?

ADDITIONAL READING

Brock TD, Smith DW, Madigan MT. *Biology of microorganisms.* Englewood Cliffs, N.J.: Prentice-Hall, 1984.

Curtis H. *Biology.* 4th ed. New York: Worth, 1983.

Gottschalk G. *Bacterial metabolism,* New York: Springer-Verlag, 1979.

Stryer L. *Biochemistry.* 2nd ed. San Francisco: W.H. Freeman, 1981.

Ferguson SJ, Sargato MC. Proton electrochemical gradients and energy transduction processes. *Annu Rev Biochem* 1982; 51:185–217.

Haddock BA, Jones CW. Bacterial respiration. *Bacterial Rev* 1977; 41:47–99.

Harold FM. Ion currents and physiological functions in microorganisms. *Annu Rev Microbiol* 1977; 31:181–203.

Zubay GL. *Biochemistry.* Reading, Mass.: Addison-Wesley, 1983.

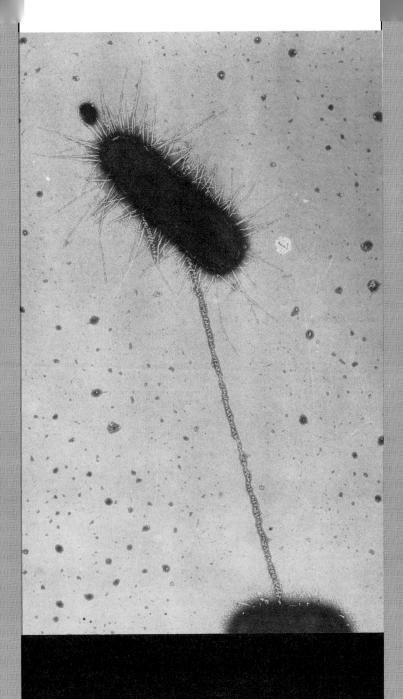

MICROBIAL GENETICS

CHAPTER OUTLINE

Just like human offspring, microbial progeny resemble their "parents" because they inherit their characteristics. Over the past half century we have made exhilarating progress in our understanding of inheritance, most of it first obtained through experiments on microorganisms. A few species of bacteria, a couple of yeasts, and a handful of algae have served as the indispensable "guinea pigs" of modern genetic research. As time passed, we also discovered that procaryotic and eucaryotic inheritance share many unifying principles. But there are also sharp differences.

A **gene** is a unit of heredity. Chemically, a gene is a sequence of nucleotides in a nucleic acid molecule (Figure 8.1). A gene has two functions: one is to carry the information to direct a particular cellular behavior; the other is to replicate itself. **Genetics** is the study of genes and their functions. Our approach will be that of **molecular genetics,** focusing on the nucleic acid and protein molecules and their chemical uniqueness. You might find it useful to refresh your memory by reviewing the sections of Chapter 2 that deal with the chemistry of the nucleic acids.

In this chapter we will discuss the genetics of both bacterial and eucaryotic cells. We will analyze how their genetic information is replicated, and how cells use this information to assemble their proteins. We will enlarge our understanding of how cells control their functions by turning genes on and off. Another objective will be to examine the natural mechanisms, mutation and recombination, that cause genetic change by altering nucleic acid sequences. We will see how recently developed laboratory techniques allow us to tailor new gene combinations, with an expanding range of practical and ethical implications.

In Chapter 7, we looked at biosynthesis as a metabolic process that creates order by assembling macromolecules from atoms and molecules. Biosynthesis, as we saw, consumes a large amount of metabolic energy from ATP generation. However, it requires more than energy — it also depends on a genetic blueprint, a source of information. In this chapter, our primary concern will be with this blueprint, and how the cell uses it.

FIGURE 8.1   A Gene
This transmission electron micrograph shows a gene called the *lac* operon, responsible for lactose metabolism. The gene was isolated from the bacterium *E. coli*. It is actually about 1.4 μm long. (×191,800. Courtesy of Lorne A. MacHattie, University of Chicago, from Shapiro, J. et al., *Nature* (1969) 224:768–774.)

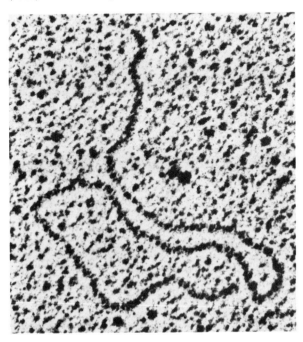

HOW IS INFORMATION CODED?

We first need to come to an understanding of what we mean by **information** when we are talking about a cell. People use languages as their primary information sources, for tasks such as translating French novels or assembling model airplanes from the instructions. We pass on large amounts of accumulated information to our children as we teach them, using language. It may seem to be stretching a point to suggest that any such human activities could be compared to events going on inside a bacterium, which doesn't (so far as we know) talk! At the most basic level, though *information equals order.* Human languages contain order because of the letters in their alphabets and the defined meanings of their words. The order in proteins and nucleic acids is contained in the sequences of variable units. Just like words in a sentence, these molecular subunits have to be placed in order.

DNA is the central repository of all coded information in cellular life. However, as we will see later, both RNA (in certain viruses) and certain very odd protein molecules (in infective particles called prions) can encode information and act as genes.

Nucleic acids and proteins cannot be correctly assembled without reference to a pattern or mold that physically guides their formation. Such a mold is called a **template.** Thus not only do genes contain information in the specified sequence of their units, but a second detailed source of information, the template, is needed to replicate genes. To complete our earlier comparison to human parents and children, unless a cell or virus can pass on its genetic information to offspring, the information is of no lasting significance.

INFORMATION COPYING AND EXPRESSION

Cells have a basic pattern of information flow centering on DNA. When the cell prepares to reproduce, it replicates or copies all of its DNA, which you will recall is a double-stranded double-helix structure. After this, each progeny cell may have a complete set of instructions. On the other hand, when a portion of the information coded in the DNA is to be used to make things happen in the cell, only the appropriate sections of the DNA information are decoded. This involves transferring the DNA information to single RNA copies. The RNA copies then apply the information to the task of assembling a protein:

$$\text{DNA} \underset{\substack{\text{gene}\\\text{replication}}}{\longleftarrow} \text{DNA} \underset{\substack{\text{gene}\\\text{expression}}}{\longrightarrow} \text{RNA} \longrightarrow \text{protein.}$$

We can understand the two different but compatible things that genes do. One function is **gene replication,** the process of supplying information to the next generation of cells. The second function is **gene expression,** which is the process of putting the stored information to work (Figure 8.2). Only some of the genes in a cell are being expressed at any one time. Gene expression is selective, and adjusted to the cell's immediate needs.

Let me caution you that although molecular genetics has a lot of rules, almost every one of them is regularly broken. In later chapters we will encounter viruses that don't need DNA at all, enzymes that copy RNA "backward" into DNA, and the protein bodies called prions, which appear to replicate themselves quite effectively with no nucleic acids at all.

When you are trying to picture what a gene can do, think about a floppy diskette for a microcomputer. You can copy it, so that someone else can have the information to use on his or her machine. Or you can load its program into the computer and run it, putting the stored information to work. Just as a cell does, you put only a certain program to work at any particular time, and you choose the program you use for the job you want done.

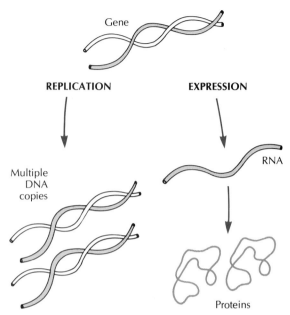

FIGURE 8.2 The Two Functions of DNA
DNA can either be replicated to make additional DNA copies for new cells, or its genes can be expressed to govern the activity of the existing cell.

DNA — ITS STRUCTURE, SYNTHESIS, AND ORGANIZATION

Deoxyribonucleic acid (DNA) is the material of which genes are made in all known cellular life forms. We shall become familiar with its physical structure, how it is replicated, and how it is organized into chromosomes.

THE BACTERIAL GENOME

The **genome** of an organism is the total of all its genes. The **genotype** is an itemized designation of some or all of an organism's genes, some of which may be "visible" through their effects on the organism, and others of which may be "invisible" since they are not being expressed.

Phenotype is a term that designates only the visible characteristics of the organism, the results of the action of its genes.

A procaryote's genome encodes all its essential genetic information in one circular chromosome, a closed double helix of DNA. In its natural state, the bacterial chromosome is often supercoiled into loops. There is no nuclear membrane separating it from the rest of the cell contents. Chemically pure, physically intact DNA can be extracted from the bacterial cell far more readily than it can from eucaryotic cells. For this reason, bacterial DNA is widely used for research.

DNA REPLICATION. A bacterium will normally replicate its chromosome once in each division cycle. The first step in replicating DNA is to unwind the helix (Figure 8.3). Unwinding takes energy and is catalyzed by an enzyme. Next the hydrogen bonds holding the base pairs together must be broken by another enzyme, separating the two strands of the helix. Unwinding and helix separation create **replication forks.** Replication forks are the site of the actual copying of the DNA. The copying process is complex; in *E. coli*, at least 13 enzymes are associated with the fork.

The actual replication of DNA is asymmetrical. You will recall that the backbones of DNA strands consist of alternating sugars and phosphates. Each DNA strand has two ends, one bearing a free (unbonded) phosphate group and the other bearing a free hydroxyl group. Since the paired strands run in opposite directions, and since enzymes add new nucleotides only to hydroxyl ends, the two strands must be copied in opposite directions. Some molecules of the copying enzyme, **DNA polymerase,** go "up" one side of the fork, while other molecules of the enzyme go "down" the other.

As you learned in Chapter 2, the bases in the nucleotides pair only with their complementary partners. Thus on each DNA strand the enzymes insert the nucleotide that is complementary to

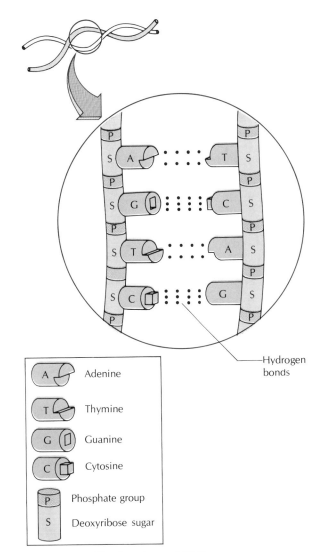

FIGURE 8.3 The Structure of DNA
A DNA double helix has two strands. The backbones of the strands consist of alternating covalently bound sugars and phosphates. One of four nitrogenous bases — adenine, guanine, cytosine, or thymine — is attached to each sugar. Each base pairs with a specific, complementary partner, holding the strands together by hydrogen bonds. Base pairs are the cell's means for transferring and decoding genetic information.

the exposed base on the original strand. That is, where an adenine is exposed on the old strand, a thymine is put into the growing new strand (Figure 8.4). The new DNA is synthesized in sections; these sections are then linked together by **DNA ligases,** enzymes that attach cut or free ends of DNA strands. DNA ligase also detects and "erases" any sections where the bases do not pair correctly because of errors, and fills in the spaces with the correct nucleotides.

The replication fork begins at an initiation point, then moves bidirectionally around the circular chromosome until there are two complete double-stranded daughter helixes. DNA synthesis is **semiconservative,** meaning that each daughter helix ends up with one "old" strand plus one "new" strand synthesized as its complement.

After the new DNA strands are synthesized, some bases may be enzymatically modified, by adding methyl groups, for example. These chemical changes, called **modification,** mark the DNA as unique to the bacterial species in whose cells they occur. Modification makes the DNA of *Escherichia coli* different from that of *Streptococcus pneumoniae*. Modification protects the DNA from being destroyed by the cell's own DNA-digesting enzymes.

DNA ORGANIZATION

There are at least four functional types of DNA, only two of which code for proteins.

- The first type, in procaryotes the largest portion, contains **structural genes.** These are the genes that give rise to enzymatic and structural proteins for cell growth.
- There are genes that yield protein products, **regulatory proteins** such as the repressors, whose cellular role is to influence the expression of other genes.
- The third type of DNA is composed of genes for ribosomal RNA and transfer RNA. The RNA products of these genes are used to assist in assembling proteins.

- The fourth type of DNA is not copied into RNA at all. This DNA forms the regulatory regions containing spacer, starter areas, and stop signals.

It is important to understand that in DNA, only one of the two strands in a region functions as a source for specific protein-assembling information. We call this the "sense" strand. The other strand, the complementary DNA "nonsense" strand, has an equally important function; it exists as a template for replicating the sense strand of the DNA.

DNA is not just an unpunctuated string of nucleotides, or a "string of beads." Although it has been hard to find out how the genes are organized on the chromosome, it now seems clear that most if not all genes are sandwiched in between "start" and "stop" signals. These are the punctuation on the chromosome. In bacteria, sometimes the structural genes for a group of proteins that cooperate in a function — several enzymes in a pathway, for example — will be arranged together in the genome, and operate under the same control signals. This type of grouped arrangement of control signals plus structural genes is called an **operon** (Figure 8.5).

PLASMID DNA. Bacterial cells often contain a great deal of DNA that is not part of their chromosome. This DNA is found in **plasmids,** the small closed circles of DNA found in the cytoplasm of procaryotes and eucaryotes. Plasmid DNA is copied in the same way as the main chromosome, but not necessarily in synchrony. A host cell will treat any gene that is part of one of its plasmids as one of its own, to replicate and express.

Plasmid DNA is chemically like the host DNA, but its information is different. The genetic material plasmids carry is not essential to the host's survival, so host strains may be found both with and without plasmids. However, plasmids may alter the host's behavior in many important ways. For example, one plasmid in *Bacillus anthracis* genetically enables its bacterial host to produce its capsule, while another in-

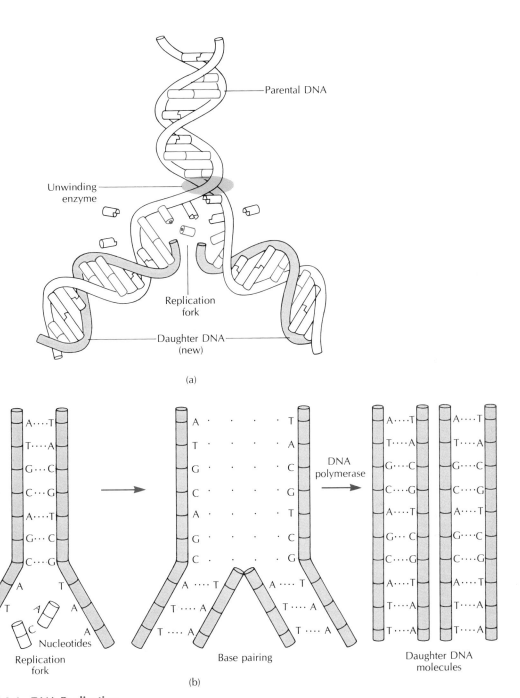

FIGURE 8.4 DNA Replication
(a) At the growing point or replication fork, enzymatic reactions supply energy to unwind the helix. (b) Each strand is copied to form a complementary strand by specific base pairing, followed by linking the nucleotides by DNA polymerase.

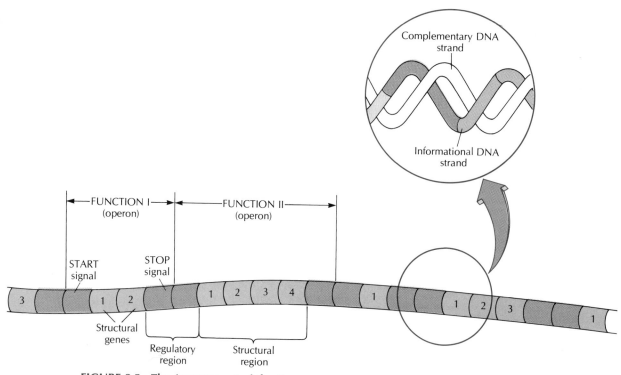

FIGURE 8.5   The Arrangement of the Genome
In bacteria, the genome is arranged in functional regions called operons. Each operon contains a regulatory region and a structural region. The regulatory genes control the one or more structural genes arranged in a sequence. Of the two strands of DNA, one is informational — it bears the protein-assembly code, and is both replicated and transcribed. The other is complementary, and is used only in replication, not in gene expression.

creases its frequency of sporulation (Figure 8.6). Other plasmids enable hosts to produce toxins or resist the lethal effects of an antibiotic. Some are **transmissible** — they have the necessary genes to direct their successful transfer to a new host cell. Others are not transmitted.

THE EUCARYOTIC GENOME

Eucaryotic cells, you will recall, have nuclei enclosed in nuclear membranes containing linear chromosomes (usually in pairs). Their chromosomal material contains a large amount of protein in addition to the DNA. The DNA of the eucaryotic chromosome is organized in regular repeating units called **nucleosomes.** A nucleosome contains a length of DNA of about 140 base pairs, and this is coated by two molecules each of four histones, basic proteins. A nucleosome does not equal a gene because it is too small. The chromosomal material is supercoiled into many large loops. During most of the cell cycle the chromosomal material is extended, and too slender to be seen with a light microscope. During cell division, however, it winds up compactly into rod-shaped structures that can be readily seen and studied with the light microscope.

We have learned that eucaryotic cells reproduce by mitosis. Their cell cycle alternates periods of division with periods during which the cell grows and carries out its normal functions. DNA replication occurs during the latter period. Some portions of the DNA, containing the genes that are most active in the cell, replicate early, while portions containing inactive genes replicate later. The replication process is semiconservative, and uses enzyme systems similar to those of procaryotes.

The genetic organization of eucaryotic DNA is quite complex. A eucaryotic gene may be split; it will contain coding segments (exons) that carry information needed to assemble a particular protein. However, the gene may be repeatedly interrupted by noncoding DNA (introns). After the gene is copied into RNA, the extraneous information (the introns) is enzymatically removed. Eucaryotic genomes also contain extensive repetitive sequences that may serve a regulatory function; they also contain much DNA that serves no known function. Eucaryotic genetic regulation does not appear to use an operon system. It is more complex and variable, and less well understood, than the bacterial regulatory system. Researchers are currently making

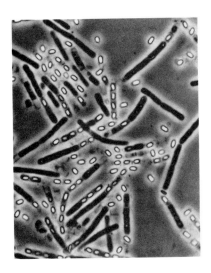

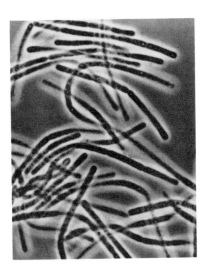

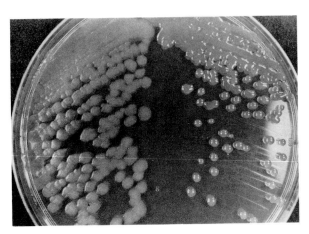

FIGURE 8.6 Phenotypic Effects of Plasmids in *Bacillus anthracis*

These micrographs show two examples of the phenotypic effects of plasmids on the microorganism *Bacillus anthracis*. (a) *B. anthracis* cells after sporulation at 42°C for 20 hours. (b) In contrast, a strain of *B. subtilis* bearing the plasmid pX01 shows no signs of sporulation after an identical incubation period. (c) Colonies of *B. anthracis* growing on an enriched medium on 20% CO_2 to encourage capsule formation: The colonies on the left, showing a granular appearance, are *B. anthracis* without plasmid; the colonies on the right, showing a smooth, mucoid appearance, are *B. anthracis* bearing the plasmid pX02. (Courtesy of Curtis B. Thorne, University of Massachusetts at Amherst.)

rapid headway in figuring out some of the regulatory puzzles, which are keys to understanding major human illnesses such as cancer, birth defects, and immunological diseases.

RNA — ITS STRUCTURE, SYNTHESIS, AND ORGANIZATION

There are several types of RNA, or **ribonucleic acid.** All are transcribed or copied from DNA. You will recall that RNA differs from DNA in that it contains ribose in the backbone instead of deoxyribose, uracil substitutes for thymine, and the molecules are smaller. Much RNA is single-stranded. RNAs differ in their size and in the number of copies a cell has of each. All RNAs act as intermediates in protein synthesis.

THE CLASSES OF RNA

The three major classes of RNA are messenger RNA, transfer RNA, and ribosomal RNA.

- *Messenger RNA (mRNA)* provides direct sequencing instructions to select and line up the individual amino acids that will form specific proteins. mRNA contains many unique molecules, but is only about 5% of the total cell RNA. Each structural gene in DNA, when copied, yields a unique mRNA message. The mRNAs vary in molecular size, averaging 1500 nucleotides in length. They are mostly single-stranded. The nucleotide sequence of one message differs from all other mRNAs except those copied from the same gene.
- *Transfer RNAs (tRNAs)* carry amino acids, which they deliver to the appropriate coded locations in the growing protein. tRNAs are the smallest RNAs, averaging 75 nucleotides in length. The number of types of tRNA present in a cell vary, but there are multiple molecules of each type. Transfer RNA is the second most abundant RNA. Each tRNA molecule has a three-

dimensional secondary structure containing some base-paired regions (Figure 8.7). At one end of the tRNA molecule is a three-base sequence called the **anticodon.** The anticodon is a crucial part of the tRNA's structure. It is the part of the molecule that recognizes the spot to which the amino acid is to be delivered.

- *Ribosomal RNAs (rRNAs)* form part of the structure of the ribosomes, the cellular structure on the surface of which proteins are assembled. There are three different types of rRNAs, but there are many thousands of copies of each per cell. rRNA is always the most abundant type of RNA in the cell. Each cell has multiple DNA se-

FIGURE 8.7   tRNA Structure
A transfer RNA has a primary structure, a sequence of between 70 and 80 nucleotides. It also has a base-paired secondary structure that produces a clover-leaf profile. There are several functional sites. The tRNA binds the carboxyl group of the amino acid to adenine. The right-hand tRNA arm binds to ribosomes. The anticodon, a three-base sequence unique to each different tRNA, binds to a specific codon on the messenger RNA.

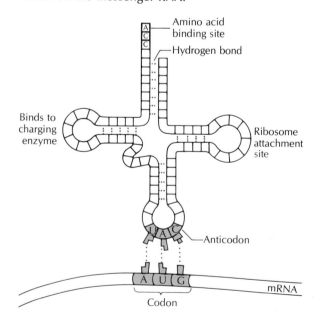

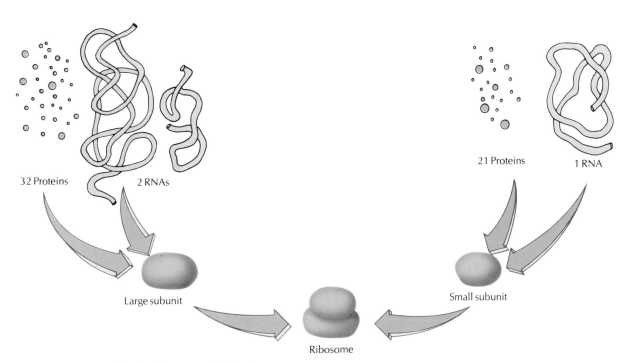

FIGURE 8.8   Ribosome Substructure
Each ribosomal subunit is a complex of a long, linear rRNA molecule in association with numerous proteins.

quences, all of which produce rRNA. A single, long RNA molecule is transcribed from the gene. The large RNA is then cleaved to yield three parts, and each of the three parts becomes a portion of a ribosomal subunit by associating with specific ribosomal proteins (Figure 8.8). A rapidly growing cell of *E. coli* may contain around 180,000 ribosomes.

RNA SYNTHESIS: TRANSCRIPTION

RNA is made by a process called **transcription.** To transcribe, in common terms, means to make a copy of something. Normally, you copy only that portion of a document for which you have an immediate need. RNA synthesis occurs at localized sites on the DNA genome, and single genes or groups of genes are transcribed. In cellu-

lar transcription, RNA copies are made only of those parts of the DNA genome necessary for current activities. The RNA transcripts will be of the mRNA, rRNA, or tRNA type, depending on which gene was copied to make them.

The DNA molecule serves as the template to line up RNA nucleotides, to produce a complementary sequence of nucleotides in the copy. Only the informational strand or "sense" strand of the DNA is transcribed. The transcribing enzyme is **RNA polymerase,** composed of six protein subunits. One subunit in the RNA polymerase binds to the start region on the DNA strand (Figure 8.9). The DNA unwinds locally, then RNA nucleotides base-pair to the exposed DNA nucleotides on the unwound sense strand. Note that since RNA contains uracil rather than thymine, an adenine in the DNA template will call for a uracil in the RNA transcript. The RNA

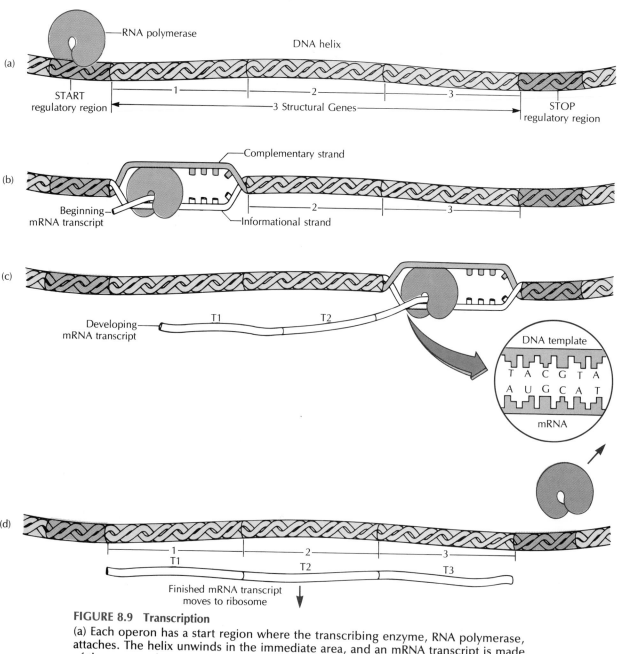

FIGURE 8.9   Transcription

(a) Each operon has a start region where the transcribing enzyme, RNA polymerase, attaches. The helix unwinds in the immediate area, and an mRNA transcript is made of the structural genes on the informational DNA strand (b,c). The stop region prevents further transcription. The DNA helix closes up after the transcribing enzyme has passed through (d), and the finished mRNA transcript moves away to begin directing protein assembly.

polymerase then catalyzes the formation of co-valent bonds between the RNA nucleotides, per-manently linking them together. Another RNA polymerase subunit recognizes the stop signals at the end of the gene and halts transcription.

THE GENETIC CODE. As DNA is being tran-scribed into RNA, information is being trans-mitted. What is the actual content of that infor-mation? Let's remind ourselves that the encoded data contains instructions for sequencing the specific amino acids to be incorporated into a protein. The information is encoded in formal nucleotide patterns. These make up the **genetic code,** the dictionary of the cell. The base se-quences in the DNA form the "words" in the dictionary. Each word, as transcribed into mRNA, is a three-base sequence called a **codon.** Each codon identifies one specific amino acid as the one and only choice to be used next in the protein assembly process. Nucleic acids have a choice of four bases available with which to code. These must provide unequivocal in-structions for 20 amino acids. If the bases are arranged in all the possible different groups of three, there will be 4^3, or 64, possible three-letter "words," more than enough to identify the 20 amino acids. A sequence of experiments has clar-ified the correspondence between codons and amino acids (Table 8.1).

First Position

TABLE 8.1 The genetic code

Second Position

		U		C		A		G		Third Position
U		UUU	Phe	UCU	Ser	UAU	Tyr	UGU	Cys	U
		UUC		UCC		UAC		UGC		C
		UUA	Leu	UCA		UAA	Stop	UGA	Stop	A
		UUG		UCG		UAG	Stop	UGG	Trp	G
C		CUU	Leu	CCU	Pro	CAU	His	CGU	Arg	U
		CUC		CCC		CAC		CGC		C
		CUA		CCA		CAA	Gln	CGA		A
		CUG		CCG		CAG		CGG		G
A		AUU	Ile	ACU	Thr	AAU	Asn	AGU	Ser	U
		AUC		ACC		AAC		AGC		C
		AUA		ACA		AAA	Lys	AGA	Arg	A
		AUG	Met	ACG		AAG		AGG		G
G		GUU	Val	GCU	Ala	GAU	Asp	GGU	Gly	U
		GUC		GCC		GAC		GGC		C
		GUA		GCA		GAA	Glu	GGA		A
		GUG		GCG		GAG		GGG		G

A messenger RNA is an **unpunctuated sequence** of three-base codons. It has a **start codon,** AUG, marking the point at which translation will begin, then a large number of protein-assembly codons, and ends with a **stop codon** marking the point at which translation will terminate, which may be UAG, UAA, or UGA (Figure 8.10). As each codon in the linear message is translated, the equivalent amino acid will be incorporated into a linear protein. Thus the information in the message is said to be colinear with the information in the protein product. If a message contains 1500 bases, it will code a protein of about 500 amino acids. Any change in the sequence or number of mRNA bases will produce a corresponding change in the sequence of amino acids in the protein product.

TRANSLATING THE GENETIC CODE

Translation is protein synthesis, the process in which the mRNA dictates the sequence of amino acids to be incorporated into the growing protein. Translation poses two tactical problems. First, the information encoded in the nucleotide code of DNA and RNA must be decoded into a quite different chemical form, the amino acid code of proteins. This translation process, as with verbal languages, will utilize a "dictionary," the genetic code. Second, the message, the amino acids, and the synthesizing enzymes must be brought and held together so peptide bonds can be formed. In the very simplest terms, we say that protein synthesis lines up the amino acids, then links them together.

Protein synthesis, like all biosynthesis, requires energy. As a preliminary step, amino acids must be activated before they can be incorporated into proteins. Activation is a two-step process. In the first step each amino acid molecule becomes covalently bonded to an ATP molecule. In the second step, each amino acid is loaded on its corresponding transfer RNA. This loading is very exact; that is, a molecule of alanine will be attached only to an alanine tRNA, or a molecule of leucine to a leucine tRNA.

ASSEMBLING THE PROTEIN. Figure 8.11 illustrates the steps in the translation process. To initiate protein synthesis, the messenger RNA first binds to a small ribosomal subunit. Then the first transfer RNA, which corresponds to the start codon AUG, binds to a large ribosomal subunit. The first transfer RNA brings the amino acid formylmethionine. This is always the first amino acid placed in the protein; in some proteins formylmethionine is deleted later, so it is not part of the finished product.

Ribosomes have two tRNA binding sites; they can hold only two tRNAs at a time. Each incoming tRNA always binds in one site first. Once the peptide has been attached to it, it moves over into the second site. Last, after it in turn transfers the peptide to the next tRNA, it leaves the ribosome entirely.

A peptide chain begins to grow as the amino acid from the first tRNA is enzymatically transferred to the amino acid on the second tRNA. A peptide bond secures the two amino acids together, forming a chain of two. Next, the third amino acid is added to the chain after the second. This is how a peptide chain grows. As elongation progresses, the message moves along the ribosome, continuously bringing new codons into the translation area. More energy, supplied by

FIGURE 8.10 How mRNAs Are Arranged
The bases of an mRNA are read in groups of three, called codons. A start codon signals the beginning point for translation. Some protein assembly code follows. Each group of three bases codes for one amino acid. A stop codon at the end terminates translation.

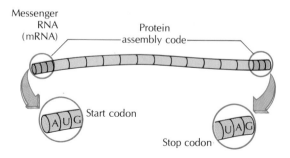

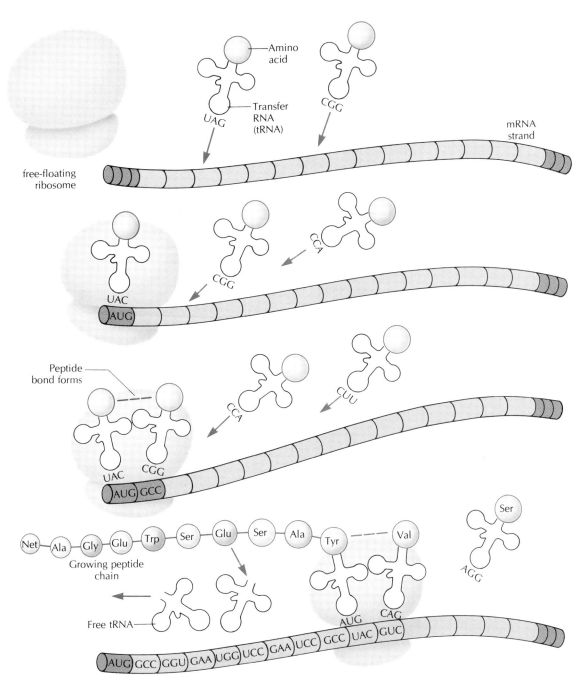

FIGURE 8.11   Protein Synthesis
Assembling a protein requires a messenger RNA, ribosomal subunits, and tRNAs
bearing amino acids. The mRNA binds to the small ribosomal subunit, while
tRNAs are held in place by the large subunit. An mRNA codon attracts and binds to a
tRNA anticodon with a complementary base sequence, holding it in place so
ribosomal enzymes can transfer and bond the growing polypeptide chain to the free
amino acid.

GTP, is required for these movements. The assembly of the protein continues in this fashion until the end of the message to be translated is reached.

The final codon is a stop signal, a triplet that binds no tRNA. When this signal codon is reached, the ribosomal complex dissociates. The polypeptide is released. It may then be modified — it may be split, disulfide bonds may be formed, and carbohydrates or lipids may be added to yield the biologically active final product.

SPECIAL FEATURES OF EUCARYOTES. Eucaryotic protein synthesis is generally similar to procaryotic protein synthesis. However, there are structural differences between procaryotic and eucaryotic ribosomes. These are great enough for substances such as chloramphenicol and tetracycline to selectively inhibit procaryotic protein synthesis. Thus, these antibiotics are extremely helpful in treating bacterial infections. Eucaryotic cells make a much wider range of modified proteins — proteins to which other substances are added, such as lipoproteins and glycoproteins. Thus, after synthesis, many eucaryotic proteins go through extensive modification primarily occurring in the Golgi bodies.

HOW MICROORGANISMS ADAPT

All organisms are able to adapt to their changing environments by making changes in the pattern of expression of their genes. This may alter their phenotype. In fact, many phenotypic changes are changes in gene expression. Let's consider some everyday examples of phenotypic change due to adaptation. For example, sun exposure affects the color and texture of our skins, because it modifies how our genetic potential for skin pigmentation is expressed. Plants do not express their full genetic potential for growth if they are deprived of light or exposed to environmental pollutants. In phenotypic changes such as these the genetic-information content is unchanged; what is altered is the degree to which genes are expressed as the organism seeks to survive.

Microorganisms respond to their environments in ways that visibly change their morphology and staining patterns. During different growth phases, bacteria may exhibit various forms. Some types, such as the myxobacteria and the stalked bacteria, pass through simple life cycles. In response to nutrient deprivation, aging, and other events, most bacteria may lose motility, develop irregular cell shapes, and their cell wall may weaken, changing the Gram reaction. These phenotypic changes are temporary; if we transfer the bacteria to new media they may be reversed. Altered colony characteristics reflect the moisture content of the media and the availability of suitable substrates from which the bacteria may make capsules (see Figure 3.9). Crowding, of course, markedly changes the appearance of colonies, as a result of nutrient shortage.

CONTROLLING GENE EXPRESSION

If a cell is to express a gene, that gene must first be transcribed. The most efficient way for a cell not to express a gene is not to transcribe it. Genetic regulation centers on controlling the activity of RNA polymerase, the transcribing enzyme. The cell can vary the supply of most gene products from none to moderate to maximum.

We noted earlier that genetic functions in bacteria are arranged in operons on chromosomes. Let us look at the operon in more detail. If we read along the DNA in the direction of transcription ("downstream"), we come first to a **promoter,** a portion of the gene to which RNA polymerase attaches. Then comes an **operator region.** This is the section of the gene that functions as a type of on/off switch. **Regulatory genes,** which are not within the operon, produce regulatory proteins or repressors, which may attach to the operator. By doing so they block the forward movement of the RNA polymerase. The structural gene or genes are downstream from the control units. Structural genes are successfully transcribed only if RNA polymerase gets past the operator. This linear sequence — control elements plus "controlled" structural gene(s) — comprise the unit called an operon.

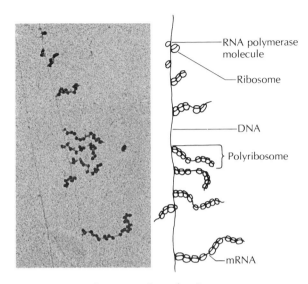

RNA polymerase molecule

Ribosome

DNA

Polyribosome

mRNA

FIGURE 8.12 The Expression of a Gene
This electron micrograph shows a single gene. Nine molecules of RNA polymerase (not all visible) have attached themselves to this gene one after another. The RNA transcripts that are being produced stick out from the gene at approximately right angles. The first to begin synthesis is now the longest. Even as the mRNA is being transcribed off the gene, ribosomes attach and protein synthesis begins. The "oldest" and longest mRNAs have the most ribosomal complexes. Each ribosome, in turn, has a growing polypeptide on it, which is not visible in this preparation. (×47,100. From O.L. Miller, Jr., B.A. Hamkalo, and C.A. Thomas, Jr., Visualization of bacterial genes in action, *Science* (1970) 169:392–395, Fig. 24. Copyright 1970 by the AAAS.)

Operons may produce either inducible or repressible responses to the environment.

THE INDUCTION RESPONSE. Inducible operons are turned on for transcription when a cell encounters a specific substrate (Figure 8.13). They usually regulate catabolic pathways and respond to extracellular conditions. Let us look at one highly studied inducible system, the lactose operon in *E. coli*. The presence of lactose signals induction. Lactose is thus the **inducer.** The lactose operon has a regulator, a protein that binds to the operator region when lactose is not present, preventing transcription. When lactose

is added to the medium, however, it combines with the specific regulator protein. In binding, lactose alters the regulator so that it falls off the operator. Now the RNA polymerase is no longer blocked. Transcription begins immediately (Figure 8.13d), and the cell promptly makes the enzymes for metabolizing lactose. The enzymes function until the lactose (inducer) molecules are consumed. Once lactose is gone, the regulator proteins are again free, and they will once again prohibit transcription.

THE REPRESSION RESPONSE. Repression is a different type of gene-regulation pattern. It turns transcription off when the end product of the enzymatic process it controls builds up. Repressible systems generally regulate biosynthetic pathways and respond to intracellular signals. For example, in *E. coli*, the amino acid tryptophane appears as the last product in a sequence of enzymatically catalyzed reactions. When tryptophane levels are high, the amino acid (the repressor) combines with the regulator protein for the tryptophane operon. The combined regulator–repressor unit binds to the operator and blocks further gene expression.

EUCARYOTIC REGULATION. Nothing exactly like an operon system has been demonstrated in eucaryotic cells. However, eucaryotic genes are tightly regulated. Many regulatory events occur during development. Developing embryos have an intricate timetable in which some genes are turned on, while others are turned off, sometimes permanently. These shifts correlate in still unclear ways with alterations in the coiling patterns of the chromatin and the appearance or disappearance of visible bands in the chromosomes.

Other regulatory events occur throughout life, on a day-to-day basis. Circulating hormones and protein growth factors act as chemical signals having their major influence on gene regulation in the various tissues of higher organisms. We will spend some time in Chapter 12 exploring the curious question of how blood cells "turn on" certain genes used to make protein

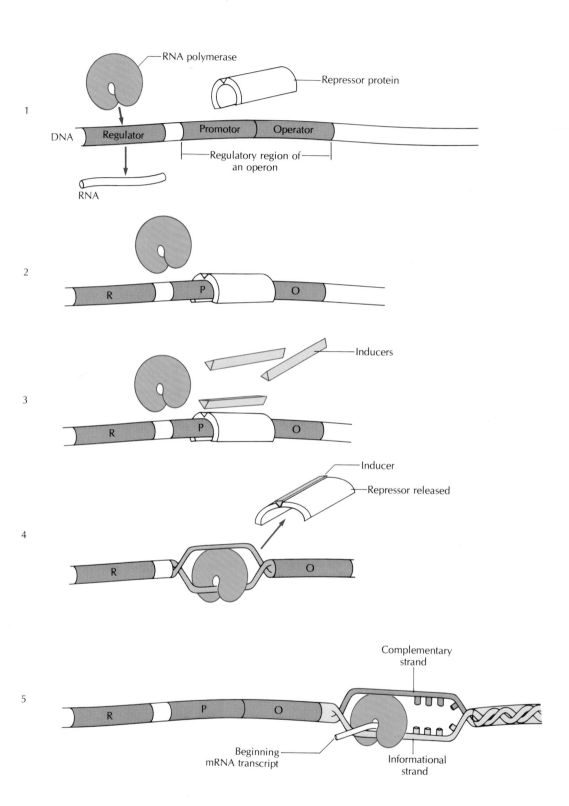

FIGURE 8.13 Induction
An inducible operon has a regulatory region containing an operator and a promoter. In addition, a regulator gene may produce repressor proteins, which can block transcription by binding to the start signal in the operator region. When the repressor protein is in place, RNA polymerase cannot work. At any time, however, an inducer (wedge) may enter the cell and bind to the repressor protein, changing its shape so it no longer binds to the operator. Once the barrier to transcription is removed, RNA polymerase can again bind at the promotor. Structural genes are transcribed, producing enzymes for using the inducer substance.

antibodies. We will also explore evidence that some forms of cancer are primarily a result of genes that have escaped from their normal framework of regulatory signals.

HOW MUTATION CHANGES MICROORGANISMS

There are two basic methods of genetic change, mutation and recombination. **Mutation** is basically a chemical event, a change in the nucleotide sequence of the DNA. Mutation is genetically significant if the DNA alteration causes the gene to code for a dysfunctional protein, or if it alters how the gene is regulated. In fact, mutations cause the formation of **alleles,** alternate forms of a gene. Mutational changes are permanent, barring a statistically unlikely **back mutation,** that is, a mutation that exactly reverses the original change. Mutation is something that happens to an individual cell.

Spontaneous mutations are those for which there is no immediate, apparent cause. Other mutations, the ones we are most concerned about, are induced. **Induced mutations** are changes in the base sequence of DNA provoked by known contact with a chemical or with radiation. The contrast between spontaneous and induced mutations is partly a false distinction. Mutation-causing chemicals, cosmic radiation,

and natural radioactive isotopes are now and have always been part of our natural environment. They are probably the undetected source of most spontaneous mutation. The rest can be attributed to the rare random errors in DNA replication and cell division. Mutation rates have become a matter of pressing social concern in the era of nuclear weaponry and toxic wastes. Researchers once again turn to the bacteria to study how mutation occurs and what conditions increase or decrease mutation rates.

The normal DNA replication process is extremely accurate. Copying errors occur spontaneously at the rate of about one per 100 million bases copied. Changes in a single DNA base are called **point mutations.** There are several types (Figure 8.14).

- **Substitution mutations** occur when an incorrect base is substituted for the normal base within a pair. An example would be the insertion of a G for an A. Substitution mutations occur as a result of temporarily incorrect base pairing. Such an error has the potential for changing one codon. Genetically, the change will be significant if this results in coding for a different amino acid (Figure 8.14b). The rest of the protein will be unchanged unless the mutation creates a stop signal. In that case, called a **nonsense** mutation, an abbreviated protein will be made (Figure 8.14c).

- **Addition mutations** are ones in which an uncoded, extra base is inserted into the DNA. **Deletion mutations** occur when a coded base in the DNA is skipped. Both seem to occur from spatial disruptions of the DNA helix, stretching or contorting. If the number of bases added or lost is anything other than three or a multiple of three, the mutation will cause a frame shift.

- **Frame-shift mutations** are addition or substitution mutations that have the effect of distorting the way in which codons are read. The results are far-reaching. Remember that the message is always read in

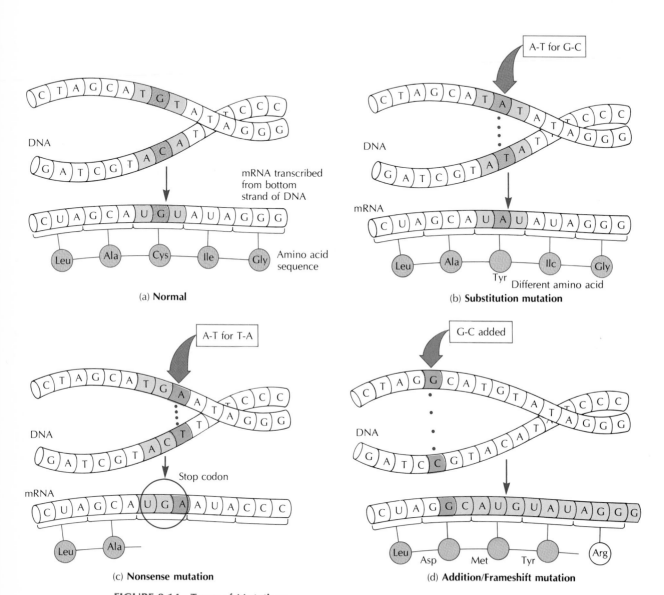

FIGURE 8.14 Types of Mutations
(a) A normal gene segment transcribed into mRNA codes for a sequence of five amino acids. (b) A substitution of adenine for guanine in the DNA leads to a substitution of tyrosine for cysteine in the peptide. (c) A substitution of adenine for thymine creates a nonsense mutation, coding a stop signal inappropriately in the middle of the peptide, which will now not be completed. (d) An addition of an extra guanine into the DNA creates a shift in the reading frame, so that from the second codon onward the code is translated incorrectly.

three-base units and that there is no punctuation. From the error point to the end of the message all codons are read off-center, a frame shift (Figure 8.14d). This gives a different message and a useless protein. Try taking any sentence and separating it into random "words" of five letters long. That's shifting the reading frame, and it completely destroys the sense, without altering a single letter. The translation machinery of the cell will faithfully translate any message, whether it "makes sense" or not.

MUTAGENIC AGENTS

A **mutagen** is a chemical or physical condition that increases the observed rate of mutation. Mutation is entirely random, and all portions of the genome are susceptible. Mutagenesis is a technique used by scientists to produce mutant organisms with desired traits. These techniques have been used on fruit flies, yeasts, bacteria, and mice.

Some chemical mutagens are **base analogs,** which are chemical modifications of the normal purine and pyrimidine bases (Figure 8.15). Base analogs pair incorrectly and the cell may synthesize them into DNA with errors. If the dose of mutagen is high enough, the cell's mutation load will be lethal. This is why base analogs are widely used as drugs in human medicine to treat rapidly growing tumors. Their effect is to kill or inactivate rapidly reproducing cells. Unfortunately, they also injure normal tissues. Other base analogs such as idoxuridine are useful in treating infections with DNA viruses such as Herpes.

All electromagnetic radiation with wavelengths shorter than visible light will cause mutation (review Figure 3.1); so also will high-energy subatomic particles such as protons and neutrons. High-energy radiations have an ionizing effect, creating ions in the cell that may break covalent bonds and actually snap DNA helixes. There are cellular mechanisms that re-

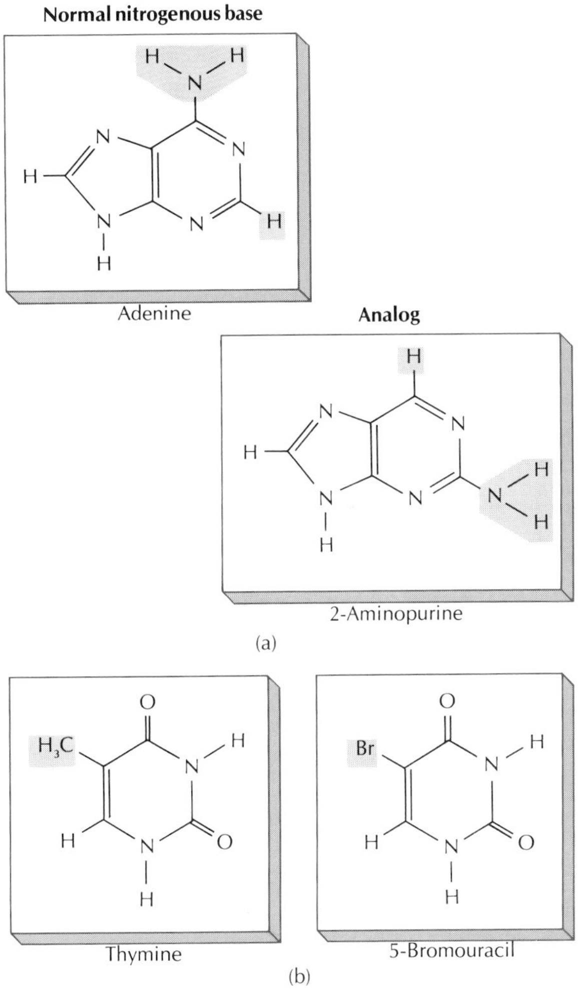

FIGURE 8.15 Base-analog Mutagens
(a) The normal nitrogenous base adenine is converted into a mutagen by the addition of an amino group. (b) The normal nitrogenous base thymine is converted into a mutagen by the substitution of a bromine atom for its original methyl group.

pair this damage, but they are very error-prone, so incorrect bases may be linked into the DNA strands, causing mutations. Ultraviolet radiation forms covalent linkages between adjacent thymines on a strand of DNA, and these interfere with DNA replication. As ultraviolet radiation is a component of sunlight, this explains the lethal effect of sunlight on some bacteria as well as sunlight's ability to cause skin cancer in overexposed humans.

MEASURING MUTATION RATES

Bacteria are particularly useful in testing the potential mutagenesis of a chemical compound. The **Ames test** is widely used to evaluate industrial chemicals, dyes, new drugs, and similar chemicals whose potential for cellular damage must be determined. The bacterium *Salmonella typhimurium* normally does not require the amino acid histidine for growth, since it produces histidine itself. A strain of a "normal" genotype, such as this, is called a **prototroph.** The Ames test uses a mutant strain of *S. typhimurium* that has lost the ability to make its own histidine, and thus requires histidine in the medium for growth. This type of nutritionally dependent mutant strain is called an **auxotroph.** The testing procedure evaluates the chemical's potency in causing the auxotrophic strain to **back-mutate** to the prototrophic state where the organism can again grow in a medium without amino acid because it produces it itself.

The test bacteria are exposed to the suspect chemical in a range of concentrations, then bacterial samples are plated out on media lacking histidine. Only back-mutants form colonies, so the more colonies formed, the higher the mutagenic potential of the chemical tested.

A major and somewhat controversial use of the Ames test is as a rapid screening test for **carcinogens** — cancer-causing chemicals. The logic for using the Ames test in this way is that much work has shown that compounds that are mutagenic are also apt to be carcinogenic. The Ames test is rapid (24 hours) and inexpensive, factors that recommend it for quick screening of suspected carcinogens. Carcinogenicity studies in small mammals such as mice, by comparison, may require two or more years and tens of thousands of dollars per chemical tested. At present,

the Ames test is used first, then if there is anything in the results to arouse suspicion, the compound may be tested in mice or other animals. It has been suggested, however, that the Ames test may miss some potential carcinogens entirely.

PRODUCING AND SELECTING MUTANTS

We may choose to produce mutations in bacteria as an aid to analyzing a chemical's mutagenic potential, evaluating the number of genes involved and the order in which they participate in metabolic pathways, or introducing phenotypic markers that identify a cell and its progeny, thus permitting the analysis of how genes behave. On the practical side, mutagenesis has been widely used to improve the productivity of industrial microorganisms. It is often possible to discern much by comparing a mutant with the normal organism — the **wild-type** strain. The choice of mutagenic agent is dictated by the results sought. The bacteria usually receive a high dose of the mutagenic agent that kills most of them and induces mutation in almost all of the survivors. The lab worker then plates out the exposed population under conditions that select the desired mutants.

A sample of the exposed population contains a huge number of cells, with mutants that have been altered in all conceivable ways. **Selection** techniques are designed to permit growth and identification of the sought-after mutant strains, while suppressing wild-type survivors and other mutants. Certain types of selection are straightforward. For example, you can select for a penicillin-resistant mutant of a penicillin-sensitive microorganism simply by plating the exposed population on media containing the drug. You know that any colonies that develop must be derived from resistant mutants. Other selection procedures are very complex. A useful screening method for large numbers of potential mutants is the **replica-plating** technique (Figure 8.16), shown here being used to isolate auxotrophs that have lost the ability to synthesize amino acids.

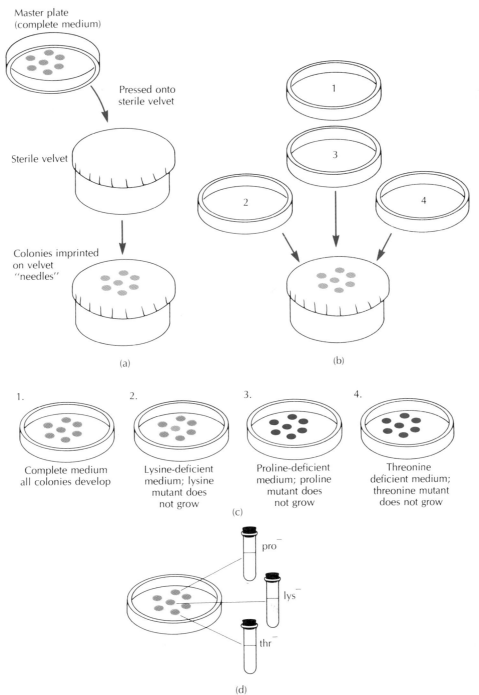

Master plate
(complete medium)

Pressed onto
sterile velvet

Sterile velvet

Colonies imprinted
on velvet
"needles"

(a)

1

3

2

4

(b)

1.
Complete medium
all colonies develop

2.
Lysine-deficient
medium; lysine
mutant does
not grow

3.
Proline-deficient
medium; proline
mutant does
not grow

4.
Threonine
deficient medium;
threonine mutant
does not grow

(c)

pro⁻

lys⁻

thr⁻

(d)

FIGURE 8.16 Selection of Mutant Bacteria
Replica plating uses velvet fabric bearing thousands of miniature inoculating probes
on its surface. A culture is exposed to a mutagenic treatment, then plated out on a
complete medium so all organisms, both normal and mutant, form colonies. Prints
of the colonies are then transferred to the sterile velvet (a). The velvet pad is then
used to inoculate a number of selective media. In our example the media lack one
amino acid, and test for mutants that have lost the ability to make that amino acid
themselves (b). Mutants can be identified by the selective media on which they
do or do not grow (c). They can then be isolated to pure culture on complete
media (d).

NATURAL GENETIC EXCHANGES

Mendel, who we acclaim as the father of genetics, had a view of genes as highly stable units of heredity that followed set rules. He worked out several important laws for inheritance (gene exchange), which every biology student memorizes in his or her first biology class. We tend to get the idea that this is the only way genetics really operates — that all the world behaves like pea plants. However, the last 10 years in genetics have been rather a shock. We are learning rapidly that not only do genes undergo continuous internal change and reshuffling, but also that almost any gene may be transferred to almost any organism, by any of a half-dozen previously unthought-of mechanisms. Microorganisms gave us the first hints about this genetic flexibility.

Recombination is any process by which genes from two separate organisms are brought together into one cell, or even one chromosome. At the simplest level, recombination includes steps for gene exchange between two organisms, followed by steps that prepare the genetic material from the two sources to work together in the recipient cell. Our attention in the remainder of this chapter will be focused primarily on recombination in some of its many forms.

Both eucaryotes and procaryotes undergo recombination at times. The best-understood gene exchanges occur just as Mendel studied them, during **sexual reproduction.** As we saw in Chapter 5, sexual reproduction results in the formation of a zygote in which the genetic material is equally derived from each of two parents (review Figure 5.2). The zygote is **diploid;** that is, it has at least two copies of each piece of genetic information. Each new individual has genes derived from two sources, and is not genetically the same as either parent. By its very nature, therefore, sexual reproduction yields recombination.

Asexual reproduction produces new individuals that are exact copies of their single parent. This can be observed in bacterial populations, where a pure culture normally consists of vast numbers of genetically identical individuals. The cell produced by asexual reproduction is normally **haploid,** having only one copy of each information unit. When single cells with new traits do appear in individuals in a pure culture, we may ascribe them to mutation. Asexual reproduction then, in itself, does not yield recombination. Recombination in asexually reproducing species occurs, but by means not directly associated with reproduction itself, to be discussed in the pages that follow.

When new traits appear in individuals in a mixed population, they can often be traced to recombination of the DNA of two unlike individuals, with the resulting cell having some traits of both. Clearly, to get this kind of effect, we need two different sources of DNA. If the two recombinants are genetic variants of the same species, the process is called **legitimate recombination.** If the two DNAs originate from different species, we call the process **illegitimate recombination.**

In the 1940s, researchers identified three intriguing genetic exchange processes in bacteria. These were named transformation, conjugation, and transduction. The three processes differ in the source of the introduced DNA, and in the mechanics of getting it into the bacteria. All result in the formation of recombinants. Studying these procaryotic phenomena provided a basis of information about genetic recombination. We are now realizing that eucaryotes have an equally large repertoire of both legitimate and illegitimate recombination mechanisms of their own.

INTEGRATING NEW GENES

In bacterial recombination, the donor genetic material becomes a permanent part of the recipient cell's genome in a process called **integration.** At the chemical level, integration occurs when the donor DNA either replaces or is added to the recipient DNA, becoming covalently bonded with the rest of the original DNA. Integrated

BOX 8.1
Who Are You Calling Illegitimate?

Hemoglobin is found in vertebrate animals' red blood cells. But legume plants such as soybeans have a gene to produce a form of hemoglobin. Now, soybeans don't have red blood cells; they don't even have blood. They do use the gene, however, as part of their nitrogen-fixation symbiosis. Where did they get that gene? We certainly are not used to thinking of ourselves as closely related to soybeans!

Legitimacy, socially, has always meant that you know who your parents are. Genetically, it means that recombination occurs between two organisms of the same species, with compatible DNA sequences that can line up side by side before recombining.

Illegitimacy, socially, doesn't have much meaning anymore. In genetics, however, it has increasing importance. An illegitimate recombination is one occurring between DNA segments from unrelated sources, such as humans and bacteria. These segments have little or no base-sequence similarity. They join end to end without a preliminary base-pairing step. These are the type of experiments that are done in the laboratory with engineered plasmids.

Illegitimate recombination also seems to occur freely in nature. The causes are transposable genetic elements or **transposons.** They were first discovered in corn. A transposon is a DNA sandwich — one or more transferable genes, sandwiched between two insertion sequences. The insertion sequences (IS) are inverted repeats, unusual DNA base sequences. A unique enzyme, the **transposase,** recognizes the insertion elements and cleaves the DNA at these points. Thus the transposon is freed to move to another DNA site. That site may be anywhere. There is no apparent prohibition against its inserting itself in the middle of any piece of DNA it comes across — animal, vegetable, or bacterial. All it needs is a compatible insertion site.

Transposons may locate in either chromosomal or plasmid DNA. They move antibiotic-resistant genes (R) from a plasmid to the chromosome, to a virus, or from one plasmid to another. Transposons assemble clusters of resistance factors making the host bacteria **multiple-drug resistant.** They also transfer drug resistance between unrelated species.

In higher organisms, transposons may serve as switches, turning existing genes in the DNA on or off, or they may promote chromosome rearrangements. These events may be an essential part of gene function. An example is antibody synthesis, where gene rearrangement is necessary before antibodies can be made.

We clearly don't know all that transposons can do yet. We also don't know the limits of our own "genetic illegitimacy." Where did that soybean get that gene, anyway?

genes are passed on with the rest of the genome to the cell's progeny, by whatever means of reproduction is used.

How DNA is integrated during recombination varies among the species and the different processes; there is still considerable debate about details. What follows is a highly generalized picture of the steps in integration, as shown in Figure 8.17.

- New or "donor" DNA enters the recipient cell. Normally, donor DNA must be from the same species to be compatible.
- Recombination is based on homology; that is, donor and recipient strands pair up, matching complementary base sequences. The pairing need not be perfect, allowing for some gene differences between the recombinants.
- The strands of both donor and recipient DNA are broken or "nicked" enzymatically by **endonucleases,** enzymes that break DNA strands internally to yield uneven, single-stranded ends.
- The free ends of donor DNA are rejoined covalently to those of the recipient DNA by **DNA ligases,** the all-purpose repair enzymes.

The recipient cell in this example remains a haploid, having but a single copy (the donor's) of the relevant information. It has simply had some of its genes replaced by genes from another source. In other types of recombination a length of donor DNA is added directly into the recipient chromosome. If this is a second copy of the information, the recipient cell becomes a **partial diploid,** a cell having duplicate genes for certain functions. In a third strategy, the new genes neither replace old genes nor duplicate existing ones; they add entirely new genetic capabilities to the recipient's repertoire. In any case, once integrated the new genetic material will be replicated right along with the rest of the genome, and passed on to all progeny in cell division.

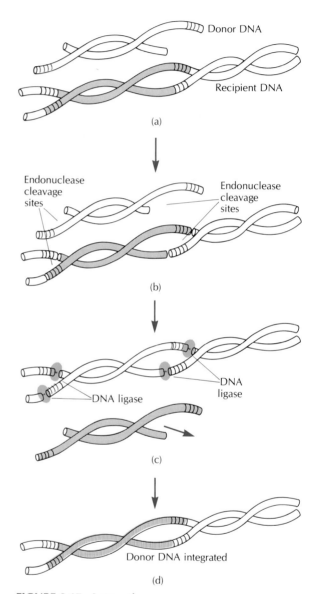

FIGURE 8.17   Integration
In legitimate recombination, entering donor DNA lines up to the site on the host chromosome bearing specific homologous genetic material (a). The host genome is excised by an endonuclease enzyme, leaving short, single-stranded ends (b). A DNA ligase enzyme then repairs the broken strands (c), inserting donor DNA in place of recipient DNA (d).

TRANSFORMATION AND GENE TRANSFER

Transformation is a process in which short pieces of DNA from a dead donor cell are taken up by a living recipient cell called a **competent cell** and integrated into its chromosome (Figure 8.18). It is the simplest example of gene recombination between bacteria. The transformation process is of historical interest. It was first observed in the "transformation" of unencapsulated nonlethal strains of *Streptococcus pneumoniae* into lethal strains, which appear encapsulated, with smooth colonies. This was the first demonstration of recombination in bacteria. Later the "transforming factor" was identified as DNA, providing the first clear proof that DNA was the stuff genes were made of; with this discovery the era of molecular genetics was launched.

The transforming process can transfer roughly 20 genes at a time; these may be from any portion of the donor chromosome. All donor genes are transmitted with equal frequency. Transformation has been observed in the laboratory in at least five major bacterial groups. We assume that it can also occur in nature, when two strains of bacteria are growing simultaneously in a natural environment. Transformation could, and probably does, also occur when a human is infected by more than one strain of a bacterium.

Transformation is an important tool of genetic engineering. It is used to introduce artificially constructed segments of DNA into bacterial hosts.

PLASMIDS AND GENE TRANSFER

Plasmids, you will recall, are small rings of extrachromosomal DNA found in most cells. Transmissible plasmids are those that contain genetic information that directs their host cell to take steps that will allow a special kind of gene transfer — the transfer of the plasmid from the host bacterium to another recipient bacterium.

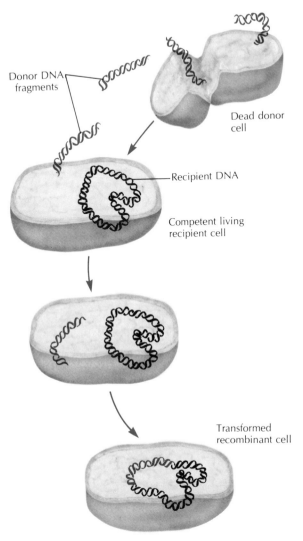

FIGURE 8.18 Transformation
In transformation, a portion of double-stranded donor DNA is released from a dead cell and taken up by a competent living recipient cell. Once donor DNA is integrated, the recipient cell has a transformed genome.

We can perhaps understand the plasmids and their importance by considering the resistance factor plasmids found in *Shigella dysenteriae, Serratia marcescens,* and other Gram-negative bacteria of humans. **Resistance factor plasmids** contain two types of genes. They have genes for their own transfer (the **RTF** or **resistance transfer factors**), plus **R genes** that confer resistance to antibiotics and heavy metals. One resistance plasmid that has been studied has, in addition to the RTF segment, R genes for up to eight different antibiotics, sulfonamide drugs, and four heavy metals (Figure 8.19). Possession of such a plasmid obviously renders a bacterial strain very difficult to control by antibiotics and some disinfectants. What is worse, R factors are freely transferred between strains of the same species and between related species.

How are plasmids transferred? The process typically involves the formation of pili, hollow tubular structures, by the donor cell. The donor cell binds via these pili to a recipient. The plasmid DNA passes out of the donor and into the recipient. Typically, plasmid DNA does not integrate with host DNA either before or after transfer. The two segments of DNA, chromosome and plasmid, remain independent and coexist within the host. The F factor, discussed shortly, is an example.

BACTERIAL CONJUGATION AND GENE TRANSFER

Bacterial conjugation is a special type of plasmid-mediated mating between two bacterial cells. In this mating, some or all of the bacterial chromosomal DNA is transferred. One bacterium serves as a donor or "male" and the other as a recipient or "female." "Maleness" is conferred by the presence of a special segment of DNA containing at least 11 genes.

In its simplest form, this DNA segment exists as a free plasmid, the **F factor.** Cells that have the plasmid are designated F^+. F^+ cells can transmit the F factor to female (F^-) cells that then also become F^+. These cells do not transmit any DNA from their bacterial chromosome.

The strains of bacteria that conjugate, however, are ones in which the DNA segment (the plasmid) has become integrated into the chromosome. Integrated F factor converts the bacteria into **Hfr** males, named for their *h*igh *f*requency of *r*ecombination. Hfr males don't transmit just the plasmid DNA; they also transfer attached portions of their own bacterial DNA. Conjugation occurs between Hfr and F^- bacteria (Figure 8.20). The steps in conjugation are as follows:

- A conjugation tube is formed by the male partner.
- The donor chromosome opens at the point next to the Hfr site. The donor replicates its DNA, passing one copy through the tube to the recipient while retaining the other.
- As long as the cells remain attached, transfer continues, and genes are transmitted in order. If mating is uninterrupted, the entire

FIGURE 8.19   Resistance Factor Plasmids
An electron micrograph of an R factor from *Hemophilus influenzae* reveals that some portions of the circle are double-stranded while others, imperfectly paired, form single-stranded loops. The interpretive drawing shows Ap is the ampicillin-resistance region, and Tc is the tetracycline-resistance region. This organism's resistance to ampicillin is posing an increasing clinical problem. (Courtesy of Dr. R. Laufs, Universität Hamburg, from the *Journal of Bacteriology*.)

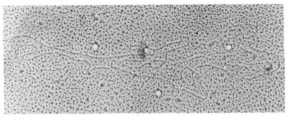

0.5 μm

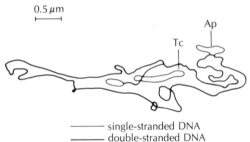

Ap

Tc

——— single-stranded DNA
——— double-stranded DNA

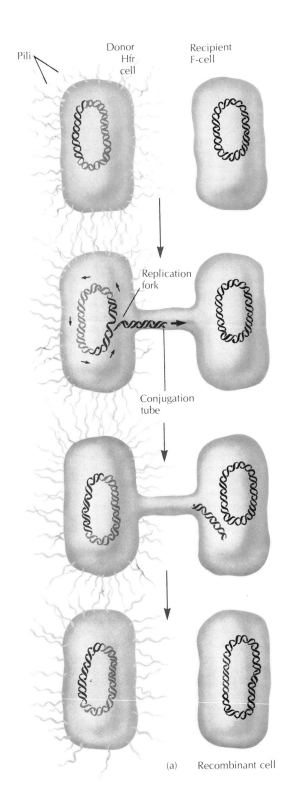

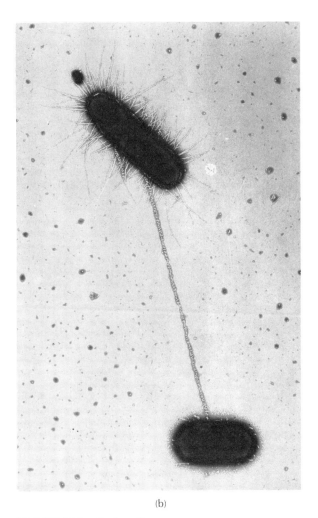

(b)

FIGURE 8.20   Conjugation
(a) The Hfr donor cell forms a pilus or conjugation tube to an F⁻ recipient cell. During conjugation, the donor copies its DNA and one copy enters the recipient. The longer conjugation lasts, the more genes are transferred. Only if the entire circle was transferred would the recipient become Hfr itself (not shown here). When conjugation is over, the cells separate and the donor DNA is integrated. (b) An electron micrograph of conjugation. The donor cell is covered with pili, one of which forms the conjugation tube. (Courtesy of Charles Brinton, Jr., and Judith Carnahan.)

donor chromosome will be transferred. The Hfr genes are the last to be transferred.

- The introduced donor DNA is integrated into the host DNA, replacing it. Both donor and recipient survive the process and end up with a complete genome.

VIRUSES AND GENE TRANSFER — TRANSDUCTION

Viruses live as parasites on cells, including bacterial cells. Viral life cycles will be discussed in detail in Chapter 9. For now, mainly we need to know that viruses infect a host cell, replicate inside it, then move on to another host cell, much like any other parasite. Viruses are important in a gene-transfer process called **transduction.** In transduction, bacterial DNA is carried from donor bacterium to recipient bacterium as a passenger molecule within a virus particle. The gene transfer actually occurs as an accidental side effect of virus infection of the host and recipient.

In general terms, when a transducing virus infects its host bacterium, the viral DNA integrates with the host chromosome. Later, the virus DNA leaves the host chromosome, which fragments. The virus genes direct the cell to make more viruses. The newly produced virus particles, either in addition to or in place of their own DNA, contain host bacterial DNA. After the virus leaves the first host, it infects a second host, injecting the first bacterium's DNA. That "donor" DNA may become integrated into the second host, the recipient (Figure 8.21).

FIGURE 8.21 Transduction
In generalized transduction, a virus first enters a susceptible host bacterium. After interaction, the host DNA fragments, and pieces of host DNA are packed into virus coats. Then the bacterium lyses, viruses are released, and they go on to infect again. The virus injects bacterial DNA from its last host into its new one. This transduced DNA integrates with the second host's own DNA.

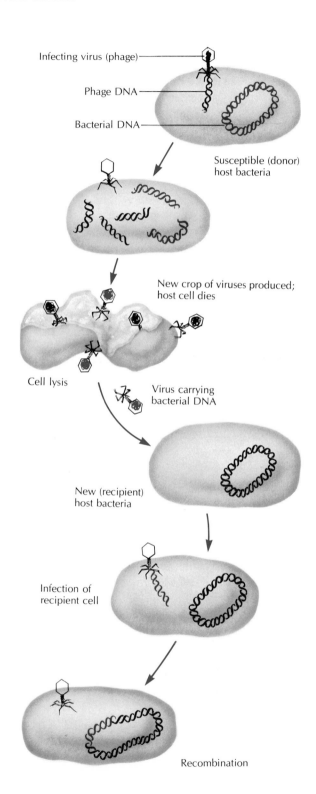

Infecting virus (phage)

Phage DNA

Bacterial DNA

Susceptible (donor) host bacteria

New crop of viruses produced; host cell dies

Cell lysis

Virus carrying bacterial DNA

New (recipient) host bacteria

Infection of recipient cell

Recombination

Nature has clearly provided any number of ways in which genes may be recombined. We don't know the limits (if any) to this genetic behavior. We do know that it gets increasingly hard to be sure if our genes are "our own." Perhaps a major force in evolution is that we all — bacteria and humans — evolve more rapidly by sharing any new genetic "ideas" that arise by mutation.

GENETIC ENGINEERING

We have just examined some of the natural recombination mechanisms as they occur in procaryotes. We saw that the enzymes necessary in integration, those to nick DNA and to splice it back together, are found in all bacterial cells. Researchers have adopted these natural tools and learned how to recombine DNA from diverse sources into hybrid molecules.

Genetic engineering is the technology of creating and using recombinant DNA in the production of desirable products. A **recombinant DNA** molecule is a gene sequence containing DNAs from two or more different sources enzymatically spliced together into a recombinant molecule. A commonly used phrase for this technique is **gene splicing.**

RECOMBINANT DNA RESEARCH

Plasmids play a key role in recombinant DNA research, having characteristics that make them exceptionally useful. Plasmid DNA is easily isolated and purified. By using DNA cutting and splicing enzymes, groups of foreign genes can be introduced into the plasmids. Recombinant plasmids can then be readily reintroduced into their host bacteria, either with their own transfer factors or by transformation. Then, as the bacterial population grows, the plasmids will be replicated too.

To make recombinant DNA, you need purified plasmid DNA and purified DNA from the source organism. Both are treated with nicking enzymes that chop them into large fragments. Let's look more closely at this class of naturally occurring enzymes, which are called **restriction endonucleases.** These enzymes are restricted in their site of action. They break the DNA strand internally, but each different restriction endonuclease recognizes a particular nucleotide sequence, such as AATT, as the point at which to break the DNA strand. An endonuclease breaks a DNA strand unevenly, leaving one strand a few bases longer than the other. These single-stranded ends are called "sticky ends" because of their ability to base-pair. If the same restriction endonuclease is used on two different DNAs, each will be left with "sticky ends" with the same base sequences.

Then the fragments of DNA are mixed together. The fragments bind to each other by the complementary sticky ends. The mixture is treated with DNA ligase, the enzyme that repairs breaks in DNA. The DNA ligase covalently bonds the matched fragments together in a manner very similar to natural integration (Figure 8.22).

DNA CLONING.   Researchers can construct plasmids containing the DNA sequences of which they wish to have large amounts for biochemical or genetic research. They then insert the recombinant plasmid into its bacterial host. As the plasmids' hosts grow and multiply, they will mass-produce plasmid DNA, including the desired DNA sequences, in bulk. This is DNA **cloning.** In effect, the researcher has a bacterial culture operating as a factory making a specific DNA to order.

Here is an example of a recent valuable application of DNA cloning. Human chromosomes contain DNA sequences called **oncogenes.** These are genes that seem to have some role in causing cancer. There are dozens of researchers who need to have oncogene DNA available for different research projects. Rather than laboriously extracting the tiny amounts of oncogene DNA obtainable from cancerous tissues, they can obtain much larger amounts more easily by

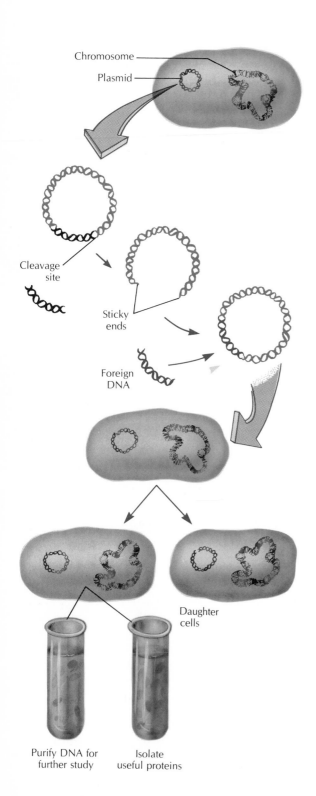

cloning the oncogenes in bacterial plasmids. With more material available, more experiments can be done, and research on the genetic basis of cancer progresses more rapidly.

PROTEINS FROM RECOMBINANT BACTERIA.   Often, we are primarily interested in the gene product — the protein the gene encodes — not the gene itself. In such cases, we wish to have the bacterial host express the gene. Perhaps the desirable gene is one that originates in a plant or animal cell and makes plant or animal protein. To get host bacteria to make plant or animal proteins, we must incorporate into a plasmid not only the gene itself, but also the bacterial regulatory sequences that will allow transcription.

Insulin is a protein hormone that regulates blood sugar levels in humans and other mammals. Diabetic people often lack insulin, and must take it daily to remain healthy. For more than 50 years, we have treated diabetes with insulin obtained from animal pancreas tissue. It is a costly product, and the supply is barely adequate to meet a growing demand. Many diabetics also become allergic to the "foreign" animal protein.

FIGURE 8.22 Constructing a Recombinant Plasmid
A plasmid is isolated in pure form from its host bacterium then treated with a restriction endonuclease. Fragments with "sticky" single-stranded AATT ends are obtained. The plasmid fragments are mixed with similarly treated fragments of DNA from the foreign organisms. The complementary sticky ends readily hybridize and are joined by DNA ligase. The hybrid plasmids are reinserted into a culture of receptive host bacteria. If DNA cloning is desired, quantities of bacteria can be harvested and the plasmid DNA isolated. If a protein product is desired, additional genes (regulatory signals) will be added to the plasmid so that it will be able to express the gene and make the protein. The product is harvested from the culture fluid and purified.

FIGURE 8.23 Commercial Insulin Production
Human insulin, produced by host bacteria containing plasmids with insulin genes, is being manufactured on a large scale. It is now sold under prescription by the trade name Humulin. (Courtesy of Johnson/Lilly.)

Insulin genes isolated from human pancreatic cells have been recombined with plasmid genes and regulatory elements. The plasmids, in growing *E. coli* cells, express the insulin genes and cause their host bacteria to produce human insulin. The first human insulin from recombinant DNA technology was licensed for sale in 1983 under the trade name Humulin (Figure 8.23). Other human insulins, hormones, and immunity factors are also now available.

IMPLICATIONS OF GENETIC CHANGE IN MICROORGANISMS

When a bacterial strain acquires altered characteristics it may have major effects on human health. Genes may actually make it more difficult for our systems to resist disease, by enhancing the ability of pathogenic bacteria to resist antibiotics, form capsules, create toxic cell-surface chemicals, grow faster, and produce toxins. The fact that these traits may be transferred between bacterial strains, and even among species, increases their negative impact.

The use of bacteria and fungi in industrial production of valuable biochemicals constitutes a multi-million-dollar business. Microbial productivity can be greatly enhanced by induction and selection of high-yielding mutants. For example, genetic and cultural selection techniques have been used with *Penicillium notatum*, the fungus that yields penicillin. With the original wild-type fungal strains first used in 1941, only four units of penicillin per milliliter of culture fluid could be obtained. Step by step, higher- and higher-yielding mutant strains were selected. Present strains produce up to 50,000 units per milliliter. The recombinant microbes that are being developed produce entirely new products, such as synthetic vaccines.

There are environmental implications for genetic change, too. Many microorganisms, notably the bacterium *Pseudomonas* and its relatives, decompose some of the simpler hydrocarbons in crude oil. Some strains of pseudo-

BOX 8.2
The Gleam in Wall Street's Eye

Since 1980, when the U. S. Supreme Court ruled that patent rights could be awarded for recombinant microorganisms, many corporations have viewed genetic avenues of investigation as especially inviting and potentially profitable. Several hundred patent applications have been granted or are pending for gene-splicing applications. Over 100 biotechnology corporations have sprung up, some small, and some large enough to interest Wall Street investment analysts. Giant pharmaceutical firms are also deeply involved.

The action has shifted from the ivy-covered halls of the university to the corporate boardroom. University professors have been founders and major stockholders of many biotechnology firms. Biotechnology stocks are among Wall Street's most speculative and hotly traded.

What does this all mean for us? Research projects that promise eventual profits are amply supported and make rapid progress. Other worthy projects that are less likely to have a cheerful effect on corporate earnings go begging, or are left to the ever-shrinking support of Federal research funds. There is much more secrecy in research, since investigators seek to keep their successes to themselves in hopes of a patent, rather than publishing them for the information of the scientific community. In any case, for better or worse, the commercialization of biology will not now be turned back.

Did you see what Supergene Corporation's stock did on the Market yesterday? Somebody made a bundle. . . .

monads now have been given recombinant plasmids that greatly increase their rate of catabolic activity against hydrocarbons. They metabolize actively in cold and saline ocean environments in which most oil spills of concern occur. In agriculture, bacteria are used to control insect pests, and efforts are under way to transfer bacterial nitrogen-fixation genes to other bacterial and plant strains. Mutants and recombinants are being sought for specialized toxic-waste-treatment applications, such as to break down dioxin in soil and industrial wastes.

Recombinant DNA techniques are revolutionizing biological research. It is not an exaggeration to say that humans can now direct some aspects of evolution. Genetic engineering is being optimistically directed toward making "super" plants, animals, and bacteria. Although the profit motive lies behind much of this activity, there's no question that the pace of scientific advance has speeded up. For instance, we have achieved as much progress in cancer research in the last two years as in the previous 20 years. Genetic tools promise to enhance significantly the already major contributions microorganisms can make to human welfare. They may, and probably will, also be misdirected to the production of biological-warfare agents. It remains to be seen what the ultimate social and economic costs of such advances will be. For more applications of genetic engineering technology, check your daily newspaper almost every day.

SUMMARY

1. Cells use raw materials from metabolism plus the energy in ATP for growth, repair, and reproduction.

2. Protein synthesis is the means by which the genetic information of the cell is expressed.

3. DNA is the blueprint to make RNA, and RNA is the blueprint to make proteins. These steps transfer information.

4. DNA can be replicated — to provide the genetic material for new cells — or transcribed — to direct the expression of genes.

5. The bacterial genome is a single closed circle of double-stranded DNA. It is replicated semiconservatively in fragments that are hitched together. In eucaryotes, with a linear chromosomal structure, replication is discontinuous.

6. Transcription of the DNA results in the appearance of three classes of RNA molecules, mRNA, rRNA, and tRNA. The synthesis of RNA is a highly regulated copying of one of the two DNA strands in specific regions only.

7. Translation of the genetic code, contained in mRNA codons, results in the assembly of polypeptide. Amino acids are carried by specific tRNAs that base-pair with message codons on ribosomal surfaces. Enzymes simultaneously link the amino acids together and release the tRNAs.

8. Temporary and reversible variations in enzyme levels are caused by regulation of gene function at the level of transcription. By controlling the access of RNA polymerase to genes, the cell regulates the formation of mRNA and thus the synthesis of the protein in question. The genetic controls are induction and repression.

9. Bacterial genomes may be permanently changed through mutation and recombination. Mutation may be induced by exposure to chemicals or radiation; it usually reflects a single-base change in the DNA sequence for a particular gene.

10. Recombination in bacteria occurs by various

means. All follow a basic pattern. Donor DNA is introduced into a recipient cell. If the DNA is compatible, portions will become integrated into the host genome.

11. Transformation occurs when short pieces of DNA, liberated from a dead donor cell, are taken in and integrated.

12. Plasmids induce gene transfer. The transmissible plasmids transfer their own genes. When they are integrated with the host genome, they mediate conjugation, a true mating between two bacteria, in which the donor or male, with the integrated transfer factor, transmits some or all of its genome.

13. Transduction is virus-mediated gene transfer.

When a virus moves from one host bacterium to another, it may carry segments of bacterial DNA.

14. Plasmids may be enzymatically reconstructed in the laboratory by joining several types of genetic material with different information contents. The host cell may then replicate the hybrid plasmid DNA, cloning the added genes, or express those genes, producing the desired protein.

15. There are practical consequences to genetic exchanges. They determine many aspects of bacterial pathogenicity as well as responses to antibiotics. By careful selection and/or engineering of recombinants or mutants, improved microbial tools for industrial and antipollution applications are being found.

DISCUSSION QUESTIONS

1. Identify the building blocks for biosynthesis of DNA, RNA, and proteins.

2. What is the ATP cost of synthesizing a polypeptide containing 5000 amino acids? (Two ATPs are required to bind the amino acid to the tRNA, and one ATP equivalent per codon to move the message on the ribosome).

3. What do the following enzymes do? a) DNA modification enzymes, b) DNA polymerase, c) restriction endonucleases, d) DNA ligase.

4. Describe the steps in polypeptide synthesis.

5. Differentiate between induction and repression. How do feedback inhibition and genetic regulation work together?

6. Why are regulatory mechanisms essential for a cell's successful exploitation of its environment?

7. Differentiate between reproduction and recombination.

8. What are the common factors as well as the differences among transformation, transduction, and plasmid-mediated conjugation?

9. Compare the process of using a gene-splicing technique for DNA cloning or for protein production.

10. As a joint project with others, prepare a collection of current press reports on successful or promising developments in genetic engineering.

ADDITIONAL READING

Birge EA. *Bacterial and bacteriophage genetics.* New York: Springer-Verlag, 1981.

Brock TD, Smith DW, Madigan MT. *Biology of microorganisms.* Englewood Cliffs, N.J.: Prentice-Hall, 1984.

Chambon P. Split genes. *Sci Am* 1981; 244:60–71.

Dressler D, Potter H. Molecular mechanisms in genetic recombination. *Annu Rev Biochem* 1982; 51:727–761.

Kornberg RD, Klug A. The nucleosome. *Sci Am* 1981; 246:52–64.

Novick RP, Plasmids. *Sci Am* 1980; 245:103–127.

Watson JD, *Molecular Biology of the Gene.* 4th ed. Menlo Park, Calif.: Benjamin/Cummings, in press.

Wittman H G. Components of bacterial ribosomes. *Annu Rev Biochem* 1982; 51:155–183.

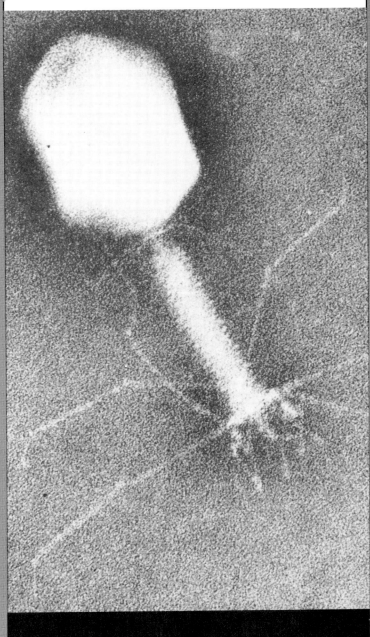

VIRUSES: PASSENGERS AND PARASITES

This chapter introduces viruses, extremely small self-replicating infectious agents very much tinier and simpler even than bacteria. We will discuss the chemistry and structure of virus particles and explain the strategy of virus replication within the host cell. We will then consider the genetics of viruses; describe virus life cycles in bacteria, plants, and insects; and discuss the importance of viruses in the natural world.

WHAT IS A VIRUS?

Early in this century, accumulated evidence led scientists to conclude that there must be infectious agents smaller than bacteria that could cause disease in animals and plants, and even destroy bacterial cells. Searching for these suspected life forms with the light microscope was unproductive; they are too small to see. The development of the electron microscope, along with other advances in biochemistry and cell culture, have allowed us to see and study the viruses. Viruses have been shown to infect every form of life, from the simplest bacteria to the most highly evolved plants and animals.

Viruses are not microorganisms because they aren't cells. They have none of the structural features that make up a cell, and they cannot replicate themselves independently. The only means by which viruses propagate is to invade a host cell. The virus particle apart from its host cell has no metabolism and is biologically inert.

If viruses are not cells, what are they? It is difficult for us to think of anything as living that is not cellular. However, remember that the entire potential of an organism is encoded in its nucleic acid genome. Where there is a genome, there can be a type of life form.

Getting down to basics, a **virus** is simply a genome surrounded by one or more exterior coats. The genome itself contains a minimum of information, genes enabling the virus to

- attach to and get into a host cell,
- redirect the host cell's metabolism to produce virus-specified nucleic acids and proteins to make more virus,
- escape from the host cell.

The viral particle doesn't maintain any systems of its own for energy production and biosynthesis. In other words, when thinking about viruses, put aside all of what you learned about cell structure, metabolism, and protein synthesis. The virus does none of this. As a parasite, it does no useful work, but simply takes advantage of the resources provided by its host, taking what we might call a "free ride."

We call viruses **obligate intracellular parasites** because they *must* be inside a host cell in order to reproduce. A practical consequence of this feature is that viruses cannot be cultivated on cell-free media. We have to provide an appropriate host-cell system in order to grow and study them.

When the virus takes over the host cell, the cell's ATP and building-block molecules are diverted to the synthesis of viral particles. The cell's nucleic acids and proteins may be broken down. This is why the infected cell is often significantly harmed or dies. Furthermore, you can see how, if millions of cells in essential tissues were destroyed, a viral infection might kill a human.

THE BIOLOGICAL IMPACT OF VIRUSES

At some point during evolution, this strategy arose for a virus replicating its genes. It was different and it was daring. Was it successful? If we measure by how widespread viruses are, unquestionably yes. All forms of life probably have viral parasites. Although the viruses of bacteria and mammals were the first to be studied, further investigation demonstrated that viral infection is ubiquitous in the five kingdoms. Viral genes are scattered through the chromosomes of most of the eucaryotes studied so far.

Research directions in virology have been driven by economic and health considerations. Most of our basic knowledge was initially gained by studying bacterial viruses, that is, viruses that infect bacteria. Another name for these viruses is **bacteriophages,** meaning "bacteria eaters." More recently, the most intensive applied virus research has been directed toward viruses that destroy agricultural crops and infect domestic animals, and toward those that cause human diseases. Some of the most common human viral diseases include measles, chickenpox, the common cold, hepatitis, and herpesvirus infections.

In addition to killing things, viruses are a force to be reckoned with in the biological world in a number of other ways, all having to do with genes and their effects. Viruses, as we saw in the last chapter, can introduce their genes into a host cell, where the viral genes may be integrated into the host genome. When this happens to bacteria, the host bacteria may acquire the capacity to produce new proteins, including deadly toxins. Animal cells bearing integrated viral genes may undergo growth changes that cause them to become cancerous. Some viruses may carry host genes with them when they move from one host cell to the next. They thus serve as a means of genetic recombination, and may be an important driving force in evolution.

VIRAL STRUCTURE

An extracellular or free virus particle is called a **virion.** Pox virions, which cause smallpox, are the largest. They are 0.2 μm in diameter, just barely visible with a light microscope. All structural observations of viruses must be done by electron microscopy (Figure 9.1).

VIRAL NUCLEIC ACIDS

Cellular organisms contain both DNA and RNA. Virions, on the other hand, contain only one of the nucleic acids. The nucleic acids may be double-stranded DNA (D2), single-stranded DNA (D1), double-stranded RNA (R2), or single-stranded RNA (R1). Since we are used to the idea that genes must be made of double-stranded DNA, we will have to do some rethinking to understand how each of these other genetic systems replicates and expresses itself.

Each virus has a fixed amount of nucleic acid corresponding to a fixed number and sequence of genes. Some tiny bacterial viruses contain only four genes while large animal viruses may contain hundreds of genes.

FIGURE 9.1 The Relative Sizes of Viruses
In the background is a portion of a yeast cell; drawn to scale against its surface is a bacterial cell and progressively smaller virus particles.

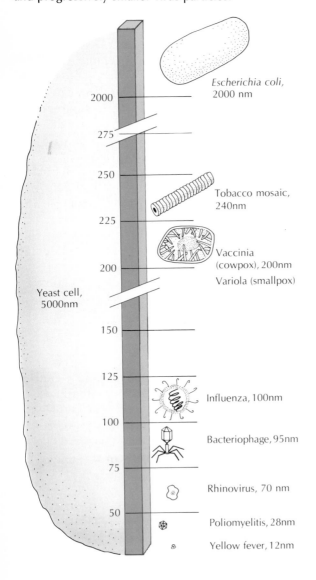

NUCLEOCAPSIDS

In the virus particle, the negatively charged nucleic acids are associated with a surrounding coat of positively charged proteins, the **capsid.** Together, the nucleic acid and its enclosing proteins may be called the **nucleocapsid.** Capsids are hollow shells or coats assembled from proteins. These proteins are products of some of the viral genes. The coat proteins self-assemble into characteristic structures as the virus is being replicated. The simplest structure has a regular **polyhedral** (many-sided) symmetry (Figure 9.2a). The most frequent type is a 20-sided polyhedral figure called an **icosahedron.**

In another type of arrangement, the nucleic acid may be coiled up inside a hollow cylinder of coat proteins (Figure 9.2b). This **helical** arrangement seen in the tobacco mosaic virus (TMV), also occurs in some bacterial and animal viruses.

ENVELOPES

Some viruses also have accessory structures (Figure 9.3) Viruses with such structures are said to be **complex.** In many animal viruses, the nucleocapsid is surrounded by an outer layer, which is called an **envelope.** These envelopes are membranelike lipoprotein bilayers. The complex virus acquires its envelope material directly from the host-cell membrane while leaving the host cell. Special proteins or enzymes coded for by viral genes may be manufactured by the protein-

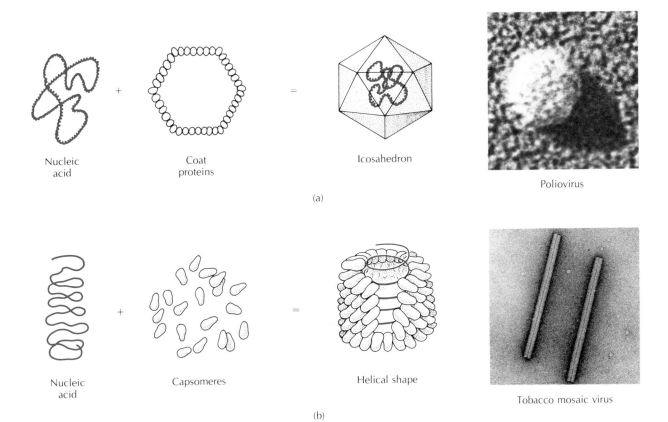

Nucleic
acid

Coat
proteins

Icosahedron

Poliovirus

(a)

Nucleic
acid

Capsomeres

Helical shape

Tobacco mosaic virus

(b)

FIGURE 9.2 The Nucleocapsid
Together, the nucleic acid and its enclosing proteins form the nucleocapsid. (a) The
simplest structure has a regular, polyhedral shape. (×726,250. Courtesy of J.J.
Cardamone, Jr. and B.A. Phillips, University of Pittsburgh School of Medicine/BPS.)
(b) In the helical arrangement, the nucleic acid may be coiled inside a hollow
cylinder of coat proteins. (Courtesy of Robley C. Williams, University of California at
Berkeley.)

synthesizing machinery of the host cell and inserted into the envelope as it is formed.

TAILS, FIBERS, AND SPIKES

Viruses may have elaborate attachment structures to bind to the host. In bacterial viruses,
protein **tails** are commonly found. They make it

possible for the viruses to inject their nucleic
acid forcibly into the host cell. **Tail fibers** specifically bind to receptor sites on the host surface.
The hollow, contractile shaft of the tail is the
tube through which nucleic acid can be injected.

In enveloped animal viruses, additional viral
proteins may form spikes called **peplomers** on
the envelope surface. These bind the virus to the
specific host tissue, such as respiratory or gastrointestinal linings, facilitating infection.

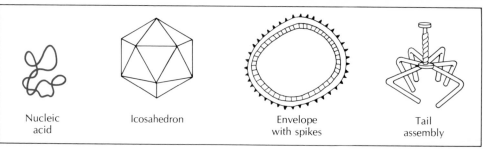

| Nucleic acid | Icosahedron | Envelope with spikes | Tail assembly |

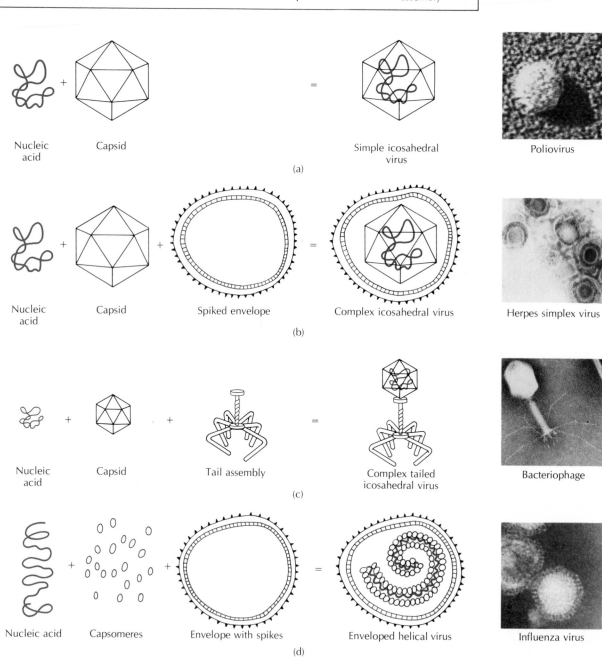

Nucleic acid + Capsid = Simple icosahedral virus

Poliovirus

(a)

Nucleic acid + Capsid + Spiked envelope = Complex icosahedral virus

Herpes simplex virus

(b)

Nucleic acid + Capsid + Tail assembly = Complex tailed icosahedral virus

Bacteriophage

(c)

Nucleic acid + Capsomeres + Envelope with spikes = Enveloped helical virus

Influenza virus

(d)

272

FIGURE 9.3 Viruses and Their Accessory Structures
The structures of viruses range from the simple to the complex. (a) Poliovirus is an example of a nucleic acid genome inside an icosahedron. (×472,500. Courtesy of J.J. Cardamone, Jr. and B.A. Phillips, University of Pittsburgh School of Medicine/BPS.) (b) The herpesvirus adds a membranelike envelope outside the capsid. (Courtesy of G.W. Gary, Jr., Viral Pathology Branch, Centers for Disease Control, Atlanta.) (c) The T4 bacteriophage has a complex tail structure with fibers and a contractile shaft. (×95,000. Courtesy of Robley C. Williams, University of California at Berkeley.) (d) In complex viruses such as influenza, the nucleocapsid is folded up and enclosed in an envelope with external protein spikes. (×113,900. Courtesy of F.A. Murphy, Viral Pathology Branch, Centers for Disease Control, Atlanta/BPS.)

VIRUS CLASSIFICATION

As yet there is no natural taxonomy for the viruses. New viruses are frequently being discovered and familiar viruses are being continuously reexamined. A partial classification appears in Table 9.1, and more complete tables for the animal viruses will be found in Chapter 10. The single most important characteristic of any virus is its nucleic acid content — whether it contains DNA or RNA; whether the genome is single- or double-stranded; its genome size; and whether the nucleic acid is linear, circular, or a group of small segments. The helical or polyhedral symmetry, total size, and presence of an envelope or a tail is also considered.

For plant and animal viruses, the **means of transmission** — the method by which the virus particles are spread from one host to another — is yet another important factor in classification. Many significant virus pathogens are spread by **arthropods,** the group of invertebrate animals that includes insects such as mosquitos, fleas, and lice, and arachnids such as ticks. These animals serve as **vectors,** the factors that phys-

ically move or transmit an infectious agent. An arthropod may carry the virus on its mouthparts or internally, and transmit it by biting, chewing, or defecating.

HOW VIRUSES INTERACT WITH HOSTS

A virus introduces its genome into a host cell. By doing so, it initiates a series of new events that will be under *viral* control, and changes the behavior of the host cell. Some of the host cell's normal activities may be bypassed, inactivated, or altered. Its growth rate or cellular morphology may change. Viral genes will direct the host cell to manufacture viral nucleic acids and proteins. The host cell may become a factory to assemble and release new virions.

Any differences between the normal activities of the uninfected host cell and the strange behavior of the infected host cell are due to the influence of the new group of genes. The host cell duplicates, transcribes, and translates the viral genes exactly as it does its own. In studying viral replication, we will draw on and further elaborate the understanding of gene expression that we developed in Chapter 8. We will use bacterial models to present general principles. We'll look first at how the viral nucleic acids are replicated, then at what types of viral proteins there are and how they are made. Then we'll examine the stages by which a bacteriophage replicates itself in a bacterial cell.

VIRAL NUCLEIC ACID REPLICATION

As we noted previously, viral nucleic acid may be one of four types — D2, D1, R2, or R1. The genome of DNA viruses, like genomes in cellular organisms, is both replicated and expressed. DNA copying mechanisms use the cell's own DNA polymerase enzymes. The host cell makes multiple copies of the viral genome, as many as 400 per infected bacterial cell (Figure 9.4a). In order to provide the huge supply of DNA nucleo-

				Examples of		Examples of
Nucleic Acid	Capsid Symmetry	Size of Capsid (nm)	Envelope Present	Bacterial Viruses	Examples of Animal Viruses	Plant Viruses
RNA	Helical	17.5 × 300	−			Tobacco mosaic virus
		9	+		Myxovirus,	
		18	+		paramyxovirus	
	Polyhedral	20–25	−	f₂		
		28	−		Enterovirus,	Bushy stunt
		70	−		reovirus	virus
DNA	Helical	5 × 800	−			
		9–10	+	Filamentous phage fd	Poxvirus	
	Polyhedral	22	−	φX 174		
		45–55	−		Polyomavirus	
		60–90	−		Adenovirus	
		140–200	+		Herpesvirus	
	Complex	Head, 95 × 65	−	T-even series		
		Tail, 17 × 115				

TABLE 9.1 A general scheme for classification of viruses

tides needed, the host's DNA may be broken up by **viral enzymes** — enzymes coded for by viral genes. The viral DNA is expressed by transcribing the information to mRNA, then using the mRNA code to assemble the protein.

When a viral genome is RNA, it carries out both replication and expression (review Figure 8.2). It serves as a template for replication of more copies of the RNA genome and directly as a messenger, coding the assembly of viral proteins.

Normal uninfected host cells do not appear to have an enzymatic mechanism for making RNA copies of RNA templates. The special type of **RNA polymerase** enzyme needed for this task is encoded in the viral genome. Let's trace the steps by which an RNA genome may be copied. First, the viral RNA enters the host cell. Next,

acting as a messenger, it directs the host to make the new RNA polymerase protein. Then the new RNA polymerase makes multiple copies of the viral genome (Figure 9.4b). Some viruses have several different pieces of RNA composing their genomes. One of each must somehow be assembled into each virion.

Those viruses that integrate their genomes with their hosts' genomes have yet another method of replicating their genomes. Once integration has occurred, their genomes are replicated right along with their host's, and are passed to the daughter cells at cell division. We have looked at how DNA fragments integrate in Chapter 8 (review Figure 8.17). Let us here look at how the genome of some single-stranded RNA viruses is copied into DNA and becomes integrated. In this process, the standard direction of

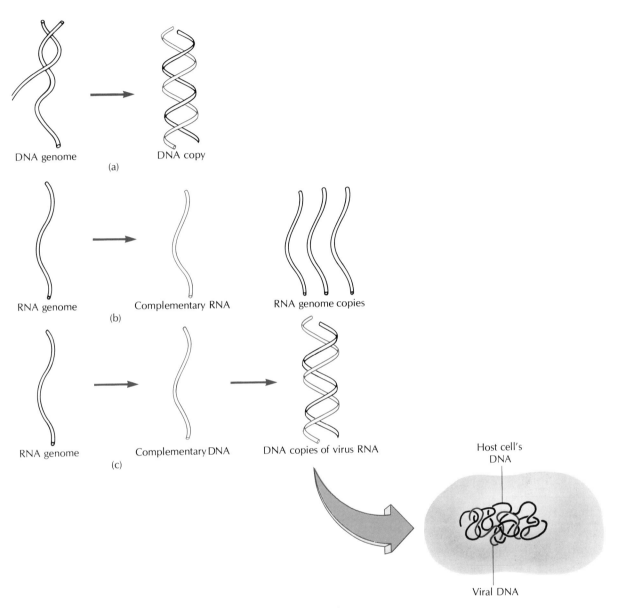

FIGURE 9.4 How Virus Genomes Are Replicated
(a) A DNA virus genome is replicated by the host's own DNA polymerases by semiconservative replication. (b) An RNA virus genome is replicated by a unique RNA-copying enzyme that is able to use RNA as a template, forming a complementary RNA. The gene for this enzyme is usually found in the virus genome. The complementary RNA then makes multiple copies of the virus genome. (c) Some RNA viruses integrate their genomes with the host DNA. To do this, they use the enzyme reverse transcriptase, which makes a DNA copy from the RNA of the virus. The DNA then integrates into the host cell's chromosomes.

information flow (from DNA to RNA) is reversed. When it was first discovered, this process provided a revolutionary challenge to generally accepted concepts of molecular biology.

The group of RNA viruses that replicate this way are called **retroviruses**, because they work "backward." Their copying enzyme is called the **reverse transcriptase.** The enzyme takes an RNA genome and makes a DNA complementary strand. This DNA strand in turn is the template for its proper DNA complement, and normal double-stranded DNA is formed. Then this viral DNA, indistinguishable from host DNA, may integrate into the host chromosomes and become a permanent part of its genome (Figure 9.4c) With integration, suddenly the distinction between viral genome and host genome has been removed. The human genome, on examination, proves to be full of scattered segments of DNA of viral origin, of largely unknown function and importance. Once again, we see that we don't really know where our genes come from! Retroviruses often provoke the cellular changes that transform a normal cell into a cancer cell.

VIRAL PROTEINS

What kinds of proteins do viruses produce? To answer this question, we need to think through what proteins viruses might need in order to replicate. Viral genes are arranged on the genome in the order of the time sequence in which they will be expressed (Figure 9.5).

Replication proteins are those that assist directly in the process of viral replication. Some replication proteins are **repressors** that block host genes. Other repressors may maintain the virus in a latent nonreplicating or nondestructive state. The viral enzymes may include endonucleases that hydrolyze host DNA, the special RNA polymerases, reverse transcriptases, or modification enzymes. Yet other proteins regulate the timing of viral gene expression so that activities occur in sequence. These **early proteins** appear near the start of viral replication; they do not become part of the finished virion.

Once replication is complete they have no further function.

Structural proteins, on the other hand, are the viral gene products that will compose the capsid and accessory structures of the new crop of virions. Animal viruses with envelopes commonly carry genes for proteins to be inserted into their envelopes. Two of these, seen in the influenza virus, are the **hemagglutinins** (red blood cell–aggregating proteins) and **neuraminidases,** enzymes that aid in invading the respiratory tract. Structural proteins are largely **late proteins.** They appear when the replication process is well advanced. They are then incorporated into the virions, assembling around the completed genomes.

PRODUCTIVE INFECTIONS

A **productive infection** is a type of viral infection that leads to the production and release of many new free virus particles. When they destroy the host cell, productive infections are also called **lytic** or **cytolytic** infections.

Let's use a bacteriophage infection as an example. The Gram-negative bacillus *Escherichia coli* may be infected by many different types of bacteriophages, with different outcomes. We'll first describe the replication of the T-even phages (T2, T4, and T6), which are large, complex viruses with tail structures attached to an icosahedral "head." The head of a T-even phage contains double-stranded DNA encoding approximately 75 genes.

The **replication cycle** of a productive virus is a sequence of three phases, initiation, replication, and release (Figure 9.6). **Initiation** requires, as a first step, the firm attachment or adsorption of the virus to the bacterium's exterior surface. Host-cell surfaces have many receptor sites, and bacterial viruses have only one attachment site. Many virions can bind to one cell. The host–parasite relationship is quite specific in that one strain of virus will often bind only to one species of bacteria, or even to just a few strains within the species.

After the T-even phage attaches itself, its tail pin pierces the bacterial cell wall. The contractile proteins of the virus tail coil and inject the viral DNA into the bacterial cytoplasm. The viral capsid, tail, and fibers remain outside and serve no further function. Once its nucleic acid is inside the cell, the remaining parts of the bacteriophage are no longer infective for other cells.

Replication begins after the viral DNA enters the host. Some of the viral genes are transcribed and translated, making early proteins.

Viral enzymes break down the host DNA, thus destroying its genes. Viral DNA is replicated — several hundred copies per infected cell — using nucleotides scavenged from the degraded host DNA. The viral late genes are then expressed, and viral structural proteins accumulate. Capsids self-assemble around viral genomes. Immature viral particles are visible using the electron microscope. Tails self-assemble in stages, joining to the heads. The viruses are mature about 25 minutes after initiation.

FIGURE 9.5 A Generalized Bacteriophage Genome
The many separate genetic functions of the phage are arranged in the order in which they are to be expressed. Each group of functions has a promotor region (solid color) controlling transcription. The early genes inactivate the host and permit the copying of viral DNA. The later genes for the head, capsid, and assembly proteins are then expressed. Maturation proteins are necessary to assemble the viral parts.

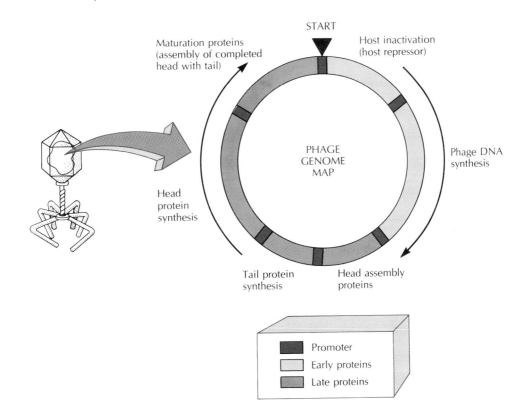

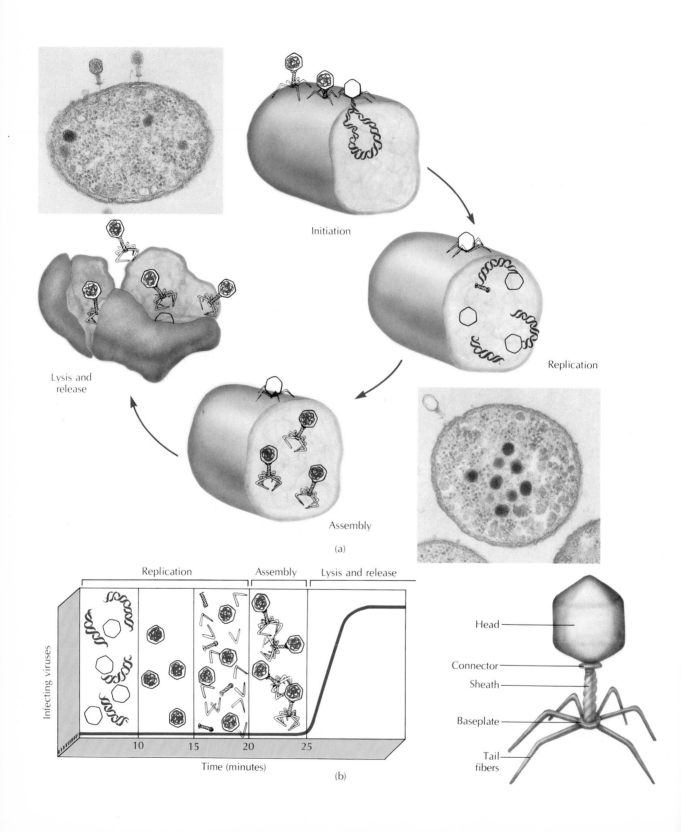

Initiation

Replication

Assembly

Lysis and
release

(a)

Replication   Assembly   Lysis and release

Infecting viruses

Time (minutes)

10   15   20   25

(b)

Head

Connector

Sheath

Baseplate

Tail
fibers

◀ FIGURE 9.6 **Replication of a Cytolytic Virus**
(a) A bacteriophage attaches to specific chemical
sites on the surface of the bacterium and penetrates
its cell wall and membrane. Once inside, its
genome redirects bacterial metabolism. First viral
DNA, then coat proteins appear; all components
are shown together for convenience. In the
scanning electron micrograph, the empty capsid can
be seen attached to the outer surface of the cell
wall. After the components have self-assembled, the
host cell lyses, releasing them to reinfect any
susceptible host bacteria that may be present.
(b) This graph depicts the various viral components
during the times that they are produced and
assembled. The detailed drawing of a T4 phage,
parasitic on *Escherichia coli*, shows the polyhedral
head attached to a contractile tail. (Photo courtesy
of Professor Jonathan King, Director, MIT Biology
Electron Microscope Facility.)

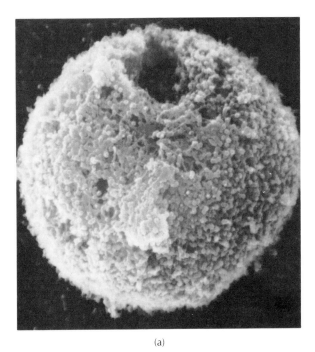

(a)

(b)

FIGURE 9.7 **Influenza Virus Budding from the Host
Cell**
(a) In this scanning electron micrograph, an infected
monkey kidney cell has rounded up and detached
from its growth surface. (b) A closer view shows
multiple fingerlike buds containing influenza virus
leaving the cell. Eventually the cell will die from
stress. (From Fumio Uno, *J. Electron Microscopy*
(1979) 28:83-92, Figs. 7a and 8a. Reproduced by
permission of the American Society for
Microbiology.)

The virions must now be **released** from the
host. In a T-even infection, the host cell now
contains little but mature viruses. The weakened cell wall ruptures and the cell lyses, releasing the virions. All that remains of the host
cell is an empty wall-membrane ghost. Viral replication (from 1 virion to 200 virions in 25
minutes) can rapidly outdistance bacterial replication (from one cell to two cells in 20 minutes).
A cytolytic virus will devastate a population of
susceptible bacteria.

Some productive infections do not cause lysis of the host cell. For example, the filamentous
bacteriophage fd, also an *E. coli* parasite, exhibits
a different, nonlytic pattern of productive infection. In contrast to lytic viruses such as the T-
even phages, fd produces virions in small numbers over a long period of time. They weaken the
host cell somewhat but do not kill it or lyse it.

There are many productive animal viruses,
as we will see in Chapter 10. Some may cause
host-cell lysis (poliovirus is an example). Influenza and measles viruses, on the other hand, are
produced and released without destroying the
cell. Like many animal viruses, they leave their
host cell by budding, which is how they acquire
their envelope (Figure 9.7). Budding does not directly destroy the host cell, though it weakens it.

FIGURE 9.8   Viral Plaques
This Petri dish of nutrient agar was inoculated with *E. coli*. Then a suspension of bacteriophage was spread over the surface. As each virus replicated, destroying even more bacteria, its progeny cleared areas where all bacteria were destroyed. These areas show up as dark, irregular spots in the photograph. (Courtesy of R. Kavenoff, University of California, La Jolla/BPS.)

BACTERIOPHAGES IN THE LABORATORY

When a suspension of a cytolytic virus is dropped on confluent growth of sensitive bacteria on a nutrient agar plate, each virus and its progeny clears a visible hole or **plaque** in the cloudy growth. By counting plaques, we can determine the number of infective viruses or **plaque-forming units** (PFUs) in a sample.

Because virus–host attachment is highly specific, we can use phages as a bacterial identification tool. For example, each *Staphylococcus aureus* subgroup is lysed by different strains of bacteriophages. If we want to identify an unknown staphylococcal strain, we confluently plate it onto nutrient agar, then place drops of different phage suspensions on areas of the growth. After incubation, we observe which phages have lysed the underlying bacteria, forming plaques (Figure 9.8). With this data, the staphylococcal strain can be assigned a precise subgroup designation. The procedure just described,

called **phage typing,** is used in tracing hospital epidemics of staphylococcal infection. It is also a useful technique in tracing outbreaks of *Salmonella* food poisoning. By pinpointing the exact strain, epidemiologists can determine the source of the outbreak.

VIRAL INFECTIONS THAT INDUCE GENETIC CHANGE

Not all virus infections immediately produce a crop of new virions. In **temperate** viruses, the viral genome integrates with the host genome, becoming latent for an indeterminate time. The primary outcome of a temperate viral infection is genetic changes for the host.

Bacteriophage **lambda** is a temperate virus of *E. coli.* Lambda initiates infection much as the T-even phages do, but once in the cell the viral DNA integrates. An integrated viral genome is called a **prophage.** In the prophage state, most of the lambda viral genes are repressed, and new phages are not produced. A prophage is replicated along with the bacterial chromosome (Figure 9.9). Each bacterial daughter cell receives a chromosome that includes the prophage. The bacterial progeny make up a clone of latently infected (also called **lysogenic**) cells. This is bacteriophage lambda's primary means of replication.

FIGURE 9.9 Replication of a Temperate Virus
When a temperate bacteriophage infects, the usual result is lysogeny — integration of the virus genome. The viral genome (colored closed circle) and the bacterial genome (black closed circle) become enzymatically connected. As the lysogenic cell continues to grow and divide normally through binary fission, it also generates multiple copies of the viral genome. When the virus is induced to deintegrate its DNA, a productive or lytic infection (left) is carried out.

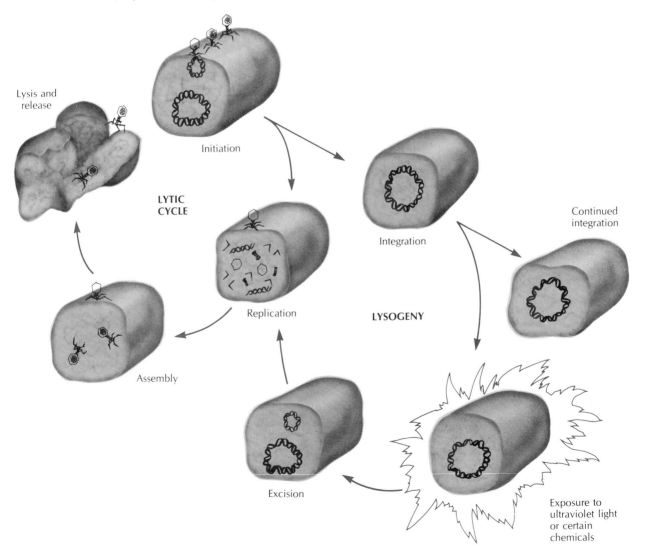

Lysis and release

Initiation

LYTIC CYCLE

Replication

Integration

Continued integration

LYSOGENY

Assembly

Excision

Exposure to ultraviolet light or certain chemicals

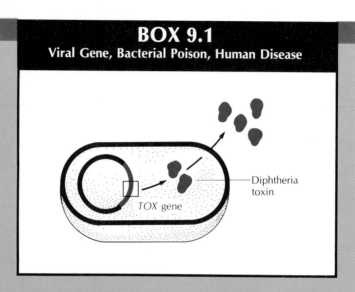

BOX 9.1
Viral Gene, Bacterial Poison, Human Disease

In diphtheria, botulism, and scarlet fever, the worst disease symptoms are due to the physiological effect of toxins released by the infecting bacteria. In each of these diseases, toxin is produced only by strains of bacteria bearing a temperate phage.

Diphtheria has been intensively studied. Toxigenic (toxin-producing) *Corynebacterium diphtheriae* ex-crete a protein that enters human cells and blocks protein synthesis. Clinically, the toxin hits the heart, kidneys, and nerve tissue hardest; fatalities are usually the result of heart failure.

The bacterial virus corynephage B is parasitic on *C. diphtheriae*. It resembles the lambda phage of *E. coli*. Its double-stranded DNA genome integrates with the bacterial ge-nome. The toxin is coded for by the viral **tox** gene. Toxin production is repressed by small amounts of iron, because the host bacterium makes a protein repressor that combines with iron and blocks transcription of the **tox** gene. Unfortunately, human plasma has so little free iron that toxigenic *C. diphtheriae* area able to make their virally coded toxins unimpeded in the human body.

Under certain circumstances, a small fraction of the latently infected cells may undergo a cytolytic infection producing infective virions. This happens when the viral DNA deintegrates from the host chromosome, and the viral genome is no longer under the control of repression. The lambda viral genes become active, mature infective viruses are produced, and the host cell lyses to release them. Exposing the host bacteria to mutagenic chemicals or radiation enhances the otherwise slight possibility that a productive infection will develop.

To summarize, a temperate phage may replicate in two ways. It usually integrates as a prophage. Then the host copies the viral genome and transfers it to all the numerous bacterial progeny. Much less commonly, the virus may carry out cytolytic productive infection.

The genetic changes come about when genes controlled by prophages are expressed by the host. This effect, in which the host's phenotype is changed, is called **lysogenic conversion.** For example, there are several major bacterial pathogens of humans that produce toxins only when carrying a prophage. Bacterial infections in which the prophage factor is important include diphtheria, tetanus, scarlet fever, and botulism.

Investigators have not yet formed a clear idea of the function of integrated viral genomes in animal and human cells. The herpesviruses integrate their DNA; they cause a range of skin and genital infections. The RNA-containing retroviruses integrate by using the reverse transcriptase mentioned earlier. Both contain one or more of the group of **oncogenes,** genes found in cancer-causing viruses that may be found also in cancerous human cells. But oncogenes are found in normal cells too, so their exact function remains unclear.

Transduction is another means by which viruses change the genetics of their hosts. In the last chapter, when we discussed the process of transduction, we saw that lysogenic viruses can serve as the mechanism for gene transfer from one host to the next (review Fig. 8.21). As a prophage detaches from the host chromosome, a small error by the excising enzymes may remove with it some adjacent bacterial DNA. This segment of DNA will be replicated many times, and eventually genomes containing a little bacterial DNA in addition to viral DNA will be placed in all the virions produced. When these virions infect again, the second host's chromosomes will acquire transduced genes. If the transduced gene is different from the host-cell gene, the bacterium may exhibit a new phenotype. For example, a bacterium unable to metabolize galactose may be transduced by a virus carrying a bacterial gene conferring the ability to use galactose. The second host will then acquire the genes to use galactose.

To summarize, viruses change the genetics of hosts in two ways. Viral genes integrated in the form of a prophage may cause lysogenic conversion. Bacterial genes, transmitted by a temperate phage, may cause transduction.

THE ECOLOGICAL ROLE OF VIRUSES

PLANT VIRUSES

Viral infection is widespread in the plant kingdom (Table 9.2) and the resulting damage has a significant impact on human welfare. Crop failure caused by virus damage can be rapid and

TABLE 9.2 Representative plant viruses			
Name	Symmetry	Size (Å)	Envelope
Tobacco etch	Helical	200 × 7000	−
Soybean mosaic	Helical	200 × 7000	−
Alfalfa mosaic	Ellipsoid	180 × 300	−
Potato yellow dwarf	Helical	750 × 3800	+
Southern bean mosaic	Icosahedral	280	−
Turnip yellow mosaic	Icosahedral	280	−

overwhelming. It can lead to widespread hunger and economic disaster.

Almost all plant viruses contain RNA as their genetic material. They are frequently helical and rodlike, as typified by tobacco mosaic virus (TMV) (Figure 9.10). However, other agents, such as tobacco necrosis virus, are icosahedral. Some plant viruses such as potato yellow dwarf virus have envelopes.

Healthy plant cells have rigid, tough cell walls, and viruses can't penetrate them. The virus needs mechanical damage to the plant cell to gain entry. Plants in the field are usually well "dusted" with virus particles, but their tissues will remain healthy as long as they remain un-

damaged. If plant leaves are injured by hail or dust storms, the jaws of chewing insects, or rough cultivation methods, viruses may invade.

Once inside plant cells, viruses migrate through the tiny pores that connect the cytoplasm of adjacent plant cells. Viral particles may also enter the plant's vascular system and be distributed with the sap. Plants, in contrast to animals, don't appear to have immune mechanisms to combat the spread of infections. Thus plant infections usually progress unchecked until the plant dies or is harvested.

Viruses interfere with plant photosynthesis and respiration; block water and nutrient transport; and alter flowering, fruit, and seed devel-

FIGURE 9.10   Plant Viruses
(a) Short segments of TMV virus are seen both from the side and in cross-section. The pattern of protein coils around the dark, central core is clearly visible. (b) A plant cell infected with TMV shows huge, regularly arranged aggregates of the developing virus. The linear viral structures here occupy a large portion of the cell's area. (From Katharine Esau, *Viruses in Plant Hosts.* Madison: The University of Wisconsin Press. © 1968 by the University of Wisconsin Press, page 92.)
(c) An electron micrograph of tomato bushy stunt virus. (×205,900. Courtesy of Robley C. Williams, University of California at Berkeley.)

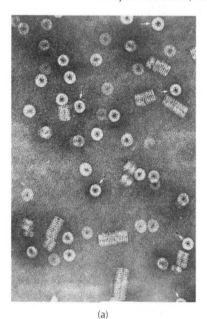

(a)

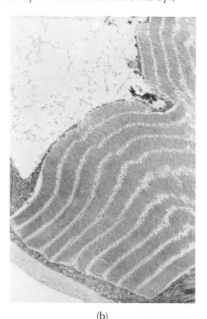

(b)

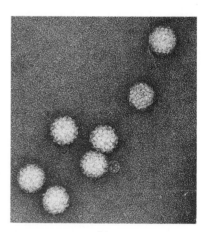

(c)

opment. Leaves become spotted, streaked, thickened, or curled, and their tissues appear distorted (Figure 9.11). Plants survive virus infections for long periods, but the crop's growth rate and productivity will be severely reduced.

The virion yield in plant infections is very high, up to 60 million virions per cell. Tobacco mosaic virus may be as much as 10% of the dry weight of tobacco leaf. Since TMV may also infect tomato plants, it isn't a good idea to drop tobacco scraps in the garden.

Viroids. A group of plant pathogens called **viroids** are even simpler than viruses, among the simplest self-reproducing agents in the biological world. They consist of short single strands of RNA, with extensive secondary structure of base-paired regions and hairpin loops. Viroids have no external structures such as capsid or envelope. They cause such plant diseases as spindle-tuber disease, chrysanthemum stunt disease, and a devastating disease of coconut trees. Viroids have not yet been unequivocally demonstrated as causative agents in animal diseases.

ANIMAL VIRUSES

Virus diseases are found in all branches of the animal kingdom. We will study them in detail in Chapter 10 and later in the book. Some important human viral diseases include measles, mumps, poliomyelitis, rabies, infectious hepatitis, and genital herpes. Viruses play an important role in the population control of animal species. There have been many examples of the rapid growth of a population, followed by an explosive outbreak of viral infection that brought population numbers down to a more supportable level. In Australia, for example, the introduction of rabbits (never previously found in Australia) led to an explosive growth of the rabbit population. Waves of rabbits overran the grazing country and stripped the grasslands to dust. Then a rabbit virus disease, myxomatosis, began to flourish, and quickly reduced the rabbit population to a low level.

FIGURE 9.11 Virus Infection in Plants
A tobacco leaf showing characteristic symptoms of mosaic disease. (Courtesy of the USDA Photographs.)

There are several slowly progressing neurological diseases, such as scrapie in sheep, the agents of which had remained unknown. Recently, however, researchers have found the scrapie agent and shown that it is composed of protein only. Yet it is both self-replicating and infectious. This unique disease agent — neither a virus nor a viroid — has been christened a **prion,** which stands for *p*rotein *i*nfective *a*gent. Researchers are now searching for prions in tissues affected with other mysterious, related human diseases, including Alzheimer's disease.

INSECT VIRUSES

Insects and viruses have two forms of interaction, important to both them and us. First, insects may act as vectors to spread virus diseases among plants and animals. Important viral insect-borne diseases of humans include yellow fever, encephalitis, dengue, and hemorrhagic fevers. In many of these cases, the virus is also using insect tissues to replicate. We will discuss insect vectors in greater detail in Chapter 14.

The second interaction is that insects themselves are parasitized by viruses. Viruses are particularly important in the natural ecology of insect populations. For example, an outbreak of gypsy moths deforested large areas of New England in 1979 to 1981. The gypsy moth population was later quickly brought under control by an outbreak of virus that destroyed the larvae in the summer of 1982 (Figure 9.12). In 1983 the gypsy moth population in most areas stood at less than 5% of its peak, pre-virus-epidemic level.

Insect viruses and other pathogens hold promise for use as biological controls of insect pests in agriculture and forestry. Insect viruses can be grown in mass production to spray or dust on pest-infested crops. The nucleopolyhedrosis viruses, tipulaviruses, and baculoviruses are pathogenic insect viruses. They are highly specific for their insect hosts, and do not parasitize helpful pollinating insects.

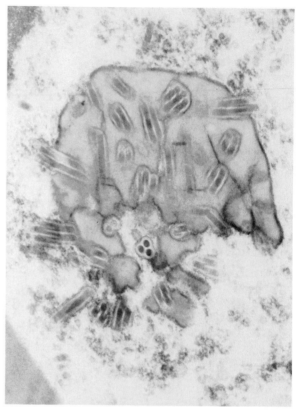

FIGURE 9.12   Nucleopolyhedrosis Virus
This virus causes a fatal disease of gypsy-moth larvae. The rodlike virus lie in groups of two or three in the cytoplasm of a gypsy-moth cell. (Courtesy of USDA Forest Service.)

SUMMARY

1. Viruses are ubiquitous, highly specialized life forms that complete a life cycle by means of obligate intracellular parasitism.

2. The free virus particle, or virion, consists of a nucleic acid genome, either DNA or RNA, enclosed in a protein capsid. Exterior to the capsid may be other structures essential to the infective process, such as the tail of bacterial viruses and the envelope of animal viruses.

3. The viral genome contains a minimum of genetic material. Only those genes needed to specify its own replication and to direct the host cell to make new viruses are present. The host cell provides the raw materials, the enzymatic machinery, and the metabolic energy. Viral replication may completely destroy the host cell, may cause reversible damage, or may cause no apparent ill effect.

4. The cytolytic bacteriophages, such as the T-even group, carry out productive lytic cycles of infection. Infection is initiated. Then early proteins are synthesized and viral DNA is manufactured. Later, capsid proteins are made and the virus particles are

BOX 9.2
A Plague on Them

For every human being on the globe there are about one billion insects. At least 3000 insect species are in direct competition with people for food crops, annually reducing human food supplies about 30%. They spoil or devour millions of tons of food and destroy vast tracts of valuable woodland and orchard.

Chemical pesticides have problems. They kill beneficial insects and are toxic to animals and human beings. It takes costly technology and oil to produce them. Detailed training is needed to administer pesticides safely and legally. Worst of all, most bugs readily develop resistance. Yet, in some circumstances they are absolutely essential if anything is to be salvaged from those billions of tiny jaws. How, then, to use chemical pesticides wisely?

Just as we do, insects have bacterial, fungal, protozoan, helminthic, and viral pathogens. The gypsy moth, *Lymantria dispar,* has its share. *L. dispar* was introduced to the United States in 1889, but its natural population controllers were not, and it spread widely. A full-scale search for parasites and pathogens to import is going on. Researchers hope to find some that can be used in a program of **integrated pest management (IPM)**. IPM is the combined use of chemicals, improved agricultural meth-

Gypsy Moths
Normal gypsy moth larvae. These voracious eaters defoliate valuable hardwood trees throughout the Northeastern and Middle Western United States. (Courtesy of USDA Forest Service).

ods, and infective pathogens to control an insect pest.

Two microbial pathogens are receiving field testing. The bacterium *Bacillus thuringensis* (BT) kills the larval or caterpillar stage of *L. dispar.* BT is also useful against tent caterpillar, spruce budworm, cabbage worm, and many other pests that have a chewing larval stage maturing to a moth adult. BT cells, mass-produced commercially as Dipel or Thuricide, contain toxic protein crystals. The voraciously feeding larvae ingest the crystals, which destroy the insects' in-

testinal tracts and kill them. The protein is not toxic to other insects or vertebrates. Recombinant DNA research seeks increases in the yield and toxicity of the crystals. BT is already a major part of spruce-budworm- and gypsy-moth control programs over forested areas of the Northern United States and Canada.

The nucleopolyhedrosis virus (NPV) is a naturally occurring pathogen of gypsy moth. The virus is grown on gypsy-moth larvae in the laboratory, then separated from the tissues of the dead larvae and dried. The virus powder (trade name, Gypchek) can be sprayed on shade trees and affected woodlands. Larvae contract the virus and die. Adults are also infected and pass the virus through mating and from generation to generation in the eggs. (Even insects can have venereal diseases!)

Microbial insect parasites are generally safe for beneficial pollinating insects, animals, or humans. They are easily stored; users don't need extensive protective gear or sophisticated instruction. When used in combination with small amounts of chemical pesticides, biological insect controls have an important part in pest control programs. (Photo courtesy of USDA Forest Service.)

assembled. The cycle terminates with digestion of the bacterial cell wall and the release of viruses.

5. Lysogenic viruses, as exemplified by phage lambda, normally do not destroy the host. Instead the viral genome integrates into the circular bacterial chromosome to form a prophage. Most or all of the viral genes are repressed. The prophage is replicated along with the bacterial genome and is passed to all the host cell's progeny. The prophage may cause new phenotypic traits to appear in the infected strain.

6. On occasion the lysogenic viral genome directs a productive lytic cycle yielding free virus. Portions of the bacterial DNA may be included in the new virions and transmitted to new bacterial hosts. This is called transduction.

7. Plant viruses almost always contain RNA. They infect plant cells following mechanical injury to the plant. The infections are slowly progressive.

8. Insects serve as vectors for virus infections.

9. Virus diseases are an important population-control factor in animal and insect populations.

DISCUSSION QUESTIONS

1. What aspects of cellular function must have evolved to a working level prior to the evolution of viruses?

2. In what ways are viruses different from plasmids? Refer to Chapter 8.

3. Summarize the genetically coded functions a virus might be expected to have.

4. How do viruses kill a host cell?

5. How do viruses cause their host cell to become genotypically altered?

6. Under what conditions is infection of a host cell by a virus very difficult to detect?

7. Describe how viruses transduce genes.

8. What are the general characteristics of viral infection of plants?

9. In what two ways are the interaction of insect and viruses important?

ADDITIONAL READING

Brock TD, Smith DW, Madigan MT. *Biology of microorganisms.* 4th ed. Englewood Cliffs, N.J.: Prentice-Hall, 1984.

Diener TO. Viroids. *Sci Am* 1981; 244:66–73.

Doane CC, McManus ML, eds. *The gypsy moth: research toward integrated pest management.* USDA Technical Bulletin 1584, Washington, D.C., 1981.

Fraenkel-Conrat H, Kimball PC. *Virology.* Englewood Cliffs, N.J.: Prentice-Hall, 1982.

Hughes SS. *The virus: a history of the concept.* London: Heinemann, 1977.

Prusiner SB. Prions. *Sci Am* 1984; 251:50–59.

Watson JD. *The molecular biology of the gene.* 4th ed. Menlo Park, Calif.: Benjamin/Cummings, in press.

10

ANIMAL VIRUSES: INFECTIOUS DISEASE AND CANCER

CHAPTER OUTLINE

Animal viruses cause many of the major unconquered infectious diseases in humans. In this chapter we will first study the animal viruses: their structure; the unique aspects of their interaction with animal cells; and the special methods used to grow, study, and identify them. We will look at patterns of viral disease in human beings. Finally, we will examine the cancer-causing oncogenes, the biology of cancer cells, and viruses that cause or are associated with cancer.

The general characteristics of viruses were described in Chapter 9. Let's review a few of the important points. The viral genome may be double-stranded DNA, single-stranded DNA, double-stranded RNA, or single-stranded RNA.

Some RNA virus genomes are made up of several unconnected segments of RNA. Viruses have unique enzyme mechanisms for copying their genomes. The DNA of animal viruses closely resembles the chromosomes of the host cell both in the chemistry of their DNA and in the nucleosome sructure. The genomes of some DNA and RNA viruses may become integrated with the host chromosomes. In this integrated state, the viruses are latent but their genes may alter the host cell's behavior. The larger animal viruses all possess lipoprotein envelopes derived from the host-cell membrane, which also contains viral proteins. Enveloped viruses always leave their host cells by the process of budding.

A BRIEF SURVEY OF THE ANIMAL VIRUSES

There are seven groups of DNA animal viruses and 15 groups of RNA animal viruses. An eighth DNA virus group, hepadnavirus, has recently been proposed. We will feature here some of the most important features of several major groups. Tables 10.1 and 10.2 contain more detailed data.

SOME IMPORTANT DNA VIRUSES

Four of the DNA virus groups are especially significant in human health.

- **Adenoviruses** occur in many distinct groups. They have spikes on the capsid.

They cause respiratory diseases of the "common-cold" variety. Certain strains cause tumors in newborn research animals and cancerous changes in cultured cells.
- **Herpesviruses** cause a wide range of infections, from a malignant tumor (Marek's disease of chickens) to a contagious skin disease (chickenpox). Their most important feature is their ability to cause latent infection, as shown by the intermittent symptoms of cold sores or genital herpes. Herpesviruses have been implicated in four different types of human cancer.
- **Poxviruses** are large and complex. They may cause skin eruptions and respiratory disease (smallpox). Some cause tumors

			Nucleocapsid	
Genus	**Genome**	**Shape**	**Size (nm)**	**Examples**
Papovavirus	D2, circular	Icosahedral	55	Rabbit papilloma Human papilloma (warts)
Polyomavirus	D2, circular	Icosahedral	45	Monkey virus SV40 Mouse polyomavirus BK virus, human brain disease
Adenovirus	D2, linear	Icosahedral	70–80	Human adenovirus of "cold"; oncogenic viruses
Herpesvirus	D2, linear	Icosahedral	100	Herpes simplex, types I and II Varicella–zoster (chickenpox) Epstein–Barr virus
Iridovirus	D2, linear	Icosahedral, some enveloped	190	Tipula virus
Poxvirus	D2, linear	Brick-shaped, enveloped	$300 \times 240 \times 100$	Smallpox Cowpox Molluscum contagiosum
Parvovirus	D1, linear	Icosahedral	20	Viruses of rabbits, dogs
Hepadnavirus	D1, circular	Icosahedral, enveloped	42	Adenovirus-associated viruses Hepatitis B

TABLE 10.1 The DNA viruses

(rabbit fibroma) and a human sexually transmitted disease (molluscum contagiosum).

- **Hepadnaviruses** are very tiny and contain single-stranded DNA. They cause hepatitis B, called serum hepatitis.

SOME IMPORTANT RNA VIRUSES

Six RNA virus groups are highly significant in human disease.

- **Picornaviruses** include very small naked polyhedral viruses. **Enteroviruses** are trans-

TABLE 10.2 The RNA viruses

Group	Genome	Shape	Nucleocapsid Size (nm)	Examples
Enterovirus	R1	Icosahedral	20–30	Poliovirus Coxsackievirus
Rhinovirus	R1	Icosahedral	20–30	Human "colds" Foot-and-mouth disease
Calcivirus	R1	Icosahedral	20–30	Swine disease
Alphavirus	R1	Spherical	50–60	Equine encephalitis viruses
Flavivirus	R1	Spherical, with envelope	40–50	Dengue Yellow fever
Myxovirus	R1, 7 segments	Spherical, with envelope	80–120	Influenza types A, B, and C
Paramyxovirus	R1	Spherical, with envelope	100–300	Mumps Parainfluenza Measles Distemper
Coronavirus	R1	Spherical, with envelope	80–120	Respiratory virus ("colds")
Arenavirus	R1	Spherical, with envelope	85–120	Lymphochoriomeningitis; Lassa virus
Bunyaweravirus	R1	Spherical, with envelope	90–100	Insect-borne group
Retrovirus	R1, 4 segments	Spherical, with envelope	100–120	C-type leukosis– leukemia–sarcoma virus B-type mammary tumor virus
Rhabdovirus	R1	Bullet-shaped, with envelope	175 × 70	Rabies
Reovirus	R2, 10 segments	Icosahedral	70–80	Reovirus Type I
Orbivirus	R2, 10 segments	Icosahedral	50–60	Bluetongue virus
Rotavirus	R2	Spherical	<36	Gastroenteritis agents

mitted via foods and liquids to the gastro-intestinal tract. They include polioviruses, echoviruses, and coxsackieviruses. The last of these may cause influenza-like symptoms, myocarditis, or neurological disease. **Rhinoviruses,** with more than 100 types, cause most cases of the common cold. Their great variability is the main barrier to controlling colds.

- **Togaviruses** include the **alphaviruses** and **flaviviruses.** They are arthropod-borne, transmitted by insect and arachnid vectors. They cause encephalitis, yellow fever, and dengue.

- **Myxoviruses** and **paramyxoviruses** are two closely similar groups of viruses that have envelopes. They cause influenza, para-influenza, measles, and mumps. Their envelopes contain hemagglutinin and neur-aminidase, proteins important in disease, synthesized under viral genetic control. Influenza A virus frequently changes its exterior surface chemistry, rendering our immune defenses obsolete. Another group of respiratory viruses is the **coronaviruses.**

- **Retroviruses** possess a unique enzyme called reverse transcriptase. This enzyme allows them to integrate. Many are oncogenic, causing leukemias, sarcomas, mammary tumors, and degenerative nerve diseases. One retrovirus causes the acquired immunodeficiency syndrome (AIDS).

- **Rhabdoviruses** are bullet-shaped and enveloped. The most important is the rabies virus.

- **Rotaviruses** are a major cause of diarrheal illness in infants. Several of the most important viral groups are illustrated in Figure 10.1.

INTERACTIONS OF ANIMAL VIRUSES WITH THE HOST CELL

Our discussion of the viral replication cycles in Chapter 9 was based on the procaryotic bacterial cell as host. Viruses that parasitize eucaryotic animal cells follow the same basic strategies, but use important additional mechanisms to coexist with their more complex hosts.

ADSORBING TO AND ENTERING THE HOST CELL. An animal cell's outer surface is a flexible cell membrane. Chemically, this membrane contains an extensive variety of glycoprotein receptor molecules, genetically specific to the animal's tissue and its species. Physically, membranes are fluid, and new material is constantly being added to the membrane, while other portions of the membrane are being drawn back into the cell and disassembled. These natural movement patterns are such that anything that binds to a cell's outside via its membrane receptors will shortly be brought into the cell. These physicochemical properties of the cell's membrane are beneficially employed in such activities as acquiring hormone molecules. Viruses, however, can use this mechanism to gain access to their intended hosts.

Many viruses contain substances on their exterior surfaces designed to aid in adhesion to the intended host cell (Figure 10.2). The adenoviruses have peplomers, or protein spikes, at the intersection of the faces of their polyhedral capsids. Enveloped viruses such as influenza and other myxoviruses have hemagglutinins, envelope proteins that bind firmly to host-cell glycoproteins. The protein of the poliovirus capsid binds directly to host-cell lipoproteins.

To enter the cell, naked and enveloped viruses use different techniques (Figure 10.2b–d). In many cases, the host cell appears to actively assist in the process. A naked virus is taken in by **phagocytosis,** the process by which cells ingest solid particles. The membrane area to which the virion is attached sinks inward. The membrane closes around the particle, retaining it in a membrane-derived vesicle. Inside the vesicle, the capsid proteins of the virus are enzymatically removed, uncoating the nucleic acid.

Enveloped viruses fuse their envelopes with the cell membrane, which has the effect of placing their nucleocapsids in the cytoplasm of the

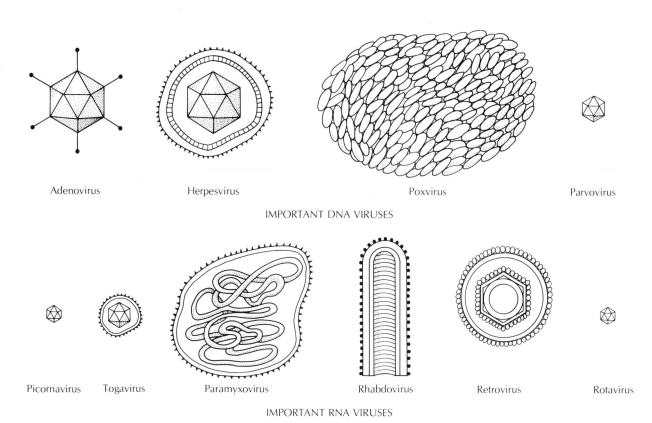

Adenovirus   Herpesvirus   Poxvirus   Parvovirus

IMPORTANT DNA VIRUSES

Picornavirus   Togavirus   Paramyxovirus   Rhabdovirus   Retrovirus   Rotavirus

IMPORTANT RNA VIRUSES

FIGURE 10.1   Some of the Important Groups of Animal Viruses

host. If the nucleic acid is coated with proteins, it will then be uncoated. The nucleic acid may stay in the cytoplasm or move to the nucleus for replication.

REPLICATION.   We outlined many of the general issues of viral replication in Chapter 9. It is important to understand that each type of nucleic acid requires a different sequence of steps and catalysts to be replicated. These vary considerably in detail from virus to virus, and a full discussion is beyond the scope of this book. However, let's take note of a few important points.

Most DNA viruses replicate their genomes in the host nucleus, as the viral proteins are being manufactured in the cytoplasm (Figure 10.3). A host-cell DNA polymerase enzyme system is usually employed for replication of the viral DNA. Somewhat later, the viral proteins move into the nucleus for the final assembly into virions. The poxviruses, however, replicate both DNA and proteins in the cytoplasm. Those DNA viruses that integrate their genomes with host chromosomes, notably the herpesviruses, do so directly, without the usual replication steps.

The active replication of RNA viruses tends to occur entirely in the host cytoplasm (Figure

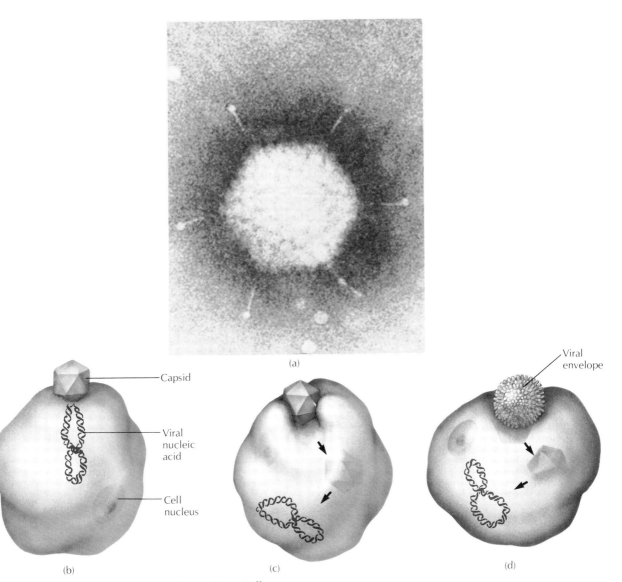

(a)

Capsid

Viral
nucleic
acid

Cell
nucleus

Viral
envelope

(b)　　　　　　　　　　　　　(c)　　　　　　　　　　　　　(d)

FIGURE 10.2 How Viruses Enter Cells
(a) Many viruses contain structures on their exterior surface designed to help them adhere to host cells. For example, the adenoviruses have protein spikes inserted in each angle of the capsid. (×500,000. Courtesy of H.C. Pereira, by permission of Academic Press, from R.C. Valentine and H.G. Pereira, *J. Molec. Biol.* (1965) 13:13-20.) (b) In some cases only the nucleic acid enters the host cell. (c) An animal cell may take in a virus that lacks an envelope by phagocytosis. Once inside, the capsid is removed. (d) If the virus has an envelope, the envelope will fuse with the host-cell membrane, releasing the nucleocapsid to enter the cell. The protein components of the nucleocapsid are removed by host enzymes, leaving the free nucleic acid.

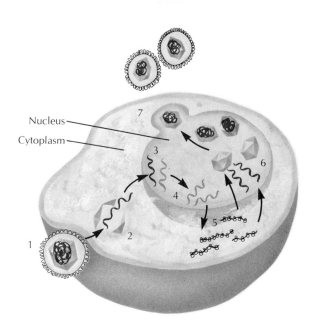

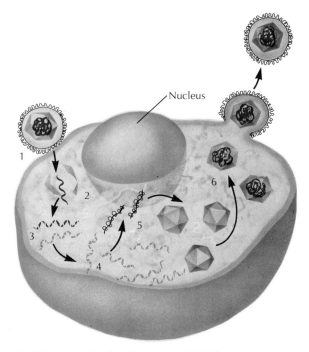

FIGURE 10.3   Replication of a DNA Virus
1) The virus particle attaches to and penetrates the cell.
2) The nucleic acids are separated from their protein coat.
3) The viral DNA enters the nucleus where one strand is transcribed into an RNA copy.
4) mRNA is processed and transported into the cytoplasm.
5) mRNA is translated and proteins are made on ribosomes.
6) Newly synthesized structural proteins and proteins that are involved in DNA synthesis are transported back into the nucleus, where viral DNA is synthesized.
7) Virus particles push through the nuclear membrane; in the process they acquire a final envelope. They then exit from the cell.

FIGURE 10.4   Replication of an RNA Virus
1) The virus particle attaches to and penetrates the cell.
2) The nucleic acids separate from the protein coat and are released in several segments.
3) First, a viral gene is translated, producing an RNA-copying enzyme. Then the viral RNA is replicated, one replicative form for each segment.
4) The replicative form of RNA in turn serves as a template for the synthesis of new genomes.
5) Capsid and envelope proteins are synthesized and begin to assemble.
6) The nucleocapsids form and the cell membrane is modified by incorporation of envelope proteins. The virions exit from the cell by budding through the membrane. In the process a segment of the cell's membrane becomes the envelope. The host cell survives and continues producing more virions.

10.4). Their RNA is transcribed as a message not only for the host cell to make the viral RNA polymerase to copy the genome, but also to make the viral proteins. The host cell's nuclear functions are less directly disturbed by this process, but the virus may shut down the cell's own protein-synthesis activities.

We see what is perhaps the most complete interaction between virus and host in the retroviruses. The single-stranded RNA of a retrovirus is copied once, by the unique retroviral enzyme called reverse transcriptase, to make a complementary DNA strand; the first DNA strand is copied once again to complete a double helix.

The "viral" DNA copy integrates into the host chromosome. Only after integration is it possible for new viral particles to be made. Most host cells with integrated retroviral genomes produce few new viral particles, however.

EXIT FROM THE HOST CELL. Cytolytic viruses, such as the poliovirus, leave the host cell when its membrane ruptures. Enveloped viruses, such as the influenza virus, leave the host by budding.

Budding appears to be a simple process. The nucleocapsid of the virus pushes out through the membrane, acquiring a segment of membrane to form its envelope (Figure 10.5). Actually, there is more to the process than appears at first glance. During viral replication, certain viral genes code for the production of envelope proteins, usually proteins used in specific adsorption of virus to host tissue. These proteins migrate to the inner surface of the host-cell membrane and are incor-

FIGURE 10.5 Budding
(a) In this electron micrograph, several copies of an enveloped Semliki forest virus bud from an animal cell. (Courtesy of M. Olsen and G. Griffiths, European Molecular Biology Laboratories.) (b) In the diagram, a nucleocapsid already formed inside the cell moves toward the cell membrane, which is modified by the insertion of virally coded proteins. The H spikes (hemagglutinin) and N spikes (neuraminidase) of influenza viruses are shown. Finally the virus leaves the cell, complete with a spiked envelope.

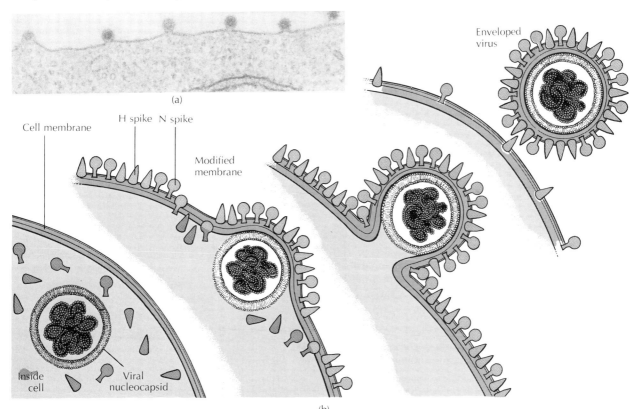

(a)

(b)

porated into it, forming modified areas of membrane. The nucleocapsid attaches to the membrane at the inner surface of these modified areas. Then, as the bud forms, the envelope material is derived from the modified part of the host-cell membrane. Up to 30% of the envelope proteins may be specifically viral.

HOW DO VIRUSES KILL CELLS?   Viruses may damage or kill cells by many means. Infective viruses often make repressors that shut down the synthesis of host-cell protein, RNA, and DNA. Mitosis is usually prevented and in some cases the host's chromosomes become markedly abnormal. The lysosomes and cell membrane become leaky, and the cell may autolyze. Certain viral proteins are directly toxic to the host cell. Some viruses cause cells to fuse, forming **syncytia** or **giant cells,** that are nonfunctional. As we will see, your body's own immune defenses, attacking the infected cell, may kill the cell as well as the viruses it harbors.

WHICH ANIMAL, WHICH TISSUE?

A virus forms a very specific chemical attachment to its host cell. As a result, it infects only certain types of cells.

Species Specificity. The polioviruses and rhinoviruses replicate exclusively in the human population in nature, although in the laboratory we are able to grow them on other types of cells and in primates. Influenza, a human virus, naturally infects rodents and swine also. These animals harbor the virus between epidemics in humans. They are a **reservoir,** a site where infectious agents remain alive, and from which the agents may spread the disease. Smallpox has no nonhuman reservoir. Consequently, we have been able to eradicate smallpox worldwide through the use of quarantine and vaccination (see Box 16.1). Since no humans now harbor it, smallpox cannot recur unless laboratory-maintained virus stocks were to escape.

The rabies virus is unusually nonselective, and has a very wide range of hosts. It infects both wild and domestic mammals, including dogs, cats, foxes, cows, skunks, raccoons, and bats. Human beings may also become infected with rabies if they are bitten by a sick animal. Rodents are the only common animals that seem to resist this virus.

Tissue Specificity. Most viruses infect one type of tissue but not others. Rhinoviruses attack only the superficial epithelium of the upper respiratory tract. Rhinovirus replication is very temperature-sensitive. They infect and replicate at 34 degrees Celsius, the temperature of the nasopharynx, but not at 37 degrees Celsius or above, the temperature of the internal tissues (Box 10.1). By contrast, the less selective Creutzfeldt–Jakob virus multiplies in the lymph nodes, liver, kidney, spleen, lung, cornea, and brain of victims.

Viral replication usually damages one type of tissue most severely. The outcome of the disease depends on whether the affected tissue is able to replace itself. Our bodies repair the destructive effects of the influenza virus within one or two days. However, we cannot replace nerve cells destroyed by poliovirus, so paralysis is permanent.

MODIFIED VIRUSES.   Although a virus has a species and tissue specificity, these may change naturally over time, or we may cause them to change by manipulations in the laboratory. Such viruses are said to be modified.

Viruses may be altered in the laboratory to grow in the tissues of a species different from their original natural host. Pasteur was the first to demonstrate this phenomenon, which he exploited to develop the first rabies vaccine. He inoculated rabies virus into the brains of rabbits, animals that do not get rabies by natural means. The rabbits did get rabies and die, and portions of their brain tissue were used to inoculate the brains of other healthy rabbits. After a series of transfers, the virus from rabbit brain became more potent for causing rabies in rabbits, but no longer caused rabies in animals other than rabbits. The virus had been modified to have a new

BOX 10.1
Creative Therapies

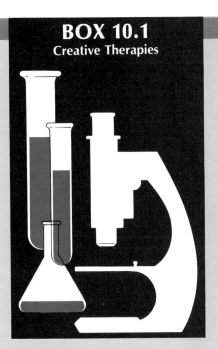

Enough about viruses is mysterious so that viral infection remains a fertile field for what we might charitably call "creative" therapeutic approaches. The common cold has stimulated its share of these. The Roman philosopher Pliny recommended kissing the hairy muzzle of a mule. New Englanders advocate hot water with honey and lemon. Hardier types go for hot rum toddies. The vitamin C enthusiasts promote the consumption of grams of ascorbic acid per day. A new device, the Rhinotherm, emits water vapor at 109 degrees Fahrenheit. The cold sufferer inhales the hot vapor for brief periods. Its use is claimed to cure colds in a day. There is at least a mild scientific rationale for most of these, although I don't know about the mule.

People suffering from warts also have an interesting choice of potential remedies. Warts, caused by papilloma viruses, appear without warning. They grow for a period, during which people try every known sort of home remedy to make them disappear. The juices of various plants, the blood of several of the smaller and more repulsive animals, nail polish and paint, and magic spells are among the choices available to the consumer of home wart remedies. They all work in time, since warts always sooner or later disappear. All that is required is patience and faith.

The same is true for colds. Any new cold treatment can make the same claim found in the old saying—untreated, a cold lasts a week, while with treatment it will go away in seven days. No one will dispute the truth of this advertising.

species preference. As a result, it was almost completely safe to use to immunize other animals and humans. Pasteur's first successful vaccines were composed of the spinal-cord tissue of rabbits that had died from rabies, in which the virus had been further weakened by varying periods of drying. He first tested his vaccine in 1885.

The Pasteur rabies vaccine was used for about 50 years, and then was replaced by a vaccine containing rabies virus modified in duck eggs, and most recently, by a vaccine containing modified rabies virus grown in human-cell cultures. Each new vaccine has been an improvement over the previous one because it reduced the risk of severe side effects. Today modification is a widely used technique in developing safe strains of infective agents for vaccine use.

Naturally occurring shifts in host specificity may be expected periodically. This phenomenon is the basis for one hypothesis about why AIDS appeared so suddenly in the 1980s as a seemingly "new" disease. It has been suggested that the causative virus, which appears to be native to the African continent, had infected only monkeys as its natural hosts until sometime in the 1970s. Then, a sudden shift in the virus's specificity allowed it to infect human beings and begin its sudden, mysterious spread.

ISOLATING, GROWING, AND IDENTIFYING VIRUSES

Viruses need living cells in which to grow. We must be able to grow viruses under controlled conditions in the laboratory in order to study them, to diagnose viral diseases, and to be able to produce viruses for vaccines. There are several ways to supply living host cells for growing viruses. These include cell culture, chick embryo inoculation, and experimental infection of laboratory animals.

CELL CULTURE

If suitable conditions are provided, cells derived from mammalian tissues can survive and divide after they are surgically removed from the animal and placed into laboratory glassware.

SOURCES OF TISSUE.   There are four types of animal tissues: epithelial, connective, muscle, and nerve. The tissues have undergone **differentiation** during embryonic life. This means that they have become different from other tissues by expressing a particular set of genetic traits. As a result, the various cell types exhibit specialized functions. Adult muscle and nerve cells have lost their ability to divide and therefore cannot readily be cultured. Cell cultures are usually taken from epithelial or connective tissues, the cells of which divide at a fixed rate, regulated by genetic and environmental signals. Our skin, cornea, and mucous membranes are all relatively thin cellular layers that shed cells from their surfaces and replace them by mitosis in underlying layers. Bone marrow, a connective tissue, matures the various types of blood cells

and releases them into the circulation. All these natural mitotic tissues can be successfully cultured in the laboratory.

Normal cells, both in the body and in the laboratory, have a control feature called **growth limitation.** That is, the cells will divide only until a certain population density is reached. As the cells in a vessel grow and divide, they undulate slowly across the glass. Whenever they touch a neighbor cell, motion ceases and cell division stops. Thus most cultured cells grow to form a **monolayer** — a sheet only one cell thick.

Tumors are solid masses of abnormal cells that may arise in any type of tissue. Tumor cells always exhibit rapid and uncontrolled growth. Tumor tissues may also be used in cell culture, and, as they lack growth inhibition, the cells multiply all over each other and form heaps many cells thick.

A **primary cell culture** is prepared by taking a sample of tissue aseptically from an animal. We mince the tissue, treat it with an enzyme to separate the cells, place them in sterile glassware, provide the cells with medium, and incubate them. A primary culture contains a mixture of cell types. Normal human amnion tissue, collected from the placenta following the delivery of a baby, makes a useful primary cell culture. Another common tissue source is human foreskin, taken at circumcision. Primary cell cultures are used to cultivate a large variety of human pathogenic viruses. These cell lines are unstable, meaning that they eventually stop dividing and die off after several transfers.

Serial cell cultures are stable cell lines that don't die off after a few generations. They are capable of an apparently unlimited number of cell divisions. These "immortal" cell clones, such as HeLa or L cells, are usually abnormal or tumorlike. However, there are some cell lines that both carry out serial growth and exhibit normal growth limitation. An example is WI-38, a cell line derived from normal human skin.

CULTURE CONDITIONS.   Cell-culture media provide amino acids and vitamins. Most cells also require blood serum, which contains protein

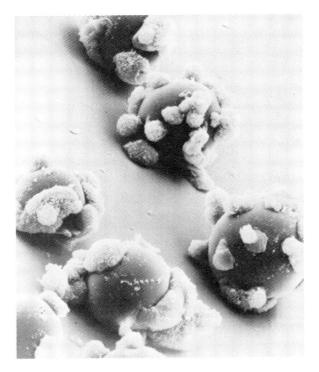

FIGURE 10.6 Large-Scale Production of Cultured Cells
(a) Human cells can be cultured in larger numbers if sufficient surface area for attachment is provided, such as by suspending inert microscopic beads in the culture fluid. Here cells of the Vero line are seen growing attached to beads (the larger spheres). (Courtesy of Bio-Rad Laboratories.) (b) A cell-culture carbon-dioxide incubator contains a roller system that supports and rolls many cell-culture vessels for circulation and aeration of media in an environment of controlled temperature and gas mixture. (Courtesy of Forma Scientific.)

growth factors to promote cell division. To prevent the growth of any microbial contaminants, antibiotics are included. Buffering and incubation in carbon-dioxide gas regulate the pH. The medium is changed frequently to replace nutrients and remove cell wastes. All manipulations are carried out under strict aseptic conditions. Bulk production techniques are highly developed, and often employ microcarrier beads suspended in the medium to increase the available surface area for cell growth (Figure 10.6). Roller bottles containing large volumes of media

and cells make it possible to manufacture the viruses used in some of the millions of doses of viral vaccines used each year.

THE CHICK EMBRYO

An intact fertilized egg is a convenient, sterile package of cells in which many viruses may be grown. As the embryo develops, the embryonic membranes resemble cell monolayers. The embryonic chicken itself contains rapidly dividing tissues, in various stages of differentiation.

The inoculum containing virus is injected into a chosen location in the egg (Figure 10.7). Viral multiplication may cause visible changes, small zones of cell destruction called **plaques** on the embryonic membranes. The embryo will eventually be killed. To harvest the virus, workers open the egg, remove the affected tissue, mince it, and separate the viral particles from the debris.

Fertilized eggs are often used to isolate and identify influenza and other viruses from patient samples. Influenza and yellow-fever vaccines are produced in embryonated eggs. Egg vaccines are hazardous to persons with allergies to eggs, however, so wherever possible they are gradually being replaced with vaccines produced in cell culture.

ANIMAL INOCULATION

Viruses can be cultivated in laboratory animals. This method is complicated and very expensive. For both humanitarian and practical reasons, it is used only when other techniques are not available or not suitable. For instance, live animals must be used to test the therapeutic effect of a new drug, or the protective effect of a vaccine. Mice, rats, rabbits, and monkeys are commonly used.

In order to render experimental results as unambiguous as possible, **inbred** animals such as rats or mice are usually used. An inbred line is the result of many generations of brother–sister mating. All individuals in an inbred line are genetically nearly identical. Experiments are easier to interpret because there is little random variation among individuals.

IDENTIFYING VIRUSES IN CULTURE

Many clinical laboratories now routinely identify viruses from patients' throat, blood, and other samples as part of the diagnostic process. One way of doing this is to observe carefully the changes the virus causes in a cell culture. The cells in a healthy cell culture will have a typical appearance. When virions enter, infect the cells,

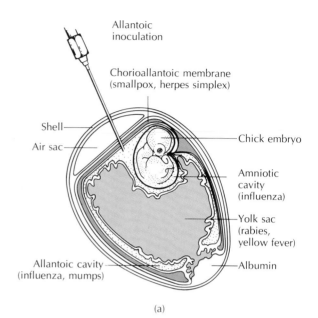

(a)

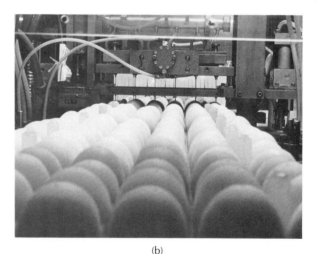

(b)

FIGURE 10.7 The Chick Embryo
An embryonated egg is a relatively cheap and convenient host used to grow viruses for vaccines. A needle passes through the air space at the blunt end of the egg, and the inoculum may be dropped on the chorioallantoic membrane, injected into the yolk sac, or introduced into the amnion. Each virus has a preferred site for replication, as shown. Viral growth is detected by the death of the embryo or by damage to embryo cells. (b) Influenza vaccine production requires mass inoculation of large numbers of fertilized eggs. This is done under strict aseptic conditions, both to keep the eggs free from contamination and to protect the workers. (Courtesy of Merck, Sharp and Dohme.)

and replicate, they cause microscopically visible **cytopathic effects** (CPE) in the cells (Figure 10.8). To a trained observer, the type of cellular damage indicates which virus is present.

For example, the vesicular stomatitis virus causes cells to come unstuck from their attachment to the glass of the culture vessel. The cells round up, and eventually wash off the glass into the culture fluid. The varicella-zoster (V-Z) virus, on the other hand, typically causes cells in culture to fuse, forming syncytia or giant cells. The influenza viruses and adenoviruses agglutinate or clump red blood cells; this is called **hemagglutination.** One virion binds to two or more red blood cells (RBCs), forms cross-connections, and thus makes clumps. Hemagglutination can be tested by adding RBCs directly to the culture flask. A positive test occurs when the red cells bind to the cultured host cells, demonstrating that they are infected with viruses that produce

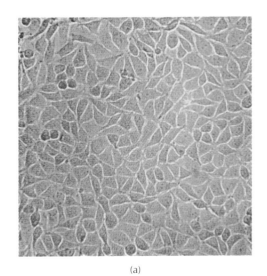

(a)

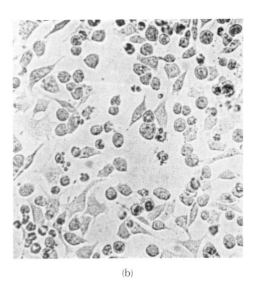

(b)

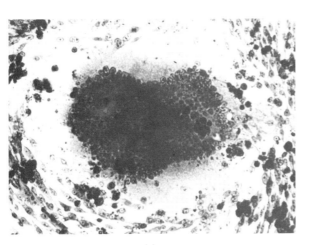

(c)

FIGURE 10.8 Characteristic Cytopathic Effects
Virus infection may cause characteristic visible cytopathic changes in cultured host cells. (a) A normal tissue monolayer, here derived from mouse L cells, is smooth and intact. (b) After infection with vesicular stomatitis virus, the cells in the monolayer have been visibly damaged and have begun to round up and detach from the glass. (c) Varicella-zoster virus infecting a culture derived from human melanoma cells has caused them to grow uncontrollably and to form a compact mound many layers thick. (Photo credits: (a) and (b), courtesy of Gail Wertz, School of Medicine, University of North Carolina, Chapel Hill; (c) from Charles Grose et al., Cell-free *Varicella-zoster* virus, *J. Gen. Virol.* (1979) 43:15–27, Fig. 1a.)

hemagglutinin, and that the protein is present on the cell's membrane. Herpes simplex, and some other viruses, cause yet another type of CPE. As herpes-damaged cells lose their attachment and detach from the glass surface, they leave cleared plaques of characteristic shapes and sizes. Microscopic observation of the internal structure of host cells may also reveal the spot where viral assembly is occurring. Although the light micro-scope cannot show the tiny individual viruses, it may reveal a dense aggregate of viral material, an **inclusion body** (Figure 10.9). Inclusion bodies are a valuable diagnostic trait for rabies infections. And finally, those viruses that integrate their nucleic acid often cause their host cells to become transformed — to take on the behavior of tumor cells and grow in an uncontrolled fashion.

USING IMMUNOLOGY TO IDENTIFY VIRUSES

Growing and isolating a virus from a patient sample is time-consuming. It is rarely possible by this approach to provide a positive identification in less than four or five days, and sometimes it takes a lot longer. However, serious illnesses may call for rapid lab results to guide the choice of therapy. As a result, immunologic diagnosis, which can often be carried out in minutes or hours, is becoming more important in clinical virology.

Antigens are substances that actively stimulate the body to produce antibodies. **Antibodies** are defensive proteins your body makes specifically to react with and neutralize antigens. We will develop these concepts further in Chapters 12 and 13.

Viruses are excellent antigens. If you have had a viral infection, antibodies will be present in your blood. When we are able to detect the antibodies, it is clear evidence that the virus is or has been there. This approach can be reversed to detect viruses (antigens) in body fluids, by treating the fluids with specific antibodies. Some of the most powerful new rapid-screening techniques quickly test samples of patients' body fluids against a selection of known antibodies. When a virus in the body fluid, such as blood serum or spinal fluid, and one of the specific antibodies combine, clumping occurs — a positive test. These tests allow a preliminary identification to be established within minutes. More detailed confirmatory tests will then follow. Inclusion bodies in tissues are often specifically identified by fluorescent-antibody and other immunologic techniques.

FIGURE 10.9   Inclusion Bodies
In this electron micrograph, clusters of rabies virus have aggregated, forming an inclusion body in the cytoplasm of the brain cell of an experimentally infected mouse. Particles in cross-section appear round, but those seen from the side show the characteristic bullet shape of rabies virus. (×340,840. From Lelio Orci and Alain Parrelet, *Freeze-Etch Histology*, New York: Springer-Verlag (1975), plate 25.)

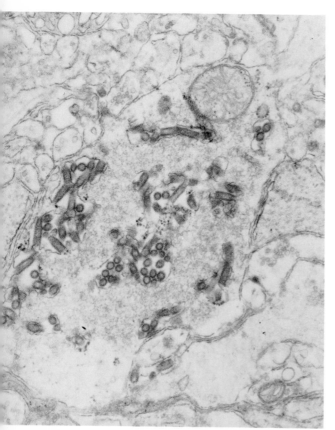

THE MANY FACES OF VIRAL DISEASE

Viral diseases come in all varieties, from brief to prolonged, from trivial to lethal. What determines the nature of the disease is the biological interaction between the host and the virus. In this section we will look at several patterns of viral disease, emphasizing biological interaction on the cellular level. In later chapters, we will be looking at specific diseases in terms of their clinical features and their effects on the whole person.

LOCALIZED INFECTIONS

As we saw, certain viruses replicate in only one type of tissue. The practical effect of this is that viral damage is restricted to one area of the body,

causing a **localized infection.** Some of the viral infections of the skin and of the respiratory and gastrointestinal tracts are of this type.

Let us use influenza as an example (Figure 10.10). We usually contract influenza by inhaling the virions, which then attach to receptors on the upper-respiratory-tract epithelium. There is a very short **incubation period** — a time during which the infective agent is multiplying in the body, but not yet causing symptoms. This period lasts for up to three days, as viruses proliferate locally in the epithelium. Many thousands of virions are released per host cell. These infect adjacent tissues, creating areas of damaged and dead cells. Tissue damage leads to the secretion of fluids, inflammation, and fever. Eventually, the surface layer of epithelial cells is lost. The

FIGURE 10.10 The Course of Localized Infection
Uncomplicated cases of influenza are localized infections. The viruses multiply in the respiratory epithelium. Replication is usually limited to the lining of the respiratory tract and thus the tissue destruction stops when no more host tissue is available and when host defenses become effective.

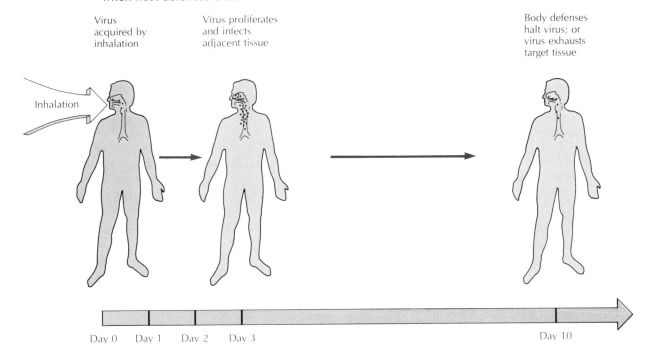

acute viral infection comes to an end when the supply of target tissue is exhausted or the body's defenses halt viral growth. Usually, the damaged tissue surface is rapidly repaired with a fresh layer of fast-growing epithelial cells. Occasionally, however, the raw, exposed surface may instead be colonized by bacteria. These bacterial invaders cause a secondary infection that may seriously delay recovery.

SYSTEMIC INFECTIONS

A **systemic** or **generalized** infection (Figure 10.11) is one that affects several tissue types and involves large areas of the body. Measles is an example. Like influenza, the measles virions are inhaled and first infect the respiratory mucosa. Unlike influenza, measles viruses then pass into lymphoid tissues (look ahead to Figure 12.2). A 10- to 12-day incubation period intervenes before symptoms appear. During this time viruses are actively multiplying in lymphoid tissue all over the body. At the end of the incubation period, viruses are carried in vast numbers to body surfaces via the blood stream. A rash suddenly appears, first on the mucous membranes, then on the skin. Simultaneously, toxic symptoms, such as fever, signal that body tissues are being destroyed.

However, at this point, when the symptoms are at their peak, the viral replication is almost at an end. The rash is actually evidence that the

FIGURE 10.11   A Systemic Infection
In a typical uncomplicated case of measles, viruses multiply as a systemic infection in several types of tissue throughout the body. The virus is inhaled and proceeds to replicate in the respiratory epithelium. It then migrates to lymph nodes and replicates further. Once large numbers of virus are produced, they migrate to the skin and the oral mucosa where they cause symptoms such as a rash and fever. As body defenses become effective, the measles virus is eliminated.

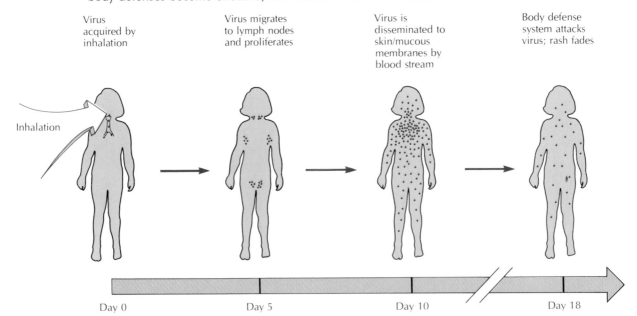

body's immune defense systems have become activated and are successfully attacking the skin cells that harbor viruses. Unfortunately, in order to destroy the viruses, the immune system must also kill their host cells, the skin cells! This is the reason for the cellular damage that is made apparent as the rash. As the immune system brings the viruses under control, the symptoms subside over a period of days. In all but a very few cases, measles viruses are completely eliminated from the body, and the disease episode ends.

PERSISTENT INFECTIONS

What happens if, as in those very few cases of measles just mentioned, the body's defenses can-

not successfully eliminate the infecting viruses? Then we have **persistent infections** — diseases in which the virus persists in the tissues for extended periods. Periods of active symptoms may occur at intervals alternating with latent periods during which the virus (and the symptoms) are not apparent. Low-level damage may occur and be cumulative over a long time. The onset of active disease may be very much delayed, appearing years after an initial viral infection occurred. Let us now look at three types of persistent infections, which are also shown in Table 10.3.

EPISODIC INFECTIONS. In an episodic infection, periods of acute cellular involvement and active

TABLE 10.3 Persistent infections				
Type	Virus	Host	Location of Virus	Disease and Symptoms
Episodic	Herpes simplex	Human beings	Epithelium, during attack; ganglion cells, between attacks	Recurrent fever blister on lip; activated by numerous stresses
	Varicella-zoster	Human beings	Epithelial cells, during attack; ganglion cells, between attacks	Shingles — widespread, very painful eruption over face and/or trunk
Chronic	Hepatitis B	Human beings	Liver	Usually none in chronic form; individuals with chronic cases act as carriers
	Epstein–Barr	Human beings	Lymphoid tissue	Usually none, occasionally causes infectious mononucleosis; associated with Burkitt's lymphoma and nasopharyngeal carcinoma
Slow	Creutzfeldt–Jakob virus ? prions	Human beings	Central nervous system and other sites	Creutzfeldt–Jakob disease, a progressive central-nervous-system degeneration
	Measles virus	Human beings	Central nervous system	Subacute sclerosing panencephalitis (SSPE)

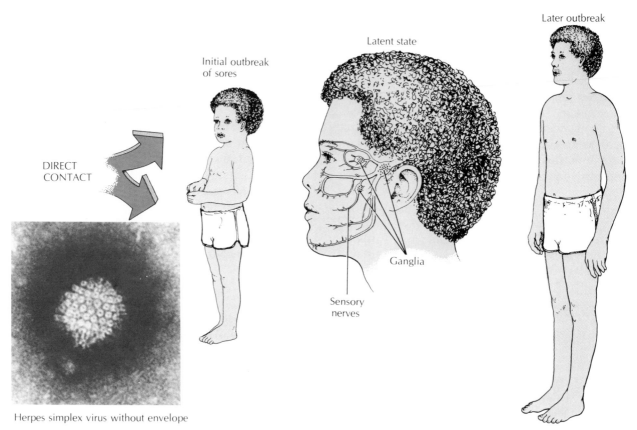

DIRECT
CONTACT

Initial outbreak
of sores

Latent state

Later outbreak

Ganglia

Sensory
nerves

Herpes simplex virus without envelope

FIGURE 10.12   An Episodic Infection
The common cold sore is caused by herpes simplex virus, which is initially
contracted (usually in childhood) by direct contact and often establishes an episodic
infection. A sore forms on the lip, and although the sore heals the virus is not
eliminated. Instead, it enters a latent stage during which it resides in sensory nerves
within the facial skin. Active and latent periods may alternate over the course of the
host's life. The virus may be reactivated periodically by minor emotional or
physiologic stress factors. (Courtesy of B. Roizman, University of Chicago/BPS.)

viral replication alternate with latent periods.
The classic example is the herpes simplex cold
sore, usually caused by herpesvirus Type I (Figure 10.12). Herpesvirus is usually contracted in
childhood by direct contact with a lesion on the
face of an adult, and causes an ulcer or sore on
the lip. This sore, the first episode, heals and a
latent stage follows. During latency the virus
resides, perhaps as a provirus, in sensory nerves
within the facial skin. Although antibodies to
herpesvirus are produced, the antibodies don't

eliminate the virus, perhaps because of its protected location inside the nerve sheath. It is currently felt that almost everyone carries latent
herpesvirus from childhood on.

What happens afterward depends on the person. In some people, occasional relatively minor
stresses such as anxiety, fever, sunburn, or common colds reactivate the virus, and another crop
of cold sores appears. In most people, the virus
remains latent for life, and they never have cold
sores. However, if their immune systems are

profoundly suppressed by cancer chemotherapy or some other severe stress, herpesvirus may reappear in an extensive, even fatal form.

CHRONIC INFECTIONS. In a chronic infection, the virus is constantly replicating at a low rate. The virus can usually be isolated from body fluids and the person is often infectious to others. Usually there are few overt disease symptoms. Virus production is limited enough that tissue damage, while cumulative, is not obvious. Evidence of an immune response can be demonstrated in the laboratory, but the immunity is unable to control the virus.

Acute hepatitis Type B (serum hepatitis) becomes established as a chronic infection in about 5% of those who recover from the first acute attack. Their serum may contain as many as a million infective viruses per milliliter. Chronically infected hepatitis patients must be ruled out as blood donors. Most patients eventually stop producing hepatitis B virus, but often not before the virus has severely damaged their livers. Chronic hepatitis also increases a person's risk of liver cancer later on. Up to 10% of persons with chronic hepatitis B eventually die of conditions related to the infection.

SLOW-VIRUS INFECTIONS. In **slow-virus infections,** the disease state develops as a continuous but slow process. An extended time — measured in years in human beings — elapses before the cumulative viral damage is sufficient to be clinically noticeable. Brain tissue is the usual target, and slow-virus diseases are degenerative, progressive, and invariably fatal.

Subacute sclerosing panencephalitis (SSPE) is a slow-virus disease that occurs very rarely some years after a person apparently recovers normally from an attack of measles. However, the measles virus has remained in the brain tissue, where its presence has caused cumulative damage. The measles virus is found in brain tissue at autopsy. SSPE patients do make abnormally high levels of antimeasles antibodies, but the antibodies clearly do no good. No one knows for sure why certain individuals develop SSPE after measles, or why the measles virus shifts both its tissue preferences and its replication pattern. With universal childhood vaccination for measles, SSPE is becoming extremely rare. Public health officials hope it will entirely disappear, along with measles.

Many researchers are now working on a degenerative neurological disease called **Alzheimer's disease.** They are investigating many hypotheses as to what causes it, including the possibility that Alzheimer's disease is caused by a slow virus. The disease is a quite common condition in older people; it causes senility. Some estimate that as many as 4 million people, up to 15% of the aging population, are affected.

CONTROLLING VIRUSES

Our bodies have many different natural means of preventing viral infection and bringing it under control. In addition, viral infections can be prevented by technical means, such as disinfection and sterilization, which kill the agents before they can infect us, and by vaccination, which raises our immune defenses.

OUR BODY DEFENSES

Our bodies have phagocytic white blood cells that can capture and digest virus particles. Our immune systems are able, when stimulated, to make specific antibodies to neutralize a virus's infectivity. We also produce sensitized blood cells that directly attack the viruses or the cells that harbor them. A detailed discussion of our body defenses is included in the next three chapters.

Another aspect of our natural defenses is **interferon,** a name used for a special class of proteins. Three types, called alpha, beta, and gamma, are produced by fibroblasts, leukocytes, and lymphocytes, respectively. It is suspected that a large number of different interferons exist, all with slightly different effects. One of the protective functions of alpha and beta interferons is to block viral replication (Figure 10.13).

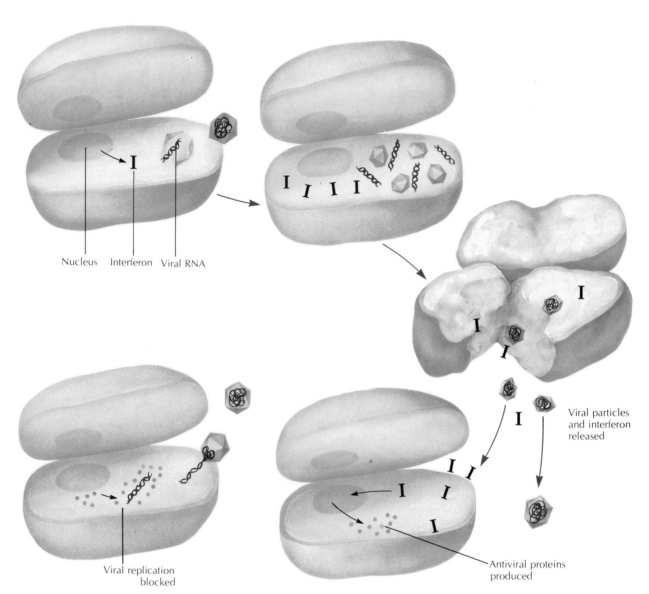

Nucleus   Interferon   Viral RNA

Viral particles
and interferon
released

Viral replication
blocked

Antiviral proteins
produced

FIGURE 10.13   How Interferon Blocks Viral Replication
When a cell is infected by a virus (top left), viral nucleic acids induce the host cell's
genes for interferon production. Interferon is synthesized and accumulates as the
virus replicates. The cell eventually releases the interferon along with new viral
particles. When interferon interacts with the cell membrane of a neighboring cell
(bottom right), that cell is induced to produce antiviral proteins. Although the
second cell may become infected, the presence of antiviral proteins blocks further
viral replication.

Unlike antibodies, interferons are not specifically directed against a particular virus. That is, they are antivirus, not anti-measles virus or anti-poliovirus. However, interferons are species-specific. Mouse or monkey interferons work only in mice or monkeys, not in humans.

When a cell is infected with virus, it starts to produce interferon because the viral nucleic acid turns on the cellular genes used for the synthesis of interferon. The protein is produced and released from the infected cells at about the same time the viral particles mature. Detectable levels of interferon appear in body fluids within 24 hours after infection occurs. The interferon then binds to the membrane surface of neighbor cells, but does not enter them. Its binding induces another set of cellular genes, altering patterns of RNA and protein synthesis within the neighbor cells. These cells are now protected. Should they be challenged by viruses attempting to infect, the interferon-induced proteins will prevent the virus from replicating.

Interferon is the primary means by which our bodies bring acute viral infections, such as the common cold and influenza, under control. Interferon's role in more protracted viral infections is not so clear. Preliminary research data show that interferon has therapeutic promise against rabies, hepatitis B, arthropod-borne encephalitis, and generalized herpes infections.

The gamma interferons appear to regulate various aspects of immunity. They are only indirectly involved in the body's defenses against acute viral infections, but may be extremely important in our defenses against cancer. Interferons of several types are now being manufactured by genetically engineered bacteria (Box 10.2).

DRUG THERAPY FOR VIRAL INFECTION

Viral chemotherapy presents several unique problems. Since viruses replicate within living cells, using many of the host cell's own systems, it is difficult (though not impossible) to find drugs that "poison" the virus without poisoning the host. We must seek enzymatic steps unique to the virus, such as its special nucleic-acid–synthesizing steps, and search for drugs that block them. Unfortunately, there are so many different viral replication processes that no one drug found so far is useful against more than one type of virus. Another problem is that, as we saw with measles infections, in many cases most of the viral replication is completed by the time symptoms appear, when we would normally start therapy. Despite these problems, we are making considerable progress in viral chemotherapy.

The biological strategy for almost all antiviral drugs is similar. They are base analogs — slightly altered chemical variants of the nucleotide subunits (Figure 10.14). They function as competitive inhibitors of nucleic-acid synthesis. Several of these chemotherapeutic drugs (Chapter 24) are licensed for human use. Examples are

FIGURE 10.14 Base Analogs
Chemicals called base analogs have structures similar to the normal nitrogenous bases of nucleic acids. Most antiviral drugs are base analogs, and act by blocking nucleic-acid synthesis. For example, guanine (left) is a normal nitrogenous base, a component of both DNA and RNA. Acyclovir (right) is a synthetic guanine analog used in the treatment of herpesvirus infections. It is incorporated into the viral DNA and hydrogen bonds normally with cytosine; it then blocks further viral replication.

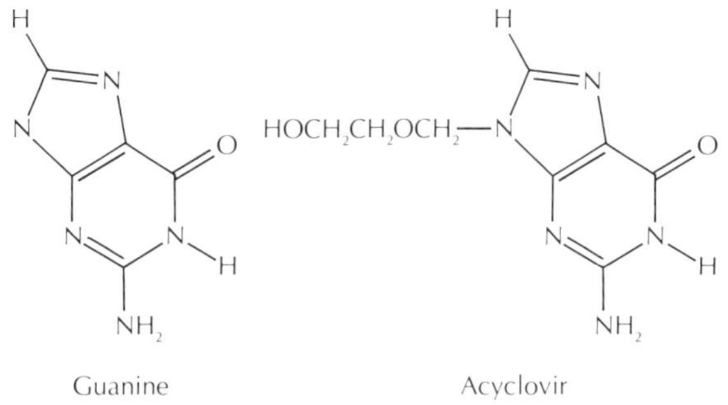

Guanine   Acyclovir

acyclovir and vidarabine, both purine derivatives, and 2-deoxy-D-glucose, a sugar derivative that mimics deoxyribose. Because these drugs interfere with nucleic-acid synthesis in the host cells — our cells, you'll recall — they are somewhat toxic for the patient if taken internally. Amantadine, which blocks the penetration of the RNA of influenza virus, is available by pre-

---

BOX 10.2
What Price Interferon?

Interferon has been, until recently, both scarce and sought after. This is because, in addition to its antiviral activity, it can affect cell motility, cell proliferation, and a variety of immunologic activities such as antibody production, allergies, graft rejection, macrophage activation, and natural lymphocyte killer activity. Interferons have great promise for a wide range of clinical problems, not the least being cancer chemotherapy.

In 1978 the American Cancer Society put up $2 million to purchase interferon to give to investigators for clinical trials. This amount of money in 1978 would buy enough interferon for only about 150 cancer patients. At that time human interferon was being produced only by culturing human leukocyte or fibroblast cells obtained from donors. The yield was low, the techniques complicated, and the price high.

Many companies began research to engineer bacterial plasmids bearing human inter-

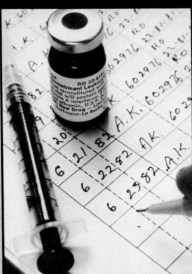

feron genes. Since 1980, a large number of *Escherichia coli* plasmids bearing genes for a variety of interferons have been cloned. Interferon genes have also been cloned in yeast plasmids. A liter of *E. coli* culture now routinely yields 10 billion units of interferon. In 1984, a huge dose of interferon could be manufactured for less than $30. Interferon is certainly no longer scarce.

As a result, human interferon from recombinant or-

ganisms is receiving extensive clinical testing. So far, tests on human volunteers are revealing a distressingly high level and wide range of side effects. Many researchers are discouraged, saying it's unlikely that interferon will be an acceptable therapy for simple viral infections such as the common cold, even though it is highly effective. Others remain hopeful that interferon will turn out to be an effective treatment for cancer, or be modified so as to be used safely for common viral infections.

Interferon may turn out to be the most important new addition to the arsenal of therapy since antibiotics were introduced, or it may end up with limited usefulness. There are many different interferons, and no one is sure that we're making and testing the "right" ones yet. It will still be a while before we know whether interferon will remain sought after.
(Photo courtesy of Hoffmann–La Roche Inc.)

scription. It prevents many cases of influenza and makes others both milder and shorter. Contrary to newspaper headlines, there is no "miracle drug" in sight for the common cold and most other virus infections.

DISINFECTION AND STERILIZATION

Most viruses are inactivated by disinfectants and sterilization. We use about the same techniques against viruses that we use to kill bacteria, fungi, and other cellular pathogens. These methods will be discussed in detail in Chapter 23. However, there are some exceptions worth noting, illustrating the dangers of being complacent about virus-disinfection procedures.

For example, the first polio vaccine was developed using killed poliovirus. The vaccine virus was grown in monkey-cell cultures, isolated, then treated with formalin (dilute formaldehyde). Enough formalin was used to kill the poliovirus, and each lot of vaccine was then tested to verify that the poliovirus was in fact nonviable. However, one batch of inadequately formalinized vaccine containing live poliovirus slipped through the testing process and was unknowingly administered to children. A number of cases of paralytic polio resulted. A more general, but fortunately less serious, slip-up was discovered later. It turned out that the monkey tissue used to cultivate the vaccine virus also was host to a monkey virus, SV(simian virus)40. SV40 was present in large amounts in all batches of vaccine. SV40 is quite resistant to formalin, so it was not killed by the level of formalin used to inactivate the poliovirus. Thus, millions of people who took the first Salk polio vaccine received live SV40 virus, too. Fortunately, it has no apparent effect in humans.

Another example of the problems of inadequate disinfection occurred in Scotland. There, sheep were vaccinated against the louping-ill virus with a formalin-treated vaccine. Unknown to anyone, the vaccine also contained the scrapie disease agent, which was unaltered by formalin treatment. As a result, 1500 sheep died of scrapie, and scrapie was introduced into a country that had previously been free from it.

Human hepatitis viruses are notorious for surviving slipshod needle- and instrument-sterilization procedures. Since viruses can neither be seen nor quickly cultured in the average laboratory, to control them we must select appropriate disinfection procedures and follow them scrupulously. There are no early warnings of failure — the first clue is illness.

VACCINATION

Unquestionably the best way to deal with virus infections is to prevent them by vaccination. Smallpox, polio, measles, rubella, yellow fever, dengue, and rabies are some of the life-threatening viral diseases prevented by timely vaccination. Vaccines are prepared from whole killed viruses, from whole live modified viruses, or from components of virions. Vaccine preparation and testing methods have been significantly improved, so that current viral vaccines are generally very safe and highly effective. Vaccination usually confers a lifelong protection against the virus used in the vaccine. See Chapter 25 for an in-depth discussion of vaccination.

VIRUSES, ONCOGENES, AND CANCER

Cancer is a very complex disease to study. Many environmental and genetic factors, such as diet, radiation, chemicals, age, ethnic group, geographic location, immune status, and virus exposures, can be shown to increase the risk of the various types of cancer. All cancers share a common underlying pattern — a genetic change in cells that causes them to grow out of control. Viruses, and alterations in certain key genes called oncogenes, are viewed as significant sources of such changes.

How do we determine what causes cancer? Recall the logic of Koch's postulates, discussed in Chapter 1. In order to say that something causes something, we must go through a logical sequence of experimental steps. In order to prove that a virus or a chemical causes cancer, we would have to administer it to human beings and observe that the disease appears. Since there are several ethical difficulties with this approach, we are not now, and won't be in the foreseeable future, in a position to say that virus X or compound Y causes cancer. We rely on statistical associations and observations of patterns of disease in populations.

Cancer occurs when a normal cell undergoes genetic change, becoming a cancer cell. All our insights at present indicate that not just one, but a sequence of events is necessary for a clinical disease state of cancer to develop. These may include a stress to a cell or changes in cellular genes, followed by the immune system's failure to do its job and take care of the problem. You can readily see that it is unwise at this time, if not impossible, to make unambiguous statements about what causes cancer.

In this section, we will become familiar with certain viruses that cause cancer in laboratory animals. We will review some of the data that suggest that specific viruses contribute to a few of the 200 or so types of human cancer. We must note that most researchers now believe that viruses are probably not the most important causes of human cancers in general. However, the study of cancer viruses has been of immense value, as it has led us to new basic understanding of the genetics of cancer.

We will investigate the nature and function of oncogenes, of which more than 20 are known at this time. The proto-oncogenes are part of the normal human genome. We find one or more oncogenes in the genome of most cancer-causing viruses. When normal oncogenes in an animal cell are activated, rearranged, or duplicated so that they are present in excess copies, they appear to cause cells to become cancerous in some studied cases. It appears that the reason some viruses can cause cancer is because they are able to disturb the host cell's oncogenes, and/or introduce extra copies.

SOME BASIC CANCER BIOLOGY

A **neoplasm** (new growth) is a group or a clone of cells with a more rapid growth rate than its surrounding tissues. A **tumor** is a solid or localized neoplasm. There are two general tumor types, benign and malignant. The term **cancer** is a general term for neoplasms, but it is most often used to refer to malignant forms. **Benign** tumors do not spread throughout the body, but they can still be dangerous. For example, a brain tumor might cause a fatal increase in intracranial pressure. However, any cells that break away from the main mass of a benign tumor lack the ability to establish themselves in other areas of tissue.

In some neoplasms, exceptionally aggressive cell lines arise from the original tumor. These are called **malignant** because they have the genetic ability to break away from the main tumor mass and to colonize other tissues and locations. Malignancies "seed" new tumor foci called **metastases** throughout the body. Metastases are a group of secondary tumors, genetically diverse even within one person. They may also be highly diverse in their response to therapy. Malignant cells are able to develop resistance to drugs used to treat them, just as bacteria develop resistance to antibiotics and insects to pesticides. As they grow, malignant tumors invade many normal organs and, if unchecked, eventually kill the host. Not all cancers are solid masses. **Leukemias** consist of both proliferating malignant cells in the bone marrow and circulating abnormal cells. The latter frequently attack normal tissue.

CELL TRANSFORMATION — THE HEART OF THE PROBLEM

Transformation is the name for the change that occurs when a normal cell develops into a cancer cell. We can cause transformation to occur in the laboratory, and watch and measure specific features of the change (Figure 10.15). Some of the transformed cell's important features are its unresponsiveness to normal growth limitation, lack of cell-to-cell communication, ability to multiply under a wider-than-normal variety of environmental conditions, and the appearance of novel antigens on its surface or in its nucleus. Although we see these changes, we still do not know exactly what causes them, although we have partial answers. Using recombinant-DNA techniques, researchers are now able to isolate genes suspected of being responsible and dissect them down to the level of individual nucleotides. We'll explore some of the available information that involves viruses directly.

ONCOGENIC DNA VIRUSES

About one quarter of the known DNA viruses will, under some conditions, cause cell transformation in a cell-culture system. Relatively fewer can do the much more significant feat of causing tumors in an intact animal.

The DNA papovaviruses cause benign tumors in human beings and rabbits. In humans, the condition is the common wart. In the rabbit, such benign growths frequently develop into malignant carcinomas later in life. In culture, these rabbit viruses have a low transforming potential. However, they will transform tissues from many different species of animals. Their DNA is integrated as a provirus. The SV40 virus of monkeys transforms not monkey cells, but cultured mouse, hamster, or human cells. However, there is no evidence at all linking SV40 to cancer in humans.

Adenoviruses are primarily respiratory pathogens, but subgroups 12, 18, and 31 are highly oncogenic. They transform hamster, rat, rabbit, and human cells in culture, and cause tumors in newborn hamsters and mice. Adenovirus DNA is integrated in transformed cells, and viral gene transcripts and virus-coded antigens are present. There is no evidence that adenoviruses cause neoplasms in humans.

The herpesviruses are the most notable oncogenic DNA viruses. They unequivocally cause naturally occurring malignancies. An example is Marek's disease virus (MDV) in chickens, which is highly infectious and can kill up to 70% of a

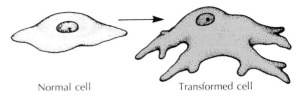

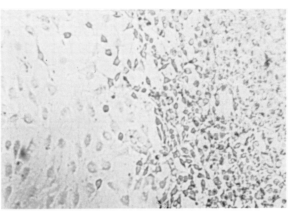

Normal cell   Transformed cell

FIGURE 10.15 Transformation
The transformation event is a genetic change during which a cell with normal controlled growth acquires traits of altered, uncontrolled growth. In the light micrograph of a monolayer of cultured 3T3 mouse cells, the left side shows an area of normal cells that are flattened and firmly attached to the glass surface on which they are growing. On the right side is an area of monolayer showing growth changes due to transformation with an oncogene. Note that the cells no longer adhere firmly to the glass and have assumed a rounded shape. (Courtesy of Dr. Robert A. Weinberg, Whitehead Institute for Biomedical Research, Massachusetts Institute of Technology.)

flock within a few weeks. Marek's disease has the distinction of being the only cancer, to date, that has been controlled by vaccination. This herpesvirus was shown to be the causative agent by Koch's classic methods; isolated MDV was introduced into healthy chickens and produced the disease.

The Epstein–Barr virus (EBV) is a human herpesvirus (Figure 10.16). It is ubiquitous — about 90% of American adults have anti-EB antibodies, denoting prior infection. Many of these

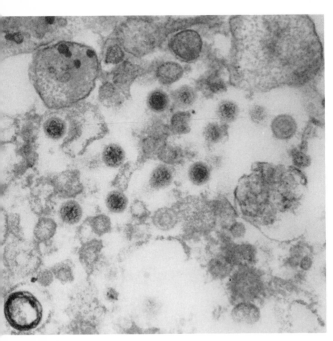

FIGURE 10.16   The Epstein–Barr Virus
EBV is a herpesvirus. It has an envelope and is polyhedral. It causes infectious mononucleosis and is strongly implicated in several types of cancer. It is shown here in tissue from a case of Burkitt's lymphoma. (Transmission electron micrograph, ×64,600. Courtesy of Alyne K. Harrison, Centers for Disease Control, Atlanta.)

with chromosome aberrations, reciprocal translocations involving chromosomes 8 and 14, 8 and 2, or 8 and 22. These translocations appear to alter the activity of a normal cellular gene found on chromosome 8. It is possible that EBV has effects of a similar nature. EBV causes infectious mononucleosis, a benign lymphoma. Infectious mononucleosis is the only one of these disease states that is common in the United States. EBV has also been implicated in Hodgkin's disease, another malignant lymphoma. All these conditions may reflect differing relationships of the EBV genome to the genomes of genetically distinct persons, living under different environmental conditions.

TRANSMISSION OF EPSTEIN–BARR VIRUS.   When we know that an infective agent causes a disease, we can discover how it is transmitted. Studies in Africa suggested person-to-person transmission of ABL. American folk wisdom labels infectious mononucleosis as the "kissing disease." EBV, in fact, is the only type of human-cancer–related virus that behaves as if it is infective. This means that the Epstein–Barr virus particles must be external to the body, or **exogenous,** at some point. Vaccination can prevent the spread of such exogenous viruses, just as it does with such "ordinary" infectious viruses as measles or polio. An EBV vaccine is being developed. Researchers expect that immunity to EBV will reduce mortality due to ABL. This would be very strong evidence that EBV is, as suspected, the primary cause of the disease.

Herpes simplex Type II, genital herpes, is generally familiar in its primary manifestation, as a sexually transmitted disease. However, studies also show that as the incidence of genital herpes has increased in the United States, the incidence of cervical cancer has increased in parallel. A similar statistical correlation has not been demonstrated in studies in Japan. With inconsistent data such as this we might suspect, but have no proof, that herpes contributes to cervical cancer.

people produce a few EB virions and are presumed to have integrated EBV DNA in the chromosomes of certain of their white blood cells called B-lymphocytes. It appears to be "normal" to have integrated EBV. However, EBV can be isolated from biopsies of malignant tissues in African Burkitt's lymphoma (ABL) and nasopharyngeal carcinoma. The evidence that EBV causes Burkitt's lymphoma is circumstantial, but very strong. The picture is clouded by the fact that EBV is not always found in ABL, and other viruses are occasionally found. Burkitt's lymphoma is, in many cases, also associated

RNA TUMOR VIRUSES

The retroviruses are the only RNA viruses that can exert a permanent effect on the genetic material of their host cell, and thus are the only ones that can cause transformation. They are frequently oncogenic, and many cause malignancy fairly freely in animals under natural circumstances. We saw in Chapter 9 that most retroviruses contain oncogenes directly associated with transformation; in fact, oncogenes were first discovered by studying the genetics of retroviruses. Not all retroviruses carry oncogenes, however. A retrovirus particle always contains some molecules of the enzyme reverse transcriptase. Retroviruses are very strongly implicated in human leukemias and breast cancer, but it has not been proven that they are directly causative.

C-TYPE RETROVIRUSES.   The leukemia, leukosis, and sarcoma viruses are a large, related group. As they bud from the cell, a characteristic **C-type particle,** with the core centered in the envelope, is formed and extruded (Figure 10.17). Unidentified viruses seen in tissue are classified as C-type if they have this appearance. Uninfected "normal" tissues in culture also often contain or produce C-type particles. We do not know where these particles come from or what they do.

The Rous sarcoma virus (RSV) is a C-type retrovirus that causes malignant disease in chickens. Most strains of RSV cannot carry out productive infection, and free virions are very rare. When infection does occur, the viral genome integrates as a provirus. This event, in cultured cells, is almost always followed by transformation. Transformed cells in turn initiate sarcomas (cancers of connective tissue) when injected into susceptible animals. The transforming potential in the viral genome is associated with the **src oncogene,** which was the first

FIGURE 10.17   Retroviruses
The retroviruses are classified in part by their shape after they bud from the host cell. (a) The C-type retroviruses, such as these in mouse brain, bud in such a way that they are spherical. (×70,414. Courtesy of Alyne K. Harrison, Centers for Disease Control, Atlanta.) (b) The D-type retroviruses, represented here by the human D-cell lymphotropic virus, form irregularly shaped envelopes and have cylindrical cores. (×71,060. Courtesy of Alkyne K. Harrison, Centers for Disease Control, Atlanta.)

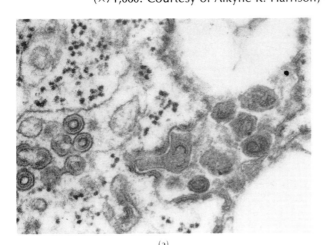

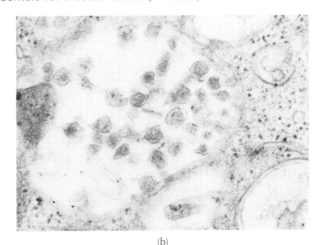

(a)   (b)

oncogene explicitly identified as such. For a period of some years it was generally assumed that the src gene was a uniquely viral gene.

How does RSV spread from cell to cell and from chicken to chicken? The virus, in the form of an oncogene-bearing provirus, is transferred from cell to cell by mitosis. The hen transfers the provirus to her eggs by meiosis. The virus, in fact, is only rarely seen as a formed virion, and is even less commonly external to its host cell. Viruses transmitted this way are called **endogenous** viruses, because they are rarely external to the body. Transmission occurs at the moment of fertilization; thus there is no practical method, such as vaccination, to prevent endogenous viruses from being transmitted. RSV has a long latent period, so relatively few birds develop sarcomas until old age. This is reminiscent of the human pattern, where most cancers develop in middle age or after. To prevent effects of endogenous virus, it may be that we must find some way to keep the provirus (and its oncogenes) permanently inactive.

Either isolated C-type murine leukemia viruses (MLV), or leukemic mouse cells containing MLV genomes, can transfer leukemia between members of inbred strains of mice. There are major differences in susceptibility among mouse strains. Mice of one strain develop leukemia many times more frequently than mice of another strain. Again, this is reminiscent of the human situation as revealed by the Epstein–Barr virus and its various diseases. It is clear that whether, and how, the oncogenic potential of a virus is expressed depends heavily on the host's genetic makeup.

T-TYPE RETROVIRUSES.   Several oddly shaped retroviruses, with a highly condensed core, have been implicated in human leukemias. These human T-cell lymphotropic viruses (HTLV) have received much attention. They are viruses that alter the behavior of one specific group of human white blood cells called T-lymphocytes. Remarkably, HTLVs are implicated in two quite opposite changes. In leukemia of the T-lymphocyte lines, caused by HTLV-I, the T cells begin to proliferate wildly, causing a malignant condition. On the other hand, HTLV-III is believed to cause AIDS, in which the fundamental problem is that the T-lymphocytes stop dividing and die out. Thus it appears that the genetic material of these viruses can either accelerate or halt cell division in the host lymphocyte.

B-TYPE RETROVIRUSES. Mammary tumor viruses (MTV) compose another retrovirus subgroup. As they bud from the host cell, they form a particle in which the core is eccentrically placed in the envelope. MTV is transmitted endogenously via egg and/or sperm in mice of certain strains. Large numbers of infectious virions are also shed in the lactating mouse's milk, and these particles can then infect their sucklings, and mice of other strains.

About one out of 25 American women will develop breast cancer in her lifetime. Evidence concerning the role of retroviruses in humans, in contrast to the mouse, is ambiguous. Milk from affected mammary glands sometimes contains viral particles; these particles contain reverse transcriptase. Tissue from malignant, but not from normal, mammary glands contains RNA sequences similar to mouse MTV ribonucleic acid. However, there is no evidence that human milk is infective.

ONCOGENES IN NATURE

Our knowledge of oncogenes is increasing daily. By the time you read this chapter, a great deal more will be known than you will find printed here. This is one of the most exciting research areas in biology.

NATURAL DISTRIBUTION OF ONCOGENES.   First, let's deal with the fact that the name "oncogene" is partly misleading. The first genes of this type that were identified came from the genomes of oncogenic retroviruses and caused transformation. The genes were also found in the abnormal cells of tissues infected by such viruses. The name **oncogene** was invented for these genes, to designate that they caused cancers, and it was generally believed that these were viral genes. To everyone's surprise, however, almost identical genes were later found to be an essential part of the genome of most, probably all, normal animals and humans. Thus, it does seem somewhat inappropriate to call them oncogenes. In summary, then, we now know that oncogenes are found in the genomes of normal animal cells, in tumor tissues, and in oncogenic viruses.

Researchers are now able to compare the forms of oncogenes found in transforming viruses (sometimes called V-oncogenes). More than 20 of these have been discovered so far. They include such intensely studied examples as **src,** the oncogene in RSV, **ras,** the oncogene of Harvey murine sarcoma virus, and **myc,** found in the virus of avian myelocytomatosis.

The normal cellular oncogenes are called **proto-oncogenes.** They are defined as normal because they are found in many different tissues of all individuals screened at any age. These genes are an essential part of the human genome. Evidence increases all the time that these oncogenes direct various crucial aspects of growth regulation.

GENETIC FEATURES OF ONCOGENES. V-oncogenes and proto-oncogenes are homologs — they have very similar coding sequences. However, the proto-oncogenes, like typical eucaryotic genes, contain both coding sequences (exons) and noncoding sequences (introns). V-oncogenes are much shorter, as they are composed of only the coding sequences. At the moment, proto-oncogene homologs to eight V-oncogenes have been mapped to the correct chromosome in the human genome. Many V-oncogenes can insert themselves seemingly anywhere in the host's chromosomes.

It is believed that the viruses probably acquired their V-oncogenes originally from the cellular genes of the host cell they parasitize as transcripts. When they infect other host cells, they then transduce these oncogenes, which have the effect of duplicating the host's already present, normal proto-oncogenes. A virus that can transduce an oncogene has the advantge that it may cause its host cell to multiply much faster, which of course means that the virus will also be more rapidly replicated.

To muddy the water still further, modified **C-oncogenes** (genes of cellular origin, with both exon and intron base sequences that are not exactly like either of the other kinds) may also cause transformation. There must be some small but critical differences in the genetic information of these closely similar genes. Researchers are comparing the various oncogenes, seeking slight variations in the hopes of discovering how these differences transform.

In a gene, small changes can be highly significant. One striking experiment compared the base sequence of a c-oncogene from a human bladder cancer with that of the proto-oncogene from the same person's normal tissues. The only difference in the DNA was the single base substitution of a thymine for a guanine (Figure 10.18). This change would have the effect of altering the 12th amino acid in the protein gene product from glycine to valine.

Studies of other tumors have revealed other mechanisms. In some cases of Burkitt's lymphoma, the c-oncogene, **c-myc,** is moved to a different chromosome than usual. In yet other tumors that have been studied, it turns out that there are duplicate copies of an oncogene. In other words, cells may go out of control either by acquiring new or altered genes from a viral infection, or through changes to existing genes (Figure 10.19a). It seems that it is not simply the presence of the gene that causes transformation, but some alteration in the way the gene interacts or is integrated with the part of our genome it is supposed to be regulating.

HOW DO ONCOGENES TRANSFORM? We can give only a preliminary answer to this question now. There is quite a bit of direct evidence supporting the hypothesis that proto-oncogene

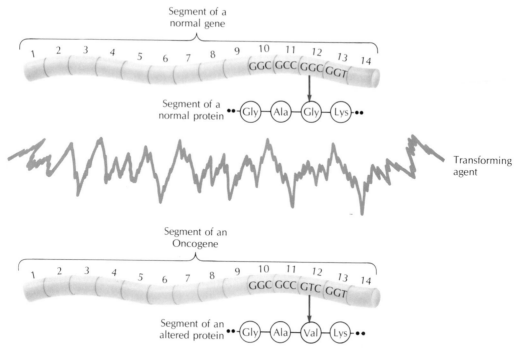

FIGURE 10.18   Molecular Variation in an Oncogene
The c-ras oncogene from normal cells and that from bladder cancer cells differ in
one nucleotide. A change from guanine to thymine has occurred in a single codon.
The gene product of the cancer cell ends up with a valine substituted for a glycine in
its amino acid sequence. It is not known why this change causes the cell to alter its
behavior.

products may act as normal growth regulatory
factors, while v-oncogene or c-oncogene products may do the same thing, just more so. About
half of the v-oncogenes have been shown to code
for enzymes of the **protein-kinase** type. These
enzymes catalyze reactions that add phosphates
to and activate membrane proteins, having the
effect of speeding up cellular activities, particularly cell division. At the same time, if the cell
has too much of a protein kinase, the enzyme
may act in an unselective and excessive fashion,
and the cell's growth patterns might become abnormal.

The following is an example in which we
understand the workings of one oncogene. The
platelet-derived growth factor (PDGF) is a human serum protein. PDGF is crucial in normal

wound healing because it encourages fibroblast
cells to divide, forming scar tissue and new connective tissue. PDGF is normally coded for by
the **sis** proto-oncogene. This gene's viral relative,
the **v-sis** gene from simian sarcoma virus, causes
an excessive production of PDGF, which in turn
causes an uncontrolled division of fibroblasts —
a sarcoma.

ENVIRONMENTAL CARCINOGENS—WHAT IS THEIR
ROLE?   Masses of data have established that environmental factors such as diet, smoking, irradiation, and chemicals increase the risk of cancer noticeably. Some researchers estimate that
as much as 90% of our cancers originate from an
environmental insult or stress. We can hypothesize that one or a series of environmental insults

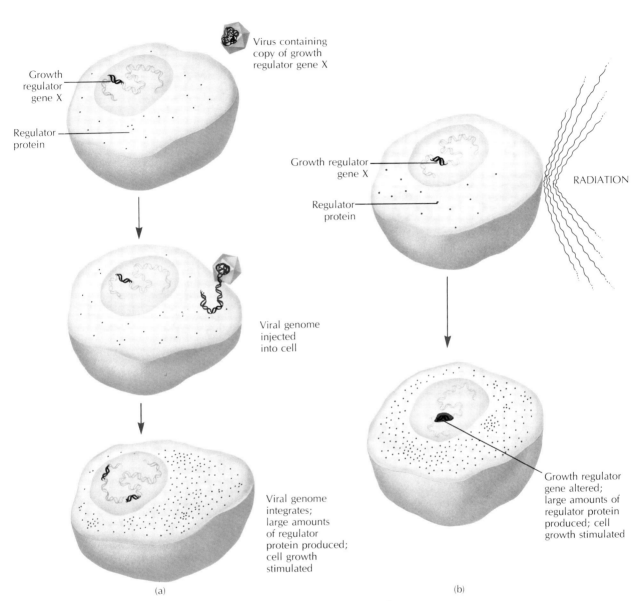

FIGURE 10.19 Two Possible Mechanisms for Oncogenesis
(a) An oncogenic virus might introduce an oncogene coding for a growth-regulator function into a host cell. After integration, the cell would possess two copies of the growth-regulator gene and would produce excessive amounts of growth protein. (b) An environmental effect such as radiation might cause a mutation in an existing regulatory gene, altering it so that it increased its output of growth protein.

could trigger an oncogene to inappropriate activity. The key event might be a mutation or chromosome rearrangement (Figure 10.19b) that could alter the expression of such a gene.

In this view, the contribution of an oncogenic virus, as part of the cell's environment, might be as just another highly efficient means of causing the crucial genetic change. Many researchers see indications that the development of cancer is a two- or even three-step process. Thus the chemical exposures, virus infections, or radiation exposures may have to be sequential and cumulative in order for clinical disease to be the final outcome. As we use our new tools to pinpoint the exact molecular basis of various cancers, we simultaneously come closer to finding ways in which the transformation event might be prevented or reversed. We are witnessing a very optimistic period in cancer research.

SUMMARY

1. The animal viruses are obligate intracellular parasites. They range from the very tiny, simple polyomaviruses to the large, complex poxviruses. Many animal viruses are enclosed in a characteristic envelope derived from host-cell membranes.

2. Viruses are classified by several features. There are eight groups of DNA viruses and 15 groups of RNA viruses; most contain serious human pathogens. Each virus has a characteristic species specificity and tissue specificity. These can be modified by artificially manipulating the conditions of viral growth.

3. Viruses are most commonly grown and maintained on cell cultures, where they can be studied in a controlled fashion. Fertilized eggs are also widely used for cultivation of viruses and vaccine production. Live animals may be inoculated with virus to study features of the viral disease and evaluate potential treatments and vaccines.

4. Animal viruses have several patterns of replication. They attach to the host cell very specifically. Then the host cell ingests the virus and enzymatically uncoats it. Cytolytic replication, integration as a provirus, or steady replication by budding may ensue, depending on the genetic and physiologic regulation systems of virus and host.

5. If new viruses are produced, they are released either by lysis of the host cell or by budding through the membrane. In the integrated state, both DNA and RNA viruses can transform cells to an uncontrolled growth pattern.

6. The replication of RNA viruses requires unconventional information transfer, either from RNA to RNA or from RNA to DNA. Unique enzymes, coded for by the viruses, execute these syntheses.

7. Viruses are identified in the laboratory by observing characteristic patterns of cell damage in culture, and by their interaction with specific antibodies. Clinical applications involve isolation and identification of virus from patient samples, or characterization of antiviral antibodies produced by the patient.

8. Some familiar viruses, such as influenza, chickenpox, and polio, usually cause acute disease. Virus multiplies to large numbers in certain types of tissue, and tissue destruction causes symptoms. The disease has a defined duration.

9. In persistent infections, the virus is present in the body for months to years. The normal defense mechanisms fail to eliminate it. Virus may be shed asymptomatically, cause sporadic episodes of cell damage, or multiply very slowly with progressive and cumulative neurologic degeneration.

10. Virus-caused cancers in animals are a special case of persistent infection. An oncogenic virus can integrate its genome and transform a cell. The cell loses its normal growth limitation, and expresses abnormal characteristics. Transformed cells, or transforming viruses, injected into susceptible animals lead to tumor growth. The integrated viral genomes that cause malignant growth, called oncogenes, may be ubiquitous in animal cells. Evidence points to viruses as contributing to several types of human cancer.

11. Body defense systems such as phagocytosis and immunity can control the spread and replication of viruses. If damaged tissue is replaced, the victim recovers. Interferon, a protein produced in response to viral infection, appears to play a major role in recovery.

12. Antiviral chemotherapy is being developed. Drugs effective against DNA viruses are available, but they are somewhat toxic.

13. Many viral infections may be prevented by the standard methods of sanitation and vaccination. Some viruses are highly resistant to disinfectants and heat, requiring special attention. We have effective vaccines against many viral diseases.

14. Proto-oncogenes in the normal human genome regulate cell growth. When altered to c-oncogenes these genes produce abnormal growth. So do the V-oncogenes carried by retroviruses. Environmental carcinogens also may alter proto-oncogenes, causing transformation.

DISCUSSION QUESTIONS

1. What are the practical implications of modifying a viral strain's range of hosts?

2. Evaluate the significance of growth limitation as a cellular characteristic both in cell culture and in the animal.

3. What are interferons? How do they differ from antibodies? What is their role in defense against viral infection?

4. Why is the enzyme RNA-dependent DNA polymerase called the reverse transcriptase?

5. Compare the means by which an oncogenic virus and an environmental carcinogen might convert a normal cell to a cancerous one.

6. [*continued*] might convert a normal cell to a cancerous one.

7. Compare the acute and the persistent disease caused by measles virus. (See also, Chapter 16.)

8. Vaccination may be a practical preventive approach for tumors caused by "contagious" cancer viruses such as Marek's disease, but not for those caused by retroviruses. Why? What preventive measures can you suggest to minimize the activation of oncogenes?

ADDITIONAL READING

Albert BA, et al. *Molecular biology of the cell.* New York: Garland, 1983

Bishop MJ. Oncogenes. *Sci Am* 1982; 247:80–93.

Croce CM, Klein G. Chromosome translocations and human cancer. *Sci Am* 1985; 252:54–73.

Dolin R. Antiviral chemotherapy and chemoprophylaxis. *Science* 1985, 227:1296–1303.

Fraenkel-Conrat H, Kimball PC. *Virology.* Englewood Cliffs, N.J.: Prentice-Hall, 1982.

Gordon J, Minks MV. The interferon renaissance: molecular aspects of induction and action. *Microbiol Rev* 1981; 45:244–266.

Hunter T. The proteins of oncogenes. *Sci Am* 1984; 251:70–79.

Kucera LS, Myrvik QN. *Fundamentals of medical virology.* 2nd ed. Philadelphia: Lea and Febiger, 1985.

Palmer EL, Mantin ML. *An atlas of mammalian viruses.* Boca Raton, Fla.: CRC Press, 1982.

Prusiner SB. Prions. *Sci Am* 1984; 251; 50–59.

Varmus HE. Form and function of retroviral proviruses. *Science* 1982; 216:812–820.

HOST AND PARASITE

11

THE HOST–PARASITE
RELATIONSHIP

CHAPTER OUTLINE

Consider how outnumbered you are by the microorganisms in your environment. Your skin carries thousands of organisms per square centimeter. Each lungful of air brings a million organisms into your upper respiratory tract. In your intestinal tract the microbial population may actually outweigh the food being digested. Humans exist in a delicate natural balance with the diverse members of the microbial world. If this balance is disrupted, the usual outcome is infectious disease. Thinking about this certainly helps us to appreciate the delicate balance we call health!

In this chapter we will begin to examine the components of the balancing act. First we will describe the general health factors that affect our ability to resist pathogenic attacks. Next, we will discover that there are various ways in which microorganisms relate to us. Many microbes are part of our normal flora. A few of the species in normal flora can become pathogens when the balance tips too far in the microorganism's favor. We will investigate what factors make one microorganism "normal" and another a pathogen. Microorganisms are more pathogenic if they have the traits of adhesiveness, invasiveness, intracellular parasitism, and toxigenicity. The human anatomy, physiology, blood cells, serum factors, the phagocytic network, and the inflammatory response help to contain infection if the microorganisms temporarily gain the upper hand.

THE STABLE STATE: HOST FACTORS IN HEALTH

Although most people can readily define "disease," they have a much harder time with "health." Good health is the most important protection we have against invading microorganisms. We will observe how a decline in physical condition clearly predisposes to infection.

NUTRITIONAL STATUS

Consistent proper nutrition is an important factor in minimizing the frequency of human infections. In particular, proteins and vitamins are directly protective. Well-nourished human beings use dietary proteins to make healthy tissues, serum proteins, and antibodies. Vitamins promote efficient metabolism and maintain the integrity of skin and membrane surfaces. On the other hand, malnourished individuals are highly susceptible to infection because their tissues may be abnormal and their serum proteins low. Once they acquire an infection, it is much more likely to be severe. Measles is rarely fatal to children in developed countries. However, in the less-developed countries of South America, measles usually hits children from one to four years of age — the age group most poorly nourished. The chance of their dying once they get measles is 100 to 400 times higher than that of a measles-infected child in the United States. In developed countries, malnutrition is encountered in the elderly poor, the mentally ill, and in chemical-dependent persons, all of whom have elevated disease rates.

RELAXATION VERSUS STRESS

When rested and stressed individuals are compared for infection risk, severity, and duration, the message is clear — stress can increase problems from disease. Severe physical or emotional stresses, including fatigue, anxiety, or depression, can produce the "stress syndrome." In this condition, enhanced production of epinephrine (adrenalin) and altered levels of hormones such as prostaglandins suppress the function of many groups of defensive cells. Stress is the major contributing factor in the development of "trench mouth" (acute necrotizing ulcerative gingivitis) in college students during exam periods. People who endure prolonged sleep deprivation, such as sailors who attempt solo oceanic crossings, almost always develop infections. A wide range of defense mechanisms are depressed during stress, but will recover when the individual becomes rested again. A new research specialty called **psychoneuroimmunology** investigates the effect of mental states on the nervous system and the body's defense.

AGE AND PHYSICAL CONDITION

Your defensive systems vary in activity over your life span. The very young and the very old have the highest risk of infection. In a young child the immune system is just gearing up, and in an older individual its effectiveness gradually wanes. However, some "childhood" diseases such as measles and mumps are less severe in children than in adults. This is partly because, as we shall see, the body defenses themselves produce many of the symptoms of an infectious disease (e.g., fever, diarrhea). Since a child's immune system is less developed, it mounts a less vigorous but sufficient attack.

Some studies have shown that physical conditioning through exercise tends to reduce the frequency and severity of infectious respiratory disease. It is unclear whether this is because conditioning strengthens respiration and circulation, or because it alters hormone levels in the body.

An underlying abnormal condition or pre-existing disease may predispose a person to infection. A person with any condition that increases his or her risk of infection is said to be **compromised.** The individual with a structurally abnormal heart, for example, has a high risk of bacterial endocarditis, a dangerous heart infection. This infection may develop following relatively minor dental procedures or minor traumas that introduce small numbers of bacteria into the blood stream. Some neoplasms can suppress the immune system and leave a patient poorly protected. A bedridden patient has a significantly increased risk of pneumonia. Hospitalized patients are frequently compromised, which is the primary reason why they sometimes develop severe infections. The complex subject of hospital infection control is covered in Chapter 26.

OCCUPATIONAL HEALTH RISKS

People working at certain jobs have a higher-than-average risk of certain infections. Public-health measures and appropriate medical treatment have eliminated most infectious occupational diseases. However, brucellosis is still occasionally diagnosed in meat-packers and leather-workers. Rabies is a risk for veterinarians, game wardens, and dog officers. Health-care personnel, who provide personal services requiring much physical contact with ill individuals, have very high occupational-disease risks. Their risk is fortunately offset by the fact that they receive information about how to protect themselves, as well as how to prevent the spread of disease to others.

NORMAL HUMAN FLORA

The body surfaces of animals are a preferred habitat for microorganisms. These areas are usually moist, warm, and bathed in nutrient-rich secre-

BOX 11.1
The Great Vitamin Controversy

The controversy over vitamins has been raging ever since they were discovered, it seems. How much is enough? How much is too much? Can vitamins help prevent diseases, especially infectious diseases and cancer?

A microbiologist would tend to quietly stress the fact that adequate vitamins are necessary for healthy tissues, and healthy tissues maintain healthy ecological balances with microorganisms. The B vitamins as a group are essential for maintaining skin structure. Vitamin C maintains the collagen fibers that reinforce connective tissue. In scurvy, the disease of vitamin C deficiency, the skin cracks and ulcerates, affording easy access for pathogens. Inadequate riboflavin intake can lead to changes in the cornea and conjunctiva that set the stage for eye infections. There's no controversy about the fact that we need these vitamins — but how much of them?

Vitamin C has also been intensely promoted as a preventive for virus diseases (the common cold) and for treatment of cancer. The available data about cancer treatment does not support any therapeutic claim. Several dozen clinical trials have tested the effectiveness of large doses (megadoses) of vitamin C in preventing or shortening common colds. The accumulated evidence indicates that on the whole, when averaging the effects on many individuals, vitamin C's effect on colds is neutral. This general conclusion does not negate the observation that a few individuals experience positive effects. What is becoming clear is that each person's requirement for each vitamin is probably a little different. What's more, your vitamin requirements change from time to time, with stress levels or life-style changes. However, megadoses of vitamins will not make up for an otherwise unhealthy life style.

(Photo courtesy of Darlene Bordwell, Boston.)

tions. The term **normal flora** describes a microbial species that is most frequently found in close benign association with human beings or other animals. Basically, the microorganism in question is well adapted to life in or on the animal. On the one hand, it is able to survive the host's defensive activities well enough to maintain an adequate population. On the other hand, both the host's defenses and its neighbor microorganisms keep its population growth and metabolic activities well under control (Figure 11.1). Thus the host isn't harmed by its presence. On the contrary, it seems our normal flora is essential to a healthy life. However, as we have recognized its beneficial roles, we have also documented its ability to cause disease. In unusual circumstances, normal flora may become parasitic. The true picture, then, is a dynamic one in which an *Escherichia coli* or *Staphylococcus epidermidis* can be truly "normal" in one place or time, but pathogenic when growing in another place in the body or at a time in the person's life when he or she is seriously compromised.

Normal flora colonize the body surfaces only. They are restricted to the skin, intestinal tract, and other areas that open directly to the outside, such as the mouth, upper respiratory tract, vagina, and lower urinary tract (Figure 11.2). Microorganisms are not "normal" in the lower respiratory tract, blood, tissues, and upper urinary and genital tracts.

A baby still in the uterus is microbiologically sterile — it has no normal flora. At birth, the newborn enters a world where all the human beings that care for it have normal flora. The baby's body is then rapidly colonized by the same organisms. Far from being undesirable, this process is essential for the baby's continued normal development. We can see this by studying animals that lack normal flora.

Experimental animals can be reared germ-free if they are surgically delivered and transferred immediately to a sterile living environment. Their immune systems do not develop at the normal rate. Since they don't encounter anything from which they need to protect them-

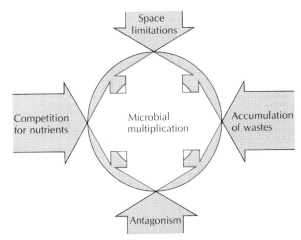

FIGURE 11.1 Microbial Population Control
Each of the many species within the normal flora of any body-surface area exerts population-control effects on the others. Controls include competition for nutrients and space, accumulation of wastes, and direct antagonism by inhibitory chemical byproducts of other organisms. These factors keep the numbers of the normal flora in check. They also help to prevent pathogens from becoming established in that area of the body.

selves, they do not express their genetic potential for immunity. If normal flora are introduced, the germ-free animals often become ill, but then, slowly, their immune systems will begin to develop. This shows us that continuous intimate contact with normal flora is important to create a state of readiness in the body defenses.

Ordinarily, the competition for nutrients and space among members of the normal flora not only controls each of them, but also excludes new colonists that are potential pathogens. Some normal flora even produce byproducts that strongly inhibit invading organisms. The normal flora, then, actually reduce the potential for infectious disease. When the normal flora is suppressed, opportunistic disease organisms frequently proliferate. These are microbial species that under normal conditions would be either

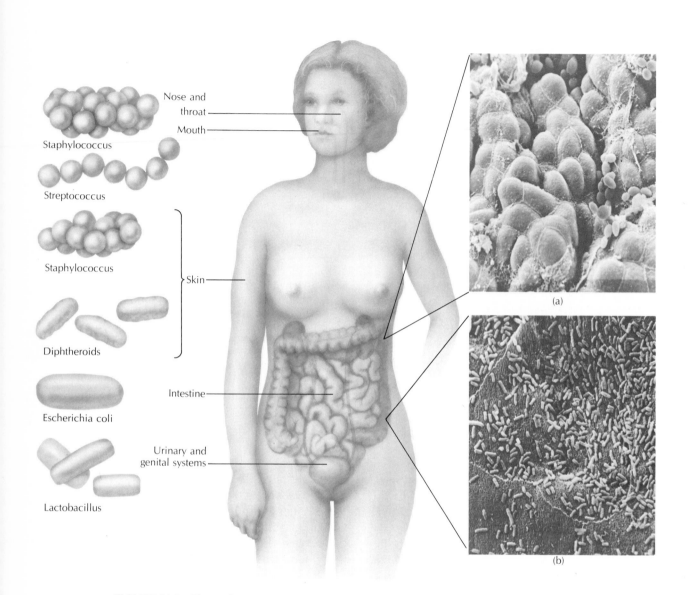

Staphylococcus

Streptococcus

Staphylococcus

Diphtheroids

Escherichia coli

Lactobacillus

Nose and throat

Mouth

Skin

Intestine

Urinary and genital systems

(a)

(b)

FIGURE 11.2   The Body's Normal Flora
Every surface of the human mucous membranes is subject to its own unique microbial population. Two dominant types are shown here. The two scanning electron micrographs show the wrinkled epithelial surface of two areas of the intestine with strands of mucous and adherent organisms, predominantly yeasts (a) and bacteria (b). (From Z. Yoshii et al., *Atlas of Scanning Electron Microscopy*. Tokyo: Igaku Shoin, Ltd. (1976).)

absent or very few in number because they are less fit and don't compete well with the normal flora. When the competition is removed, however, they take advantage of the opportunity to proliferate as a **replacement flora.** The effect can be compared with plowing a field in early spring, then leaving it uncultivated. The result will be a field full of weeds! Uncolonized skin, like bare soil, is an invitation to uncontrolled growth of whatever organism gets there first.

In the early years of the "antibiotic era," patients received very large doses of antibiotics for long periods. They frequently developed troublesome secondary infections, often with the yeast *Candida albicans*, which is resistant to the antibiotics used (Figure 11.3). This organism is a minor component of normal flora in most areas of the body. The antibiotics were repressing other portions of the normal flora, thus allowing the yeast to gain the upper hand and become troublesome.

NORMAL FLORA IN THE BOWEL.   In our gut, microorganisms are present in huge numbers — up to 100 billion organisms per gram of feces in the normal lower bowel. It is clear enough what we do for them — we provide them with warmth, moisture, and a nutritionally complete diet. What do they do for us? The picture is still incomplete.

Bowel organisms produce some vitamins, such as certain B vitamins and vitamin K. The amount of vitamin K is sufficient to augment the diet, but not enough to completely satisfy the vitamin requirements of a human. Bacterial vitamin K is very important in newborns, however, since their diet does not provide much of it. Gut bacteria in human beings do not digest cellulose to any significant extent as they do in cows and termites. Their role in supplying digestive enzymes to assist with other types of food is uncertain. They break down dietary fats and excreted bile acids. Human babies must acquire lactobacilli as the dominant member of their infant gut flora or they will have grave difficulties in digesting milk because unaided they cannot metabolize the milk sugar lactose.

It is interesting to realize that the relationship of both partners (host and flora) is designed for mutual survival. If either partner were to succeed in eliminating the other, it too would die. If the host's immune defenses were to eliminate the normal flora, he or she would be overwhelmed by opportunistic species. If the flora were to invade and kill the host, its source of

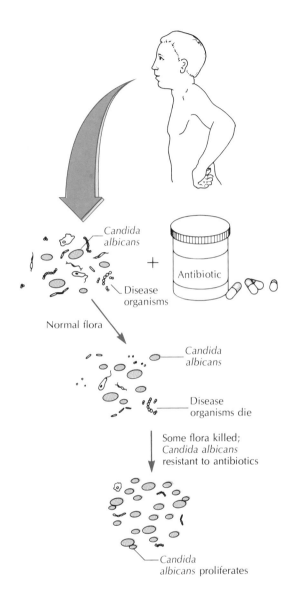

FIGURE 11.3 Regulatory Function of Normal Flora
Normal flora plays an important role in controlling the population of opportunistic pathogens. For example, the normal flora in the mouth suppress the growth of the yeast *Candida albicans*, usually present as a minor component of the flora. When a person receives antibiotics to combat a bacterial disease organism, some normal flora are also killed. The *C. albicans*, which is resistant to antibacterial antibiotics, is then free to proliferate, causing a disease of the mucous membranes, called thrush.

nutrients would be gone. For microbial flora, a dead host is not as adequate an environment as a live one. A mutual restraint on aggressiveness benefits both.

MICROBIAL PATHOGENICITY

A microbe acts as a pathogen because it possesses unique features, which it uses to establish a parasitic relationship. These may include destructive enzymes, toxins, microbial surface layers resistant to host defenses, or the ability to multiply in a location in which the host's defenses are ineffective. These features are called **virulence factors;** when they are present, some percentage of encounters (but never all) between the organism and the host will result in disease.

INFECTIOUS-DISEASE TERMINOLOGY

Infectious disease occurs when a microorganism

- gains access to a tissue or an area of the body in which it is not normally found, then
- multiplies, or, in the case of an organism of normal flora, multiplies to an abnormally large population, and
- causes a detectable change in normal tissue function.

The first two occurrences by themselves constitute **infection.** If no clinically obvious damage results, we speak of an **inapparent, subclinical,** or **asymptomatic infection.** Only at the point where normal tissue function is perceptibly altered do we have a true **infectious disease.** The course of events need not progress inevitably through these steps; in fact, host defenses usually intervene. Fortunately, we are infected and recover without ever knowing it far more often than we actually develop an infectious disease.

Pathogenicity means simply the ability to cause disease. This is not an absolute trait. That is, there is no organism so pathogenic that it

causes disease 100% of the time. The varying degrees of pathogenicity of some bacteria are shown in Table 11.1. On the other hand, we have already seen that many "nonpathogens" do cause disease in very extraordinary circumstances. A true pathogen is a microbial species that can initiate disease in a normal, healthy individual, while an opportunistic pathogen is one that can initiate disease only in a compromised individual. We often use the term **causative agent** to refer to the pathogen in a specific disease.

Virulence is a quantitative measure of pathogenicity, a number that expresses the efficiency with which the causative agent attacks. Virulence can be measured by determining the **morbidity rate** — the percentage of individuals exposed who develop the disease — or the **mortality rate** — the percentage of individuals with the disease who die. Virulence can also be measured by determining the number of microbial cells or viral particles sufficient to cause disease. A single bacterial cell of certain *Rickettsia* can initiate fatal disease, whereas a million *Salmonella typhi* bacteria may be required to cause a case of typhoid fever. By administering gradually increasing numbers of pathogens to groups of test animals the ID_{50} (infectious dose or number of particles required to infect 50% of the host group) or LD_{50} (lethal dose or number required to kill 50%) can be determined.

THE PATHOGEN'S WEAPONS

The great microbiologists of the 19th century, Pasteur, Koch, and their co-workers, recognized not only that microorganisms caused diseases, but, more importantly, that the reason that one microbial disease differed significantly from another in its symptoms and outcome was that each was caused by a different "germ," or pathogen. Each microbial pathogen is a biologically distinct species of living thing, with its own characteristic habitat, life cycle, and virulence factors. In interacting with our tissues, each species of pathogen exerts typical effects. Observing

TABLE 11.1 Varying degrees of pathogenicity of bacteria	
Pathogenicity	**Type of Bacteria**
Pathogenic — initiates disease in normal, nonimmune individuals	*Bordetella pertussis* — whooping cough *Mycobacterium tuberculosis* — tuberculosis *Neisseria gonorrhoeae* — acute genital infection *Streptococcus pyogenes* — acute pharyngitis *Yersinia pestis* — bubonic plague
Frequently opportunistic — initiates disease in compromised hosts, such as those with wounds or who are alcoholic. Can cause secondary infections	*Bacteriodes fragilis* ⎫ anaerobes that *Clostridium tetani* ⎭ infect wounds *Klebsiella pneumoniae* — may cause pneumonia principally in alcoholics *Pseudomonas aeruginosa* — colonizes burns *Streptococcus pneumoniae* — infects aged persons or persons with lung disease
Rarely opportunistic — initiates infection only in severely compromised hosts, such as premature babies, or persons with immunodeficiency or unusual structural defects	*Propionibacterium acnes* — causes infections in patients with implanted prosthetic heart valves *Staphylococcus epidermidis* — causes infections arising at the site of intravenous catheters *Streptococcus agalactiae* — subacute bacterial endocarditis in patients with abnormal heart valves *Flavobacterium meningisepticum* — meningitis in premature infants

these effects — the signs of the disease and the patient's description of the symptoms — is a key part of an accurate clinical diagnosis. Let us look now at four categories of virulence factors.

ADHERENCE FACTORS. Most pathogens must specifically adhere to host cells as a first, essential step in pathogenesis (disease production). Some bacteria use pili or fimbriae to adhere to their target tissues. *Streptococcus pyogenes,* which causes throat infections, uses fimbriae to bind to the throat lining. *Escherichia coli* uses fimbriae to bind to many surfaces, including the intestinal lining, its normal habitat (Figure 11.4). Glycocalyx and capsular substances enable oral pathogens to adhere to the tooth surface. The mycoplasmas have small round dimples with an

internal spike that help them bond to their host cell's membrane. Some spirochetes have a specialized tip structure that appears to burrow into the host-cell's surface. Viruses bind directly to glycoproteins or lipoproteins on the surface of the host cell as a prerequisite to entering.

INVASIVENESS. Although some pathogens can cause sufficient trouble on the surface of tissues, most agents must first enter some area of the body where they are not normally found, or penetrate into the tissues. An agent can invade the intestinal tract only if its exterior coating allows it to pass through the acid environment of the stomach and survive. For example, an enterovirus such as poliovirus has capsid proteins that resist acid hydrolysis. Protozoan parasites

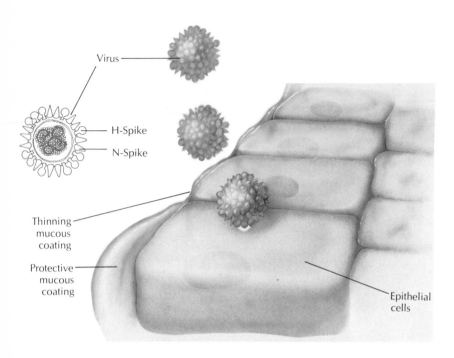

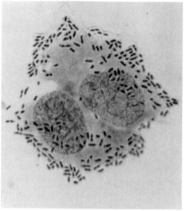

(a) VIRUSES

(b) BACTERIA

FIGURE 11.4 Invasiveness and Adherence
(a) The influenza virus possesses an envelope bearing two types of protein spikes. One is an enzyme, neuraminidase, that digests the mucous coating on epithelial cells of the respiratory membranes. Once this protective covering has been weakened, the virus can adhere to surfaces of target cells by means of the second protein in its hemagglutinin spikes. (b) Bacteria such as *E. coli* may attach to intestinal epithelial cells by means of fimbriae. Attachment sites on the surface of these cells appear to "encourage" colonization. (From G. Chabanon, C.C. Hartley and M.H. Richmond, Adhesion of *E. coli* strains, *J. Clin. Microbiol.* (1979) 10:564.)

such as *Entamoeba* enter as acid-resistant protein-coated cysts. The cysts are later uncoated by the digestive enzymes in the intestine, releasing the parasite in its active form, ready to multiply.

Some bacterial, virus, and eucaryotic pathogens produce enzymes that digest the mucous secretions that protect tissue surfaces (Table 11.2). The influenza virus carries the enzyme neuraminidase on its envelope. This enzyme attacks neuraminic acid, a key polymer in mucus, thinning the mucous coating on the surface epithelium of the respiratory membranes. The influenza virus can then establish direct contact with the cell surface so infection can be initiated. Other enzymes that enhance invasiveness hydrolyze cellular and extracellular macromolecules. Some enzymes break down membrane lipids and thus lyse cells, which is how hemolysins lyse red blood cells. Others remove barriers to their spread by digesting a path between epithelial cells (hyaluronidases) or removing fibrin clots (fibrinolysins). Two of the most invasive bacterial species, *Streptococcus pyogenes* and *Clostridium perfringens*, produce multiple invasive enzymes.

INTRACELLULAR PARASITISM. Once it has invaded, each pathogen has a characteristic **site of replication** (Table 11.3). Some pathogens multiply outside host cells, either in the spaces between cells or in body fluids. Others multiply primarily but not exclusively within host cells, as **facultative intracellular parasites.** Viruses, rickettsiae, chlamydiae and others that multiply only within host cells are **obligate intracellular parasites** (Table 11.3). Pathogens exterior to host

TABLE 11.2 Pathogenic species and their invasiveness factors

Type of Product	Specific Factor	Species or Group	Pathogenic Effect
Capsule	Type-specific polysaccharides	*Streptococcus pneumoniae* *Klebsiella pneumoniae* *Bacillus anthracis* *Yersinia pestis*	Inhibits phagocytic adherence and capture
Cell-wall component	M protein	*Streptococcus pyogenes*	Inhibits phagocytic digestion
	Waxes, mycolic acids	*Mycobacterium tuberculosis*	
Extracellular enzyme	Collagenase	*Clostridium perfringens*	Permits invasion of connective tissue
	Coagulase	*Staphylococcus aureus*	*In vivo* effect uncertain; may inhibit phagocytosis
	DNase; streptodornase	*Staphylococcus aureus* *Streptococcus pyogenes*	Reduces viscosity of purulent discharges, improving bacterial mobility
	Hemolysin	*Staphylococcus aureus* *Streptococcus species*	Lyses red blood cells. *In vivo* importance uncertain
	Hyaluronidase	*Streptococcus pyogenes*	Permits invasion of epithelial tissue
	Kinase; streptokinase	*S. pyogenes* *Staphylococcus aureus*	Activates plasma fibrinolysin and dissolves clots
	Lecithinase	*Clostridium perfringens*	Destroys cell membranes, especially red blood cells
	Leukocidin	*Staphylococcus aureus*	Lyses leukocytes; importance uncertain *in vivo*
	Lipase	*Staphylococcus species*	Breaks down lipids
	Protease	*Clostridium species*	Breaks down protein; permits attack on muscle tissue

TABLE 11.3 Site of replication of microbial parasites			
Group	Extracellular	Facultatively Intracellular	Obligately Intracellular
Bacteria	*Bacillus anthracis* *Clostridium,* all species *Corynebacterium diphtheriae* *Neisseria gonorrhoeae* *Pseudomonas aeruginosa* *Staphylococcus aureus* *Streptococcus,* all species *Vibrio cholerae* *Yersinia pestis* All spirochetes	*Brucella,* all species *Francisella tularensis* *Mycobacterium,* all species except *leprae*	*Mycobacterium leprae*
Fungi	Dermatophytes	*Candida albicans* *Coccidioides immitis* *Histoplasma capsulatum* Other systemic fungi	
Rickettsia			All species
Chlamydia			All species
Viruses			All groups

cells are comparatively vulnerable to destruction by serum proteins or phagocytes. Intracellular parasites escape phagocytosis because of their protected location. Your body can usually eradicate them only by destroying the host tissue that harbors them.

When a microorganism enters tissue, it usually is promptly contacted by a host phagocytic cell, whose function is to ingest and destroy stray particles. One of two things may then happen. Most nonpathogenic microorganisms are quickly captured and killed by the phagocyte (one reason why they are not pathogenic). Pathogens, however, usually resist either capture or digestion by the phagocyte. Some bacterial pathogens are protected by capsules or slime layers. Once inside the phagocyte, they may survive and even multiply, eventually killing the phagocytic cell. Capsular polysaccharides and cell-wall proteins, such as the M protein of *Streptococcus pyogenes,* may actively prevent phagocytic destruction. Cell-wall components may also confer resistance. The Gram-positive bacterial cell wall resists phagocytes unless it is first coated with specific antibodies. The lipids that render a *Mycobacterium* acid-fast on staining also render it so resistant that mycobacteria not only survive but multiply inside phagocytes.

TOXIN PRODUCTION. A **toxin** is a poison — a chemical substance that reversibly or irreversibly alters a key physiologic reaction. Toxins

TABLE 11.4 Pathogenic bacteria that produce exotoxins

Organism	Disease	Common Name of Toxin	Specific Activity
Bordetella pertussis	Whooping cough	—	Toxin that destroys upper-respiratory-tract epithelium
Corynebacterium diphtheriae	Diphtheria	Diphtheria toxin	Inhibits protein synthesis; damages heart, liver, nerve tissue
Clostridium botulinum	Botulism	Botulin	Blocks transmission of nerve impulses. Causes flaccid paralysis
Clostridium perfringens	Food poisoning	Enterotoxin	Acts on vomit reflex center of brain
Clostridium tetani	Tetanus	Tetanus toxin	Blocks neuromuscular junction. Causes rigid paralysis
Escherichia coli, some strains	Diarrhea	Enterotoxin	Causes water and electrolyte loss via intestinal tract
Staphylococcus aureus, some strains	Food poisoning	Enterotoxin	Activates vomit reflex center
Streptococcus pyogenes, some strains	Scarlet fever	Erythrogenic toxin	Destroys capillaries, causes rash
Shigella dysenteriae	Dysentery	—	Causes water and electrolyte loss across intestinal wall in diarrhea
Vibrio cholerae	Cholera	Choleragin	Alters enzyme activity in intestinal lining, causing rapid water and electrolyte loss by diarrhea

are a major tool for pathogenesis in bacteria. Toxins are distributed by the blood stream from the point of infection to all of the host's tissues. As a result, a small infection, caused by a noninvasive organism, may be lethal. The noninvasive pathogen *Corynebacterium diphtheriae*, which causes diphtheria, typically grows only on the surface epithelium of the nasopharynx. Yet the toxin it produces at that site is distributed widely by the circulation and can cause fatal degeneration of the heart, liver, kidneys, and nerve tissue.

Bacteria produce two types of toxins, exotoxins and endotoxins. **Exotoxins** are proteins that pathogens release into the environment where they diffuse and are free to act. Exotoxins

in tiny (microgram) doses have highly specific modes of action and have pronounced effects on host tissues (Table 11.4). A few hundred molecules of diphtheria toxin will kill a human cell. Most exotoxins, being proteins, are inactivated by heat.

Exotoxins are strongly antigenic, and the body reacts to exposure to an exotoxin by making protective neutralizing antibodies. The first immunizations to be made safe and effective enough for universal use were those that used modified toxins (**toxoids**) from diphtheria and tetanus bacteria. Toxoids are no longer toxic, but still stimulate antibody production. Antibodies produced against either a toxin or toxoid (**antitoxins**) can be collected, purified, and used to treat disease. An antitoxin arrests the disease by neutralizing the toxin, not the pathogen itself.

Endotoxins are chemically lipopolysaccharides, structural components of the cell walls of Gram-negative bacteria. They are liberated when the bacteria die and disintegrate. Endotoxins from different bacterial species all have the same nonspecific biological effects. Large (milligram) doses, usually a result of a massive Gram-negative infection, result in **septic** or **endotoxin shock** and sometimes death. Small doses cause vague general symptoms, including aches and malaise. The endotoxins are also **pyrogens,** substances that affect the temperature-regulation system of the body, causing elevated temperature, or **fever.** Pyrogens such as endotoxins come from outside the host cells and are called **exogenous pyrogens.** Other substances, released normally by our own cells, are also pyrogenic. They are called **endogenous pyrogens.**

Endotoxins are only weakly antigenic, and exposure to them gives little protective immunity. Vaccination directed against Gram-negative bacteria such as the typhoid and cholera organisms has not been very successful. There are vaccines available, but they are less protective and effective for shorter periods than those prepared against many other pathogens.

NONSPECIFIC RESISTANCE

RESISTANCE AND IMMUNITY. We mean quite different things when we speak of resistance and of immunity. **Resistance** refers to nonspecific defenses that protect the host against all microorganisms in general. Effective nonspecific resistance occurs the very first time the body meets the pathogen. It does not depend on prior exposure to the pathogen; it does not require a "learning" process by the body. Nonspecific resistance is designed to protect against first attacks by the widest possible variety of agents. We will study how it works in the remainder of this chapter.

By contrast, **immunity** refers to specific defenses directed by the body against strains of in-dividual bacteria, viruses, fungi, or other agents. It is a "learned" response, requiring the production of populations of specifically programmed genetically unique blood cells. Immunity arises only after the body's tissues have experienced contact with the agent, and then only after a five- to ten-day "learning" period. Because of this learning period, immune defenses are able to control infection only in the later stages of the first attack, although they defend immediately against second or subsequent attacks by the same agent.

Nonspecific resistance and the immune system cooperate with each other. Many of the nonspecific factors may be recruited as key components of immune responses. One way of looking at immunity is to say that it "aims" or "sharpens" the weapons of the nonspecific systems by providing a means of targeting them against the challenge of the moment. The two systems are, in fact, inseparable. Let's look now at the components of the nonspecific defenses.

ANATOMY AND PHYSIOLOGY — ADAPTATIONS FOR DEFENSE

A healthy body keeps its normal flora "in its place" and restricted to reasonable numbers. When microorganisms infect, they must either enter the body via one of its orifices or penetrate its skin. To guard against this possibility, each point of entry is shielded by anatomic and physiologic barriers to infection, which have evolved to comprise the body's first line of defense (Figure 11.5). In later chapters, we will systematically examine the anatomic and physiologic protective mechanisms of each body area. For now, let's illustrate the concept with a few examples.

SOME ANATOMIC BARRIERS. The central nervous system (the brain and spinal cord) has the most highly developed anatomic protection of any body area (look ahead to Figure 21.2). The axial skeleton (skull and vertebral column) encloses these vital structures. It not only protects

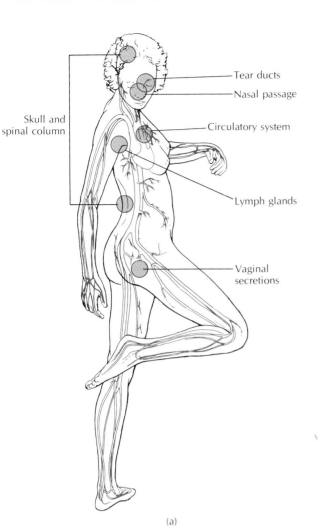

Skull and spinal column

Tear ducts

Nasal passage

Circulatory system

Lymph glands

Vaginal secretions

(a)

FIGURE 11.5 Circles of Defense
(a) In our bodies anatomic and physiologic barriers serve as the first line of defense against potential pathogens. Some examples of such barriers include the skeletal casing and meninges, the convoluted membrane of the nasal passage, the flushing action of the tears, and the low pH of the vaginal secretions. The second line of defense is provided by cells and serum factors carried by the circulatory system. The third line of defense, specific immunity, is carried out by the lymphoid organs, such as the lymph nodes. (b) One way of thinking of the body's defenses is in terms of concentric rings: General health factors — the normal anatomy and physiology of the body surfaces — form the outer barrier. If pathogens pass these barriers and get into the tissues, nonspecific blood factors and phagocytic cells are there to control them, with inflammation if necessary. Should the invader persist, specific immune mechanisms will be activated. Only rarely does an invading pathogen reach and infect a target cell.

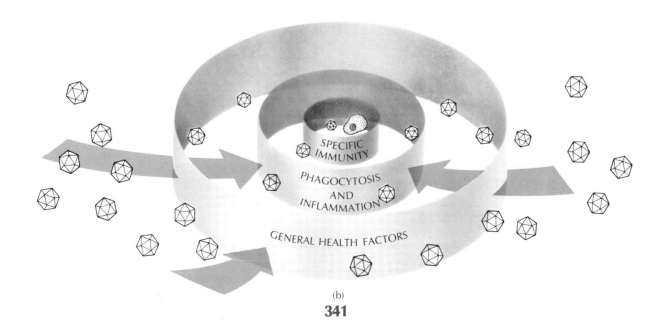

SPECIFIC IMMUNITY

PHAGOCYTOSIS AND INFLAMMATION

GENERAL HEALTH FACTORS

(b)

341

against mechanical damage but also prevents microbial invasion. Pathogens may penetrate overlying soft tissues but they rarely penetrate bone. Inside the skeletal casing, we have three layers of membrane, called meninges, that enclose the nerve tissue and present a further barrier to infection from outside. To minimize the possibility of infection from within, by pathogens carried by the blood that passes through the brain tissue, each capillary has an unusually thick covering layer. This layer, which reduces the possibility of pathogens moving from the blood into the brain tissue, is called the **blood–brain barrier.**

Anatomic barriers are important in guarding the respiratory tract. We normally inhale through our noses. The nasal passage (Figure 11.6) is not straight. Instead, it has evolved as a tortuous pathway among many highly convoluted **conchae,** bony protrusions covered with mucous membrane. Why this complexity? Entering air is forced to follow a turbulent path so that it flows over many moist, sticky surfaces with the functional characteristics of old-fashioned flypaper. Dust particles and microscopic matter are trapped by the nasal mucus. Unless the air is extremely dust-laden, or so dry as to actually dry out the mucosal surfaces, few airborne particles reach the lungs.

SOME PHYSIOLOGIC PROTECTIONS.   Our eyes are exposed to constant environmental contamination, but eye infections are quite rare. The flow of tears is a physiologic protection that gives a steady flushing action, removing invaders every few seconds. Tears also contain lysozyme, the enzyme that hydrolyzes the peptidoglycan of the bacterial cell wall. Few bacteria can infect in the face of these combined defenses. The few that do include *Chlamydia trachomatis,* the most serious eye pathogen, which is an obligate intracellular pathogen and thus is well protected.

Vaginal secretions have an acid pH of 4.4 to 4.6. The vaginal area is therefore inhospitable to microorganisms that might intrude from the

nearby skin and from the intestinal tract, where they are adapted to a pH that is near neutral. Although invaders are not usually killed by the acid, they cannot grow well, and certainly cannot compete effectively with the area's acid-tolerant normal flora.

PROTECTIVE COMPONENTS IN BLOOD

Should a microorganism happen to penetrate the anatomic and physiologic barriers, it will then be exposed to the body fluids, the blood and lymph. An adult human has about 5 liters of blood, which circulate through a closed system of arteries, capillaries, and veins (see Figure 22.1). A larger volume of tissue fluids and lymph percolate freely through the tissues and are channeled along the lymphatic ducts and lymph nodes. Lymph originates from blood, shares many of the same soluble factors as blood, and eventually flows back into the blood stream. Blood and lymph transport oxygen and nutrients, remove wastes, and regulate temperature. They contain a variety of protein factors with antimicrobial, defensive, and communication functions. Blood and lymph also transport phagocytic cells. These various elements compose the body's second, internal, line of defense. Let us become familiar with some of these factors.

BLOOD CELLS.   Blood is a tissue, and contains cellular elements suspended in a fluid matrix, the plasma. Blood cells (Figure 11.7) are of three types: red blood cells, white blood cells, and platelets. All blood cells come originally from the bone marrow, where they are derived from a single type of stem cell, the **hemocytoblast.** Red blood cells (RBCs or **erythrocytes**) are the most numerous. They circulate for up to 120 days without dividing. Their function is oxygen transport; they have no specific role in body defense.

White blood cells (WBCs or **leukocytes**) all play crucial roles in the body's defenses against infection. They are colorless unless stained.

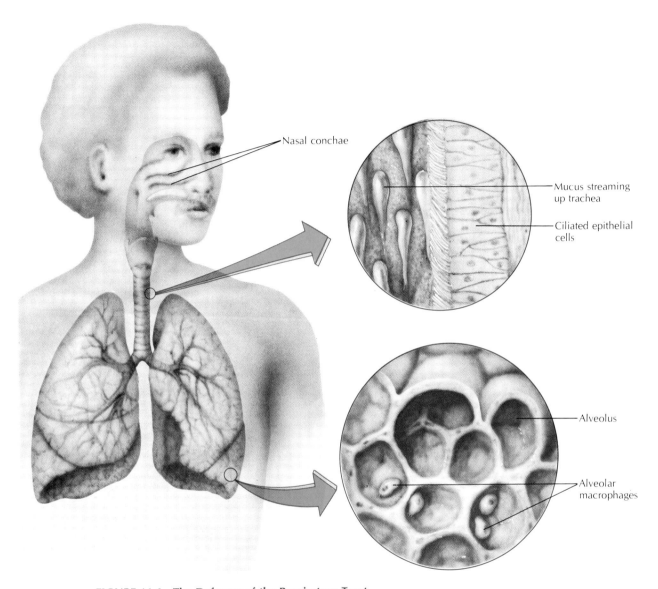

Nasal conchae

Mucus streaming up trachea

Ciliated epithelial cells

Alveolus

Alveolar macrophages

FIGURE 11.6 The Defenses of the Respiratory Tract
The nasal conchae with their moist mucous coatings trap most incoming airborne particles. Some microorganisms reach the trachea where most are trapped by more mucus and are brought back up by the movement of cilia on the epithelium. Those few particles that reach the lung are trapped and consumed by alveolar macrophages.

Blood smears are usually stained with Wright's stain, a mixture of acidic and basic dyes. This stain differentiates the leukocytes into various types. There are two general categories of leuko-cytes, the granulocytes and the agranulocytes. **Granulocytes** are of three types: neutrophils, basophils, and eosinophils. These cells have irregular, often lobed, nuclei; they get their name from

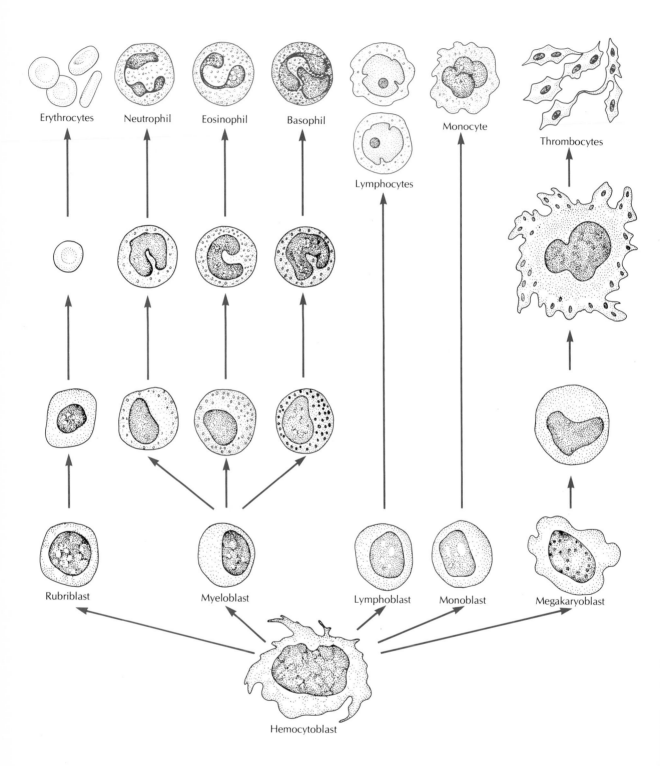

Erythrocytes   Neutrophil   Eosinophil   Basophil   Lymphocytes   Monocyte   Thrombocytes

Rubriblast   Myeloblast   Lymphoblast   Monoblast   Megakaryoblast

Hemocytoblast

FIGURE 11.7 Development of Mature Human Blood Cells

Hemocytoblast stem cells, located in the bone marrow, give rise to five partially differentiated lines of immature blood cells called blasts. The seven types of mature blood cells (top row) are formed by differentiation from the blasts and released into the blood.

their numerous, clearly visible cytoplasmic granules. With the Wright's stain, the granules in each of the three types stain different colors. The granules contain biologically active compounds.

Neutrophils, also called **polymorphonuclear leukocytes** (PMNs, PMNLs) are the most numerous white blood cells. Their granules pick up both acidic and basic dyes and stain a neutral grayish pink. Their nuclei typically have three lobes separated by constrictions. About 100 billion PMNs per day are produced in bone marrow. Your bone marrow holds about 3 trillion PMNs in reserve, to be rapidly released in case of infection. This release accounts for the great increase in circulating white blood cells (**leukocytosis**) seen in certain infections, such as acute appendicitis. In a prolonged infection, leukocyte reserves may be exhausted and **leukopenia,** an abnormally low number of circulating white blood cells, will result. Leukopenia is a diagnostic sign in typhoid fever and influenza. PMNs are mobile and phagocytic. They exhibit a behavior pattern called chemotaxis — they are able to detect substances released by damaged cells and migrate to the scene of injury or infection where they ingest the foreign particles.

Basophils contain granules that are stained deep blue-purple by the basic dyes. Although basophils are not numerous, they and the **mast cells** in connective tissue contain histamine, a chemical they release in response to injury. Histamine plays a role in coordinating the nonspecific defenses, but it also contributes to the unpleasant or dangerous effects of allergies (see Chapter 13).

Eosinophils contain granules that collect the acid stain eosin, and become a vivid orange-red. Eosinophils are few in number, except during allergic flare-ups and certain parasitic infestations. One of their roles is to counteract basophil activity. Eosinophils moderate allergic and inflammatory responses such as swelling and pain by releasing substances in their granules.

There are two groups of **agranulocytes,** the lymphocytes and the monocytes. They too contain granules, which are much less prominent than in the granulocytes. Their nuclei are round, oval, or U-shaped.

The circulating **lymphocytes** have large, darkly staining nuclei and small halos of light cytoplasm. Although all lymphocytes look much alike, they actually fall into a great number of distinct functional classes, each with a special role in defense. Some, the natural killer cells, seem to defend nonspecifically against tumors. Most lymphocytes, however, are part of the cellular network that mediates specific immunity, as discussed in the next chapter.

Monocytes are quite large compared to other white blood cells, and their nuclei are oval or U-shaped. They are actively phagocytic, like PMNs. Circulating monocytes later differentiate into the **macrophages,** larger phagocytic cells found in the tissue fluid, in body spaces, and in aggregates scattered throughout the blood and lymphatic circulations.

Platelets or **thrombocytes** are circulating cell fragments. They arise from cells called **megakaryocytes** produced in bone marrow. Platelets break down on contact with damaged tissue, releasing factors that react with plasma proteins to initiate clot formation. Platelets also release inflammatory subtances, as well as growth factors that promote cell division and healing.

SERUM AND FLUID FACTORS. The fluid portion of blood is **plasma. Serum** is plasma from which the clotting proteins have been removed. Both fluids contain a wide range of protective factors. One group of important serum proteins are the immunoglobulins, agents of specific immunity, which we will discuss in Chapter 12. **Prothrombin** and **fibrinogen** are protein clotting factors in plasma, which interact with the platelet factors to form blood clots. You've already encountered **lysozyme,** an enzyme that damages bacterial cell walls and is found in plasma and other body fluids. The interferons, proteins produced by lymphocytes, leukocytes, and fibroblasts, restrict viral multiplication and regulate immune function. **Interleukins** are interferons that have key roles in signaling white blood cells so as to coordinate the body's defenses. **Opsonins** are plasma proteins that in-

---

BOX 11.2
Why Do I Feel So Bad?

Do you have aches and pains? Here are some capsule explanations for possible common causes. Take heart — most of these symptoms are indicators that your defenses are working on your behalf.

Headache One common cause is inflammatory mediators dilating blood vessels in your brain.

Fever Pyrogens released by invading bacteria or white blood cells are persuading your hypothalamus to increase heat production. The elevated temperature will help control infection.

Swollen glands Your lymph nodes are enlarged. In an infection, cell division increases in the lymph nodes "downstream" of the infection. This may indicate that the lymphoid tissues are becoming more active in developing immunity. Also, some agents multiply in lymphoid tissue causing enlargement, as in infectious mononucleosis.

Sore throat You have swollen lymph nodes in the pharyngeal area or, less commonly, microorganisms are attacking your pharyngeal lining.

Rashes Rashes are usually an attack by your immune system on cells harboring a virus or other intracellular parasite or something you are allergic to. Some microbes or their toxins may attack skin cells directly.

Vomiting You have ingested food containing toxins, which activate the brain's vomiting reflex center. Vomiting is a protective mechanism that ejects contaminated food.

Diarrhea Toxins are causing body fluids to pour into the lumen of your intestine, accelerating peristalsis and yielding watery stools. Microbial attack on the intestinal epithelium has a similar effect. Diarrhea also protects by getting rid of the offending agents.

Coughing You are trying to rid yourself of excessive mucus. Mucus accumulates because bacterial products (or cigarette smoke) alter the permeability of your respiratory epithelium.

Runny nose You may be experiencing excessive fluid loss from the cells of a virus-damaged epithelium. The flow of fluid brings protective plasma factors to the surface of your nasal passages.

Photo courtesy of the Bettman Archive.

crease the ability of phagocytes to bind to particles. Some opsonins are nonspecific; however, certain types of antibodies are protective because they serve as specific opsonins. **Transferrins** bind free iron, thus tying up the body's iron reserves in a form unavailable to microorganisms. In infection, the host wages a war with the bacteria for the iron atoms. By increasing the amount of iron-binding proteins, the host reduces the microorganisms' growth.

THE COMPLEMENT PATHWAYS. Complement is the name given to a complex group of about 20 interacting serum proteins identified by a complicated numbering system. They are designated C1, C2, and so on up to C9. There are also B and D factors and several regulatory proteins. Once activated by an immune response or by an invading organism, they react sequentially. This sequence gives a cascade effect. Some complement proteins act as enzymes, cleaving and activating the next protein in the sequence. Others act on surrounding cells.

The complement proteins participate in all major aspects of body defense, both nonspecific and immune. The complement network is protective in three general ways. Its early steps stimulate the inflammatory response. An intermediate step releases chemotactic substances that attract phagocytes and render them highly active. Once activated, the complement pathway results in the assembly of lytic complexes that make channels through cell membranes, causing lysis of invading cells.

The key step in the complement network is the cleavage of the protein C3 into two fragments (Figure 11.8). Fragment C3a promotes inflammatory changes useful in fighting infection, and C3b opsonizes particles (coats them with opsonins) and activates the next steps in the complement cascade. C3 may be activated in two distinct ways, either by nonspecific or specific (immune) mechanisms.

The nonspecific **alternate pathway** does not require antibodies. It is triggered by microbial polysaccharides or endotoxin, contained in the cell membranes of bacteria, yeast, or protozoa. These substances cause the production of a C3 activator, C3bB, which splits C3 into fragments C3a and C3b. This action begins the cascade effect. The C3b fragment may combine with another activator and cleave the protein C5, or it may bind to receptor proteins on macrophages and leukocytes, increasing their ability to ingest the invader cells.

The **classic pathway,** however, must be triggered by specific antibody binding to antigen. The antigen–antibody combination forms a complex designated C142; its function is to activate C3. As each complement molecule in the cascade is activated, it in turn activates the next molecule in the pathway. Once activated by either pathway, C3 splits into two fragments. When C3b binds to the complex, the result is the cleavage of the protein C5. One fragment, C5a, is the chemotactic factor that attracts neutrophils, summoning these phagocytic cells to the scene of the reaction.

C5b is equally important. It initiates another stage of complement function, formation of a large functional cluster of complement proteins. The C5b fragment, followed by C6, C7, C8, and C9, bind sequentially on the surface of a target cell to form a lytic complex. The lytic complex forms a **transmembrane channel** through the target-cell membrane. Water flows in through this channel and lyses the target cell. Even one lytic complex may form a fatal hole eventually leading to the death of a bacterial or mammalian cell.

To summarize, the complement system functions both in nonspecific and immune defense. It contributes to three of the main defensive actions of the body — phagocytosis, inflammation, and lysis of invading cells, which we will now proceed to study.

PHAGOCYTES — THE BODY'S SCAVENGERS

In our bodies, certain cells have become specialized as the scavengers. Their function is to ingest

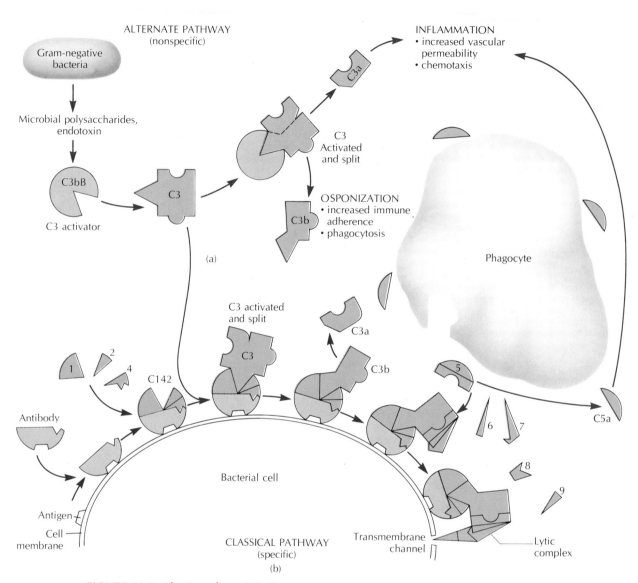

ALTERNATE PATHWAY (nonspecific)

Gram-negative bacteria

Microbial polysaccharides, endotoxin

C3bB

C3 activator

C3

INFLAMMATION
• increased vascular permeability
• chemotaxis

C3a

C3 Activated and split

OSPONIZATION
• increased immune adherence
• phagocytosis

C3b

(a)

Phagocyte

C3 activated and split

C3

C3a

C3b

Antibody

C142

1 2 4

5

C5a

6 7

8

9

Antigen

Cell membrane

Bacterial cell

CLASSICAL PATHWAY (specific)

Transmembrane channel

Lytic complex

(b)

FIGURE 11.8   The Complement System
(a) The alternate pathway may be triggered by biochemicals in the cell membrane of an invading organism. C3bB activates the C3 protein, which splits into C3a, a chemotactic stimulus for neutrophils, and C3b, an opsonin that adheres to invading organisms. (b) In the classical pathway, three proteins are triggered by the antigen–antibody complex to form C142, the C3 activator. Fragments released during activation aid in attracting neutrophils and in initiating the inflammatory response. Whether set in motion by a nonspecific stimulus or by an antigen–antibody combination, the complement pathway leads to increased vascular permeability, phagocytic chemotaxis and opsonization, and lysis of target cells.

and break down any particulate debris that might be floating in our body fluids or adhering to moist surfaces. The debris they consume includes the remains of those of our own cells that have reached the end of their usefulness. It also includes many invading microorganisms and viruses.

We have two groups of phagocytes, the circulating granulocytic PMNs and monocytes, and the more mature tissue cells that arise from them, called histiocytes and macrophages, respectively. This large family of phagocytic cells is distributed throughout the body.

Circulating phagocytes originate in and emerge from the bone marrow. Both PMNs and monocytes circulate briefly, then leave the capillaries and enter the underlying tissue. After the granulocytes enter the tissues they survive for a few days, then die. Monocytes, after leaving the circulation, become **macrophages** (literally, big eaters), which may survive for periods of years.

All the phagocytes have pronounced chemotactic abilities and are able to migrate extensively from one area in the body to another. The **wandering cells,** which migrate through tissue, police areas external to the circulation. They move actively over the moist surfaces of internal organs, such as the lung, peritoneal cavity, and uterus, covering relatively large areas (Table 11.5). These areas of the body are normally sterile, but the phagocytes are very important in dealing with any stray invading microorganisms.

TABLE 11.5 Phagocytic cells		
	Category	**Location**
Fixed	Held in spongy organ in a supporting structural network	Spleen Thymus Lymph nodes — dendritic macrophages Bone marrow Brain — microglia
	Lining cells in blood and lymph sinuses in glandular tissue	Liver — Kupffer cells Spleen Lymph nodes Adrenal gland Pituitary gland
Wandering	Macrophages and histiocytes in tissue and body cavities	Skin Peritoneal cavity Alveoli of lung Uterus Thoracic cavity Bladder
Circulating	Monocytes Granulocytes	In blood; enter tissue on maturation and in inflammation

The **fixed cells** are phagocytes that are at least temporarily immobilized in various spongy filtering organs. They include the **Kupffer cells** in the liver (Figure 11.9) and **dendritic macrophages** enmeshed in the spongy tissue of the lymph nodes. Circulating blood and lymph, respectively, pass through these organs. The fixed phagocytes carry out a routine cleansing activity by extracting floating debris as the fluid passes over them. Should small numbers of microorganisms be introduced into the blood or lymph through minor traumas (pinpricks, brushing inflamed gums) they will be removed within seconds. Unless these are the types of pathogens that are resistant to phagocytic killing, they will cause no further problem.

FIGURE 11.9   Kupffer Cells
Within the liver, phagocytic Kupffer cells are spaced along the sinusoids through which circulating blood percolates. They remove particulate debris from the blood as it flows through. (×7100.)

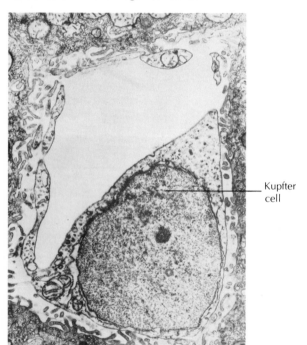

Kupffer cell

HOW PHAGOCYTOSIS WORKS.   The first step in phagocytosis is for the phagocytic cell to establish effective contact with the target particle. Phagocytes are not fussy. They will attach to living or dead microorganisms, inanimate particles such as coal dust or organic fibers; clumps of antigens bound to antibodies and complement; and whole or fragmented, living or dead human cells. Opsonization remarkably improves the phagocyte's ability to adhere to its prey. It appears that the phagocyte usually traps the target against a surface, such as the lining of a blood vessel, in order to ingest it.

Once the particle has been bound to the phagocyte, the phagocyte is stimulated to produce many ruffled extensions of its cytoplasmic membrane at its leading edge. These surround the adhering particle (Figure 11.10). Where they meet and fuse, the prey becomes enclosed in a **phagosome,** a vacuole bounded by a membrane layer derived from the cytoplasmic membrane of the phagocyte. Once in the phagosome, the target is killed and digested as follows. The numerous granules of phagocytic cells are enzyme-containing vesicles, such as lysosomes and peroxisomes. The granules fuse with the phago-

FIGURE 11.10   Phagocytosis of Microorganisms
The first step in phagocytosis is the attachment of the phagocyte's membrane to its prey; attachment is assisted by first coating the prey with complement and/or antibodies. Numerous fingerlike extensions of the cytoplasm surround the prey bacterium, their membranes fusing to form a phagosome. Enzyme-containing granules, the lysosomes and peroxisomes, fuse with the phagosome and release their contents. Once enzymes have digested the prey, the phagolysosome moves to the cell boundary and discharges its wastes. In the scanning electron micrograph a polymorphonuclear leukocyte is shown engulfing its prey, a yeast cell. (Courtesy of J. Boyles and D.F. Bainton, *Cell* (1981), 24:905–914. © MIT Press.)

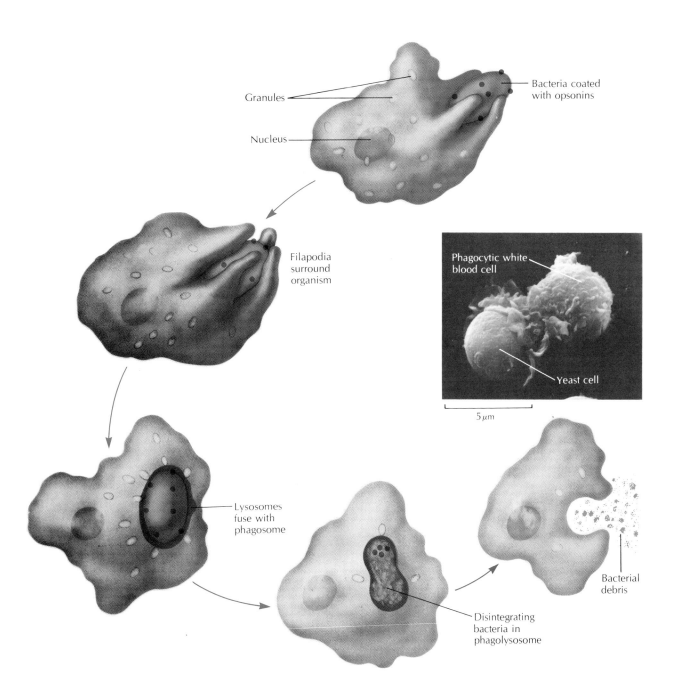

Granules

Bacteria coated with opsonins

Nucleus

Filapodia surround organism

Phagocytic white blood cell

Yeast cell

5 μm

Lysosomes fuse with phagosome

Disintegrating bacteria in phagolysosome

Bacterial debris

some, dumping their contents into the resulting **phagolysosome.** Enzymes from the granules then attack the prey. For example, the **myeloperoxidase** enzyme complex releases hydrogen peroxide and superoxides, which are rapidly lethal to many anaerobic microbes. **Acid hydrolases** create a highly acidic digestive environment, causing rapid disintegration of many polymers. Eventually, the phagolysosome moves to the cell boundary and discharges its wastes; indigestible substances may be also retained as residual bodies.

Sometimes, as we noted earlier, the phagocyte fails to kill its prey. The lysosomal enzymes cannot kill some pathogens, such as *Neisseria gonorrhoeae, Mycobacterium tuberculosis,* or the trypanosomes. In fact, these organisms use the phagocyte as a way of gaining a foothold. They survive, multiply in some cases, and are transported within the migrating phagocyte. They may then take the opportunity to parasitize other tissues.

As we shall see, one of the most important benefits of specific immunity is that it activates macrophages and vastly enhances their predatory and digestive efficiency. When immunity develops, it often enables phagocytes to destroy intracellular parasites that they had been unable to deal with unassisted.

NATURAL KILLER CELLS

Natural killer (NK) cells are large, granular lymphocytes. They are not phagocytic. Although all the other lymphocytes, so far as we know, act in specific immunity, the NK cells are nonspecific. Their function is to kill target cells, which include tumor cells and virus-infected cells. It is not clear how they recognize their targets. However, we know that NK cells kill by binding to the target cell and releasing membrane-destroying proteases and phospholipases. NK cells both produce and are activated by interferon.

Most interest in NK cells has focused on their possible role in controlling tumors by surveillance — that is, seeking for and eliminating tumor cells before they are able to establish primary tumors or metastases. Many researchers believe that this relatively recently discovered type of cell is our primary day-to-day defense against tumors. There is also evidence linking NK cells to nonspecific defense against herpesviruses, protozoan parasites, and fungal intracellular parasites.

THE INFLAMMATORY RESPONSE

Inflammation is another of our body's nonspecific responses to trauma or infection. Inflammation is a complex sequence of events characterized by the development of localized or general swelling, reddening, elevated temperature, and pain. Inflammation functions as a protective mechanism against pathogens because it tends to keep infections from spreading. However, it also has a high potential for causing tissue damage, discomfort, and, in extreme cases, even death.

The inflammatory response orchestrates the interaction of many of the body's defensive weapons at the site of a potential microbial invasion. It summons the serum proteins, the phagocytic cells, and the clotting and healing factors, while activating the fever mechanism.

HOW INFLAMMATION GETS STARTED.   Suppose your knife slips and you get a small cut on your finger (Figure 11.11). This type of mechanical

FIGURE 11.11   The Inflammatory Response
(a) A break in the skin damages cells and introduces microorganisms. (b) The damaged cells release chemotactic substances that dilate blood vessels, increase vascular permeability, cause a smooth-muscle contraction, and attract circulating phagocytes. (c) Soon PMNs migrate into the dilated capillaries and pass into the tissues by diapedesis. (d) In this close-up of diapedesis, a leukocyte adheres to the inside of the capillary wall. It then forces itself out through the gap between two cells into the tissue where it is available to ingest invading microorganisms.

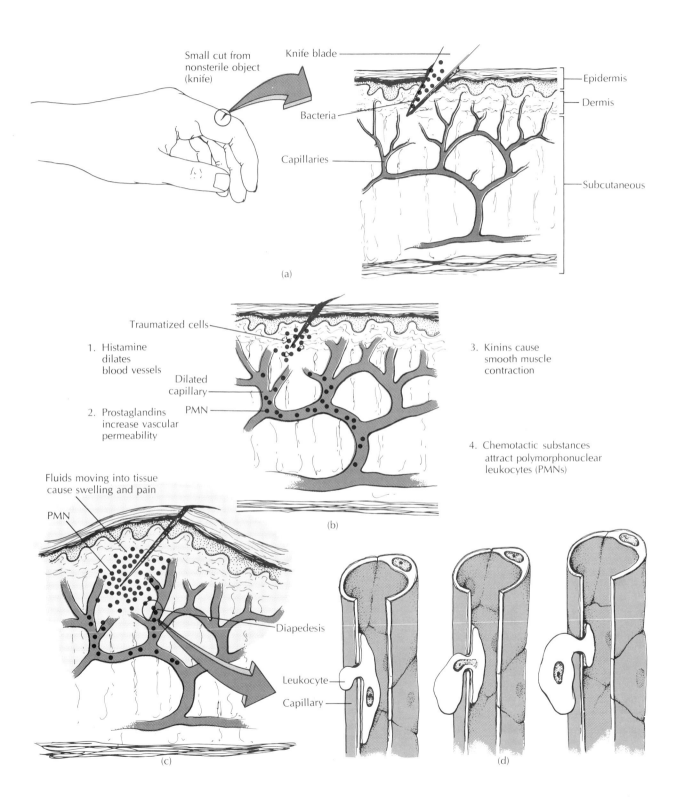

(a)

Small cut from nonsterile object (knife)

Knife blade

Bacteria

Capillaries

Epidermis

Dermis

Subcutaneous

Traumatized cells

1. Histamine dilates blood vessels

2. Prostaglandins increase vascular permeability

Dilated capillary

PMN

3. Kinins cause smooth muscle contraction

4. Chemotactic substances attract polymorphonuclear leukocytes (PMNs)

(b)

Fluids moving into tissue cause swelling and pain

PMN

Diapedesis

Leukocyte

Capillary

(c)

(d)

damage will initiate limited inflammation. Another triggering event is the presence of a foreign substance, for example a splinter. Both types of damage have the effect of traumatizing or killing a few body cells. A cytolytic viral infection, lysis of a foreign cell by complement, an allergic response to pollen, or bacterial toxin binding to cells are additional examples of events that will initiate inflammation.

Traumatized or killed cells release some of their contents — substances that are not normally found outside the cell. These chemicals are the mediators that cause the inflammatory effects. They include **histamine,** which dilates blood vessels; **prostaglandins,** which increase permeability of blood vessels; **kinins,** which cause smooth-muscle contraction; and chemotactic substances, which attract PMNs. The more cell damage, the more mediators are released. The more mediators, the more pronounced the inflammation.

HOW INFLAMMATION PROTECTS.   Let's get back to your cut. Unless you had the foresight to use a sterilized knife, we'll assume you introduced some microorganisms into your skin and you now have a tiny infection. We will see how the inflammatory response works to make sure the injury doesn't develop into a big infection.

As your damaged cells release their mediators into the surrounding area on your fingertip, the capillary bed in the vicinity becomes much dilated and more permeable. Dilated capillaries cause the reddened appearance. Fluids bearing serum proteins leak from the capillaries into the tissue, causing swelling and pain. These serum proteins contribute to opsonizing, lysing, and otherwise inactivating any microorganisms.

The damaged cells also release chemotactic substances. Within a few minutes, PMNs and monocytes begin congregating at the site. This is a fascinating process. Phagocyte membranes have receptors for each of the many leukocyte attractants. These include C5a and factors released by lymphocytes, mast cells, and eo-

sinophils in the initial stages of inflammation. Some bacterial proteins are also strong attractants. When the phagocyte's receptors detect a gradient of chemoattractant, they activate its metabolism to increase its ATP supply. The phagocyte then begins moving, using its microtubules and microfilaments to extend a ruffled border in the direction of increasing amounts of the attractant. It will begin oriented migration within minutes of the initial stimulus and continue moving in that direction as long as it perceives a gradient. Chemoattractants can quickly collect millions of phagocytes at a site.

PMNs migrate from the capillaries into tissue by **diapedesis.** In this process, PMNs adhere to the capillary linings in the inflamed area. They then wriggle between the cells of the capillary wall, which is only one cell layer thick. Once in the tissues, they migrate through the spaces between cells (Figure 11.11d). The masses of white cells may accumulate to a visible degree, forming pus, more elegantly known as a **purulent exudate.**

Let's assume your cut bled a little, then stopped. What happened to stop the blood flow? The chemical mediators released from the damaged tissue induced platelets to break down and release clotting factors that reacted with the plasma clotting proteins. Clotting occurs both where we readily see it, over the surface of a wound, and also where we are not aware of it, within the capillaries surrounding the traumatized area. The clotting mechanism is an important nonspecific defense. Clotting blocks the spread of all but the most invasive infectious agents. Surface clots form the structural framework for scab formation, and seal off raw areas while tissue replacement is carried out. Perhaps your fingertip became somewhat hot, in addition to being red, swollen, and painful. This is a very mild expression of one of the most powerful aspects of the inflammatory response — **fever.** Fever is an elevation of the body temperature. Although we tend to regard it as undesirable, moderate fever is actually an important protective mechanism. An elevated temperature of two or three degrees is helpful, because it appears to increase the efficiency of the phagocytic cells and the rate of immune reactions. At the same time, the heat hampers the growth of many pathogens. Only when the fever becomes very high is it harmful to us.

Monocytes are at least partly responsible for fever. When stimulated by inflammatory mediators, they release **endogenous pyrogens.** These substances act on the hypothalamus region of the brain, raising the temperature, and activating shivering and other warming activities. Should there be Gram-negative bacteria infecting the inflamed area, their endotoxins will also act as pyrogens.

What are the possible outcomes of an inflammatory response? In your cut finger, assuming the microorganisms are few in number and quickly subdued, the inflammation rapidly subsides. This is because when cellular damage stops, no more mediators are released. Healing follows, as the clot becomes the network for the replacement of tissue, as normal circulation is restored, and as mitosis replaces dead cells with healthy live ones.

On the other hand, suppose you had the bad luck to pick up some highly pathogenic microorganisms that resisted phagocytosis. Then not only phagocytes but also more tissue cells would be killed in the battle, releasing even higher levels of mediators. The redness, swelling, and pain would become pronounced. Capillaries would break down; surrounding tissue would die and deteriorate, a condition called **necrosis.**

If you had received a considerably deeper cut, as contact with the circulation was cut off, the area would become anaerobic and a hollowed out area filled with debris might result. This is called an **abscess.** Abscesses eventually open to the surface and drain, at which point healing may begin. Abscesses such as boils or severe acne are typical of *Staphylococcus aureus* infections; the organism and host defenses fight each other to a standstill. The host defenses usually win eventually, but at a cost, as scarring often results.

HARMFUL EFFECTS OF INFLAMMATION. As we have seen, the biological reason for inflammation is the control of infection. However, prolonged, uncontrolled inflammation is both distressing and destructive. It is the primary sign of many hypersensitivity and autoimmune diseases, as well as chronic low-grade infections. It may produce progressive tissue destruction, scarring, and loss of function in the joints, kidney, heart, lungs, and other organs. In the infectious disease tuberculosis, the effects of inflammation and related processes are what cause the actual lung damage that may lead to death.

In modern medical practice, we use **anti-inflammatory drugs** extensively to manage some diseases. Aspirin and the corticosteroid hormones can minimize the discomfort and damage in some chronic inflammatory conditions, such as arthritis. When steroids are given, they may make the patient feel better fast, as the discomforts of inflammation subside. Indiscriminate use of steroids, however, can be very dangerous, as it weakens one of the body's most important defensive responses.

There is one early study that makes this point. Clinicians were testing large steroid doses to evaluate their usefulness in treating patients with bacterial pneumonia. The patients rapidly perked up and started asking for their dinners, exclaiming how much better they felt. However, the clinicians, who were watching the patients' x-rays closely, noted that the patients' lung damage was getting worse equally rapidly. They stopped the medication. Although the patients then reported feeling worse again, their infections were arrested and they recovered. Since antiinflammatory medications are currently our only means of controlling rheumatoid arthritis and other inflammatory diseases, **nosocomial** (treatment-caused) infection is a large problem. We must remain acutely aware of the trade-off we make when we use these drugs.

SUMMARY

1. We live our lives in ecological balance with microorganisms that surround and colonize us. Most of these organisms are not harmful. Others can cause disease if our defenses are lowered, while yet another and by far the smallest group cause disease in healthy human beings.

2. Virulence can be expressed in terms of the number of organisms necessary to initiate disease, the morbidity rate, or the mortality rate.

3. Infectious disease occurs when the organism enters the host's tissues, multiplies excessively, and manufactures chemical substances that upset the host's physiology or damage tissues.

4. An organism's ability to attach itself to tissue, its extracellular or intracellular location of replication, its invasiveness, and its toxin production contribute to virulence.

5. Bacterial exotoxins have very specific effects on key functions such as neurological activity and protein synthesis. Endotoxins have the ability to cause vascular changes, inflammatory changes and, in high doses, fatal shock.

6. The host has both nonspecific resistance and specific immune defenses. The nonspecific defenses are constantly active in surveillance; they take care of most microbial invasions. The general state of the individual's health has a profound, if incompletely understood, effect on the efficiency of defense.

7. The first line of defense is made up of the anatomic features that minimize a microorganism's chances of entering tissues, and the physiologic conditions that produce a chemical environment unsuited for rapid multiplication of invaders.

8. Blood cells and plasma factors are important internal defenses. All of the white blood cells have special functions in defense. Granulocytes and monocytes are phagocytic and mediate or control inflammation, while lymphocytes mediate specific immunity. Plasma proteins include immunoglobulins, clotting factors, lysozyme, and interferon.

9. The proteins of the complement series combine to promote chemotaxis, inflammation, adherence of phagocytes to their prey, and lysis of foreign cells. Opsonins are proteins that promote phagocytosis.

10. In the living animal, phagocytosis acts to remove the organism that has passed the external barriers and established itself inside the body. The phagocyte attaches to its prey, ingests it, exposes it to lysosomal enzymes, kills it, and digests it. The normal phagocytic cells — PMNs and macrophages — work effectively only in the presence of plasma factors such as complement and the opsonins. Their efficiency is increased by immunity.

11. The inflammatory response is a generally encountered response to infection. It delivers a high concentration of plasma factors and phagocytes to the injured or infected area. Raised temperature may inhibit microbial growth; clot formation restricts microbial migration.

12. Inflammation has damaging effects that can be managed by steroids. This may be necessary in such diseases as arthritis. However, individuals receiving antiinflammatory drugs have increased susceptibility to infection.

DISCUSSION QUESTIONS

1. In what ways does the normal flora constitute a defense against disease?

2. Starvation leads to a progressive loss of blood serum proteins. How does this loss contribute to increased susceptibility to disease?

3. In what ways do pathogenic microorganisms evade phagocytic destruction?

4. What are some differences between PMNs and macrophages?

5. Determine how each of the symptoms mentioned in Box 11.2 is related not only to pathogenic effects of the microbial agent but also to defensive efforts of the host. What happens to our defenses if we give medication for "symptomatic relief" of coughing, runny noses, inflammation, fever, or vomiting?

ADDITIONAL READING

Hirsch RL. The complement system: its importance in the host response to viral infection. *Microbiol Rev* 1982; 46:71–85.

Roitt I. *Essential immunology.* 4th ed. Oxford: Blackwell, 1981.

Stephen J, Pietrowski RA. *Bacterial toxins.* Washington, D.C.: American Society for Microbiology, 1981.

Snyderman R, Goetzl EJ. Molecular and cellular mechanisms of leukocyte chemotaxis. *Science* 1981; 213:830–837.

Taylor P. Bactericidal and bacteriolytic activity of serum against Gram-negative bacteria. *Microbiol Rev* 1983; 47:46–83.

Weinberg ED. Iron and infection. *Microbiol Rev* 1978; 42:45–66.

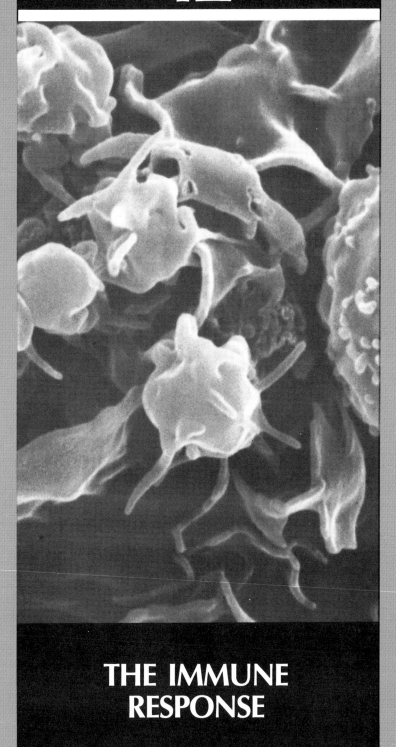

12

THE IMMUNE
RESPONSE

CHAPTER OUTLINE

Specific immunity is a powerful defensive system. It is based in the lymphoid tissues, which provide many different types of cells that have intricately interacting functions. In this chapter we will familiarize ourselves with the lymphoid system and its subsystems, examine how immune responses are controlled genetically, and study the two different types of immunity, as carried out by antibodies and by immune cells. We will see how immunity actually protects against pathogens. Finally, we will discuss techniques for using immunology in diagnosis, treatment, research, and disease prevention.

A BRIEF OVERVIEW OF IMMUNITY

The immune system is often referred to as our third line of defense. It comes into play, generally speaking, when an invading pathogen has passed our first line of defense (the anatomic and physiologic barriers) and our second line of defense (the nonspecific serum proteins and professional phagocytes). It is the essential companion to the nonspecific defenses, and a highly sophisticated means of increasing their efficiency. Immunity improves the specificity with which nonspecific defenses home in on invaders. We saw in Chapter 11 that antibodies combined with complement help phagocytes to adhere to foreign cells, and promote the lysis of the target cells. In this chapter we will see how other components of the immune system increase the aggressiveness of macrophages, enabling them to kill otherwise-resistant intracellular pathogens.

Nonspecific defenses, alone, are ineffective against infections by many pathogens — most Gram-positive bacteria, bacteria with capsules, the tuberculosis organism, and most viruses and other intracellular parasites. That is why persons born with genetic defects that prevent their immune systems from developing normally invariably have repeated and severe infections and usually have very short lives. In a person with normal defenses, but lacking specific immunity to a pathogen, the agent can sometimes penetrate the nonspecific defenses and establish an infection. It is only later, after the person has developed a specific immune response, that he or she makes a successful recovery. The acquired immunity will defend the person against subsequent invasions by the same agent.

Immune responses evolved with the vertebrate animals, reaching their highest level of development in the warm-blooded creatures — birds and mammals. Warm-blooded species have an efficient high-pressure blood circulation and a recycling lymph system, through which the cells of the immune system may travel. Their lymphoid tissues are widely distributed and have a broad range of functions.

We will be painting a picture of immune responsiveness as an intricate network of communications among cells. The **lymphocytes,** the cells that make up the immune system, do a lot of travelling around the body, and they do a lot of "talking" to each other. By this we mean that they are able to sense chemical information about all the substances they touch as they migrate through your body. They pass chemical signals to other lymphocytes and macrophages, thereby coordinating their cooperating defensive weapons.

Researchers have faced three main challenges as they have attempted to sort out this cellular communication system. The first is to discover the mechanisms by which the lymphocytes recognize the millions of different chemicals they are likely to come in contact with over your life span. The key to this puzzle lies in analyzing the genetically determined receptors that lymphocytes bear on the outer surface of their membranes.

The second puzzle is to figure out how the lymphocytes know which of the chemicals they encounter belongs there and is to be ignored, and which is a foreign entity to be attacked. The distinction between **self,** a substance that is a normal part of the body, and **non-self,** a substance that is not normally found in the body, is crucial to the immune response.

The third, and at this point the least clear issue, is to sort out how the body selectively activates, for each stimulus, the most appropriate of its several possible responses, and does so at the correct level of intensity. We will see that some forms of immunity are effective againt bacteria, while others are more effective against intracellular parasites. It is clearly important that the immune responses your body makes should fit the invader. Researchers feel that regulatory genes are at the heart of this issue.

THE TWO TYPES OF IMMUNE RESPONSE

There are two fundamentally different types of immune response, antibody-mediated immunity and cell-mediated immunity. Each type employs different subpopulations of lymphocytes that respond to the foreign material (Figure 12.1). The two types of lymphocytes are designated **B cells** and **T cells.**

Each form of immunity has **effectors,** chemical or cellular agents that carry out the response. In **antibody-mediated immunity** (AMI), the effectors are **antibodies** or **immunoglobulins,** defensive proteins that are synthesized by B cells. Antibodies are found in our blood serum, saliva, and other body secretions. Among their activities, we might note here that they neutralize toxins and kill some bacteria. We'll discuss them in detail later. The AMI response is what most people recognize or think of as "immunity." It has been known and studied for at least 100 years.

Cell-mediated immunity (CMI) has been studied only for about 30 years, and it is less well understood and less generally known. In CMI, the **T cells** are the effectors. They either acquire the capacity to kill a cellular target themselves, on contact, or may activate macrophages to do so. T cells do not secrete immunoglobulins. A major role for cell-mediated immunity is to serve as our primary protection against viruses.

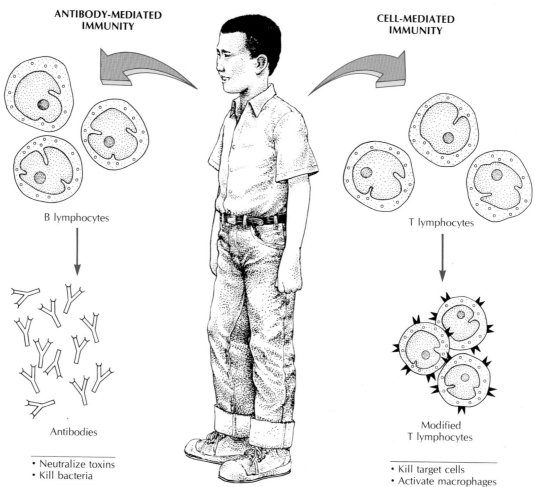

ANTIBODY-MEDIATED
IMMUNITY

CELL-MEDIATED
IMMUNITY

B lymphocytes

T lymphocytes

Antibodies

Modified
T lymphocytes

• Neutralize toxins
• Kill bacteria

• Kill target cells
• Activate macrophages

FIGURE 12.1   The Two Types of Lymphocytes
Among the body's many identical-looking lymphocytes, one group are
thymus-derived T lymphocytes, cells that have the ability to destroy cellular invaders.
Others are B lymphocytes, cells that react to antigens and initiate the production of
antibodies. Specifically activated immune cells and specific antibodies work together
to inactivate invading microorganisms and/or their products. It is essential that both
the antibody- and cell-mediated immunity systems be functional to maintain normal
resistance to disease.

When you are naturally exposed to a new immunizing agent — a microbial invader or a vaccine, for example — often both antibody-mediated and cell-mediated immune responses develop together. This means that both types of defensive weapons will appear — the immunoglobulins, effectors of AMI, and the T cells, effectors of CMI.

THE LYMPHOID TISSUES

The lymphoid system is responsible for the immune response. In an adult human, the lymphoid system weighs about 2 lbs. It consists of about 10^{12} (one trillion) lymphocytes and 10^{20} molecules of various antibodies.

LYMPHOID CELLS, TISSUES, AND ORGANS

The lymphocyte is the cell type of the lymphoid system. Lymphocytes are small, generally oval cells. They have relatively little cytoplasm; thus the nucleus appears comparatively large.

Lymphoid tissues are those that contain lymphocytes (Figure 12.2). In most lymphoid organs the lymphocytes are in close proximity or in direct cell-to-cell contact with macrophages. Thus there is intimate cooperation between lymphocytes and phagocytes. A lymphoid organ is characteristically spongy, containing large numbers of lymphocytes in various stages of maturity, very loosely retained in a meshlike mass of tissue. The lymphocytes are able to leave the spongy tissue, circulate a while in the blood and lymph, then settle elsewhere in other lymphoid tissue.

Scattered along the lymph circulation are oval-shaped structures, the lymph nodes, through which lymph flows. Antigenic particles are trapped in the spongy tissue of the nodes and phagocytosed. When a great number of microorganisms pass through the system, the nodes may become infected, tender, and swollen. The thymus and spleen are highly organized major lymphoid organs with a membranous covering. They contain both rapidly dividing immature lymphocytes and nondividing mature ones. Still other loose clusters of lymphoid tissue are found in the tonsils, in the appendix, along the respiratory tract, and in the Peyer's patches in the wall of the small intestine. Lymphoid tissue in the intestinal wall is exposed to many environmental antigens, such as microorganisms that cause gastrointestinal damage. The lymphoid patches respond by producing IgA antibodies that protect the mucosal surface from invasion. Any of the lymphoid tissues can mount an immune response singly or together.

ORIGIN AND DIFFERENTIATION OF LYMPHOCYTES

Differentiation is a general developmental process that occurs when cells, under changing patterns of genetic control, develop specialized functions. A **stem cell** is an undifferentiated cell; during development, it may give rise to several unique lines of differentiated cells. Differentiation often progresses through several stages, as certain functions are sequentially developed to their fullest, while others are gradually turned off. The signals for a cell line to differentiate may be cell-to-cell contact or diffusible chemical signals released from neighboring cells. Differentiating cells thus "learn" by the company they keep. As a result, an embryonic cell that lodges in one organ will be exposed to a specific set of signals, and thus directed along one differentiation path, while another cell, located in a different organ, will be headed toward a divergent path.

Lymphocytes arise from stem cells located in the liver in the fetus and the **bone marrow** in adults. The resulting progeny cells undergo further divisions and several stages in differentiation in other organs before they are mature enough to actually contribute to an immune response. During differentiation, some lymphocytes follow one path, becoming B cells, while others develop into T cells (Figure 12.3).

THE B-CELL LINE. Some immature lymphocytes, called lymphoblasts, having originated in the bone marrow, complete their evolution into B cells there. This stage is what commits them generally to the career of synthesizing antibodies, something no other type of mammalian cell does. At a slightly later time each B cell learns to produce a specific antibody, by a unique genetic process we will discuss shortly. Some B

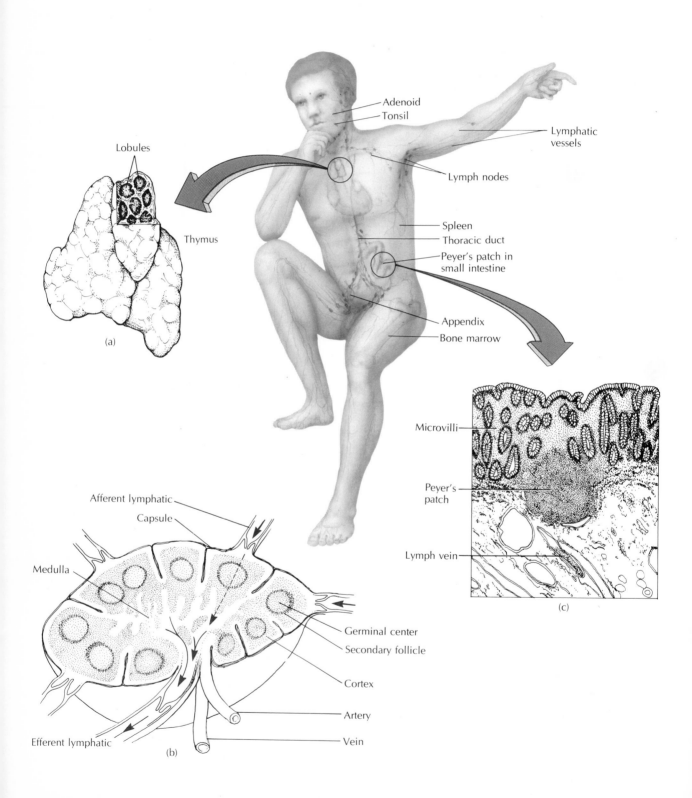

Lobules

Thymus

(a)

Adenoid
Tonsil
Lymphatic vessels
Lymph nodes
Spleen
Thoracic duct
Peyer's patch in small intestine
Appendix
Bone marrow

Afferent lymphatic
Capsule
Medulla
Germinal center
Secondary follicle
Cortex
Artery
Vein
Efferent lymphatic

(b)

Microvilli
Peyer's patch
Lymph vein

(c)

FIGURE 12.2 The Human Lymphoid System
Lymphoid organs are composed of a spongy tissue containing lymphocytes. Both lymphocytes and lymphocyte products circulate in the blood and lymph. (a) The thymus gland is active in the synthesis of thymosins, hormonelike immunoregulatory compounds. Lymphocytes fill the tissue spaces in the lobules. (b) In lymph nodes, the exterior cortex, contains B-cell follicles. When a primary follicle is activated, it develops into a germinal center where antibody-producing cells proliferate and release antibody into both lymphatic and blood circulation. Other B cells remain in the secondary follicle layer. The paracortical layers in the medulla contain T lymphocytes. (c) Peyer's patches consist of aggregates of lymphocytes beneath the microvilli that line the small intestine. Thousands of such aggregates secrete antibodies into the intestinal contents in response to antigenic stimulation.

FIGURE 12.3 Lymphocyte Differentiation
In a commonly accepted hypothesis, primitive lymphoid cells are believed to originate in fetal bone marrow. Some pass through the thymus gland and differentiate into T lymphocytes. Others mature in the bone marrow and become B lymphocytes. Antigen stimulation will cause either type (in cooperation with the other) to develop into the enlarged, functionally active T lymphocyte or plasma cell forms. The transmission electron micrograph on the left shows a small circulating lymphocyte with its prominent nucleus and relatively small amount of cytoplasm. ($\times$2996. Courtesy of T.R. Hoage, R. Jacobs, A.M. Andrews, Y. White/National Center for Toxilogical Research.) The transmission electron micrograph on the right shows a plasma cell, a fully differentiated, antibody-secreting B cell (Courtesy of Joseph Feldman).

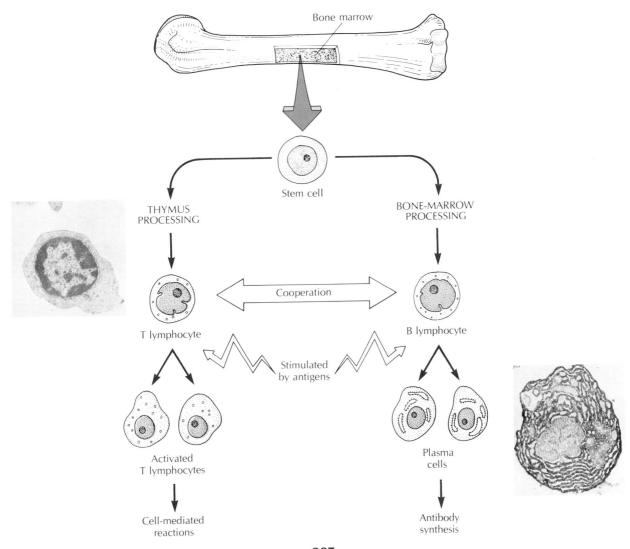

365

cells are found in special areas within all the lymphoid tissues. About 20% of circulating lymphocytes are B cells.

THE T-CELL LINE.   Those lymphocytes that are destined to become T cells mature in the thymus gland (Figure 12.4). The thymus is an unusual organ that directs the development of cell-mediated immunity. Its initial function is to establish the cellular immune response in the child, which it does by producing a large population of T cells. This work appears to be completed by adolescence. Thereafter the thymus becomes much smaller and less active. Its other

FIGURE 12.4   Changes in Thymus Tissue with Age
These micrographs show sections of thymus tissue from (a) newborn individuals, and those aged (b) 13 years, (c) 41 years, and (d) 55 years. Note that the thymus gland decreases in size with age. This decrease correlates with a decline in activity and a transfer of cell-mediated immunity to other organs, a process that ends during early adolescence. The thymus gland remains active (until middle age) in synthesis of thymosins, hormonelike immuno-regulatory compounds. (Photo courtesy of Dr. Katsuiku Hirokawa, Tokyo Metropolitan Institute of Gerontology.)

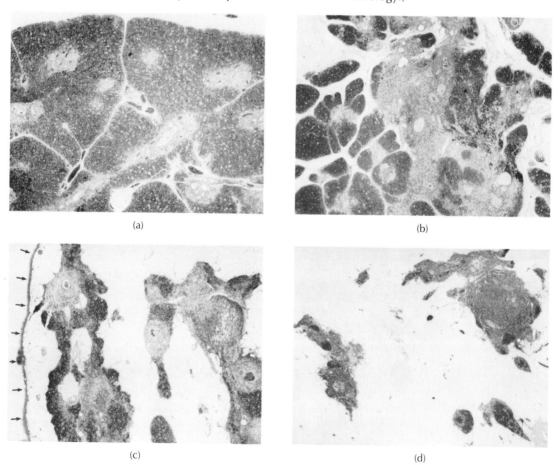

(a)

(b)

(c)

(d)

role, throughout life, is to produce signalling substances called **thymosins.** These hormones appear to stimulate various aspects of the immune system (see Box 12.2).

T cells subsequently differentiate further to acquire one or another specialized skill, such as the ability to produce **lymphokines,** which are signalling factors, the ability to activate or suppress B cells, or the ability to kill target cells. Throughout childhood, T cells proliferate rapidly and colonize T-cell areas in lymphoid organs, in close proximity to B cells. By adolescence, the lymph nodes and spleen have a full complement of T cells. Thus as the thymus ceases activity CMI functions are gradually transferred to other organs. The T cells also migrate freely. About 80% of circulating lymphocytes are T cells.

GENETICS AND IMMUNITY

Genes control all the key aspects of the immune response. Certain genes determine the surface chemistry of your body's own cells — their self antigens. Other genes direct the synthesis of the protein antibodies. Yet another group of genes regulates the intensity of response and the interaction of the various cell populations.

ANTIGENS

We have previously defined antigens as substances to which the body mounts an immune response. Chemically speaking, natural antigens are usually large, complex molecules such as proteins, glycoproteins, or polysaccharides, or cells bearing them on their surfaces. An antigen has one or more **antigenic determinants,** portions of the molecule to which an antibody or lymphocyte binds. Some antigens have multiple determinants and are "stronger" or provoke a greater response. The **valence** of an antigen is the number of its identical combining sites.

Under normal circumstances, antigens are foreign or non-self. If a substance is self, that is,

one of your own gene products, it doesn't normally work as an antigen in your body. Yet what is self to me (my blood group, for example) may be non-self to you, if you have a different blood group than I do. Let's explore this question of self and non-self in the context of tissues.

TISSUE ANTIGENS. All your cells have unique chemicals or markers embedded in the outer surfaces of their cytoplasmic membranes. Tissue cells other than red blood cells share one particular set of markers, a pattern of inherited tissue antigens called the **histocompatibility antigens.** These antigens, which are glycoproteins, are determined by the HLA genes on Chromosome 6. There are at least three different genes, and many different forms of each gene, or alleles. Each individual inherits a specific set of HLA antigens, half from each parent (Figure 12.5). Because there are so many different forms of these genes in the human population, we have less than 1 chance in 30,000 of encountering another human being, other than an identical twin, with the same marker pattern. Your HLA antigens, determined by your set of genes, unambiguously label your cells as self. In fact, your body will attack grafted tissue bearing different or non-self markers with the full force of the immune response.

BLOOD GROUP ANTIGENS. The red blood cells are the exceptions; they lack HLA antigens on their surfaces. However, they have their own class of special markers, the **blood group antigens.** These antigens are polysaccharides. Like the tissue antigens they are located on the outside of the cytoplasmic membrane (Figure 12.6). The most prominent blood group antigens are those that determine whether you are Type A, Type B, Type O, or Type AB. There are three different forms of the particular gene that determines your ABO blood group. One form, gene I^a, directs the making of the Type A polysaccharide. Another form, I^b, directs the making of a slightly different molecule, the Type B polysaccharide. The third and most common form, **i,** specifies an

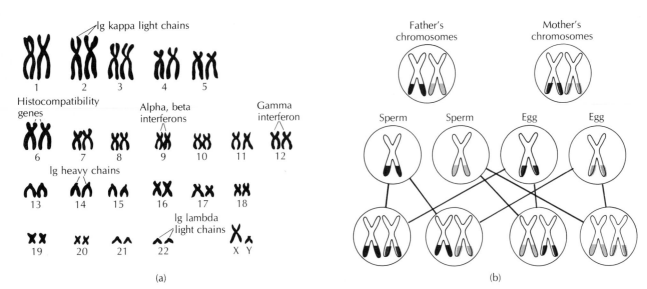

FIGURE 12.5   Genetics and Immunity
(a) The 23 pairs of human chromosomes bear genes for certain immunologic functions, which have been determined by genetic analysis. Genes for Ig kappa and lambda light chains and heavy chains are the ones that are decoded to synthesize antibody. The histocompbility region contains genes that determine "self" tissue antigens, as well as immunoregulatory genes. The interferon genes produce the three classes of interferon. (Courtesy of A.M. Winchester, *Human Genetics,* Fourth Edition. Columbus, Ohio: Charles E. Merrill (1974). Adaption of Fig. 1-7, page 11.)
(b) A human child inherits its tissue type which is determined by two sets of HLA antigens. One set is located on the copy of Chromosome 6 that the child receives from each parent. The child will then be different from either parent, with only a one-in-four chance of having the same tissue type as a sibling.

incomplete polysaccharide that is not antigenic. Since we humans are diploid, we may inherit any combination of two of these genes. As a result, our red blood cells may have one or both of the antigenic polysaccharides, or only the non-antigenic polysaccharide.

Any ABO factors you have are self to you. If you receive blood bearing the same antigens on the donor cells as you have on your cells, your body will not react to the donor red blood cells as if they were foreign. On the other hand, if you received blood of a different type, with some foreign antigens, your antibodies would destroy the donor red blood cells.

FOREIGN OR NON-SELF ANTIGENS.   To reinforce the point, an antigen must (normally) be foreign or non-self to actually provoke an immune re-

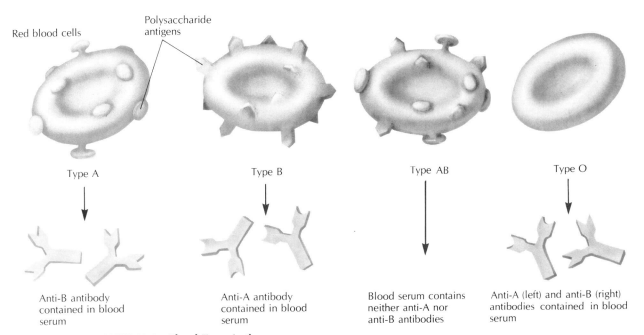

FIGURE 12.6 Blood-Type Antigens
Red blood cells have special markers, the blood-group antigens, that determine whether a person is type A, type B, type AB, or type O. Each person's blood serum contains antibody to antigens that differ from that individual's blood type. Thus, a transfusion with an incompatible blood type can lead to a dangerous attack by the recipient's antibodies on the donor's red blood cells.

sponse. We can see that this distinction is entirely relative in regard to the ABO antigens and tissue antigens, in that my "self" may be your "non-self." On the other hand, substances such as viral capsids or tetanus toxoid are foreign to all human tissue, and hence antigenic to both of us.

ANTIBODIES

Antibodies combine with specific antigens; that is, an antibody to measles virus combines with measles virus, not with poliovirus (and vice versa). Here we will see that it is the antibody's structure that is responsible for this characteristic.

STRUCTURE OF THE IMMUNOGLOBULINS. All immunoglobulins are quaternary proteins, assembled from four or more polypeptide chains held together by disulfide bonds. An immunoglobulin may be described by two features, its class and its specificity.

There are five classes of immunoglobulins, each of which has a different combination of polypeptides, described in Table 12.1 and Figure 12.7. The genes for the different antibody polypeptides are distributed on human chromosomes

TABLE 12.1 The immunoglobulins					
	IgG	IgA	IgM	IgD	IgE
Light chains	2	2(4)*	10	2	2
Heavy chains Type (number)	G(2)	A(2, 4)*	M(10)	D(2)	E(2)
Basic four-chain units	1	1(2)*	5	1	1
Antigen-binding sites	2	2(4)*	10	2	2
Percentage of total immunoglobulin in body	80	13	6	1	0.002

*Secrotary form of IgA contains double the number of units.

2, 14, and 22. Each immunoglobulin contains equal numbers of light and heavy chains. **Light chains** are shorter, contain about 220 amino acids, and are of one of two types, kappa or lambda. An antibody molecule will contain two or more identical kappas or lambdas but not both. **Heavy chains** contain about 440 amino acids. The type of heavy chain distinguishes the five separate classes of immunoglobulin.

Each immunoglobulin has a unique specificity, determined by two or more antigen-binding sites, which attach to the antigen. Both light and heavy chains contribute to forming the antigen-binding sites. Within each chain there are two types of regions, called the variable and the constant regions. **Variable regions** occur at the antigen-binding end of the molecule, and each different antibody chain has its unique variable amino acid sequence. It is the variable sequence that is responsible for the antibody's specificity. This is because the variable sequences of the polypeptides fold up into uniquely shaped grooves or pockets. The antigen binds in these grooves, very much as a substrate fits into the surface of an enzyme. Immunoglobulins may be of any of five classes. Depending on which class they belong to, antibodies may have 2, 4, or 10 binding sites per molecule.

The rest of each immunoglobulin chain is a **constant region,** in which there is little or no variation in amino acid sequence between anti-

bodies, whatever their specificity. The constant regions are important to the antibody's function, however, in that they bind complement and attach the immunoglobulins to the surfaces of phagocytes and other cells. In contrast to the variable regions, which we just saw determine antibody specificity the constant region of the heavy chains determines which class an immunoglobulin belongs to.

GENETIC CODING FOR ANTIBODIES.   There are thousands, perhaps millions of different antibodies. Each has a different specificity, because each has a different amino acid sequence. How could this happen? It certainly seems unlikely that each cell would come provided with millions of different antibody genes. In fact, they don't.

Each immature lymphocyte actually appears to have a few dozen separated immunoglobulin-gene segments. Before each lymphocyte begins to make antibody, it assembles one set of complete functional genes for a heavy chain and one for a light chain by selectively hitching together gene segments. This process involves natural gene splicing and mutation. For example, the gene for a variable region of a light chain is built by splicing one of the several hundred different variants of one gene segment to one of the four variants of another gene segment. Thus in assembling a gene for a light chain, there are at

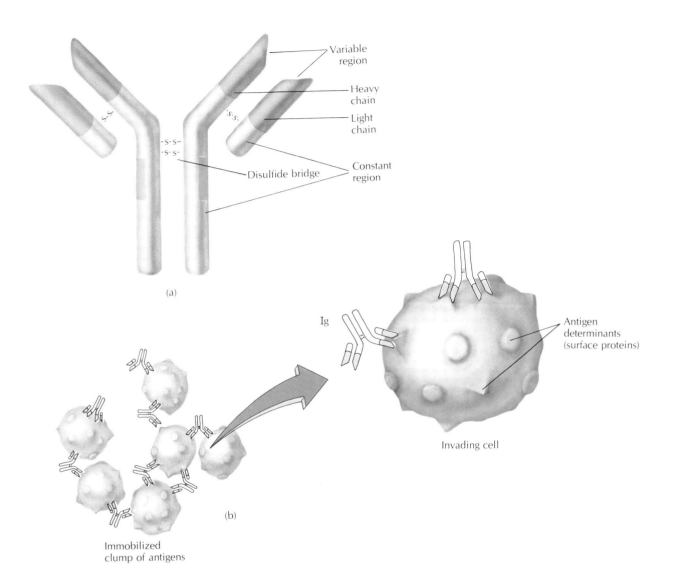

(a)

Variable region

Heavy chain

Light chain

Constant region

Disulfide bridge

Ig

Antigen determinants (surface proteins)

Invading cell

(b)

Immobilized clump of antigens

FIGURE 12.7 Immunoglobulin Structure
(a) Immunoglobulins are made up of light chains and heavy chains covalently linked by disulfide bonds. Variable regions on these chains contain amino acid sequences that form the antigen-binding sites. These sites are highly specific and bind selectively to antigens. An IgG-type immunoglobulin has four polypeptide chains. The two arms of the IgG molecule swing out at an angle, held together by the disulfide hinge. Thus they may bind to two adjacent sites on a particle or form cross-links between two particles. (b) Such cross-linking leads to agglutination, immobilizing the antigens and rendering them targets for phagocytosis.

least 2000 different combinations possible. In assembling a gene for a heavy chain, there seem to be at least 10,000 possible splicing combinations. When you then consider all the possible pairings of light and heavy chains that may occur when a complete immunoglobulin is made, you come out with 10^7, or 10 million options. Inexact splicing multiplies the possibilities further. Additionally, mutation can act on certain sensitive sites. Consequently, the number of potential different antibodies a lymphocyte could produce is huge.

Each mature lymphocyte, however, as a result of this unique customizing process, produces just one unique light chain and one unique heavy chain, making just one of the possible millions of immunoglobulins. The genetic information in its spliced genes is passed on to any and all daughter lymphocytes by mitosis. Each lymphocyte and its descendants compose a clone of cells, all of which produce a unique antibody. Many thousands of different clones exist in each person's lymphoid tissues.

DEVELOPMENT OF THE SELF/NON-SELF DISTINCTION.   There are many potential antigens in your body, but they are self antigens. Your immune system "learns" not to attack them. This is called **immune tolerance,** and it develops very early during fetal life. You will never normally mount an immune response against those antigens then or for the rest of your life. As the time of birth approaches, however, the behavior of lymphocytes begins to change. After birth, lymphocytes make contact with either self or non-self antigens. Contact with non-self antigens provokes immunization, not tolerance.

These phenomena are best explained by the **clonal selection hypothesis.** We have already seen that the evidence suggests that billions of lymphocytes are produced during fetal development, and that the genetic customizing mechanism just described renders each lymphocyte capable of producing a single immunoglobulin. From that point on, each fetal lymphocyte then incorporates some of this immunoglobulin into its outer membrane surface as an antigen receptor. By using these cell-surface–antibody molecules the lymphocyte can recognize and bind antigens. In the early fetal stage, the lymphocyte is still not mature enough to begin to form an immune response. The clonal selection hypothesis suggests that if a lymphocyte with self receptors binds self antigens it is destroyed or rendered unresponsive (Figure 12.8). This is **negative selection.** The only lymphocytes that remain active after this negative selection process are ones that attack non-self antigens.

In the second trimester of fetal life (months 4 through 6), lymphocytes begin to move into a new period of maturity. In the case of B cells, some researchers believe that the more mature B cell has become able to make a second, additional type of immunoglobulin surface receptor. When an antigen binds simultaneously to both receptors, the event stimulates the B cell to begin rapid division giving rise to a clone of daughter cells. This is **positive selection.** It is still unclear what the comparable sequence of events is for T cells, although it is known that the T cell receptors are immunoglobulinlike proteins.

Once selected, the clone of lymphocytes all mediate one single, specific immune response. Unstimulated lymphocytes remain present, potentially responsive, but inactive.

FIGURE 12.8   Clonal Selection
(a) A lymphocyte stem cell gives rise to many thousands of different fetal lymphocytes. During the negative-selection period in early fetal development, any lymphocyte that binds antigen (of necessity a self antigen) will die or become unreactive. (b) During the positive-selection period, which begins around the time of birth, lymphocytes may become exposed to non-self antigens such as microorganisms. Some contacts will positively select specific B cells, promoting proliferation, clone formation, and antibody-mediated immunity. Other lymphocytes may never contact their antigen and thus will remain unstimulated throughout life.

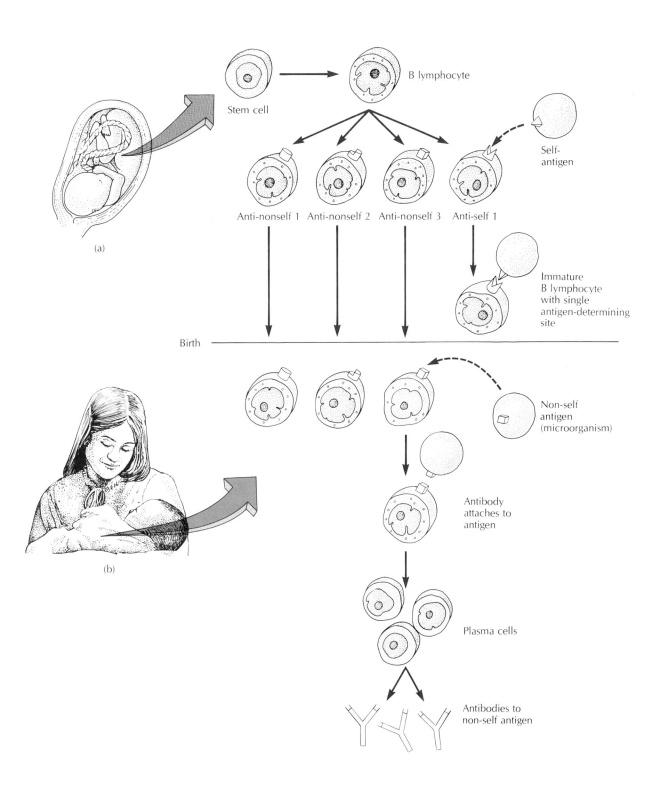

(a)

(b)

Stem cell

B lymphocyte

Self-antigen

Anti-nonself 1 Anti-nonself 2 Anti-nonself 3 Anti-self 1

Immature
B lymphocyte
with single
antigen-determining
site

Birth

Non-self
antigen
(microorganism)

Antibody
attaches to
antigen

Plasma cells

Antibodies to
non-self antigen

BOX 12.1
Monoclonal Antibodies

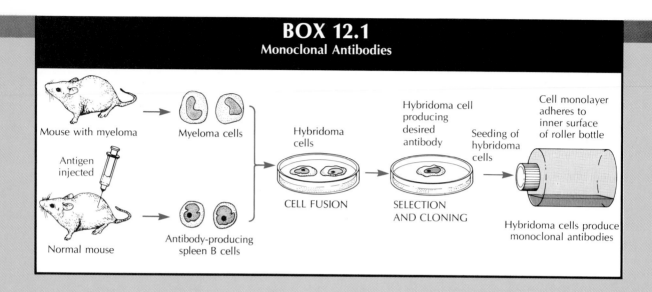

Mouse with myeloma → Myeloma cells

Antigen injected

Normal mouse → Antibody-producing spleen B cells

Hybridoma cells

CELL FUSION

Hybridoma cell producing desired antibody

SELECTION AND CLONING

Seeding of hybridoma cells

Cell monolayer adheres to inner surface of roller bottle

Hybridoma cells produce monoclonal antibodies

A sample of normal human serum contains thousands of different antibodies inextricably mixed together—the products that all the B-cell clones have pumped out from all over the body. Serum is a researcher's nightmare!

Now there is a simple, reproducible way of getting a large population of cells isolated in a pure cell culture, all of which are of one clone, producing large amounts of a single antibody: antibodies for diagnosis, pure so that tests become extremely precise; antibodies for treatment, specific so that they can be selected to target a single type of cancer cell; antibodies for research, enabling detection of any type of antigenic determinant in any part of an antigenic structure.

These pure antibodies are called **monoclonal antibodies;** they are so pure because they are the product of a single clone of genetically identical lymphocytes. Researchers obtain monoclonal antibodies by using a clever technique in which two types of cells are caused to fuse. B cells from the spleen of a mouse immunized against a specific antigen contribute the capacity to make the antibody. Mouse myeloma cells, cells of a B-cell tumor, bring the ability to proliferate freely and continuously. When the two kinds of cells are fused, the hybrid cells are called **a hybridoma.** Hybridoma cells both make a specific pure antibody and divide without limit. It is next necessary to select those hybridoma cells that produce the desired antibody. Then the chosen hybridoma cell line is propagated either in the abdominal cavities of mice or in cell culture. The hybridoma line will continue to produce its particular type of monoclonal antibodies indefinitely. This technique is a key method in thousands of research achievements every year. Monoclonal antibody products are being used in dozens of practical applications in the diagnostic laboratory.

ANTIBODY-MEDIATED IMMUNITY

We have seen how lymphocytes acquire the genetic capacity to make a specific antibody, how all those with antibodies to self antigens are possibly eliminated before birth, and how a slightly more mature lymphocyte can be stimulated to start an immune response. Now we will look at the normal child or adult, to follow how antibody-mediated immunity develops after contact with a foreign antigen.

HOW ANTIGENS ARE INTRODUCED. Antigens may get into the body by many routes, but always end up in lymphoid tissue. If antigen enters through a break in the skin or by injection, it is carried with tissue fluid to the nearest lymph node. If antigen is introduced into the blood, it is filtered out by the spleen. If it is inhaled, the lymphoid tissue in the respiratory mucosa takes it up, and if it is ingested, the antigen is picked up by the Peyer's patches in the small intestine. Once antigen is in a lymphoid organ, it is usually then phagocytosed by a polymorphonuclear leukocyte or macrophage (Figure 12.9a). The macrophage then transfers the antigen to any available lymphocytes that have surface receptors to bind and respond to it. Many aspects of this macrophage–B cell interaction have not yet been adequately explained. Additionally, it is known that certain types of T cells can either help or suppress the B cells' response.

The B cells' surface receptors (just discussed), which are the unique immunoglobulins they have been customized to make, bind the antigen, which is then drawn into the cells. This event stimulates the B cells to the final steps in their maturation. They begin rapidly dividing to form a large clone. In the lymphoid tissue, we can observe these clones as **germinal centers** of active B cells appearing within one to six days after antigen contact.

The developing clone is composed of two cell types derived from the parental B cell. These are the mature, fully functional **plasma cells,** which are large and produce antibodies, but live only about a week, and the long-lived **B memory cells,** which do not produce antibody (Figure 12.9b). All plasma cells make one class of antibody, called IgM, first. Detectable levels of specific IgM antibody appear in the blood about three to five days after the antigen has been administered. Many plasma cells within a clone then switch to making antibody of one of the other classes, but of identical specificity, and these begin to increase in 10 days to 2 weeks. Antibody molecules have an average working life measured in weeks, so the levels remain high for several months, although they then decline as most of the plasma cells age and die off and are not replaced. Some new plasma cells (and therefore some antibody) are usually produced indefinitely.

Immunizations are often given in a series of two or more doses. On second or subsequent exposures to the same antigen, the B memory cells come into play, and the immune response is both more rapid and more powerful. On antigen stimulation, some of the B memory cells promptly divide to produce a large crop of new plasma cells. The antibodies appear in a much shorter time (starting in two to three days) and in much higher concentrations. This is called the **anamnestic** (memory) response. B memory cell activity also guarantees that the elevated antibody level persists longer (Figure 12.10). This is the biological reason that a series of booster shots not only gives a high initial level of immunity but also helps to keep a person's protection active.

ANTIBODY CLASSES

As we noted earlier, plasma cells produce five different classes of immunoglobulins (abbreviated Ig). They are IgA, IgM, IgG, IgD, and IgE. This diversity is an essential feature of our protection, because each type of antibody has its own role in the defensive network.

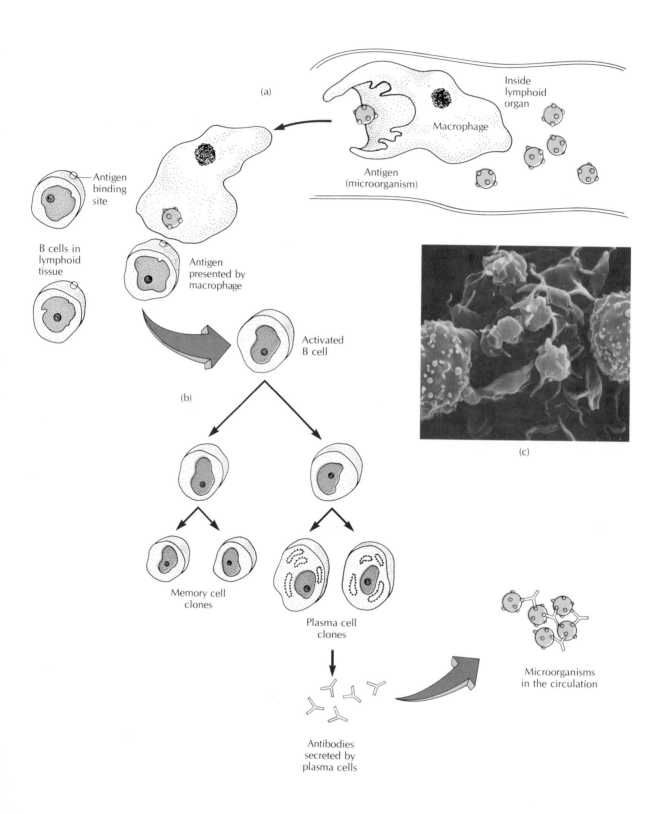

(a)

Inside
lymphoid
organ

Macrophage

Antigen
(microorganism)

Antigen
binding
site

B cells in
lymphoid
tissue

Antigen
presented by
macrophage

Activated
B cell

(b)

(c)

Memory cell
clones

Plasma cell
clones

Antibodies
secreted by
plasma cells

Microorganisms
in the circulation

FIGURE 12.9 Activation of a B-Cell Clone
(a) When an antigen such as a microorganism enters a lymphoid organ, it is phagocytosed by a polymorphonuclear leukocyte or a macrophage. The macrophage then presents the antigen to a B cell, which takes it in and becomes activated. (b) The activated B cell begins to divide, giving rise to many plasma cells and B memory cells. The plasma cells produce specific antibody, which binds to the invading microorganisms. (c) This scanning electron micrograph shows cells that may be participating in immune activation. The macrophage (center) is believed to be transferring antigen to two lymphocytes on either side. (×420. Courtesy of T.R. Hoage, R. Jacobs, A.M. Andrews, Y. White/National Center for Toxicological Research.)

FIGURE 12.10 The Anamnestic Response
When the first dose of immunization is given, a mild antibody response occurs at about six to eight days, peaks at about three weeks, then gradually declines as the first group of plasma cells ages and dies off. Memory cells, however, remain in the lymphoid tissue. When a subsequent dose of vaccine is administered, the memory cells give rise to a new population of plasma cells, and a significant increase in antibody titer occurs within about three days. The peak antibody titer achieved after a second dose may be 100 times that obtained with the first dose.

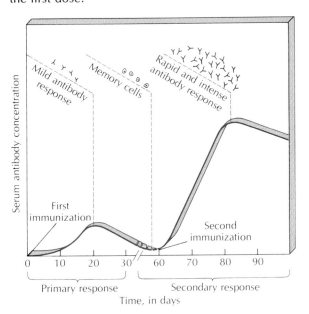

IgA is produced to some extent by clones of B cells in all the lymphoid tissues, but mainly by lymphoid tissue in the lining of the gastrointestinal and respiratory tracts. Some IgA circulates in a form composed of two light and two heavy chains, with two combining sites. Most IgA, however, is modified into a secretory form, a double-ended structure with four antigen-binding sites. Secretory IgA is found outside the body in saliva, mucus, milk, semen, and secretions, where it defends the epithelial surfaces, which are a major route of entry into the body for pathogens. For example, an individual who is immune to diphtheria secretes a type of specific IgA in his or her saliva that neutralizes diphtheria toxin. A person immune to cholera secretes a specific type of IgA that blocks cholera bacilli from adhering to the bowel lining. Secretory IgAs are exceedingly important in our day-to-day protection because they inactivate pathogens before they can infect (Table 12.2).

IgM and IgG are the two major circulating antibodies. IgM is the larger circulating antibody, the first type of antibody the infant begins to produce, and the first antibody to appear after immunization. It is a large complex formed of five of the small four-chain units linked in a circle. Because it has 10 combining sites, IgM effectively binds to the surface of antigenic cells, binding complement and initiating the cascade of events that promote complement-mediated cell lysis.

IgG is the smaller of the two major circulating antibodies. It crosses the placenta and protects the fetus. It can neutralize toxins. When IgG binds antigen, it forms complexes that activate and bind complement by the classic pathway, leading to complement-mediated phagocytosis and/or cell lysis. IgG appears later in an infection than IgM. Although IgM and IgG have somewhat overlapping functions, IgM is particularly important in promoting recovery in the early stages of a disease, while the later-appearing IgG prevents recurrence.

IgD is a four-chain antibody that is found only in trace amounts in serum. Its natural role

TABLE 12.2 Biological activities of immunoglobulin					
Characteristic	IgG	IgA	IgM	IgD	IgE
Distribution	Serum Tissue fluid	Serum Mucous secretions Glandular secretions	Serum Tissue fluid	Serum Lymphocyte surface	Serum Mast-cell surface
Complement fixation	+	−	+	−	−
Crosses placenta	+	−	−	−	−
Secreted in milk	−	+	−	−	−
Binds phagocytes	+	−	−	−	−
Roles	Neutralization of toxins, agents in body fluids Protects newborns	Defends routes of entry; preventive	First to appear; effective against organisms in blood	Lymphocyte receptor	Controls parasites; hypersensitivity

in immunity is still unclear. There is strong evidence, however, that it is a component of the surface recognition systems of lymphocytes and is essential for B cells to respond to antigen stimulation.

IgE is a four-chain antibody that attaches to the surface of basophils and mast cells and binds antigen. It is usually a minor fraction of serum globulins, although it is strikingly elevated in persons prone to allergies. IgE antibodies are responsible for allergy and anaphylaxis (a hypersensitivity reaction), and may also help to control certain helminthic infections.

CELL-MEDIATED IMMUNITY

In cell-mediated immunity not only is the agent or effector of immunity a cell, but the target is always a cell. It could be a host cell harboring an intracellular parasite or bearing new antigens of virus or tumor origin, or a tissue graft from an antigenically unmatched donor. In all cases, the antigens are chemical substances found on the surface of the target cell. The target is attacked by direct contact with either killer T cells or macrophages that have special abilities in killing other cells.

THE COMPONENTS OF CELL-MEDIATED IMMUNITY

SUBGROUPS OF T CELLS.   T cells respond to contact with antigen by dividing, just as B cells do (Figure 12.11). They give rise to several different lymphocyte subgroups, each with a specific role to play. The T-cell subgroups are distinguished by their functions and the substances they se-

BOX 12.2
Thymosins—A New Class of Signalling Substances

Glandular tissues affected by thymosins

The thymus, as we noted, is the organ that processes T lymphocytes and structures the development of cell-mediated immunity. This is a well-recognized role, although it is far from completely understood.

Recently, research groups have begun turning their attention to another function of the thymus gland, analyzing the **thymosins,** peptides secreted by the thymus. Some of these peptides function as true thymic hormones, having effects on the function of glandular tissues throughout the body. Others are immune regulators. There are two general categories, alpha and beta thymosins; there may be more than 100 different thymic peptides yet to be sorted out. Their importance is just beginning to be vigorously explored.

Thymosin alpha-1 is a potent inducer of T helper cells, while alpha-7 induces T suppressor cells. Thymopoietin influences T-cell differentiation.

Recall that the thymus shrinks with age. Decreased thymic function correlates with aging. In persons over 60, most T cells are nonfunctional "dummies" which do not respond to antigen. However, when elderly volunteers received thymosins in clinical trials, their ability to respond to antigens increased markedly. Thymosin trials have also shown promise in stimulating the defenses of children with inherited T-cell deficiency, and in promoting remissions in renal and prostate cancers.

Some beta-thymosins alter the levels of other hormones. Beta-4 stimulates the anterior pituitary gland to release luteinizing hormone, the hormone that causes women to ovulate. Other thymosins affect corticosteroid hormones. At present, we do not know what signals the onset of menopause and other hormonal changes in aging, but it has been suggested that the onset of menopause (when ovulation stops) correlates with declining thymosin levels. Some researchers suggest that thymosins may provide new tools for stimulating useful forms of immunity. Possibly they may also help in moderating some of the effects of aging.

crete. Researchers identify them by the array of chemical markers on their surfaces. T cells populate the paracortical region of lymph nodes and marginal areas in spleen. T-cell areas in a lymphoid organ usually border B-cell areas, as the two types of lymphocytes operate cooperatively. A comparison of B and T lymphocytes appears in Table 12.3.

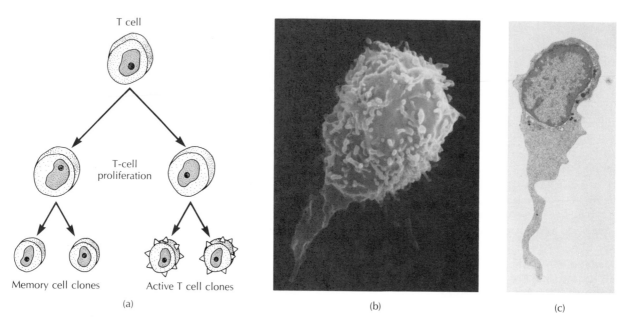

T cell

T-cell
proliferation

Memory cell clones

Active T cell clones

(a)

(b)

(c)

FIGURE 12.11 Activation of a T-Cell Clone
(a) When a T cell is activated by antigen contact, it proliferates to give rise to several types of effectors, including lymphokine producers and cytotoxic killer cells. (b) In this scanning electron micrograph, the lymphocyte is oval with surface bumps and a trailing tail of cytoplasm. (c) In the transmission electron micrograph, the dense nucleus and cytoplasm with few granules can be seen. These cells were inactive at the time they were photographed. (×514. Photo courtesy of T.R. Hoage, R. Jacobs, A.M. Andrews, Y. White/National Center for Toxicological Research.)

Two of the T-cell subgroups have regulatory roles. **T helper cells** (Table 12.4), also known as T_4 cells, produce factors that stimulate other lymphocytes. They cooperate with B cells in initiating antibody-mediated response. Helpers seem also to be essential for development of other T cells. **T suppressor cells** or T_8 cells, a second subgroup, are also regulatory, but they reduce the intensity of antibody-mediated responses. They appear to have a role in keeping other lymphocytes unresponsive to self antigens. They can also moderate cell-mediated immunity by suppressing the activity of killer cells (see below). The role of the two regulatory T cells — helpers and suppressors — is to balance the intensity of the simultaneously developing AMI and CMI responses. This ensures that immunity

works hard enough to be effective, but not so hard that it becomes destructive.

Laboratory tests can determine the ratio of suppressor to helper cells in a person's blood. This **suppressor–helper ratio** is a strong influence on how effectively the immune response is working. High suppressor–helper ratios are seen in acquired immunodeficiency syndrome (AIDS) for example, and also in cancer patients whose cancers are growing rapidly. Should the number of a cancer patient's T helper cells increase again, the patient may achieve a remission.

T cytotoxic cells, or T killer cells, are different from natural killer cells (Chapter 11) in that they are antigen-specific. Their function is, after attaching directly to an antigen-bearing cell, to secrete factors that kill the target cell.

TABLE 12.3 Comparison of B and T lymphocytes		
Site of embryonic processing	Bone marrow	Thymus
Percentage of circulating lymphocytes	20%	80%
On antigenic stimulation differentiate into:	Plasma cells Memory cells	Helper cells Suppressor cells Cytotoxic (killer) cells Delayed hypersensitivity cells Memory cells
Soluble products	Immunoglobulins	Lymphokines
Primary function	Antibody-mediated immunity	Cell-mediated immunity
Secondary function	Regulate cell-mediated immunity	Regulate antibody-mediated immunity

T delayed hypersensitivity (T_D) cells bind with a high degree of specificity to their antigens, then produce biologically active substances called lymphokines. We'll discuss these substances next. The name for this subgroup of T cells comes from the fact that delayed hypersensitivity reactions may be side effects of cell-mediated immunity.

LYMPHOKINES. Lymphokines are T-cell products that are soluble protein factors; they are completely different from antibodies in protein structure and in function. No one knows at present exactly how many different lymphokines there are, but at least 30 different functions have been ascribed to them in the laboratory. In general, we define as a lymphokine any substance that has a signalling effect on another type of lymphocyte, or some other type of cell, to increase or decrease a specific activity. Lymphokines, therefore, are intercellular communications signals. However, they are rather general signals. In an immune response involving T_D cells, it is the reaction that causes the lymphocytes to start secreting the lymphokines, not the

TABLE 12.4 Subgroups of T cells	
T helper cells	Cooperate with B cell to induce antibody synthesis Cooperate with T cell in induction of CMI
T suppressor cells	Interact with other lymphocytes to suppress aspects of immunity
T cytotoxic cells	Attach to antigens on cellular target and lyse it by membrane disruption
T delayed hypersensitivity cells	Attach to antigenic markers on cellular target, then release lymphokines that summon and activate macrophages

products themselves, that is specific. Lymphokines resemble the inflammatory mediators in that they predominantly act locally, but may be carried by the circulation. They orchestrate the combined attack of T cytotoxic cells and macrophages on antigens. Some of the lymphokines are

■ Chemotactic factor, which is a substance that attracts macrophages.
■ Macrophage inhibition factor (MIF), which reduces macrophages' movement and keeps them, once summoned, from moving away. MIF also stimulates phagocytic ability.
■ Specific macrophage activating factor, which directs macrophages against antigenic targets.
■ Lymphocyte (gamma) interferons, which are immune regulators and suppressors.
■ Interleukin-2, which is a growth factor that stimulates the growth and division of T cells.

ACTIVATED MACROPHAGES.   Nonspecific phagocytes, as previously noted, are effective only against certain types of microbial pathogens. However, when macrophages are activated by lymphokines they become much more effective. When activated, they add new surface receptors for more efficient phagocyte–prey contact. As they continue their lymphokine-triggered differentiation, they also grow larger in size, become highly motile, and accumulate more and larger enzyme granules and phagosomes (Figure 12.12). The **activated macrophage** has enhanced activity against all cellular non-self antigens. Thus a CMI response against tuberculosis may as a side benefit give a patient enhanced, nonspecific resistance against other intracellular parasites such as fungi.

PATTERN OF CMI RESPONSE

Cell-mediated immunity develops step by step. T cells bear immunoglobulinlike receptors on their surfaces to permit them to bind to antigen. Their receptor system has two parts. One recognizes the specific non-self antigen, while the other recognizes a histocompatibility antigen. To achieve a response, both these antigens must

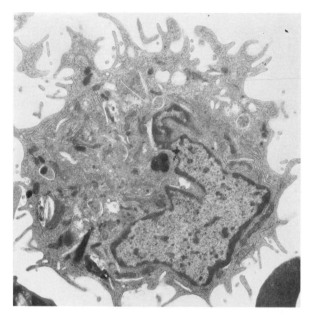

FIGURE 12.12   An Activated Macrophage
This transmission electron micrograph of an activated macrophage shows the cell extending many pseudopodia. Layers of endoplasmic reticulum studded with ribosomes are visible. The cytoplasm is full of phagolysosomes, containing unidentifiable debris from previous prey in the process of digestion. (Courtesy of T.R. Hoage, R. Jacobs, A.M. Andrews, Y. White/National Center for Toxicological Research.)

be present on the target cell, a type of "double signal." Thus CMI is only directed against our cells (or those closely enough related to have recognizable HLA antigens) that are simultaneously displaying a foreign antigen. When antigen binding stimulates T cells, they begin mitosis, and mature T cells of the various classes appear within three days, increasing for up to one week. T memory cells also develop, so that secondary responses are more rapid and stronger than primary responses. As T_D cells increase, they secrete lymphokines, which recruit macrophages to the site of the antigen. Both T cells and activated macrophages normally participate in neutralizing antigenic cellular targets.

THE EFFECTS OF CELL-MEDIATED IMMUNITY

CMI is directed against two types of targets. One type of target is the individual's own cells when they harbor intracellular infectious agents or when they are altered with tumor antigens. The other type of target is a misdirected one, and occurs in **autoimmune disease,** a condition in which our immune systems attack our own normal tissue. CMI may also be directed against the histocompatibility antigens of non-self cells, in the case of unmatched grafted tissues.

THE ANTIMICROBIAL RESPONSE. Antimicrobial CMI is primarily directed against those of our cells which harbor intracellular pathogens; these, as a result of infection, have some sort of distinctive chemical marker on their surfaces. Antimicrobial CMI depends largely on T delayed hypersensitivity cells and activated macrophages. In tuberculosis, for example, T_D cells migrate to the lung tissue where the antigenic bacteria are multiplying. There they release lymphokines that summon and activate macrophages. These macrophages ingest and destroy the tuberculosis organisms. In the process, some lung tissue is destroyed. The lesion becomes surrounded by an area of inflammation, called a **delayed hypersensitivity** response, a frequent side effect of effective antimicrobial CMI. The lesion is later enclosed by scar tissue, which walls in any surviving pathogens.

In the antiviral CMI response, cytotoxic T cells and activated macrophages zero in on tissue that contains replicating virus. In measles, for example, the infected host cell's membrane surface contains antigenic viral proteins. T cells and macrophages bind to the viral antigens on the host cell and destroy it, putting an end to viral replication in that cell.

ANTICELLULAR RESPONSES. In a cell-mediated attack against antigenic tissue, cytotoxic T cells and macrophages are also involved, destroying the grafted donor tissue in episodes of graft rejection, or destroying host tissue in autoimmune disease. Additionally, CMI is considered to be a major factor in arresting the growth of virally induced tumors, although the evidence is less clear for other forms of tumors.

PROTECTION THROUGH IMMUNITY

Let's look now at how this array of B cells, T cells, antibodies, lymphokines, and macrophages actually work together to defend you against microbial challengers. We will define **immunity** as the state of being effectively protected against challenge by an infectious agent. **Immunization** is the process of becoming protected by an immunizing material; it can be either an active or a passive process.

ACQUIRING ACTIVE IMMUNITY

Active immunity develops when your body produces an immune response after receiving an antigenic stimulus. Your immune system responds actively, and you develop protective AMI and/or CMI, usually both. You now possess many stimulated lymphocyte clones, with both active plasma and T cells, and memory cells. Your plasma cells secrete specific antibodies, usually of several different classes. Your response is, in theory, lifelong, because these cells will be replenished by cell division. In practice, the protective level of your active immunity may decline after several years, or during periods of sickness, stress, or malnutrition.

Naturally acquired active immunity is a protective state that follows an actual case of the disease. Even an inapparent disease, which causes no significant symptoms, may often supply a sufficient stimulus. The immune response will begin to develop during the incubation period and course of the disease, have valuable effects after 5 to 10 days, and promote recovery when fully developed.

Artificially acquired active immunity is obtained as a result of vaccination, a type of immunization in which one receives a vaccine. A **vaccine** is a manufactured product containing an altered harmless antigen, such as an attenuated (weakened), modified, or killed infective agent, or a toxoid. The effectiveness of vaccination depends on the amount of antigen given, how effectively it reaches lymphoid cells, how long it remains in the body as a stimulus, and how similar the vaccine antigen is to the native antigen. Researchers strive to develop safe vaccinations that provide as much protection as naturally acquired immunity; however, some vaccines are still not highly effective.

ACQUIRING PASSIVE IMMUNITY

Passive immunity is acquired when an individual receives either preformed antibodies, or presensitized lymphocytes or macrophages, from another person (an outside source). In the near future, it may be possible to induce passive immunity by administering manufactured lymphokines. Passive immunity is therefore a transferred immunity. It is also immediate; you get protection from the moment of transfer. However, passive immunity is temporary, lasting only days to weeks. This is because antibody molecules break down over time, and transferred lymphocytes or macrophages die out. Since no antigen is introduced, you are not stimulated to develop your own response to whatever you are being protected against.

Natural passive immunity is a mechanism that is highly developed in mammals, designed to protect the newborn. Newborn humans have very immature immunologic capacities of their own. Some antibody begins to appear one to two months after birth, but AMI doesn't reach full strength until age four. CMI functions are similarly slow to develop.

The newborn child's immunologic protection depends almost entirely on transfers from its mother. Maternal IgG passes through the placental membranes to the baby, who is born with IgG only to those diseases to which its mother is immune. During nursing, breast milk reinforces the baby's defenses with secretory IgA, some IgG and IgM, and lymphocytes and macrophages. Nursing, especially if prolonged beyond the first few months, bridges the gap between birth and the time, at around two years of age, when the child's own antibody production becomes adequate.

Artificial passive immunity is achieved by injecting a person exposed to a disease with serum containing preformed antibody. This approach is used to protect individuals in high-risk situations. For example, immunodeficient children risk fatal complications if they develop chickenpox. If they receive antichickenpox globulin within 96 hours of exposure, the transferred antibodies will abort the infective cycle of the virus. Rabies immunoglobulin is given to people bitten by a rabid animal, to immediately neutralize the virus and prevent the disease from developing.

AMI AND CMI—PROTECTIVE COOPERATION

CONTROLLING BACTERIA.   Extracellular bacterial infections are controlled almost entirely by AMI. Antibody neutralizes bacterial toxins, opsonizes bacteria for more effective phagocytosis, and activates complement by the classic pathway leading to lysis.

Intracellular bacterial parasites are controlled largely by CMI, because antibodies cannot enter cells. Antibodies can, however, inactivate bacteria that leave one cell to go to the next. The antimicrobial response controls acute infections, such as those by *Listeria* or *Brucella*, quickly eliminating the agent. CMI's role in chronic infections may be to contain the agent rather than eliminating it. The agent is kept from multiplying only so long as CMI levels remain high. Active disease may recur if CMI subsides.

CONTROLLING VIRUSES.   Cytolytic animal viruses that pass from one host cell to another via body fluids may be neutralized by antibody during the extracellular phase. Other viruses rarely leave the protection of the host cell. In herpes, measles, and pox virus diseases, antibody is useful in preventing infection, but relatively useless in controlling infection once it is established. CMI destroys the virus-infected tissue along with the virus. Persistent virus infections seem to be able to escape immune surveillance. Episodic herpes infections persist, for example, even though the body produces very high levels of antiherpes antibodies.

CONTROLLING FUNGI. The superficial fungi (those that infect the skin) are extracellular; they are controlled by nonspecific resistance and by cell-mediated delayed hypersensitivity. The systemic fungi, such as *Histoplasma, Coccidioides,* and disseminated *Candida,* are controlled by CMI mechanisms. Antibodies may be present but they appear to play a limited part in the control process.

CONTROLLING PARASITES. Immunity does, unfortunately, have its weak spots. It is least effective against protozoans and helminths. After most of these parasites have gained entrance, they are rarely completely eliminated. Containment is the most that immune mechanisms can achieve. Antibodies may assist in killing extracellular parasites, such as trypanosomes, which are highly resistant to phagocytosis. However, the trypanosomes are genetically able to continually change their surface antigens, thus evading any existing immunity. Extracellular forms of the malarial parasite are able to "hide" from antibodies, apparently by coating themselves with a layer of human serum proteins that conceal their own antigens. CMI is useful against *Toxoplasma,* which multiplies intracellularly. However, most parasite life cycles have highly resistant resting stages that persist despite immunity.

Immune responses, fortunately, are successful most of the time in either preventing or resolving infections. Prevention works best when the invader is met by IgA antibody at all body portals and IgG and IgM in all body fluids. The agent will then be neutralized before it can infect. How often does this happen? No one knows. But probably, for you and me, it's 100 or even 1000 times a day.

It is likely that an infective agent slips by into our tissues on only a tiny fraction of encounters. When it does, the infected cells will become the target of a cell-mediated response. CMI will destroy the agent's source of nutrients — the host cell — and expose it to destruction. This probably also happens many more times than we know it.

USING IMMUNOLOGY IN DIAGNOSIS

The antigen–antibody combination is highly specific, as we have seen. For this reason, we can use antigen–antibody reactions as clinical and research tools. **Serology** is the study of antibodies in serum, as carried out in the laboratory **in vitro** (meaning in glass, or in an artificial setting). Here we'll consider some important serologic applications.

In each immunologic reaction there is always a known and an unknown factor: the known is used to find or to quantitate the unknown. Briefly, known antigens may be used to detect unknown antibodies, or known antibodies may be used to search for unknown antigens. Some tests, or assays, are qualitative, and search only for the presence of the unknown; other assays are quantitative, and tell us how much of the unknown is present.

Antibody in an individual's body fluids is the "footprint" of a pathogen, evidence of prior or current exposure to its antigens. We may also sometimes be able to find the antigen of a pathogenic agent directly in a body fluid or tissue sample. Rapid tests for antigens seek to identify a pathogen in such samples by using extremely sensitive assays.

All serologic assays are designed to produce a visible or detectable antigen–antibody complex, allowing the positive or negative results of the test to be read. The testing method must be as free as possible from cross-reactions and other sources of error. There are literally dozens of immunologic laboratory techniques. We will concentrate on some types of tests commonly used in diagnostic applications.

AGGLUTINATION TESTS

Agglutination tests are used when the antigen is a large particle or a cell. When this type of antigen complexes with antibody, clumping or **agglutination** occurs. The clumps are large enough to see with the naked eye (Figure 12.13). We will use blood grouping, which uses red blood cells as the unknown antigen, as an example to illustrate some of the ways agglutination tests may be designed. Blood grouping is usually performed as a **slide agglutination test,** which is rapid and easy to read. In a test of this sort a drop of unknown antigen suspension and a drop of known antibody-containing antiserum are mixed together on a glass slide and observed for clumping. When clumping occurs, we know that both antigen and antibody are present, and that they have reacted. In a **tube agglutination test** we mix the reagents together in a test tube under more controlled conditions. It is somewhat more complex, but more accurately detects the weaker positive reactions that may be overlooked in a slide test. Bacterial cells are also agglutinated by their antibodies and may be used in both of these forms of agglutination tests.

Hemagglutination is the correct name for the agglutination of red blood cells. The main application for hemagglutination tests is in matching donor and recipient blood samples before transfusion. Accurate matching is essential, because red blood cells may possibly be bound, clumped, and lysed by antibodies should they be transfused into the blood of a person of a different blood type. This type of transfusion reaction may be fatal to the recipient. After the donor and recipient bloods have been grouped by a slide test to establish ABO and Rh group identity,

they are then **cross-matched.** In this process, small amounts of each pint of intended donor blood are mixed individually with small amounts of recipient blood in tube agglutination tests. A cross-match will reveal any potential clumping due to ABO, Rh, or other less-common antigen incompatibilities that might cause trouble should the blood be given to an unmatched recipient.

In some cases, antibodies against red blood cells, such as human anti-RhD antibodies, will destroy red blood cells in the body but won't agglutinate them in the test tube. We need a test to find nonagglutinating or **incomplete** antibodies against red blood cells that doesn't rely on simple hemagglutination. The **Coombs test** is used for this purpose. It is what is descriptively called a "sandwich"-type test. It uses two steps to test the compatibility of paired blood samples. In the first step, one party's red blood cells and the other's serum are mixed together and allowed to react (Figure 12.14). If incomplete antibodies are present in the serum used, they will bind (invisibly) to the red blood cells.

In the second step, **anti-human globulin** antibodies are added. These are antibodies from the serum of an animal such as a rabbit or goat. The animal has previously been immunized, using human immunoglobulins as a foreign antigen. The animal, as a result, produces antibodies against the human globulin. When these antibodies against human globulin are added in the second step of the Coombs test, they will react with any nonagglutinating antibodies that may have bound to the red blood cells in the first step. The three components together — red blood cells, incomplete human antibodies, and anti human globulin antibodies — form a "sandwich," an agglutinating complex that is readily visible (Figure 12.14b). Note that in the "sandwich" type immunologic test, the middle component of the sandwich will bind two different things. In this case, the middle of the sandwich is the incomplete antibody, which functions both as an antibody (in relation to one component) and as an antigen (in relation to the other).

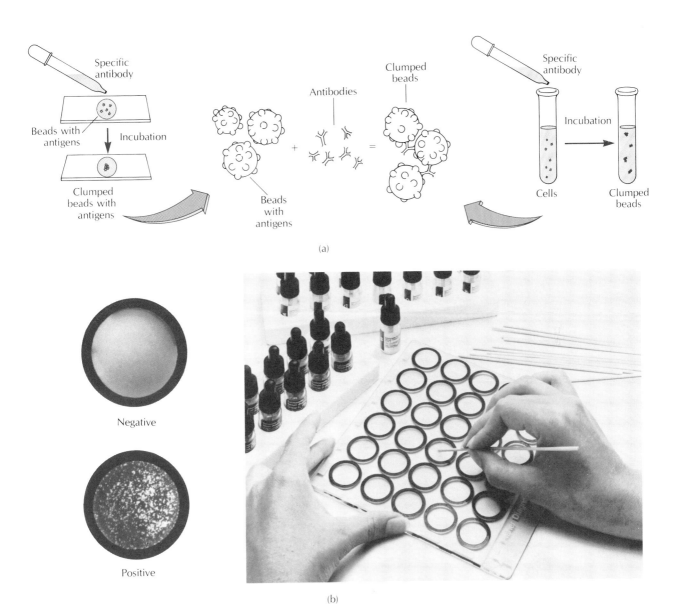

(a)

(b)

FIGURE 12.13 Agglutination Tests
(a) In a slide agglutination test, purified antigens bound to latex beads are used as a test reagent. A drop of unknown antigen suspension and a drop of known antibody-containing antiserum are mixed on the slide. In a tube agglutination test, the reagents are mixed in test tubes. In either case, when antigen binds to antibody, clumping (or agglutination) occurs. (b) In the test shown here, antibodies specific for a particular group of bacteria were bound to the beads. When a suspension of cells from a bacterial colony was mixed with the antibody-bead reagent, clumping identified the bacteria (bottom left) . A negative test (top left) remained evenly cloudy. (Courtesy of Wellcome Reagents Division, Burroughs Wellcome Company.)

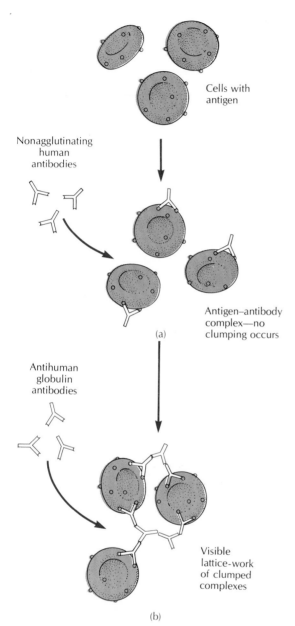

FIGURE 12.14   The Coombs Test
(a) Red blood cells from one individual and serum from another are allowed to react. Any incomplete antibodies contained in the serum will bind to the red blood cells, but agglutination does not occur. (b) Next, antihuman globulin antibodies, obtained from an animal immunized against human globulins, are added to the mixture. These react with the incomplete antibodies that have bound to red blood cells, causing agglutination to occur.

Because slide agglutination tests are the easiest of all serologic tests to set up and to read, it has been desirable to convert many clinically relevant antigen–antibody tests to an agglutination form. Soluble antigens, such as protein toxins, however, do not usually form visible agglutinating complexes with their antibodies, which are also soluble in aqueous solutions. However, it is possible to bind soluble antigens or their antibodies chemically to the surface of inert, microscopic, particulate carriers, such as latex beads, inactivated staphylococci, or tanned (phenol-treated) red blood cells. Antigens or antibodies bound to such microcarriers form relatively larger complexes that are readable in a slide agglutination–type test.

One example of this approach is found in a widely used rapid screening test for meningitis pathogens. Bacterial meningitis may be caused by any one of several different bacterial species. It is important to know as quickly as possible which species is causing any case of meningitis so the right antibiotic may be chosen. The test determines which species is present by identifying its antigen in a cerebrospinal-fluid sample from the patient. Individual drops of the patient's sample (the potential antigen source or *unknown*) are mixed on a slide with drops of four different test reagents. Each reagent contains latex beads to which are bound known specific antibody to one of the bacterial meningitis agents. If the sample contains any one of these four agents, one of the droplets will show visible agglutination. The physician may have a preliminary identification in less than 15 minutes.

The strategy of slide agglutination may be reversed so that the antigen is used to search for antibody. In one commonly used test to find out if a pregnant woman has antibody to rubella virus, latex beads carrying bound antigen-inactivated rubella virus are the known. They are mixed with patient serum, the unknown or source of potential antibody. If the patient has antibody against rubella in her serum, an agglutination forms.

As we saw previously, some viruses, including measles and rubella, will hemagglutinate red blood cells in vitro. However, if the viral envelope is first coated with specific measles or rubella antibodies, before the virus particles are placed in contact with the red blood cells, the visible hemagglutination reaction will be inhibited. Thus virology laboratories find it convenient to use a two-step **hemagglutination inhibition** (HI) test (Figure 12.15) to detect viral antibodies in patient serum. Most HI tests are quantitative, and use a technique based on titration, called the microtiter.

Titration is any technique used for determining the concentration of a substance in solution. In immunology, we start by preparing serial dilutions of the solution we want to quantitate. Serial dilution creates a series of samples containing decreasing amounts of the unknown. The procedure for setting up a 10-fold serial dilution was discussed and illustrated in Chapter 6. In serology, however, **twofold dilution** is used most often. In setting up a twofold dilution, we start with the first tube containing straight serum. In the second, we mix one part serum with one part saline or buffer solution to give a 1:2 dilution. In the third, we mix one part of the 1:2 dilution taken from Tube 2 with one part buffer, yielding one part serum in four (1:4) and so on to 1:8, 1:16, and dilutions as high as necessary (1:1024, 1:2048, etc.). **Microtiter** techniques are designed to carry out this process rapidly with a high degree of accuracy using tiny amounts of reagents. Instead of test tubes, plastic plates containing 96 tiny wells are used. The transfer of solutions is carried out by special wire loops. Microtiter techniques are used in automated immunodiagnostic systems.

PRECIPITATION TESTS

Protein and polysaccharide antigens are macromolecules. When they combine with antibody, they form macromolecular complexes, which tend to stay suspended in solution. This type of complex is called a **precipitate.** Precipitates are visible only when antigen–antibody aggregates are highly concentrated in a small area. Under those conditions the precipitate appears as a

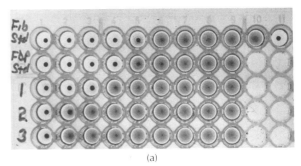

(a)

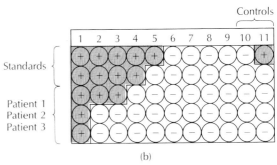

(b)

FIGURE 12.15 Hemagglutination Inhibition Test The microtiter technique is widely used for quantitative analysis of either antigens or antibodies. The test is performed in a plastic tray containing many wells, which function as tiny test tubes. (a) In this section of a plate, a known antigen (inactivated rubella virus) was added to serially diluted unknown patient sera (Rows 1 through 3). In the second step, red blood cells were added following incubation. If antibody was not present, the viruses remained capable of agglutinating red blood cells; these settled out in a layer covering the bottom of the well (top row, Tubes 6 through 10). If antibody was present, it bound the viruses and prevented them from binding to the red blood cells; these then settled to the center bottom of the well where they show up as a small, round button (Tubes 1 through 4, top row). Controls and standards are included in each plate. (Courtesy of Wellcome Reagents Division, Burroughs Wellcome Company.) (b) Compare the plus and minus signs in this drawing with the actual plate above to see how these reactions are read.

cloudy zone. Many precipitation tests use an agar gel to trap and concentrate the precipitate sufficiently so that it will be visible. The simplest precipitation method is the **immunodif-fusion plate** (Figure 12.16). First, wells are cut into highly purified agar. Then, antigen and antibody solutions are placed in different wells. The reactants are allowed to diffuse toward each other for a period of hours. At the point at which antigen meets antibody, they bind and stop diffusing, and a line of precipitate forms in the gel.

Immunoelectrophoresis, a more sophisticated precipitation test, is a multistage process. First an antigen mixture is applied to the surface of a gel slab. Then the slab is moistened with a buffer solution and an electric current is passed through it for one or more hours. This step separates the components of the antigen mixture into bands along a one-dimensional line, on the

basis of their electrical charge. Then the separated antigen mixture is exposed to antibody. Antigen bands bind antibody. Then excess unbound antibody is washed away. By carefully cutting up the gel, purified complexes can be obtained. The gel can also be stained for easier viewing, photography, or instrumental reading (Figure 12.17). Immunoelectrophoresis has clinical application in identifying the antigens of bacterial and viral pathogens amongst the mixture of proteins found in body fluids.

FIGURE 12.16   Precipitin Test
A simple precipitin test is performed by cutting wells in a plate of purified agar. Different antisera are pipetted into two wells and a mixture of antigens into another. The reactants diffuse outward from each well; where they meet, cloudy lines of precipitate form. Because each antigen has reacted with different antibodies in the serum, the precipitin lines cross instead of merging.

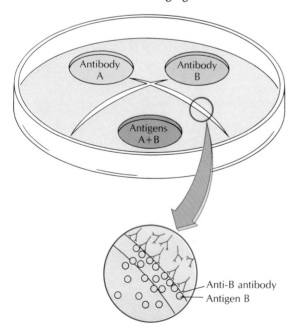

FIGURE 12.17   Immunoelectrophoresis
When a serum sample containing a mixture of immunoglobulins is placed in an electric field, the substances migrate from left to right. Next, the antibody is allowed to diffuse toward antigen. One curved precipitin line forms for each different antigen–antibody complex. The resulting pattern is made visible by a staining procedure. (Courtesy of Michael Katz.)

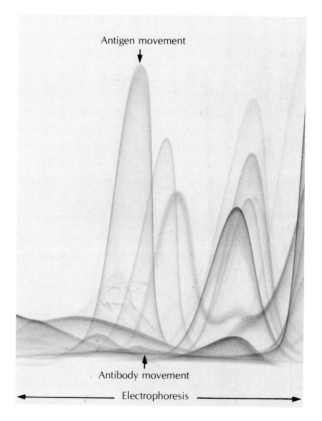

ENZYME-LINKED IMMUNOSORBENT ASSAY (EIA OR ELISA)

ELISA is a popular quantitative immunoassay method. It is based on enzyme activity, and it is simple to interpret because the reaction turns color when the result is positive.

One widely used commercial test is designed to detect rotavirus, an important agent of infant diarrhea, as an unknown antigen in a patient's stool samples. The ELISA procedure is also a sandwich-type test, this time with the antigen as the "filling" (Figure 12.18). In the first phase of the test, known antirotaviral antibody, immobilized on a plastic bead, is dropped into a tube with a small amount of suspended stool sample. If virus (antigen) is present in the stool sample, it is bound by the antibody to the surface of the bead. The bead is then washed to remove any unbound components. In the second phase of the test, a second antibody reagent is added, the **conjugate.** The conjugate also consists of antirotaviral antibody, this time bound to an enzyme. If viral antigen is present on the bead, the conjugate binds to the antigen, becoming the outer layer of the sandwich. If antigen is not present, however, the conjugate does not have anything to bind to. The mixture is again incubated, then washed.

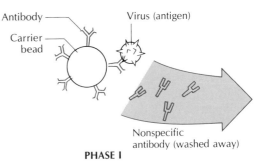

PHASE I

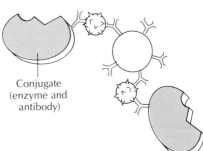

PHASE II

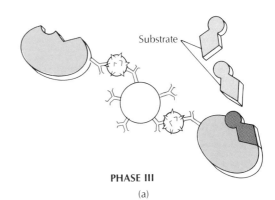

PHASE III

(a)

FIGURE 12.18 Enzyme-Linked Immunosorbent Assay (ELISA)

(a) This test works by combining three reactants to make a kind of "sandwich." First, specific antibody attached to a carrier bead binds antigen (rotavirus), and nonspecific antibody is washed away. Next, a conjugate, composed of a specific antibody conjugated to an enzyme, is added. The conjugate binds to the previously bound antigen and carrier bead, forming a three-part complex. In the final phase, the enzyme's colorless substrate is added. Presence of bound enzyme is revealed by the conversion of the substrate to a bright yellow product. (b) ELISA tests can be photometrically read to detect the formation of colored products. Automated equipment, such as this system, can also read the results of various other microbiological tests. (Courtesy of Dynatech Laboratories, Inc., Alexandria, VA.)

(b)

In the third phase of the test we finally find out what happened in the first two phases. We do this by testing to see if any enzyme activity remains bound to the bead. The enzyme's substrate, which is colorless, is added. If enzyme is present, it will convert the substrate to a bright yellow product. This is a positive test, and it indicates that there was rotavirus antigen in the stool sample. If there is no bound enzyme, the substrate remains unchanged and colorless. This is a negative test, indicating that there was no rotavirus antigen in the stool sample.

ELISA may also be adapted to the detection of antibodies. For example, it is used in testing for the antibodies formed in response to a syphilis infection. Many ELISAs have been adapted to automation by using microtiter plates, with the results read photometrically and analyzed by a computer. ELISA techniques are extremely sensitive, and can detect very tiny amounts of the unknown that is sought.

FLUORESCENT-ANTIBODY TECHNIQUES

Fluorescent dyes coupled to antibody may be used to detect antigen. These dyes, such as **fluorescein** and **rhodamine,** absorb ultraviolet (UV) light and re-emit greenish or orange visible light. They are similar to the paints used in black-light posters. We can chemically couple fluorescent dyes to antibody molecules without altering the antibody's ability to react with its specific antigen. When a fluorescent antibody binds to its antigen, we can see the complexes if we illuminate them with ultraviolet light. The reactions may be studied visually with a fluorescence microscope or measured photoelectrically with an automated UV scanner.

The **direct fluorescent-antibody (DFA) technique** is used to locate antigen in a fixed tissue section from an individual or on individual cells from a microbial culture. The sample on the slide is flooded with known fluorescent antibody. Uncombined antibody is then washed away and the slide is viewed with the fluorescence microscope. Where antigen (the unknown) has complexed with antibody, the fluorescent dye will emit visible light as the UV radiation hits it. The antigenic cell or structure may be spotted or outlined with light, revealing the distribution of antigen in the specimen (Figure 12.19). An example of this type of test is the FTA or fluorescent treponemal antibody test. The FTA test detects the presence of *Treponema pallidum*, the causative agent of syphilis. The fluid from a suspect lesion is spread on a slide, fixed, and flooded with the FTA reagent. Any *T. pallidum* present will become fluorescent.

The **indirect fluorescent-antibody (IFA) technique** is a sandwich method. It may be used when we want to detect antibody already bound to specific tissue antigens. This situation is a diagnostic feature of some immune-complex and autoimmune diseases. The essential step in making the antibody-binding sites visible is to flood the tissue specimen with fluorescent anti-human globulin antibodies. These antibodies are similar to those described under the Coombs test, but coupled to fluorescent dye. These antibodies bind to the human immunoglobulins, which are in turn bound to the antigen—a three-layered sandwich.

OTHER IMMUNOLOGIC METHODS

We will not attempt a complete discussion of all the immunologic techniques in an introductory text. However, let's make note of some others that you might encounter.

NEUTRALIZATION TESTS.   One biological role of antibody is the **neutralization** of the pathogen, that is, rendering the pathogen unable to cause disease. To test for neutralizing antibodies, we might measure how well the patient's serum does in protecting a live animal that is injected with the pathogen, whether the serum kills or damages the pathogen, or whether it detoxifies a toxin. The *Treponema pallidum* immobilization test (TPI), for example, is used to identify the syphilis spirochete in the fluid from suspected lesions. When specific antiserum is added, it stops *T. pallidum*'s undulating motion, but not the motion of unrelated spirochetes.

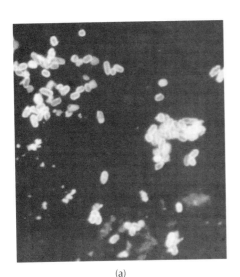

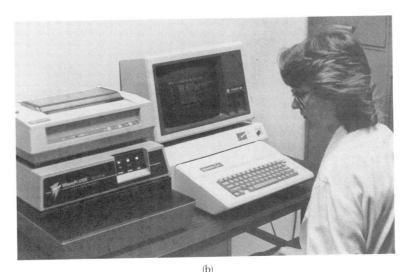

(a) (b)

FIGURE 12.19 Fluorescent-Antibody Testing
(a) *Escherichia coli* cells stained with a fluorescent antibody as viewed through a fluorescence microscope. This sample was from a fecal smear of an infant with diarrhea. The test permits rapid identification of the agent so that appropriate treatment may be chosen without delay. (Courtesy of Centers for Disease Control, Atlanta.) (b) Microtiter fluorescent-antibody assays may be read directly and quantitatively on this system. (Courtesy of Dynatech Laboratories, Inc., Alexandria, VA.)

COMPLEMENT-FIXATION TESTS. Many antigen-antibody complexes fix complement. The amount of complement fixed depends on the number of antigen-antibody complexes, and thus can indicate the amount of antibody present in a serum sample. If the antigen-antibody reaction is strong, and only a certain amount of complement is available, then all the complement will be used up. The **complement-fixation test** is a serologic test based on the competition of two antigen-antibody pairs for a limited amount of complement. The test is done in two stages, the first being a test reaction in which complement may or may not be used up, and the second being an indicator reaction that detects remaining complement by its ability to complete the lysis of red blood cells. Complement-fixation tests for many years were the most sensitive and accurate quantitative tests. They are very complex and require highly skilled technicians in order to give accurate and reproducible results. Complement-fixation tests remain the standard against which new tests are measured, but, because of their complexity, they are rarely done as routine large-volume tests.

RADIOIMMUNOASSAY. The **Radioimmunoassay (RIA)** test is an extremely sensitive method, which detects antigens on the basis of the measurement of very small amounts of radioactive iodine (^{125}I). RIA was first developed to detect minute amounts of thyroid hormones. It has since proved to be adaptable to detecting any antigenic substance to which antibody can be produced in a test animal. RIA is a competition test. Antigen, from a sample in which the antigen level is unknown, competes with a known amount of added radioactively-labeled antigen for a known limited amount of antibody. The higher the antigen level in the unknown, the lower will be the final radioactivity level in the antigen–antibody complex.

IMMUNE ELECTRON MICROSCOPY. We can make antigen–antibody complexes directly visible in the electron microscope by complexing one or the other reactant with electron-dense "dyes." This technique is called **immune electron microscopy.** The usual approach is to first complex the specific antibody with the electron-dense iron-containing protein **ferritin.** When ferritin-labeled antibody is allowed to bind to antigen-containing structures, the complexes show up as dark spots under the electron beam (Figure 12.20).

TESTS OF CELL-MEDIATED IMMUNITY

There are relatively few clinical tests that measure functional CMI. Most tests require a laboratory with cell-culture capabilities. Only the skin test is done in routine clinical settings.

SKIN TESTING. If an individual has an intracellular infection, we have seen that he or she will develop cell-mediated immunity. This response often may be detected by a skin test, in which a tiny amount of diluted antigen extracted from the organism is injected intradermally —

FIGURE 12.20   Immune Electron Microscopy
Antibodies can be chemically bound to ferritin, an electron-dense, iron-containing protein. When viewed under the electron microscope, the antigen–antibody combinations are visible as dark dots. These human cells in culture were infected with the P-3 virus and then treated with ferritin complexed anti-P-3 antibodies. The dark dots on the cell surfaces are not virus particles, but viral surface antigens bound to antibody in the host-cell membrane. (×133,400. Courtesy of Dr. John L. Carson, Infectious Disease Division, Department of Pediatrics, University of North Carolina at Chapel Hill.)

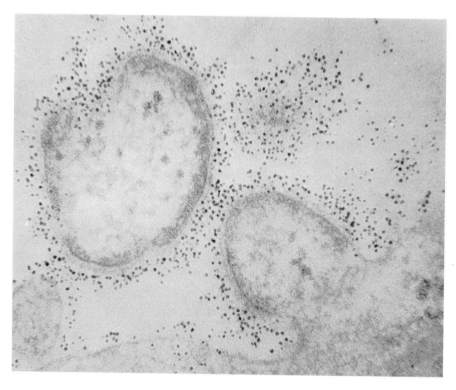

between the layers of the skin. In a positive reaction, T cells and activated macrophages respond to the antigen. Then, reddening and swelling, a delayed hypersensitivity reaction, develop around the site after 24 hours. If the individual does not have the disease or has not been immunized, there is no response. Skin tests with appropriate antigens can be used to diagnose any infection in which CMI development normally occurs. The Mantoux tuberculin test for tuberculosis is the most common example.

LYMPHOKINE ASSAYS. Increasingly, assays are being used to quantitate lymphokines produced by antigen-stimulated T cells. The lymphokine MIF (macrophage-inhibition factor) will inhibit migration of macrophages from the mouth of a capillary tube. Testing the ability of an individual's lymphocytes to produce MIF generally assesses their ability to produce CMI responses. Serum assays for interleukins, interferons, thymosins, and helper and suppressor factors may soon be useful in medical diagnosis.

SUMMARY

1. Specific immune defense is composed of antibody-mediated immunity (AMI) and cell-mediated immunity (CMI).

2. The immune responses need nonspecific defenses to complete the antigen's eradication. The two branches of the immune response work in cooperation. Antibody-mediated immunity develops when lymphocytes produce immunoglobulins that attach to antigen. Once antigen–antibody combination occurs, phagocytic cells and complement dispose of the antigen. AMI is useful against agents on epithelial surfaces, in body secretions, or in the circulation.

3. Cell-mediated immunity develops as other classes of lymphocyte attach to and recognize cellular antigens. These include host cells that harbor intracellular parasites, tumor cells, and noncompatible foreign-tissue grafts. Lymphokines produced by T cells activate macrophages, and cytotoxic lymphocytes kill cells on contact.

4. Lymphocytes are the primary functional cells of lymphoid tissue. They originate from ancestral stem cells and undergo differentiation during embryonic development. During that time, some pass through the thymus and become T cells, which carry out the CMI response and regulate the AMI response. Others undergo their development in the bone marrow. These become B cells, which produce the AMI response by giving rise to antibody-forming plasma cells. B cells also appear to regulate CMI responses. Following processing, both types of lymphocytes populate secondary lymphoid tissues in separate but adjacent areas. Lymphocytes migrate continuously from one lymphoid area to another. Both B and T cells are in a resting, nondividing state prior to antigenic stimulation.

5. Antigens are substances recognized by lymphocytes as foreign. Lymphocytes respond by rapidly dividing and differentiating into active cell clones. Antigens are large molecules containing antigenic determinants.

6. The immune response is genetically regulated. Genes determine our cellular antigens, including histocompatibility and blood group factors. Other genes produce immune regulator substances, and make immunoglobulins. This last type of gene is customized early in fetal development so that each B-lymphocyte precursor potentially makes a different antibody.

7. The human body distinguishes between "self" and "non-self." This distinction is "learned" by developing lymphocytes in the fetus. The clonal-selection theory states that, early in development, lymphoid tissues differentiate into a family of many genetically unique cells, each recognizing one antigenic determinant. During fetal life, contact of a lymphocyte with its antigen results in its dying or losing its responsiveness. All lymphocytes that could attack self do encounter their own antigen, giving permanent tolerance. After the lymphocytes have matured further, something quite different occurs. Antigen contact then exerts a positive selective influence, stimulating the lymphocyte to divide.

8. There are five classes of immunoglobulins produced by plasma cells. Each has both light and heavy chains; the identity of the heavy chain places an immunoglobulin (Ig) in its class. Each has two or more combining sites, variable amino acid sequences folded into three-dimensional grooves that hold antigen. Other portions of the Ig molecule attach the antigen–antibody complex to phagocytes, activate complement, or assist in transporting the immunoglobulin across the membrane of the secretory cell.

9. Antibody-mediated immunity develops when B cells are stimulated. B cells always have a small amount of their specific antibody incorporated as a membrane receptor. This allows them to attach and bind an antigen. The antigen–antibody complex enters the cell and promotes the production of plasma cells and memory cells. The stimulated B cells and their progeny form germinal follicles in the cortex of lymph nodes and in other lymphoid regions. Plasma cells release immunoglobulin into serum and other body fluids.

10. Cell-mediated immunity is induced when antigenic attachment to a specific T-cell receptor leads to mitosis and development of several populations of cells. T_D cells liberate lymphokines on contact with antigen. Lymphokines in turn summon and activate macrophages.

11. AMI and CMI are both induced by active immunization. AMI can be passively acquired by transfer of immunoglobulins. Passive immunity protects the fetus and newborn until its capacity for immune response develops.

12. Preventing infection depends on secretory antibodies at the body portals and circulating IgG and IgM.

13. Antibodies also control infections by extracellular parasites, and by viruses that have an extracellular phase between rounds of replication. CMI is primarily responsible for dealing with systemic fungi, and those bacteria, viruses, and parasites that multiply intracellularly. AMI and CMI must both be operating at full efficiency and in balance with each other to provide functional defense.

14. Immunodiagnosis is an important function of the clinical laboratory. Serologic tests may be structured to detect either antigen or antibody. A wide variety of qualitative and quantitative tests are available. Important considerations are speed, sensitivity, and convenience. Skin testing detects the delayed hypersensitivity side effects of CMI as evidence for the immune state. Other CMI tests detect the ability of lymphocytes to produce lymphokines.

DISCUSSION QUESTIONS

1. It has been shown that the eight-week-old, breast-fed human infant to whom oral, live-virus, polio vaccine is given often does not show an antibody response equal to that of the formula-fed infant. Why?

2. Administration of a killed-virus vaccine tends to result in development of antibody-mediated immunity, whereas an attenuated live-virus vaccine stimulates cell-mediated immunity. Why?

3. An individual genetically lacking in B-cell functions will be likely to suffer recurring infections by what type of bacteria?

4. Which portion of the immune response is compromised in the person lacking a normal complement pathway?

5. Explain how clonal selection is believed to produce both immune tolerance and immune responsiveness.

6. Compare the pathways of lymphocyte differentiation leading to the fully responding B and T lymphocytes.

7. Compare the functions of lymphokines, inflammatory mediators, and antibodies.

8. How does interferon function as a lymphokine?

ADDITIONAL READING

Albert BA, et al. *Molecular biology of the cell.* New York: Garland Press, 1983, Ch. 17.

Dausset J. The major histocompatibility complex in man. *Science* 1981; 213:1469–1474.

Leder P. The genetics of antibody diversity. *Sci Am* 1982; 246:102–115.

Mishell BB, Shiigi SM. *Selected methods in clinical immunology.* San Francisco: W.H. Freeman, 1980.

Myrvik QN, Weiser S. *Fundamentals of immunology.* 2nd ed. Philadelphia: Lea and Febiger, 1984.

Roitt I. *Essential immunology.* 4th ed. Oxford: Blackwell, 1981.

Rose NR, Friedman H, eds. *Manual of clinical immunology.* 2nd ed. Washington, D.C.: American Society for Microbiology, 1980.

Yalow RS. Radioimmunoassay: a probe for the fine structure of biological systems. *Science* 1978; 200:1236–1245.

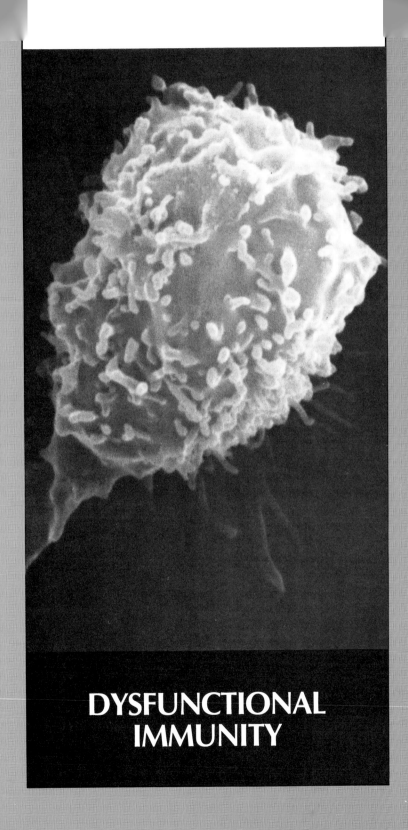

DYSFUNCTIONAL
IMMUNITY

CHAPTER OUTLINE

The immune system is a complex, powerful defensive weapon. Our survival hangs on the effectiveness of our immunity. Unfortunately, our defender may lose its potency, or may attack our own cells. In this chapter, we will discuss a number of situations in which immunity works against us, including immunodeficiencies and their effect on patients. We will survey the wide range of hypersensitivity reactions and their clinical signs. We will examine the role of immunity in transplantation of tissues, in the maternal attack against fetal antigens during Rh disease, and in the prognosis of cancer. The challenging puzzle of autoimmune disease will encourage us to think about the unsolved questions remaining in immunology.

IMMUNODEFICIENCY

Immunodeficiency is a state in which the body's ability to mount and retain an immune response is reduced or absent. Immunodeficiency may be either genetic in origin or acquired as a result of the physical stresses of life. Fortunately, most types of genetic immunodeficiency are rare. Some, however, affect a fairly large number of people. For example, about 1 in 500 individuals of European descent is genetically unable to produce secretory IgA.

The natural immunity of every person can be adversely affected by certain illnesses and high levels of stress. At these times we have reduced immunologic protection.

HOW DOES IMMUNODEFICIENCY AFFECT US?

Frequent or unusually severe infection is often a primary sign of immunodeficiency. Immunodeficient infants often have one infection crisis rapidly followed by another. In an immunodeficient patient, illnesses that are normally mild and localized may be protracted and generalized; the infective agents are frequently opportunists. Some characteristic infections in such patients include *Streptococcus pneumoniae* or *Pneumocystis carinii* pneumonia; severe, generalized herpes simplex skin lesions; and systemic *Candida albicans* infections. In the absence of functional immunity, antimicrobial drugs are much less effective. Immunodeficient children develop malignancies — often lymphoid tumors — at a rate 100 to 1000 times higher than do normal children. Adults on immunosuppressive drug regimens share a similar risk.

CONGENITAL IMMUNODEFICIENCY

Congenital or **primary immunodeficiency** is present at birth, and genetic in origin. The genetic defect causes a corresponding structural defect, and certain cell lines do not develop normally. A key lymphoid organ such as the thymus may fail to develop anatomically, lymphocytes may fail to differentiate fully and be unresponsive to antigenic stimuli, or there may be a specific metabolic defect hindering the cells' full functioning.

There are primary immunodeficiencies of B-cell function, T-cell function, or both combined. These states have often been referred to as "experiments of nature," because they provide valuable clues to help researchers understand the normal immune system. Some examples are **DiGeorge's syndrome,** a T-cell deficiency in which the thymus is rudimentary or absent; **Bruton's agammaglobulinemia,** a B-cell deficiency in which immunoglobulin levels are extremely low and lymph nodes lack germinal centers and plasma cells; and **severe combined immunodeficiency,** in which the lymphoid stem-cell line fails to differentiate at all and both B and T cells are therefore nonfunctional.

ACQUIRED IMMUNODEFICIENCY

An **acquired** or **secondary immunodeficiency** may arise in a previously normal individual following any of a number of stresses. Malnutrition, especially protein starvation, reduces serum protein levels including immunoglobulins. The stress of fighting an infection often depletes reserves of lymphocytes and granulocytes. Measles viruses have a cytotoxic effect on lymphoid cells. Syphilis, leprosy, and malaria depress cell-mediated immunity. Neoplastic diseases of lymphoid tissues (infectious mononucleosis, myelomas, leukemias, Hodgkin's disease) commonly cause immunodeficiency because lymphocytes or macrophages are no longer immunologically competent. Acquired immunodeficiency is a side effect of **immunosuppressive therapy,** which is widely used to treat autoimmune disease, to suppress graft rejection, to control inflammatory conditions, and to regulate fluid retention.

Acquired immunodeficiency syndrome (AIDS) is a recently recognized disease state marked by a progressive and irreversible loss of T-helper lymphocytes. It is accompanied by the development of a rare type of tumor and extensive and untreatable infections with multiple opportunistic viruses, fungi, bacteria, and protozoans. Current research implicates a retrovirus (human T-cell lymphotropic virus) as the primary cause. AIDS is discussed further in Chapter 22.

HYPERSENSITIVITY

Hypersensitivities are abnormal physiologic states resulting from immune reactions. They are actually reactions of antigen with immune effector at the wrong time, in the wrong place, or in the wrong proportions. The antigen binding causes inflammation and, in some cases, tissue destruction. The damage may be either transitory or chronic.

There are four categories of hypersensitivity reaction, based on the mechanism responsible for symptoms (Table 13.1). Types I, II, and III are called **immediate hypersensitivities** because symptoms appear within minutes after contact with the antigen. These reactions are all variant forms of antibody-mediated immunity. Immunoglobulins are the effectors of immediate hypersensitivities.

Type IV is **delayed hypersensitivity,** and it is cell-mediated. Symptoms appear 24 hours or more after contact with antigen. The symptoms are mediated by lymphokines from T lymphocytes.

There is no clear distinction between immunity and hypersensitivity. In fact, the mechanisms are much the same. For example, hypersensitivity, like immunity, develops only after one or more initial immunizing contacts with antigen, the process that results in **sensitization.** The term **allergen** is often used for an antigen

TABLE 13.1 The four types of hypersensitivity mechanisms

	Type I (Classic)	Type II (Cytotoxic)	Type III (Immune complex)	Type IV (Delayed)
Immune effectors	IgE	IgG, IgM	IgG, IgM	T lymphocytes
Accessory factors	Mast cells, anaphylactoid mediators, eosinophils, eosinophil mediators	Complement	Complement, inflammatory mediators, eosinophils, neutrophils	Lymphokines, macrophages
Antigen	Soluble or particulate	Cell-surface antigen	Soluble protein	Cell-surface antigen
Maximum reaction time for response in sensitized individual	30 min	Variable	3–8 hr	24–48 hr
Symptoms relieved by	Antihistamines (some), steroids (some)	Steroids	Steroids	Steroids
Clinical expressions	Anaphylaxis, asthma, hay fever, other allergy	Transfusion reaction, Rh disease, graft rejection, idiopathic thrombocytopenic purpura	Arthus reaction, serum sickness, graft rejection, rheumatoid symptoms	Granuloma, skin test, graft rejection, contact dermatitis

that elicits hypersensitivity. Over a period of time, the individual develops a hypersensitive immune response. Then, at some later time, another contact with the allergen occurs. This event is **provocative** — the immune effectors now present combine with the allergen, provoking the adverse results.

In most hypersensitivities, the allergen, although foreign, is not in itself a serious threat to health. Thus some people become hypersensitive to cat hair, pollen, poison ivy, or cosmetics (Table 13.2). People also vary widely in their individual tendency to become hypersensitive to these substances. Very simply, we call an immune response a hypersensitivity if it

is more harmful than helpful. If its protective effects outweigh its hazards, it is an immunity.

Whether a particular antigen results in immunity or hypersensitivity is not determined by the antigenic substance itself, but by the route of administration, the dosage, the timing, the individual's genetic makeup, and all sorts of unknown factors. Penicillin, for example, can be administered orally, by injection, or as a topical ointment. For about 85% of the population, contact (by any route) does not hypersensitize. However, about 15% may eventually have a reaction. Some persons develop IgE antibodies. These antibodies produce Type I reactions, possibly leading to fatal anaphylaxis if penicillin is later injec-

TABLE 13.2 Common allergens		
Ingested	**Inhaled**	**Injected**
Grains	Tree pollen	Bee venom
Nuts	Grass pollen	Wasp venom
Eggs	Weed pollen	Hornet venom
Fruits	Flower pollen	Yellow-jacket venom
Milk products	Bark dust	Penicillin
Animal protein	Mold spores	Animal insulin
Penicillin	House dust	ACTH
Hormone preparations	Insecticide	Heroin
Isoproterenol	Face powder	Cephalosporins
Aspirin	Pet dander	Other antibiotics
	Microbial spores	
	Cocaine	

ted, or to gastrointestinal upset if it is taken orally. In other individuals, IgG and IgM antibodies to penicillin develop. In these persons, penicillin injection might produce Type II or Type III hypersensitivity. Health personnel who handle penicillin preparations every day may develop contact dermatitis, a skin reaction characteristic of Type IV hypersensitivity.

IMMEDIATE HYPERSENSITIVITY

TYPE I HYPERSENSITIVITY. The effectors of Type I reactions are specific IgE antibodies. Immunologists believe that IgE plays a protective role in controlling helminth infestations. However, certain unlucky individuals produce excessive amounts of IgE, including IgE against harmless environmental antigens, thus setting the stage for Type I hypersensitivity (Figure 13.1).

IgE affects the highly granular circulating basophils and the mast cells found in connective tissue. Mast cells are especially concentrated beneath epithelial surfaces in the skin, respiratory tract, and gastrointestinal tract. The mast-cell granules contain potent inflammatory mediators. The mast-cell mediators' effects are believed to be opposed by specific "antidote" substances released by eosinophils. IgE antibodies bind strongly to the outer surface of basophils and mast cells, which is said to "prime" them.

When allergen touches the surface of an antibody-primed mast cell, it is bound by specific IgE molecules. Binding induces the mast cell to **degranulate** (Figure 13.2), releasing the mediators contained in its granules into the surrounding tissue or the blood stream. The mediators cause an immediate response. The effect does not last long, however, because the mediators are rapidly degraded in the liver or counteracted by antidote substances from eosinophils. Some of the mast-cell mediators include the inflammatory compound histamine; serotonin, a powerful blood-vessel constrictor; prostaglandins, which cause contraction of smooth muscle; factors that promote clotting; and a chemotactic substance for eosinophils.

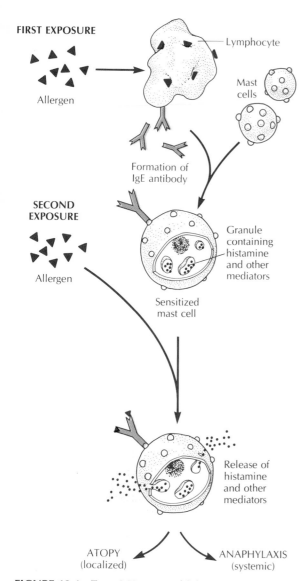

FIRST EXPOSURE

Allergen

Lymphocyte

Mast cells

Formation of IgE antibody

SECOND EXPOSURE

Allergen

Granule containing histamine and other mediators

Sensitized mast cell

Release of histamine and other mediators

ATOPY (localized)

ANAPHYLAXIS (systemic)

FIGURE 13.1   Type I Hypersensitivity
First exposure to an allergen stimulates lymphoid tissue to produce IgE antibody, which, instead of circulating freely, binds to the surface of most mast cells and basophils, priming them. Upon a second exposure to the allergen, the attached IgE molecules bind to antigen, and the binding induces the primed cell to degranulate. This process releases potent inflammatory mediators into the blood stream and surrounding tissue. A Type I hypersensitivity reaction follows, which may be either localized (atopy) or involve the entire body (anaphylaxis).

Anaphylaxis is the most intense Type I hypersensitivity reaction, involving the whole body. It usually happens when a provocative dose of antigen is injected. An insect sting (a natural hypodermic injection), or injected medication may provoke anaphylaxis. Antigen circulates rapidly to many mast-cell–containing areas, and a large mediator dose is suddenly released into body fluids. The first warning symptoms are apprehension, flushed skin, itching, and nausea. In human beings, arterioles (small arteries) dilate, causing a sharp drop in blood pressure. Simultaneously, a slow contraction of smooth (involuntary) muscle fibers occurs in the bronchioles (small air passages entering the lung) and elsewhere. The person becomes unable to breathe because of bronchial constriction. Asphyxiation and heart failure may follow. Rapid administration of epinephrine, a hormone that constricts the blood vessels and relaxes the smooth muscle, combined with artificial respiratory-support measures, will save most victims if begun soon enough. The survivor returns to normal within a short time, as the mediators are neutralized. In cases that prove fatal, penicillin injections and insect stings are the most common causes.

Atopic allergy is the localized form of a Type I reaction, and the difference between atopic allergy and anaphylaxis is one of degree. The common complaint "hay fever" is an example of an atopic allergy. The antigen, usually plant pollen, is inhaled onto the respiratory lining, where it is bound by IgE, causing degranulation of underlying mast cells. The results are excessive mucus production and localized epithelial swelling. In **asthma,** bronchial smooth-muscle spasm is added to the picture. Coughing, sneezing, and difficulty in breathing persist as long as antigen is present in the environment. Symptoms are often relieved by antihistamines. Skin and digestive atopies follow a similar mechanism. Skin testing, a process in which minute amounts of different allergens are injected into the skin, is frequently used to determine which substances are allergens for an individual (Figure 13.3). Atopic

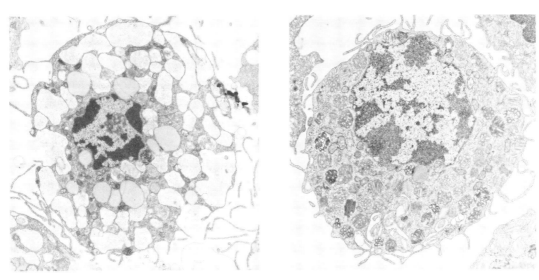

FIGURE 13.2 Mast-Cell Degranulation
(left) A normal mast cell from human lung tissue. The cytoplasm is filled with large electron-dense granules that contain histamine and other inflammatory substances. (×8050.) (right) A mast cell immediately after the allergen has been bound to its surface by specific IgE. The granules are releasing their contents into the environment, where they will cause the symptoms of a Type I hypersensitivity reaction. (×7350. Photos courtesy of John P. Caulfield and Ann Hein.)

FIGURE 13.3 Skin Testing for Allergy
Droplets of solutions containing minute amounts of different allergenic substances are injected into the skin of an allergy sufferer's forearm according to a standard pattern. If the patient is allergic to one particular antigen, a reddened area forms at the site of its injection. The responses are compared to a control spot where only sterile saline was injected. (From B.F. Feingold, *Introduction to Clinical Allergy.* Springfield, IL: Charles C. Thomas (1973), figs. 17-2 and 17-3.)

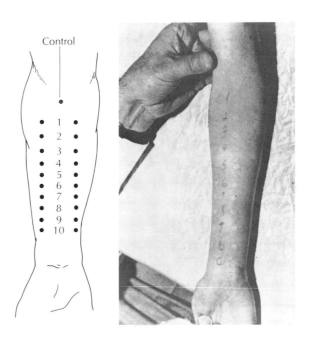

individuals can sometimes be helped by **desensitization,** a procedure in which the patient receives a series of small allergen injections. Successful desensitization is thought to depend on the development of blocking antibodies or suppressor T cells.

TYPE II HYPERSENSITIVITY.   Recall that a key aspect of protective immune responses is the mechanism in which antigen-labeled cells are destroyed by combination with specific IgG, followed by complement-mediated lysis or phagocytosis. Cytotoxic or Type II hypersensitivity is a variant of this mechanism. For example, the Type II response is the means by which incompatible red blood cells are lysed following a mismatched blood transfusion. The foreign red blood cells are coated by IgG antibodies and complement, and then they are lysed in vast numbers (Figure 13.4). The released hemoglobin and cell debris raise the blood viscosity and may

FIGURE 13.4   Type II Hypersensitivity
In a normal individual with type B blood, IgG antibody (which is anti-A) does not bind to the red blood cells, which circulate freely. However, when red blood cells from a mismatched donor (in this case, type A) are introduced into the recipient's blood stream, the IgG antibodies attach to the foreign cells, causing clumping to occur. Eventually the foreign type A cells lyse and release hemoglobin and other debris into the circulatory system, raising the viscosity of the blood to dangerous levels.

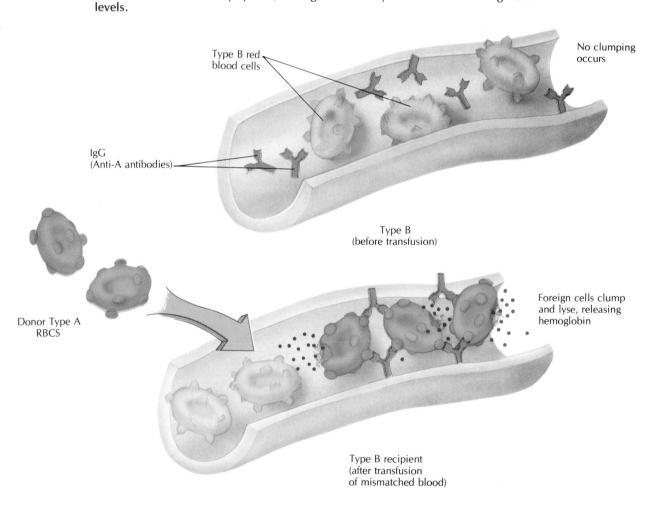

place fatal stress on the heart and kidneys. Rh disease of the newborn is another example of a Type II response in which maternal antibodies and complement severely damage the baby's red blood cells. A cytotoxic Type II response is also a major aspect of most episodes of graft rejection and some autoimmune diseases.

TYPE III HYPERSENSITIVITY. **Immune complexes** are circulating combinations of soluble antigen, antibody, and complement. Their formation is a normal outcome of protective immunity. Immune complexes are routinely scavenged by phagocytes, which then destroy the antigen. Type III hypersensitivity, also called **immune-complex disease,** occurs when more immune complexes form than the phagocytes can dispose of. The excess immune complexes then deposit in tissue, where they provoke localized inflammation and tissue damage. This situation often results from exposure to very large amounts of antigen. The antigen in an immune-complex disease is usually a protein, sometimes bearing a drug molecule. The antibody is IgG or IgM. The reaction subsides within a few days, although the lesions produced as a side effect of the inflammation may take much longer to heal.

One example of a Type III reaction is the **Arthus reaction.** This is a localized inflammation in which complexes have been deposited in small blood vessels, often causing tissue death at the site. Arthus reactions can occur in muscle tissue at the site of injection of excessive or too frequent doses of tetanus toxoid. In **farmer's lung disease,** another form of Arthus reaction is provoked when large doses of hay dust heavily loaded with microbial spores are inhaled.

Serum sickness is a Type III response that develops when serum products containing foreign protein are repeatedly administered to passively immunize an individual. The foreign proteins produce an antibody response. Both horse serum, once widely used but now rarely employed, and the human serum that has largely replaced horse serum may sensitize. About six to eight days after foreign serum proteins are injected, sufficient antiglobulin antibody is produced

to form immune complexes with the foreign globulin. The immune complexes infiltrate the walls of blood vessels. Signs of inflammation appear (fever, swollen lymph nodes, rashes, and painful joints). These signs are also typical of rheumatoid arthritis, an autoimmune disease in which Type III immune-complex formation with self antigens (not foreign serum proteins) is the underlying mechanism. An additional destructive potential of Type III disease is developed if immune complexes infiltrate the glomerular capillary bed of the kidneys, causing immune-complex **glomerulonephritis.** This condition may develop as a sequel to certain skin infections by *Streptococcus pyogenes.*

DELAYED HYPERSENSITIVITY

Recall that protective cell-mediated immunity (CMI) depends on the interaction of cell-bound antigens with T cells and activated macrophages. Type IV or delayed hypersensitivity is a destructive or misplaced form of cell-mediated response. It occurs when an allergen becomes bound to the surface of some of our own cells, which then become the target of a CMI attack. We saw in the last chapter that some degree of inflammation is an unavoidable side effect of CMI's protective role in bringing an infective agent under control. However, a delayed hypersensitivity response may also occur when your cell-mediated immune system mistakenly attempts to destroy cells to which comparatively harmless allergens such as poison ivy oil, heavy metals, or drugs such as penicillin may have bound.

Tissues to which an allergen is bound are infiltrated by T lymphocytes. These T cells release lymphokines that summon and activate macrophages. The allergen-bearing target cells are destroyed, and a **granuloma,** an area of inflamed tissue containing lymphocytes, macrophages, and giant cells, may form.

When an intracellular infective agent is eliminated by CMI, some tissue destruction is inevitable. In a persistent or widely distributed infection, however, this immune-mediated tissue destruction may be the primary source of distress, and may in fact do more harm than the pathogen itself. Although a measles skin rash is minor and quickly healed, a tubercular lung lesion, by contrast, may destroy much lung tissue and develop into a large cavity. In arrested cases, the cavity is walled off with calcium deposits and the damage is limited, although the healed lesion persists for life. If the person's CMI is not effective enough to arrest the disease, the lung destruction advances. This is not solely because of microbial invasiveness, but rather because of the continuing activity of the ineffective CMI. In fatal cases, this destruction by CMI progresses inexorably.

The **Mantoux tuberculin skin test** demonstrates local Type IV reactivity to tuberculosis antigen. In skin, the cellular changes that may be induced by introducing tuberculosis antigen are readily detected, but they are limited and quickly repaired. The antigen (a protein derivative of *Mycobacterium tuberculosis*) is injected intradermally. Within a few hours, if the person is hypersensitive to tuberculin, a reddened circle appears. Macrophages infiltrate antigen-containing tissue creating an **induration** or hardened area (Figure 13.5). The reaction peaks after 24 to 48 hours, then gradually fades. The diameter of the indurated zone is measured to indicate the reactive individual's degree of sensitivity. A positive test usually means the patient is presently harboring the tuberculosis organism or was once infected by the agent. Skin tests employing other appropriate antigens are also used to reveal hypersensitivity to the intracellular pathogens in such diseases as leprosy, histoplasmosis, and coccidioidomycosis.

Contact dermatitis is delayed hypersensitivity resulting from skin exposure to environmental antigens. A wide variety of materials may sensitize a person, usually after prolonged exposure. For example, people who work with industrial chemicals and metals such as nickel may become sensitized. Bakers may become sensitized to nutmeg or cinnamon, and photo-

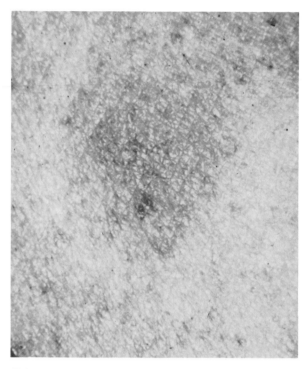

FIGURE 13.5   Mantoux Test
This highly accurate diagnostic test detects delayed skin hypersensitivity to antigens of the tuberculosis organism. A purified protein derivative (PPD) from *Mycobacterium tuberculosis* is injected intradermally. A positive reaction is the formation about 48 hours later of a large, reddened, hardened area at the site of the injection. The size of the zone increases with increasing hypersensitivity. (Courtesy of Dr. Leon J. Le Beau, Department of Pathology, University of Illinois at Chicago.)

graphers to developer chemicals. The most familiar of this unpleasant group of afflictions are poison ivy and poison oak (Figure 13.6). These plants contain in their leaves and stems oils that sensitize many people, often after several contacts. When the antigenic oil gets on the skin, it binds to epithelial cells. The area is invaded by immune T lymphocytes. The skin reddens, blisters, and breaks down. Clear serum is exuded from blisters and ulcers may form. The person suffers intense itching as a rule. Scratching is

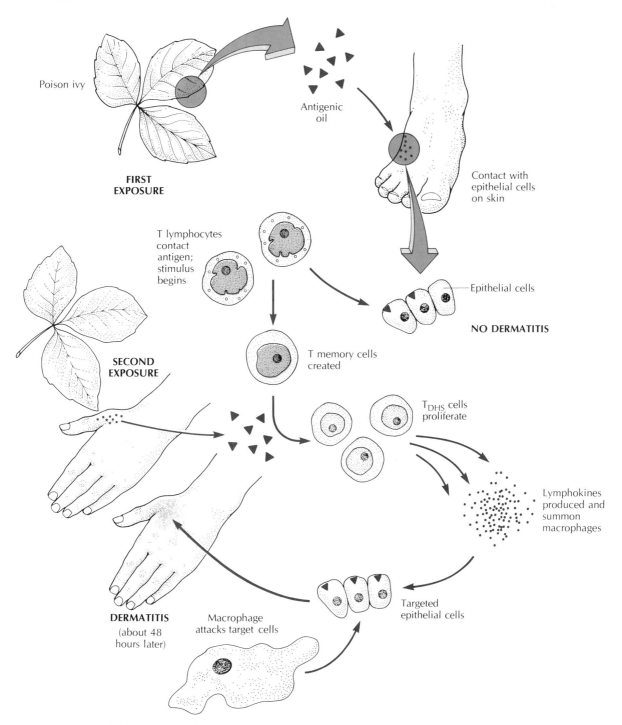

Poison ivy

Antigenic oil

FIRST EXPOSURE

Contact with epithelial cells on skin

T lymphocytes contact antigen; stimulus begins

Epithelial cells

NO DERMATITIS

SECOND EXPOSURE

T memory cells created

T_{DHS} cells proliferate

Lymphokines produced and summon macrophages

DERMATITIS (about 48 hours later)

Macrophage attacks target cells

Targeted epithelial cells

FIGURE 13.6 Contact Dermatitis
When the antigenic oil from poison ivy touches the skin, it binds to epithelial cells. If specific T lymphocytes circulating in the area come in contact with the antigen, they may be stimulated to produce T memory cells capable of initiating an immune response. Upon subsequent exposure to the same antigen, these T cells become activated; they proliferate and release lymphokines. The lymphokines cause a series of cell-mediated immune responses, including the attraction of macrophages, which attack the targeted epithelial cells. About 48 hours later, dermatitis results, with severe itching and blistering.

409

harmful not so much because it "spreads" the allergen, but because it damages more tissue, and this of course enhances the inflammation.

Anti–inflammatory drugs such as the corticosteroids reduce the symptoms of the Type IV delayed hypersensitivities. However, because steroid therapy is immunosuppressive, it is unwise to use it when the hypersensitivity reflects an active, helpful CMI response fighting an infection. Similarly, it is wise to minimize the use of these powerful drugs in the less serious contact hypersensitivities. Steroids are extremely helpful in moderating the cell-mediated aspects of graft rejection and autoimmune disease.

TRANSPLANTATION

The replacement of failing organs with healthy ones is a long-cherished medical goal. As surgical techniques advance, more graft procedures are tried. The major barrier to success has been graft rejection. Our bodies attack tissue bearing non-self antigens with the combined weapons of antibody-mediated and cell-mediated immunity. The search for methods for selectively suppressing this attack without leaving the transplant recipient totally unprotected continues.

BLOOD TRANSFUSIONS

Blood transfusions are a specialized form of tissue graft, since blood is a tissue. Why are blood "transplants" relatively simple while tissue transplants are so tricky? It is because red blood cells have a relatively small group of antigens compared to the other tissues. Red-blood-cell antigens include the ABO blood-group antigens, the Rh antigens, and a few others. These are the ones most commonly considered when transfusion compatibility is checked, and they are fairly easily matched in almost all cases. By comparison, the histocompatibility or HLA antigens are found on all other tissues; these are the factors that pose such difficulties in transplanting these tissues. Because red blood cells lack HLA antigens, one Type A person could give blood to another Type A, even if their tissue antigens were so poorly matched that a kidney donation, for example, would be impossible.

We produce **isoantibodies** from birth to any ABO antigens we do not possess. That is, a Type A person will have anti-B antibodies. Thus a mismatched transfusion will produce an almost instantaneous "graft rejection" by the Type II mechanism. In the Rh system, on the other hand, an Rh-negative person does not have anti-Rh antibodies until after contact with Rh-antigen–bearing, or Rh-positive, red blood cells. Thus a sensitization step must occur before mismatched blood will cause a reaction.

TISSUE TRANSPLANTATION

There are several categories of tissue grafts, based on the genetic relatedness of donor and recipient. An **autograft** is tissue moved from one area of an individual's body to another. Skin grafts may be obtained from an undamaged area of the body to replace burned tissue; bone transplants from the tibia are used for facial reconstruction. An **isograft** is tissue transferred from one identical twin to another. Because of the twins' identical genetic makeup, isografts are immunologically compatible. **Allografts** are tissues transferred between unrelated members of a species. Most kidney, heart, and liver transplants are allografts. Success depends largely on the degree of HLA genetic similarity between donor and recipient. Family members are often suitable donors, because they have at least some genetic relatedness. **Heterografts** are transfers of tissue between members of different species. Pigskin may be used to form a temporary covering over burn wounds until autografting can be done.

TISSUE MATCHING. In testing for histocompatibility, the HLA antigens of proposed donor or

BOX 13.1
Are You Ready for a Brain Transplant?

Jokes about brain transplants and home brain surgery are so old they have whiskers. Most people assume that the one thing you couldn't possibly be able to do, or want to do, would be a brain graft. In fact, it is surprisingly easy to graft brain tissue; it's been done experimentally in rats for a decade. The brain is a privileged site because the blood–brain barrier keeps immune cells out of the brain tissue. This makes the brain more tolerant than almost any other part of the body to grafts of foreign tissue.

No, the goal of brain grafts is *not* to transfer knowledge! We can, however, transplant segments of types of brain tissue whose role is to secrete certain chemicals that are essential to the function of other parts of the brain. This is especially effective in regard to improving the function of the brain centers that regulate such skills as balance, muscle coordination, and memory. For example, aged rats slowly lose their sense of balance. They regain their balancing skills after they receive grafts of that portion of another rat's brain that produces a chemical called **dopamine.** Dopamine is

the brain chemical that is insufficient in humans with Parkinson's disease, a condition marked by muscle tremors, weakness, and an unsteady gait. Two human Parkinson's patients in Sweden have received autografts to the brain of their own dopamine-producing adrenal gland tissue. Their symptoms became much more manageable afterward, although they were not entirely cured.

Certainly the most exciting aspect of brain-graft research is that which focuses on Alzheimer's disease, a devastating form of senile memory loss that afflicts millions of aging people. Researchers can cause an experimental Alzheimer's-like condition of total memory loss in rats. They surgically damage the area of a rat's brain that supplies a chemical called **acetylcholine** to the hippocampus, the part of the brain that plays the key role in forming memories of events as they happen. Afterward, researchers graft a segment of acetylcholine-producing rat brain tissue into the damaged area, and the forgetful rats regain up to 80% of their original memory capacities. Some hope this approach might turn out to be feasible for treating human Alzheimer's disease. Should this distant possibility come to reality, you might ask where we would get the donor brain tissue. Fortunately, it is highly possible that we will be able to grow the needed cells in cell culture by then.

No, I don't know where you can get a "microbiology final exam" graft. Don't even ask.

recipient cells can be identified by exposing them to specific antibodies. If the HLA antigens partly or completely match by this method, then other tests may be done to verify compatibility or indicate the potential severity of rejection. There are regional and national organ banks, and registries of potential donors and recipients from which immunologists try to find suitable matches.

It is rare indeed that a randomly selected donor tissue perfectly matches the recipient; the chance is less than 1 in 30,000. Tissue matching aims for the best match, but rarely achieves a perfect match. Mismatched tissue can be implanted without difficulty in **privileged sites** — sites, such as the cornea of the eye, in which the grafted tissue has no contact with lymphoid tissue and thus will not stimulate an immune rejection.

REJECTION MECHANISMS.   When implanted tissue is rejected, a combination of immune mechanisms is at work. **Acute rejection** develops several weeks after the transplant. The patient first becomes sensitized. Then T lymphocytes migrate to and accumulate in the graft, releasing lymphokines and promoting macrophage activity. Gradual destruction of graft tissue by CMI occurs. There may also be some antibody-mediated cytotoxicity. Either the graft will be irreversibly devitalized and slough off, or the response will subside so the tissue can survive (Figure 13.7). There may be repeated episodes of acute rejection during the life of a graft.

FIGURE 13.7   Skin Grafts
(a) The success of a graft depends largely on establishing and maintaining blood circulation between the graft and the graft bed. A successful graft rapidly acquires a blood flow from the graft bed into the grafted tissue. (b) A rejection response occurs if immunologic effects block circulation. The grafted tissue appears pale due to lack of blood flow. (c) Eventually, it will die and slough off.

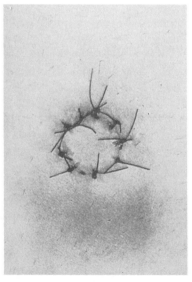

(a)

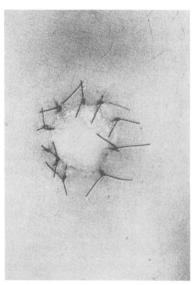

(b)

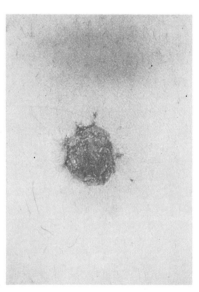

(c)

Chronic rejection results from progressive deposition of immune complexes in the arteries that feed the graft. Eventually the blood supply fails. Even though a graft has survived acute rejection it may continue to be threatened by chronic rejection. The recipient seldom develops true tolerance for the foreign antigens. Instead, the body keeps trying to get rid of the implanted material.

MANAGING REJECTION WITH IMMUNOSUPPRESSION. Immunosuppressive treatment is usually essential for graft survival. A number of different drugs are used to increase the chances of a successful graft. **Cyclosporin** is a newly discovered fungal peptide that is revolutionizing transplant medicine. It gives very reliable control of allograft rejection, with much less nonspecific immunosuppression than previous drugs. Some studies see evidence that it even seems to induce permanent tolerance in some cases. Cyclosporin has made liver transplants possible, and has increased the success rate of allograft kidney transplants from 50% to 90%. It does not appear to be powerful enough, however, to suppress the rejection of heterografts, such as baboon hearts, in humans. **Antilymphocyte globulin** (ALG) is serum containing antibodies against human T lymphocytes. ALG temporarily reduces the number and effectiveness of recipient lymphocytes. Steroids, especially **prednisone,** suppress rejection mechanisms based on cell-mediated immunity. **Antimitotic agents** halt the growth of all rapidly dividing cells, thus slowing down the development of immunity, and prolonging the period before rejection develops. An antimitotic drug called azathioprine is used in some kidney transplants.

THE MATERNAL–FETAL INTERACTION

When a mother is carrying a fetus, she is always, in a sense, the recipient of a partly unmatched "graft." This is because half the baby's HLA antigens are inherited from his or her father, not from the mother. For nine months, she normally maintains the graft without any evidence of a developing rejection process. This occurs despite the fact that maternal lymphocytes make direct contact with fetal cells, in the fetal layer of the placenta (Figure 13.8a). Perhaps the fact that fetal lymphocytes also circulate in the mother's body, crossing the placenta during pregnancy, may partly explain this special case of immune tolerance.

Erythroblastosis fetalis, or Rh disease of the newborn, is an increasingly rare exception to the rule that fetuses are immunologically tolerated. This condition could be viewed as a type of "graft rejection." It occurs when an Rh-negative mother makes anti-Rh IgG antibodies. These pass, along with other beneficial antibodies, through the placenta into the fetus. The maternal antibodies damage and destroy an Rh-positive baby's red blood cells. The infant is either stillborn or extremely ill at birth.

Cases such as these occur because the mother has earlier become sensitized to the Rh antigen. In a few instances, she may have had a mismatched Rh-positive blood transfusion. Much more often, sensitization will have resulted from an earlier pregnancy, commencing immediately after the birth of an Rh-positive child. During the birth process, some of a baby's red blood cells often escape into the mother's circulation (Figure 13.8b). If the baby's cells are Rh positive, they may cause her to become sensitized to the Rh factor. If, in a succeeding pregnancy, she again carries an infant with Rh-positive red blood cells, the maternal anti-Rh IgG will attack the baby's red blood cells. The severity of the infant's disease will depend on how early the IgG begins to destroy its red cells. Erythroblastosis fetalis is now prevented by giving Rh-negative mothers RhoGam, a product containing anti-Rh antibodies, soon after delivery. These antibodies cause any Rh-positive baby's cells in the mother's circulation to be destroyed before they have a chance to sensitize her.

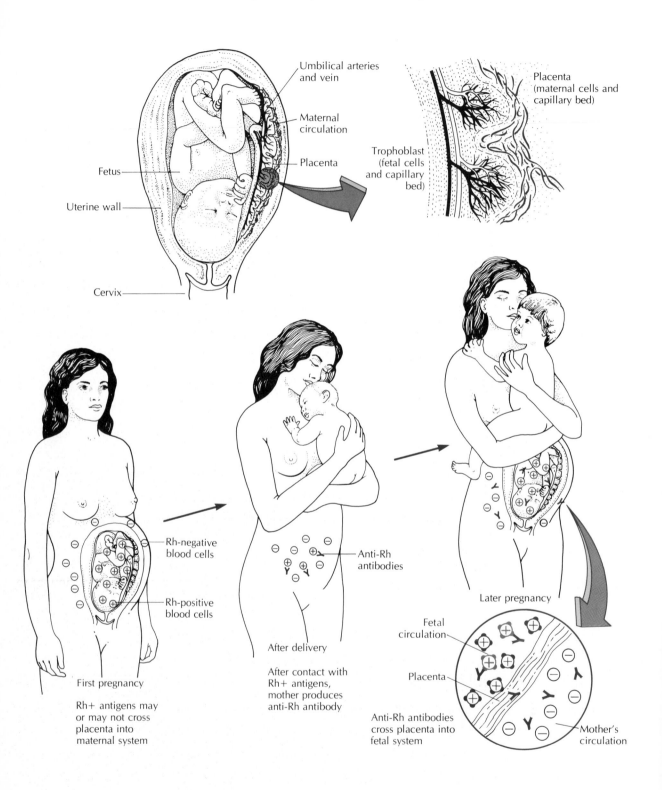

Umbilical arteries and vein

Maternal circulation

Placenta

Fetus

Uterine wall

Cervix

Placenta (maternal cells and capillary bed)

Trophoblast (fetal cells and capillary bed)

Rh-negative blood cells

Rh-positive blood cells

First pregnancy

Rh+ antigens may or may not cross placenta into maternal system

Anti-Rh antibodies

After delivery

After contact with Rh+ antigens, mother produces anti-Rh antibody

Later pregnancy

Fetal circulation

Placenta

Anti-Rh antibodies cross placenta into fetal system

Mother's circulation

FIGURE 13.8 The Fetus *in Utero*
(a) The placenta consists of both fetal trophoblast and layers of maternal cells, in intimate association. Lymphocytes from mother and fetus circulate through extensive capillary networks with minimal separation. Yet neither individual develops effective immunity against the other's mismatched antigens during a normal pregnancy. (b) During the first pregnancy of an unsensitized Rh-negative mother carrying an Rh-positive fetus, sensitization does not occur. During and after delivery, however, a few of the baby's red blood cells may be left behind in the mother's blood. The mother is sensitized by this new antigen and begins making anti-Rh antibody. Should she later carry another Rh-positive fetus, her anti-Rh antibodies can enter its circulatory system and destroy its red blood cells.

TUMOR IMMUNOLOGY

The role of immunity in cancer is a complex, and not wholly beneficial one. We are still far from understanding entirely how it operates.

TUMOR ANTIGENS

A naturally arising tumor is basically self, in that it shares genes for the histocompatibility antigens of the normal tissue from which it arose. However, many tumors do not express their HLA antigens, a trait that in at least one case has been linked to an oncogene in the tumor cells.

At the same time, during transformation, it becomes to some extent non-self, in that it expresses new tumor antigens. A tumor is an antigenic mosaic. In other words, it gives the immune system mixed signals. Furthermore, metastases of a single tumor may diversify and bear a wide range of antigens. Both antibody- and cell-mediated immunity respond to tumor antigens, but the response may be ambiguous in its effect. It frequently happens that these tumor-associated antigens do not stimulate a strong enough immune response to produce rejection of the tumor. This may in part be due to the fact that T cells must be stimulated by both a foreign antigen and an HLA antigen in order to respond.

In cases where the HLA antigens are suppressed, response will not occur.

Let's consider some of the antigens found on tumor tissue. **Oncofetal antigens** are substances that are normal and characteristic on fetal tissues, but are repressed during maturation and not normally found on adult cells. They are sometimes reactivated on tumor cells, however. **Alpha-fetoprotein** (AFP) is an antigen produced in hepatic (liver) cancers and **carcinoembryonic antigen**(CEA) appears in intestinal-tract cancers. Detection of these antigens is helpful in monitoring tumor growth after surgery or chemotherapy, and may potentially be a tool of early diagnosis.

It is generally held that antigens, such as these, alert some body mechanism which continually detects and eradicates newly arising malignant cells, or developing small clones. As long as this surveillance mechanism is alert, there is no problem. An error or a temporary lapse, however, permits the occasional tumor clone to escape and grow too large to be dealt with. Originally, immunologists proposed a form of immune surveillance directed against tumors. Although this concept still has its active supporters, most evidence shows that the immune system is really not particularly good at eliminating tumor cells. The most generally accepted idea at present holds that there is indeed primary surveillance, but that it is not based entirely on immune mechanisms. The natural killer (NK) lymphocytes, which do not appear to have specific immune functions, are currently thought to be the key components of our antitumor surveillance network (Figure 13.9).

Why do tumors still arise in the face of all our defenses, and why is immunity relatively ineffective against tumors? There are several possibilities. We know that a tumor is established in one cell or a small initial population, and stays microscopic for days or weeks. It is possible that during this time its tumor antigens may not make contact with the immune system. Alternatively, it may induce immune tolerance. The tumor, then, by one means or another,

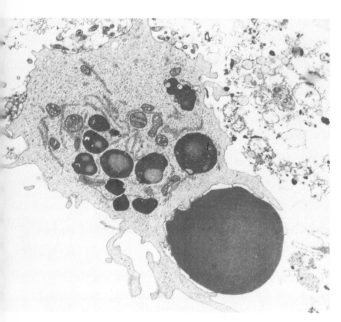

FIGURE 13.9 Natural Killer Cells
Natural killer (NK) cells are a type of nonspecific lymphocyte. They are considered to be a major factor in surveillance for newly arising tumor cells. In this transmission electron micrograph, an NK cell is seen engulfing a cell prior to destroying it. (Courtesy of Dr. Eckhard R. Podack, Scripps Clinic and Research Foundation. See E.R. Podack and G. Dennert, Cell mediated cytolysis, *Nature* (1983).)

sneaks through the defensive network. Once established, most malignancies, especially lymphoid types, are actively immunosuppressive. The activity of T-suppressor cells may rise relative to that of T-helper cells; blocking antibodies may appear. When radiation and chemotherapy are employed to treat the malignancy, both have the side effect of suppressing immune responses.

IMMUNOTHERAPY FOR TUMORS

With appropriate direction, the immune system might become more useful in curing cancer than it seems to be naturally. This idea is supported by promising recent advances in immunotherapy.

In cases where a malignancy is caused by a known virus, and transmitted by an infectious process, we are able to vaccinate to prevent the infection. Marek's disease in chickens (see Chapter 10) is completely controllable by the vaccine strategy. An Epstein–Barr virus vaccine is under development, which might reduce Burkitt's lymphoma. The hepatitis B vaccines being used now to prevent serum hepatitis may, as an additional benefit, prove to reduce some types of hepatitis B-associated liver cancers, although it is too soon to tell. The vaccination strategy is most promising for a few of the DNA oncoviruses. Unfortunately, RNA-containing retroviruses often are not transmitted by the sort of infectious process that vaccines can prevent. Additionally, it appears that viruses directly cause only a small percentage of our total cancer incidence.

CMI appears to be the body's primary way of destroying tumor tissue. Depressed CMI activity is a bad sign in a tumor patient. Tumor growth coincides with decreasing CMI; tumor regression, with increasing CMI. Substances that have an irritant or inflammatory effect when injected into tissue may stimulate localized broad-spectrum CMI. Bacillus Calmette–Guérin vaccine (BCG, a killed suspension of tuberculosis organisms) has been injected into or around melanomas or breast tumors, causing areas of delayed hypersensitivity and often promoting regression. Since the side effects are extensive, this type of treatment is limited to surface or localized tumors.

Monoclonal antibodies, described in Box 12.1, promise to be powerful new weapons in cancer immunotherapy. They are valuable because of their ability to seek out and bind specifically to antigenic tumor cells (Figure 13.10). One strategy for their use is as follows. A biopsy of tumor tissue is obtained from a patient and the tumor cells are used as a source of tumor antigens. Mice are immunized with these antigens, and their spleen B cells used to form a hybridoma-cell line. The hybridoma cells are grown and produce monoclonal antibodies against the tumor antigen, which are then harvested and purified. The antibodies by themselves would not do the tumor much harm. However, the antibody molecules can be chem-

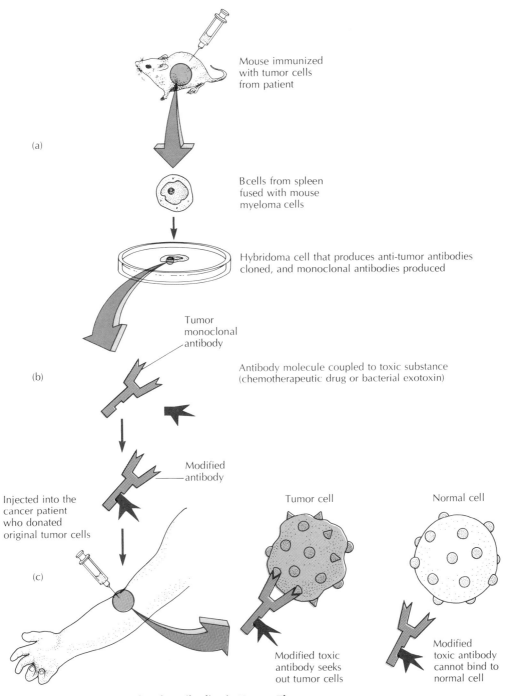

(a)

Mouse immunized
with tumor cells
from patient

B cells from spleen
fused with mouse
myeloma cells

Hybridoma cell that produces anti-tumor antibodies
cloned, and monoclonal antibodies produced

Tumor
monoclonal
antibody

(b)

Antibody molecule coupled to toxic substance
(chemotherapeutic drug or bacterial exotoxin)

Modified
antibody

Injected into the
cancer patient
who donated
original tumor cells

Tumor cell

Normal cell

(c)

Modified toxic
antibody seeks
out tumor cells

Modified
toxic antibody
cannot bind to
normal cell

FIGURE 13.10 Monoclonal Antibodies in Tumor Therapy
Specific monoclonal antibodies are receiving clinical testing in new approaches to
tumor therapy. (a) In a typical procedure, tumor cells, obtained by biopsy, immunize
a mouse against the tumor antigens. The mouse is killed and its spleen B cells used
to produce hybridoma cells. Hybridoma cells that produce highly specific anti-tumor
antibodies are selected and cloned, and their antibodies are purified. (b) Next, the
antibodies are chemically coupled to a bacterial exotoxin or a chemotherapeutic
drug with the potential to kill human cells. (c) The modified antitumor antibody is
now ready for administration to the patient from whom the original biopsy was
obtained. The antibodies bind and deliver their lethal drug specifically to the tumor,
while normal cells (to which the antibody cannot bind) are unharmed.

417

ically coupled to toxic compounds, such as chemotherapeutic drugs or bacterial exotoxins. When these modified monoclonal antibodies are then injected into the patient, they bind specifically to the tumor and deliver the toxic dose directly to the surface of the tumor cell. Normal tissue is almost entirely spared the toxic effects of the bound drug or toxin. Because they circulate, the monoclonals theoretically are also able to seek out and kill tiny unsuspected metastases. Because of the selective antibody binding, the effect of monoclonal-antibody therapy on normal tissue is very much less than that of conventional chemotherapy.

AUTOIMMUNE DISEASE

Chronic, progressive disease often involves a slow degeneration of one or more types of tissue. The actual causes of degenerative diseases can be extremely obscure. Some degenerative diseases are caused by an underlying immune mechanism; the tissue destruction results from hypersensitivity. The immunity is directed against self tissue; thus it is called **autoimmunity.** Some characteristic autoimmune diseases are shown in Table 13.3.

AUTOANTIBODIES

The clonal selection process occurring during prenatal development appears to ensure that people become permanently tolerant to all self antigens. It should therefore be impossible for autoimmunity to develop, yet it does. At least 50% of persons over 70 years of age have circulating antibodies against one or more body components, clearly suggesting that tolerance is not permanent. Antibodies against the components of the cell nucleus (antinuclear antibodies) and against the tissues of the thyroid gland are commonly encountered examples. Having autoantibodies against self components is not necessarily harmful. Only when the autoantibodies actually attack tissue does **autoimmune disease** result. Autoimmune disease develops much more commonly in females than males, for reasons that remain unclear. It is much more common in the elderly, and rare in young people,

TABLE 13.3 Some characteristic autoimmune diseases		
Disease	**Autoantibodies to**	**Clinical Picture**
Autoimmune hemolytic anemia	Red-blood-cell antigens	Hemolysis; anemia
Idiopathic thrombocytopenic purpura	Platelets	Hemorrhages in skin and mucous membranes
Systemic lupus erythematosus	DNA, nuclear antigens	Facial rash, lesions of blood vessels, heart, and kidneys
Rheumatoid arthritis	Immunoglobulin	Inflammation and deterioration of joints and connective tissue
Pernicious anemia	Intrinsic factor	Absorption of vitamin B_{12} from intestine prevented, severe anemia
Myasthenia gravis	Acetylcholine receptor on muscle	Progressive neuromuscular weakness
Addison's disease	Adrenal gland	Weakness, skin pigmentation, weight loss, electrolyte imbalance
Juvenile diabetes (some)	Islet tissue (pancreas)	Severe early diabetes

unless they have a genetic predisposition, as may occur in some diseases such as rheumatoid arthritis.

SOME EXAMPLES OF AUTOIMMUNE DISEASES

In **autoimmune hemolytic anemia,** the person produces antibodies against his or her own red blood cells, which are either lysed or opsonized. In some cases, the hemolysis is aggravated when the person gets chilled, because certain of the antibodies bind only at cooler temperatures. The Coombs test is the diagnostic test used to detect anti–red-cell antibodies.

Idiopathic thrombocytopenic purpura is a platelet deficiency. Antiplatelet IgG is produced, and it destroys the platelets as rapidly as they are formed. Clotting is markedly abnormal. Purpura (tiny hemorrhages from capillaries in the skin and tissue) occur. In advanced stages, hemorrhages from larger vessels become life-threatening. Splenectomy, or removal of the spleen, is often helpful, as the spleen is the site where many of the body's plasma cells are produced.

Systemic lupus erythematosus (SLE) is a condition marked by such symptoms as a facial rash, fever, intestinal complications, and progressive kidney failure. In SLE, antibodies against many different self antigens can be demonstrated. Fluorescent-antibody tests may be used to detect antinuclear antibodies against native and denatured DNA, histones, and nucleoprotein complexes. In an older procedure, the **LE test,** blood from a patient with SLE is drawn and allowed to stand for an hour. If the serum contains IgG antinuclear antibody, during the waiting period this antibody causes lymphocytes to swell and their nuclei to extrude. The nuclei are then ingested by phagocytic polymorphonuclear leukocytes (PMNs). Stained blood smears prepared after the waiting period is completed show "LE cells," which are the PMNs stuffed with lymphocyte nuclei they have ingested. A newer test for SLE employs a rapid-slide method (Figure 13.11). SLE may eventually

Control

Positive Test

FIGURE 13.11 SLE Test
In the disease systemic lupus erythematosus, one diagnostic sign is the formation of antinuclear antibodies. This rapid-slide screening test uses latex beads coated with nuclear antigens (top). When mixed with a patient serum sample containing antinuclear antibodies, clumping occurs (bottom). (Courtesy of Fisher Diagnostics, Fisher Scientific Company.)

be fatal. It kills through an immune-complex attack on the kidney leading to renal or cardiac failure.

Rheumatoid arthritis (RA) is an inflammatory disease, a progressive joint degeneration marked by thickening of the soft synovial tissues

BOX 13.2
Why Many Hypotheses Can Be Valuable

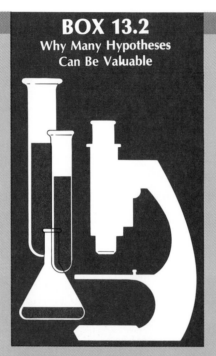

How can people become immune to themselves? Why does autoimmunity occur so frequently, when it "should not" happen at all? Let us pose five possible ways of thinking about this paradox.

(1) Autoimmunity could arise through the proliferation of a clone of lymphocytes that for some reason survived eradication during the fetal period. (2) If some self antigen, not "exposed" during the fetal period, suddenly became exposed, its responder clones would not have been eradicated and could still react. (3) Perhaps a suppressor signal fails, so that inactivated clones become active again. (4) Exposure to an environmental non-self antigen could produce antibodies that accidentally cross-react with a self antigen. (5) The body might react to a viral envelope containing a mixture of viral and host-cell antigens, and mistakenly mount an immune response against the self components of the mixture.

Investigators can become very ingenious in proposing hypotheses to explain things they do not understand. These five examples make a point about the value of hypothesizing as a scientific activity. Most theories presented to you, in this and other textbooks, seem rather self-evident. This is because they have been around for a long time and are amply bolstered by evidence. Remember, though, that when Pasteur presented his germ theory of disease, when Watson and Crick presented their hypothesis of DNA structure, or when F.M. Burnet presented his clonal selection theory, each hypothesis was just one among many theories. All were intensely questioned and even strongly resisted by some highly respected scientists. In these three cases, the competing theories have now been eliminated by experimental activity, so that those of Pasteur, Watson and Crick, and Burnet are accepted as accurate explanations of the facts.

Each of the autoimmunity hypotheses mentioned above is now being experimentally challenged. A few years from now, there may be just one generally accepted idea left. The successful hypothesis may be one of the above, a combination of them, or something completely new. In the meantime, each hypothesis will have stimulated worthwhile experiments. Some of the greatest breakthroughs in scientific research have come from attempting to verify a hypothesis that turns out to be false!

and erosion of the bony surfaces. The patient may develop an enlarged spleen, a depressed white-blood-cell count, and cardiac, pulmonary, or renal complications. The serum of individuals with rheumatoid arthritis contains **rheumatoid factor**, an autoantibody that reacts with normal or partly denatured IgG. Thus, in this disease, an immunoglobulin is the antigen. The destructive inflammatory changes in joints and organs are due to aggregates of immune complexes. Many rheumatologists feel that RA is provoked by some external cause such as an infection, but no agent has yet been demonstrated. There is a strong indication that genetic factors are important in predisposing an individual to rheumatoid arthritis.

TREATING AUTOIMMUNE DISEASE

Immunosuppressive therapy is widely used in autoimmune disease. It relieves the symptoms but does not cure. Surgical removal of the affected tissue is often not practical. Replacing a damaged kidney with another human kidney is futile; the new kidney will surely be destroyed by the same mechanism that damaged the original organs. However, synthetic replacement joints, which are not antigenic, are extremely helpful in ensuring that patients with rheumatoid arthritis can lead productive lives. However, progress in the area of autoimmune disease is slow, as we do not yet know how to turn off the basic destructive process.

SUMMARY

1. The immune system can cause diseases by the same mechanisms it uses to protect us. These mechanisms differ in the nature of some effectors, the massiveness of the response, and the target.

2. Immunodeficiency is primary if a genetic defect is the underlying cause. It is acquired if it results from stress to the host. The primary immunodeficiencies are rare failures of B-cell or T-cell differentiation. Acquired immunodeficiencies result from malnutrition, infection, or tumors. They can be induced by radiation, steroids, and cancer chemotherapy.

3. Hypersensitivity is excessive or misplaced response to an antigen. Types I, II, and III hypersensitivities are mediated by immunoglobulin. The responses are immediate and subside quickly when the antigen is removed.

4. Type I hypersensitivities occur when antigen combines with IgE-labeled mast cells. The mast cells release active mediators. These cause atopic allergy when locally released in small amounts or cause anaphylaxis when systemically released in larger amounts.

5. Type II reactions are cytotoxic; they destroy cells. IgG or IgM antibody and complement attach to antigenic cells and destroy them, as occurs in mismatched transfusions.

6. Type III immune-complex diseases result when antigen is excessive and soluble antigen–antibody complement complexes form in larger numbers than can be disposed of by phagocytes. Inflammation occurs near the site of antigen entry (Arthus reaction) or where the complexes lodge in lymphoid filters and kidneys (serum sickness, glomerulonephritis).

7. Type IV hypersensitivity is cell-mediated, and the time of onset, after antigen contact, is delayed. The allergen becomes cell-bound in the body. Manifestations are granuloma formation, induration, contact dermatitis, and cell-mediated destruction in autoimmunity and graft rejection.

8. Successful tissue transplantation from donor to recipient requires a close match between the histocompatibility antigens of the paired tissues. There are several tests to evaluate the acceptability of donor tissue to recipient lymphocytes. Imperfectly matched tissue is always a target for immunologic rejection, which may be partly controlled by immunosuppressive therapy. Fetal tissue, always unmatched, is not rejected by the mother. The Rh-negative mother may, however, produce antibodies harmful to her baby's Rh-positive red blood cells.

9. Naturally arising tumors sometimes evade destruction. Immunity is relatively ineffective in antitumor defense. Conventional radiation and chemotherapeutic treatments for malignancy are immunosuppressive, as is tumor growth itself. Monoclonal antibodies are being used in new forms of immunotherapy.

10. Autoimmune disease occurs when the individual responds to self antigens, and destroys target tissues or organs. The process can be controlled to some extent by immunosuppression. Autoimmunity is paradoxical in view of the clonal selection process. Many hypotheses have been put forth to explain autoimmune disease. They are guideposts for those searching for cures for the immunopathologic diseases.

DISCUSSION QUESTIONS

1. Compare the primary and secondary types of immunodeficiency. What approaches are possible to help people with each type?

2. Can you add some ideas of your own to the list of hypotheses for the origin of autoimmunity?

3. Explain the relation between delayed hypersensitivity and cell-mediated immunity.

4. Immunosuppressive therapy produces acquired immunodeficiency. Prepare a list of states in which immunosuppressive therapy is advisable. By reference to a pharmacology handbook such as the *Physicians' Desk Reference*, prepare a list of conditions in which the immunosuppressive drug prednisone is used. What is the relationship between your two lists?

5. What steps could be taken to prevent instances of anaphylaxis?

6. Ten years ago, any person who received a serious flesh wound was given an injection of tetanus toxoid if more than six months had elapsed since the last tetanus immunization. Now the injection is not recommended unless five years or more have passed since the previous immunization. What do you think the reason is for this change?

7. It has been observed that cancer patients are usually immunodeficient. One might say that they are immunodeficient because they have cancer, or that they have cancer because they are immunodeficient. What is the evidence for either position?

ADDITIONAL READING

Buisseret PD. Allergy. *Sci Am* 1982; 247:88–97.

Collier RJ, Kaplan DA. Immunotoxins. *Sci Am* 1981; 251:56–64.

Herberman RB, Ortaldo JR. Natural killer cells: their role in defenses against disease. *Science* 1981; 214:24–30.

Roitt I. *Essential immunology.* 4th ed. Oxford: Blackwell, 1981.

Rose NR. Autoimmune diseases. *Sci Am* 1981; 246:80–103.

Rose NR, Friedman H. *Manual of clinical immunology.* 2nd ed. Washington, D.C.: American Society for Microbiology, 1980.

14

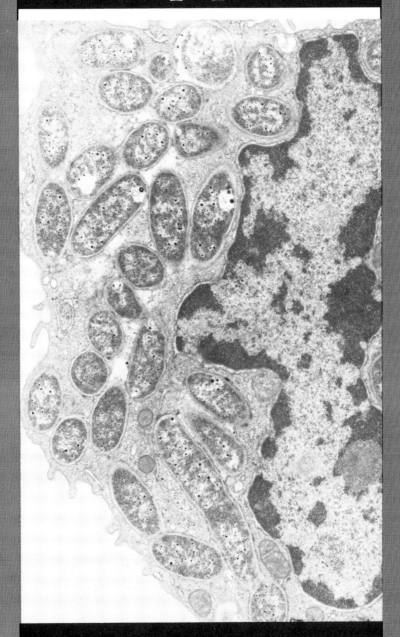

DISEASE AND
POPULATIONS

CHAPTER OUTLINE

So far, we have focussed on the relationship between one individual host and one specific parasite — a single case of infectious disease. Now we will turn to the factors that influence the frequency and spread of infectious disease within host populations. We will define epidemiology and public health and outline their major functions, which are crucial in controlling and eradicating communicable diseases. We'll become familiar with the steps in the chain of transmission of infectious agents, and we will see how disease control strategies are aimed at interrupting this chain. To conclude, we will summarize the major programs in current public health areas.

EPIDEMIOLOGY

Epidemiology is the quantitative study of the occurrence of disease in populations. The methods of the discipline were initially developed during investigations of communicable-disease outbreaks. However, epidemiologic methods quickly developed into very powerful research tools, and they were soon adapted for investiga-

ting noncommunicable diseases as well. Today's epidemiologist may be equally concerned with outbreaks of bacterial food poisoning (a communicable condition) and lead poisoning (a noncommunicable environmental disease). He or she may also search for factors contributing to diseases such as diabetes or stroke, for which the underlying cause is not known.

The epidemiologist's perspective on disease differs from that of other health-care professionals. Whereas a doctor or nurse is directly involved with individuals, an epidemiologist is concerned with groups of people. Hospital personnel see ill people and carry out therapeutic activities. Epidemiologists rarely render personal health services. They are concerned with prevention; they aim to minimize new cases of disease and, wherever possible, to eliminate a specific disease from the population.

Epidemiologists work closely with clinical practitioners, however. Doctors provide epidemiologists with essential data by reporting cases of disease as they are encountered. The epidemiologist, in turn, communicates his or her findings, alerting practitioners and the public of specific disease risks, effective therapeutic strategies, and useful preventive measures to assist in personal and public decisions about health issues.

QUESTIONS AN EPIDEMIOLOGIST CAN ANSWER

In most communicable diseases, scientists already have discovered the causative agent, using the classic methods developed by Robert Koch. These methods, described in Chapter 1, form the basis of direct approaches to control or cure. An epidemiologist analyzing an infectious disease may evaluate several therapies to determine which is most effective, compare the best ways to use different control measures, do follow-up studies to find out how well a proposed strategy has worked, and do the epidemiologic detective work to trace the origins of new epidemics and bring them under control. Examples of epidemiologic questions in infectious disease are

- Which antibiotic gives the most reliable treatment for typhoid fever, based on the outcome of many treated cases?
- In a sizable clinical trial, is a new vaccine actually effective in preventing measles? At how early an age can it be given effectively?
- What are the agents, reservoirs, and modes of transmission of a "new" disease like AIDS?
- Which is the most cost-effective means of controlling mosquito-borne diseases such as malaria — swamp drainage or aerial spraying?

In noncommunicable diseases such as arteriosclerosis and lung cancer, epidemiologists are likely to identify several contributing factors, rather than a single clear cause. Some epidemiologic questions in noninfectious diseases are

- What effect will different types of exercise programs and diet plans have on the risk of heart attack?
- What effect will the addition of beta-carotene to the diet have on reducing the development of cancer of the intestine?

- What types of public education are apt to be most effective in persuading people to undertake life-style changes such as stopping smoking?
- In terms of years of life saved, are we better off investing our limited number of cancer-research dollars in screening for industrial carcinogens or in doing research on dietary factors in cancer?

An **association** is a statistical correlation between two events. For example, cigarette smoking has a strong association with an increased risk of lung cancer. But people who have breathed sparkling clean air all their lives and have never been near a cigarette or another smoker occasionally develop the disease. So it is a matter of interpretation whether or not cigarette smoke is the **causative factor;** at the very least, it is one of several **predisposing factors** that augment the effect of the (unknown) causative agent. In arteriosclerosis, a minimum of 15 predisposing factors have been identified. Many of these factors are related to a person's life style — dietary pattern, exercise habits, emotional behavior — which makes analysis very complicated. In order to study arteriosclerosis and other noncommunicable diseases, it is necessary to study groups of factors over long periods. One such long-term research project is the Framingham study of heart disease, which has been following the health of the same group of people for more than 20 years.

Epidemiologists may also be detectives. This is the role in which they most often attract public attention. When a disease outbreak develops, epidemiologists are responsible for identifying the cause, finding where the agent is coming from, providing the key information to assist in bringing it under control, and organizing health workers and other segments of the community in a cooperative effort to break the chain of transmission. This process may be long and difficult, as it was in the case of legionnaires' disease (Box 6.1). Within hospitals, epidemiologic studies of hospital-acquired (nosocomial) disease are often carried out by an infection-control team with coordination among the hospital services. All data on infectious diseases, as well as other required information, are entered in hospital records.

PUBLIC HEALTH

Public health is day-to-day practical disease control. Some examples of public health measures include restaurant regulations and inspection, vaccination programs, and drinking water analysis. These measures are based on sound epidemiologic principles. The scope of a public health problem may be as small as a single contaminated drinking-water well or as large as an international outbreak of influenza. Public health measures are functions of local, state, federal, and international agencies, which coordinate the collection, analysis, and dissemination of epidemiologic information, design control programs, and enforce public health regulations. Individual cities frequently maintain Bureaus of Health to enforce local health regulations, provide some public health–care services, and monitor any local problems that develop. Each state has a department of health, with associated laboratories and staff epidemiologists.

The United States Public Health Service requests and collects data from cities, states, and territories and runs the Centers for Disease Control (CDC) in Atlanta, Georgia. The role of the CDC is extremely important, and has attracted public attention in recent years in regard to its investigations of such "new" diseases as legionnaire's disease, toxic-shock syndrome, and AIDS. The CDC publishes the *Morbidity and Mortality Weekly Report* (MMWR), a summary of current epidemiologic investigations. Several reports from MMWR provided the material for boxes in the next few chapters. The CDC also operates educational programs, research projects, and reference laboratories to assist local and state laboratories.

Diseases do not respect national boundaries. The international surveillance and control activities of the World Health Organization (WHO) have become increasingly important. With the availability of high-speed transportation, as well as the continuing factors of war and massive dislocations of populations, the risk of global disease outbreaks is magnified. WHO, with the cooperating national governments, has carried out successful multinational vaccination campaigns against yellow fever and cholera. In one of the great achievements of 20th-century medicine, smallpox has been eradicated (Figure 14.1). The WHO information network gives advance warning of the approach of new influenza strains so that appropriate precautionary measures can be taken.

Public health is often a matter of public policy, and the responsibility for it falls to policymakers, legislators, and ourselves as the voting public. Public health agencies provide the factual health information and practical recommendations that must be available for rational decision making. The public health agencies, and legislation, structure and enforce water and air safety regulations, food purity standards, restaurant food handling, quality-control criteria for pharmaceuticals, sewage treatment and discharge, and animal health. Many of these topics are covered in Chapters 23 through 28.

You now have a good general idea of the scope of epidemiology and public health. For the remainder of this chapter we will concern ourselves primarily with the epidemiology of communicable diseases.

HOW COMMUNICABLE DISEASES ARE TRANSMITTED

Each new case of a communicable disease is the end result of a chain of events, the **chain of transmission**. The case is also potentially the first link in a subsequent chain that will produce yet more cases. If one link in a chain can be broken, communicable-disease transmission is arrested.

SOME DEFINITIONS

Let us acquire some basic vocabulary used in epidemiology. The **prevalence** of a disease is the

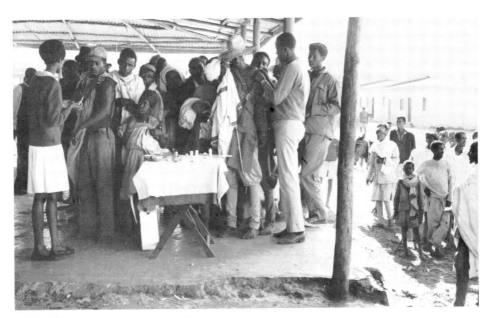

FIGURE 14.1 Smallpox Vaccination
Vaccination is one of our most powerful weapons for controlling epidemic diseases. On market day in a small Ethiopian village, a World Health Organization team administered smallpox vaccine to people of all ages, and gathered information about possible new cases. Since 1977, no new cases of smallpox have been reported, and smallpox is now considered to be eradicated. (Photo courtesy of the World Health Organization.)

actual proportion of the population that is infected at any given time. It is a "snapshot" of the health status of the population. **Endemic** diseases are those that at any time are infecting a relatively constant, small percentage of a population, sometimes called the "background" level. **Sporadic** diseases are those in which there are small, localized, unpredictable outbreaks. Typhoid fever in the United States follows this pattern. An **outbreak** or **epidemic** occurs when the number of cases rises significantly above the expected background level. Although the term outbreak means much the same thing, the word "epidemic" has much more frightening connotations, and epidemiologists do not use the term unless there is a major threat to the health of a great many people. **Pandemics** are epidemics that cross national boundaries, threatening large areas of the world. Influenza outbreaks are al-

ways pandemic in nature, often moving from the Eastern to the Western Hemisphere. A new virus strain arising in the U.S.S.R. in 1977, for instance, caused epidemics throughout Western Europe and North America within a few weeks. The great historic pandemics of cholera, smallpox, plague, and measles periodically decimated the populations of Europe.

The **incidence** of a disease is the number of new cases arising in an area in a set period of time, such as the number of cases of typhoid fever in a state per year. The incidence of infectious diseases has changed markedly over the last century. The diseases that still have a high incidence, such as gonorrhea, hepatitis, and chickenpox, are ones for which our existing control methods are only partly adequate. The now rare diseases, such as polio, rabies, and diphtheria, are ones that we effectively control.

THE CHAIN OF TRANSMISSION

The chain of transmission may be analyzed in five components. They include (1) the reservoir of the infective agent in which it multiplies and from which it passes; (2) the route by which the agent leaves the reservoir; (3) the means of transmission of the agent to the potential new case; (4) the route of entrance into the person's body; and (5) the susceptible new host (Figure 14.2).

LIVING RESERVOIRS.   A **reservoir of infection** is the site where the agent multiplies and survives between infections. The most important reservoir for human infectious diseases is other human beings, because most of our disease agents are adapted to multiply in human bodies. The smallpox and measles viruses have no other significant reservoir. Living reservoirs used by other agents include animals and arthropods. A reservoir is said to be **infective** when actively releasing viable pathogens.

A human being will often serve as a reservoir when acutely ill, but may be equally infective as a **carrier,** an infected person with no obvious disease signs. Healthy carriers are those with inapparent infections. Incubatory carriers are persons who are already shedding significant numbers of the agent before their symptoms appear. Convalescent carriers are those who no longer have symptoms but are still shedding live pathogens.

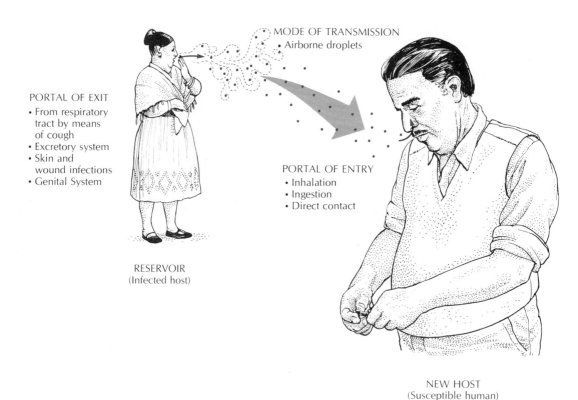

MODE OF TRANSMISSION
• Airborne droplets

PORTAL OF EXIT
• From respiratory tract by means of cough
• Excretory system
• Skin and wound infections
• Genital System

PORTAL OF ENTRY
• Inhalation
• Ingestion
• Direct contact

RESERVOIR
(Infected host)

NEW HOST
(Susceptible human)

FIGURE 14.2   Transmission of Infectious Disease
Infectious diseases are transmitted through a chain of events. In this example, an individual acting as a human reservoir, coughs droplets of respiratory secretions containing the agent, into the air. Air currents transmit the infectious droplets to the vicinity of another person, who unknowingly inhales them. If the second individual should be susceptible to the agent in question, a new active case of the disease may result.

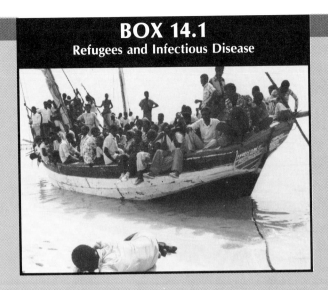

BOX 14.1
Refugees and Infectious Disease

Human history, from the Exodus of the Jews from Egypt to today, has been marked by recurring wars of conquest followed by the expulsion of entire populations from their homelands. Some of these events, like the Ethiopian famine and resettlement, the expulsion of the Indochinese "boat people" from Vietnam and the Hmong people from Cambodia, are well publicized. Others, such as recurrent local wars in distant places, receive little international attention. When and where will the next wave of displaced persons arrive? No one knows. But it is certain that there will be a long succession of these episodes of human misery, and that infectious-disease epidemics will add to the refugees' sufferings.

Epidemics almost always occur, fanned by the mal-nutrition and poor or non-existent housing and sanitation endured by the refugees on the road and in refugee camps. Often there are outbreaks of cholera and dysentery, but it is not always possible to predict what the dominant infectious challenge may be.

The hazards of flight take their worst toll on the old and the young. When Kampuchean (Cambodian) refugees fled to safety in Thailand during 1979 and 1980, at the rate of several thousand, per day, there were few survivors under 5 or over 55 years old — they hadn't survived the protracted and arduous flight to the Thai border. For those refugees who made it to safety, the dominant health problems to be surmounted were malnutrition, malaria, and upper respiratory disease.

During the same period, refugees of Chinese extraction expelled from Vietnam were being carried on the prevailing ocean currents to Hong Kong or Malaysia. In the refugee camps, these refugees required treatment for measles, tuberculosis, hepatitis B, and intestinal parasites.

Many of these refugees have resettled in the United States. Public health agencies closely monitor their health to watch for any infectious diseases they may have brought with them, and to help them become aware of the entirely different set of health risks — arteriosclerosis, obesity, substance abuse — they face in the United States.

(Photo courtesy of Miami Herald/Black Star.)

Infection may be contracted from inside or outside the body. An infection contracted by acquiring agents from an outside reservoir or source is an **exogenous** infection. However, normal flora may also serve as the reservoir of an opportunistic agent, which may, with reduced resistance, be "transmitted" to another area of the body and establish an **endogenous** infection, one in which the agent came from within.

ANIMAL RESERVOIRS.   **Zoonoses** are diseases of animals that are transmissible to human beings. In a zoonosis, the disease is typically much more prevalent in the animal population than in human beings, who are accidental hosts. For example, the major reservoir of the plague organism *Yersinia pestis* is wild rodents (Figure 14.3), while reservoirs for the equine encephalitis virus are wild birds and possibly snakes.

FIGURE 14.3   Transmission of Plague
The plague bacillus *Yersinia pestis* is found in a variety of small wild rodents in the Western and Southwestern United States. It is almost always transmitted to humans by flea bites. Hunters, trappers, and campers are at risk of infection by this route. (Courtesy of Susan C. Straley, Paula A. Harmon, and Lawrence Melsen, University of Kentucky at Lexington.)

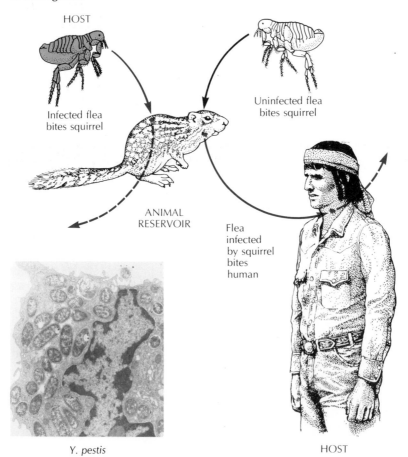

HOST

Infected flea bites squirrel

Uninfected flea bites squirrel

ANIMAL RESERVOIR

Flea infected by squirrel bites human

Y. pestis

HOST

Arthropods, including insects and arachnids, play a dual role in disease transmission. They act as reservoirs when an infective agent reproduces within them, to yield an infective stage that may then be communicated to humans. For example, as we saw in Chapter 5, the malaria parasite completes the sexual stages of its life cycle in the mosquito and the asexual stages in human beings (review Figure 5.17). Ticks, mites and fleas (Figure 14.4) are reservoirs of the rickettsial diseases. The second key role of arthropods in the chain of transmission is their ability to function as a **vector** or means of actual physical trans-

mission, as discussed shortly. Major groups of vectors are shown in Table 14.1.

NONLIVING RESERVOIRS.   Soil is a primary reservoir for some infectious bacterial agents and many pathogenic fungi. It is primary because the agents live and reproduce in the soil as their natural habitat. Many produce copious spores that are infective if inhaled. Soil is a secondary reservoir when it is fertilized or otherwise contaminated with infective agents from human or animal wastes or remains, a common practice in less-developed countries. These agents, for

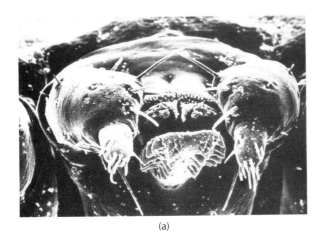

(a)

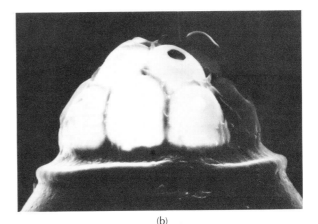

(b)

(c)

FIGURE 14.4 Arthropod Vectors
These scanning electron micrographs show the mouth parts of some common arthropod vectors: (a) The tick is notorious for transmitting rickettsial diseases such as Rocky Mountain spotted fever. (×80. Courtesy of Catherine Ellis, Science Photo Library/Photo Researchers, Inc.) (b) The louse transmits typhus fever. (Courtesy of Presbyterian University of Pennsylvania Medical Center/Photo Researchers, Inc.) (c) The mosquito is a carrier of malaria and yellow fever, as well as viral encephalitis. (Courtesy of David Scharf/Peter Arnold, Inc.)

Class	Group	Genera	Disease	Agent
TABLE 14.1 The major groups of vector insects and arachnids				
Arachnida (eight-legged)	Ticks	*Americanum* *Dermacentor*	Q fever Rocky Mountain spotted fever	*Coxiella burnetii* *Rickettsia rickettsii*
	Mites	*Haemaphysalis* *Leptotrombidium*	Tularemia Scrub typhus	*Francisella tularensis* *Rickettsia tsutsugamushi*
Insecta (wingless, six-legged)	Lice	*Pediculus*	Epidemic typhus Cholera Impetigo Trachoma	*Rickettsia prowazekii* *Vibrio cholerae* *Streptococcus pyogenes* *Chlamydia trachomatis*
Insecta (winged, six-legged)	True bugs	*Panstrongylus* *Triatoma*	American trypanosomiasis	*Trypanosoma cruzi*
	Fleas	*Xenopsylla*	Plague	*Yersinia pestis*
	Mosquitos	*Anopheles* *Culex* *Aedes* *Haemogogus*	Malaria Encephalitis Dengue Yellow fever	*Plasmodium* sp. Arbovirus Dengue agent Arbovirus
	Flies	*Chrysops* *Phlebotomus* *Glossina*	Tularemia Leishmaniasis African trypanosomiasis	*Francisella tularensis* *Leishmania donovani* *Trypanosoma brucei*
	Cockroaches	*Blatella* *Periplanata*	Amebic dysentery Hepatitis A	*Entamoeba histolytica* Type A virus

which soil is not a natural habitat, usually do not multiply, and their survival time may be short. However, while they remain viable, the soil can be a source of infection.

Water is rarely a primary reservoir, as relatively few human pathogens are naturally adapted to a water habitat. Two exceptions are the hospital opportunists *Pseudomonas aeruginosa* and *Flavobacterium meningosepticum*, which multiply well in water. However, water is frequently a secondary source. Essentially all surface waters (ponds, lakes, and rivers) in the United States receive some sewage, variously treated, and are potentially infective. Fecal organisms may survive for long periods in fresh water; in salt water they die off much faster. Water collectors in household humidifiers, vaporizers, and air conditioners, as well as in large institutional or industrial climate-control devices, frequently become colonized with microorganisms. These agents may then be dispersed into the humidified air in a cloud of tiny droplets called an **aerosol.**

Droplets of body fluids such as respiratory or rectal mucus may become airborne. Once expelled into the environment they evaporate, leaving any organisms inside the droplet protected by a dried mucoprotein film. The residual

TABLE 14.2 Airborne diseases			
Source of Airborne Contaminant	**Disease**	**Agent**	**Classification**
Spores from soil	Anthrax	*Bacillus anthracis*	Bacterium
	Aspergillosis	*Aspergillus* sp.	Fungus
	Blastomycosis	*Blastomyces dermatitidis*	Fungus
	Coccidioidomycosis	*Coccidioides immitis*	Fungus
	Cryptococcosis	*Cryptococcus neoformans*	Fungus
	Histoplasmosis	*Histoplasma capsulatum*	Fungus
Aerosols, droplets, or droplet nuclei	Chickenpox	Varicella agent	Virus
	Common cold	Many agents	Virus
	Diphtheria	*Corynebacterium diphtheriae*	Bacterium
	Influenza	Influenza agent	Virus
	Legionellosis	*Legionella pneumophila*	Bacterium
	Measles	Measles agent	Virus
	Meningitis	*Neisseria meningitidis*	Bacterium
	Mumps	Mumps agent	Virus
	Rubella	Rubella agent	Virus
	Smallpox	Variola agent	Virus
	Tuberculosis	*Mycobacterium tuberculosis*	Bacterium
	Whooping cough	*Bordetella pertussis*	Bacterium

droplet nuclei, very much lighter than the original droplets, may remain airborne and drift almost indefinitely. Droplet nuclei may be highly infective. Some airborne diseases are shown in Table 14.2.

Thanks to our public health regulations, much of the fresh, raw food available in the United States is inherently safe. A food is a primary reservoir only when the animal from which the meat or milk derived was infected with a zoonotic disease. Pork, bear, or walrus meat may be a source of trichinosis, and unpasteurized dairy products a source of tuberculosis, brucellosis, and Q fever. Raw poultry and eggs often come to market contaminated with *Salmonella* derived from the intestinal contents of the birds.

Unfortunately, safe food can be readily turned into dangerous food by mishandling. Clean foods become secondary reservoirs when they are inoculated with pathogens by dirty hands, dirty utensils, or contaminated water. Prepared foods are sometimes allowed to stand for periods of hours at incubation temperatures that permit bacterial multiplication. Food microbiology is one subject of Chapter 28.

EXIT FROM THE RESERVOIR. An infective agent usually leaves a living reservoir in body fluids. Saliva, tears, respiratory secretions, exudates, blood, urine, feces, genital secretions, and skin flakes or crusts may provide means of exit. Infections are not transmitted if no exit is available. The viable tuberculosis bacilli in a healed lung lesion, for example, are physically trapped within the cavity. As long as the lesion remains enclosed, the person is not infective. A knowledge of the means of exit of a disease agent is important because it tells us what body fluids may be hazardous, and forms the basis for hospital precautions to prevent the spread of disease.

MEANS OF TRANSMISSION.   Transmission may be either direct or indirect (Figure 14.5). A disease is horizontally transmitted when it is

FIGURE 14.5   Means of Transmission
Transmission of disease may take several routes: Vertical transmission occurs when disease is passed between generations, as during fetal development. Horizontal transmission occurs when disease spreads from one person to another within a group. The pathogenic agent may be passed by direct contact, by ingesting contaminated food or liquids, and by wounds made by infected insects or animals. Having invaded an organism through a portal of entry (such as the nose, mouth, ears, or skin), the disease usually follows a predictable pattern, beginning with a period of incubation and ending when the host's natural immune system — sometimes aided by outside treatment — brings the microorganisms under control.

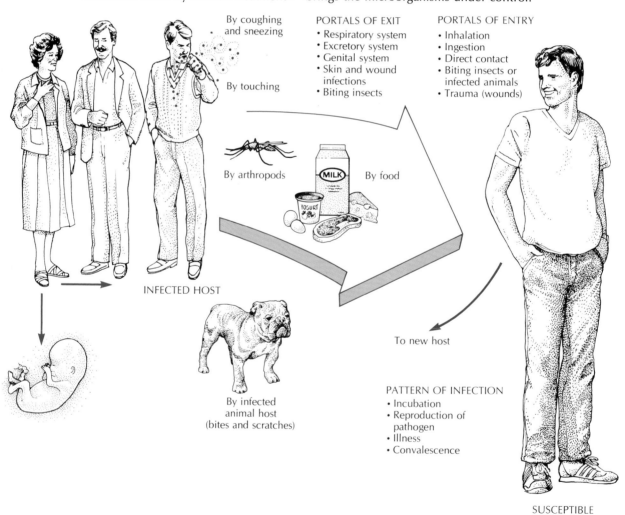

By coughing and sneezing

By touching

PORTALS OF EXIT
• Respiratory system
• Excretory system
• Genital system
• Skin and wound infections
• Biting insects

PORTALS OF ENTRY
• Inhalation
• Ingestion
• Direct contact
• Biting insects or infected animals
• Trauma (wounds)

By arthropods   By food

INFECTED HOST

By infected animal host (bites and scratches)

To new host

PATTERN OF INFECTION
• Incubation
• Reproduction of pathogen
• Illness
• Convalescence

SUSCEPTIBLE HOST

spread among the members of a group from person to person. It is vertically transmitted when it is passed between generations, from parent to child during reproduction (with the egg or sperm cell), during fetal development, or during the birth process. For some diseases, such as hepatitis, there are several means by which virus may move from one human to the next.

Direct transmission or **direct contact** occurs when the susceptible person touches, ingests, or is bitten by the reservoir. The pathogenic agent is transmitted directly, and not exposed to a hostile environment, which might inactivate it, for any significant time.

The sexually transmitted diseases (Table 14.3) are a special case of direct transmission. They may be spread both horizontally and vertically (no pun intended). Syphilis, in which the agent is extremely sensitive to cooling, is spread horizontally, almost exclusively by direct contact of epithelial surfaces during sexual activity. The spirochete may also pass across the placenta to the fetus. Herpesvirus may be transmitted horizontally through direct contact or vertically to the fetus as it comes in contact with infected cervical surfaces during childbirth.

Trichinosis is another example of a disease spread by direct means, in this case the ingestion of inadequately cooked animal flesh containing the parasite's cysts. Tuberculosis is (now rarely) spread by drinking milk from diseased cows.

Insects and arachnids are an important means of direct transmission in the infectious disease cycle; when they transmit a disease they are called **vectors.** If the insect or arachnid is the reservoir and the means of transmission it is called a **biological vector. Arthropod-borne diseases** are those transmitted primarily or exclu-

TABLE 14.3 Sexually transmitted diseases		
Disease	**Agent**	**Classification**
Candidiasis*	*Candida albicans*	Fungus
Chancroid	*Hemophilus ducreyi*	Bacterium
Gonorrhea	*Neisseria gonorrhoeae*	Bacterium
Granuloma inguinale	*Calymmatobacterium granulomatis*	Bacterium
Hepatitis B*	Hepatitis B agent	Virus
Inclusion conjunctivitis	*Chlamydia trachomatis*	Bacterium
Genital herpes	Herpes simplex Type 2	Virus
Genital warts	Papilloma viruses	Virus
Lymphogranuloma venereum	*Chlamydia trachomatis*	Bacterium
Nongonococcal urethritis	*Chlamydia trachomatis* (and other agents)	Bacterium
Pediculosis (pubic lice)*	*Phthisus pubis*	Insect
Scabies*	*Sarcoptes scabiei*	Insect
Syphilis	*Treponema pallidum*	Bacterium
Trichomoniasis*	*Trichomonas vaginalis*	Protozoan

*Also transmitted by nonsexual means.

sively by insects or arachnids. The arthropod's bite injects saliva or regurgitated blood, stored from a previous blood meal, into the person bitten. If these fluids contain infectious agents, direct transmission occurs. A specific arthropod-borne disease can usually be transmitted by only one or a very small number of arthropod species. The geographical distribution of the disease coincides with the habitat of the vectors. When we can control the insect population, we can often reduce or eradicate the disease it transmits.

Indirect transmission occurs when the susceptible person contacts food, drink, clothing, or inanimate objects previously contaminated by contact with the reservoir. An indefinite period of time may elapse between the time the **fomite,** or object bearing the agent, was originally contaminated and the moment at which the susceptible individual touches, wears, or eats it. During that time the agent is not multiplying and must be able to survive exposure to an alien environment.

For example, you can acquire a cold indirectly by picking up a tissue used by someone who has the infection, thereby transferring the agent to your hands, and then touching your mouth. Additionally, some insects have a role in indirect transmission as **mechanical vectors.** The housefly and the cockroach walk on garbage or feces, contaminate their feet, then walk on uncovered food and inoculate it with the organisms.

The **fecal–oral route** of transmission of gastrointestinal diseases is probably the most significant form of indirect transmission. It requires ingestion of fecally contaminated material. Fecal–oral transmission might occur if you ate an apple that had been washed in fecally contaminated water and had a gastrointestinal agent on its skin, or ate raw shellfish harvested from sewage-contaminated shellfish beds.

Understanding direct and indirect means of transmission is useful because it sensitizes us to the many ways a microscopic infective agent may move or be moved through a population. It is sobering to reflect that if you were the susceptible host in some of the examples given above, you would see and feel nothing to warn you of your risk. Covering food, washing your hands every time after blowing your nose or going to the bathroom, even if you think your fingers remained clean, and cooking your pork thoroughly even though you know the farmer who raised it, are plain common sense.

PORTALS OF ENTRY.   Unavoidable normal activities such as breathing, eating, and socializing constantly admit organisms to our bodies. Portals of entry include the nose, mouth, eye, ear, skin, anus, urethra, and vagina. New entrance routes may be opened by insect bites, trauma, wounds, or surgery. Tetanus, gangrene, and other wound infections follow the creation of an abnormal portal of entry. The ability to take advantage of a normal or abnormal entrance route, you will recall, depends on the infecting organism's virulence. (Chapter 11)

POPULATION SUSCEPTIBILITY.   To complete the chain of transmission, an infective agent must find a new susceptible host. An infective agent may have a different impact on disparate groups because of variations in susceptibility. You will recall that malnutrition, underlying diseases, or emotional stress predispose individual people toward disease. When such conditions are widespread in a population, the group too becomes much more susceptible. The physical hardships of inadequate or overcrowded housing, the stresses of displacement and loss experienced by victims in war-torn areas, and the acute challenges of natural calamities such as earthquakes or hurricanes invariably amplify infectious disease. Under these conditions, endemic disease flares into epidemic outbreaks.

Even in peaceful times, population susceptibility to many infectious diseases shows regular seasonal variations in incidence. These may be due either to physical factors (cold temperatures, a rainy season that encourages the breeding of mosquitos) or social factors (summer picnics, crowded indoor winter gatherings) or both (Figure 14.6). Chickenpox, meningitis, whooping cough and influenza, all contracted

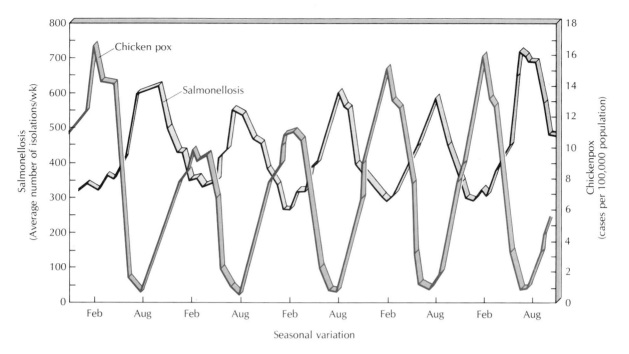

FIGURE 14.6 Seasonal Variations in Diseases
This graph compares monthly cases of salmonellosis (left) and chickenpox (right)
reported in the United States over a period of four years. Chickenpox, like most
respiratory infections acquired by inhalation, is most common during the late winter.
Salmonellosis, like many gastrointestinal infections, is most common during the
summer. (Data are from the Centers for Disease Control, Atlanta.)

primarily by respiratory routes, are most prevalent in the late winter and early spring. In the
United States, diseases such as salmonellosis or
shigellosis, which are contracted by the
fecal–oral route, or viral encephalitis, transmitted by insect vectors, are most common in
the summer months.

A seasonal increase in cases is not necessarily an epidemic. Only when the peak incidence exceeds a certain level, defined as the epidemic threshold, is an epidemic said to be
present (Figure 14.7). Thus, although the number
of influenza cases always goes up in the spring,
the number only reaches epidemic proportions
every few years. So far, we are unable to predict
in advance which years these will be.

Different diseases may affect one age group

more than others (Figure 14.8). Mumps and
chickenpox are most common in the young, and
pneumonia is more common in the elderly. The
age distribution varies in different nations, as
each country has a different percentage of its
total population in each age group. In Latin
America, children make up a very high proportion of the population, so childhood diseases
such as measles are a dominant public health
issue. In the United States and Europe, a declining birth rate, in conjunction with increased
life expectancy, have increased the proportion of
the elderly. Therefore, the health issues of the
elderly have become an increasing concern.

Immunity is a key factor in whether an individual is susceptible to an infectious agent or
not. Just as we can define the immune status of

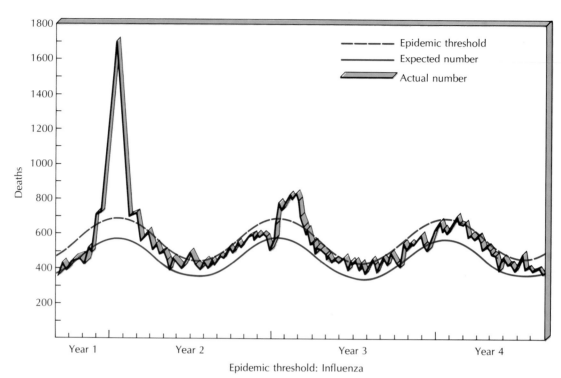

FIGURE 14.7   Annual Influenza Patterns
The number of influenza-related pneumonia deaths in the United States peaks around February each year. On the basis of past experience, an "expected number" can be calculated. An increase over this number occurs in certain years. When the deaths exceed the epidemic threshold, as they did in Years Two and Three on this graph, the conditions that define an epidemic have occurred. (Data are from the Centers for Disease Control, Atlanta.)

an individual, we can quantitate the **herd immunity** of a population. Herd immunity is the percentage of a population that is functionally immune to an agent and able to resist the challenge of exposure. The extent of herd immunity determines whether an epidemic is able to develop, and its severity if it occurs.

When the numbers of new cases in an outbreak are graphed against elapsed time, epidemics have very defined shapes (Figure 14.9). No epidemic goes on forever; it eventually runs out of new victims. The rapidity of the epidemic's development depends on the number of susceptible persons; the denser the susceptible

population, the more explosive the outbreak. Where there are few susceptible individuals in a group, on the other hand, an agent will spread in an endemic fashion, with few and isolated cases. This is because with this very high degree of herd immunity the cycle of transmission is very rarely completed (Figure 14.10). Of course, many epidemics are arrested, by forceful public health measures, before they run their "natural" course.

Herd immunity has been well studied in regard to measles. In a city of 300,000 people, it requires between 2500 and 5000 new cases of measles a year for the virus to perpetuate itself.

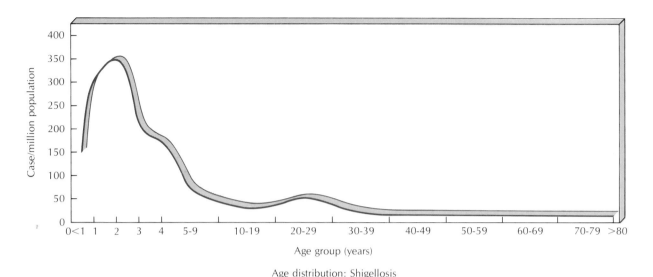

Age distribution: Shigellosis

FIGURE 14.8 The Rate of Shigellosis in Various Age Groups
Single or sporadic cases of shigellosis are most common in infants between 1 and 2
years of age. Usually, all ages are involved when there is a foodborne outbreak.
Note that the age scale at the bottom of the graph expands in the interval between
ages 0 and 10 years. (Data are from the Centers for Disease Control, Atlanta.)

In any given year, almost all susceptible people
will get measles and, as a result, become resistant. It is almost unknown for a person to get
measles more than once in a lifetime. Therefore
there must be a minimum of 2500 new susceptible people, largely children, added to the
city's population each year. If all these new
young susceptibles can be promptly made resistant by vaccination, the virus will be deprived
of its reservoirs and will die out. The U. S. Public
Health Service has set a goal of complete eradication of measles, to be attained by achieving as
close as possible to 100% herd immunity. The
estimated level at present is 94%.

Why is it really essential to immunize every
single child? If most but not all children were
vaccinated, and only a few susceptible (unvaccinated) children were added to the population per year (the current state of affairs) measles virus would gradually disappear. This
situation is an improvement from the public
health point of view. However, from the individual's point of view, it is still less than satisfactory. In this interim period, several thousand

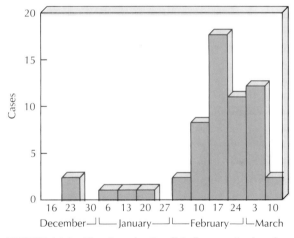

FIGURE 14.9 Kinetics of a Small Influenza Outbreak
This graph shows the pattern of influenza cases
among nursing students in a small hospital. Clinical
illness had been present at a low level for more
than a month before the real outbreak. However,
conditions necessary for transmission apparently
became optimal in February, and the agent quickly
infected a number of susceptible students.
Influenza virus was isolated in only two cases. By
the second week in March, there were no more
susceptible students to perpetuate the outbreak.

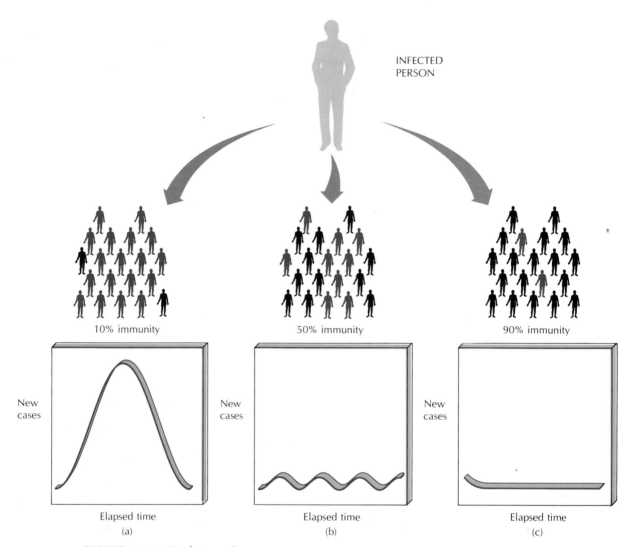

FIGURE 14.10   Herd Immunity
(a) Epidemics spread rapidly through populations in which only a small percentage of the people have resistance to the infective agent. Such an epidemic develops quickly, but it also subsides quickly once all susceptible persons have experienced the disease and become immune. (b) When about half the group is immune, the infective agent spreads more slowly and the incidence of disease is sporadic. The infectious agent persists at a low level over a long period. (c) When a high percentage of the group is immune, an introduced case of the disease is rarely transmitted, and the agent disappears from the population as soon as the original cases have recovered.

unprotected children every year become ill with measles, and a dozen or so die or suffer permanent disability. We can see, then, that public health programs can bring certain infectious diseases under some degree of control if they can achieve a herd-immunity level estimated at about 70%. The disease can be eradicated only if the immunity level approaches 100%. The 100% goal is being sought for seven preventable diseases of childhood — measles, mumps, rubella, diphtheria, pertussis, tetanus, and polio. It can probably never be attained, since no vaccine is ever 100% effective.

BREAKING THE CHAIN OF TRANSMISSION

The process of tracking down and dealing with an epidemic follows certain steps, which we will now describe. Each case poses novel challenges and difficulties for the investigator. Epidemiologic detection usually involves working backward, from the cases to the causative agent, then to the means of transmission and ultimately back to the reservoir or source. Once the chain of transmission has been fully established, we can interrupt it at the most convenient and effective point.

CASE REPORTING

More than 30 infectious diseases are designated by law as **notifiable diseases.** Some examples are gonorrhea, measles, rabies, AIDS, and typhoid fever. Health practitioners are required to report any cases of these diseases they diagnose to local and state public health officials. Notification gives a more or less accurate estimate of the prevalence of a disease. Of course, not all cases are seen by physicians, and not all treated cases are reported. Many diseases, especially sexually transmitted diseases, are underreported. Nevertheless, the notification system is a mechanism to assemble and compile the needed data on disease prevalence. This makes it possible to spot

an unexpected concentration of cases or sporadic outbreaks. A decision may then be made to investigate. In the case of a non-notifiable disease such as salmonellosis, much depends on the alertness of the individual physician who suddenly realizes that he or she has seen an unusual upsurge of gastrointestinal upsets and brings them to the epidemiologists' attention. This lead will start the epidemiologist on an investigation that may detect unsafe food-handling practices in a local restaurant, for example (Figure 14.11), or pinpoint malfunctioning water-chlorination equipment.

FIGURE 14.11
An inspector observes food-handling procedures in a restaurant kitchen. (Courtesy of U.S. Food and Drug Administration.)

IDENTIFYING THE INFECTIOUS AGENT

The causative agent must be recovered and cultured, from each patient if possible, to establish whether all the cases are really part of a single outbreak. It would be possible to have an outbreak of 52 cases of diarrheal illness, for example, in which one case could be a coincidental food allergy and another a previously undetected ulcerative colitis. Such cases should be carefully eliminated because they will confuse the issue.

Once the agent is isolated, it is subjected to the most precise identification methods possible, such as bacteriophage typing, serotyping, determining its antibiotic susceptibility pattern (its **antibiogram**), and the complete spectrum of metabolic testing. If isolates from all cases are identical, there is strong evidence that all cases had a single source. Isolation of an identical organism from the suspected reservoir or source will confirm the attribution.

SEARCHING FOR THE RESERVOIR

The identity of the infectious agent provides immediate clues to the type of reservoir to be suspected. *Salmonella* or *Shigella* gastrointestinal illnesses suggest fecally contaminated foods; staphylococcal upsets suggest foods inoculated from sores on the body of a food handler. Respiratory illnesses, transmitted by air currents, usually originate in secretions expelled by a carrier or, less commonly, in vapor sprayed from an air-conditioning device.

In some epidemics the source is never found. Following the first recorded outbreak of legionnaires' disease in Philadelphia in 1976 (see Box 6.1) the source could not be pinpointed. Three years passed, while improved techniques for isolating *Legionella* were developed. In the meantime, however, the hotel in which the outbreak centered was totally remodeled, so that the source was presumably completely removed. It has since become clear that one reservoir of *Legionella* is large-scale cooling towers and climate-control systems. This finding helped to explain why many cases occur in the summer months and cluster in large buildings, such as hotels and hospitals, in which this type of equipment is used.

BRINGING THE OUTBREAK UNDER CONTROL

Swift, directed action can often arrest an epidemic before it achieves its full destructive potential. This action may involve one or a combination of approaches.

When there is a single apparent source, the oubreak can be controlled by eliminating the source. Infective persons can be treated, trained in personal cleanliness, or switched to different occupations. Leaking water or sewer mains can be fixed, herds of infected dairy cows can be destroyed or milk from the cows kept off the market, inadequately processed batches of canned goods can be recalled, and contaminated shellfish beds can be closed to harvesting.

Where there is a large reservoir, such as a group of carriers, it is more effective to interrupt transmission. During community influenza outbreaks, nonessential personnel and visitors are frequently barred from hospitals. Schools are closed to arrest epidemics of head lice, or daycare centers to interrupt the transmission of hepatitis. Isolation and segregation procedures are useful in hospitals. When rabies in wild animals reaches epidemic proportions in an area, public health officials initiate campaigns to destroy any stray, wild dogs or cats, and veterinarians assist in control by organizing pet-vaccination campaigns.

The preferred approach to controlling epidemic disease may be to increase herd immunity. A vaccination campaign may be initiated to seek out and actively immunize all susceptible persons in epidemic zones and surrounding areas. This approach is effective only if the disease has a relatively long incubation period. Because influenza has an incubation period from one to three days, no immediate protection can be gained from immunization after an epidemic has begun. The effect of the immunization campaign

BOX 14.2
A Classic Epidemiologic Whodunit

CENTERS FOR DISEASE CONTROL

MMWR

MORBIDITY AND MORTALITY WEEKLY REPORT

Current Trends

This story has all the features of a mystery — suspense, chills, and a happy ending. A shigellosis outbreak occurred in May 1979 among employees in a children's hospital in Pennsylvania. Two hundred eighty employees (38%) and visitors became ill, with complaints of vomiting and/or diarrhea; 142 (51%) had positive stool cultures for *Shigella sonnei*. The organism was found in staff members on almost every unit in the hospital. Staff shortages during the outbreak were so severe the hospital had to be closed to new admissions for three days.

To determine the source of the epidemic, epidemiologists sent questionnaires to 1700 employees. The employees were asked to describe the symptoms of disease and recall where and what they had eaten during the period. On the basis of 1093 responses, investigators linked illness to eating tuna salad and other cold foods from the salad bar in the hospital cafeteria. Hot foods were not implicated.

The epidemiologists next carefully questioned the cafeteria workers. They found out that one employee had diarrhea on May 17, the first day of the outbreak. She had previously been exposed at home when her child had severe diarrhea. This employee was found to be culture-positive for *S. sonnei*. She had worked on May 17 and May 21, and was the person responsible for preparing all salads and sandwiches in the cafeteria. The two peaks in the outbreak occurred on May 19 and May 23,

two days after each date on which this person worked. This pattern was consistent with the one- to two-day incubation period of foodborne shigellosis.

The *Shigella* strain responsible was tested and found to be resistant to ampicillin and tetracycline, and sensitive to cotrimoxazole. As was standard practice at that time, everyone with symptoms received a five-day course of the appropriate antibiotic. Culture-positive cafeteria employees were required to have a series of three negative rectal cultures before they were allowed to return to work. The happy ending? The outbreak was controlled effectively enough so that none of the hospitalized children became ill.

on the number of new cases will only take hold two weeks or so afterward, by which time most people who were susceptible in the first place will have already experienced the disease. For good results, influenza vaccination must occur well in advance of possible exposures or epidemics. Measles, however, has an incubation period of about 10 days, so outbreaks can be arrested by crash vaccination campaigns in a threatened community. The vaccination campaign is usually combined with the exclusion of unvaccinated children from school. Passive immunization with injections of immune globulin may be used to abort epidemics of hepatitis A in children who have been exposed at a day-care center.

PUBLIC HEALTH AND DISEASE PREVENTION

The community's health is best protected in a society where there is a cooperative mixture of governmental surveillance and regulation, dedicated medical practice, and, perhaps most important, informed self–protection by individuals. Routine public health measures have been generally adopted throughout the developed world; they are seen as offering the best possible chance of preventing epidemics. We will present here a brief summary of drinking-water and sewage treatment, vaccination programs, and several other key aspects of public health as examples. All are discussed more fully in Chapters 26 through 28.

SAFE WATER SUPPLIES

A safe water source for drinking, washing, and food preparation is essential to avoid outbreaks of gastrointestinal disease. All public water supplies are subject to regulation. They supply treated water, which must be regularly analyzed for chemical and microbial contaminants. The **coliforms** are especially important. Coliforms are a group of lactose-fermenting Gram-negative enterobacteria, including *Escherichia coli* and closely related organisms, whose habitat is the human intestinal tract. These organisms are **indicator organisms,** not usually pathogenic themselves, but a sign of fecal contamination. Most laboratory tests to evaluate the safety of water or the effectiveness of sewage treatment do so by counting the number of coliforms per volume of the sample.

Almost all municipal water supplies derive from contaminated surface water, thus the water requires careful pretreatment before it goes to the consumer. Water is purified by settling, filtration, chlorination, or a combination of these methods. Individual home water supplies — wells or springs — are usually safe without treatment if local guidelines about distance of wells from sewage disposal facilities are observed. However, wells and springs should be checked periodically. Home water supplies will be tested on request by public health laboratories, usually for a minimal charge.

EFFECTIVE SEWAGE DISPOSAL

Sewage consists of three types of fluids, **black water** from toilets, **gray water** which is wash water, and **storm runoff,** rainwater from gutters and storm drains. The first two bear a heavy load of human microorganisms, often including pathogens. The accumulation of untreated sewage poses an overwhelming threat to public health, and adequate sewage treatment is therefore mandated by law. A list of waterborne diseases appears in Table 14.4. The type of treatment used is also regulated by law so as to minimize the material's negative effects on the environment.

Sewage treatment may involve up to three stages. **Primary treatment** consists of the collection of sewage through a network of pipelines, screening to remove large objects, and sedimentation in tanks. The sedimentation process separates sewage into **sludge** (the solids) and **effluent** (the liquid wastes). If only primary treatment is used, the effluent will be chlorinated and dis-

TABLE 14.4 Waterborne diseases		
Disease	**Agent**	**Classification**
Amebic dysentery	*Entamoeba histolytica*	Protozoan
Amebic meningoencephalitis	*Naegleria fowleri*	Protozoan
	Acanthamoeba sp.	Bacterium
Bacillary dysentery	*Shigella* sp.	Bacterium
Cholera	*Vibrio cholerae*	Bacterium
Acute gastroenteritis	Many	Bacterium, virus
Giardiasis	*Giardia lamblia*	Protozoan
Hepatitis A	Hepatitis A agent	Virus
Leptospirosis	*Leptospira* sp.	Bacterium
Schistosomiasis	*Schistosoma* sp.	Metazoan
Typhoid fever	*Salmonella typhi*	Bacterium

charged into a nearby body of water and sludge may be burned or dumped. This represents the minimum permissible level of treatment; more extensive measures are highly desirable.

Secondary treatment further reduces the amount of wastes. The sewage from primary treatment is transferred to tanks containing a complex microbial population. The microorganisms decompose over 90% of the remaining organic wastes, while pathogens are eliminated by ecological competition. To guarantee that all pathogens are eliminated, the effluent is chlorinated before it is released.

Tertiary treatment is a relatively new development. It incorporates additional chemical treatments to cleanse the water of all remaining organic and inorganic pollutants, returning it to an entirely drinkable condition.

FOOD SAFETY

There are many food-safety programs in operation. (As illustrated in Figure 14.12, this was not always so.) For example, inspection programs in commercial meat-packing facilities are designed to detect diseased animals, enforce standards of dressing, cutting, grading, and refrigeration, and detect antibiotic or pesticide residues, bacterial contaminants, or parasites. Refrigeration and handling of foods in supermarkets and restaurants are subject to regulation and inspection. Milk is classified by its raw bacterial count, pasteurized prior to commercial sale, and dated. Processed milk and dairy products are periodically analyzed to check that pasteurization has been adequate. Imported foods are regulated to guarantee that they are prepared by methods as stringent as those required for domestic products. Canning and preserving is carried out according to strict standards with rigid quality-control checks. On the rare occasion that a batch of contaminated food reaches the market, public health officials recall every can in warehouses or on store shelves. The public is alerted by media announcements to return, or throw out, unopened cans.

FIGURE 14.12   Food Safety

It is easy to take for granted the numerous food-safety programs in operation today. Earlier generations, however, were not so well protected. Before the turn of the century, green fruit purchased by grocers often rotted on the counters instead of ripening attractively. In 1832, a severe cholera epidemic that erupted in New York was attributed to contaminated fruit. The scare created a suspicious attitude toward fresh produce that lingered for years. The magazine *Harper's Weekly* attempted to raise public awareness of the problem through inflammatory articles and illustrations. This 1870 cartoon warned citizens about cartloads of decayed oranges and rotten bananas being sold in markets throughout the city "to partake of which was almost certain death." (Courtesy of the Bettman Archive, Inc.)

CONTROLLING HUMAN CARRIERS

In certain diseases, active surveillance of known carriers and screening to detect unsuspected carriers can pay large dividends. We know, for example, that some fraction of hepatitis patients become chronic carriers and continue to shed infectious virus indefinitely. A small percentage of those who recover from typhoid fever retain the bacterium in their gallbladders and excrete live pathogens in their feces. Systematic follow-up of patients who have had hepatitis or typhoid can identify these carriers (Figure 14.13). They can receive treatment and/or instruction in hygienic measures to avoid infecting their contacts.

Several of the sexually transmitted diseases, including gonorrhea, may produce no obvious symptoms, especially in women, so a person may be infective without being aware that anything is wrong. Routine examinations that include *Neisseria gonorrhoeae* culture detect a substantial number of positive individuals. Gonorrhea carriers and their sexual contacts may then be treated before serious clinical complications occur. Active case-finding is used to discover and treat sexual contacts of newly diagnosed syphilis patients.

VACCINATION PROGRAMS

Immunization of infants and children can prevent epidemic childhood diseases such as measles and polio, and can initiate a lifelong immunity against diseases such as tetanus. Immunization programs have helped us achieve remarkable decreases in infant mortality. One unfortunate byproduct of the success story of vaccination is public apathy about these supposedly vanquished diseases. To ensure compliance with immunization recommendations, schools now require proof of current vaccination before a child may be enrolled in kindergarten. Consequently, after declining in the last decade, immunity levels in children under age 10 are now at a satisfactory level. However, many teenagers are still unimmunized, and public health professionals are concerned about continuing outbreaks in this age group.

International immunization requirements are also helping to control outbreaks of infec-

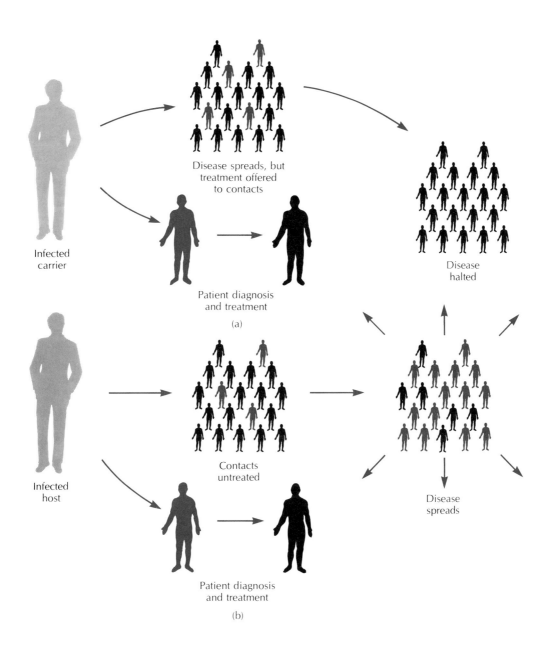

Infected carrier

Disease spreads, but treatment offered to contacts

Patient diagnosis and treatment

(a)

Disease halted

Infected host

Contacts untreated

Patient diagnosis and treatment

(b)

Disease spreads

FIGURE 14.13   Contact Follow-up
In certain diseases, detection and treatment of unsuspected carriers is essential in halting the spread of infection. (a) If a carrier receives timely, proper treatment and instruction, he can avoid infecting further contacts. (b) If the first case (index case) is diagnosed and treated without also treating the exposed contacts, the disease will continue to spread.

tious disease. Travelers abroad are urged to comply with the international vaccination recommendations prepared and updated by the World Health Organization. Most countries enforce vaccination regulations at their borders. Military personnel are routinely immunized against a wide range of potential pathogens so they will be prepared for hazardous overseas duty.

Medical and veterinary personnel have unusual occupational hazards. They often need specialized immunizations. Tables of recommendations will be found in Chapter 25.

ANIMAL HEALTH PROGRAMS

A variety of programs and requirements have been established to reduce the incidence of disease from exposure to infected animals (Figure 14.14). Dairy herds are regularly tested for tuberculosis and brucellosis. Certain types of imported pets, such as parrots, turtles, and snakes, must be tested before they are offered for sale to guarantee they are not carrying infectious agents. Almost all communities now require evidence of rabies vaccination before a dog may be licensed. Because unvaccinated and unlicensed strays are responsible for many human rabies exposures, public health officials are empowered to impound and destroy stray pets.

INSECT AND RODENT CONTROL

Rodents and their insect parasites transmit specific infections such as the bubonic plague. They also increase infection rates by the nonspecific spread of garbage and filth. Rat control is a con-

FIGURE 14.14   Animal Health Programs
Areas such as Prince Edward Island in Canada that are free of certain infectious diseases may adopt quarantine restrictions to protect livestock against the introduction of disease agents. (Courtesy of Cynthia Norton.)

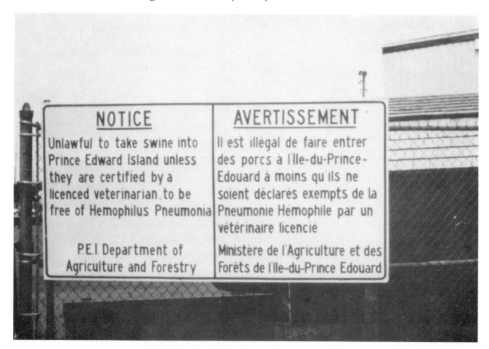

stant necessity in many densely populated urban neighborhoods. A large federally funded rat control program has been extremely successful in many large cities.

Where arthropod-borne diseases are endemic, there are usually many infected persons to serve as a continual reservoir. If immunization is not possible, as in the diseases malaria and trypanosomiasis, aggressive elimination of the vector may be the most workable control strategy (Figure 14.15). Mosquitos can be controlled by draining their breeding areas in stagnant water to deprive them of a place to deposit eggs, oiling the surface of stagnant water to kill larvae, or applying pesticides to kill adults. Houses may be sprayed with pesticides to eliminate the biting bugs that transmit trypanosomiasis. Mosquito netting, protective clothing, and insect repellent are all useful personal insect controls. Ensuring clean, adequate water supplies for bathing helps to eliminate lice.

From the examples on the past few pages, we can see clearly that public health measures are the practical applications of epidemiology. They are designed to disrupt the chain of transmission of an infectious disease by the most practical and cost-effective means. We should not take these features of our society for granted — they are too important to our health and safety.

FIGURE 14.15 Control of Insect Vectors
Killing mosquito larvae by spraying pesticides is one way to control the spread of malaria. This type of swampy area is a prime breeding ground for mosquitos. Preventing the maturation of the larvae will stop the transmission of the malaria parasite. (Photo courtesy of the World Health Organization.)

SUMMARY

1. Epidemiology is the study of the occurrence and spread of disease in populations. Its goals are the control and eventual elimination of communicable and noncommunicable disease, and the institution and maintenance of public health safeguards.

2. Communicable disease agents pass from a reservoir of infection via an exit route, carried by one or several modes of transmission. They then enter a susceptible host. Living reservoirs include human beings, animals, and arthropods; nonliving reservoirs include soil, water, air, and food. Direct means of transmission such as contact, inhalation, ingestion, and arthropod bite transmit organisms that have little capacity to survive in a hostile environment. Hardier microorganisms may be indirectly transmitted by contact with contaminated objects, food, or water. The act of transmission almost always goes undetected.

3. Population susceptibilities vary widely. The introduction of an agent may or may not cause epidemics; outbreaks may vary in their severity. Adverse physical and social conditions, certain patterns of age distribution, and low levels of herd immunity all increase population susceptibility.

4. Epidemiologists investigate outbreaks. An epidemic becomes evident when notifiable diseases or unusual clusters of illness are reported to epidemiology officials. These officials carry out field investigations, attempt to isolate an infective agent, and work to make the most complete possible identification. They search for the reservoir(s) and the mode of transmission of the epidemic. The outbreak can then be arrested by whatever means seem most practical.

5. Public health measures include provision of safe water supplies and systematic collection and treatment of sewage. Techniques in these areas are continually improving, as are the stresses on the systems.

6. Food safety is improved by meat inspection, restaurant and processor inspection, pasteurization of dairy products, testing of dairy herds, and strict controls on cannery operation.

7. Human reservoirs are detected by case follow-up and routine screening. Universal immunization of children for certain diseases is required, and special immunizations are provided for travelers, military personnel, and medical and veterinary workers.

8. The effect of animal diseases on the human population is minimized by vaccination or elimination of infected animals. Arthropod control is important in reducing arthropod-borne diseases. Rodent control has been successful in many cities.

DISCUSSION QUESTIONS

1. List some communicable diseases that are endemic, sporadic, and epidemic in your community. How might these differ from those found in an equatorial community or a research colony in Antarctica?

2. Find out what methods of treatment are used for your household water supply and your household sewage. What are their strengths and weaknesses? Consider taking a field trip to any readily accessible water- or sewage-treatment plants.

3. What arguments can be advanced for and against 100% vaccination programs?

4. Suppose you were investigating an outbreak of gastrointestinal disease that had struck a large number of individuals following attendance at a banquet. What steps would you take?

ADDITIONAL READING

Austin DF, Werner SB. *Epidemiology for the health sciences: a primer on epidemiological concepts and their use.* Springfield, Ill.: Charles C Thomas, 1982.

Benenson AS, ed. *Control of communicable diseases in man.* 14th ed. Washington, D.C.: American Public Health Association, 1985.

Evans A, Feldman HA, eds. *Bacterial infections of humans: epidemiology and control.* New York: Plenum, 1982.

Evans A, ed. *Viral infections of humans: epidemiology and control.* 2nd ed. New York: Plenum, 1982.

Friedman GD. *Primer of epidemiology.* 3rd ed. New York: McGraw-Hill, 1984.

Infection control in the hospital. 4th ed. Chicago: American Hospital Association, 1979.

Morbidity and Mortality Weekly Report. Published weekly by Centers for Disease Control, Atlanta, GA 30333.

Standard methods for the examination of water and wastewater. 16th ed. Washington, D.C.: American Waterworks Association, 1985.

State and City Epidemiology Newsletters.

PART
THREE

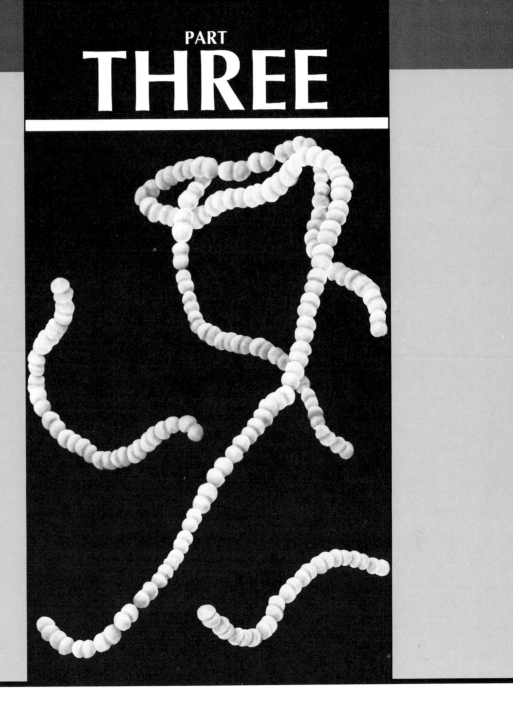

HUMAN
INFECTIOUS DISEASES

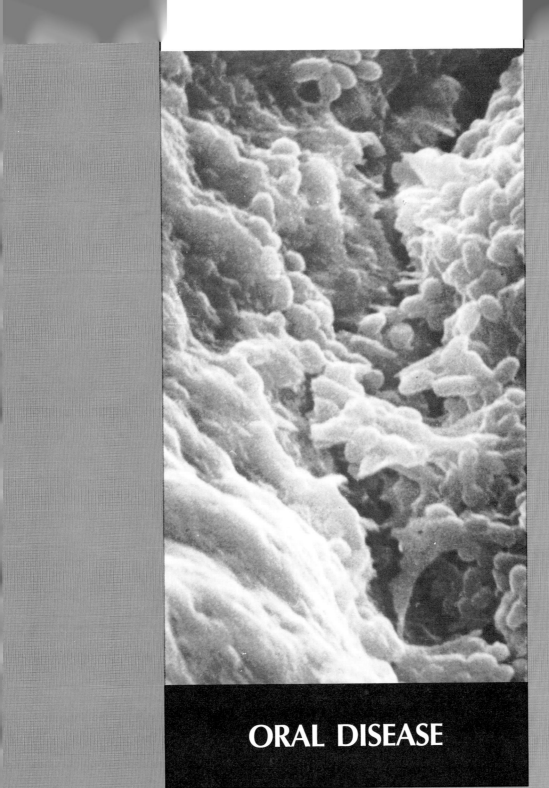

ORAL DISEASE

CHAPTER OUTLINE

INTRODUCTION TO INFECTIOUS DISEASE

With this chapter, we begin our survey of the human infectious diseases. Some of these, such as the oral infections that are the subject of this chapter, are endogenous, and are caused by overgrowth of the normal flora. The others, covered in subsequent chapters, are exogenous and are caused by a single, externally acquired pathogen.

Please note the key organizational features of our presentation of the pathogens. Each major pathogen is given an in-depth major examination in one chapter, corresponding to the part of the body in which it causes its most noteworthy or most common infections. In many cases, however, a pathogen will also cause infections in other areas of the body, so you may often meet it elsewhere as well.

In this section of the book, we will become familiar with a number of major pathogens. Each

---

Microbial diseases featured in this chapter:

Dental caries
Periodontal disease
Acute necrotizing gingivitis

---

is presented from four aspects. We discuss first the microbiology of the microbial species. Next, we examine the actual diseases it causes in the area of the body we are currently studying — in this chapter, the mouth. Next, we deal with the treatment most commonly used. Finally, we discuss the epidemiology of the microbial pathogen, and the public health measures that are useful in controlling it.

THE IMPACT OF ORAL DISEASE

About 90% of you reading this have a chronic, irreversible microbial disease. "Oh, come on," you say. But wait, that disease is dental caries, or tooth decay. The most universal infection we know is oral disease in its two main forms — tooth decay and gum disease. Most people are completely unaware that these are microbial. The good news is that oral diseases may be completely prevented, if we use our common sense. An understanding of the microbiology of oral disease may help create motivation!

The dental profession has exhibited wise leadership in a national research and education campaign to prevent and reduce oral disease. It is already taking effect. If you have children, you are almost certain to have noted that their teeth are in better shape than yours were at the same age. My dentist jokes that the dental profession is putting itself out of business.

In this chapter we will begin by familiarizing ourselves with the mouth's normal anatomy and physiology. We'll also look closely at the normal flora that live there. Next we'll examine the biological activities of these symbiotic flora. It is these activities, when uncontrolled, that cause oral disease. Then we will analyze the two main categories of oral disease, tooth decay (dental caries) and gum disease (periodontitis). We will also summarize some of the infection hazards to dental health personnel of working with patients with non-oral diseases such as hepatitis. In conclusion, we will examine the major strategies for preventing oral disease.

THE MOUTH — A HOSPITABLE HOME FOR MICROORGANISMS

Let us start by studying the part of the oral cavity that's designed for the intake and chewing of foods: the teeth and their supporting structures.

ANATOMY OF THE MOUTH

Figure 15.1 shows the major structures of the tooth and the relationship of the tooth and gum.

Note that the teeth and gums are composed of several different tissue types and that both structures provide diverse microenvironments for microbial growth.

THE SOFT TISSUES. The inner cheek, gum surfaces, and tongue are composed of soft tissues. They are all covered with a type of tissue called **epithelium.** Characteristically, epithelial tissues are surface layers that are continuously being replaced. Although epithelium is readily colonized by microorganisms, its older cell layers slough off regularly, creating an enforced turnover that places an upper limit on the number of microbial colonists. Since these epithelial surfaces are renewable, any injuries tend to heal rapidly. The tongue surface is very porous. Aerobic microorganisms accumulate on the tongue in vast numbers, forming a film or coating, especially during sleep. A protective structural protein, **keratin,** toughens the surface of the **gingiva** (gum) nearest to the tooth, where there is extensive wear and tear from chewing.

TOOTH STRUCTURE. The tooth's top or crown is coated with **enamel.** This is a transparent, crystalline substance composed of calcium and phosphate salts, plus a strengthening matrix of protein. Normal tooth enamel is the hardest tissue substance in the body. When perfectly clean and smooth, enamel effectively resists microbial adhesion and chemical attack. However, age and hard use cause scratches, cracks, and pits in teeth, providing improved opportunities for microbial adherence and growth. Tooth surfaces are basically nonrenewable; our bodies cannot repair any significant damage that occurs on a tooth surface. This is why we say that tooth decay is chronic and irreversible.

The **dentin,** the hard substance that underlies the enamel, is also calcium-based, but less hard. Dentin is arranged in tubules (visible in Figure 15.1b) that render it porous. When microorganisms enter dentin, they can migrate slowly through it. Normal dentin is opaque and gives the tooth its whitish color. Neither enamel nor dentin contains living cells. Tooth **pulp,** which

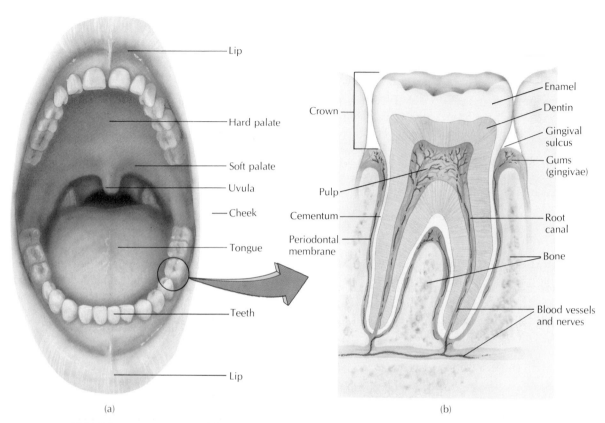

FIGURE 15.1   Anatomy of the Mouth
(a) The mouth and accessory organs. (b) Diagram of a cross-section of a molar or grinding tooth showing its layers, the blood vessels and nerves, and the gum and attachments. Note the raylike appearance of the dentin caused by the dentin tubules.

fills the hollow center of the tooth, on the other hand, is composed of living cells. Their blood supply and nerve endings enter through the tiny **apical pore** at the end of the tooth root.

The tooth is attached to the jawbone by microscopic fibers of **collagen,** a strong, flexible protein. The **periodontal membrane** lines the tooth socket, the gingiva, and the tooth surface from the point below the gum where the enamel coat ends, on down the root. The **gingival sulcus** is the pocket between the gum and the side of

the tooth. It is normally no more than 2 mm deep. This pocket tends to readily collect food debris and microbes.

PHYSIOLOGY OF THE MOUTH

The mouth is extremely hospitable to microorganisms. The temperature is warm, the tissue surfaces are moist, and nutrients are plentiful and varied. Different environmental conditions occur in various areas of the mouth, favoring distinct groups of organisms. Some of the different microenvironments include the cheek and gum surfaces, the tongue, and the tooth surfaces, all of which are mainly aerobic. The gingival sulcus, and cracks in the crown of the tooth, on the other hand, are primarily anaerobic.

The mouth has several natural defenses to control microbial growth. If we use good oral hygiene and eat an appropriate diet, these defenses are sufficient to prevent oral disease. One defense is *saliva*, the secretion of the salivary and mucosal glands. About 1500 ml of saliva are secreted and swallowed daily. The saliva has a nearly neutral pH of about 6.8, which buffers any acids produced by oral bacteria. It also contains calcium ions, which stabilize the enamel. Its dissolved oxygen inhibits growth of anaerobic bacteria. Saliva also contains secretory IgA and other immunoglobulins for immune defense against some pathogens. Leukocytes also migrate around the oral cavity, especially near areas of inflamed gum or pulp tissue. They provide for phagocytosis and inflammatory responses, thereby destroying some microorganisms.

NORMAL FLORA OF THE MOUTH

The human mouth is colonized by bacteria, fungi, and protozoa. These microorganisms are always present in large numbers and in complex mixtures. If we carefully analyzed a sample from one tiny spot we might identify 40 or more bacterial species. Using anaerobic counting methods would more than double the number of species we would find. The distribution of normal oral flora is shown in Table 15.1.

The same organisms, normal flora, are found in both normal and diseased teeth and gums. Thus, in contrast to most of the communicable diseases we'll be studying in future chapters, dental disease does not require that a pathogen be introduced. On the contrary, it is an endogenous process. When we ignore good oral hygiene, and eat a diet high in sugar, we tip the

TABLE 15.1 Distribution of normal oral flora	
Environment	**Numbers**
Saliva	40 billion/ml
	110 billion/ml
Inner cheek epithelium	5 to 25 bacteria/cell
Tongue surface	100 bacteria/cell
Gingival crevice	16 billion/g
	40 billion/g
Plaque	25 billion/g
	46 billion/g

balance in favor of very large microbial populations. Our own normal flora expand their activities, and tissue is destroyed.

THE ORAL BACTERIA

Oral bacteria predominate over the fungi and protozoa in both numbers and importance. We will look at six types of bacteria that contribute to oral disease.

Streptococci are Gram-positive, chain-forming cocci. They are facultative anaerobes and are found in all the oral microenvironments. They ferment actively, producing lactic acid from carbohydrate. Some species manufacture sticky polymers such as **dextran** from sugars. These polymers are a major component of dental plaque, the coating that forms on unclean teeth. *Streptococcus mutans* is the species considered to be the major contributor to dental caries (Figure 15.2).

Lactobacilli are anaerobic, Gram-positive rods. They ferment strongly and are highly acid-tolerant, surviving at pH levels low enough to dissolve tooth enamel. Lactobacilli are always found in cavities and probably play a role in the decay process.

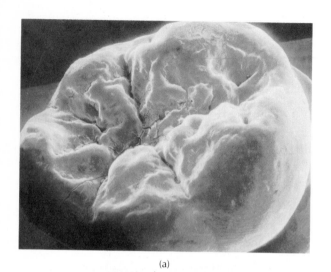

(a)

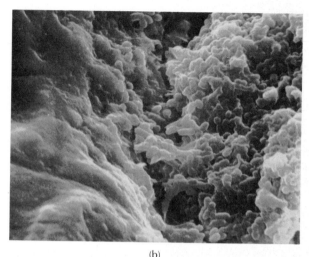

(b)

FIGURE 15.2 The Surface of the Tooth
(a) This scanning electron micrograph of a normal tooth surface reveals numerous pits and fissures. Such irregularities in the tooth enamel encourage accumulation of cellular debris. Soon after the tooth erupts, microorganisms invade and colonize these sites, thus contributing to caries of the tooth crown. (Courtesy of Dr. John Gwinnett.) (b) At a higher magnification, cells of *Streptococcus mutans* are visible clinging to eroded areas of tooth enamel. This bacterial species is considered to be a primary agent of dental caries. (×2736. Courtesy of Z. Skobe, Forsyth Dental Center/BPS.)

Actinomycetes are Gram-variable, anaerobic, often branching rods that clump together to form mats. Actinomycetes are found in the gingival sulcus and in deeper levels of surface plaque. They may produce **calculus,** a hard, abrasive material composed of plaque, which contains calcium phosphate.

Bacteroides are Gram-negative, anaerobic rods found largely in the gingival sulcus. In gum disease, they proliferate. *Bacteroides* produce endotoxin, promoting a destructive inflammatory response. Their enzymes may destroy the collagen fibers holding the tooth in place.

Fusiforms such as *Fusobacterium* are filamentous, Gram-variable, and anaerobic. Their rod-shaped cells tend to taper at both ends. These organisms proliferate in gum disease.

Spirochetes, which are flexible Gram-variable spirals, flourish in anaerobic mouth en-

vironments. *Borrelia,* in association with fusiform bacilli, grows rapidly in **acute necrotizing ulcerative gingivitis,** more commonly known as trench mouth or Vincent's angina, which we'll discuss shortly.

HOW THE ORAL POPULATION EVOLVES

During and after birth, a baby begins collecting organisms from its environment. As every mother knows, babies quickly learn to put all sorts of things in their mouths. Eventually they probably introduce just about every known species of microorganism into their mouths. However, only those microorganisms adapted to oral conditions persist. Most oral microorganisms form stable attachments to oral surfaces by means of pili or fimbriae, or by secreting a sticky polymer. The important microbial oral flora are listed in Table 15.2.

Before the first baby tooth erupts, all the oral surfaces are soft, smooth, and provide mainly aerobic environments. Once teeth appear, anaerobic environments are formed. Aerobes are then augmented by anaerobes. The baby teeth are usually widely spaced, providing few tight spots to trap food debris that microbes can use for growth.

The way the flora develop thereafter will depend somewhat on dietary habits. The healthy

TABLE 15.2 Important microbial members of the oral flora

Group	Species	Oxygen needs	Relevant Physiologic Traits	Association with Disease
Streptococci	*Streptococcus salivarius*	Facultative anaerobic	Acid, levan production	Infectious endocarditis
	S. mitis	Facultative anaerobic	Acid production	Pulp infection, infectious endocarditis
	S. mutans	Facultative anaerobic	Acid, dextran, levan production	Caries, infectious endocarditis
Lactobacilli	*Lactobacillus acidophilus*	Facultative anaerobic	Acid production, acid tolerant	Caries
	L. casei	Facultative anaerobic	Acid production, acid tolerant	Caries
Diplococci	*Branhamella catarralis*	Aerobic	Nonfermentative	Caries, tongue lesions
	Veillonella parvula	Anaerobic	Nonfermentative	Plaque formation, mixed infections
Actinomycetes	*Actinomyces israelii*	Anaerobic	Proteolytic, acid production	Actinomycosis
	A. naeslundii	Anaerobic	Mineralization, acid production	Periodontitis, caries
	Bacterionema matruchotii	Facultative anaerobic	Mineralization	Periodontitis
Bacteroides	*Bacteroides melanogenicus*	Anaerobic	Proteolytic, produces collagenase, endotoxin	Periodontitis, acute necrotizing ulcerative gingivitis
Fusiforms	*Fusobacterium*	Anaerobic	Acid production	Acute necrotizing ulcerative gingivitis
Spirochetes	*Treponema macrodentium*	Anaerobic	Nutritionally fastidious	Acute necrotizing ulcerative gingivitis
	T. vincentii	Anaerobic	Nutritionally fastidious	Acute necrotizing ulcerative gingivitis
Diphtheroids	*Corynebacterium* sp.	Species vary	Heterogeneous	Pulp infection, root-canal infection

mouth contains predominantly aerobic or facultatively anaerobic, acid-producing flora in small numbers. The diseased mouth's flora shifts toward larger numbers of bacteria including anaerobic, protein-digesting species. Eating firm or crunchy foods such as apples promotes cleansing and mechanically removes bacteria, whereas eating soft and sticky foods causes debris to adhere to the teeth. Frequent sugar consumption favors the growth of the carbohydrate-loving **cariogenic** (caries-producing) members of the oral flora, the streptococci and lactobacilli.

Microbial numbers also fluctuate with the state of oral hygiene. Brushing, flossing, and professional cleaning reduce the total microbial mass, and shift the species distribution favorably. These practices remove plaque, and temporarily eliminate some anaerobic environments that favor gum disease.

BOX 15.1
The Benefits of Civilization

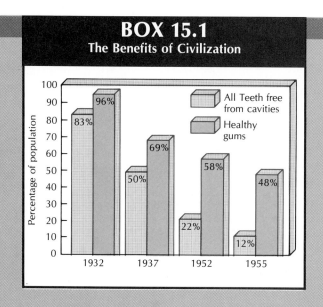

Most primitive peoples enjoyed very good dental health, although they may have suffered severely from other infectious diseases. The coming of civilization tends to reverse this picture. As primitive peoples are relieved of smallpox, cholera, or other plagues, they acquire epidemic oral disease in exchange.

This pattern was documented by dental officers of the British Royal Navy. They visited the remote, inhospitable island group of Tristan da Cunha, located in the middle of the South Atlantic. In 1932, the population of 162 inhabitants subsisted on a limited but balanced diet of potatoes, fish, eggs, and a little milk and vegetables. Only when a ship called (usually less than once a year) did the inhabitants obtain white flour or sugar. After 1932, ships docked more and more frequently. Eventually the Tristan Development Company set up a trading operation. By 1955, the staple foods were being used less, and the inhabitants were basing their diets on baked foods, consuming 4.5 lb of purchased white flour and 2 lb of sugar per week per person. This graph starkly illustrates what happened to their teeth.

PLAQUE

As we noted, microorganisms form **plaque,** a sticky, whitish coating composed of a polysaccharide/protein mixture in which microorganisms are embedded. Plaque is present to some degree on all tooth surfaces. Microbial plaque formation is a necessary first step in the development of all dental diseases.

HOW PLAQUE FORMS

There is almost no such thing as a completely clean tooth. After your teeth are carefully cleaned and polished by the dentist, within five minutes plaque formation begins again.

The clean enamel surface is so smooth that organisms can't adhere to it directly. However, the enamel has a slight negative surface charge, and positively charged salivary proteins start to coat it immediately after it is cleaned. This **acquired pellicle** is readily removed by gentle rubbing and is not harmful in itself. It does, however, provide a base to which microorganisms can adhere. The microorganisms settle out of the saliva, or emerge from pits in the enamel, to reestablish their community.

Microbial growth next forms corncob plaque, which consists of oddly shaped deposits of aerobic cocci surrounding strands of filamentous bacteria, which attach by one end to the tooth surface. These aggregates rapidly develop on the pellicle (Figure 15.3). These early aerobic colonists multiply even more rapidly if they are provided with sugars, and the plaque layer deepens with time.

FIGURE 15.3 Plaque
(a) This scanning electron micrograph of corncob plaque reveals its microbial structure. Filamentous bacteria are the "cob," adhering at one end to the tooth or plaque surface; they are surrounded by a coating of cocci, the "kernels." (Photo courtesy of Sheila J. Jones, University College, London.) (b) This electron micrograph shows the layers of microbial flora associated with plaque on the surface of human tooth enamel. (×7176. Courtesy of Dr. Max A. Listgarten, School of Dental Medicine, University of Pennsylvania.)

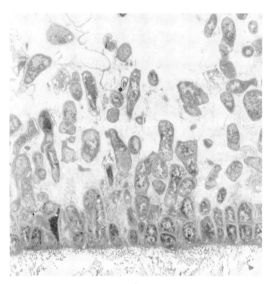

(a)

(b)

If the plaque is not brushed off, but accumulates over a period of days, the deep plaque layers nearest the tooth surface become more and more anaerobic because they don't get the benefits of fresh saliva with its dissolved oxygen, pH buffering, and calcium ions. The shift in environmental conditions causes anaerobic forms to proliferate. Over a period of weeks, "old" plaque at or under the gumline tends to calcify into calculus, which is so hard it cannot be brushed off; it must be scraped. The calcium and phosphate in calculus appear to crystallize from the saliva.

As you can see, plaque formation and its buildup are sequential and progressive. There is no practical way to prevent plaque from beginning to accumulate. Routine dental hygiene is beneficial because it regularly removes the fresh accumulation. With good home hygiene, we need never get to the anaerobic stage or the calculus stage at all.

MICROBIAL ACTIVITY IN PLAQUE

Microorganisms metabolize actively at all depths in plaque. We will look at four important aspects of their activities — the formation of dextrans, acids, inflammatory substances, and calculus.

Certain organisms, most notably *Streptococcus mutans*, are active polysaccharide producers. The most important of these polymers, *dextran*, is formed from sucrose molecules. Dextran helps microorganisms attach to teeth and forms a major component of the total plaque. Only a few bacteria synthesize dextran, but others make additional sticky substances such as glucan and levan from sugars. Without dietary sugar, the bacteria do not make these materials. You can test this for yourself. After morning brushing, on one day eat something sugary about every hour. At the end of a few hours, notice how sticky your teeth are. The next day, try to avoid all sugars (this is not easy — you'll have to read food labels for hidden sugar) for the same period. You'll note that your tooth surfaces are almost completely clean at the end of the time.

Microbes are also active in acid formation. Observe what happens in your mouth when you eat a small bit of sugary food. You may note that the sweet flavor is soon replaced by a sour, acid taste that lasts long after the sweet taste is gone. This change occurs because fermenting anaerobic bacteria and fungi in your mouth produce acid from carbohydrates. The streptococci and lactobacilli are most significant. As sugars are eaten, they seep into the plaque. Then, as the bacteria ferment the sugar molecules, releasing acids, the pH rapidly drops (Figure 15.4). Even after the sugary food has been swallowed, acid production continues as the bacteria use up the sugar that is still trapped in the plaque. The acid is strongest in the most anaerobic area — at or

FIGURE 15.4.   Acid Production from Glucose
To record these pH changes, a microelectrode was placed in the plaque on a subject's teeth. The subject took a mouthful of 10% glucose solution, swished it around a moment, and then spit it out. The pH meter recorded the sharp fall of pH as bacteria in the plaque converted the sugar to acid. A pH minimum was reached, near pH 5, close to the critical point at which calcium salts on the tooth surface begin to dissolve. Note that the pH rose again, but much more slowly than it dropped. It took almost 20 minutes for the pH to return to near normal after even this momentary contact with sugar. The thicker the plaque or more prolonged the sugar contact, the longer the pH will stay at or near the critical point with resulting damage to the tooth.

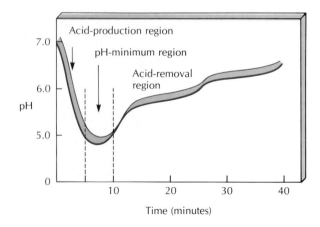

near the tooth surface. When sugary foods are eaten frequently, pH levels rarely return to normal. The tooth enamel is exposed to a continuous acid bath.

Some Gram-negative microorganisms, especially at or below the gumline, produce inflammatory substances such as endotoxins. Both *Bacteroides* and *Actinomyces* release protein-digesting enzymes such as collagenase, which may gradually loosen the tooth from the jaw.

Most researchers believe that microorganisms are also the cause of calculus formation (Figure 15.5a). Both *Actinomyces naeslundii* and *Bacterionema matruchotii* (Figure 15.5b) form calcium phosphate crystals in culture.

DENTAL CARIES

Dental caries results from microbial acid and plaque formation. Plaque sets the stage for caries because it collects the acid-forming bacteria on the tooth surface, supplies an anaerobic environment for fermentation, traps the acids, and excludes the protective saliva. Dental caries is endemic in developed nations. In the 15-to-24-year age group, almost 80% of tooth extractions are of teeth lost due to caries.

ACID DECALCIFICATION — THE FIRST STEP

Caries lesions are basically the outcome of chemical attack on the enamel and dentin, and

FIGURE 15.5 Calculus
(a) This scanning electron micrograph of calculus looks superficially like a coral growth. The structure of calculus is an accretion of calcium salts. The pores are holes in which bacterial cells were embedded before the specimen was prepared for the microscope. (Photo courtesy of Sheila J. Jones, University College, London.)
(b) *Bacterionema matruchotii* strain 14266 forms large numbers of dark, pointed calcium phosphate crystals, which are visible both intracellularly and in the medium. It is believed that this organism helps create calculus in the mouth. (×17,700. From Lie and Selvig, *Scand. F. Dent. Res.* (1974), 82:8–18.)

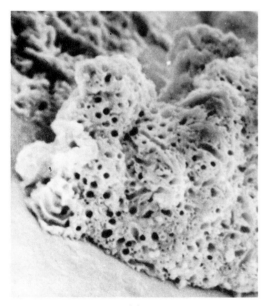

(a)

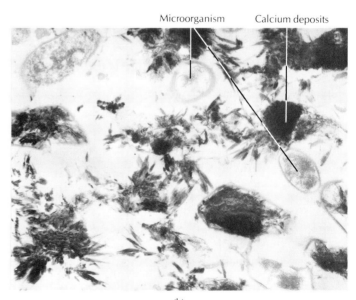

Microorganism   Calcium deposits

(b)

they develop in a stepwise fashion (Figure 15.6). Enamel's mixture of calcium salts and proteins is hard and resistant at a neutral pH, but it begins to dissolve when the pH falls below a certain level called the **critical pH.** This value varies among individuals. For most, it lies close to pH 5.0. When the critical point is reached, material dissolves from the enamel surface. This process is called **acid decalcification.** A roughened "white spot," not yet a cavity, forms to indicate a site of developing caries. The dentist tells you "We're watching it." How rapidly the lesion expands from that point on depends mainly on how much acid exposure the tooth endures.

Acid decalcification can be partially reversed by the precipitation of new calcium ions from the saliva, thus stabilizing the inorganic part of the enamel. Fluoride ions in drinking water, foods, or toothpastes also chemically stabilize calcium salts. When built into the tooth enamel, fluoride ions seem to lower the critical pH so that weaker acids no longer damage the tooth.

THE LESION ENLARGES

The original lesion enlarges slightly every time the local pH drops to the critical point. Over a period of weeks or months, the cavity penetrates into the dentin, where microorganisms and their products spread more rapidly through the dentin

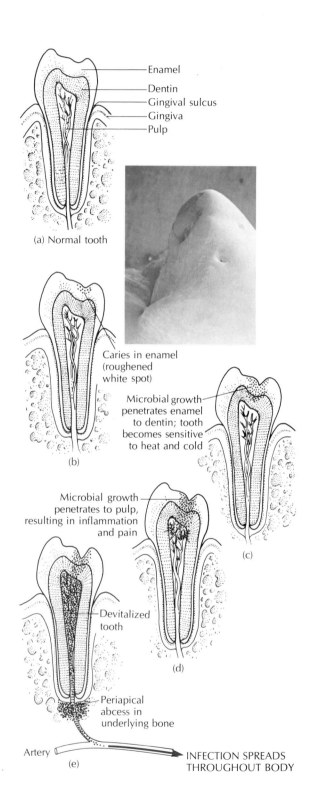

FIGURE 15.6   The Progression of Untreated Caries
(a) Normal tooth. (b) A carious lesion developing in a fissure on the crown of a bicuspid. This is an area in which plaque readily accumulates. The scanning electron micrograph shows a carious lesion developing in a fissure on the crown of a bicuspid. (Courtesy of W. Rosenberg, Iona College/BPS.) (c) The enamel has eroded and the lesion now involves the top layers of the dentin. It can be readily treated at this stage. (d) The dentin has been penetrated, and microbial invaders have reached the pulp. The inflammatory response here would be causing acute pain. This tooth can be saved only through extensive reconstruction. (e) An abscess has formed in the jaw bone. At this point, the individual's general health is also threatened.

tubules, and soften it. The microbial colony becomes larger, and so does the area that has been destroyed. The person will begin to feel a painful sensitivity to hot and cold foods and to sugars and salts at this point. It may take up to two years to get to this stage. Dentists treat caries of the enamel and dentin by removing the decayed tooth material and replacing it with a material such as the metal alloy amalgam or porcelain — the common filling. Unfortunately, the areas under the filling and around its borders remain a prime location for reactivating caries; consequently, fillings often must be replaced.

DEVITALIZATION — THE FINAL BLOW

Eventually, if a lesion is left untreated, it will penetrate into the pulp cavity. There the living tissue sets up an inflammatory response to the microbial invaders, causing acute pain. The live cells of the pulp are usually killed. The tooth is then said to be **devitalized.** In the past, such teeth were invariably pulled. Now a root canal filling may be done. This is a procedure in which the pulp cavity and root canals are emptied of cellular material, any adjacent abscesses are drained, and the cavity sterilized and filled. Such devitalized filled teeth may last indefinitely.

INFECTION BEYOND THE TOOTH

An untreated lesion may eventually spread beyond the tooth pulp to cause infections in the underlying jawbone or at the gum surface (gum boils). In some cases, spreading infections of the bone (osteomyelitis) may be the outcome. Bacteria entering the blood stream may also cause severe infections in the blood itself (septicemia) or in the heart (infectious endocarditis). It is possible, then, for tooth decay to be fatal, if it is totally neglected!

PERIODONTAL DISEASE

Periodontal disease, meaning around the tooth, is a pattern of gum inflammation and destruction of the collagen and cementing proteins that support the tooth and its roots (Figure 15.7). It is prevalent in older age groups; as much as 97% of the population is affected to some extent by age 45. From that age on, about 80% of tooth loss is

FIGURE 15.7 Healthy and Diseased Gums
(a) Normal, clean healthy teeth. The gums are smooth, not swollen. There is no accumulated material on the tooth surface. (b) A diseased region. The plaque is prominent at the gum line and between the teeth. The gingiva are swollen, malformed, and have retracted, exposing the neck of the teeth. The neglect that caused this state has been going on for a long time. (Photos: courtesy of Myron Nevins, DDS, Co-Editor, *International Journal of Periodontics and Restorative Dentistry.*)

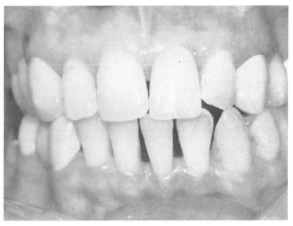

(a)

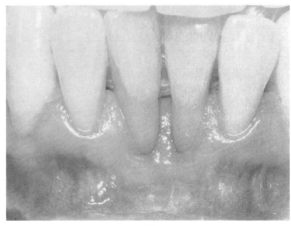

(b)

caused by periodontal disease. Like caries, periodontal disease can be well controlled by conscientious tooth care and sensible diet.

There are three common forms of periodontal disease. All are endogenous; they develop because of overgrowth of microbes normally present in the gingival sulcus (Figure 15.7c). The contributing microbial activities are the production of endotoxins, enzymes, and calculus.

GINGIVITIS

Gingivitis is the most common periodontal abnormality. It is an inflammation of the gingiva, which swell, become deeply red in color, and bleed easily. As the gums swell, they may appear to grow out over the tooth surface. However, the side of the tooth remains adequately attached to the gum and the socket.

The cause of gingivitis is microbial debris (plaque and calculus) in the gingival sulcus. The inflammation is caused by a combination of factors, including abrasion by calculus, bacterial endotoxins, and complement reactions. Inflammation subsides when the provoking microbial aggregates and calculus are removed. Flossing, which removes food particles and plaque from the gingival sulcus, is the most effective preventive measure.

Some hormonal factors increase susceptibility to gingivitis, by changing the physical condition of the gingiva. Some pregnant women notice that their gums are much more susceptible to irritation and bleeding. Psychological stress also has harmful effects on the gums, through altering levels of hormones.

PERIODONTITIS

Periodontitis is a more severe condition that develops from untreated gingivitis. You will remember from Chapter 11 that severe and persistent inflammation can kill the involved tissues. In periodontitis, tissue in the gingival sulcus is destroyed. The key event in periodontitis is the separation of the gingival epithelium from the enamel at the side of the tooth. The normal sulcus deepens into a debris-packed pocket. The pocket becomes profoundly anaerobic, and shelters anaerobic bacteria that produce enzymes. These enzymes attack the epithelium, the connective tissue, and the fibers binding the tooth to the jawbone. The neck and roots of the tooth, which are never normally exposed, are bared and become increasingly susceptible to root caries. The surrounding tissue surfaces may become ulcerated and pus may form (pyorrhea).

These events may trigger cell-mediated immunity. T lymphocytes reacting to the pocket flora may release a lymphokine that causes certain bone cells to remove calcium salts from the surrounding jawbone. This bone softening loosens the tooth in its socket, and the tooth eventually falls out.

Nutritional deficiencies are important in periodontitis in many people. Calcium, vitamin B, and vitamin C deficiencies contribute to unhealthy or ulcerated gums and tooth loss.

In treating periodontal disease, it is first necessary to remove the offending masses of microorganisms. If deep pockets have already been formed, surgery may be necessary to reconstruct the area.

ACUTE NECROTIZING
ULCERATIVE GINGIVITIS

Acute necrotizing ulcerative gingivitis (ANUG) is a specific, rapidly developing form of periodontal disease found mainly in people between 13 and 30 years of age. Many people believe it is communicable, but this is not the case. Actually, it appears to cluster among individuals of the same age who share the same stresses, not necessarily the same silverware. Patients with ANUG usually have a history of acute anxiety or stress. The condition is common in new military recruits or in students during examination periods. People who are under stress usually have elevated steroid levels, which correlate with depressed cell-mediated immunity. Thus ANUG appears to be precipitated primarily by hormonally altered host defenses, upsetting the natural symbiosis with the normal flora.

ANUG develops suddenly, appearing as necrosis (tissue death) of the gum epithelium between the teeth. Pain, bleeding, and a whitish coating or membrane are other signs. In deep lesions, large numbers of *Capnocytophaga, Eikenella*, spirochetes, and fusiforms proliferate (Figure 15.8). ANUG responds promptly to anti-

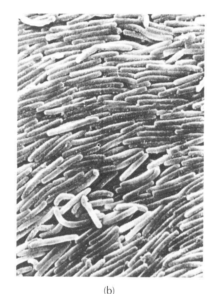

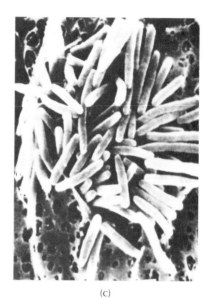

(a)

(b)

(c)

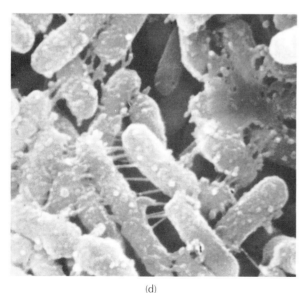

(d)

FIGURE 15.8 Periodontal Disease Organisms *Capnocytophaga:* (a) *C. gingivalis,* (b) *C. ochraceae,* (c) *C. sputigena.* (Photos: courtesy of Malcolm A. Lynch (ed.), *Burket's Oral Medicine: Diagnosis and Treatment* (7th ed.). Philadelphia: J. B. Lippincott (1977), Fig. 10.2 (a), p. 178.) These organisms are increasingly being considered key factors in periodontal disease. (d) *Eikenella corrodons:* This Gram-negative bacillus is often found in the lesions of acute necrotizing ulcerative gingivitis. (×19,800. Courtesy of Stanley C. Holt, University of Texas at San Antonio.)

biotic therapy and local cleaning. But if it is not quickly treated, there may be scarring, leaving the person especially susceptible to periodontal disease later in life.

ORAL LESIONS AND INFECTION HAZARDS

When a dental professional examines your mouth, he or she is checking primarily for tooth decay and gum disease. However, the dentist also observes whether there are signs of other infectious diseases visible in the oral cavity. About 50 infectious diseases that primarily affect such areas as the skin, respiratory tract, and urogenital tract, may also affect the oral tissues (Table 15.3). The oral lesions and infective secretions of these diseases are an important occupational hazard to dental practitioners themselves. Some examples are syphilis, meningitis, and polio. More hazardous are diseases such as hepatitis B, where there are no oral lesions to alert the practitioner of the risk; however, the saliva and any blood shed during a dental treatment are full of infective viruses.

Beyond that, contaminated instruments, tissues, and dressings may transmit the disease to the next patient. Modern high-speed drills, which are cooled and lubricated by water sprays, must be correctly used to prevent the spread of aerosols.

Current dental practice stresses the importance of two precautionary activities intended to reduce the spread of infection. First, the oral examination should include careful observation of any medical abnormalities. If any are found, the patient will then be referred to a medical doctor for follow-up. This practice promotes rapid detection and treatment. Second, dentists have adopted rigorous preventive techniques for their own and their clients' protection. These include using facial masks, using disposable products wherever possible, and autoclaving instruments rather than simply boiling or steaming them.

PREVENTING ORAL DISEASE

With present techniques we can prevent both dental caries and periodontitis. If the disease is already established, we can stop it from progressing further. There are three approaches. We can reduce the numbers of undesirable bacteria through good oral hygiene, by brushing and flossing. We can consume a diet favorable to dental health, including adequate calcium and vitamins, and limited sugars. We can improve the acid resistance of the tooth structure by incorporating fluoride. Unfortunately, only a relatively few highly motivated individuals actually carry through on such a program. Human nature is not likely to change in the near future, it seems. Practical-minded researchers hope eventually to find prevention methods that will work despite casual oral hygiene and unwise dietary habits.

CONTROLLING PLAQUE

An important means of controlling plaque is to teach children effective brushing habits early, and to retrain adults. An effective brushing technique is one that completely removes the freshly formed plaque. Brushing also toughens up the gingiva and stimulates its blood supply. Brushing the tongue (although it tickles) significantly reduces the microbial count. In older children and adults, flossing removes plaque from the inner surfaces and cleans the gingival crevice, a key to preventing periodontal disease.

PREVENTING CARIES

Dietary management and fluoridation are useful strategies in controlling caries. Perhaps the most important single preventive action is reducing sugar intake. An illustration of the contributors to dental caries is presented in Figure 15.9.

CONTROLLING DIET.   As we saw, carbohydrates nourish and encourage acid-forming bacteria. Some carbohydrates are more cariogenic than others. Many studies, in a number of countries,

TABLE 15.3 Infectious diseases with oral lesions or hazards

Disease	Agent	Primary System or Organ	Oral Manifestation	Transmission Hazard
Candidiasis	*Candida albicans*	Skin, upper respiratory tracts, urogenital, systemic	Thrush (infants) denture stomatitis, root-canal infections	Low
Chickenpox	Varicella-zoster virus	Skin	Vesicles on gingiva and mucosa	High
Diphtheria	*Corynebacterium diphtheriae*	Nasopharynx	Lesions of oral mucosa, paralysis of palate	Moderate
Gonorrhea	*Neisseria gonorrhoeae*	Urogenital tract	Stomatitis, gingivitis	High
Hepatitis B	Hepatitis B virus	Liver	None	High
Herpangina	Coxsackie A	Upper respiratory tract	Oropharyngitis, lesions of tongue and soft palate	High
Histoplasmosis	*Histoplasma capsulatum*	Lung	Ulcers of tongue	Low
Influenza	Influenza virus	Lower respiratory tract	Inflammation or ulceration of oral mucosa	Moderate
Measles	Rubeola virus	Skin, upper respiratory tract	Koplik's spots, inflamed painful mucosa	Moderate
Meningitis	*Neisseria meningitidis*	Central nervous system	Petechiae of oral mucosa	High
Mumps	Mumps virus	Salivary gland	Parotitis, stomatitis, painful chewing	Moderate
Oral herpes	Herpes simplex Type 1	Skin, oral cavity	Primary acute stomatitis, recurrent ulcerations of lip and oral mucosa	High
Pharyngitis Staphylococcosis	*Staphylococcus aureus*	Skin, any body area	Osteomyelitis, parotitis, periapical abscesses	Moderate
Polio	Poliovirus	Nervous system	Paralysis of jaw	High
Salivary-gland virus disease	Cytomegalovirus	Gastrointestinal, salivary gland, congenital	None; saliva infective	High
Scarlet fever	*Streptococcus pyogenes*	Upper respiratory tract	Rash on soft palate, raspberry tongue	High
Syphilis	*Treponema pallidum*	Urogenital tract	Primary chancre may be oral, secondary mucous patches, gummae	High
Tuberculosis	*Mycobacterium tuberculosis*	Lung	Ulcers of tongue and other mucosa	High

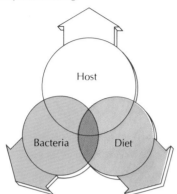

Maintenance of oral
health through
• host defenses (antibodies, tissue integrity)
• host hygiene (flossing and regular,
  frequent brushing)

Host

Bacteria   Diet

Controlling acidogenic   Encouraging strong
and plaque-synthesizing   tooth enamel through
bacteria through   • controlling intake of
• vaccination   carbohydrates,
• reducing growth   particularly sucrose
  substrate   • fluoride
• antibacterial drugs   • calcium and vitamins

FIGURE 15.9 The Multifactorial Concept of Caries
Caries develop when three factors interact: The
individual's mouth must be inhabited by certain
types of acidogenic and plaque-synthesizing
bacteria, the person's diet must supply ample
carbohydrates, and the person must fail to use good
oral hygiene or be deficient in innate defenses, or
both. If any one of these factors can be corrected,
caries will be arrested.

carrots, apples, or whole-grain breads are even
helpful in that they have a mechanical cleaning
effect on the teeth.

Many people in the United States get 30% or
more of their daily calories from sucrose, often
consumed in a dozen or more snacks. This practice sets up the precise oral environment that
most favors caries development. Two decades
ago Americans consumed an average of 50 kg of
sugar per year; at that time our 12-year-olds aver-
aged nine decayed, filled, or missing teeth. In
Japan, also a highly developed nation, annual per
capita sugar consumption was about 12 kg, and
12-year-olds had only about three damaged
teeth. The good news is that fluoridation and
improved dental care are making inroads to improve our dental health. Caries declined sharply
(32%) in the United States between 1972 and
1979 and continues to decline. The bad news is
that our sugar consumption is still increasing to
57.5 kg. (126.8 lbs.) of conventional sweeteners
each in 1984. Studies have shown clearly that
between-meal sugar snacks are far more damaging than a sugar dessert with the meal, followed by brushing. Healthy eating habits for caries prevention include minimizing sucrose
intake, avoiding frequent or prolonged contact
with sugar snacks, and rinsing the mouth with
water after eating sugar, if brushing is not practical.

One way of reducing sugar intake is to substitute some of the sugar-free foods and drinks
that are winning increased sales (Table 15.4). Replacement sweeteners such as saccharin or aspartame are used instead of sucrose or glucose,
primarily to reduce the caloric content of foods.
These sweeteners also make the foods less cariogenic. Xylitol and sorbitol are extensively used
as sugar substitutes in chewing gum and candies.
Box 15.2 provides more information about artificial sweeteners.

FLUORIDE.   The addition of the mineral fluoride
to water supplies in many communities is perhaps the most important factor in reducing tooth

age groups, and socioeconomic categories have
shown that the type of carbohydrate (starch or
sugar), the amount, and the time of day it is
eaten (with meals or between meals) are all factors. Sucrose is by far the most damaging carbohydrate because it contributes both to plaque
formation and to a drop in pH. Other sugars,
such as lactose from milk, contribute mainly
acid. Starches are much less damaging, since
they release little glucose in the mouth. Some
foods containing complex carbohydrates, such as

			TABLE 15.4 Substitute sweeteners			
Sweetener	Classification	Source	Sweetness Compared to Sucrose	Fermentation by Oral Bacteria	Plaque Formation	Metabolism by Human Beings
Sucrose	Disaccharide	Fruits, vegetables	Equal	Rapid	Strong	Rapid
Glucose	Simple sugar	Fruits, vegetables	Less	Rapid	Moderate	Rapid
Fructose	Simple sugar	Fruits, honey	More	Rapid	Moderate	Rapid
Xylitol	Five-carbon sugar alcohol	Fruits, vegetables, animal tissue	Equal	None	None	Rapid
Aspartame	Dipeptide	Synthetic	More	None	None	Rapid
Saccharin	Complex organic molecule	Synthetic	More	None	None	None

decay in children. Fluoride ions can be incorporated into the enamel of a developing tooth, or can be adsorbed to the surface of a fully formed tooth. Fluoride stabilizes enamel and dentin and increases their acid resistance. The effect of fluoride is greatest when the fluoride ions are built into a developing tooth. Some pregnant women take fluoride to make sure it is available during the period when the fetus's teeth are forming. Children receiving fluoride incorporate it into their baby teeth and developing permanent teeth. Topical applications and/or fluoride-containing toothpastes are also highly effective in reducing caries. Fluoride is also beneficial to the elderly, because it aids in calcium retention, strengthens bone structure and reduces the risk of fractures.

SEALANTS. Newly erupted molars normally have pits and fissures in their enamel. Without special precautions, the crowns of molars rapidly develop caries. Increasingly, dentists recommend sealing newly erupted, caries-free molars. This procedure involves applying a layer of tough synthetic resin to the crown of the molar. Sealants prevent the development of caries as long as they remain in place; teeth may be resealed if the resin peels off.

PREVENTING PERIODONTAL DISEASE

Two important factors in preventing the gum diseases are diet and hygiene. Dietary calcium and vitamins are necessary. Vigorous brushing of the sides of the teeth and the gums, and daily flossing are also essential.

In concluding this section on prevention, I realize that you have heard most of this before. Perhaps after examining what the oral flora are actually doing, you will see exactly why these measures are effective and will be able to use them and assist others to use them wisely.

BOX 15.2
Alternative Sweeteners

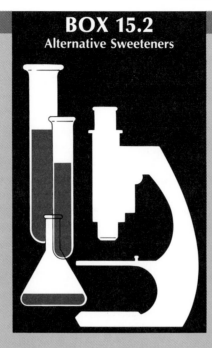

Most human beings (and some other animals) undeniably have a "sweet tooth." We love the taste of sweet foods, but we are constantly told we shouldn't eat them, for one reason or another. Therefore we search for alternatives to sucrose. Three groups of people have something to gain by avoiding sugars — diabetics, persons concerned about their teeth, and weight watchers.

In terms of dental health, the most desirable replacement sweetener is one that bacteria cannot ferment or make dextran from. For a diabetic, the sugar should be one that is not metabolized by the same pathway as glucose. For the dieter, it should not be metabolized at all, or should provide very few calories. Different sugarless sweeteners fulfill different combinations of these requirements, but each has certain drawbacks, just as sucrose does.

At the moment, aspartame, (tradename, Nutrasweet), a compound of two amino acids, is the most generally accepted. It has few calories and is not acidogenic. However, it should not be used by people with phenylketonuria, and may cause mood changes in some people. Xylitol is a possible choice for candies and gum. It is palatable, it is non-cariogenic, and it is tolerated by diabetics. However, xylitol may cause bowel upsets in some, it does have calories, and it is under review for possible cancer risks. Saccharin contains no calories, but foods containing it must bear a label warning of its possible cancer risks.

SUMMARY

1. Oral disease affects almost 100% of the population. Tooth decay and periodontal disease are a result of the growth and metabolism of endogenous microorganisms — the normal flora. Poor oral hygiene and a high-sugar diet tip the microbe–host balance in favor of enhanced bacterial growth.

2. The mouth contains both hard and soft surfaces. It provides both aerobic and anaerobic environments for microorganisms. The moist, warm, nutrient-rich oral cavity is bathed, oxygenated, and buffered by the saliva.

3. The normal flora of the mouth is huge in numbers and diverse in species. The streptococci and the lactobacilli create low pH environments in which tooth enamel will erode. *Actinomyces* and *Bacteroides* are anaerobes active in pits and fissures on the tooth surface and in the gingival sulcus. They contribute to periodontal disease when they multiply excessively. Fusiform bacilli and spirochetes are found in increased numbers in the lesions of acute necrotizing ulcerative gingivitis (ANUG). All oral disease is endogenous, related to overgrowth of normally harmless oral microflora.

4. The mouth is defended by the saliva, bacterial inhibitors, secretory IgA, and the inflammatory response.

5. Plaque is a mixture of living and dead microorganisms, salivary proteins, and the polysaccharides dextran and levan. Ingested sucrose is used by some bacteria to make dextran. As plaque builds up, the deeper layers become anaerobic, and calcium salts accumulate, turning soft plaque into calculus.

6. Within plaque, the microbial mass forms organic acids, toxins, and enzymes that may destroy the teeth or supporting structures. Dental caries develop when prolonged acid exposure demineralizes the enamel. Calcium salts and acid-soluble protein matrix are both lost. The carious lesion can eventually invade the pulp and devitalize the tooth.

7. Periodontal disease is an inflammation of the gingiva and supporting structures of the tooth, caused by irritating calculus at or below the gumline, and the destructive effects of microbial enzymes. Advanced cases — periodontitis — require radical measures. ANUG is an acute, severe form of gingivitis that appears to be brought on by stress.

8. Prevention of oral disease clearly calls for microbial control. Plaque formation can be reduced by minimizing dietary sucrose. Plaque buildup can be controlled by conscientious oral hygiene. Fluoride significantly reduces caries development. Tissues can be kept healthy by adequate dietary calcium and vitamins. Sealants can prevent caries on the crowns of molars.

DISCUSSION QUESTIONS

1. Calcium salts both dissolve and precipitate in the mouth. Outline these processes and the factors that favor or inhibit them.

2. Secretory IgA in the mouth doesn't eliminate the oral flora, so what might its role be?

3. What is the proposed role of *Streptococcus mutans* in production of caries?

4. Frequent sucrose ingestion contributes to caries production in three ways. What are they?

5. What are the microbial activities in the periodontal pocket?

6. How are teeth loosened by microbial activities?

ADDITIONAL READING

Genco RJ, Mengenhagen S E, eds. *Host–parasite interactions in periodontal disease.* Washington, D.C.: American Society for Microbiology, 1981.

Leverett DH. Fluorides and the changing prevalence of dental caries. *Science* 1982; 217:26–30.

Loesche WJ. *Dental caries: a treatable infection.* Springfield, Ill.: Charles C Thomas, 1982.

Newbrun E. Sugar and dental caries: a review of human studies. *Science* 1982; 217:418–423.

Nolte WA, ed. *Oral microbiology.* 4th ed. Saint Louis: CV Mosby, 1982.

MICROBES AND MICROBIOLOGISTS AT WORK

The working world of microbiology encompasses many areas of specialization: agrigenetics, DNA research, drug production, environmental control—the list seems endless. Microbiological processes, such as fermentation, affect your everyday life; microbiological research, such as diagnosis and treatment of cancer, may one day save your life. The microbes and microbiologists at work on the following pages represent a cross-section of the current applications of this dynamic science.

Microbiology and Agriculture

Microbiological techniques have contributed greatly to important agricultural breakthroughs. Besides developing ways to control both internal and external plant and animal parasites, microbiologists have created strains of hardier plants that are better able to withstand disease and pests.

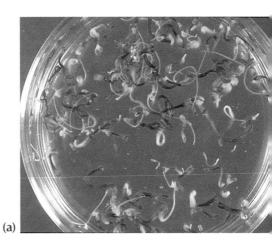

(a)

(a) Through agrigenetics, crop seedlings are regenerated from embryos in the laboratory. (b) After a period of early development in a culture dish, the seedlings are suspended in a nutrient solution. (c) The seeds are then coated with microbial treatments designed to enhance crop yield by making the plants more resistant to disease. If this genetically engineered resistance is passed on to future generations of seedlings, the technique could then be applied to nutritionally and economically important crops outside the laboratory. (Photos a and b, Plantek/Photo Researchers, Inc.; c, Cetus Corporation, photo by Chuck O'Rear.)

(b)

(c)

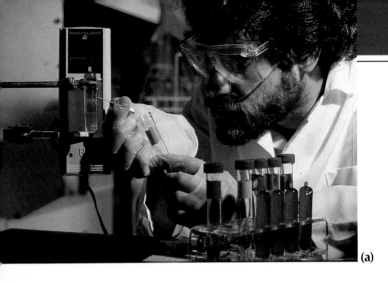

(a)

Using recombinant DNA technology, researchers have been able to combine the genetic characteristics of different species in test tubes. When the rDNA created by this process is introduced into a bacterium, the bacterium takes on the characteristics programmed by the newly acquired genes.

(b)

Recombinant DNA technology has made possible the development of a vaccine for treating infectious hepatitis. In (a), a researcher extracts plasmids from solutions; the plasmids will be cut open using restriction enzymes and will then be fused with the plasmids of microorganisms such as bacteria and yeast. (b) A researcher examines a flask of cytosine, one of the four building blocks of DNA. Such purified bases are easily assembled into whatever gene a researcher finds useful. (c) With the aid of an electron microscope, a technician examines the hepatitis B vaccine created in genetically engineered yeast (shown on monitor). The hepatitis B vaccine, manufactured in 800 liter vats (d), is closely monitored for consistency. (Photos a, c, and d, Hank Morgan/Photo Researchers, Inc.; b, © Erich Hartmann, Magnum Photos.)

(c)

(d)

Microbiology and Medicine

(a)

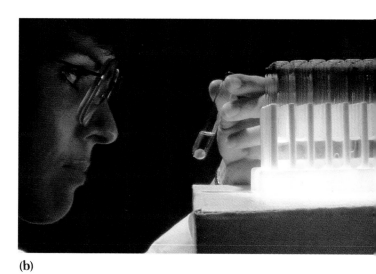

(b)

In tracing the path of a disease, microbiologists perform research in the field as well as the lab. (a) Interviews with potential or real victims often uncover crucial information about the disease's effects and route. (b) In the lab, samples taken during interviews are studied to detect the presence of cancer cells. (c) Monoclonal antibodies (bright green area), created using recombinant DNA technology, bind to breast cancer cells. These antibodies will become components of immunotoxins for the treatment of breast cancer. (d) A technician adjusts the flow of a purifier in a lab where interferon (an anticancer product) is being produced. Bottles collect the purified product after all impurities have been removed. (Photo a, Burt Glinn/Magnum Photos, Inc.; b, Cetus Corporation, photo by Chuck O'Rear; c, Cetus Corporation, photo by Sylvia Hsieh-Ma; d, Cetus Corporation, photo by Chuck O'Rear.)

(d)

(c)

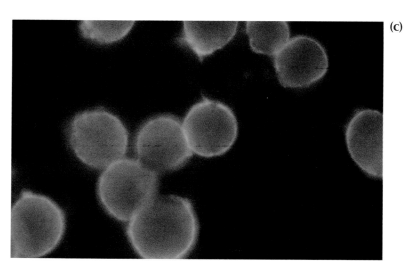

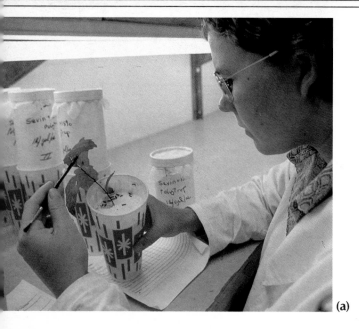

(a)

Microbiology and Environment

Microbiologists work in a variety of environmental fields ranging from oceanography to forestry. Some microbiologists are involved in developing pesticides and herbicides to protect plant life in our environment. A chief concern is the effect that such synthetic substances may have on humans and animals.

(b)

(a) A lab technician checks the mortality rate of gypsy moth caterpillars that have eaten leaves sprayed with a pesticide. (b) A field researcher examines a sample of aquatic life in a Montana creek, looking for changes in animal and plant population due to acid rain or other pollution. (c) Samples of water from rivers and lakes are returned to the lab, where they are closely studied for further signs of pollution's effects on water organisms. (Photo a, USDA, Science Source/Photo Researchers, Inc.; b, © 1981 Jack Fields/Photo Researchers, Inc.; c, © Stephen L. Feldman/Photo Researchers, Inc.)

(c)

16

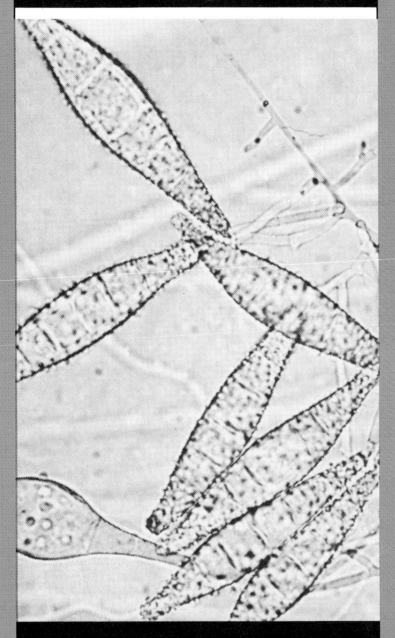

SKIN AND EYE
INFECTIONS

CHAPTER OUTLINE

The skin is our major point of contact with the microbe-filled outer world. Skin is an organ, and it is strikingly well-adapted for its role of microbial barrier. However, because of its exposed position, the skin is frequently damaged, then invaded, so skin infections are relatively common. Although these conditions are usually minor, we cannot dismiss them as unimportant. Some skin infections may spread via the circulation to become generalized and life-threatening. Many viral infections, for example, involve the skin as only part of their invasive activities.

In this chapter we will examine bacterial, viral, fungal, and protozoan skin diseases. We will study seven pathogens in detail, reviewing

Microbial pathogens featured in this chapter:

Staphylococcus aureus
Dermatophytic fungi
Herpes simplex virus
Varicella-zoster virus
Measles virus
Rubella virus
Pseudomonas aeruginosa

the organism, the diseases it causes, the appropriate therapy, and the useful control measures. We will also become familiar with the special infection hazards of burns and bites. Finally, we will look at the surface of the eye and its infections.

THE SKIN — A BARRIER TO INFECTION

Our skin is an organ containing epithelial and connective tissue as its primary components. It also contains glandular, circulatory, nervous, and muscular elements. The skin surface environment is hostile to the majority of microorganisms, yet certain well-adapted types thrive there.

ANATOMY OF THE SKIN

The thin, outermost layer of the skin is the **epidermis** (Figure 16.1), composed of epithelial tissue. Epithelial tissue consists of generally rectangular cells, cemented together by a mucopolysaccharide called **hyaluronic acid.** Epithelium is an effective barrier to most microorganisms. Its surface layers are composed of dead, hardened, and protein-rich cells. These layers are continuously flaked off and shed. The surface is constantly replenished from the deepest layer of rapidly dividing epidermal cells. The newest cell layers rise toward the surface, as new cells are added beneath them. They progressively flatten, lose water, and acquire a strong, flexible protein called **keratin.**

A somewhat thicker layer of connective tissue, the **dermis,** lies beneath the epidermis. Connective tissue is composed largely of a matrix containing protein fibers — elastin and collagen. The dermis also contains scattered fibroblasts, mast cells, lymphocytes, and macrophages. Blood capillaries, lymph sinuses, and sensory nerve endings run throughout.

The dermis also contains small **sweat glands,** whose ducts pass through the epidermis to the skin surface. **Pilosebaceous units** in the dermis (Figure 16.2) contain both a **hair follicle** and one or more **sebaceous glands** that secrete oil to the skin along the hair shaft.

The **subcutaneous tissue** is another layer of connective tissue that lies beneath the dermis and is not actually part of the skin. It anchors the skin to underlying structures. Subcutaneous tissue contains the fat-cell deposits that provide padding, insulation, and reserve energy supplies.

PHYSIOLOGY OF THE SKIN

The physiology of the epidermis is highly specialized. It is influenced by the chemical nature of the glandular secretions. Sebum is an oily secretion that coats and lubricates the skin. Sweat serves for both cooling and excretion of salts. As sweat evaporates, the surface liquids become hypertonic. The pH of the skin secretions is moderately acidic. Surface skin layers are completely aerobic, but the pilosebaceous units are moderately anaerobic.

The cells of the dermis and lower levels of epidermis receive nutrients via the capillary bed; tissue fluids are drained away via the lymphatics. This circulatory network may readily spread microorganisms that enter or invade the skin, but it also carries defensive phagocytes when they are needed (Figure 16.3).

There is little available nutrient material for microorganisms on the skin surface, but the skin microflora make use of what there is. Some metabolize the fats and oils of sebum, amino acids from sweat, and materials scavenged from surface epithelial cells for nutrients. Many skin fungi are able to use keratin.

Although moisture is continuously supplied by sweat, it is lost through evaporation, leaving salt behind. Water for microbial metabolism is therefore limited, and only salt-tolerant microorganisms do well on skin. In general, then, the skin is a poor environment for the growth of most microbial species. Only the specially adapted microorganisms find it hospitable and survive.

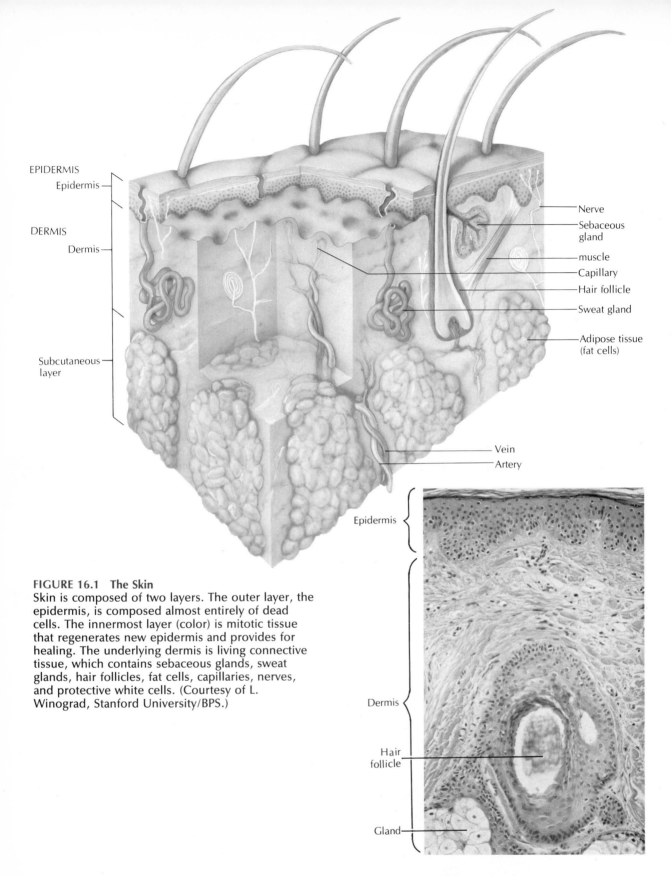

EPIDERMIS
Epidermis

DERMIS
Dermis

Subcutaneous
layer

Nerve
Sebaceous
gland
muscle
Capillary
Hair follicle
Sweat gland
Adipose tissue
(fat cells)

Vein
Artery

Epidermis

Dermis

Hair
follicle

Gland

FIGURE 16.1 The Skin
Skin is composed of two layers. The outer layer, the
epidermis, is composed almost entirely of dead
cells. The innermost layer (color) is mitotic tissue
that regenerates new epidermis and provides for
healing. The underlying dermis is living connective
tissue, which contains sebaceous glands, sweat
glands, hair follicles, fat cells, capillaries, nerves,
and protective white cells. (Courtesy of L.
Winograd, Stanford University/BPS.)

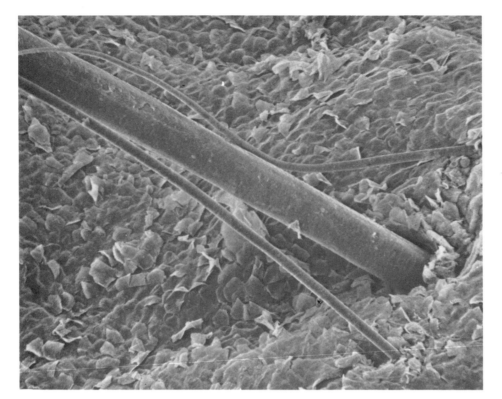

FIGURE 16.2   Scanning Electron Microscopy of the Skin
The surface of the scalp shows flattened, dead cells covering the surface. Three hairs
of different diameter are seen emerging from their follicles. Deep in the follicles,
pockets of anaerobic bacteria live and multiply. (×280. From W.H. Fahrenbach in
W. Montagna and P.F. Parakkal, *The Structure and Function of Skin*. New York:
Academic Press (1974).)

FIGURE 16.3   A Macrophage
This electron micrograph shows the numerous
projecting filipodia characteristic of macrophages.
These defensive phagocytes migrate via the
capillaries into the dermal layer of skin where they
protect it from invading microorganisms. (×660.
From W.H. Fahrenbach in W. Montagna and P.F.
Parakkal, *The Structure and Function of Skin*. New
York: Academic Press (1974).)

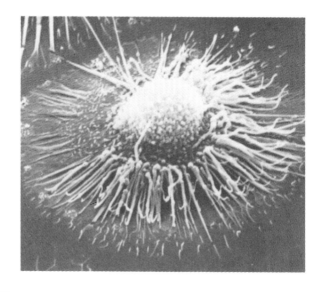

THE SKIN'S NORMAL FLORA

Specially adapted microorganisms, in particular bacteria and fungi, colonize the skin in tremendous numbers. They adhere to all exposed layers of epithelium. Rigorous scrubbing, which removes the loose outer layers, simply exposes the organisms resident on the deeper, living layers.

HOW THE MICROFLORA ARE DISTRIBUTED

There are two distinct microenvironments in skin. The surface epithelium is a dry, salty, aerobic environment. The pilosebaceous units are salty, but moist and anaerobic. Moisture levels, oiliness, and distribution of glandular units vary from one area of the body to another, and so do the numbers and species distribution of microorganisms. The total count ranges from 10^3 to 10^8 microbes per square centimeter. Climate, age, state of health, and degree of personal cleanliness also influence the normal flora. A person's normal flora may be as unique and distinctive as a fingerprint.

SOME IMPORTANT GROUPS OF SKIN FLORA

We can only begin to touch on the great diversity of microbial species found on the skin. Some of the dominant forms include:

- **Propionibacteria**   The propionibacteria are anaerobic Gram-positive bacilli that multiply in the pilosebaceous units. *Propionibacterium acnes* utilizes fats and oils for growth. Because these organisms are almost always present in the blocked pilosebaceous units characteristic of acne, it was once assumed that *P. acnes* was the cause. However, its role in acne is still unclear.
- **Staphylococci**   The staphylococci are facultatively aerobic Gram-positive bacteria found in more than 85% of skin samples. They are highly salt-tolerant and colonize the skin surface and an occasional pilosebacous unit. *Staphylococcus epidermidis* is normally nonpathogenic. The opportunistic pathogen *S. aureus* is found in 5 to 25% of samples, particularly from moist areas such as the nose, groin, and armpit.
- **Micrococci**   The micrococci are strictly aerobic, Gram-positive cocci that often produce bright pigments. They are salt-tolerant, widely distributed, but seldom numerous.
- **Diphtheroids**   The diphtheroids are aerobic to anaerobic, variably shaped, Gram-positive rods. *Corynebacterium xerosis* is numerous in moist areas such as the armpit. Diphtheroids actively digest lipids. The fatty-acid breakdown products they release help exclude some pathogens, as they are directly inhibitory to many microbial species.
- **Yeasts**   Yeasts are common on skin. They include *Pityrosporum*, a small budding yeast, actively lipophilic, generally found on the scalp, and *Candida* species. *Candida albicans* is often seen in normal skin, but is also an opportunistic pathogen.

INTERACTIONS OF BACTERIA ON THE SKIN

The skin and its flora together establish a selective environment that prevents colonization by less adequately adapted pathogens. Resident flora rapidly and efficiently consume available nutrients, providing unbeatable competition to any slower-growing potential invader. Furthermore, resident species release fatty-acid breakdown products that actively inhibit other bacteria. These fatty-acid fragments both limit the growth of normal residents and prevent pathogens from establishing themselves. These byproducts of fatty-acid metabolism are the main contributors to body odor, and a primary target of personal cosmetic products. Deodorants and antimicrobial soaps are formulated to inhibit growth of Gram-positive organisms. Their use shifts the balance of the flora in favor of Gram-negative, nonlipid-digesting species. These Gram-negative species, however, have a slightly higher potential for overstepping the thin line that divides the "normal" from the pathogenic. It appears that, whereas you cannot have too clean a mouth, it may be unwise to have a too well-scrubbed skin. This observation certainly does not apply to the systematic scrubbing of hands in the hospital, the single most important step in preventing disease transmission.

BACTERIAL SKIN INFECTIONS

We will now consider the infections caused by a well-known bacterial pathogen, *Staphylococcus aureus* (Figure 16.4).

STAPHYLOCOCCUS AUREUS INFECTIONS

THE ORGANISM. *Staphylococcus aureus* is a small Gram-positive coccus. Under the microscope it appears in grapelike clusters. It is salt-tolerant and grows in high-salt selective media. Its salt-tolerance makes it ideally suited to colonize, and sometimes invade, the skin. Colonies on agar media may have pale to golden-yellow pigment. On agar media containing

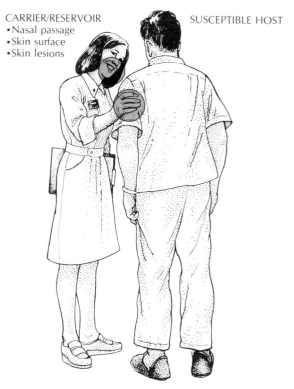

CARRIER/RESERVOIR
• Nasal passage
• Skin surface
• Skin lesions

SUSCEPTIBLE HOST

TRANSMISSION:
• Nasal secretions
• Contaminated fomites
• Hands/skin lesions
• Aerosols
• Contaminated air and food

COMMON DISEASES:
• Impetigo
• Carbuncles
• Pneumonia
• Abscesses
• Toxic shock syndrome
• Kidney infections

FIGURE 16.4 *Staphylococcus aureus*
This organism, which is found on the skin and in the nostrils of many normal people, can cause several types of infections when cellular defenses fail. The scanning electron micrograph shows the characteristic clustering arrangement of this genus, and the diagram indicates some of the key factors in the transmission of *S. aureus*. (Courtesy of Z. Yoshii et al., *Atlas of Scanning Electron Microscopy in Microbiology*. Tokyo: Igaku Shoin, Ltd. (1976).)

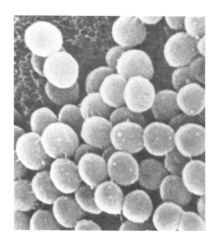

blood, *S. aureus* usually produces beta hemolysis, a zone of clearing around each colony. The less pathogenic *S. epidermidis* lacks the yellowish pigmentation, is not beta-hemolytic, and does not ferment mannitol. The **coagulase** test detects an enzyme produced by most strains of *S. aureus* that coagulates the protein fibrin found in normal plasma. *S. aureus* also produces an enzyme that hydrolyzes DNA (a DNase), that may be detected by a DNase test. The various strains of *S. aureus* may be identified by bacteriophage typing, that is, testing to see which strains of bacteriophage lyse the staphylococcal isolate. This approach is often needed in tracing epidemics.

THE DISEASES.   Many people think of *S. aureus* as very pathogenic. In fact it is only a weak pathogen, or opportunist. In a healthy person, staphylococcal infections, although relatively common, are usually effectively localized and controlled by host defenses, so systemic invasion by this agent does not occur. But *S. aureus* has the potential to cause a range of infections, from superficial skin infections in most people, to serious, even fatal problems in newborns and in compromised adults (Table 16.1).

Furuncles, commonly called pimples or boils, are individual, localized abscesses that develop when hair follicles or sebaceous glands become obstructed. Multiplication of *S. aureus* creates inflammation and local cell death. Pus forms; eventually the abscess opens and the furuncle drains. Furuncles are examples of what are called **self-limiting infections,** ones that are normally taken care of entirely by the body's defenses without outside intervention. Healed furuncles may leave scars, particularly in the axilla (armpit) or on the face.

Carbuncles are abscesses that form on the back of the neck and upper back. This particularly thick and inelastic area of skin tends to prevent normal drainage of the abscess. Instead, pressure forces the abscess to expand laterally and into underlying connective tissue, attaining great size. Surgical drainage may be required,

TABLE 16-1 Some Infections *Staphylococcus aureus* May Cause
Furuncles
Carbuncles
Staphylococcal scalded–skin syndrome
Pemphigus neonatorum
Impetigo
Pharyngitis
Pneumonia
Deep abscesses
Osteomyelitis
Meningitis
Kidney infections
Surgical infections
Food toxicity
Toxic-shock syndrome
Septicemia
Infective endocarditis

along with antibiotic treatment to control bacteremia.

Newborn babies have very little resistance to staphylococcal infection. Some strains of *S. aureus* cause **pemphigus neonatorum,** a severe infection with abscesses that spread rapidly over the entire body. Infections of newborns by another strain, phage type 2 staphylococci, yield the **staphylococcal scalded-skin syndrome.** Here an enzyme called **exfoliatin** causes the epidermis to separate and peel from the dermis. This rapidly spreading condition may progress to bacteremia and death within 36 hours.

Impetigo is a superficial infection of the skin found almost exclusively in children under ten; often it is caused by combined infection of *S. aureus* with *Streptococcus pyogenes* (see Chapter 17). The bacterial agents probably invade a small lesion such as a scratch or insect bite. They produce **hyaluronidase,** an enzyme that digests the epithelial cementing substance. The enzyme enlarges the lesion laterally, causing a spreading blisterlike rash. The lesions are covered by thin crusts; they usually heal without scarring. The organism is readily spread by contaminated fingers to other areas of the child's body and to other children.

THE TREATMENT. Staphylococcal infections may be acquired by healthy persons in the community, but they are also acquired by hospitalized persons. The treatment approach for infections from these two sources is different. Community-acquired *S. aureus* usually responds well to many antimicrobial drugs such as the penicillins and cephalosporins (see Chapter 24 for a discussion of antibiotics). However, hospital-acquired strains are highly likely to be resistant to several antibiotics. To be sure of selecting an effective drug, each strain should be tested for susceptibility. Concern over increasing resistance has led to increased use of topical (surface-applied) disinfectants, rather than antibiotics, for treating localized staphylococcal skin infections in general.

EPIDEMIOLOGY AND CONTROL MEASURES. Ten to thirty percent of individuals are nasal carriers of *S. aureus.* They disseminate the organism via nasal secretions, contaminated fomites, hands, air, or food. Another very important, and extremely common reservoir is the skin lesion. By touching or picking at pimples, large numbers of staphylococci are transmitted to the hands and fingernails. This method of spreading staphylococci is particularly important for food-handlers, who may transfer the bacteria to foods during preparation, leading to outbreaks of staphylococcal food poisoning (see Chapter 19). Preventive efforts are directed to controlling these exceedingly common reservoirs, by educating personnel in correct precautionary behavior. This includes strict aseptic procedures in dealing with high-risk populations, sanitary methods of handling contaminated materials, and personal cleanliness. The only really effective means of control is to prevent the organism (which is very common, after all) from reaching susceptible individuals. There is no available vaccine, and no other direct means of protection.

FUNGAL AND PROTOZOAN INFECTIONS

The skin is a favored location for the development of fungal parasites, most of which require aerobic environments, tolerate dry conditions, and have limited nutritional requirements that they can satisfy by utilizing keratin and other available materials. Fungi that multiply on skin are called **dermatophytes.**

DERMATOMYCOSIS

Dermatomycosis is a general term for mycotic (fungal) parasitism of the skin. The infection usually involves the superficial epithelium, but may sometimes reach the dermis, or the subcutaneous layers. There are numerous potential agents, sites, and symptom patterns. Our discussion will focus on the general aspects. One dermatomycosis that many people experience at one time or another is **tinea pedis,** or athlete's foot. We will use it as a representative example of a dermatophyte disease. Other less common manifestations are summarized in Table 16.2.

THE ORGANISMS. Several members of the genera *Trichophyton, Microsporum,* and *Epidermophyton* are able to cause dermatomycosis. In the United States, infection by *Trichophyton rubrum* is the most common. Human parasitic fungi are in the phylum Deuteromycetes. These fungi rarely reproduce sexually, but multiply asexually by forming spores. Skin fungi are saprophytes — they grow on dead tissues and their biochemicals. Thus these fungi are found growing slowly in the dead skin layers, utilizing the keratin for growth. They may also invade keratin-containing hair and nails.

Microscopic examination of affected tissue reveals reproductive structures called **microconidia** and **macroconidia** (Figure 16.5) and mycelial growth. Dermathophytes can be isolated on differential selective media containing generous amounts of glucose and antibiotics to suppress bacterial growth. They develop slowly, often requiring two to three weeks to form

Condition	Agents	Location	Description	Treatment
TABLE 16.2 The superficial and cutaneous dermatomycoses				
Tinea barbae	*Trichophyton mentagrophytes* *Microsporum canis*	Bearded areas	Localized inflammation, may become ulcerated	Griseofulvin
Tinea capitis	*Microsporum canis* *M. audouini* *Trichophyton sp.*	Scalp	Several forms; hair loss, inflammation, scaling, temporary baldness	Oral griseofulvin
Tinea corporis	*Microsporum sp.* *Trichophyton sp.*	All body areas	Flat, spreading ring-shaped lesions	Oral griseofulvin Topical anti-fungal ointment
Tinea cruris	*Candida albicans*	Groin	Flat, spreading lesions	Antifungal ointments
	Epidermophyton floccosum *Tricophyton sp.*			Gentian violet solution
Tinea pedias and manuum	*Trichophyton rubrum* *T. mentagrophytes* *Microsporum sp.*	Hands and feet	Scaling, cracking and inflammation between toes or fingers; blisters	Oral griseofulvin Topical ointments or dusts
Tinea unguium	*Trichophyton sp.* *Epidermophyton floccosum* *Candida albicans*	Nails	Thickening, deformation of nail; nail may disintegrate	Oral griseofulvin
Tinea versi-color	*Pityrosporum orbiculare*	Body surfaces	Scaly discolorations without inflammation	Topical disinfectants as above

identifiable colonies. After colonies have developed, dermatophytes may be identified by colony shape, color, diffused pigments in the medium, and by their microscopic morphology.

THE DISEASES.   Tinea pedis is one form of a general type of mycotic disease called **ringworm**. Athlete's foot is characterized by itching and burning between and around the toes, followed by the appearance of fluid-filled vesicles. Peeling, cracking, and ulceration may occur, especially on the soles. The actual symptoms seem to be due primarily to delayed hypersensitivity to fungal antigens. Other cutaneous mycoses create similar conditions on other body areas, and sometimes cause hair loss. Tinea cruris (jock itch) affects the groin, and tinea capitis the scalp; the latter occurs particularly in children.

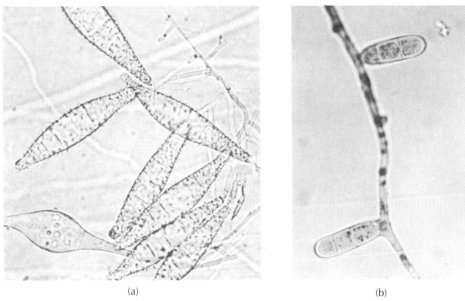

(a)   (b)

FIGURE 16.5   Dermatophytes
These fungi reproduce asexually, forming distinctive macroconidia that are useful in
identification. (a) Macroconidia of *Microsporum canis,* the most common cause of
ringworm in the United States. (b) *Epidermophyton floccosum* is one cause of
"athlete's foot." (From C.W. Emmons et al, *Medical Mycology* (3rd ed.).
Philadelphia; Lea and Febiger (1977).)

THE TREATMENT. Treatments are varied, depending on the location and severity of the infection. Topical ointments or powders containing
clotrimazole, miconazole, or tolnaftate, chemicals that inhibit fungal growth, are commonly
employed. More severe cases may require oral
griseofulvin, an antifungal antibiotic. Corticosteroid ointments relieve the symptoms (itching
and inflammation) but do not inhibit the fungi.

EPIDEMIOLOGY AND CONTROL MEASURES. Dermatomycoses are spread by fungal spores. The
spores are shed by the reservoir, usually an infected person, and contaminate floors and footwear,
from which they are picked up by the next person. Everyone is susceptible to the infection, but
actually having the fungi in the tissue is not
always significant. In the case of athlete's foot,
almost all of us do have the fungi in the skin of
our feet. Symptoms appear only when the organisms are actively multiplying. This, in turn,
usually is associated with hot, moist feet due to
inadequately ventilated footwear. Thus, although preventing transmission by disinfecting
bathing areas, wearing protective footwear, and
other common-sense measures is important,
proper foot hygiene is equally beneficial in that
it keeps those dermatophytes already present
from proliferating to the point where they cause
symptoms. In the case of tinea capitis, the
reservoir may be a dog with ringworm. Pets

with symptoms should be treated and contact with the pet limited until its lesions heal and disappear.

LEISHMANIASIS

Cutaneous leishmaniasis is a tropical protozoan disease of the skin and mucous membranes. It is caused by *Leishmania brasiliensis* or *L. tropica*, and transmitted by sandfly bites. *Leishmania* multiplies extensively inside the host's macrophages, and soon spreads far beyond the site of the original fly bite (Figure 16.6). *L. tropica* lesions tend to heal spontaneously, but cause scarring. In *L. brasiliensis* infections the initial sore heals, but then ulcers erupt on the mucous membranes of the nose, mouth, respiratory tract, or vagina. Extensive tissue destruction and death follow if untreated. Pentostam and antimalarial drugs are used in therapy.

VIRAL SKIN INFECTIONS

Viruses multiply avidly in epithelial tissues, including the skin and mucous membranes. We include here those viral diseases whose most prominent symptom is a rash or eruption on the skin. The viruses in these diseases may also be found in other tissues.

HERPES SIMPLEX INFECTIONS—COLD SORES OR FEVER BLISTERS

THE VIRUS.   The herpes simplex virus is a large, envelope-bearing DNA virus (Figure 16.7). There are two types. Type 1 predominates in the oral and skin lesions, and is the focus in this chapter. Type 2 is the usual cause of the genital tract and neonatal infections (see Chapter 20). Herpesvirus samples are readily obtained by touching a sterile swab to the lesion. Herpesvirus is

FIGURE 16.6   *Leishmania* Undergoing Phagocytosis
(a) A hamster macrophage extends slender filipodia as it prepares to engulf the protozoan parasite, *Leishmania donovani*. (b) The parasite has been completely encircled by several layers of protoplasmic extensions. Unfortunately, encirclement signals, not the destruction of the invader, but the first step in establishing disease, since *Leishmania* multiplies preferentially inside macrophages. (From K.P. Chang, *Leishmania donovani* promastigate macrophage interaction *in vitro, Exp. Parasitol.* (1979) 48:175-189, Figs. 17 and 18.)

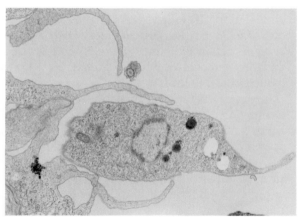

(a)

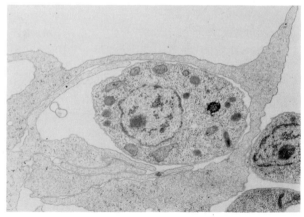

(b)

commonly identified by fluorescent-antibody staining of material from lesions. It can also be readily isolated in cell culture and identified by its distinctive effects on the host cells.

THE DISEASES. Herpesviruses have alternating periods of active viral replication and periods of latency. Almost everyone is exposed to herpesvirus Type 1 early in life and develops a primary infection, usually completely asymptomatic. However, 5 to 10% of persons experience fever, malaise, and an acute stomatitis (infection of the mouth). Most people have no further experience with oral herpes. Some, however, develop a characteristic pattern of recurrent fever blisters as the secondary phase of the disease. These lesions occur on the lip, gum, or tongue. The person recognizes a tingling or burning sensation as a warning that one or more vesicles are about to appear. The vesicles

erupt, ulcerate, then heal without scarring. Weeks or months may pass before the next recurrence. Cell-mediated immunity is crucial in healing herpes lesions. Antibodies to herpes, on the other hand, which are found in 70 to 90% of adults, do not appear to be protective. In immunosuppressed individuals, herpes lesions may spread over large areas of the skin and mucous membranes. This is the very serious condition called **disseminated herpes** (Table 16.3).

TABLE 16.3 Some infections herpes simplex virus may cause
Genital herpes
Herpes encephalitis
Corneal infection
Disseminated herpes
Oral herpes
Pharyngitis
Esophagitis

FIGURE 16.7   Herpes Simplex Virus
Herpes simplex virus (HSV) particles consist of polyhedral nucleocapsids, or cores, surrounded by envelopes, as shown in this transmission electron micrograph of an HSV-infected cell. (×33,540. Courtesy of G.W. Gary, Jr., Viral Pathology Branch, Centers for Disease Control, Atlanta.)

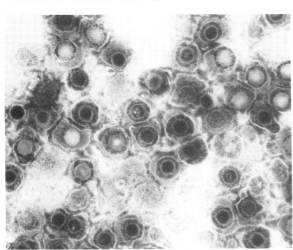

THE TREATMENT. Considerable effort has been expended to develop better drug treatments for herpesviruses. New drugs are constantly being tested and released. All current antiherpes drugs, whether for topical application or systemic therapy, are nucleoside analogs — slightly altered versions of the normal DNA building blocks, containing either altered bases or different sugars. Drug therapy is not usually prescribed for fever blisters, which heal spontaneously. Several drugs, including acyclovir, are used in treating disseminated herpes.

EPIDEMIOLOGY AND CONTROL MEASURES. The human being is the only reservoir of the herpes simplex virus. All active lesions are highly infective. The virus can be transmitted by direct contact such as kissing, by saliva (in coughs and sneezes, or on eating utensils), and by the touch of contaminated hands. Pity the poor baby who is kissed by every uncle and cousin in the family!

Control hinges on preventing transmission. Persons with active herpes lesions should be cautious around highly susceptible individuals, including young babies and debilitated people. Active lesions should be kept clean and dry. Precautions should extend to disinfection of a herpetic person's personal belongings and eating utensils. There is no preventive vaccination.

CHICKENPOX AND SHINGLES

THE VIRUS.   The varicella-zoster or herpes zoster virus causes both chickenpox (varicella) and shingles (zoster). It is a DNA virus, structurally similar to other herpesviruses (Figure 16.8). Laboratory identification is made on the basis of inclusion bodies in material from lesions, characteristic changes in cultured cells, fluorescent-antibody tests, or detection of specific antibodies in the serum of a convalescent person.

THE DISEASES.   **Chickenpox** is primarily a disease of children, in whom it is usually mild. However, people who missed having it in childhood and thus have no immunity may contract it in adulthood and have very severe cases. Following exposure, there is an incubation period, during which the virus multiplies in the respiratory epithelium. It then moves via the blood to the skin of the trunk and face, where it replicates. A vesicular rash, fever, and malaise appear at the onset of clinical illness. Successive crops of vesicles may appear at intervals. The vesicles crust over and heal, with occasional scarring. Complications such as secondary pneumonia and encephalitis are very rare in children, but more common in adults.

Shingles occurs principally in older children and adults. The varicella-zoster virus remains latent in the body after recovery from childhood chickenpox, then suddenly reactivates. Crops of vesicles appear on the skin along peripheral nerves of the face and trunk. Intense burning pain and fever is experienced. Healing may take several weeks and may be followed by transient paralysis or neuralgia (nerve pain). Shingles is frequently triggered by some kind of stress. Persons who are under severe emotional stress, receiving cancer therapy, or in an immunodeficient state are particularly susceptible.

TREATMENT.   In ordinary childhood chickenpox there is no need for treatment other than acetaminophen for fever. Aspirin is not recommended because its use is associated with an increased risk of a rare complication called Reye's syndrome (discussed in Chapter 18). Lotions are used to alleviate the itching. In the immunosuppressed child, specific passive immunization with antiserum and drug therapy with nucleoside analogs are used to control what might otherwise become a fatal infection.

EPIDEMIOLOGY AND CONTROL MEASURES.   The human being is the only host for varicella-zoster virus. Chickenpox usually stimulates lifelong immunity that prevents subsequent attacks in the chickenpox pattern of infection. However, since latent virus remains underneath the myelin sheath of nerve cells, the normal adult's immune system is all that keeps the virus in latency, and each person who has recovered from childhood infection must be considered a lifelong reservoir.

The virus is transmitted by droplets of respiratory secretions from an active case, or by contact with the vesicles or fomites contaminated with such discharges. Shingles lesions contain infectious virus that may cause chickenpox in young contacts. When there is a chickenpox outbreak in a community, usually in the early spring, sick children should be kept from school and out of contact with other susceptible children for one week or until six days after the last group of vesicles appears. Practically, though, it has been assumed that every child would get the disease in early grade-school years, and that it was better to do so earlier than later. Immunosuppressed children are an exception — they should be carefully protected from contact with infective playmates.

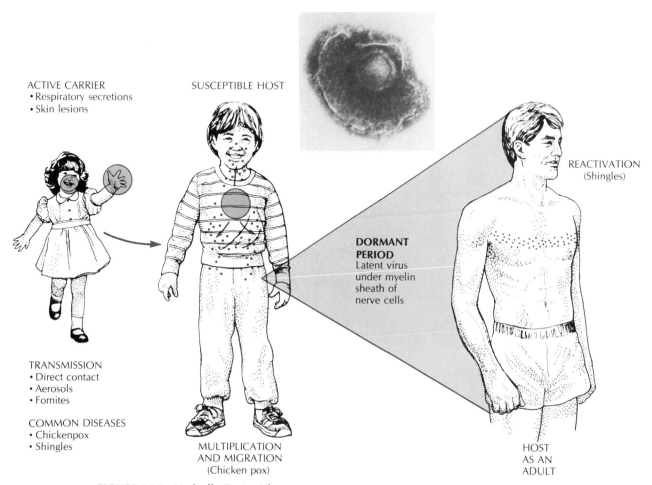

ACTIVE CARRIER
• Respiratory secretions
• Skin lesions

SUSCEPTIBLE HOST

REACTIVATION
(Shingles)

**DORMANT
PERIOD**
Latent virus
under myelin
sheath of
nerve cells

TRANSMISSION
• Direct contact
• Aerosols
• Fomites

COMMON DISEASES
• Chickenpox
• Shingles

MULTIPLICATION
AND MIGRATION
(Chicken pox)

HOST
AS AN
ADULT

FIGURE 16.8 Varicella-Zoster Virus
Chickenpox is usually a relatively mild disease of children. Following exposure to an
active carrier, there is an incubation period in which the virus first multiplies in the
respiratory epithelium and then migrates to the skin of the trunk and face. A
vesicular rash breaks out. After the child recovers from the disease, the chickenpox
virus enters a dormant period only to reactivate later in life in some individuals.
Crops of vesicles appear on the skin along the peripheral nerves of face and trunk.
Healing may take several weeks. (Courtesy of Centers for Disease Control, Atlanta.)

In 1982, there were more than 167,000 cases
of chickenpox in the United States, making it
the second most common notifiable disease.
This high rate points up the inadequacy of control measures. This situation is likely to improve
dramatically soon, because a chickenpox vaccine is in clinical testing. When it is made available to the public, we can expect that public
health measures will shift emphasis toward
reducing the susceptibility of the population
through childhood vaccination. The incidence of
varicella may then decline sharply.

BOX 16.1
The Eradication of Smallpox

How do you go about eradicating a disease, that is, getting rid of all active cases, carriers, and reservoirs? To start with, only certain diseases could be eradicated. Neither the streptococcal diseases nor the staphylococcal diseases are candidates. This is because they don't meet certain criteria. Smallpox, fortunately, did. The criteria and how smallpox met them are explained below.

(1) Smallpox exists only in humans. Its agent does not multiply in any other animal in nature, and doesn't inhabit water or soil. (2) It loses viability fairly quickly when shed from the body. (3) Once people recover, they stop being infective. Furthermore, one cannot carry the agent without being sick, so there are no apparently well people walking around shedding virus. No carriers, in other words. (4) Finally, permanent immunity is obtained by having the disease or by vaccination.

The eradication program devised by the World Health Organization and its member nations focussed on vaccination, active case-finding (rewards were offered to anyone who turned in a sick neighbor), and quarantine. International vaccination requirements stopped the spread of the disease across most boundaries. Each time an affected village was discovered, all inhabitants were vaccinated and quarantine was imposed until all active cases had disappeared. In 1967, when the campaign started, smallpox was endemic in 33 countries, with 10 to 15 million cases a year, and about 3 million deaths. Smallpox was eliminated first from the Far East, then India, the Middle East, and lastly Africa. The last cases were among nomadic tribes in Somalia and Ethiopia in 1977. (Photo courtesy of the World Health Organization/P. Almasy.)

MEASLES

THE VIRUS.　The rubeola virus, a paramyxovirus, is an envelope-bearing, RNA virus (Figure 16.9). There is only one type of naturally occurring rubeola virus. Its envelope bears hemagglutinins and a fusion protein, which causes eucaryotic cells to fuse. As a result, measles virus from patient samples can be identified in cell culture by the virus-induced formation of multinucleated giant cells. Virus-containing giant cells can also be identified by applying fluorescent anti-measles antibodies to slides of pharyngeal epithelial tissue. The virus multiplies in epithelial lymphoid, and nerve tissues.

THE DISEASE.　Measles begins with an incubation period of 9 to 11 days, during which time the virus multiplies asymptomatically in the respiratory mucosa. After this period, the virus is disseminated to the entire body, localizing in the skin and mucosa (review Figure 10.11). Usually the first overt sign of infection is a measles rash called **Koplik's spots** on the oral mucosa. One or two days later a rash also emerges over the rest of the body, and fever, malaise, sore throat, and photophobia (light sensitivity) develop. Normally, rapid improvement occurs within three days, and the virus is completely eliminated during the convalescent period.

Measles is more worrisome than many other childhood infectious diseases, however, because there may be complications. In 1 case per 1000 encephalitis develops. This condition may either leave neurologic damage or even kill the patient. Secondary bacterial pneumonias may also occur. Between these two complications, the measles mortality in the United States is about one death per 1000 cases.

The rubeola virus also appears to be responsible for **subacute sclerosing panencephalitis (SSPE)**, a very rare neurologic degenerative disease. It occurs several years later in about five of every million people who appear to recover normally from measles. Unlike measles, SSPE is noncommunicable. The rubeola virus may also cause fetal damage and congenital malformations. All in all, measles is a disease well worth preventing.

THE TREATMENT. In uncomplicated cases, one treats the symptoms only, with bed rest, good nutrition, and acetaminophen. Should a secondary bacterial pneumonia develop, antibiotics are called for. No effective treatment is available for the neurologic complications.

EPIDEMIOLOGY AND CONTROL MEASURES. Measles is a highly communicable, pandemic, children's disease. In rural Africa and Central America, where there is extreme poverty, and no vaccination, measles causes 21% of all deaths in children under four. The fatality rate is 5%, about 50 times higher than it is in the United States.

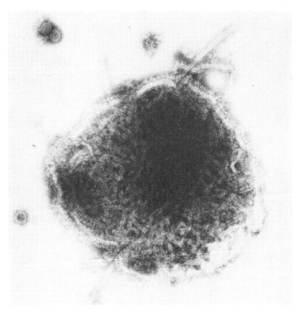

FIGURE 16.9 Measles Virus
The measles or rubeola virus is an enveloped RNA virus that causes a serious childhood disease. However, measles can be effectively controlled by vaccination. (Courtesy of Centers for Disease Control, Atlanta.)

Control of measles centers on vaccination to provide protective immunity. Since there is only one type of virus, complete natural active immunity follows a single case of clinical or asymptomatic illness. The live-virus measles vaccine, prepared from modified virus grown in chick embryo-cell culture, is recommended for all children at age 15 months. If given prior to 12 months, protection is unreliable. The vaccine is also recommended for any adults who lack a history of either effective vaccination or physician-diagnosed measles. An earlier killed-virus vaccine was proven ineffective. It has been discontinued, and persons who received it should be revaccinated with the live-virus vaccine, as should anyone who was vaccinated before the age of 15 months. (Table 16.4.)

Since measles vaccine was introduced, the incidence of measles has dropped over 90%, saving about 400 lives per year. In 1982, only 1714

TABLE 16.4 Epidemiology of skin infections

Disease	Reservoir	Transmission	Incubation Period Days
Pseudomonas infections	Water, soil, human beings	Various: airborne, water droplets, fecal contamination of hands	Short
Staphylococcal infections	Human beings	By hand or fomite; from nasal carrier or lesion	4–10
Streptococcal infections	Human beings	Direct contact with patient carrier; also milk or foodborne	1–3
Tinea pedis	Human beings	Spores on floors, washrooms, clothing	Unknown
Candidiasis	Human beings	By contact with secretions, contaminated objects	2–5; variable
Fever blister	Human beings	Contact with saliva, lesions; by contaminated hands; bite	To 14
Varicella-zoster	Human beings	By contact with or inhalation of respiratory secretions or skin lesions	13–17
Measles	Human beings	By contact with respiratory secretions and urine	8–13
Rubella	Human beings	By contact with respiratory secretions and perhaps urine and feces	14–21

cases were reported in the United States, and the number continues to decline. Like rubella (discussed next), most measles cases now occur in children 10 years of age and older. The junior and senior high school age group now contains the most susceptible individuals — persons who either were not vaccinated at all (but haven't yet had the disease) or were vaccinated by earlier, less effective methods. Almost all older people have had measles, and are immune; almost all

TABLE 16.4 Epidemiology of skin infections (continued)		
Period During Which Communicable	**Patient Control Measures**	**Epidemic Control Measures**
While symptoms persist	Disinfect soiled materials; use burn management techniques	Search for source; institute improved technique
While lesions are present or carrier state persists	Disinfect dressings; avoid contact with infants, ill persons	Search for source, treat and segregate
While discharge persists until 24 hr after penicillin therapy	Isolate until noncommunicable; disinfect discharges and soiled material	Determine source and manner of spread
While lesions persist and viable spores are present	Disinfect socks and shoes	Cleansing of floors, showers
While lesions persist	Disinfect secretions and soiled materials	Not important in skin infections
While lesions are present; saliva infectious even while asymptomatic	Those with lesions should avoid newborns, children with skin trauma, immunosuppressed persons disinfect famlies	Not applicable
From five days before eruption to six days after last lesions; while lesions are present	Exclude child from school, disinfect soiled articles	None
From onset of symptoms to four days after rash	Isolation only in hospital; disinfection: none	Reporting; universal vaccination
From one week before to four days after onset of rash; affected neonates shed virus indefinitely	Isolate only from susceptible pregnant women; disinfection: none	Reporting; universal vaccination

*Source: Benenson, A. S., *Control of Communicable Diseases in Man*. 4th ed., 1985.

younger children have had the vaccine, and are immune.

A national campaign to eradicate measles from the continental United States was begun in 1978. Such a campaign faces several difficulties. It is difficult to find all nonimmune people, there are errors in a small percentage of immunization records, and there are a few persons who refuse immunization. Returning travellers and immigrants reintroduce the virus from parts of the world where measles is still uncontrolled. Because a few unvaccinated people remain, the reintroduced virus may start small outbreaks.

RUBELLA (GERMAN MEASLES)

THE VIRUS.   The rubella virus is a relatively large, enveloped RNA virus in the togavirus group. It can be isolated from throat swabs, feces, and urine into cell culture. Laboratory diagnosis is usually serologic. There are a wide range of tests both for identifying the virus and for determining the presence of specific antibodies in the patient's serum. Sensitive techniques can distinguish the natural or "wild" virus from the virus strain used in the vaccine.

THE DISEASES.   Rubella is typically a children's disease, although susceptible adults also contract it. Usually, it is extremely mild. After an incubation period of two to three weeks, a slight rash spreads rapidly over the body. Fever, sore throat, swollen lymph nodes, and coldlike symptoms may be present although they are rarely severe. Almost 30% of cases have no symptoms whatever, and the fact that the person ever had rubella is discovered only when antibodies are detected by a lab test. The disease usually disappears in two or three days.

Rubella can have serious effects, however, if a woman in the first trimester (three months) of pregnancy contracts it (Figure 16.10). The virus crosses the placenta and invades and multiplies

---

BOX 16.2
The "Rubella Bulge"

All over the country, colleges and universities are seeking to introduce or expand programs for deaf students. Enrollment in programs for the deaf is expected to double to about 12,000 students in the first half of the 1980s, a "bulge" in the usual graphed number. This is largely because of a rubella epidemic that swept the country in 1963 to 1965, which left in its wake thousands of babies born with congenital rubella syndrome. These children, in addition to deafness, often had blindness and neurologic problems. Their intelligence is usually entirely nor-

mal, however, so many seek higher education. Their tragedy was not preventable because, in a sense, they were born too soon.

Rubella vaccine was introduced in 1969. The number of congenital rubella cases has dropped sharply, and is now around 50 per year. These current cases, unfortunately, are ones that were preventable. The largest remaining number of unvaccinated women of childbearing age is in the college years. Unless they choose vaccination, we are not yet through with these tragedies.
(Photo courtesy of Elizabeth Crews.)

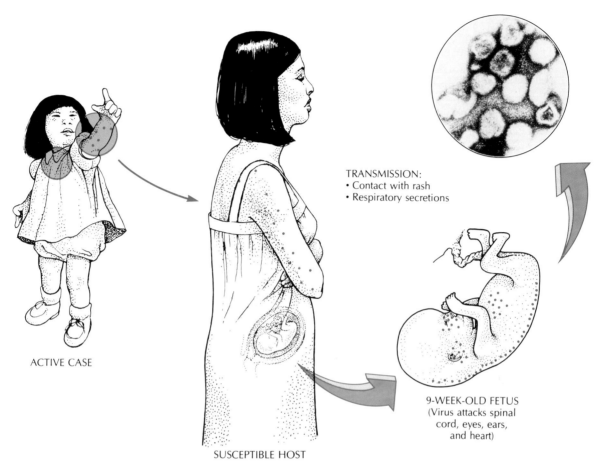

TRANSMISSION:
• Contact with rash
• Respiratory secretions

ACTIVE CASE

SUSCEPTIBLE HOST

9-WEEK-OLD FETUS
(Virus attacks spinal
cord, eyes, ears,
and heart)

FIGURE 16.10 Congenital Rubella Syndrome
If a woman in the first trimester of pregnancy contracts German measles, the virus
may cross the placenta and multiply in the fetus. Infants who survive may be born
with congenital rubella syndrome and suffer from abnormalities of the eyes, ears,
central nervous system, heart, and genitalia. (×34,117. Courtesy of Photo
Researchers/DC.)

in the fetus. In over 30% of such cases, the virus
produces severe fetal malformation or fetal
death. Surviving infants are born with **congenital rubella syndrome,** a condition characterized by abnormalities of the eyes, ears, central
nervous system, genitalia, and heart (Box 16.2).

THE TREATMENT. No specific therapy is indicated in mild childhood rubella. The symptoms
may be treated with rest, fluids, and fever con-

trol medication, if needed. There isn't any treatment for congenital rubella syndrome. All that
can be done is to provide appropriate assistance
to the child in leading as normal a life as possible.

EPIDEMIOLOGY AND CONTROL MEASURES. Humans are the only reservoir for the rubella virus.
It is transmitted by contact with the rash and by
respiratory secretions.

Vaccination is the cornerstone of control. Like natural infection, vaccination confers life-long immunity. Live-virus rubella vaccine is recommended for all normal infants over 12 months of age and any unvaccinated child up to the age of puberty. The vaccine is usually administered in a trivalent formula called MMR (measles, mumps, and rubella) at 15 months. Evidence of rubella vaccination is required for admission to public schools in most states. Hospital employees, because of the risk of exposure and subsequent transmission to pregnant clients, are usually required to have evidence of current rubella vaccination. Several recent outbreaks have occurred among hospital personnel. One interesting case history involved a male obstetrical intern who contracted rubella. Before his symptoms became apparent, he exposed almost 100 pregnant women, but fortunately no cases of rubella resulted. Any woman of reproductive age without serologic evidence of rubella immunity should be immunized. However, she must not be pregnant at the time of immunization and should avoid pregnancy for three months after vaccination. In extensive studies of inadvertent first-trimester vaccination of pregnant women, there is no evidence the rubella vaccine causes any adverse effects on the developing fetus. Nonetheless, this vaccine, like all live-virus vaccines, is contraindicated for pregnant women.

INFECTIONS IN SKIN TRAUMA

The skin's defenses are designed to protect intact skin. When the skin is injured, microorganisms may gain access to larger areas and deeper tissues. Two types of injury, burns and bites, pose special problems.

BURNS

A severe, full-thickness burn (called a third- or fourth-degree burn) poses a life-threatening infection hazard. Some experts say that burn management is the most challenging field of medical care. The burn patient undergoes not only tissue destruction but profound metabolic imbalances that compromise the immune system. The burn site is an aerobic wound that remains open until covered by grafting. Recovery may take months, because of the need to find and implant multiple grafts to reconstruct the damaged area. The burn site presents an unparalleled opportunity for microbial invasion throughout the recovery period. Infection most commonly begins two to four weeks after injury. Seriously burned patients are treated in special burn-care centers where their unique metabolic, pain control, and psychological needs may be effectively met.

CONTROL OF INFECTION IN BURNS

Protective isolation is frequently used to control infections following burns. That is, the patient is placed in an environment in which he or she is protected from exposure to environmental flora or the normal flora of the hospital staff. Prophylactic (preventive) antibiotics may be given briefly. The burn is cleansed initially and daily with silver sulfadiazine, an antibacterial drug. These methods control the common human pathogens such as *Staphylococcus aureus* and *Streptococcus pyogenes.* Burn infections are mainly caused by other agents — highly resistant opportunists, particularly *Pseudomonas aeruginosa.* Occasional epidemics are caused by *Klebsiella pneumoniae, Enterobacter cloacae, Providencia stuartii, Serratia marcescens*, and *Flavobacterium meningosepticum.* Burn pathogens are usually aerobic organisms.

A burn becomes infected first at the margin of the burn or **eschar** (the covering that forms over the burn). These areas are watched carefully to detect the early, possibly curable stages of infection. Routine tissue and blood cultures may be performed. If infection occurs, appropriate antimicrobials can be administered. However, bacteria readily move from the burned tissue into the circulation, and pneumonia and/or septicemia may develop. Survival rates are poor once these complications set in.

PSEUDOMONAS AERUGINOSA INFECTION

THE ORGANISM.   *P. aeruginosa* is a small, highly motile, Gram-negative rod (Figure 16.11). It is aerobic, nonfermenting, and extremely adaptable to varying nutritional and environmental conditions. Its normal habitat is wide-ranging. It is found in soil, water, aqueous solutions of many types, and in the human gastrointestinal tract in some people. It forms a blue-green pigment, **pyocyanin,** and a fluorescent yellow pigment, **fluorescein.** It has a strongly positive oxidase test and a strange, grapelike aroma. Serotyping, phage typing, and pyocin typing are other specialized identification tests.

Pathogenic strains of *P. aeruginosa* produce many virulence factors. For example, **exotoxin A** inhibits protein synthesis in the host. **Protease** and **elastase** exoenzyme attack tissue components and contribute to invasiveness. With its numerous virulence factors, *P. aeruginosa* is able to cause infections in many areas of the body in compromised individuals (Table 16.5).

THE DISEASE.   *P. aeruginosa* enters the burned area and proliferates in the fluid that seeps from the wound. Its proteases allow it to penetrate subcutaneous layers. The exotoxins may cause skin grafts to fail and cause scarring. Severely burned people have depressed phagocytosis and cell-mediated immunity. As a result, *P. aeruginosa* may be able to initiate a septicemia that distributes the toxic products throughout the body.

THE TREATMENT.   There is no bacterium like *P. aeruginosa* when it comes to drug resistance. All strains are resistant to many of the common, well-tolerated antimicrobial agents, such as penicillin and tetracycline. New drugs are constantly being developed and promoted primarily on the basis of antipseudomonas activity. These drugs, including gentamicin, amikacin, carbenicillin, tobramycin, and the newest cephalosporins may initially work, alone or in combination. However, after a new drug is used for a

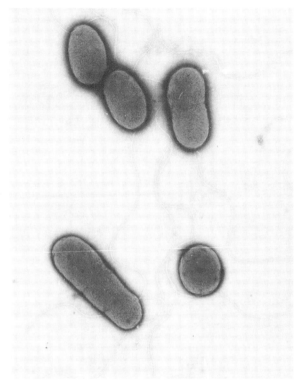

FIGURE 16.11   *Pseudomonas aeruginosa*
This motile, Gram-negative rod is very adaptable to a variety of habitats, and pathogenic strains produce many virulence factors. For these reasons, *P. aeruginosa* is able to cause infections in many areas of the body of compromised individuals. If the organisms should enter burn wounds, their exotoxins may cause skin grafts to fail, with resultant scarring. (×15,000. Courtesy of J.J. Cardamone, Jr., University of Pittsburgh/BPS.)

TABLE 16.5 Some infections
Pseudomonas aeruginosa **may cause**

Infant diarrhea
Corneal infections
Pneumonia (in cystic fibrosis)
Wound infections (burns)
Ear infections
Nosocomial infections
Septicemia

while, resistant strains steadily increase until the drug is useless. Then another drug must be adopted. Many of the antimicrobials that do work require caution because they can be toxic to the patient's liver and kidneys.

EPIDEMIOLOGY AND CONTROL MEASURES.   The reservoirs for *P. aeruginosa* include the human bowel and many aquatic habitats. It not only survives but multiplies in fresh water, some commonly used hospital disinfectants, liquid soaps, and water reservoirs in respiratory-therapy apparatus. The ways it can be transmitted seem limitless. Control measures hinge largely on strict aseptic techniques used by personnel and for all materials the patient will have contact with. New places in which *P. aeruginosa* can lurk are repeatedly discovered and disinfected. Prophylactic antibiotics must be used sparingly, to prevent the rapid buildup of microbial resistance. It is probably impossible to control *Pseudomonas* infections unless we exclude the organism from the burn unit. We cannot do that unless we also exclude human beings, the one reservoir we can't autoclave!

Vaccines against *P. aeruginosa* and/or its virulence factors are still in the research stage. Antitoxins and toxoids for the exotoxin have been developed for experimental trial. Most of the major advances in burn care of this decade are coming through advances in skin graft techniqes.

BITES

A bite is a very likely site for a serious infection, introducing a large sample of the oral flora of the biting individual. This flora is preselected for growth in the body and therefore is more likely to infect than is a random bunch of soil organisms rubbed into an ordinary cut or scrape.

Human bites can be very dangerous. Although we tend to view animal bites more seriously, our attitude should be exactly the opposite. The anaerobic normal flora of the human mouth includes such genera as *Actinomyces* and *Bacteroides*, which may cause bite infections. Some people also carry the true pathogen *Streptococcus pyogenes*. If a human bite is not adequately cleaned, or is prematurely closed by sutures or tight bandaging, infection is inevitable. Such wounds should always be professionally dressed.

Animal bites may also be a problem. However, the oral flora of a normal animal is adapted to that species, and therefore is somewhat less likely to thrive when transferred to human tissue. There are, of course, exceptions, and animal bites should receive immediate medical attention. Monkey or other primate bites are more hazardous to human beings than are those of less closely related animals. *Pasteurella multocida* is the most common pathogen in animal bites. These Gram-negative bacilli, closely related to the plague and tularemia agents, can cause severe infections in human beings. The organism responds to therapy with antimicrobials.

Rat-bite fever is another infection caused by an animal bite. It is characterized by fever, headache, chills, swollen lymph nodes, arthritis, and a rash on palms and soles. The disease responds promptly to treatment with penicillin or tetracycline. The infection can also be acquired from mouse, weasel, cat, and squirrel bites.

EYE INFECTIONS

Although the eye is a part of the nervous system, we will cover eye diseases here because the parts of the eye that usually develop infections — eyelids, conjunctiva, and corneal surface — are all composed of epithelium and modified connective tissue, similar to the skin and continous with it.

THE EYE — STRUCTURE AND FUNCTION

The eye is a complex organ, the structures of which are shown in Figure 16.12. The eyeball, which is rarely infected, is protected by an impermeable, multilayered coat of connective and epithelial tissues. Microorganisms enter the eyeball proper only via blood or gross trauma.

The surface structures of the eye are more exposed to the environment. The eyelids are thin muscular structures continuous with the facial skin. They are lined with a thin mucous epithelium, the **conjunctiva.** This membrane is fused to the surface of the cornea, forming a closed **conjunctival sac.** The **cornea** covering the exposed front of the eye is composed of connective tissue and has no blood supply; it is thin, tough, and transparent. **Lacrimal glands** supply tears via a network of lacrimal ducts, and the tears drain down the **nasolacrimal duct** into the nasopharynx.

The eye surface is washed and lubricated by mucus, sebum, and tears. Tears prevent drying and minimize mechanical damage by dust. Continual tear flow prevents microbial colonization, because most microorganisms cannot attach firmly enough to avoid being washed out. In addition, tears are a source of lysozyme. There are scanty normal flora on the conjunctiva.

BACTERIAL INFECTIONS

The microorganisms that cause eye infection are either intracellular parasites or have superior powers of attachment.

The intracellular bacterial parasite *Chlamydia trachomatis* (discussed further in Chapter 20) causes two eye infections. The first, **trachoma,** is highly communicable. The agent may

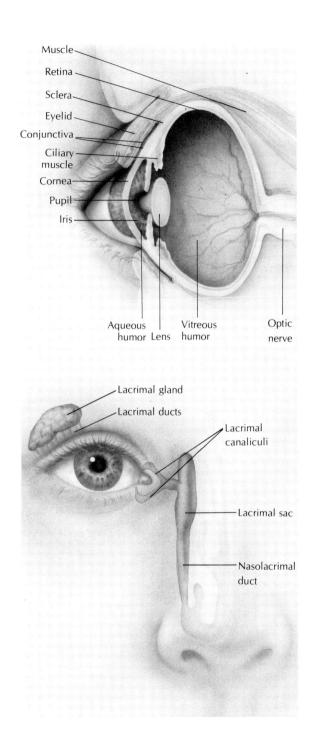

FIGURE 16.12 Structures of the Eye
(a) The eyeball is protected by several layers of connective and epithelial tissues. (b) Tears, secreted by the lacrimal gland, pass through the many tear ducts to flush the conjunctiva and cornea from all sides. Microorganisms are washed away before they have time to attach to the eye surface. The eyelids are lined with a thin mucous epithelium, the conjunctiva, which is fused with the cornea, creating a thin, transparent membrane over the exterior surface of the eyeball.

inflame the conjunctiva, scar the cornea, and cause gross deformity of the eyelids, resulting in reduced vision and possible blindness. In the endemic areas, including the Middle East, certain regions of the United States, and parts of North Africa and South America, trachoma is a major cause of preventable blindness. The agent is spread by direct contact with infected eyes or their secretions, and better sanitation methods are a key to reducing its toll. Tetracycline may be used therapeutically. There is no vaccination.

The second eye infection caused by *C. trachomatis* is **inclusion conjunctivitis.** This disease is a byproduct of chlamydial genital disease, probably the commonest sexually transmitted disease. Inclusion conjunctivitis is a much less severe condition than trachoma, caused by different strains of *C. trachomatis.* Inclusion conjunctivitis affects newborns and adults. The symptoms include inflammation of the conjunctiva, but the cornea is only superficially involved and scarring does not follow. The disease usually disappears spontaneously if it is not treated. Adults spread the chlamydiae to the eye from its reservoir, the adult urogenital tract, with contaminated hands. Neonates acquire chlamydiae during the birth process. Contaminated swimming pools have also been implicated as sources.

Gonococcal conjunctivitis, caused by *Neisseria gonorrhoeae,* the bacterial agent of gonorrhea, follows a very similar epidemiologic pattern to that described above for the relationship of inclusion conjunctivitis to chlamydial genital disease. However, an eye infection with the pathogen *N. gonorrhoeae* may cause blindness if an infant acquires it from its infected mother during a vaginal delivery. Silver nitrate or penicillin drops are placed in the eyes of all newborns to prevent this condition. Adults may also get gonococcal conjunctivitis from touching their eyes with contaminated genital secretions.

Pinkeye is a highly contagious disease caused by *Hemophilus influenza* biotype *aegyptius.* The disease is a conjunctivitis with profuse watery discharge and swelling. Children get it frequently, but pinkeye responds promptly to treatment with sulfonamide ointments.

VIRAL INFECTIONS

Viruses, as intracellular parasites, are able to infect the epithelium of the cornea and conjunctiva of the eye. Some commonly encountered eye pathogens are the adenoviruses and herpes simplex Types 1 and 2. Herpes lesions may cause blindness; they are treated with idoxuridine.

SUMMARY

1. The skin is our most important barrier to infection, covering all of the exposed parts of the body. It has a constantly renewed exterior layer of hardened, dead epithelium, with tight intercellular boundaries to block microbial entrance. The skin's surface is hostile to microorganisms; available moisture is low, and salts provide a high osmotic pressure. The nutrients, including keratin, fats, and small amounts of amino acids, do not support most human pathogens.

2. The skin hosts a large microbial population adapted to survival under these environmental conditions. The flora is mainly Gram-positive, salt-tolerant, and able to metabolize fats secreted in sebum. Inhibitory fatty acid breakdown products aid in microbial population control. Several anaerobic species inhabit the deeper dermal regions.

3. Trauma and incompletely developed or temporarily compromised host defenses lead to skin infection. Disturbance of the normal flora or alteration of the physiologic state of skin may provide a foothold for the pathogen.

4. Two species of bacteria, *Staphylococcus aureus* and *Streptococcus pyogenes,* cause most bacterial skin conditions. *S. aureus* causes abscesses; certain strains can cause spreading infections. *S. pyogenes* may also cause impetigo.

5. Certain dermatophytic fungi can invade superficial or deeper layers of skin, causing the various forms of ringworm.

6. Several human viral diseases show a skin eruption or rash as the major sign. Within the spots, viruses are present and multiplying with localized tissue destruction. Other symptoms, such as fever, malaise, and prostration, are effects of the inflammatory response and viral multiplication. This pattern, with various degrees of severity, describes chickenpox, measles, and rubella. In both measles and chickenpox, virus may persist in the latent form after recovery, causing later serious disease. It is hoped that measles will eventually be eradicated.

7. Burns expose patients to a long, difficult convalescent process during which they have a very high infection risk. *P. aeruginosa* is the primary

pathogen. Human and animal bites may result in severe anaerobic infections.

8. The surface structures of the eye, including eyelids, conjunctiva, lacrimal network, and corneal surface are occasionally subject to infection. The eye is protected principally by the constant cleansing action of the tears, which contain antibacterial substances.

9. The eye is a secondary site of infection for two venereal-disease organisms, *N. gonorrhoeae* and *C. trachomatis*. Infections occur in adults and children because of transfer of contaminated secretions to the eye. Newborns are infected during birth. Trachoma is the most serious of the eye infections. It is endemic in hot, dry areas of the world, and causes much blindness, mostly in children.

DISCUSSION QUESTIONS

1. Compare the roles of sweat, sebum, and tears as microbial control factors.

2. How does the skin obtain immunologic protection?

3. If an infective agent does colonize the skin, it will remain localized so long as it does not penetrate to which dermal structures?

4. Why would it be extremely difficult to eradicate the varicella-zoster virus from the human population over a 10-year period as

smallpox was eradicated? Why is measles a more suitable candidate for eradication, as well as a more important target?

5. What role does personal hygiene play in the effectiveness of the skin as an infection barrier?

6. Describe how dermatophytes cause the symptoms of ringworm.

7. From references, become familiar with the special techniques of burn asepsis.

ADDITIONAL READING

Henderson DA. The eradication of smallpox. *Sci Am* 1976; 235:25-33.

Lennette EH, ed., *Manual of Clinical Microbiology*, 4th ed. Washington, American Society for Microbiology, 1985.

Mims CA. *The pathogenesis of infectious disease.* 2nd ed. London: Academic Press, 1982.

Youmans GP. et al. *The biological and clinical basis of infectious diseases.* 2nd ed. Philadelphia: WB Saunders, 1980.

17

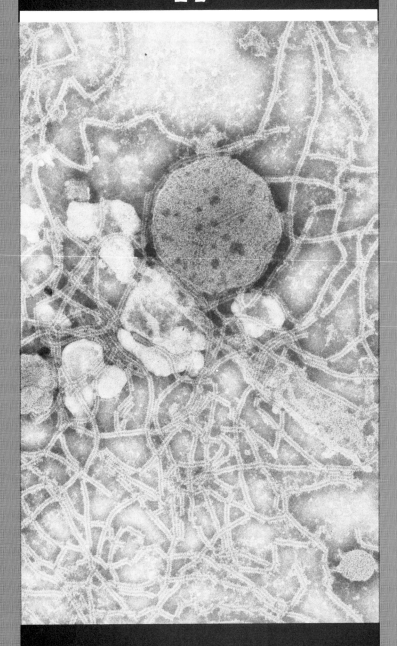

UPPER-RESPIRATORY-
TRACT INFECTIONS

CHAPTER OUTLINE

PASSAGES AND LININGS OF THE URT
Upper-Respiratory-Tract Anatomy
Physiology of the Upper Respiratory Tract

NORMAL UPPER-RESPIRATORY-TRACT FLORA
The Flora of the Nose
The Flora of the Throat
Upper-Respiratory-Tract Carriers

BACTERIAL DISEASES
Streptococcal Pharyngitis

Diphtheria
Otitis media

VIRAL INFECTIONS
Acute Rhinitis
Mumps

FUNGAL INFECTIONS

The upper respiratory tract (URT) is a network of connecting passageways, the habitat for a large, complex, and primarily protective normal flora. Many people also carry pathogens in their URT flora, making the URT a significant reservoir for the spread of infections.

About 90 to 95% of all URT infections are caused by viruses. These infections are generally mild, self-limiting, and have symptoms similar to one another. The nuisance we call the common cold may be caused by any one of about 150 different viruses.

The other 5 to 10% of URT infections are caused by bacteria and fungi. These infections are more specific in their symptoms, and should be promptly diagnosed and treated because they may be serious and give rise to complications.

PASSAGES AND LININGS OF THE URT

The upper respiratory tract is composed of the nasopharynx, oropharynx, outlets of the lacrimal ducts, Eustachian tubes, tonsils, sinuses, and mastoid cells (all shown in Figure 17.1).

Microbial pathogens featured in this chapter:

Streptococcus pyogenes
Corynebacterium diphtheriae
Common cold viruses
Mumps virus
Candida albicans

FIGURE 17.1   The Upper Respiratory Tract
The upper respiratory tract consists of a network of convoluted, connecting passageways which may harbor both normal, protective flora and pathogenic flora. When air is inhaled, it flows from the nostrils into the nasal cavity. There it is directed against the conchae and warmed, moistened, and cleansed before entering the pharynx. Embedded in the walls of the pharynx are masses of lymphoid tissue called tonsils. The opening of the sinuses, the nasolacrimal ducts, and the Eustachian tubes are all located in the nasopharynx. The sinuses are membrane-lined air spaces in the bones of the skull, which normally drain into the nasopharynx. The Eustachian tube is a slender duct that connects the pharynx to the middle ear. It equalizes atmospheric pressure on both sides of the eardrum and drains mucous secretions.

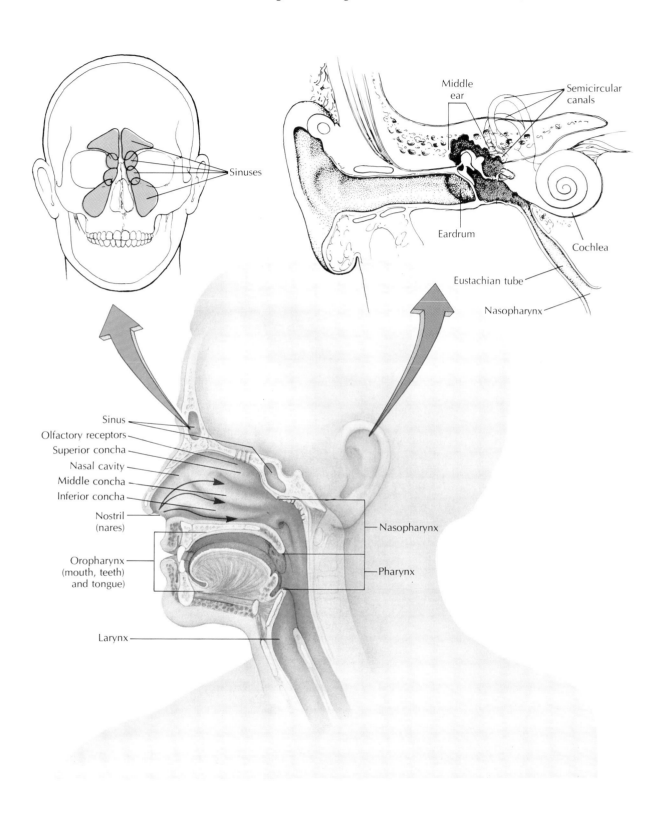

Sinuses

Middle
ear

Semicircular
canals

Eardrum

Cochlea

Eustachian tube

Nasopharynx

Sinus

Olfactory receptors

Superior concha

Nasal cavity

Middle concha

Inferior concha

Nostril
(nares)

Oropharynx
(mouth, teeth)
and tongue)

Nasopharynx

Pharynx

Larynx

UPPER-RESPIRATORY-TRACT ANATOMY

The airway, designated the **nasopharynx,** starts with the paired **nasal passages.** Air is inspired through a convoluted pathway that begins with the **nares** (nostrils). Air currents are directed against conchae, structures rather like baffles covered with moist mucous membranes. The larger airborne particles are trapped on the hairs and sticky mucous surfaces. The air is moistened, warmed, and cleansed. Air then passes through the **pharynx,** a central vestibule, in which the nasopharynx and oropharynx join, then downward into the lower respiratory tract (LRT), which begins at the **larynx.** The food passageway, which includes the teeth, tongue, and other mouth structures, is called the **oropharynx.**

The eye connects to the nasopharynx via the **nasolacrimal ducts.** These ducts drain tears from the inner margin of each eye into the nasopharynx; the tears carry microorganisms, debris, and lysozyme.

The **middle ear** is connected to the pharynx by means of the slender **eustachian tube.** This tube is a device for equalizing atmospheric pressure on both sides of the eardrum. It also drains mucous secretions. The duct can readily be compressed or collapsed, especially in small children, causing the secretions to become trapped and to accumulate.

Tonsils are masses of lymphoid tissue. There are several groups of these. The **palatine tonsils** are embedded in the sides of the oropharynx; they are the ones whose swelling is called **tonsillitis.** Swelling of the **pharyngeal tonsils,** embedded in the back of the nasopharynx, gives rise to the condition called **adenoid disease.** Swelling of the tonsils may obstruct the eustachian tubes and thereby contribute to middle-ear infection (otitis media). Often all the tonsils swell together.

Sinuses are cavities lined with mucous membrane in the cranial bones underlying the face. They are air-filled, and continuous with the nasal passages. When they are inflamed, and don't drain, **sinusitis** results. The **mastoid process,** an extension of the temporal bone beneath the ear, is a spongy bone filled with numerous tiny cavities, the **mastoid cells.** An infection that penetrates into these bony cavities is called **mastoiditis.**

Epithelium, amply supplied with mucous glands, lines all parts of the respiratory tract. Most of the tissue is ciliated epithelium, which bears a blanket of cilia on the exterior surface of each cell. The ciliary motions are coordinated so as to sweep a continuous layer of mucus toward the pharynx. This mechanism is referred to as the **mucociliary elevator** (Figure 17.2). The mucociliary elevator is also continuous down the LRT to the ends of the airway. Throughout the respiratory tract the mucociliary layer is a major defense mechanism, functioning to cleanse and to remove trapped particles and microorganisms. Mucociliary action brings mucous secretions to the mouth, where they are swallowed. The microorganisms in the mucus are then killed by stomach acid.

PHYSIOLOGY OF THE UPPER RESPIRATORY TRACT

The nasopharyngeal secretions are high in dissolved salts, and mildly acidic, normally about pH 6.5. The mucosal lining of the nasal passage responds to large changes in temperature and to low relative humidity by swelling, causing congestion. In the winter, people in northern regions experience frequent environmental changes, such as going from a hot, dry house to the cold outdoors to a heated car, and so forth. These repeated stresses seem to increase the tissue's susceptibility to infection.

In the rest of the URT, physiologic conditions are less salty and closer to a neutral pH, thus suitable for the growth of more delicate microorganisms that use the sugars and proteins of the secretions and our foods for nourishment. The temperature of the URT is usually two or more degrees lower than the interior body temperature, a factor important in the replication of some cold viruses.

The URT's main defenses, in addition to the mucociliary elevator, include secretory IgA and other immunoglobulins. These antibodies pro-

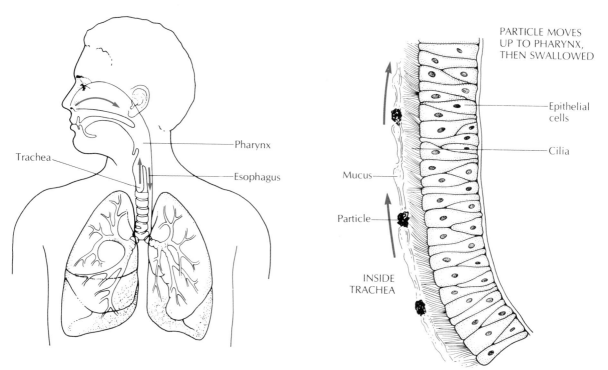

FIGURE 17.2   The Mucociliary Cleansing Mechanism
The respiratory tract is lined with mucous glands and epithelium. Many epithelial cells are ciliated, and their ciliary motions are coordinated to sweep a continuous layer of mucus toward the pharynx. This mucociliary layer is a major defense mechanism, cleansing the respiratory tract of foreign particles and microorganisms.

tect by preventing pathogens from forming an attachment to the URT lining, and to some extent by killing them. Phagocytic neutrophils and macrophages readily pass from the circulation into the mucous coat where they may capture and destroy pathogens. A respiratory-tract infection characteristically causes inflammation of the epithelial lining. The inflammatory process increases the volume of secretions. The secretions may be plentiful and thin (runny nose) or thick and viscous.

NORMAL UPPER-RESPIRATORY-TRACT FLORA

The nasopharynx and oropharynx harbor slightly different groups of microorganisms. Non-pathogenic normal flora are present in large numbers in healthy human beings. However, each genus present also contains one or more species that are either pathogens or opportunists. For example, the genus *Streptococcus* is represented in the oropharyngeal flora by several normal species, such as *S. salivarius*. However, the harmful species *S. pyogenes* is also quite at home in the oropharynx, where it causes infections, and often lives asymptomatically in a carrier state.

THE FLORA OF THE NOSE

The flora of the nasal passage is similar to the external skin flora. Gram-positive, salt-tolerant organisms such as *Staphylococcus aureus*, *Staph. epidermidis*, and the aerobic diphtheroids

TABLE 17.1 Bacterial flora of the upper respiratory tract

Normal Flora	Prevalence in Nasopharynx	Prevalence in Oropharynx	Species with Pathogenic Potential
Staphylococcus epidermidis	4	3	
S. aureus	3	2	*S. aureus*
Diphtheroids	3	3	*C. diphtheriae*
Alpha and nonhemolytic *Streptococcus* sp.	1	4	*S. pyogenes*
S. pneumoniae	1	2	*S. pneumoniae*
Neisseria sicca and *N. subflava*	1	3	*N. meningitidis* *N. gonorrhoeae*
Branhamella catarrhalis	1	3	
Hemophilus influenzae and *H. parainfluenzae*	1	2	*H. influenzae*
Enterobacteria *Klebsiella pneumoniae*	1	1	

4 = Found in almost all samples.

3 = Found in more than half of the samples.

2 = Found in less than half of the samples.

1 = Found in only an occasional sample.

predominate. From 10 to 30% of individuals carry *S. aureus* permanently or sporadically, on a day-to-day basis. The nasopharyngeal flora gradually becomes similar to and merges with the oropharyngeal flora in the pharynx.

THE FLORA OF THE THROAT

The oropharyngeal (throat) flora is composed of organisms associated with the teeth, gums, tongue, and saliva, which we studied in Chapter 15. It also includes other microorganisms that preferentially attach to the pharyngeal epithelium. Since one prerequisite for pathogenesis is attachment to the target tissue, potential pathogens are in this second group.

It is often necessary to sample the throat flora for laboratory study, by collecting a **throat swab** specimen. The proper method of collection is to obtain a sterile swab, which should be rubbed firmly over the posterior pharynx and

tonsillar area where the potential pathogens may be found (Figure 17.3). Contact with the oral surfaces and saliva should be avoided because the organisms found there are not clinically significant. They are also so numerous that they may overgrow the culture and prevent isolation of the true pathogens. After collection, the sample is carefully streaked on a blood agar plate and incubated with an elevated carbon dioxide atmosphere to simulate the conditions in human tissue. The medical technologist learns to recognize significant colony morphologies and to subculture colonies of presumed pathogens for complete indentification.

The oropharynx harbors several species of alpha-hemolytic streptococci, including the potential pathogen *S. pneumoniae*. The Gram-negative diplococci, such as *Branhamella catarrhalis* and *Neisseria sicca*, and the tiny, fastidious Gram-negative bacilli of the genus *Hemophilus* are also common. Gram-negative enterobacteria are quite uncommon. In fact, significant numbers of enteric organisms or *Pseudomonas* indicate altered body defenses and should be regarded with concern. Anaerobic organisms such as *Bacteroides*, the anaerobic streptococci, and diplococci such as *Veillonella* are a minor component of the oropharyngeal flora.

FIGURE 17.3 The Throat Culture
To culture pharyngeal flora for pathogens, a sterile swab is used to obtain a sample from the region behind the uvula. The swab should not be allowed to pick up saliva. The photograph shows a bacterial sample streaked across a blood-agar medium. Colonies of *Staph. aureus* growing on the culture medium have created clear zones where red blood cells have been destroyed. (Courtesy of Centers for Disease Control, Atlanta.)

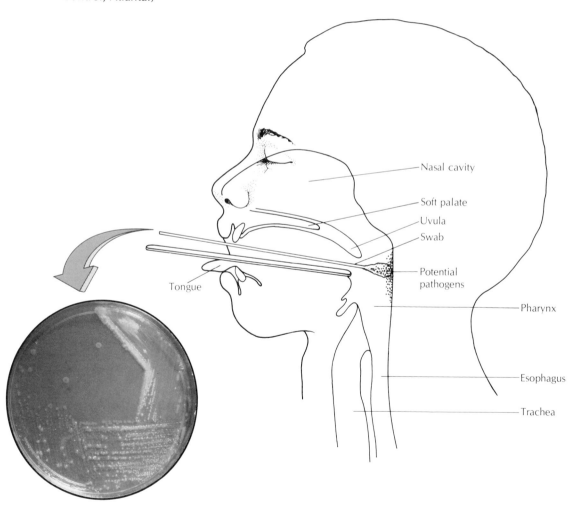

UPPER-RESPIRATORY-TRACT CARRIERS

Human carriers are the most significant reservoir in almost all respiratory infections. The infective agents are transmitted by direct contact, via airborne droplets, or by contaminated fomites. Carrier rates rise during the cold winter months or in crowded habitations, such as dormitories and military barracks. The causative agent of meningococcal meningitis, for example, may reach carrier rates approaching 90% in small populations, although the usual carrier rate for this organism appears to be about 15 to 20%.

Diphtheria is a bacterial disease in which the persistence of carriage of the organism prevents us from eradicating it. Persons immune to diphtheria can be healthy carriers of the diphtheria agent. The carrier rate for diphtheria is sporadically high in crowded urban populations and in the rural South and Southwest United States. The meningitis and diphtheria examples show us that we have no way of knowing in advance if a person carries a pathogen. Thus the respiratory secretions of any individual, sick or well, must be considered potentially hazardous. Responsible hygienic measures should be habitual.

BACTERIAL DISEASES

We will cover in detail two serious bacterial diseases of the URT, streptococcal pharyngitis and diphtheria. Both are potentially severe, and prompt differential identification is a key to successful therapy.

STREPTOCOCCAL PHARYNGITIS

THE ORGANISM. *Streptococcus pyogenes* is a Gram-positive coccus that often grows in chains, especially in liquid culture. It is facultatively anaerobic, fermentative, catalase-negative, and susceptible to the antibiotic bacitracin. Enriched media are required for successful isolation and maintenance. On blood agar tiny colonies are surrounded by a zone of beta-hemolysis. When freshly isolated, *S. pyogenes*

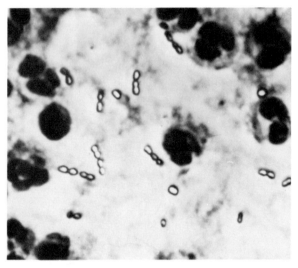

FIGURE 17.4 *Streptococcus pyogenes*
The strains of *S. pyogenes* that cause the most important form of bacterial pharyngitis carry fimbriae that preferentially adhere to the pharyngeal epithelium. Such adherent strains are associated with the risk of rheumatic fever in inadequately treated persons. (×630. Courtesy of Gene Cox, Science Photo Library/Photo Researchers, Inc.)

can be distinguished from other small beta-hemolytic colonies with a test that determines whether it is inhibited by bacitracin. A filter paper "A-disk" containing bacitracin is pressed onto the streaked area of a throat swab plate. Any beta-hemolytic colonies that do not grow in a zone around the disk are presumed to be colonies of *S. pyogenes*. They can then be subcultured from a neighboring area of the plate.

The streptococci are classified by serologic typing. The Lancefield typing scheme detects various surface carbohydrates, and on this basis divides the streptococci into about 20 lettered groups. Group A, to which *S. pyogenes* belongs, is the most important in human disease. Exterior to the carbohydrate layer the Group A streptococci bear a layer of M proteins that form their fimbriae. M proteins are important in pathogenesis because they inhibit phagocytosis and attach the bacterium to the tissue it parasitizes (Figure 17.4). Serologic typing of M proteins is

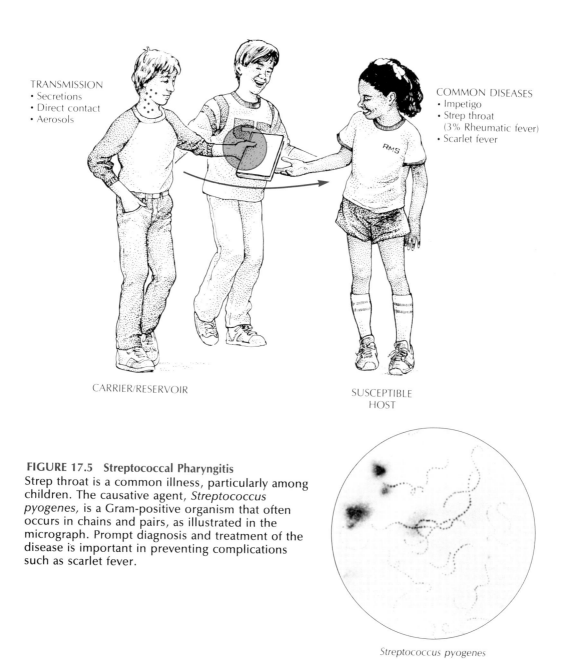

TRANSMISSION
• Secretions
• Direct contact
• Aerosols

COMMON DISEASES
• Impetigo
• Strep throat
(3% Rheumatic fever)
• Scarlet fever

CARRIER/RESERVOIR

SUSCEPTIBLE HOST

FIGURE 17.5 Streptococcal Pharyngitis
Strep throat is a common illness, particularly among children. The causative agent, *Streptococcus pyogenes*, is a Gram-positive organism that often occurs in chains and pairs, as illustrated in the micrograph. Prompt diagnosis and treatment of the disease is important in preventing complications such as scarlet fever.

Streptococcus pyogenes

used to pinpoint the source of *S. pyogenes* strains in outbreaks.

The strains that cause throat infections are different from those that cause skin and other infections. A key factor is the ability to adhere to the throat lining.

THE DISEASES. Acute **streptococcal pharyngitis** (strep throat) is a common complaint with over 200,000 cases per year in the US, particularly in children and adolescents (Figure 17.5). The symptoms include pain, reddening, and inflammation of the pharyngeal epithelium. Tonsils may swell and become coated with exudates, white patches of pus that are visible at the back of the pharynx. Fever or malaise may also be present. Other "cold-type" symptoms, such as a runny nose or cough, are not present.

Only 5 to 10% of all pharyngitis cases are caused by *S. pyogenes,* Group A. All others are caused by viruses of many types. Streptococcal sore throats are more severe than viral sore throats. A positive diagnosis of streptococcal pharyngitis requires a throat culture positive for *S. pyogenes.*

Prompt treatment is important in preventing possible complications. If untreated, about 3% of streptococcal pharyngitis patients develop **rheumatic fever.** This is a serious, often recurrent, hypersensitivity that damages heart valves and causes arthritis. Rheumatic fever is discussed further in Chapter 22.

Scarlet fever is a form of streptococcal pharyngitis in which the infecting strain produces an **erythrogenic toxin.** The toxin causes a diffuse rash, probably a hypersensitivity response. Toxin-producing strains, uncommon in the United States for the past few years, are often more generally virulent and cause more acute symptoms than strains that do not produce toxin, and thus cause only streptococcal sore throats.

Pathogenic *S. pyogenes* can also cause the skin infections **impetigo** or **pyoderma,** alone or in mixed infections with *S. aureus.* The two organisms are compared in Table 17.2. Skin infections with certain streptococcal types carry the risk of poststreptococcal kidney complications. These strains secrete a **nephritogenic antigen** that provokes an immune complex kidney disease called **glomerulonephritis,** usually about 10

TABLE 17.2 A comparison of *Staphylococcus aureus* and *Streptococcus pyogenes*		
Feature	**Staphylococcus aureus**	**Streptococcus pyogenes**
Microscopic morphology	Gram-positive coccus in grape-like clusters	Gram-positive coccus in pairs and chains
Species identification	Yellow pigment, aerobic B-hemolysis, salt-tolerant, mannitol-fermenting, catalase-positive, coagulase-positive	Nonpigmented, B-hemolysis; aerobic and anaerobic, catalase-negative, bacitracin-sensitive, Group A carbohydrate
Strain identification	Bacteriophage typing	Serological identification of M- and T-proteins
Extracellular products	Hemolysins Hyaluronidase DNAase Penicillinase Exfoliatin	Hemolysins DNAase Hyaluronidase Nephritogenic factors Rheumatogenic factors
Pathogenicity	Low for normal adults, high for the newborn and the compromised	Highly pathogenic for all individuals

TABLE 17.3 Some infections *Streptococcus pyogenes* may cause

Streptococcal pharyngitis
Scarlet fever
Impetigo
Erysipelas
Meningitis
Surgical infections
Neonatal infections
Infectious endocarditis
Puerperal fever
Rheumatic fever
Glomerulonephritis

days after the primary infection. Blood and protein in the urine, edema, and hypertension are the features of this condition. Other *S. pyogenes* infections are listed in Table 17.3.

THE TREATMENT. A mild *S. pyogenes* infection is self-limiting. However, because of the risk of serious poststreptococcal complications, streptococcal diseases should not be allowed to run their course. Prompt diagnosis and therapy are advisable. Penicillin, erythromycin, or other antibiotics, begun promptly, relieve symptoms and prevent the development of ear infections, mastoiditis, abscesses, rheumatic fever, or glomerulonephritis. Phagocytosis and circulating antibody are important in recovery.

EPIDEMIOLOGY AND CONTROL MEASURES. The reservoir for *S. pyogenes* is the human nasopharynx and, to a lesser extent, skin lesions and other orifices such as the anus. Many persons are asymptomatic carriers. The organism is spread in contaminated secretions, in aerosols, and on hands. Since *S. pyogenes* is both virulent

and readily communicable, thorough precautions must be taken to prevent its spread. Control measures focus on asepsis, disinfection, and personal hygiene. *S. pyogenes* pharyngitis is common during the winter months among children 5 to 14 years of age. There is no protective immunity after recovery, and useful vaccines have not been developed.

DIPHTHERIA

THE ORGANISM. Diphtheria is caused by *Corynebacterium diphtheriae*, a nonsporulating, Gram-positive rod. The cells often look beaded because they contain intracellular granules. Cells tend to lie together in parallel arrays called **palisades**, and are often irregular or club-shaped (Figure 17.6a). The organism can be isolated from throat or nasopharyngeal swabs by incubation on Löffler's medium, blood agar, or a medium containing potassium tellurite (Figure 17.6b).

Diphtheria toxin is the major virulence factor of this species (Figure 17.6c). It is an exotoxin, produced only by strains of *C. diphtheriae* lysogenized by a bacteriophage (see Box 8.1). Diphtheria toxin enters the host cell and inhibits an enzyme needed for protein synthesis. The blocked cell eventually dies.

THE DISEASE. *C. diphtheriae* initially infects the oropharynx, or, less commonly, the nasopharynx. Multiplication on the surface epithelium causes inflammation, which in turn leads to leakage of serum proteins and white blood cells. This exudate coagulates to form a tough, whitish **diphtheritic membrane** in the throat or nasal passage; the membrane is one major diagnostic feature of the disease. Although painful and unpleasant, the membrane itself is not life-threatening unless it obstructs breathing.

Toxin is produced in the pharynx, absorbed into the blood stream, and distributed to all the body's organs. Manifestations of the toxin's effects are variably delayed for a few days to several weeks. When the heart, kidney, and pe-

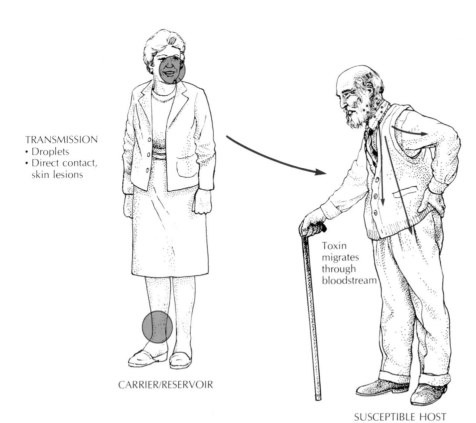

TRANSMISSION
• Droplets
• Direct contact, skin lesions

CARRIER/RESERVOIR

Toxin migrates through bloodstream

THREE AND MORE WEEKS AFTER ONSET OF DISEASE:
• Difficulty speaking and swallowing
• Demyelination
• Myocarditis
• Congestive heart failure
• 10% Mortality

SUSCEPTIBLE HOST

FIGURE 17.6   *Corynebacterium diphtheriae*
Once a disease primarily of children, diphtheria now occurs mainly in older individuals who have either never been immunized or have lost the protection of childhood immunization. The bacterium is often carried in the pharynx of immune persons and can be transmitted readily by airborne means. Once established, it multiplies rapidly in the host's upper respiratory system. While these bacteria do not invade tissues, they do produce a virulent exotoxin that is carried via the circulation to host cells where it blocks protein synthesis. The disease may be fatal to victims, particularly older individuals whose heart and kidneys become affected. The light micrograph shows the characteristic club-shaped appearance of the organism.

Corynebacterium diphtheriae

ripheral nervous system are affected by the toxin, they may undergo degenerative changes. Demyelination (loss of the myelin sheath) incapacitates nerve cells, often giving rise to another of the key diagnostic signs of diphtheria, a diffi-culty in swallowing. After a week or more, cardiac abnormalities progressing to heart failure may appear. Later neurologic effects — muscle weakness and variable paralysis — develop after a month or more.

THE TREATMENT.   One form of specific therapy is antitoxin, in the form of hyperimmune human serum, which is given so that its antibodies will neutralize any unabsorbed toxin. Antibiotics such as penicillin and erythromycin can reduce the microbial population in the pharynx, minimizing further toxin production and promoting healing of the lesions under the membrane. However, any toxin that has already been absorbed by tissues will continue to have its effects regardless of therapy. Clearly, the earlier the disease is diagnosed and therapy is begun, the better the patient's chances of rapid recovery.

EPIDEMIOLOGY AND CONTROL MEASURES.   The incidence of diphtheria in the United States has dropped dramatically since immunization became generally available in 1923. However, the mortality rate from diphtheria has remained unchanged at about 10% for the past 50 years. Diphtheria is not effectively controlled in many less-developed countries.

Diphtheria was originally primarily a disease of the first five years of life, claiming the lives of tens of thousands of children per year in the 19th century. Only five cases of diphtheria occurred in 1983. These rare cases of diphtheria now occur mainly in older individuals who have lost the protective effects of their childhood immunization or who have never been immunized.

One reservoir for the diphtheria organism is the human pharynx. Another source, especially in warmer climates, is skin lesions infected with the diphtheria organism. Insect bites occasionally become infected with *C. diphtheriae* from the pharynx and may become highly infective. The mode of transmission is droplets spread from the pharyngeal reservoir or direct contact from skin lesions.

Control of diphtheria can be achieved with universal immunization using diphtheria toxoid. The toxoid is recommended for all persons. Children receive the trivalent **DPT vaccine**, which contains antigens to immunize against three diseases, diphtheria, pertussis, and tetanus. Adults should be reimmunized every 10 years, especially if they are exposed to diphtheria. A lower dose of toxoid is administered to adults, because of the risk of a hypersensitivity reaction. A skin test called the **Schick test** may be done to verify if an individual has immunity to *C. diphtheriae* toxin or toxoid. As pointed out in Box 17.1, recovered or immunized persons may still be carriers.

Because some immune persons continue to be reservoirs, and because of uncontrolled diphtheria elsewhere in the world, diphtheria cannot be eradicated by the means at our disposal. This lethal disease remains a distant, but constant threat. Immunization will remain a necessity for the foreseeable future.

OTITIS MEDIA

Otitis media is an infection of the middle ear that primarily affects small children. It is estimated that about 70% of cases of otitis media are bacterial in origin. Of the bacteria that have been identified as agents, *Streptococcus pneumoniae* (see Chapter 18) and *Hemophilus influenzae* (see Chapter 21) are the most important. They are common "normal" throat inhabitants. Ear infection is an endogenous process resulting from overgrowth of these microbes. A key factor in flare-ups of ear infection is an obstructed Eustachian tube. This may occur as a secondary effect of almost any other upper respiratory infection in small children, due to swollen tonsils. The reason otitis media affects children so much more often than adults is because of the way the Eustachian tube is situated, so that when the tonsils swell, the tube is compressed. As children's skulls mature, the tubes become less subject to compression by swollen lymphoid tissues.

In acute otitis media there is sharp pain, and the eardrum reddens and bulges (Figure 17.7). The symptoms are the result of trapped, accumulated fluid and microorganisms causing inflammation in the enclosed middle ear. In chronic cases, exudate is present in the middle ear, but other symptoms may be absent. Aside from pain and loss of sleep, the risk of perforation and scarring of the eardrum are concerns

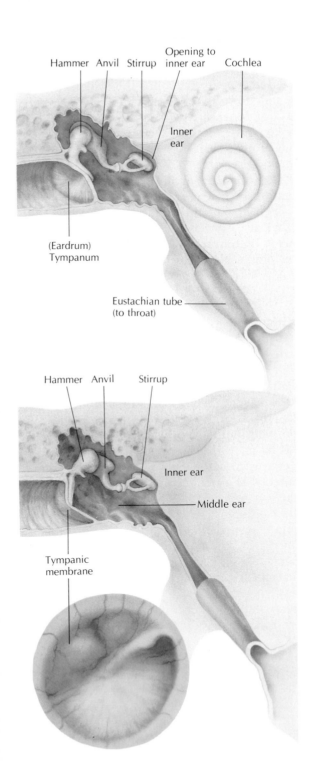

Hammer Anvil Stirrup Opening to inner ear Cochlea

Inner ear

(Eardrum) Tympanum

Eustachian tube (to throat)

Hammer Anvil Stirrup

Inner ear

Middle ear

Tympanic membrane

if the disease is not treated or does not respond to treatment. Repeated or prolonged cases of otitis media may result in hearing loss or mastoiditis.

It is not usually practical to isolate and identify the organism, because isolation requires a difficult and painful procedure — aspiration of fluid from the middle ear. Instead, usual practice is to prescribe antimicrobial agents effective against the most probable organisms. Amoxicillin, a penicillin derivative, is a popular choice. Recurrent or chronic cases are frequently treated by installing drainage tubes. This procedure helps in about 65% of cases. Surgical removal of tonsils and adenoids is done only occasionally.

Otitis media is extremely common, and it is the leading cause of preventable hearing loss. Studies of Eskimo childen in Alaska and children in rural southern Alabama both showed that about 80% of children have at least one episode of acute otitis media. Up to 20% of school-age children have chronic asymptomatic otitis media. Preventive measures for the infection are not well worked out. Vaccines against key strains of *S. pneumoniae* and *H. influenzae* could be helpful but are not yet available. Parents can prevent serious problems from developing by being alert to symptoms (signs of pain, such as rubbing or pulling on the ear, or apparent difficulty hearing) and obtaining treatment early and as often as needed.

FIGURE 17.7   The Middle Ear
The middle-ear cavity contains the auditory bones (the hammer, anvil, and stirrup) and is connected to the pharynx by the slender Eustachian tube. If masses of tonsilar material located nearby become swollen, they may press the Eustachian tube shut, preventing drainage from the middle ear. Normal microbial flora may then accumulate and cause infection. In severe cases, swelling and pressure may rupture the tympanic membrane and result in permanent damage to the delicate auditory structures.

BOX 17.1
Diphtheria-California

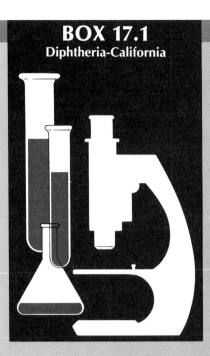

On May 10, 1979, a 68-year-old man with no history of diphtheria immunization came to a Los Angeles County Health Department clinic complaining of a sore throat. When clinicians noted a pharyngeal membrane, they took a throat culture, and the patient was referred to a hospital with a diagnosis of diphtheria. A physician in the ear, nose, and throat clinic of the hospital removed the membrane and sent it for culture, made a diagnosis of throat infection, and sent the patient home on oral penicillin. The membrane culture was positive for *C. diphtheriae*. Although his throat healed, the patient was hospitalized three weeks later with heart irregularities, but was released without further treatment. Two throat cultures taken then were reported as negative. Another three weeks later, the patient returned to the hospital with trouble swallowing and a nasal quality to his voice. He was admitted with a diagnosis of diphtheritic nerve damage and was put on tube feeding. He was left with a neurologic problem involving the muscles of the soft palate.

The county health department followed up the family members. One 9-year-old granddaughter, who had been previously immunized, turned out to be carrying toxigenic *C. diphtheriae*. The immunizations of all family members were brought up to date, and they all remained well.

This case illustrates the importance of keeping adults up to date with recommended immunizations and of being alert to the possibility of now very uncommon "childhood" diseases in unimmunized adults. Diphtheria still exists. Although the incidence of the disease has declined sharply, the pathogen lives on in thousands of healthy carriers.

Source: adapted from *Morbidity and Mortality Weekly Reports* 1979; 28.

VIRAL INFECTIONS

Viruses are the bane of the upper respiratory tract. As we said, about 90 to 95% of URT infections are caused by viruses. Our knowledge of these viral infections lags far behind our knowledge of the bacterial infections. One of the main reasons is that the viral disease patterns are overlapping. A "common cold" looks much the same, whether it is caused by rhinoviruses, coronaviruses, or echoviruses. When you consider that over half of all known human viruses infect the upper respiratory tract, usually causing much the same symptoms, you will understand the potential confusion.

ACUTE RHINITIS

Rhinitis is an inflammation of the nasal mucous membrane resulting in a profuse watery secretion, sneezing, and tear secretion. This is the all-too-familiar common cold. On any given winter day in the United States, 10% of the population — that is, in excess of 20 million people — have cold symptoms! Children under 5 average six colds per year; children from 5 to 14 have about five colds a year, and those over 14 have about three colds a year. There is a great difference, largely unexplained, in individual susceptibility.

THE VIRUSES.   The **rhinoviruses** (Figure 17.8) are RNA viruses. They replicate best in cell cultures incubated at 33 degrees Celsius rather than 37 degrees Celsius, and seem to be restricted in their infectivity to the cooler areas of the respiratory tract. There are at least 56 strains proven to infect human beings, and close to 100 additional strains of unestablished pathogenicity.

Respiratory syncytial viruses (RSV) may also cause colds. In children under one year of age, however, RSV is noted as a significant lower-respiratory-tract pathogen, causing a severe, often fatal form of pneumonia. The **adenoviruses** cause a spectrum of conditions from mild colds to severe acute respiratory disease (ARD) with sore throat and fever. ARD outbreaks occur in military camps and other densely populated areas. **Parainfluenza** viruses may cause both upper and lower respiratory diseases.

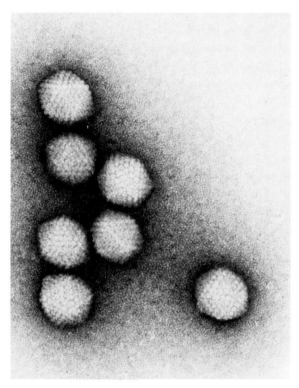

FIGURE 17.8   The Rhinovirus
This tiny icosahedral DNA virus is the commonest cause of the cold. So far, over 100 varieties have been discovered. (×5. Courtesy of Centers for Disease Control, Atlanta.)

THE DISEASE.   The common cold is normally mild and self-limiting. Respiratory epithelium becomes inflamed and hypersecretive, but there is no appreciable cellular destruction. The watery secretion carries interferon and immunoglobulins, and probably helps bring the infection under control by washing away virus, reducing viral replication, and, once an immune response appears, neutralizing new crops of virus. Colds can become serious only in small children whose incompletely developed immunity is unable to prevent viruses from advancing into the lower respiratory tract. Colds, like other primary respiratory-tract infections, appear to increase susceptibility to secondary infections.

THE TREATMENT.   There is no specific therapy for colds at this time. Medications such as decongestants and analgesics can alleviate the symptoms but do not alter the course of the disease. Remember, too, that because colds are caused by viruses, antibiotic treatment is of no use. Interferon and some drugs are being experimentally tried. Some authorities recommend high doses of vitamins such as vitamin C to shorten the acute phase, but the evidence for the value of any vitamins at dosages above their usual required dietary levels is not convincing.

Each cold confers on the individual some immunity to the agent responsible. However, there are many different agents. If each person had three colds a year, becoming immune to one agent each time, it might take 30 or more years to acquire a full spectrum of protection. So far, there are no vaccines for colds.

EPIDEMIOLOGY AND CONTROL MEASURES. The frequently recurring cold is a disease of modern urban life. In the past, cold viruses confined to small, geographically isolated populations would quickly "burn themselves out." With the advent of rapid transportation, however, cold viruses tend to be widely disseminated. Central heating exposes the nasal mucosa to hot, dry air in the wintertime and thereby places a new stress on the target tissue. Since people stay indoors more, in groups, in the winter, the cold viruses are effectively spread.

Control measures center on interrupting transmission (Figure 17.9), and protecting your-

FIGURE 17.9 Transmission of the Common Cold
The victim's respiratory secretions contain infectious viruses from about one day before symptoms appear through the first five days of symptoms. A susceptible person may catch the cold in many ways, such as by inhaling airborne droplets from an uncovered sneeze or cough, by touching contaminated objects, or by direct contact (as in a handshake).

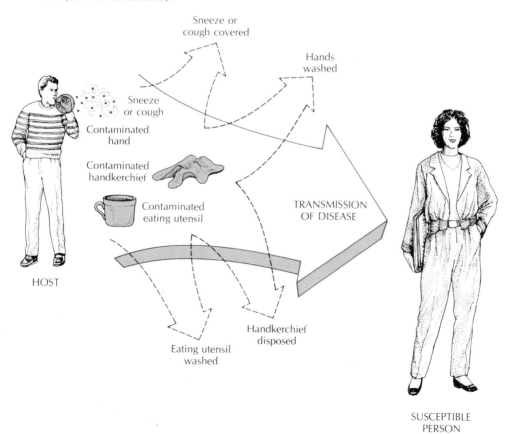

self and others from exposure. Since contaminated hands and tissues appear to be the most important means of transmission, sensible measures to prevent transmitting your cold include properly disposing of your tissues, washing your hands after you blow your nose, and avoiding close contact with other persons.

MUMPS

The primary site of mumps viral replication is the parotid salivary gland, located just below and in front of the ear. We discuss mumps among the upper-respiratory-tract diseases because the symptoms (swollen tissues) affect the throat and because the virus enters via the respiratory tract. Mumps is a disease of glandular tissue, however, rather than epithelium like the other URT diseases.

THE VIRUS. The mumps virus is a paramyxovirus (Figure 17.10), a large envelope-bearing RNA virus. It has hemagglutinin and neuraminidase units incorporated into the envelope for adhesion to its target tissues.

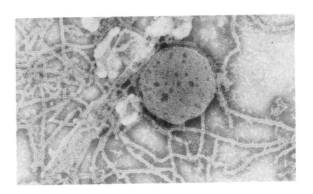

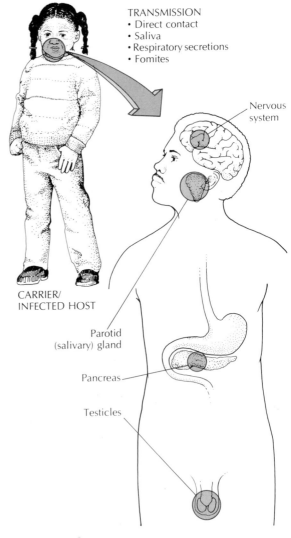

TRANSMISSION
• Direct contact
• Saliva
• Respiratory secretions
• Fomites

Nervous system

CARRIER/ INFECTED HOST

Parotid (salivary) gland

Pancreas

Testicles

FIGURE 17.10 Mumps
The mumps virus is a paramyxovirus. The electron micrograph shows many of its released helical nucleocapsids, plus one complete particle in which the nucleocapsid is coiled within an envelope. The envelope bears hemagglutinin and neuraminidase molecules. Although mumps is typically a disease of children, it can infect the parotid glands and other glandular tissues of susceptible individuals of all ages. In some adult males the pancreas and the testes may become swollen as well, causing a more severe form of the disease than that of young children. The virus may also cause a transient encephalitis. (×42,356. Courtesy of Frederick A. Murphy and S.G. Whitfield, Centers for Disease Control.)

THE DISEASE. Mumps is characteristically a disease of small children, but it may infect susceptible individuals of all ages. In typical cases one or both parotid glands become greatly swollen, and the patient has a fever. Pain on talking or eating may be pronounced. The mumps virus may attack other glandular tissues, and the pancreas and reproductive organs may also become swollen. Orchitis (testicular swelling) occurs in about 20% of adult men. In such cases sterility is rare, but possible. Mumps virus can also invade the nervous system, and transient inflammation of the brain tissue and its coverings may be present in about 10% of cases. Nerve deafness, fortunately usually in only one ear, occurs rarely. Mumps symptoms are usually transient and disappear within a defined period. Severe complications or death are rare.

THE TREATMENT. There is no specific therapy for mumps. However, medication for pain or fever reduction may be used.

EPIDEMIOLOGY AND CONTROL MEASURES. Mumps virus is spread through direct contact with saliva, respiratory secretions, or fomites. People in the incubation period, those who are actively ill, and those who have asymptomatic cases may all be infectious. The disease, either active or asymptomatic, confers permanent immunity. There is no truth to the tale that you are not immune to mumps if you "had it only on one side." The occasional occurrence of unrelated types of glandular swelling help to keep this piece of folklore alive.

Effective control depends largely on universal vaccination. There is a highly effective live-virus vaccine available, recommended for all infants and other susceptible individuals, especially men without evidence of immunity. Mumps vaccine is usually given simultaneously with measles and rubella vaccines in the trivalent MMR (measles:mumps:rubella) formulation. There are still more than 3000 cases of mumps per year in the United States, occurring mainly in those few states that do not require evidence of mumps vaccination for admission to public schools.

FUNGAL INFECTIONS

One significant fungal infection of the upper respiratory tract, called **thrush** or **candidiasis,** is caused by the yeast *Candida albicans. C. albicans* and, to a lesser extent, several other species of *Candida,* are part of the normal flora in a significant minority of individuals. They are opportunists, able to cause infections when there are predisposing environmental and constitutional factors. They erupt in a wide variety of locations, causing diseases that are merely bothersome, as well as lethal systemic infections (Table 17.4). The important factor in the development of candidiasis is not acquiring the agent itself but the predisposing factors that allow it to overgrow its "normal" limits and produce disease.

TABLE 17.4 Some infections *Candida albicans* may cause

Thrush
Urinary tract infection
Vulvovaginitis
Intertrigo
Perleché
Mucocutaneous candidiasis
Disseminated candidiasis
Fungemia
Endocarditis
Other nosocomial infections

THE ORGANISM. *Candida albicans* grows as a budding yeast under most laboratory conditions, but develops **pseudomycelia,** chains of elongated cells, under semianaerobic conditions or in tissue (Figure 17.11). When yeast cells from a colony are incubated in serum, characteristic **germ tubes** develop. *C. albicans* is identified by cultivation on differential media and microscopic examination. The organisms are facultative anaerobes. They ferment sugars, and have few nutritional requirements. Yeasts grow more rapidly than mycelial fungi, but more slowly than bacteria, a difference that is reflected in the speed with which their respective infections develop.

THE DISEASES. *Candida albicans* infects most body systems, but is most common on epithelium. We therefore see it causing problems on the skin and in the body orifices and intestinal lining.

Thrush is a *Candida* infection of the tongue, oral mucosa, and orapharynx. The mucosa becomes acutely inflamed and the reddened patches become covered with a grayish-white exudate. This exudate may be confused with the membrane seen in diphtheria. Thrush is often an indication of mildly inadequate cell-mediated immunity. It occurs in about 5% of newborns, usually those that are formula-fed and thus do not receive passive immunity from their mothers, and it develops within the first few days of life. It causes the baby discomfort and difficulty in feeding. Thrush may also be seen in as many as 10% of elderly, mildly immunodeficient patients. Thrush often develops in people receiving cancer chemotherapy or other immunosuppressive therapy. It is especially likely if the patient is also receiving antibacterial antibiotic therapy, which has the effect of substantially unbalancing the normal oral flora. Strong psychological and physical stresses, diabetes, obesity, and alcoholism, all of which suppress cell-mediated immunity, also predispose individuals to *Candida* infections.

Mucocutaneous candidiasis is an extensive form of this infection that occurs in acutely immunodeficient individuals, most commonly children. The *Candida* lesions proliferate over the mucous membranes and the skin, eventually involving the whole body. Antifungal drug treatment is often relatively useless — they cannot control the organism's growth in the absence of directed macrophage activity.

There are several manifestations of *Candida* skin infection. *Candida* causes **intertrigo** on areas where two layers of skin rub together, as between fingers, under the breasts, and between the buttocks. Vesicles appear followed by scaling, and the lesions are difficult to heal. **Perleche** is a related cracking or fissuring of the corners of the mouth. Candidiasis of the skin is experienced by people who repeatedly immerse their hands in water, such as physicians, dentists, dishwashers, and fishermen.

THE TREATMENT. Oral *Candida* infections are treated by rinses with specific antifungal antibiotics. Drugs such as amphotericin are employed for systemic infections. Skin infections are treated with antifungal ointment containing vioform (iodochlorhydroxyquin) or other antifungal agents.

EPIDEMIOLOGY AND CONTROL MEASURES. *Candida albicans* often inhabits the mouth, urogenital tract, and gastrointestinal tract of human beings. About half of the healthy population carries *C. albicans* at one or more of these sites. All normal adults appear to have effective cell-mediated immunity against *C. albicans*. Transmission is frequent and occurs through a wide variety of means, including direct contact and fomites such as tissues, eating utensils, and clothing. Control measures include particular care to keep the skin in healthy condition and attention to early signs of infection in debilitated or immunosuppressed individuals. The risk of candidiasis is reduced to the extent that immunosuppressive activities and conditions can be avoided.

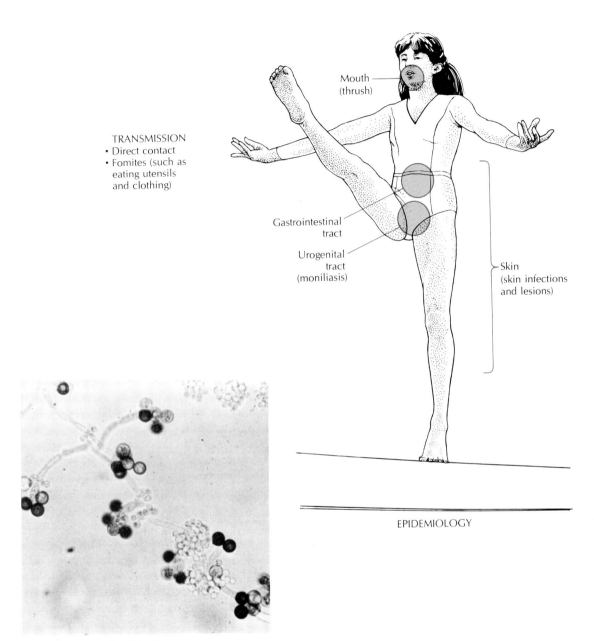

TRANSMISSION
• Direct contact
• Fomites (such as eating utensils and clothing)

Mouth (thrush)

Gastrointestinal tract

Urogenital tract (moniliasis)

Skin (skin infections and lesions)

EPIDEMIOLOGY

Candida albicans

FIGURE 17.11 *Candida albicans*

Candida albicans is commonly found on the epithelial tissue of a normal, healthy person. However, should host defenses fail or normal flora be disturbed, it can overgrow and cause problems on the skin, in intestinal linings, in the mouth, and in the urogenital tract, especially the vagina. While most cases of *C. albicans* in adults are endogenous, the microorganism can also be transmitted by direct contact and fomites. The transmission electron micrograph shows budding yeast cells. (×427. Courtesy of Centers for Disease Control, Atlanta.)

	TABLE 17.5 Epidemiology of URT infections		
Disease	Reservoir	Transmission	Incubation Period in Days
Streptococcal pharyngitis	Human beings	Direct contact, milk, water, food	1–3
Diphtheria	Human beings	Direct contact, airborne, rarely fomites, raw milk	2–5
Common cold	Human beings	Direct oral contact droplets, soiled articles, hands	1–3
Acute respiratory disease (ARD) due to adenovirus	Human beings	Direct oral contact, droplets, soiled articles, hands	Few days to week or more
Mumps	Human beings	Droplets, direct contact with saliva	12–26
Thrush	Human beings	Droplets, direct contact, fomites	2–5, variable

SUMMARY

1. The upper respiratory tract is the site of our most common infections. In order for pathogens to establish themselves, they must be able to attach to the epithelial lining and compete effectively with the normal flora. Secretory immunoglobulins in mucus may prevent infection in the resistant individual by preventing attachment.

2. Almost all viral upper-respiratory-tract infections are mild and self-limiting, and multiplication of the pathogenic agent causes little or no tissue destruction. Virus infections comprise 90 to 95% of the cases. Bacterial infections, while less common, are potentially far more serious.

3. Streptococcal pharyngitis caused by *Streptococcus pyogenes* is an acute, self-limiting disease. Since it carries a significant risk of the later development of rheumatic fever, prompt diagnosis and antibiotic treatment are needed.

4. Diphtheria caused by toxigenic strains of *Corynebacterium diphtheriae* is a rare but severe illness with significant mortality. It can be completely prevented by vaccination with diphtheria toxoid.

5. Otitis media, in severe or recurrent cases, can lead to loss of hearing. It is caused by one of several bacterial and viral agents, and is most common in children under five.

Period When Communicable	Patient Control Measures	Epidemic Control Measures
Untreated, while discharge present	Disinfect discharges, soiled materials; special attention to manual contamination risk	Trace carriers; determine source of food and/or milk contamination, if any
Two to four weeks after onset, while organism present	Isolation, minimum 14 days, disinfection of all patient-owned articles	Immunization of exposed group, especially infants and children quarantine contacts; search for carriers
From one day before onset to five days after	Cover sneezes, dispose of tissues in sealed bag, keep hands clean	None
For duration of active disease; virus may have latent period	Cover sneezes, dispose of tissues in sealed bag, keep hands clean	Avoid crowding
From six days before symptoms to nine days later. Inapparent cases may be communicable	Isolation for up to nine days after swelling develops, disinfection of soiled articles	Routine immunization of all infants
While lesions present	Disinfect secretions-soiled materials	May be needed in nursery outbreaks

TABLE 17-5 Epidemiology of URT infections (continued)

6. Nonbacterial pharyngitis, the common cold, and other viral conditions are seldom serious. Their major importance is in lost time and misery. They may also predispose the convalescent patient to secondary infections. The causative viruses are numerous and ubiquitous, and there are no practical therapeutic measures at present.

DISCUSSION QUESTIONS

1. When you go to your doctor complaining of a sore throat, what considerations might the doctor keep in mind in diagnosing and treating your case?
2. Why should you not expect the doctor to prescribe antibiotics for every case of sore throat?
3. Construct scenarios for steps in four hypothetical examples of rhinovirus transmission.
4. Under what circumstances does candidiasis occur?
5. What is the difference between controlling an infectious disease and eradicating it, epidemiologically speaking?

6. Because diphtheria is so rare, physicians may practice for an entire career without seeing a case, and they do not ordinarily suspect it when seeing "sore throat" patients. Explain how this situation relates to the fact that the mortality rate is higher among the first cases in an outbreak than among the later cases.

7. Why haven't we been able to easily develop a cold vaccine?

ADDITIONAL READING

Fenner FJ, White DO. *Medical virology.* 3rd ed. New York: Academic Press, 1985.

Giebnick GS, Quie PG. Otitis media: the spectrum of middle ear infection. *Ann Rev Med* 1978; 29:285-306.

Hoeprich PD, ed. *Infectious diseases.* 3rd ed. New York: Harper & Row (in press).

Read SE, Zabriskie JB, eds. *Streptococcal diseases and the immune response.* New York: Academic Press, 1980.

Skinner FA, Quesnel LB, eds. *The streptocci.* New York: Academic Press, 1978.

Youmans GP, Paterson PY, Sommers HM. *The biological and clinical basis of infectious diseases.* 2nd ed. Philadelphia: WB Saunders, 1980.

18

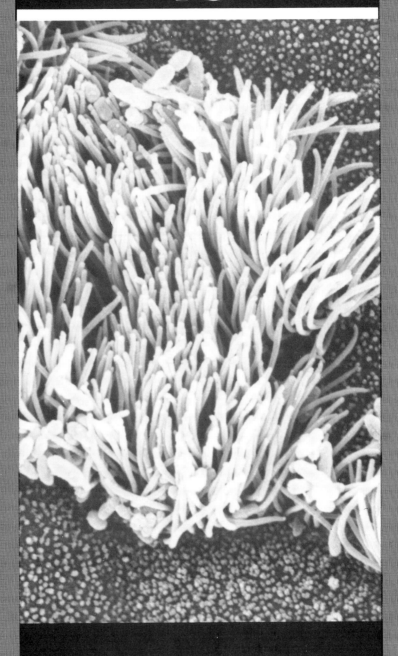

LOWER-RESPIRATORY-TRACT INFECTIONS

The lower respiratory tract (LRT) as an organ system has the duty of gas exchange, that is, taking in oxygen and giving off carbon dioxide. Any disease that seriously interferes with the lower respiratory tract threatens your life. Infections may block gas exchange by causing accumulation of fluids in the lungs or by destroying the tissues of the air-exchange sacs, the alveoli. One LRT disease, pneumonia, is the most important microbial killer in the United States.

In this chapter we will examine infections of the larynx, trachea, bronchi, and the lungs. These are well-defended parts of the body, but it will become clear that each pathogen has its own ways of evading the host's defenses.

ANATOMY AND PHYSIOLOGY

In this section, we will familiarize ourselves with the anatomy, physiology, and natural defenses of the lower respiratory tract.

ANATOMY OF THE LRT

The larynx, the trachea, and the bronchial tree comprise the lower airway (Figure 18.1). The **larynx** is the organ of speech. It connects the bot-

Microbial pathogens featured in this chapter:

Bordetella pertussis
Streptococcus pneumoniae
Mycoplasma pneumoniae
Legionella pneumophila
Mycobacterium tuberculosis
Influenza virus
Histoplasma capsulatum

FIGURE 18.1 The Lower Respiratory Tract
The lower respiratory tract includes the larynx, trachea, bronchi, and lungs, making up an organ system whose primary function is the exchange of gases. The airway (color) divides into a mass of branching tubes, the bronchi. These become finer and finer, ending up as microscopic bronchioles that are embedded in the lung tissue. Lung tissue is composed of thin-walled sacs called the alveoli, which are tightly packed and covered with a meshwork of capillaries. Within the alveoli, oxygen is transferred to the blood and carbon dioxide is released into the lung. Mucous secretions keep all tissue surfaces of the lower respiratory tract moist. The scanning electron micrograph shows tufts of cilia lining the epithelium of the trachea. The rhythmic beating of these cilia moves mucus upward along the respiratory passage. (×2900. Photo courtesy of Ellen Roter Dirksen, University of California at Los Angeles.)

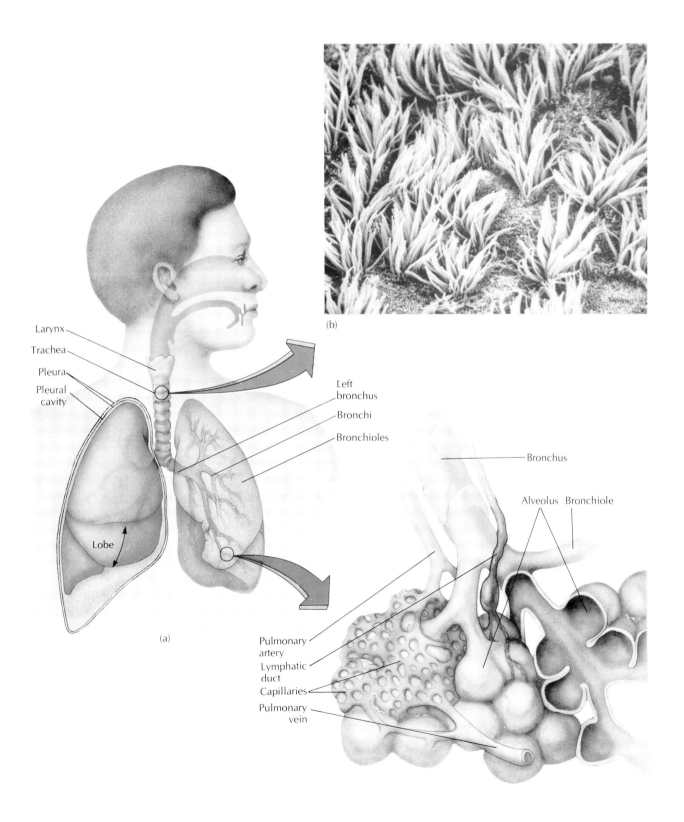

(b)

Larynx

Trachea

Pleura

Pleural cavity

Lobe

(a)

Left bronchus

Bronchi

Bronchioles

Bronchus

Alveolus Bronchiole

Pulmonary artery

Lymphatic duct

Capillaries

Pulmonary vein

tom part of the pharynx with the trachea. It is composed of cartilage, a firm but flexible connective tissue, and contains voluntary muscles, including the vocal cords. At the top of the larynx is the epiglottis, a small flap of tissue that closes off the airway, preventing food from going down it while you are swallowing. The **trachea** is a thin-walled, straight tube supported in the open position by rings of cartilage reinforced by smooth muscle. The **bronchi** are two branches of the trachea that enter the lungs, continually branching into smaller tubules that eventually have a combined diameter almost 100 times that of the trachea. The smallest of these tubules are the **bronchioles.** They have less cartilage and are less rigid than the bronchi. Spiral bands of smooth muscle fibers allow the bronchioles to dilate or constrict markedly. The structures of the airway are lined with ciliated epithelial tissue which also contains mucus-producing cells, IgA-producing lymphoid tissue, and elastic fibers.

The bronchioles are embedded in and end in the **lung tissue,** which is organized in lobes, two for the left lung and three for the right. The lobes are further subdivided into lobules, each of which is served by a particular branch of the bronchial tree. Infections tend, at least at first, to be localized in one lobe. The bronchioles end in **alveoli,** the thin-walled air sacs that comprise the lung tissue. Alveolar chambers are one cell layer thick and richly netted over by capillaries. They are highly elastic.

The heart and lungs lie in the **thoracic cavity** or chest cavity, which is lined with two layers of membrane, the **pleura.** These are moist and glide easily over each other during the breathing movements. The **pleural cavity,** or space between the pleural layers, may fill with fluid during infection.

PHYSIOLOGY OF THE LRT

SECRETIONS IN THE LOWER RESPIRATORY TRACT. Mucus production is a normal function of the epithelial lining of the airway. The mucociliary elevator, discussed in Chapter 17, operates as a continuous cleansing system bringing trapped particles from the base of the bronchioles up to the pharynx. Certain factors increase mucus production and cause the material to become thicker, more tenacious, and difficult to move. These changes may be due to irritants, such as smoke and dust, or to inflammation caused by infection. This thickened mucus is called **sputum.** We normally attempt to expel sputum by coughing. However, the coughing may not be effective, especially if the individual lacks muscular strength or if the sputum is highly viscous. When sputum accumulates, it seriously interferes with air flow and gas exchange.

THE LUNG SECRETIONS AND DEFENSES. Alveolar surfaces are designed for gas exchange. They are moist, so that oxygen and carbon dioxide cross the thin tissue layer of the alveolar sacs by first dissolving in the aqueous film on the surface. This fluid also contains a **surfactant,** a wetting agent that lubricates the lung surfaces and keeps them from sticking together. The lung secretions contain IgA antibodies and specialized phagocytes called **alveolar macrophages.** Excess fluid may accumulate in the lung because of circulatory malfunction or inflammation.

PHYSIOLOGIC RESPONSES TO LUNG INFECTION. When there are irritants in the lung, they stimulate inpouring of fluid from the blood. This fluid contains the blood-clotting protein **fibrin.** The fibrin accumulates and forms a mass of gelled sputum called a **consolidation.** This response has both harmful and beneficial effects. It blocks gas exchange in the affected area, but helps contain the infection. The fluid influx concentrates immunoglobulins in the area of inflammation. The local macrophage population is rapidly reinforced by monocytes and polymorphonuclear leukocytes migrating through the capillary walls. If the phagocytes are successful, the microorganisms will be eliminated. If the agent resists phagocytosis, as most successful lung pathogens do, infection progresses.

The lower respiratory tract has no normal flora. Any invading pathogen meets no resis-

BOX 18.1
What Does Microbiology Have To Do With Smoking?

You've heard more than you ever wanted to about smoking and cancer, smoking and heart disease. . . . How about smoking and respiratory infection? Did you know that smokers have more colds, more influenza, and more chronic bronchitis? Why should this be so?

The chemicals in tobacco smoke affect the cells lining the respiratory tract. In response, these cells increase their secretion of mucus. However, the chemicals also paralyze the cilia on the surface of the lining. This means the mucus sticks and thickens. It does not adequately carry out its function of sweeping microorganisms back up out of the lower reaches of the lower respiratory tract. Instead, the mucus provides a semi-inert medium for microbial growth.

At the same time, smoke particles are drawn into the lungs, where some stick and remain. The alveolar macrophages police the lungs. They are strategically placed to deal with foreign particles that escape the filtering mechanisms of the upper respiratory tract, such as bacterial cells, dust particles, and droplet nuclei.

The alveolar macrophages are highly efficient scavengers that attempt to digest captured particles enzymatically. If you are smoking, the macrophages must deal with the smoke particles in addition to their regular clean-up duties, and they may be unable to remove all entering pathogenic agents.

Some inhaled material (asbestos fibers, smoke particles, or food fragments) resists degradation, so the alveolar macrophage's attempts to remove it will be futile. Excessive amounts of hydrolytic enzymes will be continually released and will damage the lung tissue. One explanation for the scarring associated with emphysema, a smoking-related lung disease, is that the frustrated efforts of the macrophages to remove the smoke particles actually break down the alveolar walls, which are then repaired with inelastic scar tissue.

---

tance from microbial competitors. It must cope only with host defenses to establish itself.

BACTERIAL INFECTIONS

LRT infections can be caused by bacteria, viruses, fungi, and protozoans. Bacteria cause up to 50% of LRT infections. These infections are of-ten described clinically in general terms that bear no reference to the infective agent. For example, the term "pneumonia" indicates only a state where there is fluid in the lung; it does not indicate any one of the many possible causes for this state. Transmission of LRT infections occurs most commonly by inhaling infective agents. The agent may become airborne in the form of aerosolized respiratory secretions, dried droplet nuclei, or spores in dust clouds (Figure 18.2). Quite often it appears that an agent will first establish itself in the upper respiratory tract in a carrier state. It will then periodically be inhaled into the LRT as air is inspired — a form of self-inoculation. Only if host defenses fail will this self-inoculation take effect and clinical disease result. A less common route by which microorganisms reach the lung is from a focus of infection elsewhere in the body, via the blood stream.

It is often difficult for lab workers to identify the causative agent of a lung infection. Sputum cultures are usually the first step, but these may fail to yield the causative agent because the agent may be few in number and overgrown by salivary contaminants. Samples may be obtained by bronchial brushing or transtrachial needle aspiration, but these are invasive procedures and are not often used. Most LRT pathogens cause bacteremia, so blood cultures may reveal the organisms. All culture methods require one or more days for results to be obtained. More rapid results may be obtained by identifying antigens produced by the offending agent in body fluids such as serum, sputum, or urine; such testing methods are generally sensitive, specific, and fast.

FIGURE 18.2   Patterns of Transmission of Lower-Respiratory-Tract (LRT) Infections
Infections of the LRT occur most often through a two-step process. (a) First, airborne infectious organisms are inhaled and establish themselves in the upper respiratory tract. (b) Later, when the host draws air into his or her lungs, the agent spreads into the LRT, in a form of self-inoculation. (c) The immune system often brings the agent under control during either Step a or b. However, should body defenses fail to halt the disease, it can spread via the blood stream to other areas of the body.

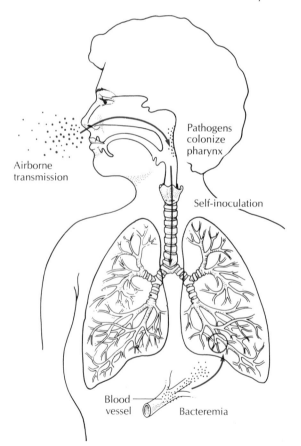

WHOOPING COUGH

THE ORGANISM.   **Whooping cough** or **pertussis** is usually caused by the small, Gram-negative bacillus *Bordetella pertussis* (Figure 18.3). *Bordetella parapertussis* and adenovirus may also cause this condition. *B. pertussis* is ovoid and has a capsule when multiplying in the host or when freshly isolated. The organism is aerobic and slow-growing. It can be isolated on a glycerin–potato agar to which blood is added. Penicillin, which does not affect this organism, is often incorporated into isolation media to suppress competing normal flora in the saliva, which often contaminate the specimen. Samples are collected by having the patient cough directly onto the open sterile plate. Fluorescent-antibody staining of secretions may provide rapid identification.

THE DISEASE.   When *B. pertussis* multiplies in the airway, it causes irritation by producing exotoxins and endotoxins. Thick, mucoid fluid is

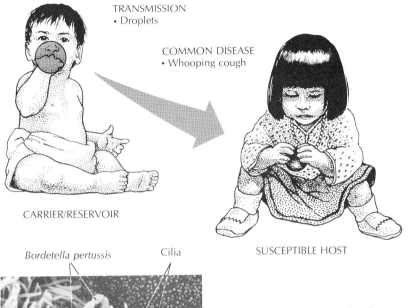

TRANSMISSION
• Droplets

COMMON DISEASE
• Whooping cough

CARRIER/RESERVOIR

SUSCEPTIBLE HOST

Bordetella pertussis   Cilia

FIGURE 18.3 *Bordetella pertussis*
The causative agent of whooping cough is a gram-negative, rod-shaped bacterium, shown in the scanning electron micrograph growing on the cilia of respiratory epithelial cells from the trachea. The reservoir of *B. pertussis* is the human respiratory tract. Transmission occurs when a susceptible person inhales droplets shed by an infected carrier. Children and the elderly are the most susceptible to whooping cough. (Courtesy of Kenneth E. Muse, Duke University Medical Center/BPS.)

secreted, forming a meshlike tangle of cilia and bacteria on the walls of the trachea and bronchi. During the **catarrhal period,** which lasts about two weeks, the symptoms are mild and resemble those caused by the common cold. Then the **paroxysmal stage** begins, marked by fits of rapid, violent, and uncontrollable coughing several times daily. These fits usually fail to expel the offending material, leave the patient exhausted and anoxic (short of oxygen) and may induce vomiting or small hemorrhages. Almost half of the small children with pertussis must be hospitalized during this stage, which also lasts about two weeks. It is followed by a long convalescence of weeks or months during which the symptoms gradually disappear. Secondary infections may occur during this stage.

THE TREATMENT. Antimicrobial therapy using erythromycin, tetracyline, or chloramphenicol stops the agent's growth and keeps the disease from worsening. However, it doesn't significantly shorten the course of the illness. Supportive measures include providing oxygen and maintaining adequate nutritional intake as needed until the paroxysmal phase ends.

EPIDEMIOLOGY AND CONTROL MEASURES.   The reservoir of *B. pertussis* is the human respiratory tract. Infants and children are most susceptible to whooping cough, although cases also occur in the elderly. Peak incidence occurs during the winter, but cases are also seen throughout the year. Transmission occurs when droplets shed from the respiratory tract of individuals in the catarrhal stage are inhaled by susceptible persons. There were under 2463 cases in the United States in 1983.

Whooping cough is controlled by required childhood immunization with the DPT (diphtheria–pertussis–tetanus) vaccine. Active disease confers lifelong immunity, but circulating antibody levels fall off rapidly. Thus even immune mothers give little or no passive immunization to their infants. This is why vaccination is needed as early as eight weeks of age. Booster shots are given later on.

The P component of the DPT vaccine is composed of killed *B. pertussis* cells or crude bacterial extracts, and it contains much endotoxin. The endotoxin material is responsible for the common side effects of fever and muscle soreness that are associated with these immunizations. However, there are occasional severe side effects, such as high fever and neurological damage. The vaccine is almost 90% effective, but we still have a significant number of pertussis cases occurring even in children who have been immunized. Because of shortages of the existing vaccine, the risk of severe side effects, and the clear need for a more effective protection, researchers are working intensively on possible new pertussis vaccines. Any new vaccine will probably not contain whole cells, but may be based on a pertussis toxoid, or on purified surface proteins from the agent.

BACTERIAL PNEUMONIA — GENERAL FEATURES

Pneumonia, an accumulation of fluid in the lungs, is often visible as an opaque area on a chest x-ray. Fever, cough, respiratory distress, and acute chest pain are the usual symptoms. The bronchial tree may or may not be involved; one or more lobes of the lung may be damaged. As fluid accumulates within a lobe, the microorganisms multiply extracellularly in the exuded fluids, which make a fine culture medium. Alveolar tissue damage varies with the agent. We'll look now at several types of bacterial pneumonia.

PNEUMOCOCCAL PNEUMONIA

THE ORGANISM.   About 90% of cases of bacterial pneumonia are caused by *Streptococcus pneumoniae* (Figure 18.4). This is a Gram-positive coccus that characteristically forms pairs or short chains. The cocci are enclosed in a smooth, thick polysaccharide capsule essential for virulence.

The organism is identified on blood agar by observation of tiny alpha-hemolytic colonies whose growth is inhibited around filter paper disks impregnated with optochin (ethylhydrocupreine hydrochloride). The cells of *S. pneumoniae* are very fragile, and lyse when suspended in solutions of bile salts. The agent is further distinguished from other alpha-hemolytic streptococci by its ability to ferment the carbohydrate inulin. There are more than 80 serotypes of *S. pneumoniae*, identified by capsular polysaccharides. Serotype 3 is the most virulent.

THE DISEASE.   Pneumococcal pneumonia begins with a severe, shaking chill, followed by fever. **Pleurisy** — inflammation of the pleura — is usually present, creating intense chest pain. The pleural cavity may fill with fluid and may itself, in extreme cases, become infected (**empyema**). Within the lung tissue, multiplication of the bacteria also causes inflammation and influx of fluids. The fibrin-bearing fluid forms a circumscribed consolidation that prevents expansion and contraction of the lobe and blocks gas exchange. Bacteremia is very common; one reliable way of establishing that *S. pneumoniae* is present is by blood cultures.

Convalescence, even when antibiotics are used, depends on the patient's ability to develop antibodies against the capsular antigen of the invading strain. Once antibodies are present, phagocytosis, previously ineffective, becomes highly efficient. As microbial multiplication is arrested, the fluid is gradually reabsorbed and the lung tissue returns to normal. Remarkably, there is usually little residual damage.

THE TREATMENT.   Almost all strains of *S. pneumoniae* are sensitive to penicillin and its derivatives. Mortality remains significant, however, because many persons contract pneumococcal pneumonia as a secondary infection; they are already severely ill from other causes and are therefore unable to develop the defenses needed to remove residual microorganisms by phagocytic action.

EPIDEMIOLOGY AND CONTROL MEASURES. Pneumococcal pneumonia is spread by human carriers. From 20 to 70% of individuals carry pneumococci in the nasopharynx some of the time. Epidemic outbreaks occur in institutional settings where there may be a high density of unusually susceptible people. Single-serotype outbreaks result when a single carrier has introduced a new strain to such a susceptible population. Often, however, there are multiple strains present in an epidemic, and it seems more likely that the outbreak is caused primarily by a dramatic increase in susceptibility in the group. Sporadic cases occur most frequently in children under five and in older persons with cardiopulmonary problems, particularly during the winter. Pneumococcal pneumonia frequently takes the life of a patient seriously ill from other causes.

Vaccination is available to control this disease to some extent. Protective immunity fol-

SUSCEPTIBLE
HOST

CARRIER/RESERVOIR

TRANSMISSION
• Airborne
• Direct contact
• Fomites

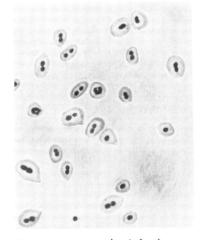

FIGURE 18.4 *Streptococcus pneumoniae* **Infection** The classic form of pneumonia is caused by a Gram-positive bacterium, which usually forms cell pairs. Often the organism is already present in the host's nasopharynx, and infection is most likely to occur when normal defense mechanisms are weakened. The streptococci have a thick capsule that makes them especially resistant to phagocytosis. The micrograph shows a quellung reaction in which exposure to specific antiserum has caused the bacterial capsules to swell. (Courtesy of the Centers for Disease Control, Atlanta.)

lows contact with pneumococcal polysaccharide, but it is type specific and not very long-lasting. A pneumococcal polysaccharide vaccine (trade name, Pneumovax) containing antigens for 14 common strains of pneumococcus, was released in 1977. It is recommended for the elderly and those with chronic respiratory or cardiac disease or other predisposing conditions. The vaccine has not been widely accepted, and many who would benefit from it do not receive it.

MYCOPLASMAL PNEUMONIA

THE ORGANISM.   Recall that *Mycoplasma pneumoniae* and other mycoplasmas are the smallest, free-living cellular organisms. Although they are considered to be bacteria, they lack peptidoglycan in the cell wall, include sterols in their cytoplasmic membrane, and have an unusual division mechanism. *M. pneumoniae, M. hominus*, and *Ureaplasma urealyticum* are pathogenic for human beings.

    Mycoplasma species are difficult to isolate and cultivate in cell-free media. They require serum or other sterol sources and prolonged incubation periods from 10 to 30 days in order to form colonies less than a tenth of a millimeter in diameter. However, mycoplasmas are readily isolated by inoculation into cell cultures and are major contaminants in cell-culture labs.

    Mycoplasmas are said to be cell-adapted; that is, they need close association with a host cell for best growth. They have specialized attachment organs for adhesion and grow intimately attached to or fused with the membrane of the living cell (Figure 18.5a). Mycoplasmas use the host-cell membrane as a source of materials for their own exterior layer (Figure 18.5b). As a result of this "borrowing," there is a high degree of antigenic similarity between host and parasite that interferes with the host's ability to destroy mycoplasmal parasites by immune mechanisms.

THE DISEASE.   In the LRT, mycoplasmas cause **primary atypical pneumonia;** it is called atypical because the agent escapes isolation by standard bacteriologic tests. It is a mild condition, of the sort sometimes called "walking pneumonia" because the patient is still able to carry on normal activities. The pneumonia is marked by slight sputum production, little consolidation, and no tissue damage. Other clinical manifestations of mycoplasma infection include upper-respiratory-tract infection, pharyngitis, bronchitis, and eye or ear infection.

THE TREATMENT.   Tetracycline is the drug of choice. Mortality is low and recovery is usually uneventful.

EPIDEMIOLOGY AND CONTROL MEASURES.   Mycoplasmal pneumonias are similar epidemiologically to pneumococcal pneumonias. However, no vaccine has been developed and there are no specific preventive measures.

LEGIONELLOSIS

Although it sometimes seems we are already familiar with all the microbial diseases, nature has periodic surprises in store. In July 1976, a large outbreak of severe pneumonialike disease occurred at an American Legion convention in Philadelphia, resulting in 29 deaths. Despite the coordinated efforts of the nation's top epidemiological experts, the causative agent eluded isolation for many months. When it became possible to detect the agent in tissue, and find specific antibodies against it in patient sera, it became clear that this was not a "new" pathogen but rather one that had been around for a long time. On serological evidence, the pathogen was retrospectively implicated in unsolved epidemics going back to 1965. **Legionnaire's disease,** as the condition was immediately named, aroused much public interest and concern. It was seen as a potential new "plague." However, the actual incidence of legionellosis is low and it is not primarily an epidemic disease.

THE ORGANISM.   The genus *Legionella* contains over 20 species. They are Gram-negative bacilli normal in structure but unusual in their surface

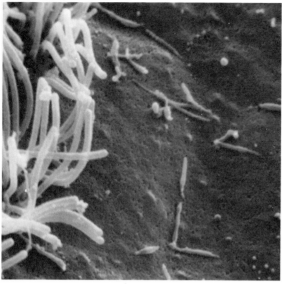

(a) *Mycoplasma pneumoniae*

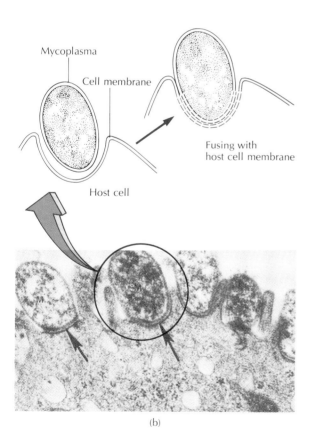

(b)

FIGURE 18.5 *Mycoplasma pneumoniae*
(a) *M. pneumoniae* organisms cling to the outer surface of ciliated epithelial cells. The background of the scanning electron micrograph shows the membrane of such a cell. Attached to its surface are many irregularly shaped *Mycoplasmas*, some spherical, some elongated. To the left, a tuft of larger cilia is visible. (×10,000. Courtesy of Michael G. Gabridge, PH.D., Bionique Laboratories, Inc.)
(b) The mycoplasmal cell has no cell wall; its outermost layer is a cell membrane. When it binds to a host cell, its membrane and that of the host cell become closely associated. Some membrane materials may be exchanged and the membranes may partly fuse. In some cases (arrows), the barrier between parasite and host cell appears to vanish. (From G.H. Cassel et al., Pathobiology of mycoplasmas, *Microbiology* (1978), American Society for Microbiology, Figures 1 and 2, p.400.)

chemistry (Figure 18.6). They are not closely related to any of the other human Gram-negative pathogens or microflora. *Legionella* can be isolated and maintained on media containing cysteine and iron salts, such as charcoal–yeast extract agar. Growth at 35 degrees Celsius in candle jars is slow, and colonies become visible in 4 to 5 days. *Legionella pneumophila* produces beta-lactamase, an enzyme that inactivates penicillin-type antibiotics. The bacilli can be identified in pleural fluid, sputum, or lung tissue by direct fluorescent-antibody microscopy. ELISA tests may be used for antigen detection.

THE DISEASE. Legionellosis is a type of pneumonia with about a 10-day incubation period, high fever, coughing, chest and abdominal pain, and diarrhea. In severe cases, shock and kidney failure may occur, causing death. There is also a much milder form of the disease, called **Pontiac fever,** in which the incubation period is about two days. The symptoms, although similar, are much milder, the kidneys and liver are never involved, and no deaths have been reported.

THE TREATMENT. Erythromycin, rifampin, and the tetracyclines have been used successfully for therapy. Convalescing patients develop antibody against the organism.

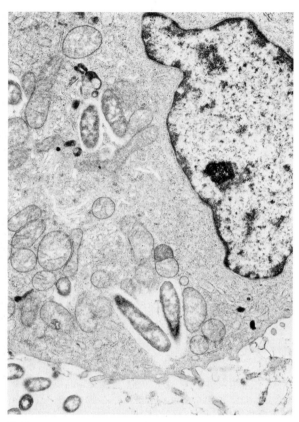

FIGURE 18.6 *Legionella pneumophila*
This electron micrograph shows a cluster of
Legionella pneumophila, a bacterium that causes
legionnaires' disease. (×17,250. Courtesy of John J.
Cardamone, Jr. and P.W. Dowling, University of
Pittsburgh School of Medicine/BPS.)

EPIDEMIOLOGY AND CONTROL MEASURES.   There
are several facets to the epidemiology of legionellosis. The normal habitat of *Legionella* appears
to be water and soil, where the organism acts as
a decomposer species. The agent also multiplies
in standing water such as puddles or in large
air-conditioning systems and water tanks. It becomes airborne in vapor or dust clouds arising
from construction sites, and is inhaled. There is
no convincing documentation of person-to-
person transmission.

The patient with legionellosis often is a person with some preexisting illness. Pontiac fever,
on the other hand, occurs in healthy people. Its
peak is during the summer months, especially
for clustered outbreaks. A key control measure is
the adequate chlorination of water supplies,
tanks, and humidification systems.

OTHER BACTERIAL PNEUMONIAS

There are many other forms of bacterial pneumonia, some of which we'll cover here briefly.
Pneumonia caused by *Hemophilus influenzae*
occurs most commonly in children. Most adults
have natural active immunity to *H. influenzae*
capsular antigen, and are not susceptible. The
clinical manifestations are similar to pneumococcal pneumonia, but chills and pain are
usually absent. Bacteremia is common, and
meningitis is a frequent complication.

Klebsiella pneumoniae is a facultatively anaerobic, Gram-negative rod, a member of the enterobacterial group. It is an important cause of
pneumonia in alcoholics, diabetics, and other
adults who have extensive metabolic disturbances or chronic cardiopulmonary disease.
Clinically, a key symptom is repeated and prolonged chills. There may be extensive tissue destruction leaving scar tissue on recovery. The
mortality rate has been quite high even with
treatment.

Pseudomonas aeruginosa, a ubiquitous opportunist, causes pneumonia particularly in
children with cystic fibrosis. The organism is
often highly resistant to most antibiotics.

TUBERCULOSIS

Tuberculosis, which used to be called the "white
plague," is being brought partly under control in
the developed countries. However, more than
20,000 new cases a year are still diagnosed in the
United States. In impoverished areas of the
world tuberculosis is largely out of control;
about 3 million people per year die of a disease
for which there are at least 10 effective drugs.

Most of these deaths occur in regions where there is neither money nor personnel to deliver the treatment.

THE ORGANISM. **Tuberculosis** is defined as the chronic progressive pulmonary disease caused by *Mycobacterium tuberculosis.* However, very similar illness can also be caused by **atypical mycobacteria,** different species such as *Mycobacterium bovis,* a pathogen of cattle. *M. tuberculosis* is by far the agent most commonly involved.

M. tuberculosis is a slender, rod-shaped organism (Figure 18.7). Its Gram-stain reaction is variable and uninformative. Its cell walls contain up to 60% lipids, which render the mycobacteria acid-fast, a property that forms the basis for the differential staining technique discussed in Chapter 3.

Mycobacteria are strictly aerobic, metabolically versatile, but extremely slow-growing, dividing only once about every 20 hours. The organism was first isolated by Robert Koch in 1882, on his 251st attempt. To isolate the organism, sputum is chemically digested to thin it and to kill flora other than the mycobacteria. The sample is then incubated for up to eight weeks on Lowenstein–Jensen or Middlebrook media in an atmosphere resembling that of the air in the alveoli, containing 5 to 10% carbon dioxide. Colonies develop extremely slowly; they are typically dry, wrinkled, and pale yellow. *M. tuberculosis* may be differentiated from other species of mycobacteria by its ability to produce niacin. *M. tuberculosis* also manufactures a lipid called **cord factor,** which makes the cells grow in coiled arrangements. Cord factor is almost always found in isolates from active disease, and is probably a virulence factor.

THE DISEASE. There is an important distinction between tuberculosis *infection* and tuberculosis *disease.* A few decades ago, infection (the presence of the organisms) was extremely common. Overt symptomatic disease was then and still is much less common.

Tuberculosis may be primary or secondary. We'll begin with the primary infection (Figure 18.7a). The causative organism is inhaled in the form of droplet nuclei from the respiratory secretions of patients in the active stages of the disease. Once in the alveoli, it is phagocytosed by the macrophages, then **primary infection** begins. Nonspecific macrophages cannot inactivate mycobacteria, so the organisms survive and actively multiply within the macrophages. Irritation of lung tissue causes an influx of fluid and polymorphonuclear leukocytes. The organism is carried by the phagocytes to lymph nodes and other organs throughout the body. During this early infection period there is little actual tissue destruction, and the lymphoid tissues receive an antigenic stimulus.

If the body's defenses respond, specific cell-mediated immunity (CMI) develops from three to four weeks after infection, as shown by the tuberculin skin test (Figure 18.7b). Infection is revealed by the conversion of the tuberculin test from negative to positive. **Tuberculin** is an extract of the organism. A form of tuberculin called **PPD (purified protein derivative)** is generally used, and it may be administered in several ways. The **Mantoux test,** which employs an intradermal injection, is the most accurate method. A positive response is revealed by a reddened, indurated (hard) area at the site of administration. A positive test may indicate either past or present infection.

As CMI develops, immune lymphocytes and macrophages arrive at the infective site. These ingest and destroy the bacteria. Some host cells fuse to form Langhans' giant cells. Host and bacterial cells also may break down to a cheesy mass in the center of the lesion, where a fibrous layer forms, sealing in the organisms. The enclosed lesion is called a **tubercle** (Figure 18.7c); its formation arrests the infection. Anaerobic conditions within the tubercle are extremely unfavorable to the mycobacteria; few survive, and they do not multiply. The tubercle gradually calcifies. If it is large enough, it may show up on a chest x-ray. In most people, all the events just

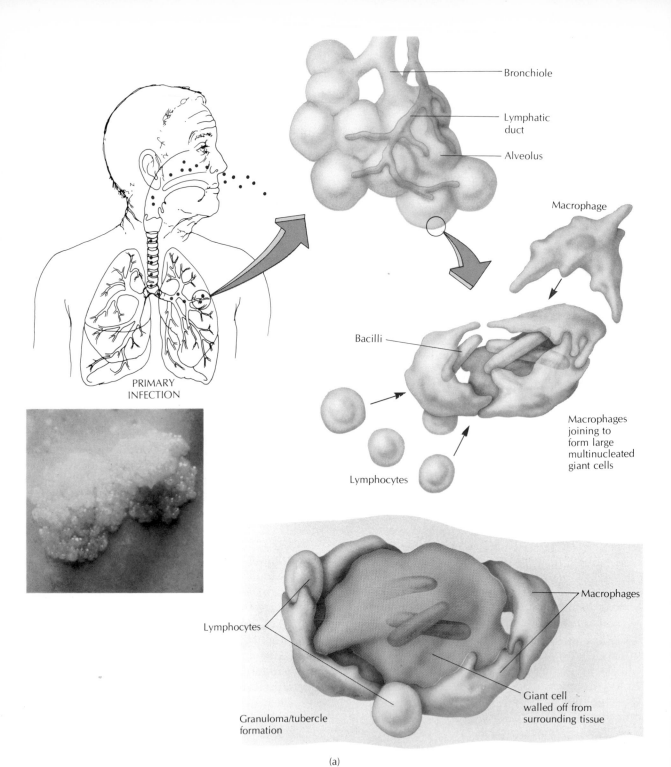

Bronchiole

Lymphatic duct

Alveolus

Macrophage

Bacilli

Macrophages joining to form large multinucleated giant cells

Lymphocytes

PRIMARY INFECTION

Lymphocytes

Granuloma/tubercle formation

Macrophages

Giant cell walled off from surrounding tissue

(a)

FIGURE 18.7 *Mycobacterium tuberculosis*
(a) In a primary infection, the organism is first inhaled in contaminated respiratory secretions or ingested in contaminated food. Bacteria spread to the alveoli where they survive and multiply within nonspecific macrophages. In response to antigenic stimuli, specific CMI develops, and lymphocytes and macrophages congregate to ingest the invaders. Host cells may join to form giant cells which then break down, along with the bacteria, to form a fibrous tubercle that seals in the pathogens. (b) A skin test can reveal the development of CMI

540

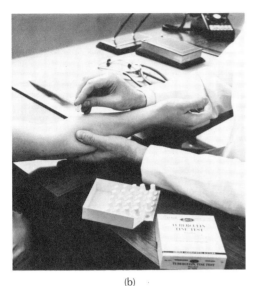

(b)

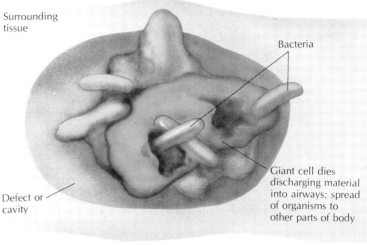

Surrounding tissue

Bacteria

Defect or cavity

Giant cell dies discharging material into airways; spread of organisms to other parts of body

(d)

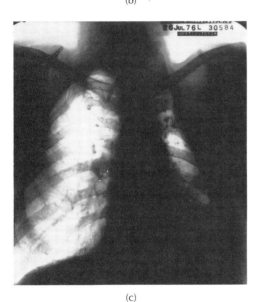

(c)

ACID-FAST STAINING

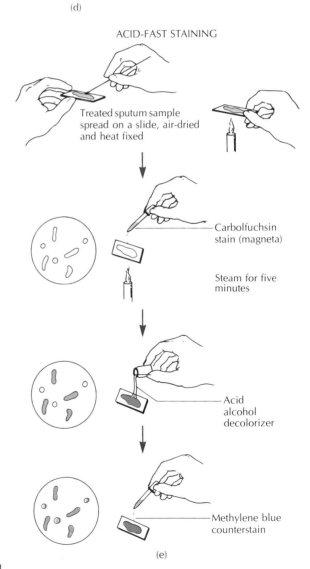

Treated sputum sample spread on a slide, air-dried and heat fixed

Carbolfuchsin stain (magneta)

Steam for five minutes

Acid alcohol decolorizer

Methylene blue counterstain

(e)

following an infection. A form of tuberculin (purified protein derivative [PPD]) is administered intradermally, and the development of a hardened, red area at the site of administration indicates a positive reaction. (c) Secondary infections occur if the "healed" lesion (the white spots on the lungs in the x-ray) should break open. (d) In secondary or uncontrolled primary infections, bacteria spread via the blood stream from the infected lung to the rest of the body. (e) Bacilli will be found in the sputum of any patient with an uncontrolled case of tuberculosis. The organisms are customarily identified by the acid-fast staining procedure depicted in the diagram. (Photo credits: (a) courtesy of Centers for Disease Control; (b) courtesy Lederle Laboratories, Wayne, New Jersey; (c) courtesy of Robert B. Morrison, M.D. and St. David's Community Hospital, Austin, Texas.)

541

described are asymptomatic; the person is not aware that anything out of the ordinary is happening. If CMI remains high, there will be no further tubercular activity during the individual's life span. However, the individual's positive tuberculin reaction will usually persist indefinitely.

Sometimes the body's CMI is unable to control the mycobacteria. **Progressive tuberculosis** occurs when the tubercle cannot be adequately walled off. The host is losing the battle to control the infection with CMI, but hypersensitivity reactions that destroy lung tissue are still occurring. A cavity opens in the lung tissue and rapidly enlarges. Bacterial multiplication is rapid. Fever, fatigue, weight loss, and cough appear, and the disease is obviously in full swing. The mycobacteria may spread via the bronchi to other areas of the lung tissue. It is still possible to arrest the disease if the patient's immune defenses are improved by rest and diet and if appropriate antimicrobial drugs are administered.

Secondary or **reactivation tuberculosis** occurs when a healed primary lesion reactivates (Figure 18.7d). The lesion breaks open with renewed microbial multiplication. Secondary tuberculosis has the same general characteristics as progressive tuberculosis, including much tissue damage. Reactivation, which correlates with declining CMI, is usually unsuspected at first. Patients who have reactivated shed the pathogen and are the major reservoir for new primary infections in susceptible infants and children.

A now rare condition called **miliary tuberculosis** occurs when mycobacteria cause bacteremia and establish multiple lesions throughout the body. It tends to develop when the pathogen is ingested rather than inhaled, for example, when *M. bovis* is ingested in unpasteurized milk from infected cows.

THE TREATMENT.   In all forms of tuberculosis, the strength of the host's immunity is a large determining factor in whether the disease advances or is contained. Therefore one objective of therapy is to improve the patient's general health and immune status. This is done by providing rest, a good diet, and improved living conditions, if necessary.

The other objective is to begin effective chemotherapy. Isoniazid, rifampin, ethambutal, and streptomycin are frequently found to be effective. Other drugs have undesirable side effects and are used only if these less toxic drugs fail. *M. tuberculosis* rapidly develops resistance to any drug used singly, so combinations of two or, preferably, three drugs are prescribed. The treatment period ranges from 12 to 36 months. Noncompliance — failure to follow the treatment — often leads to reactivation of drug-resistant strains.

EPIDEMIOLOGY AND CONTROL METHODS.   There was a rapid drop in incidence rates of tuberculosis in the early part of the 20th century, as milk pasteurization became widespread and standards of living steadily improved. A second decline started around the late 1940s as chemotherapy became available. The rate has remained relatively constant for about 10 years now. We are failing to make furthur headway, largely because today tuberculosis tends to be a disease of poverty, and the percentage of the population living in poverty is not declining very much.

Tuberculosis caused by *M. tuberculosis* is almost always contracted by inhaling droplet nuclei recently exhaled by a person with an open tubercular lesion. *M. bovis*, however, can be acquired from dairy products obtained from infected cows. Public health regulations enforcing tuberculin testing of dairy herds and pasteurization of milk prevent this, but in occasional cases these precautions are not followed.

The risk of contracting the disease by contact with an active case ranges from 24 to 46%. Each active case, if untreated, gives rise to an average of 12 new cases per year. These numbers will give you an idea of how important control measures are. Inactive cases (persons with healed lesions) should receive periodic chest x-rays to detect possible reactivation.

Control measures center on promptly diagnosing active cases, then starting antimicrobial therapy, which renders the patient noninfective within one or two weeks. Since isolation is usually impractical, and household contacts have already been exposed, patients are sent home to recuperate. However, they are instructed to control their respiratory secretions by covering the nose and mouth while coughing and carefully disposing of soiled materials to prevent any further spread.

As a simultaneous control measure, it is important to identify and tuberculin test all known contacts of the active case. If a contact is a "converter" — someone who previously tested tuberculin-negative and now tests positive — he or she has probably contracted the disease and will require antibiotic treatment. Other tuberculin-negative contacts may, at the physician's discretion, be placed on short-term preventive doses of drugs.

A tuberculosis vaccination has been available for many years, but its protective effect in the population at large has not been unambiguously established. This **BCG vaccine** (bacille Calmette–Guérin) contains a live, attenuated strain of *M. bovis*. BCG was initially developed in Europe, where it is widely used to immunize young children and others with negative tuberculin tests. The immunization process converts the recipient's tuberculin reaction to positive. However, BCG is little used in the United States, where public health policy has chosen to focus on control through antibiotics.

VIRAL INFECTIONS

Viruses infect both the airway and the lung tissues. They rarely cause serious illnesses by themselves, but often a secondary bacterial infection follows, causing further problems.

INFLUENZA

Influenza is a pandemic disease, recurring yearly throughout the world during the colder months. Periodically, new strains of influenza virus emerge to which few people have preexisting immunity, and there is a global outbreak with thousands of deaths.

THE VIRUS. The influenza viruses are myxoviruses. Their genome consists of seven separate pieces of RNA. The envelope bears spikes of antigenically distinctive hemagglutinins (H) and neuraminidases (N). Hemagglutinin assists the virus in adhering to the target cell, while neuraminidase assists it in penetrating the mucous coating over the respiratory epithelium. These proteins are inserted into the virus envelope as it buds out through the host-cell membrane (Figure 18.8). The gene for the H antigen exists in several different forms, on one of the seven gene segments. The gene for the N antigen also exists in several forms on different gene segments. Thus as influenza viruses replicate, different combinations of H and N genes may be incorporated into the virions through genetic recombination, and new strains may arise.

There are three types of influenza virus — A, B, and C. Types A and B are pathogenic; Type C is only rarely pathogenic. Type A influenza viruses usually cause the most severe and the longest-lasting cases of influenza. They are also the most genetically variable for the H and N antigens. When a new influenza strain appears, it is serotyped and designated by the type, the place and date it first appeared, and its H and N antigens. Thus we get strain designations such as A/Philippines/2/82 (H3N2) or A/Brazil/11/78 (H1N1). Usually at least one new Type A variant appears every year.

THE DISEASE. Influenza is characterized by sudden onset of fever, chills, and muscle aches, sometimes accompanied by pharyngitis. A three- to five-day period of illness is common. Recovery is slow, and overexertion may cause a relapse.

If the tissues become severely inflamed, coughing, sputum production, and chest pain may be present, and recovery will be slow. A pneumonia caused by influenza virus may develop (rarely) in the elderly or those with underlying cardiopulmonary disease. It is rapidly progressive and often fatal.

Major complications may also interfere with recovery from influenza. The most serious complication is a bacterial secondary pneumonia (Figure 18.9). This condition has a significant mortality rate.

Public health officials attempt to judge the impact of influenza by counting what are called **pneumonia–influenza (P and I) deaths.** There may be over 30,000 deaths from these causes in an epidemic year.

Another serious complication of some viral diseases, especially influenza and chickenpox, is **Reye's syndrome.** This is a condition found mainly in children. It is characterized by a history of a recent virus infection, followed by sudden loss of consciousness, central-nervous-system abnormalities, and potential failure of the liver and kidneys. About 200 cases per year of this disease are reported; the actual incidence is undoubtedly much higher. Mortality is about 30%. The actual relation of the viral infection to the development of Reye's syndrome is not clear. However, taking aspirin has been statistically associated with development of the syndrome, so many pediatricians now advise against giving aspirin to children with influenza or chickenpox.

THE TREATMENT.   Amantadine, a specific antiviral drug, reduces the severity of symptoms of influenza and shortens their duration in some

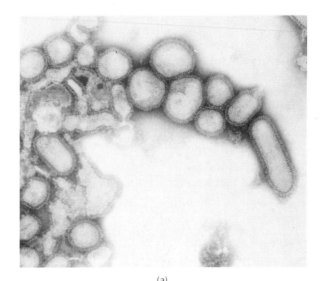

(a)

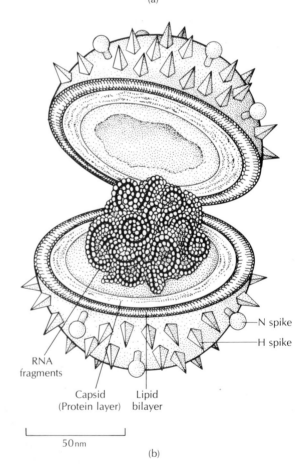

N spike

H spike

RNA fragments

Capsid (Protein layer)

Lipid bilayer

50 nm

(b)

FIGURE 18.8 The Influenza Virus
Influenza viruses are enveloped RNA viruses. Their shapes are highly variable. (a) The micrograph shows a specimen of Type A/Port Chalmers/1/73. (×36,400. Courtesy of F.A. Murphy, Centers for Disease Control, Atlanta.) (b) The surface projections on the envelope are proteins that enable the virus to adhere to the respiratory epithelium and to break down mucus. These hemagglutinins and neuraminidases undergo genetic changes in the Type A influenza viruses.

FIGURE 18.9 Secondary Bacterial Infection of Virus-Infected Cells
Viral infections often alter host tissues, making them more susceptible to secondary bacterial infections. Bacteria are able to adhere firmly to such altered cells. Here human epithelial cells in culture were infected with a virus, causing them to fuse. A large colony of *Streptococcus pneumoniae* (center) has attached to the virus-infected cells (visible as background material in the scanning electron micrograph) and are multiplying. (×3758. Courtesy of John L. Carson, PH.D., Department of Pediatrics, University of North Carolina at Chapel Hill.)

persons if it is given during the incubation period and early phase of the disease. Nonspecific treatment of the symptoms is the usual choice, however. A number of antiviral drugs are being tested, including interferon.

EPIDEMIOLOGY AND CONTROL MEASURES.   The reservoir for influenza virus is human beings, and possibly other warm-blooded animals, including pigs and ducks. Influenza is transmitted primarily by droplets of respiratory secretions. The incubation period is very short (one to three days), so epidemics spread explosively when there are many susceptible persons. New influenza strains are pandemic, and spread within weeks from one hemisphere to another. Crowds should be avoided during epidemic periods.

Prophylactic vaccination is widely practiced but not completely successful. This is because the choice of H and N antigens in the virus strains to be used to make next winter's vaccine must be made this year, about nine months in advance. These choices sometimes don't coincide with the virus' random genetic changes during the intervening months. Thus the vaccine may immunize against antigens no longer present in the prevalent viral strain. Influenza prediction, like weather forecasting, is an inexact science. Annual vaccination is beneficial enough, however, so that it is recommended for high-risk individuals — the elderly and those with underlying cardiopulmonary conditions — who would be most likely to develop pneumonia or other complications.

Universal vaccination, however, does not appear to be justifiable, because the injections are only partially effective and are not free of risk. In 1976, there was a national campaign to vaccinate everyone against a newly appeared influenza strain called the swine flu. Among people who took the vaccine, a number of cases of **Guillain–Barré disease,** a syndrome with muscular pain, weakness, and neurologic abnormalities, including paralysis, occurred. Allergic reactions to the traces of egg protein remaining in the vaccine also occur.

VIRAL PNEUMONIA

Viral pneumonia is characterized by a gradual onset, less pronounced fever than the bacterial disease, and little chilling. There is also less sputum production than there is with bacterial pneumonia. The agent, multiplying intracellularly, affects the lung tissue. Although viral pneumonia in general is less severe than the bacterial form, the respiratory syncytial virus causes a very severe pneumonia when it strikes children under a year of age. There is no specific antiviral therapy to treat it.

FUNGAL INFECTIONS

The lung is a prime site for fungal infection. Most fungi are aerobes and the lungs, of course, contain abundant oxygen. Fungi, being eucaryotic, are slow-growing and compete poorly in areas that have a large normal bacterial flora, such as the upper respiratory tract. However, since the lower respiratory tract lacks a resident flora, fungi do well once they are established. Fungal infections of deep tissues, called **systemic**

mycoses, have a slow onset and development. Fungal infection provokes strong delayed hypersensitivity and granuloma formation. The balance between multiplication of the agent and the effectiveness of the host response determines whether the infection progresses or is brought under control. If control measures fail, the fungus disseminates to other areas of the body, and in such cases the infection is often fatal.

HISTOPLASMOSIS

Histoplasmosis is a widely distributed mycosis caused by *Histoplasma capsulatum.* It is endemic in at least 30 states of the United States and in more than 60 other countries.

THE ORGANISM.   *Histoplasma capsulatum* is a dimorphic fungus of the deuteromycete group (Figure 18.10). In nature it grows in soil enriched by the droppings of bats or birds such as starlings and chickens. In soil or on laboratory media, the fungus develops in the hyphal form. The white to brown mycelial mold forms asexual microconidia and macroconidia. The conidial spores are infectious when inhaled. *H. capsulatum* in tissue develops as a budding oval yeast. It may be isolated with modified Sabouraud's medium or other formulas incubated at 30 degrees Celsius. Identification is based on specific colonial and microscopic morphology, followed by biochemical tests. A CMI response to the fungus produces a delayed hypersensitivity reaction to **histoplasmin,** an antigen derived from cells of the fungus. Histoplasmin reactivity is assessed by a skin test. The histoplasmin test is useful in diagnosis, but there is a high frequency of false positive reactions.

THE DISEASE.   Most cases of histoplasmosis are asymptomatic. The primary acute form of histoplasmosis is the most common disease pattern. After the spores are inhaled, they germinate in the lung, where the fungi are ingested by macrophages, in which they multiply. A lesion develops that resembles that of primary tuberculosis.

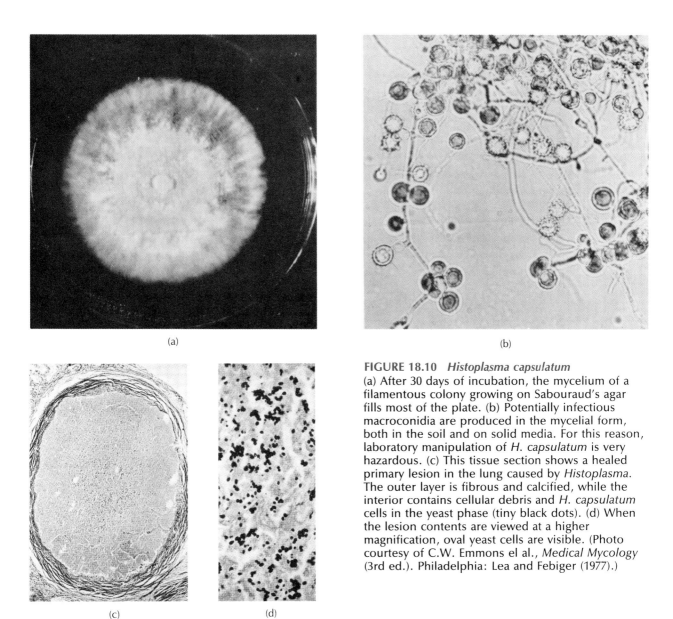

(a)

(b)

(c)

(d)

FIGURE 18.10 *Histoplasma capsulatum*
(a) After 30 days of incubation, the mycelium of a filamentous colony growing on Sabouraud's agar fills most of the plate. (b) Potentially infectious macroconidia are produced in the mycelial form, both in the soil and on solid media. For this reason, laboratory manipulation of *H. capsulatum* is very hazardous. (c) This tissue section shows a healed primary lesion in the lung caused by *Histoplasma*. The outer layer is fibrous and calcified, while the interior contains cellular debris and *H. capsulatum* cells in the yeast phase (tiny black dots). (d) When the lesion contents are viewed at a higher magnification, oval yeast cells are visible. (Photo courtesy of C.W. Emmons el al., *Medical Mycology* (3rd ed.). Philadelphia: Lea and Febiger (1977).)

Symptoms range from none to fever, respiratory distress, cough, weight loss, and joint pains. The agent disseminates to all other parts of the body, but a CMI response develops that brings the fungus under control. The pulmonary lesion may heal completely, or there may be residual scar- ring or calcification (Figure 18.10c). Should the lesion reactivate, the still-viable yeast cells can give rise to progressive disease. Effective CMI is essential to control the invading organism suc- cessfully; chronic disease indicates an under- lying CMI deficiency.

In a very few cases, principally in infants and children, histoplasmosis progresses rapidly to a severe disseminated form. The CMI response is too weak to control fungal multiplication, and lesions may develop in the liver, lymph nodes, adrenal glands, and the gastrointestinal tract, as well as in the lungs. Without prompt treatment, death is certain. A chronic, cavitary form, similar to progressive tuberculosis, may develop in adults. It advances slowly over a period of years and often causes death.

THE TREATMENT.   Therapy may not be needed in mild primary histoplasmosis. In severe cases, the antimicrobial drug amphotericin B markedly improves survival rates, and the diseased tissue is sometimes surgically removed.

EPIDEMIOLOGY AND CONTROL MEASURES.   In the endemic areas, particularly in the Ohio and Mississippi River valleys, more than 80% of the adult population have positive histoplasmin tests. This reveals the pervasiveness of inapparent or mild infection in these areas. Yearly in the United States, about 500 cases are reported and 40 to 50 deaths occur.

As we noted, in nature the fungus multiplies and sporulates in soil, and histoplasmosis is contracted by inhaling dust-containing spores. The budding form produced when the fungus multiplies in human tissues is not infectious; thus person-to-person transmission does not occur.

Roosting sites for birds, chicken coops with old accumulations of droppings, bat-infested areas such as caves and attics or dust from these areas contain infective spores. Therefore control measures include minimizing exposure to soil contaminated with bird droppings. Masks should be worn when contact is unavoidable. Soil decontamination is not practical, but periodic campaigns to clean up some of the above hazards are carried out. Lab cultures may be infective, so technicians should exercise caution in dealing with suspect cultures. There is no available vaccination.

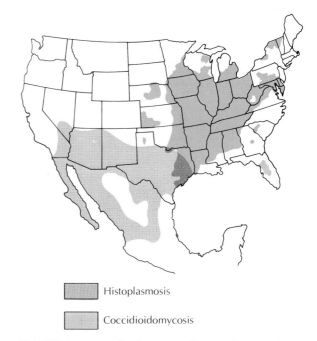

Histoplasmosis

Coccidioidomycosis

FIGURE 18.11   Endemic Areas of Systemic Mycosis
Histoplasmosis (color) is endemic in areas of the eastern central United States, while coccidioidomycosis (gray) is endemic in the dry, desert areas from Texas westward. However, because people travel frequently between regions, cases of either disease may be encountered far from the endemic area.

OTHER FUNGAL LUNG INFECTIONS

COCCIDIOIDOMYCOSIS.   **Coccidioidomycosis,** a mycosis caused by *Coccidioides immitis,* occurs commonly in desert areas in the Southwestern United States, Mexico, and South America. The endemic areas overlap only slightly with those of histoplasmosis. As much as 80% of the population of these regions may show immunologic evidence of infection. Visitors may also become infected, so isolated cases are diagnosed occasionally in nonendemic regions. *C. immitis,* when multiplying in soil or on laboratory media, produces **arthrospores** that are infectious when inhaled. Lab cultures also form arthrospores and are very hazardous to laboratory personnel.

In the host's lung tissue, the arthrospore develops into a **spherule** (Figure 18.12a). Within the spherule, thousands of sporangiospores develop. In two to three days the spherule ruptures, releasing mature spores that disseminate within the lung. Sporangiospores are not infectious, however, so person-to-person transmission does not occur.

The common form of coccidioidomycosis is acute but mild; less commonly, severe respiratory symptoms of coughing and chest pain persist for up to four weeks. Generally, the disease

FIGURE 18.12   Coccidioidomycosis
When the fungus, *Coccidioides immitis,* multiplies in soil or on laboratory media, it develops a mycelium with septate hyphae (a). These hyphae produce many arthrospores, which can cause lung infections in individuals who inadvertently inhale them. Once in the lung, an arthrospore initiates a different developmental cycle, which results in the production of spherules containing many sporangiospores (b). The mature spores may either disseminate the disease through body tissues or be exhaled to begin a new growth cycle outside the body. (Photo courtesy of C.W. Emmons et al., *Medical Mycology* (3rd ed.). Philadelphia: Lea and Febiger (1977).)

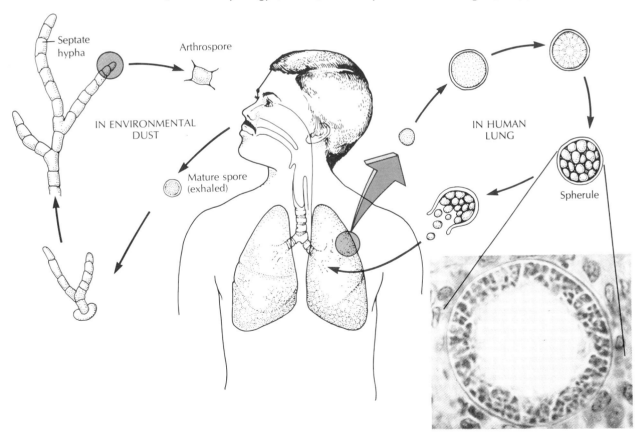

is self-limited. However, less common progressive disseminated and chronic cavitary forms are seen. Skin tests using the antigen coccidioidin and serologic tests are used in diagnosis. Treatment with amphotericin is essential in the more severe forms of the illness. Recovery gives permanent immunity. There is no specific preventive action.

PNEUMOCYSTOSIS—A PROTOZOAN DISEASE

The protozoan *Pneumocystis carinii* (Figure 18.13) causes a form of pneumonia, called **pneumocystosis,** in profoundly immunosuppressed individuals. Once considered a clinical rarity, this infection is suddenly being diagnosed with greater frequency. In acquired immunodeficiency syndrome (AIDS) it is the most common life-threatening infection, appearing in almost 60% of the cases. The parasite is found in both cyst and sporozoite form in the intraalveolar tissue. Although antiprotozoan drugs such as pentamidine or trimethoprim–sulfa combinations are used, the infection is usually unresponsive and fatal because of the total absence of normal host immune defenses. Although pneumocystosis usually occurs in adults with AIDS, it also occasionally occurs in immunosuppressed children.

FIGURE 18.13 *Pneumocystis carinii*
This protozoan parasite causes a form of pneumonia that is common in immunosuppressed persons, such as AIDS patients. The SEM shows cells of *P. carinii* joined by interparasitic fibrils (arrow). The cultured cells on which the parasites are growing show retraction (star), a characteristic cytopathic effect. (×10,873. Courtesy of Visuals Unlimited/L.L. Pifer.)

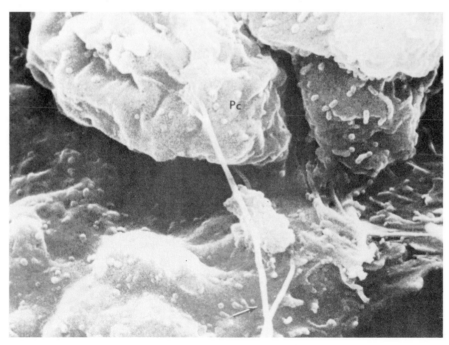

BOX 18.2
Caught In A Dust Storm

The San Joaquin Valley in California is an important endemic area for coccidioidomycosis. A violent windstorm in the valley on December 20 and 21, 1977, created extensive dust clouds that spread to many areas of California. State and local health officials became concerned that dust-borne arthrospores of *Coccidioides immitis* would expose people over a large area. They started surveillance. In fact, in January 1978, 11% of the serologic samples from persons with suspected coccidioidomycosis tested by the health department were positive for *C.*

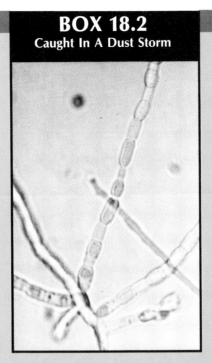

immitis, up from 2% in January 1977. During the same period, the University of California at Davis reported that 18% of samples submitted were positive compared with 4% of samples tested in January 1977. Several of these patients lived outside endemic regions of the state. The data confirmed that persons traveling through the San Joaquin Valley during the storm, or those exposed to dust clouds from the area, experienced a transitory increased risk of coccidioidomycosis.

Photo from C.W. Emmons et al., *Medical Mycology*, Philadelphia: Lea and Febiger (1977).

SUMMARY

1. Lower-respiratory-tract infections can become severe and life-threatening because they cause secretion and accumulation of tissue fluids forming sputum. Sputum may be viscous or semisolid, difficult to expel, and may obstruct the proper flow of air through the airways and exchange of gases at the alveolar surface.

2. Mycobacteria, viruses, and fungi also cause cell and tissue destruction. The agent may be cytolytic, or the delayed hypersensitivity response that accompanies cellular immunity may destroy cells harboring the antigenic target.

3. Lower-respiratory-tract infections are most common and severe in the weakened or immunologically compromised host. Risk increases with age; underlying cardiopulmonary disease; or a history of smoking, diabetes, or alcoholism.

4. In the healthy individual, a mild viral infection of either the upper or lower respiratory tract may temporarily reduce resistance, allowing a secondary pneumonia to develop.

5. Tuberculosis and the systemic fungal infections are usually asymptomatic, relatively harmless, and rapidly resolved. Progressive or chronic infections, which are serious and life-threatening, correlate with lowered resistance due to underlying disease, malnutrition, or other compromising factors.

6. Bacterial infections of the lower respiratory tract can be effectively treated with suitable antimicrobial drugs if the patient is not otherwise overwhelmingly ill. Active tuberculosis requires combined drug therapy. Systemic fungal infections are treated with amphotericin, but mortality is high if treatment is not begun early enough.

7. Vaccination is available for selected lower-respiratory-tract infections, including pertussis, pneumococcal pneumonia, tuberculosis, and influenza. Pertussis vaccination is essential for all infants. Vaccination for the other diseases is usually recommended only for high-risk individuals.

DISCUSSION QUESTIONS

1. Why does the lower respiratory tract not contain a normal flora?

2. Compare the clinical features of the different bacterial pneumonias.

3. What would be the major element of a campaign to reduce the incidence of tuberculosis in a rundown inner-city area? How might the problems differ from those encountered in a remote rural village?

4. Review the techniques for laboratory identification of the bacterial pneumonia agents.

5. Compare the treatments used for each type of lower-respiratory-tract infection.

ADDITIONAL READING

Emmons CW, et al. *Medical mycology.* 3rd ed. Philadelphia: Lea and Febiger, 1977.

Lennette, EH. *Manual of Clinical Microbiology.* 4th ed. ASM, Washington, D.C. 1985.

McGinnis, M. *Laboratory Handbook of Medical Mycology.* Orlando, Fla: Academic Press, 1980.

Robbins JB. Towards a new vaccine for pertussis. In: Leive L, Schlessinger D, eds. *Microbiology 1984.* ASM Washington, D.C.: American Society for Microbiology, 1984.

Thornsberry C, Balows A, Feeley JC, and Jakubowski W. *Legionella: Proceedings of the 2nd International Symposium.* ASM Washington, D.C., 1984.

See also, books listed at the end of Chapter 17.

19

GASTROINTESTINAL
TRACT INFECTIONS

CHAPTER OUTLINE

The gastrointestinal (GI) tract digests the food we eat, carrying out mechanical, chemical, and absorptive operations. When all goes well, you feel terrific after a good meal. However, when pathogenic microbial activities follow your meal, upsetting the function of your GI tract, you become miserable. Most of the GI diseases are not life-threatening, but they are a common source of minor illness and discomfort. Others, such as cholera, botulism, and hepatitis, are very serious indeed.

Most GI pathogens have their reservoir in the human GI tract, and are transmitted by the fecal–oral route via water, food, or fingers. Bacterial, fungal, viral, protozoan, and helminthic organisms will all be prominent in this chapter.

Microbial pathogens featured in this chapter:

Salmonella species
Shigella species
Vibrio cholerae
Enteropathogenic *Escherichia coli*
Hepatitis Viruses
Giardia lamblia
Clostridium botulinum
Helminthic parasites

We will distinguish two general categories of GI infections. In the first, the agent is present and active; in the second, it is a toxic product, rather than the organism itself, that causes the trouble. This distinction is important because the two categories of disease require different control measures.

THE GI TRACT — THE OUTSIDE THAT IS INSIDE

The GI tract is a continuous, hollow tube running through our bodies from the mouth to the anus. Visualize it as the hole in a doughnut. The hollow part or **lumen** contains a mixture of food in the process of being digested, the digestive secretions, and a complex microbial flora. As long as the wall of the intestinal tract is physically intact, this digestive mixture is actually still outside our bodies. Few pathogens have the ability to pass through the intestinal wall. This is good because real trouble starts when gut bacteria escape from the intestines and enter the normally sterile body cavities or the blood stream.

ANATOMY OF THE GI TRACT

We discussed the mouth and oropharynx earlier. In this chapter, we will deal with the rest of the GI tract, from the epiglottis to the anus, and with its accessory glands, the liver and pancreas (Figure 19.1a).

ALIMENTARY TRACT. The **esophagus** is a straight tube, about 25 cm long in the adult, running from the pharynx at the epiglottis to the stomach. The tube is muscular, flexible, and collapsible. It enters the stomach through the **cardiac sphincter,** a muscular ring that closes to prevent stomach contractions from squeezing the gastric contents back up into the esophagus.

The **stomach** itself is an expandable, muscular reservoir that collects food during a meal, and changes size as food is added or emptied out.

Partially digested food leaves the stomach via the **pyloric sphincter.** About 1 ml of the gastric contents is expelled with each stomach contraction. The stomach lining bears glands that secrete the highly acidic gastric juice, composed of hydrochloric acid, digestive enzymes, and mucus.

The main work of digestion occurs in the **small intestine,** a long (about 610 cm), much-folded tube about an inch in diameter. It is made up of three continuous regions, the **duodenum,** the **jejunum,** and the **ileum.** Stomach contents empty into the duodenum. Bile, a liver secretion that aids in the digestion of fats, and pancreatic secretions bearing hydrolytic enzymes are mixed with the fat to facilitate digestion. Muscular contractions known as **peristalsis** move the fluid mixture along, mix it, and bring it into contact with the intestinal lining where the nutrients may be absorbed into the blood stream. The lining of the small intestine is arranged in mobile fingerlike projections called **villi,** which are specialized to increase absorptive capacity. They are capable of limited movement and are supplied by many blood vessels.

The **large intestine** or bowel has a greater diameter than the small intestine and is thick-walled and muscular. It reabsorbs water from the residue left after the food has been digested. The ileum joins the large intestine at a right angle. The main portion of the large intestine beyond this junction is the **colon.** A twisted, coiled tube, the **appendix,** branches off at the junction point, and is a potential site of obstruction and infection. The colon ends at the **rectum,** which in turn connects with the **anal canal.** Wastes pass out through the anus when the anal sphincter is voluntarily relaxed, or involuntarily when propelled by unregulated peristalsis.

All the organs of digestion lie in the peritoneal or abdominal cavity. Like the thoracic cavity, this hollow space is normally sterile. It is lined with a membrane, the **peritoneum,** and the organs are suspended by membrane folds called **mesenteries.**

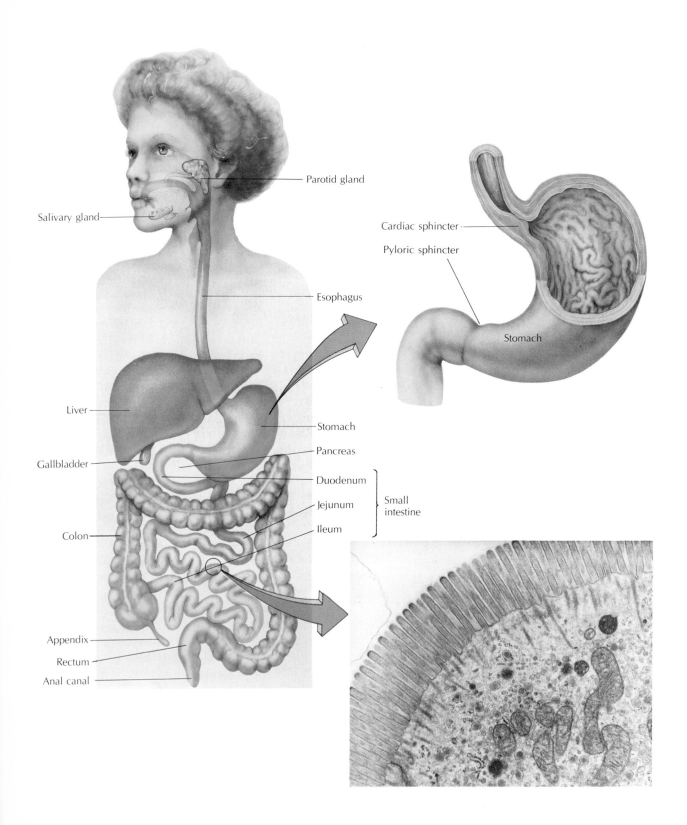

Parotid gland

Salivary gland

Esophagus

Liver

Gallbladder

Colon

Appendix

Rectum

Anal canal

Stomach

Pancreas

Duodenum

Jejunum

Ileum

Small intestine

Cardiac sphincter

Pyloric sphincter

Stomach

FIGURE 19.1 The Gastrointestinal System
The GI tract is a continuous, hollow tube, running between the mouth and the anus. Enzymatic secretions produced by the digestive glands break down food as it passes through. Although a huge microbial population exists in the lower GI tract, few invasive pathogens are able to pass through the intestinal wall into the blood stream or body tissues. The stomach itself is an expandable, muscular reservoir that secretes highly acidic gastric juices from glands in the mucosal lining. The bulk of digestion, however, occurs in the small intestine. This micrograph of the lining of the small intestine shows a surface epithelial cell bordered with villi. Several bean-shaped mitochondria are visible in the cell cytoplasm. The villi, which increase the absorptive surface, are permeated with abundant blood vessels. (×11,000. From Lelio Orci and Alain Perrelet, *Freeze-Etch Histology*. New York: Springer-Verlag (1975), plate 142.)

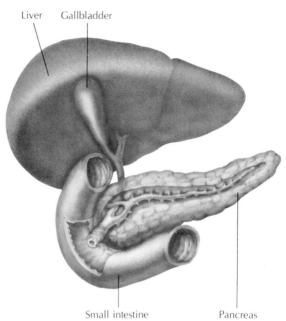

Liver Gallbladder

Small intestine Pancreas

FIGURE 19.2 The Digestive Glands
The liver supplies bile to assist in the digestion of fat molecules. The bile, in turn, is stored in the gallbladder and released into the intestinal tract as needed. Most digestive enzymes are produced in the pancreas. The enzymatic secretions pass to the small intestine through pancreatic ducts.

DIGESTIVE GLANDS. The digestive glands include the liver and pancreas (Figure 19.2). The **liver** lies on the right side of the body immediately under the diaphragm. It has many functions. Its role in digestion is to supply bile to assist in fat digestion. Internally, the liver is a mass of **lobules.** In a lobule, blood flows in one direction through spongelike sinuses and is cleansed of the bile components. The **bile,** a solution of waste products, collects and flows in the opposite direction to bile ducts and the **gallbladder.** The gallbladder stores the bile and releases it to the intestinal tract as needed; it is also frequently a site of infection.

The **pancreas** produces most of the digestive enzymes, as well as hormones such as insulin and glucagon, which regulate blood sugar levels. The enzymatic secretions of the pancreas travel to the intestine by means of the pancreatic ducts and are mixed with food shortly after it enters the small intestine.

PHYSIOLOGY OF THE GI TRACT

We will limit our overview of the physiology of digestion to those normal functions with which

infection most frequently interferes, and to the host defenses against such infections.

MOTILITY. As mentioned, food is moved by rhythmic waves of involuntary smooth-muscle contraction called peristalsis (Figure 19.3). About three contractions per minute advance the contents about 1 cm. This rate is carefully adjusted by the body to allow for optimal digestion and absorption. The soluble building-block molecules released during food digestion must be adequately absorbed from the GI tract into the blood, or they will just pass through and be wasted. The normal rate may be altered by infection, emotional stress, diet, or numerous other factors. Slowed or arrested peristalsis produces constipation, while accelerated peristalsis

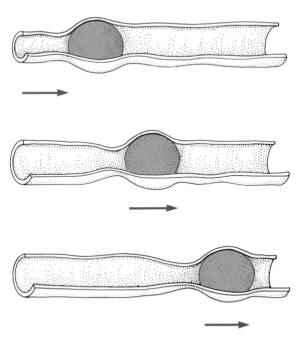

FIGURE 19.3 Peristalsis
Smooth, rhythmic waves of involuntary muscle contraction move food along the GI tract at a rate that is adjusted for optimal digestion and absorption.

causes diarrhea. Microbial products affect the peristaltic rate by irritating or inflaming the bowel or by affecting the portion of the central nervous system that controls peristalsis.

WATER AND ELECTROLYTES. The various digestive glands secrete a large volume of water into the GI tract. Daily, about 8 liters of water move from the body's internal fluid reserves into the GI tract. Almost all of this fluid is recovered by reabsorption in the large intestine if the intestinal contents are moving through at the normal, slow rate. In diarrhea, rapid passage permits little reabsorption, and there is a significant water loss. In moderate cases, **dehydration** occurs. If fluid loss continues, **hypovolemia** (abnormally low blood volume) develops. This may culminate in shock and death. The normal adult blood volume is about 10 liters, so the uncompensated loss of several liters of water per day in severe

diarrhea cannot be sustained for long. In small children, severe dehydration develops quickly; it is the major threat to life in childhood GI infections.

When water is lost, physiologically important inorganic ions (**electrolytes**) such as sodium (Na^+), potassium (K^+), chloride (Cl^-), and bicarbonate (HCO_3^-) are lost with it. A proper electrolyte balance is essential to regulate the function of nerve and muscle tissue and the internal pH of body fluids. Massive electrolyte loss associated with GI infection thus compromises all body functions. Diarrhea, for example, promotes loss of HCO_3^- ion, leading to acidosis, a condition where the blood pH is abnormally low. Alternatively, persistent vomiting, with loss of hydrochloric acid, may lead to alkalosis, a condition of elevated blood pH.

DEFENSES. The GI tract has many defenses (Figure 19.4). First, its epithelium is constantly shed and replaced, limiting the length of time a microorganism can adhere. Its mucous lining also inhibits microbial penetration into the tissue of the tract. Additionally, stomach acid may be lethal to some microorganisms and may chemically neutralize some toxic materials. However, stomach acid is unreliable as a sterilizing mechanism. It works well in the empty stomach, but in the presence of a full meal — which is, after all, how pathogens often enter — the stomach pH may be diluted and not be sufficiently acidic. Bile destroys the cell walls of some delicate bacteria.

Peristalsis, diarrhea, and vomiting are also protective mechanisms. Diarrhea and vomiting empty the intestinal tract of toxic or indigestible contents; normal peristalsis keeps the bowel contents moving along, which holds down the total number of microorganisms.

The GI tract is well protected immunologically. Many antigens make their first contact with the immune system via the aggregates of lymphoid tissue in the mucosa of the GI tract. These lymphoid aggregates, called **Peyer's patches,** produce large amounts of IgA, which neutralize many pathogens (refer to Figure

12.2c). Intestinal macrophages are plentiful, and are particularly important in protecting infants. Peritoneal macrophages police the abdominal cavity for possible microbial escapees from the tract itself.

NORMAL FLORA OF THE GI TRACT

While the normal human body is composed of about 10^{13} cells, it supports around 10^{14} microbial colonists. Thus "they" outnumber "us" by 10 to 1! Most of these microorganisms inhabit the GI tract, predominantly the mouth and the large intestine. In thinking of the distribution of the GI tract's flora, it is helpful to think of the tract as a tube containing a flowing stream. The rate of flow is fast in some regions, and slow in others. On the average, food traverses the 10 to 13 ft of the small intestine in less than 5 hours, but then takes an additional 18 hours to pass through the last 4 ft of large intestine. The chemistry of the stream varies with the secretions that are added to it at various points, the diet, and the individual's physiologic state. In the fast-flowing areas, microbial colonization probably requires attachment to the surface epithelium. This restricts the microbial numbers because of the limited available space on the surface of the intestinal wall. In the colon, where the fecal matter moves slowly, organisms live both attached to the wall and free in the contents.

The microbial flora is as varied as the habitats. Over 300 species of microorganisms have been recovered by microbial sampling. Some of these organisms were probably not true colonists, but were just passing through at the moment of sampling.

MICROBIAL FLORA OF DIGESTIVE ORGANS

The normal esophagus has no appreciable flora; some bacteria may attach temporarily to its epithelium. However, damaged esophageal surfaces often become colonized.

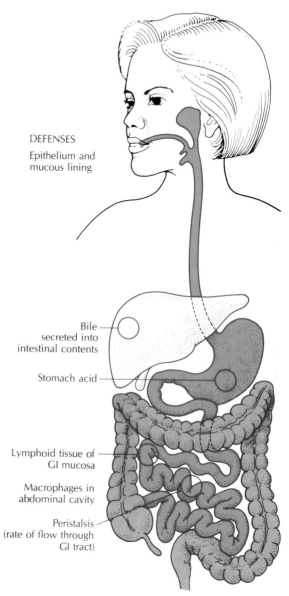

DEFENSES
Epithelium and mucous lining

Bile secreted into intestinal contents

Stomach acid

Lymphoid tissue of GI mucosa

Macrophages in abdominal cavity

Peristalsis (rate of flow through GI tract)

FIGURE 19.4 Defenses of the GI Tract
The GI tract is well protected against pathogens. Its epithelium and mucous lining inhibit microbial penetration, and numerous enzymatic secretions are hostile to invading organisms. The rate of flow through the tract keeps the number of microorganisms manageable. Note that flow is fastest in the small intestine (darker area) and slowest in the colon (light area), where a large microbial population flourishes.

BOX 19.1
Diet, Normal Flora, And Intestinal Cancer

Do the foods we eat put us at risk of developing cancer of the digestive tract? It has been suspected for years that they do, as studies of the people of different countries turned up widely varying rates of the different cancers. For example, the Japanese diet contains little red meat, leans heavily on vegetables, but includes very large amounts of pickled foods. Native Japanese have high rates of stomach cancer but low rates of bowel cancer. Their relatives who have moved to the United States and adopted the typical American diet, which is high in animal protein and fats but low in vegetable fiber, develop the American cancer pattern — frequent colon cancers, but few stomach cancers.

Most studies confirm the hypothesis that fiber and fats are key factors in colon and breast cancers. Diets that are high fat–low fiber statistically appear to promote cancer most effectively. Diets with low fat–high fiber are the least cancer-promoting.

A partial picture is emerging of one mechanism whereby an apparently "harmless" food substance becomes a carcinogen. The hypothesis is as follows. First, fats ingested in the diet are partly digested into another type of lipid by the microflora. Apparently, different people's floras vary in terms of whether they produce this potentially harmful lipid. Next, this lipid, made soluble by the bile salts, is acted upon by bowel anaerobes such as *Bacteroides* to form an active carcinogen. The carcinogen may or may not have any effect, depending on other components of the diet, some of which appear to counteract carcinogens directly. It seems that in the low fat–high fiber diet little of the harmful product is made; in addition, because of the fiber, any carcinogens present move through the intestinal tract more rapidly and have less chance to do harm.

Favored anticarcinogenic foods right now include the plants of the cabbage family — cabbage, cauliflower, brussels sprouts, broccoli, and others. In countries where these foods form a large part of the diet, people enjoy low colon cancer rates. These plants contain chemicals that appear to be anticarcinogens. I think I'll have broccoli for supper tonight!

(Photo courtesy of Darlene Bordwell, Boston.)

The stomach is essentially sterile when empty because its pH drops to about 2. Microbial activity in the normal stomach is not significant.

The upper small intestine has a small number of normal flora attached to the lining, consisting predominantly of facultative anaerobes such as *Lactobacillus, Streptococcus,* and the coliforms. Little is known about their contributions to intestinal function. The number of organisms increases rapidly in the jejunum and ileum. At the same time, intestinal secretions neutralize the stomach acid, raising the pH of the intestinal contents to near neutrality. **Diverticula** (pouches that develop in the intestinal wall of many older individuals), ulcerated areas, and obstructions promote extensive growth of normal flora. Bad effects of such growth may include vitamin depletion, gas production leading to distention, and tissue breakdown.

After the fluid mixture enters the colon, its movement slows down even further, and microbial numbers soar. Feces contain about 10^{12} bacteria per gram, about 40% by weight of the total fecal mass. These bacteria are predominantly obligate anaerobes such as *Bacteroides.* They live off of food residues, epithelial cells that have been shed, and mucus.

ROLE OF THE NORMAL FLORA

In animals such as cows, gut bacterial enzymes help to make usable otherwise indigestible fibrous foods. There is also evidence that bacteria break down some portion of the fiber in the human diet, fermenting it into small fatty acids, which we then can utilize as a source of food energy. Gut bacteria also produce certain vitamins, notably vitamin K. On the negative side, gut bacteria produce toxic end products (such as ammonia) that pass into the blood and are either detoxified by the liver or excreted by the kidneys. Anaerobic digestion of dietary proteins yields the gases hydrogen, carbon dioxide, methane, and hydrogen sulfide. These gases are partially reabsorbed, but the remainder is passed as **flatus.** Amino acid byproducts such as indole and skatole, short-chain fatty acids, and amines are also produced by the gut bacteria. In concentrated amounts, these not only inhibit the bacteria but are also toxic to the person. Such toxicity occurs when a person is constipated. Prolonged retention of feces leads to excessive microbial growth and accumulation of these normal microbial products to higher than normal levels, where they cause systemic toxicity and general discomfort. Recent evidence shows that gut bacteria under certain conditions may produce carcinogens. For example, amines found in food, and food additives such as nitrite, can be converted by bacteria to carcinogenic nitrosamine compounds.

The intestinal microflora's main positive value appears to be its competitive exclusion of pathogens. Many enterobacteria produce substances lethal or inhibitory to other bacteria. For example, some species produce H_2S, which represses the growth of *Escherichia coli;* other species produce short-chain fatty acids that inhibit *Salmonella, Shigella, Pseudomonas,* and *Klebsiella.*

Antibiotic therapy disturbs the normal bowel flora. The antibiotic-resistant members of the resident flora are able to multiply freely without competition. *Candida albicans* or *Staphylococcus aureus,* normally very minor components of the flora, can become predominant. This disturbance of the flora is only temporary, and the population balance returns to normal when antibiotic therapy stops.

KEY GROUPS OF MICROORGANISMS

STRICT ANAEROBES. Up until a few years ago, if you had asked any microbiologist to name the dominant gut bacterial species, the answer would likely have been *Escherichia coli.* Improved methods of anaerobic culture have shown us that *E. coli* and other facultative anaerobes are just a minor component, the tip of the iceberg. The strict anaerobes — *Bacteroides, Fusobacterium,* and others — outnumber the facultative enterobacteria by perhaps 1000 to 1.

LACTIC ACID BACTERIA.   These fermentative organisms produce lactic acid from carbohydrates. They include *Lactobacillus, Streptococcus,* and *Bifidobacterium.* In the newborn they are the first microbial colonists of the GI tract. Lactic acid bacteria are the dominant organisms of the stomach and upper sections of the small intestine in adults.

THE ENTEROBACTERIACEAE.   Bacteria of the family Enterobacteriaceae (the enterobacteria) are a small but important part of the total flora. These Gram-negative facultatively anaerobic rods contain many closely related species, including most of the bacterial GI pathogens. The family is subdivided into six tribes, each containing one or more genera and species (Figure 19.5). Species are differentiated by biochemical testing. The genera usually considered normal ferment lactose; the serious pathogens such as *Salmonella* and *Shigella* do not. A variety of rapid identification methods, usually automated, have been developed for clinical identification of the enterobacteria.

Plasmids and chromosomal genes are freely exchanged among the enterobacteria, promoting relatively rapid changes in characteristics. A notable effect of this genetic exchange is that enterobacteria rapidly acquire antibiotic resistance. **Enteropathogenic** strains have the ability to produce enterotoxins, chemicals that induce severe diarrhea. As just mentioned, the genera *Salmonella* and *Shigella* are pathogens, and *Proteus, Klebsiella,* and *Serratia* species are opportunists that sometimes cause gastrointestinal disease or urinary-tract infection. *Escherichia coli,* so often named as a normal flora organism, has the dubious distinction of being the bacterial species that is most often isolated in hospital infections in the United States.

Coliform organisms is a term used in sanitary microbiology to designate the lactose-

FIGURE 19.5   Classification of the Enterobacteria
The family Enterobacteriaceae is separated into six tribes, groups that generally share key biochemical and genetic traits. These are further subdivided into genera, and the genera into species. Biochemical tests such as lactose fermentation, H$_2$S production, and urease are important in identifying these.

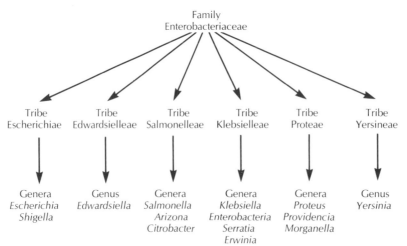

fermenting enterobacteria, such as *E. coli* and *Enterobacter*. They are universal in human feces. Since coliforms are readily isolated and identified, they are used as **indicator organisms** to check for fecal contamination of food, water, and other samples.

BACTERIAL INFECTIONS

Bacterial GI infections produce a wide spectrum of symptoms and severity, reflecting two features — the invasiveness and the toxigenicity of the infecting organism. The susceptibility of the person infected also makes a big difference in the severity of the disease. The epidemiology of the bacterial GI infections that we will cover is shown in Table 19.1 on pp. 564–565.

SALMONELLOSIS

Salmonellosis is a form of GI infection that is rapidly increasing in incidence (Figure 19.6a). **Typhoid fever,** the most severe form of salmonellosis, was historically a major killer, but it is now rare in the developed nations. The less severe forms of salmonellosis, however, are increasingly common, favored by our changing food-production methods and eating habits.

THE ORGANISM. The genus *Salmonella* contains three species, *typhi, enteritidis,* and *choleraesuis. Salmonella* is motile and can survive for limited periods outside the body in moist environments, such as food, water, and soil. It can be isolated from the mixed flora of the gut by plating on highly selective media to eliminate lactose-positive enterobacteria. *Salmonella* itself is typically lactose-negative and H_2S-positive. After initial identification, serologic tests are essential to determine which of the more than 2100 serotypes of *Salmonella* has been recovered (Figure 19.6b). Typhoid fever is always caused by *S. typhi.* The other forms of salmonellosis are caused by the other species.

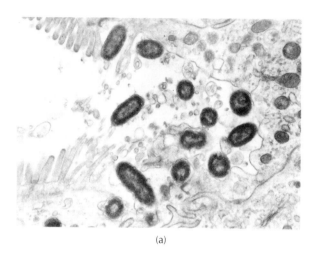

(a)

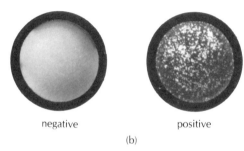

negative   positive

(b)

FIGURE 19.6 *Salmonella*
(a) The micrograph shows *Salmonella* organisms attacking the lining of the intestine between two villi. Apparently, there is extensive cell damage, and the microvilli are being detached. (b) In a slide agglutination test in which a small amount of pure culture is mixed with matching antiserum, the occurrence of agglutination confirms the organism's serotype (right). A control containing *Salmonella* cells and serum without the type-specific antibody (left) remains evenly cloudy without visible clumping. (Reprinted by permission of Harper & Row, Philadelphia, PA; from A. Takeuchi, *American Journal of Pathology* (1967), 50:125, 127.)

THE DISEASES. In typhoid fever the bacilli pass through the lining of the upper small intestine into the underlying lymphoid tissue. They are phagocytosed but not killed, then disseminated within the migrating phagocytes, multiplying

TABLE 19.1 Epidemiology of bacterial GI infections

Disease	Reservoir	Transmission	Incubation Period (Days)
Typhoid fever	Human beings	Oral–fecal; contaminated food and water. Urine may be infective	7–21
Salmonellosis	Human beings Poultry, other animals	Oral–fecal, contaminated meat, poultry, milk, and eggs	$\frac{1}{2}$–$1\frac{1}{2}$
Bacillary dysentery	Human beings, primates	Oral–fecal, direct contact, water, food, and flies	1–7
Cholera	Human beings	Oral–fecal by water, flies, and hands	2–3
Enteropathogenic Escherichia coli	Human beings	Oral–fecal, during birth, by direct contact, or fomites	$\frac{1}{2}$–3

Adapted from Benenson, AS, *Communicable Diseases of Man,* 4th ed. 1985.

during an incubation period of about two weeks. Then symptoms begin. During the first week there is bacteremia, with fever, headache, and general discomfort due to bacterial endotoxin. In the second week, if the disease is untreated, lesions of the intestinal mucosa and Peyer's patches develop. Rosy spots develop on the skin of the abdomen. Fever and prostration become more severe. In the second or third week, as the patient's condition worsens, intestinal hemorrhage and perforation, or pneumonia, may appear. Convalescence begins in the fourth week, as effective cell-mediated immunity develops. About 3% of patients develop chronic infections of the gallbladder or urinary bladder. Antibiotics may eradicate the carrier state, but surgical removal of an infected gallbladder is occasionally required.

Salmonellosis is a **gastroenteritis** or inflammation of the stomach and intestine, marked by nausea and/or diarrhea and varying degrees of damage to the intestinal lining. The infections are usually self-limiting; symptoms disappear after a few days and antibiotic therapy is not usually needed. However, some patients require hospitalization, and a few who are already ill with other diseases may die.

THE TREATMENT.   The antibiotic chloramphenicol is used in treating typhoid fever, but there are some resistant strains. Steroids and extensive life-supportive therapy may be essential in more severe or advanced cases. In salmonellosis, the treatment needed may range from little or none through antibiotics such as ampicillin to hospitalization and fluid replacement.

TABLE 19.1 Epidemiology of bacterial GI infections (continued)

Period When Communicable	Patient Control Measures	Epidemic Control Measures
As long as agent is in excreta. Ten percent carry three months; 2 to 5% are permanent carriers.	Isolate; disinfect wastes and soiled articles	Intensive search for case or carrier and vehicle
Variable — days to weeks; temporary carriers are common	Remove from food handling; Patient or child care. Enteric precautions	Report. Investigate food sources and handlers.
During acute illness or during short carrier state	Isolate; rigid personal and enteric precautions	Report. Investigate food sources and handlers
While stool is positive, usually a few days, rarely months or years	Isolation not necessary if staff practices strict cleanliness and disinfection of wastes and soiled articles	Boil or treat water, supervise food preparation, investigate transmission
Not known; during fecal colonization	Isolate infected infants and suspects. Disinfect discharges and soiled articles	Admit no more babies to nursery. Provide separate staff for exposed babies. Investigate source of outbreak

EPIDEMIOLOGY AND CONTROL MEASURES (FIGURE 19.7). *S. typhi* is found only in human beings. It is spread by the oral–fecal route. The reservoir is the convalescent or chronic carrier, and the vehicle is water, food, or soil contaminated with untreated sewage. Toilets located close to wells, and the use of polluted drinking water are risk factors. Improved sanitation and carrier control are the corresponding major control measures. In the United States, sporadic cases (about 400 a year) are either acquired abroad or traced to contact with chronic carriers, often elderly persons who have been carriers for decades. Systematic follow-up measures now ensure that recuperating patients do not continue to excrete *S. typhi*. Recovery usually confers a lifelong immunity. Available killed-organism vaccines produce discomfort and relatively short protection.

The other *Salmonella* species are primarily animal pathogens. Their reservoirs include cows, pigs, dogs, turtles, fish, chickens, and numerous other animal species. Transmission to humans is accidental. The increasing incidence relates to changes in food production and consumption patterns, mass production of eggs and meat, and the increasing trend toward consuming "fast" and convenience foods. Chicken and other domestic fowl are perhaps the most significant reservoir of *Salmonella* strains. A large proportion of chicken meat is inoculated during butchering with organisms accidentally released from the birds' intestines. Laying hens also may contaminate eggs. *Salmonella* is killed by cooking; however, surfaces and utensils that have touched contaminated raw food may transfer an inoculum to other food that will be served

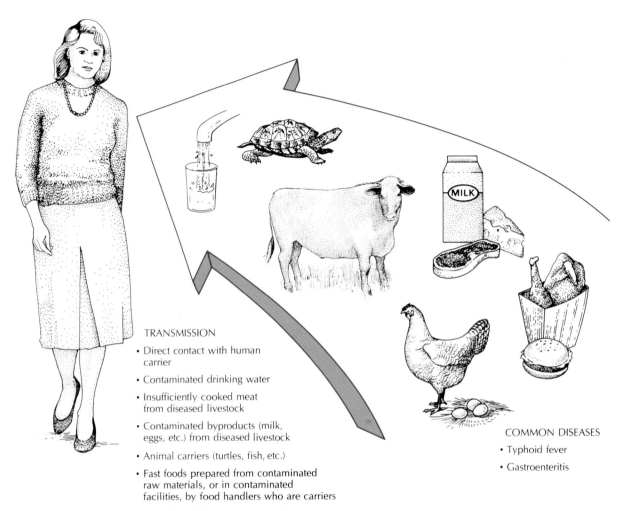

TRANSMISSION

- Direct contact with human carrier
- Contaminated drinking water
- Insufficiently cooked meat from diseased livestock
- Contaminated byproducts (milk, eggs, etc.) from diseased livestock
- Animal carriers (turtles, fish, etc.)
- Fast foods prepared from contaminated raw materials, or in contaminated facilities, by food handlers who are carriers

COMMON DISEASES

- Typhoid fever
- Gastroenteritis

FIGURE 19.7   Epidemiology of Salmonellosis
Salmonellosis is a form of GI infection that has increased in incidence. *S. typhi*, the cause of typhoid fever, is found only in human beings. Once a major killer, it is now relatively rare in developed nations, although some people may remain chronic carriers. Other *Salmonella* species may be transmitted to humans via animal carriers. Transmission is related to the increasing trend toward mass production and convenience foods. The organism may pass from animals to humans via contaminated milk, eggs, insufficiently cooked meat, or poultry. Contaminated drinking water and sloppy sanitation habits of food handlers may also result in an outbreak of the disease.

uncooked, such as salads. Human food handlers are also a common reservoir.

The market feed cycle for domestic livestock raised for meat re-uses slaughterhouse wastes as an inexpensive protein source for animal feeds.

Inadequately decontaminated fish meal may be fed to chickens. Chicken or pig wastes may be fed to healthy beef animals which in turn become *Salmonella* carriers. Thus it spreads from chickens to other domestic animals.

Convenience foods and restaurants have a role in *Salmonella* transmission. Americans now eat more than 50% of their meals away from home. Precooked turkeys and rare roasts of beef sold at delicatessen counters are consumed without further cooking. These meats have been the source of many *Salmonella* outbreaks. Inadequate heating, which fails to cook the interior of the meat during initial preparation, is at fault. Gravies held on steam tables for prolonged periods have infected customers of fried-chicken take-out chains. An outbreak traced to a restaurant may be due to contaminated food sources, inadequate preparation, or sloppy bathroom habits on the part of kitchen help, some of whom may be *Salmonella* carriers. When an outbreak originates at a restaurant, there are usually many more cases than if a similar situation arises in the home. Despite intensive surveillance, the incidence of salmonellosis has increased by 50% in the last 10 years, with almost 41,000 cases reported in the United States in 1982. Improved food handling practices are the key to control (Chapter 28).

SHIGELLOSIS

Shigellosis or **bacillary dysentery** is similar in its disease pattern to salmonellosis. There are both severe and mild forms.

THE ORGANISM. *Shigella* is a Gram-negative enterobacterial genus, all of whose members are associated with human disease. Initial isolation of this agent can be accomplished on a differential medium such as MacConkey's agar or a mildly selective formula such as Salmonella-Shigella or Hektoin agar. The bacteria are lactose- and H_2S-negative, and differ from the salmonellas in being nonmotile. *S. flexneri* and *S. sonnei* in the United States and *S. dysenteriae* in the Far East are primary agents of shigellosis.

THE DISEASE. The characteristic bacillary dysentery is a diarrhea with frequent watery stools, pain, and passage of blood and mucus. *Shigella* is invasive and penetrates the large intestine, causing inflammation and local microabscesses. The lesions remain shallow, but may cover a large area. In the healthy adult, shigellosis resolves spontaneously in about a week. In debilitated adults or small children, fluid and electrolyte loss may become life-threatening. Mortality is very low — about 0.5% — when adequate medical care is provided.

THE TREATMENT. Tetracycline, chloramphenicol, and ampicillin can all be used against susceptible strains. But resistance is a major problem. In fact, *Shigella* is the organism in which the genetic elements called resistance transfer factors (RTFs) were first demonstrated. RTFs, discussed in Chapters 8 and 24, have been responsible for a worldwide spread of resistance to sulfa drugs and tetracycline in *Shigella* species.

EPIDEMIOLOGY AND CONTROL MEASURES. Human beings are the only reservoir for the shigellae. Shigella is shed by actively ill and convalescent persons for one to three months. Prolonged carriage is rare. Shigella persists for up to three days in sea water and for 30 days in milk, eggs, and seafoods. As few as 200 bacilli may be enough to cause infection. Control measures are based on procedures to eliminate fecal contamination of food, water, and fomites. There were over 18,000 cases of shigellosis in the United States in 1982. Children under five years old have the highest case rates. Institutionalized populations have very high rates of infection, and very severe manifestations. The disease has historically been more severe in the Far East, where more virulent *Shigella* strains are found and water contamination is widespread. Recovery confers partial protective immunity. There is a vaccine, but it is not generally used.

CHOLERA INFECTIONS

Cholera was pandemic in the 19th century, with major outbreaks in Europe and the United States as well as elsewhere. However, with the introduction of water and sewage treatment, we thought we had vanquished cholera in the Western world. However, sporadic new cases have recently appeared in the United States, around the Mississippi delta region, showing that an old foe is not dead after all.

THE ORGANISM.   *Vibrio cholerae* is a Gram-negative, curved rod (Figure 19.8a). Vibrios have the epidemiologically important ability to survive and thrive outside the body in cool water; thus they can be spread and transmitted easily. The El Tor strain is most common in the Western hemisphere; fortunately, it mainly causes mild or inapparent infections. More pathogenic strains are prevalent in the Far East. *V. cholerae* can be isolated from stool specimens on selective media containing tellurite and those with an alkaline pH. Further identification depends on microscopic observation of the bacterial cells, which are typically C-shaped or S-shaped. Biochemical differentiation and serologic typing are also used. The vibrio's key virulence factor is a toxin called **choleragin.** The toxin activates the cells of the intestinal lining to greatly increase their secretion of fluid and electrolytes into the lumen of the gut (Figure 19.8b). The end result of toxin production is voluminous diarrhea.

THE DISEASE.   *Vibrio* colonizes the small intestine and attaches to the mucosa but does not penetrate or damage its surface. As it multiplies, it produces choleragin, which then causes a severe diarrhea. The outcome depends on the degree of fluid-electrolyte imbalance and the

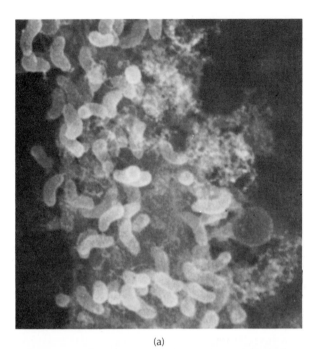

(a)

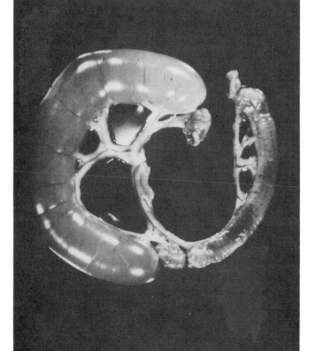

(b)

FIGURE 19.8   *Vibrio cholerae*
(a) In this scanning electron micrograph, masses of curved vibrios are seen attached to strands of intestinal mucus. (×11,600. Courtesy of M. Neal Guentzel, D. Guerrero, and T.V. Gay, Division of Life Sciences, University of Texas at San Antonio.) (b) A rabbit ileal loop test demonstrates the effect of cholera toxin on the secretion of fluid into the lumen of the bowel. Segments of an anesthetized rabbit's ileum were tied off. A control segment (right) was perfused with saline solution, while cholera toxin was introduced into the segment to the left. The toxin-exposed segment became distended with secreted fluid, which would, in turn, give rise to an acute diarrhea. (From ASM Slide Collection.)

ability of the patient to stand the stress. If the patient is strong, the process subsides naturally in a few days. Otherwise, loss of blood volume (hypovolemia) leads to apathy and shock, and death may occur. From 50 to 70% of those who become actively ill die rapidly if they do not get adequate therapy. It has been observed in India that an adult infected with a highly pathogenic strain may be dead within three hours after the diarrhea begins. On the other hand, in the United States with less pathogenic strains, a healthy adult may have a mild or even subclinical case. Cholera is hardest on the very young and very old. Malnutrition increases the mortality risk.

THE TREATMENT. Fluid-electrolyte replacement is all that is needed to reduce the mortality rate sharply. In Western countries we take for granted the availability of sterile intravenous fluids and personnel to administer them. However, in less-developed countries, these may be totally unavailable, especially in amounts sufficient to deal with an epidemic of thousands of cases. A massive relief effort may be necessary to handle the situation. For example, in the cholera outbreak that followed the India–Pakistan war of 1971, support teams sent by the World Health Organization were able to save most of the patients they could get to. They pioneered the emergency field use of oral or nasogastric fluids, supplied through a tube down the nasopharynx to the stomach. In contrast to intravenous fluids, these need not be sterile and can be mixed on the spot. Antibiotics are not of much use in treating cholera, because it is the effect of the toxin, not bacterial growth, that must be addressed.

EPIDEMIOLOGY AND CONTROL MEASURES. Recent studies suggest that the primary source of *V. cholerae* may be certain small aquatic invertebrates. Asymptomatic human carriers are also very important in the chain of transmission of cholera. The vibrios are spread largely via contamination of water and (to a lesser extent) food by human feces. Recent cases in the United States have been traced to inadequately cooked

crayfish and shellfish harvested from fecally contaminated water. Sewage-treatment methods are the most important means of prevention and control. Proper cooking of shellfish harvested from areas where sewage contamination is a possibility is also important. There is a cholera vaccine that may be taken by those anticipating exposure through travel to countries where the organism is endemic.

CAMPYLOBACTER INFECTIONS

Campylobacter jejuni (Figure 19.9) causes intestinal upsets or abortion in cattle, sheep, goats,

FIGURE 19.9 *Campylobacter jejuni*
The scanning electron micrograph reveals the rigid helical shape of this spirillum, causes a significant percentage of gastroenteritis cases. It is microaerophilic and requires special media and incubation conditions for isolation. There are many means of transmitting the agent, notably via contaminated poultry products, dairy products, and water. (Courtesy of Dr. Dorothy L. Patton, Department of Medicine, University of Washington, Seattle.)

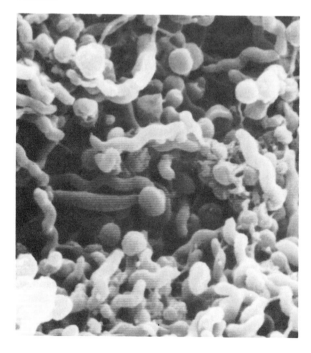

dogs, pigs, and fowl. Although data on its prevalence are incomplete, it appears that this widely distributed bacterial species is one of the most important causes of human gastroenteritis, particularly of children. In the United Kingdom epidemiological surveys suggest that *Campylobacter* infections outnumber *Salmonella* infections. In one study with infants under eight months old, *C. jejuni* caused 31% of the cases of gastroenteritis. The symptoms include diarrhea, abdominal cramps, fever, headache, and vomiting. The organism responds to several antibiotics, but drug therapy is usually not necessary.

Reservoirs include human beings and some domestic animals, especially puppies and dogs with diarrhea. The agent may be contracted by contact with their feces. Additionally, the agent is frequently foodborne. Raw chicken and other poultry products frequently have surface contamination with *Campylobacter*. Raw milk from infected cows has been shown to spread the disease within farm communities. Waterborne transmission is also a factor; transmission of *C. jejuni* through a defective branch of the town water supply affected 20% of the population of a small city.

ENTEROPATHOGENIC *ESCHERICHIA COLI*

As you know, most *E. coli* strains are harmless normal flora. There are, unfortunately, also some pathogenic strains that cause acute illness. Ordinary *E. coli* strains may also act as opportunistic pathogens, especially in the urinary tract and in newborns. Table 19.2 lists some diseases caused by *E. coli*.

THE ORGANISM.   *Escherichia coli* is a facultatively anaerobic Gram-negative rod that normally inhabits the large intestine of all human beings and most warm-blooded animals. Certain serotypes of *E. coli* have special virulence factors that enable them to cause two types of GI infection, one a severe diarrhea similar to cholera and

TABLE 19.2 Some diseases caused by *Escherichia coli*
Cystitis
Hemorrhagic colitis
Infant diarrhea
Neonatal meningitis
Neonatal septicemia
Turista

the other an invasive **hemorrhagic colitis,** a condition of colon inflammation with bleeding from the intestinal wall. The role of *E. coli* in these diseases was noted only about 15 years ago. The diarrhea syndrome is caused by strains that produce a **heat-labile enterotoxin,** that is, a toxin readily destroyed by heat. Adherence factors (Figure 19.10) are also important in virulence. Hemorrhagic colitis is caused by a serotype that is highly invasive, a characteristic not found in "normal" strains of *E. coli*.

THE DISEASES.   The diarrhea syndrome appears in infants, and also in adults as travelers' diarrhea, or **turista.** Infant diarrhea is a scourge of hospital nurseries. In the diarrheal form, heat-labile enterotoxin interacts with the lining of the small intestine, in much the same way as does cholera toxin, although heat-labile toxin has a much lower potency. Outpouring of fluid and electrolytes leads to diarrhea and dehydration. The condition can be rapidly fatal to newborn infants, in whom the case fatality rate is about 16%. The disease is mainly just an uncomfortable nuisance to adults.

Hemorrhagic colitis patients frequently suffer from a bloody diarrhea that lasts from 5 to 12 days. Hospitalization is often necessary. The transmission of the invasive *E. coli* strains responsible has been positively linked to undercooked hamburger, and less conclusively to other foods.

THE TREATMENT. Fluid replacement is essential, as in all severe diarrheas. Antimicrobial drugs may be of value, but multiple-drug resistance is widespread in *E. coli*, especially in hospital-acquired strains.

EPIDEMIOLOGY AND CONTROL MEASURES. Infant diarrhea occurs more often in hospital-delivered babies and hospitalized infants than in home-delivered children. It appears that the newborn's GI tract rapidly becomes colonized with *E. coli* strains from its attendants; in a hospital setting these strains may be pathogenic, antibiotic-resistant strains that, although harmless to the adult carrier, are dangerous to the newborn. Breast-feeding confers some protection against *E. coli* (and other intestinal pathogens) because human milk transfers large amounts of secretory IgA.

Sanitation and nutritional standards have a profound effect on the frequency and severity of this illness, as they do on so many others. In Latin America the death rate from infant diarrhea (for which *E. coli* is not the only agent) is high. In Nicaragua almost 2000 of every 100,000 children under age five die of infant diarrhea each year. In the United States, only 5 out of 100,000 children die of this illness.

Travelers' diarrhea, or turista, is often contracted by Westerners visiting less-developed countries. Scattered epidemiological studies show that enteropathogenic *E. coli* strains are the primary villains. Local residents are spared because they have acquired immunity to the bacterial strains present in their environment.

VIRAL INFECTIONS

The intestinal tract is a common entry point for viruses. Many viruses then move on to other tissues where they cause disease, and we discuss them in other chapters. Here we will look at viruses that cause gastroenteritis and hepatitis.

VIRAL GASTROENTERITIS

Almost everyone has had the "24-hour flu," "intestinal flu," "green death," or whatever the local nickname may be. Calling this illness "flu" is misleading; it has nothing to do with the influenza virus.

Viral gastroenteritis has a rapid onset, with repeated vomiting and/or diarrhea; it differs from some of the bacterial infections just discussed in that in adults it is much briefer — it is usually over within 12 to 24 hours. The lining of the small intestine is affected, undergoing mild, reversible tissue changes. Infants, however, may have more prolonged symptoms and more severe dehydration. Recent advances in virus identification have pinpointed certain of the responsible agents.

FIGURE 19.10 The Adherence Factor
The ability of certain *E. coli* strains to adhere to epithelial tissue plays an important role in the organism's virulence. In this scanning electron micrograph, enteropathogenic *E. coli* (strain 505) are shown adhering by capsular material to normal calf villi. (Courtesy of J.J. Hadad and C.L. Gyles, University of Guelph, Ontario.)

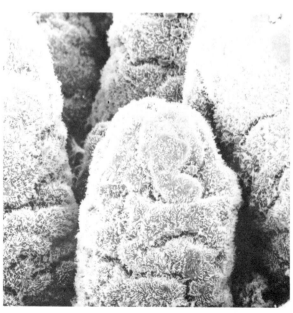

In infants the **rotavirus** (Figure 19.11) is the most common cause of viral diarrhea. It is a relatively large RNA virus. The incubation period of rotavirus infection is 48 to 72 hours. Vomiting is the main symptom, and illness lasts two to seven days. Rotaviruses are frequently carried by adults, but rarely cause illness in persons older than two years. Recovery stimulates immunity, and the immunity of nursing mothers protects their infants. Breast-fed children have less rotaviral diarrhea than formula-fed children.

The **Norwalk agent** is perhaps the commonest GI virus affecting adults. It is a parvovirus, a tiny DNA virus. The incubation period of the disease is 24 to 36 hours, nausea is the major symptom, and the illness lasts from 8 to 24 hours. Recovered individuals may shed virus in the stool for one month. There is a rather short-term, type-specific immunity. Adenoviruses, coxsackieviruses, polioviruses, and echoviruses also cause adult viral gastroenteritis.

FIGURE 19.11   The Rotavirus
The rotavirus is a relatively large RNA virus and a common cause of infant diarrhea. There are many different strains, which affect infants of various mammalian species. (From Jose Esparza and Francisco Gil, A study on the ultrastructure of human rotavirus, *Virology* (1978), 91:141–150.)

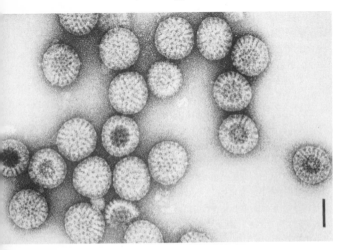

VIRAL HEPATITIS

Hepatitis is liver inflammation, often marked by **jaundice** — an accumulation of bile pigments giving a yellow tinge to the skin and eyes. A few hepatitis cases are caused by noninfectious processes such as chemical exposure, but most are caused by several unrelated viruses (Figure 19.12). The two best-studied viral agents are Types A and B, and at least two other "non-A, non-B" agents are known to exist. Very little is known about the causative agents of non-A, non-B hepatitis. They have never been cultured. The only practical way to tell which hepatitis agent is which is by serologic testing. The three types of hepatitis are compared in Table 19.3.

THE VIRUSES.   **Hepatitis A** (commonly known as infectious hepatitis) is the disease caused by the hepatitis A virus (HAV). It is a **short-incubation disease.** The agent is a small RNA virus, classed as enterovirus type 72. It is exceptionally resistant to heat and cold inactivation.

Hepatitis B or serum hepatitis is the traditional, if incompletely accurate, term for the **long-incubation disease** caused by hepatitis B virus (HBV). The DNA-containing agent has recently been placed in a new viral classification, the hepadnaviruses. The virion, about 40 to 50 nm in diameter, is composed of a surface coat and a viral core, about 27 nm in diameter. Its surface coat contains a specific antigen, HBsAg (also called Australia antigen). Immunity to

FIGURE 19.12   Hepatitis Viruses
(a) Hepatitis A virus, an RNA-containing enterovirus. (b) Hepatitis B virus, a DNA-containing hepadnavirus; note the three distinct types of particles shown in the photograph. (c) The viruslike particles associated with non-A, non-B hepatitis. (d) The diagram depicts some of the factors in the epidemiology of hepatitis. (Photo credits: a, ×156,600, courtesy of D.W. Bradley; b, reproduced from C.R Madelay, *Virus Morphology*, by permission of the author and Churchill Livingstone, Edinburgh; c, ×136,422, reprinted from the Centers for Disease Control's June 1978 *Morbidity and Mortality Report,* 27(24):Fig. 1.)

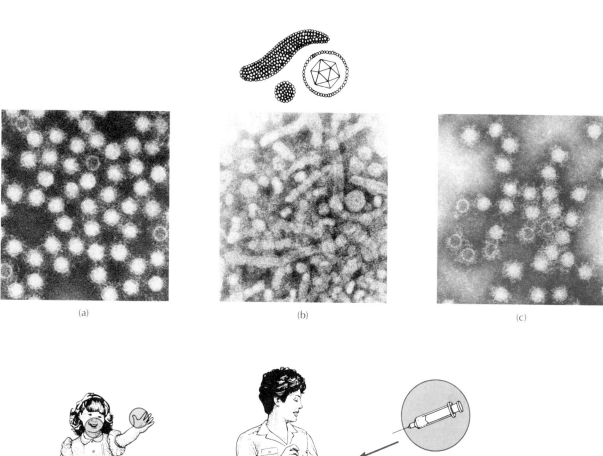

(a) (b) (c)

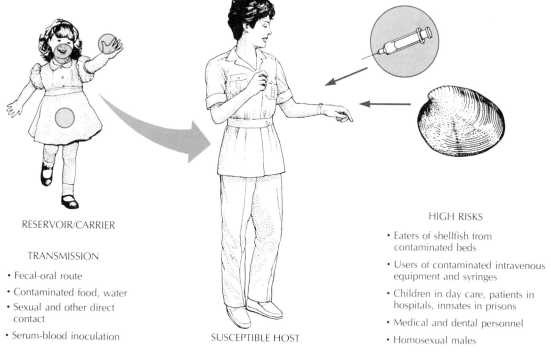

RESERVOIR/CARRIER

TRANSMISSION

- Fecal-oral route
- Contaminated food, water
- Sexual and other direct contact
- Serum-blood inoculation

SUSCEPTIBLE HOST

HIGH RISKS

- Eaters of shellfish from contaminated beds
- Users of contaminated intravenous equipment and syringes
- Children in day care, patients in hospitals, inmates in prisons
- Medical and dental personnel
- Homosexual males

(d)

TABLE 19.3 Comparison of types of hepatitis				
Non-A, Non-B	Common Name	Most Common Principal Sources	Transmission	Clinical Onset
Type A	Infectious hepatitis	Food and water	Fecal–oral	Acute
Type B	Serum hepatitis	Blood	Serum, secretions	Gradual injection
Non-A, Non-B	None	Probably blood	Probably injection	Probably gradual

this antigen is protective. Incomplete spherical forms, empty coats composed only of surface material, are found in the serum of most patients. The viral core contains the circular viral DNA and a core antigen, HBcAg, which can be detected serologically. Antibodies against this core material are made, but are not protective. A second virus, the RNA-containing **delta virus,** has been found in conjunction with HBV in some cases. The two viruses acting together appear to cause a more serious disease.

THE DISEASES. The incubation period of hepatitis A is about 30 days. It is followed by a **preicteric phase** (before jaundice) of about a week, marked by fever, abdominal tenderness, nausea, and loss of appetite. The **icteric phase,** when actual jaundice is present, coincides with increasing liver damage, as indicated by rising levels of liver enzymes in the serum. Hepatitis A is diagnosed by detecting either HAV antigen or antibodies. The clinical symptoms usually disappear fairly rapidly, and recovery is complete after about three months. Liver regeneration

may take up to a year and occasional relapses occur. Mortality is about 2%.

Hepatitis B has a much longer incubation period, averaging 60 days, and a gradual onset. The characteristics of the preicteric and icteric symptoms are generally similar to those of hepatitis A. However, hepatitis B varies widely in its severity. Some people have a subclinical disease detectable only if serum liver enzymes happen to be monitored. However, the viral antigens are detectable in the serum. A variety of tests are available to detect surface or core antigens or antibodies for diagnostic purposes.

People with poor cell-mediated immunity, however, may suffer massive liver necrosis that is rapidly progressive and fatal. Some individuals, particularly those with mild or subclinical cases, retain the virus in the liver tissue for periods of months or years and experience a persistent infection. Five to ten percent become permanent HBV carriers — as many as 200 million worldwide. The serum of volunteers with this condition is used as a source of coat antigen from which hepatitis B vaccine is prepared.

		TABLE 19.3 Comparison of types of hepatitis (continued)		
Incubation Period	Chronic cases	Prevention by Immune Serum	Serological Diagnosis	Vaccination Available
15–50 days	Rare	Effective	Immune electron microscopy of virus in stool. Anti- hepatitis A virus detected by radioimmunoassay	No
50–180 days	Rather common	Ineffective	HbsAg or anti-Hbs antibody detected by many means	Yes
Unknown	Probably common	Unknown	Not yet available. Diagnosis suggested when A and B tests are negative	No

The chronically infected state has serious implications for the patient. Liver damage may continue and be cumulative. In addition, there is much evidence that hepatitis B virus is oncogenic. It is believed to play a key role in primary hepatocellular carcinoma, a relatively common and rapidly fatal form of liver cancer.

Non-A, non-B hepatitis (NANB) is a clinically heterogeneous entity, sometimes indistinguishable from Type B. However, analysis of antibody response pinpoints the existence of at least two distinct virus agents, one of which has a shorter and one a longer incubation period. It is currently believed that the NANB viruses are the primary cause of transfusion-related hepatitis, now that serologic screening of donor blood effectively eliminates hepatitis B carriers as blood donors. At this time, there is no comparable direct test to screen donor blood for the NANB viruses. Non-A, non-B hepatitis is diagnosed indirectly, by what clinicians call a diagnosis of exclusion; that is, it is clinical hepatitis where neither anti-HAV or anti-HBV immunoglobulins can be found.

THE TREATMENT. There are no specific therapeutic measures for hepatitis. Supportive therapy, including bed rest and special diet is used as needed. An exposed person may be given immune serum globulin (ISG) as a passive immunization in the hope of moderating the disease. ISG is also used as a preventive measure, and should be given immediately if HAV exposure has occurred. If it is given within two weeks of exposure, 80 to 90% of the cases may be prevented or converted to inapparent infections. Following HBV exposure, either specific hepatitis B immunoglobulin (HBIG) or ISG should be given within one week.

EPIDEMIOLOGY AND CONTROL MEASURES. The various hepatitis viruses are spread by partly separate, partly overlapping patterns (Figure 19.12d). In general, hepatitis viruses may be transmitted by the fecal–oral route, accidental or intentional administration of blood or blood products, saliva, other direct contact, or sexual intercourse.

The fecal–oral route is the most common means of transmission for HAV, with contaminated water or food as the vehicle. Shellfish from sewage-flooded ocean beds also carry an HAV risk. Control measures depend on sewage and shellfish monitoring for the general population. For hospitalized patients, a particular set of procedures called enteric precautions are used by attendants. Small children are common targets. Recent outbreaks have focussed public attention on the risks of hepatitis A and other infections in day-care centers. Not surprisingly, pathogens that are transmitted by the fecal–oral route spread freely among dense populations of diapered and only recently toilet-trained children. Where water and sewage systems are primitive, mild childhood hepatitis is common, and there are few severe adult cases. In more-developed areas, most adults are susceptible, so although there are fewer total cases, a larger proportion of cases are serious.

The serum inoculation method is only one means of transmission of HBV; fecal–oral and sexual transmission also occur. There are several identified high-risk groups. Because infection occurs from contaminated blood products, intravenous equipment, and syringes, hepatitis is a particular problem in the hospital environment. Medical personnel may be infected by a minor accidental needle scratch. Dental personnel are at risk due to saliva and blood exposure from either active cases or chronic carriers. Other groups with especially high risks are, again, small children in day-care or nursery-school situations. Residents of institutions for the mentally retarded, prison inmates, and drug abusers who share equipment are other high-risk groups. Homosexual males have a very high risk due to sexual transmission.

In controlling hepatitis B, precautionary measures can be taken to disinfect the serum and wastes of active cases. Blood-donor screening is carried out by the American Red Cross and other organizations that collect and use blood for transfusion or the preparation of blood products.

Carriers can be excluded from occupations such as food-handling in which they pose a risk to others.

A killed-virus hepatitis B vaccine (Heptavax) became available in 1982 to immunize high-risk groups against hepatitis B. It is composed of the inactivated viral particles from the serum of volunteer carrier donors. It is a "first-generation vaccine"; that is, it is effective but has problems of supply, expense, and potential infection hazard. By the time you read this, it will probably have been supplemented or replaced by an improved synthetic vaccine, containing hepatitis B viral antigens produced by recombinant bacteria, that began its final clinical testing in 1984.

PROTOZOAN INFECTIONS

Protozoans both invade and colonize the intestine. They are frequent members of the normal intestinal flora of nonhuman animals, but in human beings it is not clear if they can be benign. Giardiasis and amebic dysentery are surprisingly common intestinal infections. The epidemiology of both viral and protozoan infections covered here is shown in Table 19.4 on pp. 578–579.

GIARDIASIS

Diarrheal infection due to *Giardia lamblia* (refer to Figure 5.16) is the most common form of intestinal parasitism in the United States, affecting an estimated 3.8% of the population, with about 15,000 cases reported in 1982.

THE ORGANISM.   *Giardia lamblia* is a member of the Class Mastigophora. It has a two-stage life cycle, alternating between the vegetative stage and the resting cyst stage. The trophozoite bears several flagella and has two nuclei positioned so that the cell appears to have two large, owlish eyes. The cyst stage, by which *Giardia* is transmitted, is oval and is marked by a central, rod-shaped reinforcing structure called an axostyle.

THE DISEASE. **Giardiasis,** the active disease state, affects the small intestine. The disease has an acute onset, with symptoms of cramps, diarrhea, bloating, flatulence, and, occasionally, constipation. Complications are rare and recovery occurs within a few days. Giardiasis may affect persons of any age.

THE TREATMENT. Metronidazole (trade name, Flagyl) is the usual therapeutic drug. While metronidazole is being taken, alcoholic drinks are highly toxic and must be avoided.

EPIDEMIOLOGY AND CONTROL MEASURES. *Giardia* cysts are spread primarily by water, and also by hand-to-mouth inoculation and (rarely) food. Giardiasis is increasing in prevalence in day-care centers. It has traditionally been associated with sewage contamination in urban settings (Figure 19.13). However, recent surveys found that its incidence in the United States was highest in the sparsely populated states of Maine, Minnesota, and Colorado. Many cases have developed following wilderness camping, where contamination of the water by human sewage was extremely unlikely, so some other source must have been responsible. It is now clear that beavers and muskrats may serve as a wild-animal reservoir. Beaver populations across the United States have recovered strongly from their brush with extinction during the early 1800s when beaver hats were all the rage. Now their dams and colonies are common in many rural areas. Their feces add to the ever-increasing

FIGURE 19.13 Giardiasis outbreak, Vail, Colorado
This Giardiasis epidemic, totaling about 55 cases, was caused by a sewerline obstruction that allowed untreated sewage to back up and leak into the town's water supply. After the problem was found and corrected, the outbreak gradually subsided.

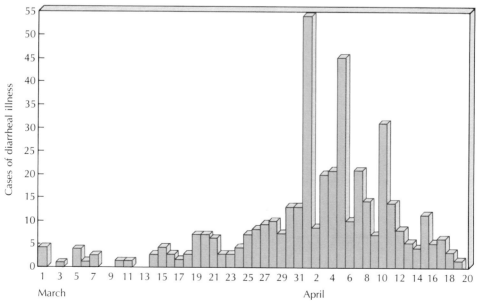

TABLE 19.4 Epidemiology of viral and protozoan GI infections			
Disease	Reservoir	Transmission	Incubation Period (Days)
Viral gastroenteritis, infant	Human beings, animals (?)	Probably oral–fecal, water, food	1–2
Viral gastroenteritis, adult	Human beings	Probably oral–fecal	1–2
Hepatitis A	Human beings, primates	Oral–fecal, direct contact, injection	15–50
Hepatitis B	Human beings, primates	Parenteral inoculation of blood, personal contact	50–180
Amebiasis	Human beings	Water contaminated with cysts, hand-to-mouth	Variable, 2–4 wk
Giardiasis	Human beings, animals (?), beavers	Oral–fecal; water and hand-to-mouth	6–22

Adapted from Benenson, AS, *Communicable Diseases of Man*, 14th ed. 1985.

volumes of human sewage to guarantee that giardiasis will be a large public health problem. One issue to be faced is that *Giardia lamblia* is relatively resistant to chlorination. To deactivate it successfully, slightly more chlorine must be used than may be needed to kill other organisms. This poses a problem in a time when the chlorination of public water supplies is under review because of indications that it poses some previously unsuspected health hazards.

AMEBIC DYSENTERY

Entamoeba histolytica is an ameboid protozoan in the Class Sarcodina, and moves by means of pseudopodia. It has a two-stage life cycle (Figure 19.14). The resting stage, or cyst form, is enclosed in a resistant coating that allows it to survive in the exterior environment and to pass unscathed through the acid stomach. Once in the intestine, the cysts germinate and *E. histolytica* multiplies in the trophozoite form, a typical shapeless ameba. Trophozoites continue to be produced as the organisms multiply within the intestinal wall, but in the lumen of the gut, cysts are formed. Both may be shed in the stool. Diagnosis is based on demonstrating the presence of the microorganism in specially prepared stool specimens.

As the trophozoites invade the intestinal lining, recurrent episodes of inflammation and diar-

TABLE 19.4 Epidemiology of viral and protozoan GI infections (continued)

Period When Communicable	Patient Control Measures	Epidemic Control Measures
Up to eighth day after onset	Isolate infants; enteric precautions	Search for vehicle and source
During acute stage, possibly shortly after	Enteric precautions; no disinfection	Search for vehicle and source
Maximum infectivity; last half of incubation period into icteric phase	Enteric precautions first two weeks; disinfect blood, wastes, and contaminated objects	Determine source, provide immune serum globulin for exposed contacts
Blood infective long before symtoms appear, through acute and chronic phases. Chronic state may persist for years	Enteric precautions; disinfect materials contaminated with blood	Search for cases among patients attending same clinic or service, strict control of blood products; preventive vaccination
While cysts are passed, perhaps for years	Carriers excluded from food handling, sanitary disposal of feces	Determine source and vehicle
Entire period of infection	Sanitary disposal of feces	Determine source and vehicle

rhea occur. Sometimes ulcers develop, probably due to bacterial attack on tissues weakened by the amebas. Bleeding and intestinal perforation develop in severe, untreated cases. Liver, lung, or brain abscesses are rare. The usual treatment is metronidazole given by mouth. Surgical treatment of abscesses is needed if drug therapy alone is not effective. Repeat infections occur, and there is no vaccination.

The reservoir for *E. histolytica* is human beings, and the agent is spread via fecal–oral transmission. Water, foods, and flies are often implicated. Cysts are infective, whereas trophozoites probably are not. The chronic carrier, who sheds mainly cysts, is potentially very hazardous. The disease is endemic throughout the world, with the case incidence high where sanitary standards are low. About 4% of the population are infected in urban industrialized areas and from 5 to 30% in rural areas. There were over 6000 cases of this disease reported in the United States in 1983.

CRYPTOSPORIDIOSIS

A little-known protozoan, *Cryptosporidium*, is found in persistent, intractable diarrhea lasting for more than one year and apparently untreatable. This rare condition is one of the opportunistic infections that may accompany acquired immunodeficiency syndrome.

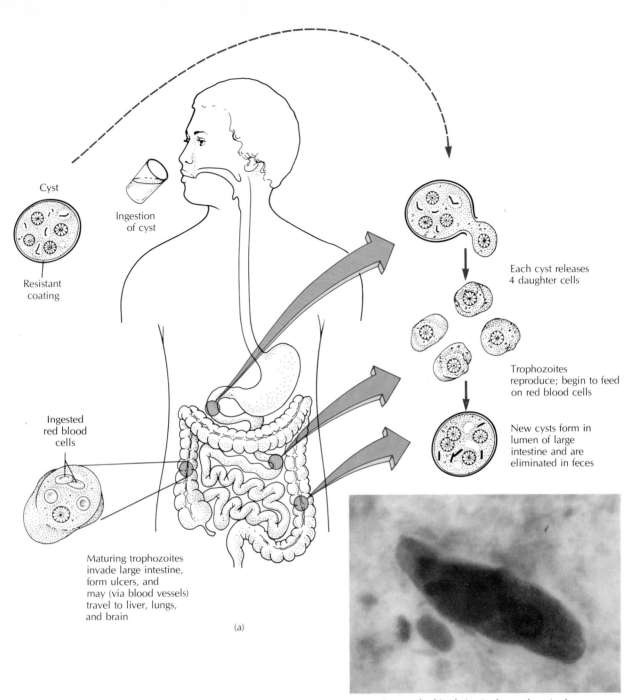

Cyst

Resistant
coating

Ingestion
of cyst

Ingested
red blood
cells

Maturing trophozoites
invade large intestine,
form ulcers, and
may (via blood vessels)
travel to liver, lungs,
and brain

(a)

Each cyst releases
4 daughter cells

Trophozoites
reproduce; begin to feed
on red blood cells

New cysts form in
lumen of large
intestine and are
eliminated in feces

Entamoeba histolytica in the trophozoite form.

(b)

FIGURE 19.14 *Entamoeba histolytica*
(a) *Entamoeba histolytica* is an ameboid protozoan with a two-stage life cycle. In its
cyst form, it can pass unharmed through the stomach to the intestine. Once in the
intestinal wall, the cysts germinate and multiply in the trophozoite form. At the same
time, infectious new cysts are formed in the lumen of the gut. These may be shed in
the stool. (b) A trophozoite of *E. histolytica* showing its nucleus and several ingested
red blood cells. (×1000. Courtesy of the Centers for Disease Control, Atlanta.)

HELMINTHIC INFECTIONS

Many multicellular helminthic worms use the intestinal tract as a preferred site for growth, especially of the adult forms. We will familiarize ourselves with four of the more common parasites.

PINWORM DISEASE

Pinworm disease is very common in small children. It is caused by the nematode, or roundworm, *Enterobius vermicularis*. This parasite has a simple life cycle in which humans are the only host (Figure 19.15). The infection is acquired by ingesting the eggs, which may be acquired from the hands, clothing, or objects such as doorknobs or bathroom fixtures. The eggs develop into adult worms in the large intestine. The female, which is about 1 cm long, migrates at night to the anal opening, where she lays thousands of eggs. This is very irritating to the anal skin, and the usual sign of pinworm disease is an intense itching. There may also be diarrhea. The person's sleeping patterns are disturbed and his or her behavior may change. When the infected person scratches, the eggs are liberally spread around the home and transferred under the fingernails. The diagnostic method is simple. A piece of plain adhesive tape is applied to the skin right around the anal opening. Eggs, if present, will be picked up by the tape, which is then applied to a blank microscope slide. Any eggs will be readily visible under the microscope. The drug pyrivinium will eradicate the worms. However, the home must be carefully cleaned to remove residual eggs, or reinfection will occur.

HOOKWORM DISEASE

Hookworm disease, caused by the nematodes *Necator americanus* and *Ancyclostoma duodenale*, is a worldwide problem in warmer climates, affecting as many as 400 million people. It is a debilitating disease, but it can be successfully treated with drugs such as mebendazole.

The hookworm life cycle (Figure 19.16) alternates between the human small intestine and

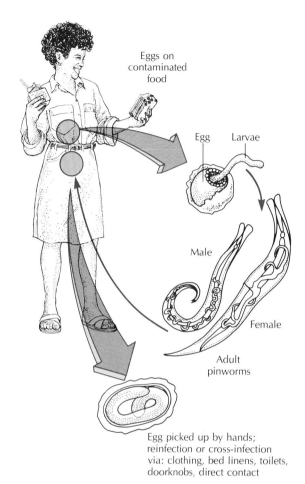

FIGURE 19.15 The Life Cycle of the Pinworm
Enterobius vermicularis adult males and females reproduce in the human intestine. At night, the female emerges and lays eggs on the skin just outside the anus. These eggs may be unintentionally spread, then ingested by the same or a different person. They soon hatch into larvae, which mature into the adult form and continue the cycle.

feces-contaminated soil. Eggs hatch in the soil and develop into larvae, which infect by entering through the skin. They take a roundabout path to the gut, moving through the blood stream to the lungs, passing into sputum, and moving to the mouth where they are finally swallowed. As many as a thousand adult worms may infest one person. The worms cling with their hooks to the intestinal wall, feeding on blood. The females lay millions of eggs, which return to the environment and will continue the cycle if feces are not adequately disposed of.

A healthy, well-nourished human can tolerate the parasitism with few ill effects. However, a malnourished child will develop a severe anemia with possible physical and mental retardation. Poorly nourished adults suffer profound

FIGURE 19.16   The Life Cycle of the Hookworm
Hookworm larvae enter the body through the skin and migrate via the blood to the lungs and then to the intestine. There they mature into adults, sucking blood and causing anemia while they shed millions of eggs that will pass back into the soil to continue the cycle. Hookworm is controllable if feces are adequately disposed of.

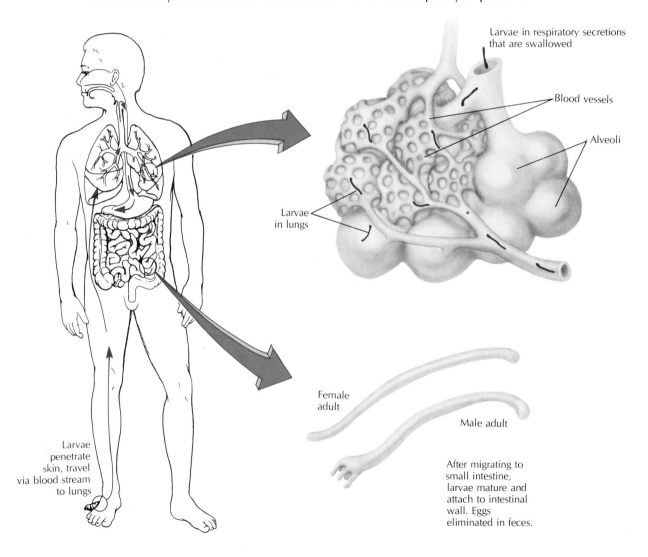

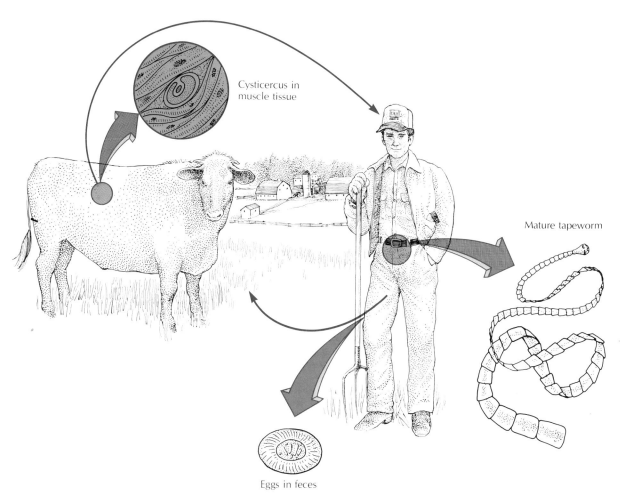

FIGURE 19.17 The Life Cycle of the Beef Tapeworm
Taenia saginata larvae form cistercercae in the muscle of cows. When we eat
inadequately cooked contaminated beef, the larvae emerge in our intestines. There
they attach to the intestinal lining and mature into adult hermaphroditic worms that
shed eggs. If human feces contaminated with tapeworm eggs in turn contaminate
grass on which cows graze, the cycle completes itself.

weakness and fatigue. The cycle may be broken
by a combination of drug therapy, an adequate
nutritional program, and improved sanitation.

TAPEWORM DISEASE

There are a great many different tapeworms with
different life cycles and hosts. These parasites
infect food and domestic animals, which are the
usual reservoir for human infection. We will use

the beef tapeworm *Taenia saginata* as an example (Fig. 19.17). The stages of the parasite's life
cycle alternate between the cow and the human.
The human is the definitive host, which, you
will recall, is the host in which the adult parasite
reproduces. Human symptoms are usually so
mild that we may be unaware of the parasite's
presence, even though the worm may grow to be
6 m long. Diarrhea and bowel obstruction occur

rarely. The tapeworm eggs are shed in the affected human's feces. If the feces are deposited on soil, the eggs hatch and develop. Larvae may then be ingested by cows grazing in the contaminated area. In the cow's muscle, larvae form cystercerci. When we eat undercooked meat from an affected cow, the cycle begins again. Fortunately today less than 1 in 1000 cattle in the United States have tapeworm.

The pork tapeworm *Taenia solium* is even less common, but it is potentially more dangerous. The eggs are ingested and hatch in the human intestine, and larvae invade various organs. In another tapeworm disease called **hydatid disease,** the human is an intermediate host for the canine tapeworm *Echinococcus granulosus.* This disease is a problem in parts of Alaska where sled dogs are an intimate part of the human community. Hydatid cysts can be very damaging to human muscle and organs.

TRICHINOSIS

Trichinosis is a disease that is much feared. It is now rare (about 100 cases per year occur in the United States) and readily preventable. Furthermore, most human infestations are mild or asymptomatic. The nematode parasite *Trichinella spiralis* is acquired from several animals, including the domestic pig and some wild species such as the bear and walrus (Figure 19.18). Cross-contamination from pork to beef sometimes occurs through contaminated work surfaces and meatgrinders in the butcher shop. When we ingest the cysts, they mature to adults in our intestines. The adult parasites produce larvae that migrate through the intestinal wall into the blood stream, then leave the blood stream to encyst in the muscles. The cysts provoke inflammation, and often break down, releasing toxins into the circulation. Symptoms include fever, puffiness around the eyes, and sometimes collapse and death in very heavy infestations. Treatments include thiabendazole, which kills the worms, and steroids to reduce the inflammation. However, nothing is very effective in dealing with extensive cyst populations.

Trichinosis is almost unknown today in commercially raised pigs. Adequate cooking of meats is essential, however. An internal temperature of 137 degrees Fahrenheit is just sufficient to kill trichinae. Many more people get trichinosis from eating foods raw, or have been infected by touching meat before it was cooked, than have gotten it from eating pork with a slight tinge of pinkness. Home-raised pigs should still, even with the declining frequency of this disease, never be fed uncooked garbage. If you buy locally raised pork, a few questions about the farmer's feeding practices might not be a bad idea.

FOOD INTOXICATION

We can distinguish between two types of foodborne disease. In the **foodborne infections** such as salmonellosis, the agent itself is ingested and multiplies in the gut, causing disease. Transmission of these diseases is largely related to poor sanitation. In a **food intoxication,** on the other hand, the microbial agent produces a toxin in the food prior to the food's ingestion. It is the ingested toxin, independent of microbial growth in the gut, that causes the disease state. In this case, we may have the odd situation of having a microbial *disease* without an *infection.* Food intoxications are invariably associated with bad food-handling and preservation practices.

TOXINS THAT CAUSE GASTROENTERIC UPSETS

Some toxins upset gut function by acting on the brain centers that control vomiting and peristaltic rate. Their symptoms are practically indistinguishable from those of some of the infections previously discussed. Although unpleasant, these toxicities are rarely serious.

For example, some strains of *Staphylococcus aureus* make an enterotoxin. It is a neurotoxin that, when absorbed from food, activates the brain center controlling the vomiting reflex.

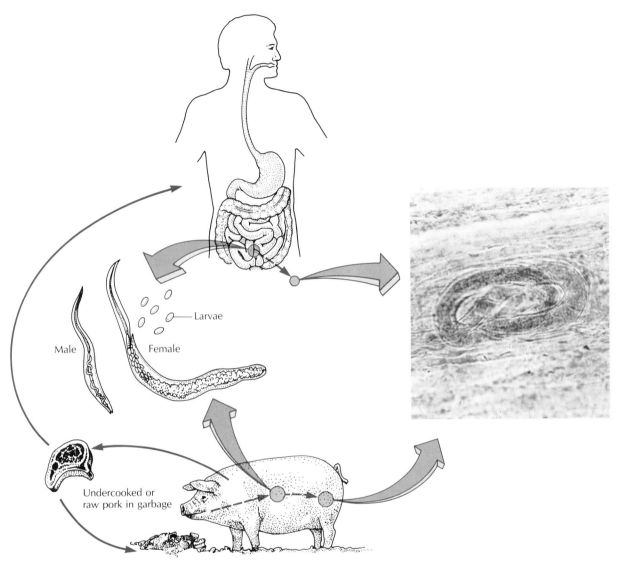

Male

Female

Larvae

Undercooked or
raw pork in garbage

FIGURE 19.18 The Life Cycle of the Trichinosis Parasite
Trichinella spiralis is a parasite that matures inside several animal species, including
the pig and the bear. Eggs develop into small worms that encyst in the animal's
muscle, as shown in the micrograph. When someone eats contaminated,
undercooked meat, the worms emerge, mate, and produce more larvae, which form
new cysts in the human's muscle, producing local and systemic inflammatory
symptoms. (Courtesy of Centers for Disease Control, Atlanta.)

Two to eight hours after ingestion of toxin, the
person begins to vomit forcefully and repeatedly
(you'll recall that vomiting is a protective mechanism). Diarrhea occasionally develops as well.
Symptoms usually disappear in 24 hours.

The problem often originates with a food
handler who is a nasal carrier or has open skin
lesions bearing enterotoxigenic strains of *S. aureus*. Less commonly, the source may be milk
from a cow with staphylococcal mastitis (infec-

tion of the udder). There are two improper food-handling steps involved. First, foods are accidentally inoculated during preparation with large numbers of *S. aureus.* Then the food is held for some hours without refrigeration at temperatures warm enough so the introduced bacteria are able to multiply and produce toxin. Certain foods are frequently incriminated because they are good staphylococcal media and because they are habitually mishandled. These foods include cream-filled baked goods, salads made with mayonnaise, cured meats such as ham, fish dishes, and dairy products.

Clostridium perfringens Type A causes an intoxication very similar to staphylococcal food poisoning. The organism produces a neurotoxin, and the symptoms differ only in the longer (approximately 18 hours) incubation period. *C. perfringens* is a spore-former that is exceptionally widely distributed in soil and the intestines of warm-blooded animals. *C. perfringens* spores from dust are found on the surface of most fruits and vegetables. The types of food that are potential problems can be deduced from the fact that *C. perfringens* is both anaerobic and thermophilic. Pots of soups or stews are first cooked, then often held in a warm condition on the back of the stove, or reheated briefly for subsequent meals. Turkeys may be stuffed hours in advance of cooking, then cooked by slow methods so their interiors reach only moderately warm temperatures. Crock-pots or other slow cookers cook foods over long periods at temperatures well below boiling. All of these practices enable the spores to germinate and the bacteria to produce toxin. These examples also demonstrate the reason for the most important rule of food handling: *Keep cold foods cold and hot foods hot.* No moist food should ever be allowed to sit for any length of time at temperatures above 4 degrees Celsius or below 100 degrees Celsius.

TOXINS WITH SYSTEMIC EFFECTS

Some microbial toxins affect the nervous or circulatory systems, causing potentially life-threatening infections. The most serious is **botulism.**

THE ORGANISM.   *Clostridium botulinum* is an anaerobic, Gram-positive, spore-forming rod that is widely distributed in soil. The spores are a natural contaminant of many foodstuffs. Although we ingest the spores frequently, they do not survive in the human gut, except in infants under one year of age. With this exception, humans do not develop disease by ingesting spores or cells of *C. botulinum*, only its toxin. Infant botulism is discussed in Chapter 20.

Vegetative cells of *C. botulinum* require highly anaerobic environments. Growth is inhibited by low pH, but not by mild salting. In the course of growth, the bacilli excrete toxin. When toxin is ingested, it is activated, not destroyed, by stomach acid. **Botulin,** the active toxin product, blocks transmission of nerve impulses (Figure 19.19).

THE DISEASE.   In botulism, the botulin is absorbed from the intestine and enters nerve fibers. Double vision, dizziness, lack of coordination, nausea, and vomiting are followed by inability to swallow, a flaccid paralysis of voluntary muscles, and respiratory failure.

THE TREATMENT.   Immediately upon diagnosis, an antitoxin (containing antibodies to the three most common types of toxins) is administered. Life-supporting measures to assist respiration are usually essential. Recovery is slow and depends on neutralizing the toxin within the nerve tissue.

EPIDEMIOLOGY AND CONTROL MEASURES.   The interior of a sealed can of food may provide superior anaerobic growth conditions. That is why commercial canning processes are designed specifically to kill *C. botulinum* spores. Inspection and quality-control checks are frequent, and failures are extremely rare. Canning is similar to autoclaving, using both elevated heat and pressure. Unfortunately, many home canners do not take seriously the processing instructions given in cookbooks and instruction manuals. We often hear of grandmothers who have been feeding

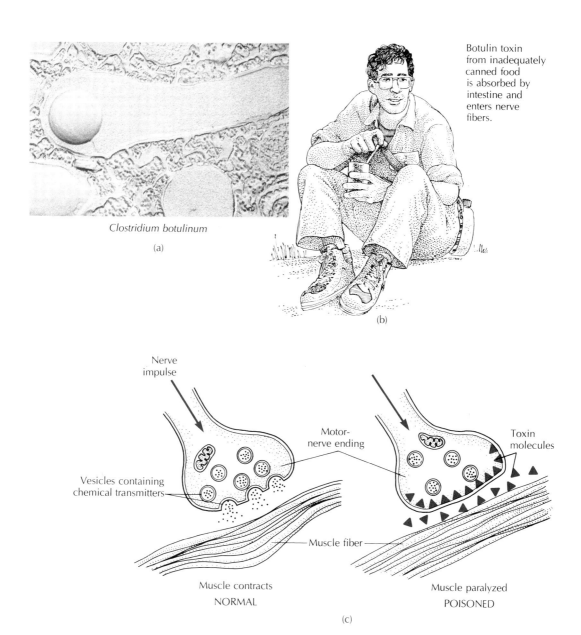

Botulin toxin
from inadequately
canned food
is absorbed by
intestine and
enters nerve
fibers.

Clostridium botulinum

(a)

(b)

Nerve
impulse

Motor-
nerve ending

Toxin
molecules

Vesicles containing
chemical transmitters

Muscle fiber

Muscle contracts
NORMAL

Muscle paralyzed
POISONED

(c)

FIGURE 19.19 The Effect of Botulinum Toxin
(a) Freeze etching of cells of *Clostridium botulinum* (type A), showing subterminal to
terminal endospore. (×23,650. Courtesy of T.J. Beveridge, University of Guelph,
Ontario/BPS.) (b) Spores that survive inadequate cooking and canning procedures
will multiply and produce an exotoxin, which accumulates in food. When a person
eats the contaminated food, the toxin is absorbed from the intestine and enters
nerve fibers. (c) In the normal neuromuscular control mechanism, a nerve impulse
from the central nervous system releases chemicals that cause muscle fibers to
contract. However, in botulin-poisoned individuals, the toxin blocks transmission of
the impulse from the nerve to the muscle, and contraction does not occur. The
person's muscles become flaccid and nonfunctional.

their families inadequately canned foods (boiled, not pressurized) for many years without incident.

Sodium nitrite and sodium nitrate are added to commercially prepared processed meats such as bacon and hams, hot dogs, and bologna. These chemicals make the interior of the product more aerobic, inhibiting clostridial growth.

Fewer than 40 cases of foodborne botulism are reported per year in the United States, and the mortality rate is less than 10%. Of the 11 outbreaks reported in 1983, 10 were traced to home-canned foods, and only 1 to restaurant-prepared food in an outbreak involving 28 people. Recent outbreaks have been traced to home-processed olives, beans, or fish. Uncooked cold-cured or processed meats and sausages can also be sources. Vaccination is theoretically possible, but not practiced because the actual risk is very low.

BOX 19.2
Recipe For An Epidemic

One November, a Thanksgiving dinner in which ham and turkey were the main course was shared by approximately 350 students, faculty, and guests at a college in Florida. It resulted in an outbreak of staphylococcal food poisoning in which more than 50 people developed nausea and vomiting, followed by diarrhea. None reported fever. Most people got sick four to five hours after eating; the range was two to eight hours. Two persons were hospitalized.

Food histories obtained from 64 persons who ate the meal incriminated ham as the vehicle of transmission. Bacterial cultures of leftovers revealed 10 million or more *Staphylococcus aureus* cells per gram in the ham and turkey. No specimens from patients or food handlers were cultured, so the original source was never found.

Because of the holiday rush, the dinner was prepared by a large number of students instead of the usual kitchen personnel. Interviews with these students indicated that hams and turkeys had been partially thawed at room temperature prior to cooking, and that the temperature controls on the ovens used for cooking were unreliable. After cooking, the ham and turkey were sliced and stored at temperatures favorable for the growth of *S. aureus* (less than 60°C [140° Fahrenheit]) for as long as eight hours before serving.

In this incident we see many of the ingredients for an epidemic — improper thawing of meat, prolonged holding of prepared foods at room temperature, unreliable equipment, and a lack of trained, supervised personnel to ensure safe preparation and handling.

Source: Adapted from *Morbidity and Mortality Weekly Report* 1979; 28:153.

FUNGAL TOXINS. Fungi are significant agents in food spoilage. Grains, nuts, or beans are relatively stable foods, if they are stored dry. However, fungi will grow rapidly if these foods are incompletely dried or stored in a damp place. Fungi produce **mycotoxins,** a diverse and poorly understood group of chemicals that includes carcinogens, mutagens, and hemorrhagic agents. Fungal contamination may not be obvious, and the food may still be eaten by hungry people or superficially cleaned up and marketed. Illness and death may follow.

Aspergillus flavus, a fungus that grows on peanuts and stored animal feeds, can cause abortion and death in sheep and cattle. Its toxin, **aflatoxin,** is one of the most potent liver carcinogens known. Aflatoxin finds its way into the human diet in peanut butter and on grains such as corn that have been stored in damp surroundings. It is not clear how significant its effects are as a human poison.

Ergotism develops in persons who have consumed grains on which the fungal pathogen *Claviceps purpurea* has multiplied. This fungus produces **ergot,** a complex mixture of toxins. Ergotism causes a hemorrhagic breakdown of capillaries, gangrene of the extremities, hallucinations, and death. One of the active compounds in ergot was once used as an abortion-inducing drug.

There is intense political controversy over whether or not certain **tricothecene** mycotoxins have been used as biological warfare weapons in the Middle East and Southeast Asia. Tricothecenes are found on animal feeds produced from corn in this country. There have been widespread outbreaks of toxicity in chickens and pigs fed such contaminated grains. The symptoms include internal bleeding, stunted growth, erratic behavior, and immune system failure. It is hypothesized that this condition is due to proliferation of *Fusarium moniliforme* on corn that was stored while inadequately dried, but this is not universally accepted.

ALGAL TOXINS. The algae rarely find their way into a discussion of medical microbiology. However, certain algal species may poison fish, livestock, and poultry as well as human beings. Certain marine algal blooms, called **red tides,** are composed of the dinoflagellates *Gonyaulax tamarenis* (refer to Figure 1.1) or *Gymnodinium.* These algae produce a **paralytic shellfish poison.** As clams, mussels, and oysters feed, their digestive organs become packed with the toxic algae. Such shellfish cannot be made safe to eat. People who do eat shellfish with a high concentration of toxin develop symptoms similar to those of botulism. Paralytic shellfish poison or saxitoxin is a neurotoxin, somewhat less potent than botulin but the most potent algal toxin known. The dinoflagellate algae bloom erratically, their growth being correlated with warm periods when heavy rainfall brings ample supplies of nutrients washed off the land. Monitoring programs along portions of the Atlantic, Gulf, and Pacific coastlines where red tides occur warn the public during periods when harvesting shellfish is unsafe.

SUMMARY

1. The gastrointestinal tract is readily disturbed by microbial activity. GI upset may be caused by the introduction of a pathogenic organism or its toxic products.

2. Infectious disease may disturb normal peristalsis, produce excessive fluid loss, ulcerate the intestinal mucosa, or invade tissues beyond the intestine.

Short-term, nonspecific gastroenteritis may be caused by bacterial parasitism, bacterial toxins in food, protozoa, or viruses. On the basis of symptoms alone it is usually not easy to distinguish among these causes.

3. Typhoid fever, bacillary dysentery, and cholera are the most serious bacterial GI infections. Typhoid and cholera have been controlled in the developed countries by the installation of modern water- and sewage-treatment systems. The diseases are now prevalent only in less-developed areas.

4. Other bacterial diseases, such as salmonellosis, shigellosis, and enteropathogenic *E. coli* infections are increasing problems in advanced nations because available control measures are not effective.

5. Hepatitis is caused by at least two and probably four or more different viruses. They are spread by a wide range of means, including sewage contamination, blood and serum products, and sexual activity.

6. Protozoans such as *Giardia lamblia* are common pathogens. Helminthic infections include pinworm, hookworm, tapeworm, and trichinosis.

7. Food toxicities are due to ingestion of preformed bacterial, fungal, and algal toxins. Staphylococcal food poisoning is common but mild. Other forms are less common.

DISCUSSION QUESTIONS

1. In what ways does the body keep micro-organisms safely inside the GI tract? What barriers must be passed in order for an organism to become disseminated in tissue?

2. Give reasons why the large intestine supports different and a larger number of flora than the small intestine.

3. Describe how the range of symptoms of GI infections are caused by or related to the activities of pathogens.

4. Many of the infectious agents discussed in this chapter, such as enteropathogenic *E. coli*, the non-A, non-B hepatitis viruses, and the rotaviruses, have only recently been identified. Why, do you suppose, have some of these extremely common agents escaped detection so long?

5. As you look back over the chapter, you will note that there is an almost complete lack of effective vaccination against GI diseases. Speculate on some possible causes.

6. If you experienced a short, acute bout of nausea, vomiting, and diarrhea, from which you recovered after one or two days, name three possible microbial causes. How would the microbiology laboratory proceed to find out which it was?

ADDITIONAL READING

Collins M. Algal toxins. *Microbiol Rev* 1978; 42:725-746.

Cukor G, Blacklow NR. Human viral gastroenteritis, *Microbiol Rev* 1984; 48:157-179.

Draser BS, Barrow, PA. *Intestinal Microbiology*, ASM Washington, DC, 1985.

Hentges DJ, ed. *Human intestinal microflora in health and disease*, New York: Academic Press, 1985.

Riemann H, Bryan FL. *Food-borne infections and intoxications*. 2nd ed. New York: Academic Press, 1979.

Savage DC. Microbial ecology of the gastrointestinal tract. *Ann Rev Microbiol* 1977; 31:107-133.

Sussman M, ed. The Virulence of *Escherichia coli*. Orlando, Fla.: Academic Press, 1985.

Tiollais P, Charnay P, Vyas GN. Biology of hepatitis B virus. *Science* 1981; 213:406-411.

Tzipori S. Cryptosporidiosis in animals and humans. *Microbiol Rev* 1983; 47:84-96.

Wolin MJ. Fermentation in the rumen and large intestine. *Science* 1981; 213:1463-1468.

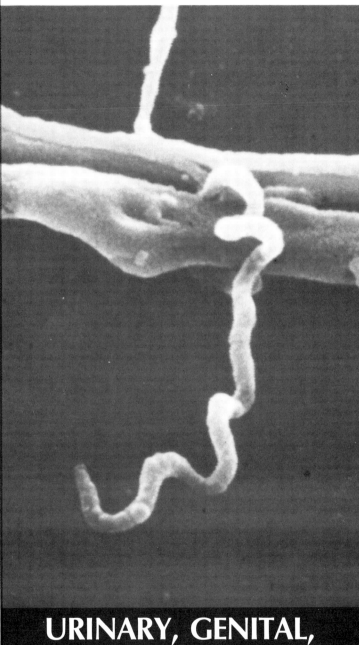

20

URINARY, GENITAL, AND REPRODUCTIVE INFECTIONS

CHAPTER OUTLINE

The urinary and reproductive systems are anatomically very close to each other, and, in the male, share some structures. However, their physiology and their functions are vastly different. Both contain some normal flora in the portions nearest their outer orifices, and both may become infected by microorganisms.

The function of the urinary system is to collect, concentrate, and excrete wastes, while helping to regulate the body's supplies of water and electrolytes. Urinary-tract infections involve several organs, particularly the bladder and the kidneys. These infections are especially common in women.

The function of the reproductive system in men is the production and distribution of sperm, and in women, the production and distribution of eggs and the prenatal development, birthing,

Microbial pathogens featured in this chapter:

Neisseria gonorrhoeae
Treponema pallidum
Chlamydia trachomatis
Toxoplasma gondii

and nourishing of infants. Sexual activity is the most frequent source of genital infections. The sexually transmitted diseases (STDs) are becoming increasingly prevalent and diverse. Sexually active adults may be exposed to up to 14 different infections, most of which are relatively minor. Other STDs, however, may be severe and a few are incurable so far. We will look at a few of the most frequently encountered STDs and other diseases of the genital organs.

Pregnancy, delivery, and the first months of life carry special risks of infectious disease for both mother and child. We will examine some important infections of the period before and immediately after birth.

THE URINARY TRACT

The urinary tract in both male and female is normally sterile starting about 1 cm above the external opening. It consists of the kidneys, the ureters, the bladder, and the urethra (Figure 20.1a).

ANATOMY OF THE URINARY TRACT

THE KIDNEYS. The kidneys, which produce urine, are paired organs about 10 to 12 cm long, located behind the peritoneum. Their outer layer is the **renal capsule,** under which lies the **renal cortex.** Beneath the cortex is the **medulla,** which contains a network of **calices** and the **renal pelvis** (Figure 20.1b). Urine is formed in the renal cortex, concentrated in the renal medulla, collected in the renal pelvis, then conducted down the **ureters** to the **urinary bladder.**

The kidney is made up largely of individual filtering units called **nephrons,** composed of a **corpuscle** and a **tubular network.** Most of the corpuscular portions of the nephrons lie in the renal cortex and most of the tubular portions lie in the medulla, although there is some overlap. The corpuscle is the actual filtering unit in which urine is formed, containing a **glomerulus** (a tuft of capillaries) and a **Bowman's capsule** (a tiny collecting vessel) (Figure 20.1c).

THE URETERS. The **ureters** are slender tubules about 26 to 28 cm long. Their walls contain smooth-muscle fibers and work with a type of peristalsis. This action moves urine toward the bladder, even when you are lying down, and also helps to stop backflow of urine from the bladder

to the kidney. The ureters enter the bladder through slitlike openings.

THE BLADDER. The **bladder** is a highly elastic receptacle for urine, surrounded by a strong, muscular coat with a mucosal lining. Urine passes from the bladder through a canal called the **urethra** and is voided. Urination is controlled by an internal sphincter at the base of the bladder.

THE URETHRA. In females the urethra is short, from 2.5 to 2.8 cm, and exits between the clitoris and the vaginal opening. Because the female urethral opening is close to both anal and vaginal orifices, there is a high probability of transfer of flora from the GI tract to the urethra. In the male, the urethra is much longer, opening at the end of the penis. Because the opening is distant from the anal opening, incidental contamination of the urethra with GI flora is less likely.

PHYSIOLOGY OF THE URINARY TRACT

THE URINE. Blood pressure forces dilute fluid filtrate bearing wastes out of the plasma, through the capillary wall, and into the Bowman's capsule. As the filtrate moves along the tubular network of the nephron, much of the useful water and electrolytes are reabsorbed into the blood, while the wastes remain in the filtrate and become concentrated into urine. In an average day, an adult produces about 1.5 liters of urine. Your kidneys adjust for your daily water, salt, and protein intakes. The urine may vary in volume, pH, and amounts and kinds of dissolved substances.

The dissolved solids in urine include wastes, such as urea, that serve as nutrients for many species of bacteria. Urine is normally sterile when it is produced, but it can be converted into a microbial culture medium both in the bladder and after voiding. Any condition that causes urine to accumulate in the bladder or tends to block the urethra predisposes to microbial growth in the retained urine, and thus encourages urinary-tract infection.

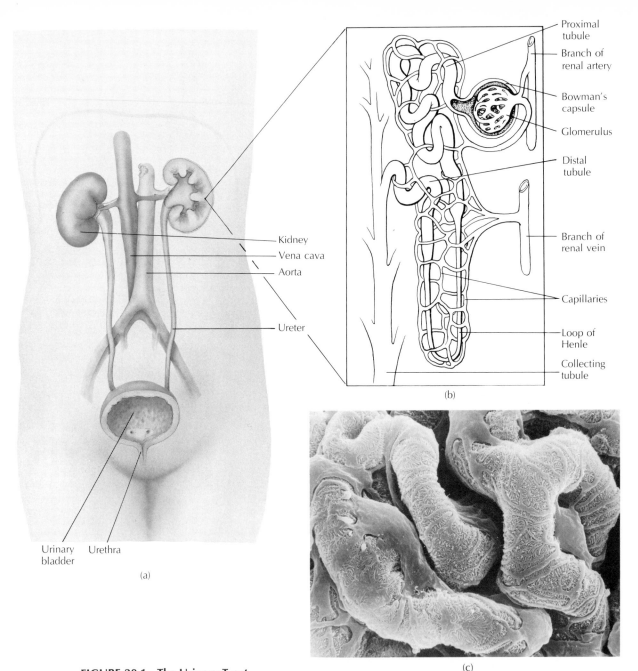

Proximal
tubule

Branch of
renal artery

Bowman's
capsule

Glomerulus

Distal
tubule

Branch of
renal vein

Capillaries

Loop of
Henle

Collecting
tubule

(b)

Kidney

Vena cava

Aorta

Ureter

Urinary
bladder

Urethra

(a)

(c)

FIGURE 20.1 The Urinary Tract
The mechanical flushing action of urine is the urinary system's primary defense. (a)
Urine is produced and collected in the kidneys, then flows down the ureters to the
bladder where it is stored until voided via the urethra. The kidney itself is enclosed
in a membranous capsule, and its tissues comprise an outer cortex and an inner
medulla. Each kidney contains thousands of individual filtering units called
nephrons. (b) Within the nephron, blood pressure forces dilute fluid bearing wastes
out of a tuft of capillaries called the glomerulus. As the fluid, or urine, passes along
the looped tubules, water is reabsorbed and the urine becomes more concentrated.
This fluid is stored in a hollow Bowman's capsule until discharged into the renal
pelvis. (c) The scanning electron micrograph of rat kidney tissue shows a glomerulus
within its capsule. (×1440. Courtesy of Dr. P.M. Andrews, Georgetown University
School of Medicine.)

DEFENSES OF THE URINARY SYSTEM. There are several levels of defense for the urinary system. The mechanical flushing action of urine appears to be the most important factor in preventing microbial colonization. An acid pH mildly inhibits growth of the most likely bacterial invaders, the intestinal flora. Thus the urine itself is the major barrier to prevent the growth of stray microorganisms that may ascend to the bladder from the urethra.

Some cells of the bladder lining are phagocytic. When bladder infection occurs, additional polymorphonuclear leukocytes enter the bladder. Lysozyme is found in urine and it increases during infection. In addition, the bladder lining produces secretory IgA. Immunoglobulins are also produced in the kidney, where they not only protect, but may also on occasion contribute to immune complex diseases such as post-streptococcal glomerulitis.

ANATOMY AND PHYSIOLOGY OF THE REPRODUCTIVE TRACTS

The reproductive tracts of males and females are very different in structure and function. Consequently, they develop rather different patterns of symptoms when infected, and also have some infections particular to one sex.

ANATOMY OF THE REPRODUCTIVE SYSTEMS

THE MALE. The reproductive organs in the male are the **testes.** These produce reproductive cells called **sperm,** which are collected in the **epididymis** and conducted via ducts across the exterior bladder surface (Figure 20.2). The sperm is lubricated with **seminal fluid,** the secretion of the **prostate gland** and other glands, and stored as **semen** in the seminal vesicles. On ejaculation, semen enters the urethra within the tissue of the prostate, then is delivered outside the body through the **penis.** The urethra, the prostate, and, less commonly, other internal structures may become infected.

THE FEMALE. The reproductive organs in the female are the **ovaries.** These produce **ova** (eggs) during the fertile years. The ova pass out of the ovary into the peritoneal cavity, then are swept up by fingerlike projections at the end of the **fallopian tubes** and carried down to the uterus (Figure 20.3). Because the ends of the fallopian tubes are open to the peritoneal cavity, invading microorganisms may move from the vagina up into the uterus, then from the fallopian tubes out of the reproductive tract into the peritoneal cavity.

The **uterus** is a thick-walled, muscular organ specialized to support the growth and nourishment of a fetus. Its lining, the **endometrium,** thickens in preparation for receiving a fertilized ovum. If the ovum is not fertilized, the functional layer of the endometrium is sloughed off in menstruation. When conception occurs, the endometrium receives the ovum, which grows into the top tissue layers. These layers become modified to form the maternal half of the **placenta,** the organ that nourishes the fetus.

The **cervix** is the muscular cuff at the opening of the uterus. Its surface epithelium is readily colonized by certain pathogens. The **vagina** is the elastic, ribbed tube that connects the uterus to the exterior of the body. Its opening is immediately adjacent to the anus. The vagina undergoes minor mechanical trauma during intercourse, and infections may be readily established. There are few pain-sensitive nerve endings within the vagina, so inflammation and infection may not be promptly detected.

The **vulva** is made up of the external genitalia, including the **labia, clitoris,** and adjacent tissues. The vulva is often infected simultaneously with the vagina in **vulvovaginitis.**

PHYSIOLOGY AND DEFENSES OF THE REPRODUCTIVE TRACTS

THE MALE. In the male, the urethra, which serves both the urinary and reproductive tracts, is partially protected by urine flow as previously noted. However, this defense does not extend to the accessory glands, which have few defenses.

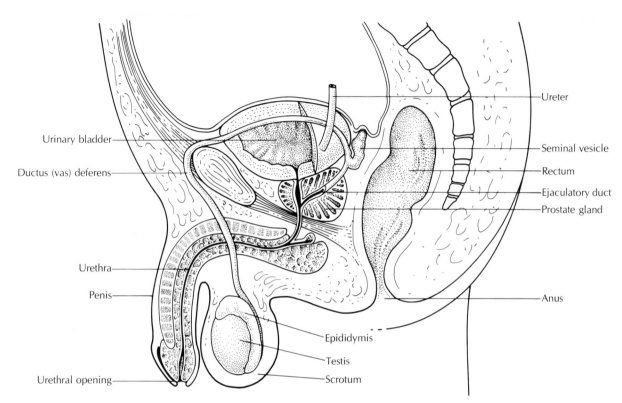

FIGURE 20.2   The Male Reproductive System
The testes produce sperm cells, which mix with the secretions of the prostate and other glands to form semen. The urethra serves as a common duct for both semen and urine. Because of the length of the male urethra, urinary-tract infection is uncommon in men. Genital infections in men are usually limited to the urethra and prostate gland.

Once microorganisms reach the prostate gland or seminal vesicles, the body has great difficulty eradicating them, and they frequently cause chronic problems.

THE FEMALE.   The vagina produces secretions that vary in chemical composition with age. Prepubertal girls secrete a mucus of nearly neutral pH. At puberty (sexual maturity) the vaginal secretions become more acidic, protecting against colonization by intestinal flora. A continuous shedding and renewal of the vaginal epithelial lining limits colonization by normal flora. The uterus contains a population of macrophages to deal with organisms that ascend through the cervical opening. Immunoglobulins and lysozyme are secreted in the vaginal mucus.

Changes in the metabolic activity and growth rate of the vaginal surface tissues occur at different phases of the menstrual cycle. These

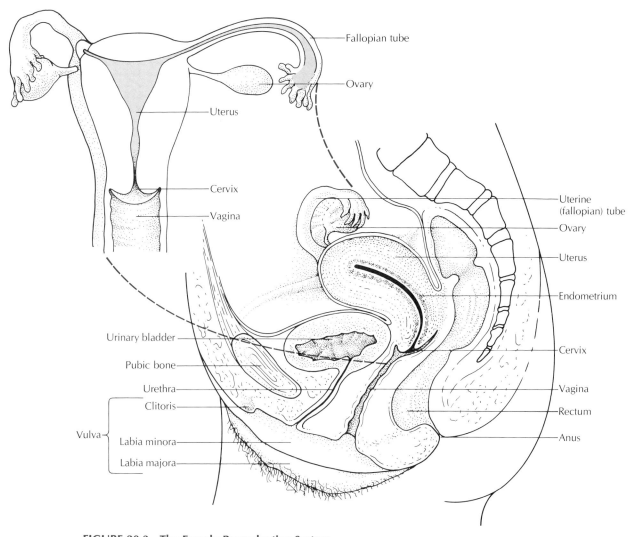

Fallopian tube

Ovary

Uterus

Cervix

Vagina

Uterine (fallopian) tube

Ovary

Uterus

Endometrium

Cervix

Vagina

Rectum

Anus

Urinary bladder

Pubic bone

Urethra

Clitoris

Vulva

Labia minora

Labia majora

FIGURE 20.3   The Female Reproductive System
In females the reproductive tract is completely separate from, but adjacent to, the urinary tract. The surfaces of both the vulva and the vagina contain a normal flora. These may migrate to infect the uterus and ultimately the peritoneal cavity via the open ends of the fallopian tubes.

normal changes, or the use of female hormones for birth control or to control menopausal discomfort, may increase the susceptibility of the vagina to infection. The use of tampons during menstruation appears to increase the risk of infections such as toxic-shock syndrome, probably by altering the vaginal habitat to favor toxin production by the causative agents.

NORMAL FLORA OF THE URINARY AND REPRODUCTIVE SYSTEMS

The normal flora of the urogenital tracts are listed in Table 20.1. In both sexes, the normal flora of the urinary tract is restricted to about the first centimeter of the urethra. The female reproductive tract possesses a large vaginal microflora throughout.

URETHRAL FLORA

Many organisms may be cultured from the external opening of the urethra. Often these organisms are derived from adjacent skin flora and normal fecal flora. *Staphylococcus epidermidis* is usually the most prominent. Both aerobes and anaerobes are represented. Not surprisingly, the great majority of bacteria that reach the bladder and establish themselves are motile. Microorganisms are not normally found in the upper urethra, bladder, ureters, or kidneys.

VAGINAL FLORA

The vaginal flora of women of reproductive age (Figure 20.4) is a mixture of lactobacilli, anaerobic *Bacteroides* and *Clostridium*, aerobic diphtheroids, staphylococci, and yeasts. This mixture shifts in response to hormonally induced changes in the physiology of the vaginal mucosa. The lactobacilli acidify the vaginal fluids by fermenting the sugars in the secretions, lowering the pH to 3.5 to 4.5 and, as we mentioned previously, inhibiting the growth of many invaders.

TABLE 20.1 Normal flora of the urogenital tracts

Group	Distal Urethra, Both Sexes	Vagina of Adult	Percentage of Individuals
Bacteria	*Staphylococcus epidermidis*	*Staphylococcus aureus*	5–15
	Anaerobic micrococci	α-hemolytic streptococci	47–50
	Group D streptococci	Enterococci or Group D streptococci	30–90
	Nonpathogenic *Neisseria*	*Streptococcus pyogenes*	5–20
	Lactobacillus sp.	Anaerobic streptococci	30–60
		Nonpathogenic *Neisseria*	
	Aerobic diphtheroids		
	Escherichia coli	*Lactobacillus*	50–75
	Klebsiella sp.	Aerobic diphtheroids	45–75
	Proteus sp.	*Clostridium* sp.	15–30
	Pseudomonas sp.	*Actinomyces* sp.	25–75
		Enterobacteriaceae	15–40
		Hemophilus sp.	
Fungi	Yeasts (irregularly)	*Candida albicans*	30–50
Protozoa	*Trichomonas vaginalis*	*Trichomonas vaginalis*	10–25

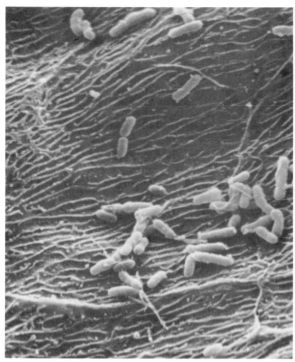

FIGURE 20.4 Normal Human Vaginal Flora
The vaginal flora of a woman of reproductive age consists of a mixture of bacteria and yeasts. This scanning electron micrograph shows a segment of vaginal epithelium to which numerous bacilli have adhered.

Before puberty, and after menopause, lactobacilli are less active.

The normal vaginal flora causes problems only if there is a tissue trauma or a physiologic imbalance. Antibiotics may change the microbial distribution, wiping out the beneficial, susceptible lactobacilli, and allowing the yeast *Candida albicans* to overgrow and produce an endogenous infection. *Streptococcus, Clostridium,* and *Bacteroides* introduced into the uterus during a septic abortion, or a difficult or badly managed childbirth, have the potential to cause an infection fatal to the new mother.

URINARY-TRACT INFECTION

Primary urinary-tract infections are almost always bacterial caused by organisms of GI tract origin. Viral or fungal infections may be brought by the circulation to the urinary system secondarily from other areas of the body.

CYSTITIS

Cystitis, or bladder infection, is a common condition. Almost every woman has at least one episode during her lifetime. One feature of cystitis is **bacteriuria** — bacteria in the urine.

Surveys of healthy persons show that asymptomatic bladder infection is quite common. About 1 to 2% of school-age girls, 5 to 6% of young, pregnant women, and up to 10% of elderly women have bacteriuria. Affected females outnumber males by about 30:1 in each age group, except for newborns and for elderly men with prostate obstruction. It is a good idea to review again the distinction between infection and disease in this context. Many people are infected, with microbial multiplication, apparently for prolonged periods. However, episodes of disease, with active tissue damage, are rare and sporadic, occurring when the bacterial population reaches a certain uncontrollable size.

Symptomatic cystitis results when there is microbial invasion and inflammation of the bladder wall and urethra. Red and white blood cells are frequently seen in the urine. The infected person experiences a burning sensation, pain on urination, and the need to void frequently. Left untreated, cystitis usually resolves spontaneously in time. However, antibiotics relieve the symptoms and reduce the risk of ascending kidney disease, particularly important in pregnant women. Ampicillin and the sulfa drugs are common choices for treatment of the infection.

Overt cystitis in women often follows sexual intercourse and is common during pregnancy. Other predisposing factors, in both sexes, include local tumors, bladder stones, prolonged bedrest, spinal-cord injuries, diabetes, and insertion of catheters or medical instruments.

Accurate laboratory diagnosis of urinary-tract infection requires quantitative determination of the number of bacteria present. A clean urine sample — a "clean catch" — should be obtained as follows: The patient first cleanses the external urogenital area with soap or a disinfectant solution and rinses carefully. Then he or she partially empties the bladder before collecting a **midstream sample** in a sterile container.

If the urine cannot be cultured within one hour of the time the sample is taken, it should be refrigerated until laboratory services are available. If it is not, organisms will multiply, giving misleading results. Some urine-collection vessels contain a bacteriostatic agent to stop multiplication between collection and laboratory analysis.

Bacterial counts of less than 10,000 per milliliter are considered negative, since this many organisms could be introduced into sterile urine simply by passage through the urethra. Counts between 10,000 and 100,000 per milliliter are suspicious, and repeat cultures may be ordered. Counts greater than 100,000 per milliliter indicate active infection.

In the otherwise healthy person, about 90% of cystitis episodes are due to *Escherichia coli* or other related enterobacteria such as *Proteus mirabilis*. *Streptococcus faecalis* causes most of the remaining 10% of cases. In hospitalized patients a larger variety of agents may be found, and antimicrobial resistance is a problem. *Serratia marcescens* and *Klebsiella pneumoniae* may be especially troublesome. *Proteus* and *Pseudomonas* predominate in paraplegic patients.

KIDNEY INFECTIONS

Bacteria reach the the kidney by either the ascending or the bloodborne route (Figure 20.5). Ascending infections are most common. At least 50% of women with acute clinical cystitis also

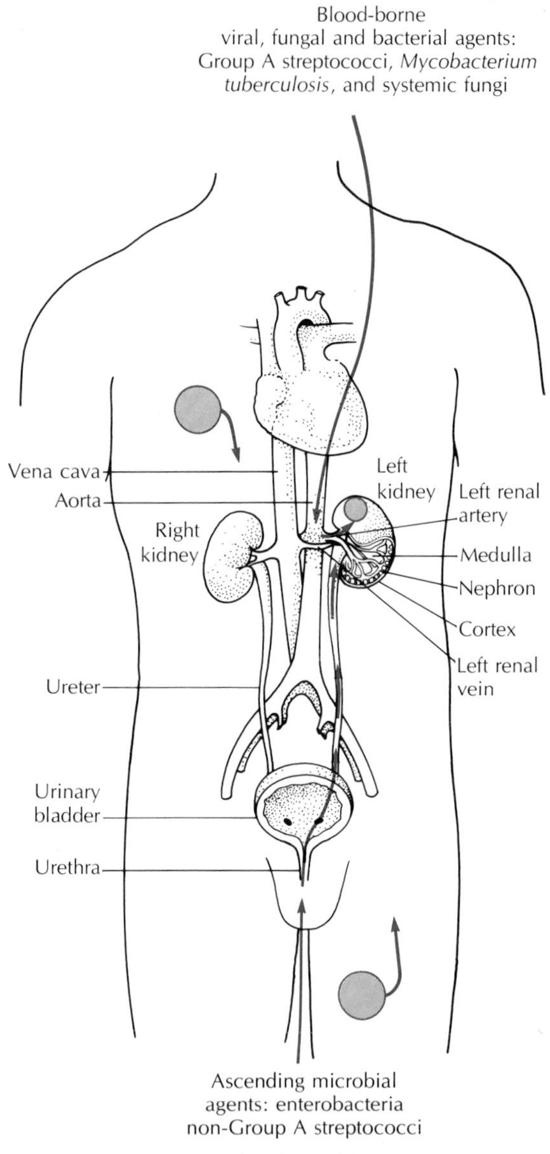

FIGURE 20.5   Routes of Kidney Infection
The kidney may become infected by two routes. Most commonly, it is by an ascending infection (color) in which motile bacteria migrate up the urinary tract. Less often, kidney infection is bloodborne (black) with the agent originating from an infection located elsewhere in the body and being transported by the blood stream to the glomeruli.

have some degree of kidney infection. Gram-negative organisms in the urine produce endotoxin, which weakens the smooth muscle at the slitlike openings of the ureters permitting the bacteria to migrate upward to the kidneys. The bacterial pili allow them to bind firmly to the epithelium in the ureter so they are not flushed out.

Other microorganisms, such as staphylococci, Group A streptococci, *Mycobacterium tuberculosis*, and systemic fungi may be transported by the blood to the kidney. However, microorganisms brought to the glomeruli by the blood rarely infect, because there is efficient phagocytosis occurring in the blood.

Acute **pyelonephritis,** or inflammation of the renal pelvis, is characterized by high fever, chills, severe pain, and spasm of the small intestine. Prompt antibiotic treatment usually brings the pathogen under control so that residual kidney damage will be minimal. Chronic pyelonephritis, often asymptomatic, may escape early

notice and treatment. When this happens, the ducts in the kidney pelvis may be obstructed and lose elasticity; eventually the affected kidney becomes nonfunctional and shrinks.

Glomerulonephritis, or inflammation of the renal cortex, is predominantly an immunologic diseases. Immune complexes on the glomerular membranes provoke complement-mediated cell damage. Glomerulonephritis may follow infections with *Streptococcus pyogenes* (see Chapter 17). The condition is also one aspect of certain autoimmune diseases, such as systemic lupus erythematosus.

Leptospirosis is a kidney infection caused by spirochetes of the genus *Leptospira* (Figure 20.6). These bacteria may cause infections in domestic animals, particularly dogs, in rodents, and in human beings. Contaminated urine is a primary means of spreading leptospirosis.

GENITAL-TRACT INFECTION

Sexual activity is one of the most effective means of transmitting microorganisms from one person to another. In fact, sexual contacts may transmit not only genital infection, but also skin, respiratory, and gastrointestinal diseases. The type of disease spread is determined in part by the sexual practices of the individuals. Infections transmitted predominantly by sexual activity are called **sexually transmitted diseases (STDs)** (replacing the less precise term **venereal disease**). However, there is no absolute distinction between STDs and non-STDs. As we have already seen in other contexts, an infectious agent can often be transmitted in several ways. Some diseases formerly thought to be nonsexual, such as hepatitis B, are now included in the sexually transmitted list. It works the other way too, as most STDs can be transmitted by means other than sexual contact under certain highly specialized circumstances. It seems clear that any genital infection, and many gastro-

FIGURE 20.6 *Leptospira interrogans*
This species of spirochetes causes a kidney infection (leptospirosis) in dogs, cattle, swine, rodents, and human beings. The agent is transmitted by urine or direct contact with infected animals or carcasses. Note the hooked end, a characteristic feature in preparations of this organism. (Courtesy of Dr. Dorothy L. Patton, Department of Medicine, University of Washington, Seattle.)

intestinal and other infections, may be spread sexually. The epidemiology of some major STDs is shown in Table 20.2.

GONORRHEA

Worldwide gonorrhea appears to be the most frequently diagnosed and reported STD. Perhaps the most unfortunate characteristic of the gonorrhea bacterium is that humans don't develop immunity from contact with it. People can have it over and over — practically on a weekly basis. So far, all research approaches have been unsuccessful in the search for a vaccine.

THE ORGANISM.   **Gonorrhea** is the sexually transmitted disease caused by the Gram-negative aerobic diplococcus *Neisseria gonorrhoeae* (Figure 20.7). This species grows only in a warm atmosphere with elevated levels of carbon dioxide, thus it survives poorly outside the body. Therefore specialized collection measures are necessary so that it can be successfully recovered and cultured from a clinical sample. The organism is grown on a chocolate agar, which contains hemolyzed blood and antibiotics. To ensure survival **transport cultures** are used. These are sealable bottles or plates containing a chocolate-agar medium and a gas mixture rich in carbon dioxide. Transport cultures are designed to be inoculated in the doctor's office or clinic, then shipped to a laboratory. After incubation, individual colonies are oxidase tested, since all neisserial species are oxidase-positive. Species identity can be determined by the use of sugar-fermentation tests, which require overnight incubation. *N. gonorrhoeae* cannot ferment fructose, sucrose, or maltose. Immunologic methods can be used for a more rapid identification, requiring only a few minutes.

The gonococci bear many pili on their cell envelopes. The pili are one key to pathogenesis

TABLE 20.2 Epidemiology of sexually transmitted diseases			
Disease	**Reservoir**	**Transmission**	**Incubation Period (Days)**
Gonorrhea	Human beings	Direct sexual contact or contact with fresh infectious exudates	2–5
Syphilis	Human beings	Direct sexual contact, contact with contaminated blood, transplacental	21
Nongonococcal urethritis and vulvo-vaginitis; lympho-granuloma venereum	Human beings, other animals (?)	Direct sexual contact (others?)	5–7
Herpes	Human beings	Direct sexual contact	Variable

Source: Adapted from Benenson, AS *Control of Communicable Diseases in Man.* 14th ed. 1985.

because they attach the coccus to the mucosal surface. The genome of the neisseriae is arranged so that the organism can frequently alter the antigenic nature of its pili to evade the host's immune system.

For many years, gonorrhea was quickly and effectively cured with penicillin. Now, however, all the *N. gonorrhoeae* strains in worldwide circulation have become progressively more resistant to penicillin since the drug's introduction in the 1940s. At first, penicillin resistance was relative, and the gonococcus strains continued to respond to penicillin in ever-increased doses. In 1976, however, new strains appeared that had absolute resistance to penicillin because they produced a **penicillinase** enzyme. These strains are referred to as **PPNG** (penicillinase-producing *N. gonorrhoeae*). PPNG strains have spread worldwide. Now it is wise to perform a laboratory penicillinase test on each new isolate of the gonococcus, to determine whether penicillin will be effective for that strain, before therapy is begun.

THE DISEASE. In the male, gonococci invade the submucosal tissues of the urethra, and less commonly those of the rectum and pharynx. After a short incubation period of 2 to 7 days, pain and burning on urination appear, along with a whitish discharge. Most men with symptoms promptly seek therapy, but some either are asymptomatic or choose to ignore the discomfort.

In untreated gonorrhea, the symptoms will subside after about eight weeks, but chronic inflammation of the epididymis or constriction of the urethra commonly remain. There is a 1 to 2% chance of developing disseminated gonorrhea, marked by fever, arthritis, and skin lesions. Sterility, myocarditis, hepatitis, and meningitis are less common complications.

In the female, the gonococcus invades the

TABLE 20.2 Epidemiology of sexually transmitted diseases (continued)

Period When Communicable	Patient Control Measures	Epidemic Control Measures
Months or years, especially in untreated female. Ends within hours of specific therapy	Avoid sexual contact until noncommunicable; disinfect soiled articles	Search for contacts within 10 days before onset; treat
Variable; during stages 1 and 2 and in relapses; possibly in latency; becomes noncommunicable 24 hr after specific therapy	Avoid sexual contact until noncommunicable; disinfect articles soiled with exudate	Search for contacts within 3 months prior to appearance of primary stage, 6 mo prior to secondary stage, 1 yr prior to latent stage; treat
Unknown	Appropriate drug therapy if agent is identified; avoid sexual contact until symptoms clear	Concurrent treatment of spouses, sexual contacts; active contact search not required
As long as lesions are present; virus may be excreted even when visible lesions appear healed	Avoid sexual contact; disinfect soiled articles	None

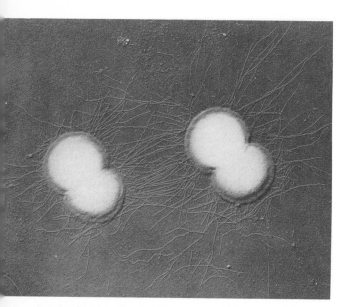

FIGURE 20.7 *Neisseria gonorrhoeae*
In this transmission electron micrograph, gonococci are seen in typical pairs. Pili on the bacteria aid it in attaching to the host it parasitizes. (Courtesy of Cheun-mo To and Charles C. Brinton, Jr.)

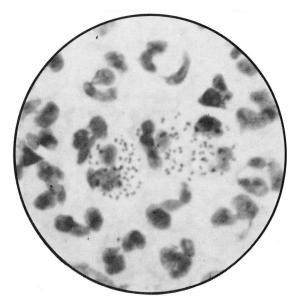

FIGURE 20.8   Intracellular *Neisseria gonorrhoeae*
This photomicrograph shows a Gram stain prepared from exudate from the urethra of a male patient with gonorrhea. The field contains many neutrophils, indicating an inflammatory response. Three of these polymorphonuclear leukocytes (center) ingested the gram-negative diplococci, which are visible as paired dark spots. This type of is a common, quick diagnostic test for Gonorrhea. (Courtesy of Dr. Nicholas J. Fiumara, Director, Division of Communicable and Venereal Diseases, State Department, Massachusetts Department of Public Health.)

cervical lining, and less commonly the urethra, rectum, pharynx, and associated glands. Cervical gonorrhea is largely asymptomatic, the only clue being the appearance of a whitish discharge. In both male and female, examination of a Gram-stained smear of the discharge will show the typical Gram-negative diplococci within polymorphonuclear leukocytes (Figure 20.8). This finding is diagnostic.

If left untreated in the female, about 20% of cases of gonorrhea will progress to pelvic inflammatory disease (PID), which we'll discuss later. This risk is multiplied from four to seven times if the woman is wearing an intrauterine device (IUD) for birth control. Chronic PID and sterility are the major risks of untreated gonorrhea in women. About 3% of females develop bacteremia and disseminated infection.

As we noted in Chapter 16, infants' eyes may be infected with the gonococcus during birth.

This was once a major cause of infant blindness, but it is now prevented by the obligatory use of silver nitrate or penicillin eye drops at birth. It is also increasingly being recognized that children are at risk of developing gonorrhea (and the other STD's) as a result of sexual abuse.

THE TREATMENT.   Gone are the days when a diagnosis of gonorrhea could be automatically followed with an injection of penicillin that always worked. Penicillin is still the drug of choice, in combination with probenecid, a substance that prolongs the penicillin's action. However, penicillin is chosen only after the strain's susceptibility to penicillin has been established. Furthermore, the dose needed even for susceptible strains has increased almost 100-fold since the days when penicillin was first used for treatment. When penicillinase-positive strains are identified, an ampicillin–probenicid combination or spectinomycin is used. Spectinomycin resistance has also appeared in some gonococcus strains. With all these variables, it is essential to carry out follow-up cultures to confirm that the person is in fact cured. In up to 50% of cases in some regions, removal of the gonococcus does not completely alleviate the symptoms. This is because a secondary STD is also present, requiring a different form of treatment.

EPIDEMIOLOGY AND CONTROL METHODS. The human genital tract (and to a lesser extent the throat and anus) are the only reservoirs of *N. gonorrhoeae*, and, as we noted, the organism dies very rapidly when outside the body. Thus the spread of gonorrhea by other than sexual means is very rare. It is often communicated by asymptomatic males or females. Although not every contact results in disease, repeated exposure to an infected sex partner will eventually do so.

Control centers on the identification and treatment of exposed contacts (review Figure 14.14). This includes all persons with whom the diagnosed individual had sexual contact within 10 days prior to the onset of symptoms. A person receiving treatment becomes noninfective within 24 hours after the antimicrobial drug is first administered. The short incubation period of gonorrhea, sometimes only 48 hours, makes it difficult to identify and treat sexual contacts before they spread the disease further. Contact tracing is formally the responsibility of STD clinics and public health officials, but it is also a personal responsibility of the infected individual. The incidence of gonorrhea in the United States hit an all-time high in 1975; vigorous control efforts have now brought it down somewhat 12%. However, with over 900,000 new cases in 1983, gonorrhea is still far from being "under control."

SYPHILIS

Syphilis, although one of the less common STDs, is one of the most serious, as it can kill. Fortunately, it can be readily cured in its early stages.

THE ORGANISM. **Syphilis** is a sexually transmitted disease caused by *Treponema pallidum*, a species of spirochete bacteria (Figure 20.9). They take the Gram stain poorly and are too slender to be seen with ordinary bright-field microscopy. They can be seen if silver stains, which coat the cell with a thick metallic layer, are used. Live spirochetes in the fluid from lesions are visible by darkfield microscopy because of their undulating movements. *T. pallidum* has never been grown *in vitro*. Therefore, little is known of its metabolism. The organisms are rapidly killed by heat, cold, drying, soap and water, and a large variety of disinfectants and antimicrobial agents.

A variety of serologic techniques are employed in the laboratory diagnosis of syphilis. Rapid-slide tests (Figure 20.9b) are generally used as screening tests. They are useful for checking large numbers of apparently normal people, as in the routine prenatal and (in some states) premarital testing. Their usefulness is limited by the fact that they give a small number of false positive reactions, indicating that a person has syphilis when he or she in fact does not. When a person's serum gives a positive result on a rapid-slide test, or he or she has a suspicious clinical picture, more specific tests using *T. pallidum* antigen, which must be obtained from animal tissues, are used to confirm the diagnosis.

THE DISEASE. The spirochetes enter the body through minute abrasions on the genitals or any other mucosal surface that touches a lesion borne by the sexual partner. Once in the body, the agents migrate via the lymphatics and blood stream, from which they infect all tissues. The development of symptoms is slow and proceeds in three stages.

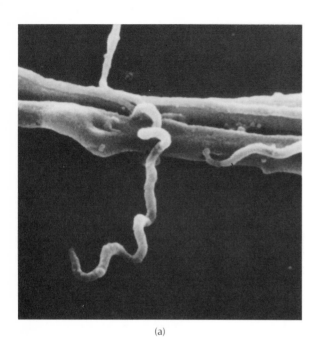

(a)

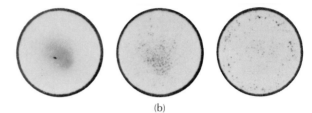

(b)

FIGURE 20.9 *Treponema pallidum*
(a) This scanning electron micrograph shows a slender loosely curled spirochete adhering by its tip to a rabbit testes cell. (Courtesy of N.S. Hayes, K.E. Muse, A.M. Collier, and J.B. Baseman, Parasitism by virulent *Treponema pallidium* of host cell surfaces, *Infection and immunity* (1977), 27:174–186.) (b) In the rapid slide test for syphilis, a nonreactive or negative test (left) remains evenly suspended; a minimally reactive test (center) shows small clumps of antigen–antibody clusters, and a positive serum (right) yields large peripheral clusters. (Courtesy of Joel B. Baseman)

The **primary stage** occurs about three weeks after infection (Figure 20.10a). A sore called a **hard chancre** appears at the original site where the spirochetes entered. This form of lesion is typically an open, painless ulcer with a hard base and some clear serumlike fluid. In the male, the penis is the usual site; but in the female the chancre most often develops within the vagina or on the cervix, where it is not readily detectable. The fluid leaking from any hard chancre is swarming with live spirochetes and is extremely infective. Darkfield examinations of the fluid will show the spirochetes, permitting an immediate positive diagnosis. The chancre heals spontaneously in three to six weeks, leaving little scarring. A latent period without symptoms follows, during which the spirochetes continue to multiply throughout the body.

About six to eight weeks later, if there has been no treatment, the **secondary stage** develops. In this stage the wide distribution of the spirochetes in the body becomes obvious. Many skin and mucous-membrane lesions appear, all highly infective (Figure 20.10b). Hair loss, swollen lymph nodes, and general symptoms such as fever, malaise, and sore throat are also seen. This phase also regresses spontaneously. A second latent or asymptomatic phase follows, and lasts months or years. In the first four years, relapses to the infective secondary stage may occur. In the later years, the disease does not relapse and is infective only by blood transfer.

The third stage, **tertiary syphilis,** will eventually appear in about 35 to 40% of untreated individuals who do not die first of other causes. There are three manifestations of tertiary syphilis. All include degenerative, destructive changes caused primarily by hypersensitivity, and with little apparent activity by the spirochetes. The benign form features large ulcers called **gummas** that may occur in any organ; these are destructive but localized. The cardiovascular form consists of an erosion of the aorta, eventually leading to a fatal hemorrhage. Neurologic forms (tabes dorsalis or paresis) may result in psychosis, paralysis, and loss of sensory functions.

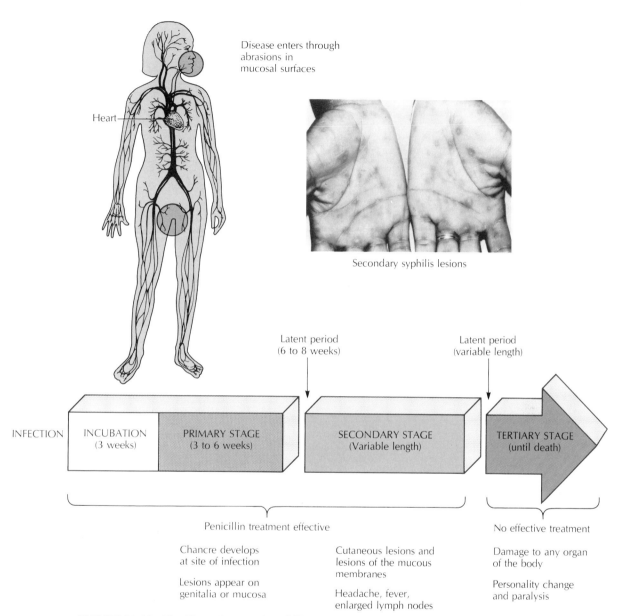

Disease enters through abrasions in mucosal surfaces

Heart

Secondary syphilis lesions

Latent period (6 to 8 weeks)

Latent period (variable length)

| INFECTION | INCUBATION (3 weeks) | PRIMARY STAGE (3 to 6 weeks) | SECONDARY STAGE (Variable length) | TERTIARY STAGE (until death) |

Penicillin treatment effective

No effective treatment

Chancre develops at site of infection

Lesions appear on genitalia or mucosa

Cutaneous lesions and lesions of the mucous membranes

Headache, fever, enlarged lymph nodes

Damage to any organ of the body

Personality change and paralysis

FIGURE 20.10 The Three Stages of Syphilis
Syphilis is a sexually transmitted disease in which the causative agent, *Treponema pallidum*, enters the body through tiny abrasions on the genitals or other mucosal surfaces. Once in the host, it quickly migrates via the circulatory system, infecting all tissues. About three weeks after infection, a hard chancre appears at the site where the organism entered; this is the beginning of the primary stage of the disease, which lasts from three to six weeks and ends with the disappearance of the lesion. A latent period without symptoms is followed by the secondary stage, marked by the appearance of numerous infectious lesions on the skin and mucous membranes. A second latent period may last for months or even years. The third stage manifests itself by large ulcers, erosion of the aorta, and/or neurologic damage. (Photo courtesy of Leonard Winograd, Stanford University/BPS.)

THE TREATMENT.   Syphilis is customarily treated with penicillin, but several other drugs are
equally effective in penicillin-intolerant people.
Drug resistance has not been a problem so far.
Cure is possible in all but the tertiary stage,

where the infection can be arrested but the tissue destruction cannot be reversed.

EPIDEMIOLOGY AND CONTROL MEASURES.   The
sole reservoir of *T. pallidum* is human beings. It

BOX 20.1
Current Trends in STDs

Diagnosis	Cases per 100 visits to STD clinics	
	Men	Women
Gonorrhea	24.0	15.7
Nongonococcal urethritis	24.8	—
Genital herpes	3.4	1.5
Venereal warts	4.3	3.0
Syphilis	1.7	1.4
Trichomonas vaginitis	—	10.4
Candida vaginitis	—	6.1
Nonspecific vaginitis	—	11.3

Periodically, national surveys
attempt to rank the true incidence of the various STDs. The
data are by nature subject to
considerable error. The error,
however, is entirely on the low
side. Many STD cases are
never reported, and many STD
patients actually have two or
more simultaneous STDs. The
report usually lists only the
most obvious one.

The accompanying table
shows disease incidence in a
six-month study involving six
STD clinics. Since these were
people who went to the clinic
because they knew they had a
problem, the results do not

necessarily reflect the incidence in the population as a
whole.

Gonorrhea is arguably the
single most common STD.
Close to 1 million new cases
per year are reported to the
Centers for Disease Control. A
large culture-testing program
evaluated close to 9 million
cultures obtained at random
from healthy women. A total
of 4.7% were culture-positive
for gonorrhoea.

The nongonococcal
urethritis/vaginitis complex,
which includes the chlamydial-
and ureaplasma-type infections, is probably even more

common, but since the condition is not notifiable, the available data is poor. One study
estimated that each year nearly
3 million Americans suffer chlamydia infection, and it costs
more than $1 billion to treat
them.

Genital herpes is a rapidly
increasing STD problem. There
are about 200,000 to 500,000
new cases annually. Since
there is no cure, the number
of total cases is cumulative,
with perhaps 5 million U.S.
cases. From 1965 to 1979, visits
to doctors' offices for genital
herpes increased by 830% (that
is not a misprint).

is spread by serum, largely via the serous fluid from the lesions during sexual contact, but also transmitted across the placenta to a fetus, or by other contacts with infective blood. The second and third methods are very rare, so syphilis is as close as any disease gets to being an absolute STD. Any open lesion is infective and the probability of contracting disease on any single exposure is about 10%.

Syphilis has a long (six to eight weeks) incubation period, providing enough time so that contacts can usually be traced and cured before their hard chancre appears and they in turn become infective. For this reason, case-finding methods are much more successful in breaking the chain of transmission of syphilis than that of gonorrhea. The public fear of syphilis helps to ensure that infected people seek treatment. Antibiotic therapy arrests the infection and renders the person noninfective within 24 hours. Sexual contacts of primary syphilitics within three

months prior to diagnosis, and of secondary syphilitics within six months prior to diagnosis, should be traced and receive penicillin therapy. To date, this public-health strategy has been only partially effective. Syphilis is still a major problem, with about 33,000 new cases reported in the United States per year (Figure 20.11). Although there are slight variations from year to year, syphilis incidence has not changed significantly since the early 1960s. We probably will not make further headway against syphilis until a vaccine against it is available.

THE NONSPECIFIC STDS

There are many cases of urethral or vulvovaginal infection in which neither the gonococcus nor the syphilis spirochete are present. They are referred to at present as **nongonococcal urethritis** and **nonspecific vulvovaginitis.**

There is, of course, always an infectious

FIGURE 20.11   Reported Cases of Sexually Transmitted Disease in the United States.
The incidence of syphilis reached its peak around 1942, then declined with the introduction of penicillin. Gonorrhea also declined but rebounded as the causative agent became penicillin-resistant. The introduction of birth control pills around 1965 had two effects: these medications both increased a woman's susceptibility to gonorrhea and provided an impetus for increased sexual freedom.

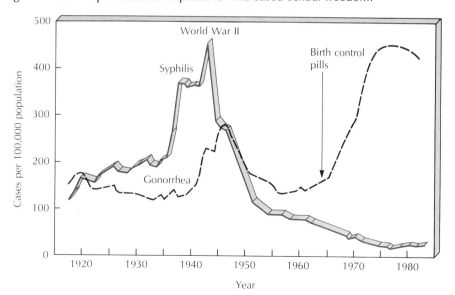

agent, but it may be bacterial, viral, fungal, or protozoan, one of many different species. Chasing down the actual agent in any case may involve a considerable laboratory workup and take quite a few days. None of the agents involved have yet been made legally notifiable, so we have relatively little data on their prevalence. By the available measures, however, these infections are more widespread in the United States than either gonorrhea or syphilis, particularly in college students and middle and upper income groups. Nonspecific, therefore, does not imply that the disease is insignificant.

CHLAMYDIAL INFECTIONS

The chlamydiae appear to be the leading cause of nonspecific STDs. Convenient and reliable monoclonal antibody and EIA methods for their identification have recently become available.

THE ORGANISM.   *Chlamydia trachomatis* is an obligate intracellular parasite that forms characteristic inclusion bodies in the host cell. Its life cycle alternates between an elementary body, adapted to extracellular survival, and a reticulate body, which is the form that multiplies inside host cells. Chlamydiae actually form microcolonies with membrane-bound vacuoles in the host cell, where they are well protected from the host's normal defense mechanisms.

THE DISEASES.   *C. trachomatis* appears to cause some 30 to 50% of cases of nongonococcal urethritis in males, and an equivalent amount of nonspecific vulvovaginitis in females. This agent is also responsible for a wide range of other infections (Table 20.3), including another relatively uncommon sexually transmitted disease called lymphogranuloma venereum. Most chlamydial STDs produce no significant complications or after-effects, but some cases in females lead to pelvic inflammatory disease, contributing to an increased risk of ectopic pregnancy and sterility.

THE TREATMENT.   Chlamydial infections cannot be effectively treated with penicillin, as can

TABLE 20.3 Some diseases caused by *Chlamydia trachomatis*
Arthritis
Inclusion conjunctivitis
Lymphogranuloma venereum
Neonatal pneumonitis
Nongonococcal urethritis
Nonspecific vulvovaginitis
Pelvic inflammatory disease
Trachoma

syphilis and most cases of gonorrhea. However, tetracycline and the sulfas are highly effective against the chlamydias. This distinction underscores the need for making an accurate laboratory diagnosis of the agent that is causing the problem before ordering the antibiotic.

EPIDEMIOLOGY AND CONTROL MEASURES.   The reservoir of *C. trachomatis* is human beings, and the agent may be found in and transmitted by many body secretions; many cases are asymptomatic. Although our understanding of the extent of chlamydial infection and its epidemiology is still incomplete, the new rapid means of diagnosing *C. trachomatis* should not only be helpful in selecting appropriate therapy, but will also begin to provide the data needed to design epidemiologic control strategies.

OTHER AGENTS OF GENITAL INFECTION

The bacterial agents called mycoplasmas frequently cause urogenital infection. Recall that mycoplasmas lack a cell wall, and parasitize by means of a firm attachment to the host-cell membrane (review Figure 18.5). *Ureaplasma urealyticum* (Figure 20.12) is frequently isolated from the inflamed urethra. In women, *Ureaplasma urealyticum* may be carried in the cervix or the endometrium. This organism appears to be transmitted by males to females, tightly bound to sperm. If a pregnant woman is infected, *Ureaplasma* is a major contributor to fetal death, recurrent miscarriage, prematurity, and low birth weight (the most common cause of neonatal death). Ureaplasmas are identified by culture from urethral and vaginal discharges, and from placental surfaces. Tetracycline is usually an effective treatment, but about 15% of strains are resistant. Because tetracycline also controls *Chlamydia*, it is most often used. Erythromycin and spectinomycin are alternative antibiotic selections.

Gardnerella vaginalis is a tiny Gram-negative rod, found both in normal vaginal flora and in the nonspecific genital infections. First identified as a nonspecific vulvovaginitis agent in females, it is now clear that it also causes inapparent infections in males and is transmitted

sexually. It responds to ampicillin therapy.

Certain agents are resistant to cold and drying, and are readily transmitted nonsexually by fomites. The yeast *Candida albicans*, which causes the infection **moniliasis** (after its older generic name *Monilia*) is one of these. *Candida* vulvovaginitis is a troublesome infection characterized by itching or burning sensations and the secretion of a copious exudate. Although it is transmitted sexually, it is most frequently an endogenous infection, a simple overgrowth of this normally minor component of the vaginal flora; the yeast can readily be seen in stained smears. It is treated with topical nystatin.

Trichomonas vaginalis is a flagellate protozoan that causes a common vulvovaginitis called **trichomoniasis.** The protozoan can be detected in freshly prepared, unstained, wet mounts of the vaginal discharge (Figure 20.13).

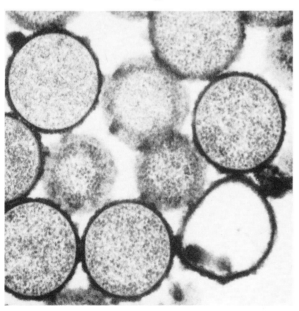

FIGURE 20.12 *Ureaplasma urealyticum*
This mycoplasma is a major cause of nonspecific sexually transmitted diseases (STDs) in both sexes. It is also a cause of spontaneous abortion, low birth weight, and premature births. (From J. Robertson and E. Smook, *J. Bacteriology* (1976), 128:658–660.)

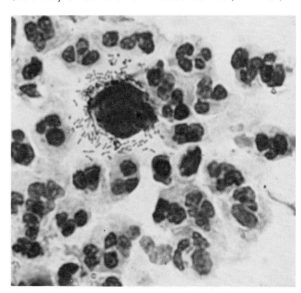

FIGURE 20.13 *Trichomonas aginalis*
This protozoan flagellate is a common causative agent of nonspecific vulvovaginitis in women. It may be transmitted both sexually and by indirect means. It is seen here in a Gram stain from a urethral discharge. The trichomonad is surrounded by rod-shaped bacteria and white blood cells. (Courtesy of Centers for Disease Control, Atlanta.)

The infection is treated with oral metronidazole (Flagyl), or with topical povidone–iodine creams. It can be transmitted sexually, or asexually by contact with contaminated objects.

GENITAL HERPES

Despite the fact that our society is just "discovering" genital herpes, to our sorrow, the disease is not new. The Romans knew a lot about venereal diseases; after all, they are named after Venus, the Roman goddess of love. Rome had significant herpes outbreaks during some of their later, more liberal empires. Rigid laws to control behavior were enacted, with as little effect as such laws usually have.

The virus, herpes simplex, was discussed in Chapter 16. Herpes simplex Type II usually, but not always, is responsible for lesions of the genital area. In women, lesions may appear on the vulva, in the vagina, or around the cervix. In men, the lesions form on the skin of the penis. In both sexes, the vesicular lesions appear sporadically, alternating with latent periods. The fluid from the active lesions is infective. It appears that a person is infective if an active lesion is present, and possibly up to 24 hours before the appearance of a lesion. It is becoming clearer that (as with oral herpes caused by the Type I virus) many persons have an initial lesion but then never have a second eruption. The true incidence of genital herpes infection is hard to estimate, but there are certainly more than 5 million persons who carry the Type II virus in the United States—between 2 and 4% of the total population.

Herpes cannot be cured at present. Some drugs are used to treat the active lesions. A topical cream containing the drug idoxuridine is prescribed to shorten the duration of the lesions and reduce the discomfort. Oral acyclovir has also been approved to assist in controlling the eruptions. Given the magnitude of the research attack on genital herpes, there is hope that ways to control or cure this infection will soon be found. A vaccine is currently being tested in an-

imals, but it is too early to predict its usefulness in humans.

Although it may be extremely painful and psychologically distressing, genital herpes by itself is not an immediate threat to life and limb. However, it has two extremely worrisome features. If a pregnant woman has an active lesion, her child may be infected during normal vaginal delivery. This infectious contact can produce a frequently fatal herpes encephalitis in the newborn. Herpes lesions may also develop in the baby's eyes that may cause blindness. To prevent infection of the infant, cesarean deliveries are used in women who have active herpes lesions at the time of delivery. Chronic herpes infection of the cervix is also a serious concern because it has been statistically associated with increased risk of cervical cancer.

PELVIC INFLAMMATORY DISEASE

Pelvic inflammatory disease (PID) is an inflammation of the pelvic organs, especially the uterus, fallopian tubes, and ovaries. It is one of the serious complications of sexually transmitted diseases such as gonorrhea or chlamydial infection. However, it may also be an ascending infection caused by the normal vaginal flora. Anaerobes, enterobacteria, alpha-hemolytic streptococci, and diphtheroids, are the most common agents. PID may develop spontaneously, but often follows surgery or abortion. The woman may have lower abdominal pain severe enough to mimic acute appendicitis. Peritonitis and septicemia may follow. PID may result in sterility through scarring of the fallopian tubes. Thus although PID usually subsides without treatment, antimicrobials are always given to prevent more serious trouble. This condition is considered a major public health problem, with an estimated 1.75 million cases per year in the United States. As previously mentioned, intrauterine devices significantly increase the risk of acquiring PID. About 25% of women with PID are hospitalized, and PID is a frequent reason given for performing hysterectomies.

ENDOMETRITIS

Endometritis is an infection of the uterine lining that may occur after abortion, surgery, or cesarean section. After a difficult vaginal delivery, it may appear as **puerperal fever.** The causative agents are usually the normal vaginal species, introduced to the uterus by some means. The most threatening feature of puerperal fever is the fatal septicemia that often follows. Such deaths were frequent in the era of illegal abortions, and they still may occur if abortions are carelessly performed. Even in these cases, no one need die, as antibiotics can prevent death, if started soon enough.

TOXIC-SHOCK SYNDROME

Toxic-Shock Syndrome (TSS) is a generalized toxemia caused by certain strains of *Staphylococcus aureus.* It was first described in 1977 as a disease of children, and it occurs sporadically in persons of all ages and both sexes. However, its most well-publicized and well-studied expression is in women of childbearing years who use tampons to absorb menstrual flow. The disease symptoms are caused by *S. aureus* toxins including enterotoxin F and pyrogenic (fever-inducing) toxin C. The bacteria are probably endogenous, and there is no evidence that sexual transmission is a particularly significant factor. Symptoms may include fever greater than 104 degrees Fahrenheit, abnormally low blood pressure, dizziness, and fainting. The toxin may cause GI symptoms such as diarrhea and vomiting. The skin may be involved with a rash and later desquamation (shedding of the outer layers). Some cases are fatal (mortality rate from 2.5 to 4.8%) and others leave the victims with neurologic problems and gangrene of the extremities. In 1980 the role of tampons as a risk factor, especially high-absorbency types, was documented by the Centers for Disease Control. Widespread publicity prompted a decline of about 30% nationwide in tampon use, but since then, tampon use has gone back up. Certain types of tampons were removed from the market. TSS diagnosed in menstruating women has declined markedly, to around 300 cases per year. TSS in other groups — nonmenstruating women, men, and children — is still occurring and it remains a rather mysterious disease. The risk of TSS to menstruating women is minimized by wearing tampons intermittently, if at all, and with frequent changes.

PROSTATITIS

Prostatitis, inflammation of the prostate gland, is a very common condition in older men. This acute or chronic inflammation is usually caused by *E. coli,* although other enterobacteria, enterococci, and intracellular organisms are found. In the acute form of prostatitis there is swelling and intense pain, as well as fever and urinary obstruction. The condition usually responds promptly to antimicrobials effective against Gram-negative bacteria. Chronic bacterial prostatitis, on the other hand, may have less pronounced symptoms such as lower back pain, urinary frequency, and very slow urinary flow. The urine may or may not show bacteriuria at any given time. However, the bacteria in the prostate serve as a focus to initiate repeated bouts of cystitis, otherwise usually rare in the male. Treatment is complicated by a physiologic diffusion barrier between the plasma and the prostatic fluid, preventing drugs from getting from the circulation into the tissues where they are needed. Trimethoprim–sulfa combinations have shown promise.

INFECTIONS OF THE PRENATAL PERIOD

Particular infections threaten the mother and her child before, during, and after birth. In the concluding sections of this chapter, we will look at these infectious hazards of reproduction.

In the prenatal period (before birth) when the fetus is in the uterus (in utero) it is protected, fed, and insulated from most traumas of independent life. However, it is not totally isolated from infection. Many agents, predominantly viruses, circulate in the blood, and are able to cross the thin placental barrier from the mother's blood to that of the fetus. Some of these agents can then multiply in the fetal tissues and kill or deform the fetus. The agents that cause intrauterine infection are listed in Table 20.4.

THE FETAL MEMBRANES AND PLACENTA

In the early days of the fertilized egg's development, while it moves down the fallopian tube, it acquires an outer **trophoblast** layer. At seven or eight days this layer becomes known as the **chorion,** and it is covered with fingerlike villi. On reaching the uterus, the egg starts to bury itself in the uterine wall. The placenta then develops as a double organ: half of the tissues are derived from the mother's uterine lining, the other half from the fetal chorion. The fetal circulation is connected to its half of the placenta by the **umbilical vessels.** The placenta is well formed by the third week of pregnancy, and by the fifth month it occupies one half of the uterine surface.

The fetus has its entire chemical contact with the world via the placenta. Plasma components including water, nutrients, electrolytes, and serum proteins such as immunoglobulins pass through from maternal placenta to fetal placenta (Figure 20.14). Alcohol, anesthetics, antimicrobial agents, and other drugs move across the placenta too. The fetal wastes are transferred to the maternal circulation. Maternal red blood cells cannot move across the placenta, however, and the fetus develops its own red and white blood cells.

FETAL INFECTIONS

Virus particles, many times smaller than blood cells, can cross to the fetal circulation from the mother's blood when the mother has viremia, as do a few cellular bloodborne pathogens. The result is an **intrauterine infection.** Some agents upset the fetus's developmental timetable and create birth defects. These are called **teratogens.** Infections occurring during the first trimester carry a higher risk of producing malformation because the basic differentiation of fetal tissues and construction of organ systems is occurring then. Intrauterine infection can cause spontaneous abortion, stillbirth, or premature delivery. Alternatively, the fetus may survive to term, but be critically ill at birth, as appears to be the case with congenital AIDS.

When intrauterine infection occurs, the fetus may be stimulated to begin making antibody early. If an infant shows a high level of IgM antibody at birth, this is strong evidence of prenatal infection.

CYTOMEGALOVIRUS INFECTION. Infection by cytomegalovirus (CMV) is the most common form of intrauterine infection. CMV may be

TABLE 20.4 Agents that cause intrauterine infection

Group	Organism
Viruses	Rubella virus*
	Herpes simplex Types I and II*
	Cytomegalovirus*
	Varicella-zoster virus
	Poliovirus
	Coxsackievirus
	Influenza viruses
	Hepatitis viruses
	Measles virus
	Variola-vaccinia
	AIDS virus*
Bacteria	*Treponema pallidum* *
	Listeria monocytogenes
	Mycobacterium tuberculosis
	Ureaplasma urealyticum sp.
Protozoa	*Toxoplasma gondii* *
	Plasmodium sp.
	Trypanosoma sp.

*Most significant.

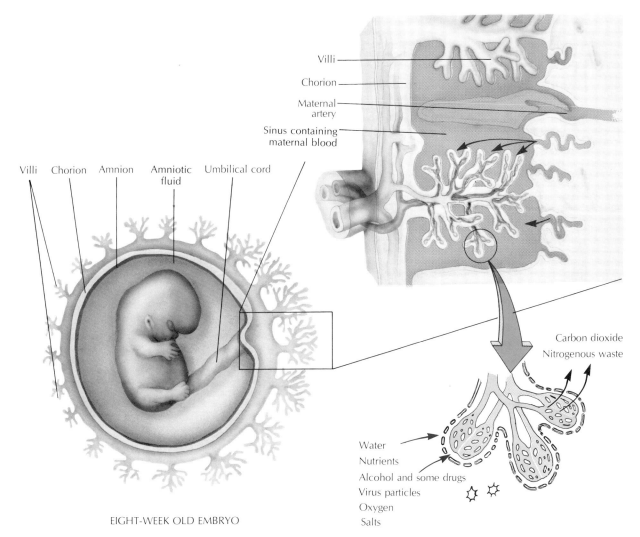

Villi

Chorion

Maternal artery

Sinus containing maternal blood

Villi · Chorion · Amnion · Amniotic fluid · Umbilical cord

Carbon dioxide
Nitrogenous waste

Water
Nutrients
Alcohol and some drugs
Virus particles
Oxygen
Salts

EIGHT-WEEK OLD EMBRYO

FIGURE 20.14 Fetal and Maternal Circulation
The placenta is formed from tissues in the mother's uterine lining and from the fetal chorion. Numerous villi covering the chorion project into the wall of the uterus. Here pockets of tissue are kept rich in blood by branches of the uterine arteries. Although fetal and maternal circulatory systems do not directly connect, materials move by diffusion and active transport across capillary membranes in both directions.

found in about 1% of all live-born babies. Most of these infants display no apparent ill effects, but at a later time about 10% of these reveal some mental retardation. These figures imply that intrauterine CMV infection is an important cause of mental retardation in children. A smaller number of infants with CMV will have clinical disease with liver and spleen enlargement, jaundice, and other neurologic defects. Few of these severely affected infants survive.

The epidemiology of CMV is complex and poorly understood. The agent (Figure 20.15) is one of the herpesviruses, and has their typical behavior pattern in its ability to establish a latent state. Its exact prevalence is uncertain. One study found that over 80% of a sample group of people over 35 had antibodies to CMV. CMV infection appears in 50 to 90% of organ trans-plant recipients, presumably reactivating because they have been immunosuppressed. The virus is found in the vaginal secretions of 10% of healthy women, and it can be transmitted either across the placenta or at birth.

CONGENITAL RUBELLA SYNDROME. As we saw in Chapter 16, when the mother contracts rubella during the first trimester of pregnancy, the virus is transmitted to the fetus and causes diminished tissue growth, tissue necrosis, and chromosomal damage. The most severely affected fetuses die and are miscarried. Damage in survivors shows up in the cardiovascular and nervous systems, especially the sense organs. Due to the widespread use of live rubella virus vaccine, the incidence of congenital rubella syndrome is now less than 10 cases per year in the United States. The vaccine virus is harmless if inadvertently given to a pregnant woman, but intentional administration is avoided as a precaution.

TOXOPLASMOSIS. *Toxoplasma gondii* (Figure 20.16) is an obligate intracellular protozoan parasite with a worldwide distribution. It infects many domestic and wild animals and human beings, causing a disease called *toxoplasmosis*. The parasite reproduces sexually in the intestine of cats, which shed oocysts in the feces. These oocysts may infect other animals and possibly human beings. In the infected nonfeline animal,

FIGURE 20.15 Cytomegalovirus
CMV, a herpesvirus, is the commonest virus to cross the placenta and infect the fetus. It causes many congenital infections, some very serious. Infected infants may retain the virus, which often crops up later in life when immune defenses are low. In the transmission electron micrograph, a number of cytomegaloviruses are maturing within the nucleus of a human embryo fibroblast. At top left and right, two virus particles are acquiring envelopes as they push through the inner layer of the nuclear membrane. (×38,000. Courtesy of Michael Fons and Thomas Albrecht; See Albrecht et al., *Lab. Invest.* 42:1,1980; Cavallo et al., *J. Gen. Virol.* 56:97, 1981.)

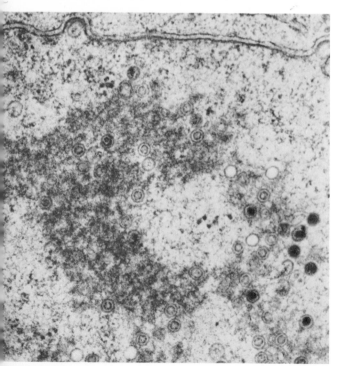

FIGURE 20.16 Toxoplasmosis
Toxoplasmosis affects many mammalian species. The cat is a definitive host, and in its intestine *Toxoplasma gondii* reproduces sexually to produce sporozoites. Humans may contract the disease after handling contaminated cat litter or sand. Children or adults usually contract a mild case. However, if a pregnant woman becomes infected, her fetus will develop congenital infection that causes severe damage to the brain and sense organs, even death. The pathogen also infects cows, and undercooked beef may transmit the disease to humans. (From E. Scholtyseck, *Fine Structure of Parasite Protozoa.* New York: Springer-Verlag (1979), Fig. 41, pp 120–121.)

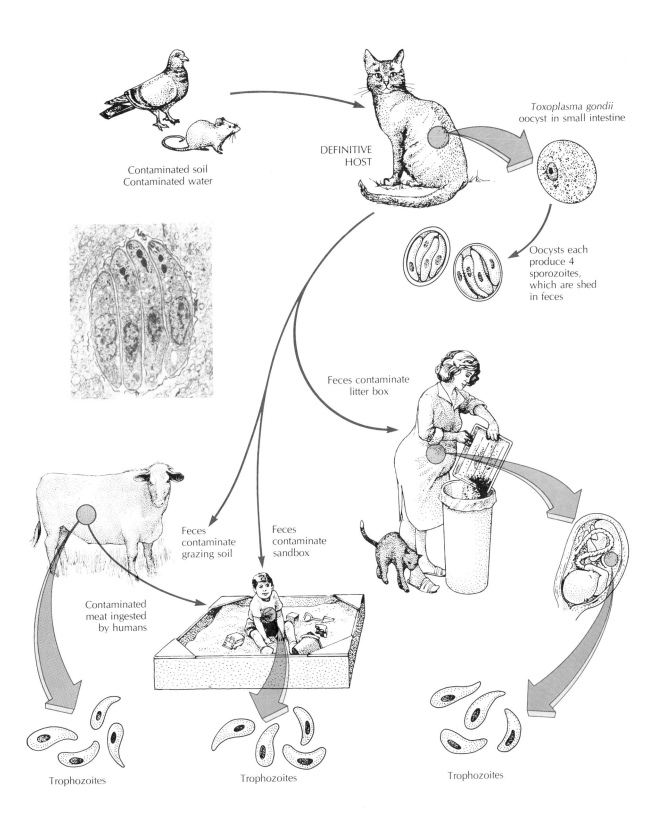

Contaminated soil
Contaminated water

DEFINITIVE
HOST

Toxoplasma gondii
oocyst in small intestine

Oocysts each
produce 4
sporozoites,
which are shed
in feces

Feces contaminate
litter box

Feces
contaminate
grazing soil

Feces
contaminate
sandbox

Contaminated
meat ingested
by humans

Trophozoites

Trophozoites

Trophozoites

cysts give rise to active trophozoites that invade all tissues. Once a sufficient immune response develops, the trophozoites stop multiplying and form tissue cysts that last for life. Such tissue cysts are found in 10% of lamb or beef samples and 25% of pork samples. Eating inadequately cooked cyst-containing meat, or coming into contact with infective cat feces, may transmit the infection to human beings.

Human toxoplasmosis is usually inapparent, though occasionally there is mild lymphoid tissue involvement. Serious infection and overwhelming disease is rare, and usually correlated with immunodeficiency. Positive serologic evidence of past infection is found in from 25 to 45% of persons of childbearing age, depending on the area of the United States.

The woman who has had toxoplasmosis earlier in life and recovered will not transmit the now inactive tissue form of the parasite to a fetus when she becomes pregnant. However, if she contracts the disease for the first time during pregnancy, particularly in the last trimester, the fetus will be infected during the period when the trophozoites are actively multiplying, and tissue cysts will form in the fetal tissues as well as in the mother. Most infected newborns are asymptomatic at birth. However, cysts may later reactivate. Then trophozoites will be released, and can destroy the retina of the eye or the central nervous system. Infants showing active disease at birth may have damage in many organs, and about 80% are mentally retarded.

Active disease may be detected by testing a pregnant woman's serum. The TORCH test series includes serologic tests for Toxoplasma, Rubella, Cytomegalovirus, and Herpes. It is widely recommended for prenatal screening. If the Toxoplasma test is positive, pyrimethamine plus sulfonamide can then be given. This treatment may prevent neonatal toxoplasmosis. The same drug combination is used to treat affected neonates.

CONGENITAL SYPHILIS. *Treponema pallidum* can cross the placenta after the 16th week of pregnancy. Twenty-five percent of affected fe-

tuses die before birth and another 30% die shortly after birth. Signs of infection include lesions of the skin and mucous membranes, bone deformity, and liver and kidney diseases. Some deformities may not become apparent until later. If routine prenatal blood screening is carried out, syphilis can be detected early in pregnancy. If penicillin therapy is then administered before the 16th week, the mother will be cured and fetal infection prevented. Despite prenatal screening, there is still a significant incidence of congenital syphilis; 239 cases were reported in 1983.

MATERNAL INFECTIONS

Pregnancy is a healthy state. An otherwise healthy, well-nourished pregnant woman is not overly susceptible to infectious disease, but there are some risks we might note. First, influenza and other lower-respiratory-tract infections contracted in late pregnancy are more hazardous than for nonpregnant women. Second, the urinary system undergoes mechanical compression due to the pressure of the fetus, but it is required to work overtime to excrete for the fetus too. These stresses increase the risk of cystitis or of inapparent chronic pyelonephritis. Some women experience outbreaks of gingivitis or of vaginal candidiasis due to hormonal changes associated with pregnancy.

Maternal infections are significant not only in themselves, but also because they may lead to neonatal infections. Infants born to mothers with uncontrolled urinary-tract infection have six times the normal infection risk themselves. This is yet another pressing argument for thorough prenatal medical care.

INFECTIONS OF THE NEONATAL PERIOD

At birth the newborn child leaves the sterile world of the uterus and enters one filled with microorganisms. This stress could potentially overwhelm the baby's incompletely developed immune defenses, unless the child continues to be assisted by its mother. The mother's lymphoid system is designed to serve as an accessory immunologic organ for the baby. From a biological point of view, mother and child are only partially separated by the birth process. The remaining physical attachment is breast-feeding.

NEONATAL IMMUNITY

An infant fully developed at birth possesses the same lymphoid cells, tissues, and organs as the adult, and has a larger and more active thymus. However, the baby's lymphocytes, macrophages, inflammatory responses, and cell-mediated immune responses are not very active, as they have not yet been stimulated or programmed. Any immunoglobulins in the infant's serum have been acquired by transfer from the mother. The fetus has the capacity to make some IgG and IgM as early as two months before birth, but this potential will not have been used unless there was intrauterine infection. Complement levels are only half those of the infant's mother, opsonizing activity is weak, and phagocytosis is only partially effective. On the other hand, serum lysozyme and interferon levels are elevated.

In the normal scheme of things, the infant's immune system will be triggered by exposure to the challenge of the normal flora and environmental microbes. One of the ironies of nature is that on the one hand these contacts are a great threat to the baby's survival and on the other are an essential initiation to life, without which he or she wouldn't develop a defensive capacity.

During the later months of pregnancy, large amounts of maternal IgG are passed to the fetus. These antibodies will partially protect the infant from agents to which the mother is immune. If the mother had no antibodies against a specific disease, let's say polio, the infant won't either. The role of IgG is to control established infections, not to prevent them. Thus these immunologic gifts will ameliorate, not prevent, the baby's infections. Protection continues for only as long as the antibodies last, which is no more than six months.

Human milk plays a key role in the infant's immunity. It contains a wide range of immune factors that supplement and stimulate the infant's immature defense capabilities. In some societies, infants are nursed for two years or even longer. This two-year nursing period extends a sort of immunologic "umbrella" over the child until the time, between two and three years, when his or her own immune system becomes adequate.

In our own society, breast-feeding fell out of favor for a time because of the availability of commercially prepared infant formulas. Formula-feeding gained acceptance because of its convenience, and because children fed formula put on more weight. Today, however, we know more about human milk's immunologic benefits, the infant's nutritional requirements, and the psychological and sociological role played by breast-feeding. In 1979, the American Academy of Pediatrics adopted a resolution strongly encouraging all mothers to breast-feed their infants. The labels on commercial formulas now bear a statement cautioning purchasers that breast-feeding is preferable unless there is a medical contraindication.

NEONATAL INFECTIONS

The neonatal period is the first month after birth. A neonate is highly susceptible to infections. Although infection is often contracted in the hospital nursery, this situation is seldom due to carelessness or negligence. Rather, it is because the first few days of a baby's life are the riskiest, and the nursery just happens to be where the infant is. High standards of care, constant observation, and rapid, effective evaluation of any adverse signs can minimize the danger and help keep minor problems from escalating into major ones. The epidemiology of some neonatal infections is outlined in Table 20.5.

A leading cause of neonatal death is **neonatal sepsis** or septicemia, the general multiplication of microorganisms in the blood. *Streptococcus agalactiae*, Lancefield Group B, is a leading cause of neonatal sepsis. This organism, four

TABLE 20.5 Epidemiology of neonatal infections			
Disease	Reservoir	Transmission	Incubation Period (Days)
Diarrhea, enteropathogenic *Escherichia coli*	Human beings	Oral–fecal; water, food, weighing scales, dressing tables, contaminated hands	1–3
Neonatal inclusion conjunctivitis	Human beings	Contact with infected vaginal discharges during birth	5–12
Neonatal gonorrheal ophthalmia	Human beings	Contact with infected birth canal	1–2
Group B streptococcosis	Human beings (?)	Contact with adult carriers, usually maternal vaginal secretions at birth	1–3
Neonatal staphylococcosis	Human beings	Spread on hands of hospital staff; airborne fomites	4–10
Thrush	Human beings, others	Contact with secretions of mouth, skin, vagina, and feces of patients and carriers	2–5

Source: Adapted from Benenson, AS *Control of Communicable Diseases in Man*. 14th ed. 1985.

the vaginal flora of 12.5% of normal mothers in one study, causes septicemia in infants up to several weeks old. The initial symptoms may be mild irritability and failure to nurse, without fever. It can progress rapidly to septicemia and death, and early intervention may be of critical importance to the outcome. Staphylococcal infections in neonates are also a major hazard, carrying a risk of fatal invasion of the lungs or of septicemia.

Bacterial meningitis, an inflammation of the membranes covering the brain and spinal cord, poses a great risk in newborns and in infants. Predisposing factors include traumatic labor and delivery, prematurity or low birth weight, and primary immunodeficiency. Early antimicrobial therapy for other infections also tends to disrupt the development of the normal flora and select for opportunistic microbial strains that are antibiotic-resistant. The Group B streptococci, discussed above, commonly cause meningitis, as does *Listeria monocytogenes*, a small, motile, facultatively anaerobic, Gram-positive rod, which is widespread in nature (in nonhuman animals and in human beings). *Listeria* may be acquired at birth, just prior to birth, or earlier in fetal development. The earlier exposures almost always lead to abortion or stillbirth, but neo-

TABLE 20.5 Epidemiology of neonatal infections (continued)

Period When Communicable	Patient Control Measures	Epidemic Control Measures
For duration of fecal colonization, more than one month	Isolate infected infants and suspects; disinfect discharges and soiled articles	Suspend maternity service unless another, clean, nursery and separate staff are available
While exudate is produced; while genital infection persists in female	Isolate for 48 hr; aseptic techniques prevent nursery transmission	Search for sexual contacts; treat maternal infection
For 24 hr following specific treatment or until discharges cease	Isolate for 24 hr; carefully dispose of conjunctival discharges	Routine administration of silver nitrate eye drops at birth prevents disease
While infection is present	Isolate; disinfect all discharges and soiled articles	None
As long as purulent lesions continue to drain or carrier state persists	Isolate all cases and suspects; carefully dispose of dressings and soiled articles	Determine characteristics of epidemic strain; search for source. Group isolation and quarantine
For duration of lesions	Segregate afflicted infants, disinfect secretions and contaminated articles	Segregation; increased emphasis on cleanliness

natal infection can be treated successfully with erythromycin if diagnosed in time. An epidemic in California in 1985 was caused by eating Mexican-style cheeses. Of the 86 cases, 58 were mother-infant pairs, and 21 babies were stillborn or died after birth. Gram-negative bacilli such as *E. coli*, strain K1, and other enterobacteria are important agents of meningitis in the infant under two months of age. The mortality rate is 40 to 80%. *Flavobacterium meningosepticum*, found in tapwater, has also caused some nursery outbreaks of meningitis.

Pneumonia is a great risk in the premature infant whose lungs may not be fully developed.

Chlamydia trachomatis is the leading cause of neonatal pneumonias, but most of the organisms already mentioned in this section are also occasionally implicated.

Gastroenteritis and diarrhea due to enteropathogenic strains of *E. coli* and other enterobacteria have already been mentioned in Chapter 19. These infections, although not life-threatening in adults, are leading causes of infant death in less-developed nations. Breast-feeding reduces the risk.

Thrush, an oropharyngeal infection caused by *Candida albicans*, is quite common in neonates. It is contracted by exposure to a maternal

BOX 20.2
The Immunologic Gift

Lactation — the mammalian way of feeding the young with milk secreted from mammary glands — evolved as a biological strategy over a period of 200 million years. The bottle-feeding of infants has been widespread for less than a century. We tend to think of human milk as a food, and have discovered that its nutritional role can be adequately performed by a number of formulas based on cow's milk, soybean hydrolysates, and other substances. However, evolution designed lactation as far more than a feeding mechanism. Microbiologists became interested in lactation in considering its immunologic benefits.

Human milk, as we note in the text, contains a variety of immunoglobulins. Secretory IgA is abundant in the **colostrum,** the first thin, lipid-free fluid produced by the breast. Secretory IgA is continually secreted in milk throughout the nursing period, along with lesser amounts of IgG and IgM. The source of secretory IgA is the mother's gut lymphoid tissue; secretory IgA is transported by the blood to the breast tissue, then concentrated in the milk. Recently, it has been shown that when a new microorganism colonizes

the infant's GI tract, it normally colonizes the mother's GI tract also. She is stimulated to produce specific secretory IgA antibodies. These are then transferred to the baby to assist it in controlling the new threat.

Maternally produced milk antibodies also appear to protect the infant from hypersensitivities. The infant gut lining can be readily sensitized by potential allergens in the diet. However, antibodies in the mother's milk combine with these antigens and appear to block allergic sensitization.

Food allergies are thus very uncommon in the breast-fed child.

Human milk also contains white blood cells — as many per milliliter as are found in the mother's blood. The nursing mother donates about 25% of her daily monocyte production to her infant. Once in the baby's intestine, these milk monocytes develop into large, active macrophages. They destroy ingested microorganisms and secrete complement. Other white blood cells in milk include B and T lymphocytes. Once in the gut, these lymphocytes may secrete antibodies. It is thought that maternal lymphocytes may play a key role in stimulating the infant's gut lymphoid tissue to become operational.

Of course, cow's milk also contains these factors, which are extremely beneficial for calves. But antibodies against calf diseases have no apparent value for human babies. The cow's white cells are killed by pasteurization and refrigeration. If they did survive, (as they do in raw milk) it is doubtful they could communicate with human lymphoid cells to stimulate them. (Photo courtesy of Janice Fullman/The Picture Cube.)

vaginal infection or contaminated fomites. Thrush is seen as a reddening of the oral membranes, with grayish-white patches of exudate. Thrush is not a severe condition, and usually disappears quickly, as the neonate's defenses improve and the competitive normal flora establishes itself. The discomfort, however, may cause the baby to feed poorly, sleep fitfully, and gain weight slowly.

Infant botulism or the "floppy baby syndrome" is characterized by lack of muscle tone, inability to hold up the head, and weak or absent suckling motions. It results when *Clostridium botulinum* Type B becomes established as part of the infant's intestinal flora. The botulinum toxin is produced in the intestine, absorbed into the blood stream and causes muscle weakness. Antimicrobial therapy and supportive measures will cure infant botulism. Many samples of honey contain *C. botulinum* spores, and honey has been implicated in many infant botulism cases. Many experts now recommend that raw honey not be fed to infants under one year of age. The rest of us may continue to enjoy honey without fear, because *C. botulinum* is incapable of establishing itself in the adult GI tract.

MATERNAL INFECTIONS

Apart from losing sleep worrying about all of the possible problems her baby might have, what risks does the new mother herself face?

At birth, the placenta separates from the uterine wall, leaving an area of raw, exposed tissue. This is a prime site for microbial invasion if bacteria are introduced to the uterus. As mentioned before, the resulting endometritis is called **puerperal sepsis** or **childbed fever.** This outcome was horribly common a century or so ago, after the invention of obstetrical forceps led doctors to start introducing instruments into the uterus of the delivering woman, and before aseptic technique was developed suggesting that they should sterilize their instruments first. As many as one in eight women having a hospital (their idea of "modern") delivery then died of septicemia. Before antibiotics, puerperal fever was almost invariably fatal. Now, prompt therapy saves most victims. With modern aseptic techniques, in uncomplicated deliveries, this disease should be only a historical curiosity. But, sadly, techniques are used by fallible human beings, and cases (and deaths) do still occur.

Mastitis is a very common infection of the mammary tissue of the nursing mother. *Staphylococcus aureus* is the usual pathogen. Swelling, reddening, and pain in the affected area of mammary tissue develop, along with fever. Early cases can often be treated with hot compresses, but antibiotics may be administered. In extreme cases, abscess formation may occur, sometimes requiring surgical drainage. Even if this happens, antibiotics usually work well and quickly, and nursing may be resumed.

SUMMARY

1. The male urinary and reproductive systems use the urethra as a common passage to the exterior of the body. Infections of the male urethra may involve both systems. Bladder and kidney infections are far less common in males than in females.

2. In the female, the urinary and reproductive systems are anatomically separate. The urinary-tract and genital/reproductive infections of women often stem from abnormal colonization by normal flora from other body areas.

3. Both the urethra and the vagina have normal flora. The vaginal flora illustrates the protective effect of resident organisms. In women of reproductive years, the flora creates an acidic physiologic environment hostile to the most likely invaders, the bowel flora.

4. Cystitis is usually caused by bowel enterobacteria. Many symptomatic infections develop from preexisting, inapparent infection. Women have a much greater risk of cystitis than men.

5. Bladder infections may ascend to affect the renal pelvis, collecting tubules, and medulla. Acute pyelonephritis is readily detected and cured; if treated, it has little residual effect. Chronic pyelonephritis may be present for months or years before being detected, causing cumulative, irreversible damage.

6. Male genital infections primarily involve the penis and urethra. In untreated disease, the accessory glands may be attacked. In the female, such infections involve the vulva, vagina, and cervix, progressing, if untreated, to the endometrium of the uterus, to the fallopian tubes and ovaries, and potentially to the peritoneal cavity.

7. The sexually transmitted diseases are those transmitted primarily by sexual acts. They are regarded as serious if they lead to sterility or life-threatening complications. Gonorrhea and syphilis fall into this category. Nonspecific STDs appear to be self-limiting and nondestructive but may be transmitted to the unborn child. Gonorrhea, the nonspecific STDs, and genital herpes are the most common STDs.

8. Pelvic inflammatory disease may be caused by STD agents or by normal flora. Endometritis is usually caused by vaginal flora, and chronic prostatitis by *E. coli* or other enterobacteria. These infections can be causally associated only rarely with any specific sexual contact.

9. Infections can be transmitted across the placenta to the fetus when the agent is present in the mother's blood. Infection of the fetus in early pregnancy may lead to spontaneous abortion or severe malformation. When infection occurs later in pregnancy, the fetus may survive but be born with acute systemic infection. Many intrauterine infections cause mental retardation that is not apparent until later in childhood.

10. Immediately after birth and for its first month, the newborn is especially susceptible to normal pathogens and opportunists. It is protected by maternal IgG passed through the placenta that slowly wanes in effectiveness. A wide range of maternal immune factors are continually supplied in the milk for as long as nursing continues. These minimize but do not prevent infection. Some infective agents commonly pathogenic to newborns, but less so to mature individuals, include the Group B streptococci, *Listeria monocytogenes*, and *Clostridium botulinum* Type B.

11. During the later months of pregnancy, maternal risks include increased incidence of pneumonia and urinary-tract infection. After birth, introduction of organisms into the uterus may lead to puerperal fever. The nursing mother may develop mastitis.

DISCUSSION QUESTIONS

1. Trace the movement of an ascending urinary-tract infection from its source to the kidney. Compare the patterns seen in males and females.

2. How does the pH of vaginal secretions affect infection susceptibility?

3. It is a common statement that increased sexual freedom has contributed to the increase in STDs. This is certainly true, but it is not the only reason. List some others.

4. Intrauterine infections have different results depending on the period of pregnancy in which they occur. Why?

5. Referring to Chapter 12, if necessary, review the specific immunologic functions to

which breast milk might contribute in the newborn.

6. We employ certain epidemiologic control measures against syphilis, yet the disease is still rampant. Why does the disease resist control, and what additional control measures would you suggest?

ADDITIONAL READING

Cannon JG, Sparling PF. The Genetics of the Gonococcus. *Ann Rev Microbiol;* 38:111–134, 1984.

Chesney, PJ et al. The Disease Spectrum, Epidemiology, and Etiology of Toxic-Shock Syndrome. *Ann Rev Microbiol* 1984; 38:315–338.

Felman YM, Nikitas JA. Nongonococcal urethritis: a clinical review, *JAMA* 1981; 245: 381–386.

Fitzgerald TJ. Pathogenesis and immunology of *Treponema pallidum. Ann Rev Microbiol* 1981; 35:29-54.

Harwood CS, Canale-Parola E. Ecology of Spirochetes. *Ann Rev Microbiol;* 38:161–192, 1984.

Jelliffe DB, Jelliffe EFP. Recent trends in infant feeding. *Ann Rev Public Health* 1981; 2:145-58.

Schachter J, Caldwell HD. Chlamydiae. *Ann Rev Microbiol* 1980; 34:285-309.

South MA, Alford CA Jr. Congenital intrauterine infections. In: Stiehm ER et al., eds. *Immunological disorders in infants and children.* 2nd ed. Philadelphia: WB Saunders, 1980.

21

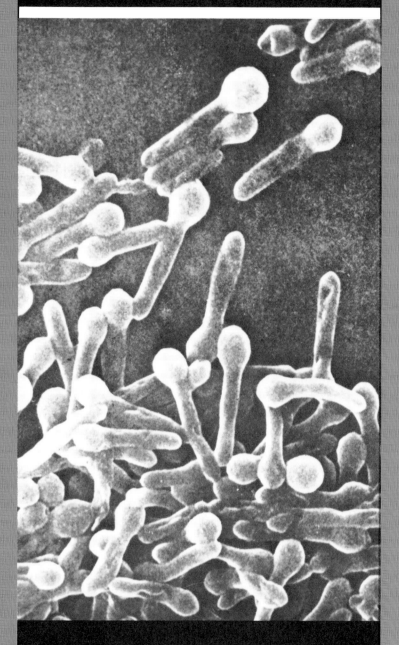

NERVOUS-SYSTEM
INFECTIONS

CHAPTER OUTLINE

THE NERVOUS SYSTEM

Anatomy of the Nervous System

Physiology of the Nervous System

BACTERIAL INFECTIONS

Hemophilus Meningitis

Meningococcal Meningitis

Leprosy

Tetanus

FUNGAL INFECTIONS

VIRAL INFECTIONS

Viral Meningitis

Arthropod-Borne Encephalitis

Rabies

Poliomyelitis

Slow-Virus Infections

PROTOZOAN INFECTIONS

Primary Amebic Meningoencephalitis

The nervous system is the body's communications network, designed to collect, process, and utilize information. It is a highly complex system, with structures that extend throughout the body. The nervous system is normally sterile. Most components are extensively protected and comparatively invulnerable to infection. Infection, when it does occur, may be devastating. In fact, one definition of death is irreversible loss of central-nervous-system function. In nonfatal infections, death of nerve cells may cause permanent neurologic impairment.

In this chapter we will examine some infectious agents that cause central-nervous-system disease. We will pay particular attention to how they succeed in invading this protected tissue. We will also study leprosy, a disease of the peripheral nerves, and tetanus, a wound infection that affects nerve impulses, causing paralysis.

THE NERVOUS SYSTEM

Let us first examine the basic architecture of the nervous system. Our interest, as usual, is in how it is defended and how infection affects its function.

Microbial pathogens featured in this chapter:

Arthropod-borne viruses

Clostridium tetani

Hemophilus influenzae

Neisseria meningitidis

Poliovirus

Rabiesvirus

ANATOMY OF THE NERVOUS SYSTEM

The basic units of the nervous system are differentiated nerve cells, designed to carry nerve impulses. The nerve cell is called a **neuron** (Figure 21.1a). It is composed of a **cell body,** containing the nucleus and organelles, and one or more slender extensions or **processes** that extend from it, in some cases for long distances. Processes that conduct impulses toward the cell body are called **dendrites;** those that conduct impulses

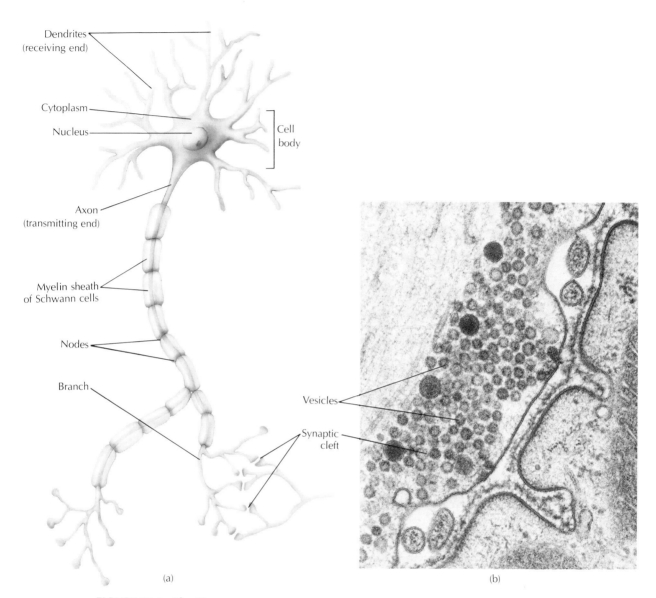

FIGURE 21.1 The Neuron

(a) The neuron, the basic conducting cell of nerve tissue is typically composed of a cell body and two processes. A nerve impulse is received by the dendrite, passes through the cell body, and moves down the axon to the next receptor. The axons of many nerves are protected by a myelin sheath, which is made up of complex lipids secreted by the Schwann cells. (b) In this micrograph, the synaptic cleft is visible as a thin space between the axonal end of one neuron and the receptor membrane of another. Nerve impulses are chemically transmitted across the synapse upon release of the neurotransmitter molecules, which are contained in many saclike vesicles within the axonal ending. (×56,400. Micrograph prepared by John Heuser, M.D.)

away from the cell body and communicate with receptor cells are called **axons.** Many nerve-cell processes are **myelinated,** that is, covered by an insulating sheath of **myelin,** a lipoprotein. The **gray matter,** found in both the brain and the spinal cord, contains cell bodies and unmyelinated nerve fibers. **White matter** is composed of bundles of myelinated nerve fibers. Demyelination causes loss of function or death of the neuron.

A nerve impulse passes between the axon of one neuron and the dendrite of the next at the **synapse,** a region at which the two nerve fibers come very close to each other but do not actually touch (Figure 21.1b). Some microbial toxins (e.g., botulin and tetanospasmin) act at the synapse, causing abnormal neurologic function.

For descriptive purposes, the nervous system is separated into two divisions, the central nervous system and the peripheral nervous sytem. The **central nervous system (CNS)** is composed of the brain and the spinal cord (Figure 21.2a). It is the integrative and control center of the nervous system. Its structures are completely enclosed by elements of the skeleton (the skull and the spinal column), except for the openings through which blood vessels, the spinal cord, and the cranial and peripheral nerves pass. These openings, and the very thin, porous bone underlying the cranial sinuses, are routes by which infective agents may enter.

The brain's anatomy is exceedingly complex. We will summarize the main areas (Figure 21.2b) and their functions. The **cerebrum** is concerned with sensory perception, voluntary motion and its coordination, and abstract thought. The **cerebellum** is responsible for motor coordination. The **medulla** and **pons,** which are directly connected to the spinal cord, regulate involuntary vital functions such as respiration, peristaltic action, and heartbeat.

The **spinal cord** receives and coordinates nerve impulses from the peripheral spinal nerves. It is composed of a central area of gray matter, surrounded by a layer of white matter arranged in structures called **horns.** At the point where the spinal nerves connect with the spinal cord, they divide into a dorsal and a ventral foot. The **dorsal root** contains sensory nerve fibers, while the **ventral root** consists of motor nerve fibers. Incoming signals typically enter the dorsal root, where they are referred to other nerve cells (Figure 21.2c).

The CNS is covered by connective-tissue membranes called **meninges.** The outer covering, the **dura mater,** is very tough. Beneath the dura mater lies the thinner **arachnoid membrane,** separated from the innermost layer, the **pia mater,** by the subarachnoid space. All three layers follow the shape of the nervous tissue and follow the cranial and spinal nerves as they exit from the skull and spinal column, forming a covering that is continuous with the nerve-sheath material.

The **peripheral nervous system,** which transmits sensory input to and from the CNS, is composed of sensory and motor neurons. **Sensory neurons** have the function of receiving and relaying sensations. Their dendrites lie in the skin, sense organs, thoracic and abdominal organs, and muscle (Figure 21.3). Their cell bodies lie in **ganglia** (enlarged bodies along nerves) at some distance from the dendrite end, and from there their axons continue to the CNS. Sensory neurons detect environmental stimuli such as heat and cold. **Motor neurons,** which have the function of controlling muscle contractions, usually have their dendrites and cell bodies within the central nervous system. Their long axons extend to individual muscle fibers.

A **nerve** serving a particular area consists of the processes of a large number of both sensory and motor neurons. A nerve contains no cell bodies. The neuron processes are arranged in bundles covered with a protective myelin sheath. On the one hand, the nerve sheath is a major factor in protecting the nerve, since it helps keep bloodborne pathogens out. On the other hand, the sheath also renders the nerve tissue relatively impermeable to blood components such as antibodies, and excludes many antimicrobial drugs. As a result, should a pathogen establish itself within a peripheral nerve, it

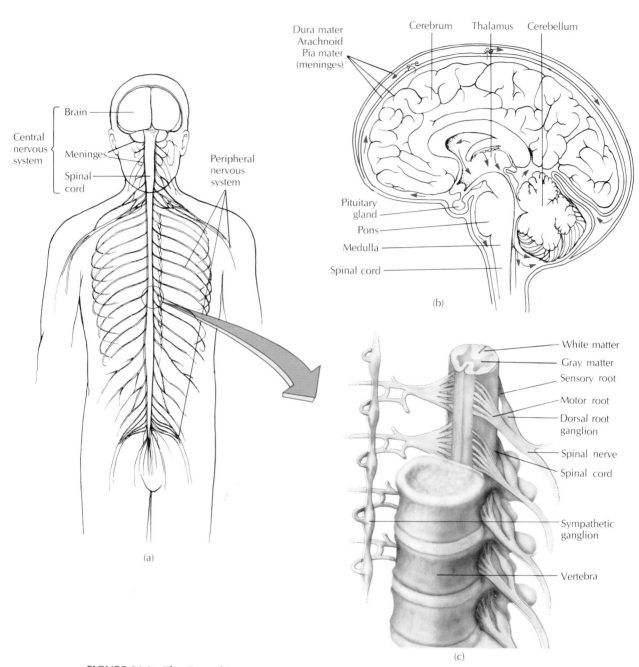

FIGURE 21.2 The Central Nervous System
(a) The central nervous system consists of the brain and spinal cord. It is covered and protected by three membranous layers, the dura mater, arachnoid, and pia mater. (b) The three major areas of the brain are the cerebrum, the cerebellum, and the medulla. Cerebrospinal fluid, originating in the choroid plexus, circulates around the brain and its coverings (arrows).

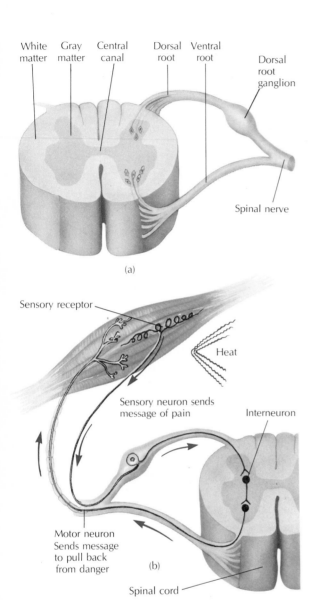

White matter | Gray matter | Central canal | Dorsal root | Ventral root | Dorsal root ganglion

Spinal nerve

(a)

Sensory receptor

Heat

Sensory neuron sends message of pain

Interneuron

Motor neuron Sends message to pull back from danger

(b)

Spinal cord

FIGURE 21.3 Relaying Sensory Messages
(a) Incoming nerve impulses enter the spinal cord along the dorsal root and are transmitted to other neurons. The messages are relayed to appropriate brain centers and/or to outgoing neurons contained in the ventral root. (b) The spinal cord serves as a reflex center. When sensory neurons detect environmental stimuli, such as heat and pain, the message is transmitted from one neuron to another within the gray matter. The message leaves the cord through a motor neuron and stimulates the appropriate muscle fibers.

may then be effectively shielded from many host defenses.

PHYSIOLOGY OF THE NERVOUS SYSTEM

The **cerebrospinal fluid (CSF)** is a clear, watery secretion that fills the spaces between the brain and spinal cord and the skeletal enclosures, and circulates slowly within the hollow areas of the brain. CSF is formed by filtration of fluid from blood plasma through capillary walls. Blood cells are not normally found in CSF. The total quantity of CSF is regulated to give a constant, slight pressure. Thus the CSF functions as a cushion for the brain tissue; to some extent, it also mediates nutrient exchange. Any increase in the intracranial-fluid pressure is abnormal. Examination of the CSF provides a means of determining the presence of infectious agents in the central nervous system.

The brain requires a large blood circulation. It demands a major portion of the body's nutrient and oxygen supply, which are delivered by capillary networks in both gray and white matter. There is no lymphatic circulation within the brain. An important physiologic feature of brain tissue is the so-called blood–brain barrier. As we noted previously, this is an anatomic and physiologic trait of brain tissue, which limits the diffusion of many chemicals from the blood into the brain tissue. Nutrients and oxygen are able to diffuse out of the capillaries and into the brain cells at the normal rate. However, many drugs are prevented from diffusing efficiently from the blood into the brain tissue or the CSF. The concentration of an antibiotic in normal brain tissue may be only 1/20 of its concentration in the serum, making it difficult to achieve therapeutic drug levels. Fortunately, inflammation increases the permeability of the brain tissue, thereby reducing the barrier effect. This allows drugs increased access during infection, when they are most needed. Even so, extraordinarily high dosages of antimicrobials are usually required to

BOX 21.1
Do Brain Cells Replace Themselves?

Science is full of dogmas — unchallengable, well-known "facts." One of these is the "fact" that no more brain cells are formed by higher animals after infancy. This presumption of the inability of brain tissue to regenerate is based on many, many observations of the permanent devastation that a brain injury or a brain infection leaves. This chapter is full of descriptions of diseases that, at present, kill or leave the recovered person severely handicapped.

Contrary to this dogma, however, a group of researchers studying learning behavior in birds have established that there is substantial production of new neurons in normal adult birds — in some brain areas, $1\frac{1}{2}$% of the neurons are replaced with new ones every day. The new neurons are formed in many different brain areas. These results, which go far beyond some less detailed suggestions of new neuron formation in fish and in rodents, have caused much questioning of the conventional wisdom. Of course, we can't automatically assume that this discovery means that human brains can also regenerate. Yet for the first time it seems possible that just a few years from now we might be investigating ways to promote human brain repair, and that the diseases we discuss in this chapter may have more hopeful prognoses.

control CNS infections. The blood–brain barrier also partially protects against microbial invasion from the blood.

The nervous system is also protected by phagocytic activity. Brain tissue contains cells called **microglia,** which are actively phagocytic. In the case of a CNS infection, circulating phagocytic white cells enter the cerebrospinal fluid. The presence of white blood cells in a CSF sample indicates CNS inflammation and infection.

There are several routes of infection (Figure 21.4). One primary route to the central nervous system is by extension of an upper-respiratory-tract infection or a skin infection in the head area. It is possible for some agents to penetrate through the cranial sinuses, middle ear, or mastoid bone, especially in neonates. Agents in the oropharynx, skin, lungs, or GI tract may also become bloodborne and escape across the blood–brain barrier to the CNS. Other local infections become established first in peripheral nerve endings, then migrate to the CNS under the nerve sheath. Trauma such as a skull fracture or surgery is yet another pathway for invasion.

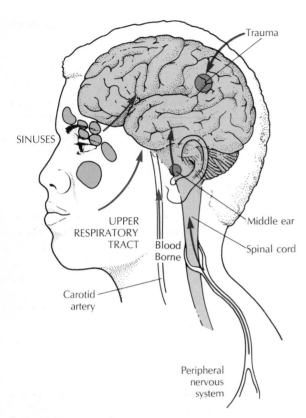

FIGURE 21.4 Routes of Infection
Infectious agents may reach the brain from the upper respiratory tract, via the blood, via the middle ear, by moving along a peripheral nerve, and through trauma.

BACTERIAL INFECTIONS

Bacteria usually infect the meninges, a condition called **meningitis;** less commonly, they attack the underlying brain tissue, causing **encephalitis.** Some of the meningitis agents are summarized in Table 21.1. Bloodborne organisms occasionally establish abscesses beneath the meninges, or in brain tissue. *Staphylococcus aureus* is the most common abscess-former, but other agents, including anaerobes, are seen. CNS infections, as previously noted, are difficult to

treat because of the diffusion barrier. Speed is of the essence in diagnosis and antimicrobial therapy.

HEMOPHILUS MENINGITIS

The most common bacterial meningitis agent is *Hemophilus influenzae* type B, a tiny, fastidious, Gram-negative coccobacillus with rounded ends. It causes a wide range of infections, especially in children.

THE ORGANISM. *H. influenzae* is facultatively anaerobic and requires two blood components, the X-factor (hematin) and the V-factor (the coenzyme NAD [nicotinamide adenine dinucleotide]). These are provided in a chocolate-agar medium to which the coenzyme NAD has been added.

Pathogenic *Hemophilus* strains have an antigenic polysaccharide capsule. There are six capsular types, designated A through F, but only Type B is commonly pathogenic (Figure 21.5). Isolates may be typed by reaction with specific typing serum. The polysaccharide capsule swells in the presence of the matched serum (the Quellung test). Direct microscopic tests on cerebrospinal fluid, latex-bead slide tests, and countercurrent immunoelectrophoresis are techniques to identify capsular antigens.

Encapsulated *H. influenzae* type B is a versatile pathogen and an obligate, though extracellular, parasite. However, its unencapsulated strains are normal in the oral flora and rarely cause disease. *H. influenzae* got its name because it caused a great deal of lethal secondary pneumonia during the 1918 influenza epidemic. The name is misleading, since the bacterium, although it does cause a variety of diseases (Table 21.2), does not cause influenza.

THE DISEASE. Meningitis often develops after the organism has traveled from a middle-ear infection to the meninges. Often there has been a recent viral respiratory infection to weaken the defenses. One early sign of meningitis is a painfully stiff neck. The symptoms include high fe-

TABLE 21.1 Bacterial meningitis agents

Disease	Number of Cases Per Year	Percent of Total Cases	Case Fatality Rate (%)
Hemophilus influenzae	1885	46	7.1
Neisseria meningitidis*	1095	27	13.5
Streptococcus pneumoniae	456	11	28.2
Group B Streptococcus	130	3	22.4
Listeria monocytogenes	68	2	29.5
Other	235	6	36.6
Unknown	212	5	16.7
Total bacterial meningitis	4081	100	13.6
Meningococcemia	289		25.1

*Excludes cases of meningococcemia alone.
Source: *Morbidity and Mortality Weekly Reports* 1979; 28: 277.

ver and convulsions, which may lead to collapse and death. The disease may develop rapidly. In a typical case, a child died less than eight hours after his temperature started to climb. Recovering children face an increased chance of learning disability. Mortality is high if the disease is untreated or treated too late.

THE TREATMENT.   The key factor in treatment is speed; antibiotics must be started without delay. Ampicillin was formerly the drug of choice, but because at least 15% of *H. influenzae* strains are now resistant, chloramphenicol–ampicillin combination therapy is often started pending a laboratory report on drug susceptibilities. Sulfa drugs are also sometimes used.

EPIDEMIOLOGY.   *H. influenzae* is the most important agent of meningitis in children under five, but it is not commonly pathogenic in adults. Nonpathogenic *H. influenzae* is widely distributed in the normal pharyngeal flora of both adults and children. The pathogenic type B is carried by 5 to 7% of persons. Most meningitis cases appear to be endogenous. We all are frequently exposed to the agent from aerosols, direct contact, and contaminated objects. A vaccine containing *H. influenzae* Type B purified polysaccharide antigen was licensed in 1985. Its

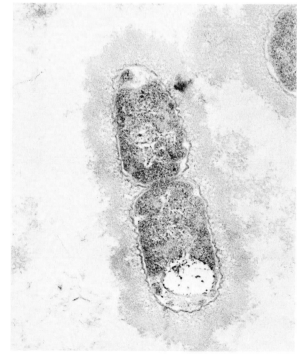

FIGURE 21.5   *Hemophilus influenzae*
H. influenzae type B is the most common cause of infant meningitis. It also causes ear infections and other diseases. In this electron micrograph the thick layer of capsular material around the cells has been emphasized by preparation with antibody to the type B capsular polysaccharide. (×42,630. Courtesy of F.L.A. Buckmire, Medical College of Wisconsin.)

**TABLE 21.2 Some diseases caused by
*Hemophilus influenzae***

Arthritis
Epiglottitis
Laryngitis
Otitis media
Meningitis
Osteomyelitis
Pericarditis
Cellulitis
Pneumonia
Sinusitis
Chronic conjunctivitis

use is recommended for all children at 24 months of age. Children at high risk may be immunized earlier, at 18 months of age.

Contact with capsular antigens stimulates a protective immunity. Thus the normal early childhood exposure to the relatively small number of antigenic types is sufficiently immunizing so that *Hemophilus* infections are not common past early childhood.

MENINGOCOCCAL MENINGITIS

The meningococcus *Neisseria meningitidis* causes the "classic" form of meningitis, which becomes the most prevalent form from about three years of age on.

THE ORGANISM.   *N. meningitidis* is a fastidious, Gram-negative diplococcus. Its cultural requirements and biochemical behavior are quite similar to those of its close relative the gonococcus. It may be isolated on Thayer–Martin or other chocolate agars in a 5% carbon dioxide atmosphere. Species identification depends heavily on sugar-fermentation tests. Pili assist the organism in adhering to target tissue (Figure 21.6). The organism has a polysaccharide capsule that

is a major factor in pathogenesis because it inhibits phagocytosis. Serologic typing of the capsular substance allows identification of at least seven major lettered groups (A, B, C, D, X, Y, and Z) and a variety of numbered strains. The organism produces an endotoxin that provokes hemorrhages, intravascular coagulation (random clotting in small blood vessels), and systemic shock.

THE DISEASE.   Infection starts with oropharyngeal colonization. Within 7 to 10 days after colonization occurs, immunocompetent individuals will have formed specific anticapsular antibodies. These antibodies in serum prevent further invasion, leading to a healthy carrier state. However, if the immune response is inadequate, bacteremia develops. The organisms may then multiply in the blood to cause **meningococcemia,** or may invade the meninges to cause meningitis. In the worst cases, both conditions prevail. Development of the disease is often explosively rapid. Unless treatment is both prompt and appropriate, the risk of mortality is high.

The primary signs of meningococcal meningitis are a severe headache, an immovably stiff neck, and many polymorphonuclear leukocytes in the cerebrospinal fluid. The organism itself cannot always be seen or successfully cultured from the CSF. **Petechiae** (small skin hemorrhages) are often present. In the meningococcemic form, large hemorrhagic lesions develop from which the causative organism can be demonstrated in a direct Gram stain. In both forms, circulatory failure due to endotoxin shock may be rapid and often irreversible.

THE TREATMENT.   Penicillin, ampicillin, or penicillin plus sulfonamide are the usual drugs of choice, given in massive, intravenous doses. Chloramphenicol may be substituted in individuals allergic to penicillin-class drugs. Supportive measures are often necessary to sustain life.

EPIDEMIOLOGY.   Meningococcal meningitis occurs principally in late winter and early spring.

Immunity develops to types A and C but not to type B. Vaccines for types A and C have been licensed for use in high-risk populations. Type C vaccine is given routinely to U.S. military recruits. Vaccine may also be of benefit during epidemics, for travelers, and for exposed contacts of cases. Currently, around 3000 cases of meningococcal infection are reported annually in the United States.

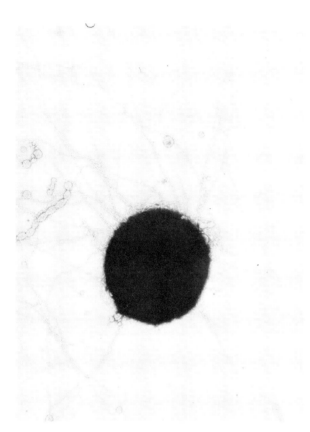

FIGURE 21.6 *Neisseria meningitidis*
This scanning electron micrograph shows the coccus with many long, hollow pili. The pathogen uses these pili to adhere to host tissue. (×36,387. From McGee, Zell A. et al., *J. Infect. Dis.* (1979) 24:194, with permission from the authors and the *Journal of Infectious Diseases.*)

Military recruits and school groups have a high risk of epidemic outbreaks, largely because of crowded living conditions. Oral rifampin may be given for protection to exposed contacts of an active case. The healthy carrier is the primary reservoir, and the agent is spread by droplets and direct contact. Since the meningococcus is readily inactivated outside the body, indirect spread is unlikely. The carrier rate depends on the population surveyed and the season.

LEPROSY

Leprosy is a slowly progressive peripheral-nervous-system infection caused by the acid-fast bacillus *Mycobacterium leprae,* which initially involves the peripheral nerve endings in the skin. The bacilli multiply within Schwann cells — the cells that produce myelin sheath material for peripheral nerve cells. The area supplied by the affected nerve loses sensation and becomes much swollen, and hands and feet are often damaged and disfigured. The disease is concentrated mainly in tropical and subtropical areas (Figure 21.7).

Leprosy occurs in two forms. In **tuberculoid leprosy,** the milder form of the disease, host cell-mediated immune defenses, although weakened, are able to restrict the spread of the lesion. The leprosy bacilli are prevented from penetrating the central nervous system. The disease is self-limited, although there may be the severe local damage just described. A positive skin test with *M. leprae* antigen (**lepromin**) is diagnostic.

Lepromatous leprosy, the more severe form of the disease, develops if cell-mediated immunity (CMI) is highly inadequate. Multiple spreading lesions form; eventually the entire skin surface may be affected. Bacteremia develops, and the organism invades internal organs. Blindness and severe deformity result; death is due to secondary infection or other conditions. The lepromin skin test may even be negative if CMI is sufficiently depressed.

Leprosy remains a mysterious disease in several ways. The agent has still not been cul-

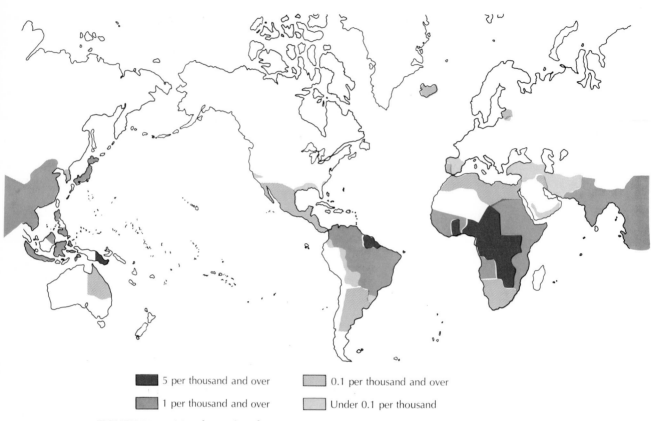

| 5 per thousand and over | 0.1 per thousand and over |
| 1 per thousand and over | Under 0.1 per thousand |

FIGURE 21.7 *Mycobacterium leprae*
Worldwide distribution of leprosy. There are at least 12 million cases of leprosy
worldwide, concentrated in tropical and subtropical countries.

tured in bacteriologic or cell culture media; thus little is known about it. However, the bacillus can be cultivated in certain live animals, including the armadillo and one species of monkey. Furthermore, although it is now clear that leprosy is not particularly communicable, its exact means of transmission is not known. The current working hypothesis is that it is spread by nasal secretions. There is a suspicion that armadillos may be a natural reservoir of the microorganism.

In treating leprosy, dapsone is the antimicrobial drug of choice. Rifampin is also becoming well accepted. Three to six months of

therapy renders active patients convalescent and noninfective; prolonged isolation is no longer required. Therapy usually must be continued for many years or for life. Resistance to the available drugs is cropping up, however, spurring the search for new means of treatment.

Although there are only about 5000 cases of leprosy in the United States, there are about 12 million cases worldwide. This includes 3.5 million in India alone, which uses 50 tons of dapsone a year. In many less-developed countries, few patients receive any kind of regular treatment. Preliminary trials of a killed bacterial vaccine began in 1982.

TETANUS

Much feared, but entirely preventable, **tetanus** is very uncommon now. However, because the causative bacterial spores are found everywhere, it remains a constant threat.

THE ORGANISM. Tetanus is a neuromuscular disease caused by *Clostridium tetani*, a Gram-positive, strictly anaerobic spore-forming bacillus that is difficult to isolate. It is unable to ferment sugars or produce indole. It has pronounced swarming motility. Its major pathogenic factor is the neurotoxin **tetanospasmin.** This extremely potent exotoxin causes muscles to go into prolonged rigid contractions or spasms. This happens because tetanospasmin blocks the neurologic inhibitory impulses that normally act to prevent random muscular contractions and terminate completed actions. Muscles innervated by toxin-saturated nerve tissue contract uncontrollably.

THE DISEASE. *C. tetani* spores are introduced into a wound with soil or fecal contamination. However, they germinate only under very anaerobic conditions. Thus tetanus most characteristically follows a puncture wound (Figure 21.8). This wound may be a very small one, such as a thorn prick, and it may completely heal without evidence of infection. However, the organism remains within the wound, multiplying and producing toxin during the incubation period, which is highly variable.

Clinical tetanus occurs after sufficient growth of *C. tetani* gives rise to damaging levels of toxin. The toxin is absorbed by peripheral nerves. The nerves that control the jaw muscles are often affected first; the spasm called lockjaw may be the first clear symptom. As toxin levels increase, other voluntary muscles become progressively affected, and death may occur, due to spasms of the thoracic muscles causing respiratory failure and exhaustion.

THE TREATMENT. Treatment begins with certain steps taken to prevent the situation from worsening. Tetanus vaccine should be administered immediately to stimulate renewed production of antibodies if the patient has a history of previous tetanus vaccination. Antitoxin of human origin may be given to any patient who has never been immunized. Antimicrobials, usually penicillin, are given to kill the vegetative bacteria in the wound, interrupting the toxin production.

However, toxin already present and fixed to nerve cells will not be inactivated by these measures. It will continue to have its effect until it has been gradually metabolized. Other therapeutic measures are required to support the patient through this period, which may be two to six weeks long. Muscle relaxants, sedatives, and intensive artificial supports such as respirators may be required. In special treatment centers, the mortality rate may be held to 10%, but overall it remains around 60%. Death is most likely in persons over 50.

EPIDEMIOLOGY AND CONTROL MEASURES. The natural habitat of *C. tetani* is the intestinal tract of warm-blooded animals. It is spread by feces to soil, where the spores remain and may be transferred to almost any surface or object. Little can be done to avoid contact with the spores except to lead an indoor, sedentary life. In the United States, the typical tetanus patient is a male over 50, who develops the disease during the summer, after getting an accidental puncture wound during outdoor farming, gardening, or construction activities. He is susceptible because many years have elapsed since his childhood vaccinations.

Control is based on vaccination. The toxin is a potent antigen. It is converted to a nontoxic but still antigenic toxoid for vaccine use. The toxoid is given to infants in the DPT (diphtheria–pertussis–tetanus) series, and to adults or accident victims in the form of single injections. Tetanus toxoid stimulates a protective immunity lasting from 10 to 20 years in normal individuals. Toxoid booster shots used to be given routinely to persons receiving care for wounds, but it became clear that hypersensitivity could develop. Now toxoid is generally included in wound treatment only if more than 10 years have passed since the last toxoid administration.

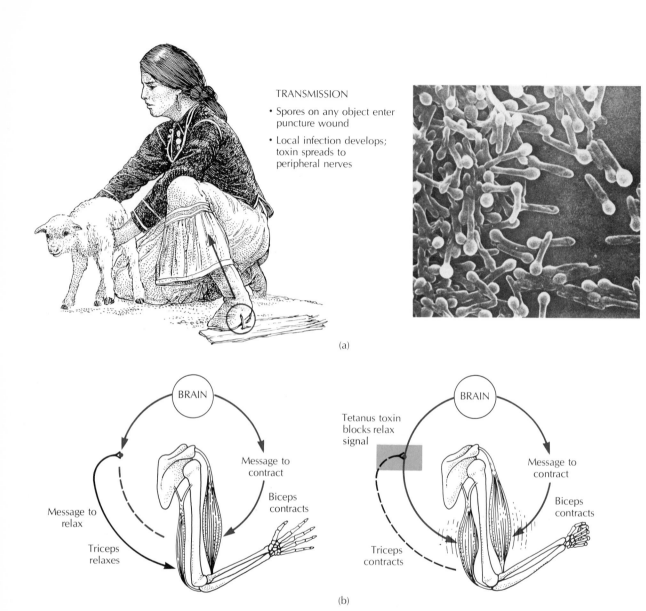

TRANSMISSION
- Spores on any object enter puncture wound
- Local infection develops; toxin spreads to peripheral nerves

(a)

BRAIN

Message to contract

Biceps contracts

Message to relax

Triceps relaxes

Tetanus toxin blocks relax signal

BRAIN

Message to contract

Biceps contracts

Triceps contracts

(b)

FIGURE 21.8 *Clostridium tetani*
(a) This scanning electron micrograph shows the characteristic drumstick shape of the organisms, which are swollen at one end of the cell by a terminal spore. (×2730.) (b) In a healthy person, a nerve impulse from the brain sends a message to contract (color) to the biceps. Other muscles in the triceps are prevented from contracting by inhibiting nerves (black lines). Tetanus toxin, however, blocks the relaxation pathway, and muscles contract and remain in a state of tension.

FUNGAL INFECTIONS

Fungi occasionally form brain lesions. The most notable infection is **cryptococcosis,** a disease caused by the yeast *Cryptococcus neoformans* (Figure 21.9). The fungus multiplies in bird feces worldwide, most significantly (for city-dwellers) in the droppings of the common pigeon. Transmission probably involves inhaling dust from dried bird droppings. The initial infection occurs in the lung. The vast majority of infections remain trivial and inapparent; in this respect, it

FIGURE 21.9 *Cryptococcus neoformans*
A negative, stained preparation uses India ink to create an opaque background on the microscopic slide. The yeastlike cells of *C. neoformans* show up surrounded by a clear zone formed by their capsules (×1744. Courtesy of Norman L. Goodman, University of Kentucky Medical School (published in *Diagnostic Medicine*, February 1985.))

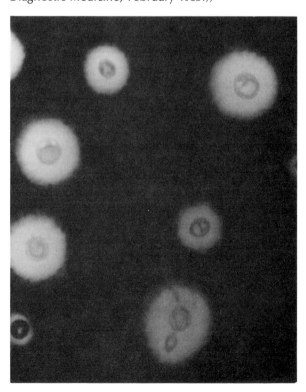

closely resembles coccidioidomycosis and histoplasmosis (Chapter 18).

Dissemination from the lungs to the brain is rare, and is always associated with immunodeficiency. Deep lesions occur in both gray and white matter. Exudates accumulate on the brain surface; meningitis is also present. Amphotericin B may be used in treatment, but the prognosis in such cases is very poor. Diagnosis is based on identification of the thick polysaccharide capsule of the yeast or direct examination of spinal fluid, culture of the organism from cerebrospinal fluid, and use of latex agglutination tests to identify fungal antigen in body fluids.

VIRAL INFECTIONS

Nerve tissue, in the lab, is a good medium for viral growth, but fortunately viruses are usually excluded from the brain by its defenses. We will examine some of the uncommon but extremely hazardous diseases in which viruses invade the central nervous system.

VIRAL MENINGITIS

From 30 to 40% of meningitis cases are caused by viruses. The term **aseptic meningitis** is applied to such cases, because nothing is found on bacteriologic examination of the cerebrospinal-fluid sample. The term is a misnomer, however, because an infective agent is indeed present. It is often the mumps virus, coxsackieviruses types A and B, or echovirus.

The disease usually has a gradual onset, a short duration, and is relatively benign with few late complications. It is most common in children and young adults. The agents enter by their typical respiratory or intestinal routes, and meningitis follows viremia. Treatment is supportive, and recovery is usually complete. There are almost 10,000 cases reported per year in the United States, but fatalities are rare.

ARTHROPOD-BORNE ENCEPHALITIS

THE AGENTS.   More than 350 viruses are classified as **arboviruses — *ar*thropod-*bo*rne viruses.** An arbovirus can multiply in both arthropod and vertebrate tissues. The arthropod is both host and vector for the virus, which it transmits to vertebrates by biting. The arboviruses are RNA viruses, currently placed in the togavirus group. Five of these viruses are important in the North American continental area (Table 21.3). They are encephalitis viruses, and are spread by mosquito bites.

THE DISEASES.   The arthropod bite introduces virus into capillaries, allowing the agent to migrate; it eventually enters neurons and supporting tissues in the CNS. Encephalitis ensues (Figure 21.10). Early signs are headache and fever; then sensory disturbances, confusion, muscular weakness, and convulsions may follow. Swelling, brain hemorrhage, and neuron death may occur. The severity of neuron loss and quantity of destroyed tissue vary. The nerve cells that die are not replaced, so permanent mental impairment typically follows severe cases. Mortality rates, especially for Eastern Equine Encephalitis (EEE), are high.

THE TREATMENT.   There are no specific antiviral drugs for this group of RNA viruses. Supportive therapy is all that is possible.

EPIDEMIOLOGY AND CONTROL MEASURES.   Arboviruses are transmitted by mosquitos among horses, wild birds, possibly snakes, and humans. Roughly 200 cases per year are recorded in non–epidemic years. In 1975, a year in which there was an epidemic of Saint Louis Encephalitis, there were about 1500 cases.

Mosquito control is one very important public-health measure for preventing viral encephalitis. Seasonal flooding or heavy rains, and the accumulation of water-holding receptacles such as discarded tires or metal cans around dwellings, increase mosquito populations. Removing breeding areas and taking precautions to minimize contact with mosquitos are wise preventive measures.

TABLE 21.3 Characteristics of arthropod-borne encephalitis in North America		
Disease	**Geographical Range**	**Arthropod Vector**
Eastern equine encephalitis	New Hampshire to Texas, Eastern United States	*Culiseta, Aedes, Culex*
Western equine encephalitis	All states west of Mississippi River	*Culex*
Venezuelan equine encephalitis	Northern South America, Central America, Florida, and Texas	*Psorophora, Aedes*
Saint Louis encephalitis	West Central and Southern United States, Panama, Trinidad	*Culex*
California encephalitis	North Central and Southwest United States	*Culex, Aedes*

Source: Youmans GP, et al. *Biologic and clinical bases of infectious diseases, 1980.*

FIGURE 21.10 Eastern Equine Encephalitis virus
These viruses, seen here in brain tissue, cause the most virulent form of arthropod-borne encephalitis. The disease is transmitted by mosquitos from infected horses to human beings.

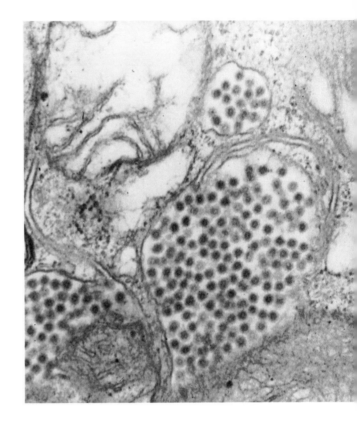

Since equines are major hosts, horses, mules, and donkeys in endemic areas should be vaccinated. Proof of vaccination for Venezuelan equine encephalitis (VEE), Eastern equine encephalitis (EEE), and/or Western equine encephalitis (WEE) may be required in order to ship horses from one state to another.

Because of the comparative rarity of the disease, human vaccination is not routine. Killed-virus vaccines are available. Laboratory workers who handle arbovirus may receive appropriate experimental vaccines. The vaccine for Japanese B encephalitis, a very severe disease common in the Far East, is manufactured in Japan and widely used to control epidemics in India.

TABLE 21.3 Characteristics of arthropod-borne encephalitis in North America (continued)

Animal Host	Age Group	Mortality Percentage	Aftereffects
Birds, horses	Children	60–75	90% of survivors
Birds, horses, snakes (?)	Infants adults < 50	5–15	Moderate
Horses, rodents	All ages	0.6	Low
Birds	Adults < 50	2–11	Low
Rabbits, squirrels, field mice	Children	1	Low

RABIES

Only two individuals have ever survived a documented case of rabies once symptoms developed. The public has a great horror of rabies, a horror that is useful in that it motivates essentially all exposed persons (several thousand a year) to seek prophylactic treatment. Antirabies treatment is clearly effective, as shown by the fact that in the United States there are usually fewer than five human cases diagnosed per year.

THE AGENT.   The rabies virus is a large, complex, envelope-bearing RNA virus, readily inactivated by sunlight, drying, or chemical agents. It can be cultivated in cell culture and duck eggs. Laboratory diagnosis is based on direct microscopic study of brain tissue from rabid animals or deceased patients. In brain tissue, the virus forms characteristic inclusion bodies called **Negri bodies** (refer to Figure 10.9) These are unequivocally identified by a fluorescent rabies antibody test on brain tissue.

THE DISEASE.   Rabies is contracted as the result of the bite of a rabid animal (Figure 21.11), which introduces saliva containing infective virus particles. The virus rapidly adsorbs to the nearest peripheral nerve endings. It enters the nerves and moves by unknown means toward the CNS. The length of the incubation period appears to depend on the location of the bite, the density of nerve endings in the vicinity of the bite, and the distance from the bite area to the CNS. Thus bites on the face, where the tissue is richly supplied with nerve endings in close proximity to the brain, are considered to be the riskiest.

Rabies virus also multiplies in the salivary gland, kidney, pancreas, and the cornea. However, the worst manifestations of the disease are caused by destruction of gray matter. Lesions develop in the cerebellum, medulla, and other areas of the brain. Once viruses are actively replicating in the CNS, **hydrophobia,** a painful spasm caused by an uncoordinated swallowing reflex, develops. Later the victim experiences paralysis and coma. In some cases, extreme excitability develops. When this does not happen, rabies may not be suspected or diagnosed until after the victim has died and an autopsy has been performed. Nerve destruction is progressive and irreversible.

THE TREATMENT.   Once rabies symptoms have developed, very little can be done. In 1970, a young boy survived rabies with prolonged use of advanced life support technology, but this treatment has not been successfully repeated. A number of persons have been maintained for some weeks, but have eventually succumbed to the disease.

EPIDEMIOLOGY AND CONTROL MEASURES.   Both domestic pets and wild animals can serve as reservoirs and transmit the disease to human beings. Where pet vaccination is widely used (the United States, Western Europe, and Canada), wild animals have replaced dogs as the most significant source of rabies. The skunk and fox account for the majority of documented animal cases (about 3000 a year) in the United States. Rabies is also seen in all other mammals (but is very rare in rodents). Transmission has also been documented by inhalation of dust in caves inhabited by rabid bat colonies, and in two cases by corneal transplants. Since saliva is infective, persons who have had direct, unprotected contact with a rabies patient may be considered exposed, especially if cuts, scratches, or lesions were present on their skin or mucous membranes.

Prophylactic immunization is a key control measure. Exposed individuals are given rabies vaccine and rabies immune globulin. No treatment is required unless the exposed person's skin is broken or a mucosal surface has been contaminated by the animal or human patient's saliva. The vaccine is produced in human diploid cell culture and has minimal side effects. Veterinarians and animal control officers usually receive prophylactic vaccination. More information on postexposure prophylaxis is given in Table 21.4.

Some preventive measures can be used by anyone. Immunize all pets, especially pet skunks and raccoons. Avoid wild animals that,

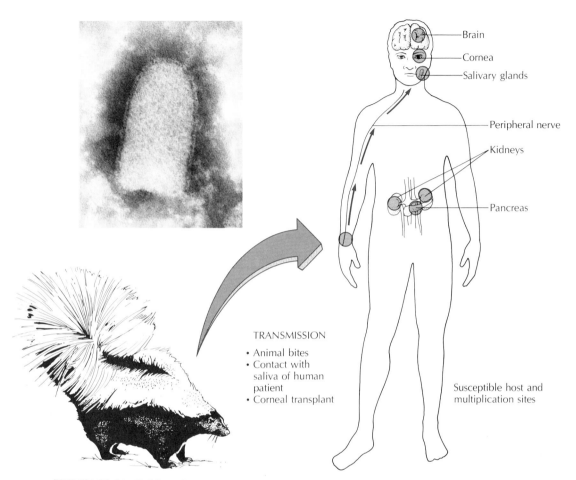

FIGURE 21.11 Rabies Virus
Rabies is contracted through the bite of an infected animal. Virus contained in the animal's saliva rapidly enters the victim's nerves and moves slowly toward the CNS. Viruses may multiply in the salivary glands, kidney, pancreas, and cornea and eventually cause widespread destruction of gray matter and death. The micrograph shows the typical bullet shape of a single rabies virus. (×199,013. Courtesy of Alyne K. Harrison, Viral Pathology Center, Centers for Disease Control, Atlanta.)

because they are ill and sluggish, seem unusually "friendly," putting out of your mind all those cute cartoon creatures. If you are bitten, be sure to wash the wound immediately and thoroughly, and swab it with soap and water for at least 20 minutes. This first-aid measure gives a good chance of inactivating and removing the virus before it enters the nerve tissue. Seek medical attention promptly.

POLIOMYELITIS

Poliomyelitis, or polio, is a viral infection of the motor neurons. It is one of those diseases that we think we've beaten, but which can always crop up again, because the virus is still around.

THE AGENT. The polioviruses are small RNA viruses belonging to the enterovirus group. There

	TABLE 21.4 Postexposure prophylaxis in rabies	
Nature of Exposure	**Status of Animal at Time of Exposure**	**Recommended Treatment**
Contact, no lesions; indirect contact	Rabid	None
Licks of skin; scratches or abrasions; minor bites (covered areas or arms, legs, trunk)	Suspected*	Start vaccine. Stop if animal is healthy for five days or proved negative.
	Rabid; animal unavailable for observation	Start serum and vaccine[†]
Licks of mucosa; major bites (multiple or on face, head, neck, or finger)	Suspect or unavailable	Start serum and vaccine. Stop if animal remains healthy.
	Rabid	Complete course of vaccination.

*All unprovoked bites are suspect until animal is proved negative by fluorescent-antibody test.
[†]Health professionals who have had inadvertent, unprotected contact with a rabid patient, and had mucosal or skin lesions at the time of exposure, should receive this treatment.

are three types, I, II, and III, present in nature both as wild-type pathogenic strains and as attenuated vaccine strains. The viruses can be propagated in monkeys and in cell cultures of tissues from monkeys and others. They are readily identified antigenically.

THE DISEASE.   If a person has some degree of immunity to polio, as with a baby nursing from a mother immune to polio, he or she may have a subclinical infection. In the person with no immunity to poliovirus, subclinical cases also often occur, but severe paralytic polio becomes more common.

Infection begins when the virus is ingested or possibly inhaled (Figure 21.12). It moves from its original site via the lymphoid tissue to the blood, and then across the capillary wall to the neurons of the CNS. In particular, the virus attacks the motor neurons of the spinal cord and brain stem. Meningitis also develops; this is the predominant feature of nonparalytic po-

lio. In paralytic forms, the muscles activated by the destroyed neurons no longer work. In fatal cases, involuntary functions of the medulla are lost. Paralytic effects, once developed, are irreversible.

THE TREATMENT.   Therapy is nonspecific, designed to minimize chances of secondary infection, and to prepare for possible rehabilitation.

EPIDEMIOLOGY AND CONTROL MEASURES.   The human gastrointestinal tract is a natural reservoir for poliovirus. Both wild and vaccine viruses may be transmitted, primarily by the oral–fecal route. Contaminated swimming areas and shellfish have been identified as particularly important sources of the virus. Before vaccination was introduced, the peak incidence of polio was during the hot summer months. Vaccination is the primary control measure in the United States, responsible for a decline of paralytic polio from more than 18,000 cases in 1954 to only 8 in

1982. At present, polio is still prevalent in countries with neither vaccination nor adequate sanitation. However, even in these countries, very mild infant infection is the rule, and paralytic cases the exception. In developed countries, 70% or more of cases occur in nonvaccinated older children and adults and are almost always paralytic. Both inactivated (Salk) polio vaccine (IPV)

FIGURE 21.12 Poliovirus
This small, polyhedral RNA virus may be ingested or inhaled by a susceptible host. The human gastrointestinal tract is a natural reservoir. From the mucosa of the small intestine, the virus moves into the lymph nodes and then across capillary walls to the neurons of the CNS. This virus preferentially attacks the cell bodies of motor neurons. By destroying them, the virus makes it impossible for nerve impulses to activate the affected muscle, and paralysis results.

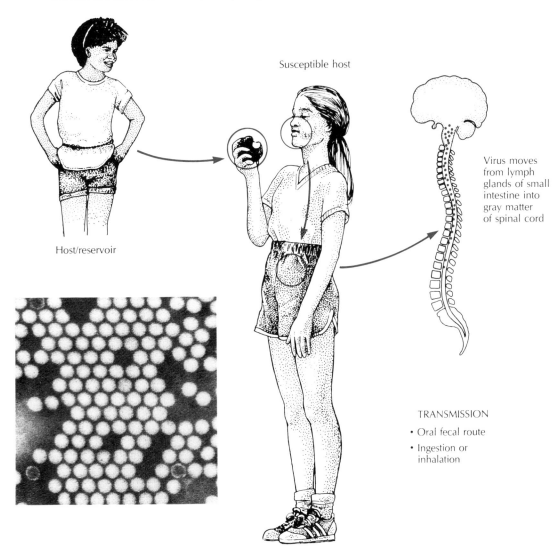

Susceptible host

Host/reservoir

Virus moves from lymph glands of small intestine into gray matter of spinal cord

TRANSMISSION
• Oral fecal route
• Ingestion or inhalation

and oral attenuated live-virus (Sabin) vaccine (OPV) are available. OPV is generally recommended in the United States. All normal infants receive a series of doses. The breast-fed infant may not respond fully to the earliest doses because maternal antibody may inactivate the vaccine

BOX 21.2
An International Polio Outbreak

Since the widespread adoption of polio vaccination, outbreaks of polio have been rare, because outbreaks require many susceptible individuals in a dense cluster. These conditions were fulfilled recently, resulting in a significant outbreak among members of the Amish community, a group that for religious reasons has traditionally refused vaccination.

The viral agent was a type I strain that was originally spotted in Kuwait in 1977. By some means, the virus was transported to the Netherlands. The outbreak started there in April 1978. The first victim was a 14-year-old girl from a village near Utrecht. She attended a large regional school attended by many Amish children from neighboring communities. An outbreak rapidly developed, all cases occurring in unvaccinated persons who were members of the Amish communities.

That summer, Amish visitors from the Netherlands traveled to Canada. Six Canadian Amish, in British Columbia, Alberta, and Ontario, developed polio, with the same type I wild-type viral strain. Each of these individuals had had contact with the visitors from the Netherlands. As of October 12, 1978, there had been 110 cases, 80 of which were paralytic.

At that time, United States health officials were saying cautiously that there was no evidence that the virus had spread to the United States. However, in late summer 1978, an Amish family from Ontario moved to a town in Pennsylvania. In January 1979, the first American polio case appeared in that town. After a three-month lag, other cases, totaling 16, appeared. These occurred in four states and Canada, and all except two were Amish persons.

An immunization program was quickly developed to serve the Amish communities in the United States and Canada, and the majority of nonvaccinated persons accepted the vaccine. Note that the persons who transported the virus from one area to another were all perfectly healthy, reflecting the fact that most persons experience polio as an asymptomatic infection. At the most only 1 in 100 develop paralytic disease. Extrapolating from these figures, the roughly 100 paralytic cases would suggest the occurrence of up to 10,000 asymptomatic cases over the same period. The long lag between the summer 1978 and spring 1979 phases of the epidemic can be accounted for by a series of asymptomatic cases. This story illustrates forcefully why there is a continuing need for polio vaccination.

virus before it can stimulate the child's immune system. Thus it is important that all children complete the series later in childhood.

A small risk is associated with OPV. The live virus is an attenuated mutant strain with greatly reduced affinity for nerve cells. It does multiply in the recipient's intestine, and is shed in the stool for a period. Immunodeficient persons may acquire full-blown paralytic polio from vaccine virus. There is also a slight possibility of the virus's reverting to a virulent form. Most of the few cases of polio each year in the United States are caused by vaccine virus.

Because of these risks of OPV, the public-health policy of some European countries is to use IPV exclusively. IPV is a slightly less effective antigenic stimulus because there is no live virus to stimulate lymphoid tissue. However, IPV poses no risk for immunodeficient persons, and the virus cannot revert to wild type.

SLOW-VIRUS INFECTIONS

There is a small group of poorly understood neurologic diseases classed as **slow virus infections.** They are marked by a gradual development and progress occurring over years, even decades, a pathology of irreversible neurologic degeneration, and the presence of a virus agent. One of these, subacute sclerosing panencephalitis (SSPE) is caused by measles virus. Others include Creutzfeldt–Jakob disease, kuru, and progressive multifocal leukoencephalopathy.

PROTOZOAN INFECTIONS

Few protozoans invade nervous tissue directly, although some protozoan infections such as trypanosomiasis (sleeping sickness) have neurologic symptoms. The rare ameboid infection described here is an exception.

PRIMARY AMEBIC MENINGOENCEPHALITIS

Primary amebic meningoencephalitis (PAM) is a degenerative neurologic infection, caused by free-living aquatic amebas. About 10 cases of

PAM per year are documented in the United States, almost all fatal. Most cases are caused by *Naegleria fowleri*, a small ameba found in brackish or stagnant warm water. The amebas live in hot springs, puddles, and cracks in the walls of concrete swimming pools.

The motile trophozoite form of the organism probably is inhaled during swimming, passes through the nasal mucosa and the cranial bone, gaining access to the brain. Such a sequence must be due to an unlikely combination of events because cases are so rare. Once present, the ameba attacks and feeds on brain tissue (Figure 21.13). Diagnosis is often made by viewing the motile amebas in microscopic examination of spinal fluid. There is no reliable treatment, and the disease is almost always fatal.

Table 21.5 reviews the epidemiology of some of the major nervous-system infections.

FIGURE 21.13 Amebic Meningoencephalitis
The ameba, *Naegleria fowleri* (left) can be seen snipping off a fragment of a mouse epithelial cell (lower right). When this trophozoite gains access to the brain, it feeds on brain tissue. Drugs have generally not been effective in stopping its relentless munching, and only one person is known to have survived this rare disease. (Courtesy of Dr. Thomas Brown and the editors of the *Journal of Medical Microbiology*.)

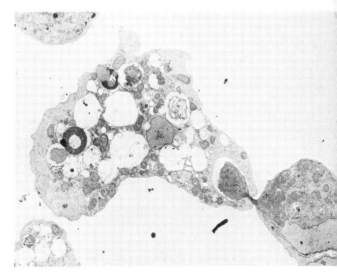

TABLE 21.5 Epidemiology of nervous-system infections

Disease	Reservoir	Transmission	Incubation Period
Meningococcal meningitis	Human beings	Direct contact and droplets from carriers, occasionally cases	2–10 days
Leprosy	Human beings	Not established; probably via skin lesion or respiratory tract	Average, 3–5 yr
Cryptococcosis	Soil, bird excreta	Presumably by inhalation	Unknown
Rabies	Many wild and domestic mammals (rarely rodents)	Saliva to bite, scratch or mucosa; airborne; tissue-graft	Usually 2–8 wk; variable
Poliomyelitis	Human beings	Direct contact with pharyngeal secretions or feces; milk	7–12 days
Tetanus	Soil, feces	Contaminated wound	3 wk, variable

Source: Benenson AS. *Control of Communicable Diseases in Man.*

SUMMARY

1. In contrast with respiratory or gastrointestinal infections, nervous-system infections are uncommon, affecting only a small percentage of the population. They are usually very severe. Most have high mortality rates and tend to leave residual handicaps.

2. The major protections of the nervous system are anatomic. The CNS is sheathed in the meninges and encased in the axial skeleton. Peripheral-nerve fibers are individually insulated by a myelin sheath. Bundles of fibers (nerves) have impermeable connective-tissue coverings. Restricted diffusion between circulation and brain tissue reduces the migration of microorganisms and drugs to brain tissue.

3. Meningitis is a bacterial infection of the meningeal space. It is most common in neonates and infants, but is found in all age groups. Different agents, however, predominate in different groups of people. *Hemophilus* is most common in infants. Meningococcal meningitis is the typical, very severe form in adults.

4. Cryptococcosis is a rare, often fatal, systemic fungal infection in immunodeficient adults.

5. Viruses have a particular affinity for nervous tissue, as they do for epithelial and glandular tissue. Viral multiplication in the cell bodies of neurons may destroy them. Neuron loss is not repaired, and

TABLE 21.5 Epidemiology of nervous-system infections (continued)		
Period Communicable	**Patient Control Measures**	**Epidemic Control Measures**
Until bacteria are no longer present in nasopharyngeal discharges; until 24 hr after therapy is initiated	Isolate until 24 hr after start of chemotherapy; disinfect discharges and soiled articles	Early diagnosis; relieve crowding of target group; prophylaxis with sulfonamide or rifampin
While bacilli are present in skin; until three month's dapsone therapy completed	If patient is on dapsone, isolation not required; disinfect nasal discharges and dressings	Search for possible family or contact cases
Not transmitted from person to person	Disinfect discharges and dressings	None
For 3–5 days before onset of symptoms and during course of disease	Isolate for duration of illness; disinfect saliva and materials soiled with it. Attendants wear protective gown and rubber gloves	Control disease by pet vaccination; control stray dog population
As early as 36 hr after infection; virus in feces for 3–6 wk or longer. Most infections 7–10 days before and after onset	Isolate for not more than seven days if hospitalized; disinfect throat, fecal discharges, and articles of clothing	Mass vaccination with oral vaccine at earliest sign of outbreak; search for sick persons among contacts
Noncommunicable	No special precautions	Universal childhood vaccination; boosters

leads to permanent neurologic defects. Encephalitis and poliomyelitis are two examples. Partially effective drug therapy is available against the DNA viruses, but there is none for the RNA-containing rabiesvirus, poliovirus, and togavirus groups.

6. Surveillance and vaccination have led to the almost complete control of rabies and polio in the United States. However, there appears to be no possibility of eradicating either.

DISCUSSION QUESTIONS

1. What is the difference between a meningitis and an encephalitis in terms of the types of tissue attacked? What does this difference mean in terms of the patient's making a complete recovery?

2. Compare the relative advantages of inactivated and live-virus polio vaccination.

3. Several viruses studied in earlier chapters seem able to establish latency in peripheral

nerves. Review these and discuss the mechanisms that allow such long-term survival.

4. Poliomyelitis and primary amebic meningo-encephalitis have a similar case incidence of about 10 cases per year. However, one is treated as a serious public-health threat while the other is considered an unfortunate natural accident. Why?

ADDITIONAL READING

DeVoe IW. The meningococcus and mechanisms of pathogenicity. *Microbiol Rev* 1982; 46:162-190.

Kaplan MM, Koprowski H. Rabies. *Sci Am* 1980; 242:120-134.

Youmans GP, et al. *Biologic and clinical bases of infectious diseases.* 2nd ed. Philadelphia: WB Saunders, 1980.

22

CIRCULATORY INFECTIONS

CHAPTER OUTLINE

In this chapter we will examine infections of the cardiovascular and lymphatic systems — the body's circulatory network. The body's major defenses — its phagocytic network, circulating antibodies, and lymphoid cells — are all centered in the blood and lymphatic circulations and their associated organs. For an agent to infect here, in the body's central stronghold, requires high virulence or very specialized mechanisms.

As we examine the circulatory pathogens we will pay special attention to the ways in which they evade the host defenses. Each pathogen has a way of "hiding out" in a sheltered microenvironment. Some inhabit red and white blood cells. Others multiply inside macrophages or lymphocytes. Still others cover themselves with coatings of coagulated blood.

Circulatory infections are very common nosocomial diseases. Many of the procedures carried out in hospitals, such as surgery, intra-

Microbial pathogens featured in this chapter:

Non-Group A streptococci
Rickettsia rickettsii
Epstein- Barr virus
Plasmodium species
HTLV/LAV virus

venous feeding, and injections, may introduce microorganisms into the patient's blood.

CIRCULATORY SYSTEM

The circulatory system consists of the cardiovascular system and the lymphatic system (Figure 22.1). The **cardiovascular system** is com-

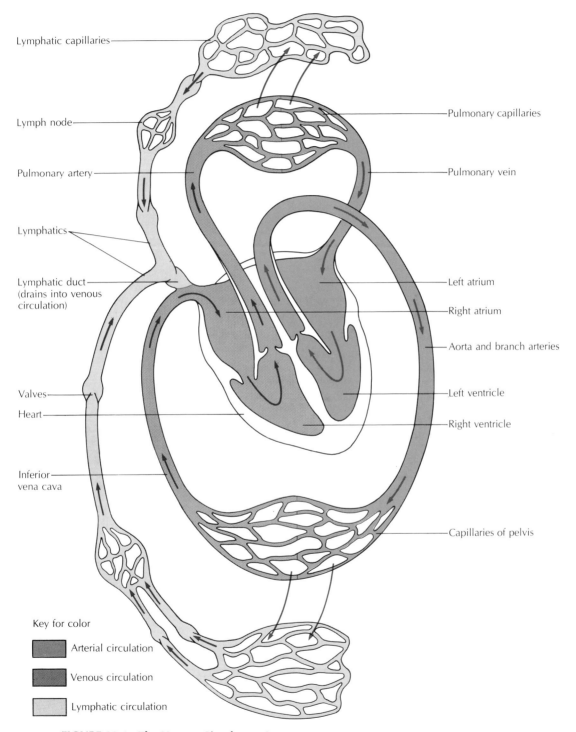

FIGURE 22.1 The Human Circulatory System
The body's major defenses are centered within the blood and lymphatic circulation.

posed of the heart and blood vessels, which circulate the blood through the body. The **lymphatic system** includes the lymphatic vessels, lymphoid organs, and lymph nodes. This system collects fluid from the tissues and conducts it back to the cardiovascular system; it also is responsible for immunity. The normal circulatory anatomy and physiology is complex, and you might find it useful to refer to a good, general anatomy textbook. We will restrict our coverage to structures and functions commonly affected by pathogenic agents.

CARDIOVASCULAR ANATOMY

THE HEART.   The heart, a compact muscular organ, is specialized for efficient blood pumping (Figure 22.2). The heart is suspended in the thoracic cavity (see Chapter 18) and covered by the pleura, forming a membranous sac called the **pericardium.** The heart itself consists of a mass of cardiac muscle, the **myocardium,** and it is lined inside, as is the entire circulation, with a delicate, smooth network of epithelial cells, the **endothelium.** The heart contains four chambers, the right and left atria and the right and left ventricles. The flow of blood into, through, and out of these chambers is unidirectional. Backflow is prevented by a system of **valves.** Valves are leaflets or folds of endothelium reinforced with dense fibrous connective tissue. Infectious disease may attack the endothelium, the valves, or in a few instances the myocardium itself.

THE VASCULAR NETWORK.   Blood leaving the heart is transported through the vascular network, which is composed of arteries, capillaries, and veins. **Arteries** conduct blood away from the heart. They are rigid, with thick, muscle-reinforced walls, and lined with endothelium. The intact arterial wall is almost impermeable to microorganisms and diffusible chemicals.

Given effective heart action, the most important factor affecting fluid flow is the condition of the vessel walls. A perfectly smooth lining is very important, because it offers minimal resistance to the blood flow and no rough spots on which clots might form. Irritations, erosions, or surgical procedures can create rough spots or scar lines, leading to turbulent blood flow, like rapids in a brook. Turbulence promotes clotting. Surface irregularities on the arterial walls also provide niches in which bloodborne bacteria can attach and escape phagocytosis, and may collect clotted blood that further impedes blood flow.

Arteries subdivide into progressively smaller branches in the tissues. The final division is into **capillaries** (Figure 22.3). A capillary wall is only one endothelial cell thick. Lymphocytes and polymorphonuclear leukocytes can move from capillaries into surrounding body tissues. This movement is much increased in inflammation. Red blood cells, however, do not move from intact capillaries. In spongy organs, such as the spleen, liver, and placenta, the circulation divides into irregular passageways called **sinusoids** rather than into capillaries.

Veins, which are thinner-walled than arteries, collect blood from the capillaries and return it to the heart. The returning blood is under little pressure because of its distance from the heart. It is prevented from flowing backward by valves in the vein wall. These valves, and the vein endothelium, are susceptible to inflammation, a condition called **phlebitis.**

THE LYMPHATIC ANATOMY

Lymph is fluid that has been forced out of capillaries into tissues under the pressure of the heart's pumping action. Lymphatic vessels collect lymph and carry it toward the heart, merging with the blood circulation at the **vena cava,** the large vein that brings returning blood to the right atrium. **Lymph capillaries** are composed of endothelium, and are highly permeable. They are readily penetrated by many infective agents, which are then transported in the lymph. As you will recall from Chapter 12, lymph is filtered through aggregates of lymphoid tissue, the **lymph nodes,** which trap infective agents. Most invaders are promptly destroyed by phagocytic

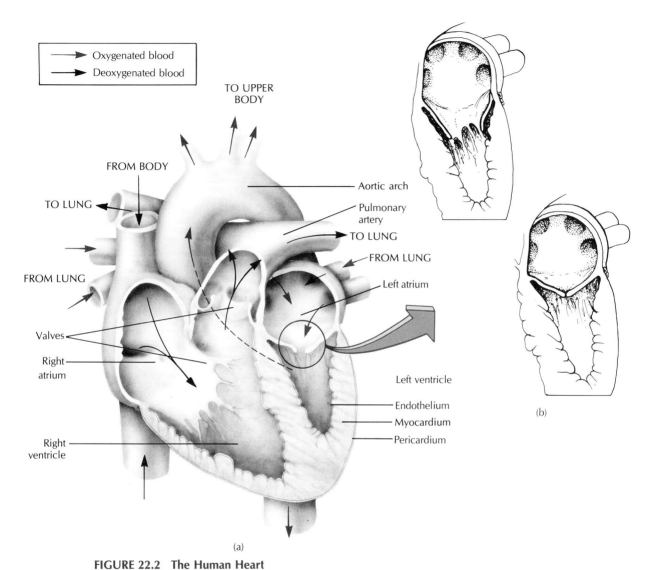

FIGURE 22.2 The Human Heart
(a) Blood flows in and out of the heart's four chambers, which are lined with a smooth endothelium. Blood returning from body tissues and the lungs enters the right and left atria, respectively. The right ventricle expels blood into the lungs via the pulmonary arteries, while the left ventricle forces blood out through the aorta. (b) Blood flow is regulated by cardiac valves that prevent backflow under the pressure of contracting heart muscle.

cells. Resistant agents, however, may survive and multiply in the lymph nodes, pending the development of an immune response. To the ex-

tent that they are not subsequently controlled by immunity, a serious or fatal disease state may develop.

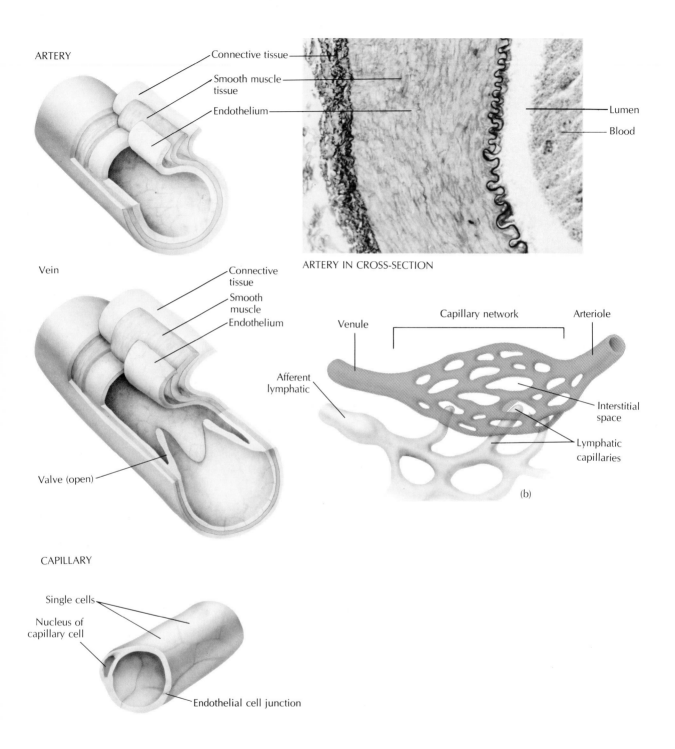

ARTERY

Connective tissue

Smooth muscle tissue

Endothelium

ARTERY IN CROSS-SECTION

Lumen

Blood

Vein

Connective tissue

Smooth muscle

Endothelium

Valve (open)

Venule

Afferent lymphatic

Capillary network

Arteriole

Interstitial space

Lymphatic capillaries

(b)

CAPILLARY

Single cells

Nucleus of capillary cell

Endothelial cell junction

(a)

FIGURE 22.3 Structure of Arteries, Veins, and Capillaries

The arterial and venous vessels and the lymphatic system serve all parts of the body. (a) Arteries have thick, muscular walls with an endothelium that is impermeable to blood cells and other substances. The micrograph shows a cross-section of an artery in which this endothelium is plainly visible. The walls of veins are thinner and less elastic; these vessels contain valves that prevent the backflow of blood. Because capillary walls are only one cell thick, materials such as pathogens and phagocytic cells may readily pass through. Note that the vessels are not drawn to scale; capillaries are much smaller than veins or arteries. (b) Blood and lymph capillaries intertwine within the tissues. Fluid leaves the blood circulation and drains into the lymph circulation, passing through the lymph nodes, where infectious agents are filtered out. Arrows indicate the direction of blood flow through the vessels and of fluid flow across interstitial spaces into either blood or lymphatic capillaries.

PHYSIOLOGY OF THE CIRCULATORY SYSTEM

The normal vascular network maintains a specific state of permeability, allowing certain substances to pass through to nourish the tissues, plus sufficient fluid to transport them. Capillary permeability is affected by disease, because it is significantly increased by inflammatory changes. When fluid seeps into tissue faster than it can be drained off by the lymphatics, the result is swelling or **edema.**

Disease may also cause capillaries and other vessels to rupture. Ruptured capillaries form tiny dark spots called petechiae on the skin, gums, and other tissues throughout the body. The rupture of a larger blood vessel results in hemorrhage.

Coagulation, or blood clotting, is a primary life-saving mechanism. Clots not only arrest blood loss but also wall off areas of potential infection. However, clotting can also be initiated by inappropriate stimuli. We noted that one such stimulus is an irregularity in a vessel lining. Bacterial products may also initiate inappropriate

clotting. Anaerobic bacteria in particular produce extensive **intravascular coagulation** — clotting within capillaries — near a focus of infection.

A **thrombus** (blood clot within a blood vessel) or a fragment of a clot is dangerous because it may break away and then get stuck downstream, cutting off blood flow. An **embolus** is a thrombus that has completely occluded a small vessel. Because an embolus cuts off the nutrient and oxygen supply, the tissue normally served by the blocked vessel may die. Thrombosis is a grave matter in the brain, lungs, or kidneys.

When an infective agent has been trapped in the lymphoid tissue, antigenic stimulation increases the mitotic rate of lymphocytes, as we saw in Chapter 12. This causes the lymph node to swell, a condition called **lymphadenitis** or **lymphadenopathy.** Swollen lymph nodes may also reflect recruitment of additional lymphocytes from the circulation. The swelling will usually persist as long as the pathogen is present. A cellular immune response may develop within the lymphoid tissue containing the pathogen. Inflammation, granuloma formation, and ulceration may result before the agent is finally destroyed.

ROUTES OF INFECTION

The usual route for circulatory infection is from surface tissue to lymphatics to blood (Figure 22.4). It is easier for microorganisms to invade the lymphatic system than the cardiovascular system because the lymphatic system's capillaries are thinner-walled. Many infectious agents entering the lungs (e.g., *Mycobacterium tuberculosis*) or the gut (e.g., *Salmonella typhi*) move promptly into the lymphatics and are disseminated.

Trauma allows microorganisms to enter the blood stream directly. Most escape phagocytosis only if they chance to settle on dead tissue or an abnormal, rough surface. These two conditions tend to occur in conjunction in many hospitalized persons (the trauma of an intravenous catheter and the rough surface around its point

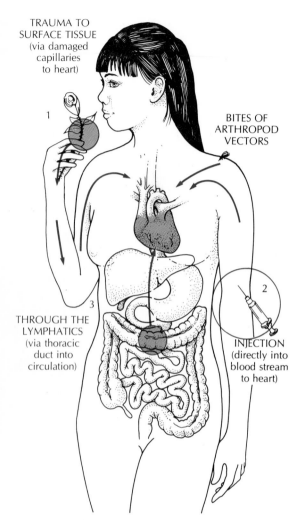

TRAUMA TO
SURFACE TISSUE
(via damaged
capillaries
to heart)

1

BITES OF
ARTHROPOD
VECTORS

2

3

THROUGH THE
LYMPHATICS
(via thoracic
duct into
circulation)

INJECTION
(directly into
blood stream
to heart)

FIGURE 22.4   Routes of Circulatory Infection
The blood circulation may become infected by
several routes: (1) Trauma to surface tissue, such as
a thorn prick or an insect bite, may allow organisms
to enter the blood directly via damaged capillaries.
(2) Nosocomial infections occur through activities
such as dentistry or injections that pierce tissue.
(3) Most common is the route by which organisms
enter the lymphatics (shown here from the
intestinal lining) and pass via the thoracic duct into
the circulation.

of entry, for example). Thus many circulatory
infections are **iatrogenic** (caused by a therapeutic
measure).

Circulatory infections may be transmitted
by some of the many **arthropod vectors** (insects
and arachnids) that feed on vertebrate blood. The
arthropod may withdraw infective blood from
one host and regurgitate a portion to infect its
next host. Or the infective agent may multiply
in the vector and be excreted in its saliva or feces. Commonly as an arthropod bites it also def-
ecates. The bite itches, and the host scratches,
thereby inoculating the puncture site with the
insect's feces.

BLOOD-CULTURE TECHNIQUES

In most circulatory infections, correct diagnosis
and treatment depend on isolating the agent
from a blood sample. Blood culture has proved to
be a most challenging task. Various commercially developed systems have been marketed for
collection, incubation, and detection of microorganisms from the blood.

Because of contamination, almost every
known microorganism has been isolated from
the blood at some time. Collection procedures
must avoid introducing misleading skin flora.
Multiple cultures from separate sites are advisable. Initial samples should be obtained before
antimicrobial therapy is begun, since the therapy may decrease the microbial numbers to the
point that isolating them can prove difficult or
impossible.

Even in a full-scale circulatory infection, the
bacterial count of a blood sample will probably
be less than 50 cells per milliliter. Three 10-ml
samples of intravenous blood collected one or
two hours apart from different sites are recommended. Normal blood components may also
have inhibitory effects on the growth of the
pathogens. Such effects may be minimized by
fivefold dilution of the blood sample in medium.
Culture methods must be flexible enough for the
possible detection of aerobes, anaerobes, micro-
aerophils, and wall-deficient bacterial L forms.

Growth of microorganisms in a blood culture may be detected microscopically by the Gram stain or by staining with acridine orange dye. This dye causes any bacteria present to fluoresce. Subculture to fresh plated media allows isolation and identification. Growth of free-living bacteria can also be monitored by detecting the release of radioactive carbon dioxide from labeled nutrients in the medium. Intracellular parasites such as rickettsiae and viruses cannot be detected by these means.

BACTERIAL CARDIOVASCULAR INFECTIONS

One characteristic of cardiovascular infection is **bacteremia,** a state where bacteria are present in the blood. It may be a transient state or a side effect of a localized infection in various body organs. It becomes life-threatening if it develops into **septicemia** — bacteria multiplying in the blood. This will occur if the agent is an extracellular pathogen of high virulence, if the host's defenses are severely depressed or compromised, or if there is a host site available for microbial attack, such as a damaged heart valve. We will look at some examples of bacterial infections of the cardiovascular system.

INFECTIVE ENDOCARDITIS

Infective endocarditis is an inflammation of the endocardium, particularly in the area of the heart valves. It exhibits a spectrum of severity with the **acute** and the **subacute** forms as the two extremes (Table 22.1). In its most serious, or acute form, endocarditis is a rapidly progressive,

TABLE 22.1 Comparison of acute and subacute forms of infective endocarditis		
Feature	**Acute Form**	**Subacute Form**
Important agents	Pathogenic bacteria include: *Staphylococcus aureus*, *Streptococcus pyogenes*, *S. pneumoniae*	Relatively nonvirulent bacteria include: Viridans streptococci, Group E streptococci, Enterococci, Enterobacteria
Underlying condition	Normal heart or implanted prosthesis	Congenital or acquired heart abnormality
Provoking incident	Trauma; surgical infection	Minor trauma, toothbrushing, dental work, gynecological procedures, other surgery or catheterization
Onset	Moderately rapid	Insidious
Pathogenesis to heart valve	Vegetation, destruction of underlying tissues	Vegetation, slow destruction
Potential for secondary or general infection	Usually progresses to general infection or septicemia	Little potential for extension to other normal tissues

frequently fatal process in which normal heart valves are destroyed (Figure 22.5). *Staphylococcus aureus, Streptococcus pyogenes,* or other true pathogens are the agents. The infection involves large areas of the vascular system. It occurs following mechanical invasion of blood vessels and fortunately is quite rare.

The more common but less immediately life-threatening subacute form is more slowly progressive because the agent is of low virulence. It occurs only in people who already have abnormal heart valves. Invariably fatal if left untreated, subacute endocarditis can almost always be cured if the organism(s) are isolated and

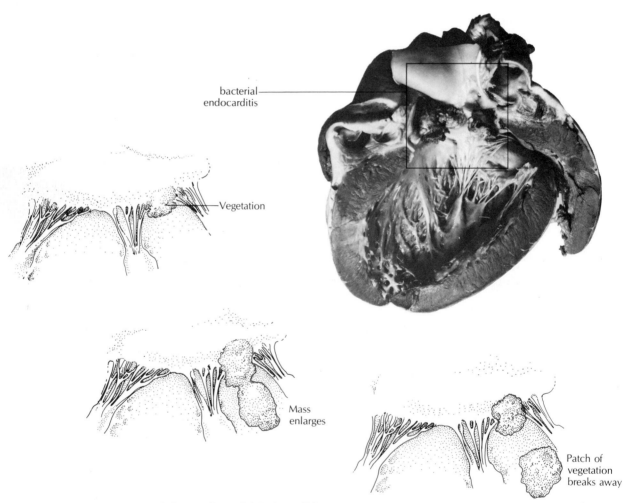

bacterial endocarditis

Vegetation

Mass enlarges

Patch of vegetation breaks away

FIGURE 22.5   Subacute Bacterial Endocarditis
SBE occurs when normal flora of low virulence enter the blood. The bacteria adhere to the surface of a damaged heart valve, forming a vegetation at the site. This growing aggregate of fibrin and bacterial colonies may periodically break off and form clots in smaller blood vessels. (Courtesy of Armed Forces Institute of Pathology/BPS.)

TABLE 22.2 The non–Group A streptococci			
Lancefield Group	**Representative Species**	**Hemolysis**	**Diseases Caused**
B	*Streptococcus agalactiae*	Beta	Neonatal meningitis Neonatal septicemia Suppurative arthritis
C	*S. equii* *S. equisimilis* *S. dysgalactiae*	Beta	Respiratory infection Infective endocarditis (IE) and bacteremia
D	Enterococcus group *S. faecalis* *S. faecium*	None (or variable)	Abdominal abscesses IE and bacteremia Urinary tract infection
	Nonenterococcus group *S. bovis, equinus*	None (or variable)	IE and bacteremia
F	*S. anginosius*	Variable	Abscesses Sinusitis (rare) Meningitis
H	*S. sanguis*	Alpha	IE and bacteremia
K	*S. salivarius*	Alpha	IE and bacteremia
Ungrouped			
Aerobic	*S. mitis* *S. mutans* *S. pneumoniae*	Alpha Alpha Alpha	IE and bacteremia Dental caries, IE Pneumonia, IE
Anaerobic or microaerophilic	*Peptococcus magnus* *P. saccharolyticus* *P. prevotii* *Peptostreptococcus*	Variable	Necrotizing pneumonia Abscesses Wound infections Gynecologic infections

an appropriate antimicrobial therapy instituted. Although there are many possible agents, we will discuss only the non–Group A streptococci, which cause 60 to 70% of all cases.

THE ORGANISMS.   You will recall the Lancefield classification of the streptococci from Chapter 17. As we saw, the most effective way of identifying the various groups is on the basis of their surface antigens. Latex agglutination or coagglutination techniques are frequently employed because they are rapid, inexpensive, and easy to perform. The characteristics of the non–Group A streptococci are summarized in Table 22.2.

Streptococcus mutans, Streptococcus sanguis, and the enterococci are prominent in infective endocarditis.

THE DISEASE.   **Subacute bacterial endocarditis (SBE)** develops on heart valves damaged by rheumatic fever, hypersensitivities, syphilitic lesions, congenital heart deformities, atherosclerosis (lipid plaques in the arterial walls), and surgical scars. SBE begins when normal flora of low virulence enter the blood, usually from the mucosa. Dental and gynecologic procedures are a common means of introducing organisms, as the oral and vaginal floras are rich in streptococci. The bacteria attach to the valve surface as the blood passes through the heart. They persist, and a **vegetation** forms at the site. The vegetation is a loose, irregular aggregate of fibrin containing microcolonies of bacteria. The mass enlarges slowly, in a downstream direction. Periodically, fragments break off and lodge in smaller blood vessels. Fragmented clots lodging in the spleen, liver, kidney, or cerebral circulation may be fatal. Classic symptoms of systemic infection, such as fever and toxicity, are accompanied by heart murmurs, fatigue, and signs of impaired cardiac function.

THE TREATMENT.   SBE cannot be cured without antimicrobial therapy. However, choice of the drug(s) and dosages to use may be quite a challenge to the therapist. The vegetation protects the bacteria because neither antimicrobials nor immune factors penetrate it very effectively. The drug(s) used must be bactericidal (they must kill the agent) and the dosage high. Therapy extends over many weeks, and follow-up blood cultures are done to ensure complete clearance of the agent. Immunity has a limited effect on the progress of the disease.

EPIDEMIOLOGY AND CONTROL FACTORS.   Certain individuals have a clearly identifiable risk of SBE because of their medical history. Before they undergo any procedure likely to cause bacteremia, such as having their teeth cleaned, a brief course of prophylactic antibiotics should be given.

TABLE 22.3 Some human rickettsial diseases			
Group	**Disease**	**Agent**	**Geography**
Spotted fever	Rocky Mountain spotted fever	*Rickettsia rickettsii*	Western hemisphere
	Rickettsialpox	*R. akari*	North America, Europe
Typhus	Epidemic typhus	*R. prowazekii*	Worldwide
	Brill–Zinsser disease	*R. prowazekii* latent in tissues, reactivates	South America, Europe
	Endemic typhus	*R. typhi* *R. mooseri*	Worldwide
Q fever	Q fever	*Coxiella burnetii*	Worldwide

THROMBOPHLEBITIS

Thrombophlebitis accounts for about 10% of hospital-acquired infections. It is an irritation of the venous endothelium, developing at or near the site of a surface lesion. Most often, it appears at the site of an improperly maintained intravenous catheter. Inserting a needle through the skin layers always provides a potential for infection. Irritation is caused by the continued presence of the needle, movement, and mechanical stress. Daily care of the site is crucial, and frequent rotation to a new site is recommended. All parts of the intravenous equipment, except the catheter itself, require daily changing.

SEPTICEMIA

Septicemia is rarely a primary disease. It may appear as a later development of almost any infection. Uncontrolled multiplication of bacteria in blood can occur only when phagocytic action is depressed. Septicemia characteristically develops in immunosuppressed or debilitated individuals. It kills via **septic shock.**

When bacteria multiply in blood, they often release toxins. Thus there is some truth behind the old-fashioned name for septicemia — **blood poisoning.** Some exotoxins of the Gram-positive organisms are destructive to blood cells. Exotoxins such as streptokinase, coagulase, and the fibrinolytic enzymes probably play a role in intravascular coagulation. After clotting has occurred in numerous small blood vessels, the supply of available clotting factors in the blood may be so depleted that widespread hemorrhages may develop. This is the hemorrhagic form of septic shock.

Endotoxin is released by dying Gram-negative organisms. In large amounts endotoxin causes loss of smooth muscle tone in the venous walls, weakening of the cardiac muscle, and capillary breakdown. As septic shock develops, the patient first becomes feverish and dry. Then there is a sudden, unexplained drop in blood pressure as the veins dilate and blood pools in the tissue. The patient becomes cold and clammy, and cardiac output drops off. The mortality rate from septic shock is estimated to be

		TABLE 22.3 Some human rickettsial diseases (continued)		
Vector	Reservoir	Potential Severity	Rash	Weil–Felix Reaction
Tick	Tick, wild rodent, dog	Severe	Yes	+
Mite	Rodent	Mild	Yes	−
Body louse	Human being	Severe	Yes	+
Body louse	Human being	Moderate	Yes or no	+
Flea	Rodents	Moderate	Yes	+
Ticks; non-vector transmission	Domestic livestock	Mild	No	−

70%. Therefore it is critically important to catch the earliest indications of septicemia. Antimicrobials may then be used before irreversible damage has been done, somewhat improving the patient's chances of survival.

ROCKY MOUNTAIN SPOTTED FEVER

The obligate intracellular parasites called rickettsiae cause a number of serious cardiovascular infections (Table 22.3). Rickettsiae multiply in vascular endothelial cells. They are currently divided into the spotted fever group, the typhus group, and the scrub typhus group. **Q fever** is a pneumonia caused by the related agent *Coxiella burnetii*. In the United States, the most common cardiovascular rickettsial disease is **Rocky Mountain spotted fever.** We will use this disease as a model for discussing rickettsial infection.

THE ORGANISM.   *Rickettsia rickettsii* and its variants cause the group of diseases called spotted fever. This group also includes typhus, boutonneuse, and rickettsialpox. Rickettsiae are typical bacteria in structure, with a Gram-negative cell wall. They divide by binary fission. An intracellular location is essential for them to grow (Figure 22.6a), but it is not clear why. In any case, if a rickettsia leaves its host cell for long, it will die. Thus a living vector is essential for transmission of this group.

Rickettsiae may be cultivated for study in the yolk sac of eggs or in tissue culture, multiplying freely in the cellular cytoplasm. Laboratory diagnosis of rickettsial disease takes advantage of the fact that antibodies formed during rickettsial disease also cross-react with antigens of the enterobacterial species *Proteus vulgaris*. This phenomenon is called the **Weil–Felix reaction** (Figure 22-6b).

THE DISEASE.   The agent of Rocky Mountain spotted fever, *Rickettsia ricketsii*, is introduced by a tick bite. After about two to six days, symptoms appear. There is a very high fever (up to 105

FIGURE 22.6   Rickettsial Infection
(a) A normal, uninfected chick embryo cell in monolayer culture. Visible are the nucleus (n), mitochondrion (m), endoplasmic reticulum (er), and Golgi apparatus (g). In the second photo, 96 hr after infection with *Rickettsia rickettsii*, spaces have appeared in the cytoplasm and the nuclear membrane has begun to break down. In the third photo, 120 hr after infection, the nuclear membrane has completely disintegrated. The remains of the cell contain membrane-enclosed rickettsiae. No other host-cell structures are identifiable. (Photo courtesy of David J. Silverman and Charles L. Wissman, Jr., *R. rickettsii* cytopathology. *Infection and Immunity* (1979) 26:714. By permission of ASM Publications.) (b) Laboratory diagnosis of some rickettsial diseases, such as typhus, utilizes the Weil–Felix reaction. This test is based on the observation that antibodies formed during infection also react with antigens of *Proteus vulgaris*. Therefore when a suspension of *Proteus vulgaris* cells is mixed with patient serum, a clumping reaction signals that the person has, or has had, typhus.

to 106 degrees Fahrenheit), extreme muscle tenderness, and a severe headache that does not respond to analgesics such as aspirin. On the second or third day, a rash appears on the patient's ankles and wrists. The location of the rash is a valuable diagnostic clue. The organism attacks endothelium throughout the cardiovascular network, with its most pronounced effects on the kidneys and other major organs. Widespread intravascular coagulation occurs, obstructing blood flow. Neurologic abnormalities may appear. If untreated, up to 20% of victims die from vascular collapse or kidney failure.

THE TREATMENT.   Prompt therapy with chloramphenicol or tetracycline is life-saving. Death from Rocky Mountain spotted fever is frequently the result of a missed diagnosis or delayed therapy. Supportive measures maintain kidney function and counteract the effects of toxicity. Specific antibodies appear between the 8th and 12th days and may aid in recovery. These antibodies protect against reinfection.

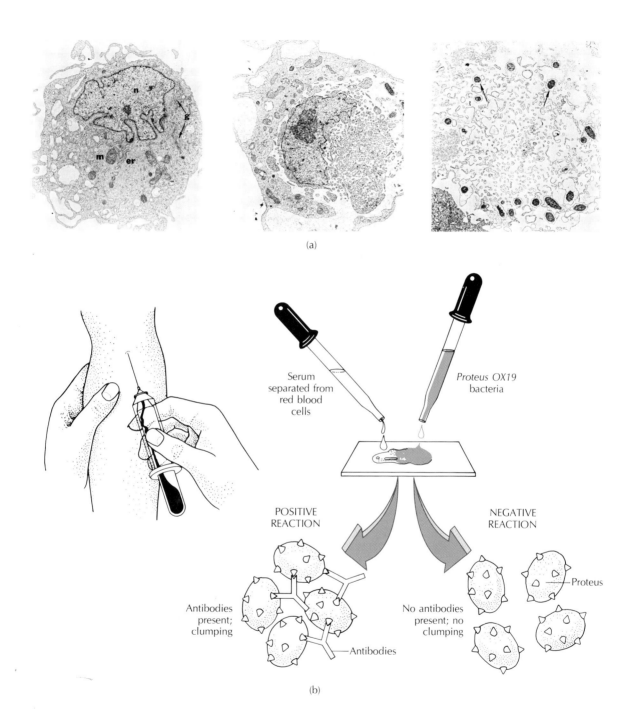

(a)

Serum
separated from
red blood
cells

Proteus OX19
bacteria

POSITIVE
REACTION

NEGATIVE
REACTION

Antibodies
present;
clumping

No antibodies
present; no
clumping

Proteus

Antibodies

(b)

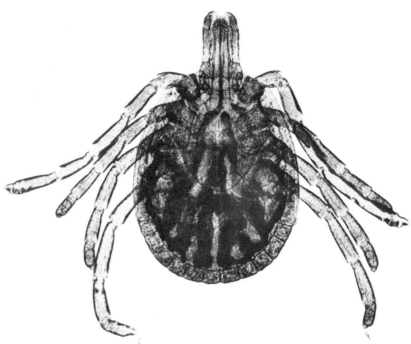

FIGURE 22.7   A Tick
Ticks are important vectors of circulatory infections, such as the rickettsial infections and tularemia. When the tick bites, it fastens its mouth parts into the tissue, injects a natural anticoagulant, and proceeds to ingest blood for several hours. During this painless feeding process, pathogens may be introduced into the unsuspecting victim's circulatory system. (Courtesy of Carolina Biological Supply Co., Burlington, NC.)

EPIDEMIOLOGY AND CONTROL MEASURES.   The agent of Rocky Mountain spotted fever is carried by wood ticks and dog ticks, multiplying in the tick's salivary glands (Figure 22.7). It is transmitted to a mammalian host during the tick's prolonged blood meal, which will last for several days unless the tick is discovered and pulled off. Many different kinds of mammals can be infected. Often a small geographic area will contain a highly infected tick population. Only a few miles away, the ticks may not be infected. It is wise, when hiking, working outdoors, or camping out in tick country to wear a head covering and clothing that will prevent tick attachment, and to examine your skin and scalp frequently for ticks. A tick removed within six hours is unlikely to infect you. Domestic pets should also be de-ticked, as they too may contract the disease. Afterward hands should be washed well, as all parts of the tick may be contaminated. A killed-bacteria vaccine is available, but it is recommended only for persons with high occupational risks, such as lab workers.

Most Rocky Mountain spotted fever in the United States occurs in warm months in the Eastern states, particularly along the Atlantic coast and in the southern mountains. It is uncommon in the Rockies at this time, although a few years ago it was a serious concern there. About 1000 cases per year can be expected.

OTHER RICKETTSIAL DISEASES

The typhus group of diseases is represented by three conditions, epidemic typhus fever, Brill–Zinsser disease, and endemic or murine typhus.

EPIDEMIC TYPHUS FEVER. *Rickettsia prowazekii* causes epidemic typhus fever and Brill–Zinsser disease. **Epidemic typhus fever** was once widespread throughout the world, but it is now found mainly in regions of Asia and Africa. Typhus is transmitted by the human body louse, which thrives under conditions of severe poverty, intense crowding, or mass migration. The disease is characterized by abrupt onset, extremely high temperature, severe headache, and a rash that appears first on the upper trunk. In severe cases, the patient's spleen and kidneys become involved, and there may be myocarditis or neurologic changes. In some persons, the agent persists in an inactive form after recovery. Chloramphenicol or tetracycline and supportive measures are used to treat typhus. Vaccines are available, and vaccination is highly recommended for travelers to endemic areas. Insecticides are widely used as a control measure to destroy the vector.

As we mentioned, some persons who experience epidemic typhus fail to completely eliminate the agent, which remains latent in their tissues. **Brill–Zinsser disease** is a mild form of typhus that reappears in someone who has had epidemic typhus first, some years before. There is often no rash. Since it is an endogenous infection, a vector is not needed to initiate the disease. If lice are present, however, the person with Brill–Zinsser disease may be bitten and serve as a reservoir to infect the lice with typhus; the lice, in turn, may cause new cases in other susceptible persons.

ENDEMIC TYPHUS. **Endemic** or **murine typhus** is a mild infection caused by *Rickettsia typhi.* It infects rats and mice and is transmitted to human beings by the feces of the rat's fleas being rubbed into a bite. In the United States, less than 100 cases per year occur, many of them in Texas.

RHEUMATIC FEVER

Unlike the diseases discussed above, **rheumatic fever** is not an infection. It is a sequel to an infection, a tissue-destroying hypersensitivity. Rheumatic fever will develop in about 3% of individuals as a sequel to untreated Group A streptococcal pharyngitis, discussed in Chapter 17. In untreated (that is, lengthy) cases of streptococcal pharyngitis, some people's immune systems respond by making antistreptococcal antibodies that cross-react with components of the heart, joints, and skin.

Within one to two weeks, a group of symptoms, collectively called rheumatic fever, will appear. The heart valves become acutely inflamed in 40 to 50% of first attacks, and about two thirds of these result in residual valve damage. Other symptoms — arthritis, rash, and chorea (muscle tremors) — may not cause permanent damage. An antibody called ASO (**antistreptolysin O**) appears in the patient's serum and is used as a diagnostic clue. Each additional attack of streptococcal pharyngitis spurs renewed antibody production and another attack of rheumatic fever. Damaged heart valves are often damaged further. Antiinflammatory drugs are useful to suppress certain aspects of the immune responses.

Penicillin or long-acting penicillin analogs are usually given as a preventive measure over a period of years. This approach prevents additional incidents of streptococcal pharyngitis and stabilizes the condition of the patient with rheumatic fever.

VIRAL CARDIOVASCULAR INFECTIONS

Viruses attack the endothelium too. They can cause diseases in which toxicity, hemorrhage, and circulatory collapse are the main features.

DENGUE

Dengue is a mosquito-borne viral disease endemic in many tropical regions. It is caused by a single-stranded RNA togavirus (Figure 22.8).

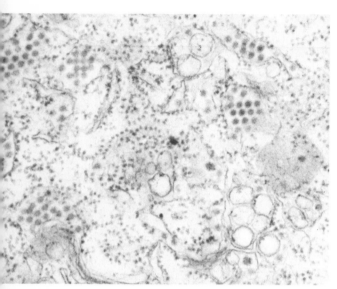

FIGURE 22.8   Dengue Virus
The dengue virus is an enveloped flavivirus that is transmitted by the bites of mosquitos. Dengue is endemic in most of the world's tropical belt. The viruses are shown here replicating in the cytoplasm of a host cell. (×51,200. Courtesy of Alyne K. Harrison, Viral Pathology Division, Centers for Disease Control, Atlanta.)

There are four serotypes, so a person may contract the disease several times.

In classic dengue the person experiences fever, muscle aches, rash, and headache but the course is benign and recovery is prompt. In hemorrhagic dengue, found principally in small children, the outcome may be fatal. About two to six days after the onset of symptoms the patient develops generalized hemorrhage, bleeding gums, hypotension, and intravascular coagulation. The mortality rate from this form of dengue is high. There is no specific therapy, only supportive measures, and no vaccine at present.

Dengue is restricted to the latitudes 25° north to 25° south. It occurs on some Caribbean Islands and has been brought back to the United States by returning tourists. The virus is spread by the mosquito *Aedes aegypti* and is predominantly encountered in urban areas. The reservoir is human beings and possibly monkeys. Mosquito control is the only useful public-health measure.

YELLOW FEVER

Yellow fever is a viral infection caused by the yellow-fever virus. The disease was once a scourge of U.S. coastal cities. In 1793, for example, a yellow fever epidemic carried off almost 10% of the population of Philadelphia. Now yellow fever is not seen in the United States, but it is endemic in much of tropical South America and sub-Saharan Africa. In endemic regions, wild monkeys serve as a reservoir, while in urban areas human beings are the reservoir. The yellow-fever virus, a flavivirus (Figure 22.9) is transmitted by mosquitos. Epidemic (or urban) yellow fever is transmitted by the insect from person to person. Endemic (or jungle) yellow fever is transmitted to people from monkeys when they enter forests in which infected monkeys dwell.

Yellow fever may have several manifestations. In its most severe form, there is massive hemorrhage, with bloody vomit and urine. The liver is affected, causing the jaundice that yellows the skin and gives the disease its name. In most cases, however, the disease is mild or asymptomatic. Control of yellow fever is based on vaccination and on controlling the mosquito vector. An effective live-virus vaccination is available and recommended for persons traveling to areas in which yellow fever is endemic.

EPIDEMIC HEMORRHAGIC FEVERS

A number of acute viral hemorrhagic diseases have been described but have received limited study to date. They all share the symptoms of vascular infection — high fever, muscle ache, hemorrhage, and coagulation. Lassa fever, Ebola virus, and Marburg fever are notorious for their high mortality rates and have aroused global concern. Yet all these epidemics, so far, have remained contained. As yet there is little systematic information available about this disease group. Rodent reservoirs and arthropod transmission are involved in several of these diseases.

PERICARDITIS

Viruses may cause **pericarditis** — inflammation of the pericardium. The major symptoms are chest pain and fluid accumulation in the pericardial sac. The pressure exerted by this fluid may severely compress the heart. There is no specific therapy, and cases usually clear up spontaneously.

MYOCARDITIS

Myocarditis is an inflammation of the heart muscle, producing impaired heart contractions.

FIGURE 22.9 Yellow-Fever Virus
(a) The flavivirus responsible for yellow fever. (×102,850. Courtesy of Erskine Palmer, Centers for Disease Control, Atlanta.) (b) This virus can reproduce either in mosquitos, in monkeys, or in human beings. In endemic tropical regions, wild monkeys act as the reservoir, and human beings working in the jungle may catch the disease through mosquito bites. In urban areas, mosquitos transmit the disease directly from person to person in an epidemic form.

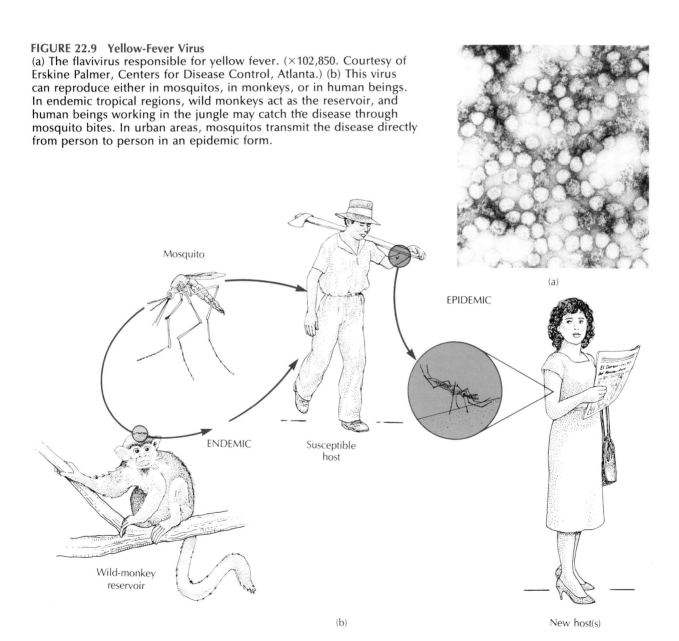

The coxsackieviruses are major agents of myocarditis; echoviruses, polioviruses, and influenza viruses may also cause myocarditis. Many patients with viral myocarditis are only mildly affected. They become easily fatigued and experience malaise, but recover spontaneously. With more severe cardiac symptoms, the patient may or may not recover full heart function.

Myocarditis may also be bacterial. Myocarditis due to the action of diphtheria toxin on the heart is the cause of most diphtheria mortality. Endotoxins released in Gram-negative infections also weaken the myocardium. Myocarditis develops in advanced stages of tuberculosis, syphilis, and many parasitic diseases.

BACTERIAL LYMPHOID INFECTIONS

Certain bacteria survive and replicate in lymphoid tissues. In doing so they provoke an injurious inflammatory process. Plague, brucellosis, and tularemia are the most notable bacterial lymphoid infections.

PLAGUE

The term **plague** once referred to any fearsome or serious epidemic. Now we reserve it for the disease caused by *Yersinia pestis* — called **yersiniosis,** and synonymous with the historic **bubonic plague.** It has been many decades since there has been a large-scale epidemic of plague, although scattered cases continue to occur.

Yersinia pestis is a small, Gram-negative rod. It can multiply extracellularly in plasma or tissue fluid, but prefers the intracellular environment of a macrophage phagosome. *Y. pestis* has virulence factors that include a capsule, high levels of catalase, and endotoxin. *Y. pestis* causes two different disorders, depending on how it is transmitted. If introduced by a flea bite, *Y. pestis* migrates to the regional lymph nodes, commonly in the armpit or groin. An extremely painful, swollen, ulcerated lymph node called a **bubo** develops, which explains why the disease is called bubonic plague. Fever and severe endotoxin poisoning develop. Peripheral circulation is blocked, producing blackened and necrotic fingers and toes (hence the name Black Death). If airborne bacteria are inhaled in person-to-person transmission, the lungs become infected and **pneumonic plague** results. This form progresses inevitably to septicemia, and approaches 100% mortality if not treated promptly. Streptomycin and tetracycline are the drugs of choice for the treatment of plague.

Plague is now not seen in urban regions of the United States, largely because of improved sanitation and much reduced rat populations in cities. However, in some Western states, wild rodents such as the ground squirrel and prairie dog are a permanent reservoir (review Figure 14.3). Fleas transmit the disease among animals. Human beings can contract plague either by capturing and skinning infected animals or by being bitten by an infected flea. Cats and dogs, by temporarily harboring infected fleas, can also bring the disease to human beings. There are about 20 cases per year, mostly in New Mexico, Arizona, and California. A heat-killed bacterial vaccine is available for persons at high risk. Persistence of a wild reservoir of the agent, as well as the rat–flea transmission system, prevents its eradication. However, should a major outbreak of plague occur now, it could in all probability be quickly contained through quarantine measures, rat eradication, and mass vaccination.

BRUCELLOSIS

Several members of the genus *Brucella*, including *B. melitensis, B. suis, B. abortus,* and *B. canis,* cause **brucellosis.** The brucellae are slow-growing, fastidious, gram-negative rods, requiring 10% carbon dioxide for incubation. They infect a wide range of mammals, both domestic and wild, causing various diseases. Spontaneous abortion is the most common manifestation in domestic herd animals. The placenta, fetus, urogenital secretions (including semen), and the meat or milk derived from the infected animal may all be infectious. Farmers, veterinarians, and meat packers are the persons most commonly infected. The agent remains viable for 21 days in refrigerated meat; in milk, it is destroyed by pasteurization. Routine testing of dairy and meat animals is practiced in areas in which the disease is prevalent. A live, attenuated vaccine is administered to calves.

In humans, *Brucella* produces abscesses of the liver, spleen, bone marrow, and lymph nodes. Brucellosis has vague, undefined symptoms, including aches, chills, sweating episodes, and fever. The fever goes up and down — thus the disease's common name, **undulant fever.** Mortality is low if tetracyclines or trimethoprim–sulfa combinations are given.

TULAREMIA

Tularemia is a highly infectious bacterial lymphoid infection. Fewer than 50 bacilli can cause

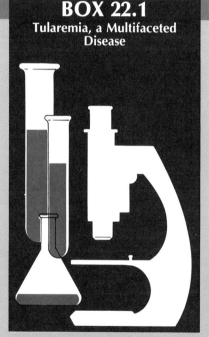

BOX 22.1
Tularemia, a Multifaceted Disease

Tularemia is remarkable for the surprising variety of circumstances in which you can acquire it. For example:

■ While hunting, a 19-year-old Washington State man found a dead rabbit. With his bare hands, bruised and scratched from his work as an automobile mechanic, he cut off the front paws for good luck charms and gave them to a friend. The hunter developed tularemia. His friend fortunately threw the rabbit paws away, and remained well.

■ Another Washington State man shot, killed, and skinned an apparently healthy black bear, cutting his left hand in the process. Tularemia developed within two days.

■ Eleven cases of tularemia developed on an Indian reservation in Montana during early summer, the peak of the tick season. The dogs on the reservation almost all had antibodies to tularemia, and most ticks removed from them contained the bacterium.

■ Four sheep shearers in Colorado developed tularemia after shearing sheep covered with wood ticks. The initial lesions appeared on their left hands. This is because shearers part the fleece with their bare left hands, while shearing with the right. Engorged ticks are often destroyed in the process, and the bacteria-containing gut contents contaminate the left hands.

■ An Alaskan man became ill after dressing a rabbit killed by his dog. A number of dead rabbits had recently been observed in the area, suggesting an active outbreak of rabbit tularemia.

■ Two young Georgia boys became ill after handling a dead rabbit they had found.

■ A laboratory technician in Colorado developed severe tularemia after working with a laboratory isolate of *Francisella tularensis*.

Source: *Morbidity and Mortality Weekly Reports* 1979; 27(12); 1979; 28(44); and 1980; 29(5).

disease, sometimes even entering through un-broken skin. The causative agent, *Francisella tu-larensis*, is closely related to *Brucella*. It infects more than 100 different animal species and many insects and arachnids. Modes of trans-mission include tick bite, contamination of the skin or conjunctiva, inhalation, or ingestion. It is surprising and fortunate that, despite this great virulence and host versatility, there are just over 200 cases per year in the United States. Tu-laremia is commonly acquired during rabbit-hunting season by hunters skinning infected an-imals. A lesion forms at the site of entry, and buboes develop in lymph nodes. Should the agent enter the blood, the disease may resemble typhoid fever. Diagnosing tularemia is difficult, and cultivating it is exceptionally hazardous to laboratory personnel. Streptomycin will arrest the disease. Recovery confers a protective im-munity; there is also an experimental vaccine. Sensible preventive measures involve avoiding tick or deerfly bites and not handling sick or dead wild rabbits.

VIRAL LYMPHOID INFECTIONS

Viruses frequently replicate in lymphoid tissue during the incubation period of an illness. A child's swollen lymph nodes and tonsils are warnings every parent recognizes — an infec-tious disease is about to erupt. Sometimes the swelling recedes with no further symptoms be-cause the body's defense mechanisms win, and destroy the invaders. At other times viruses seem to establish a prolonged latency. We have previously mentioned the role of retroviruses in cancers of the lymphoid tissues (Chapter 10). Here we will cover two other important viral infections of lymphocytes, infectious mono-nucleosis and AIDS.

INFECTIOUS MONONUCLEOSIS

Infectious mononucleosis is one of several dis-eases believed to be caused by the Epstein–Barr

virus (EBV). The others include nasopharyngeal carcinoma and Burkitt's lymphoma (discussed in Chapter 10).

THE VIRUS.   Epstein–Barr virus, a member of the herpesvirus group, is a DNA virus (Figure 22.10). It multiplies in B lymphocytes. The virus is iden-tical to that isolated from tumor tissues in Burkitt's lymphoma; antibodies produced in Burkitt's disease and in nasopharyngeal car-cinoma are identical to those found in infectious mononucleosis. EBV is being intensely studied as an oncogenic agent. It is of particular interest because it may be one cause of the specific chro-mosomal rearrangements seen in cancerous cells in Burkitt's lymphoma.

THE DISEASE.   In infectious mononucleosis, the virus multiplies during a long incubation period; then a state of general fatigue, relatively mild fever, sore throat, and swollen lymph nodes de-velop. Moderate enlargement of the spleen is common, and liver-function tests are abnormal. Rashes or jaundice are less common symptoms.

The disease results in proliferation of circu-lating lymphocytes and monocytes; many of the lymphocytes are atypical. Some researchers con-sider infectious mononucleosis to be a self-limiting form of cancer of the B-cell producing lymphoid tissues, which the body is able to cure itself. Diagnosis is made by demonstrating the presence of **heterophil** antibodies, which are not specific for EBV but which agglutinate specially treated sheep, horse, or ox red blood cells. It is also possible to test directly for anti-EBV anti-bodies. The disease is usually benign, with mild symptoms clearing up in two to three weeks. Complications develop in less than 5% of cases, probably associated with the failure of the lym-phoid tissues to restrict viral multiplication.

THE TREATMENT.   There is no specific therapy for infectious mononucleosis. Bed rest and analge-sics for the sore throat may be advised. Recovery confers immunity, not only against mono-nucleosis but apparently also against the other EBV diseases.

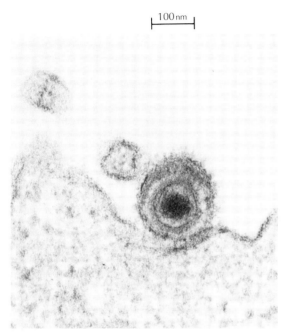

⊢ 100 nm ⊣

FIGURE 22.10   Epstein–Barr Virus
The E-B virus is a DNA-containing, enveloped herpesvirus. It causes infectious mononucleosis and has been implicated in three different malignant cancers. (Courtesy of Jack D. Griffith, University of North Carolina, Chapel Hill.)

EPIDEMIOLOGY AND CONTROL MEASURES.   The patient excretes EBV in his or her saliva. Droplets of airborne saliva, or exchanges of saliva during direct contact such as kissing, are the most likely means of transmission. This disease affects only human beings, although closely related viruses have been found in monkeys.

The virus is globally distributed, but its disease manifestations are localized. Infectious mononucleosis is largely limited to the United States, Canada, Europe, and Scandinavia. It is quite rare in tropical or less-developed nations. Burkitt's lymphoma and nasopharyngeal carcinoma are predominantly but not exclusively found in Africa and Southern China, respectively. There is no widely accepted explanation for this curious distribution pattern.

As we noted, EBV is believed to be oncogenic. For this reason, vaccine research has been pursued cautiously. A purified capsular protein preparation potentially may be safe and effective, but this line of research is still in an early stage.

ACQUIRED IMMUNODEFICIENCY SYNDROME

Acquired immunodeficiency syndrome (AIDS) is a "new" disease, in that it was first described in 1981. Efforts to track its origin indicate that an infection of the AIDS type in humans can be traced back only a few years. AIDS gives all the indications of being a transmitted disease, and of being caused by an infectious agent. Researchers have evidence that the human T-cell lymphotropic virus type III (HTLV-III) and a similar, perhaps identical viral strain designated LAV (lymphadenopathy-associated virus) cause this disease (Figure 22.11). HTLV belongs to the family of T-type retroviruses; the Type I strains have also been associated with a relatively uncommon type of human leukemia.

In AIDS, the agent appears to multiply in and kill T_4 or helper T-lymphocytes, the number of which declines steadily. T suppressors soon outnumber T helpers, which causes the B lymphocytes to become inactive. Progressively, there are fewer and fewer functional lymphocytes of any type. Eventually the patient is totally without immunity. All individuals diagnosed with AIDS in the United States have either died, are slowly worsening, or are stable at this time; none have recovered. Lymphadenopathy appears to be the first symptom; some cases do not progress beyond this point. Others develop Kaposi's sarcoma, a dermal cancer. The mortality rate is highest in those (over half) who develop opportunistic infections with otherwise unusual agents such as *Pneumocystis carinii*. In one recent case, the patient acquired seven different conditions cumulatively. He had Kaposi's sarcoma, two systemic viral infections (cytomegalovirus and generalized herpes simplex), one bacterial infection (*Mycobacterium avium*), two fungal infections (*Cryptococcus neoformans*

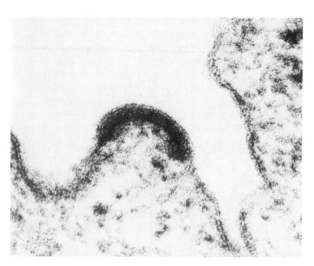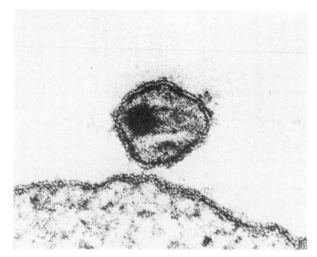

FIGURE 22.11   Human T-Cell Leukemia Virus
These micrographs show a single HTLV virus emerging from the host cell. HTLV Type I has been associated with an uncommon form of human leukemia. Antibodies to HTLV Type III appear in many cases of AIDS. (Courtesy of Syed Zaki Salahuddin.)

and *Candida albicans*), and one protozoan infection (*Pneumocystis carinii*). He had received vast doses of eight different antibiotics to no effect. At the time of his death generalized infections had taken over every organ in his body.

AIDS was first recognized in homosexual males in the United States, with cases clustering in the New York and Los Angeles areas. It was hypothesized that AIDS was a sexually transmitted disease. Sexual transmission, however, turned out to be only one aspect of its epidemiology. Shortly, cases appeared among intravenous-drug abusers and hemophiliacs. They shared a common factor — exposure to other people's blood — drug users through shared needles, and hemophiliacs through receiving clotting factors prepared from donor whole blood. Other affected groups included Haitians, for whom the route of exposure is unclear, and the mates and infants of heterosexual AIDS cases. At the moment, the facts can best be explained by proposing an infective agent that is transmitted by blood and/or intimate physical contact. It

now appears that the HTLV viruses are endemic in parts of Africa, where they affect, almost exclusively, heterosexual adults and their infants. The HTLV-III virus, the most virulent strain, was transported from there to Haiti, and was brought back and introduced into the U.S. homosexual population by returning vacationers. From there it has spread in a limited fashion among the other affected groups mentioned earlier.

Some of the preventive measures suggested by public-health officials are for homosexual males to protect themselves by limiting their sexual contacts, for persons with AIDS to refrain from exposing others through sexual contacts, and for homosexuals and people of other high-risk groups to abstain voluntarily from giving blood.

The number of new cases diagnosed each year doubles the total previous number of known cases. It is believed that over 400,000 people in the United States may have been exposed, including possibly 94% of all hemophiliacs. Serological screening tests became available in 1985 to detect presence of antibodies to HTLV-III in the blood of potential donors. However, it is still undetermined what the risk or prognosis is for a person with a positive test and no current symptoms. The incubation period for AIDS may be two or three years. It is also highly likely that many exposed persons have a milder disease and recover.

Vaccines to stimulate antibodies against the coat and envelope proteins of HTLV-III are under development.

PROTOZOAN INFECTIONS

Two protozoan infections of the blood and tissues are among the world's greatest unsolved microbiological problems. Malaria, endemic through much of the semitropical and tropical zones, may infect in excess of 200 million persons per year. American trypanosomiasis (Chagas' disease) infects about 15 million persons annually, largely in South America. The total number of cases of both diseases has already been greatly reduced by insect-control measures, but major problems still remain in the areas of effective therapy and vaccination.

MALARIA

Because of its worldwide importance, effective malaria control is one of the World Health Organization's highest priorities. Many research groups the world over are searching for a safe malaria vaccine.

THE ORGANISM. The disease **malaria** is caused by any of the four species of the sporozoan parasite genus *Plasmodium*. The disease varies in severity depending on the species of parasite responsible (Table 22.4) *Plasmodium* has a complex life cycle that requires two hosts, the female *Anopheles* mosquito and the human being (Figure 22.12). The parasite undergoes three developmental cycles. The first cycle, which occurs in the mosquito, is sexual **sporogony.** Male and female gametes, acquired from a meal of human blood, fuse in the mosquito's stomach. The zygote then multiplies actively for several days to produce many **sporozoites.** Sporozoites migrate to the insect's salivary glands. When the infected mosquito bites its next human victim, sporozoites are injected into the new host's blood stream.

Natural infection in human beings is made up of the second and third cycles, together re-

TABLE 22.4 Comparison of human malarial diseases				
Feature	*Plasmodium falciparum*	*Plasmodium vivax*	*Plasmodium ovale*	*Plasmodium malariae*
Incubation period	12 days	14 days	14 days	30 days
Persistence in liver	No	Yes	Yes	No
Duration of erythrocytic cycle	48 hr	48 hr	48 hr	72 hr
Mortality	High if untreated	Low	Low	Low

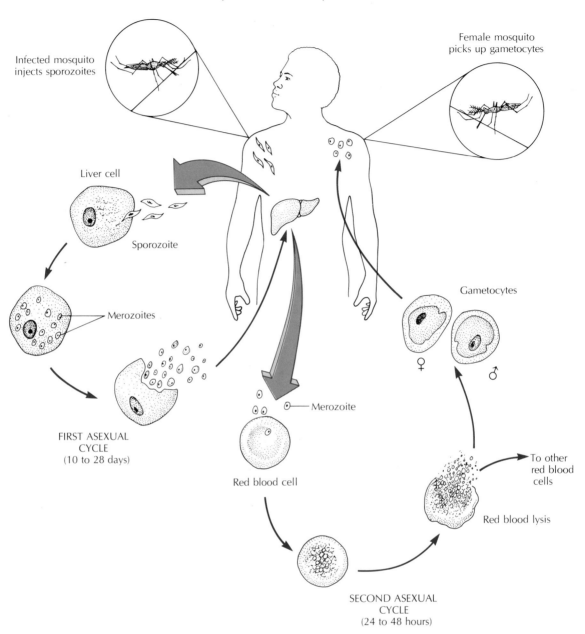

Infected mosquito injects sporozoites

Female mosquito picks up gametocytes

Liver cell

Sporozoite

Merozoites

Gametocytes

♀   ♂

FIRST ASEXUAL
CYCLE
(10 to 28 days)

Merozoite

Red blood cell

To other
red blood
cells

Red blood lysis

SECOND ASEXUAL
CYCLE
(24 to 48 hours)

FIGURE 22.12   The Life Cycle of *Plasmodium* in Human Beings
The malarial parasite completes two asexual phases in human beings. When an
infected mosquito bites a human, sporozoites pass from its salivary glands into the
person's blood stream. These sporozoites migrate to the liver where they undergo
several divisions to produce merozoites. The first phase of the disease occurs as the
merozoites multiply. During the second phase, the merozoites reproduce asexually
in red blood cells. As the red cells lyse, releasing more merozoites into the
circulatory system, the host may experience cyclic chills. Some of the parasites
produced enter new asexual replication cycles, while others mature into male and
female gametocytes. The gametocytes may then be ingested by another mosquito,
permitting the parasite to complete its life cycle by sexual replication within the
insect host.

ferred to as **schizogony.** In the second or **pre-erythrocytic cycle,** the injected sporozoites enter liver cells and for 10 to 28 days undergo further division, yielding 5000 to 10000 **merozoites** per liver cell. The merozoites are then released into the blood. This starts the third or **erythrocytic cycle.**

Merozoites are highly invasive. After they enter erythrocytes, each rapidly multiplies asexually into from 6 to 24 daughter merozoites, depending on the species of parasite. When the daughter merozoites are mature, in 48 to 72 hours, the red blood cell ruptures and releases them. These enter new red cells and multiply again. New waves of merozoites are released at regular two- or three-day intervals. Some merozoites, instead of replicating asexually, carry out meiosis in red cells, producing gametes, the male **microgametocytes** and female **macrogametocytes.** These forms have no future if they remain in the human host. However, if they are taken in by a mosquito of the correct species, they will start the whole cycle over again.

Each of these *Plasmodium* developmental states — sporozoites, merozoites, and gametocytes — infects a distinct target cell and has a distinct set of surface antigens. Each stage is hidden from the immune system of its host for all but brief periods through intracellular replication. Each stage undergoes a series of antigenic changes that aid it in evading or deflecting the host's immune responses.

A person can also become infected by blood transfusion or any needle puncture bearing merozoites. When this happens, the recipient will immediately develop erythrocytic-stage malaria without a pre-erythrocytic phase. In *P. vivax* and *P. ovale* malaria, inactive parasites may persist indefinitely in the liver, reactivating years after the initial episode. They not only cause the individual to relapse but also serve as a source of the parasite from which to infect others.

The four *Plasmodium* species are differentiated primarily by careful examination of blood films stained with Wright's or Giemsa's stain. A highly trained microscopist can identify the different species of *Plasmodium* by the appearance of various forms in the erythrocytic stage. Neither culture nor animal inoculation is useful. Indirect malarial fluorescent-antibody techniques have been used occasionally.

THE DISEASE. The symptoms of malaria reflect wholesale destruction of red blood cells. Successive crops of merozoites tend to mature in synchrony and are released by hemolysis at about the same time. The hemolysis causes a sudden onset of fever and chills that usually lasts for no more than six hours. Anemia develops; hemoglobin from the destroyed red cells is excreted in the urine as a blackish pigment, thus the descriptive name sometimes used for malaria — "blackwater fever." In *P. falciparum* malaria (the most severe type) intravascular coagulation, liver and kidney necrosis, and brain congestion may kill about 10% of those infected, mainly young children. Deaths from the other three types are uncommon.

THE TREATMENT. Primary chemotherapy is directed against the merozoite stage of the protozoan. Chloroquine is the drug of choice because it has the greatest effectiveness and least toxicity. Unfortunately, *P. falciparum* in many areas of Africa and Asia has become resistant to this drug. In such areas, Fansidar, a combination of pyrimethamine and sulfonamide, is recommended. Drugs should be used for a period of several weeks. In *P. vivax* and *P. ovale* malaria, the parasites must also be eradicated from the liver to achieve a cure. Primaquine is used for this purpose.

EPIDEMIOLOGY AND CONTROL MEASURES. Both susceptible mosquito and human populations are required for transmission of the disease. Control measures are focussed on both hosts. Large-scale mosquito control with DDT (dichlorodiphenyltrichlorethane) was introduced after World War II, with a dramatic reduction of malaria incidence. However, insects resistant to the pesticide soon emerged, and these gains were partially wiped out. After DDT was found to be

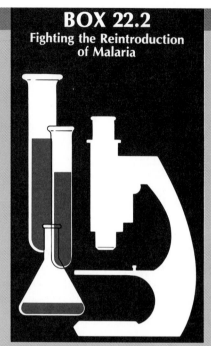

BOX 22.2
Fighting the Reintroduction of Malaria

The United States has enough mosquitos, goodness knows, but luckily we have almost no *Anopheles* mosquitos. Although the other kinds do plenty of biting, they cannot serve as hosts for malaria parasites if they do chance to suck them in. The Sutter and Yuba counties of California, however, are a rich agricultural area containing 120,000 acres of rice paddies. These wetlands provide a prime breeding ground for *Anopheles freeborni*, a mosquito species that serves as a vector for *Plasmodium vivax* malaria. The climate, which resembles that of the Punjab region of India, has attracted more than 5000 Punjabi immigrants. They often import cases of vivax malaria in its chronic liver form. Thus the area has human reservoirs of the parasite, the host mosquito species, breeding grounds — everything needed for real trouble. This unique situation presents a serious threat of reintroducing a complete cycle of endemic mosquito-transmitted malaria.

The challenge is to detect infected persons quickly, and to treat them to eradicate the parasite, while preventing them from being bitten by any mosquitos. The following measures are being taken. Unused rice paddies are kept drained. Fish are stocked in flooded paddies to eat the mosquito larvae. All newly arrived Punjabis are blood-tested, and any with positive vivax malaria are started on antimalarial drugs. Insecticides are fogged over a one-half-mile radius around the person's household and work place, starting within 12 hours of the detection of any new case and repeated weekly. So far, these measures have been successful.

Source: *Morbidity and Mortality Weekly Reports* 1980; 29(5).

harmful to animals, its use was halted in the United States, although it and other even more toxic pesticides are widely used in other countries. Drainage and oil-filming of standing water are widely used to eradicate mosquito breeding places. People are protected directly by household screening, protective clothing, and effective repellents.

Malaria is also prevented by taking drugs prophylactically. Either chloroquine or Fansidar should be taken regularly by those living in or visiting a malarial area. On leaving the area, one should take primaquine to clear the liver of parasites. The armed services use these precautions, which were largely responsible for the low incidence of malaria in Korean and Vietnam war veterans. Unfortunately, in many of the worst-affected countries, there is no money to buy drugs and no health service to distribute them.

Malaria was once endemic in the United States, but mosquito control eradicated it a number of years ago. Occasionally people return from endemic areas as "one-case epidemics." The pathogen spreads no further unless they become blood donors or share syringes in the injection of drugs. This imported malaria can be prevented in travelers by the systematic use of prophylactic drugs as just mentioned.

In some less-developed areas of the world, malaria is largely out of control, and will remain so until a vaccine becomes available. Under the direction of the Agency for International Development (AID), a network of universities and other research institutions is carrying out an international effort to develop a malaria vaccine. One current possibility is a vaccine containing a protein called the CS (circum–sporozoite) antigen, which is the major surface antigen of the sporozoite stage. The gene for the CS protein has been cloned, making possible quantity prod-

uction of the protein. Testing of a sporozoite vaccine is scheduled to begin in 1986.

TRYPANOSOMIASIS

The trypanosomes cause severe illnesses both in Central and South America and in Africa. The trypanosome diseases have many similarities.

Trypanosoma cruzi and *T. rangeli* cause **American trypanosomiasis,** or **Chagas' disease.** It is seen from Texas to Argentina and is a leading cause of cardiovascular death in males in endemic areas. *T. rhodesiense* and *T. gambiense* cause **African trypanosomiasis** or **sleeping sickness,** an equally devastating disease.

In both diseases, the flagellar parasite (Figure 22.13) is transmitted by biting insects that infest

FIGURE 22.13 *Trypanosoma*
(a) A micrograph of *Trypanosoma cruzi,* the causative agent of Chagas' Disease, or South American trypanosomiasis. (×4897. Courtesy of Stephen G. Baum, MD, Albert Einstein College of Medicine, New York.) (b) A cross-section of the organism, revealing the internal structures; note the section of the flagellum at the bottom. (From Erich Scholtyseck, *Fine structure of Parasitic Protozoa,* New York: Springer-Verlag (1979), Fig. 2, page 42.)

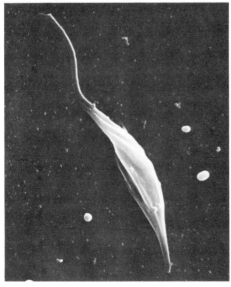

(a)

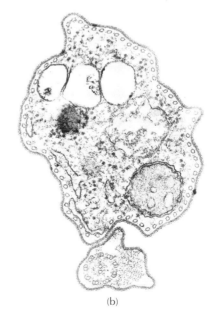

(b)

open, wood-frame, and mud-floor dwellings. The agent is passed in a motile **trypomastigote** form in the insect's saliva or feces, and these are rubbed into the itching bite wound. After an incubation period, the acute phase develops. During this phase, the motile forms are numerous in the blood plasma; nonmotile organisms are multiplying in the myocardium, smooth and skeletal muscles, and sometimes the central nervous sytem. Generalized toxic symptoms are present.

In Chagas' disease, about 10% of individuals die during the acute phase, usually of heart failure. The rest recover spontaneously. The vector is the triatomine or reduvid bug.

In Africa, sleeping sickness takes two forms. The vector for both is the tsetse fly. Rhodesian sleeping sickness, found in Eastern Africa through Ethiopia, Zambia, and Zimbabwe, is often rapidly fatal. The disease is so lethal that it has prevented people from inhabiting and utilizing some of the most fertile lands and river valleys on the continent. The Gambian form of sleeping sickness, in Senegal through west central Africa to the Sudan and Uganda, is much milder. If significant tissue damage has been sustained, the disease passes into a chronic phase marked by cardiac disease and/or enlargement of the esophagus and colon due to smooth muscle destruction. These conditions may eventually prove fatal. There is no effective specific therapy. Suramin, a nitrofuran drug, has been used with variable or incomplete success.

The trypanosomes have a highly developed ability to alter their surface antigens or **variable surface glycoproteins (VSGs)** through continual genetic variation. Researchers estimate that *T. brucei* has the genetic potential to produce hundreds, possibly thousands of different antigen combinations. Although a population of trypanosomes all make only one VSG at any given time, they periodically shift in synchrony to create a new antigen combination every few days during the course of the disease. Although the patient produces antibodies against VSGs, the antigenic target constantly shifts so that the patient's immunity is never effective. By the same token, the immunity that develops following infection is not protective. With this remarkable ability of the parasite to change antigens, the prospects for vaccination are understandably dim at this time.

Epidemiologic control is based on improving the housing conditions of the target population and spraying to eliminate the vector (Figure 22.14). The economic and social impacts of these

FIGURE 22.14   Control of the Tsetse Fly
A laboratory technician in Nigeria examines and dissects captured tsetse flies in an effort to determine the type of trypanosome present in a given area at a given time. (Photo courtesy of World Health Organization/R. daSilva.)

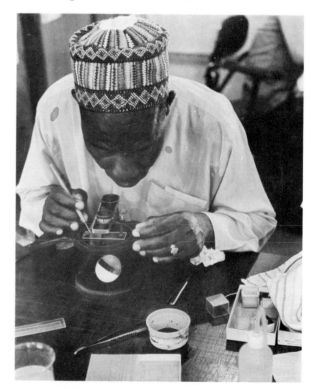

diseases in Central and South America, where in some areas up to 50% of the people are affected, are great. In Africa, the parasites also infect domestic animals and make large areas of rich grazing lands in Uganda and other countries lethal for cattle. All in all, the trypanosomal diseases are a major barrier to economic progress in large areas of the world.

HELMINTHIC INFECTIONS

We encountered the helminths earlier in Chapter 5. Here we'll note some major helminthic parasites found around the world in semitropical and tropical climates.

FLUKE DISEASES

Schistosomiasis and other diseases caused by the many species of flukes are found in Asia, Africa, South America, and the Central American and Caribbean areas, afflicting about one in twenty of the world's population — about 300 million people. U.S. citizens contract them occasionally during travel, and immigrants to this country often need treatment on arrival for fluke infestation. In the United States, the only endemic fluke disease is the very mild skin disease, called **swimmer's itch,** caused by a schistosome (Figure 22.15).

The flukes, you'll recall, are flatworms. Their complex life cycle (review Figure 5.19) involves several free-swimming larval stages and an intermediate stage in which they develop in snails. Fluke diseases, then, are intimately linked to standing or slow-moving bodies of water, such as rice paddies or impoundment areas above dams. The more serious fluke diseases are not endemic in the United States only because we are lucky enough to lack the appropriate snail host species.

Fluke larvae called cercaria often enter human skin through the bare feet of people work-

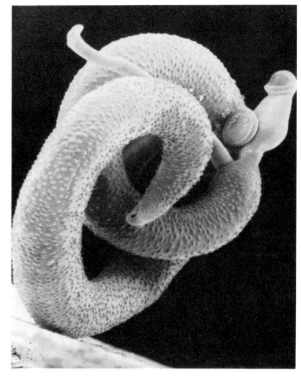

FIGURE 22.15 A Schistosome
This SEM shows the entwined bodies of male and female worms of *Schistosoma mansoni*, one of three species that causes schistosomiasis in humans. The slender body of the female is almost hidden by the larger body of the male. Adult worms, which live from 3 to 30 years, remain intertwined except when the female moves away each day to lay her eggs. (Courtesy of Ming Wong.)

ing in rice paddies or netting fish. The larvae then quickly mature and multiply in the veins of various body organs. There are many different kinds, such as lung flukes, liver flukes, and blood flukes. Tissue damage includes abscesses, ulcers, and allergic reactions. Chemotherapy with praziquantel, oxamniquine, or imidazole drugs is used, and may eradicate the parasites if the infection is treated early enough. Once the flukes are well established, however, therapy is usually not able to bring about a cure.

Control measures hinge on eliminating the snails, and preventing the contamination of water with human feces. Periodic drainage of

TABLE 22.5 Epidemiology of some circulatory and lymphoid infections			
Disease	Reservoir	Transmission	Incubation (Days)
Rocky Mountain spotted fever	Ticks, dogs, rodents	Tick bite, tick feces	3–10
Typhus	Human beings	Body louse feces	12
Dengue	Human beings, monkeys, mosquitos	Mosquito bite	5–6
Yellow fever	Human beings, monkeys, mosquitos	Mosquito bite	3–6
Plague	Wild rodents, other animals	Flea bite, handling diseased animals, droplet inhalation	2–6
Brucellosis	Many domestic and wild animals	Contact with tissues, discharges, infected meat or dairy products	Variable
Tularemia	Many domestic, wild animals	Almost any means, commonly handling diseased animals	3
Malaria	Human beings, monkeys	Mosquito bite, transfused blood, contaminated syringe	Variable
American trypanosomiasis	Human beings, animals	Feces of triatome bug	5–14
Infectious mononucleosis	Human beings	Person-to-person by pharyngeal secretions	2–6 wk

Adapted from Benenson AS, *Control of Communicable Diseases in Man*. 14th ed., 1985.

rice paddies and ponds helps disrupt the life cycles of both the parasite and the snails. Research is underway to develop both better therapeutic drugs and preventive vaccines. For review purposes, the epidemiology of some circulatory and lymphoid diseases are outlined in Table 22.5.

TABLE 22.5 Epidemiology of some circulatory and lymphoid infections (continued)		
Communicable	**Patient Control Measures**	**Epidemic Control Measures**
Not directly communicable from person to person	Remove all ticks from patient	Identify ticks: clear people from high-risk areas
Not directly communicable	Apply insecticide to delouse patient, especially hair, clothing, and bedding	Apply insecticide to all contacts
Not directly communicable	Patient housed in screened quarters	Mosquito destruction
Not directly communicable	Patient housed in screened quarters	All houses in vicinity should be sprayed, vaccinate susceptible persons
Not directly communicable except when patient has pneumonic form	Rid patient and effects of fleas; strict isolation until 48 hr after antibiotic therapy begins; disinfect discharges	Intensive case-finding; careful management of contacts flea and rodent destruction
Has evidence of communicability from person to person	Disinfect discharges	Search for vehicle of infection such as infected dairy herd
Not directly communicable	Disinfect discharges	Search for source of infection
Not directly communicable	Patient housed in screened quarters; disinfect materials contaminated with blood	Mass spraying, case-finding, chemoprophylaxis
Not directly communicable	None	Spray houses, improve dwellings
Unknown, presumably long	Disinfect articles soiled with nasal or throat discharges	None

SUMMARY

1. The circulatory diseases develop after specific events introduce the agent into the circulation. The causative agents also must have mechanisms that promote their continued survival in the face of the host defense mechanisms. Some agents accumulate a protective layer of clotted blood, while others multiply intracellularly, in the endothelium, lymphocytes, or red blood cells. Unchecked multiplication of an unprotected extracellular organism (septicemia) is an indication that host defenses are paralyzed or absent.

2. The symptoms of the circulatory diseases reflect the modes of pathogenesis. The circulation can distribute bacterial toxins and inflammatory substances. These lead to fever, malaise, and chills. When the circulation is blocked by intravascular coagulation or when oxygen-carrying red blood cells are destroyed, tissues become oxygen-starved. Muscle aches and liver, kidney, and neurologic damage result. Fragility of red cells or capillary hemorrhage produces petechiae, bloody vomit, or darkened urine. Hyperactivity of lymphoid tissues provoked by bacterial parasitism leads to swelling and ulceration.

3. Insect vectors transmit many circulatory infections. An insect bite is a very effective way for the organism to be introduced to the circulation. The bite mechanism allows the organisms to bypass the surface defenses of skin and mucosa, and to enter the connective tissues and their capillary beds directly.

4. The endemic area of a vector-transmitted disease corresponds to the habitat of the vector species. When pesticides first became available, it was hoped they would eradicate malaria and other vector-borne tropical diseases. However, vectors have become increasingly resistant. Pesticide use is being supplemented with other means of reducing vector populations, such as drainage, and protecting persons from their bites.

5. The antibody-mediated immune response is not particularly helpful in combating circulatory infection. Most of the agents survive in microenvironments in which they are not exposed to antibody. We do not have a clear understanding of how the body actually does bring these infections under control in convalescence. Vaccines have been largely ineffective.

DISCUSSION QUESTIONS

1. What preexisting conditions place a person at risk for acquiring infective endocarditis? What types of experiences are likely to introduce the necessary microorganisms?

2. Continuous penicillin prophylaxis is recommended for persons with rheumatic fever but not for persons with heart-valve abnormalities, who receive penicillin only just before a scheduled dental procedure or other trauma. What reasons can you advance for this difference?

3. Review previous chapters and prepare a list of infectious diseases that can culminate in septicemia.

4. Discuss personal actions an individual might take to minimize the chance of being infected by (a) tickborne, and (b) mosquito-borne agents.

5. Identify the target tissues for (a) the rickettsial diseases, (b) plague, (c) brucellosis, (d) Epstein–Barr virus, and (e) the malaria parasite.

6. Identify diseases described in this chapter for which effective drug control is lacking. Also list those for which vaccination is sorely needed but unavailable.

7. By reference to the most current *Health Information for International Travelers*, determine what special vaccinations and precautions you would require before visiting (a) Portugal, (b) Kenya, (c) Indonesia, and (d) Jamaica. Note recommendations for malaria, yellow fever, dengue, and cholera.

ADDITIONAL READING

Balows A, Sonnenwirth AC, eds. *Bacteremia: laboratory and clinical aspects.* Springfield, Ill.: Charles C Thomas, 1983.

Burgdorfer W, Anacker RL, eds. *Rickettsiae and rickettsial diseases.* New York: Academic Press, 1981.

Donelson JE and Turner MJ. How the trypanosome changes its coat. *Sci. Am.* 252:44–51, 1985.

Godson, GN. Molecular approaches to malaria vaccines. *Sci. Am.* 252:52–59, 1985.

Health information for international travelers. Atlanta: U.S. Department of Health and Human Services (Centers for Disease Control). Published annually.

Kolata G. Scrutinizing sleeping sickness, *Science* 226:956–959, 1984.

Kolata G. Avoiding the schistosome's tricks, *Science* 227:285–287, 1985.

Noble ER and Noble GA. *Parasitology: The biology of animal parasites,* 5th ed. Philadelphia: Lea and Febiger, 1982.

Skinner FA, Quesnel LB. *The streptococci.* New York: Academic Press, 1978.

Yekutiel P. *The eradication of infectious diseases: a critical study.* New York: S. Karger, 1979.

PART
FOUR

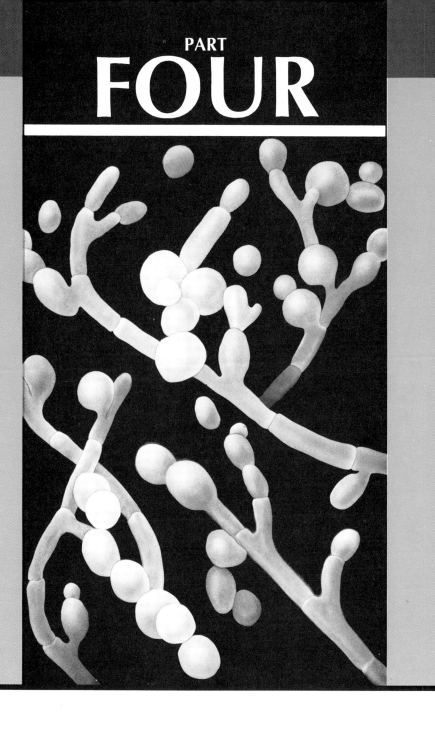

CONTROLLING
MICROORGANISMS

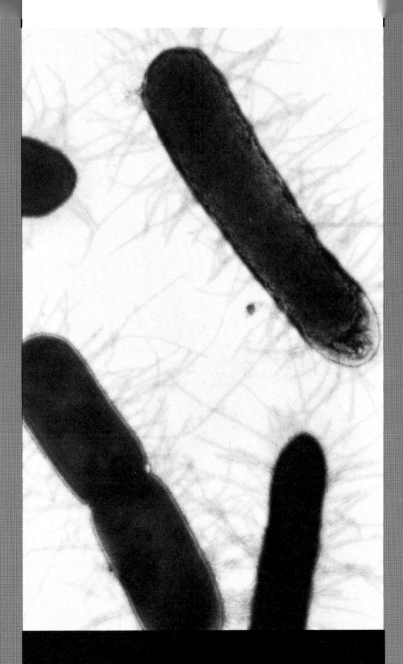

DISINFECTION AND STERILIZATION

CHAPTER OUTLINE

Disinfection and sterilization are everyday techniques that are used to ensure the safety of foods, medications, and household and work environments. The goal of these procedures is to reduce the transmission of infection. In hospitals, vast amounts of time and money are spent on these functions. In our daily lives, we are deluged with advertisements for personal and home products that promise to make us healthier and to protect us from disease.

In this chapter, we will introduce you to the basic concepts of disinfection and sterilization and explore both their usefulness and their limitations. These concepts are relatively new, having been introduced only in the past 100 years (Figure 23.1). We will not be concerned with presenting detailed instructions for the use of equipment and commercial products. These instructions may be obtained from the manufacturers, and should be carefully followed. Each health-care institution has its own sanitation procedures, which you may learn as needed. We will also discuss isolation and precaution techniques used when caring for persons with various infections, examine some infectious disease issues in occupational safety for health personnel, and look briefly at surgical asepsis.

BASIC CONCEPTS OF MICROBIAL CONTROL

Disinfection and sterilization is an applied science. It draws on basic concepts of microbial growth and survival and the means of transmission of infectious disease. You may find it useful to spend a few minutes looking back through Chapters 6 and 14 before proceeding.

Disinfection and sterilization may have any or all of three objectives:

■ Adopt general precautionary and aesthetic methods, such as washing and sanitizing eating implements after use.

FIGURE 23.1 Surgery a Century Ago
Not so many years have passed since Pasteur proposed the Germ Theory of Disease; in 1878 he suggested the sterilization of surgical instruments. Yet only a century ago surgery was performed without any attempt to control microorganisms. In *The Gross Clinic* by Thomas Eakins, the professor and his students demonstrate surgical procedures in a lecture hall without wearing gloves or masks and dressed only in ordinary street clothes. (Courtesy of the Jefferson Medical College of Thomas Jefferson University, Philadelphia.)

- Establish protective barriers for particularly susceptible body areas, persons, or populations, such as sterilizing before surgery everything that might make contact with the exposed tissues.
- Contain and inactivate pathogenic microorganisms that might be transmitted from infective individuals, such as using isolation procedures in the treatment of a person with hepatitis.

SOME TERMINOLOGY

A number of terms are commonly used (and not uncommonly misused), in designating the various levels of microbial control.

- **Sterilization** is any process that completely destroys all forms of microbial life. Sterilization implies a very high confidence level that the procedure worked — a vanishingly small probability that any viable microorganisms will be left. In commercial canning, for example, processing should reduce the possibility of *Clostridium botulinum*·contamination to one surviving spore per 10^{12}(one million million) containers. The term "sterilize" cannot correctly be applied to skin cleansing; it is not possible to sterilize skin without killing the tissue.
- **Disinfection** is the removal or destruction of pathogens. It reduces the numbers of viable organisms in the material. Some more resistant microorganisms may not be completely eliminated. The term "disinfectant" is correctly used to describe a chemical treatment used on an inanimate object — linens, a scalpel, the air, or a floor.
- **Antiseptic** refers to a treatment applied to living skin or tissue to kill microorganisms. Often the same chemical preparation may be used in concentrated solution as a disinfectant and in a dilute solution as an antiseptic. Many disinfectants cannot be used on living tissue at all.

- **Decontamination** refers to a process in which both living pathogens and their toxic products are removed.
- **Degerming** is the term used for careful tissue cleansing, using an antiseptic. It may reduce the total bacterial population 1000-fold.
- **Sanitizing** is the systematic cleansing of inanimate objects to reduce the microbial count to a safe level. Examples are the usual procedures for cleaning lavatories and toilets.

The following terms designate whether the organisms are temporarily or permanently neutralized.

- **-Cidal** is a suffix referring to a treatment that causes irreversible cellular changes so that microbes cannot reproduce — the functional definition of death. A prefix indicates the type of organism affected, e.g., bactericidal, fungicidal.
- **-Static** is a suffix referring to treatments that halt microbial growth for as long as the inhibitory substance or state is present. If the inhibition is removed, growth begins again. Thus spoilage may begin soon after a frozen food is defrosted. As in -cidal treatment, a prefix indicates the type of organism affected, for example, bacteriostatic or fungistatic.

KINETICS OF MICROBIAL DEATH

When microbial-control measures are all applied, microbial death is not instantaneous. All microbes are not equally susceptible, and therefore do not all die at the same time. Suppose you expose a microbial population to a lethal treatment for a period of time. In the first time interval, a certain fraction dies. In the next equal time interval, the same fraction of the remaining population dies. When you graph the number of organisms surviving over a long period of exposure to the lethal treatment, you get a **logarithmic death curve** (Figure 23.2). Practically speaking,

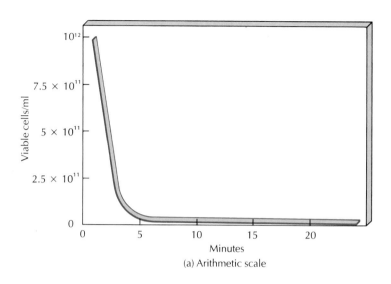

(a) Arithmetic scale

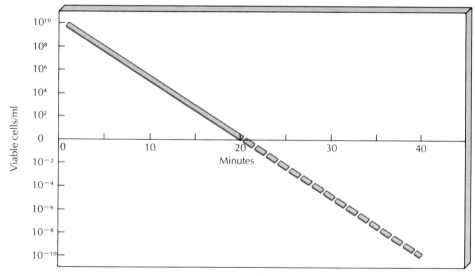

(b) Logarithmic scale

FIGURE 23.2 Microbial Death Curves
Microbial killing is exponential. The graphs illustrate a hypothetical situation in which 90% of the viable microbial population dies every two minutes. (a) The arithmetic scale (black) does not show negative numbers. Thus, it looks like complete death has been achieved at about six minutes. This is not strictly true because of the prolonged possibility of a few survivors. (b) The starting concentration on the logarithmic scale (color) is 10^{10} (10 billion) viable cells per milliliter. At 20 minutes, it has dropped to 10^0, or one viable cell per milliliter. The concentration continues to drop at the same rate, so at 40 minutes there is 10^{-10} viable cell per milliliter — which translates to one viable cell in 10 billion milliliters. In other words, out of every 10 billion 1-ml samples, all except one probably would be sterile. This might be considered an acceptable risk.

this means you must prolong the treatment period until the probability of encountering a single survivor is acceptably slight. Shortening the exposure time may leave a number of survivors.

The time required to sterilize depends on the initial number of microorganisms in the material, which is called the **microbial load.** The larger the contaminating population, the longer the sterilization time. This is because you start higher up on the death curve. If objects to be sterilized or disinfected are thoroughly sanitized first, the microbial load will be reduced. Thus treatment begins farther down on the death curve and will be more effective more quickly. To kill the organisms, the treatment must be strong enough and long enough so the cells cannot afterward repair the damage and recover.

HOW MICROBES ARE KILLED

The destructive effects of sterilizers and disinfectants are largely nonspecific. They damage all

---

BOX 23.1
Laboratory-Associated Infections

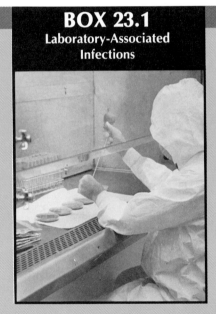

In a modern microbiology laboratory, sophisticated aseptic techniques, disinfection, and containment procedures are practiced. Employing these procedures, however, is a relatively recent development. Many microbiologists can recall a time, only a few years ago, when tuberculosis culture work was done right out on the bench in the main laboratory. We used no special precautions and had all sorts of people milling about. Few of us were ever tuberculin-tested, so there are no data on the outcome of this sloppy procedure. Today, however, samples submitted for isolation and identification of *Mycobacterium tuberculosis* are handled with extreme care. Workers use a separate restricted area, biological safety cabinets, and protective cloth-

ing (see photograph). They are regularly tuberculin-tested to detect infection.

We have recognized that diagnostic laboratory work is definitely hazardous. The samples submitted to a clinical laboratory may contain anything or nothing. The technologist

does not know if the sputum just received contains normal flora or plague bacilli. Blowing out pipettes by mouth, flaming heavily loaded loops, popping off test-tube caps, mouth pipetting and careless uses of needles and syringes are hazardous acts that are now frowned upon.

There are about 4000 laboratory-associated infections per year in the United States, and there have been at least 173 deaths in this century, most of which occurred before 1950. In recent years, deaths have accompanied the first laboratory isolations of newly discovered viral pathogens such as Lassa fever and Marburg disease. The last recorded smallpox victim (1978) was an "innocent bystander," a laboratory photographer.

forms of life. This is why disinfectants are toxic if taken internally, and can't be allowed to contaminate foods.

Microbial proteins may be destroyed by **denaturation,** a destructive change in the secondary and tertiary structures. Moist heat coagulates cellular proteins, and alcohol causes them to form precipitates (Figure 23.3). **Alkylation** is a type of chemical reaction that replaces hydrogen atoms in the amino acids with other functional groups, altering the protein's function. Ethylene oxide gas and the aldehydes are alkylating agents.

Cell membranes may be destroyed by the action of agents that interfere with the weak bonds that hold membranes together. Some examples of these agents are detergents, other surface-active agents, alcohols, and phenolic compounds; some of these will be discussed later in this chapter.

Radiation attacks nucleic acids and proteins. Ultraviolet light alters DNA bases. Other forms of radiation such as x-rays and gamma rays ionize most cellular constituents, randomly breaking covalent bonds.

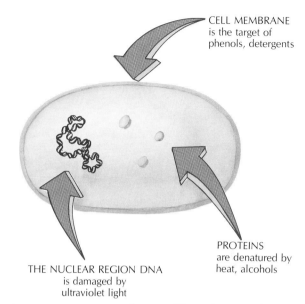

CELL MEMBRANE
is the target of
phenols, detergents

PROTEINS
are denatured by
heat, alcohols

THE NUCLEAR REGION DNA
is damaged by
ultraviolet light

FIGURE 23.3 Approaches to Killing Microorganisms The microbial cell can be attacked on several fronts. Its membrane may be disrupted, its proteins denatured, or its nucleic acids inactivated.

TESTING AND EVALUATION

Sterilizing agents and disinfectants are tested by the manufacturer, using standardized methods. Recommendations for use are then based on the manufacturer's testing.

There are over 8000 registered disinfectants. Disinfectants for use on inanimate objects are regulated by the U.S. Department of Agriculture, and chemicals used in foods to inhibit microbial growth are regulated by the Food and Drug Administration. The Association of Official Analytical Chemists (AOAC) Procedures Manual is the standard reference for evaluation procedures.

INITIAL TESTS

Not all microorganisms are equally susceptible to killing by any one technique; some are particularly resistant. These include the tuberculosis organism, viruses (especially hepatitis B), and bacterial endospores. Disinfectants or sterilizers are tested against a battery of pure cultures to define their range of usefulness. Testing that uses pure cultures is called **initial testing.**

In initial testing, two variables of exposure may be tested: the intensity of exposure and the time of exposure. Killing can be accomplished using different combinations of these two factors. Applying a high temperature for a short time may have the same final effect as applying a lower temperature for a longer time. The choice of conditions depends on practical considerations such as the time available, how resistant the materials are to heat, and cost.

If the time of exposure is fixed, for example, at 10 minutes, you can then test to find out the intensity of exposure (disinfectant concentration, temperature, radiation intensity) needed to kill a pure stock culture in that period of time.

The **thermal death point** (TDP) is the temperature that kills all members of a 24-hour-old broth culture of a given species of bacteria. The **Z value** is the temperature needed to kill 90% of the organisms in a fixed time period (Figure 23.4a). It is important to realize that the TDP or the Z value may be different for *Salmonella*, for example, than it is for *Staphylococcus*, because the two microbial species are not equally sensitive to heat.

Suppose now that it is the intensity of exposure that is fixed to a temperature of 100 degrees Celsius, and the time is the variable to be tested. The **thermal death time** (TDT) is the time needed to kill all organisms in a 24-hour broth culture at the given temperature. The **D value** is the time needed to kill 90% of the organisms (Figure 23.4b).

Z values and D values are used to compare control measures. They may be extrapolated to any desired degree of lethality. In practice, we use Z values to find the most effective treatment to do a job when only a certain amount of time can be allocated: for example, when we need to know how intense a radiation beam must be to sterilize disposable syringes on an assembly line when the syringes are under the beam for only one second. D values are used when the intensity of exposure is fixed, but the time can be extended within limits, for example, to decide how long it takes to sterilize a rubber object at 80 degrees Celsius (the upper temperature limit for rubber to keep its elasticity).

Chemical disinfectants are evaluated by comparing their effectiveness to that of phenol (carbolic acid) by a ratio called the **phenol coefficient** (PC). Phenol, a cyclic organic alcohol, has an honorable history as the first disinfectant used in surgery. Modern disinfectants are generally much more effective than pure phenol, which is rarely used by itself now, but phenol is still the standard used for measuring disinfectant effectiveness. The PC value is highly dependent on environmental variables such as species of microorganism, growth phase, temperature, and pH.

IN-USE TESTS

New sterilizers and disinfectants are also tested in conditions that simulate actual use, such as with mixed cultures, on actual sample surfaces and equipment, and with typical application and handling methods. Later, field trials are conducted in actual clinical settings. Many of the larger hospitals carry out their own disinfection testing programs. All hospitals also routinely monitor the effectiveness of their control programs and housekeeping. This permits prompt detection of equipment failure, ineffective lots or dilutions of disinfectant, and human error.

PHYSICAL CONTROL AGENTS

Temperature is the most commonly used physical method for killing and inhibiting microorganisms. In addition, there are some useful applications for radiation. Filtration and controlled air flow are increasingly being used not to kill but to physically exclude microorganisms from materials or areas.

HEAT

Moist or dry heat are extremely common agents for microbial control. Moist heat is more effective than dry. When hot air contains moisture, it readily coagulates protein. Converting water to steam gives the water molecules additional energy. When steam touches a cooler object, it condenses (turns back to water) giving up the additional heat energy to the object. The temperature of steam can vary, depending on the atmospheric pressure. At sea level, in an unsealed vessel, steam can be no hotter than 100 degrees Celsius. In a pressurized vessel, steam may be heated to much higher temperatures, depending on the pressure imposed. The steam in turn will heat the objects it touches to the elevated temperature.

One application of moist heat is plain unpressurized boiling. This method kills most vegetative microorganisms, but is useless against

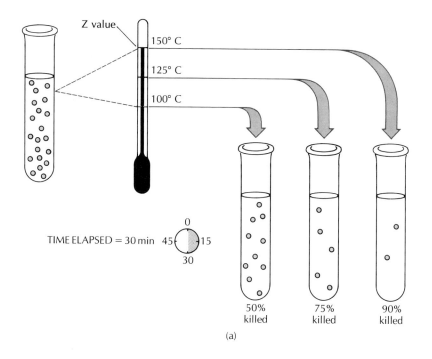

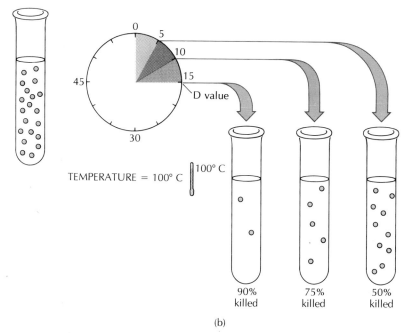

FIGURE 23.4 Determining Exposure Conditions
(a) If the time of exposure is constant at 30 minutes, different percentages of the organism will be killed by exposure at different temperatures. Ninety percent of the initial population is killed at 150 degrees Celsius. This is the Z value. (b) If the temperature is held at 100 degrees Celsius, the number of organisms killed depends on the time of exposure. Ninety percent are killed at 15 minutes; this is the D value. Note that complete sterility is not achieved in any of these situations.

hepatitis virus and most bacterial spores. Flowing unpressurized steam has a similar range of effectiveness, although its heating is more rapid. Neither boiling nor exposure to unpressurized steam can be relied on for complete sterilization.

AUTOCLAVING.   The most effective use of heat is **autoclaving,** the use of pressurized steam within a sealed chamber (Figure 23.5a). Materials to be sterilized are placed in a chamber, which is then sealed. Steam flows in from a generator and pushes out the room-temperature air. Once all air has been displaced, temperature-sensitive valves close, and steam continues to fill the chamber typically until the pressure is 15 pounds per square inch (psi). At this pressure, the temperature of the pressurized steam will reach 121 degrees Celsius. If all the air is not pushed out, the desired pressure may be obtained, but the temperature will not be high enough. Timing

FIGURE 23.5   The Autoclave
(a) The autoclave is a chamber for holding steam under pressure. Once the chamber has been loaded with materials to be sterilized, air is drawn from the chamber and steam flows in. When all the air has been exhausted, valves close while steam continues to enter, building pressure to the desired setting. (Courtesy of 3M.)
(b) A masking tape product printed with a heat-sensitive dye may be used in packaging materials to be autoclaved. (c) Biological controls such as the Attest system are used to test the effectiveness of autoclaving.

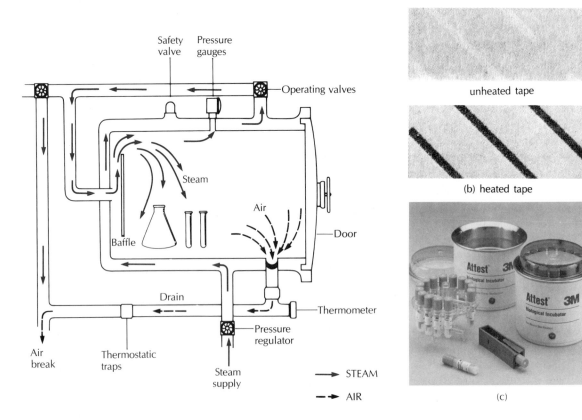

should begin only when a temperature gauge verifies a chamber reading of 121 degrees Celsius. Many autoclaves now have digital automatic controls and vacuum pumps to speed the exchange of air for steam.

To kill all microorganisms, the steam must be able to penetrate throughout the load and touch all the surfaces. Equipment, glassware, and linens should be packed and wrapped so that the movement of the steam is not obstructed. For example, in pipettes or other long thin containers steam enters well only if they are lying horizontally. Similarly, bottles must not be sealed completely, screw-cap tops should be loosened, wrappings permeable to steam should be used, and syringes should be disassembled.

It may take longer than 15 minutes for the interior of a wrapped package of linens or the contents of a large vessel to reach the necessary temperature. True sterilization begins only when *all* the contents of the autoclave have reached 121 degrees Celsius. When running large volumes of liquid, or dense objects, it may be necessary to autoclave for 30 minutes to an hour.

How do you know if the autoclave has reached the sterilizing temperature? Indicator cultures composed of a vial of spores and a vial of sterile nutrient broth may be processed with each load, preferably at its center. Afterward, the spores are mixed with the broth, and the indicator culture is incubated. The sterilization procedure is judged adequate if no growth appears (Figure 23.5b). An immediate check can be obtained with tapes imprinted with stripes of heat-sensitive dyes; these are used to package materials to be autoclaved. They develop a visible color only when their temperature reaches 121 degrees Celsius. On completion of a sterilizing run, the dyes should have turned color in all areas of the load.

Liquid and dry materials should be autoclaved separately because they require different treatment in the final steps. If the load contains liquids, the steam pressure must be released gradually after autoclaving so the liquids will not boil violently, causing breakage and loss.

However, if dry materials are processed it is important that the steam be rapidly evacuated so the materials will dry off. A soggy sterile wrap is completely useless because once removed from the chamber it is rapidly penetrated by airborne microorganisms. Wrapped materials should not be removed until they have dried.

The Attest system (Figure 23.5c) is widely used for testing the effectiveness of autoclave techniques. Test ampules containing spores of *Bacillus stearothermophilus* are placed in a large package in the most inaccessible part of the load; the purpose is to measure the greatest "challenge" the autoclave is being asked to meet. When the run has been completed, the ampule is squeezed in order to moisten the spores with medium. The ampule is then incubated in a special 56-degree-Celsius incubator and checked after 24 hours for signs of growth.

Autoclaving is not a suitable treatment for all objects and materials. Sharp-edged instruments may be rusted or dulled by autoclaving; solvents, oils, or greases may be evaporated; powders may be wetted; and equipment with tight-fitting surfaces that cannot be disassembled cannot be penetrated completely. All of these items are suitable for dry-heat sterilization, however.

OTHER HEAT TECHNIQUES. There are several other methods of destroying microorganisms with heat. **Pasteurization** is the partial disinfection of a fluid by heating. The process is designed to kill pathogens and reduce total microbial count while leaving the taste and consistency of the milk or other beverage unchanged. After regular pasteurization, some percentage of the harmless microorganisms survive. Pasteurization is applied not only to milk but also to other beverages produced commercially, such as beer and cider.

Tyndallization or **intermittent sterilization** may be carried out with minimal equipment. A solution or food preparation in a closed container is first boiled for one hour to kill all vegetative cells present. (Spores will not be killed in this first step.) The substance is then incubated overnight to encourage spore germination. It is again boiled for one hour, which will eliminate the spores that have germinated during the incubation. It is incubated and boiled for one hour a third time, after which it is presumed to be sterile. An alternative schedule, used for materials that are destroyed by boiling, specifies heat treatments of 60 degrees Celsius for one hour on five successive days (Figure 23.6).

Dry heat is useful for sterilizing empty glassware by placing it in an oven at 180 degrees Celsius for one to three hours. The effectiveness of dry heat is increased if the oven has a circulating fan. Ovens may be used for all nonvolatile materials. Some types of glassware must be allowed to cool gradually to prevent cracking. A home oven set at 330 degrees Fahrenheit will sterilize materials if they are heated for three hours.

COLD

Cold is largely bacteriostatic; microbial growth is inhibited by cold but few organisms are actually killed by it. Refrigeration slows food spoilage because relatively few bacteria are able to grow at 4 degrees Celsius, the recommended temperature for food refrigerators.

Freezing to −10 degrees Celsius arrests microbial growth and completely kills some food pathogens — the agents of trichinosis and toxoplasmosis are destroyed, for example, but tularemia bacilli are not. Slow freezing in an inefficient freezer causes the formation of relatively large intracellular ice crystals, which rupture some microbial cells. It also has an undesirable effect on the consistency of foods, since it destroys the structure of the plant or animal tissue being preserved. With flash freezing — rapid, very-low-temperature treatment — the ice crystals formed are much smaller. This is more desirable in terms of the palatability of the foods, but it eliminates fewer of the microorganisms. Superfast freezing in liquid nitrogen (−196 degrees Celsius) does not kill microorganisms, and is one of the preferred ways of

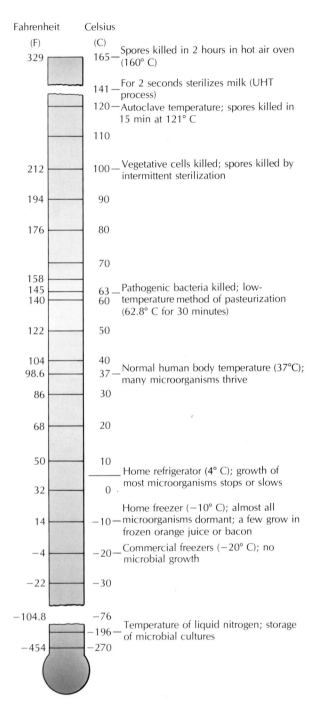

FIGURE 23.6   Temperature Ranges and Microbial Control
Heat and cold may be used to kill microorganisms or to arrest their growth.

preserving viable microbial stock cultures, as is lyophilizing (freeze-drying). In general, refrigeration or freezing cannot be counted on to eliminate microbes. In fact, when refrigerated or frozen material is returned to warm temperatures, microbial growth may begin again immediately.

OSMOTIC CONTROL

One of the oldest human food preservation strategies is that of depriving microorganisms of the water they need for growth. This is accomplished either directly by drying the food, or by adding generous amounts of salt or sugar to cause unfavorable osmotic conditions. These approaches cannot be considered disinfection or sterilization, thus we will not discuss them further here. However, let us stress once again that microorganisms thrive in moist environments. Therefore we should keep a watchful eye on moist environments when we are attempting to control microbial growth.

RADIATION

Because the lethality of radiation is well known, it has been much investigated as a microbial-control mechanism. To date, however, its applications remain limited.

Ionizing radiation is high-energy radiation of wavelengths shorter than ultraviolet (review Figure 3.1). This form of energy ionizes molecules, a property it shares with beams of subatomic particles such as protons and neutrons. The reactive ions formed by the ionization nonspecifically alter cellular proteins and nucleic acids.

Material to be irradiated usually must be positioned very close to the energy source. Only very high-energy radiations, such as gamma rays, have enough penetrating power to sterilize the interior of large objects. Gamma rays emitted from the radioactive isotope cobalt-60 are used industrially to sterilize mass-produced disposables such as syringes, petri dishes, and surgical dressing materials. The great expense of installing and safely operating radiation equipment limits its use. Unfortunately, sterilizing doses of gamma rays make foods unpalatable and reduce nutritional value.

Ultraviolet rays are a lower-energy, nonionizing form of radiation found in sunlight and emitted by special mercury-vapor lamps. Nucleic acids strongly absorb ultraviolet radiation, which may cause mutations that prevent normal gene expression and DNA replication (Figure 23.7), as discussed in Chapter 8. Repair enzymes may reverse the damaging effects of ultraviolet light, especially if the irradiated organisms are exposed to visible light shortly afterward. Ultraviolet lamps are used to kill organisms in air or within relatively small, closed areas such as microbiological transfer cabinets. The lethal effect occurs only at close range, in clean, dry air. Ultraviolet is screened out by water vapor, glass, plastic, and opaque coverings. Ultraviolet radiation must be used with care as it is hazardous to human skin and eyes.

ULTRASOUND. **Ultrasonic vibrations,** sound waves of an energy level too high for humans to hear, are a source of energy that can be used to rupture microbial cells under controlled conditions. When a liquid is bombarded with ultrasound, tiny bubbles are produced. These bubbles collapse, leaving tiny holes or cavities that refill with fluid. Bubble formation and collapse (**cavitation**) happens in microseconds, causing intense localized pressure changes. The pressure variations may rupture microbial cells. Ultrasound is not a reliable sterilizer, however. Ultrasonic devices are more commonly used to cleanse items such as intricate equipment, dentures, and jewelry because the scrubbing action caused by cavitation can clean microscopic fissures. Such objects, once thoroughly cleansed, may then be effectively sterilized by other means.

FILTRATION

Rather than killing organisms, it is sometimes more desirable to simply remove them from air or a solution by filtration. Two types of filters may be used.

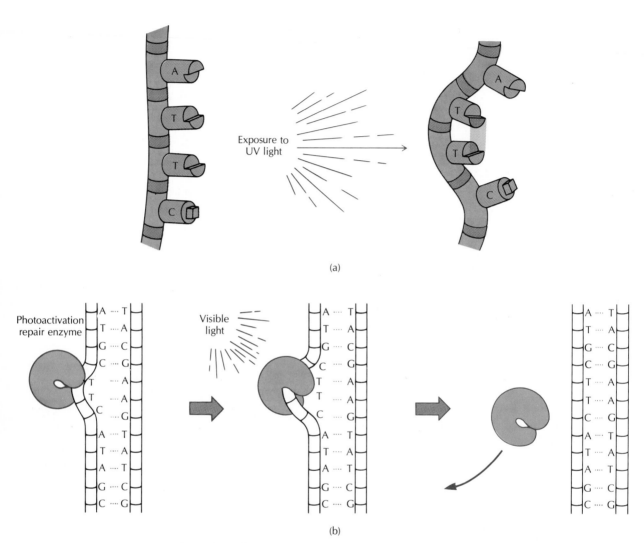

FIGURE 23.7   Ultraviolet Light Mutations
(a) When ultraviolet light strikes DNA, its energy may form a covalent bond between adjacent thymines, called a thymine dimer. This dimer interferes with normal gene function and may even kill the cell. (b) Cells have several natural repair mechanisms for thymine dimers. In one, a photoreactivation repair enzyme attaches to the distorted sequence, absorbs visible light, and breaks the covalent bond. The DNA strand is then as good as new.

Depth filters are designed like sponges, with a thick layer of porous material. Liquid passes through many tortuous channels, in which microorganisms will eventually become stuck. The liquid emerges into a sterile vessel. Filter materials include porcelain, asbestos, or porous glass. Depth filtration is a slow process. Because the particles are retained, the filter eventually clogs and requires long and tedious cleaning. The filter may also absorb too much of the fluid, retain

BOX 23.2
Cleanliness or Compulsion?

In the grand old days of King Henry VIII of England or King Louis XIV of France, sanitary standards were rather relaxed. People (even kings) were bathed three times in their lives — when they were born, before their marriage, and after they died. Even Henry probably only had eight baths! Foods were heavily spiced in an effort to hide the taste of advanced spoilage. The chamber pots and slop jars were emptied out the upstairs windows into the street or lugged to an open cesspool somewhere all too near at hand.

In the 20th century, Western culture has made a complete turnaround. We now pursue what may be completely unreasonable standards of hygiene. TV advertisements show us wild-eyed housewives stalking through their shining homes with cans of spray disinfectant. Should we find ourselves next to a person who clearly hasn't bathed recently, we shy away with real fear as well as repugnance. We refuse to use milk that is a little sour even for cooking. The thought of using an outhouse disgusts us. We have developed a new fear — microbiophobia.

Realistically, how much cleanliness is enough? Remember the grounds for susceptibility to infection. Extreme youth, extreme age, and underlying ailments are what make one highly vulnerable. If you are healthy, you can basically be infected only by pathogens. Those hairy little "germs" of the advertising media — normal flora (yours or someone else's) and environmental organisms — are extremely unlikely to cause you any problems. They can cause problems for your newborn infant or your elderly grandmother, however.

It is necessary, however, to take certain essential precautions, because when a pathogen such as a wild-type poliovirus or a *Shigella* bacillus happens by, you may pay the price for lax hygiene standards. You won't know in advance when that is going to occur, either. Clean drinking water, adequate sewage disposal, updated vaccinations for all ages, and properly handled food are the basic requirements. On the other hand, from a health standpoint, it doesn't matter very much whether you shower once a day or once a week. Those air freshener sprays can be promoted only for their aesthetic values. That sink full of unwashed dishes isn't going to kill anyone. How much grubbiness is too much? If you're a healthy adult, you may be as grubby at home as your personal tastes permit, and you'll probably get away with it. But when there is a high-risk person around, you owe it to them to adopt higher standards. And watch out for certain types of exposure to people with infectious disease. Remember that Henry VIII gave each of his six wives syphilis!

(Photo courtesy of Historical Pictures Service, Chicago.)

important substances such as enzymes, or introduce chemical contaminants.

Because of these difficulties, **sieve filters** such as the **membrane filter** have almost entirely replaced the depth filter. A sieve or membrane filter is made of a thin sheet of a synthetic polymer such as cellulose acetate. The polymer sheet contains millions of very tiny, uniformly sized holes. Fluid to be filtered is passed through under vacuum. If membranes of a sufficiently small pore size are used, all rigid bacterial cells may be removed from the fluid and retained (Figure 23.8). However, there is no filter that reliably removes the tiny, highly flexible mycoplasmas, chlamydias, and rickettsias, or the viruses. Membrane filters are sterilizable, inexpensive, and disposable.

AIR-CONTROL SYSTEMS

A controlled flow of air can be used as a barrier to infection, by directing the movement of airborne microorganisms. Air-control systems may be used in several ways.

High-efficiency particulate air (HEPA) filters have a pore size so small that the air that passes through is rendered sterile as well as clean. HEPA filters are widely used in fans for air control. You would find HEPA filters cleaning the air supplied to operating rooms, to patients in protective isolation units, and to the "clean rooms" in which sensitive electronic equipment is assembled. HEPA filters also may be used to remove pathogens in contaminated air. The exhaust from facilities where extremely hazardous microorganisms are handled is vented to the outside only after HEPA filtration.

Air will always flow from an area of higher pressure to one of lower pressure. **Positive pressure** in a room — that is, air at a pressure slightly higher than that of the surrounding area — directs air flow outward. In a surgical unit, the operating room may be provided with a constant flow of sterilized air to maintain an atmospheric

FIGURE 23.8   Membrane Filter and Bacteria
These scanning electron micrographs show two types of bacteria trapped on the surface of a membrane filter: (a) a spirochete; (b) mixed bacteria removed from a water sample. (Courtesy of Dr. Glenn Howard, Pall Corporation, Glen Cove, NY.)

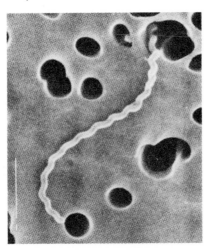
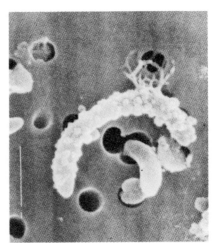

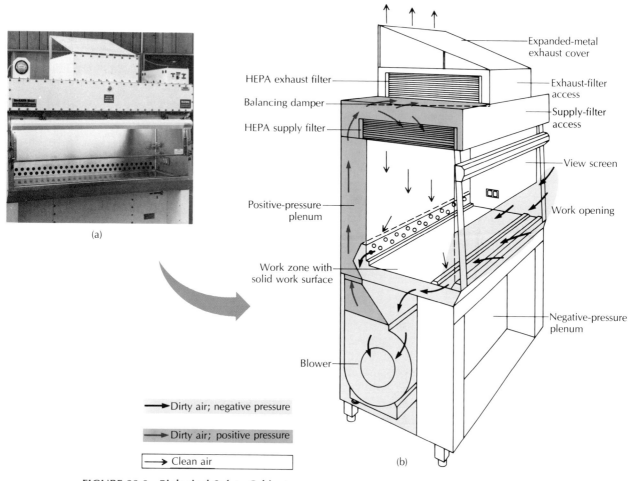

FIGURE 23.9 Biological Safety Cabinet
(a) This laminar flow hood is designed to protect the experimenter working on low-to-moderate-risk materials. (b) Room air (dirty) is drawn in at the front opening, where the worker's hands enter. It flows under and in back of the work zone (white area) to the fan. A supply of HEPA-filtered clean air is directed down over the work area to prevent contamination of cultures. All air vented to the outside (at top) has been sterilized by passage through a HEPA filter.

pressure slightly higher than that of the exterior environment. All air flows are from the inside (clean) to the outside (unclean). Unsterilized air cannot enter because the interior pressure is higher. Random airborne contamination is thereby excluded.

Negative pressure in a room draws air inward. This technique may be used to isolate patients with highly communicable diseases.

Negative pressure within the unit prevents air from moving out and blocks airborne dissemination of the patient's pathogen. The air may be HEPA-filtered before discharge.

Laminar flow transfer hoods are found in many laboratories. They are work areas in which a unidirectional layer of air is drawn over the work surface. A negative-pressure type of laminar flow hood is used for transferring *My-*

cobacterium tuberculosis cultures and other biohazards (Figure 23.9). A constant inward flow of air protects the worker sitting outside from accidental airborne contamination from the cultures he or she is working with. The exhaust air is vented through a HEPA filter before leaving the building. Positive pressure transfer hoods are used when the material being worked with is not hazardous, but it is imperative to protect it from airborne contaminants. Applications include cell culture work and the formulation of sterile intravenous fluids in pharmacies.

CHEMICAL CONTROL AGENTS

We frequently use chemicals to control organisms, often with some degree of misplaced faith. It is important to understand just what a disinfectant can and cannot do.

In the past, any disinfectant or antiseptic worth its salt had to smell bad and sting like fire. Many newer disinfectants do not have these disadvantages. A good disinfectant should effectively destroy all pathogenic microorganisms in a reasonable period of time, be readily soluble, and actively coat surfaces and penetrate into cracks and fissures. It should not evaporate too rapidly. The chemical should be nontoxic (not irritate or blister, kill living tissue, or cause allergic responses); it should not corrode metals, dull blades, degrade rubber or plastics, or leave toxic residues. If it is inexpensive and if it does not stain or have an objectionable smell, all the better.

The chemical-disinfectant groups we will discuss include surfactants (detergents and soaps), halogens, phenolics, alcohols, gases, metals, and dyes. Many commercial disinfectant preparations contain combinations of these substances for a stronger disinfectant effect. The properties of some common disinfectants are given in Table 23.1.

EFFECTS OF ENVIRONMENTAL VARIABLES

A disinfecting treatment that may be completely effective in one environment may be relatively ineffective in another. Many things can interfere with the chemical's proper operation. For example, many contaminated materials contain or are coated by organic matter such as food, saliva, blood, or excretions. Many disinfectants are readily inactivated by such organic matter. Disinfectants bind nonspecifically, so the organic matter absorbs some of the chemical, reducing the amount available for antimicrobial action. In order to compensate for losses due to nonspecific binding, we can either reduce the amount of organic matter by prewashing or increase the concentration of disinfectant to compensate for losses due to nonspecific binding. In chlorinating drinking water, for example, the amount of disinfectant is adjusted to the amount of organic matter in the water source. Cities that draw their water supplies from rivers with a heavy organic load must use a great deal of chlorine. Their treated drinking water may be safe but it is not very tasty.

Most disinfectants operate best at a neutral or slightly basic pH. Also, most disinfectants work better at higher temperatures. Thus, treatment should be carried out at room temperature or above, or should be prolonged if cold temperatures are unavoidable.

Certain disinfectants are ionic (carry an electrical charge). When the charge is positive the ion is called a cation; when it is negative, it is an anion. Ionic disinfectants bind to the microbial surface by electrostatic interaction. Cationic detergents, for example, bind to anionic groups in the bacterial membrane. If other anions, such as those found in soaps, are present, however, they compete with the cell surface in binding the cations, and neutralize the disinfectant.

A chemical used in disinfection must have free access to all contaminated surfaces and areas. Improper application, failure to take equipment apart before immersing it, or residual coatings of grease or wax may prevent effective action.

SURFACTANTS

Surfactants or **surface active compounds** are wetting agents that lower the surface tension of

TABLE 23.1 Characteristics of common disinfectants			
Class	Disinfectant	Antiseptic	Other Properties
Gas			
Ethylene oxide	+3 to +4	0	Toxic; good penetration; requires relative humidity of 30% or more; bactericidal activity varies with apparatus used; absorbed by porous materials. Dry spores highly resistant; moisture must be present, and presoaking is desirable
Liquid			
Glutaraldehyde, aqueous	+3	0	Sporicidal; active solution unstable; toxic
Formaldehyde + alcohol	+3	0	Sporicidal; noxious fumes; toxic; volatile
Formaldehyde, aqueous	+1 to +2	0	Sporicidal; noxious fumes; toxic
Phenolic compounds	+3	0	Stable; corrosive; little inactivation by organic matter; irritates skin
Chlorine compounds	+1 to +2	0	Flash action; much inactivation by organic matter; corrosive; irritates skin
Alcohol	+1	+3	Rapidly -cidal; volatile; flammable; dries and irritates skin
Iodine + alcohol	0	+4	Corrosive; very rapidly -cidal; causes staining; irritates skin; flammable
Iodophors	+1 to +2	+3	Somewhat unstable; relatively bland; staining temporary; corrosive
Iodine, aqueous	0	+2	Rapidly -cidal; corrosive; stains fabrics; stains and irritates skin
Quaternary ammonium compounds	+1	0	Bland; inactivated by soap and anionics; absorbed by fabrics
Hexachlorophene	0	+2	Bland; insoluble in water, soluble in alcohol; not inactivated by soap; weakly -cidal
Mercurial compounds	0	±	Bland; much inactivated by organic matter; weakly -cidal.

Source: Favero, MS, *Manual of Clinical Microbiology,* 3rd ed., Washington, D.C.: American Society of Microbiology, p. 955, 1980.

Maximal practical usefulness in the hospital environment is indicated by 4; little or no usefulness by 0.

water. As a result, a disinfectant solution containing a surfactant more readily wets all available surfaces. Although few surfactants have significant antimicrobial effects by themselves, they are frequently incorporated into mixtures with more active disinfectants. They give the product much improved penetrating or wetting powers, thus augmenting the other active ingredients' access to the microorganisms.

Natural **soaps** consist of sodium ions bonded to anionic long-chain fatty acids derived from plant oils or animal fats. Soaps were the first surfactants used. They are weakly charged and have little -cidal activity, but are household standbys in routine degerming. The term **detergent** indicates a varied group of synthetic surfactants and cleansers that have largely replaced natural soaps. Home laundry and dishwashing detergents are also anionic, but more strongly charged than soaps. Thus they have some lytic effect on cell membranes.

Some detergents, however, are cationic because they contain a positively charged ion, to which is attached one or more large hydrophobic chains. Examples are the **quaternary ammonium compounds,** or **quats** (Figure 23.10). Cationic quats bind avidly to the naturally anionic cell

membranes of bacteria and viral envelopes, and in some cases will destroy them. Cationic detergents are useful as sanitizers.

Quats are -cidal, however, only against vegetative organisms, not against spores or acid-fast bacilli. Weak quat solutions even support growth of *Pseudomonas aeruginosa* and some enterobacteria. Anionic detergents or soaps neutralize and completely inactivate cationic detergents, so the two types should not be mixed.

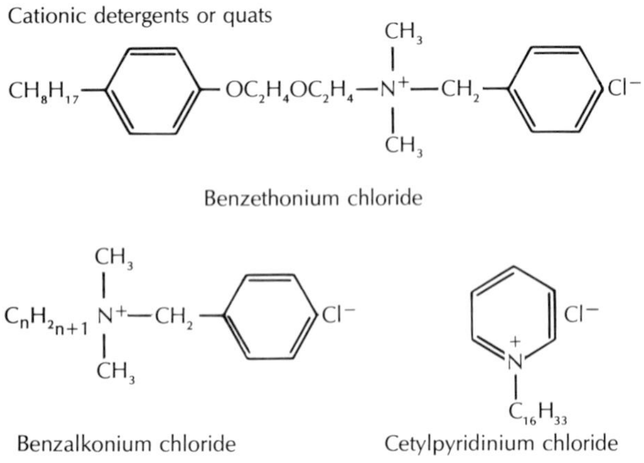

HALOGENS

Chlorine, bromine, iodine, and fluorine are the **halogens** — highly reactive elements that characteristically exist as anions with single negative charges. Compounds containing halogens have a long history of usefulness in microbial control.

Chlorine gas and chlorine compounds, such as chloramines and hypochlorites, are widely used to sanitize drinking-water supplies, swimming facilities, and dairy- and food-processing equipment. Chlorine compounds are inexpensive and lethal to a wide range of microorganisms. However, they are corrosive to certain materials and irritating to living tissues. They are also volatile and for that reason they have a short-term effect. When large amounts of organic matter are present, it absorbs much of the added chlorine, and the amount of chlorine used must therefore be increased to compensate for the organic load. A margin of **residual chlorine** — that is, active chlorine left over after absorption by the organic material in the water — is required for complete safety. This means that more chlorine must be added than will be immediately consumed to react with the organic matter. Inadequate chlorination has led to outbreaks of legionellosis, salmonellosis, giardiasis, and other waterborne infections.

Iodine was one of the first antiseptics to be used, dissolved in an alcoholic solution, or **tincture.** Tincture of iodine stung, stained, and often became overly concentrated as the alcohol evaporated, causing serious burns. Nowadays iodine

FIGURE 23.10   Some Surface-Active Disinfectants

Cationic detergents or quats

Benzethonium chloride

Benzalkonium chloride

Cetylpyridinium chloride

is commonly used in an **iodophor** form, that is, chemically bound to a surfactant carrier. The common iodophor **povidone–iodine** does not stain, sting, or harm tissue, but it is somewhat less active than unbound iodine. Iodophor formulations are widely used clinically for skin degerming before insertion of needles, for preparing the site of surgical incisions, and for hand scrubbing. Iodine hypersensitivity occurs in some people and should be watched for.

PHENOLICS

Phenolics are derivatives of phenol, containing six-carbon rings (Figure 23.11). A 5 or 10% phenol solution destroys the lipid components of membranes and kills all vegetative bacteria and enveloped viruses. Phenol's main use today is as the starting material from which other phenolic compounds are made. Very strong -cidal activity can be attained by adding chemical groupings containing chlorine to the phenol molecule.

Three examples of such phenolics are ortho-phenylphenol, chlorophene, and hexachlorophene. These compounds are long-acting and stable. They are not active against spores. One phenolic (hexachlorophene) was once indiscriminately included in over-the-counter disinfectant lotions and in soaps, until it was shown to cause brain damage in baby rats. Since then, products containing hexachlorophene have been controlled and can be obtained only with a prescription in situations where such a powerful disinfectant is truly needed. Extensive skin absorption of phenolics should be avoided.

ALCOHOLS

Ethyl and isopropyl alcohol are widely used disinfectants (Figure 23.12). In solutions of 70 to 95% in water, alcohols are effective against numerous agents including the tuberculosis organism and enveloped viruses, but not against spores and hepatitis virus. They coagulate cellular proteins, an effect that occurs most efficiently when there is some water present. Alcohols have the disadvantage of evaporating rapidly and having no residual action. They are often mixed with iodophors and cationic detergents for a combined bacteriocidal effect.

FIGURE 23.11 Some Phenolic Disinfectants

Phenolics

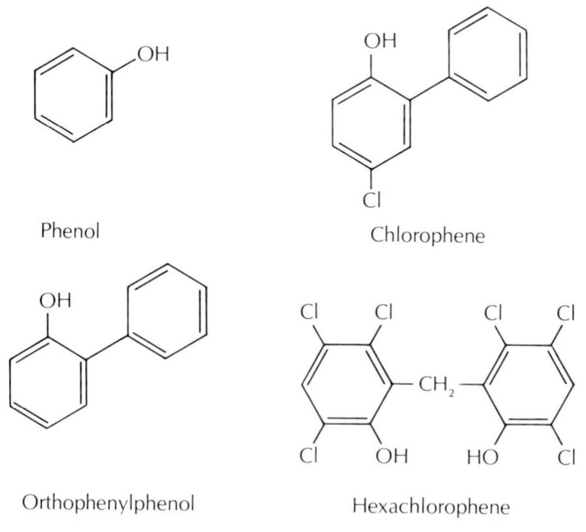

Phenol

Chlorophene

Orthophenylphenol

Hexachlorophene

FIGURE 23.12 Some Alcohol and Aldehyde Disinfectants

Alcohols and aldehydes

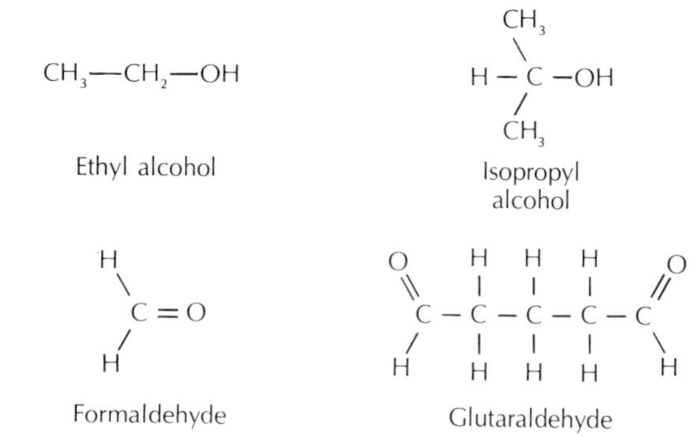

$CH_3\text{—}CH_2\text{—}OH$

Ethyl alcohol

Isopropyl alcohol

Formaldehyde

Glutaraldehyde

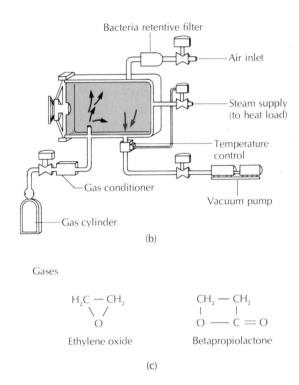

(a)

(b)

Gases

Ethylene oxide

Betapropiolactone

(c)

FIGURE 23.13   Gas Sterilization
(a) A gas sterilization unit. (Courtesy of American Sterilizer Co., Erie, Pa.)
(b) A diagram of a simplified EtO gas-sterilization system: Once the chamber is loaded and sealed, air within is replaced by the gas sterilant. The gas is moistened and heated by steam, then removed by a vacuum pump after sterilization is complete. (c) Ethylene oxide is a protein-altering gaseous agent commonly used in sterilizers. Betapropiolactone is an alternative gas sterilant.

ALDEHYDES

The **aldehydes** are highly reactive compounds that alter the chemistry of proteins, changing their conformation and rendering them inactive. Formaldehyde is one example, all too familiar to many of us as a preservative in dissection specimens. A 37% solution of formaldehyde in water is called **formalin.** When placed overnight in a closed container with contaminated inanimate objects, formalin gives off a vapor that is a very effective sterilant. Bacteria or viruses to be used in vaccines are often killed by formalin treatment. There is some evidence that formaldehyde is a carcinogen in human beings, and it should be

used with care. Another aldehyde is glutaraldehyde. Two percent glutaraldehyde solutions are used for sterilizing endoscopes and respiratory therapy equipment.

GASES

Gas sterilization is also called **fumigation.** It may be carried out in instruments similar to autoclaves. Ethylene oxide gas (EtO) is the most commonly used. EtO sterilizers are found along with autoclaves in most hospital central supply areas (Figure 23.13). EtO is a protein-altering agent that readily penetrates paper or plastic wrappings. It may be used to sterilize pre-wrapped plastic disposables such as syringes and culture-collecting devices, respiratory therapy mouthpieces, and wrapped instruments.

The exposure conditions for gas sterilization must be carefully controlled. Ethylene oxide gas, which by itself is explosive, is supplied in non-explosive mixtures with carbon dioxide or Freon. During sterilization, the temperature must be hot, the relative humidity must be maintained between 30 and 50%, and the gas concentration must be fixed. From 4 to 12 hours of exposure are required. EtO gas blisters living tissues, so sterilized articles must be aired from 12 to 24 hours afterward to allow the gas to dissipate. EtO gas cannot be used to sterilize biochemicals because it chemically alters many of them. It is hazardous to work with and careful ventilation is required.

OTHER MISCELLANEOUS CHEMICALS

METAL IONS. The ions of heavy metals such as mercury, silver, arsenic, and copper have long been known to be toxic (Figure 23.14). They were used in insecticides, rat poisons, and the earliest disinfectants, but they are used less now. Some exceptions are the 1% silver nitrate solution that is infused into the eyes of neonates to prevent ophthalmic gonorrhea and the silver sufadiazine used in the routine cleansing of severe burns. Copper sulfate, a fungicide and algicide, is used to control microbial growth in swimming pools and lakes; it also has some agricultural applications. The industrial and agricultural uses of heavy metals, once widespread, are now restricted by law to reduce the environmental hazards. For example, mercury compounds were once used in vast quantities to control microbial growth in industrial processes such as those at pulp and paper mills. The mercury ended up in the bottom sediments of rivers or bays. There, bacteria converted it to methyl mercury, a more soluble form that is highly toxic to mammals. Eating fish from such waters causes severe neurologic disease due to mercury poisoning.

PEROXIDES AND OZONE. **Peroxides** are unstable compounds that decompose to yield chemically reactive free atomic forms of oxygen, called **nascent oxygen.** Nascent oxygen is highly toxic to anaerobes. Hydrogen peroxide solutions are used occasionally to clean superficial wounds. A medicinal zinc peroxide preparation may be used to cleanse deep wounds. **Ozone** is a gaseous oxygen compound generated by electrical discharges passing through air. It has been used for small-scale water purification.

ACIDS AND ALKALIS. Strong acids, such as sulfuric or nitric acid, and strong alkalis, such as lye, were once used somewhat in disinfection

FIGURE 23.14 Metal Ions
The ions of heavy metals are toxic to many microorganisms. Here, inhibition of bacterial growth is visible in the clear zones around both the silver coin and heart. A nonsilver coin (top) has been moved aside to show microbial growth beneath. (Courtesy of Richard Humbert/BPS.)

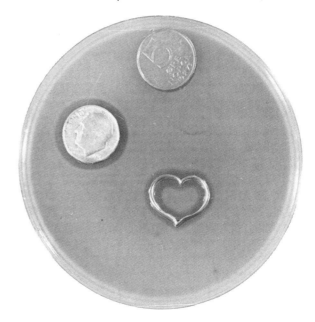

but are no longer because of their harsh and destructive effects. Milder organic acids may have static effects on microbial growth. A mixture of benzoic and salicylic acids (Whitfield's ointment) is used to control superficial mycosis. Undecylenic acid is the active ingredient of a preparation used for athlete's foot. Benzoic, propionic, sorbic, and acetic acids are widely used as food preservatives.

DYES.   Some of the aniline dyes, derived from phenol, are inhibitory to certain Gram-positive bacteria and superficial fungi. Gentian violet may be used to treat ringworm and *Candida* infections. Dyes are an integral part of many selective microbiological media because they inhibit unwanted species of microorganisms.

CONTROL OF INFECTIOUS DISEASE

Disinfection and sterilization are intensively used in clinical facilities to reduce the general population of microorganisms, to restrict the spread of pathogens from a patient with infectious disease to other patients or the staff, and to prevent microorganisms from infecting patients, especially compromised individuals.

As you know, each infectious disease has its own specific chain of transmission. Certain of the patient's body surfaces and secretions bear and transmit the agent, while others do not. In each case, therefore, the selection of the specific precautionary measures to be employed must be based on an understanding of the epidemiology of the patient's disease.

CONCURRENT DISINFECTION

Concurrent disinfection is the immediate and ongoing disinfection of a patient's contaminated secretions plus any articles the secretions may have soiled. Disinfection procedures are most often applied to discharges from the mouth and nose — soiled tissues, and eating utensils — as well as to feces, urine, urinals, and bedpans. An attendant's hands must also be cleaned after giving patient care or handling soiled articles. Concurrent disinfection is the primary component in the various categories of precautions (see below).

TERMINAL DISINFECTION

Terminal disinfection is the cleaning procedure that occurs after the patient has vacated the unit. It may include thorough disinfection of walls, woodwork, floors, and furniture. All containers or utensils used by the patient are removed and sterilized. This routine, between-patient, cleaning procedure should be especially thorough when the last patient had a communicable disease.

ISOLATION AND PRECAUTIONS

Isolation is a situation in which a patient is kept in a private room and special protective measures are imposed on all persons and materials entering and leaving the room. An isolated environment may be used to prevent the dissemination of a communicable disease or to protect a severely compromised patient. Different types of isolation are recommended for various conditions. The system and techniques proposed by the U.S. Department of Health and Human Services include three categories of isolation.

Strict isolation is applied when there is a need to contain a highly communicable, dangerous disease such as diphtheria or rabies. Gowns, masks, and gloves are required for all persons entering the room. Articles leaving the room must be protectively wrapped before being sent elsewhere in the hospital for processing.

Respiratory isolation is used to contain highly communicable, hazardous diseases such as tuberculosis, in which the agent is spread by respiratory secretions. The procedures are much like strict isolation except that they apply only to articles apt to be contaminated with saliva or sputum.

Protective isolation is used to protect extremely vulnerable or immunosuppressed patients from exposures. In this case, staff members and visitors wear gowns, masks, and gloves to prevent their flora from reaching the patient.

Strategies called **precautions** are designed to exercise a very high level of control over the patient's excretions, dressings, or other materials, which may be loaded with the infective agent, without requiring a private room or the use of gowns and masks. **Enteric precautions** call for the disinfection of feces and any fomites that may potentially be contaminated with feces, and are used in cases of typhoid fever, cholera, hepatitis, and other gastroenteric conditions. **Wound and skin precautions** are used to control leakage of fluids from skin infections and infected surgical wounds, burns, puerperal sepsis, and other infectious states where the risk is localized. Other types of precautions are applied to blood and respiratory secretions.

The physician or infection-control officer may order the adoption of one of these types of precautions or isolations. In the **card system,** a color-coded outline of the required procedures is placed on the door of the patient's unit. These should be checked by staff members every time they prepare to enter. Gowns, gloves, and other special equipment are also placed at the door. Visitors are required to observe the same procedures as staff. They are requested to report to the nurses' station before entering, in order to receive instructions. Each unit should have available complete and current procedures manuals. Newcomers to the unit should receive full orientation.

The primary problem in implementing isolation is noncompliance. Doctors and nurses are often hurried, and the time required to put on gloves and masks may seem excessive at the moment. However, if infectious disease is to be controlled, it is crucial that all persons — including doctors, visitors, and housekeepers — entering an isolation room use the indicated precautions. Some diseases requiring isolation measures and their color-coding in the card system are shown in Table 23.2.

SURGICAL ASEPSIS

Sepsis is the uncontrolled presence of infectious agents or their toxins in the blood or tissues. In early and primitive medicine, because people were not aware of the existence of microorganisms, they used septic techniques — that is, techniques in which microbial control was not considered. The mental picture of a surgeon sharpening his dull scalpel on his boot heel before operating is a sufficient example of the deficiencies of this approach.

When the role of microbes in disease became apparent, septic techniques were supplanted by antiseptic ones. **Antisepsis** as an approach assumed that the presence of pathogens in the environment was inevitable, and that they should be controlled chemically. Surgeons operated in a cloud of disinfectant spray designed to prevent the development of infection. This technique was a significant improvement, but still far from perfect.

Asepsis is a set of techniques used to exclude potential infective agents, rather than relying on killing them once they are present. In asepsis, medical practitioners maintain a barrier to infection around the uninfected patient, surgical field, injection site, or whatever, based on the scrupulous sterility or cleanliness of all materials coming in contact with them. Asepsis applied to an infected patient, on the other hand, uses the barrier approach to confine the infective agent to that patient, and exclude it from the larger environment.

A modern surgical suite is very complex, not only in its architecture and equipment but also in the aseptic techniques practiced. The primary thrust is always to minimize the number of microorganisms at large in the environment prior to and during surgery (Figure 23.15). Let's look at how asepsis is achieved.

HUMAN FLORA.   Both patient and surgical team bring their flora into the operating suite. These organisms must be contained so they cannot reach the incision. Traffic flow is restricted and unidirectional; staff are not allowed to walk back and forth among the rooms. Sterile gown or

TABLE 23.2 Some precautionary measures and the card system		
Type of Procedure	**Strict Isolation**	**Respiratory Isolation**
Card Color	Yellow	Red
Diseases	Anthrax, inhalation	Measles
	Burn wound, major, infected with *Staphylococcus aureus* or Group A *Streptococcus*	Meningococcal meningitis Meningococcemia
	Congenital rubella syndrome	Mumps
	Diphtheria	Pertussis
	Disseminated neonatal herpes simplex	Rubella Tuberculosis
	Disseminated herpes zoster	
	Lassa fever	
	Marburg disease	
	Plague, pneumonic	
	Pneumonia, *Staphylococcus aureus* or Group A *Streptococcus*	
	Rabies	
	Skin infection, major *Staphylococcus aureus*	
	Smallpox	
	Vaccinia	
	Varicella	

Source: Compiled from *Isolation Techniques for Use in Hospitals,* Washington, D.C.: Government Printing Office, 1975.

*Exceptions are in columns to the left.

TABLE 23.2 Some precautionary measures and the card system (continued)		
Protective Isolation	Enteric Precautions	Wound and Skin Precautions
Blue	Brown	Green
Agranulocytosis	Cholera	Infected burn wounds, other*
Dermatitis, noninfected vesicular bullous, or eczematous disease when severe and extensive	Diarrhea, acute with suspected infectious etiology	Gas gangrene Herpes zoster, localized
Extensive noninfected burns (some)	Enterocolitis, staphylococcal	Melioidosis (draining sinuses)
Lymphomas and leukemias (some)	Gastroenteritis caused by: *Escherichia coli* *Salmonella* sp. *Shigella* sp. *Yersinia enterocolitica* Hepatitis, viral, all	Plague, bubonic Puerperal sepsis Wound or skin infections where discharge not adequately contained by dressings (other)
	Typhoid fever	

wash suit, gloves, mask, complete hair covering, and foot coverings replace contaminated street clothes. These are referred to as **occlusive clothing** because they enclose the body, preventing surface microflora from becoming airborne.

The surgical team degerm their hands and forearms by means of a systematic scrub using a disinfectant soap. Then they put on sterile gloves. It is estimated that at least 25% of surgical gloves are punctured during operations, so the hand degerming process is an important backup measure.

The patient's flora is also controlled by covering all exposed skin surfaces with sterile occlusive drapes. The incision site is exactingly degermed and shaved, then sterile drapes are

FIGURE 23.15   Surgical Asepsis
A modern operating room is set up to accommodate highly technological equipment, all of which is designed for efficient disinfection or sterilization. The surgical staff wears occlusive clothing and scrubs their hands before gloving. The patient is covered with a sterile drape. (Courtesy of National Institutes of Health.)

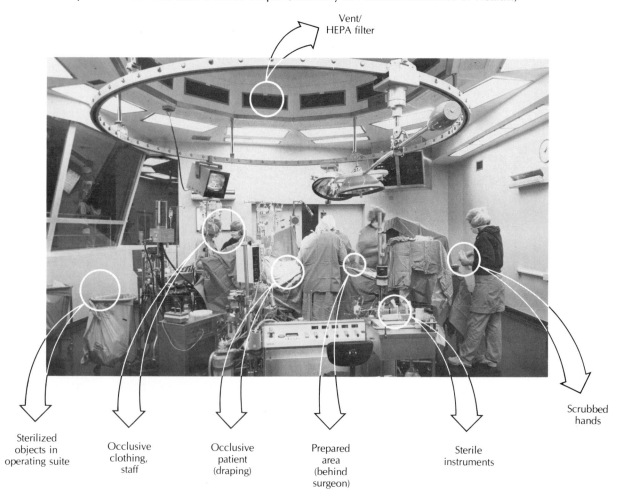

Vent/
HEPA filter

Sterilized objects in operating suite

Occlusive clothing, staff

Occlusive patient (draping)

Prepared area (behind surgeon)

Sterile instruments

Scrubbed hands

placed around the incision area. Shaving should be delayed until just prior to the surgical procedure. If skin is shaved too far in advance, in the interim the inevitable tiny nicks may develop small infections. Should the scalpel pass through one of these clusters of bacteria, the underlying tissues will be contaminated.

When an incision must pass through a mucous membrane or into an existing focus of infection, contamination of the surgical site is inevitable. Infection may be prevented by the use of prophylactic antimicrobial agents.

Aseptic techniques are also applied to the air supply of the surgical suite. HEPA-filtered air is supplied, at positive pressure, and "used" air is conducted to the exterior. Ultraviolet light may be used to kill airborne microorganisms when the surgical suite is unoccupied. A "greenhouse," which is basically a large plastic isolation unit with glove ports, may be placed over the operating table in the highest-risk situations.

All materials brought into the operating suite must be sterile or be free of transmissible microorganisms. This includes drapes, dressing materials, sutures, instruments, rinsing solutions, and medications. Many of these supplies are purchased as sterilized disposables. Others are wrapped and sterilized by the surgical aides or by the hospital's central supply service. Labels confirming sterility and shelf life are required on each package.

Following surgery, materials are cleaned and resterilized, usually by the central supply service. All surfaces are scrupulously disinfected. Anesthesia and x-ray equipment receive special cleaning. Patients with infections are normally placed last in the day's operating schedule so that there will be plenty of time for disinfection. Postsurgical infections now occur after less than 2% of operations in most hospitals.

SUMMARY

1. Disinfection and sterilization reduce the total number of microorganisms in the environment. They may be used with any of three goals: to adopt general precautionary and aesthetic measures; to establish protective barriers for particularly susceptible body areas, persons, or populations; or to contain and inactivate pathogenic microorganisms from infective patients so they can spread no farther.

2. If the destruction of a particular pathogen is required, a disinfectant effective against that pathogen may be selected. If complete destruction of all microorganisms is required, then sterilization is employed.

3. Microbial death is exponential. Treatments are continued until the probability of survivors is acceptably small.

4. Degerming, disinfection, and sterilization are different in the degree of aggressiveness with which the microorganisms are assaulted.

5. A desired result can usually be obtained by many different physical or chemical treatments. Time, cost, and simplicity are important deciding factors.

6. Disinfection and sterilization nonspecifically destroy a cell's structure and its biochemical integrity. As a result, they all harm human tissue to some extent, unless carefully used. Disinfectants cannot be taken internally as therapeutic agents. Their role is to interrupt the transmission of disease.

7. All disinfectant and sterilization treatments are effective only within certain defined limits of time, temperature, and concentration. If significant variation is introduced in their use, the treatments may become ineffective.

8. Some treatments are -cidal, meaning that they kill. Their effect is irreversible. Others are -static, meaning that they arrest growth; they are effective only for as long as they are used.

9. The physical means of sterilization include heat and irradiation. Cold, drying, and ultrasound partially kill or inactivate most organisms. Filtration removes the microorganisms from fluids or air.

10. Many different chemical disinfectants are available. They include the phenolics, surfactants, alcohols and aldehydes, halogens, gases, metals, peroxides, and acids. The most widely used preparations contain several different substances whose activities complement each other.

11. Disinfection, precautions, and isolation procedures are systematically used in hospitals. They all combine a variety of disinfection and sterilization methods with good aseptic technique. The type of isolation and the procedures used are selected according to the infection's mode of transmission.

12. In the surgical suite, we attempt to block all avenues of microbial access to the incision. Particular attention is paid to occluding human surface flora, providing sterile equipment and supplies, and working in a flow of clean or sterile air.

DISCUSSION QUESTIONS

1. Compare the effects on microbial cells and endospores of moist heat, steam under pressure, and dry heat.

2. It is not possible to say with absolute certainty that an object has been rendered sterile. Why?

3. A disinfectant was added to a test tube containing a growing culture of bacterium. Growth rapidly stopped. How can you tell if the disinfectant was bacteriocidal or bacteriostatic?

4. Design an in-use test to evaluate the effectiveness of a new degerming preparation for the skin.

5. Describe how a positive-pressure air supply could be incorporated into the protective isolation of a burn patient.

6. Referring to the epidemiology tables in Chapters 16 through 22, give examples of diseases that require concurrent disinfection of some sort; isolation; respiratory precautions.

7. What sources of microorganisms remain in the scrupulously clean operating room?

ADDITIONAL READING

Benenson AS, ed. *Control of communicable diseases in man.* (14th ed.) Washington, D.C.: American Public Health Association, 1985.

Block SS, ed. *Disinfection, sterilization, and preservation.* 3rd ed. Philadelphia: Lea and Febiger, 1983.

Castle M. *Hospital infection control.* New York: John Wiley, 1980.

Centers for Disease Control. *Isolation techniques for use in hospitals.* 2nd ed. Washington, D.C.: Government Printing Office, 1975. (DHEW publication no. 75-8043.)

Collins CH, Allwood MC, Bloomfield SF, Fox A, eds. *Disinfectants: their use and evaluation of effectiveness.* London: Academic Press, 1981.

Miller BM, ed. *Laboratory Safety Principles and Practices.* Washington, D.C.: ASM, 1984.

Physicians Desk Reference to Pharmaceutical Specialties and Biologicals. Oradell, N.J.: Medical Economics. Published annually.

Pike RM. Laboratory-associated infections: incidence, fatalities, causes, and prevention. *Ann. Rev. Microbiol* 1979; 33:41-66.

24

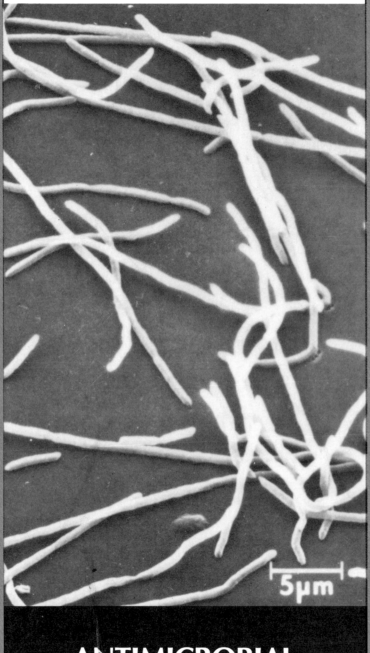

5μm

ANTIMICROBIAL DRUGS

CHAPTER OUTLINE

We use a wide range of naturally occurring and synthetic drugs to treat infectious diseases. The value of antimicrobial drugs lies in their specificity. They have an ability to damage or kill microbial cells without an equally toxic effect on our cells. Used under carefully controlled conditions, these drugs offer comparative safety for the patient but swift elimination of the pathogens.

The search for useful chemotherapeutic drugs began in the 19th century, with the insights of Paul Ehrlich, who envisioned a "magic bullet" — a drug that would have the specificity described above. At first, chemists concentrated on synthetic compounds, but most were much too toxic. The sulfonamides (sulfa drugs) were produced in the 1930s and provided a breakthrough. Later, researchers discovered that nature had been millions of years ahead of us. Natural microbial inhibitors were produced by soil fungi and bacteria, and they could be purified and used for treating humans. In rapid succession, Alexander Fleming made the observations leading to the discovery of penicillin; Selman Waksman identified streptomycin; and tetracycline and chloramphenicol were discovered.

By the 1950s we were firmly launched into the antibiotic era. The annual global production of penicillin alone is over 100 million pounds. On the one hand, we can now cure many once-lethal diseases. On the other hand, we now have sobering evidence of the drugs' potential toxicity and the organisms' ability to develop resistance.

Successful antimicrobial therapy requires that host, drug, and microorganism factors all be taken into consideration. We will study the chemical characteristics of antimicrobial drugs, their modes of action and their spectra of activity. We will examine the problems of microbial resistance and the laboratory methods for determining susceptibility. We will discuss how drugs get into and out of the body, and the types of toxicity problems that may arise. Then we will examine the various categories of drugs, looking at substances used against bacteria, mycobacteria, fungi, viruses, protozoa, and helminths.

CHARACTERISTICS OF AN ANTIMICROBIAL DRUG

KILLING OR INHIBITION?

An **antimicrobial drug** is any substance that can be used therapeutically to treat microbial disease by killing or inhibiting a disease agent. **Antibiotics** are naturally occurring substances produced by microbes. A **semisynthetic drug** is a natural antibiotic somewhat modified by chemists. An entirely **synthetic drug** is made by chemists from simple organic materials.

You'll recall that disinfectants can be classified as -cidal or -static. The same distinction can be made for antimicrobials. Some antimicrobials have irreversible effects that render the pathogen nonviable. Others inhibit growth, but the organisms recover once the drug is discontinued. A -cidal drug should be chosen in preference to a -static one whenever possible. However, a -cidal drug given in an inadequate dosage may prove only -static. The term **minimum inhibitory concentration** (MIC) denotes the concentration that arrests growth. The **minimum lethal concentration** (MLC) is the term for the concentration that kills.

SPECTRUM

The term **spectrum** refers to a drug's range of effectiveness(Table 24.1). A **broad-spectrum** drug is useful against many different types of organisms, such as bacteria, rickettsia and chlamydia, or fungi and protozoa. A **narrow-spectrum** drug is useful against a limited group of organisms (Figure 24.1). No drug safe for human use is effective against all true bacterial species, much less against all microbial pathogens.

DIFFERENTIAL TOXICITY

A useful antimicrobial drug must have differential toxicity. That is, it must damage the microbial cell at concentrations that do not significantly damage the human cell. At levels that are too low, it will not damage the pathogen; at levels that are too high, it will probably injure the patient.

Dosages for each drug must be evaluated individually against each pathogen. The dosage level necessary for clinical control of a particular pathogen is called the **therapeutic dose**. The dosage level at which unacceptable toxicity ensues is called the **toxic dose**. The ratio of the two is the **therapeutic index**. Between the two levels lies the safety margin. The penicillins have a wide safety margin. This allows considerable flexibility in dosage. Amphotericin B, an antifungal drug, has no safety margin — any effective dose will also produce some toxicity. These safety margins may be narrowed by patient illnesses such as kidney disease, which potentiate the drug's toxic effects.

THE IMPORTANCE OF HOST DEFENSES

Antimicrobial drugs alone do not eradicate infections. Generally, when an infectious disease has developed, the host defenses are temporarily overwhelmed, but they do not quit. Antibody

TABLE 24.1 Spectrum and effect of selected antimicrobial drugs			
Drug	Lethality in Therapeutic Dosage	Spectrum	Usable Against
Penicillin G	-cidal	Narrow	Gram-positive bacteria
Ampicillin	-cidal	Broad	Gram-positive and some Gram-negative bacteria
Cephalosporins	-cidal	Narrow	Gram-positive and some Gram-negative bacteria
Erythromycin	-static	Narrow	Gram-positive
Clindamycin	-static	Narrow	Gram-positive, especially anaerobes
Streptomycin	-cidal	Broad	Gram-negative, mycobacteria
Gentamicin	-cidal	Narrow	Gram-negative
Tetracycline	-static	Broad	Most bacteria, rickettsia, chlamydia
Chloramphenicol	-static	Broad	Most bacteria, rickettsia, chlamydia
Sulfonamides, trimethoprim	-static	Broad	Bacteria, protozoans
Isoniazid	Dose-dependent	Narrow	Mycobacteria
Amphotericin B	-static	Narrow	Fungi
Metronidazole	-static	Narrow	Protozoans
Chloroquine	-static	Narrow	Protozoans
Amantadine	-static	Narrow	Influenza virus
Acyclovir	-static	Narrow	DNA viruses

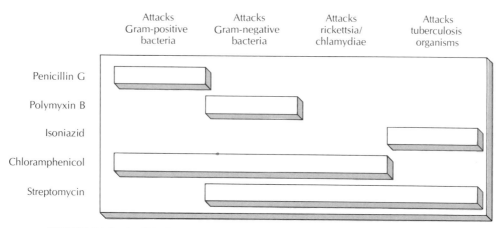

FIGURE 24.1 Antibiotic Spectrum
The spectrum of an antibiotic is the range of microorganisms against which it is effective. Penicillin G is an example of a narrow-spectrum drug, while chloramphenicol is a broad-spectrum drug.

and cell-mediated immunity continue to operate at some level. A drug's role is either to reduce the microbial population or to keep it in check. This buys time until the host defenses are able to dispose of the remainder of the invaders.

In the immunosuppressed patient, immune defenses are ineffective. Therefore -cidal drugs must be given in large doses. Even then, drug treatment often fails because of the difficulties of maintaining high dosage levels throughout the tissues, and because resistant strains of microorganisms appear during therapy. In the acquired immunodeficiency syndrome, patients die of overwhelming multiple infections despite massive doses of many different drugs.

ANTIMICROBIAL DRUG ACTIONS AND MICROBIAL RESPONSES

In this section we will examine the ways in which drugs attack microorganisms. We will also discuss the mechanisms of microbial resistance. We will then compare various laboratory methods for evaluating susceptibility or resistance of clinical isolates to individual drugs.

MODES OF ACTION OF ANTIMICROBIAL DRUGS

Therapy against bacteria is comparatively easy, because the targets are procaryotic cells, while the host cells are eucaryotic. As we saw in Chapter 3, the two cell types have fundamental structural and functional differences, which may be exploited in designing a specific therapy. We don't have this advantage with eucaryotic fungal or protozoan parasites. In the worst situation of all, a virus infection, we must try to inhibit or eliminate a replicating virus that is virtually impossible to target distinct from its host cell. Any drug that inactivates the virus is all too likely to arrest the host cells as well. This explains why we have more and better drugs in the antibacterial category than in any other (Table 24.2).

TABLE 24.2 Selected modes of antimicrobial action	
Cell-wall–synthesis inhibitors	Penicillins Cephalosporins Bacitracin
Protein-synthesis inhibitors	Aminoglycosides Tetracyclines Chloramphenicol Erythromycin Clindamycin
Membrane-active inhibitors	Amphotericin B Polymyxin
Nucleic-acid–synthesis inhibitors	Actinomycin D Rifampin
Enzyme inhibitors	Sulfonamides Diaminopyrimidines Isoniazid

CELL-WALL INHIBITORS.   The peptidoglycan layer in the cell wall of the procaryotes is unique to them and vital to their survival. Neither the peptidoglycan components nor the enzymes for peptidoglycan synthesis are found in eucaryotic organisms. Thus drugs that attack the bacterial cell wall operate with complete specificity (Figure 24.2). They are the only group that does so. Examples of cell-wall inhibitory drugs are the penicillins and the cephalosporins, which block formation of the peptide bridges that link together the carbohydrate chains of the peptidoglycan.

PROTEIN-SYNTHESIS BLOCKERS.   The proteins of procaryotes are synthesized on smaller ribosomes than are eucaryotic proteins. The bacterial ribosomes therefore provide a target for many antimicrobial drugs. Drugs such as the aminoglycosides, erythromycin, and chloramphenicol bind to ribosomal subunits, blocking translation (Figure 24.3).

Protein synthesis inhibitors are broad-spectrum and inhibit all procaryotes to varying extents. These drugs are -static; translation resumes once drug treatment stops. Some protein synthesis inhibitors are toxic to certain human organs in high doses or on prolonged exposure.

MEMBRANE-ACTIVE DRUGS.   Certain drugs are able to cause destructive changes in the cell membranes of microorganisms. The membranes of different categories of organisms have different lipid and sterol compositions. For example, fungal cell membranes contain a sterol called ergosterol. The polyenes are antifungal agents that bind to ergosterol, and cause cell leakage. They also bind somewhat to mammalian membrane sterols, causing toxicity when used in large doses.

NUCLEIC ACID INHIBITORS.   Certain drugs interfere with the infectious agent's ability to synthesize nucleic acids. Rifampin is a drug that inhibits RNA synthesis. It is widely used for treating tuberculosis. DNA synthesis inhibitors, however, are too nonspecific and too toxic to be useful in the treatment of microbial pathogens. They are, however, used in a limited way for treating life-threatening viral infections and to kill tumor cells in cancer chemotherapy.

ANTIMETABOLITES.   Certain drugs act as competitive inhibitors that block enzyme reactions. They thus arrest microbial metabolism of an essential compound, and are called **antimetabolites.** The enzymatically blocked cell then stops growing; it will eventually die if the blockade continues. The best-understood antimetabolites are the sulfonamides and diaminopyrimidines, which interfere with the synthesis and utilization of the vitamin tetrahydrofolic acid (folic acid for short). This vitamin is essential for nucleic acid synthesis. Most bacteria need to make their own folic acid internally because they are impermeable to external folic acid. Thus their metabolism may be conveniently shut off by folic acid antimetabolites. The sulfonamides block the conversion of para-aminobenzoic acid to dihydrofolic acid, and tri-methoprim blocks the conversion of dihydro-folic acid to tetrahydrofolic acid (Figure 24.4). Because human cells don't synthesize folic acid (they must receive their supply from dietary sources), they are not affected.

MICROBIAL RESISTANCE

MECHANISMS OF RESISTANCE.   Certain antimi-crobials have no effect whatever on some mi-crobes. There are at least four mechanisms by which microorganisms may evade the toxic ef-fects of antimicrobials. One is simply that the drug cannot penetrate into the target cell. An-other is that the pathogen may get around a blocked enzyme by shifting to alternate meta-bolic pathways. The pathogen may make an en-zyme that destroys the drug. Or it may produce a "drug inhibitor" that binds the drug and inac-tivates it.

HOW MICROBIAL RESISTANCE DEVELOPS.   We have documented many instances in which mi-croorganisms have become resistant to anti-biotics. Perhaps the best known is the devel-opment of penicillin resistance by the gonorrhea agent. Resistance develops when an initially sus-ceptible species of microorganism changes gen-etically over time. How does this happen? An antimicrobial presented to a vast microbial pop-ulation acts on a wide variety of genetically het-erogeneous individual cells. Regardless of pre-vious drug exposure, a few individual cells will

FIGURE 24.2 Antibiotic Mode of Action
Individual drugs interfere with specific aspects of a microorganism's structure or function. Penicillin blocks cell-wall synthesis; as a result the walls of actively growing cells may be fatally weakened. (a) Twenty minutes after treatment with penicillin, this bacterial cell wall has formed a definite bulge (×84,800). (b) An *E. coli* cell lyses as its cell wall ruptures. (×20,640. Courtesy of M. E. Bayer, M.D., Fox Chase Cancer Center, Philadelphia.)

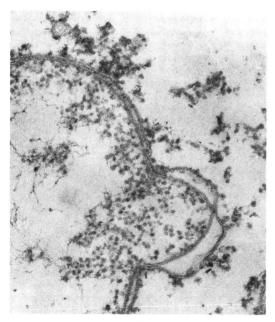

(a)

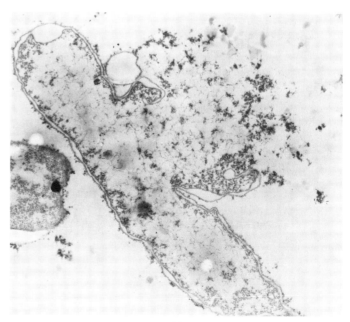

(b)

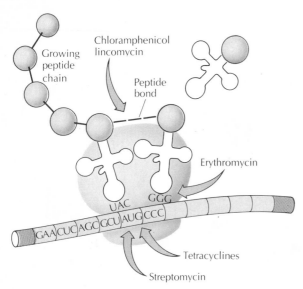

FIGURE 24.3   Protein-Synthesis Inhibitors
Antibiotic drugs may bind to ribosomal subunits, blocking initiation or preventing formation of peptide bonds. The arrows point to specific areas where the antimicrobial drugs chloramphenicol, lincomycin, erythromycin, tetracyclines, and streptomycin may inhibit protein synthesis.

FIGURE 24.4   Folic-Acid Inhibitors
In making DNA, bacteria make tetrahydrofolate, a precursor of thymine, from para–aminobenzoic acid. One of the two steps in the process is blocked by sulfonamides and the other by trimethoprim. The two chemicals together — called cotrimoxazole — block both steps.

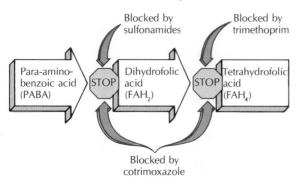

possess genes rendering them less susceptible to the drug. These genes are of no value to the organism in the absence of the drug. In fact, studies show that, in the normal flora, drug-resistant strains disappear when drug use is discontinued. However, in the presence of the drug, it is the few resistant cells that are favored because they are the only ones that can continue to multiply. The resistant strain rapidly emerges as dominant. Thus antibiotic resistance is selected, not created, by exposing microorganisms to antibiotics. Table 24.3 shows some examples of acquired resistance.

In some cases, the acquisition of resistance may be a multistep process. Over a period of months or years, successive mutant populations emerge that can tolerate increasingly high drug levels. The early history of gonorrhea therapy was marked by a gradual increase in the dose of penicillin needed for cure.

In other cases, one-step mutations confer absolute resistance. In recent years, some gonococcus strains have acquired a gene for an enzyme that breaks down penicillin molecules. These strains are now resistant to any dosage level. Other one-step mutations have modified

TABLE 24.3 Significant examples of acquired resistance	
Organism	Significant Percentage of Isolates Now Resistant to:
Hemophilus influenzae	Ampicillin
Neisseria gonorrhoeae	Penicillin, spectinomycin
Pseudomonas aeruginosa	Gentamicin, carbenicillin
Plasmodium falciparum	Chloroquine
Staphylococcus aureus	Penicillin, methicillin
Mycobacterium leprae	Dapsone

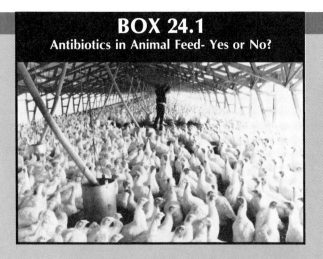

BOX 24.1
Antibiotics in Animal Feed- Yes or No?

Almost half the antibiotics manufactured every year are put into animal feed. According to a 1979 government report, essentially all commercially raised poultry, 90% of the pigs and veal calves, and 60% of the beef cattle receive antibiotics as part of their daily feed. Animals receiving antibiotics in their feed put on weight faster than those that do not. The antibiotics also keep the animals healthy under the intense crowding of the modern beef feedlot, pig barn, or chickenhouse, where each animal is allotted scarcely more room than it is standing on. Thus animals fed antibiotics are more economical to raise. The farmers are able to make some profit in a time when it is becoming very difficult just to break even in farming; the consumer is able to buy meat at lower prices. The growth stimulus caused by antibiotics remains, paradoxically, even though the animals' normal flora becomes highly resistant to the antibiotics being fed them.

Some of the drugs most heavily used in animal feed are the ones we count on in human medicine, such as penicillin, chlortetracycline, and sulfamethazine. If the animals receiving these drugs develop resistant bacteria, what does that mean for us? This question had been kicked around for many years, but the debate suddenly intensified when plasmid-borne resistance factors were discovered. It became clear that plasmids passed readily from one genus of human bacteria to another genus of human bacteria. Wasn't it then likely that they could pass from a genus of animal bacteria to a genus of human bacteria? By feeding animals antibiotics and developing resistance on a wholesale basis in the name of profitability, weren't we acting with a shortsightedness for which society as a whole might pay later? Yet, until 1984, no one had been able to document any specific case in which antibiotic-resistant bacteria had spread from antibiotic-fed animals to humans. Then, CDC investigators showed that an outbreak of a very virulent type of salmonellosis, which hospitalized a number of people, could be traced to their having eaten hamburg meat from a South Dakota cattle herd infected with antibiotic-resistant *Salmonella newport*.

This issue is surrounded by a climate of conflicting academic, medical, economic, and political pressures. Despite the CDC findings, which now go far beyond the South Dakota incident, the agriculture and pharmaceutical industries are unified in their commitment to continuing the practice. As of this writing, it is unclear what the outcome will be.

(Photo courtesy of Junebug Clark/Photo Researchers Inc.)

ribosomes and enzymes so they are no longer "fooled" by a competitive inhibitor.

Resistance may be conferred by either chromosomal or plasmid-borne genes. The plasmid R (resistance) factors readily recombine under the control of transposons (refer to Box 8.1), which group together individual resistance genes to assemble multiple-drug–resistant genetic combinations. Plasmids bearing four or more resistance genes are not uncommon and groupings of 14 are known. Imagine a bacterium resistant to 14 different drugs! The spread of R factors among strains, species, and genera of normal flora can be demonstrated to occur in the bowels of healthy hospital workers.

One especially significant one-step resistance mechanism is bacterial production of **beta-lactamase.** This enzyme, a penicillinase, cleaves and inactivates the **beta-lactam ring,** a chemical feature of penicillins and cephalosporins. Clavulinic acid, oxacillin, and cloxacillin can block some beta-lactamases.

TESTING SUSCEPTIBILITY IN THE LABORATORY

A pure culture of a patient's pathogen is routinely tested for its susceptibility to antimicrobial drugs in the clinical laboratory. This is the "sensitivity" part of the standard culture and sensitivity test package physicians often order when an infection is suspected. Laboratories seek to combine accuracy and standardization with speed and low cost. Here we'll consider some types of susceptibility testing procedures.

DISK DIFFUSION. The **Kirby–Bauer disk diffusion method** is an older accepted standard method. A plate of a nutrient medium called Mueller–Hinton agar is spread with a smooth, even inoculum of the agent. Then paper disks, impregnated with standard concentrations of antimicrobial drugs, are placed on the moist surface of the inoculated agar (Figure 24.5). The drugs diffuse out of the moistened disks into the surrounding agar. Drug concentration is highest near a disk and decreases with distance from the disk. The inoculum is thus exposed to a drug concentration gradient. If the strain is susceptible, a **zone of inhibition** appears following incubation (Figure 24.6). This is a circular area around the disk in which there is no microbial growth. A small zone usually means the organism is inhibited only by the highest drug levels, while a large zone indicates susceptibility to lower levels. The agar, the plates, the inoculum, the disks, and the incubation are standardized. Therefore, the diameter of each final zone can be measured and compared with standard evaluation charts. The size of the zone is translated into the concentration of the drug that would need to be obtained in the patient's serum to be effective against the pathogen. Just because a drug produces a zone of inhibition does not necessarily mean that the organism can be treated with that drug, since the serum levels required might be unacceptably toxic. An isolate is considered **susceptible** only when its zone is large enough to indicate inhibition at a clinically attainable and safe concentration. If the zone is borderline, it is **intermediate.** If the zone is of a smaller size, it means that the organism is only inhibited by very high concentrations of the drug, which would not be safe or sometimes even possible to use, and it is reported as **resistant.** The Kirby–Bauer technique is semiquantitative, meaning that it gives approximate answers.

MINIMAL INHIBITORY CONCENTRATION (MIC) DETERMINATIONS. The clinician is increasingly demanding to know not only which drug to prescribe but also exactly how much drug will be needed to achieve control. For example, in endocarditis, the organism is deeply entrenched, and therapy will be useless unless a high enough dosage is given. The MIC is determined by a serial dilution technique in which samples of the microbial isolate are exposed to a series of concentrations of the drug (Figure 24.6). Most laboratories use partly or wholly automated microdilution, inoculation, and readout systems.

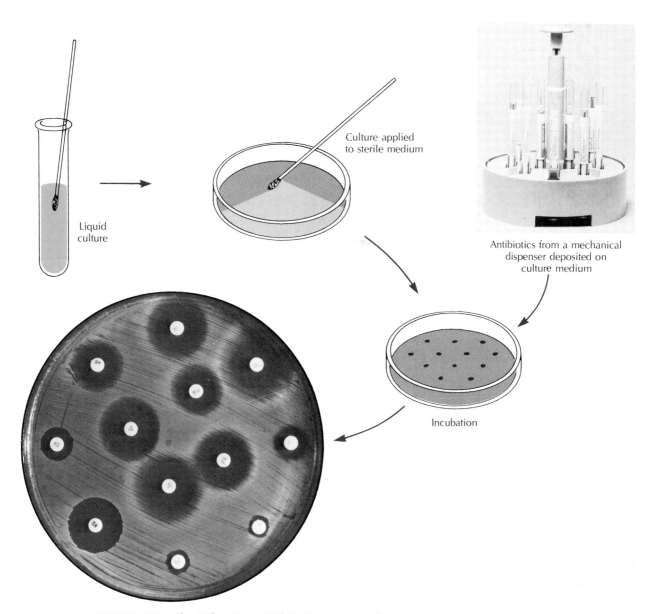

FIGURE 24.5   The Kirby–Bauer Disk Diffusion Method
In this test for susceptibility to antimicrobial drugs, a plate of nutrient medium is streaked with an even inoculum of the agent. Next, drug-impregnated paper disks are placed on the agar surface. The various drugs diffuse outward into the inoculum. Where the agent is susceptible, a zone of inhibition, in which there is no microbial growth, forms a circular area around the disk. The diameter of each zone is measured and compared with a table to determine whether susceptibility is great enough for the drug to be effective if given to the patient. (Photo credits: top right, courtesy of Pfizer, Inc.; bottom left, courtesy of Dr. Leon J. LeBeau, Pathology Department, University of Illinois at Chicago.)

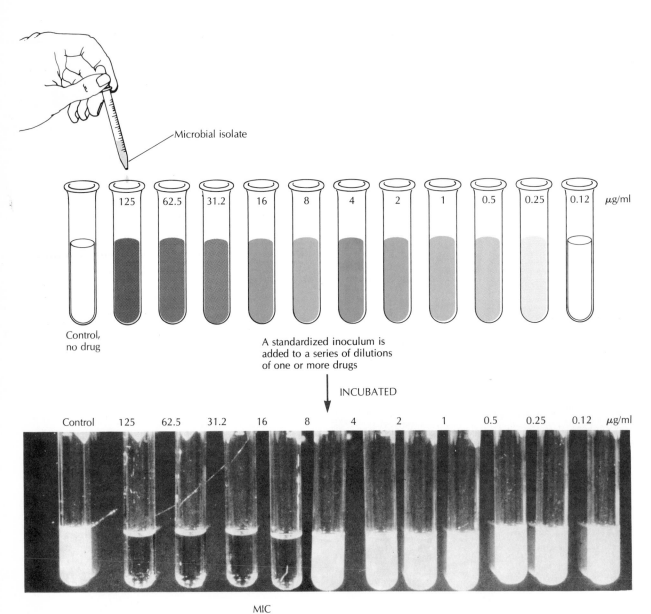

FIGURE 24.6 Minimal Inhibitory Concentration Testing
The technician prepares a series of dilutions of one or more drugs. Next, a standardized inoculum of the microbial isolate is added to each dilution. Below, a series of tests for sensitivity of *Staphylococcus aureus* to ampicillin: cloudiness indicates microbial growth. According to this series, a minimum concentration of 16 mg/milliliters of ampicillin is needed to inhibit the growth of the bacteria. (Courtesy of Dr. Leon J. LeBeau, University of Illinois at Chicago.)

In determining the MIC, a series of dilutions of one or more drugs are prepared, then a standardized inoculum of the microbial isolate is added to each dilution. The MIC is read based on the lowest concentration of drug in which no growth appears after incubation. If desired, these dilutions can next be subcultured to drug-free media to determine if the organisms are truly dead. This step allows the determination of the minimum lethal concentration (MLC).

BETA-LACTAMASE TESTING.   When the laboratory is dealing with isolates of organisms such as *Neisseria gonorrhoeae* or *Hemophilus influenzae*, which are commonly treated with a penicillin or cephalosporin-type drug, it is important to screen quickly for beta-lactamase activity. The clinician needs the information to decide whether to commence treatment with a penicillin or cephalosporin, or to choose some other drug that is not degraded by beta-lactamase. There are many rapid testing methods available.

SERUM DRUG CONCENTRATIONS.   Human beings are provokingly individual. Although two people may each receive equivalent standard drug doses, they may end up with quite different serum concentrations, for reasons that are discussed below. This individuality has proved especially troublesome with aminoglycoside drugs such as gentamicin and amikacin. Recall that drugs are toxic above certain serum levels, yet are ineffective below their MIC. These two levels are relatively close in the aminoglycosides, thus their therapeutic range is quite narrow. Aminoglycosides are usually reserved for seriously ill, hospitalized patients. In the first days of drug therapy, the patient's serum drug level is monitored frequently. This allows the dosage to be fine-tuned to give an acceptable serum level.

COMBINATION THERAPY

If one antimicrobial is good, will two be better? As you might expect, the answer is sometimes yes, sometimes no. Some poorly designed multiple-drug combinations do very little, or are worse than either drug individually. However, drug combinations are frequently successful against resistant organisms that neither drug could control alone. This is called a **synergistic effect.** Table 24.4 shows some accepted synergistic combinations. An example is the combination of a penicillin-class drug (which increases the cell wall permeability) with a protein-synthesis inhibitor (which sometimes encounters difficulty in penetrating the microorganism's cell wall) (Figure 24.7).

Combining drugs is also useful in treating organisms in which resistance may appear during prolonged therapy. It is extremely unlikely that a single microbial cell would simultaneously experience two different favorable mutations in order to survive. Cotrimoxazole is a combination of sulfisoxazole and trimethoprim, two different folate-synthesis inhibitors, useful in treating chronic infections such as prostatitis.

TABLE 24.4 Some accepted synergistic combinations	
Combination	**Used Against**
Ampicillin/gentamicin	Enterococci, *Streptococcus* Groups B, D
Carbenicillin/gentamicin	*Pseudomonas aeruginosa*
Cotrimoxazole (sulfisoxazole/trimethoprim)	Enterobacteria, especially in prostate *Pneumocystis carinii*
Sulfonamide/pyrimethamine	*Plasmodium falciparum*
Isoniazid/rifampin	*Mycobacterium tuberculosis*

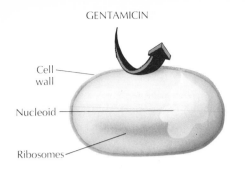
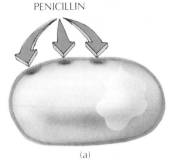
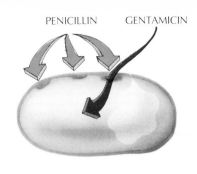

Cell wall

Nucleoid

Ribosomes

(a)

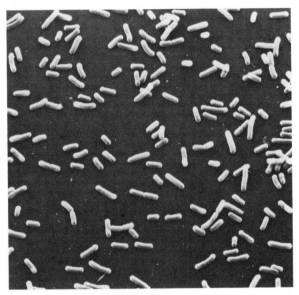

(b) *E. coli*, untreated

Treated with cefamandole

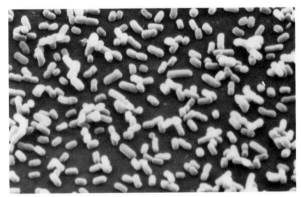

Treated with mecillinam

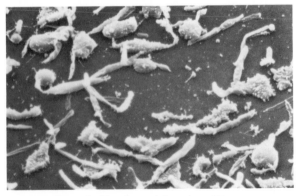

Treated with cefamandole and mecillinam

FIGURE 24.7 Synergistic Effects
In some instances a combination of drugs may be more effective against resistant organisms than a single drug. (a) For example, penicillin alone cannot kill this Gram-negative species, although the drug may weaken the cell wall. Gentamicin by itself cannot penetrate the cell, so it cannot have a detectable effect on growth. However, when the two drugs are administered together, the penicillin increases the bacterium's permeability, allowing the gentamicin to enter and block protein synthesis. (b) The Gram-negative bacterium *E. coli* is only moderately affected by either of two cell-wall–synthesis inhibitors, cefamandole and mecillinam, when they are administered separately. When both are used together, however, severe cellular disruption results. (Courtesy of M.J. Kramer, Y.R. Mauriz, M.D. Timmes, T.L. Robertson, and R. Cleeland, Morphological changes produced by Amdinocillin alone and in combination with Beta-lactam antibiotics: in vitro and in vivo, presented at "An International Review of Amdinocillin," and published in *AM. J. Med.,* August 23, 1983 (pp 30-41).)

Tuberculosis treatment, which usually takes at least 18 months, is most commonly carried out with combinations of two or more drugs, such as isoniazid plus rifampin.

THERAPY AND PROPHYLAXIS

Antimicrobial drugs are primarily used for therapy, that is, to treat obvious infectious disease. Their prophylactic use — to prevent infection — is very controversial. It is the consensus of the medical community that prophylactic use should be very limited, to minimize toxic effects and the spread of resistant strains. Antibiotics for prevention are generally accepted in rheumatic fever and congenital heart disease, for protecting people who have come into contact with tuberculosis, and in preoperative situations where there is a high degree of risk or where the introduction of microorganisms is unavoidable.

DYNAMICS OF DRUG ADMINISTRATION

Most of us have a television-induced, overly simplistic view of what happens when we take a pill or receive a shot. We view our body as an open vessel and visualize the drug surging immediately to the affected area for "instant relief." We are unaware that the drug reaches many parts of our body other than the part that needs it, that there are areas of the body where the drug may become concentrated, and that there are yet others where the drug may penetrate poorly or not at all. We also forget that the drug is a foreign substance. Our ungrateful body will immediately start to do what it does with all foreign chemicals — get rid of it. We will look here at how a drug becomes distributed in the body, how long it remains active, and how the body excretes it.

ROUTE OF ADMINISTRATION

The choice of an appropriate route of administration is influenced by many factors. Some drugs are supplied in various forms suitable for administration by several different routes, while others are restricted to a single route. There are four main routes of administration.

In **topical administration** the drug is applied to the skin or mucous membrane surfaces. Ointments, salves, or soaking solutions are used. The topical route has the advantage of eliminating unnecessary exposure of the rest of the body to the drug. Many drugs such as neomycin, bacitracin, or nystatin, although too toxic or too poorly absorbed for other routes, are extremely useful in topical forms.

Oral administration (*per os* or p.o.) is favored by both patient and practitioner because of its convenience and comfort. It has its limitations, however. A lot depends on the patient's receiving the drug according to a fixed time schedule to ensure that a certain minimum level of the drug is present in the body at all times. Even so, serum drug levels vary considerably from hour to hour. Penicillin G is an example of a drug that cannot be given orally; it is inactivated by stomach acid. The polymyxins are not absorbed from the GI tract and are excreted unchanged in the feces. Patient factors such as intestinal irritation or ulceration may alter absorption of any oral medication. The very ill, unconscious, or vomiting patient cannot take oral medication.

In **intramuscular** (IM) administration an injection is given **parenterally,** meaning directly into the tissues. From the muscle tissue, the drug is gradually absorbed into the circulation. The absorption period may range from minutes to days depending on the drug. A mixture of procaine and penicillin, administered intramuscularly, is absorbed much more slowly than penicillin G alone and thus gives a steady, prolonged effect. An injection of the long-acting benzyl penicillin will be continuously effective for a month. However, some drugs are extremely painful given intramuscularly, and cause tissue damage.

Intravenous (IV) **infusion,** administration directly into a vein, may be the best choice for a person who is seriously ill. IV administration provides constant, readily adjustable serum levels. A few drugs, however, such as erythromycin and certain cephalosporins, are unsuitable because they may cause thrombophlebitis at the needle site.

SERUM DRUG LEVELS

Effective therapy requires that the MIC of the drug be achieved at the site of microbial attack and usually also in serum. When a new drug is readied for licensing and marketing, dosage recommendations are prepared based on studies of the drug's behavior in human volunteers. Manufacturers' recommendations take into account the initial dose needed to achieve the desired serum level — the **loading dose** — and the frequency with which it must be replenished.

After administration, serum drug concentration increases to a peak level (Figure 24.8). Then the serum drug level begins to fall as the body eliminates more and more of the drug. The next dose is timed so that the falling drug level does not dip low enough to permit new microbial growth. Extreme fluctuations also place unnecessary stress on the kidneys and liver, which are responsible for eliminating the drug. Medication schedules, therefore, are not just rituals, they fulfill biological needs.

Each drug localizes somewhat differently in the body's various tissues and fluids. Most antibiotics administered parenterally enter tissues well and are also found in pleural fluid, the peritoneum, and the synovial fluid of the joints. Certain drugs are effectively secreted in mucus and, as a result, are readily available to fight infection on oral and vaginal surfaces. Not all body areas are equally accessible to all bloodborne substances. The central nervous system is one of the most restricted areas. Because of the blood–brain barrier, the cerebrospinal fluid concentration of most medications will be only a small fraction of the serum concentration. The interior of the eyeball and of the prostate gland are also physiologically isolated areas, which are difficult for drugs to enter. Devitalized tissue, vegetations, abscesses, and granulomas, all of which are key sites for infections, are also, unfortunately, poorly penetrated by most drugs.

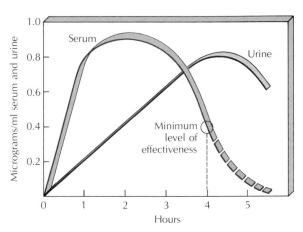

FIGURE 24.8 Serum and Urine Drug Concentrations This graph shows a generalized time sequence for the transfer of a drug from the circulation into the urine. Following administration, serum concentration rises rapidly and stays high for at least two hours. However, the kidneys immediately start to excrete the drug, bringing the serum level down after the third hour, while the urine level continues to climb. In order to maintain an effective level of the drug in serum, repeat doses must be given at least every four hours.

HOW DRUGS ARE ELIMINATED

As soon as a drug is administered, the body starts eliminating it. The **half-life** of a drug is the time needed for the serum level to be reduced by 50% from its peak. The half-life may differ, from the norm observed in healthy adults, in newborns, in obese people, or in people with renal or hepatic disease.

Most drugs are excreted unchanged via the kidneys. Those drugs with short half-lives, such as sulfisoxazole (Gantrisin), are so rapidly excreted that they go directly from the site of absorption (usually the intestine) via the blood to the urinary tract. This means that the drug's effect is concentrated in the urine and it is good for treating cystitis. Various factors may alter the time it takes the kidneys to excrete a drug. Probenicid is a chemical that may be given with a drug; it has the effect of altering the patient's renal function and prolonging the companion drug's residence time in the body. Poor kidney function hampers drug excretion. A patient with kidney disease will tend to retain medications abnormally, and may accumulate toxic serum concentrations unless the dose is reduced. Furthermore, the task of excreting some drugs imposes heavy stress on the tissue of both normal and damaged kidneys. This stress, referred to as **nephrotoxicity,** may produce temporary or permanent kidney damage.

A few drugs are excreted into the bile by the liver. The body eliminates such drugs via the feces rather than the urine. This type of drug may therefore be preferable when a patient has poor kidney function. Examples of drugs that are excreted in the feces include erythromycin, some of the semisynthetic penicillins, and chloramphenicol.

ADVERSE REACTIONS

Every medication has some side effects, and antimicrobial drugs are no exception. Antimicrobial therapy is continuously evolving, motivated by the demand for safer drugs and the need to define the safest conditions for their use. There is growing awareness of the need to use these drugs only when they are truly indicated, thus avoiding preventable toxic reactions. Toxicity reactions fall into several categories, as shown in Table 24.5.

TABLE 24.5 Major side effects of some commonly used antimicrobials		
Antimicrobial	**Common Side Effects**	**Occasional Side Effects**
Penicillin	Hypersensitivity	
Ampicillin	Hypersensitivity, GI upset	
Clindamycin	GI upset	Antibiotic-associated colitis
Chloramphenicol		Aplastic anemia
Gentamicin		Renal injury, inner-ear dysfunction
Tetracycline	GI upset	Renal and hepatic injury, tooth discoloration
Sulfonamides	Hypersensitivity	Renal and hepatic injury, anemia
Isoniazid		Hypersensitivity, GI upset, hepatic injury, peripheral neuropathy
Rifampin		Hepatic injury, hearing problems
Chloroquine	GI upset	Dizziness and headache, anemia, eye and ear problems
Metronidazole		GI upset

LIVER AND KIDNEY TOXICITY

The liver and kidney, as we have seen, detoxify and excrete foreign substances, including antibiotics. Thus, while a piece of tissue in your arm, for example, comes in contact with only a portion of the total drug administered, your liver or kidneys will eventually cope with the whole dose. In the process, they may suffer damage.

Damage to the liver shows up as increased serum levels of liver enzymes or bile pigments (jaundice). Liver damage is usually reversible after short exposures, but it is cumulative with increasing dosage and duration of therapy. Isoniazid, used in treating tuberculosis, is occasionally associated with liver dysfunction.

Kidney damage may be either glomerular or tubular. It is detected by monitoring the serum levels of retained waste products, which go up if the kidneys are not excreting at normal efficiency. Amphotericin B, neomycin, and kanamycin are drugs that are nephrotoxic at the usual dosage levels. Other drugs tend to cause trouble only if serum levels become excessively high.

BONE-MARROW TOXICITY

Antimicrobial drugs may also cause damage to the rapidly multiplying stem-cell populations in the bone marrow, causing numbers of blood cells to decline. Aplastic anemia is a rare outcome of chloramphenicol treatment. It is an irreversible, fatal disease in which the production of all circulating blood cells ceases. The polyene drugs used in treating fungal infections also weaken and hemolyze red blood cells. **Leukopenia** (decreased white-blood-cell count) is a common side effect of systemic treatment with the antiviral drug idoxuridine. Persons with a certain recessive genetic trait may react to sulfonamides by developing a hemolytic anemia, in which the red blood cells are fragile and lyse spontaneously.

NEUROMUSCULAR TOXICITIES

Neuromuscular symptoms such as dizziness, confusion, and muscular weakness may occur with some drugs. Penicillin in extreme doses may cause seizures. Streptomycin may damage the auditory nerve and cause deafness. Quinine and its derivatives often cause ringing in the ears (tinnitus).

GASTROINTESTINAL DISORDERS

Nausea, vomiting, or diarrhea are relatively common, minor side effects following administration of some oral drugs; these side effects can often be prevented by taking the drug with food. However, food may also interfere with drug absorption. For example, tetracycline should not be ingested with milk products because calcium ions prevent its absorption.

HYPERSENSITIVITY

Some drugs may cause hypersensitivity reactions. Penicillins, cephalosporins, and sulfonamides are the most notorious offenders. This is why it is crucial for health-care personnel to determine if a patient has ever had a reaction to these drugs. Parenteral administration may lead to fatal anaphylactic shock. Over 300 persons a year in the United States suffer penicillin-induced shock. Oral and topical administration in a hypersensitive person produce GI upset and hives, respectively. A person sensitized to one penicillin-class drug is likely to have a reaction to any other, but is less likely to react to cephalosporins. Penicillin hypersensitivity may be assessed by a scratch test or intradermal injection. The tester should be fully prepared to treat a possible anaphylactic response.

SUPERINFECTION

Superinfection is the overgrowth of an antibiotic-resistant microbial strain, often displacing the normal flora. In hospitals, these superinfecting opportunists may be some of the hospital "super-bugs" — multiple-drug–resistant enterobacteria. The yeast *Candida albicans*

BOX 24.2
Antibiotic Misuse

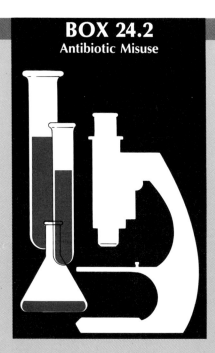

Although there have been some improvements in the last decade, antibiotics are extensively and persistently overused in human medicine. In 1960, three million pounds of antibiotics were produced in this country. By 1972 production for human use had jumped to 9.6 million pounds. This represents a growth of 360%; but over the same period the population increased only 11%. The average person has an infection requiring antibiotics once every 5 to 10 years, but in 1972 enough doses of antibiotics were produced to treat two illnesses per year for every man, woman, and child in the United States. Antibiotics are given for viral infections, without positive cultures or evidence of infection, or prescribed over the telephone. Physicians may overprescribe because of the pressure of patient expectations. Many patients aren't "patient" at all; they demand immediate relief from a disease that will cure itself. Some feel that something is missing if they leave a doctor's office without a prescription, and they will shop around until they find another physician who will fulfill their expectations. Practitioners also overprescribe to play it safe, or as a time-saving short-cut. Outside the medical-care system, antibiotics are freely available as "street drugs" right along with heroin, cocaine, and other illegal substances. We have no idea how extensively they are abused.

In the short term, antibiotic overuse is poor practice because 5% of antibiotic recipients experience side effects, and more than 2% develop superinfection. These outcomes may lengthen their illnesses. In the long term, overuse slowly but steadily promotes drug resistance, particularly in hospital-acquired strains of bacteria. This hampers our ability to deal forcefully with infections when we really need force.

Fortunately, the public is becoming more aware of the limitations of antibiotics. This could help to relieve the pressure on physicians to prescribe against their own good judgment. Detailed guidelines and professional peer review allow evaluation of drug prescription practices. Pharmacists provide expert consultation services to clinicians. The epidemic of misuse peaked around 1970; it is gradually receding, but remains a worry.

also causes troublesome mucous membrane superinfections. Another superinfection is **antibiotic-associated colitis,** a severe intestinal ulcerative disease in which *Clostridium difficile* is the causative agent. This condition is most often encountered when the patient has had prolonged ampicillin, clindamycin, or cephalosporin therapy.

ANTIBACTERIAL DRUGS

In treating bacterial infections, we have available to us a wide range of different antibacterial drugs. They are grouped in several chemical families, which we'll consider next.

THE PENICILLINS

The **penicillins** are a family of antibacterial drugs that all share a central chemical feature, the beta-lactam ring. This ring is the source of their antimicrobial activity, and they remain active only as long as the ring is intact. The main practical limitation on penicillin effectiveness is that this beta-lactam ring is vulnerable to degradation by stomach acid and attack by microbial enzymes. Because of this problem, organic chemists have been working for many years, progressively improving on the natural penicillins. They start with the 6-aminopenicillanic acid (6-APA) molecule, a metabolic product of the fungus *Penicillium chrysogenum* (Figure 24.9a). Then they add various functional groups, or side-chains, to 6-APA to produce compounds with such useful properties as acid stability, penicillinase resistance, and expanded spectrum (Figure 24.9b, also see Figure 28.8). Roughly a dozen penicillins, out of the hundreds that have been experimentally developed over the years, have shown significant value and are in general use.

Penicillins, as we noted, interfere with cell-wall synthesis in actively growing cells. In adequate dosage, they are bactericidal. However, they have no effect on inactive bacteria. Penicillins can work only if they successfully penetrate the outer cell-wall layer of the bacteria to reach the inner layers where peptidoglycan is assembled. In the human body, penicillins penetrate tissues well, except for those of the central nervous system. However, since high dosages may be used safely, effective levels in the central nervous system can be achieved. Penicillins are mainly excreted in the urine. True toxicity is extremely uncommon, which is one reason why the penicillins are the most widely used of all antimicrobials. However, hypersensitivity develops quite commonly, and ranges from mild to fatal. If a person is hypersensitive to one penicillin, no other penicillins should be used.

As we noted, there are many different penicillins available. Penicillin G is one of the original types; it is a natural fungal product, a highly potent, narrow-spectrum drug effective largely against Gram-positive bacteria. It must be given parenterally, since it does not survive passage through the stomach. It may be used for gonorrhea on a "one-dose" basis in the long-acting procaine-penicillin form. Methicillin, although penicillinase-resistant, is broken down by stomach acid. Large parenteral doses are used to treat hospitalized patients with penicillinase-positive staphylococcal infections. Combined acid and penicillinase resistance is found in oxacillin, cloxacillin, and nafcillin. They are all useful narrow-spectrum drugs with little effect against Gram-negative bacteria.

Ampicillin is an extremely popular **extended-spectrum penicillin,** meaning that it is effective not only against Gram-positive bacteria, but also against most strains of certain Gram-negative species, including *Hemophilus, Salmonella*, and other enterobacteria. Ampicillin is given orally. Amoxicillin is also an extended-spectrum drug, which may be preferred for oral administration because higher serum levels are achieved. Carbenicillin is effective against many strains of *Pseudomonas aeruginosa*. Resistant strains arise freely, so the drug is frequently combined with aminoglycosides such as gentamicin or tobramycin. Ticarcillin, piperacillin, mezlocillin and azlocillin are among the newest extended-spectrum penicillins.

THE CEPHALOSPORINS

The **cephalosporins** are produced by fungi of the *Cephalosporium* genus, and act against the bacterial cell wall. Their active chemical structure is 7-aminocephalosporanic acid (7-ACA), which contains a beta-lactam ring (Figure 24.10). Like

FIGURE 24.9 Penicillins

(a) A growing colony of *Penicillium chrysogenum,* the mold that supplies a large portion of the world's penicillin supply. (Courtesy of Pfizer, Inc.) (b) All penicillins contain the 4-sided beta-lactam ring (color type), which is essential for penicillin's activity. This ring is inactivated by penicillinase enzymes. 6-Aminopenicillanic acid is the starting material from which semisynthetic penicillins are made. Penicillins differ from one another in their side-chains, or R groups, (color screen). In penicillin G, a naturally occurring antibiotic, the side chain confers activity against Gram-positive bacteria; however, the molecule remains sensitive to both acid and enzymatic attack. The substitution of a bulky side-chain in semisynthetic methicillin renders this compound more penicillinase-resistant. Ampicillin is acid-resistant due to the substitution of an amino group on the side-chain. It can be given by mouth and has activity against some Gram-negative bacteria.

(a)

6-amino penicillanic acid

Penicillin G

Methicillin, penicillinase resistant

Ampicillin Acid-resistant expanded spectrum

(b)

741

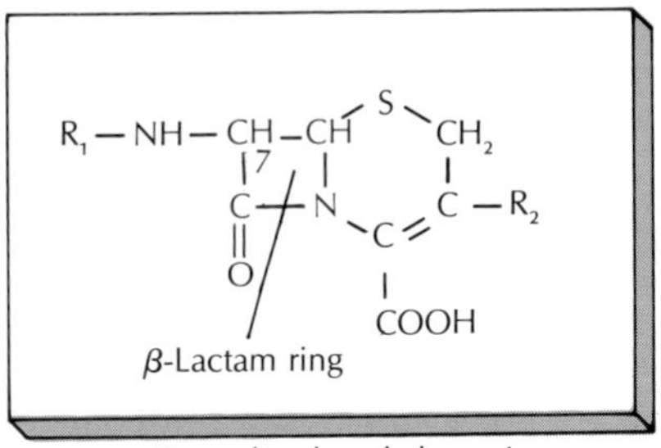

Generalized cephalosporin

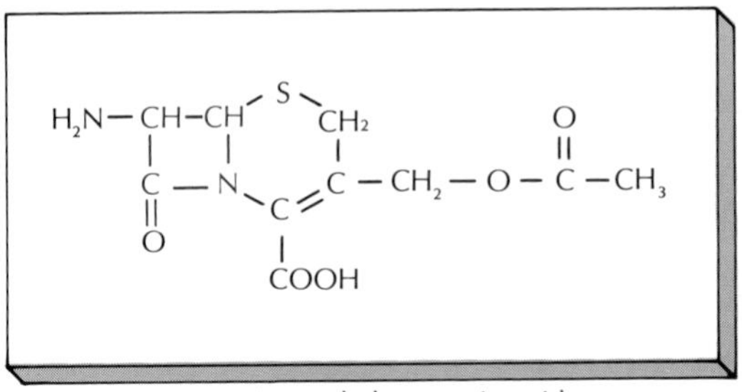

7-aminocephalosporanic acid

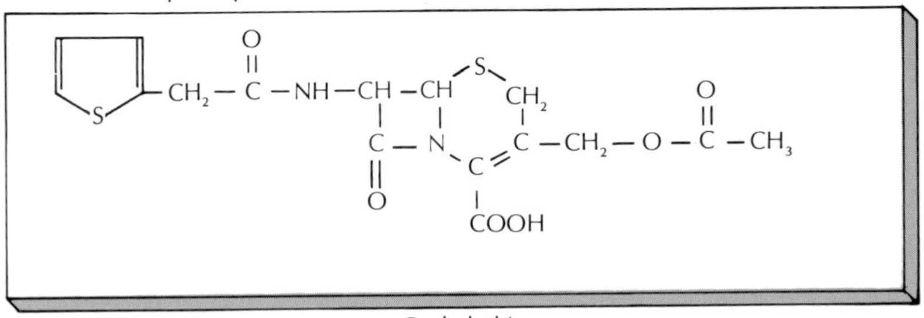

Cephalothin

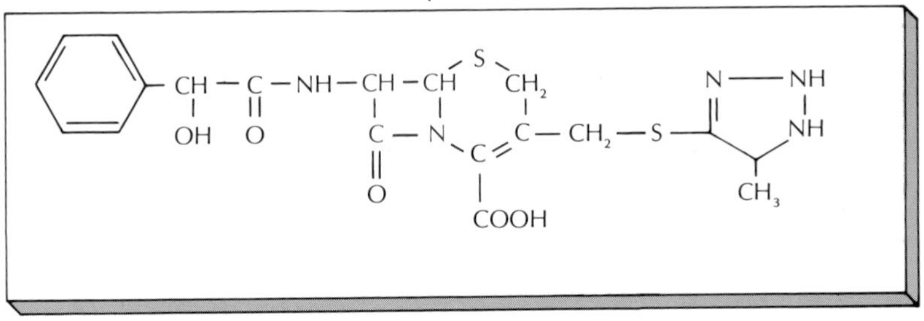

Cefamandole

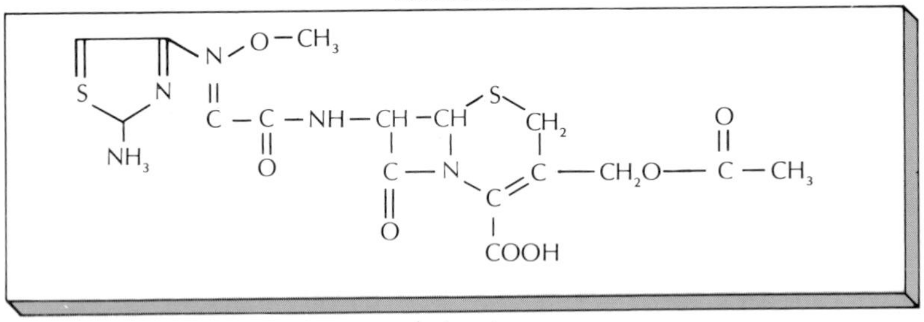

Cefotaxime

FIGURE 24.10   The Cephalosporins
All cephalosporins contain the beta-lactam ring fused to a six-membered ring
containing sulfur. They have two R groups, which may be modified for altered
activity. The 7-aminocephalosporanic acid molecule is the starting material for
antisynthetic cephalosporins. Cephalothin is a first generation cephalosporin;
Cefamandole is a second-generation cephalosporin; Cefotaxime is a third-generation
cephalosporin.

the penicillins, the cephalosporins vary in their stability against gastric juice and bacterial beta-lactamases.

Chemical modification has produced a succession of cephalosporins; the original "first-generation" cephalosporins are no longer used. The "third-generation" cephalosporins are appearing, fruits of much research and corporate investment. Cephalosporins are most often used as first alternatives to the penicillins. They are costly drugs; the cost of a 10-day course of therapy can be as much as 1000 dollars.

Cephalosporins are bactericidal in adequate dosage. Toxicities are rare with the third-generation drugs, but may include GI upset, nephrotoxicity, or thrombosis on intravenous administration. Hypersensitivity occurs less commonly than with the penicillins. Hives are the usual manifestation of a mild hypersensitivity, and anaphylaxis is very rare. Most of these drugs have "ceph-," "cef-," or "kef-" in their names. A few examples include Keflex, cefotaxime, moxalactam, and cefoperazone.

ERYTHROMYCIN AND CLINDAMYCIN

Erythromycin and clindamycin are used as alternates to the penicillins. Their spectra are similar to that of penicillin, but their modes of action are different. Erythromycin (Figure 24.11) is effective against common Gram-positive pathogens, as well as mycoplasmas and chlamydiae. It is particularly useful in diphtheria, *Listeria* infections, and *Legionella* pneumonias. Erythromycin kills bacteria by inhibiting their protein synthesis. The drug can be administered by various routes. It is excreted after breakdown by the liver, and rarely may cause liver toxicity. The related drug tylosin is widely used in animal feed.

Clindamycin is valuable because of its unusually effective action against anaerobes. It may also be used when there is poor kidney function, as it is excreted in the feces. Oral absorption is rapid and complete, and clindamycin diffuses well into deep tissues and anaerobic foci.

GI upset is a common side effect, and antibiotic-associated colitis resulting from superinfection by clindamycin-resistant anaerobes in the bowel flora may become severe.

THE AMINOGLYCOSIDES

The **aminoglycosides** include several powerful chemotherapeutic agents containing linked amino sugars, primarily valued for their effectiveness against Gram-negative bacteria. The more recently developed aminoglycosides are less toxic than the earlier ones.

The aminoglycosides inhibit protein synthesis, and also affect bacterial membranes. They are bactericidal, but one-step genetic resistance develops rapidly, and is readily transmitted among bacterial strains. A bacterial strain resistant to one aminoglycoside may also be resistant to some or all of the others.

FIGURE 24.11 Erythromycin
Erythromycin inhibits protein synthesis. It is useful against diphtheria, legionellosis, and other infections, and may be used in neonatal eye drops to prevent gonorrhea infections.

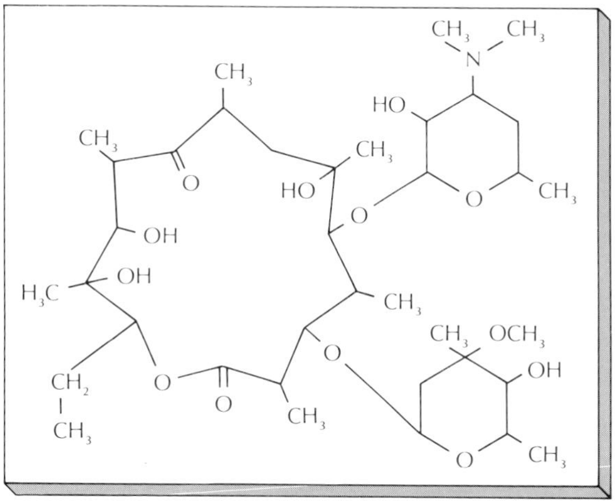

Erythromycin

The first aminoglycosides — streptomycin, neomycin, kanamycin, and paromomycin — were natural products of the filamentous bacteria *Streptomyces* (Figure 24.12). They have only limited use at this time. Streptomycin, discovered in 1944, immediately became the first life-saving drug for tuberculosis treatment. It is still included in some tuberculosis, plague, and tularemia treatments where its value outweighs

FIGURE 24.12 Aminoglycoside Antibiotics
These drugs get their group name from their structures, sugar molecules that carry amino groups. Streptomycin and gentamicin are well-known examples.

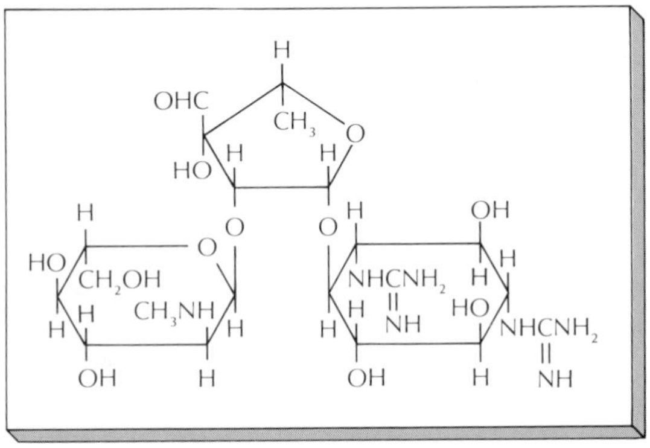

Streptomycin

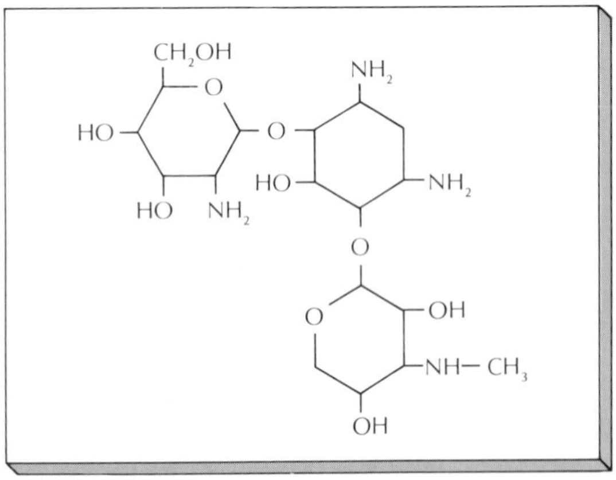

Gentamicin

its hazards. Its main drawbacks are side effects such as deafness, and the frequency of resistant strains. Neomycin is widely used topically to treat skin infections.

The newer aminoglycosides are produced by other filamentous bacteria of the genus *Micromonospora*. They are similar to the original aminoglycosides in mode of action and spectrum, but are somewhat less toxic. Gentamicin, tobramycin, and amikacin are poorly absorbed from the gut and are usually given intravenously. Serum levels, which are highly variable, must be closely monitored, especially when kidney disease is among the patient's problems. These drugs are indicated only in severe infections, such as bacteremia, necrotizing pneumonia, neonatal sepsis and meningitis, and burn sepsis. Aminoglycosides are frequently combined with a penicillin or cephalosporin. Amikacin has an expanded activity against *Pseudomonas aeruginosa*. Most anaerobes are not at all susceptible.

THE TETRACYCLINES

Tetracyclines are widely used broad-spectrum antimicrobials chemically based on a structure of four linked rings (Figure 24.13). They inhibit protein synthesis, and are bacteriostatic. Since they do not actually kill the microorganisms, tetracyclines work well only when host defenses are strong. Toxicities include nausea and photosensitivity; nephrotoxicity may be caused by outdated, deteriorated preparations.

Tooth discoloration is a distressing side effect of the use of tetracycline. When tetracycline is taken by pregnant women or young children, the infant's first or child's second teeth may be visibly, permanently discolored with grayish, brownish, or mottled markings. For this reason, tetracyclines are not recommended for children under eight years of age.

The tetracyclines are useful in treating sexually transmitted nonspecific urethritis or vaginitis because they are effective against my-

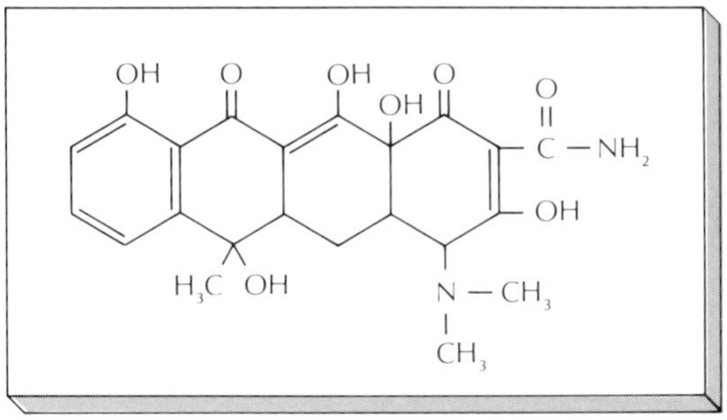

FIGURE 24.13 Protein-Synthesis Inhibitors
Tetracycline, as its name implies, has four rings in its structure. Chloramphenicol is a highly potent antibacterial drug that must be used with great caution.

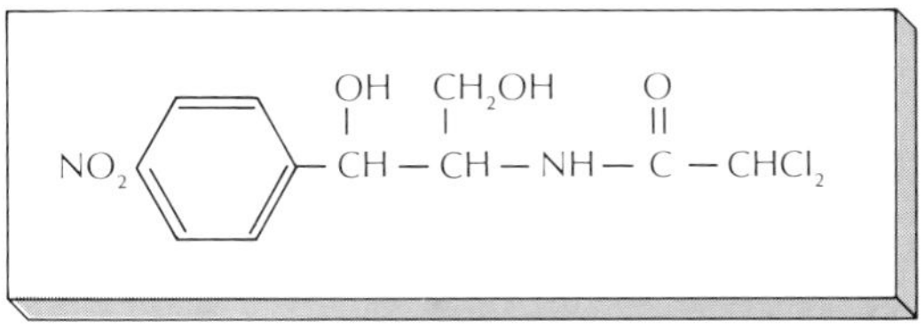

coplasmas and chlamydiae. They are the drug of choice for many rickettsial infections. Oxytetracycline is useful in urinary-tract infections. Doxycycline is a form of tetracycline that can be used to treat patients who also have renal problems.

CHLORAMPHENICOL

Chloramphenicol is a broad-spectrum bacteriostatic drug (Figure 24.13). It can be administered by any route, is readily absorbed, and distributes to all body compartments. Unlike most other antimicrobial drugs, it becomes concentrated in the central nervous system. Chloramphenicol is broken down by the liver and excreted by the kidney.

Chloramphenicol clearly has many useful qualities, but they are offset by its unpredictable

tendency to produce either a reversible dose-related anemia or a rare, irreversible, fatal aplastic anemia. In newborns the **gray-baby syndrome,** which is a fatal anemia, may occur if the dose is excessive. Lesser side effects such as GI upset, neurologic disturbances, and hypersensitivity are also seen.

The use of chloramphenicol is justified in life-threatening infections, including typhoid fever, Rocky Mountain spotted fever, typhus, and some forms of bacterial meningitis. It is also widely used in veterinary medicine, as many other mammals are less susceptible to its side effects than we are. In poorer countries its low cost makes it one of the few affordable drugs. Like many antibiotics, chloramphenicol can be purchased over the counter (without prescription) in much of Central and South America.

OTHER ANTIBACTERIAL DRUGS

The **polymyxins** are cationic peptides that destroy bacterial membranes. They are too toxic for general systemic use. Polymyxin B is occasionally used in *Pseudomonas* infections. Topical preparations, sometimes in combination with neomycin and bacitracin, are used for external ear and eye infections. Bacitracin is a related peptide, also for topical use only.

The **sulfonamides** and **diaminopyrimidines,** as described earlier, inhibit different steps in folic acid synthesis (Figure 24.14). Their effect is broad-spectrum but only bacteriostatic. Resistance develops rapidly, as does host hypersensitivity. However, these drugs diffuse well into body fluids, including cerebrospinal fluid. The short-acting sulfonamides such as sulfisoxazole are widely used in uncomplicated bladder infections. Sulfamylon (Mafenide) and silver sulfadiazine (Silvadene) are used for burn treatment. The diaminopyrimidines include trimethoprim and pyrimethamine. Cotrimoxazole combines sulfamethoxazole and trimethoprim (Figure 24.14b). It is recommended in chronic prostatitis, in some urinary tract and pulmonary

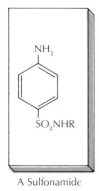

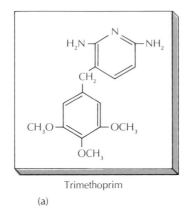

A Sulfonamide   Trimethoprim

(a)

FIGURE 24.14   Folate Inhibitors
(a) Sulfonamide drugs and trimethoprim inhibit different steps in folate synthesis. (b) The four bacterial isolates in the top row show susceptibility to a disk combining sulfa and trimethoprim. In the bottom row, each isolate was tested against individual disks of trimethoprim (left) and sulfamethoxazole, a sulfonamide, (right). Isolate 1 shows individual susceptibility to both, plus synergy (fused zone). Isolates 2 and 3, susceptible to only one drug, show no added inhibition from the presence of the other drug. Isolate 4, while susceptible only to sulfa, does show some synergy, as evidenced by the skewed zone of inhibition. (Reproduced with permission from *Culture*, September 1978, p.1. Published by Media Medica Limited, U.K.)

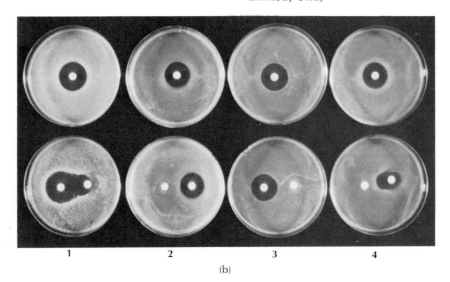

1   2   3   4

(b)

infections, and in salmonellosis. It is the preferred agent against *Pneumocystis carinii*. Fansidar, a combination of pyrimethamine and sulfadiazine, is effective against chloroquine-resistant strains of the malaria parasite.

OTHER ANTIMICROBIAL DRUGS

Mycobacteria, fungi, viruses, protozoans, and helminths are treated with drugs that are generally less familiar than those used against the common bacteria. The effectiveness of these drugs depends on their ability to interfere with the unique biological functions of each group of agents.

ANTIMYCOBACTERIAL DRUGS

Mycobacterium tuberculosis and *M. leprae*, although they are bacteria, possess unusual lipids that render their cell walls impermeable to the more commonly used antibacterial drugs. Furthermore, mycobacteria are very slow-growing and not much affected by substances that kill rapidly growing bacteria. Because long periods of treatment are needed to control tuberculosis and leprosy, the pathogens readily develop resistance to any drug offered singly. In treating active cases, therefore, two or three drugs are combined. Laboratory susceptibility tests are used to select the drugs.

There are several effective antimycobacterial drugs. **Isoniazid,** a synthetic chemical, is the single most widely used antitubercular drug (Figure 24.15). It can be administered orally, and toxicity, including neuritis and hepatitis, is uncommon. It acts bacteriostatically by a competitive inhibition mechanism involving vitamin B_6. Patients usually take vitamin B_6 supplements to help protect the liver and prevent side effects. Isoniazid is used alone for prophylactic treatment of tuberculin-positive contacts. It is combined with other drugs to treat active cases.

FIGURE 24.15 Drugs Used Against Mycobacteria
Isoniazid is the first choice both for prophylaxis and therapy of tuberculosis. Rifampin, an RNA inhibitor, is usually given with isoniazid in treating tuberculosis cases. Para-aminosalicylic acid or ethambutol may be substituted for rifampin. Dapsone is the drug of choice in leprosy.

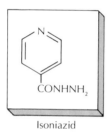

Rifampin is a drug possessing a large, complex ring structure (Figure 24.15). Its mode of action is to inhibit bacterial RNA polymerase. It is now the most commonly used companion drug to isoniazid in treating mycobacterial infection, but it is also effective against many other bacteria. Rifampin can be given orally, is well absorbed, and has low toxicity. Side effects may include jaundice, bone-marrow toxicity, and a flulike syndrome. **Streptomycin** is still occasionally used, but it requires intramuscular injection, which makes it impractical for home care. **Ethambutol,** an RNA synthesis inhibitor, can be administered orally and is widely used for primary treatment of tuberculosis (Figure 24.15).

Dapsone, a synthetic chemical of the type called a sulfone, is the mainstay of leprosy treatment because it is bacteriostatic to *M. leprae*. However, resistant strains have started to appear in many areas of the world. Rifampin, acedapsone and cofazimine are used as alternatives.

ANTIFUNGAL DRUGS

The eucaryotic fungal cell is most effectively controlled by **polyenes,** drugs whose chemical structure contains many double bonds alternating with single bonds. Polyenes such as nystatin and amphotericin B destroy cell membranes. All antifungal drugs are incompletely specific and thus have appreciable human toxicity. Superficial fungal infections, which are minor, are treated with topical drugs. So long as the drug is not absorbed from the skin, it will not be significantly toxic. Progressive systemic mycoses, on the other hand, are life-threatening, and in these infections the use of toxic drugs is unavoidable.

TREATMENT OF SUPERFICIAL MYCOSIS.   The dermatomycoses (tinea capitis, tinea pedis, etc.), candidiasis of the mucous membranes, and fungal infections of the eye and external ear are frequently treated topically. Synthetic chemicals of the class called **imidazoles,** such as clotrimazole and miconazole, are supplied as creams, solutions, or vaginal applications. They are broad-spectrum antifungals, and their only adverse effect is local irritation in some patients. Clotrimazole is also useful against *Trichomonas.*

Griseofulvin is a fungistatic drug that blocks mitosis. It is effective against every dermatophyte except *Candida albicans,* and is given orally to treat extensive or persistent dermatomycosis. Although griseofulvin is poorly absorbed, sufficient drug is deposited in the keratin-forming cells to arrest fungal spread on new growth of skin, hair, and nails. Headaches are a frequent side effect. **Nystatin,** a polyene, is widely used for candidiasis. Creams and vaginal suppositories are available for topical use. Nystatin is used prophylactically in combination with gentamicin in leukemia patients to control superinfection with bowel flora. Nausea and vomiting are the main side effects. **Tolnaftate** is useful topically against dermatophytes but not against *Candida.*

TREATMENT OF SYSTEMIC MYCOSIS.   The polyene **amphotericin B** is the most effective choice for treatment of severe systemic mycosis. It causes irreversible leakage of fungal cell membranes. It is effective against *Blastomyces, Histoplasma* (Figure 24.16), *Cryptococcus, Candida,* and *Coccidioides.* Amphotericin B is given intravenously. Most of the drug molecules bind to body fat, from which they are slowly released. Side effects are inevitable; they increase with both dosage and duration. Fever, chills, weight loss, and GI upset are common. Some degree of renal damage and anemia are usually seen.

5-flucytosine, a nucleoside analog, is effective against most systemic fungi, but unreliable with *Candida.* It may be given orally, and though side effects may occur (bowel dysfunction, leukopenia, aplastic anemia), they are much less frequent than with amphotericin B.

ANTIVIRAL DRUGS

Viral replication is highly dependent on host-cell structures and enzymes. A few of the replicative

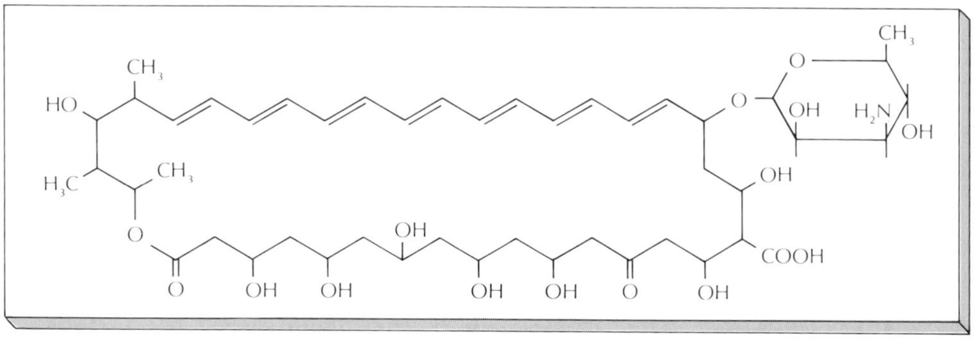

Amphotericin B

FIGURE 24.16 The Polyene Amphotericin B
Amphotericin B is often used to treat serious systemic fungal infections. It acts by
destroying the semipermeability of fungal cell membranes.

enzymes, however, are specific for viral replication, but not essential to the host's survival.
The challenge of research in antiviral therapy is
to find drugs sufficiently specific to block only
these enzymes. All the drugs mentioned below
inhibit nucleic acid synthesis in some way.

Most drug research on DNA viruses has been
directed toward the herpesviruses, particularly
the virus of genital herpes. The base analog **idoxuridine** (Figure 24.17a) is useful in some herpesvirus conditions. It has recently been shown
to reduce significantly the rate of recurrences of
genital herpes per year per person, although it
does not cure the underlying disease. It is licensed for topical use in herpes eye infections. A
sugar analog, **2-deoxy-D-glucose,** is being tested
for effectiveness in herpes eye infections and has
recently been approved for treatment of cervical
herpes. **Vidarabine** (Figure 24.17b) has shown
promise in herpes encephalitis. In a carefully designed clinical study, ara-A reduced the mortality rate from 70% to 28%. It is currently being
evaluated against progressive mucocutaneous
and neonatal herpes. **Acyclovir** (Figure 24.17c),
9-(2-hydroxyethoxymethyl) guanine, is useful
against several DNA viruses. It is less toxic than

ara-A, and has been licensed for treatment of
genital herpes.

Amantadine hydrochloride (Figure 24.17d)
and the related drug **rimantadine** are useful
against an RNA virus, the influenza virus. They
are approximately 70% effective in preventing
influenza A when taken during an epidemic by a
person who is at risk of exposure and wishes or
needs to prevent illness. These drugs also moderate the symptoms of influenza and may
shorten its duration by several days. They may
cause transitory nervousness, insomnia, or hallucinations in 2 to 5% of patients.

The interferons, you'll remember, are proteins that are part of our defense network. Some
inhibit viral replication and have been the subject of small-scale, inconclusive clinical trials
against many viral infections, including the
common cold. There is now a large supply of
human interferon produced by recombinant bacteria, and the material available for clinical investigation is plentiful. It is still unclear what
the side effects of treatment with pure interferon
may be, but patients in early trials have been
plagued with problems such as cardiac illness
and upper respiratory irritations. The next few

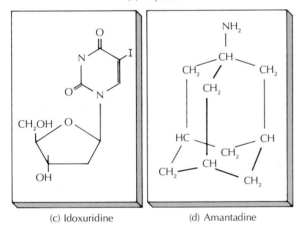

FIGURE 24.17   Antiviral Drugs
Vidarabine (a), acyclovir (b), and idoxuridine (c) are all nucleoside analogs that mimic natural starting materials for DNA synthesis. They are used to treat infections caused by DNA viruses, especially the herpesviruses. Amantadine (d) is an unusual cyclic compound that is partly effective in treating influenza virus infections.

years may or may not see the emergence of interferon as the mainstay of viral chemotherapy.

ANTIPROTOZOAN AND ANTIHELMINTHIC DRUGS

We have highly effective treatments for giardiasis, amebic dysentery, trichomoniasis, and most forms of malaria, but none for trypanosomiasis. Some antiprotozoan drugs are highly toxic, containing elements such as arsenic or antimony, and they are not available through the usual pharmaceutical network. They are provided by the Centers for Disease Control only on an experimental basis, with skilled consultation.

Metronidazole (Figure 24.18a) (Flagyl), is a generally available, safe synthetic drug widely used in the United States to treat three common protozoan diseases — *Entamoeba histolytica* dysentery, *Giardia lamblia* gastroenteritis, and *Trichomonas vaginalis* vulvovaginitis. The drug may cause nausea, drowsiness, a skin rash, or a bad taste in the mouth. There is a severe antagonistic reaction with alcoholic beverages. Metronidazole should not be used in early pregnancy or while the mother is breast-feeding because of toxicity to the fetus or infant.

Chloroquine (Figure 24.18b), a derivative of quinine, is the mainstay in malaria treatment. Primaquine or quinine may also be used in certain cases. Where chloroquine resistance is feared, pyrimethamine combined with sulfadiazine (Fansidar) is substituted.

Helminths are treated with a wide range of synthetic drugs. The less complex infestations, such as pinworm, are easily eradicated with pyrivinium. Organisms with complex life cycles, including resting or cyst forms, such as *Trichinella* or *Echinococcus* species, can be brought under control if treated soon enough, but the parasite can only be eradicated from the patient's body if treatment is begun before the cysts have formed. The **ivermectins** are a group of new antihelminthics with a high potency and a comparatively low toxicity. They are being developed for both human and animal use.

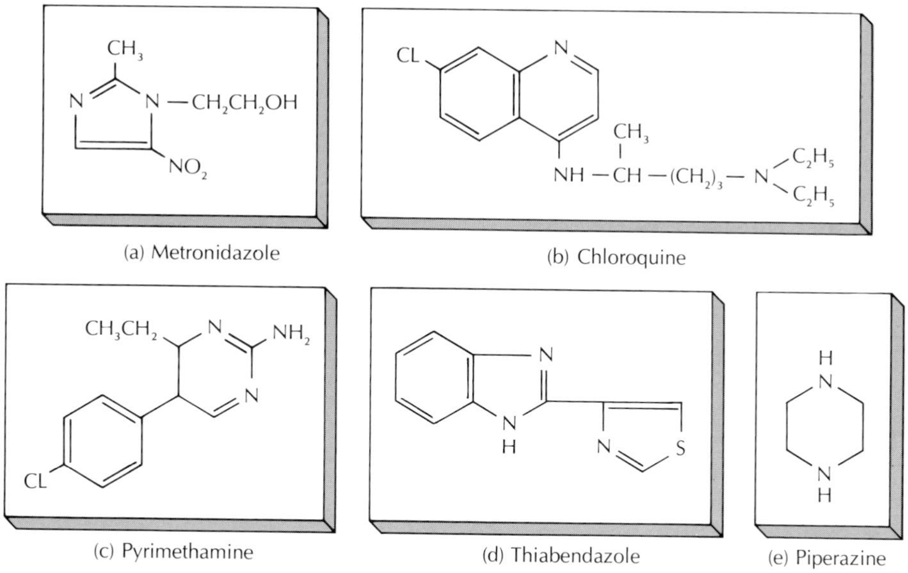

FIGURE 24.18   Antiprotozoan and Antihelminth Drugs
(a) Metronidazole is used to treat giardiasis and trichomoniasis; (b) Chloroquine is the drug most often used against malaria; (c) In cases where malarial parasites are resistant to chloroquine, pyrimethamine may be substituted; (d) Thiabendazole is an antihelminthic used against roundworms such as hookworms, whipworms, and threadworms; (e) Piperazine is highly effective against pinworms.

SUMMARY

1. Antimicrobial drugs have lowered the case fatality rates of most bacterial, fungal, and protozoan diseases.

2. All antimicrobials occasionally have some degree of toxicity and hypersensitivity. Microorganisms also readily develop and transfer drug resistance.

3. Antimicrobials should never be used unnecessarily.

4. The causative agent(s) must be isolated from the patient and antimicrobial susceptibilities determined. Both semiquantitative agar diffusion and quantitative microdilution techniques are in current use. The minimum inhibitory concentration (MIC) test is the most effective means of determining susceptibility.

5. Antimicrobials that are -static inhibit microorganisms' growth while -cidal antimicrobials irreversibly inactivate them; -cidal drugs are considered preferable.

6. Each compound has a specific mode of action. Human toxicity is low when the mode of action is specific for the target pathogen, and spares human tissue. Toxicity is higher when the drug is relatively nonspecific.

7. Toxic effects are most commonly manifested in the gastrointestinal tract, liver and kidney, nervous system, and bone marrow. Hypersensitivity may also occur.

8. When a drug is administered, it follows a specific pattern of absorption, distribution, and excretion. This influences the choice of dosage, timing, and route of administration. The drug must reach the anatomic site of the infection. Effective drug levels at that site must be continually maintained.

9. Different groups of antimicrobials are recommended for different types of pathogens. Effective, relatively nontoxic therapy is available for almost all bacterial diseases. The penicillins, cephalosporins, aminoglycosides, and tetracyclines are a few of the best-known groups of antibacterial drugs.

10. Choices are fewer and attended by significantly more risk for mycobacterial, fungal, viral, protozoan, and helminthic pathogens. Mycobacteria are often treated with isoniazid, rifampin, or dapsone.

Superficial fungi are treated with imidazoles, griseofulvin, nystatin, or tolnaftate. Systemic fungi are treated with amphotericin B. Antiviral therapy is just beginning; some useful drugs are idoxuridine, acyclovir, and amantadine. Protozoans and helminths are treated with quinine derivatives, imidazoles, and other more toxic drugs.

11. Because microorganisms continually become resistant to newly introduced drugs and because many major diseases still lack specific therapeutic agents, we are far from the conquest of microbial pathogens by chemical means.

DISCUSSION QUESTIONS

1. What is the meaning of minimal inhibitory concentration? Spectrum? Differential toxicity? What restrictions do each place on therapy?

2. What is the argument in favor of using bactericidal rather than bacteriostatic antimicrobial treatments? Are there any exceptions?

3. Give a number of reasons why antimicrobial therapy may fail.

4. Describe the disk and MIC methods for determining susceptibility. What does each tell you? Which seems preferable?

5. Why does antimicrobial therapy have to be chosen with careful consideration of the recipient's liver and kidney function?

6. In what areas of infectious-disease management are more and better antimicrobials most urgently needed?

ADDITIONAL READING

Abraham EP. The beta-lactam antibiotics. *Sci Am* 1981; 244:76-86.

Galasso GJ, Merigan TC, Buchanan RA, eds. *Antiviral agents and viral diseases of man.* New York: Raven Press, 1979.

Gale EF, et al. *The molecular basis of antibiotic action.* 2nd ed. New York: John Wiley, 1981.

Goodman LS, Gilman A, eds. *The pharmacological basis of therapeutics.* New York: Macmillan, 1979.

McDermott W. Pharmaceuticals: their role in developing societies. *Science* 1980; 209:240-245.

Ogawara H. Antibiotic resistance in pathogenic and producing bacteria, with special reference to beta-lactam antibiotics. *Microbiol Rev* 1981; 45:591-619.

Physician's Desk Reference. Oradell, N.J.: Medical Economics Co. Published annually.

Warren KS. The control of helminths: nonreplicating infectious agents of man. *Ann Rev Public Health* 1981: 2:101-116.

IMMUNIZATION

The ultimate goal of medicine is to prevent disease from occurring. But often, the needs of the sick are so much more immediate and pressing than the needs of the healthy that it is hard to keep this goal in view. One common complaint about modern health care is that it isn't focussed on health — it concentrates on disease. There are clear signs, however, that health is becoming a more "interesting" issue. With the revival of family-practice medicine and home-health-care services, we are beginning to pay attention to providing healthy individuals with the information and help they need to stay healthy. Immunization, the process of introducing substances into the body to render it either temporarily or permanently immune, is a vital tool in preventing infectious disease. It is receiving more public attention and more research funds than it has in several decades. In this chapter we will survey the current practices in immunization and their value in disease prevention and control.

IMMUNE SERA AND VACCINES

The objective of this section is to examine the substances we have available with which to immunize. While doing so, we'll review some relevant terms from Chapter 12, and add a few new ones. **Immunoprophylaxis** refers to preventive immunization. This is by far the most widely used approach. Examples include all the common childhood immunizations. However, **immunotherapy,** that is, immunization offered after disease develops in hope of moderating symptoms, is also available for some diseases. An example is the use of antiserum in cases of tetanus.

Active immunity, such as the protection we get from a measles vaccination, follows exposure to specific antigens, stimulates the body to produce humoral or cellular immune mediators, and is long-lasting. **Vaccination** is the process of inducing active immunity, by administering microorganisms or their products in a nonpathogenic form, as a vaccine (Figure 25.1).

One type of **vaccine** is a liquid biological product that contains infective agents; these

FIGURE 25.1 Conferring Active or Passive Immunity to an Exotoxin
(a) Some bacteria produce proteins called exotoxins that destroy parts of the host cell or inhibit specific metabolic functions. These exotoxins may be converted to toxoids by heat or chemical treatment. In this inactivated form, they can no longer cause disease but may be used to stimulate the body's antibody production. (b) Active immunity occurs when a toxoid, introduced in vaccine form, stimulates the body to produce antibody. (c) Passive immunity results when immune serum containing antibody from another source neutralizes the toxin of a specific infectious agent.

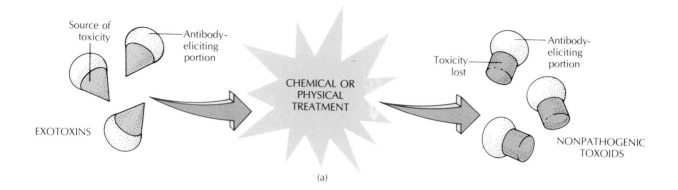

Source of toxicity
Antibody-eliciting portion
EXOTOXINS
CHEMICAL OR PHYSICAL TREATMENT
Toxicity lost
Antibody-eliciting portion
NONPATHOGENIC TOXOIDS

(a)

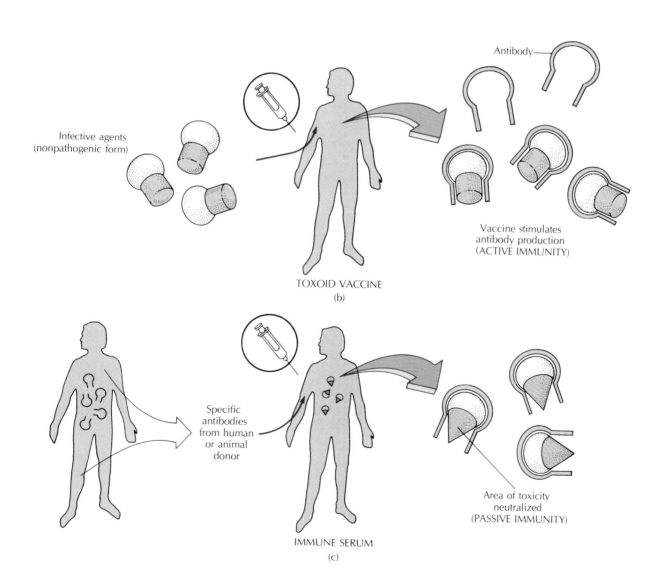

Infective agents (nonpathogenic form)
Antibody
Vaccine stimulates antibody production (ACTIVE IMMUNITY)
TOXOID VACCINE

(b)

Specific antibodies from human or animal donor
Area of toxicity neutralized (PASSIVE IMMUNITY)
IMMUNE SERUM

(c)

agents may be killed, or they may be **attenuated,** cultivated under conditions that remove their pathogenicity. Thus a specific attenuated vaccine virus is used in the polio vaccines.

A **toxoid** is another form of vaccine containing a microbial exotoxin, chemically treated to destroy its toxicity while retaining its ability to immunize. For example, we use tetanus and diphtheria toxoids for immunization. **Combined vaccines** contain two or more different immunizing substances. The most familiar is the DPT vaccine that contains two different toxoids (diphtheria and tetanus) and killed *Bordetella pertussis* bacteria. **Polyvalent vaccines** contain two or more strains of the same organism. For example, the trivalent oral polio vaccine (TOPV) contains live, attenuated virus of the three pathogenic types of poliovirus.

Passive immunity results when immunoglobulins or other immune products are transferred to the individual who needs them. It affords only temporary protection. We passively immunize by giving an **antiserum,** or, more accurately, **immune serum,** which is a product prepared from the blood of immune animals or human donors. An immune serum contains antibody against specified infectious agents or their toxins. Currently available human immune serum globulin (ISG) products are derived from normal human donor sera. Other antisera of historical interest or for research purposes have been made by immunizing horses, cows, or rabbits with appropriate vaccines. We use antisera to protect persons exposed to a disease such as hepatitis, or to alleviate their symptoms once they have the disease, as in botulism.

ROLE OF IMMUNIZATION IN PUBLIC HEALTH

Universal immunization, that is, immunization of a whole population, is considered to be a justifiable public-health measure wherever a large number of people are at risk for an infectious disease, either because the disease is endemic, or because it is predictably epidemic. In addition, the disease must be serious enough for at least some individuals that the risk from the disease is significantly greater than the risk of side effects arising from the immunization. In the United States, we enforce immunization for only seven infectious diseases — measles, mumps, rubella, diphtheria, whooping cough, tetanus, and poliomyelitis. **Routine immunization** is the systematic vaccination of all individuals, as they reach the optimal age, at a mutually convenient time. The American Academy of Pediatrics, working closely with the Centers for Disease Control, formulate and update recommendations for immunizing preschool children. **Individual immunization** is that received by someone who needs special protection not needed by the general population, such as veterinarians who need rabies vaccination, or travelers who need inoculation against foreign disease agents.

PASSIVE IMMUNIZATION

In the 1890s, it was discovered that, at least in some cases, immunity was passively transferrable. That is, if serum obtained from a person who had recovered from a dangerous disease was administered to a patient with the same disease, the patient sometimes received an immediate therapeutic benefit. Serum treatments became widely used, but much of this therapy was useless. Some of it was lethal, as when serum was grossly contaminated. Then it was discovered that horses could be immunized to produce larger amounts of serum, removing the need for human donors. This period was still the era of horse-drawn transportation, however, so many people who had extensive daily contact with horses were often allergic to horse antigens. When horse serum was injected, serum sickness and fatal anaphylactic shock were quite common side-effects. Then antibiotics were discovered in the 1940s, and they quickly replaced serum therapy. Now passive immunization, employing human serum, is used only in special instances (listed in Table 15.3). The switch to human sera and improved production methods

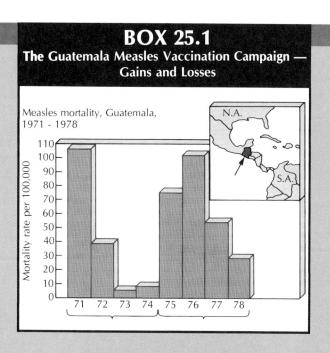

BOX 25.1
The Guatemala Measles Vaccination Campaign — Gains and Losses

Measles mortality, Guatemala, 1971 - 1978

Guatemala, like all of Central America, has had a high measles mortality rate. However, from 1972 through 1974, a vigorous vaccination campaign reduced measles deaths 90%, to less than 4 per 100,000. Vaccine coverage reached 94% in 1974. But after 1975, the campaign fizzled, and the measles death rate rebounded, reaching 102.1 per 100,000 in 1976. The highest death rates occurred in the highland areas populated largely by Mayan Indians.

What happened? Investigators traced some problems to lost or inaccurate vaccination records, and others to a collapse of the network of health-care personnel. Beyond that, the vaccine itself seemed to have failed too — it protected only a little more than half the children who received it. Investigators then looked at how the vaccine had been refrigerated, shipped, and stored at various points from its manufacture to its ultimate destination. Storage of vaccine in the capital city and in the regional health departments was generally satisfactory, as was shipment from the capital. However, vaccine was shipped to rural health centers in poorly insulated containers that could not keep it cool. Aged kerosene refrigerators in health centers often lacked replacement parts or fuel, and power for electric refrigerators was often interrupted. Personnel often received no instructions in vaccine handling or refrigerator maintenance.

This study points out several potential difficulties in mass-vaccination campaigns. One problem is maintaining public interest and momentum, especially in politically unstable areas. Another is actually achieving good vaccine coverage and records, which depends on maintaining a stable network of committed personnel. A third is protecting the perishable vaccine, since live-virus vaccines must be refrigerated until administration. In those areas of Central America now torn by revolution, vaccination campaigns are practically non-existent, and any gains made in the '70s have evaporated.

have vastly increased the safety of passive immunization. Research efforts focus on the use of monoclonal antibodies; these are often linked to toxins as **immunotoxins** and investigated in cancer chemotherapy.

Human immune sera are produced by pooling carefully selected donor sera, from individuals exposed to the agent or toxin we wish to protect against. These sera are then purified, concentrated, and packaged aseptically. Lyophilized (freeze-dried) preparations are normally dispensed with a vial of sterile fluid with which to resuspend them. Careful processing has largely eliminated the risk of transmitting infection via serum. Most antisera of human origin are expensive; many are in extremely short supply.

Some examples of available immune sera include the following.

- **Immune serum globulin (ISG)** is a nonspecific product containing antibodies against measles, diphtheria, polio, and hepatitis A. It may be used to protect infants exposed to these diseases.
- **Hepatitis B immune globulin (HBIG)** contains a very high level of anti-hepatitis B antibodies. It is used for prophylaxis in patients in institutions and hemodialysis units (if hepatitis B is endemic), of intimate contacts of hepatitis B patients, and of infants born to mothers with acute hepatitis B.
- **Rabies immune globulin (RIG)** is made from serum of immunized human volunteer donors. Rabies immune globulin may be administered following the bite of a rabid animal.
- **Tetanus immune globulin (TIG)** is a human product used in patients with massive or dirty wounds without previous tetanus immunization.
- **Antivenins** are antisera containing antitoxins, used to treat bites of poisonous snakes and spiders. There are three types: for coral snakes; for rattlers, copperheads, and moccasins; and for black widow spiders.

HISTORY OF ACTIVE IMMUNIZATION

The early history of active immunization was presented in Chapter 1. Smallpox, rabies, and anthrax vaccines were the first developed; in each case, a weakened infective agent was used. Then reliable chemical means of converting toxins to toxoids were developed, leading to safe universal immunization against diphtheria and tetanus. Vaccines using killed bacteria, such as the typhoid vaccine, were also developed (Figure 25.2). Once egg and cell-culture techniques for cultivating viruses were worked out, in the early 1950s, it became possible to carry out bulk production of viruses. First killed-virus, then live-virus vaccines were made. Now, researchers are

FIGURE 25.2   The First Typhoid Fever Inoculation In 1909, Major Frederick Russell administered the first typhoid fever inoculation in the United States to Army medical officers. By 1910, some 31,000 doses of the serum had been administered at the Army Medical School. (Courtesy of Culver Pictures, Inc.)

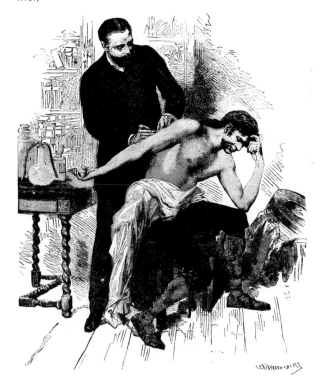

working on vaccines that contain synthetic antigens, made by recombinant bacteria. The pathogenic agents themselves are not present in the vaccine at all. Other experimental vaccines use the comparatively harmless vaccinia virus as a vector to introduce single antigen-coding genes of viruses such as herpes (Figure 25.3).

PRODUCTION AND STANDARDIZATION

Let's look now at how various vaccines are produced, and survey some of those that are available. In the United States, licensed vaccines are produced from approved stock cultures, under control of the Bureau of Biologicals of the Food and Drug Administration. An acceptable vaccine should be easily reproducible, and convenient to assay for potency. It should have few if any side effects, immediate or delayed. On administration, the recipient's antibody titer should rise and he or she should show definite protection when exposed to the pathogen. Any living agent used in a vaccine should be incapable of causing fetal death or malformation when given to a pregnant woman. These criteria are all ensured as far as possible by research and clinical testing before a new vaccine is licensed.

All immunizing materials are subject to regulations establishing the standardized dosage and requiring lot testing, numbering, and labeling. Package inserts must be provided as with other licensed pharmaceuticals. All packages must carry an expiration date.

The storage instructions on the package must be carefully followed. Vaccines lose potency rapidly if held at room temperature; they should be stored frozen or refrigerated. Allowing a liquid vaccine to stand at room temperature also carries the hazard that contaminants introduced by poor technique will multiply and render the vaccine dangerous.

Bacterial vaccines are made with killed bacterial cells or purified cell-free bacterial products. First, the stock strain of the agent is grown in bulk on standardized media. The bacteria are

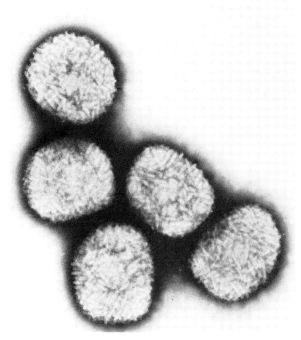

FIGURE 25.3 Vaccinia Virus
It was a major breakthrough when scientists learned that it was possible to use the vaccinia virus, a nonpathogenic relative of the smallpox virus, to induce immunity to smallpox. Now vaccinia viruses are being investigated as vectors to induce immunity against other antigens. (Reproduced from C.R. Madeley, *Virus Morphology,* by permission of the author and Churchill Livingstone, Edinburgh.)

washed to remove the medium, then killed by exposure to heat, to chemicals such as formalin, or to ultraviolet light. After careful checks for sterility and toxicity are done, the material is dispensed into sterile sealed vials. Companion vials containing the amount of sterile fluid needed to dilute the vaccine to the correct dosage are often supplied with the vaccine. Normally, each vaccine vial contains a small number of doses, each containing between 10^8 and 10^9 bacteria or the equivalent amount of material derived from them.

Toxoids are derived from the toxins produced by toxigenic strains of bacteria grown on high-yield media. After growth the bacterial cells are removed and discarded, while the culture medium is retained and chemically treated to concentrate and purify the toxin. The toxin is then treated with formalin until it has completely lost its toxicity. Finally, the toxoid is adjusted to a standard dosage, tested, and dispensed. Vaccines made with rickettsias must be cultivated in the yolk sacs of embryonated eggs. After growth, the microorganisms are separated from the yolk material and formalin-inactivated.

Some of the widely used bacterial vaccines include the following.

- **Diphtheria toxoid** is a formalin-treated toxin. It can be obtained alone, in combination with tetanus toxoid (Td), or with both tetanus toxoid and pertussis vaccine (DPT). Older children and adults need a much lower dose than infants because of the risk of severe reactions. DPT vaccines for infants should never be administered to adults because the dosage is too high.
- **Pertussis vaccine** may be either killed, whole bacteria or bacterial subunits, depending on the manufacturer. The material contains endotoxin and has caused side effects such as fever, soreness, and neurologic damage. The vaccine is very effective in preventing pertussis when properly administered. It has been in very short supply recently, because several manufacturers have ceased making it due to concerns about its toxicity.
- **Tetanus toxoid** is a formalinized toxoid. It provides long-lasting protection, and hypersensitivity reactions are uncommon. Reimmunization is limited to every 10 years.
- **Pneumococcal polysaccharide vaccine** (Pneumovax) is polyvalent, with antigens from the 14 most prevalent strains of *Streptococcus pneumoniae*. It is recommended for high-risk groups, including the elderly, those with abnormal spleen function, chronic cardiopulmonary disease,

or any condition likely to increase the risk of pneumonia.
- *Hemophilus influenzae* **Type B vaccine** is composed of purified Type B capsular material. It is recommended for all children of 24 months of age

Bacterial vaccines are also available to immunize against anthrax, tuberculosis, cholera, meningitis, plague, Rocky Mountain spotted fever, tularemia, typhoid fever, and typhus fever. All are infrequently used.

Some **viral vaccines** are produced in eggs. Increasingly, human-cell–culture lines are being substituted. The virus strains used in **live-virus vaccines** are attenuated. **Killed-virus vaccines** are usually formalinized. In **split-virus vaccines,** only the major antigens of the virus, usually from the capsid or envelope, are retained and concentrated, while the genetic material is removed and discarded.

Some of the currently available viral vaccines include the following.

- **Measles vaccine** is a live-viral vaccine, cultivated in chick-embryo–cell culture. It is marketed alone or in combination with rubella vaccine (MR), or with rubella plus mumps vaccine (MMR). All children should be immunized against measles at 15 months of age. Previously unimmunized children and adults, persons who received the earlier killed-virus vaccines (now known to be ineffective), and children immunized before the age of 15 months should also be vaccinated. It should not be given to pregnant women, or to persons with fevers, tuberculosis, or immunodeficiency. About 90% of recipients develop protective immunity, and the effect seems to persist for more than 15 years. A moderate fever follows vaccination in 5 to 15% of recipients.
- **Mumps vaccine** is prepared from live virus grown in chick-embryo–cell culture and is sold alone or in combination with measles and rubella vaccines. It is recommended for all children. The duration of immunity

is more than 10 years. Side effects are minimal.

- **Rubella vaccine** is a live, attenuated cell-culture vaccine universally recommended for all children. The success rate with the vaccine is 95%, and the immunity is long-lasting. Rash, swollen glands, joint pain, and neuritis are possible side effects. Pregnant women should not be vaccinated, even though there is no indication that the vaccine virus is capable of fetal damage.

- **Polio vaccines** are either live or killed virus types. In the United States, the live trivalent oral polio vaccine (TOPV) is universally recommended for all normal children. It should be given in three doses in the first year of life. It confers long-lasting immunity in over 90% of recipients. TOPV can be administered to pregnant women. When the vaccine is given to the breast-fed infant whose mother has active immunity, more doses are essential after breast-feeding has been concluded. In rare cases (approximately 12 per year in the United States) paralytic polio caused by the vaccine virus follows vaccination, either in the vaccinated person or in contacts. The injectable inactivated polio vaccine is recommended for immunodeficient individuals. The killed virus cannot cause paralytic disease, but it does not promote as complete or as lasting an immunity.

- **Influenza vaccines** are usually made from virus grown in eggs (Figure 25.4). Adminis-

FIGURE 25.4 Production of Influenza Virus Vaccine
(a) Embryonated eggs about to be inoculated with influenza virus are placed on a conveyer belt. (b) Trays of virus-laden eggs (background) pass to a manifold where their contents are sucked out under vacuum and collected in a sealed vessel (foreground). This material containing influenza virus will be extensively purified to remove egg proteins and standardized before packaging as vaccine. (Courtesy of Merck & Co., Inc.)

(a)

(b)

tration to a certain group of persons is readily justified. In a moderate influenza epidemic year, between 10,000 and 20,000 related secondary pneumonia deaths occur in the elderly and chronically ill. Severe epidemics, even in the antibiotic era, have claimed 70,000 lives in one winter. Influenza vaccination can prevent most such fatalities. Whole inactivated virus is recommended for adults and split-virus is used in children. The antigen content of influenza vaccines is adjusted from year to year based on forecasts of probable epidemic strains of virus.

- **Rabies vaccine** is a human diploid cell-culture vaccine. It is given to veterinarians, dog officers, and game wardens (high-risk occupations). It is also given to about 30,000 exposed persons/year in the United States, most of whom have been bitten by a rabid or suspect animal. Its side effects are local reactions, fever, and malaise.

- **Hepatitis B vaccine** in its first form (Heptavax) contains antigenic particles that are removed from the serum of volunteer chronic hepatitis B patients and inactivated. It is used by high-risk individuals, such as health workers, hemophiliacs, male homosexuals, and institutionalized persons.

Viral vaccines are also available against yellow fever, some forms of encephalitis, and adenovirus infection. New vaccines are under development against chickenpox, hepatitis B (a synthetic vaccine), Epstein–Barr virus, certain other herpesviruses, and the virus(es) of AIDS.

PRACTICAL ISSUES IN IMMUNIZATION

Immunization is safe and effective when it is carried out with attention to some basic rules. We will now look at some of the practical aspects of immunization, including the antigen to be used, the route of administration, contraindications to vaccination, and the risks of contamination, toxicity, and hypersensitivity.

ANTIGEN STIMULUS. When we have an infectious disease, the pathogen presents its surface antigens to the body's lymphoid tissue. Some pathogens contain strong antigens, and provoke a strong response, and these usually can be readily converted to vaccines. Other pathogens, such as the gonorrhea organism, do not elicit a strong response, and we have generally been unsuccessful at developing vaccines from these.

A vaccine is at its best when it most closely simulates the "natural" strong antigenic stimulus of disease, without causing any symptoms. The antigen needs a period of intimate contact with lymphoid tissue. The duration of the stimulus can be prolonged by mixing antigens with inert materials such as aluminum salts or organic compounds that delay absorption and elimination.

Live-agent vaccines in general are much more effective than killed-agent vaccines (Figure 25.5a). The value of a live-agent vaccine derives from the fact that the agents multiply in the body for a period until the body's developing immunity eliminates them. This gives a long, intimate stimulus that very closely simulates natural infection. Inactivated or nonliving antigens, by contrast, are eliminated fairly rapidly from the body and don't replace themselves. To maintain a continuing antigenic stimulus, a series of injections must be given at intervals to take advantage of the memory function in immunity (Figure 25.5b).

ROUTES OF ADMINISTRATION. The only practical route for most immunizations is the intramuscular route (Figure 25.6). Deposit of a small volume of vaccine into muscle tissue leads to relatively slow, steady absorption. If larger volumes must be used, they will cause greater pain and muscle irritation.

Live-polio vaccine is an exception; it can be given orally because it contains an enterovirus that survives passage through the stomach. No matter how much children would prefer it, most other vaccines cannot be given orally because the stomach acid would denature them. Polio-

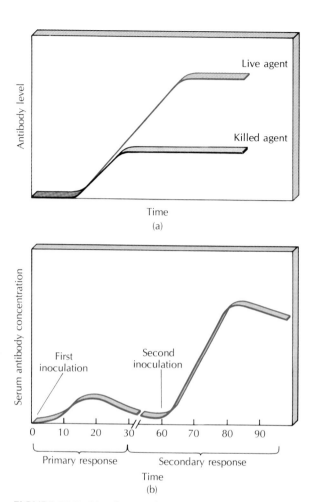

FIGURE 25.5 Vaccines and Antigen Stimulation
(a) Live vaccines are preferable because a live agent multiplies in the body and gives a stronger stimulus to lymphoid tissue. The resulting immune response builds to higher levels over a longer period of time. (b) When several doses of antigen are given in sequence, the body's memory response comes into play, giving higher final antibody levels than if just one immunization was used.

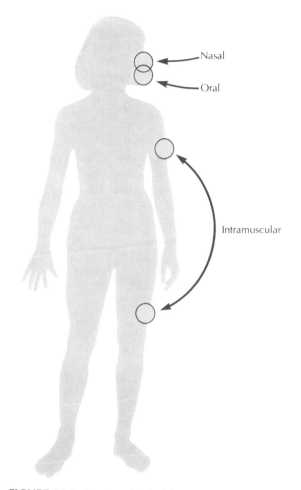

FIGURE 25.6 Routes of Administration
Vaccines may be administered orally, if the agent is not affected by stomach acidity, in an aerosol that is inhaled, especially useful for respiratory agents, or by intramuscular injection.

virus, on reaching the intestine, stimulates the lymphoid tissue in the intestinal wall, which produces most of the protective response.

A few vaccines may be administered as inhaled aerosols. They stimulate lymphoid tissue in the respiratory passage, provoking local secretion of antibodies. This approach is being investigated for vaccination against respiratory viruses such as influenza.

DURATION OF IMMUNITY. After vaccination, there is a period of one or more weeks during which immunity builds to protective levels. Thereafter, the period of protection varies with the antigen used and the individual recipient. Some vaccines, such as typhoid vaccine, protect for a relatively short time and require boosters at yearly intervals to maintain protection. Smallpox vaccine, on the other hand, gives very-long-term protection, measured in years.

CONTRAINDICATIONS.   There are many **contraindications** — specific health reasons why a medicine should not be administered to a certain individual. For vaccines, the contraindications are always detailed on the package insert, and they should be reviewed before the responsibility of administering a vaccine is assumed. For example, live-virus vaccines should not be given to pregnant women because the viruses might cause fetal damage. Persons who are receiving steroids or who are known to have immunosuppression resulting from malignancy, drug therapy, or other causes should not be immunized. Table 25.1 presents other situations in which particular vaccines are contraindicated.

CONTAMINATION.   In the past, batches of vaccine have been inadvertently released in which the vaccine agent was not completely killed or in which there were bacterial or viral contaminants. Gross contamination in commercially produced vaccines is now essentially unknown. However, vaccine vials are occasionally contaminated during use, by faulty aseptic techniques in offices and clinics.

TABLE 25.1 Contraindications for vaccination	
Condition	**Vaccines Not Given**
Illness with fever	Mumps, measles, rubella, diphtheria–pertussis–tetanus
Congenital immunodeficiency	Mumps, smallpox, yellow fever, polio (TOPV), measles, rubella
Leukemia, lymphoma, or other malignancy	Mumps, smallpox, yellow fever, polio (TOPV), measles, rubella, bacillus Calmette–Guérin
Immunosuppression	Mumps, smallpox, yellow fever, rabies, polio (TOPV), measles, rubella, bacillus Calmette–Guérin
Pregnancy	Mumps, smallpox, yellow fever, measles, rubella, bacillus Calmette–Guérin
Egg hypersensitivity	Yellow fever, influenza, typhus, Rocky Mountain spotted fever
Antibiotic hypersensitivity	Yellow fever, measles
Previous severe reaction	Cholera, diphtheria toxoid, tetanus toxoid, pertussis
Positive tuberculin test	Bacillus Calmette–Guérin
Unstable neurologic condition, e.g. uncontrolled epilepsy	Pertussis vaccine; diphtheria–pertussis–tetanus

BOX 25.2
What Is an Acceptable Risk?

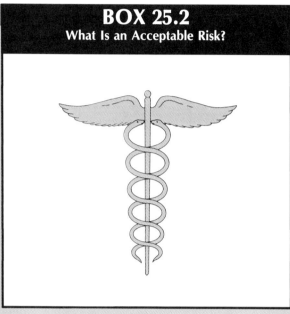

Vaccination, like other medical actions, must conform to the primary rule of the Hippocratic oath: "First, do no harm." Yet, as has been amply documented, it is possible for vaccination to do far more harm than good. For example, there is no reason whatever for American civilians to get smallpox vaccinations, since smallpox has been eradicated worldwide. Yet, up to 1983, thousands of smallpox vaccinations were being given per year, with life-threatening complications and fetal deaths. Responding to this problem, federal agencies ordered the vaccine withdrawn from the market.

In coming to terms with the issue of "acceptable" risks, smallpox is a comparatively easy case. It is clear that, when smallpox was still a major world problem, seven or eight deaths annually due to smallpox vaccination in the United States were "acceptable." This was because there was still a definite possibility that we might have had a severe imported smallpox epidemic, and if we did, thousands might have died. It is equally clear that, now that smallpox has been eradicated, the death of even one fetus is unacceptable. The problem of acceptable risk becomes more difficult in the gray areas in which neither benefits nor risks can be satisfactorily quantitated. For example, the risk/benefit analysis is very complicated when we discuss the use of the new live-virus chickenpox vaccine. Chickenpox is an almost universal childhood disease, but it isn't very harmful. Is it worth protecting against with a vaccine that has risks?

Take the issue one step further. Suppose someone developed a vaccine for the common cold. How great a risk (mortality per million doses administered) would be acceptable for this vaccine? Colds cost the country billions of dollars per year and untold misery, but rarely kill anyone. If a cold vaccine caused the death of even 1 person in 10 million, would it be worth it? In considering this, remember there are risks inherent in injecting or swallowing anything, even sterile distilled water, not to mention in driving or taking the subway to the clinic!

TOXICITY AND HYPERSENSITIVITY. Vaccines prepared from Gram-negative bacterial cells contain much endotoxin. The endotoxin can cause soreness, fever, and headache that may last for several days, for example after receiving a typhoid vaccination. Recipients should be made aware of what to expect. In the case of pertussis vaccination in infants, a very strong reaction on the first injection of the series usually suggests reducing the dosage of future injections or discontinuing the series.

Vaccines may provoke allergic responses. Viral vaccines are commonly produced in chicken or duck eggs. Many persons are allergic to egg proteins and react strongly to the traces of egg remaining in the vaccine. Other viral vaccines are produced in cell cultures, which always contain antibiotics to keep down chance bacterial contaminants. Neomycin allergy is a relatively common problem with vaccines prepared in cell culture. Before receiving vaccines of these types, the patient should always be questioned to find out if he or she has a history of these allergies.

There are certain rare neurologic disturbances, such as the Guillain–Barré syndrome, for which the true cause is unknown. Some cases of Guillain–Barré syndrome may be autoimmune conditions that are provoked either by the process of vaccination or by antigenic substances in the vaccine.

CURRENT RECOMMENDATIONS

In the United States, several professional organizations cooperate to formulate and update immunization recommendations. They include the American Academy of Pediatrics, the American Academy of Internal Medicine, the American Public Health Association, and the U.S. Public Health Service Advisory Committee on Immunization Practices. The recommendations of these groups then may be made legally binding by state or federal regulations. Internationally, immunization recommendations differ widely from country to country.

RECOMMENDATIONS FOR INFANTS AND CHILDREN

In the United States, it is required that all normal children be immunized against seven potentially common and serious diseases. A suggested timetable for administering the vaccines is shown in Table 25.2. The timing is dictated by the need to offer initial protection at the ages when the child is most likely to contract the disease and/or when the disease would have the most serious consequences. However, since the infant's immune system develops slowly, some vaccinations, such as that against measles, do not become effective until the child is at least 15 months of age. Later, boosters are provided at regular intervals to heighten immunity and take advantage of the child's increased ability to respond to antigenic stimulus. When immunization has not been started at the recommended time, or when it is uncertain that the child will be brought in to complete the sched-

TABLE 25.2 Recommended childhood immunizations	
Age	Immunizations
2 months	DPT, TOPV
4 months	DPT, (TOPV optional)
6 months	DPT, TOPV
15 months	MMR, tuberculin test
$1\frac{1}{2}$ years	DPT, TOPV
4 to 6 years	DPT, TOPV
14 to 16 years	Td

Legend:
DPT = Diphtheria and tetanus toxoids combined with pertussis vaccine, childhood dosage
TOPV = Trivalent oral polio vaccine
MMR = Combined measles, mumps, and rubella live-virus vaccine
Td = Combined tetanus and diphtheria toxoids, adult type.

uled series of immunizations, the sequence may be altered.

The greatest problem with childhood vaccination programs is children's resistance and parents' apathy. What would happen if we let kids choose whether or not they got their immunizations? Universal vaccination has been re-emphasized as a key public health policy since the late 1970s. In almost all states, proof of vaccination is now required by law for a child to enter public school. When students are actually excluded from school for nonvaccination, their parents' motivation to complete the children's immunizations often improves rapidly. The percentage of children with adequate, documented immunizations has risen sharply in the past few years, and approaches 95% for most of these seven diseases. Mumps is an exception — vaccination is not required in a few states.

It is not a coincidence that these seven diseases, for which universal vaccination has been mandated, are all either rapidly declining as public-health problems (measles) or are now almost unknown (diphtheria). These diseases illustrate clearly how potent a force vaccination can be in controlling the spread of infectious disease.

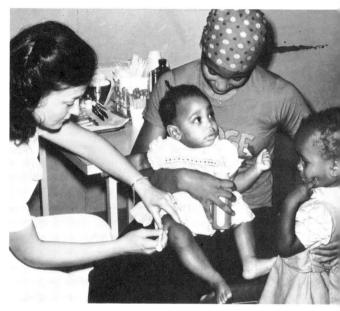

FIGURE 25.7 Vaccination Programs
All healthy infants in the United States are immunized against seven infectious diseases. Here is a child receiving a booster shot. (Courtesy of Centers for Disease Control, Atlanta.)

ADULT IMMUNIZATION

Vaccination is for kids — that's the prevailing public opinion. But adults also require vaccination. Many adults are inadequately protected. A recent study of 1900 persons showed that only 39% of persons over 20 years of age had adequate levels of tetanus immunity, even though all were members of a health-maintenance organization from which tetanus immunizations could be obtained free of charge. In fact, essentially all cases of diphtheria and tetanus in this country occur in older adults, many of them with childhood immunizations far in their past. Adults need booster vaccinations, but they usually exercise the "right to choose" that they didn't have when they were kids, and avoid getting them. Adults in specialized occupations are more conscientious about receiving necessary

immunizations, and many adults do obtain immunizations after exposure to an infectious disease such as hepatitis. Table 25.3 show immunizations that certain adults may need. Note that adults should receive tetanus–diphtheria toxoid (adult dosage) every 10 years.

IMMUNIZATION FOR TRAVELERS

Elsewhere on the globe, there are endemic diseases not found in the United States. When you travel, you must realize that you have less resistance than the local residents. If you plan to travel to these endemic areas, you may need special vaccination. The need depends largely on your itinerary. Updated lists of suggested immunizations for each foreign country or geographic

TABLE 25.3 Immunization for adults

Immunization Before Disease Exposure	Comments
All adults	
Tetanus–diphtheria toxoid	Must be given every 10 years
Certain adults	
Influenza vaccine	Different types available each year
Women of child-bearing age	
Rubella vaccine	Only if shown susceptible by antibody testing and if pregnancy can be prevented for at least three months after vaccination
Some postpubertal men	
Mumps vaccine	Most adults have natural immunity
Travelers to foreign countries	
Typhoid vaccine	Two injections plus booster every three years
Cholera vaccine	Needed within two months of departure to endemic area
Yellow fever vaccine	Endemic areas in Africa, South America
Typhus vaccine	Only for extended stay in endemic area
Polio vaccine	Tropical areas or developing countries
Plague vaccine	Endemic in Vietnam, Kampuchea, Laos
Immune serum globulin for viral hepatitis	Extended stay in tropical areas and developing countries
Persons with unusual occupational exposure	
Smallpox vaccine	Certain laboratory personnel
Plague vaccine	Certain laboratory personnel, field workers
Rabies vaccine	Certain laboratory personnel, veterinarians
Anthrax vaccine	Certain laboratory personnel, textile workers, veterinarians
Rocky Mountain spotted fever vaccine	Certain laboratory personnel, persons repeatedly exposed to ticks in endemic areas
Tularemia vaccine	Certain laboratory personnel, repeated contacts with enzootic areas
Equine encephalitis vaccine	Certain laboratory personnel, field workers
Special epidemiologic situations	
Bacillus Calmette–Guérin (BCG) vaccine	Some hospital personnel, children of tubercular parents
Meningococcal vaccine	Only available for serogroups A and C
Adenovirus vaccine	Some military personnel

Immunization After Exposure to Disease But Before Onset of Symptoms

Rabies vaccine and rabies immune globulin	Most animal cases now in wildlife
Immune serum globulin for viral hepatitis	Probably indicated for type B as well as type A
Immune serum globulin for measles	Only for high-risk susceptibles with exposure

TABLE 25.3 Immunization for adults (continued)	
Immunization After Onset of Clinical Illness	
Tetanus immune globulin (and tetanus–diphtheria toxoids)	Human preferable over horse antitoxin
Diphtheria antitoxin	Give immediately after clinical diagnosis
Botulism antitoxin	Give immediately after clinical diagnosis

area are prepared by the Centers for Disease Control. Information can be obtained by contacting local or state public-health agencies. It is wise to take care of this well in advance of your departure date, since there may be long delays in obtaining the less commonly used vaccines. The requirements of the countries to be visited should be carefully followed and all necessary certificates obtained. Otherwise, you may experience serious difficulty in crossing borders.

SUMMARY

1. We can systematically utilize the body's natural defenses for its protection. Immunization against many infectious diseases has become practical. The general use of vaccination has contributed, along with sanitary engineering, antibiotics, and other factors, to the great 20th-century increase in human life expectancy.

2. Immunization is passive when immune sera bearing desirable antibodies are administered. Such protection is immediate but transitory.

3. Antisera from human donors minimize the chance of hypersensitivity reactions. Such sera are expensive and in short supply. Passive immunization is normally reserved for those faced with an immediate threat of serious infection not readily controlled by antimicrobials.

4. Vaccination is actively immunizing by administering an altered, nonpathogenic antigen. Immunization products may include toxoids, live attenuated organisms, killed organisms, and subunits or chemical derivatives of organisms.

5. After administration, there is a delay before measurable antibody appears. It may be several weeks before full protection is established. The immunity lasts for a period that varies with both the nature of the antigen and the peculiarities of the individual recipient.

6. Extensive research, field testing, and clinical testing go into the development of a new vaccine before it can be licensed. Careful production controls are designed to guarantee the safety and effectiveness of each vaccine lot. Conscientious attention to the health status and history of each recipient, as well as scrupulous aseptic technique during administration, make vaccination maximally safe.

7. In the United States, routine childhood vaccination for three bacterial and four viral diseases is recommended and, to varying degrees, required by law.

8. In adulthood, periodic vaccination against certain pathogens or toxins is wise. Everyone should receive diphtheria and tetanus boosters every 10 years. Travelers, persons with certain occupational hazards, and contacts of persons with certain communicable diseases may require immunization.

DISCUSSION QUESTIONS

1. Under what conditions are immune sera, in contrast to vaccines, particularly useful?

2. What problems remain with human-origin immune sera?

3. By reviewing previous chapters, identify other bacterial diseases in which toxoid vaccines could be useful.

4. The pneumococcal polysaccharide vaccine and the influenza vaccine are recommended for about the same group of recipients. What types of persons are in this group and why do they need these vaccines?

5. The live-virus vaccines are generally preferred to killed-virus vaccines. Why? Also, why do live-virus vaccines have a larger number of contraindications?

6. Explain why each of the seven childhood vaccinations is recommended, based on the characteristics of the disease it prevents and the expected benefits and risks.

ADDITIONAL READING

Anderson RM, May RM. Directly transmitted infectious diseases: control by vaccination. *Science* 1982; 215:1053–1060.

Benenson AS. *Control of communicable diseases in man.* 14th ed. Washington, D.C.: American Public Health Association, 1985.

Collier RJ and Kaplan DA. Immunotoxins. *Sci Am* 1984; 251:56–64.

Germanier R. *Bacterial Vaccines.* Orlando, Fla.: Academic Press, 1984.

Godson GN. Molecular approaches to malaria vaccines. *Sci Am* 1985; 252:52–59.

Gulginetti VA, ed. *Immunization in clinical practice: a useful guide to the vaccines, sera and immune globulins in clinical practice.* Philadelphia: JB Lippincott, 1982.

Health Information for International Travel. Atlanta: Centers for Disease Control. Published annually.

Lerner RA. Synthetic vaccines. *Sci Am* 1983; 248:66–74.

Physicians Desk Reference. Oradell, N.J.: Medical Economics Co., 1984.

Recommendations of the Public Health Service Advisory Committee on Immunization Practice (ACIP). Atlanta: Centers for Disease Control. Published periodically in *Morbidity and Mortality Weekly Report.*

Roitt IM, ed. *Immune Intervention 1: New Trends in Vaccines.* Orlando, Fla.: Academic Press, 1984.

26

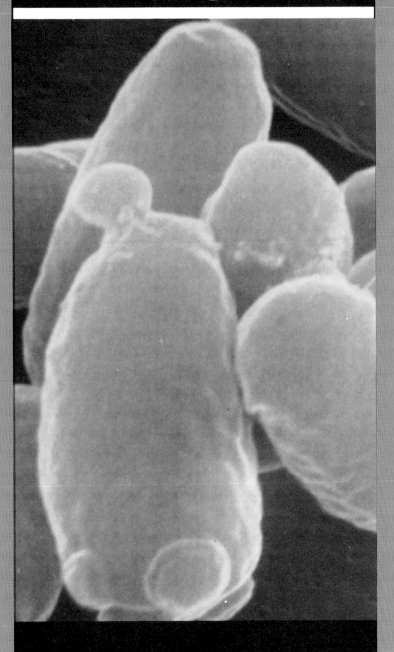

HOSPITAL INFECTION
CONTROL

In the community, infectious diseases tend to occur when a virulent pathogen penetrates the normal defenses of a basically healthy individual. Epidemiologists call these **community-acquired diseases** to differentiate them from **hospital-acquired** or **nosocomial infections,** which patients develop in hospitals, nursing homes, or other health-care facilities. In a typical nosocomial infection, a relatively nonvirulent opportunist penetrates the reduced defenses of a person who is already seriously ill. Nosocomial infections are more difficult to prevent, more unpredictable to treat, and tougher to cure than community-acquired infections.

In this chapter we will examine the three components of the hospital infection control problem — the opportunistic microorganisms, the compromised hosts, and the unique features of the hospital environment. We will also become familiar with the key aspects of hospital infection control programs.

THE SCOPE OF THE PROBLEM

Why do hospital-acquired infections occur? Is this a new problem, and if so, why? How big is the problem?

HOW HOSPITALS HAVE DEVELOPED

Until the last century, hospitals were places of last resort. They served only the destitute who were dying. In medieval hospitals, called "pesthouses," several patients — sick, dying, and dead — often shared the same bed (Figure 26.1). Medical practice was primitive, but the patients either died (or got well and left) so fast that disease transmission occurring within the hospital escaped notice. Furthermore, until the middle of the 19th century, medical practice did not include most of the techniques that we now know can cause nosocomial infection.

The first worker to recognize that hospital-acquired infection was occurring was the Viennese physician Ignatz Semmelweis. In 1850, he published documentation of the fact that physician-attended childbirth in a hospital actually reduced the mother's chances of survival. The reason for this was that physicians had recently begun to use obstetrical forceps, but that their instruments (and their hands) were grossly contaminated with microorganisms. Semmelweis proved for the first time (and certainly not the last) a highly unpopular point — that hospitals could be hazardous to your health. He attempted to introduce improved techniques in

FIGURE 26.1   A Typical Hospital before the Germ Theory of Disease
Medieval hospitals served as places to shelter the sick — if they were
poverty-stricken and had no where else to go — until they either recovered or died.
There was no awareness of how infectious diseases spread, and no attention was
directed toward preventing transmission. (Courtesy of the Bettmann Archive.)

the hospital in which he worked, but his approach was so unpopular that he was finally forced to resign.

In the past it was assumed that you went to a hospital to die. Our assumptions now are quite different. Today we expect, usually correctly, that our stay in the hospital will cure us or improve our health. With few exceptions, hospitals today follow reasonable standards of care and their staffs are conscientious in carrying them out. Why, then, do nosocomial infections continue to occur, little diminished by an "epi-demic" of scrubbing, disinfecting, and surveillance? The reason is, paradoxically, that modern medicine has improved so dramatically. Many persons who just a few years ago would have died rapidly from cancers, burns, heart attacks, diabetes, or kidney failure now survive, return to their homes and enjoy additional years of useful and happy life. But first they must endure a prolonged period of complex, highly technical therapy during which infection is a constant threat. For example, almost 80% of all childhood leukemia patients can now be effectively treated. But the treatment, which may take months, severely depresses their immune systems, making them highly susceptible to infection for the duration of treatment. In an imperfect world, no benefit is without cost. Hospital infections are one price paid for medical advances.

About 3.3% of persons discharged from a representative group of hospitals in the United States during the period 1980-1982 had developed nosocomial infection during their stays. The rate may be about 8% in nursing homes. Large university-associated research and teaching hospitals tend to have high rates (above 6%) because they treat the sickest patients and offer the most complex and highest-risk forms of therapy.

The personal and economic costs of nosocomial infections are staggering. Data from a 1977 CDC survey showed that about 1.84 million patients in the United States developed nosocomial infections in that year. The infections added an average of three days to their hospital stays, at an average cost of $174 per day. The total bill borne by insurers, individuals, and the government was $966 million. In the 1980-1982 CDC survey, about 1.2% of the infections were considered to have caused the patient's death directly, while an additional 3.5% were contributory factors.

SOME DEFINITIONS

A nosocomial infection, as we saw, is one that is acquired in the hospital. If the infection was present on admission, it is not nosocomial; if it appears after discharge but the incubation period coincides with the hospital stay, it is nosocomial. If an infection is hospital-acquired, the hospital has medical responsibility for its treatment, ethical responsibility for tracing and eliminating the source and means of transmission, and legal responsibility if there is a question of negligence or malpractice.

Many nosocomial infections are **iatrogenic diseases,** that is, diseases that are the direct re-

sult of a diagnostic or therapeutic procedure. An example is a bacteremia resulting from implantation of a replacement heart valve.

Nosocomial infections require compromised hosts. You'll recall that a person becomes compromised when any or all of the three layers of defense — anatomic integrity, phagocytic network, or immune system — are altered. Many hospital patients are already compromised when they enter the hospital (for example burn patients). In addition, modern medical practice provides many new ways of worsening the compromised state, such as using immunosuppressive drugs. At the same time, modern techniques facilitate many entirely new means of transmission. For example, nosocomial infection can be transmitted by inhaling vapor from a contaminated respiratory therapy apparatus or by receiving an injection with a nonsterile needle. In the hospital, these and many other new or "unnatural" means of transmission supplement their "natural" analogs, which in these two examples would be inhaling air contaminated by someone else's sneeze and being bitten by an insect.

Nosocomial infections are most commonly caused by opportunistic microorganisms, which, you'll recall, are ones that do not cause infection in healthy individuals. The usual habitats of opportunists are on body surfaces, in the throat or gastrointestinal tract, or in air, liquids, or food. Such organisms tend to be very hardy and tolerate heat, cold, and drying conditions well. Lacking the usual virulence factors of invasiveness and toxicity, opportunists require lowered bodily defenses to enter and survive in human tissues. They therefore infect only compromised hosts.

INTERACTIVE FACTORS IN HOSPITAL INFECTIONS

Nosocomial infections occur because of the interaction of three factors (Figure 26.2). As we noted, hospitals contain opportunistic or pathogenic microorganisms and serve patients who

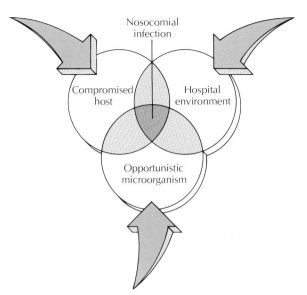

FIGURE 26.2 **Interactive Factors in Nosocomial Infection**
Three factors interact to produce nosocomial infections: the compromised host has minimal resistance to microbial invasion; opportunistic microorganisms (often highly drug resistant) are present within the hospital; and the hospital environment provides numerous opportunities (such as surgery and intubation) for disease transmission within a dense population.

are compromised hosts. The third factor is the hospital environment — its architecture, its population density, its staffing patterns, and the special types of intimate, physical contact that occur within its walls. The environment may provide unique means of transmission.

OPPORTUNISTIC MICROORGANISMS

We will consider any organism to be an opportunist if it either causes disease only in compromised hosts, attacks them more often than nor-

mal hosts, or causes markedly more severe and generalized infection in compromised individuals. A very large number of microbial agents fall in this category. These include bacteria, fungi, protozoan parasites, and viruses (Tables 26.1 and 26.2).

THE MOST SIGNIFICANT OPPORTUNISTS

The enterobacteria are the most numerous and bothersome group of opportunists. *Escherichia coli* is the most common opportunist, reported in 19.7% of all nosocomial infections. In a nursery it can cause neonatal meningitis, and in other hospital areas it is the leading cause of bladder infection. *Klebsiella*, *Proteus*, and *Enterobacter* are also frequently encountered. *Serratia marcescens* hospital strains usually lack the characteristic red pigment of this species. All enterobacteria readily acquire and exchange antibiotic resistant plasmids. These five enterobacterial genera combined account for over 40% of all nosocomial infections, including conditions such as urinary tract infections, pneumonias and septicemias.

Nosocomial *Staphylococcus aureus* infections were the first to be seriously studied, and they have been partly controlled. Yet they still are the second most common form of nosocomial infection, accounting for about 11% of the total, primarily in skin lesions and surgical wound sites. Other staphylococci cause about 5% of the infections, with one important category being bacteremias. The decline in *S. aureus* prevalence is a result of specific infection control techniques using disinfectants aimed at eliminating staphylococci.

The **enterococci** (streptococci that are normal inhabitants of the gastrointestinal tract) may, as opportunists, cause urinary tract and surgical wound infections, accounting for about 10% of all nosocomial infections.

We are familiar with *Pseudomonas aeruginosa* as the primary opportunistic invader of burn wounds (Chapter 16). This organism can

TABLE 26.1 Some opportunistic bacteria

Organism	Microscopic Appearance	Source	Infections in Compromised Hosts
Acinetobacter calco-aceticus	Gram-negative diplococcus	Soil, water, skin	Meningitis, septicemia
Bacteroides	Gram-negative rod	Upper respiratory, gastrointestinal, and genitourinary tracts	Abdominal
Cardiobacterium hominis	Gram-negative rod	Upper respiratory; gastrointestinal tracts	Endocarditis
Hemophilus spp.	Gram-negative rod	Upper respiratory tract	Septicemia, pneumonia, meningitis, endocarditis
Mycobacterium spp.	Acid-fast rod	Human beings, soil	Chronic or acute lesions; lung, disseminated
Neisseria spp.	Gram-negative diplococcus	Upper respiratory tract	Meningitis, endocarditis, septicemia
Pseudomonas aeruginosa	Gram-negative rod	Colonized human, moist items in hospital environment	Pneumonia, burn infection, septicemia, any other organ
Staphylococcus aureus	Gram-positive coccus	Colonized human, moist items in hospital environment	Urinary and respiratory infections, endocarditis
Staphylococcus epidermidis	Gram-positive coccus	Ubiquitous	Wound infection, septicemia, any other organ
Streptococcus spp.	Gram-positive coccus	Ubiquitous, human or animal upper respiratory or Gastrointestinal tract	Wound infection, septicemia, endocarditis
Enterobacteria: *Escherichia coli* *Enterobacter cloacae, aerogenes* and *agglomerans* *Klebsiella* spp. *Proteus* spp. *Serratia* spp. *Providencia* spp. *Citrobacter freundii* *Citrobacter diversus*	Gram-negative rod	Ubiquitous	Urinary-tract infection, neonatal meningitis, peritonitis, bacteremia, pneumonia, septicemia, gastrointestinal upsets

also invade other types of wounds or any body orifice. In the absence of active phagocytosis it can invade and destroy any organ. Recurrent *P. aeruginosa* infections are common in cystic fibrosis and in immunosuppressive conditions (Figure 26.3); they account for about 9% of all nosocomial infections.

Candida albicans and other *Candida* species cause fungemia in patients who have immunosuppressive malignancies or who are undergoing immunosuppressive therapies. Fungal infections in general are responsible for about half of the deaths following kidney and other

organ transplants. Yeasts cause 4.5% of reported nosocomial infections.

Viruses of the herpes group, as you know, often establish latent states following the initial infection. Many of us, perhaps most, harbor latent herpesviruses. In the immunologically incompetent infant, an initial infection itself may be overwhelming. Furthermore, when a normal adult becomes immunodeficient, latent herpes simplex, herpes zoster, and/or cytomegaloviruses may reactivate into severe, generalized infections. Accurate data on the prevalence of nosocomial virus infection is not available.

TABLE 26.2 Some opportunistic fungi, protozoans, and viruses

Group	Organism	Predisposing Factor	Form of Disease
Fungi	*Candida albicans, tropicalis, parapsilosis*	Prosthetic heart valve; immunosuppression; allograft	Mucocutaneous candidiasis, dissemination, candidemia
	Aspergillus oryzae, fumigatus	Prosthetic heart valve; immunosuppression; allograft	Pulmonary infection, infection of prosthetic heart valves
	Cryptococcus neoformans	Hodgkin's disease and other lymphomas	Severe disseminated cryptococcosis; central nervous system involvement
Protozoans	*Pneumocystis carinii*	Immunosuppression; allograft	Rapidly progressive pulmonary lesions
Viruses	Herpes simplex	Immunosuppression; allograft	Generalized cutaneous eruption; pneumonia; encephalitis
	Varicella-zoster	Hodgkin's disease, other malignancies; immunosuppression; allograft	Shingles, pneumonia, disseminated disease, hemorrhage
	Cytomegalovirus	Immunosuppression; allograft; blood transfusion	Fever, pneumonitis, hepatitis

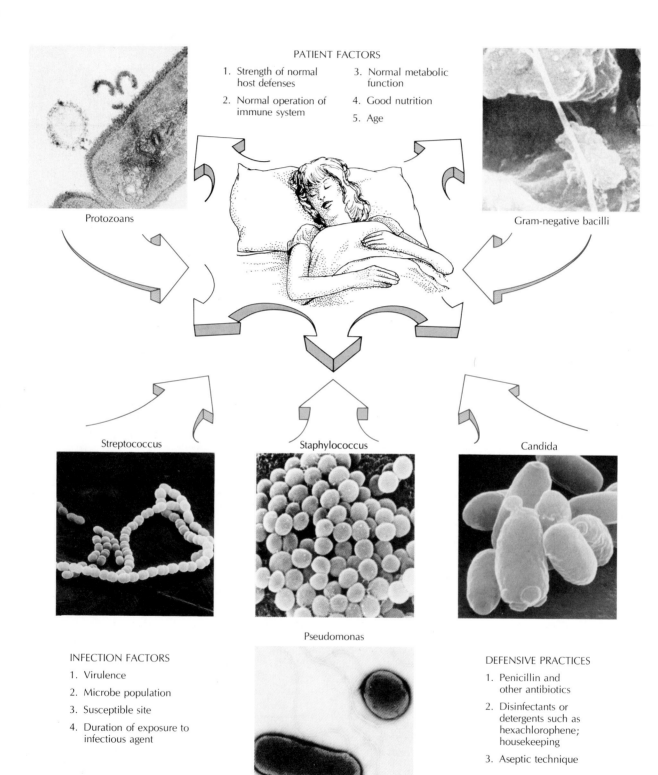

PATIENT FACTORS

1. Strength of normal host defenses
2. Normal operation of immune system
3. Normal metabolic function
4. Good nutrition
5. Age

Protozoans

Gram-negative bacilli

Streptococcus

Staphylococcus

Candida

Pseudomonas

INFECTION FACTORS

1. Virulence
2. Microbe population
3. Susceptible site
4. Duration of exposure to infectious agent

DEFENSIVE PRACTICES

1. Penicillin and other antibiotics
2. Disinfectants or detergents such as hexachlorophene; housekeeping
3. Aseptic technique

FIGURE 26.3 Significant Opportunists
A large number of microorganisms cause disease in compromised hosts, such as hospital patients. Several of the more common opportunists are shown here. In general, the organisms that cause nosocomial infections tend to be hardy, tolerant to heat and cold, and resistant to antibiotics. (Photo credits: Protozoan (×13,260), courtesy of Visuals Unlimited/L. L. Pifer; Gram-negative bacilli (×15,000), courtesy of J. J. Cardamone, Jr., University of Pittsburgh/BPS; Staphylococcus, from Z. Yoshii et al., *Atlas of Scanning Electron Microscopy in Microbiology.* Tokyo: Igaku Shoin, Ltd. (1976); Streptococcus, from D. L. Shungu, J. B. Cornett, and G. D. Shockman, Autolytic mutants of *S. faecium, J. Bacteriol.* (1979) 138:601, Fig. 1a; Pseudomonds (×14,060), SEM by Gary Gaard, in Diane A. Kuppels and Arthur Kelman, Isolation of pectolytic flourescent pseudomonads from soil and potatoes, *Phytopathology* (1980) 70:1110–1115.)

MULTIPLE DRUG RESISTANCE

Multiple drug resistance is present when a microbial strain is simultaneously resistant to many antibiotics. It arises by natural genetic means. Various strains of bacteria carrying different plasmid resistance factors (Chapter 8) may coexist in the intestines of hospitalized patients and healthy staff members. Their plasmids recombine, producing new gene groupings containing an ever-increasing number of resistance factors. Multiple-drug resistance provides a strong survival advantage for bacteria in an environment, such as the hospital, filled with antibiotics. Multiple resistant strains, called (very respectfully) "superbugs," become members of the flora of patients and staff and become permanent features of the hospital environment (Figure 26.4).

Thus in nosocomial infections it is sometimes not the name of the organism, but its origin that matters. A strain of *Serratia marcescens* that is endemic among patients and staff in a hospital is apt to have a totally different and far more extensive pattern of antibiotic resistance than will a strain of *S. marcescens* isolated in the community. A hospital-acquired *Staphylococcus aureus* infection is automatically more serious than a community-acquired one because it is highly likely to be resistant to many antibiotics. Resistance develops rapidly with use of an antibiotic; the percentage of *S. aureus* isolates resistant to methicillin in teaching hospitals went from 5.9% in 1980 to 10.6% in 1981 and to 16.1% in 1982. Similar patterns are seen in other bacteria, revealing steadily declining effectiveness of specific drugs against them.

COLONIZATION

When a patient enters the hospital, his or her normal patterns of life are altered. The diet is different from that at home and the timing and extent of daily activity shifts. These changes alone are enough to alter the normal flora. At the same time, the patient encounters the microorganisms indigenous to the hospital. Within a brief period the new patient's skin, mucous membranes, and gut become colonized with the hospital microorganisms. In general, these are just different strains of the species that are his or her normal colonists — different, that is, in their enhanced antibiotic resistance. The patient becomes a new part of the reservoir for these strains and can pass them on to others.

The new colonists enter into a close relationship with the host that is not, initially at least, a harmful one. However, if the patient's defenses sag, the colonists can easily take over and produce disease. The disease, if and when it occurs, may be highly resistant to antibiotic treatment.

New hospital staff members are also promptly colonized. If you get a new job at a health care facility, you usually acquire the hospital flora sooner than you can get your permit for the hospital parking lot! You also become a part of the reservoir for these strains, and have the potential for transmitting them to others.

FIGURE 26.4   A Modern Hospital Environment
The mixed population of microorganisms that persist in hospitals must survive heavy selection pressures from disinfectants and antibiotics. Those that succeed may colonize staff and patients or survive in water, air, or food systems. Hospitals are crowded, busy environments, with people constantly coming and going. New patients and staff members are colonized within hours of entry and soon acquire the potential for transmitting them to others. (Courtesy of Sepp Seitz/Woodfin Camp & Associates.)

THE COMPROMISED HOST

The list of factors that may compromise a hospital patient is long. It grows longer with each new technical advance in medical intervention. Some compromising factors increase susceptibility by thwarting natural defenses, while others create new means of transmission. See Table 26.3 for selected examples.

SOME INDICATORS OF SUSCEPTIBILITY

The health-care professional assessing a new patient can form a fairly accurate estimate of the patient's degree of susceptibility by observing certain key indicators. The first four can be readily noted at the bedside. The other two can be readily spotted by examining the patient's laboratory reports.

■ Age — Most people begin and end their lives in hospitals. The very young or very old are always most susceptible because their immunologic defenses are not fully functional. Neonates, especially if they are premature or suffered birth trauma, and the very aged are to be considered automatically at risk.

■ Anatomical damage — Wounds, burns, skin ulcerations, and surgical incisions create abnormal portals of entry for microbial invaders. These conditions always dictate special infection control considerations.

- Intubation — When a tube or catheter is introduced into a body orifice or the tissues, it disturbs normal surface defenses and adds to the risk of infection. The risk increases the longer the tube remains in place.
- Nutritional Problems — Patients who are very thin or severely wasted may have been inadequately nourished for a long time. Invariably, their antimicrobial defenses will be depressed.
- Leukopenia — Sometimes the number of circulating white blood cells is abnormally low. This indicates depressed blood cell formation, a condition called Leukopenia. When polymorphonuclear cells fall below 500 per milliliter, the nosocomial infection risk becomes very great.
- Metabolic disease — Signs of metabolic diseases including elevated blood sugar, high blood urea nitrogen, and acidosis tend to cause deficient phagocytosis, complement activity, or cell-mediated immunity. When the primary disease state is successfully brought under control, the threat of infection diminishes.

The admitting diagnosis, history, and the orders for necessary diagnostic procedures and therapy will provide a complete picture of the patient's risk. Of course, this picture changes as the patient's condition improves or deteriorates. A high-risk patient requires careful observation so that a developing infection can be detected as early as possible.

SPECIAL MEANS OF TRANSMISSION IN HOSPITAL UNITS

Patients with certain high-risk conditions such as burns are often grouped in units that offer

TABLE 26.3 Factors predisposing to infection in the compromised host	
Mechanism	Predisposing Factor
Circumvents external anatomic barriers	Burns, dermatitis Intravenous and urinary catheters Injections, diagnostic procedures Surgery Trauma
Impairs cellular immunity	Steroid therapy Alcoholism Antineoplastic and immunosuppressive drugs Burns Congenital defects Diabetes, especially with acidosis Radiation therapy Leukemia, lymphoma, Hodgkin's disease, other malignancies Malnutrition Removal of spleen Kidney failure, especially with acidosis
Impairs antibody defenses	Cirrhosis Malnutrition Removal of spleen

BOX 26.1
The Progressively Compromised Patient

The failing patient may undergo a cumulative sequence of compromising events. An initial organ failure or a malignancy may lead to hospitalization; that, in turn, may lead to a primary nosocomial infection. These problems may progress, in time, to secondary infections or organ failures. The odds on dealing successfully with each additional stress get poorer.

The pattern of progressive compromise was revealed by a recent study of hospital-acquired fungemia, a secondary infection. The predisposing factors in the cases studied overlapped markedly. For example, at the time fungemia developed, quite late in the patients' illnesses, they all had received prior antibiotic therapy and were receiving fluids by intravenous catheter. (Two thirds were getting total parenteral nutrition.) All except one had urinary catheters. Thus this was a group of people that was also more or less immobilized in bed. Almost 90% had concurrent bacterial infections despite the antibiotic therapy, suggesting that the antibiotics weren't working anyway because of poor immune defenses. Almost 70% of the study group had had recent surgery. The time sequence of these cumulative factors, of course, differed from patient to patient. But each therapy, by attempting to solve one immediate and pressing problem, was seemingly setting the stage for another problem.

The infective agent in most of these cases was a *Candida* species, and almost all the patients died. As you can see, in cases such as these, it might be very difficult to assign an exact cause of death. It is also clear that, whatever we attempt to do using all the tools of modern medicine, we are frequently helpless to save a dying person.

specialized care. The disadvantage is that grouping these individuals may set the stage for epidemic spread of nosocomial infections. Clusters of compromised patients occur in the newborn nursery, intensive care units, hemodialysis unit, surgical suite and recovery areas, burn unit, oncology service, and other areas.

Certain procedures also increase the patient's risk of nosocomial infection. Anesthesia may contribute to the development of pneumonia after surgery by altering the respiratory reflexes. Inhalation or respiratory therapy poses a hazard because it takes great care to keep the equipment from becoming colonized with bacteria that may then be blown forcefully into the patient's lungs during treatment. A tracheostomy, a surgical incision made into the trachea through which a tube is introduced to permit the flow of air, poses a double threat. The abnormal airway bypasses the normal anatomic defenses located in the upper respiratory tract, while the surgical wound may allow infection of the surrounding tissues.

The urinary or Foley catheter, used to allow urine to drain from the bladder, is a clearly documented risk. About 15% of all hospital-acquired infections involve the urinary tract; most follow urinary catheterization.

The tips of intravenous catheters, which pass into a vein through a skin puncture, readily become colonized with skin flora. If left in place for more than 48 hours, the risk of overt infection increases sharply, especially if a nutrient fluid such as dextrose-saline or total parenteral nutrition fluids are given. These fluids may be contaminated during preparation if appropriate aseptic techniques are not used. Alternatively, skin microorganisms from the catheter tip may migrate up the tubing to the reservoir and there multiply to a high population density, using the nutrients for growth. Fluids for intravenous infusion are preferably prepared in a centralized pharmacy that has appropriate equipment for aseptic manipulations.

Surgical dressings, too, pose an infection hazard. Changing procedures are designed to protect the patient, other patients, and staff members. When the wound is known to be infected, wound precautions (Chapter 23) are used.

Finally, immunosuppressive therapy to relieve inflammation, reduce pulmonary edema, permit graft acceptance or treat malignancy markedly increases the patient's infection risk. Recent or current use of steroids (such as prednisone), cytotoxic cancer drugs, antilymphocyte serum, or radiation therapy add to the compromised state of the patient (Table 26.4).

TYPES OF NOSOCOMIAL INFECTION

Urinary tract infections are the most common nosocomial problems; as we noted, they are strongly correlated with catheterization. Drug-resistant enterobacteria produce a stubborn urinary tract infection that may eventually lead to kidney disease or septicemia. Surgical wound infections are the second most prevalent type of nosocomial infection. Lower respiratory tract in-

---

TABLE 26.4 Iatrogenic effects — How therapy predisposes to infection	
Therapy	**Effect**
Antibiotics	Alter normal flora and select resistant types May encourage overgrowth of some fungi
Steroids	Depress antibody formation Depress phagocytic function Suppress interferon production Suppress lymphocytes and cellular immunity
Radiation and antineoplastic drugs	Depress antibody formation Depress bone-marrow production of leukocytes Suppress the phagocytic system Injure tissues
Surgery, insertion of foreign material, intubation	Provide portal of entry to highly susceptible tissues Provide a site for adherence and persistence of organisms

fections, the third most common, occur in both medical and surgical patients, especially if the patients are immobilized. These types of nosocomial infections are shown in Figure 26.5.

Anesthesia or prolonged unconsciousness create special risks of LRT infection. In the debilitated person, lower respiratory tract infections are more likely to progress to septicemia.

FIGURE 26.5 High-Risk Conditions
Situations that make the patient particularly vulnerable to infection include (a) units such as the newborn nursery where dense grouping may encourage epidemics, (b) procedures such as respiratory therapy that might subject a patient to contaminated equipment, (c) treatment involving catheters or surgical instruments that puncture skin and veins, and (d) extensive injuries such as burns. (Photo credits: a, courtesy of Baron Wolman/Woodfin Camp & Associates; b, courtesy of William Thompson © 1982; c, courtesy of Frank Siteman/Stock, Boston; d, courtesy of Russ Kinne © 1979/Photo Researchers, Inc.)

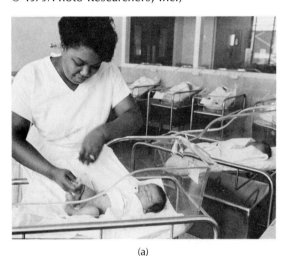

(a)

(b)

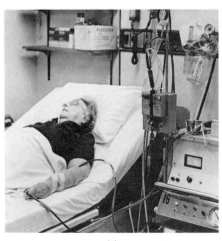

(c)

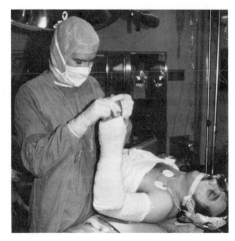

(d)

Gastroenteritis, which is most common in the newborn nursery or the pediatric ward, is often caused by enteropathogenic *Escherichia coli*. Unsatisfactory sanitation in the kitchen can produce outbreaks of *Salmonella* and other common gastrointestinal pathogens. Skin infections are the outstanding problem in the nursery, where *Staphylococcus aureus* may be endemic. Generalized herpetic infections appear most often tissue transplantation units. All severe burns become colonized. The risk of infection depends on the size and depth of the burn and the difficulties experienced in covering the wound. Upper respiratory tract infections are common in pediatric patients, paralleling the frequent occurrence of these conditions in healthy children. Any of these initially localized infections may develop into bacteremia or septicemia and eventually cause death. This is why early detection is so crucial.

THE HOSPITAL ENVIRONMENT

In studying the chain of transmission of infectious disease (Chapter 14), we learned that we must first identify the source or reservoir of the agent and then the means by which it is transmitted to the new victim. The same rules apply to nosocomial infection, but as we noted earlier there are a number of variations on the usual routes (Figure 26.6).

RESERVOIRS, SOURCES, AND MEANS OF TRANSMISSION

In a modern hospital the number of potential reservoirs and sources of infection is somewhat restricted as the patient usually has no contact with soil, animals, or insects. Four basic categories of reservoirs and sources are present (Figure 26.7).

HUMAN RESERVOIRS. People are, as always, the most significant reservoir. There is no doubt that direct person-to-person contact is the main means of spreading nosocomial infections. Car-

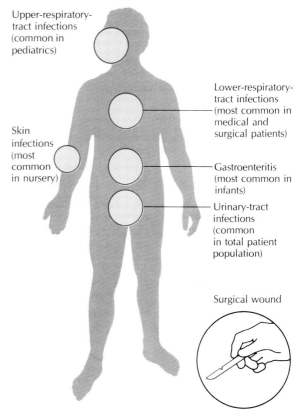

FIGURE 26.6 Types of Nosocomial Infection
Urinary tract infections outnumber all other types and may occur in patients of any type, ususally when they are wearing urinary catheters. Other infections tend to affect specific services and types of patients.

riers among the staff exchange microorganisms with colonized patients; personnel carry the organisms from one bed to the next. But general patient care would be impossible without continuous close contact. Therefore, interrupting the chain of transmission is entirely dependent on personal cleanliness and aseptic technique. All authorities agree that *frequent and conscientious hand-washing* is the most important single practice in breaking the chain of infection.

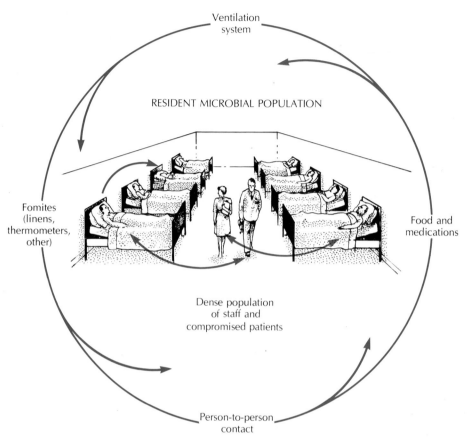

FIGURE 26.7   Sources and Means of Transmission in the Hospital
The hospital environment provides numerous opportunities for the spread of
nosocomial infections. Common pathways for the transmission of disease include
exchange of microorganisms between patients and staff, airborne spread of
organisms, and contact with fomites such as linens, food, equipment, and other
contaminated objects.

Patient-to-patient transmission may occur
directly by droplets, as when a person with un-
diagnosed, active tuberculosis infects a room-
mate. More typically there is an intermediate,
such as shared equipment, and the route is indirect. In nursery staphylococcal outbreaks, the
agent spreads rapidly, usually via nurses' hands.
Patient-to-staff transmission also occurs, either
directly or indirectly. For example, direct trans-

mission occurred when a physician in an obstetrical clinic contracted rubella after treating a
patient. An example of indirect patient-to-staff
transmission is when a nurse contracts hepatitis
from patient serum introduced into the hand
through an accidental needle stick.

Staff-to-patient transmission may also be direct or indirect. A physician may directly transmit his or her flora to a patient when changing a

dressing. Indirect staff-to-patient transfer occurred, for example, when a kitchen worker, an intestinal carrier of *Salmonella*, infected food supplies that were later served to a number of patients. Because of the 24-hour schedule in a hospital, each patient is exposed to three sets of care-givers every day, and thus the range of possible staff/patient contacts is large.

FOMITES.   Inanimate objects in the patient's environment are called **fomites** when they serve as reservoirs and sources of infection. Nonsterile dressing materials, contaminated linens, and unclean equipment may all cause problems. Eating utensils contaminated by patients may remain a hazard if they are not adequately disinfected during dishwashing. Soiled linens, waste paper, improperly discarded disposable syringes, needles, and broken glass, all are hazardous to housekeeping personnel.

It is usually the occasional slip-up in an otherwise satisfactory procedure that causes trouble. Thus human error often has a substantial role in hospital infection.

FOODS AND MEDICATIONS.   Contaminated foods and medications are often reservoirs that cause outbreaks among patients. Furthermore, food scraps contaminated by patients (for example, in hepatitis, where the patient's saliva is infective)

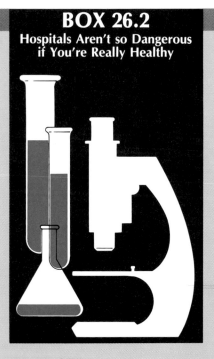

BOX 26.2
Hospitals Aren't so Dangerous if You're Really Healthy

In 1944, an Englishman named McIlroy began a long career as an imaginary invalid. His ability to magnify or fake symptoms was linked to a pathological compulsion to be hospitalized, a psychological disorder called Munchausen's syndrome. McIlroy was treated for acute respiratory failure, a collapsed lung, neurologic disorders, severe abdominal pain, and urine retention. He had 207 documented hospital admissions, under many fictitious names, and the British National Health Service spent several million pounds trying to cure his "ailments."

McIlroy underwent surgical exploration of the skull, abdominal and orthopedic operations, hundreds of x-rays and blood tests, at least 48 spinal taps, three encephalograms, numerous myelograms, and innumerable intravenous and urinary catheterizations. He would seem to have been the world's likeliest candidate for death from nosocomial infection. However, he walked out of his last hospital in 1978. At last report, he was residing peacefully in an old-age home.

Source: Excerpted from *Scientific American* 1979; 241:80.

may be a hazard to the kitchen workers if they are returned to the kitchen without disinfection.

AIR CURRENTS AND SUPPLIES.   Ventilation is a major factor in some nosocomial diseases, such as surgical infections and legionellosis. Airborne transmission of disease becomes significant in any setting where there are high population densities, as there are in a modern hospital. Air flows along corridors, up warm air currents in laundry chutes, and through air-conditioning systems, carrying organisms from one person to another. Beds in multi-bed rooms are rarely more than a few feet apart, facilitating airborne transmission. These and other environmental factors that can spread infection are presented in Table 26.5.

PROGRAMS TO CONTROL HOSPITAL INFECTION

CONCEPTUAL APPROACHES

No method of control is likely to completely eradicate nosocomial infections in the immediate future. However, there are two conceptual approaches, which, when implemented simultaneously, can improve the standard of care significantly. The first is to reduce the number of organisms to which the patients are exposed. The means to this end include improving overall cleanliness, reducing the impact of staff carriers by education and/or reassignment, and using asepsis and isolation in a sophisticated way. The other basic approach is to maintain and improve patients' resistance. This approach often means

TABLE 26.5 Some sources of infection in the hospital			
Type of Source	Examples	Agent Cultured	Case(s) Documented
Personnel	Surgeon's hair	*Staphylococcus aureus*	Yes
	Nurse's hands	*Citrobacter diversus*	Yes
	Obstetrics-clinic staff	Rubella virus	No
	Surgical assistant	*Streptococcus pyogenes*	Yes
Medications	Intravenous solution	*Candida albicans*	Yes
	Corneal transplant	Rabies virus	Yes
	Blood fraction	*Pseudomonas cepacia*	Yes
Equipment	Stethoscope	*Staphylococcus aureus*	No
	Oxygen-humidifying bottle	*Pseudomonas aeruginosa*	No
	Endoscope	*Salmonella typhimurium*	Yes
	Breast-milk pump	*P. aeruginosa*	Yes
	Anesthesia equipment	*P. aeruginosa*	Yes
	Elastoplast wound dressing	*Rhizopus oryzae*	Yes
Environment	Floor-scrubbing machine	*P. aeruginosa*	No
	Carmine dye in food	*Salmonella cubana*·	Yes
	Hand lotion	*Klebsiella pneumoniae*	Yes
	Sink traps, nursery and surgery	*P. aeruginosa*	Yes
	Tomato	*P. aeruginosa*	No
	Chrysanthemum leaf	*P. aeruginosa*	No

that certain therapeutic options are to be avoided unless absolutely essential, such as unnecessary immunosuppression, indiscriminate use of antibiotics, catheterization done simply for the convenience of the attendants, and use of invasive procedures when noninvasive ones would suffice.

These principles are the basis of infection-control programs, which now reach into all phases of hospital operation. Certain stringent requirements for infection control must be met before a hospital can obtain accreditation. However, infection-control requirements for other types of health care and residential facilities such as nursing homes are generally much less stringent, and they vary greatly from place to place.

INFECTION-CONTROL PROGRAMS

The Joint Committee on Accreditation of Hospitals requires that each institution:

- provide definitions of nosocomial infections, for uniform surveillance, reporting, and identification;
- collect and review infection records for patients and personnel;
- publish and update its aseptic, sanitation, and isolation procedures;
- make a commitment to equal standards of care for the isolated patient;
- safely operate service areas such as laundry and waste disposal;
- guarantee laboratory support for the infection-control program;
- provide an employee health program, orientation, and in-service education relative to infection control; and
- audit antibiotic usage by medical staff.

Most of these responsibilities are assumed by the hospital's infection control committee, which includes representation from a wide range of departments. A hospital epidemiologist, usually either a doctor or the microbiology director, provides overall supervision and acts as a liaison

with the medical staff. The infection control nurse or infection control clinician does the day-to-day work of surveillance, using a case-finding approach to ensure that all nosocomial infections are discovered, treated, and analyzed. In case-finding, the infection control nurse may review patient charts and microbiology laboratory reports daily. He or she may also prepare daily and monthly summaries of nosocomial infections. This type of systematic data collection provides an early warning of developing situations and objective information as to how well each service is doing. Surveillance of employee health and follow-up of discharged patients may also be necessary.

The employee health service provides pre-employment physical examination to detect communicable diseases such as tuberculosis and provide vaccination, if needed, for diphtheria, tetanus, measles, mumps, polio, and rubella. When an employee comes to work ill, he or she may need to be transferred to tasks that require no patient contact or sent home with treatment, if indicated.

Hospital visitors frequently have little understanding of infectious disease or its prevention. All visitors should be required to check in at the nursing station for any necessary instructions before entering a patient's room. If you are visiting, you should be aware that it is a dubious favor to visit a friend or family member if you bring an infectious disease along with the flowers!

A new employee's orientation should include education on his or her role in the control of infection. Procedure manuals giving details of infection control procedures must be available and kept up to date. Review material or new information is presented through in-service programs. Supervisors should educate by informal discussion of failures in technique, and teach colleagues by good example.

Antimicrobial drugs are still widely over-prescribed, and inappropriate use of broad-spectrum drugs instead of narrow-spectrum drugs is common. Antibiotic prescriptions of questionable value can be discovered and corrected by systematic auditing of pharmacy records.

In hospitals in which the infection control staff enjoys the full cooperation of all departments, the attack rate for nosocomial infection has been rapidly lowered from the 5% average of the early 1970s to below 2% in community hospitals. This translates to a nationwide reduction of several million cases per year.

SUMMARY

1. Hospital-acquired infections occur because the hospital houses and groups together patients whose antimicrobial defenses are less than adequate. Ill health, in whatever form, increases susceptibility to infection.

2. The hospital environment has a high population density, extensive use of common facilities, and a large amount of human traffic — visitors, shift changes of staff, and support workers.

3. The widespread use of antimicrobial drugs in clinical settings has led to the rapid selection of multiply-resistant hospital microflora. These hospital strains colonize the hospital staff, and transfer to patients within a few hours after the patients' admission.

4. Opportunistic microorganisms cause disease in the compromised host. Many of these agents are not exclusively restricted to a life within host tissues. They survive indefinitely in or on equipment and supplies in the hospital.

5. A helpful approach to the issue of host susceptibility is to evaluate the degree of compromise. Such analysis may be based on the number of compromising factors present and their severity. This allows the identification of patients for whom extra precautions are warranted. Compromising factors are cumulative in effect.

6. All accredited hospitals in the United States follow infection-control guidelines established by the Joint Committee on Accreditation of Hospitals. The guidelines require the formation of an interdisciplinary infection-control committee and the appointment of two or more infection-control officers.

7. A successful infection-control program will include active case-finding, preparation of daily and monthly reports on observed nosocomial infections, surveillance of employee health, orientation and inservice education on the control of infectious disease, the preparation and updating of procedures manuals for all areas of the hospital, and antimicrobial drug audits.

8. When all constituencies within the hospital work together, the incidence of nosocomial infections can be markedly reduced.

DISCUSSION QUESTIONS

1. Why are nosocomial infections particularly difficult to prevent?

2. List aspects of clinical therapy that are directly linked to increased infection risk. Pair them with suggestions to minimize their effect.

3. What might be the role of each of the following staff persons in spreading nosocomial infection — dietician, laundry worker, laboratory technician, physical therapist?

4. In case-finding, what might an infection-control nurse look for as clues to infection?

5. Discuss the contributions of antimicrobial drug therapy to increasing nosocomial infections; to controlling them.

ADDITIONAL READING

American Hospital Association. *Infection control in the hospital.* Chicago: American Hospital Association. Revised frequently.

Bennett JV, Brachman PS. *Hospital infections.* Boston: Little, Brown, 1979.

Larson, E. *Clinical Microbiology and Infection Control.* Boston: Blackwell Scientific, 1984.

Nosocomial Infection Surveillance 1980-1982. *Morbid Mortal Weekly Rep* 1984; 32:1SS-16SS.

Joint Commission on Accreditation of Hospitals. *Infection-control standards adopted by the Board of Commissioners.* Washington, D.C.: Joint Commission on the Accreditation of Hospitals. Revised frequently.

FIVE

MICROORGANISMS: THE ENVIRONMENTAL AND ECONOMIC IMPACT

27

SOIL AND WATER
MICROBIOLOGY

CHAPTER OUTLINE

Microorganisms are a diverse and fascinating group of life forms. In the last two units, we had ample opportunity to observe how microorganisms affect humans in the intimate situations of individual health and disease. It is now time to expand our perspective and acknowledge the microorganisms' fundamental roles in the activities that maintain our planet as a uniquely favorable environment for life.

In this chapter we will examine how microorganisms contribute to the cycling of essential nutrients. We'll discuss the water cycle, and how it distributes nutrients and microorganisms within and between soil and water. We'll analyze soil, its structure, its microflora, and microbes' role in soil fertility. We'll describe the chemical and microbial components of fresh water and of the marine environment. Then we'll review the major classes of pollutants and be-come familiar with how water is purified and tested for safety. We'll also become familiar with the methods currently used for treating sewage.

NUTRIENT CYCLES

A fundamental physical law, the Law of Conservation of Matter, states that *matter is neither created nor destroyed*. This means that all biochemical processes, which endlessly move atoms from one compound to another, use and reuse a fixed, limited pool of atoms. Four biologically critical elements — carbon, nitrogen, sulfur, and phosphorus — are comparatively scarce in the Earth's crust. Therefore, their atoms must be constantly shuffled from one compound to another. For example, different nitrogen compounds are needed by plants than are needed by animals. But all nitrogen compounds are potentially useful to some organism. In the biosphere, no biologically vital element is ever left unused for long.

Nutrient cycles are sequences of biologically mediated chemical conversions of elements. Thus, there is a carbon cycle, a nitrogen cycle, and so on. Nutrient cycles transfer elements from one molecular form to another, and from forms useful to one type of organism to forms useful to another.

The basic biological fact underlying nutrient cycling is that one organism's waste product is usually another organism's nutrient. Each living organism participates in the nutrient cycles. An organism takes an element from the environment in one molecular form, uses it for growth or energy, and releases it back to the environment in another molecular form, as a waste product. The next organism in the cycle takes up this waste product, uses it as a nutrient, then returns it to the environment in yet a third form. These sequential changes form a cycle because the element is eventually regenerated in its original form.

Nutrient cycling is global, and occurs in soil, fresh water, and the oceans. Microorganisms are an indispensable link in all the cycles. From a global perspective, if carbon, nitrogen, sulfur, and phosphorus did not cycle, but remained tied up in unavailable forms, life would soon cease to exist. From a practical perspective, when we compost organic wastes or treat sewage, we are deliberately encouraging nutrient cycling, to get rid of waste products that have temporarily accumulated at one step of a nutrient cycle.

THE WATER CYCLE

Water performs many life-sustaining functions. As a solvent, it transports nutrients, including compounds of carbon, nitrogen, sulfur, phosphorus, and metals. Also as a solvent, it carries harmful compounds — pollutants — that threaten living things. Chemically, water (H_2O) serves as the ultimate source of oxygen, to renew the atmospheric pool, and of hydrogen, from which organisms build carbohydrates.

The water cycle consists of the physical movement of water from one environment to another, powered by solar energy (Figure 27.1). About 98% of the earth's water is in liquid form, in the streams, rivers, lakes, oceans, and underground deposits. Only 2% is found elsewhere — in atmospheric vapor, in ice or snow, in the soil, or in the cells and bodies of living things.

Solar energy evaporates pure water from all surfaces, but most importantly from the oceans, which cover 70% of the planet. Any dissolved materials such as mineral salts are left behind. The movement of warm and cool air masses creates a global atmospheric circulation, from which we get the weather patterns. The water vapor eventually falls back to earth as rain or snow, bringing soluble chemicals picked up from the atmosphere. Most of this precipitation accumulates and runs once again to the oceans as **surface water.** The rest percolates down through the soil into the underlying rock and seeps more slowly back to the oceans. This subsurface **groundwater** may remain in underground deposits called **aquifers** for thousands of years.

As rainwater runs over the land surface and through the soil, it picks up all sorts of chemicals, both nutrients and pollutants. Much of this burden accumulates, century after century, in the oceans. The oceans, in consequence, not only become constantly saltier, but also become the ultimate repository, or dump, for many of the toxic wastes of our society.

All of the nutrient cycles (such as the carbon and nitrogen cycles) are directly linked to the water cycle, since water is the carrier of most of the intermediate compounds. Our society's problems with inadequate or contaminated drinking water and sewage and toxic-waste disposal are also intimately connected to the indispensable, unstoppable water cycle.

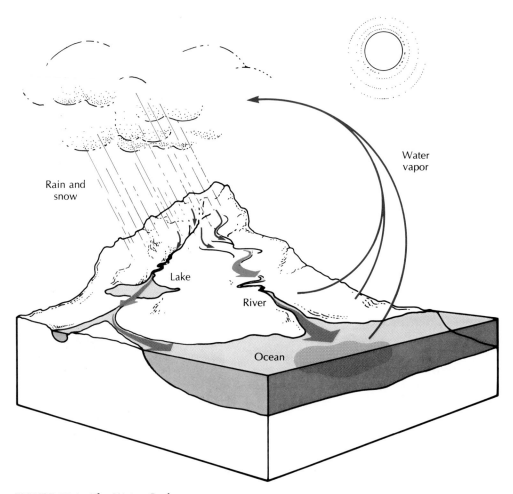

FIGURE 27.1 The Water Cycle
The energy of the sun drives the water cycle on our planet. Water evaporates from the earth's surface, especially the oceans, and is carried by air currents into the upper atmosphere. Eventually, it returns to earth as snow or rain. As rainwater collects and runs toward the ocean, it dissolves and carries with it a variety of suspended chemicals and minerals.

THE CARBON CYCLE

As you know, atoms of carbon make up the backbone of all organic molecules. No organism can manufacture a single new organic molecule without a carbon source. You'll also recall that organic compounds are the energy source for all heterotrophs.

Most of the earth's carbon is in the form of atmospheric carbon dioxide (CO_2) (Figure 27.2). **Carbon fixation** is the term we use for the incor-

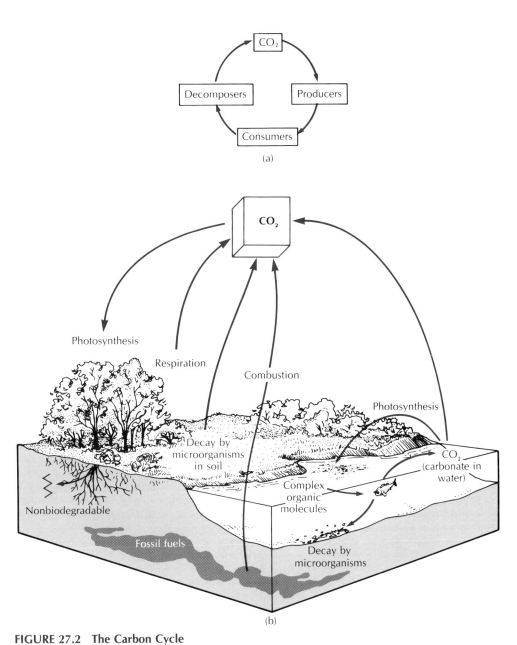

FIGURE 27.2 The Carbon Cycle
(a) In its simplest outline, the carbon cycle is the passage of carbon from
atmospheric CO_2 to producers, to consumers, to decomposers, and finally back to
atmospheric CO_2. (b) The reality is more complex. Carbon dioxide is fixed both on
land and in water, and producers give off some CO_2. Decay processes may be slow
or even halted by fossilization — as with the organic matter that formed fossil fuels.
Furthermore, some organic materials are nonbiodegradable.

poration of carbon atoms from atmospheric CO_2 into organic carbon compounds. **Producers,** which are the photosynthesizers and chemolithotrophs, are the organisms that are able to fix carbon dioxide into carbohydrates and other organic molecules. Their **primary production** is the metabolic activity that increases their own biomass.

It is helpful to view the rest of the carbon cycle as a progressive "unfixing" of carbon atoms. Unfixing is a consequence of metabolic reactions, primarily respiration, which break down carbohydrates and other biochemicals to form CO_2. It is the biomass of the producers that is broken down, directly or indirectly, as food for all the heterotrophic species. These species are of two sorts, the **consumers,** which graze on plants, prey on other consumers, or are parasites; and the **decomposers,** which metabolize the wastes and dead tissues of other organisms. As the biomass of the producers is used, each user in turn releases some of the CO_2. Consumers, including grazers, predators, and parasites, break down the more readily metabolized compounds. Residual fixed carbon is left in the parts of the dead tissues and animal wastes that resist decomposition, collectively called **detritus.** Most bacteria and fungi find their prime ecologic role in utilizing these leftovers. Communities of decomposers complete the unfixing of carbon.

Some carbon atoms stay fixed for a very long time in naturally occurring compounds that resist microbial enzymes. We call these **refractory** or **persistent compounds.** For example, under anaerobic conditions plant lignins, found in wood, persist for a long time without being decomposed. **Peat** is an accumulation of such undecomposed plant material, preserved by overlying bog water. Once dried, peat is used as a home-heating fuel in some areas. Coal, oil, and natural gas derive from organic detritus that was covered with water and rock before microorganisms had a chance to decompose it. When we burn these fossil fuels we complete the car-

bon cycle, returning the carbon to the atmosphere as carbon dioxide.

Some human-made chemicals, such as certain plastics and pesticides, are also refractory; these we call nonbiodegradable. Nonbiodegradable compounds have molecular structures so unlike anything found in nature that there are no decomposers with enzymes able to attack them. These chemicals are broken down (extremely slowly) by nonbiological processes, such as exposure to the ultraviolet radiation in sunlight. If they accumulate, they may pose severe environmental problems.

Research suggests that genetic changes such as mutation or recombination of existing genes may enable a microorganism to acquire the enzymatic tools needed to attack a previously refractory compound. One of the promises of genetic engineering is that it may be able to provide custom-designed decomposers. We have already put together, by genetic engineering, bacterial strains that can rapidly break down some of the components of spilled crude oil. Industrial research groups are involved in crash programs to engineer microorganisms that will decompose dioxin, an extremely toxic and refractory byproduct of several manufacturing processes.

THE NITROGEN CYCLE

Nitrogen atoms are required by all organisms for incorporation into proteins and nucleic acids. Both inorganic and organic forms of nitrogen can be used by various organisms. Like carbon, nitrogen also moves from a free to a fixed status in its cycle. Plants most often use nitrate ion (NO_3^-), and can use ammonium ion (NH_4^+). Animals need organic nitrogen, mostly as the amino ($-NH_2$) group of amino acids. There is a certain group of microorganisms that uses each form of nitrogen (Table 27.1). In fact, microbes have the primary responsibility for interconverting the nitrogen compounds in the environment.

The nitrogen cycle is a five-step process. It is shown in Figure 27.3. The only bountiful nitro-

TABLE 27.1 Microorganisms in the nitrogen cycle		
Process	**Some Microbial Participants**	**Comments**
Nitrogen fixation*	*Rhizobium, Azotobacter, Beijerinckia, Frankia*	Symbiotic
	Bacteria such as *Azotobacter, Beijerinckia*	Free-living aerobic
	Cyanobacteria such as *Anabaena, Nostoc,*	
	Clostridium	Free-living anaerobic
Assimilation	All microorganisms	
Ammonification	*Bacillus*, most aerobic species	
	Clostridium, most anaerobic species	
	Proteus	Breaks down urea
Nitrification*	*Nitrosomonas, Nitrosospira*	Forms nitrite
	Nitrobacter, Nitrococcus	Forms nitrate
Denitrification	*Pseudomonas, Bacillus*	Some species

*Microorganisms are essential for these processes.

gen compound on Earth is nitrogen gas, which, ironically, is chemically inert and useless to plants and animals alike! As the first step in the cycle, nitrogen gas must be removed from the atmosphere and fixed into nitrogenous organic compounds. This step is carried out by **nitrogen-fixing** microorganisms. The free-living cyanobacteria fix the nitrogen for aquatic environments. On land, particularly in cultivated agricultural areas, bacterial nitrogen-fixers form **nodules** on the root tissues of host plants. *Rhizobium* nodulates leguminous plants (peas, beans, clover, alfalfa), and binds gaseous nitrogen into the amino groups found in glutamic acid and related molecules. The amino acid is absorbed by the plant cells and incorporated into protein for growth. A significant amount of nitrogen fixation in grasslands, swamps, and forests is carried out by free-living bacteria such as *Azotobacter* and *Beijerinckia* and by actinomycete symbionts such as *Frankia.*

Much smaller amounts of fixed nitrogen also become available to plants when nitrogen gas is oxidized to nitrate by lightning or combustion, and falls to earth dissolved in rain. Industrial processes also manufacture inorganic nitrate fertilizer. Even so, microorganisms contribute over 90% of the annual global nitrogen fixation required for plant growth.

In a second process, which works in opposition to nitrogen fixation, some nitrogen is lost to the atmosphere as nitrogen gas, as a result of microbial reduction of nitrates. This **denitrification** seems wasteful from a narrow perspective because it depletes the available supply of fixed nitrogen. Yet the magnitude of the nitrogen loss is quite small. Denitrification occurs only in anaerobic environments, such as waterlogged soil, where the microorganisms in question are reducing nitrates as an energy-yielding mechanism. In the total nitrogen economy, the amount of nitrogen fixation going on seems to

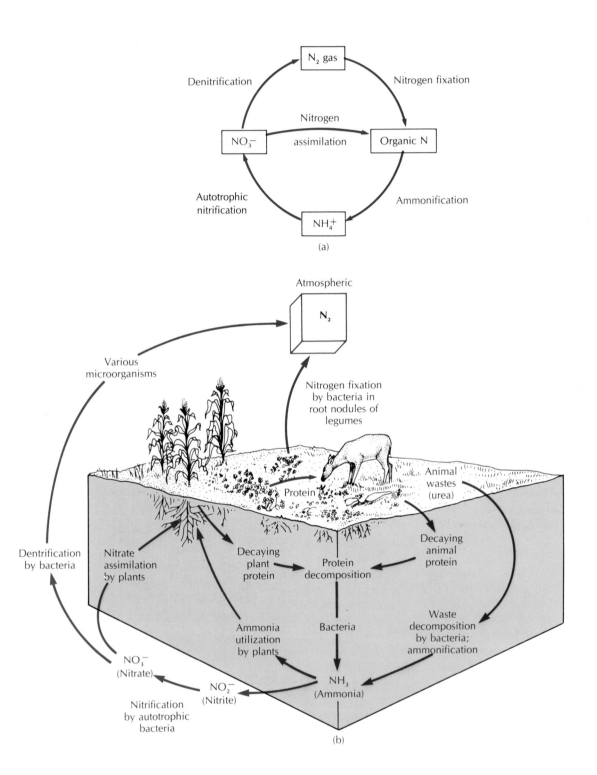

(a)

(b)

FIGURE 27.3 The Nitrogen Cycle
(a) In its simplest schematic form, the nitrogen cycle consists of five basic steps. (b) A more complex view shows nitrogen fixation capturing nitrogen from the atmosphere and converting it to usable form. The cycle is interconnected by the metabolic activities of microorganisms, plants, and animals. At certain points, some nitrogen gas escapes back into the atmosphere.

effectively balance the amount of denitrification that is occurring; thus the global supply of fixed nitrogen remains essentially stable.

The fixed nitrogen, in the form of organic amino acids, is passed from producers to consumers to decomposers in the food chain. As animals produce wastes and decomposers work on them, amino acids are broken down and ammonium ion (NH_4^+) is released. This is the third step in the nitrogen cycle, called **ammonification.** As some of this vaporizes as ammonia gas, a small amount of nitrogen is lost into the atmosphere.

The fourth step in the nitrogen cycle is **assimilation,** the process whereby plants and microorganisms make new amino acids, using inorganic nitrogen compounds. Both nitrate and ammonium are readily assimilated, but nitrate is usually preferred by plants and microorganisms to make amino acids and build tissue.

We now know where ammonium comes from, but where does nitrate come from? Nitrate is produced by the fifth step in the nitrogen cycle, called **nitrification,** a strictly microbial activity that oxidizes ammonium to nitrate. The chemolithotrophic bacteria *Nitrosomonas* and *Nitrobacter* derive energy from nitrifying, while regenerating a critical plant nutrient.

In summary, the nitrogen cycle has five major steps. Microorganisms are essential to three — nitrogen fixation, nitrification, and de-

nitrfication. Microorganisms contribute with plants to assimilation, and with animals to ammonification. This is yet another example of how microbial activities are essential for the productivity of the biosphere.

THE SULFUR CYCLE

Sulfur is an essential element in the amino acids cysteine and methionine. Living things need relatively small amounts of sulfur, and the natural geologic reserves of sulfur in the form of sulfate are usually adequate. The chemical interconversions of the sulfur cycle are generally similar to those of the nitrogen cycle. However, the elemental form of sulfur (S) does not physically leave the biosphere by escaping into the atmosphere, as does nitrogen gas (Figure 27.4a).

Sulfate ion ($SO_4^=$) is the form of sulfur used by plants, animals, and most microorganisms (Table 27.2). Organic sulfur compounds (predominantly cysteine and methionine) pass along the food chain. Microbial decomposition releases sulfur in the form of a bad-smelling gas, hydrogen sulfide (H_2S). H_2S is sequentially reoxidized by a number of bacterial species, first to elemental sulfur and then to sulfate. These oxidations are the energy-yielding mechanisms used by chemolithotrophs such as *Beggiatoa* and *Thiobacillus. Rhodospirillum* (Figure 27.4b) oxidizes H_2S and elemental sulfur during anaerobic photosynthesis.

Sulfate reduction occurs in anaerobic environments. Anaerobic bacteria such as *Desulfovibrio* use sulfate in place of oxygen as an electron acceptor for respiration, yielding H_2S. Sulfate reduction is analogous to the denitrification step of the nitrogen cycle.

OTHER ELEMENTAL CYCLES

Phosphorus, calcium, and metals such as iron, magnesium, nickel, and manganese are also of prime biological importance. They may be transformed by microorganisms by oxidation or re-

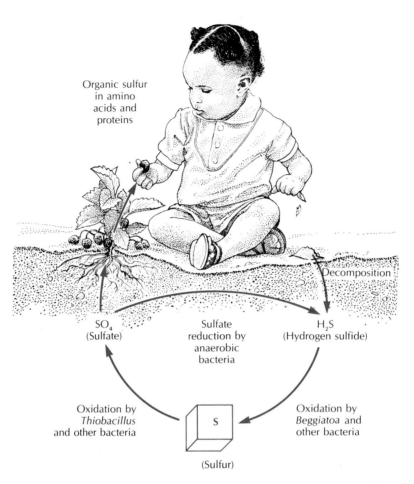

Organic sulfur in amino acids and proteins

Decomposition

SO₄ (Sulfate)

Sulfate reduction by anaerobic bacteria

H₂S (Hydrogen sulfide)

Oxidation by *Thiobacillus* and other bacteria

S

(Sulfur)

Oxidation by *Beggiatoa* and other bacteria

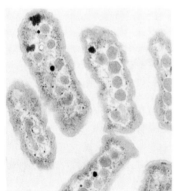

Thiobacillus

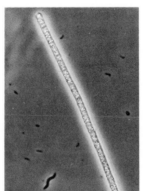

Beggiatoa

Desulfovibrio

FIGURE 27.4   The Sulfur Cycle
Microorganisms are able to change the oxidation states of sulfur within both organic and inorganic compounds through oxidation-reduction reactions. During decomposition, anaerobic bacteria release sulfur from organic compounds in the form of hydrogen sulfide. This compound is then oxidized by lithotrophic bacteria into elemental sulfur and sulfate. Hydrogen sulfide may also be produced by sulfate-reducing bacteria, such as *Desulfovibrio*, during anaerobic respiration. Some species such as *Beggiatoa* apparently derive energy from the oxidation of hydrogen sulfide even though they need organic carbon for growth. Chemolithotrophic species, such as *Thiobacillus*, actually oxidize sulfur as their source of energy. Photosynthetic species such as *Rhodospirillum* use hydrogen sulfide as a source of reducing power. Plants use sulfur by reducing sulfate and incorporating it into the amino acids cysteine and methionine for the formation of proteins. (Photo credits: *Desulfovibrio*, courtesy of J. Eckstein, R. Apkarian, and R. M. Atlas; *Beggiatoa* (×624), courtesy of P. W. Johnson and J. McN. Sieburth, Univ. of Rhode Island/BPS; *Thiobacillus* (×15,360), courtesy of Dr. Jessup M. Shively.)

duction into forms that are used by plants. Mineral deposits and ocean sediments are the primary sources of these elements.

SOIL AND TERRESTRIAL ENVIRONMENTS

The land supports the majority of higher plants and animals. Soil is the essential constituent of all terrestrial habitats. Plants obtain many of their nutrients from the soil. Organisms may live either in the upper layer of the soil or directly on the soil surface. Terrestrial producers manufacture, on the average, from 1 to 10 g dry weight of new organic matter per square meter of earth surface per day.

SOIL COMPOSITION

Soils contain both inorganic and organic components. The mineral (inorganic) portion con-

TABLE 27.2 Microorganisms in the sulfur cycle		
Process	**Some Microbial Participants**	**Comments**
Sulfate assimilation	Almost all organisms	
H₂S release*	Enterobacteria	Facultative anaerobes
	Clostridium, Bacteroides	Strict anaerobes
Sulfur oxidation*	*Beggiatoa, Thiothrix*	Accumulate sulfur
	Thiobacillus, Thiomicrospira,	Produce sulfate
	Rhodospirillum, Chlorobium	Photosynthetic, diverse products
Sulfate reduction*	*Desulfatomaculum, Desulfovibrio*	Strict anaerobes

*Microorganisms are essential for these processes.

sists of weathered rock. Soil particles in descending order of size are called rock, gravel, sand, silt, and clay. These particles, when loosely arranged rather than compacted, leave spaces for water and air to penetrate soil. The mineral content of any soil depends on that of the rock from which it is derived. Soil is built up either from below, as the underlying rock crumbles, or from above as waterborne sediment is deposited in valleys and riverbeds. Plant cover protects soil from erosion and maintains its thickness.

Most soils have definite layers or zones (Figure 27.5) with varying amounts of organic matter. **Subsoil,** which is principally inorganic, contains few nutrients, and supports little life. **Topsoil,** the fertile zone, has organic matter and a significant microbial population that recycles nutrients. Subsoil matures into topsoil as it accumulates humus. **Humus** is amorphous nutrient-rich organic matter composed of undecomposed plant residues. In undisturbed temperate grasslands, plant growth will annually add more organic matter to topsoil than the decomposers will break down. The topsoil layer thus becomes thicker. Humus lightens up the soil structure so plants are able to form larger root systems. It also improves soil's water-holding capacity, and supplies organic matter to support a rich and varied population of microbial decomposers (Table 27.3).

SOIL MICROORGANISMS

When microbial populations in soil expand, soil fertility improves. This relationship is explained by what we already know of the carbon, nitrogen, and sulfur cycles. Microorganisms, by recycling wastes, provide plant nutrients. The greater the number of microorganisms, the stronger the plants will be. The more plants produce on the soil, the more wastes will be left to nourish the recyclers.

In a fertile garden soil bacteria may number up to 2.5 billion per gram. These bacteria include

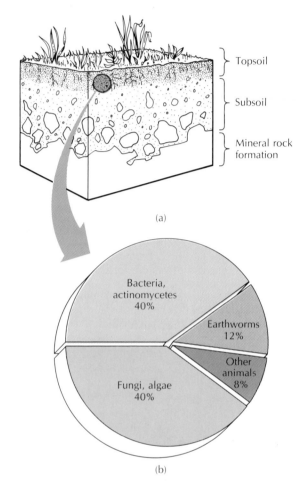

FIGURE 27.5   The Soil
(a) A typical soil profile shows a thin layer of topsoil that contains decomposing organic matter and supports all terrestial forms of life. Beneath the topsoil lie layers of infertile subsoil and rock formations. (b) This diagram shows the proportion of various life forms present in a typical soil sample once plants have been removed.

aerobic spore-formers, anaerobic organisms, heterotrophs that decompose a variety of substrates, and lithotrophs (Figure 27.5). Hundreds of thousands of fungi are also present and aid in decom-

TABLE 27.3 Microorganisms in soil			
Type of Organism		Roles	Numbers/Gram
Bacteria	Autotrophs/heterotrophs Aerobes/anaerobes Cellulose digesters Protein digesters Sulfur oxidizers Nitrifiers, denitrifiers Nitrogen fixers	Decomposers, mineral cycles, plant symbiosis	2.5 billion (direct count) 15 million (dilution plate)
	Actinomycetes	Decomposers	700,000
Fungi	Mostly mycelial forms	Decomposers, mycor- rhizal associations	400,000
Photosynthetic	Algae, diatoms, cyano- bacteria	Producers — minor importance; limited to lighted surface	50,000
Protozoa	Flagellates and amebas	Consumers — mainly consume bacteria	30,000

position. Algae inhabit the surface of bare soil. Protozoans in soil feed on the soil bacteria.

THE RHIZOSPHERE

The zone immediately surrounding the root network of a plant is termed its **rhizosphere.** Dense populations of microorganisms utilize the organic substances the plant secretes from its roots. **Root nodules** formed by bacteria on legumes are a major site of nitrogen fixation in the soil (refer to Figure 1.3). Legumes excrete organic nitrogen into the soil to nourish surrounding organisms. Also, if left to decompose in the soil, they make a long-range contribution to its nitrogen reserves.

The soil microorganisms in turn nourish the plant by carrying out their part in the nutrient cycle right on the spot. The **mycorrhizae,** symbiotic associations of fungal mycelia with the roots of certain plants, are a part of the rhizosphere (see Box 27.1). So are nitrifiers, sulfur bacteria, and the general group of decomposers.

Soil productivity is dependent on soil type, soil usage, temperature range, and rainfall. The most productive areas in the world are all now close to maximum human utilization. Many soils are farmed with no regard for maintaining humus to support the microbial populations on which sustained fertility depend. Rapidly growing human populations create pressure to develop and farm marginal lands, often leading to catastrophic erosion. Agricultural methods that preserve and enhance microbial populations can help. To continue to disregard nutrient and topsoil conservation would be disastrous.

BOX 27.1
Mycorrhizae

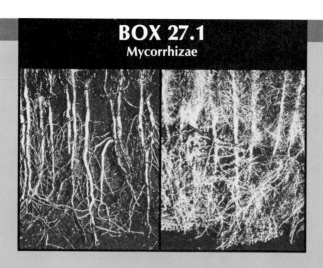

The vast majority of economically important crop plants benefit by a symbiotic relationship with soil fungi. These symbioses are called mycorrhizae or fungus roots, and they arise from colonization of the root hairs of plants by beneficial soil fungi. Plants with mycorrhizae show increased vigor and more rapid growth. The reason is as follows: As plants grow, the limiting factor in their growth is often their ability to absorb minerals such as phosphate, zinc, copper, molybdenum, and ammonium, as well as water. The larger the root surface area a plant has, the more effectively it will absorb nutrients. Root hairs associated with fungal hyphae have a greatly increased surface area. The hyphae, in fact, serve to expand the plant's root system.

Thriving pine, hemlock, oak, birch, and poplar trees develop a fungus mantle over the surface of their feeder roots. To prevent various microbial diseases, tree nurseries often start seedlings in sterilized soil. Later, they inoculate the soil with the seedlings' appropriate mycorrhizal species. Inoculated seedlings develop larger root systems and grow several times as fast as uninoculated ones. The advantage continues after the seedlings are moved to their final planting site in natural soil, as the mycorrhizal fungus is transplanted along with the tree.

In some annual crop plants such as corn or wheat, the fungal hyphae actually penetrate the roots, helping the plant to be better able to extract water and scarce nutrients such as phosphate from poor soil. The practical impli-

cations of this relationship were largely ignored as a means of improving crop fertility until recently. However, population pressures are forcing agriculture to marginal soils. At the same time, inorganic fertilizers are becoming scarcer and more expensive. Agriculturalists are now attempting to maximize crop yields by encouraging the beneficial fungi. Field experiments on marginal, nutrient-poor soil, using no artificial fertilizers, have demonstrated 20% or greater yield increases in potatoes, corn, wheat, and barley when either the soil or the plants were inoculated with fungi. Working *with* microorganisms, not against them, may provide one part of the answer to the world's food crisis.

(Photo courtesy of D. H. Marx, USDA.)

THE FRESHWATER ENVIRONMENT

As we saw in Chapter 1, the first life forms evolved in water. Water masses currently cover approximately 70% of the planet's surface. Thus aqueous environments, both fresh and saltwater, are dominant features in global ecology. Micro-organisms make key contributions to all phases of aquatic biology.

FRESHWATER ORGANISMS

Microorganisms inhabit the water at all depths (Figure 27.6). The microscopic floating organisms are called **plankton.** They are bacteria,

FIGURE 27.6   Freshwater Environments
The warm, sunlit limnetic zone of a lake supports photosynthetic microorganisms and the creatures that feed on them. As detritus from above settles to the dark bottom, it becomes food for various animals, as well as for many heterotrophic bacteria and fungi.

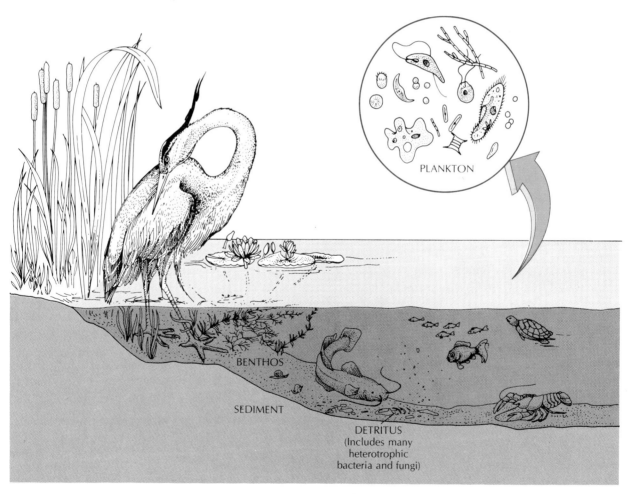

PLANKTON

BENTHOS

SEDIMENT

DETRITUS
(Includes many
heterotrophic
bacteria and fungi)

cyanobacteria, algae, protozoans, and the larvae of multicellular animals. Swimming animals are called **nekton.** Plants and animals either attached to the bottom or found in the sediments are classed as **benthos.** Organic wastes and bodies of dead organisms (detritus) settle slowly toward the bottom. This detritus is the prime nutrient source for bottom-dwelling organisms.

There is an enormous variety of freshwater microbial species. A brief review of Chapters 4 and 5 will remind you of the many bacteria, cyanobacteria, and eucaryotic microorganisms that live in water. Budding and filamentous bacteria coexist with the more familiar pseudomonads and vibrios. Fungi live on the surfaces of the aquatic plants. Protozoans — ciliates, flagellates, and amebas — either swim freely or attach to surfaces. At or near the surface cyanobacteria and algae thrive. At greater depths, where the light is dim, photosynthetic bacteria live in limited numbers. Sulfate-reducing bacteria, methane producers, and clostridia live in anaerobic sediments.

PRODUCTIVITY

The primary production of biomass in water is effectively restricted to shallow depths. Photosynthesizers require light, and light penetrates only the top layers of deep bodies of water. The light is absorbed even by clear water, while suspended particles such as silt further reduce water's transparency. Productivity is greater along shorelines (the **littoral zone**) and in surface waters (the **limnetic zone**).

Temperature also affects productivity. In freshwater, particularly in the temperate zones, the coming of summer spurs high productivity, while winter reduces productivity to near zero. Shallow ponds or pools are often warmer than deeper waters, and are thus more productive. The water of a river entering the sea will often be warmer than the ocean water, and productivity will be high in the mixing zone.

Photosynthesizers also need dissolved carbon dioxide, nitrogen, and phosphate for growth.

Water dissolves small but usually adequate amounts of carbon dioxide from the air. In freshwater the nitrogen-fixing cyanobacteria may supply organic nitrogen. Fixed nitrogen compounds are also washed into streams or lakes as rainwater runs off the land. Fertilizers, agricultural animal wastes, and sewage are major nitrogen sources.

Phosphate enters water largely from mineral sources such as weathered rocks. Biological processes tie up much of the phosphate in highly stable calcium salts forming shells, bones, and teeth. These structures decompose very slowly. The phosphate tends to accumulate in sediment, "trapped" because it is so slowly released. Phosphate is frequently the most scarce nutrient in water; total primary production is usually limited by the amount of phosphate available. Inputs of phosphate, such as might occur due to runoff from a heavy rainstorm, promote sudden bursts of planktonic growth called **blooms.** Some phosphates enter water from fertilizer runoff, and domestic sewage contributes phosphates via detergent products and human waste.

EUTROPHICATION

The word eutrophic literally means "well-fed." **Eutrophication** is the nutrient enrichment of natural waters such as rivers and lakes, usually from artificial sources such as sewage or agricultural fertilizers and manure. Eutrophication frequently results in excessive growth of algae. If a body of water is very well supplied with nitrogen compounds and phosphates, rapid microbial growth will in turn speed up such natural events as accumulating mud, decreasing depth, and changing fish population that typically occur very slowly in all lakes. A eutrophic lake will be rapidly, possibly irreversibly altered (Figure 27.7).

In freshwater, as we saw, excess nitrogen and phosphorus may come from either agricultural or domestic wastes. When these nutrients are plentiful, algae multiply; when the nutrients are used up, the algae die suddenly. A wilderness

(a)

(b)

FIGURE 27.7 Lakes and Pollution
(a) A wilderness lake contains a relatively small amount of dissolved nutrient
material. Algal growth is slight, and the water stays clear during the summer for
there is little organic matter to undergo decomposition. (Courtesy of Jen and Des
Bartlett/Photo Researchers, Inc.) (b) A lake undergoing eutrophication, however,
contains windrows of accumulated dead algae and froth. This material will eventually
sink and be decomposed by microorganisms, thereby exhausting the dissolved
oxygen supply of the lakewater. (Courtesy of Ron Curbow/Photo Researchers, Inc.)

lake, as you might guess, is unlikely to receive
excess nutrients. Its algal growth is slight, and
its water stays very clear. The few algae are effectively consumed or decomposed when they die,
so the lake's bottom sediments build up slowly
if at all, and it stays deep.

A lake receiving runoff from rich soils will
have a different pattern from that of a wilderness
lake. Each rainfall will bring a surge of both nutrients and sediments. Algae will bloom, then
die. There may be too many dead algae to be
consumed or decomposed in the upper layers, so
a deposit of dead algae will accumulate with the
sediments. There they will decompose very

slowly. Year by year, the lake will become shal-
lower. In this type of natural eutrophication, in
which the time frame for change is measured in
centuries, the lake does not usually lose its animals or become aesthetically objectionable.

Now consider a lake receiving heavy supplements of runoff from chicken farms, beef feed-
lots, or a city. Its eutrophication is rapid (Figure
27.7b). Warm-season algal blooms are massive,
converting the top few feet of the water to a fluid
resembling pea soup. When all the algae die, they
sink, forming a thick, brownish scum on the bottom. Vast numbers of bacteria and fungi work to
decompose this scum, and their metabolism

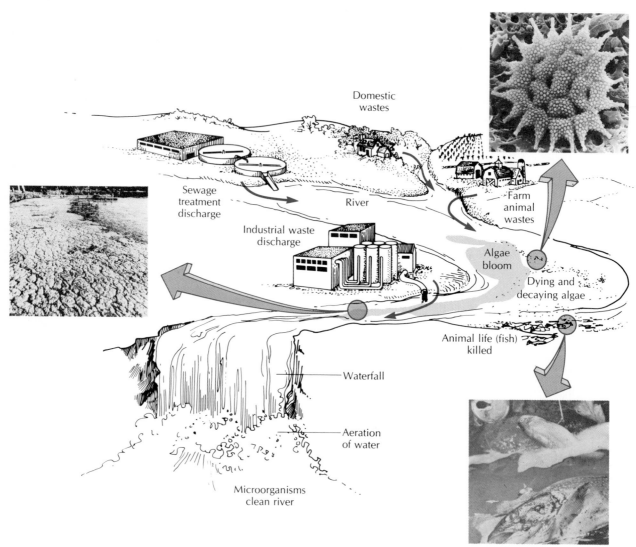

FIGURE 27.8   Pollution and Purification of a River
The water quality of a rapidly moving river may change markedly over the course of a few miles. Nutrient-bearing waste discharges promote a massive algal bloom, while further downstream the decomposition of dying algae deplete the water and may cause fish kills — especially in broader, slowly moving reaches of the river. Yet should the river have an opportunity to reoxygenate by passing over rapids or waterfalls, the increased oxygen supply will enable microorganisms to cleanse the water. (Photo credits: left, Environmental Protection Agency; upper right, Jeremy Pickett-Heaps, University of Colorado; lower right, Grant Heilman, Inc.)

strips the lake water of all its dissolved oxygen. Aquatic animals suffocate and die. Even so, much of the scum remains undecomposed after the microbes exhaust the oxygen. The lake rap-idly becomes unattractive, malodorous, and in-creasingly shallow. Disastrous changes in such a lake can be observed in periods as short as 20 years.

BIOCHEMICAL OXYGEN DEMAND. The more organic matter to be decomposed in a body of water, the more oxygen the microorganisms will need to complete the decomposition process. This oxygen-consuming property of a body of water can be measured; it is called the **biochemical oxygen demand** (BOD). The BOD is determined by taking a water sample, aerating it by shaking to guarantee that it is fully charged with dissolved oxygen, then sealing it and incubating it for 5 days at 20 degrees Celsius. During this period, the microorganisms in the sample will be decomposing the organic material. At the end of the period, the amount of oxygen left is determined. If there was little BOD, most of the oxygen will still be present; but if there was much BOD, little or no oxygen will be left. The BOD method is widely used to evaluate the degree of eutrophication of natural waters, as well as in drinking-water and wastewater technology.

Running water — streams and rivers — become eutrophic too, but because they are more able than lakes to pick up additional oxygen from the air, they cleanse themselves naturally, and may recover. When nutrients are piped into a stream, algae proliferate immediately downstream, then die. If the water flows slowly, the dissolved oxygen supply will be severely depleted and there may be a fish kill. However, if and when the stream becomes re-aerated, by passing over rapids or falls, the oxygen supply will be replenished. The decomposers will then be able to resume their work and the water quality will be somewhat improved downstream (Figure 27.8). In a later section, we will discuss methods of removing contaminants from water.

MARINE MICROBIOLOGY

Some areas of the oceans teem with life. Much of the oceans' vast volume, however, contains few nutrients and supports few living things. Our evolving knowledge of the oceans produces many surprises. One is the discovery of unique, dense, ecosystems around the volcanic rifts deep in the Pacific Ocean.

OCEAN ZONES

Oceanographers consider the oceans as consisting of several zones, as shown in Figure 27.9. Coastal regions are bordered by the **intertidal zone,** or area between the high-tide line and the low-tide line. All types of microorganisms proliferate in this zone, which is covered with water some of the time, and exposed to the air the rest of the time. The macroscopic algae (seaweeds) are a major nutrient source for intertidal microbes.

From the shore, the ocean bottom drops off slowly on the continental shelf. The intertidal zone merges with the **neritic zone,** that portion of the ocean's waters that lie over the continental shelf. The waters of neritic zones contain dense populations of photosynthesizers, which become the food for marine animals of all sorts, including large schools of fish and marine mammals. The neritic zones receive most of their nutrients from freshwater runoff — the rivers that flow into the oceans. Other nutrient supplements come from the sediments as deep ocean waters rise to the surface.

The **oceanic zone** encompasses the deeper waters over the continental slope and deep sediments. Some microbial forms are present and active, but they are much less numerous than in inshore waters.

Photosynthetic activity is limited to the upper 200 meters or less of the oceanic zone, the lighted **euphotic zone.** Here surface organisms live, then die. The detritus sinks very slowly, passing through the dark **aphotic zone** on its way to the bottom, and nourishing consumers and decomposers along the way. Detritus contains much calcium phosphate. When it reaches the bottom sediments or **benthic zone,** it nourishes a population of bacteria and simple invertebrates. Any inorganic components of detritus form sediments that are rich in the essential nutrient phosphorus.

Over periods of years, slow-moving ocean currents transport water masses from one area to another. Rising currents bring water from great depths back to the surface, a process called **upwelling.** This bottom water has spent thou-

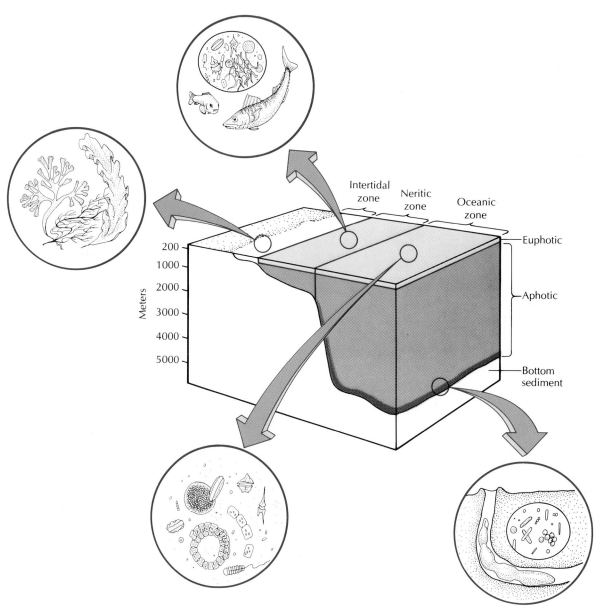

FIGURE 27.9   Microorganisms in the Ocean
Single-celled organisms are the dominant form of life in the oceans. Countless
diatoms, dinoflagellates, and cyanobacteria comprise the major food supply for fish
and marine animals. Bacteria and, to a lesser extent, fungi proliferate throughout the
water column.

sands of years flowing slowly across the ocean bottom, dissolving phosphate from the phosphate-rich sediments. When this enriched water rises to the euphotic zone, it supports dense algal blooms, which in turn nourish vast schools of fish. Upwelling areas are the world's richest fishing zones.

Volcanic fault lines and **rifts** — cracks in the earth's crust — are found in many areas of the ocean floor. At a rift, ocean water sinks deep into the underlying earth's crust. Within the crust, the water is heated to temperatures of 500 degrees Celsius or higher, and dissolves minerals such as iron and sulfur from the crust. It cannot boil because of the tremendous pressures at these depths. At places where this superheated water rises to the surface of the crust (which is the ocean floor), **rift communities,** composed of unique life forms, thrive.

MARINE ORGANISMS

Evidence up till now suggests that the oceans were the cradle of life, and that the first life forms were microorganisms. Microorganisms still remain a predominant feature more of oceanic than of terrestrial life. Essentially all oceanic productivity is microbial (Table 27.4). The major primary producers are cyanobacteria, diatoms, and dinoflagellates. The few true plants and the seaweeds are minor contributors when you consider the oceans as a whole, important though they may be in the salt marsh or tide pool.

TABLE 27.4 Microorganisms in the major oceanic zones	
Zone	**Microbial Contributors**
Intertidal	Producers — cyanobacteria, unicellular and multicellular algae Consumers — protozoa, bacteria Decomposers — aerobic bacteria and fungi
Neritic euphotic zone	Producers — photosynthetic diatoms, dinoflagellates Consumers — foraminiferans, radiolarians Decomposers — aerobic bacteria and fungi
Oceanic euphotic zone	Producers — photosynthetic diatoms, dinoflagellates Consumers — foraminiferans, radiolarians Decomposers — aerobic bacteria
Aphotic zone	Producers — probably none Consumers — protozoa Decomposers — aerobic bacteria
Sediments	Producers — chemolithotrophs, especially in hot springs Consumers — detritus feeders Decomposers — aerobic and anaerobic bacteria

Photosynthetic microorganisms in turn nourish swarms of small marine invertebrates, such as copepods, amphipods, and krill. Then the small fish (and baleen whales) consume these. The bacteria are represented by some producers (the methanogens, photosynthesizers, and other chemolithotrophs) and by many consumers and decomposers (including pseudomonads, vibrios, and photoluminescent bacteria). The deep sediments contain **barophiles,** bacteria that are adapted to intense pressures. All marine organisms either tolerate or require more salts than do terrestrial or freshwater forms.

BIOLOGICAL ACTIVITIES IN THE OCEANS

PRODUCTIVITY.   The photosynthesizers thrive in the euphotic zone: their population density is determined by the availability of nitrogen and phosphorus and by temperature. The richest marine habitats are estuaries (shallow areas, such as Chesapeake Bay, where nutrient-laden freshwater from rivers mixes with saltwater) coral reefs, upwelling zones, and shallow areas on the continental shelf such as the Grand Banks off the New England coast. Oceanic productivity averages 0.5 g of dry organic matter per square meter of ocean surface per day worldwide. Marine ecosystems are responsible for about 44% of the gross annual global production of organic matter.

You may recall from Chapter 7 that in photosynthesis producers not only make carbohydrates but also release oxygen as a byproduct. The cyanobacteria and algae regenerate almost half of the oxygen that supplies the globe's aerobes. In addition, the cyanobacteria fix nitrogen and are the only significant source of "new" amino acids in the vast oceanic zone.

GEOCHEMICAL ACTIVITIES.   The activities of marine microorganisms may result in the accumulation of geologic deposits containing economically important minerals. Deposits of diatom frustules compose diatomaceous earth, and foraminiferan deposits form chalk and limestone. Sulfur domes are formed of elemental sulfur, possibly produced by chemolithotrophic sulfur-cycle bacteria. Petroleum deposits are accumulations of hydrocarbons formed from the oil droplets of innumerable marine algae.

Many areas of the ocean floor are strewn with **manganese nodules.** These are irregularly shaped, metallic globules made up of iron, manganese, and other, rarer, metals. It is believed that microorganisms precipitate the metals from seawater, although the exact mechanism remains to be discovered. Because they contain extremely valuable metals, manganese nodules have significant commercial potential.

RIFT COMMUNITIES.   The discovery of the rift communities is comparatively recent. Scientists have determined that the rift communities are based on a food chain in which chemolithotrophic bacteria are the primary producers. One type of volcanic hot spring typical of the rifts is diagrammed in Figure 27.10. Superheated water rising out of the crust carries dissolved hydrogen sulfide (H_2S) and carbon dioxide (CO_2) gases, which are used by sulfur bacteria in a carbon-fixing reaction:

$$6CO_2 + 6H_2S \longrightarrow C_6H_{12}O_6 + 6S$$

The carbohydrate formed serves as the basic organic nutrient for a food chain. The bacteria nourish dense fields of giant clams, tube worms, crabs, sea anemones, and some fish at depths of 2000 to 3000 meters below the surface. These rift communities are sustained solely by chemolithotrophy. Their uniqueness stems from the fact that they are believed to be the only self-contained communities on Earth that are totally independent of sunlight as an energy source. Recent fossil discoveries suggest that similar communities have existed previously. They developed on earlier sites of sea-floor volcanic activity, now extinct, as long as 95 million years ago. Some researchers compare the environmental conditions in these extraordinary hydro-

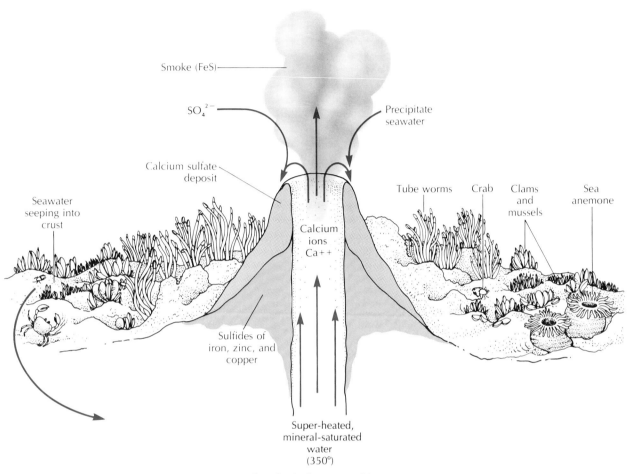

FIGURE 27.10 Volcanic Rifts and Biological Communities
Along a rift zone, seawater seeps through cracks in the newly formed crust and is
heated by hot basaltic rock to temperatures of 350 degrees Celsius or more. Sulfate
from the seawater reacts with iron in the rock to produce hydrogen sulfide and iron
oxides. This superheated solution, heavily charged with sulfides, calcium, and other
dissolved minerals, rises through a vent back toward the ocean floor. As water from
the hot spring meets cold seawater, many inorganic materials precipitate out as
calcium sulfate and ferrous sulfide, building chimneylike structures.
Chemolithotrophic bacteria, both free-living and dwelling in animal tissue, fix carbon
from dissolved carbon dioxide in the seawater. These producers support a rich
variety of bacterial decomposers and animal life, including giant tube worms,
barnacles, clams, and filter-feeding crabs.

thermal communities to the conditions believed
to have been those of the primitive earth. From
studying the rift communities, we may discover
much about the earliest forms of life.

WASTES IN THE OCEANS. Any chemical released
into the soil or freshwater will find its way into
the oceans, if it is not degraded first. Steadily
increasing amounts of human industrial and do-

mestic waste are being dumped into the oceans. Seawater rapidly kills human fecal organisms, and marine organisms decompose many of the simpler organic wastes. Other pollutants of prime concern include persistent materials such as crude oil, nonbiodegradable pesticides, and radioactive wastes. Our waste-disposal practices are based on the assumption that the vast volume of seawater will dilute these wastes until their effect is insignificant. We don't know if this is a safe assumption. We probably won't know until the ocean biosphere shows signs of disruption that we can't ignore.

WATER TREATMENT

A safe source of water is something Americans take very much for granted. We have been able to enjoy clean water because of municipal water treatment, one of the major public-health advances of the century (Figure 27.11). Treated water for drinking, cooking, and washing protects us against outbreaks of waterborne diseases that were once a leading cause of death (Table 27.5).

Water supplies originate from either surface water (lakes, reservoirs and streams) or ground water (wells driven into aquifers). Initially most water supplies are at least slightly contaminated, and require some purification before use.

You will notice that the concepts and techniques covered in this section on water treatment and the next section on sewage treatment tend to overlap. This is inevitable, because one of the things our society is particularly good at is turning pure water into sewage. In fact, the flush toilet has been described as a device that instantaneously turns five gallons of drinking water into five gallons of sewage. Another reality at issue is the fact that one city's discharged sewage, a few miles down the river, may become part of another city's drinking-water supply. Thus, we often have the job of reversing our previous actions, and trying to turn sewage back into pure water.

There are three aspects to water safety. One is selecting and maintaining a clean water source. Another is using effective treatment techniques. A third is running consistent, accurate water-quality tests. Together, these functions guarantee safe water. The safety records of most municipal systems to date have been exemplary.

FIGURE 27.11   Advances in Sanitation, 1890s Style In 1872, New York City introduced an improved method of garbage disposal. Instead of tossing wastes from dumping platforms into local waterways, the city used barges to tow the garbage out to sea. It is clear from this cartoon of the period that bathers did not appreciate the benefits of the new system. Even today, wastes that we believe have been removed a safe distance away may still come back to haunt us — for example, the current toxic-waste contamination of many local water supplies that often originates from supposedly "safe" landfills. (Courtesy of the Bettmann Archive.)

Coney Island

TABLE 27.5 Some waterborne pathogens	
Bacteria	*Salmonella typi*
	Shigella dysenteriae
	Vibrio cholerae
	Campylobacter jejuni
	Escherichia coli
Viruses	Poliovirus
	Hepatitis viruses
	Rotaviruses
	Norwalk virus
Protozoans	*Giardia lamblia*
	Entamoeba histolytica
	Balantidium coli
	Naegleria fowleri
Helminths	Roundworms
	Tapeworms
	Schistosomes

WATER POLLUTANTS

Let's now consider the pollutants that may render water unsafe or unsuitable for drinking. There are four major classes of water pollutants: infectious agents, biochemical oxygen demand, nitrogen and phosphorus, and toxic chemicals. Each has its own effects on our health and that of the organisms with which we share our environment. Municipal water supplies — lakes, reservoirs, and deep wells — are selected because they have low levels of these pollutants. They are usually legally closed to recreation and development to protect their initial water quality as much as possible.

INFECTIOUS AGENTS. Infectious agents enter water supplies almost exclusively with fecal (sewage) contamination. Human feces may carry bacteria that cause typhoid fever, dysentery, and cholera, the viruses that cause hepatitis and polio, and the parasites that cause giardiasis, amebiasis, and the helminthic diseases. Infectious agents were the type of pollutants that water treatment was originally designed to control and those with which it still deals best.

BIOCHEMICAL OXYGEN DEMAND. We saw earlier that decaying organic matter imposes a biochemical oxygen demand (BOD) on a body of water. This occurs because decomposer organisms consume dissolved oxygen in the water as they break down the organic compounds. The actual organic pollutant may be wood fibers from a paper mill, potato peelings from a French-fried potato plant, organic residues from human feces or any other biodegradable material. Water with a high BOD will usually be turbid and may smell or taste objectionable. Since dissolved organic matter absorbs chlorine, the heavier the BOD, the more chlorine will be needed to render the water safe and the worse the water will taste. Municipal water services prefer to use as their sources waters with little BOD, and remove as much of it as possible by water-treatment steps, prior to adding chlorine.

NITROGEN AND PHOSPHORUS COMPOUNDS. The plant nutrients nitrogen and phosphorus occur as pollutants in water supplies receiving agricultural runoff or sewage (treated or untreated). In general, we do not consider these pollutants to be directly harmful to human beings, with the exception that very high levels of nitrate are toxic and render a water supply unusable. However, nitrogen and phosphorus promote algal blooms that increase the BOD. Water dense with algae tends to look, smell, and taste objectionable. Conventional water treatment was not designed to do anything to remove plant nutrients.

TOXIC CHEMICALS   The size of the chemical industry in this country has vastly increased over the past three decades. As a result, more and more chemical contaminants, most of which are byproducts of manufacturing processes and organic solvents, find their way into dumps, and from there they leach into surface and ground waters. Heavy metals such as mercury, cadmium, and chromium are toxic. Organic solvents such as benzene, chloroform, and carbon tetrachloride are both toxic and carcinogenic. Gasoline and fuel oil leaking from rusted, buried tanks create fumes that contaminate wells and are potentially explosive. Persistent organic compounds such as chlorinated pesticides and PCBs (polychlorinated biphenyls) have been shown to have a wide range of toxic effects on the skin and organs.

These contaminants may become serious public-health hazards if they leach through the soil and eventually reach wells and municipal water supplies. Leaching and underground movement of contaminated water is a slow process, difficult to detect and, at the moment, impossible to stop. In fact, we have only recently started to monitor our water supplies for toxic wastes. Conventional water treatment was not designed to remove these substances. Researchers are scrambling to develop methods to cope with these problems, which seem likely to increase at a rapid rate.

WATER-TREATMENT METHODS

As we stated, conventional water treatment is directed at infectious agents and BOD. Removal of nitrates, phosphates, and toxic chemicals requires special techniques (Figure 27.12). These techniques are not often used because they are difficult and costly, or, in some cases, have not yet been developed to the point where they are practical to use on large volumes of water.

The first step in all forms of water treatment is to collect the water into a reservoir where it is allowed to stand so that some silt and BOD will settle out. If the water remains very cloudy, it may then be drawn into a settling basin and mixed with alum. The alum binds the particulate matter to form a jell-like precipitate, called a **flocculation,** in which clay particles, organic

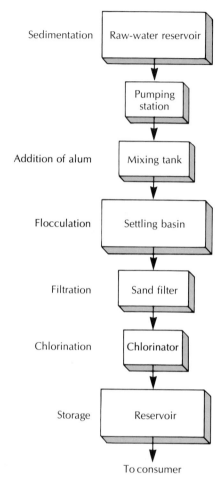

FIGURE 27.12 Water Purification
In most municipal water-treatment systems, the water is cleansed by flocculation and filtration and finally chlorinated before being pumped to consumers.

debris, bacteria, and viruses are trapped and dragged to the bottom. The clarified water is then passed through a sand filter that removes almost all microorganisms that escaped the flocculation treatment. Sand filters are periodically backflushed to rinse out accumulated detritus. The last step in water treatment is chlorination, which kills any infectious agents that escaped removal in the earlier steps. The amount of chlorine must be actively monitored to provide a slight excess of **residual chlorine** (Chapter 23). An adequate residual chlorine level is particularly important to control *Giardia lamblia* cysts, which are more resistant to chlorine than the ordinary fecal bacteria. The fully treated water is now free of pathogenic agents and is routed to the users through a distribution system of water mains. Unfortunately, in some older cities, water mains have become a very unreliable link in the system (see Box 27.2).

BOX 27.2
Crisis Under the Streets

In the older major cities of the United States, such as New York, Boston, and Chicago, the water and sewerage systems were built over 100 years ago. They are not only old, but are deteriorating rapidly.

Let us use New York City as an example, with the understanding that the situation is equally bad in a number of other cities. New York City has 6200 mi of water mains of varying sizes and materials. Located under busy streets, they are subjected to continuous stresses from traffic vibration, soil compaction, and excavation of foundations. Aging cast iron and concrete mains crack, and steel mains corrode. Water mains break daily in New York. Breaks are also increasing — from 415 in 1969 to almost 600 in 1979.

Because water within the mains is supplied at high pressure, these breaks are not likely to permit contaminated ground water or sewage to enter and mix with the drinking water. Thus they are more of an inconvenience than a health threat. However, another class of plumbing accident is more hazardous. Cross-connections are illegal plumbing connections between water lines and cooling or wastewater lines. These illegal conditions occur in large numbers because there are not enough inspectors to enforce plumbing codes and to make rigorous inspections. A cross-connection often lets wastewater siphon back into the drinking-water source. In 1975, poisonous chromates from an air conditioner siphoned into a drinking-water source and caused 20 people to be hospitalized. In 1978, public concern over legionnaires' disease in the Murray Hill area of Manhattan provoked a systematic plumbing inspection of all the buildings in the area. Although *Legionella* was not found, plenty of other problems were. The area drinking water was found to be heavily contaminated with *E. coli*, indicating sewage leakage. Over 50 illegal cross-connections were discovered within a two-block area. One wonders what would be found if this detailed inspection were extended throughout the city.

Deteriorating equipment, broken mains, and illegal plumbing connections — a quiet crisis, but a massive one. These are all correctable problems, but unfortunately, correcting them will require huge sums of money. It is difficult to foresee a source for these sums in today's hard-pressed urban budgets.

WATER-QUALITY TESTS

Many tests for water safety are directed at detecting infectious agents. These tests detect fecal contamination by looking for, not the rare pathogens, but the common fecal coliform bacteria. Fecal coliform bacteria are reliable indicators of untreated sewage contamination. Municipal water supplies must conform with standards set by the Environmental Protection Agency for testing methods and drinking-water quality. Most laboratories use a **membrane-filter technique** to enumerate coliforms (refer to Figure 6–14). A 100-ml water sample is drawn through the sterile filter, which is then incubated on an absorbent pad soaked with a differential medium. After bacterial growth has occurred, coliform colonies are a deep purple color and have a greenish sheen. No more than one coliform per 100 ml of water after treatment is acceptable. Excessive counts usually indicate a system failure, or a temporary need to increase the chlorine because of a heavier organic burden in the water source. An older method, the multistep tube fermentation technique or **most probable number** (MPN) **method** may also be used. Special tests to detect other fecal indicator species, the fecal streptococci and clostridia, are used in some laboratories. Some experts feel that these give a more accurate reflection of the degree of fecal contamination.

WASTEWATER TREATMENT

Wastewater treatment is an industrial technology, developed to collect sewage or wastewater and remove as many pollutants as possible from it, before it is discharged back into the environment. Sewage treatment is a form of applied microbiology, in that it uses microorganisms to do a useful job. In fact, people can't treat sewage, only microbes can!

Sewage consists of several components from different sources. Human urine contains few bacteria; fecal matter, in contrast, contains high BOD and billions of organisms per gram, many of which are potentially pathogenic. Toilet wastewater is called **black water;** it is highly hazardous. The second component of sewage is **gray water,** consisting of bathwater, dishwater, and water used for washing clothes; it contains some organisms of human origin, but its potential hazard is less. The third component of sewage is storm runoff from gutters, storm drains, and culverts. It carries few human pathogens but may be high in plant nutrients and BOD. A fourth component of sewage may be industrial wastes from diverse sources such as electroplating plants, paint factories, canneries, and fish-processing plants, to give a few examples. Some industries contribute high BOD loads, others discharge salts or metals, and yet others release toxic organic chemicals. Industrial wastes may place severe strains on the functioning of the treatment facility. Some municipalities refuse to handle industrial wastes, requiring that industries assume the responsibilities and costs of treating their own wastewater.

PRIMARY TREATMENT

Let us now look at the steps in sewage treatment. Sewage treatment may include up to three stages, primary, secondary, and tertiary (Figure 27.13). Primary treatment first screens collected sewage to remove large floating objects from the fluid. Then the raw sewage passes to primary settling tanks. In these tanks, over a period of about one hour, half the organic matter settles out as primary sludge. This sludge is removed and dewatered. Afterward, this primary sludge may be handled in several ways. It may be directly dumped into the ocean from barges or buried in a sanitary landfill. Alternatively, it may be placed in an **anaerobic sludge digester** for further treatment. In the digester, anaerobic bacteria decompose the organic materials, releasing methane gas as a major end product. The methane gas is usually collected and burned as a source of energy to run portions of the treatment facility.

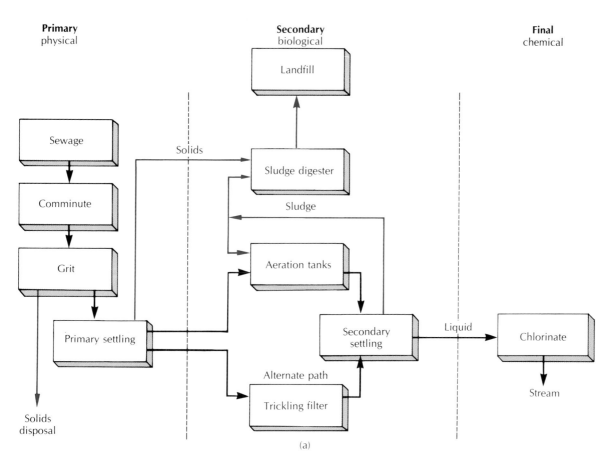

Primary
physical

Secondary
biological

Final
chemical

(a)

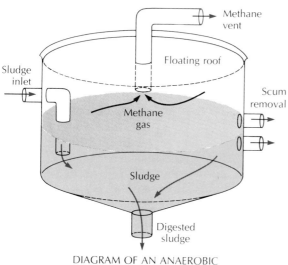

DIAGRAM OF AN ANAEROBIC
SLUDGE DIGESTER
(b)

FIGURE 27.13 Sewage Treatment
(a) Primary treatment removes as much solid material as possible from the sewage. The effluent, which still contains dissolved or suspended matter, is then biologically treated during secondary treatment. Microorganisms growing in aeration tanks either break wastes down to carbon dioxide or grow at their expense, flocculating out in large masses that can be easily removed by settling. The secondary effluent is chlorinated and discharged. Sludge and solids are burned, digested, composted, or used as landfill. (b) Diagram of an anaerobic sludge digester.

The fluid from primary treatment, called **effluent,** still contains about 60% of the original BOD, microorganisms, and other pollutants initially present. Where primary treatment is the only treatment, as in many systems in older cities, the effluent is chlorinated to kill microorganisms, then discharged. Although the chlorinated effluent is no longer an infectious hazard, it is still heavily loaded with BOD. Primary effluent has a negative aesthetic impact, and it has severe eutrophying effects on the body of water into which it is discharged. The significance of these problems can be appreciated if we note that a large municipal sewage-treatment plant produces many millions of gallons of effluent and tons of sludge per day.

SECONDARY TREATMENT

Across the country, federal grant money has helped communities to upgrade their existing primary-treatment programs by adding secondary treatment. The advantage of secondary treatment is that it rapidly removes dissolved organic matter, eliminating up to 90% of the BOD remaining after primary treatment. In secondary treatment, aerobic growth conditions are created in the sewage. This encourages the development of a dense mixed microbial flora, which carries out an accelerated decomposition of the BOD. Most microorganisms of human origin are eliminated by the brisk microbial competition. Decomposition can occur both aerobically and anaerobically, but it is more rapid and more complete when oxygen is available. In addition, the end products of aerobic decomposition are not particularly smelly. This is important because aerobic secondary treatment requires large holding tanks exposed to the air.

The **activated sludge process** is a method of secondary treatment that is now favored over the older **trickling filter** approach, because it can handle larger volumes of sewage faster. In this technique the primary effluent is piped to a sequence of aeration tanks (Figure 27.14a) where it is mixed with activated sludge, composed of sludge from previous batches containing floc-

cules (gelatinous aggregates) of the decomposer population — bacteria, fungi, and protozoans (Figures 27.14b–e). The sludge microflora grows rapidly on the organic molecules in the effluent with a minimum of odor. After only a few hours, almost all the BOD has either been broken down to carbon dioxide or incorporated in microbial biomass. The now much larger volume of activated sludge is separated from the secondary effluent in a secondary settling tank. The sludge is relatively odorless and noninfective. If it does not contain large amounts of salt or toxic substances, it may be dried and sold as fertilizer. Alternatively, most of the activated sludge may be disposed of by whatever means are used for the primary sludge. A portion of the sludge is reserved as a "stock culture" and piped back to the aeration tanks to inoculate subsequent batches of effluent.

The secondary effluent is low in BOD, and most of the microorganisms have either been removed by protozoan predation or entrapped in the floccules and retained in the sludge. The effluent still, however, requires chlorination before discharge. It will still contain most of any plant nutrients and toxic wastes it had before treatment. Its environmental effect on the receiving water is much less disruptive than that of primary effluent, but still may be significant if a relatively large amount of effluent is discharged into a relatively small river or lake.

Secondary treatment is sensitive to disruption by large amounts of toxic industrial wastes, sudden reductions in the volume of waste, or heavy rainstorms. These disruptions unbalance the process by washing out or killing off the activated sludge "stock culture." When this happens, the treatment plant's waste-consuming efficiency will fall off temporarily until the population re-establishes itself.

TERTIARY TREATMENT

Some communities choose to spend the money to carry out complete sewage treatment, yielding drinking-quality water as the end product. To achieve this goal, a group of additional chem-

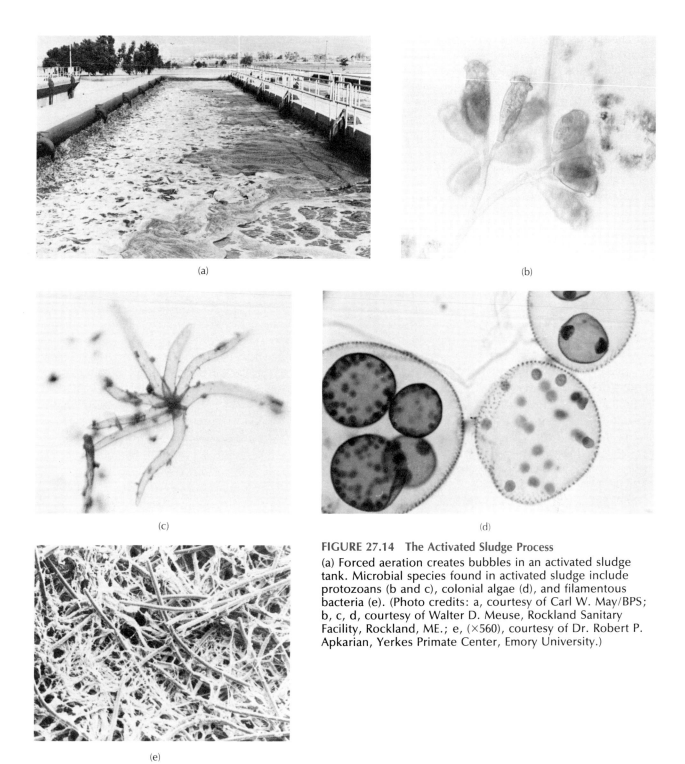

FIGURE 27.14 The Activated Sludge Process
(a) Forced aeration creates bubbles in an activated sludge
tank. Microbial species found in activated sludge include
protozoans (b and c), colonial algae (d), and filamentous
bacteria (e). (Photo credits: a, courtesy of Carl W. May/BPS;
b, c, d, courtesy of Walter D. Meuse, Rockland Sanitary
Facility, Rockland, ME.; e, (×560), courtesy of Dr. Robert P.
Apkarian, Yerkes Primate Center, Emory University.)

ical treatments, collectively referred to as tertiary treatment, must be used to remove all
chemical contaminants. To remove the nitrogen
and phosphorus contamination, algae may be allowed to grow. After they have incorporated all
the plant nutrients available into their cells, the
algae are then harvested. Activated charcoal filters may be used to remove traces of metals or
nonbiodegradable organics. Resins that absorb
ions are used to remove inorganic ions, especially sodium and metals. In some desert communities where water supplies are critical, tertiary treatment is essential so that the final
effluent can be directly recycled for domestic
use. Other communities invest in tertiary treatment because they value the environmental
quality of the lake or stream that is used to receive the effluent, and wish to avoid any ecological damage or eutrophication.

SMALL-SCALE SEWAGE TREATMENT   Many rural
homes have septic tanks and drainage fields to
handle their sewage (Figure 27.15). The **septic
tank** is a primary treatment receptacle in which

the sludge settles out and decomposes anaerobically. The effluent flows on out of the tank
into the **drainage** (leaching) **field,** a network of
buried pipes laid in gravel, where it is filter-
purified. Self-contained composting toilets have
also been designed. They are used in areas where
homes are built on bedrock or where, for other
reasons, there is inadequate drainage to install a
septic system.

MINES AND METAL RECOVERY

A special problem in wastewater treatment often develops around mines. The scrap heaps or
tailings left where coal or metals have been
mined can be a serious source of water pollution.
Rainwater and microbial action slowly leach
strong acids or metals out of scrap piles in toxic
amounts. The contaminated water, if uncontrolled, may kill all life in nearby surface waters.
Microbial mining is a form of controlled leaching that employs the mineral-cycling abilities of
microorganisms. It has been developed as a way
of recovering valuable metals from low-grade

FIGURE 27.15   The Septic Tank and Leaching Field
The septic tank system is often used for sewage disposal in rural areas. Anaerobic
decomposition in the tank breaks down sludge, while effluent is purified by trickling
through a leaching field of gravel and sand into the soil.

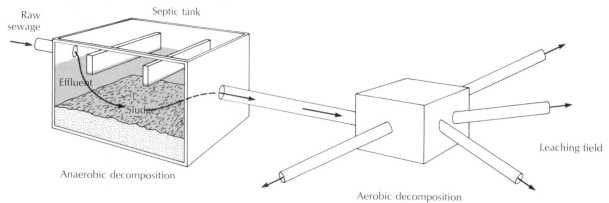

ores and wastes in which the metals are found as insoluble metal sulfide compounds. Bacteria, such as *Thiobacillus thiooxidans,* carry out chemical reactions that convert sulfide ions to sulfate ions. This process solubilizes the metal ions, which are leached out in the form of sulfate compounds. The leachate is collected and the metals purified. Copper (about 10% of current U.S. production) and uranium are currently being profitably extracted in this way, starting with low-grade ores that otherwise would not be economical to process.

SUMMARY

1. Microorganisms occupy all possible ecological niches. They perform unique environmental roles in soil and water. The planetary life-support system for higher plants and animals is completely dependent on microbial contributions.

2. Elements move in cycles, in which they are converted from inorganic to organic form (fixed) and then returned to the inorganic form. Producer organisms fix carbon derived from atmospheric carbon dioxide into new biomass. This material in turn feeds consumers. The leftovers are used by decomposers so that all carbon returns eventually to the atmosphere as carbon dioxide.

3. Nitrogen, as N_2 gas, is removed from the atmosphere by nitrogen-fixers and converted into amino acids, useful to all forms of life. Ammonification, the breakdown of nitrogenous wastes, yields ammonia. Certain chemolithotrophic bacteria use ammonia as an energy source for growth and, in the process (nitrification), oxidize it to nitrate, the prime nitrogen source for plants. Nitrogen returns to the atmosphere as nitrogen gas when certain bacteria metabolize under anaerobic conditions.

4. Sulfur is used by most organisms in the form of sulfate. In metabolism, sulfur is converted to a reduced form. Decomposition releases sulfur as free hydrogen sulfide gas. This gas is reoxidized to sulfur and sulfate by chemolithotrophic bacteria.

5. Terrestrial environments are characterized by an interaction between soil and air. Thirty percent of the globe is covered with soils of varying degrees of fertility. Soil fertility depends on the same elements needed in water, and these elements are converted to useful forms largely by microorganisms. A soil's agricultural potential improves with increasing numbers of microorganisms and humus supplies to feed them.

6. The greatest biological activity is found in the rhizosphere, the soil immediately around the roots of plants, in which mutually beneficial symbiosis among plants and soil microorganisms contributes to fertility.

7. Freshwater and saltwater constitute the largest ecosystems. The main producer organisms are the floating microscopic plankton. Some consumers are microscopic; they become food for swimming animals such as fish and whales. Much life in deeper water is found in the bottom sediments. The benthic organisms break down all undigested detritus, completing the cycling of nutrients.

8. Aquatic productivity varies greatly from area to area. Photosynthetic organisms need light, nitrogen, and phosphorus, and are limited by whichever requirement runs out first. Human activities contribute extra nutrients that overstimulate the productivity of some water masses, with unfortunate results. Lakes become eutrophic when too much productivity occurs.

9. Over long periods of time, aquatic microorganisms have produced deposits of important minerals and petroleum.

10. The hydrothermal vents on the ocean floor support unique rift communities independent of solar energy and based on the primary production of sulfur-oxidizing chemolithotrophs.

11. Water may be polluted by infectious agents, organic matter, nitrogen and phosphates, and toxic wastes. Water-treatment systems remove pollutants and prepare water for safe use. Water-quality tests evaluate the purity of water supplies.

12. The goal of sewage treatment is to remove pollutants from used water before it is released into other surface waters. Primary treatment separates the solids from the fluids. Secondary treatment re-

moves most of the dissolved organic matter. The sludge and the effluent are treated to eliminate any chance of their spreading infection. Tertiary treatment produces drinkable water. Microbial mining is used to recover valuable minerals from low-grade ores and wastes.

DISCUSSION QUESTIONS

1. List the inorganic forms of carbon, nitrogen, and sulfur; list as many organic forms as you can.

2. Explain the contributions of microorganisms to soil fertility.

3. Where does a human being fit in the carbon, nitrogen, and sulfur cycles?

4. Describe the steps in the eutrophication of a lake; the repurification of a contaminated stream.

5. Explain the ecological limitations on aquatic productivity. What direct effect do these have on human interests?

6. Describe how the hydrothermal rift communities develop using the nutrient and energy sources available.

7. Become familiar with your community's water- and sewage-treatment facilities. How modern and complete are they? What, if anything, should be done to update them?

ADDITIONAL READING

Alexander M. *Introduction to soil microbiology.* 2nd ed. New York: Wiley, 1977.

Atlas M, Bartha R. *Microbial ecology: fundamentals and applications.* Reading, Mass.: Addison-Wesley, 1981.

Edmond JM, Von Damm K. Hot springs on the ocean floor. *Sci Am,* 1983; 247: 78–93.

Ehrlich H. *Geomicrobiology.* New York: Marcel Dekker, 1981.

Greenberg AE, Trussell RR, and Clesceri LC, eds. *Standard methods for the examination of water and wastewater,* 16th ed. Washington, DC: Am. Public Health Assoc., 1985.

Laws EA. *Aquatic pollution.* New York: John Wiley, 1981.

Ourisson G, Albrecht P, and Rohmer M. The microbial origin of fossil fuels. *Sci Am* 1984; 251: 44–51.

Reinheimer G. *Aquatic microbiology.* 2nd ed. London: John Wiley, 1980.

Rueble JL, Marx DH. Fiber, food, fuel, and fungal symbionts. *Science,* 1979; 206: 419–422.

Sieburth JM. *Sea microbes.* New York: Oxford University Press, 1979.

28

FOOD, DAIRY, AND INDUSTRIAL MICROBIOLOGY

CHAPTER OUTLINE

MICROBIOLOGY OF FOODS

Microbial Roles in Food Production
Food Handling
Food Spoilage

MICROORGANISMS IN INDUSTRY

Microorganisms Used in Industry
Production Methods in Industrial Microbiology
Industrial Microbiology at Work

One of humanity's great achievements has been learning to manipulate our environment. Nowhere is our success more evident than in the ways we work with and control microorganisms. If we were to choose just one area to demonstrate the pervasiveness of microbiology in our daily lives, it might be in what we eat and drink. We have learned, through trial and error, to understand the ways in which microorganisms interact with foods. Desirable microorganisms are encouraged to grow, with beneficial effects. Harmful species are excluded, to prevent spoilage and disease. In the first half of this chapter we will examine food production techniques that use beneficial microbial species, food preservation methods, food spoilage due to uncontrolled microbial growth, and proper food handling and food storage methods.

In the 20th century, we have also awakened to the extraordinary versatility of microorganisms as miniature "chemists." We have developed methods that employ microbes for industrial production of many highly desirable chemical products, including medicines. The second half of this chapter is devoted to an overview of industrial microbiology.

MICROBIOLOGY OF FOODS

It would be possible, but certainly not easy or pleasant to do without the foods that microorganisms help produce. Let's see. . .no toast; no butter; no coffee, tea, or cocoa for breakfast; no cheese, salami, pickles, or olives for lunch; no yogurt for a snack; no alcoholic beverages (or even root beer); no soy products; no sour cream; nothing flavored with vanilla or chocolate. . . . Well, you could survive, no doubt, but eating would be mighty dull!

Fresh fruits and vegetables, fresh meats and fish, and fresh milk are among our most wholesome and nutritious foods. Microorganisms like our foods as well as we do, and grow rapidly on the nutrients in them, producing waste products. Some of these waste products are useful, either because they act as natural food preservatives or because they are flavoring essences. For example, the lactic acid produced by lactobacilli in cultured sour cream or yogurt acts as a natural preservative because it stops the growth of other, potentially more harmful organisms. The microbes that grow in cheeses and on ripening vanilla beans contribute strong flavors most people find delicious. Of course, we may hold differing opinions on whether the flavor of a very strong cheese is pleasant or not. There is a very fine line between microbial ripening and microbial spoilage, which produces very unpleasant flavors and potential toxicity. Let's first consider the beneficial roles of microorganisms in food production.

MICROBIAL ROLES IN FOOD PRODUCTION

Several groups of microorganisms are employed in food production. One major group is the lactic acid bacteria, which produce lactic acid during fermentation. This group includes genera such as *Lactobacillus, Leuconostoc, Streptococcus,* and *Pediococcus.* Also important are the yeasts, primarily *Saccharomyces,* which produce ethyl alcohol as a fermentation byproduct. Certain molds, such as *Penicillium, Aspergillus,* and *Mucor* have a role in curing ham and cheeses and in making fermented foods in the Orient. Some common microbial food products are listed in Table 28.1.

CEREAL PRODUCTS AND BREADS. Plants yield foods in the form of starchy, nutritious seeds and roots. Raw starch, however, is hard for our di-gestive systems to handle. Thus, we must alter raw starches to render them more digestible, usually by cooking. The earliest grain-raising cultures roasted their whole grains. Later, they ground the grain into flour, moistened the flour, and baked it. Some batches of moist dough must have been left in a warm spot a little longer than usual, for they were found to bubble and "raise." People observed that this raising or leavening vastly improved both the flavor of the mixture and its texture, although they did not know that the leavening was caused by microorganisms. Leavening became a general practice and caught

TABLE 28.1 Some common microbial food products		
Food	Raw Ingredient	Microorganism
Bread, rolls	Wheat flour	*Saccharomyces cerevisiae*
Sourdough bread	Wheat flour	*Lactobacillus sanfrancisco*
Pumpernickel	Rye flour	Lactic-acid bacteria
Poi	Taro root	Lactic-acid bacteria
Butter	Butterfat	*Streptococcus cremoris*
Sour cream	Fresh cream	*Streptococcus lactis*
Yogurt	Milk	*Lactobacillus bulgaricus*
Acidophilus milk	Milk	*Lactobacillus acidophilus*
Cheeses	Milk	Lactic acid bacteria, molds
Pickles	Cucumbers	*Pediococcus cerevisiae*
Sauerkraut	Cabbage	*Leuconostoc mesenteroides*
Olives	Olives	*Lactobacillus plantarum*
Kimchi	Cabbage, other	Lactic-acid bacteria
Cocoa	Cocoa beans	*Candida krusei*
Coffee	Coffee beans	*Saccharomyces* spp.
Tea	Tea leaves	Mixed bacteria, yeasts
Vanilla	Vanilla beans	*Leuconostoc mesenteroides*
Soy sauces	Soy beans	*Aspergillus orzyae,* others
Dry sausage	Pork, beef	*Pediococcus cerevisiae*
Semi-dry sausage	Pork, beef	*Pediococcus cerevisiae*
Ham, countrycure	Pork	Molds
Beer, ale	Grains	*Saccharomyces* spp.
Wine	Grapes, fruits	*Saccharomyces ellipsoideus*
Sake	Rice	*Saccharomyces sake*
Cider	Apples	*Saccharomyces* spp.
Tequila	Agave	Lactic-acid bacteria
Scotch	Barley	*Saccharomyces cerevisiae*
Vodka	Potatoes	*Saccharomyces cerevisiae*

on so well that eating unleavened bread came to be considered a hardship or a penance in certain religions.

Modern commercial baking differs from primitive methods largely in scale, but little in fundamentals. In a large bakery (Figure 28.1) industrial engineering is utilized for rapid, uniform, and economical production. The flavor of the final product depends not only on the choice of flours or seasonings, but also on the organism used to leaven it, because each species produces different byproducts. Yeast bread, for example, is leavened with *Saccharomyces cerevisiae.* Because yeast utilizes native starch very slowly, often some sugar is added to start the yeast growing. From the sugars released from the starch, the yeast produces carbon dioxide, the bubbles of which make the bread rise. Ethyl alcohol and other flavoring essences are also produced. Yeast bread retains little or no residual acid or sugar after rising, and the alcohol evaporates during baking. Sourdough bread, by contrast, is leavened by a yeast, *Tomia,* and the bacterium *Lactobacillus sanfrancisco,* which excrete lactic acid during the rising of the dough, giving the final product a sharper flavor.

If starch foods aren't cooked, they are fermented for digestibility. Fermented starches are eaten extensively throughout Africa, Southeast Asia, and on the Pacific islands. Hawaiian poi, made from taro root, is an example. Raw starch is made up into a paste and allowed to stand for several days. During this period, bacteria and fungi enzymatically convert the starch into a digestible form.

DAIRY PRODUCTS.   Fresh milk is a fine natural microbial medium. It provides a fermentable sugar (lactose) and other essential nutrients. Milk's most abundant protein is **casein.** Casein coagulates (curdles) either when the milk is acidified by microbial fermentation, or when it is treated with rennin, an enzyme that may be obtained either from calf stomach or made by microorganisms. Fermented milk products are all either acidified or thickened or both. You may not have realized that butter, made from butterfat separated from cream, derives its delicious flavor from a fermentation step. It is inoculated with microbial cultures that produce the flavored acidic compound diacetyl during a curing process. Without curing, the butter is almost flavorless. Farm butters undergo more curing, and have a stronger flavor that some people like and others do not. Sour cream is cream fermented slightly so that a small amount of lactic acid is formed. The acid thickens the milk by altering the casein. Yogurt is a similar product, but made from whole or skimmed, partially evaporated milk so that its butterfat content is lower. To make yogurt, the milk is inoculated with *Lactobacillus bulgaricus* and/or *Streptococcus thermophilus.* The mixture is then incubated at a warm temperature while the microbes produce lactic acid and the milk casein thickens.

Cheeses require microbial activity in order to develop their distinctive flavors. The steps in cheese production are as follows. First, milk or cream is inoculated with a starter culture, which begins the fermentation, producing some flavor and promoting curd formation. Rennin is added to many cheeses to speed the curdling. Curds sold promptly, without further microbial ripening, are called cottage cheeses or farmer's cheeses; they illustrate the mild flavor of unripened cheese products.

In the second stage, the curd may be pressed to varying degrees of firmness. There are soft cheeses, such as Camembert, semihard cheeses such as Muenster, and hard cheeses such as cheddar or parmesan (Figure 28.2). Semihard and hard cheeses are made from curds that are cooked and pressed and then receive a second inoculation of microbial culture to carry out the curing. Pressing forms the cheese into blocks or wheels, which are coated or wrapped and allowed to age. During aging, the microbial culture carries out further activity. The longer the ripening, the more flavor is developed. The texture of the cheese may become dry and crumbly, or soft and creamy, as ripening progresses.

During ripening, changes occur both in the interior and on the surface of the cheese. If surface growth of molds and yeasts is permitted, a

(a)

(b)

(c)

(d)

FIGURE 28.1 Bread-Making

(a) In breadmaking, mixing machines incorporate flour, yeast, and other ingredients into dough. (b) Dough is then formed into masses which will be placed in pans (c) to rise into loaves. (d) Finished loaves emerge from the oven. (e) The metabolic products of the yeast *Saccharomyces cerevisiae* causes the bread dough to rise. (Photo credits: a, b, c, courtesy of Continental Baking Company; d, courtesy of Pepperidge Farm, Inc.; e, (×4600), courtesy of A.T. Pringle.)

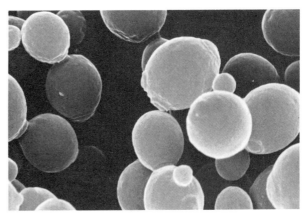

(e)

(a)

(b)

(c)

FIGURE 28.2   Cheese-Making
Microbial activity produces the distinctive taste of various cheeses. (a) Milk or cream is inoculated with a starter culture to start fermentation, and rennin is added to speed the formation of curds. (b) Next, the curds are weighed before being packed into blocks or wheels. Cheddars will be flipped by hand for over an hour before being set aside to age. (c) Finally, a clear wax coat is applied to each cheese to prevent contact with air and microbial spores; even a pinhole will allow mold to form. (Courtesy of Cabot Farmers' Co-op Creamery, Cabot, VT.)

flavorful crust develops, as in Brie and Camembert. Growth of *Penicillium roqueforti* is encouraged in the interior of the blue cheeses (Roquefort, Stilton, Gorgonzola) by poking holes into, or spiking, the ripening cheese. This permits air to enter, fostering the aerobic growth of mold. When *Propionibacterium shermanii* grows within a hard curd cheese, it releases carbon dioxide. The carbon dioxide bubbles are trapped, forming the "holes" that are characteristic of Swiss and Gruyere cheeses. There are over 400 different cheeses, made possible by combining different types of microbial cultures with different types of milk or cream (from cows, sheep, goats, water buffaloes), in a variety of shaping, firming, ripening, and aging techniques. Some varieties and the microorganisms involved are shown in Table 28.2.

VEGETABLES AND FRUITS.   Most of us are familiar with pickled vegetables such as olives, cucumber pickles, and cabbage (sauerkraut). Many other vegetables, and even fruits are also pickled, especially in Japan and other Far Eastern countries. To produce these, the foods are placed in a strong salt brine. Brining inhibits the growth of some microorganisms, while extracting the sugars and other nutrients from the vegetables to provide a medium in which other microorganisms will flourish. Members of the genera *Lactobacillus*, *Leuconostoc*, and *Pediococcus* are used in pickling. As they grow, they ferment sugars found in vegetables and brine to produce lactic, propionic, and other organic acids. These have flavor-enhancing and preservative effects.

In recent years, fermented soybean products, used for thousands of years in the Far East, have been widely adopted in Western countries. Soy sauce, also called shoyu, tamari, and other names, is made by a long, complex fermentation process. Soybeans are first inoculated with the mold *Aspergillus oryzae* and incubated briefly. They are then moistened with a strong salt solution and incubated for 6 to 18 months. During fermentation, acids and alcohols form, and the soybean proteins are hydrolyzed into free amino acids. The final product has a pH of around 4.8 and a salt concentration of 18%. The clear,

TABLE 28.2 Microbiology of cheeses			
Type of Cheese	Name	Microorganism	Origin
Soft, unripened	Cottage, cream	*Streptococcus lactis* *Streptococcus cremoris* *Streptococcus diacetilactis* *Lactobacillus citrovorum*	United States
Soft, ripened	Brie, Camembert	*Streptococcus* spp. *Penicillium* spp. *Brevibacterium* spp.	France
Semisoft, ripened	Roquefort, blue	*Streptococcus* spp. *Penicillium* spp.	France
	Brick, Monterey	*Streptococcus* spp. *Brevibacterium* spp.	United States
Hard, ripened	Cheddar, colby	*Streptococcus* spp. *Lactobacillus* spp.	Britain United States
	Edam, Gouda	*Streptococcus* spp.	Netherlands
	Swiss, Gruyere	*Streptococcus* spp. *Lactobacillus* spp. *Propionibacterium* spp.	Switzerland
Very hard, ripened	Parmesan, Romano	*Lactobacillus* spp. *Streptococcus* spp.	Italy

brown liquid is separated from the bean mash, which is sold as animal feed. The amino acid salt monosodium glutamate (MSG) is present in high concentration in the soy sauce. Some MSG is purified as a separate product, widely used as a flavor enhancer. Other fermented soy products, less known in the West, include miso, a dark-brown bean paste, sufu, a cheeselike food made by aging tofu (bean curd), and tempeh, a thin soybean cake. Just as with starch foods, microbial processing of soybean foods is desirable because it increases not only their flavor and digestibility, but also their protein content.

Most people are unaware that microbes play a key role in developing the latent flavors in fresh coffee beans, cocoa beans, vanilla beans, and tea leaves. In each case, the fresh food has little flavor. Once picked, it is allowed to dry and ripen. During ripening, the complex microbial flora naturally found on the surface of the food appears to convert flavorless substances into strong flavor essences. The ripening processes are not well understood at the scientific level, thus these key flavor-development steps are rather more of an art than a science. Chocolate, for example, contains over 1200 different natural chemicals, results of a microbial synthesis that research chemists have so far been unable to duplicate.

MEAT PRODUCTS.   Microbial curing is a process in which we allow microorganisms to act on a food for a relatively long period under controlled drying conditions. Curing is the key factor in the flavor development of hard sausages and of some hams and other meats. A dry sausage, such as Genoa salami, is prepared by adding curing and seasoning agents to ground meat. Sugar is included as a substrate for fermentation. Most commercial manufacturers also add nitrites to inhibit the growth of *Clostridium botulinum* and other anaerobes. After the sausage meat is stuffed into the casing, there is an incubation period, then a drying and curing period that may last for up to six months. A semidry sausage, such as Lebanon bologna, is heated to a pasteurizing temperature during a smoking step, arresting the curing process.

ALCOHOLIC BEVERAGES.   The great variety available in cheeses is not nearly as impressive as the variety of alcoholic beverages that are produced. It seems that every available sugary or starchy food material, be it rice, potatoes, palm sap, agave juice, or plain old grape juice, is pressed into service somewhere, by someone, to make alcohol. We will limit our discussion to beer- and wine-making.

Beer and ale are called **malt** beverages. Their starting material is grain, usually barley, that has undergone a malting process — that is, it has been allowed to sprout. When the barley seeds sprout, they produce the plant enzyme amylase. Crushing the seeds releases this enzyme into the resulting mash, where it continues to break down the seed starch into sugar. The mash may be placed with other substances such as corn, rice, or wheat into a large vat and heated in order to increase the conversion from starch to sugar. The barley starch in this mash is largely digested into sugars over a period of a few hours. Boiling produces a clear liquid, called the **wort,** which contains the fermentable carbohydrate. The wort is drained off and hops are added to contribute stability and a slightly bitter flavor.

The wort is then inoculated with *Saccharomyces cerevisiae* or *Saccharomyces carlsbergensis* (brewer's yeast) and placed in huge fermenting vats. Fermentation may take from 5 to 12 days. During this process, the yeasts convert the sugars to alcohol and carbon dioxide, and proteins and lipids are converted to higher alcohols which add to the flavor of the brew. Afterward, the spent brewer's yeast, rich in proteins and vitamins, is removed and sold for human and animal food supplementation. The beer or ale may be aged in lagering tanks before being distributed in kegs or bottles. The process is shown in Figure 28.3.

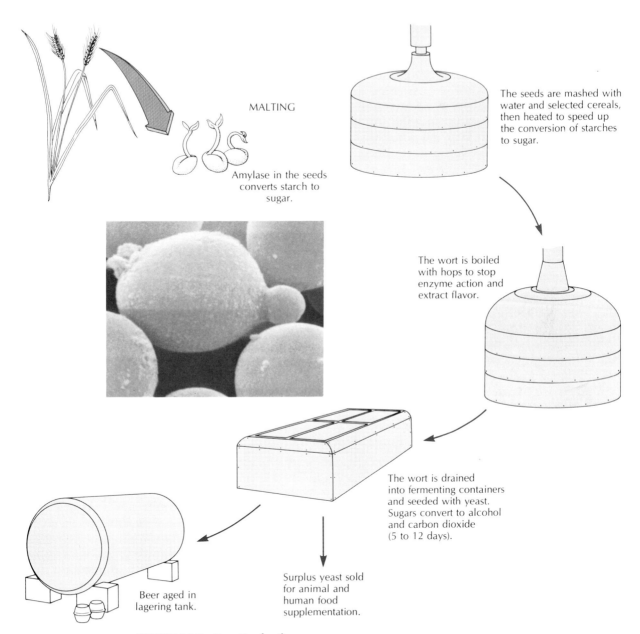

MALTING

Amylase in the seeds
converts starch to
sugar.

The seeds are mashed with
water and selected cereals,
then heated to speed up
the conversion of starches
to sugar.

The wort is boiled
with hops to stop
enzyme action and
extract flavor.

The wort is drained
into fermenting containers
and seeded with yeast.
Sugars convert to alcohol
and carbon dioxide
(5 to 12 days).

Surplus yeast sold
for animal and
human food
supplementation.

Beer aged in
lagering tank.

FIGURE 28.3 Beer Production
To make malt beverages such as beer, grain is used as a source of fermentable
sugars. The first steps convert grain starches to sugars and add flavors. When yeast
is added, fermentation occurs and the sugars are converted to carbon dioxide and
alcohol. After a fixed period, the beer is drained off and the spent yeast is sold as a
food supplement. (×5200. Courtesy of R.G. Kessel, University of Iowa.)

Wine-making remains a challenging art, but scientific methods are rapidly improving the technical aspects of wine-making and taking the guesswork out of making quality wine. In general, wine-making starts with the crushing of ripe grapes. The crushed grapes are always naturally inoculated by the bacteria and molds that adhere to the grapeskins. In a few cases fermentation is allowed to proceed with this natural inoculum, but more often the pressed juice, called **must,** is pasteurized, then reinoculated with stock cultures of *Saccharomyces ellipsoideus.* Fermentation occurs in a vat; if the skins of red grapes are left in, a red wine results. White wines come from either white grapes, or the juice of red grapes minus skins. The alcoholic strength depends on the initial sugar content of the grapes — the more sugar, the more alcohol, to an upper limit around 13%. The "dryness" of the wine is more pronounced if there is little or no sugar left over after fermentation is terminated.

The wine is removed from the sediment after several days, and placed in casks to age. During aging, other, slower changes occur, which develop additional flavor. Finally, the wine may be bottled. Each step in the process, from the initial selection of the grapes to be used, to the final decisions about when to bottle, and, possibly, the blending of different batches, requires a very practical knowledge of microbiology.

FOOD HANDLING

We can preserve the wholesomeness and safety of a perishable fresh food only if we handle it correctly at each step from the time it is harvested until the time it is consumed. Food preservation, quality-control programs in food industries, and correct food preparation and food service techniques are all integral parts of the picture.

FOOD PRESERVATION. Each of the microbial food fermentations discussed in the beginning of this chapter may actually be viewed as a preservation method. In fact, people developed these techniques, in times gone by, as much for their food-keeping properties as for their flavor-enhancing ones. They are still the only food preservation methods available in many less-developed areas of the world. In industrialized countries, we use other, newer methods of food preservation on a daily basis. In all cases, one principle behind successful food preservation is to inhibit microbial growth by establishing unfavorable environmental conditions. Alternatively, we may preserve food by killing all existing organisms, then excluding any new ones. Other objectives are to preserve the food's maximum nutritive value and keep it palatable, at reasonable costs.

Drying foods inhibits microbial growth by creating unfavorable osmotic conditions, thus depriving the microorganisms of the water they need to be active. Foods that are frequently preserved by drying include fruits (raisins, apricots, pears, pineapples, prunes, papayas, apples), herbs and spices, meats (chipped beef, jerky), and fish. When grains and seed crops are harvested they must also be carefully dried before they are stored. Milk, dairy products, and eggs may be converted to a dry powder form for long-term storage.

Salting may be used, as we saw earlier, in a moist pickling process, or it may be incorporated into a microbial curing procedure. Most microorganisms cannot grow in brine, a concentrated salt solution. However, there are some halophilic organisms that may eventually attack the food. Before refrigeration was introduced, brine salting (or "corning") was the major method of preserving meat. When a large animal was slaughtered, as much as possible was immediately eaten or sold. The rest of the meat would be placed in casks or crocks of brine. We still use brine salting for salt pork and corned beef. Salting combined with drying has limited use. Up until about 30 years ago dried salt cod was a staple product, especially in New England.

Dried, salt-preserved fruits and vegetables are widely used in Japan.

The technique of preserving fruits with sugar or sugar–vinegar combinations yields jams, jellies, and preserves. Like salt, sugar inhibits microbial growth by decreasing the available water due to osmotic pressure. Some fungi, however, can grow in strong sugar solutions. To eliminate fungal spores so a preserve or syrup will have long-term keeping qualities, it must be heat-processed and sealed, or else frozen.

Smoking (Figure 28.4) is used in combination with drying or a brine cure for preserving bacon, sausages, hams, turkeys, and fish. When a food is heated in the smoke chamber some, but not necessarily all of the microbial flora is killed. Additionally, some of the chemical components of smoke remain in the food and have antimicrobial effects. However, for complete protection against trichinosis, smoked pork products must be completely cooked.

Canning involves cooking a food within a metal or glass container, which is then completely sealed while the food is still hot. This process, correctly executed, excludes both air and microorganisms. The cooking step is carried out under pressure, in a canner, autoclave, or commercial canning retort, at a temperature at which all thermoresistant bacterial spores are destroyed. However, canners keep a food's heat exposure to the minimum compatible with producing a safe product, to minimize changes in flavor and color and loss of vitamins.

In commercial canning, a safe heat exposure is one that kills the spores of *Clostridium botulinum*. Remember that microbial killing is a logarithmic process (review Figure 23.1). Each filled can must be processed long enough to reduce to a negligible level the chance of survival of any spores of *C. botulinum*. The time of processing is increased for larger-volume cans. As a safety and quality-control measure, test cans within each batch are sampled after processing. The test cans contain spores of the thermoresistant indicator species *Bacillus stearothermophilus*. If any

FIGURE 28.4 Meat Curing
Sausages hang in the curing room for a period of time while their flavor develops. Here an inspector is checking to determine if curing is progressing properly. (Courtesy of Oscar Mayer Foods Corporation, Madison, Wisconsin.)

spores have survived the processing, the entire batch must be discarded. Because of these quality controls, incidents of botulism associated with commercially canned foods are extremely rare.

Home canning can also be entirely safe, but only if the canning directions are carefully followed. Acidic foods, such as tomatoes, apple sauce, and some other high-acid fruits, are sufficiently processed by boiling, as long as they are well sealed after processing. Those acid-tolerant organisms that survive are largely aerobes, and will not grow in the sealed vessel. The dangerous clostridia will not grow under highly acid conditions. However, low-acid foods such as meats, fish, beans, corn, and the new varieties of "low-acid" tomatoes require pressurized processing to a higher temperature. The cook must pay attention to the pressure gauge, thermometer, and timer, and take no shortcuts, to be entirely sure that all clostridial spores are killed.

When canned foods are not boiled long enough or are sealed improperly, microorganisms may survive, then grow, and cause spoilage within the can. In most cases, spoilage is an anaerobic fermentation process that results in gas formation, so the affected can will swell, bulge, or even explode. The food within will be obviously decomposed. In other cases gas is not produced; changes in the food may be undetectable, or the food may show only slight acid or "off" flavors and colors. It is these almost normal-appearing contaminated foods that are the most dangerous to eat. When you use canned foods, especially home-processed products, it is important that you keenly observe the can and contents before using them and assure yourself that they appear normal. Swollen or bulging cans should be thrown away unopened.

Pasteurization is the process of mild heating of a fluid to kill pathogens. In milk and other dairy products, the target pathogens are *Mycobacterium tuberculosis* and *Brucella abortus*, which are killed at relatively low temperatures. Heating milk to 145 degrees Fahrenheit (62.8 degrees Celsius) for 30 min (batch method) or to 161 degrees Fahrenheit (71.7 degrees Celsius) for 15 sec (flash method) are effective. Unfortunately, certain other pathogens such as *Campylobacter jejuni*, which causes diarrhea, sometimes survive this treatment. Recently **ultrapasteurization** for cream and **ultrahigh temperature (UHT) treatment** for milk are finding increased use in the United States. These techniques employ much higher temperatures and yield products that stay fresh much longer. In fact UHT milk is sterile; it has been heated under pressure to 285 degrees Fahrenheit for 15 sec, then to 300

degrees Fahrenheit for 0.5 sec. It is packaged in cardboard containers and can be kept on the shelf without refrigeration for months as long as the container remains unopened and intact. UHT milk, long available in Latin America and Europe, is now being introduced in the United States.

Refrigeration and freezing are used in almost every home in the industrialized countries. Much of the fresh food we buy has been refrigerated or frozen to preserve it until purchase. Microbial growth slows down as temperature drops. Some species do die, but more, including most pathogens, just become inactive. The **psychrophiles,** however, enjoy their most rapid growth in cool temperatures. Unlike pure water, microbial cells may not actually freeze solid until the temperature is well below 0 degree Celsius. Psychrophilic growth at − 5 degrees Celsius is common, but fortunately the rate of spoilage is comparatively slow. It is important that home refrigerators and freezers be properly set and operating efficiently. The aging refrigerator with its leaking seal and its freezer compartment door half off will probably not prevent spoilage! Some foods that support microbial growth at subfreezing temperatures are fruit juice concentrates, bacon, and ice cream. Foods that have been defrosted should not be refrozen, since microbial growth may have occurred during the interval.

Radiation as a means of food preservation presents two potential advantages, which are short exposure time and indefinitely long shelf life. Despite extensive research, few commercial irradiated food applications have reached the marketing stage. The reason is that irradiation causes a decline in palatability in some foods; in others, it is too expensive to be competitive with other means of preservation.

Chemical preservatives are extensively used in foods to inhibit or kill microorganisms, as we saw in Chapter 23. Chemicals approved for use in foods by the Food and Drug Administration (FDA) are those on the GRAS (generally regarded as safe) list. Periodically, controversy arises over the safety of these chemical preservatives and food additives for human consumption. Many of the present food preservatives (Table 28.3) are in fact metabolized by the usual cellular pathways in the human body. An exception is sodium nitrite ($NaNO_2$), an inorganic salt added to hams, hotdogs, and cold cuts because it inhibits germination and growth of any botulism organisms. Additionally, nitrite ion interacts with the globins in meat to retain its attractive red color. However, there is evidence that our normal intestinal flora alters the nitrite ion to carcinogenic substances called **nitrosamines.** Because of this problem nitrites have been reduced or eliminated in many meat products, and some consumers avoid buying nitrite-containing foods.

FOOD QUALITY CONTROL. Commercial food producers operate under guidelines created by federal and state agencies. These guidelines mandate certain testing procedures to be carried out on each lot of some foods, such as dairy products and canned foods. Lot numbering allows for easy identification and recall should any food

items be discovered to be substandard or hazardous. Designation of the last date of fresh sale and shelf-life codes are used to make sure that foods are not sold past a reasonable period.

The major concern of microbiological testing in fresh foods is to determine if a food has been contaminated with bacteria of human origin. Quantitative testing for the presence of fecal coliforms and fecal streptococci is regularly carried out by the manufacturers on such foods as dairy products, fresh and frozen meats, and frozen convenience meals. Foods may also be analyzed for specific pathogens, such as salmonellae, staphylococci, toxigenic sporeformers, and viruses. Cultural, biochemical, serologic, and physical testing methods are used.

FOOD HANDLING AND SANITATION. After a food product has been properly harvested, its continuing safety depends on many factors, including proper packaging, shipping, storing, refrigeration, cooking or preparation, and serving (Figure 28.5). The packaging materials that protect a dried, canned, or frozen food must remain intact until use, requiring careful handling at all

TABLE 28.3 Some common chemical preservatives

Preservative	Organisms Inhibited	Foods Used
Propionates; propionic acid	Molds	Breads, cakes, cheeses
Sorbates; sorbic acid	Molds	Jellies, salad dressings, cheeses, syrups
Benzoates; benzoic acid	Molds, yeasts	Catsup, salad dressing, pickles, margarine, cider, soft drinks
Sulfur compounds	All organisms	Molasses, dried fruits
Sodium nitrite	Bacteria	Cured meats, hotdogs, cold cuts, some fish

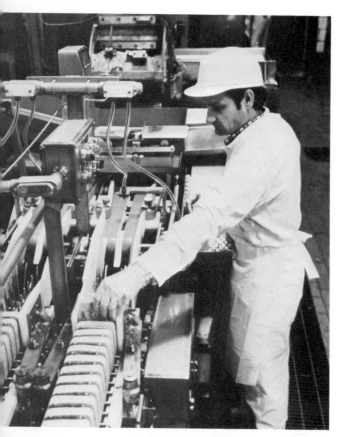

FIGURE 28.5   Food Handling and Sanitary Measures Commercial food producers must follow certain testing, packaging, and storage procedures to prevent contamination of their product by microorganisms. This Oscar Mayer employee wears gloves while operating a computerized sliced-meat packaging machine. (Photo courtesy of General Foods Corporation.)

storage. After preparation, hot foods must stay very hot, and cold foods very cold, to prevent them from being rapidly taken over by microbial colonists. The temperature range in which microorganisms proliferate most rapidly is from about 60 to 120 degrees Fahrenheit (15 to 50 degrees Celsius). Perishable foods should be kept out of that zone except for the very short periods of time required for service and consumption. Most outbreaks of food intoxication occur from such food items as roasted turkeys that have been kept warm rather than hot, or potato salads that have been kept cool rather than cold (see Box 28.1).

In the home, restaurant, or institutional kitchen, high standards of cleanliness must be maintained. It is essential to avoid introducing human microflora into the food. Common sources of significant human contamination are nasal flora, pathogens from skin lesions, and fecal flora. Careful personal hygienic habits are essential. In many restaurants, personnel are not allowed to serve foods with bare hands, but must use plastic gloves or tongs. Restaurant and institutional kitchens are operated under state and local regulations, and are subject to periodic inspections. Unfortunately, inspections are infrequent, and they may or may not be thorough. The events that lead to outbreaks of food-related diseases often arise quickly and without warning. We hear through the news media about an outbreak if a public eating place is involved, but the far more common family outbreaks never come to our attention, unless it's our family. Only recently, a small outbreak of typhoid fever was traced to a restaurant worker who carried *Salmonella typhi* in the intestinal tract, reminiscent of the famous Typhoid Mary outbreak early in this century. You may find it helpful to review Chapter 19 to refresh your memory about foodborne diseases.

steps. Punctured cans or unsealed bottles admit air, allowing aerobic growth. Plastic or foil envelopes, or cardboard boxes, are readily punctured and may admit microorganisms.

Temperature control is perhaps the most significant factor in food safety. Refrigerated or frozen foods must be maintained at the designated temperature during transportation and

FOOD SPOILAGE

About 20% of the harvest of fresh fruits and vegetables and up to 25% of the grain harvest is lost

BOX 28.1
Who Cooked Those Giblets, Anyway?

Even very large, well-run restaurants can get into big trouble when a rush period such as Thanksgiving Day comes along. November 25, 1982, turned out to be a disastrous day for a restaurant in Central Maine.

On that day, the restaurant served 1094 dinners over a period of about eight hours. A total of 160 patrons later developed salmonellosis. An analysis of the foods eaten by those who became sick and cultures from leftovers taken home in "doggy bags," suggested that the giblet gravy that was served with both turkey and chicken dinners was the culprit. Of course, since turkey gravy tends to get mixed in with everything else on your plate, and since lots of people sample their dinner companions' meals, the picture was a little messy, both literally and figuratively.

Epidemiologists reconstructed the events leading up to dinnertime on Thanksgiving. The preparations began four days before, when 42 large frozen turkeys were set out to defrost and were left at room temperature for 36 hr. The turkeys, believed to have been the original source of the *Salmonella*, were then unwrapped. The giblets were set aside in a large pan in the refrigerator, earmarked for later use in the gravy. The chief cook gave orders for the giblets to be cooked. Two days before Thanksgiving, all the turkeys were roasted in batches, then boned, some on wooden tables that were supposed to be used only for raw meats. The still warm turkey meat was bagged and, after considerable delay, refrigerated.

The day before Thanksgiving, the chief cook prepared the stuffing and the giblet gravy. The giblets were minced in a blender, then added to thickened turkey stock. The gravy was warmed but not boiled, for fear of scorching it. However, when each kitchen worker was later questioned separately, *not one* had either cooked the giblets or had seen anyone else cooking them. The inescapable conclusion is that the giblets were still raw when added to the gravy! On the big day, both the gravy and the stuffing sat on the back of the steam table at only slightly above room temperature for the entire dinner period of about 10 hours.

The cost of the outbreak included 292 inpatient hospital days, for a total of $93,108, plus outpatient fees of $20,925 and fees for medication, transportation, and baby-sitting totaling $7,045. There were 783 missed work days, a salary loss estimated at $41,428. The Maine taxpayers spent $17,754 for the investigation and control measures of the Bureau of Health. The restaurant owner's business declined sharply for several months, with losses conservatively estimated at $300,000. These were the costs of the confusion and rush that led a normally exceptionally clean and efficient operation to lose track of a few small, key details, like cooking the giblets.

annually to spoilage worldwide (Figure 28.6). The economic impact is immediately apparent. Such losses in the grain crop are especially devastating in poorer countries where every bit of the harvest is needed in order to avert starvation. Much of this loss, however, can be prevented with appropriate food-storage techniques.

SPOILAGE MICROORGANISMS.   Microorganisms are the agents of food spoilage wherever it occurs. Fortunately, few are pathogenic. Their normal habitats are foods — fruits, vegetables, and meats — at room or refrigerator temperatures, but not human bodies at 37 degrees Celsius. Their adverse effects on us are primarily economic and social, and only secondarily health-related. Rarely, those few pathogenic exceptions produce toxins in the food, but more often food losses have indirect health effects because they may lead to malnutrition. Common spoilage groups are summarized in Table 28.4.

HOW FOODS SPOIL.   Fruits and vegetables are highly perishable. Their spoilage is usually caused by the growth of soil organisms left on them at harvest time. Insect or mechanical damage assists the microorganisms to penetrate the food and accelerates spoilage. Both fruits and vegetables have some available sugars, but the higher sugar content of fruits tends to promote fungal attack while the acidity of most fruits inhibits many aerobic bacteria.

Meats are the most perishable of foods, since they contain an abundance of nutrients. Meats may acquire a diverse surface flora from the animal's skin and mucosa during slaughter, and

FIGURE 28.6   Food Spoilage
Fruits, vegetables, and meats are highly perishable items. The sugar in fruits and vegetables tends to promote fungal growth, while microbial slime accumulates on the exposed surface of meat cuts such as steaks and roasts. (Courtesy of Darlene Bordwell, Boston.)

TABLE 28.4 Microorganisms that cause food spoilage		
Food	**Bacteria**	**Fungi**
Vegetables	*Erwinia*	*Rhizopus*
	Pseudomonas	*Botrytis*
	Xanthomonas	*Geotrichum*
	Corynebacterium	*Phytophora*
		Fusarium
		Phytophora
Fruits		*Botrytis*
		Rhizopus
		Penicillium
		Aspergillus
		Cladosporium
Meats	*Clostridium*: enterobacteria	*Thamnidium*
	Pseudomonas	*Mucor*
	Alcaligenes	*Rhizopus*
	Acinetobacter	*Candida*
	Moraxella	*Torulopsis*
Fish, shellfish	Similar to meats	Little fungal activity
Eggs	*Pseudomonas*: enterobacteria	*Penicillium*
		Cladosporium
Bakery products	*Bacillus*	*Penicillium*
		Rhizopus
		Neurospora
Grains		*Aspergillus*
		Claviceps
Milk	*Streptococcus*: lactobacilli	

from the tools and cutting surfaces used during cutting. This flora simplifies to one or more types as spoilage occurs. With solid cuts, such as roasts, spoilage develops only on the exposed surface, resulting in microbial slime, and discoloration due to changes in the muscle globin pigments. The grinding of hamburger or sausage introduces microorganisms, so spoilage is more rapid and can occur all the way through. In fish, spoilage often begins around the gill region; the organisms involved are psychrophilic bacteria from the waters where the fish was caught.

Dairy products, if pasteurized, will be soured or spoiled primarily by heat-resistant lactic acid bacteria that survived pasteurization. Butter is attacked by pseudomonads and some fungi, and cottage cheese and yogurt by various bacteria and yeasts.

Prior to canning and refrigeration, people routinely ate, without ill effects, meat that was slightly "off," soured milk, rancid butter, and other foods we would now reject as substandard. This should alert us to the fact that taste and smell are not wholly reliable indicators of food safety. Foods that have dangerous levels of accumulated toxins rarely show, either by taste or smell, much sign of microbial action.

MICROORGANISMS IN INDUSTRY

The origins of **industrial microbiology**—the controlled use of microorganisms to make saleable products—go back to the earliest bakeshops and alehouses, roughly 4000 to 6000 BC. Food manufacture is the oldest of the industrial processes in which microbial activities are central. In the 20th century another form of industrial microbiology, a multi-billion-dollar manufacturing network, has developed, producing nonfood items such as drugs, solvents and other organic chemicals, enzymes, and dietary supplements. We will now cover some of the biotechnical advances that have paved the way for this extensive use of microorganisms in industry.

Bacteria have been used to manufacture bulk chemicals since the beginning of the century. In World War I, the Germans developed a fermentation process for making glycerol, while the British pioneered the production of acetone. Both chemicals were employed in making explosives. During both world wars, the Germans relied heavily on the production of bulk food yeast to maintain an adequate protein supply for the German population.

It often happens that a product such as ethanol can be manufactured either biologically, by

microbial fermentation or chemically, by fractionating crude petroleum. Until fairly recently, oil-derived products were cheaper than microbially derived ones. But as global crude oil prices and supplies have become less and less predictable, microbial processes are once again becoming economically attractive.

The microbial fermentations employ as starting materials, or **feedstocks,** such inexpensive substances as molasses and corn syrup byproducts, cottonseed meal, sulfite liquor from papermaking plants, and similar leftovers from other industries. A feedstock for a process is chosen on the basis of cost, availability, and the yield of the desired product obtained with its use. The starting price of feedstock materials varies with crop conditions. Economics, then, largely determines the proportion of a given product made by straight chemical means or by biological means.

MICROORGANISMS USED IN INDUSTRY

Of the more than 100,000 species of microorganisms believed to exist in nature, only a few hundred have so far found uses in industry. They include certain bacteria, molds, algae, and yeasts. We make use of both anaerobes and aerobes, carrying out all types of metabolism. They all share one feature, the ability to produce large amounts of a useful chemical end product. In some cases, such as penicillin production, theirs is a natural ability, which we have capitalized on by selecting those fungus strains that are the highest yielding. Increasingly, however, we are creating entirely new industries based on engineered microorganisms. These are strains that have been genetically altered so that they are able to produce for us a substance, such as insulin, normally found only in plants or animals.

GENETICS OF INDUSTRIAL MICROORGANISMS. Genetic techniques are essential in the development of a microbial strain for an industrial use

(Figure 28.7). The primary requirements for a producer strain are that it be metabolically stable and high-yielding. Sluggish producers or strains with unpredictable variations in behavior are unsuitable. Initially, most wild-type (directly from nature and genetically unaltered) microorganisms have one or both of these undesirable traits. The geneticist's task is to adapt the organism to its intended use. This often means inducing the microbial strain to make vast amounts of one of its hundreds of products, often at the expense of others. We can view the resulting organisms as "freaks." Like the chicken, the dairy cow, and certain other highly bred domestic animals with one exaggerated characteristic, most industrial microorganisms would be completely incapable of survival in the wild.

Mutation and selection are the major traditional genetic methods for improving the yield of a microbial strain. For example, these techniques have been used repeatedly over a period of 40 years with *Penicillium chrysogenum*, the fungus used to produce penicillin. Initial wild-type strains yielded only 60 milligrams of the drug per liter of media. As the result of many succeeding rounds of genetic manipulation, we now have *Penicillium* strains that yield over 20 grams of penicillin per liter (Figure 28.8).

This type of genetic "reprogramming" often requires a mutation to inactivate the organism's own natural metabolic or genetic controls. You may recall the feedback inhibition mechanism discussed in Chapter 7, whereby the accumulation of an end product causes the biosynthetic pathway to that end product to shut down. For example, this type of feedback inhibition prevents the synthesis of excessive amounts of the amino acid lysine. In order to use a bacterial strain for bulk lysine production, the feedback inhibition effect must be blocked. In one strain, an enzyme has been altered by mutation so that it is no longer sensitive to lysine's allosteric inhibition. Lysine biosynthesis in this strain is now no longer under metabolic control — it is a runaway reaction.

FIGURE 28.7 Uses of Genetic Engineering Industry relies heavily on genetic techniques to develop stable and high-yielding strains of microorganisms. Here a lab technician examines a flask of cytosine, one of the four building blocks of DNA. After purified bases are synthesized in the lab, they can be assembled into whatever gene a researcher wishes to create. (Courtesy of Erich Hartmann/Magnum.)

In yeasts such as *Saccharomyces*, sexual recombination leads to genetic hybrids, which may have more desirable behavior than either parental strain. For example, genetic crosses in *Saccharomyces carlsbergensis* have produced yeast strains that leave no carbohydrate in the wort after fermentation, resulting in what are called "light" beers (see Box 28.2)

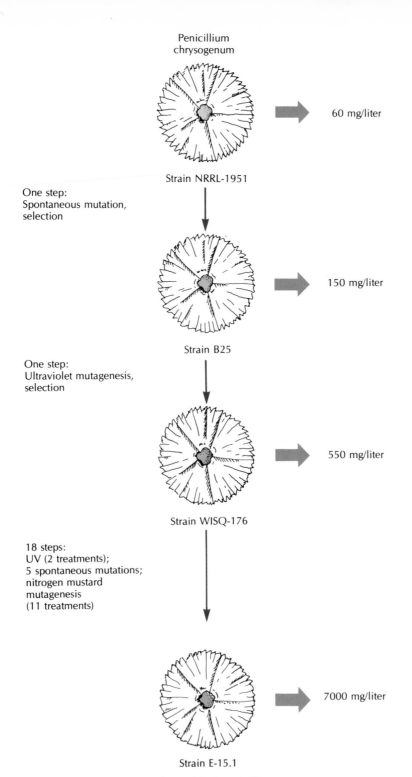

Penicillium
chrysogenum

60 mg/liter

Strain NRRL-1951

One step:
Spontaneous mutation,
selection

150 mg/liter

Strain B25

One step:
Ultraviolet mutagenesis,
selection

550 mg/liter

Strain WISQ-176

18 steps:
UV (2 treatments);
5 spontaneous mutations;
nitrogen mustard
mutagenesis
(11 treatments)

7000 mg/liter

Strain E-15.1

FIGURE 28.8 Increasing the Yield of Penicillin
Over the years, *Penicillium chrysogenum*, the mold that produces penicillin, has
been modified by classic genetic techniques until its yield is many times greater than
that of the original strain. Some higher-yielding strains resulted from spontaneous
mutations, while others were the result of induced mutations caused by ultraviolet
radiation or nitrogen mustard, a chemical mutagen.

The techniques of DNA hybridization, gene cloning, plasmid transfer, and manipulation of microbial gene expression were discussed in Chapter 8. These techniques make it possible to get microorganisms to produce entirely new substances, often ones that it has never before been practical to produce industrially. The potential for a spectrum of genetically engineered products has caught the imagination of entrepreneurs, industrial managers, and the investment industry. This area of microbiology is suffused with a high degree of excitement.

Using DNA hybridization·techniques, scientists may create plasmids bearing genes that code for any desirable bacterial, viral, animal, or plant protein (review Figure 8.20). In addition to these genes, the plasmids must contain transcriptional controls (operator and promoter sequences) appropriate to their host bacterial species. When these plasmids are reintroduced to host bacteria, the new genetic sequences may be expressed, directing bacterial bulk synthesis of the compound. Among bacteria, *Escherichia coli* is the most commonly used host. Its plasmid

BOX 28.2
Make Mine a Light Beer

The beer yeast *Saccharomyces carlsbergensis* ferments sugars provided in the wort to carbon dioxide and alcohol, as you would expect. However, it leaves 19% of the sugars (mainly maltose and its relatives, which don't taste sweet) untouched when growth stops. The beer thus has quite a few remaining sugar calories, in addition to its alcohol calories, totalling an average of 160 calories in 12 ounces. In the United States, beer has developed an "image problem," because people feel that it is more fattening than other alcoholic beverages. The beer industry has been making an effort to attract younger, more appearance-conscious consumers. Thus, the push to find a way to make and market a "low-calorie" beer.

Several ways of modifying the yeast's fermentative behavior genetically have been tried.

One successful sequence involved first hybridizing *S. carlsbergensis* with *S. diastaticus*. The first hybrid yeast fermented all the sugars, but produced a bad-tasting brew. The next crosses produced a yeast strain that consumed 90% of the sugars and tasted good. A third round of crosses with a wild *S. carlsbergensis* strain produced the final yeast (a closely guarded corporate treasure) that consumes 100% of the sugar and makes palatable beer with about 100 calories in 12 ounces. The "light" beers, now made by all leading breweries, currently account for a major share of all beer sales. Let's drink to the yeast geneticists and to our diets!

(Photo courtesy of Darlene Bordwell, Boston.)

pBR 322 (Figure 28.9), which was artificially constructed, is the most common cloning vehicle. Bacteriophage lambda may also be used to clone large gene fragments in *E. coli*. Other microorganisms used as hosts are the bacterum *Bacillus subtilis* and the yeast *Saccharomyces cerevisiae*. The bacterial plant pathogen *Agrobacterium tumefaciens* carries a large plasmid designated Ti, which has been used successfully for introducing foreign genes into the cells of some species of broad-leaved plants, such as tobacco.

Bacteria bearing viral genes may be used to produce viral proteins in bulk, free from the viral nucleic acid that would be found in the complete virus. Since animals develop their immune responses against the coat proteins of a viral infective agent, these protein products make excellent vaccines. In contrast, with vaccines made from complete virus particles they have no infective or oncogenic potential. Genetic engineering has produced a vaccine against the animal disease foot-and-mouth disease, and against the human diseases hepatitis B, influenza, and cytomegalovirus.

PRODUCTION METHODS IN INDUSTRIAL MICROBIOLOGY

Once a high-yielding, stable microbial strain is available, engineering techniques are brought to bear to create the optimal growth conditions. The growth conditions and timing are fine-tuned to the process and the organism. It is interesting to note that although these processes are traditionally called fermentations and the vessels fermenters, the growth conditions within the vessel are most often aerobic, and as close to saturated with oxygen as can be achieved. We will look at three techniques for producing biological products — the batch culture, the continuous culture, and the immobilized enzyme method (Figure 28.10).

FIGURE 28.9   Manufacturing Gene Products
Scientists use DNA hybridization techniques to join plasmids, such as this one from *Escherichia coli*, with foreign genes that code for a particular protein product. Once the altered plasmid is reintroduced to its bacterial host, it begins directing the bulk synthesis of the desired compound. (Courtesy of Stanley N. Cohen, Stanford University.)

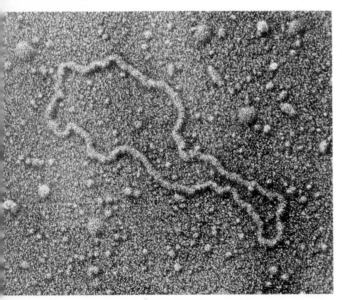

FIGURE 28.10   Three Production Methods
(a) In a batch culture, the fermenter vessel is filled with a sterile medium, then inoculated and maintained at optimal growth conditions until a high yield of microorganisms is obtained. When the yield peaks, the fermenter is emptied and the product purified. (b) In a continuous culture, fresh medium is supplied and used culture medium harvested at a constant rate. The cells remain in exponential growth. Although yields are lower than batch culture yields at any given time, a continuous culture fermenter may produce for weeks. (c) The immobilized enzyme method uses active enzyme molecules bound to an inert carrier, which is often packed in a column. As substrate solution flows through the column, substrate is progressively changed to product.

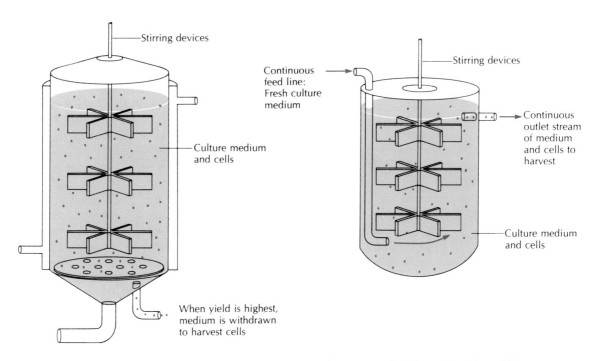

Stirring devices

Culture medium
and cells

When yield is highest,
medium is withdrawn
to harvest cells

(a) BATCH CULTURE

Continuous
feed line:
Fresh culture
medium

Stirring devices

Continuous
outlet stream
of medium
and cells to
harvest

Culture medium
and cells

(b) CONTINUOUS CULTURE FERMENTOR

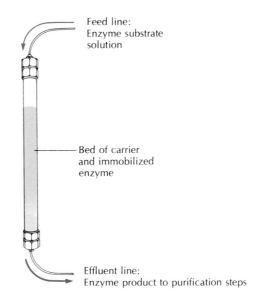

Feed line:
Enzyme substrate
solution

Bed of carrier
and immobilized
enzyme

Effluent line:
Enzyme product to purification steps

(c) IMMOBILIZED ENZYME METHOD

A **batch culture** is a closed system, much like a laboratory broth culture but on a huge scale. The very large culture vessel, called a reactor or fermenter, is made of glass or stainless steel and designed so that all parts are easy to clean and accessible to steam for sterilization. The empty vessel is filled with the desired volume of a growth medium containing the feedstock. The growth medium contains the kinds and concentrations of nutrients to support optimal yield at minimum cost. The medium may either be presterilized or sterilized once in place.

The inoculum is prepared separately, its growth timed so that it contains the optimal number of microbial cells in logarithmic growth phase. Once the inoculum is pumped in, the cells start to grow immediately in a continuation of their log phase. The cells' aeration needs (if any) are supplied by providing filter-sterilized air or other gases, and by vigorous agitation. Stirring also promotes rapid and complete use of nutrients and diffusion of waste products. Heater coils and refrigeration loops within the vessel regulate the temperature.

An array of electronic sensors within the fermenter detect and record the dissolved oxygen concentration, temperature, and concentrations of nutrients and products. Microprocessors activate the environmental controls as needed. Once the product yield has stopped increasing, the fermenter is shut down. The cells are then separated from the fluid, and the desired product is concentrated and purified.

Some microbial syntheses are most productive during the log phase, while others reach their highest yields later, in the stationary phase after microbial replication has largely stopped. This difference reflects the fact that the metabolic products of cells may be separated into two classes. One class, the **primary metabolites,** are substances the microorganism makes to use itself, for its own growth. Some examples are vitamins, amino acids, and enzymes. The maximum yield of a primary metabolite is usually achieved by culturing log phase cells. Little additional accumulation of the desired substance occurs once the culture stops growing in stationary phase.

The other class of cellular products are **secondary metabolites.** These substances are end products of metabolism (not directly used in growth) such as antibiotics. The maximum production of a secondary metabolite usually occurs after active growth levels off, in the stationary phase. In this case, a batch culture can be allowed to run as long as the rate of accumulation is significant.

Continuous culture is another industrial process, often preferable to batch culture for the production of primary metabolites. In this technique, the microbial cell population is kept in a perpetual log phase. The culture vessel is designed to be operated on a flow-through basis, that is, new sterile medium is continually fed in, while an equal amount of used medium is simultaneously withdrawn. The continuous inflow of medium provides nutrients for optimal constant growth, and the outflow of fluid bearing some cells provides a continuous source of products to be harvested. This technique eliminates most of the down time for emptying, cleaning, and refilling culture vessels. The disadvantage of continuous culture is that it is a delicately balanced process, sensitive to potentially disastrous problems of contamination, genetic instability, and equipment failure.

Immobilized enzyme techniques are a new approach to industrial biochemistry. You'll recall that using whole, growing microbial cells to carry out an enzymatic reaction requires that the microbes be fed. It is sometimes possible, as a less costly alternative, to use just their purified enzymes. The microbial enzymes are bonded onto an inert supporting surface, often a mesh of cellulose fibers. A solution of the substrate to be acted on is passed over the immobilized enzymes, and the product is recovered from the effluent.

INDUSTRIAL MICROBIOLOGY AT WORK

The products of industrial microbiology come in various categories, such as antibiotics, enzymes, industrial chemicals, and so on (Table 28.5). We

TABLE 28.5 Main microbial products in industry

Classification	Product
Industrial chemicals	Ethanol, butanol, acetone, citric acid
Amino acids	Lysine, glutamic acid
Vitamins	Riboflavin, vitamin B_{12}
Enzymes	Amylases, glucamylase, cellulase, invertase, lactase, pectinase, proteases, rennet
Polysaccharides	Dextran, xanthan, capsular polysaccharides for vaccine use
Antibiotics	Penicillins, cephalosporins, amphotericin, neomycin, kanamycin, streptomycin, tetracycline, chloramphenicol, bacitracin, polymyxin, etc.
Other pharmaceuticals	Steroids, interferon, insulin, human growth hormone and somatostatin, carotenoids
Bioinsecticides	*Bacillis thuringensis*, polyhedrosis viruses

can best illustrate the versatility of our "microbial biochemists" by giving some examples.

MICROBIAL CELL MASS. Microbial cells in bulk, such as baker's and brewer's yeasts, are manufactured at the rate of tons per day, for use both in industry and at home. Spent brewer's yeast, a byproduct of beer-making, has its own value as a protein and vitamin rich dietary supplement consumed by people and added to animal feeds. Microorganisms other than *Saccharomyces cerevisiae* have also been much studied for their potential as food producers. The term **single cell protein** (SCP) is commonly applied to microorganisms grown specifically to be used for human and animal foods. Scientists have been working intensively to develop SCP sources for many years. The advantage of SCP lies in the extremely rapid growth rates of microorganisms, and their ability to make use of organic waste products that are completely unusable by higher organisms.

The most intensive study has focussed for reasons of economy on photosynthetic organisms, such as the algae. They grow very rapidly in shallow ponds using only cheap inorganic starting materials and needing no externally supplied energy source other than sunlight. One algal species, *Spirulina*, has been used as a food in Africa and Central America for centuries. However, SCP prepared from other algal species contains compounds that cause gastrointestinal upsets in humans and, understandably, is not well accepted.

Current SCP research is focussed on heterotrophs, particularly those bacteria that are able to metabolize crude oil, oil refinery byproducts, methane, or methanol. Some of the early optimism about ending world hunger with oil fed SCPs subsided when petroleum prices began to climb. However, the answer to the question of where to find inexpensive feedstocks for bacterial SCP production is found by looking to other microbial fermentations. For example, one potential starting material, methane, can readily be made from microbial composting of plant and animal wastes. Methanol can be obtained by bacterial fermentation of wood wastes, corn stalks, and other plant wastes, or cheaply produced from plentiful coal. Then the methane or methanol can be used by the SCP bacterium *Methylophilus methylotrophus* for growth. This type of SCP is sold in England (Figure 28.11) as a pelleted animal feed called Pruteen.

At present, SCP development is primarily aimed at expanding food sources for domestic animals, and market forces are currently limiting investment in the SCP industry. A continuing problem with SCPs for human use is that the high nucleic-acid content of SCP products,

compared with their dietary equivalents of plant or animal tissue, may induce gout in sensitive individuals. Yet despite such remaining problems, the potential of SCPs for helping to feed the rapidly expanding populations of the next few decades has been clearly established. It is not just the science-fiction writers, but also the serious space-exploration scientists, who project that onboard SCP production will be the major renewable food source for interplanetary travelers on long missions in the future.

ENZYMES, PROTEINS, AND HORMONES. Enzymes have extensive commercial uses. They may originate from plants, animals, or microorganisms. In almost every case, the cost structure favors obtaining the enzymes from microbes. About 1270 tons of enzymes per year are manufactured. The proteases, which hydrolyze proteins, are the most widely used. They are obtained from *Bacillus* and *Aspergillus* species, and are used as meat tenderizers, aids to digestion, and cleaning aids such as spot removers. They may also be used as catalysts to modify other proteins, as they are in commercial baking, for example. Proteases alter the gluten protein in the flour, which shortens the mixing time to make dough and improves the texture of the finished product. Another class of enzymes, the amylases (also from species of *Bacillus* and *Aspergillus*), predigest starches into smaller carbohydrate feedstocks for fermentation (Figure 28.12), are used to prepare starch sizings for fabrics in the textile industry, and are used to modify the texture of thick syrups. Two other closely related enzymes, glucamylase and glucose isomerase, are used together with amylase in a three-step process that converts corn starch into high-fructose corn syrup (Figure 28.13). This relatively new product has now replaced much of the (more costly) cane sugar in many popular soft drinks and in some other sweets. *Aspergillus* also produces pectinase, which is used in producing canned applesauce and to clarify fruit juices before bottling. *Mucor*, a mold, now produces most of the rennin for cheese-making, re-

FIGURE 28.11   Microbial Food Production
Large-scale cultivation of the bacterium *Methylophilus methylotrophus* has expanded the food supply for domestic animals. Imperial Chemical Industries in Britain converts vast quantities of the microorganism into a pelleted animal feed called Pruteen. The scanning electron micrograph shows several specimens of *M. methylotrophus* trapped on a filter. (Photo courtesy of Imperial Chemical Industries, Agricultural Division, Billingham, U.K.)

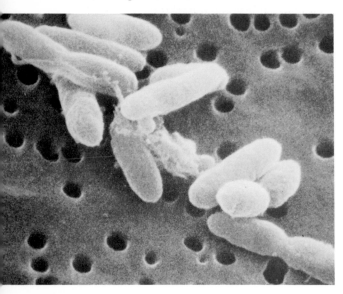

placing the enzyme derived from calves' stomachs. There is great interest in using fungal cellulases to digest cellulose to glucose. You'll recall that only microorganisms can digest cellulose, which makes up more than 50% of all the organic matter on Earth. If the cellulose could be rapidly digested to glucose, a primary energy source for animal life, this would open up a vast new source of food energy for us.

One promising new application of microbial enzymes is in the plastics industry. The small subunit molecules from which plastics are made, such as ethylene oxide and propylene oxide, are currently obtained from petroleum. The price of plastics, then, is determined largely by the price of petroleum. Petrochemical manufacturing is both increasingly costly and highly polluting. Now under development is a microbial process that utilizes three different oxidase enzymes in sequence, each from a different microorganism. The starting materials are glucose and other simple organic molecules. The end products are plastic subunit molecules and fructose, both of which are valuable. All of the techniques of genetic engineering are being called upon to create microbial strains with improved efficiency in this process. The doors of the $50-billion-a-year plastics industry are opening to microbiologists.

We have already mentioned the role that genetic engineering has played in the development of new biological products. Among the human proteins being produced in large-scale, commercial operations are alpha, beta, and gamma interferons; the peptide hormones insulin, somatostatin, and parathyroid hormone; serum albumin; and blood-clotting factors.

In addition to making some peptide hormones, microorganisms also play a role in the synthesis of the steroid hormones. They are used to carry out certain particularly difficult steps in the multi-stage synthesis of these complex organic structures. For example, the fungus *Cunninghamella* is used to convert the steroid cortexone to hydrocortisone, a drug that is widely used to treat inflammation.

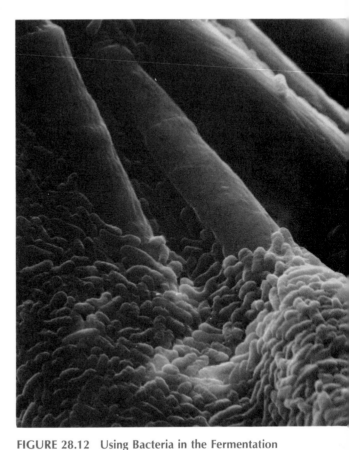

FIGURE 28.12 Using Bacteria in the Fermentation Process
Bacteria of the species *Zymomonas mobilis* are more efficient than yeast at converting carbohydrates into ethanol and are commonly used to make fermented beverages. In this micrograph, bacteria have been immobilized on cotton fibers for the industrial manufacture of ethanol. The inert fiber mesh was contained in a glass chamber 22 in. long. While nutrients and glucose flowed in one end, spent medium and ethanol passed out the other. (Courtesy of Carl E. Shively, Department of Biology, Alfred University.)

PRIMARY METABOLITES.   Proteins are made up of 20 different amino acids. Humans require eight of them as essential amino acids (amino acids that can't be made in the body and must be consumed in the diet). Animal proteins normally

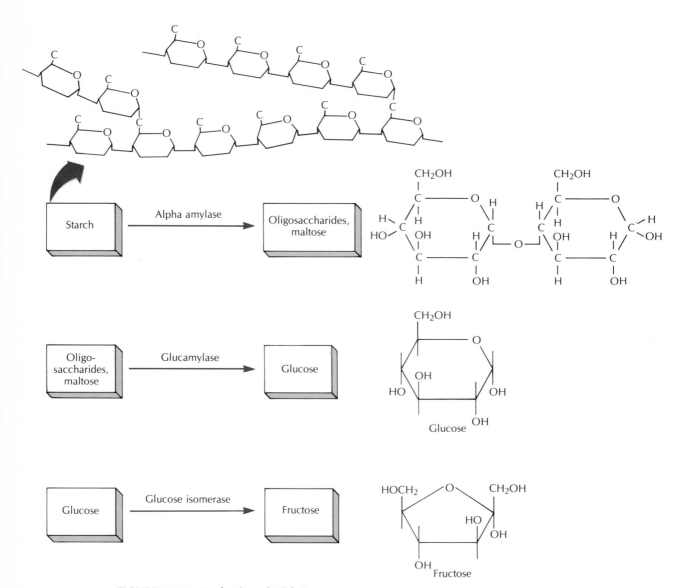

FIGURE 28.13   Production of High-Fructose Corn Syrup
The availability of high-fructose corn syrup as an inexpensive substitute for cane sugar is revolutionizing the soft-drink industry. The corn syrup is made from corn starch by a three-step process, using three enzymes of microbial origin. The carbohydrates of the final product exist almost entirely in the form of fructose, one of the sweetest-tasting sugars.

supply a balanced mix of all 20, but plant proteins, almost all that is available to a large portion of the world's population, are often seriously deficient in one or more essential amino acids. Corn proteins, for example, are very low in lysine. Therefore, feeds for domestic animals, when produced from corn, must be supplemented with lysine so the animal will have a complete diet. About 80% of the 40,000 tons of lysine produced annually are produced by the bacterium *Corynebacterium glutamicum*. One unsolved problem, which so far has not yielded a microbiological solution, is an economical synthesis of the sulfur-containing essential amino acid methionine, also much employed as a feed supplement, but still made entirely by costly chemical synthesis.

There is also a large market for the nonessential amino acids glycine and alanine as flavor enhancers. Monosodium glutamate (MSG), a salt of glutamic acid, is a well-known flavoring substance, consumed at the rate of over 30,000 tons per year in the United States.

Microorganisms are generous producers of vitamins. Of the vitamins being manufactured and sold today, microbial biosynthesis contributes most to our supplies of vitamin B_{12} and riboflavin. Vitamin B_{12} may be produced as a second product in the fermentation that yields the antibiotic streptomycin. The two most common producer organisms are *Propionibacterium shermanii* and *Pseudomonas denitrificans*. Riboflavin can be produced directly by several fungal species, or obtained as a second product from acetone–butanol fermentations.

The organic acids (formic, acetic, lactic, propionic, gluconic, butyric, and citric) can be obtained as fermentation products. Acetic acid, some of which is used to make white vinegar (Figure 28.14), is made by trickling an ethanol solution through a bed of wood chips coated by growth of aerobic acetic acid bacteria, the same organisms responsible for spoiled homemade beer and wine (recall Box 7.2). Citric acid, used as a flavoring and preservative agent, is produced by

Aspergillus. Gluconic acid, used in such diverse applications as preventing spots on dishes washed in automatic dishwashers and as a calcium salt in pharmaceuticals, is produced by several species of bacteria and fungi.

Alcohols such as ethanol, isopropanol, butanol, and glycerol can be made microbiologically, but only a small amount of ethanol is currently made this way in the United States because chemical means are currently more cost effective. However, it is reassuring to know that we are not totally dependent on foreign petroleum as a raw material for these essential compounds, because we have microbiological alternatives available.

SECONDARY METABOLITES. We noted previously that antibiotics are a significant group of secondary metabolites. The antibiotics industry was founded on the basis of microbial synthesis, and developed in conjunction with the rest of the U.S. chemical industry during and after World War II. As adequate supplies of antibiotics came on the market, they made possible a complete transformation of medical practice, as discussed extensively in preceding chapters. All of the first group of antibiotics — penicillin, chloramphenicol, the tetracyclines, streptomycin, and many others — are of direct bacterial or fungal origin. The techniques developed for bulk production of antibiotics served as the training ground in which industry acquired the tools for vastly improved production of other primary and secondary metabolites. The worldwide sales value of antibiotics in recent years has topped $5 billion, and is increasing steadily.

The major antibiotic-producing organisms are the filamentous fungi (*Penicillium, Cephalosporium*), the filamentous bacteria (*Streptomyces*), and certain nonfilamentous bacteria (*Bacillus*). The naturally occurring penicillins and cephalosporins have increasingly been supplemented by semisynthetic drugs derived from a fungal product by chemical modification. For example, *Penicillium* produces penicillin G when grown for about seven days on a wheat bran–nutrient medium. The penicillin G may be used in its original state as a drug, but it may also be modified by bacterial enzymes to 6-amino-penicillanic acid. This compound is then modified chemically to make ampicillin, carbenicillin, azlocillin, and dozens of other drugs. The cephalosporins in current use are produced by modification of the 7-alpha-cephalosporanic acid molecule, itself produced by enzymatic mod-

ification of cephalosporin C, the natural antibiotic made by *Cephalosporium acremonium* (Figure 28.15).

Industrial microorganisms now contribute vastly to our economic and physical well-being. Yet the full benefits of industrial microbiology await the further application of genetic engineering, which will undoubtedly enable us to produce a large repertoire of additional essential chemicals.

FIGURE 28.14 Acetic Acid Production
To produce vinegar, a diluted solution of ethyl alcohol is trickled over a loosely packed bed of wood chips, on which a mixed culture, predominantly *Acetobacter aceti*, is growing. The bacteria oxidize the ethyl alcohol to acetic acid.

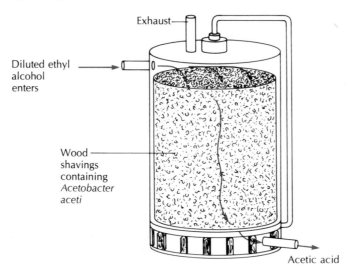

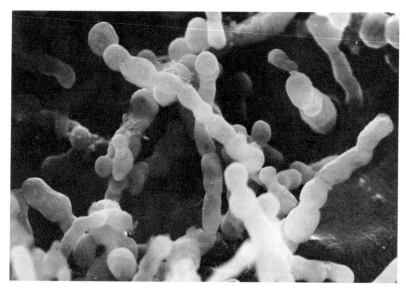

FIGURE 28.15 Industrial Microorganisms
The natural antibiotic, cephalosporin C, is manufactured by the mold *Cephalosporium acremonium*. This scanning electron micrograph shows the swollen filamentous hyphae, typical of the high-production phase of the organism. (×2355. Courtesy of Erika A. Hartwieg, Electron Microscopy Facility of the MIT Biology Department.)

SUMMARY

1. Microorganisms play an important role in human food production. They enhance the flavors of spices and flavorings and provide the tanginess of cheeses. Also, by fermenting sugars in foods, they lower the pH and preserve the food.

2. Some important food groups produced by microbial action include raised breads, alcoholic beverages, cottage cheese, yogurt, hard sausages, pickles and sauerkraut, and soy sauce.

3. Foods are safe only if handled so as to control unwanted microbial growth. Drying, salting, canning, refrigeration, freezing, and chemical preservation are the main methods used to keep food from spoiling.

4. Proper methods of storage, shipping, and food service are an essential part of public health.

5. Fresh fruits, vegetables, meats, fish, and dairy products all spoil easily. Food spoilage may be prevented or delayed by sensible food handling.

6. Microbial processes may produce thousands of tons per year of useful chemical products. Industrial microbiology uses large batch or continuous fermenters to make hundreds of different products.

7. Primary metabolites are substances the producer microorganism uses for growth. These include amino acids, vitamins, and enzymes. Secondary metabolites are end products of metabolism, such as antibiotics.

8. Recombinant DNA technology is being used in conjunction with older genetic techniques to expand the productivity and versatility of industrial microorganisms.

DISCUSSION QUESTIONS

1. How do microorganisms alter food flavors? How do they act as food preservers?

2. Investigate further the role of microorganisms in the preparation of such dairy products as butter, yogurt, and sour cream.

3. There is a thin line between useful microbial growth and spoilage. What is the difference between sour milk and milk that has been converted to yogurt?

4. As a class project, have a cheese-tasting lab. As you taste, present the specifics of the organisms and processes used to achieve each result.

5. What are the major food-handling problems and how does each contribute to a decline in food quality?

6. Become familiar with a variety of chemical products produced industrially by microorganisms.

7. How is the science of genetics employed in industrial microbiology?

8. The next time you go food shopping make a mental note of the types of food preservation used in each product you buy.

ADDITIONAL READING

Brill WJ. Safety concerns and genetic engineering in agriculture. *Science* 1985; 227:381–384.

Burg RW. Fermentation products in animal health. *ASM News* 1982; 48:460–463.

Crueger W, and Crueger A (T.D. Brock ed, English edition). *Biotechnology: A textbook of industrial microbiology.* Sunderland, Mass: Sinauer Assoc. Inc., in association with Science Tech Inc., 1984.

Demain AL. Industrial microbiology. *Science* 1981; 214:987–994.

Industrial microbiology. *Sci Am* 1981; 245(3). (The entire edition is devoted to review articles on key aspects of industrial microbiology.)

Richardson GH, ed. *Standard methods for the examination of dairy products,* 15th ed.

Washington, DC: Am Public Health Assoc, 1985.

Rose AH, ed. *Food microbiology* (Volume 8 of the Economic Microbiology Series). New York: Academic Press, 1983.

Speck ML, ed. *Compendium of methods for the microbiological examination of foods,* 2nd ed. Washington, DC: Am Public Health Assoc, 1984.

Swaminathan MS. Biotechnology research and Third World agriculture. *Science* 1982; 218:967–972.

Watson JD, Tooze J. *The DNA story: a documentary history of gene cloning.* San Francisco: W.H. Freeman, 1981.

GLOSSARY

Abscess   An anaerobic hollowed-out infection site filled with debris.

Acetyl group   The two-carbon fragment that remains after pyruvic acid is oxidized and a carbon-dioxide group is split off.

Acid   A compound that dissolves in water, dissociates, and increases the free-hydrogen-ion concentration.

Acid decalcification   A roughened "white spot" on a tooth indicating a site of developing caries.

Acid-fast stain   A common differential staining technique, used in diagnosing tuberculosis infections.

Acid hydrolases   Enzymes that create a highly acidic digestive environment, causing rapid disintegration of many polymers.

Acidic solution   A solution that contains more hydrogen ions than pure water and has a pH of less than 7.

Acquired immunodeficiency   Immunodeficiency resulting from malnutrition, illness, or other stress.

Acquired immunodeficiency syndrome (AIDS)   A disease state involving irreversible loss of lymphocytes and accompanied by a rare type of tumor and extensive infections; it is thought to be caused by the HTLV-III virus, which is transmitted by blood and/or intimate physical contact.

Acquired pellicle   The coating of positively charged salivary proteins on the tooth enamel, providing a basis for microorganism growth.

Activated macrophage   A nonspecific phagocyte that has been activated by lymphokines, becoming much more efficient at capturing cellular non-self antigens.

Activated sludge process   A method of secondary sewage treatment that can handle large volumes of sewage quickly.

Activation energy   The minimum energy that must be supplied to increase collisions between molecules to the point at which bonds break and reaction occurs; without it there is no chemical reaction.

Active immunity   The immune response created after receiving an antigenic stimulus; it may be naturally or artificially acquired.

Active site   The point at which the substrate binds an enzyme.

Active transport   The process by which the cell's energy is used to move sugars, amino acids, and ions into the cell against a concentration gradient.

Acute   The most serious and rapidly progressive form of an illness.

Acute necrotizing ulcerative gingivitis (ANUG)   (. . . NEH-kruh-ty-zing UHL-sur-ah-tihv jihn-juh-vy-tuhs)   A rapidly developing form of periodontal disease (also known as trench mouth or Vincent's angina).

Acute rejection   The destruction of graft tissue within several weeks after transplantation.

Acyclovir   A chemotherapeutic purine derivative that acts as a competitive inhibitor of nucleic-acid synthesis; used to treat herpes virus infections.

Adaptation   The process by which living organisms adjust to changing environmental conditions.

Adenoid disease   Swelling of the pharyngeal tonsils.

Adenosine diphosphate (ADP) (uh-DEHN-uh-seen dy-FAWS-fayt)   The substance formed when adenosine triphosphate is hydrolyzed, releasing energy.

Adenosine triphosphate (ATP) (. . . try-FAWS-fayt)   The nucleotide that is the primary energy-carrying molecule of the cell.

Adenovirus (ad-uhn-oh-VY-rahs)   A virus that causes respiratory diseases of the "common-cold" variety.

Aerobic (ai(uh)r-ROH-bihk)   Growing in the presence of oxygen.

Aerobic respiration   Respiration that occurs when the terminal inorganic electron acceptor is oxygen; water is the chemical end product.

Aerosol   A cloud of tiny droplets in air.

Aflatoxin (af-lah-TAHK-suhn)   A potent liver carcinogen produced by *Aspergillus flavus*, a fungus that grows on peanuts and stored animal feeds.

African trypanosomiasis (. . . trih-pan-uh-soh-MY-ah-sihs) Also known as sleeping sickness, a cardiovascular disease caused by *Trypanosoma gambiense* and *T. rhodesiense* organisms transmitted by the tsetse fly.

Agglutination (uh-gloot-uhn-AY-shuhn) The clumping of cells, as occurs when antigen complexes with antibody.

Agranulocyte (ay-GRAN-yuh-loh-syt) A leukocyte without granules in the cytoplasm; includes lymphocytes and monocytes.

Aldehyde (AL-duh-hyd) A highly reactive compound that alters the chemistry of proteins, changing their conformation and rendering them inactive; one example is formaldehyde.

Algae Seaweeds and microscopic forms of organisms that use light energy; important as food for small aquatic animals; they are members of the kingdom Protista.

Alkylation (al-kuh-LAY-shuhn) Destruction of microbial proteins by chemical reaction that replaces hydrogen atoms in the amino acids; this alters the protein's function.

Allergen (AL-ur-jehn) An antigen that elicits hypersensitivity.

Allograft Tissue transferred between unrelated members of a species.

Allosteric interaction (al-oh-STEHR-ihk . . .) The inhibition of enzyme activity that occurs when a substance combines with an enzyme at an allosteric site.

Allosteric site The site on an enzyme at which an inhibitor binds; anything other than the active site.

Alpha hemolysis (AL-fuh hee-MAHL-luhs-uhs) The designation used when a greenish zone is seen marking the area of hemolyzed red blood cells around a bacterial colony.

Alpha virus (AL-fuh VY-ruhs) One type of the RNA virus togavirus; transmitted by insect vectors.

Alternate pathway A mechanism of activating complement that does not require antibodies but is triggered by microbial polysaccharides or endotoxin.

Alternation of generations The life cycle in which diploid forms alternate with haploid forms.

Alveolar macrophages Specialized phagocytes found in lung secretions.

Alzheimer's disease (ALTZ-hy-murz) A degenerative disease affecting older people; it may be caused by a slow virus.

Amantadine (uh-MAN-tah-deen) A drug that blocks the penetration of the RNA by influenza virus.

Ameba (uh-MEE-buh) The best-known of the subgroup of protozoans known as Sarcodinia, they use two or more pseudopodia to capture prey organisms.

American trypanosomiasis A serious cardiovascular disease (also known as Chagas' disease); it is caused by *Trypanosoma cruzi* and *T. rangeli*, parasites transmitted by biting insects prevalent in Central and South America.

Ames test A procedure that uses bacteria to identify potential carcinogens.

Amino acid A molecule that contains both an amino group and a carboxyl group.

Aminoglycosides (uh-mee-noh-GLY-kuh-sydz) Powerful chemotherapeutic agents containing linked amino sugars, primarily valued for their effectiveness against gram-negative bacteria.

Ammonification (uh-mon-nuh-fih-KAY-shuhn) The third step in the nitrogen cycle, involving the breaking down of amino acids as animal waste decomposes, releasing ammonium ion.

Amphitrichous (am-FIHT-ruh-kuhs) Having tufts at both ends.

Anaerobes (an-AY(uhr)-rohbs) Organisms that grow in the absence of oxygen.

Anaerobic sludge digester One stage of primary sewage treatment involving the decomposition of organic materials with the release of methane gas as a major end product.

Anamnestic response (an-am-NEHS-tikh . . .) The response that enables B memory cells to achieve a more rapid and powerful immune response after several exposures to an antigen.

Anaphylaxis (an-uh-fah-LAK-suhs) A possibly fatal hypersensitivity reaction involving the whole body; asphyxiation and heart failure may occur.

Anions (AN-y-ahns) The negatively charged particles that result when atoms gain electrons.

Antenna A group of light-harvesting pigments that absorb light, augmenting the action of the chlorophylls in photosynthesis.

Antibiogram The antibiotic susceptibility pattern of an infectious agent.

Antibiotic An antimicrobial substance (that is, one

that acts against microorganisms) produced naturally by a microorganism.

Antibiotic-associated colitis A severe intestinal ulcerative superinfection in which *Clostridium difficile* is the causative agent.

Antibiotic susceptibility testing Techniques for determining which antibiotics have an effect on a specific microbe being cultured.

Antibody A defensive protein that the body makes to react with and neutralize antigens; found in blood serum, saliva, and other body secretions; antibodies neutralize toxins and kill some bacteria.

Antibody-mediated immunity (AMI) The part of the body's defense system that involves the action of antibodies.

Anticodon A three-base sequence representing the part of the molecule that recognizes the spot to which the amino acid is to be delivered.

Antigen A substance (for example, a virus) that actively stimulates the body to produce antibodies.

Antigenic determinants Portions of the molecule to which an antibody or lymphocyte binds.

Antihuman globulin antibodies Antibodies from the serum of an animal that has been immunized with human immunoglobulins as a foreign antigen.

Antiinflammatory Acting against inflammation.

Antilymphocyte globulin (ALG) (an-tee LIHM-fo-syt GLAH-byoo-lihn) Serum containing antibodies against human T lymphocytes.

Antimetabolites Drugs that act as competitive inhibitors, which block enzyme reactions, arresting microbial metabolism of an essential compound.

Antimicrobial drug Any substance that can be used therapeutically to treat microbial disease by killing or inhibiting the disease agent.

Antimitotic agents (an-tee-myt-AH-tikh . . .) Drugs that halt the growth of all rapidly dividing cells, thus slowing the development of immunity and prolonging the period before rejection develops.

Antisepsis The attempt to control pathogens in the environment by using disinfectants to keep their number as low as possible (as opposed to asepsis, where the attempt is to exclude pathogens).

Antiseptic Pertaining to the treatment of living skin or tissue with chemical agents to kill microorganisms.

Antistreptolysin O (ASO) (an-tee strehp-toh-LY-suhn OH) An antibody that appears in the serum of a patient with rheumatic fever.

Antitoxin Antibodies produced against either a toxin or toxoid.

Aphotic zone (ah-FOHT-ihk . . .) The middle, dark level of the oceanic zone, between the lighted (euphotic) zone and the bottom (benthic) zone.

Aquifers The undergound deposits into which ground water percolates through the soil.

Arboviruses See arthropod-borne viruses.

Archaebacteria (ahr-kee-bak-TEE-ree-ah) A newly classified group of microbes that seem to be closer to the earliest forms of life than most present-day bacteria; their cell walls lack peptidoglycan and they differ in other important ways from other bacteria.

Arthropod An invertebrate animal, for example, mosquito, flea, or louse, or an arachnid such as a tick.

Arthropod-borne disease A disease transmitted primarily or exclusively by insects or arachnids.

Arthropod-borne viruses RNA viruses that transmit disease to vertebrates via the bite of the arthropod, which is both host and vector for the virus; the most important in North America are encephalitis viruses, which are spread by mosquito bites.

Arthropod vector An arthropod that carries a disease-causing agent from one host to another.

Arthrospore An asexual fungal spore.

Arthus reaction A localized inflammatory hypersensitivity reaction in which complexes have been deposited in small blood vessels, often causing tissue death.

Artificial passive immunity The protective state that results when a person exposed to a disease is injected with serum containing preformed antibody.

Artificially acquired active immunity The protective state that results from vaccination.

Ascomycetes (as-koh-my-SEE-tees) Fungi that form sexual ascospores; some are mycelial fungi, others are yeasts, and a few are fleshy fungi.

Ascospore (AS-koh-spawr) Sexual fungal spore produced in an ascus, formed by ascomycetes.

Ascus (AS-kuhs) A saclike structure containing ascospores; it eventually ruptures to allow the release of the ascospores when they are mature.

Asepsis (ay-SEHP-suhs) The attempt to exclude potential pathogens from entering a patient's room, the site of an injection, or a surgical field, rather than relying on killing the organisms once present.

Aseptic meningitis (ay-SEHP-tikh mehn-ihn-JY-tihs)   Meningitis caused by a virus; the term is misleading because an infective agent is present, but is not found on bacteriologic examination of cerebrospinal fluid.

Asexual reproduction   Reproduction in which new cells or organisms arise from just one parent.

Asexual spores   Fungal spores that arise from portions of the coenocytic cell containing one nucleus.

Assimilation   The fourth step in the nitrogen cycle, the process in which plants and microorganisms make new amino acids, assimilating nitrate and ammonium.

Association   A statistical correlation between two events, for example, between cigarette smoking and lung cancer.

Asthma   An atopic allergy with bronchial smooth-muscle spasm, excessive mucus production, and localized epithelial swelling occuring in response to an antigen.

Asymptomatic infection   An infection with no clinically obvious damage.

Atom   The smallest unit of matter within an element.

Atomic number   The number of protons in the nucleus of an atom.

Atomic weight   The total number of protons and neutrons in the nucleus of an atom.

Atopic allergy (at-AW-pikh . . .)   A localized type of anaphylaxis, such as asthma or hay fever.

Atypical mycobacteria   Species other than *Mycobacterium tuberculosis* that can cause illnesses similar to tuberculosis.

Autoclaving (AW-toh-klayv-ihng)   The use of pressurized steam within a sealed container as a method of heat sterilization.

Autograft   Tissue moved from one area of an individual's body to another.

Autoimmunity   The possession of circulating antibodies against one or more body components; only when the autoantibodies attack normal tissue does autoimmune disease result.

Autolysis (aw-TAW-luh-sihs)   Enzymatic self-destruction carried out by some microorganisms in old cultures.

Auxotroph (AWK-suh-trohf)   A mutant bacterial strain with nutritional needs it cannot fill itself.

Axial filaments   Groups of internally located flagella in spirochetes.

B cells   The lymphocytes that synthesize antibodies or immunoglobulins, the defensive proteins involved in antibody-mediated immunity.

B memory cells   Long-lived cells that do not produce antibodies, derived from parental B cells.

Bacillus   A rod-shaped or sausage-shaped bacterium.

Back mutation   A mutation that exactly reverses the original change in the nucleotide sequence of the DNA.

Bacteremia (bak-tee-REE-mee-uh)   The presence of bacteria in the blood, which is one characteristic of cardiovascular infection.

Bacteria   Simple one-celled procaryotic organisms, most of which live off of dead plant and animal materials and wastes.

Bacterial chromosome   The long double helix of DNA that bears all the essential genes in a bacterium.

Bacterial conjugation   A type of plasmid-mediated mating between two bacterial cells in which some or all of the bacterial chromosomal DNA is transferred.

Bacterial meningitis   Bacterial inflammation of the membranes covering the brain and spinal cord; this disease is a great risk in newborns.

Bacteriophage (bak-TEER-ih-oh-fayj)   A virus that infects bacteria (the term literally means "bacteria eater").

Bacteriorhodopsin (bak-TEER-ih-oh-roh-DAHP-sihn)   A light-reactive substance chemically related to the pigment in the human retina used by some bacteria to capture light energy.

Bacteriuria (bak-teer-ee-(Y)OOR-ee-uh)   Bacteria in the urine, which occurs in cystitis.

Barophiles (BA-rah-fihls)   Bacteria adapted to intense pressure; found in deep ocean sediments.

Basal body   The portion of the flagellum that is composed of rings bound to a central rod; it anchors the flagellum to the cell wall and cell membrane.

Base   A substance that dissociates in water and causes a decrease in the hydrogen-ion concentration.

Base analog   Chemical mutagens that are

modifications of the normal purine and pyrimidine bases; they pair incorrectly and the cell may synthesize them into erroneous DNA.

Basic solution A solution that has fewer hydrogen ions than water and has a pH greater than 7.

Basidiospore (buh-SIHD-ee-uh-spawr) A type of fungal spore formed on a basidium.

Basidium (buh-SIHD-ee-uhm) A microscopic structure formed in the sexual reproduction of certain fleshy fungi.

Basophil (BAY-zuh-fyl) A granulocyte that readily takes up basic dye.

Batch culture A closed system involving a reactor or fermenter used as a culture vessel to incubate the desired volume of a growth medium containing the feedstock.

BCG (Bacille Calmette–Guérin) vaccine (bah-SEEL kal-MEHT GEHR-uhn . . .) Vaccine used to protect against tuberculosis in Europe and elsewhere.

Benign tumors Tumors that do not spread throughout the body.

Benthic zone The lowest part of the oceanic zone; the bottom sediments of that zone nourish a population of bacteria and simple invertebrates.

Benthos (BEHN-thohs) Plants and animals that are attached to the bottom of bodies of water or are found in ocean sediments.

Beta hemolysis (BAY-tuh hee-MAW-lys-uhs) The designation used when a clear zone is seen where red blood cells were hemolyzed around a bacterial colony.

Beta-lactam ring A chemical feature of penicillins and cephalosporins responsible for their antimicrobial activity.

Beta-lactamase (. . . LAK-tah-mayz) An enzyme that cleaves the beta-lactam ring, rendering penicillins and cephalosporins inactive.

Binary series A geometric progression in which the constant factor by which the number increases is 2.

Binary fission The simple reproductive process by which a procaryotic cell characteristically enlarges and divides in two.

Binomial system The system that designates an organism by a two-part name giving its genus and species.

Biochemical oxygen demand (BOD) The amount of oxygen the microorganisms in a body of water

need to decompose the organic matter that is present.

Biological vector A vector that is also the reservoir of infection.

Biosphere The part of the global surface composed of all types of living things.

Biosynthesis The formation of organic compounds by living organisms from elements or simple compounds.

Black water Water from toilets, heavily contaminated.

Blood–brain barrier Unusually thick cell membranes that allow some substances to pass from the blood to the brain but restrict the possibility of pathogens moving from the blood to the brain tissue.

Blood group antigens Polysaccharides that determine whether an individual has type A, B, O, or AB blood.

Blooms Sudden bursts of plankton growth after added amounts of phosphate accumulate in a body of water.

BOD See biochemical oxygen demand.

Bone marrow A source of lymphocytes (from stem cells), contributing to the immune system.

Botulin (BAHCH-uh-lihn) The active toxin product involved in botulism; it blocks transmission of nerve impulses.

Botulism A potentially life-threatening illness caused by the toxin of *Clostridium botulinum*, an anaerobic, gram-positive, spore-forming rod present in soil; most outbreaks result from inadequately processed home-canned foods.

Bright-field microscope A microscope that uses visible light for illumination; the specimens are viewed against a white background.

Brill–Zinsser disease A mild form of typhus in a person who previously had epidemic typhus.

Broad-spectrum Term applied to a drug useful against many different types of organisms.

Brucellosis (broo-suh-LOH-suhs) An infection, also known as undulant fever because the patient's temperature goes up and down, caused by a *Brucella* organism, which commonly infects domestic herd animals.

Bubo (B(Y)OO-boh) A painful, swollen, ulcerated lymph node seen in yersiniosis (bubonic plague).

Bubonic plague The former term for yersiniosis, a disease transmitted by animals harboring fleas infected with *Yersinia pestis*; the peripheral circulation is blocked, causing necrosis of fingers and toes, and buboes may form.

Budding A variant of binary fission in which bacteria undergo unequal cell divisions; the smaller of the two progeny, the daughter cell, originates from growth of one end of the cell.

Buffer A substance that tends to stabilize the pH of a solution.

Calculus A hard abrasive material composed of plaque containing calcium phosphate that forms on teeth.

Calvin cycle The set of reactions leading to the reduction of carbon dioxide to carbohydrates in the dark reactions of photosynthesis.

Cancer The general term used for neoplasms, but most often to refer to malignant forms.

Candidiasis (kan-dih-DY-uh-suhs) A fungal infection of the upper respiratory tract, caused by the yeast *Candida albicans*.

Capillary action The movement of water or other liquid over solid surfaces.

Capsid The coat of positively charged proteins surrounding the negatively charged nucleic acids in a virus particle.

Capsule A dense, structured layer coating some procaryotes.

Carbohydrate An organic compound composed of carbon, oxygen, and hydrogen, including both sugars and starches.

Carbon cycle Process by which green plants and other organisms remove carbon dioxide from the air and capture the energy of the sun by photosynthesis.

Carbon fixation The incorporation of carbon atoms from atmospheric carbon dioxide into organic carbon compounds.

Carboxysomes (kahr-BAHKS-uh-sohms) The sites in the cytoplasm of some photosynthetic microorganisms where carbohydrates are manufactured.

Carbuncles Abscesses on the back of the neck and upper back.

Carcinogen A chemical that causes cancer.

Card system A color-coded outline of the required precautions of isolation procedures, placed on the door of a patient's hospital room.

Cariogenic (kair-ee-oh-JEHN-ihk) Producing caries, or dental disease.

Carotenoids (KAR-ah-tehn-oyds) Yellow-to-red accessory pigments that expand the light-absorbing potential in cyanobacteria.

Carrier Infected person with no obvious signs of disease.

Casein (KA-see-uhn) The most abundant protein in milk; its coagulation results from fermenting the milk or treating it with rennin, leading to such products as sour cream, yogurt, and cheese.

Catabolism (kuh-TAB-uh-lihzm) The process by which energy is extracted from organic nutrients by breaking them down.

Catalase (KAT-uh-layz) The enzyme that breaks down hydrogen peroxide to oxygen and water.

Catalyst A substance that can lower the activation energy needed for a chemical reaction and thus speed up that reaction.

Catarrhal period (kah-TAHR-uhl . . .) The initial stage of whooping cough, with mild symptoms resembling those of a common cold.

Cations (KAT-y-uhns) The positively charged particles that result when atoms lose electrons.

Causative agent The microbial organism responsible for a specific infectious disease.

Causative factor The chemical, environmental agent, or aspect of life style responsible for a noncommunicable disease.

Cavitation Bubble formation and collapse, which occurs due to intense localized pressure changes when ultrasound is used as a cleansing mechanism.

Cell The primary unit of life, a self-reproducing microscopic structure enclosed by an outer layer called a membrane.

Cell-mediated immunity (CMI) The part of the body's defense system that protects against viruses; T cells act as its effectors.

Cell membrane The membrane that encloses the cytoplasm of a cell.

Cell pool The cell's supply of the compounds needed for growth and reproduction.

Cell wall The rigid outer layers enclosing most bacteria, fungi, algae, and plant cells.

Cellular ATP assay A technique for precise measurement of metabolic rate by measuring the intensity of light emitted, which is proportional to ATP levels.

Cephalosporin (sehf-ah-loh-SPAWR-ihn) Antibiotic produced by fungi of the *Cephalosporium* genus, which acts against the bacterial cell wall.

Cercaria (sur-KA-ree-ah) One of the larval stages of the fluke.

Cestodes See tapeworms.

Chagas' disease See American trypansomiasis.

Chain of transmission The chain of events leading to each new case of a communicable disease.

Chemical bond The force of attraction between atoms in a molecule.

Chemical reactions The rearrangements of atoms, involving the formation and breakage of bonds between the atoms.

Chemiosmosis (keh-mee-ahs-MOH-suhs) The process by which it is thought that electrons give off energy to make ATP in the electron transport system, using a proton gradient between the interior of the cell and the medium outside the membrane.

Chemolithotrophs (kee-moh-LIH-thah-trohfs) Organisms that get their energy by chemical reactions in which they oxidize certain inorganic compounds.

Chemotaxis (kee-moh-TAKS-uhs) The ability to move in response to chemicals.

Chickenpox A childhood disease involving rash, fever, and malaise, caused by varicella-zoster virus.

Chlamydial infections Infections with *Chlamydia trachomatis*, which may be responsible for a large percentage of nongonococcal urethritis and nonspecific vulvovaginitis.

Chlamydias (klah-MIHD-ee-uhs) The obligate intracellular bacterial parasites *Chlamydia trachomatis* and *C. psittaci.*

Chloramphenicol (klawr-am-FEHN-ih-kahl) A broad-spectrum bacteriostatic drug sometimes used to treat typhoid fever.

Chlorophyll A light-absorbing pigment closely related to heme and the cytochromes, used in photosynthesis.

Chlorophyta (KLAWR-uh-fy-tuh) Green algae, which may have been ancestors of the land plants.

Chloroplasts Specialized organelles in eucaryotic cells that contain chlorophyll as well as enzymes required in photosynthesis.

Cholera A bacterial disease that was pandemic before the introduction of water and sewage treatment, involving severe diarrhea that can lead to serious loss of blood volume.

Choleragen (KAL-ur-uh-jehn) The toxin produced by *Vibrio cholerae*, the organism that causes cholera.

Chromatophores (kroh-MAT-oh-fawrz) Folded membranes and sacs of membrane material filled with liquid in anaerobic photosynthetic bacteria.

Chromosomes The long strands of DNA that form the structure for the genes of an organism and contain hereditary information.

Chronic rejection Destruction of implanted tissue after progressive deposition of immune complexes in the arteries feeding the graft, leading to failure of the blood supply.

Chrysophyta (KRIHS-uh-fuh-tuh) A type of green algae also known as diatoms, with single cells encased in silica.

-cidal Causing irreversible cellular changes that prevent microbes from reproducing.

Cilium (SIHL-ee-uhm) Short, hairlike cellular projections that move in a wavelike manner.

Citric-acid cycle See Krebs cycle.

Classic pathway A mechanism for activating complement that must be triggered by specific antibody binding to antigen.

Clonal-selection hypothesis The theory that if a lymphocyte with self-receptors binds self-antigens it is destroyed or rendered inactive; the remaining lymphocytes are the ones that attack non-self antigens.

Cloning The creation of a population of cells identical to the parent cell; DNA cloning involves the use of recombinant plasmids to make many copies of a specific DNA.

Coagulase test (koh-AG-yoo-layz . . .) A test that detects an enzyme produced by most strains of *Staphylococcus aureus.*

Coagulation Blood clotting.

Coccidioidomycosis (kahk-sihd-ee-oy-doh-my-KOH-suhs) A fungal lung disease caused by *Coccidioides immitis.*

Coccus (KAHK-uhs) A spherical bacterium.

Codon (KOH-dahn) A group of three nucleotides that specifies the insertion of an amino acid into a protein.

Coenocyte (SEE-nah-syt) The type of structure that is formed when a cell undergoes nuclear division several times without the occurrence of cell division.

Coenzymes Small, complex organic molecules that

assist enzymes; they frequently consist of nucleotides bound to vitamins.

Cofactors Inorganic ions, such as iron, magnesium, calcium, or zinc, which may form the reactive center of an enzyme and assist it to combine correctly with its substrates.

Coliform organisms (KOH-luh-fawrm . . .) Lactose-fermenting, gram-negative enterobacteria, including *Escherichia coli* and related organisms, that inhabit the intestinal tract.

Collision theory An explanation of chemical reactions in terms of collisions between atoms and molecules.

Colonial morphology The visible appearance of the colonies of a microbial species.

Colony A visible heap of microbial cells on a solid culture medium.

Colony-forming unit Visible units counted in a plate count; each unit may be a group of cells.

Community-acquired diseases Diseases developed by basically healthy individuals when a virulent pathogen invades their normal defenses.

Competent cell A living recipient bacterial cell that takes up DNA in the transformation process.

Competitive inhibitors Blocking substances that compete with the normal substrate for the active site of an enzyme.

Complement The complex series of about 20 interacting serum proteins, which participate in all major aspects of bodily defense.

Complement-fixation test A serologic test; the amount of complement fixed depends on the number of antigen–antibody complexes and thus indicates the amount of antibody present.

Compound A chemical substance that contains two or more kinds of atoms.

Compound light microscope An instrument that has two sets of lenses and that uses visible light for illumination.

Compromised At greater risk for infection, because of an underlying abnormal condition or preexisting disease.

Concentration gradient The difference in concentration that causes a substance to move by diffusion from an area where it is more concentrated to an area where it is less concentrated.

Concurrent disinfection The immediate and ongoing disinfection of a patient's contaminated secretions plus any articles they soiled.

Condensation reactions Reactions in which water molecules are split off during bond formation.

Condenser A lens system located below the microscope stage that directs light rays through the specimen.

Congenital immunodeficiency Immunodeficiency present at birth and of genetic origin.

Congenital rubella syndrome A condition involving abnormalities of the eyes, ears, central nervous system, genitalia, and heart, and found in infants affected by a maternal rubella virus infection during pregnancy.

Conidia (kuh-NIHD-ee-uh) Structures, primarily reproductive in function, formed by filamentous bacteria; also asexual fungal spores.

Conidiospores (kuh-NIHD-ee-uh-spawrs) Asexual fungal spores of mycelial ascomycetes.

Conjugate (KAHN-juh-guht) A type of antibody reagent bound to an enzyme.

Conjugation A process in which DNA is transferred from one bacterial cell to another.

Consolidation A mass of gelled sputum that occurs in response to lung infection.

Constant region The part of an immunoglobulin chain in which there is little or no variation in amino acid sequence, and which determines the class of the immunoglobulin.

constitutive enzymes Enzymes (for example those that make ATP) that are essential under all conditions and are always present.

Consumers In the carbon cycle, the species that graze on plants, prey on other consumers, or are parasites.

Contact dermatitis Delayed hypersensitivity resulting from skin exposure to environmental antigens.

Continuous culture An industrial process used as an alternative to the batch culture; the microbial-cell population is maintained in constant growth, with new sterile medium continually fed in and an equal amount of used medium withdrawn.

Contractile vacuole A simple excretory organelle that eliminates unwanted water from certain unicellular organisms.

Coombs test A test for incomplete or nonagglutinating anti–red-blood-cell antibodies.

Cord factor A lipid found in cultures of *Mycobacterium tuberculosis*; it makes the bacterial cells grow in coiled arrangements.

Coronavirus (kawr-oh-nuh-VY-ruhs) A type of respiratory RNA virus that causes mild respiratory infections.

Cortex The first coat of peptidoglycan deposited on the forespore as part of bacterial sporulation.

Coryneform group (kuh-RIHN-uh-fawrm . . .) A group of club-shaped, irregular small rods that includes the bacterial species responsible for diphtheria, *Corynebacterium diphtheriae*.

Counterstain A stain of a contrasting color, used as the last step in a differential stain process.

Coupled reaction A reaction in which an exergonic and an endergonic process are linked by sharing an enzyme and/or its products and reactants.

Covalent bond A chemical bond in which the electrons of one atom are shared with another atom.

Crista A folding of the inner membrane of mitochondria.

Critical pH The pH level at which tooth enamel begins to dissolve, leading to acid decalcification.

Cross-matching The process of mixing donor and recipient blood samples to check for ABO, Rh, or other antigen incompatibilities prior to blood transfusion.

Cryptococcosis (krihp-tuh-kah-KOH-suhs) A fungal infection caused by the yeast *Cryptococcus neoformans*; the infection can spread from the lungs (where it usually causes only trivial illness) to the brain.

c-type particle A particle with the core centered in the envelope that is formed and extruded as certain viruses (leukemia, leukosis, and sarcoma viruses) bud from the cell.

Cuvette (kyoo-VEHT) An optically clear test tube used to hold a sample in spectrophotometry.

Cyanobacteria (sy-an-oh-bak-TEER-ee-uh) Large, oxygen-producing photosynthetic bacteria with a typical gram-negative cell wall.

Cyclosporin A (sy-kloh-SPAW-ruhn AY) A fungal peptide used in controlling allograft rejection; cyclosporine.

Cyst Resistant cell formed by soil bacteria, which differ from endospores in that they lack dipicolinic acid and are less heat resistant; also a resistant hard-coated cell type in some protozoa.

Cystitis Bladder infection, especially common in women.

Cytochrome (SY-tuh-krohm) A protein containing heme groups; it is chemically similar to chlorophyll and uses oxidized iron atoms as electron acceptors.

Cytochrome oxidase A respiratory enzyme found in heterotrophic bacteria that do not ferment.

Cytolytic infection (sy-toh-LIH-tikh . . .) A productive viral infection that destroys the host cell.

Cytopathic effect (CPE) (sy-to-PATH-ihk . . .) A visible change that occurs in cells when virions enter, infect the cells, and replicate.

Cytoplasm (SY-toh-plaz-uhm) Everything inside the cell membrane in a procaryotic cell.

Cytoplasmic enzymes Enzymes that are retained within the cell to catalyze the metabolic processes.

Cytoplasmic membrane See cell membrane.

Dark field microscope A microscope with a device that scatters light from the source of illumination so that the specimen appears white against a black background.

Dark reactions One part of the reaction of photosynthesis; they fix carbon dioxide into carbohydrate, depending on the products of light reaction but not directly requiring light energy.

Deamination (dee-am-uh-NAY-shuhn) The removal of an amino group from a molecule.

Death phase The period following the stationary phase when cells in a culture can no longer find energy sources to maintain life.

Decarboxylation (dee-kahr-BAHKS-ih-LAY-shun) The removal of carbon dioxide from a molecule.

Decolorizing agent An agent that contains alcohol and is used to remove the dye complex from some types of organisms.

Decomposers In the carbon cycle, the species that metabolize the wastes and dead tissues of other organisms.

Decomposition The process by which bacteria and fungi break down the indigestible parts, waste products, and dead tissues of all the other life forms.

Decontamination The removal of both living pathogens and their toxic products.

Definition The clarity of an image.

Definitive host The species in which adult parasites mature and reproduce sexually.

Degerming   Careful tissue cleansing using an antiseptic.

Degranulation   Release by the mast cell of mediators contained in its granules after contact with an allergen.

Dehydration   Loss of water from the body, which can result from moderate diarrhea.

Delayed hypersensitivity   A cell-mediated immune reaction in which symptoms appear 24 hours or more after exposure to the antigen; it is mediated by lymphokines from T lymphocytes.

Denaturation   The process by which an overheated enzyme becomes nonfunctional; the destruction of microbial proteins by heat, alcohol, or chemical treatment.

Dendritic macrophages   Phagocytes that are immobilized in the lymph nodes and attack microorganisms in the circulating lymph.

Dengue (DEHN-gee)   A mosquito-borne viral disease that causes fever, muscle ache, rash, and headache.

Denitrification   A part of the nitrogen cycle involving the loss of nitrogen to the atmosphere as nitrogen gas, resulting from microbial reduction of nitrates.

Deoxyribonucleic acid (dee-ahcks-see-ry-boh-noo-KLAY-ihk . . .)   See DNA.

Depth filter   A spongelike filter with a thick layer of porous material; liquid passes through many channels that trap microorganisms.

Dermatomycosis (dur-muh-to-my-KOH-suhs)   Mycotic (fungal) parasitism of the skin.

Dermatophyte (dur-MAT-ah-fyt)   A fungus that multiplies on skin.

Desensitization   A treatment for atopic allergy involving a series of small allergen injections.

Detergent   One of a varied group of synthetic surfactants and cleansers that have largely replaced natural soaps.

Detritus (duh-TRY-tuhs)   Parts of dead tissues and animal wastes that remain after decomposition.

Devitalization   The destruction of the live cells in the tooth pulp.

Dextran (DEHK-stran)   Sticky polymer manufactured from sugars by oral bacteria, contributing to dental plaque.

Diapedesis (dy-ah-peh-DEE-suhs)   A process by which polymorphonuclear leukocytes adhere to the capillary linings in the inflamed area of an infection and then migrate through the spaces between cells.

Dicaryon (dy-KA-ree-uhn)   A cell with two separate haploid nuclei.

Differential media   Substances that contain pH indicators, color reagents, or other materials that act to make the growth of one organism look clearly different from the growth of other neighboring colonies.

Differential stain   A stain that distinguishes objects on the basis of reactions to the staining procedure.

Differentiated tissues   Tissues of organisms with specialized structures such as digestive tubes, male and female reproductive systems, and nervous systems.

Differentiation   A process by which tissues come to express particular sets of genetic traits, leading to specialized functions of the various cell types.

Dimorphic (dy-MAWR-fihk)   Having two forms, as in certain fungi.

Dioecious (dy-EE-shuhs)   Having separate male and female individuals in a species.

Diphtheritic membrane (dihf-thuh-RIHT-ihk . . .)   The tough whitish coagulated exudate that forms in the throat or nasal passage in patients with diphtheria.

Diploid   Having at least two copies of each piece of genetic information.

Direct count   The technique for counting individual cells or colonies by placing a sample under the microscope.

Direct fluorescent-antibody (DFA) technique   A method for detecting antigen in a fixed tissue section or on individual cells from a microbial culture.

Direct transmission   Spread of disease when the pathogenic agent is conveyed to the susceptible person by direct contact (by touching, ingesting, or being bitten by the reservoir).

Disaccharide (dy-SAK-ah-ryd)   A sugar that consists of two monosaccharides.

Disinfection   The removal or destruction of pathogens, reducing the number of viable organisms in the material.

Disseminated herpes   A condition that occurs in immunosuppressed people, with herpes lesions spread over large areas of the body.

Dissociation The process of splitting a compound into ions surrounded by water.

Diverticula (dy-vur-TIHK-yuh-luh) Pouches that develop in the intestinal wall, with ulcerated areas and obstructions.

DNA The double-stranded macromolecule with the information (or "blueprint") pertaining to a living organism that directs its growth and reproduction; it contains the bases adenine, guanine, cytosine, and uracil.

DNA ligase (. . . LY-gayz) An enzyme that links single-stranded ends of free DNA; its function is DNA repair.

DNA polymerase (. . . puh-LIHM-er-ayz) The enzyme that enables DNA to be copied and new strands to be synthesized.

DNA replication The formation of an exact copy of the DNA double helix.

Double helix The configuration of the two chains of covalently linked nucleotides that constitute a DNA molecule.

Doubling time The amount of time the cells of a species take to complete their division cycle.

DPT vaccine Vaccine containing antigens that immunize against diphtheria, pertussis, and tetanus.

Drainage field The network of buried pipes laid in gravel into which effluent flows from a septic tank; there the effluent is filter-purified.

Droplet nuclei Dried residue of droplets of body fluid, with microorganisms protected by a dried mucoprotein film; they can remain airborne and spread infection.

Dry heat Sterilization by the use of heat without water vapor or steam.

Dry weight The weight of a biological sample after the water has been evaporated, as a way of estimating the number of cells.

D value The time needed to kill 90% of the organisms in a 24-hour broth culture at a given temperature.

Early proteins Proteins that appear early in viral replication; some regulate the timing of viral gene expression so that activities occur in sequence.

Edema (eh-DEE-mah) Swelling caused by fluid accumulation in the tissue in an amount greater than can be drained off by the lymphatic system.

Effluent The liquid wastes that remain after sludge is separated from sewage.

Egg The smallest, least complex stage of the life cycle of certain organisms.

Elastase (ih-LAS-tays) A microbial exoenzyme that attacks tissue components.

Electrolytes Inorganic ions such as sodium (Na^+), potassium (K^+), chloride (Cl^-), and bicarbonate (HCO_3^-).

Electron A negatively charged particle; its mass is negligible in relation to that of protons and neutrons.

Electron acceptor A molecule that serves to receive electrons metabolized from food materials.

Electron donor The substance being oxidized when electrons are removed from a substrate.

Electron microscope A microscope that uses beams of electrons instead of light to produce an image.

Electron-transport coenzymes Coenzymes that act as temporary electron acceptors, enabling transfer of electrons from one compound to another, generating ATP by oxidative phosphorylation.

Electron-transport system A sequence of electron-transport coenzymes that pass electrons from one to the next in a defined order.

Element A substance that cannot be broken down into simpler substances by ordinary chemical means.

Elementary body One of the cell types of *Chlamydia*, representing the infectious form of the microbe.

Embolus (EHM-bah-luhs) A thrombus that has completely occluded a small blood vessel.

Empirical Based on experience or direct observation.

Empyema (ehm-py-EE-muh) Infection of the pleural cavity.

Encephalitis (ehn-sehf-ah-LY-tuhs) A bacterial infection of the brain tissue.

Endemic diseases Diseases that affect a relatively constant, small percentage of a population at a time.

Endemic typhus A mild infection caused by *Rickettsia typhi*.

Endergonic reaction (ehn-der-GAH-nihk . . .) A

type of chemical reaction that requires energy.

Endogenous Internal to the body; caused by an agent from within.

Endogenous pyrogens Substances released normally by the body that can cause fever.

Endometritis (ehn-duh-meh-TRY-tus) Infection of the uterine lining, which may occur after abortion, surgery, or cesarean section; after vaginal delivery it is called puerperal fever.

Endonuclease (ehn-doh-N(Y)OO-klee-ayz) Enzyme that breaks DNA strands internally to yield uneven, single-stranded ends.

Endoplasmic reticulum (ER) (ehn-doh-PLAZ-mihk reh-TIHK-(y)oo-luhm) A membrane network in eucaryotic cells connecting the plasma membrane with the nuclear membrane.

Endospore A resting structure formed inside some bacteria.

Endosymbiosis The evolution of specialized cell structures through a process in which some free-living procaryotic cells ingest others in a mutually beneficial relationship.

Endotoxin A lipopolysaccharide that forms part of the outer portion of the cell wall of most gram-negative bacteria; it is liberated when the bacteria die and disintegrate.

Endotoxin shock A cardiovascular collapse resulting from the release of large amounts of endotoxin in the circulation.

Energy The ability to work or to drive a process.

Enteric precautions Strategies for disinfection of feces and any fomites that may be contaminated with feces; used in patients with gastrointestinal conditions.

Enterobacteria (ehn-tur-oh-bak-TEER-ee-ah) A cluster of species of gram-negative facultatively anaerobic rods that inhabit the large intestine of mammals, including *Escherichia coli*, the most familiar of all bacteria.

Enterococci (ehn-tur-oh-KAHK-see) Streptococci that are normal inhabitants of the gastrointestinal tract.

Enteropathogenic Producing enterotoxins, chemicals that induce severe diarrhea.

Enterovirus (ehn-tur-oh-VY-ruhs) An RNA virus that is transmitted via foods and liquids to the gastrointestinal tract.

Entropy (EHN-truh-pee) Energy that is not available to do work.

Envelope The outer layer of membranelike lipo-

protein bilayers surrounding the nucleocapsid in many animal viruses.

Enzyme Large globular protein molecules that act as specific catalysts, speeding up chemical reactions without being changed themselves.

Enzyme–substrate complex A combination of an enzyme and its substrate that initiates an enzymatic reaction.

Eosinophil (ee-oh-SIHN-uh-fihl) A granulocyte that moderates allergic and inflammatory responses.

Epidemic The situation that occurs when the number of cases of a particular disorder rises significantly above the expected prevalence.

Epidemic typhus fever A once-widespread disease caused by *Rickettsia prowazekii* and transmitted by the human body louse; it is characterized by abrupt onset, high temperature, headache, and rash.

Epidemiology The quantitative study of the occurrence of disease in populations.

Epstein–Barr virus (EBV) A human herpesvirus that is thought to be the cause of infectious mononucleosis and Burkitt's lymphoma.

Equilibrium point The point at which the number of conversions from reactants to products is exactly equal to that of products to reactants.

Ergot (UR-guht) A complex mixture of toxins produced by the fungal pathogen *Claviceps purpurea*.

Ergotism A disorder involving hemorrhagic breakdown of capillaries, gangrene of the extremities, hallucinations, and death; it develops in persons who have eaten grains containing ergot.

Erythroblastosis fetalis (ih-rihth-roh-blas-TOS-suhs fee-TAL-uhs) A type of graft rejection that occurs when an Rh-negative mother's anti-Rh IgG antibodies pass through the placenta, causing damage to an Rh-positive fetus.

Erythrocyte (ih-RIHTH-roh-syt) Red blood cell.

Erythrocytic cycle The third cycle in the development of the parasite *Plasmodium*, the agent that causes malaria; the merozoites created in the pre-erythrocytic cycle multiply asexually in the red blood cells.

Erythrogenic toxin The exotoxin produced by *Streptococcus pyogenes*, which causes the diffuse rash seen in scarlet fever.

Ester linkage An -O-CO- group.

Eucaryote (yoo-KAR-ee-oht) An organism whose cells have nuclear membranes, as well as other small internal structures called organelles.

Euphotic zone (yoo-FAW-tihk . . .) The part of the oceanic zone that has photosynthetic activity, which is limited to the top 200 m reached by light.

Eutrophication (yoo-troh-fih-KAY-shun) The nutrient enrichment of natural waters, usually from sources such as sewage or fertilizers and manure, often leading to the growth of excessive amounts of algae.

Exergonic reaction (ehks-er-GAWN-ihk . . .) A type of chemical reaction that gives off energy.

Exfoliatin (ehks-FOH-lee-uh-tihn) An enzyme that causes the epidermis to separate from the dermis in staphylococcal scalded-skin syndrome.

Exoenzymes Enzymes that are excreted by microorganisms to function outside the cell's membrane to hydrolyze external nutrient sources.

Exogenous External to the body; from an outside source.

Exogenous pyrogens (ehk-SAH-jeh-nuhs PY-roh-jehnz) Substances such as endotoxins that come from without and can cause fever.

Exotoxin A protein toxin released from bacterial cells into the surrounding medium.

Exotoxin A An exotoxin produced by *Pseudomonas aeruginosa* that inhibits protein synthesis in the host.

Exponential growth The increase in size of a population by a constant factor; for example, in a binary series the population doubles at each round of division.

Extended-spectrum penicillin Penicillin effective not only against gram-positive bacteria but also against most strains of certain gram-negative bacteria.

Facilitated transport Movement of small molecules across the membrane assisted by specific membrane proteins.

Facultative anaerobes Anaerobic bacteria that can grow either with or without oxygen; however, their metabolism is more efficient in the presence of oxygen.

Facultative intracellular parasites Pathogens that multiply primarily within host cells.

Farmer's lung disease A form of Arthus reaction that is provoked by inhalation of hay dust loaded with microbial spores.

Fastidious organism A microorganism that has complex nutritional requirements.

Fatty acids Long hydrocarbon chains, usually with an even number of carbons.

Fecal–oral route Method of indirect transmission of gastrointestinal disease involving ingestion of fecally contaminated material.

Feedback inhibition Inhibition of an enzyme in a particular pathway by end-product accumulation.

Feedstocks In industrial microbiology, substances such as molasses or corn syrup byproducts, cottonseed meal, or other nutrient materials used as a basis for microbial fermentation.

Fermentation An anaerobic process of enzymatic decomposition (for example, the production of wine by microorganisms); it is the metabolic method for producing ATP in cells that cannot respire.

Fermentative bacteria Bacteria that use organic compounds, such as organic acids, as electron acceptors.

Fever Elevated body temperature.

F factor A free plasmid found in the donor cell in bacterial conjugation.

Fibrin (FY-brihn) A blood-clotting protein; it accumulates in response to a lung infection.

Fibrinogen (fy-BRIHN-uh-juhn) A protein-clotting factor in plasma, which interacts with platelet factors to form blood clots.

Fimbriae (FIHM-bree-y) See pili.

First law of thermodynamics The law stating that whenever energy is converted from one form to another, it all must be accounted for; energy is neither created nor destroyed.

Fixed cells Phagocytes that are at least temporarily immobilized in various spongy filtering organs.

Fixing The process of attaching the specimen to a slide for observation.

Flagella (fluh-JEHL-uh) Long slender protein fibrils attached to one end of a cell; used for propulsion.

Flagellin A protein with molecules that aggregate by a self-assembly process into a tubular form to form flagella.

Flatworms One of the major types of helminth, including the two subgroups—trematodes and cestodes.

Flavivirus (FLAY-vuh-vy-ruhs) One of the types of togavirus; it causes yellow fever.

Fleshy fungi Mushrooms, puffballs, bracket fungi, and other solid fungal structures formed from highly organized hyphal threads.

Fluid mosaic model A way of describing the arrangement of phospholipids and proteins that make up the plasma membrane.

Flukes One of the types of flatworms; these parasites, also known as trematodes, have two suckers that they use to adhere to the host or to suck its fluids.

Fluorescein (floo(u)r-EHS-ee-uhn) A yellow-green fluorescent dye coupled to antibody and used to detect antigens.

Fluorescence The ability to give off light of one color when exposed to light of another color.

Fomite (FOH-myt) The contaminated object involved in indirect transmission of disease.

Food intoxication Illnesses caused by ingested toxin, independent of microbial growth in the gut; the toxin is produced in the food by the microbial agent before the food is ingested.

Foodborne infections Infections such as salmonellosis, in which the disease agent itself is ingested and multiplies in the gut, causing illness.

Formalin (FAWR-muh-lihn) A 37% solution of formaldehyde in water, sometimes used for disinfection; the aldehydes sterilize by altering the chemistry of proteins.

Frame-shift mutations Addition or substitution mutations that distort the way codons are read.

Free energy Energy that is available to do work.

Freeze-etching A technique in electron microscopy that gives views of inner layers of cell walls and membranes to allow the study of their architectural features.

Fruiting body A vegetative colony of myxobacteria that converge as protection against predatory bacteria or scarcity of nutrients; also, a specialized fungal structure that produces spores.

Frustule (FRUH-stool) The silica-impregnated casing of *Chrysophyta* or diatoms.

Fumigation Gas sterilization, which may be carried out in instruments similar to autoclaves.

Functional groups Arrangement of elements in organic molecules that are responsible for most of the chemical properties of those molecules.

Fungi (FUHN-gee) Organisms, including yeasts, molds, and fleshy fungi (mushrooms), that get their nourishment from dead organic matter.

Furuncle (FOO-ruhn-kuhl) An infection such as a pimple or boil, resulting from obstruction of hair follicles or sebaceous glands and multiplication of staphylococcus aureus.

Gamete (GAM-eet) Male or female sexual cell.

Gamma hemolysis The designation used when no changes are seen in the color of a blood agar medium in the area of a bacterial colony because the microorganisms are nonhemolytic.

Gas exchange Cellular respiration, which is measured either in terms of the amount of oxygen taken up or the amount of radioactive carbon dioxide given off by a culture.

Gas vesicles Membranous gas-filled sacs found in cyanobacteria and photosynthetic bacteria, allowing movement in water by changes in buoyancy.

Gastroenteritis (ga-stroh-ehn-ter-Y-tuhs) Inflammation of the stomach and intestine.

Gene A DNA segment that specifies a particular protein or polypeptide and represents the basic unit of hereditary information.

Gene expression The process of applying the information stored in the genes.

Gene replication The process of supplying information to the next generation of cells.

Gene splicing A term used to describe the assembly of recombinant DNA.

Generalized infection An infection (such as measles) that affects several tissue types and involves large areas of the body.

Generation time The length of time cells of a species take to complete their division cycle.

Genetic code The cell's "dictionary" of the base sequences in its DNA, representing the instructions for sequencing the specific amino acids to be incorporated into a protein.

Genetic engineering The manipulation of genetic material using recombinant DNA in the laboratory setting.

Genetics The study of genes and their functions.

Genome (JEE-nohm) The sum of all the genes of an organism.

Genotype The itemized designation of some or all of an organism's genes.

Genus The first part of the binomial designation of an organism, roughly equivalent to a person's family surname.

Geometric progression A number series in which each number is derived from the previous one by multiplying it by a constant factor.

Germ   A term, meaning "seed," that was originally applied to microorganisms.

Germ-free   Lacking normal flora, as in experimental animals delivered by surgery and raised in sterile conditions.

Germ theory of disease   The theory that states that infectious diseases are caused by living microorganisms and that one type of organism causes one specific type of disease.

Germ tubes   Structures formed by yeast cells of *Candida albicans* when incubated in serum.

Germinal center   The site at which active B cells appear in the lymphoid tissue after antigen contact.

Germination   The process by which an endospore returns to the vegetative state.

Giant cells   Nonfunctional host cells that are fused by the action of certain viral proteins.

Giardiasis (jee-ar-DY-uh-sihs)   Diarrheal infection due to the intestinal parasite *Giardia lamblia*.

Gingivitis   Inflammation of the gingiva, which become swollen and red, and bleed easily.

Gliding   Type of movement seen in cyanobacteria and other bacteria that have no external structures for motility.

Glomerulonephritis (gloh-mur-(y)oo-luh-neh-FRY-tuhs)   A disease that results from invasion of the glomerular capillary bed of the kidneys by immune complexes.

Glycocalyx (gly-koh-KAL-ihks)   The capsule or slime layer that forms the coating of procaryotes.

Glycolysis (gly-KAHL-uh-sihs)   The metabolic pathway for the first stage of the cell's oxidation of glucose; it involves the conversion of glucose to pyruvic acid.

Golgi body (GAHL-jee . . .)   A stack or complex of flattened oval membrane disks in a eucaryotic cell.

Gonococcal conjunctivitis (gahn-uh-KAHK-uhl kuhn-juhnk-tuh-VY-tuhs)   An eye infection caused by *Neisseria gonorrhoeae*; it may cause blindness in an infant acquiring it from an infected mother during vaginal delivery.

Gonorrhea   The sexually transmitted disease caused by the Gram-negative aerobic diplococcus *Neisseria gonorrhoeae*.

Gram-negative bacteria   Bacteria that lose the purple color but show the red counterstain in the Gram-stain test; they typically have a multilayered cell wall.

Gram-positive bacteria   Bacteria that retain the first purple stain in the Gram-stain test; their cell walls contain a single thick layer of peptidoglycan.

Gram stain   A stain that differentiates among different types of bacteria on the basis of their reaction to the staining material.

Grana (GRAY-nuh)   Interconnected stacks of thylakoids in chloroplasts.

Granulocyte   A leukocyte with granules in the cytoplasm; includes neutrophils, basophils, and eosinophils.

Granuloma   An area of inflamed tissue containing lymphocytes, macrophages, and giant cells.

Gray-baby syndrome   A fatal anemia that can occur in newborns given excessive doses of chloramphenicol.

Gray water   Wash water, or water from the bathtub or shower, dishwater, and water used for washing clothes.

Ground water   The water that percolates down through the soil into the underlying rock, some of it remaining in underground aquifers and some seeping back to the oceans.

Growth limitation   The mechanism that makes cell division cease after a certain population density is reached.

Guillain–Barré syndrome (gee-YAHN bar-RAY . . .)   A neurologic syndrome marked by muscular weakness and paralysis; some cases occur following vaccination for a strain of influenza virus.

Gummas   The large ulcers that may occur in any organ in the benign form of tertiary syphilis.

Half-life (of a drug)   The amount of time needed for the serum level of a drug to be reduced by 50% from its peak.

Halobacteria (hal-oh-bak-TEER-ee-uh)   Members of the archaebacteria group that inhabit highly saline environments such as the Dead Sea and pickling brines.

Halogens (HAL-uh-juhnz)   Highly reactive elements such as fluorine, chlorine, and iodine that characteristically exist as anions with single negative charges.

Halophile (HAL-uh-fyl)   A salt lover; an organism that lives in environments containing 15 to 30% dissolved salts.

Haploid   Having only one copy of each piece of genetic information.

Hard chancre (. . . SHANG-kur)   An open, painless ulcer that represents the first sign of syphilis and occurs at the site where the spirochetes entered.

Heat-labile enterotoxin (. . . LAY-byl . . .)   A toxin readily destroyed by heat, formed by some pathogenic strains of *Escherichia coli* that cause diarrhea.

Heavy chain   A chain of about 440 amino acids contained in an immunoglobulin molecule; the type of heavy chain distinguishes among the classes of immunoglobulins.

Helical   Having a coiled arrangment.

Helminths   Simple invertebrate animals (rather than microorganisms) that are multicellular and have differentiated tissues; the two basic types are roundworms and flatworms.

Hemagglutination (hee-muh-gloot-uhn-AY-shuhn)   Clumping or agglutination of the red blood cells; used to match donor and recipient blood samples before transfusion.

Hemagglutination-inhibition (HI) test   A two-step test to detect antiviral antibodies in patient serum.

Hemagglutinins   Red-blood-cell–aggregating proteins.

Hemolysin (hee-mah-LYS-uhn)   An enzyme that destroys red-blood-cell membranes.

Hemolysis (hee-MAHL-uh-sihs)   The ability to destroy red blood cells.

Hemorrhagic colitis (heh-maw-RAHJ-ihk koh-LY-tuhs)   Colon inflammation with bleeding from the intestinal wall.

Hepadnavirus   A tiny virus with single-stranded DNA that causes hepatitis B.

Herd immunity   The percentage of a population that is functionally immune to an agent.

Herpesvirus (HUR-peez-VY-ruhs)   A virus that causes a wide range of infections, including a malignant tumor, a contagious skin disease (chickenpox), and the latent infections involved in cold sores and genital herpes.

Heterocysts   Specialized cells that are involved in nitrogen fixation in cyanobacteria.

Heterograft   Tissue transferred between members of different species (for example, the use of pigskin as a temporary covering over a burn wound in a human being).

Heterophil antibodies   Antibodies that agglutinate specially treated sheep, horse, or ox red blood cells; their presence is diagnostic or infectious mononucleosis.

Heterotrophs   Organisms that require organic carbon as their energy source.

Hfr   High frequency of recombination, a term applied to bacteria that transmit not only the plasmid DNA but attached portions of their own bacterial DNA in conjugation.

Hierarchical classification   The arrangement of groups of closely related species in genera, groups of similar genera in families, and groups of families in orders.

High-efficiency particulate air filters (HEPA)   Filters with a pore size so small that the air that passes through is rendered sterile as well as clean; such filters clean the air supplied to operating rooms or protective-isolation units.

Histamine (HIHS-tah-mihn)   A chemical mediator that dilates blood vessels.

Histiocyte (HIHS-tee-uh-syt)   A phagocytic cell that is a more mature form of polymorphonuclear leukocyte.

Histocompatibility antigens (hihs-toh-kuhm-pat-uh-BIHL-uht-ee . . .)   Glycoproteins determined by the HLA genes on chromosome 6, which unambiguously label an individual's cells as self.

Histones (HIHS-tohnz)   Clusters of basic proteins associated with DNA in eucaryotic cells.

Histoplasmin (hihs-toh-PLAZ-mihn)   An antigen derived from cells of *Histoplasma capsulatum.*

Histoplasmosis (hihs-toh-plaz-MOH-suhs)   A widely distributed mycosis caused by *Histoplasma capsulatum*, a dimorphic fungus of the deuteromycete group.

Homologs (HOH-muh-lahgs)   Portions of genetic material with very similar base sequences.

Homopolymer (hoh-moh-PAH-lee-muhr)   A substance composed of only one type of monomer.

Hook   A portion of a flagellum attached to the filament and connected to a basal body.

Hospital-acquired infections   Also called nosocomial infections; disorders developed in hospitals or other health-care facilities, often caused by opportunist organisms in patients who are already ill.

Host   An organism infected by a pathogen.

Humus   Nutrient-rich organic matter, with undecomposed plant residues; this substance enriches the soil below, increasing the depth of the topsoil.

Hyaluronic acid (hyl-yoo-RAHN-ihk . . .) A mucopolysaccharide found in the epidermis.

Hyaluronidase (hyl-yoo-RAHN-uh-days) A microbial enzyme that digests the epithelial cementing substance.

Hybridoma (hy-brih-DOH-muh) The hybrid result of the fusion of two kinds of cells, producing monoclonal antibodies.

Hydatid disease (HY-duh-tihd . . .) A tapeworm disease in which the human beings are intermediate hosts for the canine tapeworm *Echinococcus granulosus.*

Hydrocarbon A compound that contains only carbon and hydrogen.

Hydrogen bond The attraction between neighboring molecules that have a hydrogen atom bonded to either oxygen or nitrogen; the bond forms when the positive end of one molecule is attracted to the negative end of the other molecule.

Hydrolases Digestive or degradative enzymes that hydrolyze covalent bonds linking monomers into polymers.

Hydrolytic reaction A process in which a molecule of water is added to the reacting molecule as it splits apart; the reverse process from condensation.

Hydrophobia The painful spasm caused by an uncoordinated swallowing reflex, which occurs in rabies.

Hydrophobic compounds Compounds that are nonpolar and will not dissolve in water.

Hydroxyl group (hy-DRAWKS-uhl . . .) A functional group that consists of an oxygen attached to a hydrogen—the –OH combination.

Hypersensitivity An abnormal physiologic state resulting from an immune reaction.

Hypertonic A solution that contains more dissolved materials than the cell's cytoplasm.

Hyphae (HY-fee) The long microscopic filaments formed by molds, which may form the visible mat called a mycelium.

Hypotonic A solution that is more dilute than cytoplasm.

Hypovolemia (hy-poh-voh-LEE-mee-ah) Abnormally low blood volume.

Iatrogenic (y-a-trah-JEHN-ihk) Caused by a therapeutic measure; for example, an infection following the traumatic insertion of an intravenous catheter is an iatrogenic infection.

Icosahedron (y-koh-suh-HEE-druhn) A 20-sided geometric structure.

Icteric phase (ihk-TEHR-ihk . . .) The part of a hepatitis illness that is accompanied by jaundice.

Identification of isolates Technique for transferring pure cultures from colonies on primary isolation streak plates and comparing the results with standard microbial test profiles.

Ig See immunoglobulin.

IgA Immunoglobulin A, which is mainly produced by lymphoid tissue in the lining of the gastrointestinal and respiratory tracts; it is secreted on body surfaces and can inactivate pathogens before they infect us.

IgD A four-chain antibody that is produced only in trace amounts; it is believed to be a lymphocyte cell-surface receptor.

IgE A four-chain antibody that binds antigen on the surface of basophils and mast cells; it is responsible for allergic reactions and may also help control helminthic infections.

IgG The smaller of the two major circulating antibodies, acting to neutralize toxins; it crosses the placenta and protects the fetus.

IgM The larger of the two major circulating antibodies, and the first produced in the infant.

Illegitimate recombination Recombination of DNA from different species.

Immediate hypersensitivity An immune reaction involving immunoglobulins in which symptoms occur within minutes after contact with the antigen.

Immobilized-enzyme technique An industrial microbiological process in which the microbial enzymes are bonded onto an inert supporting surface and a solution of the substrate to be acted on is passed over the enzymes; the product is recovered from the effluent.

Immune-complex disease A condition that occurs when more immune complexes form than the phagocytes can dispose of.

Immune complexes Circulating combinations of soluble antigen, antibody, and complement that form as a normal outcome of protective immunity.

Immune-electron microscopy A process for visualizing antigen–antibody complexes by complexing one of the reactants with an electron-dense substance.

Immune tolerance   The learned response of the body's immune system in recognizing self-antigens.

Immunity   The state of being effectively protected against infection from strains of individual bacteria, viruses, fungi, or other agents.

Immunization   The active or passive process that protects individuals by increasing their resistance to an infectious disease.

Immunodeficiency   The condition in which the body's immune defenses are reduced or absent.

Immunodiffusion plate   A simple precipitation method involving an agar-gel medium.

Immunoelectrophoresis (ihm-yuh-noh-uh-lehk-troh-fuh-REE-suhs)   A multistage precipitation test that separates antigen mixtures and exposes them to antibody to obtain purified complexes.

Immunoglobulin (Ig) (ihm-yuh-noh-GLAH-b(y)oo-lihn)   An antibody; a defensive protein synthesized by B cells.

Immunology   The study of the immune system and its reactions.

Immunosuppressive therapy   A treatment for autoimmune disease; used also to suppress graft rejection, control inflammatory conditions, and regulate fluid retention.

Impetigo (ihm-pah-TY-goh)   A superficial infection of the skin found in young children; it is caused by bacterial agents that invade a scratch or insect bite and is spread by scratching.

Inapparent infection   An infection with no clinically obvious damage.

Incidence   The number of new cases of a disease arising within an area in a set period of time.

Inclusion bodies   Stored substances, such as glycogen, starch, oils, or other lipids, that the cell can withdraw and use for energy; also dense aggregates of viral material in an infected host cell.

Inclusion conjunctivitis   An eye infection caused by chlamydial genital disease.

Incomplete   Nonagglutinating; a term applied to anti–red-blood-cell antibodies that will destroy red blood cells in the body but will not agglutinate them in the test tube.

Incubation period   The period of time during which the infective agent is multiplying in the body but not yet causing symptoms.

Indicator organisms   Organisms that are not usu-

ally pathogenic themselves but represent a sign of fecal contamination.

Indirect fluorescent-antibody (IFA) technique   A method for detecting antibody already bound to specific tissue antigens in the diagnosis of some immune complex and autoimmune diseases.

Indirect transmission   Spread of disease when the pathogenic agent is conveyed to the susceptible person from food, drink, clothing, or objects contaminated by contact with the reservoir.

Induced mutation   A change in the base sequence of DNA provoked by contact with a chemical or with radiation.

Inducer   A substrate that brings about increased transcription, resulting in production of an increased amount of an enzyme.

Inducible enzymes   Enzymes that are produced only when the cell has a specific need or opportunity to use them.

Induration (ihn-d(y)uh-RAY-shuhn)   A hardened area in tissue caused by inflammation.

Industrial microbiology   The controlled use of microorganisms in the manufacture of saleable products.

Infant botulism   Also known as the "floppy-baby syndrome," A condition caused by botulinum toxin, most commonly from ingestion of raw honey containing *Clostridium botulinum* spores.

Infection   The condition that occurs when a microorganism gains access to a tissue or an area of the body in which it is not normally found and then multiplies, or, in the case of a normal flora organism, multiplies to an abnormally large population.

Infectious disease   The disease that results when an infection caused by a microorganism interferes with our normal body processes.

Infectious hepatitis (hepatitis A)   The short-incubation disease involving liver inflammation, often with jaundice, caused by the hepatitis A virus.

Infectious mononucleosis (. . . mahn-oh-n(y)oo-klee-OH-suhs)   A debilitating but self-limited disease caused by the Epstein–Barr virus.

Infective   Actively releasing viable pathogens.

Infective endocarditis (. . . ehn-doh-kahr-DY-tuhs)   Inflammation of the endocardium, which may occur in acute or subacute forms; in the acute form, rapidly progressive disease may destroy normal heart valves.

Inflammation   A nonspecific response of the body

to trauma or infection; a sequence of events with swelling, reddening, fever, and pain.

Initial body The larger, thin-walled type of cell developed in *Chlamydia* from the elementary body.

Initial testing Testing method for standardizing a disinfectant that uses pure cultures.

Initiation The first stage of protein synthesis.

Inoculum (ihn-AHK-yuh-luhm) A sample of microorganisms introduced into a culture medium.

Inorganic compounds Compounds that do not contain carbon.

Integrated pest management The combined use of chemicals, improved agricultural methods, and infective pathogens to control an insect pest.

Integration The process through which donor genetic material in bacterial recombination becomes a permanent part of the recipient cell's genome.

Interferon (ihnt-uhr-FIH(UH)R-ahn) A special class of proteins involved in our natural defenses; they do not act specifically against a particular virus but are species-specific.

Interleukin (ihnt-uhr-L(Y)OO-kuhn) A substance produced by lymphocytes, which has a key role in signaling other white blood cells to coordinate the body's defenses.

Intermediate host The species in which immature forms of a parasite develop.

Intermittent sterilization A process in which boiling and incubation periods are alternated to eliminate spores.

Intertidal zone The area between the high-tide line and the low-tide line; an area rich in microorganisms, as well as macroscopic algae.

Intertrigo (ihn-tur-TRY-goh) Dermatitis on areas where two layers of skin rub together, as between the fingers, under the breasts, and between the buttocks.

Intracellular parasites Parasites that live and reproduce within host cells.

Intramuscular (IM) administration The injection of a drug directly into the tissues, or parenterally.

Intrauterine infection (ihn-trah-YOO-ter-uhn) An infection of the fetus with virus particles that came from the blood of a mother who has viremia; a few cellular bloodborne pathogens can also have this effect.

Intravascular coagulation Clotting within capillaries that occurs near the focus of an infection with anaerobic bacteria.

Intravenous (IV) infusion The administration of a drug or other substance directly into a vein.

In vitro (ihn VEE-troh) "In glass," or in an artificial, laboratory setting.

Iodophor (y-OHD-uh-faw(uh)r) A form of disinfectant in which iodine is chemically bound to a surfactant carrier.

Ionic bond (y-AHN-ihk . . .) The attraction between positive and negative ions.

Ionizing radiation High-energy radiation of wavelengths shorter than ultraviolet, which can be used as a method of sterilization.

Ions (Y-ahns) Electrically charged particles derived from atoms.

Isoantibodies Antibodies to any ABO antigens that an individual does not possess.

Isograft Tissue transferred from one identical twin to another.

Isolation A technique of patient care to prevent the spread of infection involving a private room, with special protective measures imposed on all persons and materials entering and leaving the room.

Isomerases Enzymes that reshuffle atoms within molecules without adding or subtracting any.

Isotope A form of a chemical element in which the number of neutrons in the nucleus differs from the other forms of that element.

Jaundice An accumulation of bile pigments giving a yellow tinge to skin and eyes, a feature of hepatitis.

Kinetoplast (kih-NEH-toh-plast) An organelle embedded in the cell membrane of some protozoa; chemical reactions in it power the organism's flagella.

Kinins (KY-nihns) Chemical mediators that cause smooth-muscle contraction.

Kirby–Bauer disk diffusion method A method of testing the sensitivity of microorganisms to particular antimicrobial agents.

Koch's postulates The logical steps that are used to demonstrate the cause-and-effect relationship between a specific microorganism and a specific infectious disease.

Koplik's spots Lesions on the oral mucosa that represent the first sign of measles.

Krebs cycle A cyclical series of metabolic reactions that free up electrons for electron transport, so that their energy can be used to make

ATP; also known as the tricarboxylic-acid cycle or the citric-acid cycle.

Lag phase The period of time during which the inoculated cells adjust to the culture medium; only minimal growth and no cell division occurs.

Lambda A bacteriophage that is a temperate virus of *Escherichia coli*.

Lamellae (lah-MEHL-ee) Layers of parallel-folded membrane material containing light-capturing apparatus in procaryotic cyanobacteria.

Laminar-flow transfer hoods Work areas with a unidirectional flow of air drawn inward over the work surface, protecting the worker sitting outside from airborne contamination by the cultures being worked on.

Larva (plural, **larvae**) Immature, rapidly growing form of certain parasites and other invertebrate animals.

Late proteins Proteins that appear when the viral replication process is well advanced.

LE test A blood test to determine whether a patient has systemic lupus erythematosus.

Legionnaire's disease Also known as legionellosis; a severe pneumonialike disease caused by gram-·negative bacilli identified after an outbreak at an American Legion convention in 1976.

Legitimate recombination Recombination of DNA from genetic variants of the same species.

Lepromatous leprosy (lehp-ROH-muh-tuhs . . .) The more severe form of leprosy, which develops when cell-mediated immunity is inadequate; there are multiple spreading lesions, bacteremia develops, and the organism invades internal organs.

Lepromin (LEHP-roh-mihn) *Mycobacterium leprae* antigen, used as a skin test for leprosy.

Leprosy A slowly progressive peripheral-nervous-system infection caused by *Mycobacterium leprae* and initially involving the peripheral nerve endings in the skin.

Leptospirosis (lehp-to-spy-ROH-suhs) A kidney infection caused by spirochetes and spread by contaminated urine.

Leukemia A cancer that involves proliferating malignant cells in the bone marrow and circulating abnormal blood cells.

Leukocyte White blood cell.

Leukocytosis Greatly increased numbers of circulating white blood cells, which occurs in infections such as acute appendicitis.

Leukopenia An abnormally low number of circulating white blood cells, as in typhoid fever and influenza.

L forms Irregularly shaped mutant variants of normal bacteria, with defective cell walls.

Lichen (LY-kuhn) A symbiotic partnership between a fungus and a cyanobacterium or alga.

Ligases Enzymes that condense monomers together to form polymers (the opposite action to that of hydrolases); they are also called synthetases and polymerases.

Light chain A chain of about 220 amino acids contained in an immunoglobulin molecule.

Light reaction One of the reactions of photosynthesis; they occur only while light is provided and only take place in membrane structures, which ensure the correct spatial orientation of the components.

Limnetic zone (lihm-NEH-tihk) The surface waters of a body of water.

Lipases Enzymes that separate fatty acids from glycerol.

Lipids Organic compounds that will not dissolve in water; the simplest lipids (fats and oils) are the triglycerides.

Lithotrophs Organisms that use inorganic carbon sources such as carbon dioxide or carbonate ions for growth.

Littoral zone (LIHT-uh-ruhl . . .) The shoreline area of a body of water, where productivity is greatest.

Loading dose An initial dose of a drug that is intended to achieve a desired serum level of that medication.

Localized infection An infection with the damage restricted to one area of the body, such as a viral infection of the skin or the respiratory tract.

Logarithmic death curve A description of the relationship between equal time intervals and the number of deaths that occur in that time interval when the same fraction of the remaining population dies in each time interval.

Log phase The period during which cells in a culture medium grow at an exponential rate.

Long-incubation disease A disease in which a relatively long time elapses between exposure to the infectious agent and the onset of symptomatic illness; term used to describe Hepatitis B.

Lophotrichous (loh-FAH-trih-kuhs) Having two or more flagella at one pole.

Luciferase (loo-SIHF-uh-rayz) The light-generating enzyme that makes *Photobacterium* luminescent.

Luminescent Producing yellowish light when actively respiring.

Lyases Enzymes that remove groups from compounds.

Lymphadenitis (lihm-fuh-deh-NY-tuhs) Swollen lymph nodes, a sign of infection; the infective agent trapped in the lymphoid tissue causes antigenic stimulation and increases the mitotic rate of lymphocytes.

Lymphocyte A type of agranulocyte; lymphocytes transmit chemical signals to help coordinate the immune system.

Lymphokines (LIHM-fuh-kynz) Signaling factors produced by the T cells.

Lyophilizing (ly-AHF-uh-ly-zing) Freeze-drying, or freezing a substance and evaporating the ice in a vacuum.

Lysis (LY-suhs) Irreversible disruption of the plasma membrane of a cell.

Lysogenic conversion The process by which the host's phenotype is changed when genes controlled by prophages are expressed by the host.

Lysogenic viruses See temperate viruses.

Lysosomes Organelles that act as internal reservoirs for digestive enzymes in eucaryotic cells.

Lysozyme An enzyme found in plasma that catalyzes the breakdown of peptidoglycan.

Lytic infection A productive infection that destroys the host cell.

Macroconidia Larger conidia, or fungal spores.

Macrogametocyte (mak-ro-gah-MEET-uh-syt) The female gamete of merozoites that carry out meiosis in red blood cells of an individual with malaria.

Macromolecules Extremely large organic molecules.

Macrophage (MAK-roh-fayj) A phagocytic cell; an enlarged monocyte.

Magnification The increase in the diameter of the image, depending on the curvature of the lenses.

Malaria A disease caused by one of the sporozoan parasites of the genus *Plasmodium*; symptoms of the disease reflect serious destruction of red blood cells.

Malignant tumors Tumors that tend to "seed" new tumor foci, called metastases, throughout the body.

Malt beverages Beverages such as beer and ale, which are produced from grain that has been allowed to sprout; the resulting mash is boiled and fermented by yeasts.

Manganese nodules Nodules of metals precipitated from seawater by microorganisms, found on the ocean bottom.

Mantoux tuberculin skin test (man-TOO . . .) A test for reactivity to tuberculosis antigen.

Mast cell A cell found in connective tissue; it contains histamine, which is released in response to injury.

Mastitis (ma-STYT-uhs) Infection of the mammary tissue of the nursing mother.

Mastoiditis (mas-toyd-YT-uhs) Infection of the mastoid cells in the temporal bone beneath the ear.

Means of transmission The method by which an organism is spread from one host to another.

Mechanical vector An arthropod involved in indirect transmission of disease.

Megakaryocyte (mehg-uh-KAR-ee-oh-syt) A cell produced in bone marrow, which is involved in the production of platelets.

Meiosis (my-OH-suhs) The cell division in which only one of each pair of parental chromosomes is placed into each gamete.

Membrane filter A sieve filter made of a thin sheet of a synthetic polymer such as cellulose acetate, used to remove microorganisms from fluids.

Membrane-filter technique The technique of counting bacteria colonies by passing the sample through a membrane filter under vacuum; the microorganisms on the filter surface are then placed on a medium where they can develop into colonies.

Meningitis (meh-nihn-JY-tuhs) A bacterial infection of the meninges; two of the more common forms of this serious illness are hemophilus meningitis and meningococcal meningitis.

Meningococcemia (muh-nihn-guh-kuhk-SEE-mee-uh) Invasion of the blood with meningococcal organisms.

Merozoite (mehr-uh-ZOH-yt) A form derived by the division of sporozoites in human beings infected with malaria.

Mesophile (MEHZ-uh-fyl) An organism that grows best at moderate temperatures, between 20 and 50 degrees Celsius, including the normal flora and pathogens of humans and other mammals.

Mesosome (MEHZ-uh-sohm) Infolding of the

membrane of many Gram-positive bacterial cells.

Metabolic pathway   The sequences of events in which a compound that results from one enzymatic reaction may be further modified when it becomes the substrate for a different enzyme in the next reaction.

Metabolism   The sum of all the chemical reactions that occur in the cell, including condensation, hydrolysis, and oxidation–reduction.

Metacercaria (meh-tuh-ser-KA-ree-uh)   One of the larval stages of the fluke.

Metal ions   The ions of mercury, silver, arsenic, and copper, used as pesticides and disinfectants.

Metastases (muh-TAS-tuh-seez)   New tumor foci that have broken away from the main tumor mass.

Methanogens   Anaerobic archaebacteria that produce the gas methane.

Michaelis–Menton equation (muh-KAY-lees . . .)   The equation describing the stepwise process by which an enzyme combines with its substrate, acts on it, and releases the product.

Microaerophile (my-kroh-AY-uhr-oh-fyl)   An organism that prefers lowered oxygen concentrations and increased carbon dioxide, a combination found in the tissues and body fluids of animals.

Microbial antagonism   Destruction or inhibition of some microbes by other microorganisms.

Microbial load   The initial number of microorganisms in a particular population.

Microbiology   The study of microorganisms.

Microconidia (my-kroh-kuh-NIHD-ee-ah)   Smaller, usually single-celled conidia or fungal spores.

Microfilaments   Slender contractile protein fibrils in eucaryotic cells; they are active in movement and in cell division.

Microgametocyte (my-kroh-gah-MEET-ah-syt)   The male gamete of merozoites that carry out meiosis in red blood cells of an individual with malaria.

Micrographs   Photographs taken through the microscope.

Microorganisms   Living things so small they can only be seen through a microscope; examples include bacteria, fungi, algae, and protozoa.

Microscopy (my-KRAHS-kuh-pee)   The study of small objects or details by the formation of magnified images.

Microtiter techniques (MY-kroh-ty-tuhr . . .)   Highly accurate methods of titration using plastic plates with tiny wells rather than test tubes.

Microtubules   Hollow protein cylinders in eucaryotic cells.

Midstream sample   A urine sample taken by having the subject partially empty the bladder and then urinate into a sterile container.

Miliary tuberculosis (MIHL-ee-ay-ree . . .)   A now rare condition in which mycobacteria cause bacteremia and multiple lesions throughout the body; it commonly resulted from drinking unpasteurized milk from infected cows.

Minimum inhibitory concentration   The concentration of an antimicrobial drug that arrests growth.

Minimum lethal concentration   The concentration of an antimicrobial drug that kills the organisms.

Miracidia (mih-ruh-SIH-dee-(y)uh)   One of the larval stages of the fluke.

Mitochondria (myt-uh-KAHN-dree-uh)   Specialized organelles that trap the energy released from the breakdown of food molecules by forming ATP.

Mitosis (my-TO-suhs)   The process by which eucaryotic cells divide.

Mixed cultures   Cultures that contain many species.

Modification   The enzymatic changes that mark the DNA as unique to the bacterial species in whose cells it occurs.

Molds   One of the basic types of fungi, forming hyphae that may result in the visible mat called a mycelium.

Molecular genetics   The study of nucleic acid and protein molecules.

Molecule   A combination of atoms forming a specific chemical compound.

Moniliasis (moh-nih-LY-ah-sihs)   Vulvovaginitis caused by the yeast *Candida albicans*, characterized by itching or burning sensations and the secretion of a copious exudate.

Monoclonal antibodies   Antibodies that result from a single clone of lymphocytes, used to obtain a large population of cells in a pure cell culture.

Monocyte (MAHN-uh-syt)   Large phagocytic agranulocyte with oval or U-shaped nuclei; circulates in the blood.

Monolayer   A sheet of cells in a culture in which the layer is only one cell thick.

Monomers   Small molecules; the units that combine to form polymers.

Monosaccharides (mah-nuh-SAK-uh-rydz)   Simple sugars that contain three to five carbon atoms.

Monotrichous (muh-NOHT-rih-kuhs)   Having a single flagellum.

Morbidity rate   The percentage of exposed individuals who actually develop the disease.

Mordant   A substance added to dye to make it stain more intensely.

Morphology   The physical appearance of an organism or object.

Mortality rate   The percentage of individuals with a disease who die.

Most probable number (MPN) method   The multistep fermentation technique used to test water or food for contamination by coliform bacteria.

Motility   The ability of a cell to move.

Mucociliary elevator (myoo-koh-SIH-lee-ayr-ee)   The defense mechanism by which coordinated ciliary motions sweep mucus toward the pharynx.

Mucoid colonies   Colonies of organisms that produce glycocalyx.

Multicellular organisms   Organisms with groups of cells that work together.

Multiple-drug resistance   An effect that occurs when clusters of genes conferring resistance to several antibiotics are moved by transposons among bacteria of the same or different species, transferring drug resistance.

Murine typhus   A mild infection caused by *Rickettsia typhi.*

Mutagen (MYOO-tuh-jehn)   A chemical or physical condition that increases the observed rate of mutation.

Mutation   A chemical event that involves a change in the nucleotide sequence of the DNA.

myc   The oncogene found in the virus of avian myelocytomatosis.

Mycelium (my-SEE-lee-uhm)   Long branching bacterial or fungal filaments that grow over a solid surface to form a tangled mat.

Mycology (my-KAHL-oh-jee)   The study of fungi.

Mycoplasmas   Naturally occurring procaryotes that lack a cell wall entirely.

Mycorrhizae (my-kuh-RY-zee)   Symbiotic associations of fungal mycelia with the roots of certain plants.

Mycoses (my-KOH-seez)   Human fungal diseases.

Mycotoxins   Toxins produced by fungi, including carcinogens, mutagens, and hemorrhagic agents.

Myeloperoxidase (my-uh-loh-pah-RAHK-suh-dayz)   An enzyme complex produced by phagocytic cells that releases hydrogen peroxide and superoxide, which are rapidly lethal to many anaerobic microbes.

Myocarditis   Inflammation of the heart muscle, leading to impaired heart contractions; most cases are caused by viruses, but it may also result from bacterial infections.

Myxobacteria   Gliding bacteria with a complex developmental cycle; they live primarily by breaking down other bacteria and metabolizing their contents.

Myxospores   Resistant resting cells within the fruiting body of myxobacteria.

Myxovirus   A type of RNA virus that causes influenza, parainfluenza, measles, and mumps; they have envelopes containing hemagglutinin and neuraminidase.

Narrow-spectrum   Term applied to a drug useful against a limited group of organisms.

Nascent oxygen   The chemically reactive free forms of oxygen, yielded by unstable peroxides.

Natural-killer (NK) cells   Large granular lymphocytes, which are not phagocytic; they kill target cells including tumor cells and virus-infected cells.

Natural passive immunity   The protective mechanism passed from the mother to the fetus and nursing baby, representing the main immunologic protection of the newborn child.

Natural selection   The tendency for the most efficient organisms to be more likely to survive and reproduce themselves, while the less adaptable die out.

Natural taxonomy   A classification that groups organisms descended from a common ancestor.

Naturally acquired active immunity   The protective state that follows an actual case of the disease, even if it did not involve symptoms.

Necrosis (nuh-KROH-sihs)   Death or deterioration of tissue.

Negative pressure   In a transfer cabinet or a room, air at a pressure slightly lower than that of the surrounding area, directing air flow inward.

Negative selection   The process by which an undesirable tendency is eliminated by removal of inappropriate members of the population.

Negative stain   A procedure that colors the background of the slide and leaves the organisms uncolored so that they stand out in contrast.

Negri bodies   The viral inclusion bodies found in nerve cells, diagnostic of rabies.

Nekton   Swimming animals, as opposed to plankton, which are floating animals.

Neonatal sepsis   The multiplication of microorganisms in the blood in infants in the first weeks of life; a leading cause of neonatal death.

Neoplasm   A group or clone of cells with a more rapid growth rate than its surrounding tissues; a tumor is a solid or localized neoplasm. The term literally means "new growth."

Nephritogenic antigen   An antigen that provokes glomerulonephritis after certain streptococcal infections.

Nephrotoxicity   The capacity of a substance to place stress on or damage the kidneys.

Neritic zone   The portion of the ocean's waters that lies over the continental shelf, between the intertidal zone and the oceanic zone.

Neuraminimidase (n(y)ur-uh-MIHN-uh-days)   An enzyme that aids in invading the respiratory tract.

Neutral solution   A solution that contains an equal mixture of hydrogen ions and hydroxyl ions.

Neutralization test   A test of the ability of antibodies to incapacitate pathogens.

Neutron   Electrically neutral particle with approximately the same mass as a proton.

Neutrophils (N(Y)OO-trah-fihlz)   Also called polymorphonuclear leukocytes, these phagocytic granulocytes are the most numerous white blood cells; large reserves of them are released in certain infections.

9 + 2 arrangement   The configuration involving a circle of nine double microtubules surrounding a pair of single microtubules in a eucaryotic cell.

Nitrification (ny-trah-fuh-KAY-shuhn)   The fifth step in the nitrogen cycle, involving the production of nitrate by a microbial activity oxidizing ammonium to nitrate.

Nitrogen cycle   The combination of events involved in the conversion of nitrogen to organic substances and back to nitrogen; as one part of this process, bacteria remove nitrogen from the air and provide it to plants.

Nitrogen-fixing   Removing nitrogen from the atmosphere and fixing it into nitrogenous organic compounds as part of the nitrogen cycle.

Nitrosamine (ny-trohs-AH-meen)   The carcinogenic substances that result when our normal intestinal flora reduces the nitrite ion in food we digest.

Nodules   In the nitrogen cycle, the swellings on the root tissues of host plants caused by bacterial nitrogen-fixers.

Noncompetitive inhibitors   Substances that do not compete with the substrate for the active site of an enzyme, but instead bind the enzyme irreversibly at another location.

Nongonococcal urethritis (NGU) (nahn-gah-nuh-KAH-kuhl yoo-ree-THRY-tus)   Urethral infection with organisms other than the gonococcus or the syphilis spirochete.

Nonpolar bond   A type of covalent bond in which all parts of the bond are electrically neutral.

Non-self   An antigenic substance not normally found in the body.

Nonspecific vulvovaginitis (NSV) (. . . vuhl-voh-va-jihn-NY-tuhs)   Vulvovaginal infection with organisms other than the gonococcus or the syphilis spirochete.

Nonviable   Not capable of reproducing.

Normal flora   The microorganisms that are found on our skin, in our noses, throats, and other body openings, and in our digestive tract; they rarely make us sick and create an environment that competes with harmful microorganisms.

Norwalk agent   A parvovirus that is perhaps the most common gastrointestinal virus affecting adults.

Nosocomial (nahz-uh-KOH-mee-uhl)   Caused by treatment.

Nosocomial infections   Disorders developed in hospitals or other health care facilities, often caused by opportunist organisms in patients who are already ill.

Notifiable disease   An infectious disease such as gonorrhea, measles, rabies, AIDS, or typhoid fever, which must by law be reported to public-health officials.

Nucleic acid   A large molecule that consists of nucleotides; examples are DNA and RNA.

Nucleocapsid (n(y)oo-klee-oh-KAP-sihd)   The nucleic acid and its enclosing proteins in a virus particle.

Nucleoid (N(Y)OO-klee-oyd) A diffuse, shapeless area that contains the chromosome in a bacterial cell.

Nucleolus (n(y)oo-KLEE-ah-luhs) Dense, roughly spherical area within the nucleus of a eucaryotic cell; it contains DNA and is concerned with the manufacture of ribosomes.

Nucleosomes Small, regular repeating units of DNA in eucaryotic chromosomes; each one contains a length of DNA of about 140 base pairs.

Nucleotide A compound made up of a purine or pyrimidine base, a five-carbon sugar, and phosphates.

Nucleus Spherical structure in a eucaryotic cell that contains most of the cell's DNA.

Nutrient cycle A sequence of biologically mediated chemical conversions of elements.

Obligate intracellular parasites Parasites that can live and multiply only within and at the expense of an intact eucaryotic animal cell.

Obligate parasites Parasites that can survive only in living organisms.

Occlusive clothing Clothing such as gown or wash suit, gloves, mask, hair covering, and foot coverings, completely enclosing the body and preventing microflora from becoming airborne.

Oceanic zone The portion of the ocean that includes the deeper waters over the continental slope and deep sediments.

Oncofetal antigens Substances that are characteristic in fetal tissue but are not normally found on adult cells; they are sometimes found on tumor cells.

Oncogenes (AHN-kuh-jeens) DNA sequences in human chromosomes that have a role in causing cancer.

Oomycetes (oh-uh-my-SEE-tees) Lower fungi responsible for causing mildew and blight.

Oospores (OH-uh-spawrs) Fungal spores resulting from fusion of nonmotile gametes; they mature into heavy-walled survival spores.

Operator region The section of the gene that functions as a type of on/off switch.

Opportunists Organisms capable of causing disease only in an already weakened host.

Opsonin (AHP-suh-nihn) A plasma protein that increases the ability of phagocytes to bind to particles.

Opsonization A part of phagocytosis, the process of coating the particle with complement and/or antibody.

Oral administration The administration of medications by mouth.

Organelles Small internal structures within eucaryotic cells; each is specialized for a single cellular process.

Organic compounds Compounds containing carbon that are found in all living organisms and form the substance of cells and tissues.

Organic electron acceptor An electron acceptor in fermentation processes, which uses a cellular organic compound as the terminal electron acceptor.

Osmosis (ahz-MOH-suhs) The movement of water between two solutions separated by a semipermeable membrane.

Osmotolerance The ability of organisms, such as halobacteria and certain yeasts and molds, to grow in concentrated solutions of salts or sugars.

Otitis media (oh-TYT-uhs MEED-ee-uh) Infection of the middle ear.

Outbreak The situation that occurs when the number of cases of a particular disorder rises significantly above the expected prevalence.

Ova The eggs produced in the reproductive organ of the female.

Oxaloacetic acid (AHK-suh-luh-SEE-tihk . . .) The four-carbon compound that is a starting material in the Krebs cycle.

Oxidase positive organisms Organisms that have the respiratory enzyme cytochrome oxidase.

Oxidation The removal of electrons from a substance.

Oxidation–reduction reactions The transfer of electrons between atoms.

Oxidative phosphorylation (. . . fahs-FAWR-uh-lay-shuhn) The production of ATP in respiring cells, requiring a membrane-bound coenzyme structure.

Oxidoreductases Enzymes that catalyze redox reactions.

Ozone (O_3) A gaseous oxygen compound generated by electrical discharges passing through air and used for water purification.

Palisades Parallel sprays of cells lying together.

Pandemic   An epidemic that crosses national boundaries, threatening large areas of the world.

Parainfluenza virus   An RNA virus that causes both upper and lower respiratory disease.

Paralytic shellfish poison (PSP)   The neurotoxin produced by the algal blooms known as red tides; shellfish affected by this toxin cannot be made safe to eat.

Paramyxovirus (pahr-ah-MIHK-soh-vy-rihs)   A type of virus that causes influenza, parainfluenza, measles, and mumps; they have envelopes that contain hemagglutinin and neuraminidase.

Parasites   Organisms that live by invading other cells or the tissues of a living organism.

Parentaral administration (pah-REHNT-uh-rahl)   An injection made directly into the tissues, or intramuscularly.

Paroxysmal stage (par-ahk-SIHS-muhl . . .)   The second stage of whooping cough, in which there are fits of violent, uncontrollable coughing.

Partial diploid   A bacterial cell having duplicate genes for certain functions as a result of recombination.

Passive immunity   The protective state acquired when an individual receives preformed antibodies or presensitized lymphocytes or macrophages from another person.

Pasteurization   Partial sterilization by heat, to destroy the microorganisms that cause fermentation and decay.

Pathogenic bacteria   Bacteria that cause disease.

Pathogenicity   The ability to cause disease.

Peat   An accumulation of undecomposed plant material, preserved by overlying bog water.

Pelvic inflammatory disease   Inflammation of the pelvic organs, especially the uterus, fallopian tubes, and ovaries; a serious complication of sexually transmitted diseases, it may also occur as an ascending infection caused by normal vaginal flora.

Pemphigus neonatorum (PEHM(P)-fih-gus nee-oh-nuht-AWR-uhm)   A severe staphylococcal infection of newborns, with abscesses over the entire body.

Penicillin   One of a family of antibacterial drugs that all share a central chemical feature, the beta-lactam ring.

Penicillinase   An enzyme produced by penicillin-resistant bacteria that breaks down the beta-lactam ring.

Peplomers (PEHP-luh-mur)   Spikes of viral proteins on the envelope surfaces of animal viruses.

Peptide bond   The covalent bond that results from the condensation reaction between the carboxyl group of one amino acid and the amino group of another, forming a protein.

Peptidoglycan (PEHP-tih-doh-GLY-kuhn)   A three-dimensional macromolecule consisting of sugars and chains of amino acids; it provides structural strength to the cell walls of procaryotes.

Pericarditis   Inflammation of the pericardium, with chest pain and fluid accumulation in the pericardial sac surrounding the heart.

Periodontal membrane   The material that lines the tooth socket, the gingiva, and the tooth surface from the point below the gum where the enamel coat ends.

Periodontitis (pehr-ee-oh-dahnt-YT-uhs)   Destruction of tissue in the gingival sulcus, resulting from untreated gingivitis.

Periplasmic space   A thin space that separates the cell wall from the cell membrane in Gram-negative bacteria.

Peritrichous (peh-RIHT-rah-kuhs)   Having flagella over the entire bacterial surface.

Perlèche (pehr-LEHSH)   Crackling or fissuring of the corners of the mouth; one form of candidiasis.

Peroxides   Unstable compounds that decompose to yield chemically reactive free atomic forms of oxygen called nascent oxygen.

Peroxisomes   Organelles containing enzymes that produce peroxides and other toxic chemicals to kill captured prey.

Persistent infections   Diseases in which the agent persists in the tissues for extended periods of time.

Pertussis   See whooping cough.

Petechiae (pah-TEE-kee)   Small skin hemorrhages.

Peyer's patches   Lymphoid aggregates in the mucosa of the gastrointestinal tract, producing large amounts of IgA, which neutralizes many pathogens.

pH   The term used to describe hydrogen-ion concentration, which reflects the relative acidity of a solution.

pH gradient   The difference between the pH level within a cell and that outside.

Phaeophyta (fee-AH-fuh-tah)   True macroscopic seaweed, also known as brown algae, with two parts, a thallus and a holdfast; one example is the giant kelp.

Phage typing The process of identifying bacteria using specific strains of bacteriophages.

Phagocytosis The process by which a cell surrounds and engulfs particles.

Phagolysosome The structure that results when granules of phagocytic cells fuse with the phagosome.

Phagosome A vacuole bounded by a membrane layer derived from the cytoplasmic membrane of the phagocyte; it encloses a microorganism or other prey particle and digests it.

Pharyngeal tonsils Masses of lymphoid tissue embedded in the back of the nasopharynx; their swelling is termed adenoid disease.

Phase-contrast microscope A compound light microscope that allows examination of structures inside cells through the use of a special condenser.

Phenol coefficient (FEE-nohl . . .) A measure of a chemical compound's effectiveness as a disinfectant compared with phenol.

Phenolic (fih-NOH-lihk) A derivative of phenol, containing six-carbon rings, used as a disinfectant.

Phenotype The visible characteristics of an organism that result from the action of its genes.

Pheromones Chemical sex attractants produced by fungi to lead to mating of hyphae of opposite types.

Phlebitis (fluh-BY-tuhs) Inflammation of the valves and endothelium of the veins.

Phospholipid (fahs-foh-LIHP-uhd) A complex lipid composed of glycerol, two fatty acids, and one organic molecule containing a phosphate group; major component of all types of cellular membranes.

Photolithotrophs Organisms that get their energy from light and their carbon from CO_2.

Photosynthesis The process by which plants, algae, and some bacteria capture the energy of the sun and use it to make sugars and other food materials.

Photosystems Clusters of light-capturing chemicals found in photosynthetic organisms; they are embedded in membranes that contain chlorophyll pigments, electron-transport systems, and antennae.

Phycobilins (fy-koh-BIHL-uhns) Red or blue accessory pigments that expand the light-absorbing potential of cyanobacteria.

Phytoplankton Floating, microscopic, photosynthetic plant life of a body of water.

Picornavirus (pih-kawr-nuh-VY-ruhs) A very small naked polyhedral RNA virus.

Pili (PY-ly) Small tubular surface projections on bacteria that serve for attachment and conjugation.

Pilosebaceous unit The hair follicle and one or more sebaceous glands.

Pinkeye An eye infection with profuse watery discharge and swelling, caused by *Hemophilus ducreyi.*

Plague A term once used to describe any serious epidemic, but now reserved for yersiniosis, or bubonic plague, a disease transmitted by animals harboring fleas infected with *Yersinia pestis;* the peripheral circulation is blocked, causing necrosis of fingers and toes.

Plankton Microscopic floating organisms, including bacteria, cyanobacteria, algae, and protozoans.

Plaque (PLAK) **(1)** The sticky substance generated by slime-producing bacteria from sucrose, coating tooth surfaces and contributing to tooth decay; **(2)** a hole caused by a virus in the cloudy growth of a bacterial suspension.

Plaque-forming units (PFU) Visible plaques counted; each plaque may be caused by one or more viruses.

Plasma cells Large, antibody-producing, short-lived cells derived from parent B cells.

Plasmids Small closed circles in the cytoplasm of procaryotes and eucaryotes; they contain information that is useful but not essential for growth.

Plate count The technique for determining the number of cells or colonies by counting the colony-forming units in a culture medium.

Platelets Circulating cell fragments that arise from megakaryocytes; they break down on contact with damaged tissue and promote cell division and healing.

Platelet-derived growth factor (PDGF) A human serum protein that is crucial in wound healing because it encourages fibroblast cells to divide, forming scar tissue.

Pleomorphic Having a highly variable shape.

Pleurisy Inflammation of the pleura, the two layers of membrane that line the thoracic cavity.

Pneumocystosis (n(y)oo-muh-sihs-TOH-sihs) A form of pneumonia that occurs in very immunosuppressed people; it is commonly seen in patients with acquired immunodeficiency syndrome (AIDS); caused by the protozoan parasite *Pneumocystis carinii*.

Pneumonia   An accumulation of fluid in the lungs, usually accompanied by fever, cough, respiratory distress, and acute chest pain.

Pneumonia–influenza (P and I) deaths   A measure that public-health officials use to assess the impact of influenza (counting deaths from it and pneumonia).

Polar covalent bond   The unequal sharing of electrons between atoms.

Polar pattern   The configuration in which bacteria flagella are located at one or both ends of a cell.

Poliomyelitis (poh-lee-oh-my(uh)-LY-tuhs)   Infection of the motor neurons with the poliovirus, a small RNA virus belonging to the enterovirus group; severe cases involve varying degrees of paralysis.

Polyhedral   Symmetric and many-sided in structure; applied to the capsid symmetries of certain viruses.

Polymer   A macromolecule composed of small molecules held together by covalent bonds.

Polymerases   See ligases.

Polymyxins   Cationic peptide antibiotics that destroy bacterial membranes; used predominantly in topical preparations.

Polymorphonuclear leukocyte (PMN) (pahl-ih-mawr-fah-N(Y)OO-klee-ur LOO-koh-syt)   Phagocytic granulocyte that represents the most numerous of white blood cells; large reserves of them are released in certain infections.

Polypeptides   Long chains of amino acids forming macromolecules.

Polysaccharides (PAHL-ih-SAK-ah-rydz)   Carbohydrates composed of eight to hundreds or thousands of monosaccharides.

Polysome (PAHL-ih-sohm)   A group of ribosomes actively reading and working on one genetic message and assembling the coded protein molecules.

Polyunsaturated fatty acids   Fats or oils that have two or more unsaturated bonds; they tend to be liquid at room temperature.

Pontiac fever   A milder form of Legionnaire's disease.

Porins   Clusters of protein molecules forming pores in the outer membranes of gram-negative cell walls.

Positive pressure   In a room or transfer chamber, air at a pressure slightly higher than that of the surrounding area, directing air flow outward.

Positive selection   The process by which a desirable tendency is encouraged by increasing the members of the population with that tendency.

Pour-plate method   An isolation culture technique in which bacteria are mixed in the melted nutrient medium and the mixture is poured into a petri dish to solidify.

Povidone–iodine   An iodophor used as an antiseptic.

Poxvirus   A large and complex DNA virus that may cause skin eruptions and respiratory disease (smallpox).

Ppd (purified protein derivative)   A form of tuberculin used in testing for reactivity to tuberculosis.

PPNG (penicillinase-producing *Neisseria gonorrhoeae*)   Strains of *N. gonorrhoeae* that have absolute resistance to penicillin because they produce the enzyme penicillinase.

Precautions   Strategies for controlling a patient's excretions, dressings, or other materials that might be contaminated, without using a private room or gowns and masks.

Precipitates   Macromolecular complexes such as protein and polysaccharide antigens suspended in solution.

Predisposing factor   A disease, chemical, environmental agent, or other factor that augments the effect of the causative agent for a particular disorder in causing the risk that exposure will be followed by development of disease.

Prednisone (PREHD-nuh-sohn)   A steroid drug; one of its uses is for the suppression of rejection mechanisms.

Pre-erythrocytic cycle   The second cycle in the development of the parasite *Plasmodium*, the agent that causes malaria; the sporozoites are injected in the blood stream of the human victim (via a mosquito bite), enter the liver cells, and further subdivide into merozoites.

Preicteric phase (pree-ihk-TEHR-IHK . . .)   The portion of a liver disease before the onset of jaundice.

Prevalence   The actual proportion of the population affected by a particular disorder at a given time.

Primary amebic meningoencephalitis (PAM) (. . . muh-nihng-goh-uhn-sehf-uh-LYT-uhs)   A degenerative neurologic infection caused by free-living aquatic amebas such as *Naegleria fowleri.*

Primary atypical pneumonia Pneumonia caused by mycoplasmas, which escape isolation by standard bacteriologic tests.

Primary cell culture A sample that contains a mixture of cell types prepared from tissue taken aseptically from the animal; these cell lines are unstable and will eventually stop dividing.

Primary immunodeficiency Immunodeficiency present at birth and genetic in origin.

Primary infection The form of tuberculosis that occurs after the causative organism is inhaled and phagocytosed by alveolar macrophages.

Primary metabolites The substances the microorganism makes to use for its own growth.

Primary production In the carbon cycle, the metabolic activity of the photosynthesizers and chemolithotrophs, which increases their own biomass.

Primary stage In syphilis, the initial stage that occurs about three weeks after infection, when a hard chancre is noted.

Primary stain The purple stain crystal violet, which is the first step in the Gram stain.

Primary structure The simple linear sequence of amino acids in a polypeptide, held together by peptide bonds.

Primary treatment The first stage of sewage treatment, involving separation of sewage into sludge and effluent; the effluent is chlorinated if no further treatment is done.

Prion (PRY-ahn) A unique disease agent that is neither a virus nor a viroid and contains only protein; the term is derived from the fact that it is a protein infective agent.

Privileged sites Body areas, such as the cornea of the eye, where mismatched grafted tissue will not be rejected.

Procaryotae The kingdom of organisms that includes four divisions of the several thousand species of bacteria.

Procaryotes (proh-KAR-ee-ohts) Organisms, such as bacteria, that have a simple cell structure and contain no organized nucleus.

Producers In the carbon cycle, the photosynthesizers and chemolithotrophs, which are able to fix carbon dioxide into carbohydrates and other organic molecules.

Productive infection A type of viral infection that leads to the production and release of many new free virus particles.

Proglottids Segments growing from the neck of tapeworm larvae; these segments grow to form the body of the mature worm.

Progressive tuberculosis A form of tuberculosis that occurs when the tubercle is not adequately walled off or when the lesion opens up again, resulting in an open cavity.

Promoter A portion of a gene to which RNA polymerase attaches.

Prophage A viral genome that has been integrated into the chromosome of its bacterial host cell.

Prostatitis Inflammation of the prostate gland, which occurs commonly in older men.

Protease An exoenzyme that attacks tissue components.

Protective isolation A form of isolation used to protect extremely vulnerable or immunosuppressed patients from exposure to microorganisms.

Protein kinases Enzymes that catalyze reactions that activate membrane proteins, speeding up cellular activities.

Proteins Macromolecules made up of a sequence of amino acids; some form a helix and then fold into a globular structure.

Proton gradient The difference in concentration of H^+ between the interior of a cell and the surrounding medium.

Proton motive force (PMF) The energy potential involved in the proton gradient.

Protons Subatomic particles with a positive charge.

Proto-oncogenes Normal cellular genes, an essential part of the genome and believed to regulate the rate of cell growth and division; altered forms of these genes may cause cells to become cancerous.

Protoplast A Gram-positive bacterium without a cell wall.

Prototroph A strain of bacteria with a normal or wild-type genotype.

Protozoa Unicellular heterotrophs that move rapidly and live by capturing and eating smaller microbes.

Provocative Producing adverse results when immune effectors combine with an allergen.

Pseudohyphae (s(y)oo-doh-HY-fee) Tubular extensions from fungi consisting of sequences of buds that do not fully separate.

Pseudomycelia (s(y)ood-oh-my-SEE-lee-uh)   Chains of elongated cells.

Pseudopodia (s(y)ood-oh-POH-dee-uh)   Flexible extensions of a cell that help capture food and aid in motion.

Psychoneuroimmunology (sy-koh-noo-roh-ihm-yoo-NAH-luh-jee)   The research specialty that investigates the effect of mental states on the nervous system and the body's defense.

Psychrophiles (SYK-roh-fylz)   Cold-loving organisms, which multiply within the temperature range 10 to 35 degrees Celcius, including bacteria, fungi, algae, and protozoa.

Puerperal fever (pyoo-UR-p(uh)-ruhl . . .)   Endometritis, or infection of the uterine lining, after a difficult vaginal delivery.

Pulmonic plague   Infection with airborne *Yersinia pestis*, from person-to-person transmission; the resulting lung involvement is fatal if not treated promptly.

Pure culture   A culture with one kind of microorganism, growing in the absence of all other organisms.

Purple membrane   The outer layer of halobacteria, which incorporates a light-sensitive pigment.

Purulent exudate (PYOO-roo-lehnt . . .)   The masses of white cells that accumulate on the surface of an infection; pus.

Pyelonephritis (py(uh)-loh-nih-FRYT-uhs)   Inflammation of the renal pelvis, with high fever, chills, severe pain, and spasm of the small intestine in the acute illness.

Pyocyanin (py-oh-SY-uh-nihn)   A blue-green pigment formed by *Pseudomonas aeruginosa*.

Pyoderma (py-oh-DUHR-muh)   A skin infection caused by *Streptococcus pyogenes*.

Pyrogen (PY-ruh-jehn)   A substance that affects the temperature-regulation system of the body, causing fever.

Pyrrophyta (puh-RAH-fah-tuh)   A type of green algae also known as dinoflagellates; they are unicellular and biflagellate.

Q fever   A pneumonia caused by *Coxiella burnetti*, an organism related to rickettsiae.

Quaternary ammonium compound   Also known as quat; a cationic detergent with four organic groups attached to a central positively charged nitrogen atom, used as a disinfectant.

Quaternary structure   The structure taken by some globular proteins when several polypeptides combine into one large molecular complex.

Quinone (kwy-NOHN)   A compound related to vitamin K, involved in the electron-transport system.

Rabies   A disease contracted as a result of the bite of an animal infected with rabies virus, a large, complex RNA virus; lesions in the central nervous system cause hydrophobia, paralysis, coma, and death if prophylactic treatment is not obtained before symptoms develop.

Radioactive isotopes   Isotopes with atomic nuclei that are unstable and spontaneously emit energy or radiation.

Radioimmunoassay (RIA)   A sensitive test that measures antigens by ascertaining the amount of radioactive iodine in an antigen–antibody complex.

Ras   The oncogene of Harvey murine sarcoma virus.

Rat-bite fever   An infection with fever, headache, and rash transmitted by the bite of a rat, mouse, weasel, cat, or squirrel and caused by either *Spirillum minus* or *Streptobacillus moniliformis*.

Reaction rate   The speed at which a chemical reaction occurs.

Reactivation tuberculosis   See secondary tuberculosis.

Recombinant DNA   An experimental gene sequence containing DNAs from two or more sources enzymatically spliced together into a recombinant molecule.

Recombination   Any process by which genes from two separate organisms are brought together into one cell or one chromosome.

Red tides   Marine algal blooms that produce a paralytic poison; the algae are eaten by shellfish, which themselves become toxic. The affected shellfish cannot be made safe to eat.

Redia   One of the larval stages of the fluke.

Reduction   The addition of electrons to a substance.

Reflection   An effect that occurs when light strikes an object and bounces off.

Refraction The bending of light rays as they move from a transparent medium, such as air, to another, such as a glass microscope lens, causing blurring of images.

Refractive index The relative velocity at which light passes through different parts of a specimen.

Refractory compounds Also called persistent compounds; carbon atoms that stay fixed for long periods in naturally occurring or synthetic compounds that resist microbial enzymes.

Regulatory genes Genes that produce regulatory proteins or repressors; control the expression of other genes.

Regulatory proteins Repressors and other protein products whose cellular role is to influence the expression of other genes.

Replacement flora Microbial species on body surfaces that proliferate only in the absence of normal flora.

Replica-plating A technique used in microbial genetics to select chosen strains, such as recombinants that express new genetic combinations or auxotrophs that have lost the ability to synthesize amino acids.

Replication Formation of new copies of a structure, for example the copying of new strands of DNA.

Replication cycle The sequence of initiation, replication, and release of a virus in its host cell.

Replication forks The site of the actual copying and separation of DNA, where new strands of DNA will be synthesized.

Replication proteins Proteins that assist directly in the process of viral replication.

Repressible enzymes Enzymes that the cell can stop making if it has no immediate need for the enzymes' product.

Reproductive mycelium The portions of the mycelium of a fungus that produce asexual or sexual spores.

Reservoir A site in nature where an infectious agent remains alive, and from which the agent may spread to cause new cases of the disease.

Residual chlorine Active chlorine left over after absorption by the organic material in the water.

Resistance factor plasmids Bacterial plasmids that contain genes for their own transfer (RTF) as well as R genes that confer resistance to antibiotics and heavy metals; these are readily passed from one strain of bacteria to another.

Resistant Term applied to a microorganism, incapable of being inhibited or killed by a particular drug at a safe or attainable concentration.

Resolving power A measurement of how close together two objects can be and still be perceived clearly as separate under the microscope.

Respiration The cellular process by which ATP is generated; the process used by organisms to produce carbon dioxide. A respiring cell transfers electrons via some type of electron-transport system from organic compounds to an inorganic electron acceptor.

Respiratory bacteria Bacteria that use oxygen or another inorganic electron acceptor (see respiration).

Respiratory isolation A form of isolation used to contain highly communicable, hazardous diseases that are spread by respiratory secretions.

Respiratory syncytial virus (RSV) An RNA virus that can cause colds; in infants it can cause a severe, often fatal form of pneumonia.

Restriction endonucleases Naturally occurring enzymes that break the DNA strand internally; they are restricted in action because each recognizes and cleaves only at a particular nucleotide sequence.

Retroviruses RNA viruses that work "backward" in that they use a copying enzyme to take an RNA genome and make a DNA complementary strand.

Reverse transcriptase The copying enzyme used by retroviruses.

Reye's syndrome (RAYZ . . .) A condition found mainly in children involving sudden loss of consciousness, central-nervous-system abnormalities, and potential failure of liver and kidneys, often developing following a virus infection such as influenza or chickenpox.

Rhabdovirus (RAB-duh-vy-ruhs) A bullet-shaped, enveloped RNA virus; one of its types is the rabies virus.

Rheumatic fever A serious, often recurrent hypersensitivity that damages heart valves and causes arthritis; it is a sequel to untreated Group A streptococcal pharyngitis.

Rheumatoid arthritis (RA) An inflammatory disease with progressive joint degeneration, thickening of soft tissues, and erosion of bony surfaces.

Rheumatoid factor (RF) An autoantibody that reacts with normal or partly denatured IgG; found in serum of individuals with rheumatoid arthritis.

Rhinitis (ry-NY-tuhs) Inflammation of the nasal mucous membrane, with profuse watery secretion, sneezing, and tear secretion (the common cold).

Rhinovirus (ry-nuh-VY-ruhs) An RNA virus, which has more than 100 types and causes most cases of the common cold.

Rhizosphere (RY-zoh-sfeer) The soil zone that surrounds the root network of a plant.

Rhodamine (ro-DAH-mihn) A red fluorescent dye that may be coupled to antibody and used to detect antigen.

Rhodophyta (roh-DAH-fuh-tah) Red algae, some of which produce carrageenan, alginic acid, and agar, substances used for thickening food and gelling laboratory media.

Ribosome (RY-boh-sohm) The major cell structure participating in protein synthesis.

Rickettsiae (rihk-EHT-see-ee) Obligate intracellular bacterial parasites that are typically transmitted by the bites or feces of arthropods such as ticks or lice and cause circulatory infections.

Rift communities The marine life forms that develop where the warm water dissolves minerals from the earth's crust in ocean rifts.

Rifts Cracks in the earth's crust, which are found in many areas of the ocean floor.

Ringworm A mycotic disease such as tinea cruris (jock itch).

RNA A nucleic acid that resembles DNA but is single-stranded; it contains adenine, guanine, cytosine, and uracil. It acts as an intermediate in protein synthesis in cellular life and forms the genome of the RNA viruses.

RNA polymerase The enzyme, composed of six protein subunits, that enables transcription of RNA from the DNA template.

Rocky Mountain spotted fever The most common cardiovascular rickettsial disease in the United States, caused by the bite of a tick that is host for the organism *Rickettsia rickettsii*; the disease involves a high fever, a rash on the ankles and wrists, and cardiovascular effects.

Root nodules Structures formed by bacteria on the roots of legumes; the bacteria within the root nodules fix nitrogen from the atmosphere into amino acids and excrete them into the soil.

Rotavirus The RNA virus that is the major cause of diarrheal illness in children.

Rough colonies Colonies of bacteria that do not produce glycocalyx and appear dryish.

Rough ER Endoplasmic reticulum with associated ribosomes.

Roundworms One of the two groups of helminths, with a more complex structure and a complete digestive system, compared with flatworms.

Salinity A measure of the concentration of dissolved salts in a fluid.

Saliva The secretion of the salivary and mucosal glands, which represents one of the natural defenses of the mouth against microbial growth.

Salmonellosis (sal-muh-nehl-OH-suhs) A gastrointestinal infection caused by organisms of the genus *Salmonella*; the gastroenteritis is marked by nausea and/or diarrhea and possible damage to the intestinal lining.

Sanitation Public health techniques used to protect individuals and communities by cleaning and sterilizing eating or drinking utensils and by treating food, water, and sewage to prevent transmission of the microorganisms that cause disease.

Sanitizing The systematic cleansing of inanimate objects to reduce the microbial count to a safe level.

Saprophytes (SAP-roh-fyts) Heterotrophic organisms that use organic compounds they obtain as wastes or residues from dead organisms.

Saturated fatty acid A fatty acid in which each of the carbon atoms is attached to the maximum number of hydrogen atoms.

Scanning electron microscope (SEM) An electron microscope that produces highly magnified images of the surface structures of specimens.

Scarlet fever A form of streptococcal pharyngitis in which the toxin from the infecting strain causes a rash.

Schick test A skin test for immunity to *Corynebacterium diphtheriae* (the organism that causes diphtheria).

Schizogony (skihz-AHG-uh-nee) The combination of the second and third cycles of the life of the sporozoan parasite *Plasmodium,* the agent that causes malaria; these stages take place in the human host.

Scolex   The head of the larva of the tapeworm, which attaches its suckers to the intestinal wall.

Second law of thermodynamics   The law stating that in any energy-converting process, some of the energy will be changed into a form that is unavailable for work.

Secondary immunodeficiency   Immunodeficiency resulting from malnutrition, illness, or other stress.

Secondary metabolites   Substances such as antibiotics, which are end products of metabolism, not directly used in growth.

Secondary stage   The form of syphilis that develops about six to eight weeks after the onset of the primary stage if there has been no treatment; there is wide distribution of the spirochetes in the body with many skin and mucous-membrane lesions, as well as generalized symptoms.

Secondary structure   The repeated coiling or zigzag pattern that results from the formation of hydrogen bonds between amino acids.

Secondary treatment   The second stage of sewage treatment, reducing the amount of waste; microorganisms are added to decompose most of the remaining organic waste, and the effluent is chlorinated.

Secondary tuberculosis   The condition found when a healed primary lesion reactivates.

Selection   The process by which one identifies and encourages the growth of the preferred mutant strains in a cell population.

Selective media   Chemicals added to media used for culturing microorganisms that allow only certain types to grow.

Self   An antigen that is a normal part of the body.

Self-limiting infection   An infection that is taken care of by the body's defenses without outside intervention.

Semiconservative   The outcome of DNA synthesis in which each daughter helix ends up with one "old" strand and one "new" one.

Semipermeable membrane   A membrane that permits certain molecules and ions to pass through but excludes others.

Semisynthetic drug   A natural antibiotic somewhat modified by chemists.

Sensitization   The process in which hypersensitivity develops after one or more initial immunizing contacts with antigen.

Sepsis   The uncontrolled presence of infectious agents or their toxins in the blood or tissues.

Septa   The crosswalls perforated by pores found in mold hyphae.

Septicemia (sehp-tuh-SEE-mee-uh)   Bacteria multiplying in the blood, which is the life-threatening form of bacteremia; it can lead to septic shock if not treated quickly.

Septic shock   Formerly called blood poisoning, a massive microbial infection that results when bacteria multiply in blood and release toxins that are destructive to blood cells.

Septic tank   A primary sewage-treatment receptacle in which the sludge settles out and decomposes anaerobically, while the effluent is released into the soil.

Serial cell cultures   Stable cell lines that do not die off after a few generations but are capable of seemingly limitless cell divisions.

Serial dilution   A laboratory technique in which counts are made of the bacterial colonies in successive dilutions of a sample; is also applied to quantitating antigen or antibody, or to measuring antimicrobial susceptibility.

Serology   The study of antibodies in serum, as carried out in the laboratory in vitro.

Serum hepatitis (hepatitis B)   The long-incubation liver disease caused by hepatitis B virus.

Serum sickness   The immune-complex disease that develops when serum products containing foreign protein are repeatedly administered to induce passive immunization, but produce hypersensitivity.

Severe combined immunodeficiency   A condition in which the fetal lymphoid stem cell line fails to differentiate at all and both B and T cells are therefore nonfunctional.

Sewage   Waste materials from various sources and needing various forms of treatment before it can be discharged back into the environment.

Sex pili   Pili that attach two bacteria to each other prior to conjugation.

Sexual reproduction   Reproduction in which a new individual is formed by fusion of sex cells with genetic material from two parents.

Sexual spores Fungal spores that are formed by fusion of haploid nuclei from two different parental mating types.

Sexually transmitted diseases Infections transmitted predominantly by sexual activity; replaces the previous term venereal disease.

Shigellosis (sheh-gehl-LOH-suhs) A condition also known as bacillary dysentery and involving a diarrhea with frequent watery stools, pain, and passage of blood and mucus, caused by bacteria of the genus shigella.

Shingles A condition caused by varicella-zoster virus that remains latent in the body after childhood chickenpox and then reactivates.

Short-incubation disease A disease with a relatively short period elapsing between the time of exposure to the infectious agent and the onset of the illness; term used for hepatitis A.

Sieve filter A filter made of a thin sheet of a synthetic polymer such as cellulose acetate.

Simple stain A process involving the bathing of a fixed slide of microbial cells with a single dye.

Single-cell protein (SCP) A food substitute used in animal feed; it is produced by pelleting microorganisms grown on industrial waste.

Sinusitis Inflammation of the sinuses.

Sinusoids The irregular circulatory passageways in spongy organs such as the liver and spleen.

Site of replication The site in a tissue or organ at which a specific pathogen multiplies.

Site-specific The tendency in recombination for donor and recipient strands to pair up, matching complementary base sequences of their attachment sites, then join end to end.

Sleeping sickness See African trypanosomiasis.

Slide agglutination test A test in which a drop of unknown antigen and a drop of known antibody-containing antiserum are mixed together and observed for clumping; used to test for blood grouping.

Slime layer A loosely associated layer coating some procaryotes.

Slow-virus infection A disease process that develops slowly (over years in humans) before the cumulative viral damage is clinically noticeable.

Sludge The solid portion of sewage after separation from the effluent.

Smear A suspension of material containing microorganisms to be studied; the material is spread over a slide for observation.

Smooth colonies See mucoid colonies.

Smooth ER Endoplasmic reticulum that lacks ribosomes.

Soap A compound of fatty acids and sodium ions, used as surfactant and in routine de-germing.

Solute The substance that is dissolved by the solvent in the formation of a solution.

Solution Two or more substances that are completely mingled, so that their molecules are evenly dispersed.

Solvent The liquid that is used to dissolve another substance to form a solution.

Species name The second part of a binomial designation of an organism, roughly equivalent to a person's first or given name.

Spectrum (1) The band of wavelengths of electromagnetic radiation; (2) the range of effectiveness of a drug.

Spheroplast A gram-negative bacterium lacking a complete cell wall.

Spherule A structure that develops in the host's lung tissue from a fungal arthrospore; when it ruptures, mature spores disseminate in the lung.

Spirillum (spih-RIHL-uhm) A helical, rigid rod-shaped bacterium.

Spontaneous generation The hypothetical concept that living organisms can arise directly from nonliving matter.

Spontaneous mutation A mutation for which there is no immediate, apparent cause.

Sporadic diseases Diseases that have small, localized, unpredictable outbreaks.

Sporangium (spah-RAN-jee-uhm) The sac in lower fungi in which asexual spores are produced.

Spore A general term for a microscopic resting or reproductive cell formed by certain bacteria, fungi, and protozoa. Spores have resistant coatings and reduced or absent cell activity.

Spore septum The structure that separates the developing spore from the rest of the cell contents.

Sporogony (spuh-RAHG-uh-nee) The first cycle of the development of the parasite *Plasmodium*, the agent that causes malaria; this cycle occurs in the mosquito and involves the fusion of male and female gametes ingested from the blood of a human bitten by the mosquito.

Sporozoite (spawr-uh-ZOH-yt)   The form that results from fusion of male and female gametes of the parasite *Plasmodium* in the mosquito.

Sporulation   Spore formation, or the process of forming endospores.

Sputum   The thickened mucus secreted by irritated tissues in the lower respiratory tract and expelled by coughing.

Src gene   The first identified oncogene, which is involved with the transforming potential in the Rous sarcoma viral genome.

Staining   The use of some form of dye in the preparation of microorganisms or tissues for observation with a microscope.

Staphylococcal scalded-skin syndrome   A rapidly spreading condition in neonates in which the enzyme exfoliatin causes the epidermis to peel from the dermis.

Start codon   The codon AUG that represents the point at which RNA translation begins.

Static treatments   Treatments that halt microbial growth for as long as the inhibitory substance or state is present.

Stationary phase   The period during which the population of cells in a culture does not change after a period of growth.

Stem cell   An undifferentiated cell, which may develop into one of several unique types of differentiated cells.

Sterilization   A process that completely destroys all forms of microbial life.

Steroids   Lipids that have a characteristic multiring structure.

Sterols   Steroids that contain hydroxyl groups in certain positions; the best known sterol is cholesterol.

Stock cultures   Pure cultures that are maintained and subcultured for future use.

Stop codon   The codon that represents the point at which RNA translation ends.

Streak-dilution method   A technique for isolating a specific organism in a culture by spreading microorganisms over the surface of a solid medium.

Streptococcal pharyngitis   "Strep sore throat," with pain, reddening, and inflammation of the pharynx, and usually fever, caused by *Streptococcus pyogenes*.

Strict aerobes   Aerobic bacteria that require oxygen for growth, although some may survive without growth in the absence of oxygen.

Strict anaerobes   Anaerobic bacteria that do not use oxygen and tolerate exposure to it poorly.

Strict isolation   A form of isolation used to control the spread of a highly communicable, dangerous disease such as diphtheria or rabies.

Strong acid   A compound such as hydrochloric acid, in which all the hydrogen ions are free and contributing to the acidity of the solution.

Structural genes   The most prevalent of the four functional types of DNA sequences, these code for enzymatic and structural proteins for cell growth.

Structural proteins   The viral gene products that will compose the capsid and accessory structures of the new crop of virions; these are largely late proteins.

Subacute   A slowly progressive yet life-threatening form of an illness.

Subacute bacterial endocarditis (SBE)   The more common form of endocarditis, which happens only in people with abnormal heart valves and is most frequently caused by the non-Group A streptococci.

Subacute sclerosing panencephalitis (SSPE)   A rare fatal degenerative disease that occurs in people who appeared to have recovered from measles.

Subclinical infection   An infection that does not result in clinically obvious damage.

Subcultures   Portions of cultures that have been transferred to fresh media for future use as stock cultures.

Subsoil   Inorganic material, with few nutrients and supporting little life; the lowest layer of soil.

Substrate   A compound with which an enzyme interacts.

Substrate-level phosphorylation   The direct process by which energy is released by oxidizing a substrate at the reaction site to phosphorylate ADP to ATP.

Sulfate reduction   The use by anaerobic bacteria of sulfate in place of oxygen as an electron acceptor for oxygen.

Sulfonamides (suhl-FAHN-uh-mydz)   Broad-spectrum bacteriostatic drugs that inhibit steps in folic-acid synthesis; short-acting sulfonamides are commonly used in uncomplicated bladder infection.

Superinfection   The overgrowth of an antibiotic-resistant microbial strain, often displacing the normal flora.

Suppressor–helper ratio   The ratio of suppressor to helper cells among the T cells in a person's blood; this ratio indicates how effectively the immune response is working.

Surface tension   The capacity of water or other liquid to spread over solid surfaces.

Surface water   The water that accumulates and runs into the ocean, as opposed to that which percolates down through the soil.

Surfactant   Surface active compound, or wetting agent that lowers the surface tension of water; also the substance that lubricates the lung surfaces and keeps them from sticking together.

Susceptible   Applied to a microorganism, capable of being inhibited by a particular drug at a clinically attainable and safe concentration.

Swarmer   A term applied to a type of bacteria that migrate rapidly over the whole surface of a culture plate.

Swimmer's itch   A mild skin disease transmitted by flukes.

Symbiosis (sihm-by-OH-suhs)   An interaction between two species of organism.

Syncytia (sihn-SIHSH-(ee)uh)   Nonfunctional oversized host cells that have been caused to fuse by the action of certain viral proteins.

Synergistic effect (sihn-ur-JIH-stihk . . .)   The effect found when a combination of drugs is successful in treating conditions that neither could control individually.

Synthetases   See ligases.

Synthetic drug   A drug made by chemists from simple organic starting materials.

Syphilis   The sexually transmitted disease caused by *Trepenoma pallidum*, a species of spirochete bacteria.

Systemic infection   An infection (such as measles) that affects several tissue types and involves large areas of the body.

Systemic lupus erythematosus (SLE)   A condition involving antibodies against many different self antigens and marked by symptoms of facial rash, fever, intestinal complications, and progressive kidney failure.

Systemic mycoses   Fungal infections of deep tissues.

T cells   Lymphocytes that act as effectors in cell-mediated immunity.

T cytotoxic cells   Antigen-specific killer cells; they attach directly to an antigen-bearing cell.

T delayed hypersensitivity (T_D) cells   T cells that bind to their antigens with a high degree of specificity and then produce lymphokines.

T helper cell   A type of T cell that stimulates immune responses; it cooperates with B-cells in initiating antibody-mediated responses.

T suppressor cell   A type of T-cell that has a regulatory function in reducing the intensity of antibody-mediated responses.

Tail   Protein accessory structure of a bacterial virus that makes it possible for the virus to inject its nucleic acid forcibly into the host cell.

Tail fiber   The part of the virus tail that specifically binds to receptor sites on the host surface.

Tapeworms   One of the types of flatworms; these parasites, also known as cestodes, have immobile adult stages adapted to life as intestinal parasites.

Taxonomy   The science of biological classification, involving groupings by similarities.

Temperate viruses   Viruses causing infections that do not immediately produce a crop of new virions; the viral genome integrates with the host genome and becomes latent for some time.

Template   The molecular pattern or mold that physically guides the assembly of nucleic acids and proteins.

Teratogen (tuh-RAT-uh-juhn)   An agent that crosses from the mother's blood into the fetal circulation and upsets the fetus' developmental timetable and/or causes birth defects.

Terminal disinfection   The cleaning procedure that occurs after the patient has vacated the unit, involving disinfection of walls, woodwork, floors, and furniture.

Terminal electron acceptor   An inorganic compound that combines with electrons in the last stage of the electron-transport system to form a low-energy waste product.

Tertiary structure   The three-dimensional shape created by the folds and bends of a long amino acid chain.

Tertiary syphilis   The untreatable third stage of syphilis, which eventually appears in some untreated individuals and involves irreversible degenerative and destructive changes.

Tertiary treatment   The third stage of sewage treatment, involving additional cleansing of the water with chemicals that remove remaining organic and inorganic pollutants, yielding a drinkable final product.

Tetanus A neuromuscular disease caused by the gram-positive, strictly anaerobic spore-forming bacillus *Clostridium tetani*; the organism lives in soil or fecal matter and enters the body usually through a puncture wound.

Tetracyclines Widely used broad-spectrum antimicrobials that inhibit protein synthesis; because they are bacteriostatic, they are best used only when the host defenses are strong.

Thallus (THAL-uhs) The sheetlike body of the sea lettuce or other multicellular protists.

Therapeutic dose The dosage level necessary for clinical control of a particular pathogen.

Therapeutic index The ratio of the therapeutic dose and the toxic dose.

Thermal death point The temperature that kills all members of a 24-hour-old broth culture of a given species of bacteria.

Thermal death time The time needed to kill all organisms in a 24-hour broth culture at the given temperature.

Thermophiles (THUR-muh-fylz) Heat-loving organisms, which grow between 40 and 85 degrees Celsius; these organisms can rapidly spoil cooked foods that are held at warm temperatures for several hours.

Throat swab Method for sampling microbial throat flora for laboratory study, using a sterile swab rubbed over the posterior pharynx and tonsillar area.

Thrombophlebitis (thrahm-boh-fleh-BY-tuhs) An irritation of the venous endothelium, developing at or near a surface lesion, frequently at the site of an improperly maintained intravenous catheter.

Thrombus A blood clot within a blood vessel, or a fragment of a clot, which is dangerous because it may break away and get stuck downstream, cutting off blood flow.

Thrush A fungal infection of the upper respiratory tract caused by the yeast *Candida albicans*.

Thylakoids (THY-luh-koydz) Flattened membrane disks that contain chlorophyll and divide by binary fission; they carry out the "light" reactions of photosynthesis.

Thymosins Signaling substances or hormones produced by the thymus.

Tincture An alcoholic solution.

Tinea pedis (TIHN-ee-uh PEE-duhs) The dermatomycosis commonly known as athlete's foot.

Tissues Groups of differentiated cells with the same structure and function; examples include nerve and muscle tissues.

Titration A technique to determine the concentration of a substance in solution, using serial dilutions.

Togavirus (toh-gah-VY-ruhs) An arthropod-borne RNA virus, transmitted by insect and arachnid vectors, which causes encephalitis, yellow fever, and dengue.

Tonsillitis Swelling of the palatine tonsils.

Tonsils Masses of lymphoid tissues, of which there are several groups: the palatine tonsils and the pharyngeal tonsils.

Top soil The fertile layer of soil above the subsoil, containing organic matter and a microbial population that recycles nutrients.

Topical administration The use of a drug applied to the skin or mucous-membrane surfaces.

***Tox* gene** The viral gene that codes for toxins, which are released by the infecting bacteria in diphtheria.

Toxic dose The dosage level at which an unacceptable toxicity occurs.

Toxic-shock syndrome (TSS) A generalized toxemia caused by strains of *Staphylococcus aureus*, most commonly in women using tampons to absorb menstrual flow.

Toxin A poisonous chemical substance that alters a key physiologic reaction; bacteria can produce toxins that are spread by the blood stream from the point of infection.

Toxoid Modified toxins used to stimulate antibody production in the process of immunization.

Toxoplasmosis (tahk-suh-plaz-MOH-suhs) A disease caused by the protozoan parasite *Toxoplasma gondii*; the infection is usually mild but can affect the fetus if it occurs in pregnant women.

Trachoma (truh-KOH-muh) A highly communicable eye infection caused by *Chlamydia trachomatis*.

Transamination (trans-am-uh-NAY-shuhn) The enzymatic transfer of an amino group from one compound to another.

Transcription   The process by which RNA is made; RNA synthesis occurs at localized sites on the DNA genome, and single genes or groups of genes are copied.

Transduction   The mechanism by which bacterial DNA is carried from donor to recipient bacterium as a passenger molecule within a virus particle.

Transferases   Enzymes that move functional groups such as amino or phosphate groups from one molecule to another.

Transferrins   Proteins that bind free iron, thus tying up the body's iron reserves in a form unavailable to microorganisms.

Transformation   **(1)** In bacteria, a process in which short pieces of DNA from a dead donor cell are taken up by a living recipient cell and integrated into its chromosome; **(2)** in mammalian cells, a change that occurs when a normal cell develops into a cancer cell.

Translation   Protein synthesis using RNA as a template.

Transmembrane channel   A passage formed through the cell membrane by aggregates of specialized proteins such as complement proteins, leading to lysis of the target cell.

Transmissibility   The characteristic in some plasmids of having the necessary genes to direct their successful transfer to a new host cell.

Transmission   An effect that occurs when light strikes a transparent object at a right angle and passes through without affecting the object.

Transmission electron microscope (TEM)   An electron microscope that allows the viewer to see details of thin sections of the internal structures of specimens.

Transport cultures   Sealable bottles or plates containing a chocolate agar medium and a gas mixture rich in carbon dioxide; the technique is used to culture organisms such as *Neisseria gonorrhoeae*, which grows only in an atmosphere with elevated carbon dioxide.

Transposase   The enzyme that recognizes the insertion elements for transposons and cleaves the DNA at those points.

Transposon   A transposable genetic element, or a DNA sandwich—one or more transferable genes sandwiched between two insertion elements.

Trematodes (TREEM-uh-tohds)   See flukes.

Tricarboxylic acid cycle (tri-kahr-BAWK-sihl-ihk . . .)   See Krebs cycle.

Trichocysts   Sticky threadlike harpoons attached to some protozoa that entangle the organism's prey.

Trichomoniasis (trihk-uh-muh-NY-ah-sihs)   A vulvovaginitis caused by the flagellate protozoan *Trichomonas vaginalis.*

Trickling filter   An older method of secondary sewage treatment.

Tricothecene (truh-KOH-thuh-seen)   A form of mycotoxin found on animal feeds produced from corn, which have been allowed to stay damp and have developed fungal growth.

Triglycerides (try-GLIH-suhr-ydz)   The simplest lipids, formed by combining three fatty acid molecules with a single glycerol molecule.

Trophozoite (troh-fuh-ZOH-yt)   The vegetative, often motile, cell form of a protozoan.

Trypanosomiasis   See American trypanosomiasis; African trypanosomiasis.

Trypomastigote (try-poh-MAS-tih-goht)   The form of the parasite in trypanosome diseases that is passed in the insect's saliva or feces; these are rubbed into the itching bite wound, leading to infection in human beings.

Tube agglutination test   An immunologic test in which the reagents are mixed in a test tube; more accurate but more complex than the slide agglutination test.

Tubercle   An enclosed lung lesion produced by infection with *Mycobacterium tuberculosis.*

Tuberculin   An extract of *Mycobacterium tuberculosis,* used in testing for tuberculosis (see ppd).

Tuberculoid leprosy   The milder form of leprosy, in which the leprosy bacilli do not penetrate the central nervous system.

Tuberculosis   The chronic progressive pulmonary disease caused by *Mycobacterium tuberculosis.*

Tularemia (t(y)u-luh-REE-mee-uh)   A highly infectious bacterial lymphoid infection with the organism *Francisella tularensis,* which resembles *Brucella;* it is transmitted by tick bite, contamination of the skin (often by skinning an affected rabbit), inhalation, or ingestion.

Tumor   A solid mass of abnormal cells, or neoplasm, which may arise in any type of tissue.

Turbidity Cloudiness of a suspension, which increases when more organisms are present.

Turista Travelers' diarrhea, or diarrhea caused by pathogenic strains of *Escherichia coli*.

Twofold dilution Titration using successive serial dilutions in which each sample is mixed with an equal quantity of saline or buffer; the first tube is a 1:2 dilution, the second 1:4, the third 1:8, and so on.

Tyndallization (tihn-dahl-ih-ZAY-shuhn) A process of intermittent sterilization in which boiling and incubation periods are alternated to eliminate spores.

Typhoid fever The most severe form of salmonellosis, now rare in developed nations; caused by *Salmonella typhi*.

Ultra-high temperature (UHT) treatment The process for treating milk at a temperature much higher than used in regular pasteurization, so that the milk can be stored for several months at room temperature while the container is unopened.

Ultrapasteurization A technique in which cream is processed at a much higher temperature than is involved in regular pasteurization, yielding a longer-lasting product.

Ultrasonic vibrations Sound waves of an energy level too high for humans to hear; these can rupture microbial cells and may be used as a method of cleansing and sterilization.

Ultraviolet microscopes Microscopes that use an ultraviolet light source, often used in conjunction with fluorescent dyes to absorb ultraviolet light and emit orange, yellow, or greenish light, making cell outlines more visible.

Ultraviolet rays Invisible, non-ionizing radiation used to kill organisms in air or within small areas; alters DNA and causes mutations.

Unpunctuated sequence A linear message of three-base codons; as each codon is translated, the equivalent amino acid will be incorporated into a linear protein.

Upwelling The process by which water from great depths is brought back to the surface by rising ocean currents.

Vaccination A procedure in which an altered disease agent or its products are administered to confer immunity.

Vaccine A manufactured product containing an altered harmless antigen, such as an attenuated, modified, or killed infective agent, or a toxoid.

Valence (1) The number of electrons an atom must gain or lose in order to achieve a filled outer shell; the atom's combining capacity; (2) the valence of an antigen is the number of its identical combining sites.

Valence electrons The electrons in the outermost shell of a molecule, which are directly involved with bond formation.

Variable region The sequence within a light or heavy chain in an immunoglobulin molecule that is responsible for the antibody's specificity.

Variable surface glycoproteins (VSGs) The changeable surface antigens of trypanosomes; since the antigenic target shifts, the patient's immunity is ineffective.

Vector A living carrier (for example, a flea or tick) that physically moves or transmits an infectious agent from one host to another.

Vegetation A loose, irregular aggregate of fibrin-containing microcolonies of bacteria, found in subacute bacterial endocarditis.

Vegetative cell A cell that is actively growing, compared with cells that are reproducing or resting.

Vegetative mycelium The portions of the mycelium of a fungus that gather nutrients and then extend.

Venereal disease Older term for a sexually transmitted disease.

Viable Living; able to reproduce.

Vidarabine A chemotherapeutic purine derivative that acts as a competitive inhibitor of nucleic-acid synthesis; used to treat herpesvirus infections.

Viral enzymes Enzymes coded for by viral genes.

Virion An extracellular or free virus particle, containing only one of the nucleic acids.

Viroids Plant pathogens that are simpler than viruses, consisting of short single strands of RNA and having no external structures such as a capsid or an envelope.

Virology The study of viruses.

Virulence A quantitative measure of pathogenicity, determined by the morbidity rate or the mortality rate.

Virulence factors Features that enable a microbe to establish a parasitic relationship.

Virus   A submicroscopic, parasitic filterable genome consisting of a nucleic acid surrounded by a protein coat.

Vulvovaginitis   Simultaneous inflammation of the vulva and vagina.

Wastewater treatment   An industrial technology aimed at transforming sewage or wastewater into a form that can safely be discharged back into the environment by removing as many pollutants as possible from it.

Wavelength   Measurement of electromagnetic radiation in terms of the distance from one peak to the next.

Weak acid   A compound, such as acetic acid, that only partly dissociates.

Weil–Felix reaction   The cross-reaction of antibodies formed during rickettsial disease with antigens of the enterobacterial species *Proteus vulgaris*; this phenomenon is used as the basis for laboratory diagnosis of rickettsial disease.

Whooping cough   An infectious disease caused by *Bordetella pertussis* and characterized by a peculiar type of paroxysmal coughing.

Wild-type   The genetic variant of a bacterial species that is most commonly encountered in nature; the most fit.

Wort   The soluble portion of the mash used in the process of making beer.

Wound and skin precautions   Precautions used to control leakage of fluids from skin infections, wounds, or burns, or other localized infectious states.

Yeasts   Unicellular fungi, usually oval in outline, that reproduce by budding.

Yellow fever   A viral infection caused by the yellow fever virus transmitted by mosquitos; severe cases involve massive hemorrhage and liver damage.

Yersiniosis (yuhr-sihn-ee-OH-suhs)   The disease formerly called bubonic plague and transmitted by animals harboring fleas infected with *Yersinia pestis*.

Z value   The temperature needed to kill 90% of the organisms in a 24-hour broth culture in a fixed time period.

Zone of inhibition   The circular area around the disk with no microbial growth in the Kirby–Bauer disk diffusion test for antimicrobial susceptibility.

Zoonosis (zoh-AHN-uh-suhs)   A disease of animals that can be transmitted to humans.

Zooplankton (zoh-uh-PLANG(K)-tuhn)   Tiny aquatic animals such as protozoa.

Zoospore   Motile, reproductive spore propelled by a single flagellum.

Zygomycetes (zy-goh-my-SEE-teez)   Terrestrial fungi that may reproduce through both sexual and asexual process; one example is *Rhizopus nigricans*, the common black bread mold.

Zygospore   A fungal spore that forms at the point of fusion in the sexual reproductive cycle of zygomycetes.

Zygote   The resulting diploid cell after the fusion of two haploid gametes.

APPENDIX ONE
DESCRIPTIONS OF
BACTERIAL GENERA

This appendix provides brief summary descriptions of one hundred of the more significant bacterial genera mentioned in the text. The listing is alphabetical, rather than taxonomic, for easy reference. The descriptions are condensed from either *Bergey's Manual of Systematic Bacteriology*, volume I, 1984, designated Reference A, or *Bergey's Manual of Determinative Bacteriology*, 8th edition, 1974, designated Reference B. The more recent source was used wherever a genus was covered in both sources. As these are intended as summary descriptions, no effort has been made to take note of all of the many exceptions to most generalizing statements. Thus, for the most complete and detailed information, the references and the original publications cited in their bibliographies should be consulted. Another very useful concise guide to the groups of bacteria is Chapter 19 in *Biology of Microorganisms*, 4th edition, by T. D. Brock *et al.*, Prentice-Hall, 1984.

Each description includes the microscopic morphology, Gram reaction, motility, oxygen requirement, and key metabolic patterns of the organism. Note that, in order to maintain consistency with the text, the terms lithotroph and heterotroph have been substituted for the terms autotroph and organotroph, respectively, whenever they were used in the reference descrip-

tions. The optimum temperature and any unusual growth conditions are noted. The habitat or sources, significance in nature, and importance to humans are summarized. The type species, that is, the species most characteristic of the genus, is listed; the total number of valid species given in the reference is noted. In many cases, brief descriptions of key species follow the main listing.

Acetobacter Cells ellipsoidal to rod-shaped, straight or slightly curved, 0.6–0.8 μm by 1.0–4.0 μm; Gram-negative. Usually motile with peritrichous or lateral flagella. Obligately aerobic. Metabolism is respiratory, never fermentative. Chemoheterotrophic. Oxidize ethanol to acetic acid; oxidize acetate and lactate to CO_2 and H_2O. Some species synthesize cellulose microfibrils. Optimum temperature 25–30° Celsius. Found in many forms of alcoholic beverage, on flowers and fruits, and in soil and water. May rot apples, pears, and pineapples. Important in the industrial production of vinegar. Four species; type species is *Acetobacter aceti*, first described by Pasteur in 1864. Reference A:268.

Acholeplasma Cells spherical, 0.3 μm minimum diameter, and filamentous, usually 2–5 μm in length. Cells are bounded by a plasma membrane containing sterols and have no cell wall. Stain Gram-negative. Nonmotile. Facultative anaerobes; metabolism is fermentative. Chemoheterotrophic; have limited metabolic abilities. Optimum temperature for growth 20–40° Celsius; growth is slow and colonies tiny. Parasites of a wide range of tissues and organs of vertebrates, although pathogenicity has not been established. Also found in sewage, compost, and soil, and on plant surfaces. Although they may be troublesome contaminants of cell cultures, they have been extremely useful in basic research on plasma membranes. Eight valid species; type species *Acholeplasma laidlawii*. Reference A:775.

Acinetobacter Rods 0.9–1.6 μm in diameter and 1.5–2.5 μm in length, becoming spherical in stationary phase, occuring in pairs and chains. Gram-negative. Usual swimming motility absent; no flagella: cells may display twitching motility. Aerobic; metabolism is versatile and strictly respiratory with oxygen. Chemoheterotrophic. Optimum temperature for most strains 33–35° Celsius. Found commonly in soil, water, and sewage. May be involved in meat spoilage. Can cause nosocomial infections in compromised humans. Single and type species *A. calcoaceticus*. Reference A:303.

Actinomyces Cells irregular, diphtheroid, or branching rods; filaments 1 μm or less in diameter with true branching may predominate. Gram-positive but irregularly staining. Nonmotile. Facultative anaerobes; most preferentially anaerobic. Metabolism fermentative; produce acid from sugars, require organic nitrogen; some CO_2 required for optimum growth. Optimum temperature 35–37° Celsius. Symbionts and pathogens of mammals, especially their oral cavities. Five species; type species *Actinomyces bovis*, which causes a disease of cattle. *A. israelii*, a normal inhabitant of the mouth, may cause actinomycosis in humans; *A. naeslundii* and *A. viscosis* are implicated in periodontal disease. Reference B:660.

Aeromonas Cells straight, rod-shaped to coccoid, 0.3–1.0 μm in diameter and 1.0–3.5 μm in length. Gram-negative. Generally motile by a single polar flagellum; one species nonmotile. Facultative anaerobes. Metabolism both respiratory and fermentative. Chemoheterotrophic, using a variety of sugars and organic acids. Optimum temperature 22–28° Celsius. Normal habitat water, sludge, and sewage. May be pathogenic. Four species; type species *A. hydrophila* may be pathogenic for frogs and fish and may infect human wounds. *A. salmonicida* causes an economically significant disease of salmon and trout. Reference A:545.

Agrobacterium Rods, 0.6–1.0 μm by 1.5–3.0 μm, occurring singly or in pairs. Gram-negative. Motile by 1–6 peritrichous flagella. Aerobic; metabolism respiratory using oxygen and, in some species, nitrate as terminal electron acceptor. Chemoheterotrophic, using a wide range of carbohydrates, amino acids, and organic acids, and producing copious polysaccharide slime. Optimum temperature 25–28° Celsius. Invade the crown, roots, and stems of a wide variety of plants via wounds. Oncogenic, causing the transformation of the plant cells into tumor cells in diseases such as crown gall. Oncogenesis correlates with the possession of the plasmid Ti in the bacterial cells. Occur almost worldwide in the soil and rhizosphere of plants and can also be isolated from diseased plant tissues and from human clinical specimens, probably as incidental contaminants. Significant as an agent of disease on peach, almond, and cherry trees; on grapevines; and in greenhouses. Recently, it and its plasmids have been widely employed in experiments leading toward gene manipulation in plants. Four valid species with many different strains; type species *A. tumefaciens*. Reference A:244.

Alcaligenes Cells are rods, coccal rods, or cocci, 0.5–1.0 μm in diameter, 0.5–2.6 μm long. Gram-negative. Motile with 1–8 peritrichous flagella. Obligately aerobic; metabolism is strictly respiratory with oxygen, some strains using nitrate. Chemoheterotrophic, rarely using carbohydrates, but using a variety of organic and amino acids and producing alkali from several substrates. Optimum temperature 20–37° Celsius. Have been isolated from soil, water, dairy products, rotten eggs, and the intestinal tract of many animals. Can cause meat spoilage, may cause opportunistic infection in humans; have been recovered from a variety of clinical specimens. Two valid species and eight less well-defined species; type species *A. faecalis.* Reference A:361.

Alteromonas Cells are straight or curved rods, 0.7–1.5 μm by 1.8–3.0 μm. Gram-negative; motile by means of single polar flagella. Aerobic; metabolism is strictly respiratory, using oxygen but not nitrate. Chemoorganotrophic, utilizing a wide range of organic compounds. Many produce extracellular polysaccharide-degrading enzymes, which allow them to utilize nutrients common in their environment. One species is bioluminescent. All require a seawater base medium for growth. Temperature range wide; all grow at 20° Celsius, and some are mildly psychrophilic. Habitat is coastal seawaters and the open oceans. Eleven species; type species is *A. macleodii.* Reference A:343.

Aquaspirillum Cells rigid, generally helical, 0.2–1.4 μm in diameter; however, one species is vibrioid, another straight rods. A polar membrane underlies the cell membrane at the cell poles. Gram-negative. Motile generally by bipolar tufts of flagella. Aerobic to microaerophilic; metabolism is respiratory using oxygen, with some strains using nitrate. Usually chemoheterotrophic, using amino or organic acids as the usual carbon sources, not usually carbohydrates. Some strains fix nitrogen; one is a hydrogen lithotroph. No growth occurs in presence of 3% NaCl. Optimum temperature 30–32° Celsius. Usually occur in stagnant freshwater environments but not saline areas. Eleven species; type species *A. serpens. A. magnetotacticum* responds to magnetic stimuli. Reference A:72.

Arthrobacter Cell shape undergoes characteristic marked change in form during growth in complex media. Cells in fresh media rod-shaped; cells in older (2–7 days) media become larger and coccoid. Gram-positive but readily decolorized. Generally nonmotile or feebly motile. Strict aerobes; metabolism is respiratory using oxygen. Chemoheterotrophic, using a wide range of substrates. Optimum temperature 20–30° Celsius. Among the dominant bacteria in soil. Seven valid species; type species *A. globiformis.* Reference B:618.

Azotobacter Cells are large, ovoid, and 1.5–2.0 μm or more in diameter. They are pleomorphic and exist in highly variable groupings. Gram-negative. Form resting structures called cysts. Motile by peritrichous flagella or nonmotile. Aerobic, tolerating decreased oxygen tensions. Metabolism is respiratory. Chemoheterotrophic, nutritionally versatile. Free-living fixers of atmospheric nitrogen. Optimum temperature 32–37° Celsius. Found in fertile, slightly acid to alkaline soils, where the ability to form cysts insures survival during dry periods; also found in water. Significant contributor to soil fertility. Six valid species; type species *A. chroococcum.* Reference A:220.

Bacillus Cells rod-shaped, straight or nearly so, 0.3–2.2 μm by 1.2–7.0 μm, often in chains. Gram-positive, although reaction varies. Form heat-resistant endospores. Mostly motile with many lateral flagella. Strict aerobes, some facultative anaerobes. Metabolism strictly respiratory, strictly fermentative, or mixed, depending on species. Respire with oxygen and in some cases nitrate. Chemoheterotrophic; nutritional patterns very diverse. Temperature range wide. Found in soil, water, and animal and plant tissues. Numerous industrial applications. Forty-eight species: type species *B. subtilis.* Some species are pathogenic; *B. thuringensis* kills certain species of chewing insect larvae, and *B. anthracis* causes a severe disease of domestic animals and humans. *B. cereus* and other bacilli cause spoilage of bakery products and other foods. *B. licheniformis* and *B. polymyxa* produce useful polypeptide antibiotics. The spores of *B. stearo-*

thermophilus are used to test canning and auto-claving processes. Reference B:529.

Bacteroides Cells are pleomorphic rods. Gram-negative. Usually nonmotile; some are motile with peritrichous flagella. Obligate anaerobes; metabolism usually fermentative. Chemohet-erotrophic, using carbohydrates, peptides, or metabolic intermediates. Optimum temperature near 37° Celsius. Sources are the oral and other body cavities and intestinal contents of humans, other mammals, and insects; infections of soft tissues; and sewage. Some species are patho-genic. Thirty-nine species recognized; type spe-cies *B. fragilis*, a member of the human normal bowel flora that has been isolated in human ap-pendicitis, peritonitis, and septicemia. *B. mela-nogenicus* and other species have been isolated from diseased gingival crevices. Reference A:604.

Bdellovibrio Cells are comma-shaped rods, 0.2–0.5 μm in diameter and 0.5–1.4 μm in length. Gram-negative. Motile by a single polar flagellum. Predatory; exhibit a remarkable bi-phasic life cycle, alternating between a non-growing, free-living predatory phase and an in-tracellular, parasitic reproductive phase. Prey on other species of Gram-negative bacteria, which are killed and their cellular substance used to support bdellovibrio reproduction. Obligate aerobes. Metabolism strictly respiratory. Che-moheterotrophic; all wild-type strains require prey, but some prey-independent strains have been isolated on normal lab media. Optimum temperature 28–30° Celsius. Habitats include soil, sewage, freshwater, and marine environ-ments, wherever prey bacteria are plentiful. Three named species, one unnamed marine spe-cies; type species *B. bacteriovorus*. Reference A:118.

Beggiatoa Cells colorless, in filaments, 1–30 μm by 4–20 μm. Gram-negative. Motile by gliding. Aerobic to microaerophilic; metabolism is respi-ratory using oxygen. Grow mixotrophically, ox-idizing hydrogen sulfide to sulfur as a litho-trophic energy source, but requiring organic growth factors. Form intracellular sulfur gran-ules. Found in fresh and salt water, where they participate in the sulfur cycle. Six species; type species *B. alba*. Reference B:113.

Beijerinckia Cells are straight or slightly curved rods, about 0.5–1.5 μm in width and 1.7–4.5 μm in length; some large, misshapen cells. Gram-negative. Distinctive, refractile intracellular granules; cysts and capsules seen in some. Mo-tile by peritrichous flagella or nonmotile. Aero-bic; metabolism is strictly respiratory with oxy-gen. Chemoheterotrophic; utilize carbohydrates actively, produce copious polysaccharide slime. Free-living nitrogen fixers. Optimum tem-perature 20–30° Celsius. Found in usually acidic soils in the tropics and in irrigation water over flooded rice paddies. Contribute to soil fertility. Four species; type species *B. indica*. Reference A:311.

Bifidobacterium Cells are rods, highly variable in size and appearance, with branched or club-shaped forms, affected by cultural conditions. Gram-positive, with irregularly staining gran-ules; nonmotile. Anaerobic; metabolism is fer-mentative. Chemoheterotrophic, utilizing car-bohydrates without producing gas. Optimum temperature 36–38° Celsius; difficult to main-tain in culture. Normal habitats are the intes-tinal tract of humans and other mammals and the human vagina. No evidence of human patho-genicity; has been suggested that the organisms are helpful in the development of normal di-gestive function in the human infant. Eleven rec-ognized species; *B. bifidum* is the type species. Reference B:669.

Bordetella Cells are minute coccobacilli, 0.2–0.5 μm by 0.5–2.0 μm, singly or in pairs, rarely chains. Gram-negative, often bipolar stained. Nonmotile, two species; one species motile with peritrichous flagella. Strict aerobes; metab-olism respiratory. Chemoheterotrophic, fastid-ious. Optimum temperature 35–37° Celsius. Found on epithelium of mammalian respiratory tract as a parasite and pathogen. Three species; type species is *B. pertussis*, the agent of whooping cough. Reference A:388.

Borrellia Cells are helical, 0.2–0.5 μm by 3–20 μm, composed of 3–10 loose coils. Gram-negative. Actively motile by means of 15–20 axial filaments. Only a few have been cultivated in vitro; they are microaerophilic and chemoheterotrophic with complex nutritional requirements. Most species in the genus appear to be obligate parasites. Pathogenic in humans, other mammals, and birds, and parasitize and are transmitted by ticks and lice. Cause septicemic relapsing fevers in humans. There are 19 species, classified according to their arthropod vectors; *B. anserina*, which is pathogenic for birds, is the type species. *B. recurrentis* causes a louse-borne relapsing fever in humans. Reference A:57.

Brevibacterium Cells are short, unbranched rods. Gram-positive; motile with peritrichous flagella. Aerobic; fermentative. This is a genus of uncertain validity, appearing to be a catchall for a group of organisms of similar traits, many of which probably belong in other genera. The genus name is still used, however, despite its ambiguity. A large number of species have been described at various times. Many are of industrial importance and are extensively described in the patent literature. *B. linens* is used in the manufacture of soft, ripened cheeses. Reference B:625.

Brucella Cells are cocci, coccobacilli, or short rods, 0.5–0.7 μm by 0.6–1.5 μm. Gram-negative; nonmotile. Aerobic; metabolism is respiratory using oxygen or nitrate. Chemoheterotrophic, most strains requiring complex media with amino acids, vitamins, or serum factors. Optimum temperature 37° Celsius. Found worldwide as obligate intracellular parasites and pathogens, in a wide range of animal species; may cause disease in humans. Cause serious economic losses due to disease of domestic animals. Six valid species; type species is *B. melitensis*, which usually infects sheep and goats. Reference A:377.

Campylobacter Cells are slender, spirally curved rods, 0.2–0.5 μm by 0.5–5.0 μm, also appearing S-shaped or gull-winged. Cells have a multilayered polar membrane lying under the cytoplasmic membrane at both ends of the cell. Gram-negative. Motile with a corkscrewlike motion by a single polar flagellum at one or both ends. Microaerophilic; metabolism is respiratory. Chemoheterotrophs; do not utilize carbohydrates, depending on amino acids or metabolic intermediates. Optimum temperature near 37° Celsius; best isolation temperature 42–43° Celsius. Can be found in the reproductive organs, intestinal tract and oral cavity of humans and animals. Cause abortion, reproductive infection, and enteritis. Five species; type species is *C. fetus*. *C. jejuni* is increasingly recognized as a major human intestinal pathogen, often contracted by consuming contaminated poultry and dairy products. Reference A:111.

Caulobacter Cells are rod-shaped, fusiform, or vibrioid, 0.4–0.5 μm by 1–2 μm. Typically with a stalk, about 0.15 μm in diameter and varying in length, extending from one pole. The stalk consists of the cell wall layers, the cell membrane, and a core of twisted membranes. A small mass of adhesive material is present at the end of the stalk. In dense populations, may adhere to each other in rosettes. Binary fission is assymmetrical. A bud forms at the end of the stalk; after fission, one daughter cell retains the stalk, while the other, the bud, develops a flagellum becoming a swarmer. Gram-negative; swarmer cells are motile with a single polar flagellum. Strictly aerobic; metabolism is respiratory. Chemoheterotrophs, able to grow in very dilute media on a variety of organic compounds as sole carbon and energy sources. Optimum temperature 20–25° Celsius. Habitat is fresh and salt water, particularly in organic films at the surface, and in soil. Ten species; type species is *C. vibrioides*. Reference B:153.

Cellulomonas Cells are irregular rods, 0.5 μm by 0.7–2.0 μm; some V formations and occasional rudimentary branching. Gram-staining reaction is ambiguous. Motile by one polar or a few lateral flagella. Aerobes and facultative anaerobes; metabolism primarily respiratory. Chemoheterotrophic; most important trait ability to attack and hydrolyze cellulose. Optimum temperature 30° Celsius. Found in soil. Under investigation for possible industrial uses. Single type species *C. flavigenes*. References B:629.

Chlamydia Coccoid organisms, 0.2–1.5 μm, multiplying only within membrane-bound vacuoles in the cytoplasm of host cells. In their unique developmental cycle, small elementary bodies develop into larger reticulate bodies that divide by fission yielding new elementary bodies. Gram-negative; nonmotile. Have never been grown outside of host cells; their metabolism remains unknown. The chlamydiae are obligate intracellular parasites of humans, other mammals, and birds, infecting notably the epithelia of the eye, respiratory tract, and genital tract. Cause a wide range of diseases. Two species; type species is *C. trachomatis*, which causes trachoma, lymphogranuloma venereum, nonspecific sexually transmitted diseases, and neonatal pneumonias. *C psittaci* causes respiratory infections in birds and farm animals, and psittacosis in humans. Reference A:729.

Chlorobium Cells are rod-shaped, ovoid or vibrioid. Gram-negative; nonmotile; natural cell color green. Strictly anaerobic; can only use light energy. Photolithotrophic or photoheterotrophic. Have only Photosystem I; pigments located in unique ovoid chlorobium vesicles underlying and attached to cell membrane. Oxidize hydrogen sulfide to sulfur, forming extracellular globules. Optimum temperature 25–30° Celsius. One of the genera of so-called green bacteria. Found in mud and stagnant water where there is little or no dissolved oxygen, but plenty of hydrogen sulfide and some available light. Four species; type species *C. limicola*. Reference B:53.

Chondromyces Vegetative cells are cylindrical, untapered rods. Gram-negative. Cells are able to swarm over moist surfaces. Have life cycle in which vegetative cells aggregate to form stalked sporangia or fruiting bodies in which resting cells called myxospores are formed. Aerobic; chemoheterotrophic, predatory, obtaining nutrients by lysing other bacteria. Optimum temperature 28–30° Celsius. Found in soil, on dung, decaying wood, and fruits. Have been extensively studied as models of cell-to-cell communication and cell differentiation. Five species; type species is *C. crocatus*. Reference B:96.

Citrobacter Cells are straight rods, about 1.0 μm by 2.0–6.0 μm, singly or in pairs. Conforms to the general description of the Family Enterobacteriaceae. Gram-negative. Motile by peritrichous flagella. Facultative anaerobe; metabolism both respiratory and fermentative. Chemoheterotrophic. Use citrate as sole carbon source, decarboxylate lysine, and give a negative Voges-Proskauer reaction. Found in soil, water, sewage, and food. Isolated from feces of humans and other animals; probably a normal intestinal inhabitant. Also isolated from feces in opportunistic infections. Two species; type species *C. freundii*. Reference A:458.

Clostridium Cells are rods, 0.6–1.2 μm by 3.0–7.0 μm (type species). Gram-positive; form ovoid to spherical endospores that usually distend the bacilli. Most strains strictly anaerobic; metabolism is fermentative. Chemoheterotrophic, utilizing many substrates; some species produce active proteolytic enzymes, and some fix nitrogen. Optimum temperature of type species is 25–37° Celsius. Commonly found in the soil, marine and fresh water sediments, intestinal contents of humans and other animals, and sewage. Several species cause human diseases. Others are useful in industrial applications. Sixty-one species; type species is *C. butyricum*. *C. acetobutyricum* is used to produce the solvents acetone and butanol. *C. pasteurianum* is a vigorous nitrogen-fixer, contributing to the fertility of anaerobic soils. *C. tetani*, *C. botulinum*, *C. perfringens*, and *C. difficile* cause, respectively, the wound infection tetanus, the food toxicity botulism, the wound infection gas gangrene, and the intestinal disease antibiotic-associated colitis. Reference B:551.

Corynebacterium Cells are straight to slightly curved rods with irregularly stained segments, sometimes granules, club-shaped swellings. Size 0.3–0.8 μm by 1.0–8.0 μm (type species). Snapping division produces angular and palisade (picket fence) arrangements of cells. Gram-positive, but staining may vary. Generally nonmotile. Aerobic and facultatively anaerobic; metabolism is mixed respiratory and fermentative. Chemoheterotrophs. Wide temperature range. Divided into two groups: Section I includes nine species of human and animal parasites and pathogens found in diverse anatomic sites including skin, nasopharynx, and other body orifices. Section II includes 12 species of plant pathogens infecting wheats, grasses, potatoes, and many other species; some of these are also spoilage agents in vegetable crops. Type species for Section I is *C. diphtheriae*, which causes human diphtheria. *C. xerosis* and *C. pseudodiphtheriticum* are commonly encountered in normal human skin and nasal cultures. Reference B:602.

Coxiella Cells are short rods, usually 0.2–0.4 μm by 0.4–1.0 μm. Gram-negative; nonmotile. Obligate intracellular parasites, growing in vacuoles of the host cell. Have not been grown in cell-free culture; grow well in the yolk sac of fertilized eggs. Little is known of their metabolism. Worldwide distribution in ticks and mammals. Infection particularly in cattle, sheep, and goats. In humans cause Q fever, an aerosol-borne respiratory disease. The single type species is *C. burnetii*. Reference A:701.

Cytophaga Cells are single short or elongate rods or filaments 0.3–0.7 μm by 5–50 μm. Gram-negative. Motile by gliding. Strict aerobes or facultative anaerobes. Metabolism is respiratory, using oxygen or in one case nitrate; or mixed respiratory and fermentative. Chemoheterotrophic, often with the ability to decompose polysaccharides such as cellulose, agar, or chitin.

Brightly pigmented yellow, orange, or red. Optimum temperature 30° Celsius. Found in soil and in both fresh and salt water, where they live as decomposers. Eight species; type species is *C. hutchinsonii*. Reference B:101.

Desulfovibrio Cells are curved or occasionally straight rods, sometimes S-shaped or spiral, 0.5–1.5 μm by 2.5–10.0 μm. Gram-negative. Motile by a single or lophotrichous polar flagella. Obligate anaerobes; metabolism is mainly respiratory. Sulfate or other sulfur compounds are reduced, serving as terminal electron acceptors, yielding H_2S. Usually chemoheterotrophic, using lactate or in some cases acetate. Diverse metabolic patterns. Optimum temperature usually 25–35° Celsius. Habitats include waterlogged soil; anaerobic mud of fresh, brackish, and marine environments; and intestinal contents of animals. One of several genera of sulfate-reducing, H_2S-producing bacteria. Nine species; type species is *D. desulfuricans*. Reference A:666.

Ectothiorhodospira Cells are spiral to slightly bent rods, 0.7–1.0 μm by 2.0–2.6 μm (type species). Gram-negative. Motile by means of polar flagella. Anaerobic; obligate phototrophs, containing chlorophylls and other pigments in lamellar membrane stacks. Use hydrogen sulfide as electron donor for photosynthesis and produce sulfur globules exterior to cells. One of the photosynthetic genera referred to as purple sulfur bacteria. Also may use molecular hydrogen. Require NaCl for growth; some are extreme halophiles. Habitat is salt lakes, salt flats, and estuaries, where there are H_2S and light available. Three species; type species is *E. mobilis*. Reference B:47.

Eikenella Cells are straight rods, 0.3–0.4 μm by 1.5–4.0 μm, occasionally in short filaments. Gram-negative. Nonmotile; however, a twitching movement may occur on agar surfaces. Facultative anaerobes; metabolism is respiratory. Chemoheterotrophic, usually requiring enriched media with hemin for aerobic growth. Optimum temperature 35–37° Celsius. Occur, probably as normal flora, in the human mouth and intestine. May be opportunistic pathogens and are implicated in periodontal disease. Single and type species *E. corrodens*. Reference A:591.

Enterobacter Cells are straight rods, 0.6–1.0 μm by 1.2–3.0 μm. Conform to the general description of the Family Enterobacteriaceae. Gram-negative; motile by peritrichous flagella. Facultative anaerobes; metabolism is both respiratory and fermentative. Chemoheterotrophs; ferment lactose, produce gas from sugars, give positive Voges-Proskauer and citrate tests, and negative indol and methyl red tests. Optimum temperature 30° Celsius. Widely distributed in nature; common in intestinal contents of normal humans and animals, in soil and water, and on plant materials. Saprophytic but may be opportunistic pathogens. Five species; type species *E. cloacae*, the most frequently isolated enterobacter. Reference A:465.

Erwinia Cells are straight rods, 0.5–1.0 μm by 1.0–3.0 μm. Members of Family Enterobacteriaceae. Gram-negative; motile by peritrichous flagella. Facultative anaerobes; metabolism both respiratory and fermentative. Chemoheterotrophic, utilizing some sugars and organic acids. Optimum temperature 27–30° Celsius. Associated with plants as pathogens or saprophytes or as constituents of the flora of the plant surface. Fifteen species; type species *E. amylovora*, which causes fireblight disease on pears, apples, and related fruit trees. Reference A:469.

Escherichia Cells are straight rods, 1.1–1.5 μm by 2.0–6.0 μm. Members of the Family Enterobacteriaceae. Gram-negative; motile by peritrichous flagella. Facultative anaerobes; metabolism both respiratory and fermentative. Chemoheterotrophic; ferments lactose, produces acid and gas from sugars, gives positive indol and methyl red tests, gives negative Voges-Proskauer and citrate tests. Optimum temperature 37° Celsius. Occur in the lower intestine of warm-blooded animals and cockroaches; in their feces; and in soil, water, and foods contaminated with feces. Enumerated in tests for water quality to measure degree of fecal contamination. Two species; type species *E. coli* has been a primary tool of basic molecular genetic research during the last 40 years; it is used as a host for recombinant DNA methodologies. While most strains are harmless, some strains of *E. coli* are pathogenic, causing gastroenteritis, hemorrhagic colitis, neonatal infections, and a variety of opportunistic conditions. Reference A:420.

Flavobacterium Cells are rods with parallel sides and rounded ends, typically 0.5 μm by 1.0–3.0 μm. Gram-negative; nonmotile. Aerobic; metabolism is strictly respiratory. Chemoheterotrophic. Growth on solid media usually yellow pigmented. Temperature range 5–37° Celsius. Widely distributed in soil and water; also found in raw meats, milk, and other foods. Not part of human normal flora, but sometimes isolated from clinical specimens. In the hospital environment may cause opportunistic infections. Six species; type species is *F. aquatile*. *F. meningosepticum* infrequently causes neonatal meningitis. Reference A:353.

Francisella Cells are minute rods, 0.2 μm by 0.2–0.7 μm or larger. Faintly Gram-negative. Nonmotile. Obligate aerobes; chemoheterotrophic and fastidious. Cysteine is either required or strongly stimulatory for growth. Optimum temperature 37° Celsius. Found in a wide range of warm-blooded animals, in meats and animal products, and in water samples. Two species; type species *F. tularensis* is a highly pathogenic organism, causing tularemia in humans and in other domestic and wild animals. Reference A:394.

Fusobacterium Cells are rods with tapered or pointed ends, 0.4–0.4 μm by 3–10 μm (type species). Gram-negative; nonmotile. Obligate anaerobes; metabolism is usually nonfermentative. Chemoheterotrophs, utilizing peptone and converting threonine and lactate to propionate. Optimum temperature 37° Celsius. Found in the body cavities of humans and other animals. Some species are pathogenic, isolated from abscesses, necrotic lesions, upper respiratory tract infections, and wounds. Ten species; type species is *F. nucleatum*. Reference A:631.

Gardnerella Cells are pleomorphic rods, 0.5 μm by 1.5–2.5 μm. Gram-negative to Gram-variable; have an atypical cell wall structure. Nonmotile. Facultative anaerobes; metabolism is fermentative. Chemoheterotrophic, having fastidious growth requirements. Optimum temperature 35–37° Celsius. Found worldwide in the human genital and urinary tracts. Considered to be a major cause of bacterial nonspecific vaginitis. Single type species *G. vaginalis*. Reference A:587.

Gluconobacter Cells are ellipsoidal to rod-shaped, 0.5–0.8 μm by 0.9–4.2 μm. Gram-negative to occasionally Gram-variable; motile by polar flagella or nonmotile. Intracytoplasmic membranes at some growth stages. Obligate aerobes; metabolism is strictly respiratory with oxygen. Chemoheterotrophic, oxidizing ethanol to acetic acid, producing ketogluconic acids from glucose. Optimum temperature 25–30° Celsius. Occur in flowers, fruits, honey bees, cider, beer, fermented juices, and soft drinks, wherever sugars are plentiful. May rot fruits; no known pathogenic effects. May be used industrially to produce sorbose, tartaric acid, dihydroxyacetone, gluconic, and ketogluconic acids. Single and type species *G. oxydans*. Reference A:275.

Haemophilus Cells are minute to medium-sized coccobacilli or rods, generally less than 1 μm wide, variable in length, and very pleomorphic. Gram-negative; nonmotile. Aerobic or facultatively anaerobic; metabolism is fermentative. Chemoheterotrophic, attacking carbohydrates, reducing nitrates. Require preformed growth factors X and/or V found in blood. Optimum temperature 35–37° Celsius. Occur as obligate parasites on mucous membranes of humans and other animals. Cause a variety of diseases, including pneumonia, meningitis, middle ear infection, septicemic conditions, conjunctivitis, and opportunistic infections. Sixteen valid species; type species is *H. influenzae*, which is the

leading cause of meningitis in children. *H. ducreyi* causes the human sexually transmitted disease chancroid. *H. aegyptius* causes infectious conjunctivitis. Reference A:558.

Hafnia Cells are straight rods, about 1.0 μm by 2.0–5.0 μm. Conform to the general description of the Family Enterobacteriaceae. Gram-negative; motile by peritrichous flagella. Facultative anaerobes, having both respiratory and fermentative metabolism. Chemoheterotrophic; may be distinguished by biochemical tests. Optimum temperature for accurate testing is around 25° Celsius. Occur in the feces of humans and other animals; also found in natural environments such as sewage, soil, water, and dairy products. Often isolated from fecal specimens of normal people; may be implicated in opportunistic mixed infections. Single and type species is *H. alvei*. Reference A:484.

Halobacterium Cell shape and size depends on growth conditions, including rod, disk, club, sphere, or spindle forms. Divide by constriction. Stain Gram-negative but have atypical cell wall structure lacking peptidoglycan. Cell membrane contains ether-linked lipids. Considered to be archaebacteria. Require high salt concentrations to prevent lysis. Some strains have gas vesicles. Strict aerobes or facultative anaerobes; metabolism both respiratory and fermentative. Chemoheterotrophic, utilizing sugars and amino acids for energy. Some species also form the light-capturing purple membrane containing bacteriorhodopsin. Strains with purple membrane are able to make ATP by capturing light energy under anaerobic conditions and thus are phototrophs. Optimum temperature 40–50° Celsius. Found in salt lakes, salt flats, and proteinaceous environments such as salt fish and salted hides; may cause spoilage. Five species; type species is *H. salinarium*. Reference A:262.

Hyphomicrobium Cells are rods with pointed ends, oval, or egg- or bean-shaped forms 0.5–1.0 μm by 1–3 μm. Produce filamentous outgrowths or hyphae. Gram reaction variable or not recorded. Multiply by budding from tips of hyphae.

Mature buds motile with polar flagella. Aerobic; chemoheterotrophic; although they fix CO_2 they require very dilute organic nutrients. Denitrify. Optimum temperature 25–30° Celsius. Widely distributed in soils, and found in sea water and ocean sediment. Single and type species *H. vulgare*. Reference B:149.

Klebsiella Cells are straight rods, 0.3–1.0 μm by 0.6–6.0 μm, singly, in pairs or in short chains. Conform to the general characteristics of the Family Enterobacteriaceae. Gram-negative. Encapsulated and nonmotile. Facultative anaerobes; metabolism is both respiratory and fermentative. Chemoheterotrophic, distinguished by biochemical testing. Widely distributed in nature, found in soil, water, and the intestinal contents of humans and other animals. Several are opportunistic pathogens. Four species; type species is *K. pneumoniae*, which causes urinary and respiratory infections in humans. It is a significant nosocomial infective agent, often carrying multiple antibiotic resistance. Reference A:461.

Lactobacillus Cells are rods, varying in size, and commonly forming chains. Motility unusual; when present, by peritrichous flagella. Gram-positive, becoming Gram-negative with increasing age and medium acidity. Anaerobic, although growth of some species may occur in air; metabolism strictly fermentative. Chemoheterotrophic, having complex nutritional requirements for amino acids, vitamins, and other organic molecules. Optimum temperature 30–40° Celsius. Habitats include dairy products, grain and meat products, water, sewage, alcoholic beverages, and pickled foods; also found as normal flora in the mouth, intestinal tract, and vagina of many warm-blooded animals, including humans. Pathogenicity is highly unusual. This genus is highly valuable in the food and fermentation industries. Twenty-seven valid species; type species is *L. delbrueckii*, used in the commercial production of lactic acid. Other lactobacilli are used in the production of sour cream, yoghurt, sour-dough bread, cured olives, and other foods. Reference B:576.

Legionella Cells are rod-shaped or filamentous, 0.3–0.9 μm by 2–20 μm. Gram-negative. Motile by one or more polar or lateral flagella. Aerobic. Chemoheterotrophic, requiring cysteine and iron salts for growth. Have proved difficult to cultivate. Use amino acids as carbon and energy sources; neither ferment nor oxidize carbohydrates. Optimum temperature is 36° Celsius. Natural habitats include bodies of fresh water, especially those receiving thermal pollution, and moist soils. Urban environmental isolates from air conditioning cooling towers, evaporative condensers, and other plumbing sources. All species potentially pathogenic for humans. Six species recognized at time of publication of reference; now over 20 species recognized. Type species *L. pneumophila* is the causative agent of legionellosis and Pontiac fever. Reference A:279.

Leptospira Cells are flat, very slender helicoidal rods 0.1 μm by 6–12 μm or more. Faintly Gram-negative. Motile by axial filaments. Obligate aerobes; metabolism is respiratory with oxygen. Chemoheterotrophic, using fatty acids and fatty alcohols as carbon and energy sources. Optimum temperature 28–30° Celsius. Two species; the type species *L. interrogans* is parasitic and may be pathogenic, causing kidney infections in humans and other animals. *L. biflexa* is free-living in soil, fresh water, and marine habitats. Reference A:62.

Leptothrix Cells are straight rods, 0.6–1.5 μm by 3–12 μm. Occur in chains within a sheath or free-swimming. Sheaths often become impregnated with iron or manganese oxides. Gram-negative. Free cells motile by single polar flagella. Strict aerobes; metabolism is respiratory with oxygen. Chemoheterotrophic, utilizing organic acids, amino acids, or carbohydrates. Optimum temperature 20–25° Celsius. Found in slow-running, iron-containing uncontaminated fresh water worldwide, where they form yellowish-brown flocculent masses. Five species; type species *L. ochracea*. Reference B:129.

Leuconostoc Cells may be spherical but often lens-shaped, 0.5–0.7 μm by 0.7–1.2 μm (type species). Usually in pairs and chains. Gram-positive; nonmotile. Facultative anaerobes; fermentative. Chemoheterotrophic; cannot grow without a fermentable carbohydrate and usually need multiple growth factors and amino acids. Some species form copious polysaccharide slime from sucrose. Optimum temperature 20–30° Celsius. Found in slimy sugar solutions, on fruit and vegetables, in milk and dairy products, and in wine. Used in manufacture of sauerkraut and curing of vanilla beans, but may cause some food spoilage. Employed industrially to manufacture dextran. Six species; type species is *L. mesenteroides*. Reference B:510.

Listeria Cells are small, coccoid rods, 0.4–0.5 μm by 0.5–2.0 μm (type species) with a tendency to produce short chains and elongated, filamentous forms. Gram-positive, losing ability to retain stain in older cultures. Motile with peritrichous flagella when grown at 20–25° Celsius. Aerobic to microaerophilic; metabolism is respiratory and fermentative. Chemoheterotrophic. Optimum temperature of type species 37° Celsius. Habitats include the feces of animals and humans, vegetation, and silage. Some are parasitic on animals. Four species; type species is *L. monocytogenes*, which is widely distributed in nature and causes meningitis, septicemia, abortion, and other conditions. Reference B:593.

Methanobacterium Cells are curved, crooked to straight rods, 0.5–1.0 μm wide, and coccoid to long and filamentous. Stain Gram-variable; have atypical cell wall composition lacking peptidoglycan. Cell membrane contains ether-linked lipids. Along with other methane producers, grouped with the archaebacteria. Very strict anaerobes, readily killed by oxygen. A highly specialized physiological group which does not utilize carbohydrates or proteins in general. Reduces formate, methanol, acetate, or CO_2 to methane using hydrogen gas as electron donor; may thus be either chemoheterotrophic or chemolithotrophic. Has unusual electron-transport co-enzymes. Mesophilic to thermophilic. Widely distributed in nature, being found in anaerobic habitats such as sediments, soils, anaerobic sludge digesters, and the rumens and intestinal tracts of animals. The methane-producing genera are subject of intensive basic research to clarify issues of microbial evolution and the origins of cellular life on earth. Also used in applied research on commercial methane production. Five species at date of reference; type species *M. soehngenii*. Reference B:473.

Methylococcus Cells are spherical, about 1.0 μm, usually in pairs. Forms capsules; resting stages called cysts. Gram-negative; nonmotile. Aerobic and strictly respiratory. Chemoheterotrophic, oxidizing only one-carbon compounds such as methane, methanol, and formaldehyde as sole carbon and energy sources. Contain unique methane-oxidizing enzymes and intracellular membranous structures. Optimum temperature 37° Celsius. Isolation from aerobic soils, water, and muds where methane is available (see preceding genus for one possible source). Methane-utilizing bacteria used to produce single-cell protein for animal feed supplements. Single and type species *M. capsulatus*. Reference A:259.

Micrococcus Cells are cocci 0.5–3.5 μm in diameter, dividing in more than one plane to form irregular clusters, tetrads, or packets. Strongly Gram-positive; usually nonmotile. Aerobic; metabolism strictly respiratory. Chemoheterotrophic, utilizing organic acids. Oxidize but do not ferment sugars. Optimum temperature 25–30° Celsius; grow in presence of up to 5% NaCl. Common inhabitants of soils and fresh water; also found in human skin. Nonpathogenic. Three species; type species *M. luteus*. Reference B:478.

Moraxella Cells are short, plump rods, 1.0–1.5 μm by 1.5–2.5 μm, usually in pairs and short chains (subgenus *Moraxella*); or smaller cocci singly or in pairs (subgenus *Branhamella*). Gram-negative, but ofter resist decolorization and appear Gram-positive. Flagella absent, but twitching motility observed. Aerobic; some strains grow weakly under anaerobic conditions.

Respiratory; chemoheterotrophic, preferring enriched media with blood. Optimum temperature 33–35° Celsius. Parasitic on the mucous membranes, especially the conjunctiva, of humans and other warm-blooded animals. Most species, although of low pathogenicity, may be opportunists. Ten species in the subgenus; type species *M. lacunata*, isolated from healthy and inflamed conjunctivae. Reference A:296.

Morganella Cells are straight rods, 0.6–0.7 μm by 1.0–1.7 μm. Conform to general description of the Family Enterobacteriaceae. Gram-negative; motile by means of peritrichous flagella but do not swarm. Facultative anaerobes; metabolism is respiratory and to a limited extent fermentative. Chemoheterotrophic; give positive urease, phenylalanine deaminase, and indol tests. Optimum temperature for testing around 36° Celsius. Found in feces of humans, other mammals, and reptiles. Low pathogenicity, but occur as secondary invaders in urinary tract infections, bacteremias, wounds, and respiratory infections. Single and type species *M. morganii*. Reference A:497.

Mycobacterium Cells are slender slightly curved or straight rods, 0.2–0.6 μm by 1.0–10 μm, sometimes branching. Gram stain reaction not valuable. Stain acid-fast with Ziehl-Neelsen staining procedure. Cell walls contain waxes and long-chain, branched fatty acids. Nonmotile. Aerobes; chemoheterotrophic, with simple nutritional requirements. Very slow growing, and form rough, wrinkled, sometimes pigmented colonies on solid media. Optimum temperature 30–40° Celsius. Many are saprophytic, found in soil and water, while others are parasitic on skin and in lesions of warm- and cold-blooded animals. Twenty-nine species; type species *M. tuberculosis* is principal cause of tuberculosis in humans. *M. bovis*, *M. kansasii*, and *M. intracellulare* are among the other species that can also cause tuberculosis. *M. leprae* causes leprosy. Reference B:682.

Mycoplasma Cells are pleomorphic, varying in shape from spherical, slightly ovoid, or pear-shaped, 0.3–0.8 μm in diameter, to slender branched filaments. No cell wall; stain Gram-negative. Cell membrane contains sterols. Usually nonmotile, but gliding has been observed. Facultative anaerobes; some are fermentative, others nonfermentative. Chemoheterotrophic, using sugars and amino acids as energy sources, and requiring complex media for their very slow growth. Optimum temperature for mycoplasmas from animals is 36–37° Celsius. Parasites of the mucous membranes, especially the respiratory and urogenital tracts and the joints. Cause disease in humans, cows, sheep, rodents, and many other species. Sixty-four species; type species is *M. mycoides*. *M. pneumoniae* is the most pathogenic species for humans, causing atypical pneumonias and other systemic infections. Reference A:742.

Myxococcus Vegetative cells are slender rods with tapering or rounded ends, 0.4–0.7 μm by 2.0–10.0 μm. Gram-negative, motile by gliding. Have complex life cycle in which cells aggregate to form brightly colored, visible fruiting bodies that contain spherical microcysts. Strict aerobes; metabolism respiratory. Chemoheterotrophs, requiring several amino acids and hydrolyzing several polymers. Predatory in natural habitat; obtain nutrients by lysing other bacteria. Optimum temperature 26–30° Celsius. Isolated from dung, soil, decaying plant material, moist bark, and lichens. Six species; type species *M. fulvus*. Reference B:79.

Neisseria Cells are cocci, 0.6–1.0 μm in diameter, single but more often in pairs with adjacent sides flattened. Capsules and fimbriae may be present. Gram-negative, but tend to resist decolorization. No swimming motility. Aerobic; metabolism is respiratory and fermentative. Species differ in ability to use sugars and to reduce nitrate. Chemoheterotrophic. Pathogenic species usually isolated on media containing hemolyzed blood and in atmosphere with added CO_2. Give positive catalase and oxidase tests. Optimum temperature 35–37° Celsius. Inhabitants of mucous membranes of mammals, some species as normal flora and others as pathogens. Eleven species; type species is *N. gonorrhoeae*, the causative agent of gonorrhea. *N. meningitis* causes meningitis in humans; *N. lactamica*, *N. sicca*, and *N. subflava* are part of the human normal flora. Reference A:290.

Nitrobacter Cells are short rods, often wedge- or pear-shaped, 0.6–0.8 μm by 1.0–2.0 μm. Gram-negative. Usually nonmotile, but may produce single polar flagella in continuous culture. Reproduce by budding. Strictly aerobic using oxygen. Some strains are obligate chemolithotrophs, which oxidize nitrite to nitrate to yield energy and fix CO_2 for carbon. Some strains can be grown heterotrophically. Optimum temperature 25–30° Celsius. Widely distributed in soils; marine strains have been isolated from seawater. Significant, with related nitrifying genera, in the nitrogen cycle. Single and type species *N. winogradskyi*. Reference B:451.

Nitrosomonas Cells are ellipsoidal or short rods, 0.8–0.9 μm by 1.0–2.0 μm, occurring singly, in pairs or as short chains. Gram-negative; may be nonmotile or motile. Strictly aerobic with oxygen. Obligate chemolithotrophs, deriving energy from the oxidation of ammonia to nitrite and fixing CO_2. Optimum temperature 25–30° Celsius. Habitat is soils, fresh water, and sea water. With related genera of ammonia oxidizers, carry out key step in nitrogen cycle immediately prior to *Nitrobacter* (see above). Single and type species *N. europaea*. Reference B:453.

Nocardia Cells are highly pleomorphic, including coccoid, rodlike, branched, and mycelial forms, depending on phase of growth and pattern of development. Gram-positive; some species are acid-fast. Heat-resistant microcysts or chlamydospores produced by some members. Nonmotile. Obligate aerobes; metabolism both oxidative and fermentative. Chemoheterotrophs, utilizing sugars and organic acids; slow growing. Optimum temperature (type species) 35° Celsius. Habitat is soil; frequently isolated from tissues of warm-blooded animals. Cause a variety of lesions and systemic diseases. There are 31

species; type species is *N. farcinica*. *N. brasiliensis* in humans causes mycetoma, an invasive foot and skin disease common in tropical regions. Reference B:726.

Oceanospirillum Cells are rigid, helical rods 0.4–1.4 μm in diameter; coccoid bodies formed in old cultures. A polar membrane underlies the cell membrane at the cell poles. Gram-negative. Motile by means of bipolar tufts of flagella or by a single polar flagellum at each pole. Aerobic; metabolism is respiratory with oxygen. Chemoheterotrophic, using amino acids or the salts of organic acids as carbon sources; sugars cannot be used. Seawater required for growth. Optimum temperature 25–32° Celsius. Isolated from coastal seawater, decaying seaweed, and marine mussels. Nine valid species; type species is *O. linum*. Reference A:104.

Pasteurella Cells are spherical, ovoid, or rod-shaped, 0.3–1.0 μm by 1.0–2.0 μm. Gram-negative; bipolar staining is common. Nonmotile. Facultative anaerobes; metabolism is fermentative. Chemoheterotrophs, utilizing sugars and amino acids and giving positive nitrate reduction, catalase, and (usually) oxidase tests. Optimum temperature 37° Celsius. Live as parasites on the mucous membranes of the upper respiratory digestive tracts of mammals, birds, and (rarely) humans. Six species; type species is *P. multocida*, which causes hemorrhagic septicemia in cattle and fowl cholera in domestic fowl. In humans, this species may infect bites or be an opportunistic pathogen. Reference A:552.

Pediococcus Cells are cocci 0.6–1.0 μm in diameter, occurring in pairs or tetrads. Gram-positive; nonmotile. Microaerophilic; metabolism is fermentative. Chemoheterotrophs, utilizing carbohydrates, have complex nutritional requirements. Optimum temperature for most species around 25° Celsius. Saprophytes, found in fermenting plant materials. In food curing processes, such as making pickles and curing dry and semidry sausages, their fermentation contributes to flavor development. Five species; type species is *P. cerevisiae*. Reference B:513.

Peptococcus Cells are spherical, usually 0.5–1.0 μm in diameter. Occur singly, in pairs, tetrads, irregular masses or short chains. Gram-positive; nonmotile. Strict anaerobes; metabolism is fermentative. Chemoheterotrophic, preferentially utilizing proteins and their breakdown products. Optimum temperature 35–37° Celsius. Isolated from normal and infected human urogenital, intestinal, and respiratory tracts, inflamed appendices or gums, and other sites of infection. Also found in anaerobic mud. Five species; type species *P. niger*. Reference B:518.

Photobacterium Cells are coccobacilli or rods, 1.0–2.5 μm by 0.4–1.0 μm. Gram-negative; motile by one or more polar flagella. Facultative anaerobes; metabolism is both respiratory and fermentative. Chemoheterotrophs, utilizing hexoses and amino acids. Most notable trait is luminescence — ability to produce visible light in suitable growth conditions. Optimum temperature usually 20–30° Celsius; most require NaCl concentration near that of sea water. Habitat is sea water, surfaces and alimentary tracts of some fishes, and the luminescent organs of some marine animals, where they contribute to a complex symbiosis. Two species; type species is *P. phosphoreum*. Reference B:349.

Propionibacterium Cells are pleomorphic, diphtheroid or club-shaped rods; sizes and groupings highly variable. Gram-positive, staining irregular; nonmotile. Anaerobic to aerotolerant; metabolism is fermentative. Chemoheterotrophic, utilizing carbohydrates, peptone, or organic acids. Products include both propionic and acetic acids, plus lesser amounts of other organic acids. Optimum temperature 30–37° Celsius. Found in dairy products, on human skin, and in the intestinal tracts of humans and animals. Several species found in infections such as abscesses. Many are useful in the food and chemical industries. Eight valid species; the type species is *P. freudenreichii*, which is used in the manufacture of Swiss cheese. *P. acnes* is a member of the normal skin flora; often contaminates diagnostic samples, especially blood drawn for culture; isolated from wound infections and abscesses. Reference B:633.

Proteus Cells are straight rods, 0.4–0.8 μm by 1.0–3.0 μm. Conform to the general definition of the Family Enterobacteriaceae. Gram-negative; motile by peritrichous flagella. Most strains swarm with periodic cycles of migration, producing concentric zones of growth, or spread in a uniform film over moist surfaces of solid media. Facultative anaerobes; metabolism is respiratory and fermentative. Chemoheterotrophic; give positive urease and phenylalanine and tryptophane deaminase tests, and produce H_2S. Occur in the intestinal contents of humans and a wide variety of animals; also occur in manure, soil, and polluted waters. Pathogenic, causing urinary tract infections; also cause septic lesions at other sites as secondary invaders. Three species; type species *P. vulgaris*. *P. mirabilis* is more commonly found in clinical specimens. Reference A:491.

Providencia Cells are straight rods, 0.6–0.8 μm by 1.5–2.5 μm. Conform to the general definition of the family Enterobacteriaceae. Gram-negative; motile by peritrichous flagella. Facultative anaerobes; metabolism is both oxidative and fermentative. Chemoheterotrophs; deaminate phenylalanine and tryptophane, ferment mannose, positive for indol and for citrate utilization. Optimum temperature 35–37° Celsius. Isolated from stool samples in human diarrhea; from urinary tract infections, usually in catheterized or compromised patients; and from wounds, burns, and bacteremias. Three species; type species is *P. alcalifaciens*. Reference A:494.

Pseudomonas Cells are straight or slightly curved rods, 0.5–1.0 μm by 1.5–5.0 μm. Gram-negative; motile by one or more polar flagella. Aerobic; metabolism is strictly respiratory with oxygen and in some cases nitrate. Chemoheterotrophic, noted for their ability to use a wide range of simple organic substrates. Metabolize sugars by the Entner-Doudoroff pathway. Some species are facultative chemolithotrophs. Temperature range is wide. Widely distributed; habitats are diverse, including soil, fresh and salt water, and plant and animal tissues. Some are plant and animal pathogens. The pseudomonads are receiving intense research investigation. Some species are potentially useful for their abilities to metabolize various components of crude oil. Recombinant pseudomonads are being studied for their ability to prevent crop losses due to frost damage. Classification is in flux; 57 species are described in Reference A, some of which are taxonomically ambiguous. Type species is *P. aeruginosa*, an extensively studied organism noted for its ability to cause opportunistic infections, especially of burns, in humans. *P. mallei*, *P. pseudomallei*, and *P. cepacea* are other human pathogens. *P. solanacearum* and *P. syringae* are important plant pathogens. Reference A:141.

Rhizobium Cells are rods, 0.5–0.9 μm by 1.2–3.0 μm, commonly pleomorphic. Gram-negative; motile by either polar or peritrichous flagella. Aerobes; metabolism is strictly respiratory with oxygen. Chemoheterotrophic, utilizing a wide range of carbohydrates, organic acids, and amino acids. Fix atmospheric nitrogen into ammonia when living symbiotically. Produce copious slime when grown on carbohydrates. Optimum temperature 25–30° Celsius. Found worldwide in soil. Able to invade the root hairs of leguminous plants and incite the production of root nodules, within which they live as intracellular symbionts, fixing nitrogen. Major contribution to soil fertility. Three species; type species *R. leguminosarum*. Reference A:235.

Rhodospirillum Cells are spiral-shaped. Size range is wide. Gram-negative; motile by polar flagella. Color reddish, purple, or brown, due to photosynthetic pigments in internal membrane vesicles. Anaerobic phototrophs, carrying out photosynthesis utilizing either hydrogen gas or simple organic compounds as a source of reducing power. Most species also function as microaerophilic or aerobic chemoheterotrophs, carrying out oxidative metabolism of simple organic nutrients in the dark, in most cases requiring vitamins as growth factors. Some strains fix nitrogen. Called the nonsulfur purple bacteria because they utilize H_2S, if at all, only at very low levels (compare with *Ectothiorhodospira*). Optimum temperature (type species) 30–35° Celsius. Habitat is stagnant water, muddy fresh water, and mud exposed to light. Five species; type species is *R. rubrum*. Reference B:26.

Rickettsia Cells are short rods, 0.3–0.5 μm by 0.8–2.0 μm. Gram-negative; nonmotile. Aerobic; have not been cultivated in absence of host cells. Obligate intracellular parasites; energy-yielding metabolism includes a portion of the Krebs cycle. They do not synthesize or degrade nucleoside monophosphates. Optimum temperature that of host cells. Ecology involves a natural cycle in which arthropods such as ticks, lice, and mites, and vertebrates such as rats, flying squirrels, and humans participate. The genus is notorious for its virulence for humans. Twelve species; the type species is *R. prowazekii*, the agent of epidemic typhus fever. *R. rickettsii* causes Rocky Mountain spotted fever. Reference A:688.

Salmonella Cells are straight rods, 0.7–1.5 μm by 2.0–5.0 μm. Conform to general description of the Family Enterobacteriaceae. Gram-negative, usually motile with peritrichous flagella. Facultative anaerobes; metabolism is both respiratory and fermentative. Chemoheterotrophic. Lactose and indol tests negative; citrate, H_2S, and nitrate reduction tests usually positive. Optimum temperature 35–37° Celsius. Classified into species, but also classified into serovars (serological varients) based on identification of the surface antigens by serological methods. There are hundreds of different serovars, many of which also have species names. Habitat is the intestinal tract of many warm and cold-blooded animals; may also be isolated from water, soil, meat and dairy products, and other fecally contaminated materials and objects. Passed from one animal species to the other in the food chain. Pathogenic; cause enteric fever (typhoid fever), gastroenteritis, and septicemia in humans. Twelve serovars described in Reference A. Type species is *S. choleraesuis*, pathogenic for pigs, humans, and others. *S. typhi* causes typhoid fever. *S. typhimurium*, *S. cubana*, and *S. agona* are frequent causes of gastroenteritis outbreaks. Reference A:427.

Serratia Cells are straight rods, 0.5–0.8 μm by 0.9–2.0 μm. Conform to the general description of the Family Enterobacteriaceae. Gram-negative, motile by peritrichous flagella. Facultative anaerobes; metabolism is both respiratory and fermentative. Chemoheterotrophic; utilize a wide range of nutrients including caprylate, L-fucose; produce several extracellular hydrolases. Often produce pink to red pigments at 20–35° Celsius. Occur in the natural environment in soil, fresh and salt water, and on plant surfaces, but may be opportunistic human pathogens. Six species; type species is *S. marcescens*. Reference A:477.

Shigella Cells are straight rods. Conform to general description of the Family Enterobacteriaceae. Gram-negative; nonmotile. Facultative anaerobes; metabolism is both respiratory and fermentative. Chemoheterotrophic; ferment sugar without gas, do not ferment lactose or use citrate or malonate, negative for H_2S production. Optimum temperature 35–37° Celsius. Habitat is the intestinal tract of humans and other animals; also found in water and contaminated foods. Pathogenic, causing bacillary dysentery. Possesses transmissible antibiotic resistance plasmids. There are four species or subgroups, each with several distinctive serovars. The type species is *S. dysenteriae*. Reference A:423.

Sphaerotilus Cells are straight rods, 0.7–2.4 μm by 3–10 μm, occurring in chains within a sheath. May be attached by means of a holdfast to submerged plants, stones, or other objects. Gram-negative; single cells usually motile by a bundle of subpolar flagella. Strict aerobes; metabolism is respiratory with oxygen. Chemoheterotrophs, using sugars, alcohols, organic acids, and amino acids. Optimum temperature 25–30° Celsius. Habitat is slowly running fresh water contaminated with nutrient-rich sewage or wastes from paper or dairy industries; also grows in activated sewage sludge. Single and type species is *S. natans*. Reference B:128.

Spirillum Cells are large, rigid helices, 1.4–1.7 μm by 14–60 μm. A polar membrane underlies the cell membrane at both poles. Gram-negative; motile by large bipolar tufts of flagella. Microaerophilic; metabolism is strictly respiratory with oxygen. Chemoheterotrophic, utilizing certain organic acids but not carbohydrates or common macromolecular nutrients. Grow poorly on solid media. Oxidase positive, catalase negative. Optimum temperature 30° Celsius. Habitat is stagnant freshwater environments. Single type species *S. volutans*. Reference A:90.

Spirochaeta Cells are helical, 0.2–0.75 μm by 5–250 μm. Gram-negative; motile by axial filaments. Obligate to facultative anaerobes; metabolism fermentative and in some species oxidative. Chemoheterotrophic, using a variety of carbohydrates and producing ethanol, acetate, and gases. Optimum temperature 25–40° Celsius. In sediments, mud, and water of ponds, marshes, and swamps, where H_2S found. Free-living, nonpathogenic spirochetes. Six species; type species *S. plicatilis*. Reference A:39.

Spiroplasma Cells pleomorphic, ranging from helical and branched nonhelical filaments to spherical or ovoid. Stain Gram-negative, but test is not relevent as cell wall is entirely lacking. Cell membrane contains sterols. Helical cells are motile with flexing, twitching, and rotating movements. The method is not clear, but intracellular fibrils have been observed. Facultative anaerobes; sugars are fermented. Chemoheterotrophs, utilizing carbohydrates and some amino acids; cholesterol is required for growth. These mycoplasmas are isolated from ticks, insects, the fluids of vascular plants and insects that feed on them, and from the surfaces of flowers and other plant parts. Pathogens of numerous plant species. Four species; type species is *S. citri*, which causes "stubborn disease" in oranges and grapefruits. Reference A:781.

Staphylococcus Cells are spherical, 0.5–1.5 μm in diameter. Occur singly, in pairs, and characteristically in irregular grapelike clusters. Gram-positive; nonmotile. Facultative anaerobes; metabolism is both respiratory and fermentative. Chemoheterotrophs, utilizing a wide range of carbohydrates, and requiring some amino acids and other growth factors. Possess wide range of extracellular hydrolytic enzymes. Tolerate exposure to salt and to bile components. Optimum temperature 35–40° Celsius. Habitat is the skin, nasal passages, and mucous membranes of warm-blooded animals. Part of normal flora of humans, but may be pathogenic, particularly in compromised individuals. Three species; type species is *S. aureus*, a pathogen that causes infections such as boils, abscesses, septicemia, wound infections, toxic shock syndrome, and food poisoning. *S. epidermidis*, although harmless in normal people, is one of the more common causes of nosocomial infections. Reference B:483.

Streptococcus Cells are spherical to ovoid, less than 2 μm in diameter, varying with species. Occur in pairs or characteristically in chains when grown in liquid media. Gram-positive; predominantly nonmotile, but with a few motile strains. Facultative anaerobes; metabolism is fermentative. Chemoheterotrophic, fermenting sugars to lactic and other acids, or fermenting organic and amino acids. Contain no heme compounds such as cytochromes; catalase test is negative. Nutrient requirements generally complex; most grow best on media supplemented with blood. Optimum temperature about 37° Celsius for most species. Have been classified by biochemical methods into species and also by serological methods, which are the basis for the Lancefield groupings. Parasitic on the mucous membranes of humans and other animals; many strains are pathogenic. Several species are used extensively in the dairy and fermentation industries, while others cause food spoilage. There are 21 species; type species is *S. pyogenes*, the highly pathogenic Group A species that causes human skin and respiratory infections, meningitis, pneumonia, endocarditis, septicemia, and other diseases. *S. pneumoniae*, a lancet-shaped organism, causes pneumonia and middle-ear infections. *S. agalactiae*, in Group B, causes neonatal infections. *S. faecium* and *S. faecalis* are part of the normal human bowel flora and are rarely pathogenic. *S. lactis* and *S. cremoris* are used in making cottage cheese. Reference B:490.

Streptomyces Microscopic structures in form of slender, coenocytic hyphae, making up a vegetative mycelium. At maturity, an aerial mycelium forms, bearing chains of three to many reproductive spores called conidia, 0.5–2.0 μm in diameter. Gram-positive; nonmotile. Aerobes; metabolism is respiratory, only rarely fermentative. Chemoheterotrophs, utilizing carbohydrates and versatile in using other nutrients for growth. Produce hydrolytic enzymes, pigments, and antibiotics, such as streptomycin, neomycin, tetracycline, erythromycin, and chloramphenicol. Optimum temperature 25–35° Celsius; some species are thermophilic. Habitat is soil, decaying plant materials, and to a lesser extent water. This group of organisms is immensely useful for the antibiotics it yields in industrial processes. Reference B lists 463 species; type species is *S. albus*. Reference B:748.

Sulfolobus Cells are spherical with lobes, 0.8–1 μm in diameter. Gram-negative; cell wall present but lacks peptidoglycan. Cell membrane contains ether-linked lipids; structure unusual. These organisms considered to be archaebacteria. Nonmotile. Resemble mycoplasmas but are more refractile. Aerobes; facultative autotrophs, oxidizing elemental sulfur as an energy source, but also using organic nutrients such as yeast extract as a carbon and energy source. Optimum temperature 70–75° Celsius, pH optimum 2–3. Found in geothermal areas where environment is both acid and hot and where elemental sulfur is available. Single and type species *S. acidocalcarius*. Reference B:461.

Thermoplasma Cells are pleomorphic, varying in shape from spherical (0.1–0.3 μm) to filamentous. Gram-negative, but lack a cell wall. Membrane contains ether lipids common to archaebacteria. Nonmotile. Strict aerobes. Chemoheterotrophic, having a strict requirement for yeast extract for laboratory cultivation. Obligate thermoacidophiles, with optimum growth at 55–59° Celsius and pH 1–2. Occur freeliving in self-heating coal refuse piles, the only habitat reported so far for this remarkable organism. Single type species *T. acidophilum*. Reference A:790.

Thermus Cells are straight rods, 0.5–0.8 μm by 5.0–10.0 μm. Filaments of from 20 to more than 200 cells may occur. Most strains form rotund bodies, structures formed from the association of individual cells. Gram-negative; nonmotile. Obligate aerobes; metabolism is strictly respiratory with oxygen. Chemoheterotrophic, using sugars, amino acids, and a wide range of nutrient sources. Many strains have yellow to red pigments. Optimum temperature 70–75° Celsius. Isolated from natural habitats such as hot springs, as well as from hot water heaters and thermally polluted waters. Single type species *T. aquaticus*.

Thiobacillus Cells are small rods, 0.5 μm by 1.0–3.0 μm (type species). Gram-negative; motile by a single polar flagellum or nonmotile. Obligate aerobes except one species. Grow lithotrophically by oxidizing reduced sulfur compounds such as elemental sulfur, sulfides, and thiosulfate as sources of energy. Often accumulate sulfuric acid as an end product. Some species can also obtain energy from the oxidation of reduced (ferrous) iron. Fix CO_2. Many species are also capable of assimilating preformed organic compounds. Optimum temperature 28–30° Celsius. Some species tolerate extreme acidity; some isolates have been reported to be halophilic or thermophilic. Habitat is sea water, marine mud, soil, fresh water, acid mine waters, sewage, sulfur springs, and in or near sulfur deposits, especially in environments where H_2S is produced. With other sulfur-oxidizing genera, play essential role in the sulfur cycle. Are employed in mining pro-

cesses for extraction of copper and uranium from low-grade ores. Eight species; type species is *T. thioparus*. Reference B:456.

Thiothrix Cells are long filaments composed of short cylindrical or ovoid cells. In nature filaments usually attached to solid substrates by means of holdfasts. Form rosettes in culture. Deposit sulfur granules within the cells. Aerobic; appear to be obligate lithotrophs, deriving energy from oxidation of H_2S. Have not been consistently cultured, so little is known of their metabolism. Significance in sulfur cycle similar to other sulfur-oxidizers. Habitats include fresh water and marine habitats, in sewage plants and sulfur springs where H_2S concentration is high. Single and type species *T. nivea*. Reference B:119.

Treponema Cells are slender helical rods 0.1–0.4 μm by 5–20 μm, with tight regular spirals. Gram-negative type of cell wall, but stain poorly with the Gram stain technique. Motile by means of one or more axial filaments; capable of both rotational and translational movements. Strictly anaerobic or microaerophilic. The forms that have been cultivated ferment sugar and amino acids. Chemoheterotrophic, using a complex variety of nutrients, including fatty acids. Human pathogenic species have not been cultivated in vitro, so little is known about their metabolism. Treponemes are host-associated to parasitic, found in the oral cavity, rumen, and intestinal and urogenital tracts of humans and animals. Several species are pathogens, while others are suspected pathogens. Thirteen species; type species is *T. pallidum*, the causative agent of syphilis and yaws. *T. carateum* causes pinta. Reference A:49.

Ureaplasma Cells from 18- to 24-hour cultures are round or coccobacillary, about 0.3 μm in diameter. Pleomorphic forms under other conditions. Gram-negative, but lack a cell wall. Cell membrane contains sterols. Nonmotile. Microaerophilic; chemoheterotrophic, hydrolyzing urea but not metabolizing the usual sugars. Complex nutritional requirements; cultivated in media containing serum and yeast extract. Optimum temperature 37° Celsius. Occur predominantly in the mouth and respiratory urogenital tracts of humans and various animals, including birds. Two species; type species *U. urealyticum* is an increasingly commonly identified pathogen that causes urethritis and is believed to cause spontaneous abortion, infertility, and neonatal infections such as pneumonia in humans. *U. diversum* causes pneumonia and urogenital disease in cattle. Reference A:770.

Veillonella Cells are cocci, 0.3–0.5 μm in diameter, appearing in pairs, masses, and short chains. Gram-negative; nonmotile. Anaerobic. Chemoheterotrophic, fermenting pyruvate, lactate, and other organic acids. Nutritional requirements are complex; CO_2 is required. Optimum temperature 30–37° Celsius. Parasitic in the mouths and intestinal and respiratory tracts of humans and other animals. Pathogenicity not clear. Seven species; type species is *V. parvula*. Reference A:681.

Vibrio Cells are straight or curved rods, 0.5–0.8 μm by 1.4–2.6 μm. Gram-negative; in liquid media, motile by means of polar flagella, which are enclosed in a sheath. On solid media, may develop numerous lateral flagella. Facultative anaerobes; metabolism is both fermentative and respiratory. Chemoheterotrophic, versatile in use of many different organic nutrient sources. Most species require, and all are stimulated by, sodium ions. Very common in aquatic environments with a wide range of salinities, especially marine and estuarine waters, on the surfaces and in the intestinal tracts of marine animals. Some species pathogenic for humans. Twenty valid species; type species is *V. cholerae*, which causes cholera in humans. *V. parahaemolyticus* causes human gastroenteritis, usually contracted by eating contaminated seafoods. *V. vulnificus* may cause septicemia. *V. anguillarum* is a pathogen of marine fish and eels. Reference A:518.

Xanthomonas Cells are straight rods, usually 0.4–0.7 μm by 0.7–1.8 μm. Gram-negative; motile by a single polar flagellum. Obligate aerobes; metabolism is strictly respiratory with oxygen. Chemoheterotrophic, using a variety of carbohydrates and organic acids and requiring several growth factors. Produce bright yellow pigments. Optimum temperature 25–30° Celsius. All species are plant pathogens and are found on and in the tissues of their specific host plants. Often categorized into pathovars, based on the host specificity of the strain. Some cause disastrous crop losses. Five species; type species is *X. campestris*, a pathogen which has 126 pathovars infecting several hundred different species of plants including trees, shrubs, legumes, grains, and garden flowers: e.g., *X. campestris* pv. *citri*, which causes Florida citrus canker. Reference A:199.

Yersinia Cells are straight rods to coccobacilli, 0.5–0.8 μm by 1–3 μm. Gram-negative. Nonmotile when grown at 37° Celsius; species other than *Y. pestis* motile with peritrichous flagella when grown at or below 30° Celsius. Facultative anaerobes; metabolism is both respiratory and fermentative. Chemoheterotrophs, fermenting sugars and otherwise similar to the other enterobacteria in their metabolism. Optimum temperature 28–29° Celsius. Occur in a broad spectrum of habitats, including soil, water, foods, and normal animals (nonpathogenic species) and humans, animals, insects, and soil (pathogenic species). Several are virulent human and animal pathogens. Seven species; type species is *Y. pestis*, the causative agent of human and animal plague. *Y. pseudotuberculosis* and *Y. enterocolitica* are other species pathogenic for humans. Reference A:498.

Zoogloea Cells are straight to slightly curved, plump rods, 1.0–1.3 μm by 2.1–3.6 μm. Gram-negative; motile by means of a single polar flagellum. Cells become embedded in a fingerlike gelatinous matrix or film at later growth stages. Aerobic; respiring with oxygen or with nitrate when oxygen absent. Chemoheterotrophic, nutritionally versatile, utilizing proteinaceous nutrients, urea, organic acids, aromatic compounds, some alcohols, and some sugars. Optimum temperature 28–37° Celsius. Occur free-living in organically polluted fresh water, and in waste water in all stages of treatment. Are a major component of the beneficial microbial community that composes the activated sludge in sewage treatment. Single and type species *Z. ramigera*. Reference A:214.

Zymomonas Cells are rods with rounded ends; 1.0–1.4 μm by 2–6 μm, usually in pairs. Gram-negative; usually nonmotile, but if motile possess polar flagella. Facultative to obligate anaerobes; metabolism is fermentative. Chemo-heterotrophic, converting glucose or fructose to ethanol and other products. Is ethanol and acid tolerant; requires amino acids and two vitamins. Optimum temperature 25–30° Celsius. A serious spoilage agent in cask beers and ciders; the fermenting agent that converts agave sap to pulqe and palm sap to palm wine. Non-pathogenic. Due to its reputed antagonistic effects against other bacteria and fungi has been tested as a therapeutic agent in chronic enteric and gynecological infections. Single and type species *Zymomonas mobilis*. Reference A:576.

APPENDIX TWO
DESCRIPTIONS OF
BACTERIAL GENERA

Bergey's Manual does not comprise a natural classification scheme for the bacteria; rather it presents an arrangement of descriptions of the bacteria according to the conventions of taxonomists. This arrangement follows certain patterns of natural relatedness (or similarity) that are considered acceptable by experts. Working microbiologists find it useful because it is an authoritative consensus reference on nomenclature.

The first part of Appendix Two lists the classification scheme that will be used in the new *Bergey's Manual of Systematic Bacteriology*. It is provided for your reference. The organization of *Bergey's Manual of Determinative Bacteriology* is listed in the second part of Appendix Two. The differences between the new manual of systematic bacteriology and the eighth edition of *Bergey's Manual* demonstrate that, as our information about organisms increases, taxonomic classifications change to reflect this new knowledge.

CLASSIFICATION OF BACTERIA ACCORDING TO *BERGEY'S MANUAL OF SYSTEMATIC BACTERIOLOGY*

VOLUME 1

Kingdom *Procaryotae*

Gracilicutes
Firmicutes
Tenericutes
Mendosicutes

The Spirochetes

SPIROCHAETALES
  Spirochaetaceae
    Spirochaeta
    Cristispira
    Treponema
    Borrelia
  Leptospiraceae
    Leptospira
Other Organisms
Hindgut Spirochetes of Termites and *Cryptocercus punctulatus*

Aerobic/Microaerophilic, Motile, Helical/Vibrioid, Gram-negative Bacteria

> *Aquaspirillum*
> *Spirillum*
> *Azospirillum*
> *Oceanospirillum*
> *Campylobacter*
> *Bdellovibrio*
> *Vampirovibrio*

Nonmotile (or Rarely Motile), Gram-negative Curved Bacteria

> *Spirosomaceae*
> > *Spirosoma*
> > *Runella*
> > *Flectobacillus*
>
> **Other Genera**
> > *Microcyclus*
> > *Meniscus*
> > *Brachyarcus*
> > *Pelosigma*

Gram-negative Aerobic Rods and Cocci

> *Pseudomonadaceae*
> > *Pseudomonas*
> > *Xanthomonas*
> > *Frateuria*
> > *Zoogloea*
>
> *Azotobacteraceae*
> > *Azotobacter*
> > *Azomonas*
>
> *Rhizobiaceae*
> > *Rhizobium*
> > *Bradyrhizobium*
> > *Agrobacterium*
> > *Phyllobacterium*
>
> *Methylococcaceae*
> > *Methylococcus*
> > *Methylomonas*
>
> *Halobacteriaceae*
> > *Halobacterium*
> > *Halococcus*
>
> *Acetobacteraceae*
> > *Acetobacter*
> > *Gluconobacter*
>
> *Legionellaceae*
> > *Legionella*
>
> *Neisseriaceae*
> > *Neisseria*
> > *Moraxella*
> > *Acinetobacter*
> > *Kingella*
>
> **Other Genera**
> > *Beijerinckia*

> *Derxia*
> *Xanthobacter*
> *Thermus*
> *Thermomicrobium*
> *Halomonas*
> *Alteromonas*
> *Flavobacterium*
> *Alcaligenes*
> *Serpens*
> *Janthinobacterium*
> *Brucella*
> *Bordetella*
> *Francisella*
> *Paracoccus*
> *Lampropedia*

Facultatively Anaerobic Gram-negative Rods

> *Enterobacteriaceae*
> > *Escherichia*
> > *Shigella*
> > *Salmonella*
> > *Citrobacter*
> > *Klebsiella*
> > *Enterobacter*
> > *Erwinia*
> > *Serratia*
> > *Hafnia*
> > *Edwardsiella*
> > *Proteus*
> > *Providencia*
> > *Morganella*
> > *Yersinia*
>
> **Other Genera of the Family**
> *Enterobacteriaceae*
> > *Obesumbacterium*
> > *Xenorhabdus*
> > *Kluyvera*
> > *Rahnella*
> > *Cedecea*
> > *Tatumella*
>
> *Vibrionaceae*
> > *Vibrio*
> > *Photobacterium*
> > *Aeromonas*
> > *Plesiomonas*
>
> *Pasteurellaceae*
> > *Pasteurella*
> > *Haemophilus*
> > *Actinobacillus*
>
> **Other Genera**
> > *Zymomonas*
> > *Chromobacterium*
> > *Cardiobacterium*
> > *Calymmatobacterium*
> > *Gardnerella*
> > *Eikenella*
> > *Streptobacillus*

Anaerobic Gram-negative Straight, Curved, and Helical Rods

Bacteroidaceae
  Bacteroides
  Fusobacterium
  Leptotrichia
  Butyrivibrio
  Succinimonas
  Succinivibrio
  Anaerobiospirillum
  Wolinella
  Selenomonas
  Anaerovibrio
  Pectinatus
  Acetivibrio
  Lachnospira

Dissimilatory Sulfate- or Sulfur-Reducing Bacteria

Desulfuromonas
Desulfovibrio
Desulfomonas
Desulfococcus
Desulfobacter
Desulfobulbus
Desulfosarcina

Anaerobic Gram-negative Cocci

Veillonellaceae
  Veillonella
  Acidaminococcus
  Megasphaera

The Rickettsias and Chlamydias

RICKETTSIALES
  Rickettsiaceae
  Rickettsieae
    Rickettsia
    Rochalimaea
    Coxiella
  Ehrlichieae
    Ehrlichia
    Cowdria
    Neorickettsia
  Wolbachieae
    Wolbachia
    Rickettsiella
  Bartonellaceae
    Bartonella
    Grahamella
  Anaplasmataceae
    Anaplasma
    Aegyptianella
    Haemobartonella
    Eperythrozoon

CHLAMYDIALES
  Chlamydiaceae
    Chlamydia

The Mycoplasmas

Division Tenericutes
Mollicutes
  MYCOPLASMATALES
    Mycoplasmataceae
      Mycoplasma
      Ureaplasma
    Acholeplasmataceae
      Acholeplasma
    Spiroplasmataceae
      Spiroplasma
    Other Genera
      Anaeroplasma
      Thermoplasma
    Mycoplasma-like Organisms of Plants and Invertebrates

Endosymbionts

ENDOSYMBIONTS OF PROTOZOA
  Endosymbionts of Ciliates
  Endosymbionts of Flagellates
  Endosymbionts of Amoebas
  Taxa of Endosymbionts
    Holospora
    Caedibacter
    Pseudocaedibacter
    Lyticum
    Tectibacter
ENDOSYMBIONTS OF INSECTS
  Blood-sucking Insects
  Plant Sap-sucking Insects
  Cellulose and Stored Grain Feeders
  Insects Feeding on Complex Diets
  Taxon of Endosymbionts
    Blattabacterium
ENDOSYMBIONTS OF FUNGI AND IN-VERTEBRATES OTHER THAN ARTHROPODS
  Fungi
  Sponges
  Coelenterates
  Helminthes
  Annelids
  Marine Worms and Mollusks

VOLUME 2

Gram-positive Cocci

Micrococcaceae
  Micrococcus

Stomatococcus
Planococcus
Staphylococcus
Deinococcaceae
  Deinococcus
Other Organisms
  "Pyogenic" streptococci
  "Oral" streptococci
  "Lactic" streptococci and enterococci
  Leuconostoc
  Pediococcus
  Aerococcus
  Gemella
  Peptococcus
  Peptostreptococcus
  Ruminococcus
  Coprococcus
  Sarcina

Endospore-forming Gram-positive Rods and Cocci

Bacillus
Sporolactobacillus
Clostridium
Desulfotomaculum
Sporosarcina
Oscillospira

Regular, Non-sporing, Gram-positive Rods

Lactobacillus
Listeria
Erysipelothrix
Brochothrix
Renibacterium
Kurthia
Caryophanon

Irregular, Non-sporing, Gram-positive Rods

Animal and saprophytic *Corynebacteria*
Plant *Corynebacteria*
  Gardnerella
  Arcanobacterium
  Arthrobacter
  Brevibacterium
  Curtobacterium
  Caseobacter
  Microbacterium
  Aureobacterium
  Cellulomonas
  Agromyces
  Arachnia
  Rothia
  Propionibacterium
  Eubacterium
  Acetobacterium
  Lachnospira
  Butyrivibrio
  Thermoanaerobacter
  Actinomyces
  Bifidobacterium

Mycobacteria

Mycobacteriaceae
  Mycobacteria

Nocardioforms

Nocardia
Rhodococcus
Nocardioides
Pseudonocardia
Oerskovia
Saccharopolyspora
Micropolyspora
Promicromonospora
Intrasporangium

VOLUME 3

Gliding, Non-fruiting Bacteria

CYTOPHAGALES
Cytophagaceae
  Cytophaga
  Sporocytophaga
  Capnocytophaga
  Flexithrix
  Flexibacter Microscilla
  Saprospira
  Herpetosiphon
LYSOBACTERALES
Lysobacteraceae
  Lysobacter
BEGGIATOALES
Beggiatoaceae
  Beggiatoa
  Thioploca
  Thiospirillopsis
  Thiothrix
  Achromatium
Simonsiellaceae
  Simonsiella
  Alysiella
Leucotrichaceae
  Leucothrix
Families and genera incertae sedis
  Toxothrix
  Vitreoscilla
  Chitinophagen
  Desulfonema
Pelonemataceae
  Pelonema
  Achroonema
  Peloploca
  Desmanthus

Anoxygenic Phototrophic Bacteria

PURPLE BACTERIA
  Chromatiaceae
    Chromatium
    Thiocystis
    Thiospirillum
    Thiocapsa
    Amoebobacter
    Lamprobacter
    Lamprocystis
    Thiodictyon
    Thiopedia
  Ectothiorhodospiraceae
    Ectothiorhodospira
  Purple nonsulfur bacteria
    Rhodospirillum
    Rhodopseudomonas
    Rhodobacter
    Rhodomicrobium
    Rhodopila
    Rhodocyclus
GREEN BACTERIA
  Green sulfur bacteria
    Chlorobium
    Prosthecochloris
    Ancalochloris
    Pelodictyon
    Chloroherpeton
    Symbiotic consortia
  Multicellular filamentous green bacteria
    Chloroflexus
    Heliothrix
    Oscillochloris
    Chloronema
GENERA INCERTAE SEDIS
    Heliobacterium
    Erythrobacter

Budding and/or Appendaged Bacteria

PROSTHECATE BACTERIA
  Budding bacteria
    Hyphomicrobium
    Hyphomonas
    Pedomicrobium
    "Filomicrobium"
    "Dicotomicrobium"
    "Tetramicrobium"
    Stella
    Ancalomicrobium
    Prosthecomicrobium
  Non-budding bacteria
    Caulobacter
    Asticcacaulis
    Prosthecobacter
    Thiodendron
NON-PROSTHECATE BACTERIA
  Budding bacteria
    Planctomyces

Pasteuria
Blastobacter
Angulomicrobium
Gemmiger
Ensifer
Isosphaera
Non-budding stalked bacteria
  Gallionella
  Nevskia
Morphologically unusual budding bacteria
involved in iron and manganese deposition
  Seliberia
  Metallogenium
  Caulococcus
  Kuznezovia
Others
  Spinate bacteria

Archaeobacteria

METHANOGENIC BACTERIA
    Methanobacterium
    Methanobrevibacter
    Methanococcus
    Methanomicrobium
    Methanospirillum
    Methanosarcina
    Methanococcoides
    Methanothermus
    Methanolobus
    Methanoplanus
    Methanogenium
    Methanothrix
EXTREME HALOPHILIC BACTERIA
    Halobacterium
    Halococcus
EXTREME THERMOPHILIC BACTERIA
    Thermoplasma
    Sulfolobus
    Thermoproteus
    Thermofilum
    Thermococcus
    Desulfurococcus
    Thermodiscus
    Pyrodictium

Sheathed Bacteria

    Sphaerotilus
    Leptothrix
    Haliscominobacter
    Lieskeella
    Phragmidiothrix
    Crenothrix
    Clonothrix

Gliding, Fruiting Bacteria

MYXOBACTERALES
  Myxococcaceae
    Myxococcus

Archangiaceae
  Archangium
Cystobacteraceae
  Cystobacter
  Melittangium
  Stigmatella
Polyangiaceae
  Polyangium
  Nannocystis
  Chondromyces
Genus incerta sedis
  Angiococcus

Chemolithotrophic Bacteria

NITRIFIERS
  Nitrobacteraceae
    Nitrobacter
    Nitrospina
    Nitrococcus
    Nitrosomonas
    Nitrosospira
    Nitrosococcus
    Nitrosolobus
SULFUR OXIDIZERS
  Thiobacillus
  Thiomicrospira
  Thiobacterium
  Thiospira
  Macromonas
OBLIGATE HYDROGEN OXIDIZERS
  Hydrogenbacter
METAL OXIDIZERS AND DEPOSITERS
  Siderocapsaceae
    Siderocapsa
    Naumaniella
    Ochrobium
    Siderococcus
OTHER MAGNETOTACTIC BACTERIA

Cyanobacteria

Others

PROCHLORALES
  Prochloraceae
    Prochloron

VOLUME 4

Actinomycetes That Divide in More Than One Plane

  Geodermatophilus
  Dermatophilus
  Frankia
  Tonsilophilus

Sporangiate *Actinomycetes*

  Actinoplanes (including *Amorphosporangium*)
  Streptosporangium
  Ampullariella
  Spirillospora
  Pilimelia
  Dactylosporangium
  Planomonospora
  Planobispora

Streptomycetes and Their Allies

  Streptomyces
  Streptoverticillium
  *Actinopycnidium**
  *Actinosporangium**
  *Chainia**
  *Elytrosporangium**
  *Microellobosporia**
  (The last five genera may be merged with *Streptomyces*)

Other Conidiate Genera

  Actinopolyspora
  Actinosynnema
  Kineospora
  Kitasatosporia
  Microbispora
  Micromonospora
  Microtetrospora
  Saccharomonospora
  Sporichthya
  Streptoalloteichus
  Thermomonospora
  Actinomadura
  Nocardiopsis
  Excellospora
  Thermoactinomyces

CLASSIFICATION OF BACTERIA ACCORDING TO *BERGEY'S MANUAL OF DETERMINATIVE BACTERIOLOGY,* 8TH EDITION

KINGDOM PROCARYOTAE

DIVISION I. THE CYANOBACTERIA

DIVISION II. THE BACTERIA

PART 1.
Phototrophic Bacteria
  Order I. *Rhodospirillales*
    Family I. *Rhodospirillaceae*
      Genus   I. *Rhodospirillum*
      Genus   II. *Rhodopseudo-monas*
      Genus   III. *Rhodomicrobium*
    Family II. *Chromatiaceae*
      Genus   I. *Chromatium*
      Genus   II. *Thiocystis*
      Genus   III. *Thiosarcina*
      Genus   IV. *Thiospirillum*
      Genus   V. *Thiocapsa*
      Genus   VI. *Lamprocystis*
      Genus   VII. *Thiodictyon*
      Genus   VIII. *Thiopedia*
      Genus   IX. *Amoebobacter*
      Genus   X. *Ectothiorhodo-spira*
    Family III. *Chlorobiaceae*
      Genus   I. *Chlorobium*
      Genus   II. *Prosthecochloris*
      Genus   III. *Chloropseudo-monas*
      Genus   IV. *Pelodictyon*
      Genus   V. *Clathrochloris*

PART 2.
The Gliding Bacteria
  Order I. *Myxobacterales*
    Family I. *Myxococcaceae*
      Genus   I. *Myxococcus*
    Family II. *Archangiaceae*
      Genus   I. *Archangium*
    Family III. *Cystobacteraceae*
      Genus   I. *Cystobacter*
      Genus   II. *Melittangium*

      Genus   III. *Stigmatella*
    Family IV. *Polyangiaceae*
      Genus   I. *Polyangium*
      Genus   II. *Nannocystis*
      Genus   III. *Chondromyces*
  Order II. *Cytophagales*
    Family I. *Cytophagaceae*
      Genus   I. *Cytophaga*
      Genus   II. *Flexibacter*
      Genus   III. *Herpetosiphon*
      Genus   IV. *Flexithrix*
      Genus   V. *Saprospira*
      Genus   VI. *Sporocytophaga*
    Family II. *Beggiatoaceae*
      Genus   I. *Beggiatoa*
      Genus   II. *Vitreoscilla*
      Genus   III. *Thioploca*
    Family III. *Simonsiellaceae*
      Genus   I. *Simonsiella*
      Genus   II. *Alysiella*
    Family IV. *Leucotrichaceae*
      Genus   I. *Leucothrix*
      Genus   II. *Thiothrix*
    Families and Genera of
    Uncertain Affiliation
      Genus   *Toxthrix*
    Family *Achromatiaceae*
      Genus   *Achromatium*
    Family *Pelonemataceae*
      Genus   *Pelonema*
      Genus   *Achroonema*
      Genus   *Peloploca*
      Genus   *Desmanthos*

PART 3.
The Sheathed Bacteria
      Genus   *Sphaerotilus*
      Genus   *Leptothrix*

      Genus   *Streptothrix*
      Genus   *Lieskeella*
      Genus   *Phragmidiothrix*
      Genus   *Crenothrix*
      Genus   *Clonothrix*

PART 4.
Budding and/or Appendaged Bacteria
      Genus   *Hyphomicrobium*
      Genus   *Hyphomonas*
      Genus   *Pedomicrobium*
      Genus   *Caulobacter*
      Genus   *Asticcacaulis*
      Genus   *Ancalomicro-bium*
      Genus   *Prosthecomicro-bium*
      Genus   *Thiodendron*
      Genus   *Pasteuria*
      Genus   *Blastobacter*
      Genus   *Seliberia*
      Genus   *Gallionella*
      Genus   *Nevskia*
      Genus   *Planctomyces*
      Genus   *Metallogenium*
      Genus   *Caulococcus*
      Genus   *Kusnezovia*

PART 5.
The Spirochetes
  Order I. *Spirochaetales*
    Family I. *Spirochaetaceae*
      Genus   I. *Spirochaeta*
      Genus   II. *Cristispira*
      Genus   III. *Treponema*
      Genus   IV. *Borrelia*
      Genus   V. *Leptospira*

PART 6.
Spiral and Curved Bacteria
    Family I. *Spirillaceae*
        Genus   I. *Spirillum*
        Genus   II. *Campylobacter*
    Genera of Uncertain Affiliation
        Genus   *Bdellovibrio*
        Genus   *Microcyclus*
        Genus   *Pelosigma*
        Genus   *Brachyarcus*

PART 7.
Gram-Negative Aerobic Rods and
Cocci
    Family I. *Pseudomonadaceae*
        Genus   I. *Pseudomonas*
        Genus   II. *Xanthomonas*
        Genus   III. *Zoogloea*
        Genus   IV. *Gluconobacter*
    Family II. *Azotobacteraceae*
        Genus   I. *Azotobacter*
        Genus   II. *Azomonas*
        Genus   III. *Beijerinckia*
        Genus   IV. *Derxia*
    Family III. *Rhizobiaceae*
        Genus   I. *Rhizobium*
        Genus   II. *Agrobacterium*
    Family IV. *Methylomonadaceae*
        Genus   I. *Methylomonas*
        Genus   II. *Methylococcus*
    Family V. *Halobacteriaceae*
        Genus   I. *Halobacterium*
        Genus   II. *Halococcus*
    Family VI. *Legionellaceae*
        Genus   I. *Legionella*
    Genera of Uncertain Affiliation
        Genus   *Alcaligenes*
        Genus   *Acetobacter*
        Genus   *Brucella*
        Genus   *Bordetella*
        Genus   *Francisella*
        Genus   *Thermus*

PART 8.
Gram-Negative Facultatively
Anaerobic Rods
    Family I. *Enterobacteriaceae*
        Genus   I. *Escherichia*
        Genus   II. *Edwardsiella*

        Genus   III. *Citrobacter*
        Genus   IV. *Salmonella*
        Genus   V. *Shigella*
        Genus   VI. *Klebsiella*
        Genus   VII. *Enterobacter*
        Genus VIII. *Hafnia*
        Genus   IX. *Serratia*
        Genus   X. *Proteus*
        Genus   XI. *Yersinia*
        Genus   XII. *Erwinia*
    Family II. *Vibrionaceae*
        Genus   I. *Vibrio*
        Genus   II. *Aeromonas*
        Genus   III. *Plesiomonas*
        Genus   IV. *Photobacterium*
        Genus   V. *Lucibacterium*
    Genera of Uncertain Affiliation
        Genus   *Zymomonas*
        Genus   *Chromobac-
terium*
        Genus   *Flavobacterium*
        Genus   *Haemophilus*
                   (*H. vaginalis*)
        Genus   *Pasteurella*
        Genus   *Actinobacillus*
        Genus   *Cardiobacterium*
        Genus   *Streptobacillus*
        Genus   *Calymmatobac-
terium*
    Parasites of *Paramecium*

PART 9.
Gram-Negative Anaerobic Bacteria
    Family I. *Bacteroidaceae*
        Genus   I. *Bacteroides*
        Genus   II. *Fusobacterium*
        Genus   III. *Leptotrichia*
    Genera of Uncertain Affiliation
        Genus   *Desulfovibrio*
        Genus   *Butyrivibrio*
        Genus   *Succinivibrio*
        Genus   *Succinimonas*
        Genus   *Lachnospira*
        Genus   *Selenomonas*

PART 10.
Gram-Negative Cocci and
Coccobacilli
    Family I. *Neisseriaceae*
        Genus   I. *Neisseria*
        Genus   II. *Branhamella*

        Genus   III. *Moraxella*
        Genus   IV. *Acinetobacter*
    Genera of Uncertain Affiliation
        Genus   *Paracoccus*
        Genus   *Lampropedia*

PART 11.
Gram-Negative Anaerobic Cocci
    Family I. *Veillonellaceae*
        Genus   I. *Veillonella*
        Genus   II. *Acidaminococcus*
        Genus   III. *Megasphaera*

PART 12.
Gram-Negative, Chemolithotrophic
Bacteria
  a. Organisms oxidizing ammonia
    or nitrite
    Family I. *Nitrobacteraceae*
        Genus   I. *Nitrobacter*
        Genus   II. *Nitrospina*
        Genus   III. *Nitrococcus*
        Genus   IV. *Nitrosomonas*
        Genus   V. *Nitrosospira*
        Genus   VI. *Nitrosococcus*
        Genus VII. *Nitrosolobus*
  b. Organisisms metabolizing
    sulfur
        Genus   *Thiobacillus*
        Genus   *Sulfolobus*
        Genus   *Thiobacterium*
        Genus   *Macromonas*
        Genus   *Thiovulum*
        Genus   *Thiospira*
  c. Organisms depositing iron or
    manganese oxides
    Family I. *Siderocapsaceae*
        Genus   I. *Siderocapsa*
        Genus   II. *Naumanniella*
        Genus   III. *Ochrobium*
        Genus   IV. *Siderococcus*

PART 13.
Methane-Producing Bacteria
    Family I. *Methanobacteriaceae*
        Genus   I. *Methanobac-
terium*
        Genus   II. *Methanosarcina*
        Genus   III. *Methanococcus*

Order I. *Mycoplasmatales*
  Family I. *Mycoplasmataceae*
    Genus   I. *Mycoplasma*
  Family II. *Acholeplasmataceae*
    Genus   I. *Acholeplasma*
Genera of Uncertain Affiliation:
    Genus   *Thermoplasma*
    Genus   *Spiroplasma*
Mycoplasma-like Bodies in Plants

Chapter Opening Photo Credits

Chapter 1: R.C. Kessel and C.Y. Shih, *Scanning Electron Microscopy in Biology*. Berlin: Springer-Verlag (1974).

Chapter 2: O.R. Miller and B.R. Beatty, cover from *Science* **64**:955–957, May 23, 1969.

Chapter 3: T.M. Williams, The Pennsylvania State University.

Chapter 4: From Noel R. Krieg, *Int. J. Syst. Bacteriol.* (1974) **24**:453–458, Fig. 1, page 455.

Chapter 5: From R.G. Kessel and C.Y. Shih, *Scanning Electron Microscopy in Biology*. New York: Springer-Verlag (1974).

Chapter 6: Courtesy of Millipore Corporation, Bedford, Massachusetts.

Chapter 7: A.T. Pringle.

Chapter 8: Charles C. Brinton, Jr., and Judith Carnahan.

Chapter 9: Robley C. Williams, University of California at Berkeley.

Chapter 10: H.C. Pereira, by permission of Academic Press, from R.C. Valentine and H.C. Pereira, *J. Molec. Biol.* (1965) **13**:13–20.

Chapter 11: From Z. Yoshii *et al.*, *Atlas of Scanning Electron Microscopy*. Tokyo: Igaku Shoin, Ltd. (1976).

Chapter 12: T.R. Hoage, R. Jacobs, A.M. Andrews, Y. White/National Center for Toxicological Research.

Chapter 13: T.R. Hoage, R. Jacobs, A.M. Andrews, Y. White/National Center for Toxicological Research.

Chapter 14: Susan C. Straley, Paula A. Harmon, and Lawrence Melsen, University of Kentucky at Lexington.

Chapter 15: Z. Skobe, Forsyth Dental Center/BPS.

Chapter 16: From C.W. Emmons *et al.*, *Medical Mycology* (3rd ed.). Philadelphia: Lea and Febiger (1977).

Chapter 17: Frederick A. Murphy and S.G. Whitfield, Centers for Disease Control.

Chapter 18: Kenneth E. Muse, Duke University Medical Center/BPS.

Chapter 19: Courtesy of J.J. Hadad and C.L. Gyles, University of Guelph, Ontario.

Chapter 20: N.S. Hayes, K.E. Muse, A.M. Collier, and J.B. Baseman, Parasitism by virulent *Trepenoma Pallidum* of host cell surfaces, *Infection and Immunity* (1977) **27**:174–186.

Chapter 22: Stephen G. Baum, M.D., Albert Einstein College of Medicine, New York.

Chapter 23: S. Abraham and E.H. Beachey, V.A. Medical Center, Memphis, TN/BPS.

Chapter 24: M.J. Kramer, Y.R. Mauriz, M.D. Timmes, T.L. Robertson, and R. Cleeland, *Am. J. Med.*, August 23, 1983 (pp. 30–41).

Chapter 25: Courtesy of Merck & Co., Inc.

Chapter 27: Jen and Des Bartlett/Photo Researchers, Inc.

Chapter 28: Carl E. Shively, Department of Biology, Alfred University.

INDEX